Table of Contents

UNIT PRICES

GENERAL REQUIREMENTS	1
SITE CONSTRUCTION	2
CONCRETE	3
MASONRY	4
METALS	5
WOOD & PLASTICS	6
THERMAL & MOISTURE PROTECTION	7
DOORS & WINDOWS	8
FINISHES	9
SPECIALTIES	10
EQUIPMENT	11
SPECIAL CONSTRUCTION	13
CONVEYING SYSTEMS	14
MECHANICAL	15
ELECTRICAL	16
SQUARE FOOT	17

ASSEMBLIES

SUBSTRUCTURE	A
SERVICES	D
BUILDING SITEWORK	G

D1307816

REFERENCE INFORMATION

| REFERENCE NUMBERS |
| CREWS |
| COST INDEXES |
| INDEX |

How the Book Is Built: An Overview

A Powerful Construction Tool

You have in your hands one of the most powerful construction tools available today. A successful project is built on the foundation of an accurate and dependable estimate. This book will enable you to construct just such an estimate.

For the casual user the book is designed to be:

- quickly and easily understood so you can get right to your estimate
- filled with valuable information so you can understand the necessary factors that go into the cost estimate

For the regular user, the book is designed to be:

- a handy desk reference that can be quickly referred to for key costs
- a comprehensive, fully reliable source of current construction costs and productivity rates, so you'll be prepared to estimate any project
- a source book for preliminary project cost, product selections, and alternate materials and methods

To meet all of these requirements we have organized the book into the following clearly defined sections.

Quick Start

This one-page section (see following page) can quickly get you started on your estimate.

How To Use the Book: The Details

This section contains an in-depth explanation of how the book is arranged . . . and how you can use it to determine a reliable construction cost estimate. It includes information about how we develop our cost figures and how to completely prepare your estimate.

Unit Price Section

All cost data has been divided into the 16 divisions according to the MasterFormat system of classification and numbering as developed by the Construction Specifications Institute (CSI) and Construction Specifications Canada (CSC). For a listing of these divisions and an outline of their subdivisions, see the Unit Price Section Table of Contents.

Estimating tips are included at the beginning of each division.

Division 17: Quick Project Estimates
In addition to the 16 Unit Price Divisions there is a S.F. (Square Foot) and C.F. (Cubic Foot) Cost Division, Division 17. It contains costs for 59 different building types that allow you to quickly make a rough estimate for the overall cost of a project or its major components.

Assemblies Section

The cost data in this section has been organized in an "Assemblies" format. These assemblies are the functional elements of a building and are arranged according to the 7 divisions of the UNIFORMAT II classification system. For a complete explanation of a typical "Assemblies" page, see "How To Use the Assemblies Cost Tables."

Reference Section

This section includes information on Reference Numbers, Change Orders, Crew Listings, Historical Cost Indexes, City Cost Indexes, Location Factors and a listing of Abbreviations. It is visually identified by a vertical gray bar on the edge of pages.

Reference Numbers: Following selected major classifications throughout the book are "reference numbers" shown in bold squares. These numbers refer you to related information in other sections.

In these other sections, you'll find related tables, explanations, and estimating information. Also included are alternate pricing methods and technical data, along with information on design and economy in construction. You'll also find helpful tips on estimating and construction.

Change Orders: This section includes information on the factors that influence the pricing of change orders.

Crew Listings: This section lists all the crews referenced. For the purposes of this book, a crew is composed of more than one trade classification and/or the addition of equipment to any trade classification. Power equipment is included in the cost of the crew. Costs are shown both with bare labor rates and with the installing contractor's overhead and profit added. For each, the total crew cost per eight-hour day and the composite cost per labor-hour are listed.

Historical Cost Indexes: These indexes provide you with data to adjust construction costs over time. If you know costs for a past project, you can use these indexes to estimate the cost to construct the same project today.

City Cost Indexes: Obviously, costs vary depending on the regional economy. You can adjust the "national average" costs in this book to 316 major cities throughout the U.S. and Canada by using the data in this section.
Location Factors, to quickly adjust the data to over 930 zip code areas, are included.

Abbreviations: A listing of the abbreviations and the terms they represent, is included.

Index

A comprehensive listing of all terms and subjects in this book to help you find what you need quickly.

The Scope of This Book

This book is designed to be comprehensive and easy to use. To that end we have made certain assumptions:

1. We have established material prices based on a "national average."
2. We have computed labor costs based on a 30-city "national average" of union wage rates.
3. We have targeted the data for projects of a certain size range.

For a more detailed explanation of how the cost data is developed, see "How To Use the Book: The Details."

Project Size

This book is aimed primarily at commercial and industrial projects costing $1,000,000 and up, or large multi-family housing projects. Costs are primarily for new construction or major renovation of buildings rather than repairs or minor alterations.

With reasonable exercise of judgment the figures can be used for any building work. *However, for civil engineering structures such as bridges, dams, highways, or the like, please refer to **Means Heavy Construction Cost Data.***

Quick Start

If you feel you are ready to use this book and don't think you need the detailed instructions that begin on the following page, this Quick Start section is for you.

These steps will allow you to get started estimating in a matter of minutes.

1 First, decide whether you require a Unit Price or Assemblies type estimate. Unit price estimating requires a breakdown of the work to individual items. Assemblies estimates combine individual items or components into building systems.

If you need to estimate each line item separately, follow the instructions for **Unit Prices.**

If you can use an estimate for the entire assembly or system, follow the instructions for **Assemblies.**

2 Find each cost data section you need in the Table of Contents (either for Unit Prices or Assemblies).

Unit Prices: The cost data for Unit Prices has been divided into 16 divisions according to the CSI MasterFormat.

Assemblies: The cost data for Assemblies has been divided into 7 divisions according to the UNIFORMAT II.

3 Turn to the indicated section and locate the line item or assemblies table you need for your estimate. Portions of a sample page layout from both the Unit Price Listings and the Assemblies Cost Tables appear below.

Unit Prices: If there is a reference number listed at the beginning of the section, it refers to additional information you may find useful. See the referenced section for additional information.

- Note the crew code designation. You'll find full descriptions of crews in the Crew Listings including labor-hour and equipment costs.

Assemblies: The Assemblies (*not* shown in full here) are generally separated into three parts: 1) an illustration of the system to be estimated; 2) the components and related costs of a typical system; and 3) the costs for similar systems with dimensional and/or size variations. The Assemblies Section also contains reference numbers for additional useful information.

4 Determine the total number of units your job will require.

Unit Prices: Note the unit of measure for the material you're using is listed under "Unit."

- Bare Costs: These figures show unit costs for materials and installation. Labor and equipment costs are calculated according to crew costs and average daily output. Bare costs do not contain allowances for overhead, profit or taxes.

- "Labor-hours" allows you to calculate the total labor-hours to complete that task. Just multiply the quantity of work by this figure for an estimate of activity duration.

Assemblies: Note the unit of measure for the assembly or system you're estimating is listed in the Assemblies Table.

5 Then multiply the total units by . . .

Unit Prices: "Total Incl. O&P" which stands for the total cost including the installing contractor's overhead and profit. (See the "How To Use the Unit Price Pages" for a complete explanation.)

Assemblies: The "Total" in the right-hand column, which is the total cost **including the installing contractor's overhead and profit.** (See the "How To Use the Assemblies Cost Tables" section for a complete explanation.)

Material and equipment cost figures include a 10% markup. For labor markups, see the inside back cover of this book. If the work is to be subcontracted, add the general contractor's markup, approximately 10%.

6 The price you calculate will be an estimate for either an individual item of work or a completed assembly or *system.*

7 Compile a list of all items or assemblies included in the total project. Summarize cost information, and add project overhead.

Localize costs by using the City Cost Indexes or Location Factors found in the Reference Section.

For a more complete explanation of the way costs are derived, please see the following sections.

Editors' Note: We urge you to spend time reading and understanding all of the supporting material and to take into consideration the reference material such as Crews Listing and the "reference numbers."

Unit Price Pages

Assemblies Pages

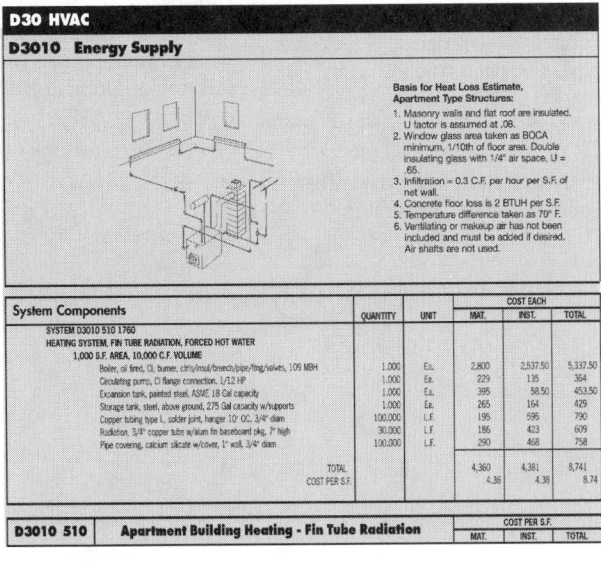

How to Use the Book: The Details

What's Behind the Numbers? The Development of Cost Data

The staff at R.S. Means continuously monitors developments in the construction industry in order to ensure reliable, thorough and up-to-date cost information.

While *overall* construction costs may vary relative to general economic conditions, price fluctuations within the industry are dependent upon many factors. Individual price variations may, in fact, be opposite to overall economic trends. Therefore, costs are continually monitored and complete updates are published yearly. Also, new items are frequently added in response to changes in materials and methods.

Costs—$ (U.S.)

All costs represent U.S. national averages and are given in U.S. dollars. The Means City Cost Indexes can be used to adjust costs to a particular location. The City Cost Indexes for Canada can be used to adjust U.S. national averages to local costs in Canadian dollars.

Material Costs

The R.S. Means staff contacts manufacturers, dealers, distributors, and contractors all across the U.S. and Canada to determine national average material costs. If you have access to current material costs for your specific location, you may wish to make adjustments to reflect differences from the national average. Included within material costs are fasteners for a normal installation. R.S. Means engineers use manufacturers' recommendations, written specifications and/ or standard construction practice for size and spacing of fasteners. Adjustments to material costs may be required for your specific application or location. Material costs do not include sales tax.

Labor Costs

Labor costs are based on the average of wage rates from 30 major U.S. cities. Rates are determined from labor union agreements or prevailing wages for construction trades for the current year. Rates along with overhead and profit markups are listed on the inside back cover of this book.

- If wage rates in your area vary from those used in this book, or if rate increases are expected within a given year, labor costs should be adjusted accordingly.

Labor costs reflect productivity based on actual working conditions. These figures include time spent during a normal workday on tasks other than actual installation, such as material receiving and handling, mobilization at site, site movement, breaks, and cleanup.

Productivity data is developed over an extended period so as not to be influenced by abnormal variations and reflects a typical average.

Equipment Costs

Equipment costs include not only rental, but also operating costs for equipment under normal use. The operating costs include parts and labor for routine servicing such as repair and replacement of pumps, filters and worn lines. Normal operating expendables such as fuel, lubricants, tires and electricity (where applicable) are also included. Extraordinary operating expendables with highly variable wear patterns such as diamond bits and blades are excluded. These costs are included under materials. Equipment rental rates are obtained from industry sources throughout North America—contractors, suppliers, dealers, manufacturers, and distributors.

Crew Equipment Cost/Day—The power equipment required for each crew is included in the crew cost. The daily cost for crew equipment is based on dividing the weekly bare rental rate by 5 (number of working days per week), and then adding the hourly operating cost times 8 (hours per day). This "Crew Equipment Cost/Day" is listed in Subdivision 01590.

General Conditions

Cost data in this book is presented in two ways: Bare Costs and Total Cost including O&P (Overhead and Profit). General Conditions, when applicable, should also be added to the Total Cost including O&P. The costs for General Conditions are listed in Division 1 of the Unit Price Section and the Reference Section of this book. General Conditions for the *Installing Contractor* may range from 0% to 10% of the Total Cost including O&P. For the *General* or *Prime Contractor*, costs for General Conditions may range from 5% to 15% of the Total Cost including O&P, with a figure of 10% as the most typical allowance.

Overhead and Profit

Total Cost including O&P for the *Installing Contractor* is shown in the last column on both the Unit Price and the Assemblies pages of this book. This figure is the sum of the bare material cost plus 10% for profit, the base labor cost plus total overhead and profit, and the bare equipment cost plus 10% for profit. Details for the calculation of Overhead and Profit on labor are shown on the inside back cover and in the Reference Section of this book. (See the "How to Use the Unit Price Pages" for an example of this calculation.)

Factors Affecting Costs

Costs can vary depending upon a number of variables. Here's how we have handled the main factors affecting costs.

Quality—The prices for materials and the workmanship upon which productivity is based represent sound construction work. They are also in line with U.S. government specifications.

Overtime—We have made no allowance for overtime. If you anticipate premium time or work beyond normal working hours, be sure to make an appropriate adjustment to your labor costs.

Productivity—The productivity, daily output, and labor-hour figures for each line item are based on working an eight-hour day in daylight hours in moderate temperatures. For work that extends beyond normal work hours or is performed under adverse conditions, productivity may decrease. (See the section in "How To Use the Unit Price Pages" for more on productivity.)

Size of Project—The size, scope of work, and type of construction project will have a significant impact on cost. Economies of scale can reduce costs for large projects. Unit costs can often run higher for small projects. Costs in this book are intended for the size and type of project as previously described in "How the Book Is Built: An Overview." Costs for projects of a significantly different size or type should be adjusted accordingly.

Location—Material prices in this book are for metropolitan areas. However, in dense urban areas, traffic and site storage limitations may increase costs. Beyond a 20-mile radius of large cities, extra trucking or transportation charges may also increase the material costs slightly. On the other hand, lower wage rates may be in effect. Be sure to consider both these factors when preparing an estimate, particularly if the job site is located in a central city or remote rural location.

In addition, highly specialized subcontract items may require travel and per diem expenses for mechanics.

Other factors —
- season of year
- contractor management
- weather conditions
- local union restrictions
- building code requirements
- availability of:
  - adequate energy
  - skilled labor
  - building materials
- owner's special requirements/restrictions
- safety requirements
- environmental considerations

Unpredictable Factors—General business conditions influence "in-place" costs of all items. Substitute materials and construction methods may have to be employed. These may affect the installed cost and/or life cycle costs. Such factors may be difficult to evaluate and cannot necessarily be predicted on the basis of the job's location in a particular section of the country. Thus, where these factors apply, you may find significant, but unavoidable cost variations for which you will have to apply a measure of judgment to your estimate.

Rounding of Costs

In general, all unit prices in excess of $5.00 have been rounded to make them easier to use and still maintain adequate precision of the results. The rounding rules we have chosen are in the following table.

Prices from . . .	Rounded to the nearest . . .
$.01 to $5.00	$.01
$5.01 to $20.00	$.05
$20.01 to $100.00	$.50
$100.01 to $300.00	$1.00
$300.01 to $1,000.00	$5.00
$1,000.01 to $10,000.00	$25.00
$10,000.01 to $50,000.00	$100.00
$50,000.01 and above	$500.00

Contingencies

Contingencies: The allowance for contingencies generally is to provide for unforeseen construction difficulties. On alterations or repair jobs 20% is none too much. If drawings are final and only field contingencies are being considered, 2% or 3% is probably sufficient and often nothing need be added. As far as the contract is concerned, future changes in plans will be covered by extras. The contractor should consider inflationary price trends and possible material shortages during the course of the job. If drawings are not complete or approved or a budget cost wanted, it is wise to add 5% to 10%. Contingencies, then, are a matter of judgment. Additional allowances are shown in Division 01250-200 for contingencies and job conditions and Division 01250-400 for factors to convert prices for repair and remodeling jobs.

Note: For an explanation of "Selective Demolition" versus "Removal for Replacement" see Reference number R15050-720.

Important Estimating Considerations

One reason for listing a job size "minimum" is to ensure that the construction craftsmen are productive for 8 hours a day on a continuous basis. Otherwise, tasks that only require 6 to 7 hours could be billed as an 8-hour day, thus providing an erroneous productivity estimate. The "productivity," or daily output of each craftsman, includes mobilization and cleanup time, break time, plan layout time, as well as an allowance to carry stock from the storage trailer or location on the job site up to 200' into the building and on the first or second floor. If material has to be transported over greater distances or to higher floors, an additional allowance should be considered by the estimator. An allowance has also been included in the piping and fittings installation time for a leak check and minor tightening.

Equipment installation time includes all applicable items following: positioning, leveling and securing the unit in place, connecting all associated piping, ducts, vents, etc., which shall have been estimated separately, connecting to an adjacent power source, filling/bleeding, startup, adjusting the controls up and down to assure proper response, setting the integral controls/valves/regulators/thermostats for proper operation (does not include external building type control systems, DDC systems, etc.), explaining/training owner's operator, and warranty. A reasonable breakdown of the labor costs would be as follows:

1. Movement into building, installation/ setting of equipment   35%
2. Connecting to piping/duct/power, etc.   25%
3. Filling/flushing/cleaning/touchup, etc.   15%
4. Startup/running adjustments   5%
5. Training owner's rep.   5%
6. Warranty/call back/service   15%

Note that cranes or other lifting equipment are not included on any lines in Division 15. For example, if a crane is required to lift a heavy piece of pipe into place high above a gym floor, or to put a rooftop unit on the roof of a four-story building, etc., it must be added. Due to the potential for extreme variation—from nothing additional required, to a major crane or helicopter—we feel that including a nominal amount for "lifting contingency" would be useless and detract from the accuracy of the estimate. When using equipment rental from Means do not forget to include the cost of the operator(s).

Estimating Labor-Hours

The labor-hours expressed in this publication are based on Average Installation time, using an efficiency level of approximately 60-65% (see item 7) which has been found reasonable and acceptable by many contractors.

The book uses this national efficiency average to establish a consistent benchmark.

For bid situations, adjustments to this efficiency level should be the responsibility of the contractor bidding the project.

The unit labor-hour is divided in the following manner. A typical day for a journeyman might be:

1.	Study Plans	3%	14.4 min.
2.	Material Procurement	3%	14.4 min.
3.	Receiving and Storing	3%	14.4 min
4.	Mobilization	5%	24.0 min.
5.	Site Movement	5%	24.0 min.
6.	Layout and Marking	8%	38.4 min.
7.	Actual Installation	64%	307.2 min.
8.	Cleanup	3%	14.4 min.
9.	Breaks, Non-Productive	6%	28.8 min.
		100%	480.0 min.

If any of the percentages expressed in this breakdown do not apply to the particular work or project situation, then that percentage or a portion of it may be deducted from or added to labor hours.

Final Checklist

Estimating can be a straightforward process provided you remember the basics. Here's a checklist of some of the items you should remember to do before completing your estimate.

Did you remember to . . .

- factor in the City Cost Index for your locale
- take into consideration which items have been marked up and by how much
- mark up the entire estimate sufficiently for your purposes
- read the background information on techniques and technical matters that could impact your project time span and cost
- include all components of your project in the final estimate
- double check your figures to be sure of your accuracy
- for more information, please see "Tips for Accurate Estimating," R01100-005 in the Reference Section.
- call R.S. Means if you have any questions about your estimate or the data you've found in our publications

Remember, R.S. Means stands behind its publications. If you have any questions about your estimate . . . about the costs you've used from our books . . . or even about the technical aspects of the job that may affect your estimate, feel free to call the R.S. Means editors at 1-800-334-3509.

Unit Price Section

Table of Contents

How to Use the Unit Price Pages

The following is a detailed explanation of a sample entry in the Unit Price Section. Next to each bold number below is the item being described with appropriate component of the sample entry following in parenthesis. Some prices are listed as bare costs, others as costs that include overhead and profit of the installing contractor. In most cases, if the work is to be subcontracted, the general contractor will need to add an additional markup (R.S. Means suggests using 10%) to the figures in the column "Total Incl. O&P."

1 Division Number/Title (03300/Cast-In-Place Concrete)

Use the Unit Price Section Table of Contents to locate specific items. The sections are classified according to the CSI MasterFormat (1995 Edition).

2 Line Numbers (03310 240 3900)

Each unit price line item has been assigned a unique 12-digit code based on the CSI MasterFormat classification.

- **03300** — Level One - CSI-MasterFormat Division
- **03310** — Level Two - CSI
- **03310-240-3900** — Means 12-digit Line Number
  - Level Four - Means
  - Level Three - CSI

3 Description (Concrete-In-Place, etc.)

Each line item is described in detail. Sub-items and additional sizes are indented beneath the appropriate line items. The first line or two after the main item (in boldface) may contain descriptive information that pertains to all line items beneath this boldface listing.

4 Reference Number Information

R03310 -010

You'll see reference numbers shown in bold rectangles at the beginning of some sections. These refer to related items in the Reference Section, visually identified by a vertical gray bar on the edge of pages.

The relation may be: (1) an estimating procedure that should be read before estimating, (2) an alternate pricing method, or (3) technical information.

The "R" designates the Reference Section. The numbers refer to the MasterFormat classification system.

It is strongly recommended that you review all reference numbers that appear within the section in which you are working.

Note: Not all reference numbers appear in all Means publications.

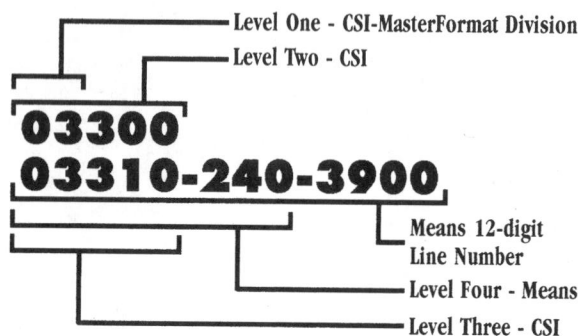

03300 | Cast-In-Place Concrete

03310	Structural Concrete	CREW	DAILY OUTPUT	LABOR-HOURS	UNIT	MAT.	2003 BARE COSTS LABOR	EQUIP.	TOTAL	TOTAL INCL O&P	
240	0010	**CONCRETE IN PLACE** Including forms (4 uses), reinforcing										240
	0050	steel, including finishing unless otherwise indicated										
	0300	Beams, 5 kip per L.F., 10' span	C-14A	15	3.804	C.Y.	227	405	50	681.50		
	0350	25' span	"	1	82		210	340	41.50	591.50		
	3800	Footings, spread under 1 C.Y.	C-14C	38.07	2.942	C.Y.	102	89	.85	191.85		
	3850	Over 5 C.Y.		81.04	1.382		93.50	41.50	.40	135.40	169	
	3900	Footings, strip, 18" x 9", plain		41.04	2.729		92	82.50	.79	175.29	232	
	3950	36" x 12", reinforced		61.55	1.820		93.50	55	.53	149.03	190	
	4000	Foundation mat, under 10 C.Y.		38.67	2.896		125	87.50	.84	213.34	276	
	4050	Over 20 C.Y.	▼	56.40	1.986		111	60	.58	171.58	217	
	4200	Grade walls, 8" thick, 8' high	C-14D	45.83	4.364		128	137	16.80	281.80	375	
	4250	14" high		27.26	7.337		152	230	28.50	410.50	560	
	4260	12" thick, 8' high		64.32	3.109		113	97.50	11.95	222.45	290	

Crew (C-14C)

The "Crew" column designates the typical trade or crew used to install the item. If an installation can be accomplished by one trade and requires no power equipment, that trade and the number of workers are listed (for example, "2 Carpenters"). If an installation requires a composite crew, a crew code designation is listed (for example, "C-14C"). You'll find full details on all composite crews in the Crew Listings.

- For a complete list of all trades utilized in this book and their abbreviations, see the inside back cover.

Crews

Crew No.	Bare Costs		Incl. Subs O & P		Cost Per Labor-Hour	
Crew C-14C	Hr.	Daily	Hr.	Daily	Bare Costs	Incl. O&P
1 Carpenter Foreman (out)	$33.55	$268.40	$52.40	$419.20	$30.20	$47.60
6 Carpenters	31.55	1514.40	49.30	2366.40		
2 Rodmen (reinf.)	35.55	568.80	59.70	955.20		
4 Laborers	24.65	788.80	38.50	1232.00		
1 Cement Finisher	30.20	241.60	44.75	358.00		
1 Gas Engine Vibrator		32.60		35.85	.29	.32
112 L.H., Daily Totals		$3414.60		$5366.65	$30.49	$47.92

Productivity: Daily Output (41.04)/Labor-Hours (2.729)

The "Daily Output" represents the typical number of units the designated crew will install in a normal 8-hour day. To find out the number of days the given crew would require to complete the installation, divide your quantity by the daily output. For example:

Quantity	÷	Daily Output	=	Duration
100 C.Y.	÷	41.04/ Crew Day	=	2.44 Crew Days

The "Labor-Hours" figure represents the number of labor-hours required to install one unit of work. To find out the number of labor-hours required for your particular task, multiply the quantity of the item times the number of labor-hours shown. For example:

Quantity	x	Productivity Rate	=	Duration
100 C.Y.	x	2.729 Labor-Hours/ C.Y.	=	272.9 Labor-Hours

Unit (C.Y.)

The abbreviated designation indicates the unit of measure upon which the price, production, and crew are based (C.Y. = Cubic Yard). For a complete listing of abbreviations refer to the Abbreviations Listing in the Reference Section of this book.

Bare Costs:

Mat. (Bare Material Cost) (92)

The unit material cost is the "bare" material cost with no overhead and profit included. *Costs shown reflect national average material prices for January of the current year and include delivery to the job site. No sales taxes are included.*

Labor (82.50)

The unit labor cost is derived by multiplying bare labor-hour costs for Crew C-14C by labor-hour units. The bare labor-hour cost is found in the Crew Section under C-14C. (If a trade is listed, the hourly labor cost—the wage rate—is found on the inside back cover.)

Labor-Hour Cost Crew C-14C	x	Labor-Hour Units	=	Labor
$30.20	x	2.729	=	$82.50

Equip. (Equipment) (.79)

Equipment costs for each crew are listed in the description of each crew. Tools or equipment whose value justifies purchase or ownership by a contractor are considered overhead as shown on the inside back cover. The unit equipment cost is derived by multiplying the bare equipment hourly cost by the labor-hour units.

Equipment Cost Crew C-14C	x	Labor-Hour Units	=	Equip.
.29	x	2.729	=	$.79

Total (175.29)

The total of the bare costs is the arithmetic total of the three previous columns: mat., labor, and equip.

Material	+	Labor	+	Equip.	=	Total
$92	+	$82.50	+	$.79	=	$175.29

Total Costs Including O&P

This figure is the sum of the bare material cost plus 10% for profit; the bare labor cost plus total overhead and profit (per the inside back cover or, if a crew is listed, from the crew listings); and the bare equipment cost plus 10% for profit.

Material is Bare Material Cost + 10% = 92 + 9.20	=	$101.20
Labor for Crew C-14C = Labor-Hour Cost (47.60) x Labor-Hour Units (2.729)	=	$129.90
Equip. is Bare Equip. Cost + 10% = .79 + .08	=	$ .87
Total (Rounded)	=	$232

Division 1
General Requirements

Estimating Tips

The General Requirements of any contract are very important to both the bidder and the owner. These lay the ground rules under which the contract will be executed and have a significant influence on the cost of operations. Therefore, it is extremely important to thoroughly read and understand the General Requirements both before preparing an estimate and when the estimate is complete, to ascertain that nothing in the contract is overlooked. Caution should be exercised when applying items listed in Division 1 to an estimate. Many of the items are included in the unit prices listed in the other divisions such as mark-ups on labor and company overhead.

01200 Price & Payment Procedures

- When estimating historic preservation projects (depending on the condition of the existing structure and the owner's requirements), a 15-20% contingency or allowance is recommended, regardless of the stage of the drawings.

01300 Administrative Requirements

- Before determining a final cost estimate, it is a good practice to review all the items listed in subdivision 01300 to make final adjustments for items that may need customizing to specific job conditions.
- Historic preservation projects may require specialty labor and methods, as well as extra time to protect existing materials that must be preserved and/or restored. Some additional expenses may be incurred in architectural fees for facility surveys and other special inspections and analyses.

01330 Submittal Procedures

- Requirements for initial and periodic submittals can represent a significant cost to the General Requirements of a job. Thoroughly check the submittal specifications when estimating a project to determine any costs that should be included.

01400 Quality Requirements

- All projects will require some degree of Quality Control. This cost is not included in the unit cost of construction listed in each division. Depending upon the terms of the contract, the various costs of inspection and testing can be the responsibility of either the owner or the contractor. Be sure to include the required costs in your estimate.

01500 Temporary Facilities & Controls

- Barricades, access roads, safety nets, scaffolding, security and many more requirements for the execution of a safe project are elements of direct cost. These costs can easily be overlooked when preparing an estimate. When looking through the major classifications of this subdivision, determine which items apply to each division in your estimate.

01590 Equipment Rental

- This subdivision contains transportation, handling, storage, protection and product options and substitutions. Listed in this cost manual are average equipment rental rates for all types of equipment. This is useful information when estimating the time and materials requirement of any particular operation in order to establish a unit or total cost.
- A good rule of thumb is that weekly rental is 3 times daily rental and that monthly rental is 3 times weekly rental.

- The figures in the column for Crew Equipment Cost represent the rental rate used in determining the daily cost of equipment in a crew. It is calculated by dividing the weekly rate by 5 days and adding the hourly operating cost times 8 hours.

01770 Closeout Procedures

- When preparing an estimate, read the specifications to determine the requirements for Contract Closeout thoroughly. Final cleaning, record documentation, operation and maintenance data, warranties and bonds, and spare parts and maintenance materials can all be elements of cost for the completion of a contract. Do not overlook these in your estimate.

01830 Operations & Maintenance

- If maintenance and repair are included in your contract, they require special attention. To estimate the cost to remove and replace any unit usually requires a site visit to determine the accessibility and the specific difficulty at that location. Obstructions, dust control, safety, and often overtime hours must be considered when preparing your estimate.

Reference Numbers

Reference numbers are shown in bold squares at the beginning of some major classifications. These numbers refer to related items in the Reference Section. The reference information may be an estimating procedure, an alternate pricing method or technical information.

Note: Not all subdivisions listed here necessarily appear in this publication.

01107 | Professional Consultant

		CREW	DAILY OUTPUT	LABOR-HOURS	UNIT	2003 BARE COSTS MAT.	LABOR	EQUIP.	TOTAL	TOTAL INCL O&P	
300	0010	**ENGINEERING FEES**									**300**
	0020	Educational planning consultant, minimum [R01107-030]				Project					.50%
	0100	Maximum				"					2.50%
	0200	Electrical, minimum				Contrct					4.10%
	0300	Maximum									10.10%
	0400	Elevator & conveying systems, minimum									2.50%
	0500	Maximum									5%
	0600	Food service & kitchen equipment, minimum									8%
	0700	Maximum									12%
	1000	Mechanical (plumbing & HVAC), minimum									4.10%
	1100	Maximum									10.10%

01200 | Price & Payment Procedures

01250 | Contract Modification Procedures

		CREW	DAILY OUTPUT	LABOR-HOURS	UNIT	2003 BARE COSTS MAT.	LABOR	EQUIP.	TOTAL	TOTAL INCL O&P	
200	0010	**CONTINGENCIES** for estimate at conceptual stage				Project					20%	**200**
	0050	Schematic stage									15%
	0100	Preliminary working drawing stage (Design Dev.)									10%
	0150	Final working drawing stage									3%
300	0010	**CREWS** For building construction, see How To Use This Book										**300**
400	0010	**FACTORS** Cost adjustments [R01250-010]										**400**
	0100	Add to construction costs for particular job requirements									
	0500	Cut & patch to match existing construction, add, minimum				Costs	2%	3%			
	0550	Maximum					5%	9%			
	0800	Dust protection, add, minimum					1%	2%			
	0850	Maximum					4%	11%			
	1100	Equipment usage curtailment, add, minimum					1%	1%			
	1150	Maximum					3%	10%			
	1400	Material handling & storage limitation, add, minimum					1%	1%			
	1450	Maximum					6%	7%			
	1700	Protection of existing work, add, minimum					2%	2%			
	1750	Maximum					5%	7%			
	2000	Shift work requirements, add, minimum						5%			
	2050	Maximum						30%			
	2300	Temporary shoring and bracing, add, minimum					2%	5%			
	2350	Maximum					5%	12%			
	2400	Work inside prisons and high security areas, add, minimum						30%			
	2450	Maximum						50%			
500	0010	**JOB CONDITIONS** Modifications to total										**500**
	0020	project cost summaries									
	0100	Economic conditions, favorable, deduct				Project					2%
	0200	Unfavorable, add									5%
	0300	Hoisting conditions, favorable, deduct									2%
	0400	Unfavorable, add									5%
	0700	Labor availability, surplus, deduct									1%
	0800	Shortage, add									10%
	0900	Material storage area, available, deduct									1%
	1000	Not available, add									2%

Important: See the Reference Section for critical supporting data - Reference Nos., Crews, & City Cost Indexes

01200 | Price & Payment Procedures

| | | **01250** | **Contract Modification Procedures** | CREW | DAILY OUTPUT | LABOR-HOURS | UNIT | 2003 BARE COSTS | | | | TOTAL INCL O&P | |
								MAT.	LABOR	EQUIP.	TOTAL		
500	1100		Subcontractor availability, surplus, deduct				Project					5%	500
	1200		Shortage, add									12%	
	1300		Work space, available, deduct									2%	
	1400		Not available, add				↓					5%	
600	0010		**OVERTIME** For early completion of projects or where	R01100 -110									600
	0020		labor shortages exist, add to usual labor, up to				Costs		100%				

01255 | Cost Indexes

200	0010	**CONSTRUCTION COST INDEX** (Reference) over 930 zip code locations in										200
	0020	The U.S. and Canada, total bldg cost, min. (Clarksdale, MS)				%					65%	
	0050	Average									100%	
	0100	Maximum (New York, NY)				↓					132.60%	
400	0010	**HISTORICAL COST INDEXES** (Reference) Back to 1953										400
500	0010	**LABOR INDEX** (Reference) For over 930 zip code locations in										500
	0020	the U.S. and Canada, minimum (Clarksdale, MS)				%		32%				
	0050	Average						100%				
	0100	Maximum (New York, NY)				↓		159.10%				
600	0011	**MATERIAL INDEX** For over 930 zip code locations in										600
	0020	the U.S. and Canada, minimum (Elizabethtown, KY)				%	92.10%					
	0040	Average					100%					
	0060	Maximum (Ketchikan, AK)				↓	146.10%					

01290 | Payment Procedures

800	0010	**TAXES** Sales tax, State, average	R01100 -090				%	4.65%					800
	0050	Maximum						7%					
	0200	Social Security, on first $84,900 of wages	R01100 -100						7.65%				
	0300	Unemployment, MA, combined Federal and State, minimum							2.10%				
	0350	Average							6.50%				
	0400	Maximum					↓		8%				

01300 | Administrative Requirements

| | | **01310** | **Project Management/Coordination** | CREW | DAILY OUTPUT | LABOR-HOURS | UNIT | 2003 BARE COSTS | | | | TOTAL INCL O&P | |
								MAT.	LABOR	EQUIP.	TOTAL		
150	0010		**PERMITS** Rule of thumb, most cities, minimum				Job					.50%	150
	0100		Maximum				"					2%	
200	0010		**PERFORMANCE BOND** For buildings, minimum	R01100 -080			Job					.60%	200
	0100		Maximum				"					2.50%	
350	0010		**INSURANCE** Builders risk, standard, minimum	R01100 -040			Job					.22%	350
	0050		Maximum									.59%	
	0200		All-risk type, minimum	R01100 -060								.25%	
	0250		Maximum				↓					.62%	
	0400		Contractor's equipment floater, minimum				Value					.50%	
	0450		Maximum				"					1.50%	
	0600		Public liability, average				Job					1.55%	
	0800		Workers' compensation & employer's liability, average										
	0850		by trade, carpentry, general				Payroll		18.57%				
	1000		Electrical						6.65%				
	1150		Insulation						16.54%				
	1450		Plumbing				↓		8.35%				

01300 | Administrative Requirements

01310 | Project Management/Coordination

			CREW	DAILY OUTPUT	LABOR-HOURS	UNIT	2003 BARE COSTS MAT.	LABOR	EQUIP.	TOTAL	TOTAL INCL O&P	
350	1550	Sheet metal work (HVAC) R01100-040				Payroll		11.81%				**350**
500	0010	**MARK-UP** For General Contractors for change R01100-070										**500**
	0100	of scope of job as bid										
	0200	Extra work, by subcontractors, add				%					10%	
	0250	By General Contractor, add									15%	
	0400	Omitted work, by subcontractors, deduct all but									5%	
	0450	By General Contractor, deduct all but									7.50%	
	0600	Overtime work, by subcontractors, add									15%	
	0650	By General Contractor, add									10%	
	1000	Installing contractors, on his own labor, minimum						47.50%				
	1100	Maximum						87.80%				
600	0010	**OVERHEAD** As percent of direct costs, minimum R01100-070				%				5%		**600**
	0050	Average								13%		
	0100	Maximum								30%		
620	0010	**OVERHEAD & PROFIT** Allowance to add to items in this R01100-070										**620**
	0020	book that do not include Subs O&P, average				%				25%		
	0100	Allowance to add to items in this book that										
	0110	do include Subs O&P, minimum				%					5%	
	0150	Average									10%	
	0200	Maximum									15%	
	0300	Typical, by size of project, under $100,000									30%	
	0350	$500,000 project									25%	
	0400	$2,000,000 project									20%	
	0450	Over $10,000,000 project									15%	

01321 | Construction Photos

			CREW	DAILY OUTPUT	LABOR-HOURS	UNIT	2003 BARE COSTS MAT.	LABOR	EQUIP.	TOTAL	TOTAL INCL O&P	
500	0010	**PHOTOGRAPHS** 8″ x 10″, 4 shots, 2 prints ea., std. mounting				Set	283			283	310	**500**
	0100	Hinged linen mounts					305			305	340	
	0200	8″ x 10″, 4 shots, 2 prints each, in color					330			330	360	
	0300	For I.D. slugs, add to all above					3.97			3.97	4.37	
	1500	Time lapse equipment, camera and projector, buy					3,575			3,575	3,925	
	1550	Rent per month					530			530	585	
	1700	Cameraman and film, including processing, B.&W.				Day	1,300			1,300	1,425	
	1720	Color				″	1,300			1,300	1,425	

01500 | Temporary Facilities & Controls

01510 | Temporary Utilities

			CREW	DAILY OUTPUT	LABOR-HOURS	UNIT	2003 BARE COSTS MAT.	LABOR	EQUIP.	TOTAL	TOTAL INCL O&P	
800	0010	**TEMPORARY UTILITIES**										**800**
	0100	Heat, incl. fuel and operation, per week, 12 hrs. per day	1 Skwk	100	.080	CSF Flr	5.20	2.58		7.78	9.80	
	0200	24 hrs. per day	″	60	.133		7.85	4.30		12.15	15.35	
	0350	Lighting, incl. service lamps, wiring & outlets, minimum	1 Elec	34	.235		2.11	8.85		10.96	15.50	
	0360	Maximum	″	17	.471		4.60	17.70		22.30	31.50	
	0400	Power for temp lighting only, per month, min/month 6.6 KWH								.75	1.18	
	0450	Maximum/month 23.6 KWH								2.85	3.14	
	0600	Power for job duration incl. elevator, etc., minimum								47	51.70	
	0650	Maximum								110	121	
	1000	Toilet, portable, see division 01590-400										

Important: See the Reference Section for critical supporting data - Reference Nos., Crews, & City Cost Indexes

01500 | Temporary Facilities & Controls

01520 | Construction Facilities

			CREW	DAILY OUTPUT	LABOR-HOURS	UNIT	MAT.	LABOR	EQUIP.	TOTAL	TOTAL INCL O&P	
500	0010	**OFFICE** Trailer, furnished, no hookups, 20′ x 8′, buy	2 Skwk	1	16	Ea.	5,700	515		6,215	7,075	500
	0250	Rent per month					148			148	163	
	0300	32′ x 8′, buy	2 Skwk	.70	22.857		8,075	735		8,810	10,000	
	0350	Rent per month					159			159	175	
	0400	50′ x 10′, buy	2 Skwk	.60	26.667		15,000	860		15,860	17,900	
	0450	Rent per month					266			266	293	
	0500	50′ x 12′, buy	2 Skwk	.50	32		17,400	1,025		18,425	20,700	
	0550	Rent per month					325			325	360	
	0700	For air conditioning, rent per month, add					36.50			36.50	40	
	0800	For delivery, add per mile				Mile	1.53			1.53	1.68	
	1000	Portable buildings, prefab, on skids, economy, 8′ x 8′	2 Carp	265	.060	S.F.	81.50	1.91		83.41	93	
	1100	Deluxe, 8′ x 12′	″	150	.107	″	88.50	3.37		91.87	103	
	1200	Storage boxes, 20′ x 8′, buy	2 Skwk	1.80	8.889	Ea.	3,375	287		3,662	4,150	
	1250	Rent per month					74			74	81.50	
	1300	40′ x 8′, buy	2 Skwk	1.40	11.429		4,025	370		4,395	5,000	
	1350	Rent per month					105			105	115	
550	0010	**FIELD OFFICE EXPENSE**										550
	0100	Field office expense, office equipment rental average				Month	139			139	153	
	0120	Office supplies, average				″	83.50			83.50	92	
	0125	Office trailer rental, see division 01520-500										
	0140	Telephone bill; avg. bill/month incl. long dist.				Month	204			204	224	
	0160	Field office lights & HVAC				″	94			94	103	

01530 | Temporary Construction

			CREW	DAILY OUTPUT	LABOR-HOURS	UNIT	MAT.	LABOR	EQUIP.	TOTAL	TOTAL INCL O&P	
700	0010	**PROTECTION** Stair tread, 2″ x 12″ planks, 1 use	1 Carp	75	.107	Tread	4.69	3.37		8.06	10.40	700
	0100	Exterior plywood, 1/2″ thick, 1 use		65	.123		1.47	3.88		5.35	7.65	
	0200	3/4″ thick, 1 use		60	.133		1.99	4.21		6.20	8.75	
900	0010	**WINTER PROTECTION** Reinforced plastic on wood										900
	0100	framing to close openings	2 Clab	750	.021	S.F.	.38	.53		.91	1.24	
	0200	Tarpaulins hung over scaffolding, 8 uses, not incl. scaffolding		1,500	.011		.18	.26		.44	.61	
	0250	Tarpaulin polyester reinf. w/ integral fastening system 11 mils thick		1,600	.010		.78	.25		1.03	1.25	
	0300	Prefab fiberglass panels, steel frame, 8 uses		1,200	.013		.74	.33		1.07	1.32	

01540 | Construction Aids

			CREW	DAILY OUTPUT	LABOR-HOURS	UNIT	MAT.	LABOR	EQUIP.	TOTAL	TOTAL INCL O&P	
750	0010	**SCAFFOLDING**										750
	0015	Steel tubular, reg, rent/mo, no plank, incl erect or dismantle	R01540-100									
	0090	Building exterior, wall face, 1 to 5 stories, 6′-4″ x 5′ frames	3 Carp	24	1	C.S.F.	24.50	31.50		56	76	
	0200	6 to 12 stories	4 Carp	21.20	1.509		24.50	47.50		72	101	
	0310	13 to 20 stories	5 Carp	20	2		24.50	63		87.50	125	
	0460	Building interior, wall face area, up to 16′ high	3 Carp	25	.960		24.50	30.50		55	74	
	0560	16′ to 40′ high		23	1.043		24.50	33		57.50	78	
	0800	Building interior floor area, up to 30′ high		312	.077	C.C.F.	2.57	2.43		5	6.60	
	0900	Over 30′ high	4 Carp	275	.116	″	2.57	3.67		6.24	8.60	
	0910	Steel tubular, heavy duty shoring, buy										
	0920	Frames 5′ high 2′ wide				Ea.	75			75	82.50	
	0925	5′ high 4′ wide					85			85	93.50	
	0930	6′ high 2′ wide					86			86	94.50	
	0935	6′ high 4′ wide					101			101	111	
	0940	Accessories										
	0945	Cross braces				Ea.	16			16	17.60	
	0950	U-head, 8″ x 8″					17.50			17.50	19.25	
	0955	J-head, 4″ x 8″					12.80			12.80	14.10	
	0960	Base plate, 8″ x 8″					14.20			14.20	15.60	
	0965	Leveling jack					30.50			30.50	33.50	

GENERAL REQUIREMENTS

1

		01540	Construction Aids	CREW	DAILY OUTPUT	LABOR-HOURS	UNIT	2003 BARE COSTS				TOTAL INCL O&P	
								MAT.	LABOR	EQUIP.	TOTAL		
750	1000		Steel tubular, regular, buy										750
	1100		Frames 3' high 5' wide R01540 -100				Ea.	58			58	64	
	1150		5' high 5' wide					67			67	73.50	
	1200		6'-4" high 5' wide					84			84	92.50	
	1350		7'-6" high 6' wide					145			145	160	
	1500		Accessories cross braces					15			15	16.50	
	1550		Guardrail post					15			15	16.50	
	1600		Guardrail 7' section					7.25			7.25	8	
	1650		Screw jacks & plates					24			24	26.50	
	1700		Sidearm brackets					28			28	31	
	1750		8" casters					33			33	36.50	
	1800		Plank 2" x 10" x 16'-0"					42.50			42.50	47	
	1900		Stairway section					245			245	270	
	1910		Stairway starter bar					29			29	32	
	1920		Stairway inside handrail					53			53	58.50	
	1930		Stairway outside handrail					73			73	80.50	
	1940		Walk-thru frame guardrail				↓	37			37	40.50	
	2000		Steel tubular, regular, rent/mo.										
	2100		Frames 3' high 5' wide				Ea.	3.75			3.75	4.13	
	2150		5' high 5' wide					3.75			3.75	4.13	
	2200		6'-4" high 5' wide					3.75			3.75	4.13	
	2250		7'-6" high 6' wide					7			7	7.70	
	2500		Accessories, cross braces					.60			.60	.66	
	2550		Guardrail post					1			1	1.10	
	2600		Guardrail 7' section					.75			.75	.83	
	2650		Screw jacks & plates					1.50			1.50	1.65	
	2700		Sidearm brackets					1.50			1.50	1.65	
	2750		8" casters					6			6	6.60	
	2800		Outrigger for rolling tower					3			3	3.30	
	2850		Plank 2" x 10" x 16'-0"					5			5	5.50	
	2900		Stairway section					10			10	11	
	2910		Stairway starter bar					.10			.10	.11	
	2920		Stairway inside handrail					5			5	5.50	
	2930		Stairway outside handrail					5			5	5.50	
	2940		Walk-thru frame guardrail				↓	2			2	2.20	
	3000		Steel tubular, heavy duty shoring, rent/mo.										
	3250		5' high 2' & 4' wide				Ea.	5			5	5.50	
	3300		6' high 2' & 4' wide					5			5	5.50	
	3500		Accessories, cross braces					1			1	1.10	
	3600		U - head, 8" x 8"					1			1	1.10	
	3650		J - head, 4" x 8"					1			1	1.10	
	3700		Base plate, 8" x 8"					1			1	1.10	
	3750		Leveling jack					2			2	2.20	
	5700		Planks, 2"x10"x16'-0", labor only, erect or remove to 50' ht	3 Carp	144	.167			5.25		5.25	8.20	
	5800		Over 50' high	4 Carp	160	.200	↓		6.30		6.30	9.85	
	6000		Heavy duty shoring for elevated slab forms to 8'-2" high, floor area										
	6010		incl. erection or dismantle labor										
	6100		1 use/month	4 Carp	36	.889	C.S.F.	29.50	28		57.50	76.50	
	6150		2 uses/month	"	36	.889	"	14.80	28		42.80	60.50	
	6500		To 14'-8" high										
	6600		1 use/month	4 Carp	18	1.778	C.S.F.	43	56		99	135	
	6650		2 uses/month	"	18	1.778	"	21.50	56		77.50	112	
755	0010	**SCAFFOLDING SPECIALTIES**											755
	1200		Sidewalk bridge, heavy duty steel posts & beams, including										
	1210		parapet protection & waterproofing										
	1220		8' to 10' wide, 2 posts	3 Carp	15	1.600	L.F.	30.50	50.50		81	113	

Important: See the Reference Section for critical supporting data - Reference Nos., Crews, & City Cost Indexes

01540 | Construction Aids

		CREW	DAILY OUTPUT	LABOR-HOURS	UNIT	MAT.	LABOR	EQUIP.	TOTAL	TOTAL INCL O&P	
755	1230	3 posts	3 Carp	10	2.400	L.F.	47	75.50		122.50	170	755
	1500	Sidewalk bridge using tubular steel										
	1510	scaffold frames, including planking	3 Carp	45	.533	L.F.	4.72	16.85		21.57	31.50	
	1600	For 2 uses per month, deduct from all above					50%					
	1700	For 1 use every 2 months, add to all above					100%					
	1900	Catwalks, 20" wide, no guardrails, 7' span, buy				Ea.	125			125	138	
	2000	10' span, buy				"	160			160	176	
	2800	Hand winch-operated masons scaffolding, no plank										
	2810	plank moving not required										
	2900	98' long, 10'-6" high, buy				Ea.	26,200			26,200	28,900	
	3000	Rent per month					1,050			1,050	1,150	
	3100	28'-6" high, buy					32,200			32,200	35,400	
	3200	Rent per month					1,275			1,275	1,425	
	3400	196' long, 28'-6" high, buy					62,000			62,000	68,500	
	3500	Rent per month					2,475			2,475	2,725	
	3600	64'-6" high, buy					85,000			85,000	93,500	
	3700	Rent per month					3,400			3,400	3,750	
	3720	Putlog, standard, 8' span, with hangers, buy					61			61	67	
	3730	Rent per month					10			10	11	
	3750	12' span, buy					92			92	101	
	3755	Rent per month					15			15	16.50	
	3760	Trussed type, 16' span, buy					210			210	231	
	3770	Rent per month					20			20	22	
	3790	22' span, buy					252			252	277	
	3795	Rent per month					30			30	33	
	3800	Rolling ladders with handrails, 30" wide, buy, 2 step					151			151	166	
	4000	7 step					510			510	560	
	4050	10 step					665			665	735	
	4100	Rolling towers, buy, 5' wide, 7' long, 10' high					1,150			1,150	1,250	
	4200	For 5' high added sections, to buy, add					190			190	209	
	4300	Complete incl. wheels, railings, outriggers,										
	4350	21' high, to buy				Ea.	1,925			1,925	2,125	
	4400	Rent/month				"	145			145	159	
800	0010	**TARPAULINS** Cotton duck, 10 oz. to 13.13 oz. per S.Y., minimum				S.F.	.48			.48	.53	800
	0050	Maximum					.57			.57	.63	
	0100	Polyvinyl coated nylon, 14 oz. to 18 oz., minimum					.47			.47	.52	
	0150	Maximum					.67			.67	.74	
	0200	Reinforced polyethylene 3 mils thick, white					.11			.11	.12	
	0300	4 mils thick, white, clear or black					.14			.14	.15	
	0400	5.5 mils thick, clear					.19			.19	.21	
	0500	White, fire retardant					.18			.18	.20	
	0600	7.5 mils, oil resistant, fire retardant					.19			.19	.21	
	0700	8.5 mils, black					.24			.24	.26	
	0710	Woven polyethylene, 6 mils thick					.47			.47	.52	
	0730	Polyester reinforced w/ integral fastening system 11 mils thick					1.05			1.05	1.16	
	0740	Mylar polyester, non-reinforced, 7 mils thick					1.15			1.15	1.27	
820	0010	**SMALL TOOLS** As % of contractor's work, minimum				Total					.50%	820
	0100	Maximum				"					2%	

01550 | Vehicular Access & Parking

		CREW	DAILY OUTPUT	LABOR-HOURS	UNIT	MAT.	LABOR	EQUIP.	TOTAL	TOTAL INCL O&P	
700	0010	**ROADS AND SIDEWALKS** Temporary										700
	0050	Roads, gravel fill, no surfacing, 4" gravel depth	B-14	715	.067	S.Y.	2.94	1.75	.31	5	6.30	
	0100	8" gravel depth	"	615	.078	"	5.85	2.03	.36	8.24	10	
	1000	Ramp, 3/4" plywood on 2" x 6" joists, 16" O.C.	2 Carp	300	.053	S.F.	1.19	1.68		2.87	3.93	

11

01550 | Vehicular Access & Parking

			CREW	DAILY OUTPUT	LABOR-HOURS	UNIT	2003 BARE COSTS				TOTAL INCL O&P	
							MAT.	LABOR	EQUIP.	TOTAL		
700	1100	On 2" x 10" joists, 16" O.C.	2 Carp	275	.058	S.F.	1.80	1.84		3.64	4.85	700
	2200	Sidewalks, 2" x 12" planks, 2 uses	1 Carp	350	.023		.78	.72		1.50	1.99	
	2300	Exterior plywood, 2 uses, 1/2" thick		750	.011		.25	.34		.59	.80	
	2400	5/8" thick		650	.012		.28	.39		.67	.92	
	2500	3/4" thick		600	.013		.33	.42		.75	1.03	

01560 | Barriers & Enclosures

			CREW	DAILY OUTPUT	LABOR-HOURS	UNIT	MAT.	LABOR	EQUIP.	TOTAL	TOTAL INCL O&P	
100	0010	**BARRICADES** 5' high, 3 rail @ 2" x 8", fixed	2 Carp	30	.533	L.F.	11	16.85		27.85	38.50	100
	0150	Movable		20	.800		11	25		36	51.50	
	1000	Guardrail, wooden, 3' high, 1" x 6", on 2" x 4" posts		200	.080		.93	2.52		3.45	4.96	
	1100	2" x 6", on 4" x 4" posts		165	.097		1.79	3.06		4.85	6.75	
	1200	Portable metal with base pads, buy					15.50			15.50	17.05	
	1250	Typical installation, assume 10 reuses	2 Carp	600	.027		1.60	.84		2.44	3.07	
	1300	Barricade tape, polyethelyne, 7 mil, 3" wide x 500' long roll				Ea.	25			25	27.50	
250	0010	**FENCING** Chain link, 11 ga, 5' high	2 Clab	100	.160	L.F.	3.58	3.94		7.52	10.10	250
	0100	6' high		75	.213		3.28	5.25		8.53	11.80	
	0200	Rented chain link, 6' high, to 500' (up to 12 mo.)		100	.160		2.26	3.94		6.20	8.65	
	0250	Over 1000' (up to 12 mo.)		110	.145		1.64	3.59		5.23	7.40	
	0350	Plywood, painted, 2" x 4" frame, 4' high	A-4	135	.178		4.47	5.40		9.87	13.25	
	0400	4" x 4" frame, 8' high	"	110	.218		8.70	6.65		15.35	19.80	
	0500	Wire mesh on 4" x 4" posts, 4' high	2 Carp	100	.160		6.95	5.05		12	15.50	
	0550	8' high	"	80	.200		10.35	6.30		16.65	21	
400	0010	**TEMPORARY CONSTRUCTION** See also division 01530										400

01580 | Project Signs

			CREW	DAILY OUTPUT	LABOR-HOURS	UNIT	MAT.	LABOR	EQUIP.	TOTAL	TOTAL INCL O&P	
700	0010	**SIGNS** Hi-intensity reflectorized, no posts, buy				S.F.	16.35			16.35	18	700

Important: See the Reference Section for critical supporting data - Reference Nos., Crews, & City Cost Indexes

01590 | Equipment Rental

			UNIT	HOURLY OPER. COST	RENT PER DAY	RENT PER WEEK	RENT PER MONTH	CREW EQUIPMENT COST/DAY	
100	0010	**CONCRETE EQUIPMENT RENTAL**							100
	0100	without operators							
	0150	For batch plant, see div. 01590-500							
	0200	Bucket, concrete lightweight, 1/2 C.Y.	Ea.	.50	19.65	59	177	15.80	
	0300	1 C.Y.		.55	26.50	80	240	20.40	
	0400	1-1/2 C.Y.		.70	32	96	288	24.80	
	0500	2 C.Y.		.75	39.50	118	355	29.60	
	0580	8 C.Y.		4.20	188	565	1,700	146.60	
	0600	Cart, concrete, self propelled, operator walking, 10 C.F.		2.05	58.50	175	525	51.40	
	0700	Operator riding, 18 C.F.		3.10	88.50	265	795	77.80	
	0800	Conveyer for concrete, portable, gas, 16" wide, 26' long		7.10	120	360	1,075	128.80	
	0900	46' long		7.50	147	440	1,325	148	
	1000	56' long		7.60	157	470	1,400	154.80	
	1100	Core drill, electric, 2-1/2 H.P., 1" to 8" bit diameter		1.68	65.50	197	590	52.85	
	1150	11 HP, 8" to 18" cores		5.93	76	228	685	93.05	
	1200	Finisher, concrete floor, gas, riding trowel, 48" diameter		4.30	88.50	265	795	87.40	
	1300	Gas, manual, 3 blade, 36" trowel		1.40	35.50	106	320	32.40	
	1400	4 blade, 48" trowel		1.45	39.50	119	355	35.40	
	1500	Float, hand-operated (Bull float) 48" wide		.08	13	39	117	8.45	
	1570	Curb builder, 14 H.P., gas, single screw		8.75	118	355	1,075	141	
	1590	Double screw		9.55	152	455	1,375	167.40	
	1600	Grinder, concrete and terrazzo, electric, floor		1.88	78	234	700	61.85	
	1700	Wall grinder		.94	39	117	350	30.90	
	1800	Mixer, powered, mortar and concrete, gas, 6 C.F., 18 H.P.		5.05	79.50	238	715	88	
	1900	10 C.F., 25 H.P.		6.25	86.50	259	775	101.80	
	2000	16 C.F.		6.55	110	330	990	118.40	
	2100	Concrete, stationary, tilt drum, 2 C.Y.		5	207	620	1,850	164	
	2120	Pump, concrete, truck mounted 4" line 80' boom		21.15	945	2,840	8,525	737.20	
	2140	5" line, 110' boom		28.60	1,325	3,950	11,900	1,019	
	2160	Mud jack, 50 C.F. per hr.		5.52	127	380	1,150	120.15	
	2180	225 C.F. per hr.		5.37	225	675	2,025	177.95	
	2190	Shotcrete pump rig, 12 CY/hr		10.65	325	980	2,950	281.20	
	2600	Saw, concrete, manual, gas, 18 H.P.		3.45	50	150	450	57.60	
	2650	Self-propelled, gas, 30 H.P.		6.70	91	273	820	108.20	
	2700	Vibrators, concrete, electric, 60 cycle, 2 H.P.		.39	19.35	58	174	14.70	
	2800	3 H.P.		.58	27.50	83	249	21.25	
	2900	Gas engine, 5 H.P.		.95	29.50	89	267	25.40	
	3000	8 H.P.		1.35	36.50	109	325	32.60	
	3050	Vibrating screed, gas engine, 8HP		1.77	54	162	485	46.55	
	3100	Concrete transit mixer, hydraulic drive							
	3120	6 x 4, 250 H.P., 8 C.Y., rear discharge		35.95	650	1,950	5,850	677.60	
	3200	Front discharge		40	705	2,120	6,350	744	
	3300	6 x 6, 285 H.P., 12 C.Y., rear discharge		38.90	660	1,980	5,950	707.20	
	3400	Front discharge		40.85	715	2,150	6,450	756.80	
200	0010	**EARTHWORK EQUIPMENT RENTAL** Without operators	R01590-100						200
	0040	Aggregate spreader, push type 8' to 12' wide	Ea.	1.55	66	198	595	52	
	0045	Tailgate type, 8' wide	"	1.50	32.50	97	291	31.40	
	0050	Augers for truck or trailer mounting, vertical drilling							
	0055	Fence post auger, truck mounted	Ea.	6.90	510	1,530	4,600	361.20	
	0060	4" to 36" diam., 54 H.P., gas, 10' spindle travel		28.05	655	1,960	5,875	616.40	
	0070	14' spindle travel		31.95	810	2,430	7,300	741.60	
	0075	Auger, truck mounted, vertical drilling, to 25' depth		118.15	2,250	6,750	20,300	2,295	
	0080	Auger, horizontal boring machine, 12" to 36" diameter, 45 H.P.		15.10	195	585	1,750	237.80	
	0090	12" to 48" diameter, 65 H.P.		21.45	485	1,450	4,350	461.60	
	0100	Excavator, diesel hydraulic, crawler mounted, 1/2 C.Y. cap.		15.20	365	1,100	3,300	341.60	
	0120	5/8 C.Y. capacity		18.50	485	1,460	4,375	440	
	0140	3/4 C.Y. capacity		21.50	520	1,560	4,675	484	
	0150	1 C.Y. capacity		23.95	555	1,660	4,975	523.60	

01590	Equipment Rental		UNIT	HOURLY OPER. COST	RENT PER DAY	RENT PER WEEK	RENT PER MONTH	CREW EQUIPMENT COST/DAY	
200	0200	1-1/2 C.Y. capacity	R01590 -100	Ea.	31.35	795	2,390	7,175	728.80	**200**
	0300	2 C.Y. capacity			41.15	1,025	3,110	9,325	951.20	
	0320	2-1/2 C.Y. capacity			53.70	1,700	5,100	15,300	1,450	
	0340	3-1/2 C.Y. capacity			88.80	2,175	6,540	19,600	2,018	
	0341	Attachments								
	0342	Bucket thumbs			2.35	205	615	1,850	141.80	
	0345	Grapples			1	202	606	1,825	129.20	
	0350	Gradall type, truck mounted, 3 ton @ 15' radius, 5/8 C.Y.			36.80	920	2,760	8,275	846.40	
	0370	1 C.Y. capacity			42.60	1,075	3,240	9,725	988.80	
	0400	Backhoe-loader, 40 to 45 H.P., 5/8 C.Y. capacity			8.20	188	565	1,700	178.60	
	0450	45 H.P. to 60 H.P., 3/4 C.Y. capacity			9.85	233	700	2,100	218.80	
	0460	80 H.P., 1-1/4 C.Y. capacity			11.90	248	745	2,225	244.20	
	0470	112 H.P., 1-1/2 C.Y. capacity			18.10	425	1,280	3,850	400.80	
	0480	Attachments								
	0482	Compactor, 20,000 lb			4	115	345	1,025	101	
	0485	Hydraulic hammer, 750 ft-lbs			1.85	135	406	1,225	96	
	0486	Hydraulic hammer, 1200 ft-lbs			3.85	190	570	1,700	144.80	
	0500	Brush chipper, gas engine, 6" cutter head, 35 H.P.			5.75	118	355	1,075	117	
	0550	12" cutter head, 130 H.P.			9.10	180	540	1,625	180.80	
	0600	15" cutter head, 165 H.P.			13.05	193	580	1,750	220.40	
	0750	Bucket, clamshell, general purpose, 3/8 C.Y.			.95	44.50	133	400	34.20	
	0800	1/2 C.Y.			1.05	51	153	460	39	
	0850	3/4 C.Y.			1.20	53.50	160	480	41.60	
	0900	1 C.Y.			1.25	73.50	220	660	54	
	0950	1-1/2 C.Y.			1.95	96	288	865	73.20	
	1000	2 C.Y.			2.05	113	340	1,025	84.40	
	1010	Bucket, dragline, medium duty, 1/2 C.Y.			.55	23	69	207	18.20	
	1020	3/4 C.Y.			.55	24.50	73	219	19	
	1030	1 C.Y.			.60	26	78	234	20.40	
	1040	1-1/2 C.Y.			.90	40	120	360	31.20	
	1050	2 C.Y.			1	45	135	405	35	
	1070	3 C.Y.			1.45	76.50	230	690	57.60	
	1200	Compactor, roller, 2 drum, 2000 lb., operator walking			6.10	48.50	145	435	77.80	
	1250	Rammer compactor, gas, 1000 lb. blow			1.50	38.50	115	345	35	
	1300	Vibratory plate, gas, 13" plate, 1000 lb. blow			1.40	33.50	100	300	31.20	
	1350	24" plate, 5000 lb. blow			2.25	58.50	176	530	53.20	
	1370	Curb builder/extruder, 14 H.P., gas, single screw			8.75	118	355	1,075	141	
	1390	Double screw			9.55	152	455	1,375	167.40	
	1500	Disc harrow attachment, for tractor		▼	.35	58	174	520	37.60	
	1750	Extractor, piling, see lines 2500 to 2750								
	1810	Feller buncher, shearing & accumulating trees, 100 H.P.		Ea.	19.65	455	1,370	4,100	431.20	
	1860	Grader, self-propelled, 25,000 lb.			17.95	425	1,280	3,850	399.60	
	1910	30,000 lb.			20.15	495	1,480	4,450	457.20	
	1920	40,000 lb.			30.60	745	2,240	6,725	692.80	
	1930	55,000 lb.			40.75	1,050	3,150	9,450	956	
	1950	Hammer, pavement demo., hyd., gas, self-prop., 1000 to 1250 lb.			18.20	405	1,210	3,625	387.60	
	2000	Diesel 1300 to 1500 lb.			25.40	625	1,870	5,600	577.20	
	2050	Pile driving hammer, steam or air, 4150 ft.-lb. @ 225 BPM			6.15	278	835	2,500	216.20	
	2100	8750 ft.-lb. @ 145 BPM			8.05	465	1,390	4,175	342.40	
	2150	15,000 ft.-lb. @ 60 BPM			8.40	500	1,500	4,500	367.20	
	2200	24,450 ft.-lb. @ 111 BPM		▼	11	545	1,630	4,900	414	
	2250	Leads, 15,000 ft.-lb. hammers		L.F.	.03	2.51	7.52	22.50	1.75	
	2300	24,450 ft.-lb. hammers and heavier		"	.05	3.33	10	30	2.40	
	2350	Diesel type hammer, 22,400 ft.-lb.		Ea.	23.15	650	1,950	5,850	575.20	
	2400	41,300 ft.-lb.			30.15	680	2,040	6,125	649.20	
	2450	141,000 ft.-lb.			56.65	1,500	4,510	13,500	1,355	
	2500	Vib. elec. hammer/extractor, 200 KW diesel generator, 34 H.P.			24.90	675	2,030	6,100	605.20	
	2550	80 H.P.		▼	42.30	1,025	3,040	9,125	946.40	

Important: See the Reference Section for critical supporting data - Reference Nos., Crews, & City Cost Indexes

01590 | Equipment Rental

			UNIT	HOURLY OPER. COST	RENT PER DAY	RENT PER WEEK	RENT PER MONTH	CREW EQUIPMENT COST/DAY	
200	2600	150 H.P.	Ea.	60.50	1,525	4,570	13,700	1,398	200
	2700	Extractor, steam or air, 700 ft.-lb.		14.20	283	850	2,550	283.60	
	2750	1000 ft.-lb.		16.30	385	1,150	3,450	360.40	
	2800	Log chipper, up to 22" diam, 600 H.P.		25.79	1,825	5,440	16,300	1,294	
	2850	Logger, for skidding & stacking logs, 150 H.P.		32.20	745	2,240	6,725	705.60	
	2900	Rake, spring tooth, with tractor		7.84	219	658	1,975	194.30	
	3000	Roller, tandem, gas, 3 to 5 ton		4.75	115	345	1,025	107	
	3050	Diesel, 8 to 12 ton		6.80	215	645	1,925	183.40	
	3100	Towed type, vibratory, gas 12.5 H.P., 2 ton		6.30	255	765	2,300	203.40	
	3150	Sheepsfoot, double 60" x 60"		2.30	103	310	930	80.40	
	3170	Landfill compactor, 220 HP		42	1,125	3,380	10,100	1,012	
	3200	Pneumatic tire diesel roller, 12 ton		8	305	915	2,750	247	
	3250	21 to 25 ton		13.15	570	1,710	5,125	447.20	
	3300	Sheepsfoot roller, self-propelled, 4 wheel, 130 H.P.		31.85	845	2,540	7,625	762.80	
	3320	300 H.P.		43.95	1,100	3,300	9,900	1,012	
	3350	Vibratory steel drum & pneumatic tire, diesel, 18,000 lb.		16.10	340	1,020	3,050	332.80	
	3400	29,000 lb.		24.20	425	1,270	3,800	447.60	
	3410	Rotary mower, brush, 60", with tractor		10.95	235	705	2,125	228.60	
	3450	Scrapers, towed type, 9 to 12 C.Y. capacity		3.82	182	545	1,625	139.55	
	3500	12 to 17 C.Y. capacity		1.51	242	726	2,175	157.30	
	3550	Scrapers, self-propelled, 4 x 4 drive, 2 engine, 14 C.Y. capacity		83.10	1,450	4,360	13,100	1,537	
	3600	2 engine, 24 C.Y. capacity		123.50	2,325	6,950	20,900	2,378	
	3640	32 - 44 C.Y. capacity		153.05	2,775	8,360	25,100	2,896	
	3650	Self-loading, 11 C.Y. capacity		42.30	830	2,490	7,475	836.40	
	3700	22 C.Y. capacity		80.75	1,675	5,060	15,200	1,658	
	3710	Screening plant 110 hp. w / 5' x 10'screen		20	370	1,115	3,350	383	
	3720	5' x 16' screen		22.25	475	1,420	4,250	462	
	3850	Shovels, see Cranes division 01590-600							
	3860	Shovel/backhoe bucket, 1/2 C.Y.	Ea.	1.65	53.50	160	480	45.20	
	3870	3/4 C.Y.		1.75	61.50	185	555	51	
	3880	1 C.Y.		1.85	71.50	215	645	57.80	
	3890	1-1/2 C.Y.		1.95	128	385	1,150	92.60	
	3910	3 C.Y.		2.30	215	645	1,925	147.40	
	3950	Stump chipper, 18" deep, 30 H.P.		3.06	133	400	1,200	104.50	
	4110	Tractor, crawler, with bulldozer, torque converter, diesel 75 H.P.		16.55	340	1,020	3,050	336.40	
	4150	105 H.P.		20.55	455	1,370	4,100	438.40	
	4200	140 H.P.		25.40	575	1,720	5,150	547.20	
	4260	200 H.P.		38.10	960	2,880	8,650	880.80	
	4310	300 H.P.		49.45	1,225	3,670	11,000	1,130	
	4360	410 H.P.		69.05	1,600	4,830	14,500	1,518	
	4370	500 H.P.		90	2,100	6,320	19,000	1,984	
	4380	700 H.P.		138.70	3,475	10,400	31,200	3,190	
	4400	Loader, crawler, torque conv., diesel, 1-1/2 C.Y., 80 H.P.		14.15	315	945	2,825	302.20	
	4450	1-1/2 to 1-3/4 C.Y., 95 H.P.		16.65	385	1,160	3,475	365.20	
	4510	1-3/4 to 2-1/4 C.Y., 130 H.P.		22.75	620	1,860	5,575	554	
	4530	2-1/2 to 3-1/4 C.Y., 190 H.P.		34.05	845	2,530	7,600	778.40	
	4560	3-1/2 to 5 C.Y., 275 H.P.		45.55	1,225	3,640	10,900	1,092	
	4610	Tractor loader, wheel, torque conv., 4 x 4, 1 to 1-1/4 C.Y., 65 H.P.		9.90	205	615	1,850	202.20	
	4620	1-1/2 to 1-3/4 C.Y., 80 H.P.		12.45	255	765	2,300	252.60	
	4650	1-3/4 to 2 C.Y., 100 H.P.		13.40	283	850	2,550	277.20	
	4710	2-1/2 to 3-1/2 C.Y., 130 H.P.		15	320	965	2,900	313	
	4730	3 to 4-1/2 C.Y., 170 H.P.		21.30	490	1,470	4,400	464.40	
	4760	5-1/4 to 5-3/4 C.Y., 270 H.P.		35.45	725	2,170	6,500	717.60	
	4810	7 to 8 C.Y., 375 H.P.		59.15	1,275	3,790	11,400	1,231	
	4870	12-1/2 C.Y., 690 H.P.		89.45	2,075	6,200	18,600	1,956	
	4880	Wheeled, skid steer, 10 C.F., 30 H.P. gas		8.50	142	425	1,275	153	
	4890	1 C.Y., 78 H.P., diesel		9.85	193	580	1,750	194.80	
	4891	Attachments for all skid steer loaders							

R01590 -100

GENERAL REQUIREMENTS | **1**

01590 | Equipment Rental		UNIT	HOURLY OPER. COST	RENT PER DAY	RENT PER WEEK	RENT PER MONTH	CREW EQUIPMENT COST/DAY	
200	4892	Auger	Ea.	.45	75.50	226	680	48.80	**200**
	4893	Backhoe		.65	108	325	975	70.20	
	4894	Broom		.66	110	329	985	71.10	
	4895	Forks		.22	36	108	325	23.35	
	4896	Grapple		.47	78.50	235	705	50.75	
	4897	Concrete hammer		1.02	170	511	1,525	110.35	
	4898	Tree spade		.81	134	403	1,200	87.10	
	4899	Trencher		.75	125	374	1,125	80.80	
	4900	Trencher, chain, boom type, gas, operator walking, 12 H.P.		2.85	94	282	845	79.20	
	4910	Operator riding, 40 H.P.		8.25	242	725	2,175	211	
	5000	Wheel type, diesel, 4' deep, 12" wide		31.70	720	2,160	6,475	685.60	
	5100	Diesel, 6' deep, 20" wide		55.40	1,075	3,210	9,625	1,085	
	5150	Ladder type, diesel, 5' deep, 8" wide		22.35	585	1,750	5,250	528.80	
	5200	Diesel, 8' deep, 16" wide		83.10	1,350	4,020	12,100	1,469	
	5210	Tree spade, self-propelled		9.60	267	800	2,400	236.80	
	5250	Truck, dump, tandem, 12 ton payload		18.25	267	800	2,400	306	
	5300	Three axle dump, 16 ton payload		25.20	415	1,240	3,725	449.60	
	5350	Dump trailer only, rear dump, 16-1/2 C.Y.		4.70	115	345	1,025	106.60	
	5400	20 C.Y.		5.20	130	390	1,175	119.60	
	5450	Flatbed, single axle, 1-1/2 ton rating		8.95	81	243	730	120.20	
	5500	3 ton rating		11.50	76.50	230	690	138	
	5550	Off highway rear dump, 25 ton capacity		39.75	995	2,980	8,950	914	
	5600	35 ton capacity		40.80	1,025	3,070	9,200	940.40	
	5610	50 ton capacity		51.20	1,275	3,815	11,400	1,173	
	5620	65 ton capacity		54.40	1,375	4,110	12,300	1,257	
	5630	100 ton capacity		77.70	1,975	5,950	17,900	1,812	
	6000	Vibratory plow, 25 H.P., walking		4.80	84.50	253	760	89	
400	0010	**GENERAL EQUIPMENT RENTAL** Without operators							**400**
	0150	Aerial lift, scissor type, to 15' high, 1000 lb. cap., electric	Ea.	2.10	58.50	175	525	51.80	
	0160	To 25' high, 2000 lb. capacity		2.45	86	258	775	71.20	
	0170	Telescoping boom to 40' high, 500 lb. capacity, gas		9.45	273	820	2,450	239.60	
	0180	To 45' high, 500 lb. capacity		10.25	320	960	2,875	274	
	0190	To 60' high, 600 lb. capacity		21.10	425	1,270	3,800	422.80	
	0195	Air compressor, portable, 6.5 CFM, electric		.43	24	72	216	17.85	
	0196	gasoline		.53	36	108	325	25.85	
	0200	Air compressor, portable, gas engine, 60 C.F.M.		3.95	41.50	125	375	56.60	
	0300	160 C.F.M.		6.80	53.50	160	480	86.40	
	0400	Diesel engine, rotary screw, 250 C.F.M.		8.45	102	305	915	128.60	
	0500	365 C.F.M.		9.70	117	350	1,050	147.60	
	0550	450 C.F.M.		12.10	135	405	1,225	177.80	
	0600	600 C.F.M.		16.65	193	580	1,750	249.20	
	0700	750 C.F.M.		19.95	203	610	1,825	281.60	
	0800	For silenced models, small sizes, add		3%	5%	5%	5%		
	0900	Large sizes, add		5%	7%	7%	7%		
	0920	Air tools and accessories							
	0930	Breaker, pavement, 60 lb.	Ea.	.35	19.65	59	177	14.60	
	0940	80 lb.		.35	25.50	76	228	18	
	0950	Drills, hand (jackhammer) 65 lb.		.40	18.35	55	165	14.20	
	0960	Track or wagon, swing boom, 4" drifter		33.45	540	1,620	4,850	591.60	
	0970	5" drifter		45.30	760	2,280	6,850	818.40	
	0975	Track mounted quarry drill, 6" diameter drill		47.95	820	2,460	7,375	875.60	
	0980	Dust control per drill		.79	12.35	37	111	13.70	
	0990	Hammer, chipping, 12 lb.		.40	22	66	198	16.40	
	1000	Hose, air with couplings, 50' long, 3/4" diameter		.03	5.35	16	48	3.45	
	1100	1" diameter		.03	5.65	17	51	3.65	
	1200	1-1/2" diameter		.05	7.65	23	69	5	
	1300	2" diameter		.10	16.35	49	147	10.60	

R01590 -100

Important: See the Reference Section for critical supporting data - Reference Nos., Crews, & City Cost Indexes

GENERAL REQUIREMENTS

1

01590 | Equipment Rental

			UNIT	HOURLY OPER. COST	RENT PER DAY	RENT PER WEEK	RENT PER MONTH	CREW EQUIPMENT COST/DAY	
400	1400	2-1/2" diameter	Ea.	.12	20	60	180	12.95	400
	1410	3" diameter		.16	27.50	82	246	17.70	
	1450	Drill, steel, 7/8" x 2'		.05	6	18	54	4	
	1460	7/8" x 6'		.05	7	21	63	4.60	
	1520	Moil points		.02	4	12	36	2.55	
	1525	Pneumatic nailer w/accessories		.40	26.50	80	240	19.20	
	1530	Sheeting driver for 60 lb. breaker		.10	8.65	26	78	6	
	1540	For 90 lb. breaker		.15	12.65	38	114	8.80	
	1550	Spade, 25 lb.		.30	5.65	17	51	5.80	
	1560	Tamper, single, 35 lb.		.44	29.50	88	264	21.10	
	1570	Triple, 140 lb.		.66	44	132	395	31.70	
	1580	Wrenches, impact, air powered, up to 3/4" bolt		.20	14.35	43	129	10.20	
	1590	Up to 1-1/4" bolt		.30	26.50	80	240	18.40	
	1600	Barricades, barrels, reflectorized, 1 to 50 barrels		.02	3.33	10	30	2.15	
	1610	100 to 200 barrels		.02	2.53	7.60	23	1.70	
	1620	Barrels with flashers, 1 to 50 barrels		.02	4	12	36	2.55	
	1630	100 to 200 barrels		.02	3.20	9.60	29	2.10	
	1640	Barrels with steady burn type C lights		.03	5.35	16	48	3.45	
	1650	Illuminated board, trailer mounted, with generator		.60	120	360	1,075	76.80	
	1670	Portable barricade, stock, with flashers, 1 to 6 units		.02	4	12	36	2.55	
	1680	25 to 50 units		.02	3.73	11.20	33.50	2.40	
	1690	Butt fusion machine, electric		21.05	430	1,290	3,875	426.40	
	1695	Electro fusion machine		8.25	173	520	1,550	170	
	1700	Carts, brick, hand powered, 1000 lb. capacity		.26	43	129	385	27.90	
	1800	Gas engine, 1500 lb., 7-1/2' lift		3.15	112	336	1,000	92.40	
	1822	Dehumidifier, medium, 6 Lb/Hr, 150 CFM		.68	41.50	124	370	30.25	
	1824	Large, 18 Lb/Hr, 600 CFM		1.36	82.50	248	745	60.50	
	1830	Distributor, asphalt, trailer mtd, 2000 gal., 38 H.P. diesel		6.70	330	985	2,950	250.60	
	1840	3000 gal., 38 H.P. diesel		7.75	350	1,050	3,150	272	
	1850	Drill, rotary hammer, electric, 1-1/2" diameter		.40	24.50	74	222	18	
	1860	Carbide bit for above		.03	5.35	16	48	3.45	
	1865	Rotary, crawler, 250 HP		80.35	1,700	5,110	15,300	1,665	
	1870	Emulsion sprayer, 65 gal., 5 H.P. gas engine		1.49	58	174	520	46.70	
	1880	200 gal., 5 H.P. engine		4.25	86.50	260	780	86	
	1900	Fencing, see division 01560-250 & 02820-000							
	1920	Floodlight, mercury vapor, or quartz, on tripod							
	1930	1000 watt	Ea.	.29	15.65	47	141	11.70	
	1940	2000 watt		.51	28	84	252	20.90	
	1950	Floodlights, trailer mounted with generator, 1 - 300 watt light		2.35	65	195	585	57.80	
	1960	2 - 1000 watt lights		3.25	110	330	990	92	
	2000	4 - 300 watt lights		2.70	76.50	230	690	67.60	
	2020	Forklift, wheeled, for brick, 18', 3000 lb., 2 wheel drive, gas		14.35	190	570	1,700	228.80	
	2040	28', 4000 lb., 4 wheel drive, diesel		11.30	252	755	2,275	241.40	
	2050	For rough terrain, 8000 lb., 16' lift, 68 HP		15.25	375	1,130	3,400	348	
	2060	For plant, 4 T. capacity, 80 H.P., 2 wheel drive, gas		7.80	127	380	1,150	138.40	
	2080	10 T. capacity, 120 H.P., 2 wheel drive, diesel		11.90	228	685	2,050	232.20	
	2100	Generator, electric, gas engine, 1.5 KW to 3 KW		1.35	27.50	82	246	27.20	
	2200	5 KW		1.85	41	123	370	39.40	
	2300	10 KW		3.25	90.50	272	815	80.40	
	2400	25 KW		5.80	113	340	1,025	114.40	
	2500	Diesel engine, 20 KW		4.95	75	225	675	84.60	
	2600	50 KW		9.35	91.50	275	825	129.80	
	2700	100 KW		13.50	118	355	1,075	179	
	2800	250 KW		38	218	655	1,975	435	
	2850	Hammer, hydraulic, for mounting on boom, to 500 ft.-lb.		1.70	63.50	190	570	51.60	
	2860	1000 ft.-lb.		3	102	305	915	85	
	2900	Heaters, space, oil or electric, 50 MBH		.93	17.35	52	156	17.85	
	3000	100 MBH		1.66	23	69	207	27.10	

01590 | Equipment Rental

		UNIT	HOURLY OPER. COST	RENT PER DAY	RENT PER WEEK	RENT PER MONTH	CREW EQUIPMENT COST/DAY	
400	3100	300 MBH	Ea.	5.33	35	105	315	63.65	**400**
	3150	500 MBH		10.71	52	156	470	116.90	
	3200	Hose, water, suction with coupling, 20' long, 2" diameter		.02	7	21	63	4.35	
	3210	3" diameter		.03	10.65	32	96	6.65	
	3220	4" diameter		.04	14.35	43	129	8.90	
	3230	6" diameter		.09	27.50	82	246	17.10	
	3240	8" diameter		.31	51.50	154	460	33.30	
	3250	Discharge hose with coupling, 50' long, 2" diameter		.01	6	18	54	3.70	
	3260	3" diameter		.02	7.35	22	66	4.55	
	3270	4" diameter		.03	10	30	90	6.25	
	3280	6" diameter		.06	23.50	71	213	14.70	
	3290	8" diameter		.34	56.50	169	505	36.50	
	3300	Ladders, extension type, 16' to 36' long		.16	26	78	234	16.90	
	3400	40' to 60' long		.19	31	93	279	20.10	
	3405	Lance for cutting concrete		2.84	109	327	980	88.10	
	3407	Lawn mower, rotary, 22", 5HP		.98	27.50	83	249	24.45	
	3408	48" self propelled		2.44	85.50	257	770	70.90	
	3410	Level, laser type, for pipe laying, self leveling		1.36	90.50	271	815	65.10	
	3430	Manual leveling		.73	48.50	145	435	34.85	
	3440	Rotary beacon with rod and sensor		.93	62	186	560	44.65	
	3460	Builders level with tripod and rod		.09	18.65	56	168	11.90	
	3500	Light towers, towable, with diesel generator, 2000 watt		2.70	76.50	230	690	67.60	
	3600	4000 watt		3.25	110	330	990	92	
	3700	Mixer, powered, plaster and mortar, 6 C.F., 7 H.P.		1.30	46.50	139	415	38.20	
	3800	10 C.F., 9 H.P.		1.55	66	198	595	52	
	3850	Nailer, pneumatic		.40	26.50	80	240	19.20	
	3900	Paint sprayers complete, 8 CFM		.64	42.50	127	380	30.50	
	4000	17 CFM		.99	65.50	197	590	47.30	
	4020	Pavers, bituminous, rubber tires, 8' wide, 52 H.P., gas		24.80	760	2,280	6,850	654.40	
	4030	8' wide, 64 H.P., diesel		37.55	1,125	3,390	10,200	978.40	
	4050	Crawler, 10' wide, 78 H.P., gas		43.80	1,325	4,010	12,000	1,152	
	4060	10' wide, 87 H.P., diesel		56.60	1,675	5,060	15,200	1,465	
	4070	Concrete paver, 12' to 24' wide, 250 H.P.		55.20	1,275	3,830	11,500	1,208	
	4080	Placer-spreader-trimmer, 24' wide, 300 H.P.		68.15	1,825	5,510	16,500	1,647	
	4100	Pump, centrifugal gas pump, 1-1/2", 4 MGPH		2.35	35	105	315	39.80	
	4200	2", 8 MGPH		3	43.50	130	390	50	
	4300	3", 15 MGPH		3.20	43.50	130	390	51.60	
	4400	6", 90 MGPH		15.05	173	520	1,550	224.40	
	4500	Submersible electric pump, 1-1/4", 55 GPM		.35	25	75	225	17.80	
	4600	1-1/2", 83 GPM		.41	28.50	85	255	20.30	
	4700	2", 120 GPM		.57	35.50	106	320	25.75	
	4800	3", 300 GPM		.95	44.50	133	400	34.20	
	4900	4", 560 GPM		6.21	110	330	990	115.70	
	5000	6", 1590 GPM		9	182	545	1,625	181	
	5100	Diaphragm pump, gas, single, 1-1/2" diameter		.72	34.50	104	310	26.55	
	5200	2" diameter		2.30	43.50	130	390	44.40	
	5300	3" diameter		2.35	45	135	405	45.80	
	5400	Double, 4" diameter		4.15	86.50	260	780	85.20	
	5500	Trash pump, self-priming, gas, 2" diameter		2.75	33.50	100	300	42	
	5600	Diesel, 4" diameter		4.35	78.50	235	705	81.80	
	5650	Diesel, 6" diameter		9.15	127	380	1,150	149.20	
	5655	Grout Pump		4.70	28	84	252	54.40	
	5660	Rollers, see division 01590-200							
	5700	Salamanders, L.P. gas fired, 100,000 B.T.U.	Ea.	1.66	14	42	126	21.70	
	5705	50,000 BTU		1.25	9.65	29	87	15.80	
	5720	Sandblaster, portable, open top, 3 C.F. capacity		.40	32	96	288	22.40	
	5730	6 C.F. capacity		.65	46	138	415	32.80	
	5740	Accessories for above		.11	17.65	53	159	11.50	

Important: See the Reference Section for critical supporting data - Reference Nos., Crews, & City Cost Indexes

01590 | Equipment Rental

		UNIT	HOURLY OPER. COST	RENT PER DAY	RENT PER WEEK	RENT PER MONTH	CREW EQUIPMENT COST/DAY	
400								**400**
5750	Sander, floor	Ea.	.72	25	75	225	20.75	
5760	Edger		.51	18.65	56	168	15.30	
5800	Saw, chain, gas engine, 18" long		.35	31.50	94	282	21.60	
5900	36" long		.50	56.50	170	510	38	
5950	60" long		.50	61.50	184	550	40.80	
6000	Masonry, table mounted, 14" diameter, 5 H.P.		1.45	63.50	190	570	49.60	
6050	Portable cut-off, 8 H.P.		1.15	35.50	107	320	30.60	
6100	Circular, hand held, electric, 7-1/4" diameter	Ea.	.18	23.50	70	210	15.45	
6200	12" diameter		.24	20	60	180	13.90	
6250	Wall saw, w/hydraulic power, 10 H.P		1.38	117	350	1,050	81.05	
6275	Shot blaster, walk behind, 20" wide		1.05	330	990	2,975	206.40	
6300	Steam cleaner, 100 gallons per hour		2.10	63.50	190	570	54.80	
6310	200 gallons per hour		2.75	78.50	235	705	69	
6340	Tar Kettle/Pot, 400 gallon		2.69	51.50	155	465	52.50	
6350	Torch, cutting, acetylene-oxygen, 150' hose		1.50	20	60	180	24	
6360	Hourly operating cost includes tips and gas		4.50				36	
6410	Toilet, portable chemical		.10	17	51	153	11	
6420	Recycle flush type		.13	21	63	189	13.65	
6430	Toilet, fresh water flush, garden hose,		.14	23.50	71	213	15.30	
6440	Hoisted, non-flush, for high rise		.12	20.50	62	186	13.35	
6450	Toilet, trailers, minimum		.21	35.50	106	320	22.90	
6460	Maximum		.64	106	318	955	68.70	
6465	Tractor, farm with attachment		9.90	228	685	2,050	216.20	
6470	Trailer, office, see division 01520-500							
6500	Trailers, platform, flush deck, 2 axle, 25 ton capacity	Ea.	4.50	90	270	810	90	
6600	40 ton capacity		6	158	475	1,425	143	
6700	3 axle, 50 ton capacity		6.55	175	525	1,575	157.40	
6800	75 ton capacity		8.30	230	690	2,075	204.40	
6810	Trailer mounted cable reel for H.V. line work		4.30	205	614	1,850	157.20	
6820	Trailer mounted cable tensioning rig		8.47	405	1,210	3,625	309.75	
6830	Cable pulling rig		53.48	2,275	6,840	20,500	1,796	
6850	Trailer, storage, see division 01520-500							
6900	Water tank, engine driven discharge, 5000 gallons	Ea.	6.10	160	480	1,450	144.80	
6925	10,000 gallons		8.55	228	685	2,050	205.40	
6950	Water truck, off highway, 6000 gallons		46.95	725	2,170	6,500	809.60	
7010	Tram car for H.V. line work, powered, 2 conductor		5.63	111	333	1,000	111.65	
7020	Transit (builder's level) with tripod		.09	18.35	55	165	11.70	
7030	Trench box, 3000 lbs. 6'x8'		.42	70	210	630	45.35	
7040	7200 lbs. 6'x20'		1.03	171	514	1,550	111.05	
7050	8000 lbs., 8' x 16'		.85	142	426	1,275	92	
7060	9500 lbs., 8'x20'		1.46	243	730	2,200	157.70	
7065	11,000 lbs., 8'x24'		1.34	224	671	2,025	144.90	
7070	12,000 lbs., 10' x 20'		2.55	385	1,150	3,450	250.40	
7100	Truck, pickup, 3/4 ton, 2 wheel drive		4.35	55	165	495	67.80	
7200	4 wheel drive		4.50	63.50	190	570	74	
7250	Crew carrier, 9 passenger		3.66	87	261	785	81.50	
7290	Tool van, 24,000 G.V.W.		5.79	112	337	1,000	113.70	
7300	Tractor, 4 x 2, 30 ton capacity, 195 H.P.		12.45	235	705	2,125	240.60	
7410	250 H.P.		17.20	325	970	2,900	331.60	
7500	6 x 2, 40 ton capacity, 240 H.P.		16.10	335	1,010	3,025	330.80	
7600	6 x 4, 45 ton capacity, 240 H.P.		19.25	370	1,110	3,325	376	
7620	Vacuum truck, hazardous material, 2500 gallon		6.60	305	919	2,750	236.60	
7625	5,000 gallon		7.16	405	1,220	3,650	301.30	
7640	Tractor, with A frame, boom and winch, 225 H.P.		13.15	230	690	2,075	243.20	
7650	Vacuum, H.E.P.A., 16 gal., wet/dry		.27	33.50	101	305	22.35	
7655	55 gal, wet/dry		.60	43.50	131	395	31	
7660	Water tank, portable		1	9.35	28	84	13.60	
7690	Large production vacuum loader, 3150 CFM		15.28	615	1,840	5,525	490.25	

01590 | Equipment Rental

			UNIT	HOURLY OPER. COST	RENT PER DAY	RENT PER WEEK	RENT PER MONTH	CREW EQUIPMENT COST/DAY	
400	7700	Welder, electric, 200 amp	Ea.	3.74	57.50	172	515	64.30	**400**
	7800	300 amp		5.24	61	183	550	78.50	
	7900	Gas engine, 200 amp		5.15	42	126	380	66.40	
	8000	300 amp		5.95	52	156	470	78.80	
	8100	Wheelbarrow, any size		.06	10.35	31	93	6.70	
	8200	Wrecking ball, 4000 lb.		1.80	68.50	205	615	55.40	
500	0010	**HIGHWAY EQUIPMENT RENTAL**							**500**
	0050	Asphalt batch plant, portable drum mixer, 100 ton/hr.	Ea.	53.05	1,350	4,060	12,200	1,236	
	0060	200 ton/hr.		58.75	1,425	4,240	12,700	1,318	
	0070	300 ton/hr.		68.50	1,675	5,040	15,100	1,556	
	0100	Backhoe attachment, long stick, up to 185 HP, 10.5' long		.29	19.35	58	174	13.90	
	0140	Up to 250 HP, 12' long		.32	21	63	189	15.15	
	0180	Over 250 HP, 15' long		.41	27	81	243	19.50	
	0200	Special dipper arm, up to 100 HP, 32' long		.85	56.50	169	505	40.60	
	0240	Over 100 HP, 33' long		1.06	70.50	212	635	50.90	
	0300	Concrete batch plant, portable, electric, 200 CY/Hr		11.80	590	1,770	5,300	448.40	
	0500	Grader attachment, ripper/scarifier, rear mounted							
	0520	Up to 135 HP	Ea.	2.70	60	180	540	57.60	
	0540	Up to 180 HP		3.20	76.50	230	690	71.60	
	0580	Up to 250 HP		3.55	88.50	265	795	81.40	
	0700	Pvmt. removal bucket, for hyd. excavator, up to 90 HP		1.30	43.50	130	390	36.40	
	0740	Up to 200 HP		1.50	65	195	585	51	
	0780	Over 200 HP		1.65	78.50	235	705	60.20	
	0900	Aggregate spreader, self-propelled, 187 HP		33.40	800	2,400	7,200	747.20	
	1000	Chemical spreader, 3 C.Y.		2	100	300	900	76	
	1900	Hammermill, traveling, 250 HP		38.35	1,650	4,920	14,800	1,291	
	2000	Horizontal borer, 3" diam, 13 HP gas driven		3.70	55	165	495	62.60	
	2200	Hydromulchers, gas power, 3000 gal., for truck mounting		8.90	245	735	2,200	218.20	
	2400	Joint & crack cleaner, walk behind, 25 HP		1.75	66	198	595	53.60	
	2500	Filler, trailer mounted, 400 gal., 20 HP		5.35	170	510	1,525	144.80	
	3000	Paint striper, self propelled, double line, 30 HP		5.10	212	635	1,900	167.80	
	3200	Post drivers, 6" I-Beam frame, for truck mounting		6.75	440	1,320	3,950	318	
	3400	Road sweeper, self propelled, 8' wide, 90 HP		21.45	345	1,030	3,100	377.60	
	4000	Road mixer, self-propelled, 130 HP		27.45	585	1,760	5,275	571.60	
	4100	310 HP		53.10	2,000	6,000	18,000	1,625	
	4200	Cold mix paver, incl pug mill and bitumen tank,							
	4220	165 HP	Ea.	63.50	1,925	5,810	17,400	1,670	
	4250	Paver, asphalt, wheel or crawler, 130 H.P., diesel		56.60	1,675	5,060	15,200	1,465	
	4300	Paver, road widener, gas 1' to 6', 67 HP		27.90	630	1,890	5,675	601.20	
	4400	Diesel, 2' to 14', 88 HP		38.20	995	2,990	8,975	903.60	
	4600	Slipform pavers, curb and gutter, 2 track, 75 HP		23.45	635	1,900	5,700	567.60	
	4700	4 track, 165 HP		32.40	1,250	3,780	11,300	1,015	
	4800	Median barrier, 215 HP		32.85	880	2,640	7,925	790.80	
	4901	Trailer, low bed, 75 ton capacity		8.95	250	750	2,250	221.60	
	5000	Road planer, walk behind, 10" cutting width, 10 HP		1.70	34	102	305	34	
	5100	Self propelled, 12" cutting width, 64 HP		4.35	380	1,140	3,425	262.80	
	5200	Pavement profiler, 4' to 6' wide, 450 HP		139.85	2,800	8,410	25,200	2,801	
	5300	8' to 10' wide, 750 HP		223.50	4,300	12,900	38,700	4,368	
	5400	Roadway plate, steel, 1"x8'x20'		.06	9.35	28	84	6.10	
	5600	Stabilizer,self-propelled, 150 HP		25.95	565	1,700	5,100	547.60	
	5700	310 HP		42.60	1,225	3,690	11,100	1,079	
	5800	Striper, thermal, truck mounted 120 gal. paint, 150H.P.		27	4.40	13.20	39.50	218.65	
	6000	Tar kettle, 330 gal., trailer mounted		2.41	38.50	115	345	42.30	
	7000	Tunnel locomotive, diesel, 8 to 12 ton		18.35	445	1,340	4,025	414.80	
	7005	Electric, 10 ton		17.50	495	1,490	4,475	438	
	7010	Muck cars, 1/2 C.Y. capacity		1.40	20	60	180	23.20	
	7020	1 C.Y. capacity		1.60	29	87	261	30.20	
	7030	2 C.Y. capacity		1.75	33.50	100	300	34	

Important: See the Reference Section for critical supporting data - Reference Nos., Crews, & City Cost Indexes

GENERAL REQUIREMENTS

1

		01590 | Equipment Rental	UNIT	HOURLY OPER. COST	RENT PER DAY	RENT PER WEEK	RENT PER MONTH	CREW EQUIPMENT COST/DAY	
500	7040	Side dump, 2 C.Y. capacity	Ea.	1.95	41.50	125	375	40.60	**500**
	7050	3 C.Y. capacity		2.60	48.50	145	435	49.80	
	7060	5 C.Y. capacity		3.65	61.50	185	555	66.20	
	7100	Ventilating blower for tunnel, 7-1/2 H.P.		1.25	34.50	103	310	30.60	
	7110	10 H.P.		1.44	36.50	109	325	33.30	
	7120	20 H.P.		2.33	39.50	119	355	42.45	
	7140	40 H.P.		4.09	61.50	185	555	69.70	
	7160	60 H.P.		6.21	91.50	274	820	104.50	
	7175	75 H.P.		7.95	123	370	1,100	137.60	
	7180	200 H.P.		17.70	160	480	1,450	237.60	
	7800	Windrow loader, elevating	▼	31.80	910	2,725	8,175	799.40	
600	0010	**LIFTING AND HOISTING EQUIPMENT RENTAL** [R01590 -100]							**600**
	0100	without operators							
	0120	Aerial lift truck, 2 person, to 80'	Ea.	16.10	605	1,820	5,450	492.80	
	0140	Boom work platform, 40' snorkel		7.90	212	635	1,900	190.20	
	0150	Crane, flatbed mntd, 3 ton cap.		12.10	178	535	1,600	203.80	
	0200	Crane, climbing, 106' jib, 6000 lb. capacity, 410 FPM		41.55	1,300	3,910	11,700	1,114	
	0300	101' jib, 10,250 lb. capacity, 270 FPM	▼	46.75	1,650	4,950	14,900	1,364	
	0400	Tower, static, 130' high, 106' jib,							
	0500	6200 lb. capacity at 400 FPM	Ea.	44.60	1,500	4,520	13,600	1,261	
	0600	Crawler mounted, lattice boom, 1/2 C.Y., 15 tons at 12' radius		4.97	380	1,136	3,400	266.95	
	0700	3/4 C.Y., 20 tons at 12' radius		27.19	595	1,790	5,375	575.50	
	0800	1 C.Y., 25 tons at 12' radius		36.25	755	2,270	6,800	744	
	0900	1-1/2 C.Y., 40 tons at 12' radius		40.20	1,025	3,070	9,200	935.60	
	1000	2 C.Y., 50 tons at 12' radius		54.45	1,300	3,930	11,800	1,222	
	1100	3 C.Y., 75 tons at 12' radius		48.15	1,350	4,030	12,100	1,191	
	1200	100 ton capacity, 60' boom		61.95	1,750	5,250	15,800	1,546	
	1300	165 ton capacity, 60' boom		86.70	2,500	7,465	22,400	2,187	
	1400	200 ton capacity, 70' boom		96.80	2,700	8,115	24,300	2,397	
	1500	350 ton capacity, 80' boom		140.95	3,875	11,620	34,900	3,452	
	1600	Truck mounted, lattice boom, 6 x 4, 20 tons at 10' radius		25.16	800	2,400	7,200	681.30	
	1700	25 tons at 10' radius		26.84	855	2,560	7,675	726.70	
	1800	8 x 4, 30 tons at 10' radius		28.52	905	2,720	8,150	772.15	
	1900	40 tons at 12' radius		30.20	960	2,880	8,650	817.60	
	2000	8 x 4, 60 tons at 15' radius		31.87	1,150	3,420	10,300	938.95	
	2050	82 tons at 15' radius		33.55	1,500	4,510	13,500	1,170	
	2100	90 tons at 15' radius		30.40	1,600	4,800	14,400	1,203	
	2200	115 tons at 15' radius		33.78	1,725	5,150	15,500	1,300	
	2300	150 tons at 18' radius		47.98	1,925	5,770	17,300	1,538	
	2350	165 tons at 18' radius		56.45	2,250	6,760	20,300	1,804	
	2400	Truck mounted, hydraulic, 12 ton capacity		33	620	1,860	5,575	636	
	2500	25 ton capacity		33.20	650	1,945	5,825	654.60	
	2550	33 ton capacity		33.85	675	2,025	6,075	675.80	
	2560	40 ton capacity		32.10	680	2,045	6,125	665.80	
	2600	55 ton capacity		47.70	1,300	3,870	11,600	1,156	
	2700	80 ton capacity		57.90	1,050	3,150	9,450	1,093	
	2720	100 ton capacity		81.75	2,125	6,375	19,100	1,929	
	2740	120 ton capacity		76.50	2,275	6,820	20,500	1,976	
	2760	150 ton capacity		85.55	2,575	7,745	23,200	2,233	
	2800	Self-propelled, 4 x 4, with telescoping boom, 5 ton		15.05	345	1,030	3,100	326.40	
	2900	12-1/2 ton capacity		22.55	520	1,565	4,700	493.40	
	3000	15 ton capacity		24.60	615	1,840	5,525	564.80	
	3050	20 ton capacity		25.30	640	1,925	5,775	587.40	
	3100	25 ton capacity		27.60	730	2,185	6,550	657.80	
	3150	40 ton capacity		49.25	1,150	3,480	10,400	1,090	
	3200	Derricks, guy, 20 ton capacity, 60' boom, 75' mast		12.20	320	957	2,875	289	
	3300	100' boom, 115' mast	▼	19.80	550	1,650	4,950	488.40	

01590 | Equipment Rental

			UNIT	HOURLY OPER. COST	RENT PER DAY	RENT PER WEEK	RENT PER MONTH	CREW EQUIPMENT COST/DAY	
600	3400	Stiffleg, 20 ton capacity, 70' boom, 37' mast	Ea.	14.11	410	1,230	3,700	358.90	600
	3500	100' boom, 47' mast		22.25	665	2,000	6,000	578	
	3550	Helicopter, small, lift to 1250 lbs. maximum, w/pilot		65.18	2,575	7,740	23,200	2,069	
	3600	Hoists, chain type, overhead, manual, 3/4 ton		.10	4	12	36	3.20	
	3900	10 ton		.56	20.50	61	183	16.70	
	4000	Hoist and tower, 5000 lb. cap., portable electric, 40' high		4.08	184	552	1,650	143.05	
	4100	For each added 10' section, add		.09	14.35	43	129	9.30	
	4200	Hoist and single tubular tower, 5000 lb. electric, 100' high		5.50	257	770	2,300	198	
	4300	For each added 6'-6" section, add		.15	24.50	73	219	15.80	
	4400	Hoist and double tubular tower, 5000 lb., 100' high		5.89	283	848	2,550	216.70	
	4500	For each added 6'-6" section, add		.16	27	81	243	17.50	
	4550	Hoist and tower, mast type, 6000 lb., 100' high		6.38	293	879	2,625	226.85	
	4570	For each added 10' section, add		.11	17.65	53	159	11.50	
	4600	Hoist and tower, personnel, electric, 2000 lb., 100' @ 125 FPM		13.02	780	2,340	7,025	572.15	
	4700	3000 lb., 100' @ 200 FPM		14.90	885	2,650	7,950	649.20	
	4800	3000 lb., 150' @ 300 FPM		16.50	990	2,970	8,900	726	
	4900	4000 lb., 100' @ 300 FPM		17.13	1,000	3,030	9,100	743.05	
	5000	6000 lb., 100' @ 275 FPM		18.54	1,050	3,180	9,550	784.30	
	5100	For added heights up to 500', add	L.F.	.01	1.67	5	15	1.10	
	5200	Jacks, hydraulic, 20 ton	Ea.	.05	10	30	90	6.40	
	5500	100 ton	"	.30	29.50	88	264	20	
	6000	Jacks, hydraulic, climbing with 50' jackrods							
	6010	and control consoles, minimum 3 mo. rental							
	6100	30 ton capacity	Ea.	1.59	106	317	950	76.10	
	6150	For each added 10' jackrod section, add		.05	3.33	10	30	2.40	
	6300	50 ton capacity		2.55	170	510	1,525	122.40	
	6350	For each added 10' jackrod section, add		.06	4	12	36	2.90	
	6500	125 ton capacity		6.65	445	1,330	4,000	319.20	
	6550	For each added 10' jackrod section, add		.46	30.50	91	273	21.90	
	6600	Cable jack, 10 ton capacity with 200' cable		1.33	88.50	265	795	63.65	
	6650	For each added 50' of cable, add		.14	9.35	28	84	6.70	
700	0010	**WELLPOINT EQUIPMENT RENTAL** See also division 02240							700
	0020	Based on 2 months rental							
	0100	Combination jetting & wellpoint pump, 60 H.P. diesel	Ea.	8.80	262	785	2,350	227.40	
	0200	High pressure gas jet pump, 200 H.P., 300 psi	"	15.70	224	671	2,025	259.80	
	0300	Discharge pipe, 8" diameter	L.F.	.01	.42	1.26	3.78	.35	
	0350	12" diameter		.01	.63	1.88	5.65	.45	
	0400	Header pipe, flows up to 150 G.P.M., 4" diameter		.01	.39	1.16	3.48	.30	
	0500	400 G.P.M., 6" diameter		.01	.45	1.36	4.08	.35	
	0600	800 G.P.M., 8" diameter		.01	.63	1.88	5.65	.45	
	0700	1500 G.P.M., 10" diameter		.01	.66	1.98	5.95	.50	
	0800	2500 G.P.M., 12" diameter		.02	1.25	3.74	11.20	.90	
	0900	4500 G.P.M., 16" diameter		.02	1.59	4.78	14.35	1.10	
	0950	For quick coupling aluminum and plastic pipe, add		.02	1.65	4.95	14.85	1.15	
	1100	Wellpoint, 25' long, with fittings & riser pipe, 1-1/2" or 2" diameter	Ea.	.05	3.29	9.88	29.50	2.40	
	1200	Wellpoint pump, diesel powered, 4" diameter, 20 H.P.		4.27	151	453	1,350	124.75	
	1300	6" diameter, 30 H.P.		5.58	187	562	1,675	157.05	
	1400	8" suction, 40 H.P.		7.59	257	770	2,300	214.70	
	1500	10" suction, 75 H.P.		10.43	300	900	2,700	263.45	
	1600	12" suction, 100 H.P.		15.58	480	1,440	4,325	412.65	
	1700	12" suction, 175 H.P.		20.76	530	1,590	4,775	484.10	
800	0010	**MARINE EQUIPMENT RENTAL**							800
	0200	Barge, 400 Ton, 30' wide x 90' long	Ea.	15.45	350	1,050	3,150	333.60	
	0240	800 Ton, 45' wide x 90' long		25.46	495	1,490	4,475	501.70	
	2000	Tugboat, diesel, 100 HP		16.95	168	505	1,525	236.60	
	2040	250 HP		32	315	950	2,850	446	
	2080	380 HP		72.70	930	2,785	8,350	1,139	

R01590 -100

Important: See the Reference Section for critical supporting data - Reference Nos., Crews, & City Cost Indexes

01700 | Execution Requirements

		01740	Cleaning	CREW	DAILY OUTPUT	LABOR-HOURS	UNIT	2003 BARE COSTS				TOTAL INCL O&P	
								MAT.	LABOR	EQUIP.	TOTAL		
500	0010		**CLEANING UP** After job completion, allow, minimum				Job					.30%	500
	0040		Maximum				"					1%	
	0050		Cleanup of floor area, continuous, per day, during const.	A-5	24	.750	M.S.F.	1.70	18.50	1.25	21.45	32.50	
	0100		Final by GC at end of job	"	11.50	1.565	"	2.65	38.50	2.61	43.76	66	
	1000		Mechanical demolition, see division 02225										

01800 | Facility Operation

		01810	Commissioning	CREW	DAILY OUTPUT	LABOR-HOURS	UNIT	2003 BARE COSTS				TOTAL INCL O&P	
								MAT.	LABOR	EQUIP.	TOTAL		
100	0010		**COMMISSIONING** Including documentation of design intent										100
	0100		performance verification, O&M, training, min				Project					.50%	
	0150		Maximum				"					.75%	

For information about Means Estimating Seminars, see yellow pages 12 and 13 in back of book

Division Notes

	CREW	DAILY OUTPUT	LABOR-HOURS	UNIT	MAT.	LABOR	EQUIP.	TOTAL	TOTAL INCL O&P

Division 2
Site Construction

Estimating Tips

02200 Site Preparation

- If possible visit the site and take an inventory of the type, quantity and size of the trees. Certain trees may have a landscape resale value or firewood value. Stump disposal can be very expensive, particularly if they cannot be buried at the site. Consider using a bulldozer in lieu of hand cutting trees.
- Estimators should visit the site to determine the need for haul road, access, storage of materials, and security considerations. When estimating for access roads on unstable soil, consider using a geotextile stabilization fabric. It can greatly reduce the quantity of crushed stone or gravel. Sites of limited size and access can cause cost overruns due to lost productivity. Theft and damage is another consideration if the location is isolated. A temporary fence or security guards may be required. Investigate the site thoroughly.

02210 Subsurface Investigation

In preparing estimates on structures involving earthwork or foundations, all information concerning soil characteristics should be obtained. Look particularly for hazardous waste, evidence of prior dumping of debris, and previous stream beds.

02220 & 02225 Selective Demolition

The costs shown for selective demolition do not include rubbish handling or disposal. These items should be estimated separately using Means data or other sources.

- Historic preservation often requires that the contractor remove materials from the existing structure, rehab them and replace them. The estimator must be aware of any related measures and precautions that must be taken when doing selective demolition, and cutting and patching. Requirements may include special handling and storage, as well as security.

02300 Earthwork

- Estimating the actual cost of performing earthwork requires careful consideration of the variables involved. This includes items such as type of soil, whether or not water will be encountered, dewatering, whether or not banks need bracing, disposal of excavated earth, length of haul to fill or spoil sites, etc. If the project has large quantities of cut or fill, consider raising or lowering the site to reduce costs while paying close attention to the effect on site drainage and utilities if doing this.
- If the project has large quantities of fill, creating a borrow pit on the site can significantly lower the costs.
- It is very important to consider what time of year the project is scheduled for completion. Bad weather can create large cost overruns from dewatering, site repair and lost productivity from cold weather.

02500 Utility Services
02600 Drainage & Containment

- Never assume that the water, sewer and drainage lines will go in at the early stages of the project. Consider the site access needs before dividing the site in half with open trenches, loose pipe, and machinery obstructions. Always inspect the site to establish that the site drawings are complete. Check off all existing utilities on your drawing as you locate them. If you find any discrepancies, mark up the site plan for further research. Differing site conditions can be very costly if discovered later in the project.

02700 Bases, Ballasts, Pavements/Appurtenances

- When estimating paving, keep in mind the project schedule. If an asphaltic paving project is in a colder climate and runs through to the spring, consider placing the base course in the autumn, then topping it in the spring just prior to completion. This could save considerable costs in spring repair. Keep in mind that prices for asphalt and concrete are generally higher in the cold seasons.

02900 Planting

- The timing of planting and guarantee specifications often dictate the costs for establishing tree and shrub growth and a stand of grass or ground cover. Establish the work performance schedule to coincide with the local planting season. Maintenance and growth guarantees can add from 20% to 100% to the total landscaping cost. The cost to replace trees and shrubs can be as high as 5% of the total cost depending on the planting zone, soil conditions and time of year.

Reference Numbers

Reference numbers are shown in bold squares at the beginning of some major classifications. These numbers refer to related items in the Reference Section. The reference information may be an estimating procedure, an alternate pricing method or technical information.

Note: Not all subdivisions listed here necessarily appear in this publication.

2

SITE CONSTRUCTION

02065	Cement & Concrete	CREW	DAILY OUTPUT	LABOR-HOURS	UNIT	2003 BARE COSTS				TOTAL INCL O&P	
						MAT.	LABOR	EQUIP.	TOTAL		
300 0010	**ASPHALTIC CONCRETE** plant mix (145 lb. per C.F.)				Ton	31.50			31.50	35	**300**
0200	All weather patching mix, hot					34			34	37.50	
0250	Cold patch					38			38	42	
0300	Berm mix				↓	34			34	37.50	

02080 | Utility Materials

		CREW	DAILY OUTPUT	LABOR-HOURS	UNIT	MAT.	LABOR	EQUIP.	TOTAL	TOTAL INCL O&P	
790 0010	**UNDERGROUND MARKING TAPE** [G1030 805]										**790**
0400	Underground tape, detectable, reinforced, alum. foil core, 2"	1 Clab	150	.053	C.L.F.	1.40	1.31		2.71	3.59	
0500	6"	"	140	.057	"	3.50	1.41		4.91	6.05	
900 0010	**PIPING VALVES** Water distribution, see also div. 15110										**900**
3000	Butterfly valves with boxes, cast iron, mech. jt.										
3100	4" diameter	B-20	6	4	Ea.	210	111		321	405	
3140	6" diameter	"	5	4.800		241	134		375	475	
3180	8" diameter	B-21	4	7		292	201	41	534	675	
3300	10" diameter		3.50	8		430	230	46.50	706.50	875	
3340	12" diameter		3	9.333		560	268	54.50	882.50	1,100	
3400	14" diameter		2	14		960	400	81.50	1,441.50	1,775	
3440	16" diameter		2	14		1,300	400	81.50	1,781.50	2,150	
3460	18" diameter		1.50	18.667		1,625	535	109	2,269	2,725	
3480	20" diameter		1	28		2,025	805	163	2,993	3,650	
3500	24" diameter	↓	.50	56	↓	2,825	1,600	325	4,750	5,950	
3600	With lever operator										
3610	4" diameter	B-20	6	4	Ea.	114	111		225	299	
3614	6" diameter	"	5	4.800		159	134		293	385	
3616	8" diameter	B-21	4	7		230	201	41	472	610	
3618	10" diameter		3.50	8		370	230	46.50	646.50	810	
3620	12" diameter		3	9.333		685	268	54.50	1,007.50	1,225	
3622	14" diameter		2	14		1,100	400	81.50	1,581.50	1,925	
3624	16" diameter		2	14		1,500	400	81.50	1,981.50	2,350	
3626	18" diameter		1.50	18.667		1,700	535	109	2,344	2,825	
3628	20" diameter		1	28		2,275	805	163	3,243	3,925	
3630	24" diameter	↓	.50	56	↓	2,450	1,600	325	4,375	5,550	
3700	Check valves, flanged										
3710	4" diameter	B-20	6	4	Ea.	415	111		526	635	
3714	6" diameter	"	5	4.800		705	134		839	985	
3716	8" diameter	B-21	4	7		1,425	201	41	1,667	1,925	
3718	10" diameter		3.50	8		2,300	230	46.50	2,576.50	2,925	
3720	12" diameter		3	9.333		3,525	268	54.50	3,847.50	4,350	
3722	14" diameter		2	14		5,625	400	81.50	6,106.50	6,900	
3724	16" diameter		2	14		7,725	400	81.50	8,206.50	9,225	
3726	18" diameter		1.50	18.667		11,400	535	109	12,044	13,600	
3728	20" diameter		1	28		13,300	805	163	14,268	16,100	
3730	24" diameter	↓	.50	56	↓	19,100	1,600	325	21,025	24,000	
3800	Gate valves, C.I., 250 PSI, mechanical joint, w/boxes										
3810	4" diameter	B-21A	8	5	Ea.	173	152	61.50	386.50	490	
3814	6" diameter		6.80	5.882		209	179	72.50	460.50	585	
3816	8" diameter		5.60	7.143		310	218	88	616	765	
3818	10" diameter		4.80	8.333		500	254	103	857	1,050	
3820	12" diameter	↓	4.08	9.804		680	299	121	1,100	1,325	
3822	14" diameter	B-21	2	14		795	400	81.50	1,276.50	1,600	
3824	16" diameter		1	28		940	805	163	1,908	2,450	
3826	18" diameter		.80	35		1,150	1,000	204	2,354	3,050	
3828	20" diameter		.80	35		1,475	1,000	204	2,679	3,400	
3830	24" diameter		.50	56		1,800	1,600	325	3,725	4,825	
3831	30" diameter	↓	.35	80	↓	2,175	2,300	465	4,940	6,475	

Important: See the Reference Section for critical supporting data - Reference Nos., Crews, & City Cost Indexes

02080	Utility Materials	CREW	DAILY OUTPUT	LABOR-HOURS	UNIT	2003 BARE COSTS				TOTAL INCL O&P
						MAT.	LABOR	EQUIP.	TOTAL	
900	3832	36″ diameter	B-21	.30	93.333	Ea.	2,600	2,675	545	5,820	7,625	900
3900	Globe valves, flanged, iron body, class 125									
3910	4″ diameter	B-20	10	2.400	Ea.	1,050	67		1,117	1,250
3914	6″ diameter	″	9	2.667		2,025	74.50		2,099.50	2,350
3916	8″ diameter	B-21	6	4.667		3,750	134	27	3,911	4,375
3918	10″ diameter		5	5.600		5,875	161	32.50	6,068.50	6,750
3920	12″ diameter		4	7		8,250	201	41	8,492	9,425
3922	14″ diameter		3	9.333		8,825	268	54.50	9,147.50	10,200
3924	16″ diameter		2	14		7,425	400	81.50	7,906.50	8,875
3926	18″ diameter		.80	35		8,150	1,000	204	9,354	10,700
3928	20″ diameter		.60	46.667		8,050	1,350	272	9,672	11,200
3930	24″ diameter		.50	56		9,550	1,600	325	11,475	13,400

02115	Underground Tank Removal	CREW	DAILY OUTPUT	LABOR-HOURS	UNIT	2003 BARE COSTS				TOTAL INCL O&P
						MAT.	LABOR	EQUIP.	TOTAL	
200	0010	**REMOVAL OF UNDERGROUND STORAGE TANKS** R02115-200										200
0011	Petroleum storage tanks, non-leaking									
0100	Excavate & load onto trailer									
0110	3000 gal. to 5000 gal. tank	B-14	4	12	Ea.		315	54.50	369.50	545
0120	6000 gal to 8000 gal tank	B-3A	3	13.333			350	243	593	805
0130	9000 gal to 12000 gal tank	″	2	20			525	365	890	1,225
0190	Known leaking tank add				%				100%	100%
0200	Remove sludge, water and remaining product from tank bottom									
0201	of tank with vacuum truck									
0300	3000 gal to 5000 gal tank	A-13	5	1.600	Ea.		50	98	148	184
0310	6000 gal to 8000 gal tank		4	2			62	123	185	229
0320	9000 gal to 12000 gal tank		3	2.667			83	163	246	305
0390	Dispose of sludge off-site, average				Gal.					4.40
0400	Insert inert solid CO2 "dry ice" into tank									
0401	For cleaning/transporting tanks (1.5 lbs./100 gal. cap)	1 Clab	500	.016	Lb.	1.40	.39		1.79	2.16
1020	Haul tank to certified salvage dump, 100 miles round trip									
1023	3000 gal. to 5000 gal. tank				Ea.				550	690
1026	6000 gal. to 8000 gal. tank								650	825
1029	9,000 gal. to 12,000 gal. tank								875	1,100
1100	Disposal of contaminated soil to landfill									
1110	Minimum				C.Y.					110
1111	Maximum				″					330
1120	Disposal of contaminated soil to									
1121	bituminous concrete batch plant									
1130	Minimum				C.Y.					55
1131	Maximum				″					110
2010	Decontamination of soil on site incl poly tarp on top/bottom									
2011	Soil containment berm, and chemical treatment									
2020	Minimum	B-11C	100	.160	C.Y.	5.60	4.58	2.19	12.37	15.60
2021	Maximum	″	100	.160		7.25	4.58	2.19	14.02	17.45
2050	Disposal of decontaminated soil, minimum									66
2055	Maximum									135

SITE CONSTRUCTION **2**

			CREW	DAILY OUTPUT	LABOR-HOURS	UNIT	MAT.	2003 BARE COSTS LABOR	EQUIP.	TOTAL	TOTAL INCL O&P	
02210		**Subsurface Investigation**										
320	0010	**DRILLING, CORE** Reinforced concrete slab, up to 6" thick slab	B-89A	28	.571	Ea.	2.35	16.25	3.33	21.93	32	**320**
	0020	Including bit, layout and set up										
	0100	1" diameter core	B-89A	28	.571	Ea.	2.35	16.25	3.33	21.93	32	
	0150	Each added inch thick, add		300	.053		.42	1.52	.31	2.25	3.17	
	0300	3" diameter core		23	.696		5.25	19.80	4.05	29.10	41	
	0350	Each added inch thick, add		186	.086		.94	2.45	.50	3.89	5.40	
	0500	4" diameter core		19	.842		5.25	24	4.90	34.15	48.50	
	0550	Each added inch thick, add		170	.094		1.19	2.68	.55	4.42	6.10	
	0700	6" diameter core		14	1.143		8.60	32.50	6.65	47.75	68	
	0750	Each added inch thick, add		140	.114		1.46	3.25	.67	5.38	7.45	
	0900	8" diameter core		11	1.455		11.75	41.50	8.45	61.70	87	
	0950	Each added inch thick, add		95	.168		1.98	4.79	.98	7.75	10.75	
	1100	10" diameter core		10	1.600		15.70	45.50	9.30	70.50	98.50	
	1150	Each added inch thick, add		80	.200		2.60	5.70	1.16	9.46	13.05	
	1300	12" diameter core		9	1.778		18.85	50.50	10.35	79.70	111	
	1350	Each added inch thick, add		68	.235		3.12	6.70	1.37	11.19	15.40	
	1500	14" diameter core		7	2.286		23	65	13.30	101.30	142	
	1550	Each added inch thick, add		55	.291		3.96	8.30	1.69	13.95	19.15	
	1700	18" diameter core		4	4		29.50	114	23.50	167	236	
	1750	Each added inch thick, add		28	.571		5.20	16.25	3.33	24.78	35	
	1760	For horizontal holes, add to above								30%	30%	
	1770	Prestressed hollow core plank, 6" thick										
	1780	1" diameter core	B-89A	52	.308	Ea.	1.56	8.75	1.79	12.10	17.40	
	1790	Each added inch thick, add		350	.046		.27	1.30	.27	1.84	2.62	
	1800	3" diameter core		50	.320		3.44	9.10	1.86	14.40	20	
	1810	Each added inch thick, add		240	.067		.57	1.90	.39	2.86	4.03	
	1820	4" diameter core		48	.333		4.58	9.50	1.94	16.02	22	
	1830	Each added inch thick, add		216	.074		.79	2.11	.43	3.33	4.64	
	1840	6" diameter core		44	.364		5.65	10.35	2.12	18.12	25	
	1850	Each added inch thick, add		175	.091		.94	2.60	.53	4.07	5.70	
	1860	8" diameter core		32	.500		7.60	14.25	2.91	24.76	34	
	1870	Each added inch thick, add		118	.136		1.31	3.86	.79	5.96	8.35	
	1880	10" diameter core		28	.571		10.25	16.25	3.33	29.83	40.50	
	1890	Each added inch thick, add		99	.162		1.41	4.60	.94	6.95	9.80	
	1900	12" diameter core		22	.727		12.50	20.50	4.23	37.23	51	
	1910	Each added inch thick, add		85	.188		2.08	5.35	1.10	8.53	11.90	
	1950	Minimum charge for above, 3" diameter core		7	2.286	Total		65	13.30	78.30	117	
	2000	4" diameter core		6.80	2.353			67	13.70	80.70	120	
	2050	6" diameter core		6	2.667			76	15.50	91.50	136	
	2100	8" diameter core		5.50	2.909			83	16.95	99.95	148	
	2150	10" diameter core		4.75	3.368			96	19.60	115.60	172	
	2200	12" diameter core		3.90	4.103			117	24	141	210	
	2250	14" diameter core		3.38	4.734			135	27.50	162.50	242	
	2300	18" diameter core		3.15	5.079			145	29.50	174.50	259	
02220		**Site Demolition**										
875	0010	**SITE DEMOLITION** No hauling, abandon catch basin or manhole	B-6	7	3.429	Ea.		92	31.50	123.50	177	**875**
	0020	Remove existing catch basin or manhole, masonry		4	6			161	54.50	215.50	310	
	0030	Catch basin or manhole frames and covers, stored		13	1.846			49.50	16.85	66.35	95	
	0040	Remove and reset		7	3.429			92	31.50	123.50	177	
	0900	Hydrants, fire, remove only	B-21A	5	8			244	98.50	342.50	480	
	0950	Remove and reset	"	2	20			610	247	857	1,200	
	1710	Pavement removal, bituminous roads, 3" thick	B-38	690	.058	S.Y.		1.62	1.27	2.89	3.89	
	1750	4" to 6" thick		420	.095			2.66	2.09	4.75	6.40	
	1800	Bituminous driveways		640	.063			1.75	1.37	3.12	4.20	
	1900	Concrete to 6" thick, hydraulic hammer, mesh reinforced		255	.157			4.38	3.45	7.83	10.55	

			CREW	DAILY OUTPUT	LABOR-HOURS	UNIT	MAT.	2003 BARE COSTS LABOR	EQUIP.	TOTAL	TOTAL INCL O&P	
02220		**Site Demolition**										
875	2000	Rod reinforced	B-38	200	.200	S.Y.		5.60	4.40	10	13.45	875
	2100	Concrete, 7" to 24" thick, plain		33	1.212	C.Y.		34	26.50	60.50	81.50	
	2200	Reinforced	▼	24	1.667	"		46.50	36.50	83	112	
	2300	With hand held air equipment, bituminous, to 6" thick	B-39	1,900	.025	S.F.		.66	.09	.75	1.12	
	2320	Concrete to 6" thick, no reinforcing		1,600	.030			.78	.11	.89	1.33	
	2340	Mesh reinforced		1,400	.034			.89	.12	1.01	1.52	
	2360	Rod reinforced	▼	765	.063	▼		1.64	.22	1.86	2.78	
	4000	Sidewalk removal, bituminous, 2-1/2" thick	B-6	325	.074	S.Y.		1.98	.67	2.65	3.80	
	4050	Brick, set in mortar		185	.130			3.48	1.18	4.66	6.65	
	4100	Concrete, plain, 4"		160	.150			4.02	1.37	5.39	7.70	
	4200	Mesh reinforced	▼	150	.160	▼		4.29	1.46	5.75	8.20	
02225		**Selective Demolition**										
730	0010	**RUBBISH HANDLING** The following are to be added to the										730
	0020	demolition prices										
	0400	Chute, circular, prefabricated steel, 18" diameter	B-1	40	.600	L.F.	27	15.20		42.20	53	
	0440	30" diameter	"	30	.800	"	36	20.50		56.50	71	
	0725	Dumpster, weekly rental, 1 dump/week, 20C.Y. capacity (8Tons)				Week					440	
	0800	30 C.Y. capacity (10 Tons)									665	
	0840	40 C.Y. capacity (13 Tons)				▼					805	
	1000	Dust partition, 6 mil polyethylene, 4' x 8' panels, 1" x 3" frame	2 Carp	2,000	.008	S.F.	.16	.25		.41	.57	
	1080	2" x 4" frame	"	2,000	.008	"	.28	.25		.53	.69	
	2000	Load, haul to chute & dumping into chute, 50' haul	2 Clab	24	.667	C.Y.		16.45		16.45	25.50	
	2040	100' haul		16.50	.970			24		24	37.50	
	2080	Over 100' haul, add per 100 L.F.		35.50	.451			11.10		11.10	17.35	
	2120	In elevators, per 10 floors, add	▼	140	.114			2.82		2.82	4.40	
	3000	Loading & trucking, including 2 mile haul, chute loaded	B-16	45	.711			18.05	10	28.05	39	
	3040	Hand loading truck, 50' haul	"	48	.667			16.95	9.35	26.30	37	
	3080	Machine loading truck	B-17	120	.267			7.05	4.37	11.42	15.70	
	5000	Haul, per mile, up to 8 C.Y. truck	B-34B	1,165	.007			.18	.39	.57	.69	
	5100	Over 8 C.Y. truck	"	1,550	.005	▼		.13	.29	.42	.52	
740	0010	**DUMP CHARGES** Typical urban city, tipping fees only										740
	0100	Building construction materials				Ton					70	
	0200	Trees, brush, lumber									50	
	0300	Rubbish only									60	
	0500	Reclamation station, usual charge				▼					85	
790	0010	**TORCH CUTTING** Steel, 1" thick plate	1 Clab	32	.250	L.F.	.18	6.15		6.33	9.85	790
	0040	1" diameter bar	"	210	.038	Ea.		.94		.94	1.47	
	1000	Oxygen lance cutting, reinforced concrete walls										
	1040	12" to 16" thick walls	1 Clab	10	.800	L.F.		19.70		19.70	31	
	1080	24" thick walls	"	6	1.333	"		33		33	51.50	
02240		**Dewatering**										
500	0010	**DEWATERING** Excavate drainage trench, 2' wide, 2' deep	B-11C	90	.178	C.Y.		5.10	2.43	7.53	10.45	500
	0100	2' wide, 3' deep, with backhoe loader	"	135	.119			3.39	1.62	5.01	7	
	0200	Excavate sump pits by hand, light soil	1 Clab	7.10	1.127			28		28	43.50	
	0300	Heavy soil	"	3.50	2.286	▼		56.50		56.50	88	
	0500	Pumping 8 hr., attended 2 hrs. per day, including 20 L.F.										
	0550	of suction hose & 100 L.F. discharge hose										
	0600	2" diaphragm pump used for 8 hours	B-10H	4	3	Day		90	14.05	104.05	152	
	0650	4" diaphragm pump used for 8 hours	B-10I	4	3			90	26.50	116.50	167	
	0800	8 hrs. attended, 2" diaphragm pump	B-10H	1	12			360	56	416	610	
	0900	3" centrifugal pump	B-10J	1	12	▼		360	67.50	427.50	625	

2

SITE CONSTRUCTION

For expanded coverage of these items see Means Heavy Construction Cost Data 2003

29

2

SITE CONSTRUCTION

		02240	Dewatering	CREW	DAILY OUTPUT	LABOR-HOURS	UNIT	2003 BARE COSTS				TOTAL INCL O&P	
								MAT.	LABOR	EQUIP.	TOTAL		
500	1000		4" diaphragm pump	B-10I	1	12	Day		360	107	467	665	500
	1100		6" centrifugal pump	B-10K	1	12	↓		360	271	631	850	
	1300		CMP, incl. excavation 3' deep, 12" diameter	B-6	115	.209	L.F.	7.45	5.60	1.90	14.95	18.90	
	1400		18" diameter		100	.240	"	9.25	6.45	2.19	17.89	22.50	
	1600		Sump hole construction, incl. excavation and gravel, pit		1,250	.019	C.F.	.68	.51	.18	1.37	1.73	
	1700		With 12" gravel collar, 12" pipe, corrugated, 16 ga.		70	.343	L.F.	12.35	9.20	3.13	24.68	31	
	1800		15" pipe, corrugated, 16 ga.		55	.436		15.75	11.70	3.98	31.43	40	
	1900		18" pipe, corrugated, 16 ga.		50	.480		18.50	12.85	4.38	35.73	45	
	2000		24" pipe, corrugated, 14 ga.	↓	40	.600		22	16.10	5.45	43.55	55.50	
	2200		Wood lining, up to 4' x 4', add		300	.080	SFCA	12.55	2.14	.73	15.42	17.90	
	9950		See div. 02240-900 for wellpoints										
	9960		See div. 02240-700 for deep well systems										
700	0010		**WELLS** For dewatering 10' to 20' deep, 2' diameter										700
	0020		with steel casing, minimum	B-6	165	.145	V.L.F.	2.30	3.90	1.33	7.53	10	
	0050		Average		98	.245		4.59	6.55	2.23	13.37	17.65	
	0100		Maximum	↓	49	.490	↓	12.25	13.15	4.47	29.87	39	
	0300		For pumps for dewatering, see division 01590-400-4100 to 4400										
	0500		For domestic water wells, see division 02520-900										
900	0010		**WELLPOINTS** For wellpoint equipment rental, see div. 01590-700	R02240 -900									900
	0100		Installation and removal of single stage system										
	0110		Labor only, .75 labor-hours per L.F., minimum	1 Clab	10.70	.748	LF Hdr		18.45		18.45	29	
	0200		2.0 labor-hours per L.F., maximum	"	4	2	"		49.50		49.50	77	
	0400		Pump operation, 4 @ 6 hr. shifts										
	0410		Per 24 hour day	4 Eqlt	1.27	25.197	Day		785		785	1,175	
	0500		Per 168 hour week, 160 hr. straight, 8 hr. double time		.18	177	Week		5,525		5,525	8,375	
	0550		Per 4.3 week month	↓	.04	800	Month		24,900		24,900	37,700	
	0600		Complete installation, operation, equipment rental, fuel &										
	0610		removal of system with 2" wellpoints 5' O.C.										
	0700		100' long header, 6" diameter, first month	4 Eqlt	3.23	9.907	LF Hdr	122	310		432	600	
	0800		Thereafter, per month		4.13	7.748		97.50	241		338.50	470	
	1000		200' long header, 8" diameter, first month		6	5.333		108	166		274	370	
	1100		Thereafter, per month		8.39	3.814		55	119		174	241	
	1300		500' long header, 8" diameter, first month		10.63	3.010		42.50	93.50		136	189	
	1400		Thereafter, per month		20.91	1.530		30.50	47.50		78	106	
	1600		1,000' long header, 10" diameter, first month		11.62	2.754		36.50	85.50		122	171	
	1700		Thereafter, per month	↓	41.81	.765	↓	18.30	24		42.30	56	
	1900		Note: above figures include pumping 168 hrs. per week										
	1910		and include the pump operator and one stand-by pump.										

		02250	Shoring & Underpinning										
400	0010		**SHEET PILING** Steel, not incl. wales, 22 psf, 15' excav., left in place	B-40	10.81	5.920	Ton	800	187	216	1,203	1,425	400
	0100		Drive, extract & salvage		6	10.667		355	335	390	1,080	1,375	
	0300		20' deep excavation, 27 psf, left in place	R02250 -450	12.95	4.942		800	156	180	1,136	1,325	
	0400		Drive, extract & salvage		6.55	9.771	↓	355	310	355	1,020	1,275	
	1200		15' deep excavation, 22 psf, left in place		983	.065	S.F.	9.30	2.05	2.37	13.72	16.20	
	1300		Drive, extract & salvage		545	.117		3.99	3.70	4.28	11.97	15.10	
	1500		20' deep excavation, 27 psf, left in place		960	.067		11.70	2.10	2.43	16.23	18.90	
	1600		Drive, extract & salvage	↓	485	.132	↓	5.20	4.16	4.81	14.17	17.70	
	2100		Rent steel sheet piling and wales, first month				Ton	190			190	209	
	2200		Per added month					19.05			19.05	21	
	2300		Rental piling left in place, add to rental					635			635	700	
	2500		Wales, connections & struts, 2/3 salvage				↓	194			194	214	
	3900		Wood, solid sheeting, incl. wales, braces and spacers,										
	3910		drive, extract & salvage, 8' deep excavation	B-31	330	.121	S.F.	1.67	3.20	.45	5.32	7.35	
	4000		10' deep, 50 S.F./hr. in & 150 S.F./hr. out		300	.133		1.72	3.52	.49	5.73	7.95	
	4100		12' deep, 45 S.F./hr. in & 135 S.F./hr. out	↓	270	.148	↓	1.77	3.92	.55	6.24	8.65	

Important: See the Reference Section for critical supporting data - Reference Nos., Crews, & City Cost Indexes

02200 | Site Preparation

02250 | Shoring & Underpinning

			CREW	DAILY OUTPUT	LABOR-HOURS	UNIT	MAT.	LABOR	EQUIP.	TOTAL	TOTAL INCL O&P	
400	4200	14' deep, 42 S.F./hr. in & 126 S.F./hr. out	R02250-450	B-31	250	.160	S.F.	1.82	4.23	.59	6.64	9.25	400
	4300	16' deep, 40 S.F./hr. in & 120 S.F./hr. out			240	.167		1.88	4.41	.62	6.91	9.65	
	4400	18' deep, 38 S.F./hr. in & 114 S.F./hr. out			230	.174		1.94	4.60	.64	7.18	10.05	
	4520	Left in place, 8' deep, 55 S.F./hr.			440	.091		3	2.40	.34	5.74	7.40	
	4540	10' deep, 50 S.F./hr.			400	.100		3.16	2.64	.37	6.17	8	
	4560	12' deep, 45 S.F./hr.			360	.111		3.34	2.94	.41	6.69	8.70	
	4565	14' deep, 42 S.F./hr.			335	.119		3.53	3.16	.44	7.13	9.30	
	4570	16' deep, 40 S.F./hr.			320	.125		3.75	3.30	.46	7.51	9.80	
	4700	Alternate pricing, left in place, 8' deep			1.76	22.727	M.B.F.	675	600	84	1,359	1,775	
	4800	Drive, extract and salvage, 8' deep			1.32	30.303	"	600	800	112	1,512	2,025	

02260 | Excavation Support/Protection

			CREW	DAILY OUTPUT	LABOR-HOURS	UNIT	MAT.	LABOR	EQUIP.	TOTAL	TOTAL INCL O&P	
850	0010	SOLDIER BEAMS & LAGGING H piles with 3" wood sheeting											850
	0020	horizontal between piles, including removal of wales & braces											
	0100	No hydrostatic head, 15' deep, 1 line of braces, minimum	B-50	545	.205	S.F.	7.80	6.20	3.18	17.18	22	
	0200	Maximum		495	.226		8.70	6.80	3.50	19	24.50	
	1300	No hydrostatic head, left in place, 15' dp., 1 line of braces, min.		635	.176		10.45	5.30	2.73	18.48	23	
	1400	Maximum		575	.195		11.20	5.85	3.01	20.06	25	
	2350	Lagging only, 3" thick wood between piles 8' O.C., minimum	B-46	400	.120		1.74	3.37	.09	5.20	7.50	
	2370	Maximum		250	.192		2.61	5.40	.15	8.16	11.85	
	2400	Open sheeting no bracing, for trenches to 10' deep, min.		1,736	.028		.78	.78	.02	1.58	2.14	
	2450	Maximum		1,510	.032		.87	.89	.03	1.79	2.44	
	2500	Tie-back method, add to open sheeting, add, minimum								20%	20%	
	2550	Maximum								60%	60%	
	2700	Tie-backs only, based on tie-backs total length, minimum	B-46	86.80	.553	L.F.	10.35	15.55	.44	26.34	37.50	
	2750	Maximum		38.50	1.247	"	18.20	35	.98	54.18	78	
	3500	Tie-backs only, typical average, 25' long		2	24	Ea.	455	675	18.95	1,148.95	1,625	
	3600	35' long		1.58	30.380	"	605	855	24	1,484	2,100	

02300 | Earthwork

02305 | Equipment

			CREW	DAILY OUTPUT	LABOR-HOURS	UNIT	MAT.	LABOR	EQUIP.	TOTAL	TOTAL INCL O&P	
250	0010	MOBILIZATION OR DEMOB (One or the other, unless noted)										250
	0015	Up to 25 mi haul dist (50 mi round trip for mob/demob crew)										
	0020	Dozer, loader, backhoe, excav., grader, paver, roller, 70 to 150 H.P.	B-34N	4	2	Ea.		51.50	112	163.50	202	
	0100	Above 150 HP	B-34K	3	2.667			68.50	199	267.50	325	
	0900	Shovel or dragline, 3/4 C.Y.	"	3.60	2.222			57	166	223	270	
	1100	Small equipment, placed in rear of, or towed by pickup truck	A-3A	8	1			25	9.25	34.25	48	
	1150	Equipment up to 70 HP, on flatbed trailer behind pickup truck	A-3D	4	2			50	41	91	121	
	2000	Crane, truck-mounted, up to 75 ton (costs incl both mob & demob)	1 EQHV	3.60	2.222			75		75	113	
	2100	Crane, tuck-mounted, over 75 ton	A-3E	2.50	6.400			190	29.50	219.50	320	
	2200	Crawler-mounted, up to 75 ton	A-3F	2	8			237	315	552	705	
	2300	Over 75 ton	A-3G	1.50	10.667			315	450	765	970	
	2500	For each additional 5 miles haul distance, add						10%	10%			
	3000	For large pieces of equipment, allow for assembly/knockdown										
	3100	For mob/demob of micro-tunneling equip, see section 02441-400										

02315 | Excavation and Fill

			CREW	DAILY OUTPUT	LABOR-HOURS	UNIT	MAT.	LABOR	EQUIP.	TOTAL	TOTAL INCL O&P	
100	0010	BACKFILL By hand, no compaction, light soil	G1030-805	1 Clab	14	.571	C.Y.		14.10		14.10	22	100
	0100	Heavy soil			11	.727			17.95		17.95	28	

SITE CONSTRUCTION 2

2 SITE CONSTRUCTION

		02315	Excavation and Fill	CREW	DAILY OUTPUT	LABOR-HOURS	UNIT	MAT.	LABOR	EQUIP.	TOTAL	TOTAL INCL O&P	
									2003 BARE COSTS				
100	0300		Compaction in 6" layers, hand tamp, add to above	1 Clab	20.60	.388	C.Y.		9.55		9.55	14.95	100
	0400		Roller compaction operator walking, add	B-10A	100	.120			3.59	.78	4.37	6.35	
	0500		Air tamp, add	B-9C	190	.211			5.25	.88	6.13	9.20	
	0600		Vibrating plate, add	A-1	60	.133			3.29	.96	4.25	6.20	
	0800		Compaction in 12" layers, hand tamp, add to above	1 Clab	34	.235			5.80		5.80	9.05	
	0900		Roller compaction operator walking, add	B-10A	150	.080			2.40	.52	2.92	4.23	
	1000		Air tamp, add	B-9	285	.140			3.52	.59	4.11	6.15	
	1100		Vibrating plate, add	A-1	90	.089			2.19	.64	2.83	4.12	
	1300		Dozer backfilling, bulk, up to 300' haul, no compaction	B-10B	1,200	.010			.30	.73	1.03	1.27	
	1400		Air tamped	B-11B	240	.067			1.91	4.41	6.32	7.80	
	1600		Compacting backfill, 6" to 12" lifts, vibrating roller	B-10C	800	.015			.45	1.36	1.81	2.18	
	1700		Sheepsfoot roller	B-10D	750	.016			.48	1.28	1.76	2.14	
	1900		Dozer backfilling, trench, up to 300' haul, no compaction	B-10B	900	.013			.40	.98	1.38	1.69	
	2000		Air tamped	B-11B	235	.068			1.95	4.51	6.46	7.95	
	2200		Compacting backfill, 6" to 12" lifts, vibrating roller	B-10C	700	.017			.51	1.55	2.06	2.48	
	2300		Sheepsfoot roller	B-10D	650	.018			.55	1.48	2.03	2.47	
130	0010		**BEDDING** For pipe and conduit, not incl. compaction										130
	0050		Crushed or screened bank run gravel	B-6	150	.160	C.Y.	18.35	4.29	1.46	24.10	28	
	0100		Crushed stone 3/4" to 1/2"		150	.160		15.60	4.29	1.46	21.35	25.50	
	0200		Sand, dead or bank		150	.160		3.93	4.29	1.46	9.68	12.50	
	0500		Compacting bedding in trench	A-1	90	.089			2.19	.64	2.83	4.12	
340	0010		**DRILLING AND BLASTING** Only, rock, open face, under 1500 C.Y.	B-47	225	.107	C.Y.	1.85	2.93	3.89	8.67	10.85	340
	0100		Over 1500 C.Y.		300	.080		1.85	2.20	2.92	6.97	8.65	
	0200		Areas where blasting mats are required, under 1500 C.Y.		175	.137		1.85	3.77	5	10.62	13.35	
	0250		Over 1500 C.Y.		250	.096		1.85	2.64	3.51	8	9.95	
	2200		Trenches, up to 1500 C.Y.		22	1.091		5.35	30	40	75.35	96.50	
	2300		Over 1500 C.Y.		26	.923		5.35	25.50	33.50	64.35	82	
440	0010		**EXCAVATING, STRUCTURAL** Hand, pits to 6' deep, sandy soil	1 Clab	8	1	C.Y.		24.50		24.50	38.50	440
	0100		Heavy soil or clay		4	2			49.50		49.50	77	
	0300		Pits 6' to 12' deep, sandy soil		5	1.600			39.50		39.50	61.50	
	0500		Heavy soil or clay		3	2.667			65.50		65.50	103	
	0700		Pits 12' to 18' deep, sandy soil		4	2			49.50		49.50	77	
	0900		Heavy soil or clay		2	4			98.50		98.50	154	
	1500		For wet or muck hand excavation, add to above				%				50%	50%	
505	0010		**FILL** Spread dumped material, by dozer, no compaction	B-10B	1,000	.012	C.Y.		.36	.88	1.24	1.52	505
	0100		By hand	1 Clab	12	.667	"		16.45		16.45	25.50	
	0500		Gravel fill, compacted, under floor slabs, 4" deep	B-37	10,000	.005	S.F.	.16	.13	.01	.30	.37	
	0600		6" deep		8,600	.006		.24	.15	.01	.40	.50	
	0700		9" deep		7,200	.007		.40	.17	.01	.58	.73	
	0800		12" deep		6,000	.008		.56	.21	.02	.79	.95	
	1000		Alternate pricing method, 4" deep		120	.400	C.Y.	11.90	10.40	.89	23.19	30.50	
	1100		6" deep		160	.300		11.90	7.80	.67	20.37	26	
	1200		9" deep		200	.240		11.90	6.25	.54	18.69	23.50	
	1300		12" deep		220	.218		11.90	5.70	.49	18.09	22.50	
	1400		Granular fill					7.95			7.95	8.70	
900	0010		**EXCAVATING, TRENCH** or continuous footing, common earth [G1030 805]										900
	0020		No sheeting or dewatering included										
	0050		1' to 4' deep, 3/8 C.Y. tractor loader/backhoe	B-11C	150	.107	C.Y.		3.05	1.46	4.51	6.30	
	0060		1/2 C.Y. tractor loader/backhoe	B-11M	200	.080			2.29	1.22	3.51	4.85	
	0090		4' to 6' deep, 1/2 C.Y. tractor loader/backhoe	"	200	.080			2.29	1.22	3.51	4.85	
	0100		5/8 C.Y. hydraulic backhoe	B-12Q	250	.064			1.98	1.76	3.74	4.94	
	0110		3/4 C.Y. hydraulic backhoe	B-12F	300	.053			1.65	1.61	3.26	4.27	
	0300		1/2 C.Y. hydraulic excavator, truck mounted	B-12J	200	.080			2.48	4.23	6.71	8.40	

02315	Excavation and Fill	CREW	DAILY OUTPUT	LABOR-HOURS	UNIT	2003 BARE COSTS				TOTAL INCL O&P
						MAT.	LABOR	EQUIP.	TOTAL	

900 0500	6' to 10' deep, 3/4 C.Y. hydraulic backhoe, 6' to 10' deep	B-12F	225	.071	C.Y.		2.20	2.15	4.35	5.70 **900**
0600	1 C.Y. hydraulic excavator, truck mounted	B-12K	400	.040			1.24	2.47	3.71	4.60
0900	10' to 14' deep, 3/4 C.Y. hydraulic backhoe	B-12F	200	.080			2.48	2.42	4.90	6.40
1000	1-1/2 C.Y. hydraulic backhoe	B-12B	540	.030			.92	1.35	2.27	2.87
1300	14' to 20' deep, 1 C.Y. hydraulic backhoe	B-12A	320	.050			1.55	1.64	3.19	4.15
1400	By hand with pick and shovel 2' to 6' deep, light soil	1 Clab	8	1			24.50		24.50	38.50
1500	Heavy soil	"	4	2			49.50		49.50	77
1700	For tamping backfilled trenches, air tamp, add	A-1	100	.080			1.97	.58	2.55	3.71
1900	Vibrating plate, add	B-18	230	.104	↓		2.64	.23	2.87	4.38
2100	Trim sides and bottom for concrete pours, common earth		1,500	.016	S.F.		.41	.04	.45	.67
2300	Hardpan	↓	600	.040	"		1.01	.09	1.10	1.68
940 0010	**EXCAVATING, UTILITY TRENCH** Common earth									**940**
0050	Trenching with chain trencher, 12 H.P., operator walking									
0100	4" wide trench, 12" deep	B-53	800	.010	L.F.		.31	.10	.41	.58
0150	18" deep		750	.011			.33	.11	.44	.62
0200	24" deep		700	.011			.36	.11	.47	.66
0300	6" wide trench, 12" deep		650	.012			.38	.12	.50	.71
0350	18" deep		600	.013			.41	.13	.54	.78
0400	24" deep		550	.015			.45	.14	.59	.85
0450	36" deep		450	.018			.55	.18	.73	1.03
0600	8" wide trench, 12" deep		475	.017			.52	.17	.69	.97
0650	18" deep		400	.020			.62	.20	.82	1.16
0700	24" deep		350	.023			.71	.23	.94	1.33
0750	36" deep	↓	300	.027	↓		.83	.26	1.09	1.55
1000	Backfill by hand including compaction, add									
1050	4" wide trench, 12" deep	A-1	800	.010	L.F.		.25	.07	.32	.47
1100	18" deep		530	.015			.37	.11	.48	.70
1150	24" deep		400	.020			.49	.14	.63	.93
1300	6" wide trench, 12" deep		540	.015			.37	.11	.48	.69
1350	18" deep		405	.020			.49	.14	.63	.92
1400	24" deep		270	.030			.73	.21	.94	1.37
1450	36" deep		180	.044			1.10	.32	1.42	2.06
1600	8" wide trench, 12" deep		400	.020			.49	.14	.63	.93
1650	18" deep		265	.030			.74	.22	.96	1.40
1700	24" deep		200	.040			.99	.29	1.28	1.86
1750	36" deep	↓	135	.059	↓		1.46	.43	1.89	2.75
2000	Chain trencher, 40 H.P. operator riding									
2050	6" wide trench and backfill, 12" deep	B-54	1,200	.007	L.F.		.21	.18	.39	.50
2100	18" deep		1,000	.008			.25	.21	.46	.61
2150	24" deep		975	.008			.26	.22	.48	.63
2200	36" deep		900	.009			.28	.23	.51	.68
2250	48" deep		750	.011			.33	.28	.61	.81
2300	60" deep		650	.012			.38	.32	.70	.94
2400	8" wide trench and backfill, 12" deep		1,000	.008			.25	.21	.46	.61
2450	18" deep		950	.008			.26	.22	.48	.64
2500	24" deep		900	.009			.28	.23	.51	.68
2550	36" deep		800	.010			.31	.26	.57	.76
2600	48" deep		650	.012			.38	.32	.70	.94
2700	12" wide trench and backfill, 12" deep		975	.008			.26	.22	.48	.63
2750	18" deep		860	.009			.29	.25	.54	.71
2800	24" deep		800	.010			.31	.26	.57	.76
2850	36" deep		725	.011			.34	.29	.63	.84
3000	16" wide trench and backfill, 12" deep		835	.010			.30	.25	.55	.73
3050	18" deep		750	.011			.33	.28	.61	.81
3100	24" deep	↓	700	.011	↓		.36	.30	.66	.87

For expanded coverage of these items see *Means Heavy Construction Cost Data 2003*

2 SITE CONSTRUCTION

02315	Excavation and Fill	CREW	DAILY OUTPUT	LABOR-HOURS	UNIT	2003 BARE COSTS				TOTAL INCL O&P	
						MAT.	LABOR	EQUIP.	TOTAL		
3200	Compaction with vibratory plate, add								50%	50%	940
5100	Hand excavate and trim for pipe bells after trench excavation										
5200	8" pipe	1 Clab	155	.052	L.F.		1.27		1.27	1.99	
5300	18" pipe	"	130	.062	"		1.52		1.52	2.37	

02320	Hauling	CREW	DAILY OUTPUT	LABOR-HOURS	UNIT	MAT.	LABOR	EQUIP.	TOTAL	TOTAL INCL O&P	
0011	**HAULING** Excavated or borrow material, loose cubic yards										200
0015	no loading included, highway haulers										
0020	6 C.Y. dump truck, 1/4 mile round trip, 5.0 loads/hr.	B-34A	195	.041	C.Y.		1.05	1.57	2.62	3.34	
0030	1/2 mile round trip, 4.1 loads/hr.		160	.050			1.28	1.91	3.19	4.06	
0040	1 mile round trip, 3.3 loads/hr. R01590 -100		130	.062			1.58	2.35	3.93	5	
0100	2 mile round trip, 2.6 loads/hr.		100	.080			2.05	3.06	5.11	6.50	
0150	3 mile round trip, 2.1 loads/hr.		80	.100			2.57	3.83	6.40	8.15	
0200	4 mile round trip, 1.8 loads/hr.		70	.114			2.93	4.37	7.30	9.30	
0310	12 C.Y. dump truck, 1/4 mile round trip 3.7 loads/hr.	B-34B	288	.028			.71	1.56	2.27	2.81	
0400	2 mile round trip, 2.2 loads/hr.		180	.044			1.14	2.50	3.64	4.49	
0450	3 mile round trip, 1.9 loads/hr.		170	.047			1.21	2.64	3.85	4.75	
0500	4 mile round trip, 1.6 loads/hr.		125	.064			1.64	3.60	5.24	6.45	
1300	Hauling in medium traffic, add								20%	20%	
1400	Heavy traffic, add								30%	30%	
1600	Grading at dump, or embankment if required, by dozer	B-10B	1,000	.012			.36	.88	1.24	1.52	
1800	Spotter at fill or cut, if required	1 Clab	8	1	Hr.		24.50		24.50	38.50	

02441	Microtunneling	CREW	DAILY OUTPUT	LABOR-HOURS	UNIT	2003 BARE COSTS				TOTAL INCL O&P	
						MAT.	LABOR	EQUIP.	TOTAL		
0010	**MICROTUNNELING** Not including excavation, backfill, shoring,										400
0020	or dewatering, average 50'/day, slurry method										
0100	24" to 48" outside diameter, minimum				L.F.					640	
0110	Adverse conditions, add				%					50%	
1000	Rent microtunneling machine, average monthly lease				Month					85,500	
1010	Operating technician				Day					640	
1100	Mobilization and demobilization, minimum				Job					42,800	
1110	Maximum				"					430,000	

02445	Boring or Jacking Conduits	CREW	DAILY OUTPUT	LABOR-HOURS	UNIT	MAT.	LABOR	EQUIP.	TOTAL	TOTAL INCL O&P	
0010	**HORIZONTAL BORING** Casing only, 100' minimum,										300
0020	not incl. jacking pits or dewatering										
0100	Roadwork, 1/2" thick wall, 24" diameter casing	B-42	20	3.200	L.F.	50.50	89	59.50	199	263	
0200	36" diameter		16	4		80.50	111	74.50	266	350	
0300	48" diameter		15	4.267		118	119	79.50	316.50	405	
0500	Railroad work, 24" diameter		15	4.267		50.50	119	79.50	249	330	
0600	36" diameter		14	4.571		80.50	127	85.50	293	385	
0700	48" diameter		12	5.333		118	149	99.50	366.50	475	
0900	For ledge, add								155	190	
1000	Small diameter boring, 3", sandy soil	B-82	900	.018		13.95	.50	.07	14.52	16.15	
1040	Rocky soil	"	500	.032		13.95	.89	.13	14.97	16.80	
1100	Prepare jacking pits, incl. mobilization & demobilization, minimum				Ea.				2,850	3,375	
1101	Maximum				"				16,100	19,300	

34 **Important: See the Reference Section for critical supporting data - Reference Nos., Crews, & City Cost Indexes**

02510	Water Distribution	CREW	DAILY OUTPUT	LABOR-HOURS	UNIT	2003 BARE COSTS				TOTAL INCL O&P
						MAT.	LABOR	EQUIP.	TOTAL	
350 0010	**DISTRIBUTION CONNECTION**									**350**
7020	Crosses, 4" x 4"	B-21	37	.757	Ea.	380	21.50	4.41	405.91	460
7040	6" x 6"		25	1.120		455	32	6.55	493.55	555
7060	8" x 6"		21	1.333		555	38.50	7.75	601.25	680
7080	8" x 8"		21	1.333		605	38.50	7.75	651.25	735
7100	10" x 6"		21	1.333		1,100	38.50	7.75	1,146.25	1,275
7120	10" x 10"		21	1.333		1,200	38.50	7.75	1,246.25	1,375
7140	12" x 6"		18	1.556		1,100	44.50	9.05	1,153.55	1,275
7160	12" x 12"		18	1.556		1,400	44.50	9.05	1,453.55	1,600
7180	14" x 6"		16	1.750		2,775	50	10.20	2,835.20	3,150
7200	14" x 14"		16	1.750		2,900	50	10.20	2,960.20	3,300
7220	16" x 6"		14	2		2,950	57.50	11.65	3,019.15	3,350
7240	16" x 10"		14	2		3,000	57.50	11.65	3,069.15	3,375
7260	16" x 16"		14	2		3,150	57.50	11.65	3,219.15	3,550
7280	18" x 6"		10	2.800		4,400	80.50	16.30	4,496.80	5,000
7300	18" x 12"		10	2.800		4,425	80.50	16.30	4,521.80	5,025
7320	18" x 18"		10	2.800		4,550	80.50	16.30	4,646.80	5,175
7340	20" x 6"		8	3.500		3,575	100	20.50	3,695.50	4,100
7360	20" x 12"		8	3.500		3,825	100	20.50	3,945.50	4,375
7380	20" x 20"		8	3.500		5,575	100	20.50	5,695.50	6,325
7400	24" x 6"		6	4.667		4,600	134	27	4,761	5,325
7420	24" x 12"		6	4.667		4,675	134	27	4,836	5,400
7440	24" x 18"		6	4.667		6,975	134	27	7,136	7,925
7460	24" x 24"		6	4.667		7,250	134	27	7,411	8,225
7600	Cut-in sleeves with rubber gaskets, 4"		18	1.556		127	44.50	9.05	180.55	218
7620	6"		12	2.333		161	67	13.60	241.60	296
7640	8"		10	2.800		219	80.50	16.30	315.80	385
7660	10"		10	2.800		305	80.50	16.30	401.80	480
7680	12"		9	3.111		360	89.50	18.15	467.65	555
7800	Cut-in valves with rubber gaskets, 4"		18	1.556		330	44.50	9.05	383.55	445
7820	6"		12	2.333		445	67	13.60	525.60	610
7840	8"		10	2.800		690	80.50	16.30	786.80	905
7860	10"		10	2.800		1,050	80.50	16.30	1,146.80	1,325
7880	12"		9	3.111		1,325	89.50	18.15	1,432.65	1,625
8000	Sleeves with rubber gaskets, 4" x 4"		37	.757		305	21.50	4.41	330.91	375
8020	6" x 6"		25	1.120		365	32	6.55	403.55	455
8040	8" x 6"		21	1.333		455	38.50	7.75	501.25	570
8060	8" x 8"		21	1.333		480	38.50	7.75	526.25	600
8080	10" x 6"		21	1.333		880	38.50	7.75	926.25	1,050
8100	10" x 10"		21	1.333		955	38.50	7.75	1,001.25	1,125
8120	12" x 6"		18	1.556		880	44.50	9.05	933.55	1,050
8140	12" x 12"		18	1.556		1,125	44.50	9.05	1,178.55	1,300
8160	14" x 6"		16	1.750		2,200	50	10.20	2,260.20	2,525
8180	14" x 14"		16	1.750		2,325	50	10.20	2,385.20	2,650
8200	16" x 6"		14	2		2,350	57.50	11.65	2,419.15	2,700
8220	16" x 10"		14	2		2,400	57.50	11.65	2,469.15	2,725
8240	16" x 16"		14	2		2,525	57.50	11.65	2,594.15	2,875
8260	18" x 6"		10	2.800		3,525	80.50	16.30	3,621.80	4,025
8280	18" x 12"		10	2.800		3,550	80.50	16.30	3,646.80	4,050
8300	18" x 18"		10	2.800		3,650	80.50	16.30	3,746.80	4,175
8320	20" x 6"		8	3.500		2,850	100	20.50	2,970.50	3,300
8340	20" x 12"		8	3.500		3,050	100	20.50	3,170.50	3,525
8360	20" x 20"		8	3.500		4,475	100	20.50	4,595.50	5,100
8380	24" x 6"		6	4.667		3,675	134	27	3,836	4,300
8400	24" x 12"		6	4.667		3,725	134	27	3,886	4,350
8420	24" x 18"		6	4.667		5,575	134	27	5,736	6,375

2

SITE CONSTRUCTION

02510 | Water Distribution

			CREW	DAILY OUTPUT	LABOR-HOURS	UNIT	MAT.	LABOR	EQUIP.	TOTAL	TOTAL INCL O&P	
350	8440	24" x 24"	B-21	6	4.667	Ea.	5,800	134	27	5,961	6,625	350
	8800	Curb box, 6' long	B-20	20	1.200		132	33.50		165.50	198	
	8820	8' long	"	18	1.333		138	37		175	210	
	9000	Valves, gate valve, N.R.S. post type, 4" diameter	B-21	32	.875		266	25	5.10	296.10	340	
	9020	6" diameter		20	1.400		340	40	8.15	388.15	445	
	9040	8" diameter		16	1.750		540	50	10.20	600.20	685	
	9060	10" diameter		16	1.750		840	50	10.20	900.20	1,025	
	9080	12" diameter		13	2.154		1,075	62	12.55	1,149.55	1,275	
	9100	14" diameter		11	2.545		2,525	73	14.85	2,612.85	2,900	
	9120	O.S.&Y., 4" diameter		32	.875		375	25	5.10	405.10	460	
	9140	6" diameter		20	1.400		485	40	8.15	533.15	605	
	9160	8" diameter		16	1.750		745	50	10.20	805.20	910	
	9180	10" diameter		16	1.750		1,075	50	10.20	1,135.20	1,275	
	9200	12" diameter		13	2.154		1,425	62	12.55	1,499.55	1,650	
	9220	14" diameter		11	2.545		2,525	73	14.85	2,612.85	2,900	
	9400	Check valves, rubber disc, 2-1/2" diameter	B-20	44	.545		295	15.20		310.20	350	
	9420	3" diameter	"	38	.632		295	17.60		312.60	355	
	9440	4" diameter	B-21	32	.875		365	25	5.10	395.10	445	
	9480	6" diameter		20	1.400		545	40	8.15	593.15	670	
	9500	8" diameter		16	1.750		800	50	10.20	860.20	970	
	9520	10" diameter		16	1.750		1,425	50	10.20	1,485.20	1,675	
	9540	12" diameter		13	2.154		2,250	62	12.55	2,324.55	2,575	
	9542	14" diameter		11	2.545		3,175	73	14.85	3,262.85	3,625	
	9700	Detector check valves, reducing, 4" diameter		32	.875		640	25	5.10	670.10	750	
	9720	6" diameter		20	1.400		985	40	8.15	1,033.15	1,150	
	9740	8" diameter		16	1.750		1,600	50	10.20	1,660.20	1,875	
	9760	10" diameter		16	1.750		3,175	50	10.20	3,235.20	3,600	
	9800	Galvanized, 4" diameter		32	.875		835	25	5.10	865.10	965	
	9820	6" diameter		20	1.400		1,225	40	8.15	1,273.15	1,400	
	9840	8" diameter		16	1.750		1,950	50	10.20	2,010.20	2,250	
	9860	10" diameter		16	1.750		3,850	50	10.20	3,910.20	4,350	
800	0010	**PIPING, WATER DISTRIBUTION SYSTEMS** Pipe, laid										800
	0020	in trench, excavation and backfill not included	R02510 -800									
	1400	Ductile Iron, cement lined, class 50 water pipe, 18' lengths										
	1410	Mechanical joint, 4" diameter	B-20	144	.167	L.F.	9.65	4.64		14.29	17.90	
	1420	6" diameter		126	.190		9.90	5.30		15.20	19.15	
	1430	8" diameter		108	.222		13.55	6.20		19.75	24.50	
	1440	10" diameter		90	.267		15.90	7.45		23.35	29	
	1450	12" diameter	B-21	72	.389		19.80	11.15	2.27	33.22	42	
	1460	14" diameter		54	.519		25	14.90	3.02	42.92	54	
	1470	16" diameter		46	.609		27.50	17.45	3.55	48.50	61	
	1480	18" diameter		42	.667		35.50	19.15	3.89	58.54	73	
	1490	24" diameter		35	.800		55.50	23	4.66	83.16	102	
	1550	Push on joint, 4" diameter	B-20	155	.155		6.95	4.31		11.26	14.40	
	1560	6" diameter		135	.178		7.85	4.95		12.80	16.40	
	1570	8" diameter		115	.209		11.20	5.80		17	21.50	
	1580	10" diameter		98	.245		17.05	6.80		23.85	29.50	
	1590	12" diameter		78	.308		17.95	8.55		26.50	33	
	1600	14" diameter	B-21	58	.483		19.75	13.85	2.81	36.41	46	
	1610	16" diameter		52	.538		27.50	15.45	3.14	46.09	57.50	
	1620	18" diameter		43	.651		30.50	18.70	3.80	53	66.50	
	1630	20" diameter		41	.683		33.50	19.60	3.98	57.08	72	
	1640	24" diameter		40	.700		44	20	4.08	68.08	84	
	1950	Butterfly valves with boxes, cast iron										
	1970	4" diameter	B-20	6	4	Ea.	360	111		471	570	

Important: See the Reference Section for critical supporting data - Reference Nos., Crews, & City Cost Indexes

02510 | Water Distribution

		CREW	DAILY OUTPUT	LABOR-HOURS	UNIT	MAT.	LABOR	EQUIP.	TOTAL	TOTAL INCL O&P	
800						**2003 BARE COSTS**					**800**
1990	6" diameter	B-20	5	4.800	Ea.	485	134		619	745	
2010	8" diameter	B-21	4	7		725	201	41	967	1,150	
2030	10" diameter		3.50	8		955	230	46.50	1,231.50	1,450	
2050	12" diameter		3	9.333		1,325	268	54.50	1,647.50	1,925	
2070	14" diameter		2	14		1,800	400	81.50	2,281.50	2,700	
2090	16" diameter		2	14		2,225	400	81.50	2,706.50	3,175	
2650	Polyvinyl chloride pipe, class 160, S.D.R.-26, 1-1/2" diameter	B-20	300	.080	L.F.	.59	2.23		2.82	4.13	
2700	2" diameter		250	.096		.89	2.67		3.56	5.15	
2750	2-1/2" diameter		250	.096		1.32	2.67		3.99	5.65	
2800	3" diameter		200	.120		1.89	3.34		5.23	7.35	
2850	4" diameter		200	.120		3.09	3.34		6.43	8.65	
2900	6" diameter		180	.133		6.65	3.71		10.36	13.15	
2950	8" diameter	B-21	160	.175		11.30	5	1.02	17.32	21.50	
8000	Fittings, ductile iron, mechanical joint										
8010	90° bend 4" diameter	B-20	37	.649	Ea.	148	18.05		166.05	191	
8020	6" diameter		25	.960		195	26.50		221.50	256	
8040	8" diameter		21	1.143		289	32		321	365	
8060	10" diameter	B-21	21	1.333		450	38.50	7.75	496.25	565	
8080	12" diameter		18	1.556		590	44.50	9.05	643.55	730	
8100	14" diameter		16	1.750		640	50	10.20	700.20	795	
8120	16" diameter		14	2		690	57.50	11.65	759.15	855	
8140	18" diameter		10	2.800		1,600	80.50	16.30	1,696.80	1,900	
8160	20" diameter		8	3.500		2,375	100	20.50	2,495.50	2,800	
8180	24" diameter		6	4.667		2,975	134	27	3,136	3,525	
8200	Wye or tee, 4" diameter	B-20	25	.960		143	26.50		169.50	199	
8220	6" diameter		17	1.412		179	39.50		218.50	259	
8240	8" diameter		14	1.714		265	47.50		312.50	365	
8260	10" diameter	B-21	14	2		455	57.50	11.65	524.15	600	
8280	12" diameter		12	2.333		570	67	13.60	650.60	750	
8300	14" diameter		10	2.800		1,175	80.50	16.30	1,271.80	1,425	
8320	16" diameter		8	3.500		1,200	100	20.50	1,320.50	1,500	
8340	18" diameter		6	4.667		2,475	134	27	2,636	2,975	
8360	20" diameter		4	7		3,150	201	41	3,392	3,800	
8380	24" diameter		3	9.333		3,600	268	54.50	3,922.50	4,425	
820	**0010 PIPING, WATER DISTRIBUTION, DUCTILE IRON** cement lined										**820**
4000	Drill and tap pressurized main (labor only)										
4100	6" main, 1" to 2" service	Q-1	3	5.333	Ea.		179		179	271	
4150	8" main, 1" to 2" service	"	2.75	5.818	"		196		196	295	
4500	Tap and insert gate valve										
4600	8" main, 4" branch	B-21	3.20	8.750	Ea.		251	51	302	445	
4650	6" branch		2.70	10.370			298	60.50	358.50	530	
4700	10" main, 4" branch		2.70	10.370			298	60.50	358.50	530	
4750	6" branch		2.35	11.915			340	69.50	409.50	605	
4800	12" main, 6" branch		2.35	11.915			340	69.50	409.50	605	
4850	8" branch		2.35	11.915			340	69.50	409.50	605	
850	**0010 PIPING, HDPE BUTT FUSION JOINTS,** SDR 21, 40' lengths										**850**
0100	4" diameter	B-22A	400	.095	L.F.	1.92	2.67	1.78	6.37	8.20	
0200	6" diameter		380	.100		3.07	2.81	1.87	7.75	9.80	
0300	8" diameter		320	.119		5.20	3.34	2.22	10.76	13.35	
0400	10" diameter		300	.127		8.10	3.56	2.37	14.03	17.05	
0500	12" diameter		260	.146		11.35	4.11	2.73	18.19	22	
0600	14" diameter		220	.173		13.70	4.85	3.23	21.78	26	
0700	16" diameter		180	.211		17.90	5.95	3.95	27.80	33	
0800	18" diameter		140	.271		22.50	7.65	5.10	35.25	42.50	
0900	24" diameter		100	.380		40.50	10.70	7.10	58.30	69	

R02510 -800

For expanded coverage of these items see *Means Heavy Construction Cost Data 2003*

02510 | Water Distribution

		CREW	DAILY OUTPUT	LABOR-HOURS	UNIT	2003 BARE COSTS				TOTAL INCL O&P
						MAT.	LABOR	EQUIP.	TOTAL	
850	1000 Fittings										850
	1100 Elbows, 90 degrees									
	1200 4" diameter	B-22B	32	.500	Ea.	29	14.25	5.30	48.55	60.50
	1300 6" diameter		28	.571		75	16.25	6.05	97.30	115
	1400 8" diameter		24	.667		159	18.95	7.10	185.05	212
	1500 10" diameter		18	.889		275	25.50	9.45	309.95	350
	1600 12" diameter		12	1.333		465	38	14.15	517.15	590
	1700 14" diameter		9	1.778		465	50.50	18.90	534.40	610
	1800 16" diameter		6	2.667		560	76	28.50	664.50	765
	1900 18" diameter		4	4		755	114	42.50	911.50	1,050
	2000 24" diameter		3	5.333		1,525	152	56.50	1,733.50	1,975
	2100 Tees									
	2200 4" diameter	B-22B	30	.533	Ea.	33.50	15.15	5.65	54.30	67
	2300 6" diameter		26	.615		88.50	17.50	6.55	112.55	132
	2400 8" diameter		22	.727		224	20.50	7.75	252.25	287
	2500 10" diameter		15	1.067		295	30.50	11.35	336.85	385
	2600 12" diameter		10	1.600		405	45.50	17	467.50	535
	2700 14" diameter		8	2		475	57	21.50	553.50	640
	2800 16" diameter		6	2.667		560	76	28.50	664.50	765
	2900 18" diameter		4	4		755	114	42.50	911.50	1,050
	3000 24" diameter		2	8		1,575	228	85	1,888	2,175

02520 | Wells

		CREW	DAILY OUTPUT	LABOR-HOURS	UNIT	2003 BARE COSTS				TOTAL INCL O&P
						MAT.	LABOR	EQUIP.	TOTAL	
900	0010 **WELLS** Domestic water										900
	0100 Drilled, 4" to 6" diameter	B-23	120	.333	L.F.		8.35	20.50	28.85	35.50
	0200 8" diameter	"	95.20	.420	"		10.55	25.50	36.05	44.50
	0400 Gravel pack well, 40' deep, incl. gravel & casing, complete									
	0500 24" diameter casing x 18" diameter screen	B-23	.13	307	Total	22,100	7,700	18,700	48,500	57,000
	0600 36" diameter casing x 18" diameter screen		.12	333	"	23,700	8,350	20,300	52,350	61,500
	0800 Observation wells, 1-1/4" riser pipe		163	.245	V.L.F.	12.15	6.15	14.95	33.25	39.50
	0900 For flush Buffalo roadway box, add	1 Skwk	16.60	.482	Ea.	33	15.55		48.55	61
	1200 Test well, 2-1/2" diameter, up to 50' deep (15 to 50 GPM)	B-23	1.51	26.490	"	495	665	1,600	2,760	3,350
	1300 Over 50' deep, add	"	121.80	.328	L.F.	13.25	8.25	20	41.50	49.50
	1500 Pumps, installed in wells to 100' deep, 4" submersible									
	1510 1/2 H.P.	Q-1	3.22	4.969	Ea.	320	167		487	600
	1520 3/4 H.P.		2.66	6.015		375	202		577	720
	1600 1 H.P.		2.29	6.987		395	235		630	790
	1700 1-1/2 H.P.	Q-22	1.60	10		1,025	335	400	1,760	2,075
	1800 2 H.P.		1.33	12.030		1,200	405	480	2,085	2,450
	1900 3 H.P.		1.14	14.035		1,375	470	560	2,405	2,850
	2000 5 H.P.		1.14	14.035		1,875	470	560	2,905	3,400
	2050 Remove and install motor only, 4 H.P.		1.14	14.035		670	470	560	1,700	2,075
	3000 Pump, 6" submersible, 25' to 150' deep, 25 H.P., 249 to 297 GPM		.89	17.978		4,075	605	715	5,395	6,200
	3100 25' to 500' deep, 30 H.P., 100 to 300 GPM		.73	21.918		4,650	735	870	6,255	7,200
	8110 Well screen assembly, stainless steel, 2" diameter	B-23A	273	.088	L.F.	36.50	2.46	8.65	47.61	53.50
	8120 3" diameter		253	.095		50.50	2.65	9.35	62.50	70
	8130 4" diameter		200	.120		57.50	3.36	11.80	72.66	81.50
	8140 5" diameter		168	.143		67.50	4	14.05	85.55	95.50
	8150 6" diameter		126	.190		80.50	5.35	18.75	104.60	118
	8160 8" diameter		98.50	.244		105	6.80	24	135.80	153
	8170 10" diameter		73	.329		132	9.20	32.50	173.70	195
	8180 12" diameter		62.50	.384		156	10.75	38	204.75	229
	8190 14" diameter		54.30	.442		177	12.35	43.50	232.85	261
	8200 16" diameter		48.30	.497		194	13.90	49	256.90	290
	8210 18" diameter		39.20	.612		242	17.10	60.50	319.60	360

Important: See the Reference Section for critical supporting data - Reference Nos., Crews, & City Cost Indexes

02520	Wells	CREW	DAILY OUTPUT	LABOR-HOURS	UNIT	2003 BARE COSTS				TOTAL INCL O&P	
						MAT.	LABOR	EQUIP.	TOTAL		
900 8220	20" diameter	B-23A	31.20	.769	L.F.	276	21.50	75.50	373	420	**900**
8230	24" diameter		23.80	1.008		340	28	99.50	467.50	530	
8240	26" diameter		21	1.143		375	32	113	520	590	
8300	Slotted PVC, 1-1/4" diameter		521	.046		1.38	1.29	4.54	7.21	8.50	
8310	1-1/2" diameter		488	.049		2.01	1.38	4.84	8.23	9.70	
8320	2" diameter		273	.088		2.79	2.46	8.65	13.90	16.35	
8330	3" diameter		253	.095		2.96	2.65	9.35	14.96	17.60	
8340	4" diameter		200	.120		3.41	3.36	11.80	18.57	22	
8350	5" diameter		168	.143		3.58	4	14.05	21.63	25.50	
8360	6" diameter		126	.190		4.97	5.35	18.75	29.07	34	
8370	8" diameter	↓	98.50	.244		7.50	6.80	24	38.30	45.50	
8400	Artificial gravel pack, 2" screen, 6" casing	B-23B	174	.138		2.01	3.86	14.85	20.72	24.50	
8405	8" casing		111	.216		2.74	6.05	23.50	32.29	38	
8410	10" casing		74.50	.322		3.44	9	34.50	46.94	55.50	
8415	12" casing		60	.400		4.65	11.20	43	58.85	70	
8420	14" casing		50.20	.478		6	13.35	51.50	70.85	83.50	
8425	16" casing		40.70	.590		8.30	16.50	63.50	88.30	105	
8430	18" casing		36	.667		9.65	18.65	72	100.30	119	
8435	20" casing		29.50	.814		11.10	23	87.50	121.60	144	
8440	24" casing		25.70	.934		12.30	26	101	139.30	165	
8445	26" casing		24.60	.976		13.75	27.50	105	146.25	173	
8450	30" casing		20	1.200		15.80	33.50	129	178.30	211	
8455	36" casing		16.40	1.463	↓	17	41	158	216	256	
8500	Develop well		8	3	Hr.	180	84	325	589	685	
8550	Pump test well	↓	8	3		48	84	325	457	540	
8560	Standby well	B-23A	8	3		46.50	84	295	425.50	505	
8570	Standby, drill rig		8	3	↓		84	295	379	455	
8580	Surface seal well, concrete filled	↓	1	24	Ea.	440	670	2,375	3,485	4,100	
8590	Well test pump, install & remove	B-23	1	40			1,000	2,425	3,425	4,250	
8600	Well sterilization, chlorine	2 Clab	1	16	↓	380	395		775	1,025	
9950	See div. 02240-900 for wellpoints										
9960	See div. 02240-700 for drainage wells										
910 0010	**PUMPS, WELL** Water system, with pressure control										**910**
1000	Deep well, jet, 42 gal. galvanized tank										
1040	3/4 HP	1 Plum	.80	10	Ea.	570	375		945	1,200	
3000	Shallow well, jet, 30 gal. galvanized tank										
3040	1/2 HP	1 Plum	2	4	Ea.	360	149		509	620	

02530	Sanitary Sewerage										
730 0010	**PIPING, DRAINAGE & SEWAGE, CONCRETE**										**730**
0020	Not including excavation or backfill										
1000	Non-reinforced pipe, extra strength, B&S or T&G joints										
1010	6" diameter	B-14	265.04	.181	L.F.	3.93	4.72	.83	9.48	12.60	
1020	8" diameter		224	.214		4.32	5.60	.98	10.90	14.50	
1030	10" diameter		216	.222		4.79	5.80	1.01	11.60	15.35	
1040	12" diameter		200	.240		5.90	6.25	1.09	13.24	17.40	
1050	15" diameter		180	.267		6.90	6.95	1.22	15.07	19.70	
1060	18" diameter		144	.333		8.45	8.70	1.52	18.67	24.50	
1070	21" diameter		112	.429		10.40	11.15	1.95	23.50	31	
1080	24" diameter	↓	100	.480	↓	12.75	12.50	2.19	27.44	36	
2000	Reinforced culvert, class 3, no gaskets										
2010	12" diameter	B-14	210	.229	L.F.	9.80	5.95	1.04	16.79	21	
2020	15" diameter		175	.274		12.55	7.15	1.25	20.95	26.50	
2030	18" diameter		130	.369		13.20	9.60	1.68	24.48	31.50	
2035	21" diameter	↓	120	.400	↓	17.15	10.40	1.82	29.37	37	

SITE CONSTRUCTION 2

02530 | Sanitary Sewerage

		CREW	DAILY OUTPUT	LABOR-HOURS	UNIT	2003 BARE COSTS MAT.	LABOR	EQUIP.	TOTAL	TOTAL INCL O&P	
730	2040	24" diameter	B-14	100	.480	L.F.	19.45	12.50	2.19	34.14	43.50	**730**
	2045	27" diameter	B-13	92	.609		24.50	16.30	7.10	47.90	60	
	2050	30" diameter		88	.636		26.50	17	7.45	50.95	64	
	2060	36" diameter	↓	72	.778		38	21	9.10	68.10	84	
	2070	42" diameter	B-13B	72	.778		52	21	16.05	89.05	107	
	2080	48" diameter		64	.875		64.50	23.50	18.05	106.05	127	
	2090	60" diameter		48	1.167		103	31	24	158	188	
	2100	72" diameter		40	1.400		145	37.50	29	211.50	250	
	2120	84" diameter		32	1.750		246	47	36	329	380	
	2140	96" diameter	↓	24	2.333		295	62.50	48	405.50	475	
	2200	With gaskets, class 3, 12" diameter	B-21	168	.167		11.75	4.78	.97	17.50	21.50	
	2220	15" diameter		160	.175		14.10	5	1.02	20.12	24.50	
	2230	18" diameter		152	.184		17.65	5.30	1.07	24.02	29	
	2240	24" diameter	↓	136	.206		26.50	5.90	1.20	33.60	39.50	
	2260	30" diameter	B-13	88	.636		35.50	17	7.45	59.95	73.50	
	2270	36" diameter	"	72	.778		53	21	9.10	83.10	101	
	2290	48" diameter	B-13B	64	.875		86.50	23.50	18.05	128.05	151	
	2310	72" diameter	"	40	1.400		224	37.50	29	290.50	335	
	2330	Flared ends, 6'-1" long, 12" diameter	B-21	190	.147		32.50	4.23	.86	37.59	43.50	
	2340	15" diameter		155	.181		38	5.20	1.05	44.25	50.50	
	2400	6'-2" long, 18" diameter		122	.230		39.50	6.60	1.34	47.44	55	
	2420	24" diameter	↓	88	.318		45.50	9.15	1.86	56.51	66	
	2440	36" diameter	B-13	60	.933	↓	82.50	25	10.90	118.40	142	
	3040	Vitrified plate lined, add to above, 30" to 36" diameter				SFCA	3.40			3.40	3.74	
	3050	42" to 54" diameter, add					3.64			3.64	4	
	3060	60" to 72" diameter, add					4.26			4.26	4.69	
	3070	Over 72" diameter, add				↓	4.54			4.54	4.99	
	3080	Radius pipe, add to pipe prices, 12" to 60" diameter				L.F.	50%					
	3090	Over 60" diameter, add				"	20%					
	3500	Reinforced elliptical, 8' lengths, C507 class 3										
	3520	14" x 23" inside, round equivalent 18" diameter	B-21	82	.341	L.F.	22	9.80	1.99	33.79	41.50	
	3530	24" x 38" inside, round equivalent 30" diameter	B-13	58	.966		39	26	11.30	76.30	95.50	
	3540	29" x 45" inside, round equivalent 36" diameter		52	1.077		49.50	29	12.60	91.10	113	
	3550	38" x 60" inside, round equivalent 48" diameter		38	1.474		76	39.50	17.25	132.75	163	
	3560	48" x 76" inside, round equivalent 60" diameter		26	2.154		116	57.50	25	198.50	245	
	3570	58" x 91" inside, round equivalent 72" diameter	↓	22	2.545	↓	166	68	30	264	320	
	3780	Concrete slotted pipe, class 4 mortar joint										
	3800	12" diameter	B-21	168	.167	L.F.	12.50	4.78	.97	18.25	22.50	
	3840	18" diameter	"	152	.184	"	19.35	5.30	1.07	25.72	31	
	3900	Class 4 O-ring										
	3940	12" diameter	B-21	168	.167	L.F.	13.20	4.78	.97	18.95	23	
	3960	18" diameter	"	152	.184	"	17.70	5.30	1.07	24.07	29	
780	0010	**PIPING, DRAINAGE & SEWAGE, POLYVINYL CHLORIDE**										**780**
	0020	Not including excavation or backfill										
	2000	10' lengths, S.D.R. 35, B&S, 4" diameter	B-20	375	.064	L.F.	1.55	1.78		3.33	4.50	
	2040	6" diameter	↓	350	.069		2.61	1.91		4.52	5.85	
	2080	8" diameter	↓	335	.072		4.40	2		6.40	7.95	
	2120	10" diameter	B-21	330	.085		6.65	2.43	.49	9.57	11.65	
	2160	12" diameter		320	.087		7.40	2.51	.51	10.42	12.60	
	2200	15" diameter	↓	190	.147	↓	11.15	4.23	.86	16.24	19.80	
790	0010	**PIPING, DRAINAGE & SEWAGE, VITRIFIED CLAY** C700										**790**
	0020	Not including excavation or backfill,										
	4030	Extra strength, compression joints, C425										
	5000	4" diameter x 4' long	B-20	265	.091	L.F.	1.66	2.52		4.18	5.75	
	5020	6" diameter x 5' long	"	200	.120		2.78	3.34		6.12	8.30	
	5040	8" diameter x 5' long	B-21	200	.140	↓	3.96	4.02	.82	8.80	11.50	

Important: See the Reference Section for critical supporting data - Reference Nos., Crews, & City Cost Indexes

02530 | Sanitary Sewerage

		CREW	DAILY OUTPUT	LABOR-HOURS	UNIT	2003 BARE COSTS				TOTAL INCL O&P	
						MAT.	LABOR	EQUIP.	TOTAL		
790	5060	10" diameter x 5' long	B-21	190	.147	L.F.	6.65	4.23	.86	11.74	14.80	790
	5080	12" diameter x 6' long		150	.187		8.80	5.35	1.09	15.24	19.25	
	5100	15" diameter x 7' long		110	.255		15.45	7.30	1.48	24.23	30	
	5120	18" diameter x 7' long		88	.318		23	9.15	1.86	34.01	41.50	
	5140	24" diameter x 7' long	↓	45	.622		42	17.85	3.63	63.48	78.50	
	5160	30" diameter x 7' long	B-22	31	.968		74.50	28	7.90	110.40	134	
	5180	36" diameter x 7' long	"	20	1.500		123	43.50	12.25	178.75	216	
	6000	For 3' lengths, add					30%	30%				
	6020	For 2' lengths, add					40%	60%				
	6060	For plain joints, deduct					25%					
	7060	2' lengths, add to above				↓	40%					

02550 | Piped Energy Distribution

		CREW	DAILY OUTPUT	LABOR-HOURS	UNIT	MAT.	LABOR	EQUIP.	TOTAL	TOTAL INCL O&P	
450	0010	**GAS STATION PRODUCT LINE**										450
	0020	Primary containment pipe, fiberglass-reinforced										
	0030	Plastic pipe 15' & 30' lengths										
	0040	2" diameter	Q-6	425	.056	L.F.	2.98	1.98		4.96	6.25	
	0050	3" diameter		400	.060		3.92	2.11		6.03	7.50	
	0060	4" diameter	↓	375	.064	↓	5.05	2.25		7.30	8.95	
	0100	Fittings										
	0110	Elbows, 90° & 45°, bell-ends, 2"	Q-6	24	1	Ea.	30	35		65	86	
	0120	3" diameter		22	1.091		31.50	38.50		70	93	
	0130	4" diameter		20	1.200		42	42		84	110	
	0200	Tees, bell ends, 2"		21	1.143		36.50	40		76.50	101	
	0210	3" diameter		18	1.333		37	47		84	111	
	0220	4" diameter		15	1.600		50.50	56		106.50	141	
	0230	Flanges bell ends, 2"		24	1		12.10	35		47.10	66.50	
	0240	3" diameter		22	1.091		15.25	38.50		53.75	75	
	0250	4" diameter		20	1.200		21	42		63	86.50	
	0260	Sleeve couplings, 2"		21	1.143		7.70	40		47.70	69	
	0270	3" diameter		18	1.333		10.95	47		57.95	82.50	
	0280	4" diameter		15	1.600		15.10	56		71.10	102	
	0290	Threaded adapters 2"		21	1.143		10.15	40		50.15	71.50	
	0300	3" diameter		18	1.333		17.80	47		64.80	90	
	0310	4" diameter		15	1.600		24	56		80	112	
	0320	Reducers, 2"		27	.889		13.50	31		44.50	62	
	0330	3" diameter		22	1.091		15.60	38.50		54.10	75	
	0340	4" diameter	↓	20	1.200	↓	20	42		62	85.50	
	1010	Gas station product line for secondary containment (double wall)										
	1100	Fiberglass reinforced plastic pipe 25' lengths										
	1120	Pipe, plain end, 3"	Q-6	375	.064	L.F.	5.15	2.25		7.40	9.05	
	1130	4" diameter		350	.069		8.40	2.41		10.81	12.90	
	1140	5" diameter		325	.074		10.95	2.59		13.54	15.90	
	1150	6" diameter	↓	300	.080	↓	11.25	2.81		14.06	16.65	
	1200	Fittings										
	1230	Elbows, 90° & 45°, 3"	Q-6	18	1.333	Ea.	37	47		84	111	
	1240	4" diameter		16	1.500		64.50	52.50		117	151	
	1250	5" diameter		14	1.714		149	60		209	255	
	1260	6" diameter		12	2		151	70		221	272	
	1270	Tees, 3"		15	1.600		54.50	56		110.50	145	
	1280	4" diameter		12	2		80.50	70		150.50	195	
	1290	5" diameter		9	2.667		163	93.50		256.50	320	
	1300	6" diameter		6	4		170	140		310	400	
	1310	Couplings, 3"		18	1.333		26	47		73	99	
	1320	4" diameter	↓	16	1.500	↓	67	52.50		119.50	153	

2

SITE CONSTRUCTION

2 SITE CONSTRUCTION

02550 | Piped Energy Distribution

		CREW	DAILY OUTPUT	LABOR-HOURS	UNIT	MAT.	LABOR	EQUIP.	TOTAL	TOTAL INCL O&P	
450	1330	5″ diameter	Q-6	14	1.714	Ea.	139	60		199	244	**450**
	1340	6″ diameter		12	2		144	70		214	265	
	1350	Cross-over nipples, 3″		18	1.333		5.90	47		52.90	77	
	1360	4″ diameter		16	1.500		6.95	52.50		59.45	87	
	1370	5″ diameter		14	1.714		10.30	60		70.30	102	
	1380	6″ diameter		12	2		10.80	70		80.80	118	
	1400	Telescoping, reducers, concentric 4″ x 3″		18	1.333		19.75	47		66.75	92	
	1410	5″ x 4″		17	1.412		51.50	49.50		101	132	
	1420	6″ x 5″	▼	16	1.500	▼	124	52.50		176.50	216	
464	0010	**PIPING, GAS SERVICE & DISTRIBUTION, POLYETHYLENE**										**464**
	0020	not including excavation or backfill										
	1000	60 psi coils, comp cplg @ 100′, 1/2″ diameter, SDR 9.3	B-20A	608	.053	L.F.	.38	1.56		1.94	2.81	
	1040	1-1/4″ diameter, SDR 11		544	.059		.69	1.74		2.43	3.43	
	1100	2″ diameter, SDR 11		488	.066		.86	1.94		2.80	3.92	
	1160	3″ diameter, SDR 11	▼	408	.078		1.80	2.32		4.12	5.55	
	1500	60 PSI 40′ joints with coupling, 3″ diameter, SDR 11	B-21A	408	.098		1.80	2.99	1.21	6	7.85	
	1540	4″ diameter, SDR 11		352	.114		4.14	3.46	1.40	9	11.40	
	1600	6″ diameter, SDR 11		328	.122		12.85	3.71	1.50	18.06	21.50	
	1640	8″ diameter, SDR 11	▼	272	.147	▼	17.55	4.48	1.81	23.84	28	
550	0010	**PIPE CONDUIT, PREFABRICATED**										**550**
	0020	Does not include trenching, fittings or crane.										
	0300	For cathodic protection, add 12 to 14%										
	0310	of total built-up price (casing plus service pipe)										
	0580	Polyurethane insulated system, 250°F. max. temp.										
	0620	Black steel service pipe, standard wt., 1/2″ insulation										
	0660	3/4″ diam. pipe size	Q-17	54	.296	L.F.	17.70	10.05	1.45	29.20	36.50	
	0670	1″ diam. pipe size		50	.320		19.50	10.85	1.57	31.92	39.50	
	0680	1-1/4″ diam. pipe size		47	.340		22	11.50	1.67	35.17	43	
	0690	1-1/2″ diam. pipe size		45	.356		23.50	12.05	1.75	37.30	46	
	0700	2″ diam. pipe size		42	.381		24.50	12.90	1.87	39.27	48.50	
	0710	2-1/2″ diam. pipe size		34	.471		25	15.95	2.31	43.26	54	
	0720	3″ diam. pipe size		28	.571		29	19.35	2.81	51.16	63.50	
	0730	4″ diam. pipe size		22	.727		36.50	24.50	3.57	64.57	81	
	0740	5″ diam. pipe size	▼	18	.889		46.50	30	4.36	80.86	101	
	0750	6″ diam. pipe size	Q-18	23	1.043		54	36.50	3.41	93.91	118	
	0760	8″ diam. pipe size		19	1.263		79	44.50	4.13	127.63	159	
	0770	10″ diam. pipe size		16	1.500		104	52.50	4.91	161.41	199	
	0780	12″ diam. pipe size		13	1.846		129	65	6.05	200.05	247	
	0790	14″ diam. pipe size		11	2.182		144	76.50	7.15	227.65	282	
	0800	16″ diam. pipe size		10	2.400		165	84	7.85	256.85	320	
	0810	18″ diam. pipe size		8	3		190	105	9.80	304.80	380	
	0820	20″ diam. pipe size		7	3.429		212	120	11.20	343.20	425	
	0830	24″ diam. pipe size	▼	6	4		259	140	13.10	412.10	510	
	0900	For 1″ thick insulation, add					10%					
	0940	For 1-1/2″ thick insulation, add					13%					
	0980	For 2″ thick insulation, add				▼	20%					
	1500	Gland seal for system, 3/4″ diam. pipe size	Q-17	32	.500	Ea.	288	16.95	2.46	307.41	345	
	1510	1″ diam. pipe size		32	.500		288	16.95	2.46	307.41	345	
	1540	1-1/4″ diam. pipe size		30	.533		310	18.05	2.62	330.67	370	
	1550	1-1/2″ diam. pipe size		30	.533		310	18.05	2.62	330.67	370	
	1560	2″ diam. pipe size		28	.571		365	19.35	2.81	387.16	430	
	1570	2-1/2″ diam. pipe size		26	.615		395	21	3.02	419.02	470	
	1580	3″ diam. pipe size		24	.667		425	22.50	3.27	450.77	505	
	1590	4″ diam. pipe size		22	.727		500	24.50	3.57	528.07	590	
	1600	5″ diam. pipe size	▼	19	.842	▼	615	28.50	4.13	647.63	730	

Important: See the Reference Section for critical supporting data - Reference Nos., Crews, & City Cost Indexes

02550 | Piped Energy Distribution

		CREW	DAILY OUTPUT	LABOR-HOURS	UNIT	2003 BARE COSTS MAT.	LABOR	EQUIP.	TOTAL	TOTAL INCL O&P	
550	1610	6" diam. pipe size	Q-18	26	.923	Ea.	655	32.50	3.02	690.52	770	550
	1620	8" diam. pipe size		25	.960		760	33.50	3.14	796.64	890	
	1630	10" diam. pipe size		23	1.043		925	36.50	3.41	964.91	1,075	
	1640	12" diam. pipe size		21	1.143		1,025	40	3.74	1,068.74	1,200	
	1650	14" diam. pipe size		19	1.263		1,150	44.50	4.13	1,198.63	1,325	
	1660	16" diam. pipe size		18	1.333		1,350	47	4.36	1,401.36	1,550	
	1670	18" diam. pipe size		16	1.500		1,425	52.50	4.91	1,482.41	1,650	
	1680	20" diam. pipe size		14	1.714		1,625	60	5.60	1,690.60	1,900	
	1690	24" diam. pipe size	↓	12	2	↓	1,800	70	6.55	1,876.55	2,125	
	2000	Elbow, 45° for system										
	2020	3/4" diam. pipe size	Q-17	14	1.143	Ea.	182	38.50	5.60	226.10	265	
	2040	1" diam. pipe size		13	1.231		187	41.50	6.05	234.55	275	
	2050	1-1/4" diam. pipe size		11	1.455		212	49	7.15	268.15	315	
	2060	1-1/2" diam. pipe size		9	1.778		240	60	8.75	308.75	365	
	2070	2" diam. pipe size		6	2.667		232	90.50	13.10	335.60	405	
	2080	2-1/2" diam. pipe size		4	4		250	135	19.65	404.65	500	
	2090	3" diam. pipe size		3.50	4.571		289	155	22.50	466.50	580	
	2100	4" diam. pipe size		3	5.333		335	181	26	542	670	
	2110	5" diam. pipe size	↓	2.80	5.714		435	193	28	656	800	
	2120	6" diam. pipe size	Q-18	4	6		490	211	19.60	720.60	880	
	2130	8" diam. pipe size		3	8		710	281	26	1,017	1,250	
	2140	10" diam. pipe size		2.40	10		910	350	32.50	1,292.50	1,575	
	2150	12" diam. pipe size		2	12		1,200	420	39	1,659	2,000	
	2160	14" diam. pipe size		1.80	13.333		1,475	470	43.50	1,988.50	2,375	
	2170	16" diam. pipe size		1.60	15		1,775	525	49	2,349	2,800	
	2180	18" diam. pipe size		1.30	18.462		2,225	650	60.50	2,935.50	3,475	
	2190	20" diam. pipe size		1	24		2,800	840	78.50	3,718.50	4,425	
	2200	24" diam. pipe size	↓	.70	34.286		3,500	1,200	112	4,812	5,800	
	2260	For elbow, 90°, add					25%					
	2300	For tee, straight, add					85%	30%				
	2340	For tee, reducing, add					170%	30%				
	2380	For weldolet, straight, add				↓	50%					
	2400	Polyurethane insulation, 1"										
	2410	FRP carrier and casing										
	2420	4"	Q-5	18	.889	L.F.	34.50	30		64.50	83.50	
	2422	6"	Q-6	23	1.043		60	36.50		96.50	122	
	2424	8"		19	1.263		98	44.50		142.50	175	
	2426	10"		16	1.500		132	52.50		184.50	226	
	2428	12"	↓	13	1.846	↓	172	65		237	288	
	2430	FRP carrier and PVC casing										
	2440	4"	Q-5	18	.889	L.F.	13.70	30		43.70	60.50	
	2444	8"	Q-6	19	1.263		28.50	44.50		73	98.50	
	2446	10"		16	1.500		41	52.50		93.50	125	
	2448	12"	↓	13	1.846	↓	52.50	65		117.50	156	
	2450	PVC carrier and casing										
	2460	4"	Q-1	36	.444	L.F.	6.75	14.95		21.70	30	
	2462	6"	"	29	.552		9.45	18.55		28	38.50	
	2464	8"	Q-2	36	.667		13.15	23.50		36.65	49.50	
	2466	10"		32	.750		18	26		44	59.50	
	2468	12"	↓	31	.774	↓	21	27		48	64	
	2800	Calcium silicate insulated system, high temp. (1200°F)										
	2840	Steel casing with protective exterior coating										
	2850	6-5/8" diameter	Q-18	52	.462	L.F.	33	16.20	1.51	50.71	62.50	
	2860	8-5/8" diameter		50	.480		36	16.85	1.57	54.42	67	
	2870	10-3/4" diameter		47	.511		42.50	17.90	1.67	62.07	75.50	
	2880	12-3/4" diameter	↓	44	.545	↓	46	19.15	1.78	66.93	82	

02550	Piped Energy Distribution	CREW	DAILY OUTPUT	LABOR-HOURS	UNIT	2003 BARE COSTS				TOTAL INCL O&P	
						MAT.	LABOR	EQUIP.	TOTAL		
2890	14″ diameter	Q-18	41	.585	L.F.	52	20.50	1.91	74.41	90	550
2900	16″ diameter		39	.615		56	21.50	2.01	79.51	96	
2910	18″ diameter		36	.667		61.50	23.50	2.18	87.18	106	
2920	20″ diameter		34	.706		69.50	25	2.31	96.81	116	
2930	22″ diameter		32	.750		96	26.50	2.45	124.95	149	
2940	24″ diameter		29	.828		110	29	2.71	141.71	168	
2950	26″ diameter		26	.923		125	32.50	3.02	160.52	190	
2960	28″ diameter		23	1.043		156	36.50	3.41	195.91	230	
2970	30″ diameter		21	1.143		166	40	3.74	209.74	247	
2980	32″ diameter		19	1.263		187	44.50	4.13	235.63	277	
2990	34″ diameter		18	1.333		189	47	4.36	240.36	282	
3000	36″ diameter		16	1.500		202	52.50	4.91	259.41	305	
3040	For multi-pipe casings, add					10%					
3060	For oversize casings, add					2%					
3400	Steel casing gland seal, single pipe										
3420	6-5/8″ diameter	Q-18	25	.960	Ea.	460	33.50	3.14	496.64	565	
3440	8-5/8″ diameter		23	1.043		540	36.50	3.41	579.91	655	
3450	10-3/4″ diameter		21	1.143		605	40	3.74	648.74	730	
3460	12-3/4″ diameter		19	1.263		720	44.50	4.13	768.63	865	
3470	14″ diameter		17	1.412		790	49.50	4.62	844.12	950	
3480	16″ diameter		16	1.500		925	52.50	4.91	982.41	1,100	
3490	18″ diameter		15	1.600		1,025	56	5.25	1,086.25	1,225	
3500	20″ diameter		13	1.846		1,125	65	6.05	1,196.05	1,350	
3510	22″ diameter		12	2		1,275	70	6.55	1,351.55	1,525	
3520	24″ diameter		11	2.182		1,425	76.50	7.15	1,508.65	1,700	
3530	26″ diameter		10	2.400		1,625	84	7.85	1,716.85	1,925	
3540	28″ diameter		9.50	2.526		1,850	88.50	8.25	1,946.75	2,175	
3550	30″ diameter		9	2.667		1,875	93.50	8.70	1,977.20	2,225	
3560	32″ diameter		8.50	2.824		2,125	99	9.25	2,233.25	2,475	
3570	34″ diameter		8	3		2,300	105	9.80	2,414.80	2,725	
3580	36″ diameter		7	3.429		2,450	120	11.20	2,581.20	2,900	
3620	For multi-pipe casings, add					5%					
4000	Steel casing anchors, single pipe										
4020	6-5/8″ diameter	Q-18	8	3	Ea.	415	105	9.80	529.80	625	
4040	8-5/8″ diameter		7.50	3.200		435	112	10.45	557.45	655	
4050	10-3/4″ diameter		7	3.429		570	120	11.20	701.20	820	
4060	12-3/4″ diameter		6.50	3.692		605	130	12.05	747.05	875	
4070	14″ diameter		6	4		710	140	13.10	863.10	1,000	
4080	16″ diameter		5.50	4.364		830	153	14.25	997.25	1,150	
4090	18″ diameter		5	4.800		935	168	15.70	1,118.70	1,300	
4100	20″ diameter		4.50	5.333		1,025	187	17.45	1,229.45	1,425	
4110	22″ diameter		4	6		1,150	211	19.60	1,380.60	1,600	
4120	24″ diameter		3.50	6.857		1,250	241	22.50	1,513.50	1,775	
4130	26″ diameter		3	8		1,425	281	26	1,732	2,000	
4140	28″ diameter		2.50	9.600		1,550	335	31.50	1,916.50	2,250	
4150	30″ diameter		2	12		1,650	420	39	2,109	2,475	
4160	32″ diameter		1.50	16		1,975	560	52.50	2,587.50	3,075	
4170	34″ diameter		1	24		2,225	840	78.50	3,143.50	3,775	
4180	36″ diameter		1	24		2,400	840	78.50	3,318.50	4,000	
4220	For multi-pipe, add					5%	20%				
4800	Steel casing elbow										
4820	6-5/8″ diameter	Q-18	15	1.600	Ea.	570	56	5.25	631.25	715	
4830	8-5/8″ diameter		15	1.600		605	56	5.25	666.25	755	
4850	10-3/4″ diameter		14	1.714		730	60	5.60	795.60	900	
4860	12-3/4″ diameter		13	1.846		865	65	6.05	936.05	1,050	
4870	14″ diameter		12	2		925	70	6.55	1,001.55	1,150	

02550	Piped Energy Distribution	CREW	DAILY OUTPUT	LABOR-HOURS	UNIT	2003 BARE COSTS				TOTAL INCL O&P
						MAT.	LABOR	EQUIP.	TOTAL	
4880	16" diameter	Q-18	11	2.182	Ea.	1,000	76.50	7.15	1,083.65	1,225
4890	18" diameter		10	2.400		1,150	84	7.85	1,241.85	1,400
4900	20" diameter		9	2.667		1,225	93.50	8.70	1,327.20	1,500
4910	22" diameter		8	3		1,325	105	9.80	1,439.80	1,625
4920	24" diameter		7	3.429		1,475	120	11.20	1,606.20	1,800
4930	26" diameter		6	4		1,600	140	13.10	1,753.10	1,975
4940	28" diameter		5	4.800		1,725	168	15.70	1,908.70	2,175
4950	30" diameter		4	6		1,725	211	19.60	1,955.60	2,250
4960	32" diameter		3	8		1,975	281	26	2,282	2,625
4970	34" diameter		2	12		2,175	420	39	2,634	3,050
4980	36" diameter	▼	2	12	▼	2,300	420	39	2,759	3,225
5500	Black steel service pipe, std. wt., 1" thick insulation									
5510	3/4" diameter pipe size	Q-17	54	.296	L.F.	14.45	10.05	1.45	25.95	32.50
5540	1" diameter pipe size		50	.320		15	10.85	1.57	27.42	34.50
5550	1-1/4" diameter pipe size		47	.340		16.85	11.50	1.67	30.02	37.50
5560	1-1/2" diameter pipe size		45	.356		18.35	12.05	1.75	32.15	40
5570	2" diameter pipe size		42	.381		20.50	12.90	1.87	35.27	44
5580	2-1/2" diameter pipe size		34	.471		21.50	15.95	2.31	39.76	50
5590	3" diameter pipe size		28	.571		24.50	19.35	2.81	46.66	59
5600	4" diameter pipe size		22	.727		31.50	24.50	3.57	59.57	75.50
5610	5" diameter pipe size	▼	18	.889		42.50	30	4.36	76.86	97
5620	6" diameter pipe size	Q-18	23	1.043	▼	46.	36.50	3.41	85.91	110
6000	Black steel service pipe, std. wt., 1-1/2" thick insul.									
6010	3/4" diameter pipe size	Q-17	54	.296	L.F.	14.65	10.05	1.45	26.15	33
6040	1" diameter pipe size		50	.320		16.35	10.85	1.57	28.77	36
6050	1-1/4" diameter pipe size		47	.340		18.35	11.50	1.67	31.52	39
6060	1-1/2" diameter pipe size		45	.356		20	12.05	1.75	33.80	42
6070	2" diameter pipe size		42	.381		22	12.90	1.87	36.77	45.50
6080	2-1/2" diameter pipe size		34	.471		23	15.95	2.31	41.26	52
6090	3" diameter pipe size		28	.571		26	19.35	2.81	48.16	60.50
6100	4" diameter pipe size		22	.727		33	24.50	3.57	61.07	77.50
6110	5" diameter pipe size	▼	18	.889		42.50	30	4.36	76.86	97
6120	6" diameter pipe size	Q-18	23	1.043		48	36.50	3.41	87.91	112
6130	8" diameter pipe size		19	1.263		69.50	44.50	4.13	118.13	148
6140	10" diameter pipe size		16	1.500		90.50	52.50	4.91	147.91	184
6150	12" diameter pipe size	▼	13	1.846		108	65	6.05	179.05	224
6190	For 2" thick insulation, add					15%				
6220	For 2-1/2" thick insulation, add					25%				
6260	For 3" thick insulation, add				▼	30%				
6800	Black steel service pipe, ex. hvy. wt., 1" thick insul.									
6820	3/4" diameter pipe size	Q-17	50	.320	L.F.	15.20	10.85	1.57	27.62	35
6840	1" diameter pipe size		47	.340		16.15	11.50	1.67	29.32	37
6850	1-1/4" diameter pipe size		44	.364		18.65	12.30	1.79	32.74	41
6860	1-1/2" diameter pipe size		42	.381		19.85	12.90	1.87	34.62	43.50
6870	2" diameter pipe size		40	.400		21	13.55	1.96	36.51	45.50
6880	2-1/2" diameter pipe size		31	.516		26	17.45	2.53	45.98	58
6890	3" diameter pipe size		27	.593		29.50	20	2.91	52.41	65.50
6900	4" diameter pipe size		21	.762		38.50	26	3.74	68.24	85.50
6910	5" diameter pipe size	▼	17	.941		54	32	4.62	90.62	113
6920	6" diameter pipe size	Q-18	22	1.091	▼	59.50	38.50	3.57	101.57	127
7400	Black steel service pipe, ex. hvy. wt., 1-1/2" thick insul.									
7420	3/4" diameter pipe size	Q-17	50	.320	L.F.	15.20	10.85	1.57	27.62	35
7440	1" diameter pipe size		47	.340		17.30	11.50	1.67	30.47	38.50
7450	1-1/4" diameter pipe size		44	.364		20	12.30	1.79	34.09	42.50
7460	1-1/2" diameter pipe size		42	.381		22	12.90	1.87	36.77	45.50
7470	2" diameter pipe size	▼	40	.400	▼	22	13.55	1.96	37.51	47

SITE CONSTRUCTION 2

02550 | Piped Energy Distribution

		CREW	DAILY OUTPUT	LABOR-HOURS	UNIT	2003 BARE COSTS				TOTAL INCL O&P	
						MAT.	LABOR	EQUIP.	TOTAL		
550	7480	2-1/2" diameter pipe size	Q-17	31	.516	L.F.	26.50	17.45	2.53	46.48	58.50	**550**
	7490	3" diameter pipe size	↓	27	.593		31	20	2.91	53.91	67.50	
	7500	4" diameter pipe size		21	.762		40.50	26	3.74	70.24	87.50	
	7510	5" diameter pipe size	▼	17	.941		56	32	4.62	92.62	115	
	7520	6" diameter pipe size	Q-18	22	1.091		61.50	38.50	3.57	103.57	130	
	7530	8" diameter pipe size		18	1.333		92.50	47	4.36	143.86	177	
	7540	10" diameter pipe size		15	1.600		110	56	5.25	171.25	212	
	7550	12" diameter pipe size	▼	13	1.846		135	65	6.05	206.05	253	
	7590	For 2" thick insulation, add					13%					
	7640	For 2-1/2" thick insulation, add					18%					
	7680	For 3" thick insulation, add				▼	24%					
	9000	Combined steam pipe with condensate return										
	9010	8" and 4" in 24" case	Q-18	100	.240	L.F.	195	8.40	.78	204.18	228	
	9020	6" and 3" in 20" case		104	.231		133	8.10	.75	141.85	159	
	9030	3" and 1-1/2" in 16" case		110	.218		99.50	7.65	.71	107.86	121	
	9040	2" and 1-1/4" in 12-3/4" case		114	.211		86.50	7.40	.69	94.59	107	
	9050	1-1/2" and 1-1/4" in 10-3/4" case	▼	116	.207	▼	75.50	7.25	.68	83.43	95	
	9100	Steam pipe only (no return)										
	9110	6" in 18" case	Q-18	108	.222	L.F.	109	7.80	.73	117.53	133	
	9120	4" in 14" case		112	.214		86.50	7.50	.70	94.70	107	
	9130	3" in 14" case		114	.211		80	7.40	.69	88.09	100	
	9140	2-1/2" in 12-3/4" case		118	.203		74.50	7.15	.67	82.32	93.50	
	9150	2" in 12-3/4" case	▼	122	.197	▼	71.50	6.90	.64	79.04	89.50	

02620 | Subdrainage

		CREW	DAILY OUTPUT	LABOR-HOURS	UNIT	2003 BARE COSTS				TOTAL INCL O&P	
						MAT.	LABOR	EQUIP.	TOTAL		
210	0010	**PIPING, SUBDRAINAGE, CONCRETE**										**210**
	0021	Not including excavation and backfill										
	3000	Porous wall concrete underdrain, std. strength, 4" diameter	B-20	335	.072	L.F.	1.86	2		3.86	5.15	
	3020	6" diameter	"	315	.076		2.42	2.12		4.54	6	
	3040	8" diameter	B-21	310	.090		2.99	2.59	.53	6.11	7.90	
	3100	18" diameter	"	165	.170		9.60	4.87	.99	15.46	19.20	
	4000	Extra strength, 6" diameter	B-20	315	.076		2.45	2.12		4.57	6	
	4020	8" diameter	B-21	310	.090		3.68	2.59	.53	6.80	8.65	
	4040	10" diameter		285	.098		7.35	2.82	.57	10.74	13.10	
	4060	12" diameter		230	.122		7.95	3.49	.71	12.15	15	
	4080	15" diameter		200	.140		8.80	4.02	.82	13.64	16.85	
	4100	18" diameter	▼	165	.170	▼	12.85	4.87	.99	18.71	23	
240	0010	**PIPING, SUBDRAINAGE, CORRUGATED METAL**										**240**
	0021	Not including excavation and backfill										
	2010	Aluminum, perforated										
	2020	6" diameter, 18 ga.	B-14	380	.126	L.F.	2.66	3.29	.58	6.53	8.65	
	2200	8" diameter, 16 ga.		370	.130		3.88	3.38	.59	7.85	10.15	
	2220	10" diameter, 16 ga.		360	.133		4.85	3.47	.61	8.93	11.40	
	2240	12" diameter, 16 ga.		285	.168		5.45	4.39	.77	10.61	13.65	
	2260	18" diameter, 16 ga.	▼	205	.234	▼	8.15	6.10	1.07	15.32	19.55	
	3000	Uncoated galvanized, perforated										
	3020	6" diameter, 18 ga.	B-20	380	.063	L.F.	4.28	1.76		6.04	7.45	

Important: See the Reference Section for critical supporting data - Reference Nos., Crews, & City Cost Indexes

02620	Subdrainage	CREW	DAILY OUTPUT	LABOR-HOURS	UNIT	2003 BARE COSTS				TOTAL INCL O&P	
						MAT.	LABOR	EQUIP.	TOTAL		
240 3200	8″ diameter, 16 ga.	B-20	370	.065	L.F.	5.90	1.81		7.71	9.30	**240**
3220	10″ diameter, 16 ga.	B-21	360	.078		8.85	2.23	.45	11.53	13.65	
3240	12″ diameter, 16 ga.		285	.098		9.25	2.82	.57	12.64	15.15	
3260	18″ diameter, 16 ga.	↓	205	.137	↓	14.15	3.92	.80	18.87	22.50	
4000	Steel, perforated, asphalt coated										
4020	6″ diameter 18 ga.	B-20	380	.063	L.F.	3.42	1.76		5.18	6.50	
4030	8″ diameter 18 ga	″	370	.065		5.35	1.81		7.16	8.70	
4040	10″ diameter 16 ga	B-21	360	.078		6.15	2.23	.45	8.83	10.75	
4050	12″ diameter 16 ga		285	.098		7.05	2.82	.57	10.44	12.80	
4060	18″ diameter 16 ga	↓	205	.137	↓	9.65	3.92	.80	14.37	17.60	
250 0010	**PIPING, SUBDRAINAGE, PLASTIC**										**250**
0020	Not including excavation and backfill										
1110	10″				Ea.	2.91			2.91	3.20	
2100	Perforated PVC, 4″ diameter	B-14	314	.153	L.F.	.70	3.98	.70	5.38	7.75	
2110	6″ diameter		300	.160		1.30	4.17	.73	6.20	8.65	
2120	8″ diameter		290	.166		1.47	4.31	.75	6.53	9.15	
2130	10″ diameter		280	.171		2.28	4.47	.78	7.53	10.30	
2140	12″ diameter	↓	270	.178	↓	3.15	4.63	.81	8.59	11.55	
280 0010	**PIPING, SUBDRAINAGE, VITRIFIED CLAY**										**280**
0020	Not including excavation and backfill										
3000	Perforated, 5′ lengths, C700, 4″ diameter	B-14	400	.120	L.F.	2	3.13	.55	5.68	7.65	
3020	6″ diameter		315	.152		3.31	3.97	.69	7.97	10.55	
3040	8″ diameter		290	.166		4.71	4.31	.75	9.77	12.75	
3060	12″ diameter	↓	275	.175		9.95	4.55	.80	15.30	18.80	
4000	Channel pipe, 4″ diameter	B-20	430	.056		1.89	1.55		3.44	4.51	
4020	6″ diameter		335	.072		3.32	2		5.32	6.75	
4060	8″ diameter	↓	295	.081		5.05	2.27		7.32	9.15	
4080	12″ diameter	B-21	280	.100	↓	11.10	2.87	.58	14.55	17.30	

02630 | Storm Drainage

		CREW	DAILY OUTPUT	LABOR-HOURS	UNIT	MAT.	LABOR	EQUIP.	TOTAL	TOTAL INCL O&P	
100 0010	**PIPING, STORM DRAINAGE, CORRUGATED METAL**										**100**
0020	Not including excavation or backfill										
2000	Corrugated metal pipe, galvanized and coated										
2020	Bituminous coated with paved invert, 20′ lengths										
2040	8″ diameter, 16 ga.	B-14	330	.145	L.F.	6.90	3.79	.66	11.35	14.20	
2060	10″ diameter, 16 ga.		260	.185		8.25	4.81	.84	13.90	17.45	
2080	12″ diameter, 16 ga.		210	.229		9.90	5.95	1.04	16.89	21.50	
2100	15″ diameter, 16 ga.		200	.240		12.05	6.25	1.09	19.39	24	
2120	18″ diameter, 16 ga.		190	.253		15.40	6.60	1.15	23.15	28.50	
2140	24″ diameter, 14 ga.	↓	160	.300		18.85	7.80	1.37	28.02	34	
2160	30″ diameter, 14 ga.	B-13	120	.467		25	12.50	5.45	42.95	53	
2180	36″ diameter, 12 ga.		120	.467		36.50	12.50	5.45	54.45	65.50	
2200	48″ diameter, 12 ga.	↓	100	.560		55.50	15	6.55	77.05	91	
2220	60″ diameter, 10 ga.	B-13B	75	.747		72	19.95	15.40	107.35	127	
2240	72″ diameter, 8 ga.	″	45	1.244	↓	107	33.50	25.50	166	198	
2500	Galvanized, uncoated, 20′ lengths										
2520	8″ diameter, 16 ga.	B-14	355	.135	L.F.	5.30	3.52	.62	9.44	12	
2540	10″ diameter, 16 ga.		280	.171		5.85	4.47	.78	11.10	14.20	
2560	12″ diameter, 16 ga.		220	.218		6.65	5.70	.99	13.34	17.30	
2580	15″ diameter, 16 ga.		220	.218		8.45	5.70	.99	15.14	19.25	
2600	18″ diameter, 16 ga.		205	.234		11.20	6.10	1.07	18.37	23	
2620	24″ diameter, 14 ga.	↓	175	.274		16.25	7.15	1.25	24.65	30.50	
2640	30″ diameter, 14 ga.	B-13	130	.431		20.50	11.50	5.05	37.05	46.50	
2660	36″ diameter, 12 ga.		130	.431		34	11.50	5.05	50.55	61	
2680	48″ diameter, 12 ga.	↓	110	.509		45.50	13.60	5.95	65.05	77.50	
2690	60″ diameter, 10 ga.	B-13B	78	.718	↓	71.50	19.20	14.80	105.50	124	

SITE CONSTRUCTION 2

2 SITE CONSTRUCTION

02630	Storm Drainage	CREW	DAILY OUTPUT	LABOR-HOURS	UNIT	2003 BARE COSTS MAT.	LABOR	EQUIP.	TOTAL	TOTAL INCL O&P
100										**100**
2780	End sections, 8" diameter	B-14	24	2	Ea.	55	52	9.10	116.10	152
2785	10" diameter		22	2.182		56.50	57	9.95	123.45	161
2790	12" diameter		35	1.371		49	35.50	6.25	90.75	116
2800	18" diameter		30	1.600		75.50	41.50	7.30	124.30	156
2810	24" diameter	B-13	25	2.240		110	60	26	196	243
2820	30" diameter		25	2.240		209	60	26	295	350
2825	36" diameter		20	2.800		277	75	32.50	384.50	455
2830	48" diameter		10	5.600		570	150	65.50	785.50	935
2835	60" diameter	B-13B	5	11.200		750	300	231	1,281	1,550
2840	72" diameter	"	4	14		1,350	375	289	2,014	2,375
3000	Corrugated galvanized or alum. oval arch culverts, coated & paved									
3020	17" x 13", 16 ga., 15" equivalent	B-14	200	.240	L.F.	20.50	6.25	1.09	27.84	33.50
3040	21" x 15", 16 ga., 18" equivalent		150	.320		26	8.35	1.46	35.81	43.50
3060	28" x 20", 14 ga., 24" equivalent		125	.384		37.50	10	1.75	49.25	59
3080	35" x 24", 14 ga., 30" equivalent		100	.480		45.50	12.50	2.19	60.19	72.50
3100	42" x 29", 12 ga., 36" equivalent	B-13	100	.560		68	15	6.55	89.55	105
3120	49" x 33", 12 ga., 42" equivalent		90	.622		82	16.65	7.25	105.90	124
3140	57" x 38", 12 ga., 48" equivalent		75	.747		91.50	19.95	8.75	120.20	141
3160	Steel, plain oval arch culverts, plain									
3180	17" x 13", 16 ga., 15" equivalent	B-14	225	.213	L.F.	11	5.55	.97	17.52	22
3200	21" x 15", 16 ga., 18" equivalent		175	.274		13.05	7.15	1.25	21.45	27
3220	28" x 20", 14 ga., 24" equivalent		150	.320		21	8.35	1.46	30.81	37.50
3240	35" x 24", 14 ga., 30" equivalent	B-13	108	.519		26.50	13.85	6.05	46.40	57
3260	42" x 29", 12 ga., 36" equivalent		108	.519		43.50	13.85	6.05	63.40	75.50
3280	49" x 33", 12 ga., 42" equivalent		92	.609		51.50	16.30	7.10	74.90	89.50
3300	57" x 38", 12 ga., 48" equivalent		75	.747		45	19.95	8.75	73.70	90
3320	End sections, 17" x 13"		22	2.545	Ea.	60.50	68	30	158.50	204
3340	42" x 29"		17	3.294	"	247	88	38.50	373.50	450
3360	Multi-plate arch, steel	B-20	1,690	.014	Lb.	.69	.40		1.09	1.38
200										**200**
0010	**CATCH BASINS OR MANHOLES** not including footing, excavation,									
0020	backfill, frame and cover									
0050	Brick, 4' inside diameter, 4' deep	D-1	1	16	Ea.	284	455		739	1,000
0100	6' deep		.70	22.857		395	650		1,045	1,425
0150	8' deep		.50	32		510	910		1,420	1,950
0200	For depths over 8', add		4	4	V.L.F.	110	114		224	296
0400	Concrete blocks (radial), 4' I.D., 4' deep		1.50	10.667	Ea.	238	305		543	725
0500	6' deep		1	16		315	455		770	1,050
0600	8' deep		.70	22.857		390	650		1,040	1,425
0700	For depths over 8', add		5.50	2.909	V.L.F.	40	83		123	171
0800	Concrete, cast in place, 4' x 4', 8" thick, 4' deep	C-14H	2	24	Ea.	355	750	16.30	1,121.30	1,575
0900	6' deep		1.50	32		515	1,000	22	1,537	2,175
1000	8' deep		1	48		740	1,500	32.50	2,272.50	3,200
1100	For depths over 8', add		8	6	V.L.F.	85	187	4.08	276.08	390
1110	Precast, 4' I.D., 4' deep	B-22	4.10	7.317	Ea.	430	212	59.50	701.50	870
1120	6' deep		3	10		555	290	81.50	926.50	1,150
1130	8' deep		2	15		645	435	122	1,202	1,525
1140	For depths over 8', add		16	1.875	V.L.F.	90.50	54.50	15.30	160.30	201
1150	5' I.D., 4' deep	B-6	3	8	Ea.	455	214	73	742	910
1160	6' deep		2	12		615	320	109	1,044	1,300
1170	8' deep		1.50	16		775	430	146	1,351	1,675
1180	For depths over 8', add		12	2	V.L.F.	101	53.50	18.25	172.75	214
1190	6' I.D., 4' deep		2	12	Ea.	745	320	109	1,174	1,425
1200	6' deep		1.50	16		970	430	146	1,546	1,900
1210	8' deep		1	24		1,200	645	219	2,064	2,550
1220	For depths over 8', add		8	3	V.L.F.	156	80.50	27.50	264	325

Important: See the Reference Section for critical supporting data - Reference Nos., Crews, & City Cost Indexes

02630	Storm Drainage	CREW	DAILY OUTPUT	LABOR-HOURS	UNIT	2003 BARE COSTS				TOTAL INCL O&P	
						MAT.	LABOR	EQUIP.	TOTAL		
200											**200**
1250	Slab tops, precast, 8" thick										
1300	4' diameter manhole	B-6	8	3	Ea.	160	80.50	27.50	268	330	
1400	5' diameter manhole		7.50	3.200		305	86	29	420	500	
1500	6' diameter manhole		7	3.429		365	92	31.50	488.50	575	
1600	Frames & covers, C.I., 24" square, 500 lb.		7.80	3.077		220	82.50	28	330.50	400	
1700	26" D shape, 600 lb.		7	3.429		365	92	31.50	488.50	575	
1800	Light traffic, 18" diameter, 100 lb.		10	2.400		119	64.50	22	205.50	255	
1900	24" diameter, 300 lb.		8.70	2.759		162	74	25	261	320	
2000	36" diameter, 900 lb.		5.80	4.138		405	111	37.50	553.50	660	
2100	Heavy traffic, 24" diameter, 400 lb.		7.80	3.077		165	82.50	28	275.50	340	
2200	36" diameter, 1150 lb.		3	8		485	214	73	772	945	
2300	Mass. State standard, 26" diameter, 475 lb.		7	3.429		370	92	31.50	493.50	580	
2400	30" diameter, 620 lb.		7	3.429		310	92	31.50	433.50	515	
2500	Watertight, 24" diameter, 350 lb.		7.80	3.077		261	82.50	28	371.50	445	
2600	26" diameter, 500 lb.		7	3.429		258	92	31.50	381.50	460	
2700	32" diameter, 575 lb.	↓	6	4	↓	530	107	36.50	673.50	785	
2800	3 piece cover & frame, 10" deep,										
2900	1200 lbs., for heavy equipment	B-6	3	8	Ea.	820	214	73	1,107	1,300	
3000	Raised for paving 1-1/4" to 2" high,										
3100	4 piece expansion ring										
3200	20" to 26" diameter	1 Clab	3	2.667	Ea.	108	65.50		173.50	222	
3300	30" to 36" diameter	"	3	2.667	"	152	65.50		217.50	270	
3320	Frames and covers, existing, raised for paving, 2", including										
3340	row of brick, concrete collar, up to 12" wide frame	B-6	18	1.333	Ea.	33.50	35.50	12.15	81.15	105	
3360	20" to 26" wide frame		11	2.182		44	58.50	19.90	122.40	161	
3380	30" to 36" wide frame	↓	9	2.667		55	71.50	24.50	151	198	
3400	Inverts, single channel brick	D-1	3	5.333		62	152		214	300	
3500	Concrete		5	3.200		47.50	91		138.50	193	
3600	Triple channel, brick		2	8		94	228		322	455	
3700	Concrete	↓	3	5.333		81	152		233	325	
3800	Steps, heavyweight cast iron, 7" x 9"	1 Bric	40	.200		9.55	6.50		16.05	20.50	
3900	8" x 9"		40	.200		14.30	6.50		20.80	25.50	
3928	12" x 10-1/2"		40	.200		13.05	6.50		19.55	24.50	
4000	Standard sizes, galvanized steel		40	.200		11.75	6.50		18.25	23	
4100	Aluminum	↓	40	.200	↓	15	6.50		21.50	26.50	

02740	Flexible Pavement	CREW	DAILY OUTPUT	LABOR-HOURS	UNIT	2003 BARE COSTS				TOTAL INCL O&P	
						MAT.	LABOR	EQUIP.	TOTAL		
300											**300**
0010	**ASPHALTIC CONCRETE PAVEMENT** for highways										
0020	and large paved areas										
0080	Binder course, 1-1/2" thick	B-25	7,725	.011	S.Y.	2.05	.31	.25	2.61	3.01	
0120	2" thick		6,345	.014		2.74	.37	.30	3.41	3.92	
0160	3" thick		4,905	.018		4.06	.48	.39	4.93	5.65	
0200	4" thick	↓	4,140	.021		5.45	.57	.46	6.48	7.40	
0300	Wearing course, 1" thick	B-25B	10,575	.009		1.51	.25	.20	1.96	2.27	
0340	1-1/2" thick		7,725	.012		2.31	.34	.27	2.92	3.37	
0380	2" thick		6,345	.015		3.10	.42	.33	3.85	4.41	
0420	2-1/2" thick	↓	5,480	.018	↓	3.82	.48	.38	4.68	5.35	

SITE CONSTRUCTION **2**

02700 | Bases, Ballasts, Pavements & Appurtenances

02740 | Flexible Pavement

			CREW	DAILY OUTPUT	LABOR-HOURS	UNIT	MAT.	LABOR	EQUIP.	TOTAL	TOTAL INCL O&P	
300	0460	3" thick	B-25B	4,900	.020	S.Y.	4.55	.54	.42	5.51	6.30	**300**
	0500	Open graded friction course	B-25C	5,000	.010	↓	1.86	.27	.33	2.46	2.82	
	0800	Alternate method of figuring paving costs										
	0810	Binder course, 1-1/2" thick	B-25	630	.140	Ton	27	3.77	3.01	33.78	38.50	
	0811	2" thick		690	.128		27	3.44	2.75	33.19	38	
	0812	3" thick		800	.110		27	2.97	2.37	32.34	36.50	
	0813	4" thick	▼	850	.104		27	2.80	2.23	32.03	36.50	
	0850	Wearing course, 1" thick	B-25B	575	.167		28	4.59	3.61	36.20	42	
	0851	1-1/2" thick		630	.152		28	4.19	3.30	35.49	41	
	0852	2" thick		690	.139		28	3.82	3.01	34.83	40	
	0853	2-1/2" thick		745	.129		28	3.54	2.79	34.33	39.50	
	0854	3" thick	▼	800	.120	▼	28	3.30	2.60	33.90	39	
	1000	Pavement replacement over trench, 2" thick	B-37	90	.533	S.Y.	3.15	13.90	1.19	18.24	26.50	
	1050	4" thick		70	.686		6.25	17.85	1.53	25.63	36	
	1080	6" thick	▼	55	.873	▼	9.95	22.50	1.95	34.40	48.50	
315	0011	**PAVING** Asphaltic concrete, parking lots & driveways										**315**
	0020	6" stone base, 2" binder course, 1" topping	B-25C	9,000	.005	S.F.	1.17	.15	.18	1.50	1.72	
	0300	Binder course, 1-1/2" thick		35,000	.001		.29	.04	.05	.38	.43	
	0400	2" thick		25,000	.002		.37	.05	.07	.49	.56	
	0500	3" thick		15,000	.003		.57	.09	.11	.77	.89	
	0600	4" thick		10,800	.004		.75	.12	.15	1.02	1.19	
	0800	Sand finish course, 3/4" thick		41,000	.001		.18	.03	.04	.25	.29	
	0900	1" thick	▼	34,000	.001		.22	.04	.05	.31	.35	
	1000	Fill pot holes, hot mix, 2" thick	B-16	4,200	.008		.42	.19	.11	.72	.88	
	1100	4" thick		3,500	.009		.61	.23	.13	.97	1.18	
	1120	6" thick	▼	3,100	.010		.83	.26	.15	1.24	1.48	
	1140	Cold patch, 2" thick	B-51	3,000	.016		.48	.40	.04	.92	1.18	
	1160	4" thick		2,700	.018		.90	.45	.04	1.39	1.74	
	1180	6" thick	▼	1,900	.025	▼	1.41	.63	.06	2.10	2.60	

02800 | Site Improvements and Amenities

02810 | Irrigation System

			CREW	DAILY OUTPUT	LABOR-HOURS	UNIT	MAT.	LABOR	EQUIP.	TOTAL	TOTAL INCL O&P	
800	0010	**SPRINKLER IRRIGATION SYSTEM** For lawns										**800**
	0100	Golf course with fully automatic system	C-17	.05	1,600	9 holes	79,000	52,000		131,000	168,500	
	0200	24' diam. head at 15' O.C. incl. piping, auto oper., minimum	B-20	70	.343	Head	17.35	9.55		26.90	34	
	0300	Maximum		40	.600		40	16.70		56.70	70	
	0500	60' diam. head at 40' O.C. incl. piping, auto oper., minimum		28	.857		52.50	24		76.50	95.50	
	0600	Maximum		23	1.043	▼	147	29		176	208	
	0800	Residential system, custom, 1" supply		2,000	.012	S.F.	.26	.33		.59	.81	
	0900	1-1/2" supply	▼	1,800	.013	"	.29	.37		.66	.90	

02815 | Fountains

			CREW	DAILY OUTPUT	LABOR-HOURS	UNIT	MAT.	LABOR	EQUIP.	TOTAL	TOTAL INCL O&P	
225	0010	**FOUNTAINS/AERATORS**										**225**
	0100	Pump w/controls										
	0200	Single phase, 100' chord, 1/2 H.P. pump	2 Skwk	4.40	3.636	Ea.	2,550	117		2,667	3,000	
	0300	3/4 H.P. pump		4.30	3.721		2,925	120		3,045	3,400	
	0400	1 H.P. pump		4.20	3.810		2,950	123		3,073	3,450	
	0500	1-1/2 H.P. pump	▼	4.10	3.902	▼	3,050	126		3,176	3,550	

Important: See the Reference Section for critical supporting data - Reference Nos., Crews, & City Cost Indexes

2 SITE CONSTRUCTION

02815 | Fountains

		CREW	DAILY OUTPUT	LABOR-HOURS	UNIT	2003 BARE COSTS				TOTAL INCL O&P	
						MAT.	LABOR	EQUIP.	TOTAL		
225	0600	2 H.P. pump	2 Skwk	4	4	Ea.	3,100	129		3,229	3,600	225
	0700	Three phase, 200' chord, 5 H.P. pump		3.90	4.103		7,475	132		7,607	8,425	
	0800	7-1/2 H.P. pump		3.80	4.211		8,450	136		8,586	9,525	
	0900	10 H.P. pump		3.70	4.324		9,425	139		9,564	10,600	
	1000	15 H.P. pump		3.60	4.444		11,200	143		11,343	12,500	
	1100	Nozzles, minimum		8	2		187	64.50		251.50	305	
	1200	Maximum		8	2		246	64.50		310.50	370	
	1300	Lights w/mounting kits, 200 watt		18	.889		310	28.50		338.50	385	
	1400	300 watt		18	.889		345	28.50		373.50	420	
	1500	500 watt		18	.889		375	28.50		403.50	460	
	1600	Color blender	▼	12	1.333	▼	292	43		335	390	

02950 | Site Restoration & Rehabilitation

02955 | Restoration of Underground Piping

		CREW	DAILY OUTPUT	LABOR-HOURS	UNIT	2003 BARE COSTS				TOTAL INCL O&P	
						MAT.	LABOR	EQUIP.	TOTAL		
100	0010	**PIPE INTERNAL CLEANING & INSPECTION**										100
	0100	Cleaning, pressure pipe systems										
	0120	Pig method, lengths 1000' to 10,000'										
	0140	4" diameter thru 24" diameter, minimum				L.F.				2.14	2.46	
	0160	Maximum				"				5.35	6.40	
	6000	Sewage/sanitary systems										
	6100	Power rodder with header & cutters										
	6110	Mobilization charge, minimum				Total				320	375	
	6120	Mobilization charge, maximum				"				750	850	
	6140	4" diameter				L.F.				1.07	1.23	
	6150	6" diameter								1.34	1.55	
	6160	8" diameter								1.60	1.87	
	6170	10" diameter								1.87	2.14	
	6180	12" diameter								2.03	2.35	
	6190	14" diameter								2.14	2.46	
	6200	16" diameter								2.41	2.78	
	6210	18" diameter								2.68	3.05	
	6220	20" diameter								2.78	3.21	
	6230	24" diameter								3.10	3.58	
	6240	30" diameter								3.75	4.28	
	6250	36" diameter								4.28	4.92	
	6260	48" diameter								4.60	5.30	
	6270	60" diameter								6	6.90	
	6280	72" diameter				▼				6.65	7.65	
	9000	Inspection, television camera with film										
	9060	500 linear feet				Total				1,075	1,230	
680	0010	**PIPE REPAIR**										680
	0020	Not including excavation or backfill										
	0100	Clamp, stainless steel, lightweight, for steel pipe										
	0110	3" long, 1/2" diameter pipe	1 Plum	34	.235	Ea.	8.05	8.80		16.85	22	
	0120	3/4" diameter pipe		32	.250		8.30	9.35		17.65	23	
	0130	1" diameter pipe		30	.267		8.90	9.95		18.85	25	
	0140	1-1/4" diameter pipe		28	.286		9.45	10.65		20.10	26.50	
	0150	1-1/2" diameter pipe	▼	26	.308	▼	9.75	11.50		21.25	28	

SITE CONSTRUCTION 2

02955	Restoration of Underground Piping	CREW	DAILY OUTPUT	LABOR-HOURS	UNIT	\multicolumn{4}{c}{2003 BARE COSTS}	TOTAL INCL O&P	
						MAT.	LABOR	EQUIP.	TOTAL		
680	0160	2" diameter pipe	1 Plum	24	.333	Ea.	10.65	12.45		23.10	30.50	**680**
	0170	2-1/2" diameter pipe		23	.348		11.60	13		24.60	32.50	
	0180	3" diameter pipe		22	.364		12.80	13.60		26.40	34.50	
	0190	3-1/2" diameter pipe		21	.381		13.75	14.25		28	36.50	
	0200	4" diameter pipe	B-20	56	.429		14.35	11.95		26.30	34.50	
	0210	5" diameter pipe		53	.453		16.50	12.60		29.10	38	
	0220	6" diameter pipe		48	.500		18.30	13.95		32.25	42	
	0230	8" diameter pipe		30	.800		21.50	22.50		44	58.50	
	0240	10" diameter pipe		28	.857		55	24		79	98	
	0250	12" diameter pipe		24	1		58.50	28		86.50	108	
	0260	14" diameter pipe		22	1.091		61	30.50		91.50	115	
	0270	16" diameter pipe		20	1.200		66	33.50		99.50	125	
	0280	18" diameter pipe		18	1.333		69.50	37		106.50	135	
	0290	20" diameter pipe		16	1.500		77	42		119	150	
	0300	24" diameter pipe		14	1.714		85.50	47.50		133	169	
	0360	For 6" long, add					100%	40%				
	0370	For 9" long, add					200%	100%				
	0380	For 12" long, add					300%	150%				
	0390	For 18" long, add					500%	200%				
	1000	Clamp, stainless steel, with threaded service tap										
	1040	Full seal for iron, steel, PVC pipe										
	1100	6" long, 2" diameter pipe	1 Plum	17	.471	Ea.	104	17.60		121.60	141	
	1110	2-1/2" diameter pipe		16	.500		110	18.70		128.70	149	
	1120	3" diameter pipe		15.60	.513		117	19.15		136.15	158	
	1130	3-1/2" diameter pipe		15	.533		122	19.90		141.90	164	
	1140	4" diameter pipe	B-20	40	.600		134	16.70		150.70	174	
	1150	6" diameter pipe		34	.706		155	19.65		174.65	201	
	1160	8" diameter pipe		21	1.143		193	32		225	262	
	1170	10" diameter pipe		20	1.200		214	33.50		247.50	288	
	1180	12" diameter pipe		17	1.412		244	39.50		283.50	330	
	1205	For 9" long, add					20%	45%				
	1210	For 12" long, add					40%	80%				
	1220	For 18" long, add					70%	110%				
	1600	Clamp, stainless steel, single section										
	1640	Full seal for iron, steel, PVC pipe										
	1700	6" long, 2" diameter pipe	1 Plum	17	.471	Ea.	53.50	17.60		71.10	85.50	
	1710	2-1/2" diameter pipe		16	.500		58.50	18.70		77.20	92.50	
	1720	3" diameter pipe		15.60	.513		61	19.15		80.15	96	
	1730	3-1/2" diameter pipe		15	.533		67	19.90		86.90	104	
	1740	4" diameter pipe	B-20	40	.600		73	16.70		89.70	107	
	1750	6" diameter pipe		34	.706		91.50	19.65		111.15	132	
	1760	8" diameter pipe		21	1.143		122	32		154	184	
	1770	10" diameter pipe		20	1.200		140	33.50		173.50	207	
	1780	12" diameter pipe		17	1.412		153	39.50		192.50	230	
	1800	For 9" long, add					40%	45%				
	1810	For 12" long, add					60%	80%				
	1820	For 18" long, add					120%	110%				
	2000	Clamp, stainless steel, two section										
	2040	Full seal, for iron, steel, PVC pipe										
	2100	6" long, 4" diameter pipe	B-20	24	1	Ea.	104	28		132	158	
	2110	6" diameter pipe		20	1.200		122	33.50		155.50	187	
	2120	8" diameter pipe		13	1.846		134	51.50		185.50	229	
	2130	10" diameter pipe		12	2		195	55.50		250.50	300	
	2140	12" diameter pipe		10	2.400		214	67		281	340	
	2200	9" long, 4" diameter pipe		16	1.500		134	42		176	214	
	2210	6" diameter pipe		13	1.846		153	51.50		204.50	249	

Important: See the Reference Section for critical supporting data - Reference Nos., Crews, & City Cost Indexes

02955	Restoration of Underground Piping	CREW	DAILY OUTPUT	LABOR-HOURS	UNIT	2003 BARE COSTS				TOTAL INCL O&P	
						MAT.	LABOR	EQUIP.	TOTAL		
680	2220	8" diameter pipe	B-20	9	2.667	Ea.	159	74.50		233.50	290	680
	2230	10" diameter pipe		8	3		220	83.50		303.50	375	
	2240	12" diameter pipe		7	3.429		244	95.50		339.50	415	
	2250	14" diameter pipe		6.40	3.750		365	104		469	570	
	2260	16" diameter pipe		6	4		385	111		496	600	
	2270	18" diameter pipe		5	4.800		395	134		529	645	
	2280	20" diameter pipe		4.60	5.217		425	145		570	695	
	2290	24" diameter pipe		4	6		730	167		897	1,075	
	2320	For 12" long, add to 9"					15%	25%				
	2330	For 18" long, add to 9"					70%	55%				
	8000	For internal cleaning and inspection, see Div. 02955-100										
	8100	For pipe testing, see Div. 15955-700										
800	0010	**CORROSION RESISTANCE** Wrap & coat, add to pipe, 4" dia.				L.F.	1.39			1.39	1.53	800
	0020	5" diameter					1.46			1.46	1.61	
	0040	6" diameter					1.46			1.46	1.61	
	0060	8" diameter					2.26			2.26	2.49	
	0080	10" diameter					3.37			3.37	3.71	
	0100	12" diameter					3.37			3.37	3.71	
	0120	14" diameter					4.93			4.93	5.40	
	0140	16" diameter					4.93			4.93	5.40	
	0160	18" diameter					4.99			4.99	5.50	
	0180	20" diameter					5.20			5.20	5.75	
	0200	24" diameter					6.40			6.40	7	
	0220	Small diameter pipe, 1" diameter, add					1.05			1.05	1.16	
	0240	2" diameter					1.15			1.15	1.27	
	0260	2-1/2" diameter					1.27			1.27	1.40	
	0280	3" diameter					1.27			1.27	1.40	
	0300	Fittings, field covered, add				S.F.	6.65			6.65	7.35	
	0500	Coating, bituminous, per diameter inch, 1 coat, add				L.F.	.26			.26	.29	
	0540	3 coat					.40			.40	.44	
	0560	Coal tar epoxy, per diameter inch, 1 coat, add					.15			.15	.17	
	0600	3 coat					.32			.32	.35	
	1000	Polyethylene H.D. extruded, .025" thk., 1/2" diameter add					.23			.23	.25	
	1020	3/4" diameter					.24			.24	.26	
	1040	1" diameter					.27			.27	.30	
	1060	1-1/4" diameter					.32			.32	.35	
	1080	1-1/2" diameter					.36			.36	.40	
	1100	.030" thk., 2" diameter					.39			.39	.43	
	1120	2-1/2" diameter					.46			.46	.51	
	1140	.035" thk., 3" diameter					.57			.57	.63	
	1160	3-1/2" diameter					.65			.65	.72	
	1180	4" diameter					.72			.72	.79	
	1200	.040" thk, 5" diameter					1			1	1.10	
	1220	6" diameter					1.04			1.04	1.14	
	1240	8" diameter					1.36			1.36	1.50	
	1260	10" diameter					1.66			1.66	1.83	
	1280	12" diameter					1.98			1.98	2.18	
	1300	.060" thk., 14" diameter					2.73			2.73	3	
	1320	16" diameter					3.20			3.20	3.52	
	1340	18" diameter					3.78			3.78	4.16	
	1360	20" diameter					4.25			4.25	4.68	
	1380	Fittings, field wrapped, add				S.F.	4.39			4.39	4.83	

For information about Means Estimating Seminars, see yellow pages 12 and 13 in back of book

SITE CONSTRUCTION 2

		DAILY OUTPUT	LABOR-HOURS	UNIT	2003 BARE COSTS				TOTAL INCL O&P
	CREW				MAT.	LABOR	EQUIP.	TOTAL	

Division 3
Concrete

Estimating Tips

General

- Carefully check all the plans and specifications. Concrete often appears on drawings other than structural drawings, including mechanical and electrical drawings for equipment pads. The cost of cutting and patching is often difficult to estimate. See Subdivision 02225 for demolition costs.
- Always obtain concrete prices from suppliers near the job site. A volume discount can often be negotiated depending upon competition in the area. Remember to add for waste, particularly for slabs and footings on grade.

03100 Concrete Forms & Accessories

- A primary cost for concrete construction is forming. Most jobs today are constructed with prefabricated forms. The selection of the forms best suited for the job and the total square feet of forms required for efficient concrete forming and placing are key elements in estimating concrete construction. Enough forms must be available for erection to make efficient use of the concrete placing equipment and crew.
- Concrete accessories for forming and placing depend upon the systems used. Study the plans and specifications to assure that all special accessory requirements have been included in the cost estimate such as anchor bolts, inserts and hangers.

03200 Concrete Reinforcement

- Ascertain that the reinforcing steel supplier has included all accessories, cutting, bending and an allowance for lapping, splicing and waste. A good rule of thumb is 10% for lapping, splicing and waste. Also, 10% waste should be allowed for welded wire fabric.

03300 Cast-in-Place Concrete

- When estimating structural concrete, pay particular attention to requirements for concrete additives, curing methods and surface treatments. Special consideration for climate, hot or cold, must be included in your estimate. Be sure to include requirements for concrete placing equipment and concrete finishing.

03400 Precast Concrete
03500 Cementitious Decks & Toppings

- The cost of hauling precast concrete structural members is often an important factor. For this reason, it is important to get a quote from the nearest supplier. It may become economically feasible to set up precasting beds on the site if the hauling costs are prohibitive.

Reference Numbers

Reference numbers are shown in bold squares at the beginning of some major classifications. These numbers refer to related items in the Reference Section. The reference information may be an estimating procedure, an alternate pricing method or technical information.

Note: Not all subdivisions listed here necessarily appear in this publication.

03100 | Concrete Forms & Accessories

		03110	**Structural C.I.P. Forms**	CREW	DAILY OUTPUT	LABOR-HOURS	UNIT	2003 BARE COSTS				TOTAL INCL O&P	
								MAT.	LABOR	EQUIP.	TOTAL		
425	0010	**FORMS IN PLACE, EQUIPMENT FOUNDATIONS** job built											**425**
	0020		1 use	C-2	160	.300	SFCA	2.51	9.20		11.71	17.15	
	0050		2 use		190	.253		1.38	7.75		9.13	13.65	
	0100		3 use		200	.240		1	7.40		8.40	12.60	
	0150		4 use	↓	205	.234	↓	.82	7.20		8.02	12.15	
620	0010	**ACCESSORIES, SLEEVES AND CHASES**											**620**
	0100		Plastic, 1 use, 9" long, 2" diameter	1 Carp	100	.080	Ea.	.53	2.52		3.05	4.52	
	0150		4" diameter		90	.089		1.56	2.80		4.36	6.10	
	0200		6" diameter		75	.107		2.75	3.37		6.12	8.30	
	0250		12" diameter		60	.133		18.05	4.21		22.26	26.50	
	5000		Sheet metal, 2" diameter		100	.080		.68	2.52		3.20	4.69	
	5100		4" diameter		90	.089		.85	2.80		3.65	5.30	
	5150		6" diameter		75	.107		1.23	3.37		4.60	6.60	
	5200		12" diameter		60	.133		2.46	4.21		6.67	9.25	
	6000		Steel pipe, 2" diameter		100	.080		1.80	2.52		4.32	5.90	
	6100		4" diameter		90	.089		6.05	2.80		8.85	11.05	
	6150		6" diameter		75	.107		12.85	3.37		16.22	19.40	
	6200		12" diameter	↓	60	.133	↓	36.50	4.21		40.71	47	

03200 | Concrete Reinforcement

		03210	**Reinforcing Steel**	CREW	DAILY OUTPUT	LABOR-HOURS	UNIT	2003 BARE COSTS				TOTAL INCL O&P	
								MAT.	LABOR	EQUIP.	TOTAL		
600	0010	**REINFORCING IN PLACE** A615 Grade 60											**600**
	0602		Slab on grade, #3 to #7	4 Rodm	4,200	.008	Lb.	.28	.27		.55	.76	

03300 | Cast-In-Place Concrete

		03310	**Structural Concrete**	CREW	DAILY OUTPUT	LABOR-HOURS	UNIT	2003 BARE COSTS				TOTAL INCL O&P	
								MAT.	LABOR	EQUIP.	TOTAL		
220	0010	**CONCRETE, READY MIX** Normal weight											**220**
	0020		2000 psi				C.Y.	68.50			68.50	75.50	
	0100		2500 psi					70			70	77	
	0150		3000 psi					70			70	77	
	0200		3500 psi					73.50			73.50	80.50	
	0300		4000 psi					76			76	83.50	
	0350		4500 psi					77.50			77.50	85	
	0400		5000 psi					78			78	86	
	0411		6000 psi					89			89	98	
	0412		8000 psi					145			145	160	
	0413		10,000 psi					206			206	227	
	0414		12,000 psi					249			249	274	
	1000		For high early strength cement, add					10%					
	2000		For all lightweight aggregate, add		↓			45%					

Important: See the Reference Section for critical supporting data - Reference Nos., Crews, & City Cost Indexes

03310	Structural Concrete	CREW	DAILY OUTPUT	LABOR-HOURS	UNIT	2003 BARE COSTS				TOTAL INCL O&P	
						MAT.	LABOR	EQUIP.	TOTAL		
240	0010	**CONCRETE IN PLACE** Including forms (4 uses), reinforcing									**240**
	0050	steel, including finishing unless otherwise indicated									
	0500	Chimney foundations, industrial, minimum	C-14C	32.22	3.476	C.Y.	142	105	1.01	248.01	320
	0510	Maximum		23.71	4.724		165	143	1.37	309.37	410
	3800	Footings, spread under 1 C.Y.		38.07	2.942		102	89	.85	191.85	253
	3850	Over 5 C.Y.		81.04	1.382		93.50	41.50	.40	135.40	169
	3900	Footings, strip, 18" x 9", plain		41.04	2.729		92	82.50	.79	175.29	232
	3950	36" x 12", reinforced		61.55	1.820		93.50	55	.53	149.03	190
	4000	Foundation mat, under 10 C.Y.		38.67	2.896		125	87.50	.84	213.34	276
	4050	Over 20 C.Y.	↓	56.40	1.986		111	60	.58	171.58	217
	4650	Slab on grade, not including finish, 4" thick	C-14E	60.75	1.449		87.50	45	.54	133.04	169
	4700	6" thick	"	92	.957	↓	84	30	.35	114.35	141
	4751	Slab on grade, incl. troweled finish, not incl. forms									
	4760	or reinforcing, over 10,000 S.F., 4" thick	C-14F	3,425	.021	S.F.	.95	.60	.01	1.56	1.96
	4820	6" thick		3,350	.021		1.39	.61	.01	2.01	2.46
	4840	8" thick		3,184	.023		1.91	.65	.01	2.57	3.08
	4900	12" thick		2,734	.026		2.86	.75	.01	3.62	4.29
	4950	15" thick	↓	2,505	.029	↓	3.59	.82	.01	4.42	5.20
700	0010	**PLACING CONCRETE** and vibrating, including labor & equipment									**700**
	1900	Footings, continuous, shallow, direct chute	C-6	120	.400	C.Y.		10.35	.54	10.89	16.65
	1950	Pumped	C-20	150	.427			11.35	5.35	16.70	23.50
	2000	With crane and bucket	C-7	90	.800			21.50	13.80	35.30	48
	2100	Footings, continuous, deep, direct chute	C-6	140	.343			8.90	.47	9.37	14.25
	2150	Pumped	C-20	160	.400			10.65	5	15.65	22
	2200	With crane and bucket	C-7	110	.655			17.55	11.30	28.85	39.50
	2900	Foundation mats, over 20 C.Y., direct chute	C-6	350	.137			3.55	.19	3.74	5.70
	2950	Pumped	C-20	400	.160			4.25	2.01	6.26	8.75
	3000	With crane and bucket	C-7	300	.240	↓		6.45	4.14	10.59	14.45

03350	Concrete Finishing										
300	0010	**FINISHING FLOORS** Monolithic, screed finish	1 Cefi	900	.009	S.F.		.27		.27	.40	**300**
	0100	Screed and bull float (darby) finish		725	.011			.33		.33	.49	
	0150	Screed, float, and broom finish	↓	630	.013	↓		.38		.38	.57	
600	0010	**PATCHING CONCRETE**										**600**
	0100	Floors, 1/4" thick, small areas, regular grout	1 Cefi	170	.047	S.F.	.07	1.42		1.49	2.18	
	0150	Epoxy grout	"	100	.080	"	3.71	2.42		6.13	7.65	
	0300	Slab on Grade, cut outs, up to 50 C.F.	2 Cefi	50	.320	C.F.	5.20	9.65		14.85	20	
	2000	Walls, including chipping, cleaning and epoxy grout										
	2100	Minimum	1 Cefi	65	.123	S.F.	3.21	3.72		6.93	9.05	
	2150	Average		50	.160		6.40	4.83		11.23	14.20	
	2200	Maximum	↓	40	.200		12.85	6.05		18.90	23	
	2510	Underlayment, P.C based self-leveling, 4100 psi, pumped, 1/4"	C-8	20,000	.003		1.35	.08	.04	1.47	1.65	
	2520	1/2"		19,000	.003		2.43	.08	.04	2.55	2.83	
	2530	3/4"		18,000	.003		3.78	.09	.04	3.91	4.34	
	2540	1"		17,000	.003		5.15	.09	.04	5.28	5.85	
	2550	1-1/2"	↓	15,000	.004		7.85	.10	.05	8	8.80	
	2560	Hand mix, 1/2"	C-18	4,000	.002		2.43	.06	.01	2.50	2.77	
	2610	Topping, P.C. based self-level/dry 6100 psi, pumped, 1/4"	C-8	20,000	.003		1.90	.08	.04	2.02	2.25	
	2620	1/2"		19,000	.003		3.42	.08	.04	3.54	3.92	
	2630	3/4"		18,000	.003		5.30	.09	.04	5.43	6.05	
	2660	1"		17,000	.003		7.20	.09	.04	7.33	8.15	
	2670	1-1/2"	↓	15,000	.004		11	.10	.05	11.15	12.30	
	2680	Hand mix, 1/2"	C-18	4,000	.002	↓	3.42	.06	.01	3.49	3.86	

CONCRETE 3

For information about Means Estimating Seminars, see yellow pages 12 and 13 in back of book

For expanded coverage of these items see *Means Concrete & Masonry Cost Data 2003*

		CREW	DAILY OUTPUT	LABOR-HOURS	UNIT	2003 BARE COSTS				TOTAL INCL O&P
						MAT.	LABOR	EQUIP.	TOTAL	

Division 4
Masonry

Estimating Tips

04050 Basic Masonry Materials & Methods

- The terms *mortar* and *grout* are often used interchangeably, and incorrectly. Mortar is used to bed masonry units, seal the entry of air and moisture, provide architectural appearance, and allow for size variations in the units. Grout is used primarily in reinforced masonry construction and is used to bond the masonry to the reinforcing steel. Common mortar types are M(2500 psi), S(1800 psi), N(750 psi), and O(350 psi), and conform to ASTM C270. Grout is either fine or coarse, conforms to ASTM C476, and in-place strengths generally exceed 2500 psi. Mortar and grout are different components of masonry construction and are placed by entirely different methods. An estimator should be aware of their unique uses and costs.
- Waste, specifically the loss/droppings of mortar and the breakage of brick and block, is included in all masonry assemblies in this division. A factor of 25% is added for mortar and 3% for brick and concrete masonry units.
- Scaffolding or staging is not included in any of the Division 4 costs. Refer to section 01540 for scaffolding and staging costs.

04200 Masonry Units

- The most common types of unit masonry are brick and concrete masonry. The major classifications of brick are building brick (ASTM C62), facing brick (ASTM C216) and glazed brick, fire brick and pavers. Many varieties of texture and appearance can exist within these classifications, and the estimator would be wise to check local custom and availability within the project area. On repair and remodeling jobs, matching the existing brick may be the most important criteria.
- Brick and concrete block are priced by the piece and then converted into a price per square foot of wall. Openings less than two square feet are generally ignored by the estimator because any savings in units used is offset by the cutting and trimming required.
- It is often difficult and expensive to find and purchase small lots of historic brick. Costs can vary widely. Many design issues affect costs, selection of mortar mix, and repairs or replacement of masonry materials. Cleaning techniques must be reflected in the estimate.

- All masonry walls, whether interior or exterior, require bracing. The cost of bracing walls during construction should be included by the estimator and this bracing must remain in place until permanent bracing is complete. Permanent bracing of masonry walls is accomplished by masonry itself, in the form of pilasters or abutting wall corners, or by anchoring the walls to the structural frame. Accessories in the form of anchors, anchor slots and ties are used, but their supply and installation can be by different trades. For instance, anchor slots on spandrel beams and columns are supplied and welded in place by the steel fabricator, but the ties from the slots into the masonry are installed by the bricklayer. Regardless of the installation method the estimator must be certain that these accessories are accounted for in pricing.

Reference Numbers

Reference numbers are shown in bold squares at the beginning of some major classifications. These numbers refer to related items in the Reference Section. The reference information may be an estimating procedure, an alternate pricing method or technical information.

Note: Not all subdivisions listed here necessarily appear in this publication.

04090 | Masonry Accessories

		CREW	DAILY OUTPUT	LABOR-HOURS	UNIT	2003 BARE COSTS				TOTAL INCL O&P	
						MAT.	LABOR	EQUIP.	TOTAL		
860	0010	**VENT BOX** Extruded aluminum, 4" deep, 2-3/8" x 8-1/8"	1 Bric	30	.267	Ea.	24	8.65		32.65	40	860
	0050	5" x 8-1/8"		25	.320		32.50	10.35		42.85	51.50	
	0100	2-1/4" x 25"		25	.320		64	10.35		74.35	86.50	
	0150	5" x 16-1/2"		22	.364		47.50	11.80		59.30	70.50	
	0200	6" x 16-1/2"		22	.364		69	11.80		80.80	94	
	0250	7-3/4" x 16-1/2"		20	.400		72.50	12.95		85.45	100	
	0400	For baked enamel finish, add					35%					
	0500	For cast aluminum, painted, add					60%					
	1000	Stainless steel ventilators, 6" x 6"	1 Bric	25	.320		92	10.35		102.35	117	
	1050	8" x 8"		24	.333		97	10.80		107.80	124	
	1100	12" x 12"		23	.348		112	11.25		123.25	140	
	1150	12" x 6"		24	.333		98	10.80		108.80	125	
	1200	Foundation block vent, galv., 1-1/4" thk, 8" high, 16" long, no damper		30	.267		18	8.65		26.65	33	
	1250	For damper, add					6			6	6.60	

04800 | Masonry Assemblies

04810 | Unit Masonry Assemblies

		CREW	DAILY OUTPUT	LABOR-HOURS	UNIT	2003 BARE COSTS				TOTAL INCL O&P	
						MAT.	LABOR	EQUIP.	TOTAL		
160	0010	**CHIMNEY** See Div. 03310 for foundation, add to prices below										160
	1800	Metal, high temp. steel jacket, factory lining, 24" diam.	E-2	65	.862	V.L.F.	174	30	18.50	222.50	264	
	1900	60" diameter	"	30	1.867		630	64.50	40	734.50	850	
	2100	Poured concrete, brick lining, 200' high x 10' diam.					5,600			5,600	6,150	
	2200	200' high x 18' diameter									4,995	
	2400	250' high x 9' diameter									3,920	
	2500	300' x 14' diameter									4,690	
	2700	400' high x 18' diameter									5,255	
	2800	500' x 20' diameter					9,800			9,800	10,800	

For information about Means Estimating Seminars, see yellow pages 12 and 13 in back of book

Important: See the Reference Section for critical supporting data - Reference Nos., Crews, & City Cost Indexes

Division 5
Metals

Estimating Tips

05050 Basic Metal Materials & Methods

- Nuts, bolts, washers, connection angles and plates can add a significant amount to both the tonnage of a structural steel job as well as the estimated cost. As a rule of thumb add 10% to the total weight to account for these accessories.

- Type 2 steel construction, commonly referred to as "simple construction," consists generally of field bolted connections with lateral bracing supplied by other elements of the building, such as masonry walls or x-bracing. The estimator should be aware, however, that shop connections may be accomplished by welding or bolting. The method may be particular to the fabrication shop and may have an impact on the estimated cost.

05200 Metal Joists

- In any given project the total weight of open web steel joists is determined by the loads to be supported and the design. However, economies can be realized in minimizing the amount of labor used to place the joists. This is done by maximizing the joist spacing and therefore minimizing the number of joists required to be installed on the job. Certain spacings and locations may be required by the design, but in other cases maximizing the spacing and keeping it as uniform as possible will keep the costs down.

05300 Metal Deck

- The takeoff and estimating of metal deck involves more than simply the area of the floor or roof and the type of deck specified or shown on the drawings. Many different sizes and types of openings may exist. Small openings for individual pipes or conduits may be drilled after the floor/roof is installed, but larger openings may require special deck lengths as well as reinforcing or structural support. The estimator should determine who will be supplying this reinforcing. Additionally, some deck terminations are part of the deck package, such as screed angles and pour stops, and others will be part of the steel contract, such as angles attached to structural members and cast-in-place angles and plates. The estimator must ensure that all pieces are accounted for in the complete estimate.

05500 Metal Fabrications

- The most economical steel stairs are those that use common materials, standard details and most importantly, a uniform and relatively simple method of field assembly. Commonly available A36 channels and plates are very good choices for the main stringers of the stairs, as are angles and tees for the carrier members. Risers and treads are usually made by specialty shops, and it is most economical to use a typical detail in as many places as possible. The stairs should be pre-assembled and shipped directly to the site. The field connections should be simple and straightforward to be accomplished efficiently and with a minimum of equipment and labor.

Reference Numbers

Reference numbers are shown in bold squares at the beginning of some major classifications. These numbers refer to related items in the Reference Section. The reference information may be an estimating procedure, an alternate pricing method or technical information.

Note: Not all subdivisions listed here necessarily appear in this publication.

		05090	Metal Fastenings	CREW	DAILY OUTPUT	LABOR-HOURS	UNIT	2003 BARE COSTS				TOTAL INCL O&P
								MAT.	LABOR	EQUIP.	TOTAL	
080	0010	**ANCHOR BOLTS**										080
	0100	J-type, incl. hex nut & washer, 1/2" diameter x 6" long	2 Carp	70	.229	Ea.	.78	7.20		7.98	12.10
	0110	12" long		65	.246		.97	7.75		8.72	13.20
	0120	18" long		60	.267		1.26	8.40		9.66	14.55
	0130	3/4" diameter x 8" long		50	.320		1.15	10.10		11.25	17.05
	0140	12" long		45	.356		1.44	11.20		12.64	19.15
	0150	18" long		40	.400		1.87	12.60		14.47	22
	0160	1" diameter x 12" long		35	.457		2.74	14.40		17.14	25.50
	0170	18" long		30	.533		3.30	16.85		20.15	30
	0180	24" long		25	.640		4.03	20		24.03	36
	0190	36" long		20	.800		5.50	25		30.50	45.50
	0200	1-1/2" diameter x 18" long		22	.727		9.75	23		32.75	47
	0210	24" long		16	1		11.60	31.50		43.10	62.50
	0300	L-type, incl. hex nut & washer, 3/4" diameter x 12" long		45	.356		1.08	11.20		12.28	18.75
	0310	18" long		40	.400		1.39	12.60		13.99	21
	0320	24" long		35	.457		1.69	14.40		16.09	24.50
	0330	30" long		30	.533		2.15	16.85		19	29
	0340	36" long		25	.640		2.45	20		22.45	34
	0350	1" diameter x 12" long		35	.457		1.88	14.40		16.28	24.50
	0360	18" long		30	.533		2.34	16.85		19.19	29
	0370	24" long		25	.640		2.89	20		22.89	34.50
	0380	30" long		23	.696		3.41	22		25.41	38.50
	0390	36" long		20	.800		3.90	25		28.90	44
	0400	42" long		18	.889		4.74	28		32.74	49
	0410	48" long		15	1.067		5.30	33.50		38.80	58.50
	0420	1-1/4" diameter x 18" long		25	.640		4.21	20		24.21	36
	0430	24" long		20	.800		5	25		30	45
	0440	30" long		20	.800		5.80	25		30.80	46
	0450	36" long		18	.889		6.60	28		34.60	51.50
	0460	42" long		16	1		7.45	31.50		38.95	57.50
	0470	48" long		14	1.143		8.50	36		44.50	66
	0480	54" long		12	1.333		10	42		52	76.50
	0490	60" long		10	1.600		11	50.50		61.50	91
	0500	1-1/2" diameter x 18" long		22	.727		6.45	23		29.45	43
	0510	24" long		19	.842		7.55	26.50		34.05	50
	0520	30" long		17	.941		8.55	29.50		38.05	56
	0540	42" long		15	1.067		11.20	33.50		44.70	65
	0550	48" long		13	1.231		12.55	39		51.55	74.50
	0560	54" long		11	1.455		15.30	46		61.30	88.50
	0570	60" long		9	1.778		16.80	56		72.80	106
	0580	1-3/4" diameter x 18" long		20	.800		9.80	25		34.80	50.50
	0590	24" long		18	.889		11.45	28		39.45	56.50
	0600	30" long		17	.941		13.35	29.50		42.85	61
	0610	36" long		16	1		15.25	31.50		46.75	66.50
	0620	42" long		14	1.143		17.10	36		53.10	75.50
	0630	48" long		12	1.333		18.80	42		60.80	86
	0640	54" long		10	1.600		23.50	50.50		74	105
	0650	60" long		8	2		25	63		88	126
	0660	2" diameter x 24" long		17	.941		14.60	29.50		44.10	62.50
	0670	30" long		15	1.067		16.50	33.50		50	70.50
	0680	36" long		13	1.231		18.10	39		57.10	80.50
	0690	42" long		11	1.455		20	46		66	93.50
	0700	48" long		10	1.600		23	50.50		73.50	105
	0710	54" long		9	1.778		27.50	56		83.50	118
	0720	60" long		8	2		29.50	63		92.50	131
	0730	66" long		7	2.286		32	72		104	148

Important: See the Reference Section for critical supporting data - Reference Nos., Crews, & City Cost Indexes

			CREW	DAILY OUTPUT	LABOR-HOURS	UNIT	2003 BARE COSTS				TOTAL INCL O&P	
05090		**Metal Fastenings**					MAT.	LABOR	EQUIP.	TOTAL		
080	0740	72" long	2 Carp	6	2.667	Ea.	35	84		119	170	**080**
	0990	For galvanized, add					75%					
340	0010	**DRILLING** For anchors, up to 4" deep, incl. bit and layout										**340**
	0050	in concrete or brick walls and floors, no anchor										
	0100	Holes, 1/4" diameter	1 Carp	75	.107	Ea.	.07	3.37		3.44	5.35	
	0150	For each additional inch of depth, add		430	.019		.02	.59		.61	.94	
	0200	3/8" diameter		63	.127		.07	4.01		4.08	6.35	
	0250	For each additional inch of depth, add		340	.024		.02	.74		.76	1.18	
	0300	1/2" diameter		50	.160		.07	5.05		5.12	8	
	0350	For each additional inch of depth, add		250	.032		.02	1.01		1.03	1.60	
	0400	5/8" diameter		48	.167		.13	5.25		5.38	8.35	
	0450	For each additional inch of depth, add		240	.033		.03	1.05		1.08	1.68	
	0500	3/4" diameter		45	.178		.16	5.60		5.76	8.90	
	0550	For each additional inch of depth, add		220	.036		.04	1.15		1.19	1.83	
	0600	7/8" diameter		43	.186		.19	5.85		6.04	9.35	
	0650	For each additional inch of depth, add		210	.038		.05	1.20		1.25	1.93	
	0700	1" diameter		40	.200		.22	6.30		6.52	10.10	
	0750	For each additional inch of depth, add		190	.042		.05	1.33		1.38	2.14	
	0800	1-1/4" diameter		38	.211		.31	6.65		6.96	10.75	
	0850	For each additional inch of depth, add		180	.044		.08	1.40		1.48	2.27	
	0900	1-1/2" diameter		35	.229		.47	7.20		7.67	11.75	
	0950	For each additional inch of depth, add		165	.048		.12	1.53		1.65	2.52	
	1000	For ceiling installations, add						40%				
	1100	Drilling & layout for drywall or plaster walls, no anchor										
	1200	Holes, 1/4" diameter	1 Carp	150	.053	Ea.	.01	1.68		1.69	2.64	
	1300	3/8" diameter		140	.057		.01	1.80		1.81	2.83	
	1400	1/2" diameter		130	.062		.01	1.94		1.95	3.04	
	1500	3/4" diameter		120	.067		.02	2.10		2.12	3.31	
	1600	1" diameter		110	.073		.03	2.29		2.32	3.62	
	1700	1-1/4" diameter		100	.080		.04	2.52		2.56	3.98	
	1800	1-1/2" diameter		90	.089		.06	2.80		2.86	4.44	
	1900	For ceiling installations, add						40%				
380	0010	**EXPANSION ANCHORS** & shields										**380**
	0100	Bolt anchors for concrete, brick or stone, no layout and drilling										
	0200	Expansion shields, zinc, 1/4" diameter, 1-5/16" long, single	1 Carp	90	.089	Ea.	.95	2.80		3.75	5.45	
	0300	1-3/8" long, double		85	.094		1.04	2.97		4.01	5.80	
	0400	3/8" diameter, 1-1/2" long, single		85	.094		1.56	2.97		4.53	6.35	
	0500	2" long, double		80	.100		1.93	3.16		5.09	7.05	
	0600	1/2" diameter, 2-1/16" long, single		80	.100		2.59	3.16		5.75	7.80	
	0700	2-1/2" long, double		75	.107		2.50	3.37		5.87	8	
	0800	5/8" diameter, 2-5/8" long, single		75	.107		3.70	3.37		7.07	9.30	
	0900	2-3/4" long, double		70	.114		3.70	3.61		7.31	9.70	
	1000	3/4" diameter, 2-3/4" long, single		70	.114		5.50	3.61		9.11	11.70	
	1100	3-15/16" long, double		65	.123		7.35	3.88		11.23	14.10	
	1410	Concrete anchor, w/rod & epoxy cartridge, 1-3/4" diameter x 15" long	E-22	20	1.200		69.50	39.50		109	139	
	1415	18" long		17	1.412		83.50	46.50		130	165	
	1420	2" diameter x 18" long		16	1.500		107	49.50		156.50	195	
	1425	24" long		15	1.600		139	52.50		191.50	236	
	1430	Chemical anchor, w/rod & epoxy cartridge, 3/4" diam. x 9-1/2" long		27	.889		10.15	29.50		39.65	57	
	1435	1" diameter x 11-3/4" long		24	1		19.45	33		52.45	73	
	1440	1-1/4" diameter x 14" long		21	1.143		37	37.50		74.50	99.50	
	1500	Self drilling anchor, snap-off, for 1/4" diameter bolt	1 Carp	26	.308		.71	9.70		10.41	15.95	
	1600	3/8" diameter bolt		23	.348		1.03	10.95		11.98	18.30	
	1700	1/2" diameter bolt		20	.400		1.59	12.60		14.19	21.50	
	1800	5/8" diameter bolt		18	.444		2.65	14		16.65	25	
	1900	3/4" diameter bolt		16	.500		4.46	15.80		20.26	29.50	

5

METALS

05090	Metal Fastenings	CREW	DAILY OUTPUT	LABOR-HOURS	UNIT	2003 BARE COSTS				TOTAL INCL O&P
						MAT.	LABOR	EQUIP.	TOTAL	
380	2100 Hollow wall anchors for gypsum wall board, plaster or tile									380
	2300 1/8" diameter, short	1 Carp	160	.050	Ea.	.23	1.58		1.81	2.72
	2400 Long		160	.050		.27	1.58		1.85	2.77
	2500 3/16" diameter, short		150	.053		.47	1.68		2.15	3.15
	2600 Long		150	.053		.51	1.68		2.19	3.19
	2700 1/4" diameter, short		140	.057		.58	1.80		2.38	3.46
	2800 Long		140	.057		.66	1.80		2.46	3.55
	3000 Toggle bolts, bright steel, 1/8" diameter, 2" long		85	.094		.21	2.97		3.18	4.87
	3100 4" long		80	.100		.32	3.16		3.48	5.30
	3200 3/16" diameter, 3" long		80	.100		.36	3.16		3.52	5.35
	3300 6" long		75	.107		.50	3.37		3.87	5.80
	3400 1/4" diameter, 3" long		75	.107		.40	3.37		3.77	5.70
	3500 6" long		70	.114		.57	3.61		4.18	6.30
	3600 3/8" diameter, 3" long		70	.114		.78	3.61		4.39	6.50
	3700 6" long		60	.133		1.36	4.21		5.57	8.05
	3800 1/2" diameter, 4" long		60	.133		2.01	4.21		6.22	8.75
	3900 6" long	▼	50	.160	▼	3.31	5.05		8.36	11.55
	5000 Screw anchors for concrete, masonry,									
	5100 stone & tile, no layout or drilling included									
	5200 Jute fiber, #6, #8, & #10, 1" long	1 Carp	240	.033	Ea.	.23	1.05		1.28	1.89
	5300 #12, 1-1/2" long		200	.040		.33	1.26		1.59	2.33
	5400 #14, 2" long		160	.050		.52	1.58		2.10	3.04
	5500 #16, 2" long		150	.053		.54	1.68		2.22	3.22
	5600 #20, 2" long		140	.057		.86	1.80		2.66	3.77
	5700 Lag screw shields, 1/4" diameter, short		90	.089		.38	2.80		3.18	4.80
	5800 Long		85	.094		.44	2.97		3.41	5.10
	5900 3/8" diameter, short		85	.094		.69	2.97		3.66	5.40
	6000 Long		80	.100		.81	3.16		3.97	5.80
	6100 1/2" diameter, short		80	.100		.96	3.16		4.12	6
	6200 Long		75	.107		1.20	3.37		4.57	6.55
	6300 3/4" diameter, short		70	.114		2.69	3.61		6.30	8.60
	6400 Long		65	.123		3.25	3.88		7.13	9.65
	6600 Lead, #6 & #8, 3/4" long		260	.031		.15	.97		1.12	1.69
	6700 #10 - #14, 1-1/2" long		200	.040		.22	1.26		1.48	2.21
	6800 #16 & #18, 1-1/2" long		160	.050		.29	1.58		1.87	2.79
	6900 Plastic, #6 & #8, 3/4" long		260	.031		.09	.97		1.06	1.62
	7000 #8 & #10, 7/8" long		240	.033		.04	1.05		1.09	1.68
	7100 #10 & #12, 1" long		220	.036		.12	1.15		1.27	1.92
	7200 #14 & #16, 1-1/2" long	▼	160	.050	▼	.07	1.58		1.65	2.55
	8000 Wedge anchors, not including layout or drilling									
	8050 Carbon steel, 1/4" diameter, 1-3/4" long	1 Carp	150	.053	Ea.	.36	1.68		2.04	3.03
	8100 3 1/4" long		145	.055		.48	1.74		2.22	3.25
	8150 3/8" diameter, 2-1/4" long		150	.053		.54	1.68		2.22	3.23
	8200 5" long		145	.055		.95	1.74		2.69	3.77
	8250 1/2" diameter, 2-3/4" long		140	.057		.83	1.80		2.63	3.73
	8300 7" long		130	.062		1.41	1.94		3.35	4.59
	8350 5/8" diameter, 3-1/2" long		130	.062		1.64	1.94		3.58	4.83
	8400 8-1/2" long		115	.070		3.48	2.19		5.67	7.25
	8450 3/4" diameter, 4-1/4" long		115	.070		1.97	2.19		4.16	5.60
	8500 10" long		100	.080		4.48	2.52		7	8.85
	8550 1" diameter, 6" long		100	.080		6.60	2.52		9.12	11.25
	8575 9" long		80	.100		8.60	3.16		11.76	14.40
	8600 12" long		80	.100		9.30	3.16		12.46	15.15
	8650 1-1/4" diameter, 9" long		70	.114		12.05	3.61		15.66	18.90
	8700 12" long	▼	60	.133	▼	15.40	4.21		19.61	23.50
	8750 For type 303 stainless steel, add					350%				

Important: See the Reference Section for critical supporting data - Reference Nos., Crews, & City Cost Indexes

05090	Metal Fastenings	CREW	DAILY OUTPUT	LABOR-HOURS	UNIT	2003 BARE COSTS MAT.	LABOR	EQUIP.	TOTAL	TOTAL INCL O&P	
380	8800	For type 316 stainless steel, add					450%					380
460	0010	**LAG SCREWS**										460
	0020	Steel, 1/4" diameter, 2" long	1 Carp	200	.040	Ea.	.07	1.26		1.33	2.05	
	0100	3/8" diameter, 3" long		150	.053		.20	1.68		1.88	2.85	
	0200	1/2" diameter, 3" long		130	.062		.33	1.94		2.27	3.39	
	0300	5/8" diameter, 3" long	↓	120	.067	↓	.64	2.10		2.74	3.99	
500	0010	**MACHINE SCREWS**										500
	0020	Steel, round head, #8 x 1" long				C	1.96			1.96	2.16	
	0110	#8 x 2" long					4.27			4.27	4.70	
	0200	#10 x 1" long					2.80			2.80	3.08	
	0300	#10 x 2" long				↓	5.25			5.25	5.75	
540	0010	**MACHINERY ANCHORS**, heavy duty, incl. sleeve, floating base nut,										540
	0020	lower stud & coupling nut, fiber plug, connecting stud, washer & nut.										
	0030	For flush mounted embedment in poured concrete heavy equip. pads.										
	0200	Material only, 1/2" diameter stud & bolt				Ea.	45			45	49.50	
	0300	5/8" diameter					50			50	55	
	0500	3/4" diameter					57.50			57.50	63	
	0600	7/8" diameter					62.50			62.50	69	
	0800	1" diameter					66			66	72.50	
	0900	1-1/4" diameter				↓	87.50			87.50	96.50	
580	0010	**POWDER ACTUATED** Tools & fasteners										580
	0020	Stud driver, .22 caliber, buy, minimum				Ea.	274			274	300	
	0100	Maximum				"	440			440	485	
	0300	Powder charges for above, low velocity				C	15.50			15.50	17.05	
	0400	Standard velocity					22			22	24	
	0600	Drive pins & studs, 1/4" & 3/8" diam., to 3" long, minimum					10.10			10.10	11.15	
	0700	Maximum				↓	39.50			39.50	43.50	
	0800	Pneumatic stud driver for 1/8" diameter studs				Ea.	1,900			1,900	2,100	
	0900	Drive pins for above, 1/2" to 3/4" long				M	415			415	460	
600	0010	**RIVETS**										600
	0100	Aluminum rivet & mandrel, 1/2" grip length x 1/8" diameter				C	4.67			4.67	5.15	
	0200	3/16" diameter					7.20			7.20	7.90	
	0300	Aluminum rivet, steel mandrel, 1/8" diameter					6.25			6.25	6.85	
	0400	3/16" diameter					5.70			5.70	6.30	
	0500	Copper rivet, steel mandrel, 1/8" diameter					5.65			5.65	6.25	
	1200	Steel rivet and mandrel, 1/8" diameter					4.89			4.89	5.40	
	1300	3/16" diameter				↓	7.35			7.35	8.10	
	1400	Hand riveting tool, minimum				Ea.	99			99	109	
	1500	Maximum					189			189	208	
	1600	Power riveting tool, minimum					715			715	785	
	1700	Maximum				↓	1,800			1,800	1,975	

METALS 5

05120	Structural Steel	CREW	DAILY OUTPUT	LABOR-HOURS	UNIT	2003 BARE COSTS				TOTAL INCL O&P	
						MAT.	LABOR	EQUIP.	TOTAL		
440	0010	**LIGHTWEIGHT FRAMING**									**440**
	0400	Angle framing, field fabricated, 4" and larger	E-3	440	.055	Lb.	.37	1.98	.18	2.53	4.21
	0450	Less than 4" angles		265	.091		.38	3.29	.30	3.97	6.70
	0600	Channel framing, field fabricated, 8" and larger		500	.048		.38	1.74	.16	2.28	3.76
	0650	Less than 8" channels	▼	335	.072		.38	2.60	.24	3.22	5.40
	1000	Continuous slotted channel framing system, shop fab, min	2 Sswk	2,400	.007		1.98	.24		2.22	2.60
	1200	Maximum	"	1,600	.010		2.23	.36		2.59	3.10
	1250	Plate & bar stock for reinforcing beams and trusses					.70			.70	.77
	1300	Cross bracing, rods, shop fabricated, 3/4" diameter	E-3	700	.034		.77	1.25	.11	2.13	3.22
	1310	7/8" diameter		850	.028		.77	1.03	.09	1.89	2.80
	1320	1" diameter		1,000	.024		.77	.87	.08	1.72	2.51
	1330	Angle, 5" x 5" x 3/8"		2,800	.009		.77	.31	.03	1.11	1.44
	1350	Hanging lintels, shop fabricated, average	▼	850	.028		.77	1.03	.09	1.89	2.80
	1380	Roof frames, shop fabricated, 3'-0" square, 5' span	E-2	4,200	.013		.77	.46	.29	1.52	1.96
	1400	Tie rod, not upset, 1-1/2" to 4" diameter, with turnbuckle	2 Sswk	800	.020		.83	.71		1.54	2.21
	1420	No turnbuckle		700	.023		.80	.82		1.62	2.36
	1500	Upset, 1-3/4" to 4" diameter, with turnbuckle		800	.020		.83	.71		1.54	2.21
	1520	No turnbuckle	▼	700	.023	▼	.80	.82		1.62	2.36
520	0010	**PIPE SUPPORT FRAMING**									**520**
	0020	Under 10#/L.F.	E-4	3,900	.008	Lb.	.85	.30	.02	1.17	1.50
	0200	10.1 to 15#/L.F.		4,300	.007		.84	.27	.02	1.13	1.44
	0400	15.1 to 20#/L.F.		4,800	.007		.83	.24	.02	1.09	1.37
	0600	Over 20#/L.F.	▼	5,400	.006	▼	.82	.21	.01	1.04	1.31

For information about Means Estimating Seminars, see yellow pages 12 and 13 in back of book

Important: See the Reference Section for critical supporting data - Reference Nos., Crews, & City Cost Indexes

Division 6
Wood & Plastics

Estimating Tips

06050 Basic Wood & Plastic Materials & Methods

- Common to any wood framed structure are the accessory connector items such as screws, nails, adhesives, hangers, connector plates, straps, angles and holdowns. For typical wood framed buildings, such as residential projects, the aggregate total for these items can be significant, especially in areas where seismic loading is a concern. For floor and wall framing, the material cost is based on 10 to 25 lbs. per MBF. Holdowns, hangers and other connectors should be taken off by the piece.

06100 Rough Carpentry

- Lumber is a traded commodity and therefore sensitive to supply and demand in the marketplace. Even in "budgetary" estimating of wood framed projects, it is advisable to call local suppliers for the latest market pricing.
- Common quantity units for wood framed projects are "thousand board feet" (MBF). A board foot is a volume of wood, 1" x 1' x 1',

or 144 cubic inches. Board foot quantities are generally calculated using nominal material dimensions—dressed sizes are ignored. Board foot per lineal foot of any stick of lumber can be calculated by dividing the nominal cross sectional area by 12. As an example, 2,000 lineal feet of 2 x 12 equates to 4 MBF by dividing the nominal area, 2 x 12, by 12, which equals 2, and multiplying by 2,000 to give 4,000 board feet. This simple rule applies to all nominal dimensioned lumber.

- Waste is an issue of concern at the quantity takeoff for any area of construction. Framing lumber is sold in even foot lengths, i.e., 10', 12', 14', 16', and depending on spans, wall heights and the grade of lumber, waste is inevitable. A rule of thumb for lumber waste is 5% to 10% depending on material quality and the complexity of the framing.
- Wood in various forms and shapes is used in many projects, even where the main structural framing is steel, concrete or masonry. Plywood as a back-up partition material and 2x boards used as blocking and cant strips around roof edges are two

common examples. The estimator should ensure that the costs of all wood materials are included in the final estimate.

06200 Finish Carpentry

- It is necessary to consider the grade of workmanship when estimating labor costs for erecting millwork and interior finish. In practice, there are three grades: premium, custom and economy. The Means daily output for base and case moldings is in the range of 200 to 250 L.F. per carpenter per day. This is appropriate for most average custom grade projects. For premium projects an adjustment to productivity of 25% to 50% should be made depending on the complexity of the job.

Reference Numbers

Reference numbers are shown in bold squares at the beginning of some major classifications. These numbers refer to related items in the Reference Section. The reference information may be an estimating procedure, an alternate pricing method or technical information.

Note: Not all subdivisions listed here necessarily appear in this publication.

6 WOOD & PLASTICS

06090		Wood & Plastic Fastenings	CREW	DAILY OUTPUT	LABOR-HOURS	UNIT	2003 BARE COSTS				TOTAL INCL O&P	
							MAT.	LABOR	EQUIP.	TOTAL		
600	0010	**NAILS** Prices of material only, based on 50# box purchase, copper, plain				Lb.	4.59			4.59	5.05	600
	0400	Stainless steel, plain					4.97			4.97	5.45	
	0500	Box, 3d to 20d, bright					1.19			1.19	1.31	
	0520	Galvanized					1.37			1.37	1.51	
	0600	Common, 3d to 60d, plain					.79			.79	.87	
	0700	Galvanized					1.20			1.20	1.32	
	0800	Aluminum					3.38			3.38	3.72	
	1000	Annular or spiral thread, 4d to 60d, plain					1.78			1.78	1.96	
	1200	Galvanized					1.70			1.70	1.87	
	1400	Drywall nails, plain					1.36			1.36	1.50	
	1600	Galvanized					1.49			1.49	1.64	
	1800	Finish nails, 4d to 10d, plain					.92			.92	1.01	
	2000	Galvanized					1.06			1.06	1.17	
	2100	Aluminum					4.95			4.95	5.45	
	2300	Flooring nails, hardened steel, 2d to 10d, plain					1.24			1.24	1.36	
	2400	Galvanized					2.21			2.21	2.43	
	2500	Gypsum lath nails, 1-1/8", 13 ga. flathead, blued					1.36			1.36	1.50	
	2600	Masonry nails, hardened steel, 3/4" to 3" long, plain					1.74			1.74	1.91	
	2700	Galvanized					1.74			1.74	1.91	
	5000	Add to prices above for cement coating					.07			.07	.08	
	5200	Zinc or tin plating					.12			.12	.13	
	5500	Vinyl coated sinkers, 8d to 16d					.56			.56	.62	
650	0010	**NAILS** mat. only, for pneumatic tools, framing, per carton of 5000, 2"				Ea.	37			37	41	650
	0100	2-3/8"					42.50			42.50	46.50	
	0200	Per carton of 4000, 3"					37.50			37.50	41.50	
	0300	3-1/4"					40			40	44	
	0400	Per carton of 5000, 2-3/8", galv.					57.50			57.50	63.50	
	0500	Per carton of 4000, 3", galv.					65			65	71.50	
	0600	3-1/4", galv.					80.50			80.50	88.50	
	0700	Roofing, per carton of 7200, 1"					35			35	38.50	
	0800	1-1/4"					32.50			32.50	36	
	0900	1-1/2"					37.50			37.50	41.50	
	1000	1-3/4"					45.50			45.50	50	
700	0010	**SHEET METAL SCREWS** Steel, standard, #8 x 3/4", plain				C	2.74			2.74	3.01	700
	0100	Galvanized					3.44			3.44	3.78	
	0300	#10 x 1", plain					3.76			3.76	4.14	
	0400	Galvanized					4.35			4.35	4.79	
	0600	With washers, #14 x 1", plain					10.20			10.20	11.20	
	0700	Galvanized					10.20			10.20	11.20	
	0900	#14 x 2", plain					10.20			10.20	11.20	
	1000	Galvanized					18.15			18.15	19.95	
	1500	Self-drilling, with washers, (pinch point) #8 x 3/4", plain					6.10			6.10	6.70	
	1600	Galvanized					6.10			6.10	6.70	
	1800	#10 x 3/4", plain					6.10			6.10	6.70	
	1900	Galvanized					6.10			6.10	6.70	
	3000	Stainless steel w/aluminum or neoprene washers, #14 x 1", plain					18.35			18.35	20	
	3100	#14 x 2", plain					25			25	27.50	
750	0010	**WOOD SCREWS** #8, 1" long, steel				C	3.36			3.36	3.70	750
	0100	Brass					11.20			11.20	12.35	
	0200	#8, 2" long, steel					3.76			3.76	4.14	
	0300	Brass					11.70			11.70	12.90	
	0400	#10, 1" long, steel					4.30			4.30	4.73	
	0500	Brass					23			23	25	
	0600	#10, 2" long, steel					7.65			7.65	8.45	
	0700	Brass					40.50			40.50	44.50	

Important: See the Reference Section for critical supporting data - Reference Nos., Crews, & City Cost Indexes

06050 | Basic Wood / Plastic Materials / Methods

	06090	Wood & Plastic Fastenings	CREW	DAILY OUTPUT	LABOR-HOURS	UNIT	2003 BARE COSTS MAT.	LABOR	EQUIP.	TOTAL	TOTAL INCL O&P	
750	0800	#10, 3" long, steel				C	11.95			11.95	13.15	750
	1000	#12, 2" long, steel					4.89			4.89	5.40	
	1100	Brass					16.30			16.30	17.95	
	1500	#12, 3" long, steel					16.20			16.20	17.85	
	2000	#12, 4" long, steel				▼	29			29	32	

06100 | Rough Carpentry

	06160	Sheathing	CREW	DAILY OUTPUT	LABOR-HOURS	UNIT	2003 BARE COSTS MAT.	LABOR	EQUIP.	TOTAL	TOTAL INCL O&P	
800	0010	**SHEATHING** Plywood on roof, CDX										800
	0030	5/16" thick	2 Carp	1,600	.010	S.F.	.43	.32		.75	.97	
	0035	Pneumatic nailed		1,952	.008		.43	.26		.69	.88	
	0050	3/8" thick		1,525	.010		.44	.33		.77	1	
	0055	Pneumatic nailed		1,860	.009		.44	.27		.71	.90	
	0100	1/2" thick		1,400	.011		.49	.36		.85	1.10	
	0105	Pneumatic nailed		1,708	.009		.49	.30		.79	1	
	0200	5/8" thick		1,300	.012		.57	.39		.96	1.23	
	0205	Pneumatic nailed		1,586	.010		.57	.32		.89	1.12	
	0300	3/4" thick		1,200	.013		.66	.42		1.08	1.39	
	0305	Pneumatic nailed		1,464	.011		.66	.34		1	1.27	
	0500	Plywood on walls with exterior CDX, 3/8" thick		1,200	.013		.44	.42		.86	1.14	
	0505	Pneumatic nailed		1,488	.011		.44	.34		.78	1.01	
	0600	1/2" thick		1,125	.014		.49	.45		.94	1.24	
	0605	Pneumatic nailed		1,395	.011		.49	.36		.85	1.11	
	0700	5/8" thick		1,050	.015		.57	.48		1.05	1.37	
	0705	Pneumatic nailed		1,302	.012		.57	.39		.96	1.23	
	0800	3/4" thick		975	.016		.66	.52		1.18	1.54	
	0805	Pneumatic nailed	▼	1,209	.013	▼	.66	.42		1.08	1.38	

06400 | Architectural Woodwork

	06440	Wood Ornaments	CREW	DAILY OUTPUT	LABOR-HOURS	UNIT	2003 BARE COSTS MAT.	LABOR	EQUIP.	TOTAL	TOTAL INCL O&P	
400	0010	**LOUVERS** Redwood, 2'-0" diameter, full circle	1 Carp	16	.500	Ea.	105	15.80		120.80	141	400
	0100	Half circle		16	.500		104	15.80		119.80	139	
	0200	Octagonal		16	.500		86	15.80		101.80	119	
	0300	Triangular, 5/12 pitch, 5'-0" at base	▼	16	.500	▼	180	15.80		195.80	223	

For information about Means Estimating Seminars, see yellow pages 12 and 13 in back of book

WOOD & PLASTICS

6

		CREW	DAILY OUTPUT	LABOR-HOURS	UNIT	2003 BARE COSTS				TOTAL INCL O&P
						MAT.	LABOR	EQUIP.	TOTAL	

Division 7
Thermal & Moisture Protection

Estimating Tips

07100 Dampproofing & Waterproofing

- Be sure of the job specifications before pricing this subdivision. The difference in cost between waterproofing and dampproofing can be great. Waterproofing will hold back standing water. Dampproofing prevents the transmission of water vapor. Also included in this section are vapor retarding membranes.

07200 Thermal Protection

- Insulation and fireproofing products are measured by area, thickness, volume or R value. Specifications may only give what the specific R value should be in a certain situation. The estimator may need to choose the type of insulation to meet that R value.

07300 Shingles, Roof Tiles & Roof Coverings
07400 Roofing & Siding Panels

- Many roofing and siding products are bought and sold by the square. One square is equal to an area that measures 100 square feet.

This simple change in unit of measure could create a large error if the estimator is not observant. Accessories necessary for a complete installation must be figured into any calculations for both material and labor.

07500 Membrane Roofing
07600 Flashing & Sheet Metal
07700 Roof Specialties & Accessories

- The items in these subdivisions compose a roofing system. No one component completes the installation and all must be estimated. Built-up or single ply membrane roofing systems are made up of many products and installation trades. Wood blocking at roof perimeters or penetrations, parapet coverings, reglets, roof drains, gutters, downspouts, sheet metal flashing, skylights, smoke vents or roof hatches all need to be considered along with the roofing material. Several different installation trades will need to work together on the roofing system. Inherent difficulties in the scheduling and coordination of various trades must be accounted for when estimating labor costs.

07900 Joint Sealers

- To complete the weather-tight shell the sealants and caulkings must be estimated. Where different materials meet—at expansion joints, at flashing penetrations, and at hundreds of other locations throughout a construction project—they provide another line of defense against water penetration. Often, an entire system is based on the proper location and placement of caulking or sealants. The detail drawings that are included as part of a set of architectural plans, show typical locations for these materials. When caulking or sealants are shown at typical locations, this means the estimator must include them for all the locations where this detail is applicable. Be careful to keep different types of sealants separate, and remember to consider backer rods and primers if necessary.

Reference Numbers

Reference numbers are shown in bold squares at the beginning of some major classifications. These numbers refer to related items in the Reference Section. The reference information may be an estimating procedure, an alternate pricing method or technical information.

Note: Not all subdivisions listed here necessarily appear in this publication.

07650 | Flexible Flashing

		CREW	DAILY OUTPUT	LABOR-HOURS	UNIT	2003 BARE COSTS				TOTAL INCL O&P	
						MAT.	LABOR	EQUIP.	TOTAL		
600	**0010**	**FLASHING** Aluminum, mill finish, .013" thick	1 Rofc	145	.055	S.F.	.35	1.53		1.88	2.99	**600**
	0030	.016" thick		145	.055		.51	1.53		2.04	3.16	
	0060	.019" thick		145	.055		.77	1.53		2.30	3.45	
	0100	.032" thick		145	.055		1.04	1.53		2.57	3.74	
	0200	.040" thick		145	.055		1.45	1.53		2.98	4.20	
	0300	.050" thick		145	.055		1.84	1.53		3.37	4.62	
	0400	Painted finish, add					.24			.24	.26	
	1300	Asphalt flashing cement, 5 gallon				Gal.	3.90			3.90	4.29	
	1600	Copper, 16 oz, sheets, under 1000 lbs.	1 Rofc	115	.070	S.F.	2.48	1.92		4.40	6	
	1700	Over 4000 lbs.		155	.052		2.91	1.43		4.34	5.65	
	1900	20 oz sheets, under 1000 lbs.		110	.073		3.90	2.01		5.91	7.70	
	2000	Over 4000 lbs.		145	.055		3.62	1.53		5.15	6.60	
	2200	24 oz sheets, under 1000 lbs.		105	.076		4.69	2.11		6.80	8.75	
	2300	Over 4000 lbs.		135	.059		4.34	1.64		5.98	7.55	
	2500	32 oz sheets, under 1000 lbs.		100	.080		6.20	2.21		8.41	10.65	
	2600	Over 4000 lbs.		130	.062		5.80	1.70		7.50	9.30	
	4300	Copper-clad stainless steel, .015" thick, under 500 lbs.		115	.070		3.16	1.92		5.08	6.75	
	4400	Over 2000 lbs.		155	.052		3.05	1.43		4.48	5.80	
	4600	.018" thick, under 500 lbs.		100	.080		4.18	2.21		6.39	8.40	
	4700	Over 2000 lbs.		145	.055		3.06	1.53		4.59	5.95	
	5800	Lead, 2.5 lb. per SF, up to 12" wide		135	.059		2.93	1.64		4.57	6	
	5900	Over 12" wide		135	.059		3.25	1.64		4.89	6.40	
	8500	Shower pan, bituminous membrane, 7 oz		155	.052		1.08	1.43		2.51	3.63	
	8550	3 ply copper and fabric, 3 oz		155	.052		1.63	1.43		3.06	4.23	
	8600	7 oz		155	.052		3.37	1.43		4.80	6.15	
	8650	Copper, 16 oz		100	.080		3.11	2.21		5.32	7.20	
	8700	Lead on copper and fabric, 5 oz		155	.052		1.62	1.43		3.05	4.22	
	8800	7 oz		155	.052		2.93	1.43		4.36	5.65	
	8850	Polyvinyl chloride, .030" thick		160	.050		.30	1.38		1.68	2.69	
	8900	Stainless steel sheets, 32 ga, .010" thick		155	.052		2.16	1.43		3.59	4.82	
	9000	28 ga, .015" thick		155	.052		2.68	1.43		4.11	5.40	
	9100	26 ga, .018" thick		155	.052		3.25	1.43		4.68	6	
	9200	24 ga, .025" thick		155	.052		4.22	1.43		5.65	7.10	
	9400	Terne coated stainless steel, .015" thick, 28 ga		155	.052		4.05	1.43		5.48	6.90	
	9500	.018" thick, 26 ga		155	.052		4.57	1.43		6	7.50	
	9600	Zinc and copper alloy (brass), .020" thick		155	.052		3.26	1.43		4.69	6.05	
	9700	.027" thick		155	.052		4.37	1.43		5.80	7.25	
	9800	.032" thick		155	.052		5.10	1.43		6.53	8.05	
	9900	.040" thick		155	.052		6.20	1.43		7.63	9.30	

07700 | Roof Specialties & Accessories

07710 | Manufactured Roof Specialties

		CREW	DAILY OUTPUT	LABOR-HOURS	UNIT	2003 BARE COSTS				TOTAL INCL O&P	
						MAT.	LABOR	EQUIP.	TOTAL		
400	**0010**	**DOWNSPOUTS** Aluminum 2" x 3", .020" thick, embossed	1 Shee	190	.042	L.F.	.74	1.56		2.30	3.22	**400**
	0100	Enameled		190	.042		.91	1.56		2.47	3.41	
	4800	Steel, galvanized, round, corrugated, 2" or 3" diam, 28 ga		190	.042		.74	1.56		2.30	3.22	
	4900	4" diameter, 28 gauge		145	.055		1.08	2.04		3.12	4.34	
	5100	5" diameter, 28 gauge		130	.062		1.40	2.28		3.68	5.05	
	5200	26 gauge		130	.062		1.36	2.28		3.64	5	

Important: See the Reference Section for critical supporting data - Reference Nos., Crews, & City Cost Indexes

07710	Manufactured Roof Specialties	CREW	DAILY OUTPUT	LABOR-HOURS	UNIT	2003 BARE COSTS				TOTAL INCL O&P	
						MAT.	LABOR	EQUIP.	TOTAL		
400	5400	6″ diameter, 28 gauge	1 Shee	105	.076	L.F.	1.98	2.82		4.80	6.55	**400**
	5500	26 gauge		105	.076		1.42	2.82		4.24	5.90	
	7500	Steel pipe, black, extra heavy, 4″ diameter		20	.400		6.05	14.80		20.85	29.50	
	7600	6″ diameter	↓	18	.444	↓	12.85	16.45		29.30	39.50	

07840	Firestopping		CREW	DAILY OUTPUT	LABOR-HOURS	UNIT	2003 BARE COSTS				TOTAL INCL O&P	
							MAT.	LABOR	EQUIP.	TOTAL		
100	0010	**FIRESTOPPING**	R07800 -030									**100**
	0100	Metallic piping, non insulated										
	0110	Through walls, 2″ diameter	1 Carp	16	.500	Ea.	9.60	15.80		25.40	35	
	0120	4″ diameter		14	.571		14.65	18.05		32.70	44	
	0130	6″ diameter		12	.667		19.70	21		40.70	54.50	
	0140	12″ diameter		10	.800		35	25		60	78	
	0150	Through floors, 2″ diameter		32	.250		5.80	7.90		13.70	18.75	
	0160	4″ diameter		28	.286		8.35	9		17.35	23.50	
	0170	6″ diameter		24	.333		11	10.50		21.50	28.50	
	0180	12″ diameter	↓	20	.400	↓	18.50	12.60		31.10	40	
	0190	Metallic piping, insulated										
	0200	Through walls, 2″ diameter	1 Carp	16	.500	Ea.	13.60	15.80		29.40	39.50	
	0210	4″ diameter		14	.571		18.65	18.05		36.70	48.50	
	0220	6″ diameter		12	.667		23.50	21		44.50	59	
	0230	12″ diameter		10	.800		39	25		64	82	
	0240	Through floors, 2″ diameter		32	.250		9.80	7.90		17.70	23	
	0250	4″ diameter		28	.286		12.35	9		21.35	27.50	
	0260	6″ diameter		24	.333		15	10.50		25.50	33	
	0270	12″ diameter	↓	20	.400	↓	18.50	12.60		31.10	40	
	0280	Non metallic piping, non insulated										
	0290	Through walls, 2″ diameter	1 Carp	12	.667	Ea.	39.50	21		60.50	76.50	
	0300	4″ diameter		10	.800		49.50	25		74.50	94	
	0310	6″ diameter		8	1		69	31.50		100.50	126	
	0330	Through floors, 2″ diameter		16	.500		31	15.80		46.80	58.50	
	0340	4″ diameter		6	1.333		38.50	42		80.50	108	
	0350	6″ diameter	↓	6	1.333	↓	46	42		88	116	
	0370	Ductwork, insulated & non insulated, round										
	0380	Through walls, 6″ diameter	1 Carp	12	.667	Ea.	20	21		41	55	
	0390	12″ diameter		10	.800		40	25		65	83.50	
	0400	18″ diameter		8	1		65	31.50		96.50	121	
	0410	Through floors, 6″ diameter		16	.500		11	15.80		26.80	36.50	
	0420	12″ diameter		14	.571		20	18.05		38.05	50	
	0430	18″ diameter	↓	12	.667	↓	35	21		56	71.50	
	0440	Ductwork, insulated & non insulated, rectangular										
	0450	With stiffener/closure angle, through walls, 6″ x 12″	1 Carp	8	1	Ea.	16.65	31.50		48.15	68	
	0460	12″ x 24″		6	1.333		22	42		64	90	
	0470	24″ x 48″		4	2		63	63		126	168	
	0480	With stiffener/closure angle, through floors, 6″ x 12″		10	.800		9	25		34	49.50	
	0490	12″ x 24″		8	1		16.20	31.50		47.70	67.50	
	0500	24″ x 48″	↓	6	1.333	↓	32	42		74	101	
	0510	Multi trade openings										
	0520	Through walls, 6″ x 12″	1 Carp	2	4	Ea.	35	126		161	236	

07840	Firestopping		DAILY OUTPUT	LABOR-HOURS	UNIT	2003 BARE COSTS				TOTAL INCL O&P	
						MAT.	LABOR	EQUIP.	TOTAL		
100	0530	12" x 24"	R07800 -030	1 Carp	1	8	Ea.	141	252		393	550	100
	0540	24" x 48"		2 Carp	1	16		565	505		1,070	1,400
	0550	48" x 96"		"	.75	21.333		2,275	675		2,950	3,550
	0560	Through floors, 6" x 12"		1 Carp	2	4		35	126		161	236
	0570	12" x 24"		"	1	8		141	252		393	550
	0580	24" x 48"		2 Carp	.75	21.333		565	675		1,240	1,675
	0590	48" x 96"		"	.50	32		2,275	1,000		3,275	4,075
	0600	Structural penetrations, through walls										
	0610	Steel beams, W8 x 10		1 Carp	8	1	Ea.	22	31.50		53.50	73.50
	0620	W12 x 14			6	1.333		35	42		77	104
	0630	W21 x 44			5	1.600		70	50.50		120.50	156
	0640	W36 x 135			3	2.667		170	84		254	320
	0650	Bar joists, 18" deep			6	1.333		32	42		74	101
	0660	24" deep			6	1.333		40	42		82	110
	0670	36" deep			5	1.600		60	50.50		110.50	145
	0680	48" deep			4	2		70	63		133	176
	0690	Construction joints, floor slab at exterior wall										
	0700	Precast, brick, block or drywall exterior										
	0710	2" wide joint		1 Carp	125	.064	L.F.	5	2.02		7.02	8.65
	0720	4" wide joint		"	75	.107	"	10	3.37		13.37	16.25
	0730	Metal panel, glass or curtain wall exterior										
	0740	2" wide joint		1 Carp	40	.200	L.F.	11.85	6.30		18.15	23
	0750	4" wide joint		"	25	.320	"	16.15	10.10		26.25	33.50
	0760	Floor slab to drywall partition										
	0770	Flat joint		1 Carp	100	.080	L.F.	4.90	2.52		7.42	9.35
	0780	Fluted joint			50	.160		10	5.05		15.05	18.90
	0790	Etched fluted joint			75	.107		6.50	3.37		9.87	12.40
	0800	Floor slab to concrete/masonry partition										
	0810	Flat joint		1 Carp	75	.107	L.F.	11	3.37		14.37	17.35
	0820	Fluted joint		"	50	.160	"	13	5.05		18.05	22
	0830	Concrete/CMU wall joints										
	0840	1" wide		1 Carp	100	.080	L.F.	6	2.52		8.52	10.55
	0850	2" wide			75	.107		11	3.37		14.37	17.35
	0860	4" wide			50	.160		21	5.05		26.05	31
	0870	Concrete/CMU floor joints										
	0880	1" wide		1 Carp	200	.040	L.F.	3	1.26		4.26	5.25
	0890	2" wide			150	.053		5.50	1.68		7.18	8.70
	0900	4" wide			100	.080		10.50	2.52		13.02	15.50

07900 | Joint Sealers

07920	Joint Sealants		DAILY OUTPUT	LABOR-HOURS	UNIT	2003 BARE COSTS				TOTAL INCL O&P	
						MAT.	LABOR	EQUIP.	TOTAL		
800	0010	**CAULKING AND SEALANTS**										800
	0020	Acoustical sealant, elastomeric, cartridges				Ea.	2.05			2.05	2.26
	0030	Backer rod, polyethylene, 1/4" diameter		1 Bric	4.60	1.739	C.L.F.	1.35	56.50		57.85	88
	0050	1/2" diameter			4.60	1.739		2.71	56.50		59.21	89.50
	0070	3/4" diameter			4.60	1.739		5.20	56.50		61.70	92
	0090	1" diameter			4.60	1.739		8.15	56.50		64.65	95.50
	0100	Acrylic latex caulk, white									
	0200	11 fl. oz cartridge				Ea.	1.83			1.83	2.01

Important: See the Reference Section for critical supporting data - Reference Nos., Crews, & City Cost Indexes

THERMAL & MOISTURE PROTECTION

7

07920	Joint Sealants	CREW	DAILY OUTPUT	LABOR-HOURS	UNIT	2003 BARE COSTS				TOTAL INCL O&P	
						MAT.	LABOR	EQUIP.	TOTAL		
800	0500	1/4" x 1/2"	1 Bric	248	.032	L.F.	.15	1.05		1.20	1.77	**800**
	0600	1/2" x 1/2"		250	.032		.30	1.04		1.34	1.92	
	0800	3/4" x 3/4"		230	.035		.67	1.13		1.80	2.47	
	0900	3/4" x 1"		200	.040		.90	1.30		2.20	2.98	
	1000	1" x 1"	↓	180	.044	↓	1.12	1.44		2.56	3.44	
	1400	Butyl based, bulk				Gal.	22			22	24.50	
	1500	Cartridges				"	27			27	29.50	
	1700	Bulk, in place 1/4" x 1/2", 154 L.F./gal.	1 Bric	230	.035	L.F.	.14	1.13		1.27	1.89	
	1800	1/2" x 1/2", 77 L.F./gal.	"	180	.044	"	.29	1.44		1.73	2.53	
	2000	Latex acrylic based, bulk				Gal.	23			23	25.50	
	2100	Cartridges					26			26	28.50	
	2300	Polysulfide compounds, 1 component, bulk				↓	44			44	48	
	2600	1 or 2 component, in place, 1/4" x 1/4", 308 L.F./gal.	1 Bric	145	.055	L.F.	.14	1.79		1.93	2.91	
	2700	1/2" x 1/4", 154 L.F./gal.		135	.059		.28	1.92		2.20	3.26	
	2900	3/4" x 3/8", 68 L.F./gal.		130	.062		.64	1.99		2.63	3.77	
	3000	1" x 1/2", 38 L.F./gal.	↓	130	.062	↓	1.15	1.99		3.14	4.33	
	3200	Polyurethane, 1 or 2 component				Gal.	47.50			47.50	52	
	3300	Cartridges				"	46.50			46.50	51.50	
	3500	Bulk, in place, 1/4" x 1/4"	1 Bric	150	.053	L.F.	.15	1.73		1.88	2.83	
	3600	1/2" x 1/4"		145	.055		.31	1.79		2.10	3.09	
	3800	3/4" x 3/8", 68 L.F./gal.		130	.062		.70	1.99		2.69	3.83	
	3900	1" x 1/2"	↓	110	.073	↓	1.23	2.36		3.59	4.97	
	4100	Silicone rubber, bulk				Gal.	34.50			34.50	38	
	4200	Cartridges				"	40.50			40.50	44.50	
	4400	Neoprene gaskets, closed cell, adhesive, 1/8" x 3/8"	1 Bric	240	.033	L.F.	.20	1.08		1.28	1.88	
	4500	1/4" x 3/4"		215	.037		.48	1.21		1.69	2.38	
	4700	1/2" x 1"		200	.040		1.40	1.30		2.70	3.53	
	4800	3/4" x 1-1/2"	↓	165	.048		2.91	1.57		4.48	5.60	
	5500	Resin epoxy coating, 2 component, heavy duty				Gal.	26			26	28.50	
	5800	Tapes, sealant, P.V.C. foam adhesive, 1/16" x 1/4"				C.L.F.	4.60			4.60	5.05	
	5900	1/16" x 1/2"					6.80			6.80	7.50	
	5950	1/16" x 1"					11.30			11.30	12.45	
	6000	1/8" x 1/2"				↓	7.65			7.65	8.40	
	6200	Urethane foam, 2 component, handy pack, 1 C.F.				Ea.	27.50			27.50	30.50	
	6300	50.0 C.F. pack				C.F.	14.05			14.05	15.45	

For information about Means Estimating Seminars, see yellow pages 12 and 13 in back of book

Division Notes

	CREW	DAILY OUTPUT	LABOR-HOURS	UNIT	2003 BARE COSTS				TOTAL INCL O&P
					MAT.	LABOR	EQUIP.	TOTAL	

Division 8
Doors & Windows

Estimating Tips

08100 Metal Doors & Frames
- Most metal doors and frames look alike, but there may be significant differences among them. When estimating these items be sure to choose the line item that most closely compares to the specification or door schedule requirements regarding:
  - type of metal
  - metal gauge
  - door core material
  - fire rating
  - finish

08200 Wood & Plastic Doors
- Wood and plastic doors vary considerably in price. The primary determinant is the veneer material. Lauan, birch and oak are the most common veneers. Other variables include the following:
  - hollow or solid core
  - fire rating
  - flush or raised panel
  - finish
- If the specifications require compliance with AWI (Architectural Woodwork Institute) standards or acoustical standards, the cost of the door may increase substantially. All wood doors are priced pre-mortised for hinges and predrilled for cylindrical locksets.

- Frequently doors, frames, and windows are unique in old buildings. Specified replacement units could be stock, custom (similar to the original) or exact reproduction. The estimator should work closely with a window consultant to determine any extra costs that may be associated with the unusual installation requirements.

08300 Specialty Doors
- There are many varieties of special doors, and they are usually priced per each. Add frames, hardware or operators required for a complete installation.

08510 Steel Windows
- Most metal windows are delivered preglazed. However, some metal windows are priced without glass. Refer to 08800 Glazing for glass pricing. The grade C indicates commercial grade windows, usually ASTM C-35.

08550 Wood Windows
- All wood windows are priced preglazed. The two glazing options priced are single pane float glass and insulating glass 1/2" thick. Add the cost of screens and grills if required.

08700 Hardware
- Hardware costs add considerably to the cost of a door. The most efficient method to determine the hardware requirements for a project is to review the door schedule. This schedule, in conjunction with the specifications, is all you should need to take off the door hardware.

- Door hinges are priced by the pair, with most doors requiring 1-1/2 pairs per door. The hinge prices do not include installation labor because it is included in door installation. Hinges are classified according to the frequency of use.

08800 Glazing
- Different openings require different types of glass. The three most common types are:
  - float
  - tempered
  - insulating
- Most exterior windows are glazed with insulating glass. Entrance doors and window walls, where the glass is less than 18" from the floor, are generally glazed with tempered glass. Interior windows and some residential windows are glazed with float glass.

08900 Glazed Curtain Wall
- Glazed curtain walls consist of the metal tube framing and the glazing material. The cost data in this subdivision is presented for the metal tube framing alone or the composite wall. If your estimate requires a detailed takeoff of the framing, be sure to add the glazing cost.

Reference Numbers
Reference numbers are shown in bold squares at the beginning of some major classifications. These numbers refer to related items in the Reference Section. The reference information may be an estimating procedure, an alternate pricing method or technical information.

Note: Not all subdivisions listed here necessarily appear in this publication.

08341 | Cold Storage Doors

		CREW	DAILY OUTPUT	LABOR-HOURS	UNIT	2003 BARE COSTS				TOTAL INCL O&P
						MAT.	LABOR	EQUIP.	TOTAL	
200	0010	**COLD STORAGE**									**200**
	0020	Single, 20 ga. galvanized steel								
	0300	Horizontal sliding, 5' x 7', manual operation, 2" thick	2 Carp	2	8	Ea.	2,300	252		2,552	2,950
	0400	4" thick		2	8		2,775	252		3,027	3,475
	0500	6" thick		2	8		2,700	252		2,952	3,375
	0800	5' x 7', power operation, 2" thick		1.90	8.421		4,500	266		4,766	5,375
	0900	4" thick		1.90	8.421		4,575	266		4,841	5,450
	1000	6" thick		1.90	8.421		5,200	266		5,466	6,150
	1300	9' x 10', manual operation, 2" insulation		1.70	9.412		3,625	297		3,922	4,475
	1400	4" insulation		1.70	9.412		3,750	297		4,047	4,600
	1500	6" insulation		1.70	9.412		4,525	297		4,822	5,450
	1800	Power operation, 2" insulation		1.60	10		6,250	315		6,565	7,375
	1900	4" insulation		1.60	10		6,400	315		6,715	7,550
	2000	6" insulation		1.70	9.412		7,250	297		7,547	8,450
	2300	For stainless steel face, add					20%				
	3000	Hinged, lightweight, 3' x 7'-0", galvanized 1 face, 2" thick	2 Carp	2	8	Ea.	1,050	252		1,302	1,550
	3050	4" thick		1.90	8.421		1,200	266		1,466	1,750
	3300	Aluminum doors, 3' x 7'-0", 4" thick		1.90	8.421		1,075	266		1,341	1,600
	3350	6" thick		1.40	11.429		1,900	360		2,260	2,675
	3600	Stainless steel, 3' x 7'-0", 4" thick		1.90	8.421		1,375	266		1,641	1,950
	3650	6" thick		1.40	11.429		2,300	360		2,660	3,125
	3900	Painted, 3' x 7'-0", 4" thick		1.90	8.421		985	266		1,251	1,500
	3950	6" thick		1.40	11.429		1,875	360		2,235	2,650
	5000	Bi-parting, electric operated									
	5010	6' x 8' opening, galv. faces, 4" thick for cooler	2 Carp	.80	20	Opng.	5,875	630		6,505	7,450
	5050	For freezer, 4" thick		.80	20		6,475	630		7,105	8,100
	5300	For door buck framing and door protection, add		2.50	6.400		480	202		682	845
	6000	Galvanized batten door, galvanized hinges, 4' x 7'		2	8		1,450	252		1,702	2,000
	6050	6' x 8'		1.80	8.889		1,975	280		2,255	2,625
	6500	Fire door, 3 hr., 6' x 8', single slide		.80	20		6,800	630		7,430	8,450
	6550	Double, bi-parting		.70	22.857		9,000	720		9,720	11,000

08344 | Industrial Doors

		CREW	DAILY OUTPUT	LABOR-HOURS	UNIT	2003 BARE COSTS				TOTAL INCL O&P
						MAT.	LABOR	EQUIP.	TOTAL	
120	0010	**AIR CURTAINS** Incl. motor starters, transformers,									**120**
	0050	door switches & temperature controls								
	0100	Shipping and receiving doors, unheated, minimal wind stoppage								
	0150	8' high, multiples of 3' wide	2 Shee	6	2.667	L.F.	268	98.50		366.50	445
	0160	5' wide		10	1.600		296	59		355	415
	0210	10' high, multiples of 4' wide		8	2		213	74		287	350
	0250	12' high, 3'-6" wide		7	2.286		221	84.50		305.50	375
	0260	12' wide		6	2.667		223	98.50		321.50	395
	0350	16' high, 3'-6" wide		7	2.286		305	84.50		389.50	465
	0360	12' wide		6	2.667		345	98.50		443.50	530
	1500	Customer entrance doors, unheated, minimal wind stoppage								
	1550	10' high, multiples of 3' wide	2 Shee	6	2.667	L.F.	268	98.50		366.50	445
	1560	5' wide		10	1.600		291	59		350	410
	1650	Maximum wind stoppage, 12' high, multiples of 4' wide		8	2		282	74		356	425
	1700	Heated, minimal wind stoppage, electric heat								
	1750	8' high, multiples of 3' wide	2 Shee	6	2.667	L.F.	550	98.50		648.50	755
	1850	10' high, multiples of 3' wide		6	2.667		550	98.50		648.50	755
	1860	Multiples of 5' wide		10	1.600		370	59		429	500
	1950	Maximum wind stoppage, steam heat								
	1960	12' high, multiples of 4' wide	2 Shee	8	2	L.F.	425	74		499	580
	2000	Walk-in coolers and freezers, ambient air, minimal wind stoppage								
	2050	8' high, multiples of 3' wide	2 Shee	6	2.667	L.F.	340	98.50		438.50	525

Important: See the Reference Section for critical supporting data - Reference Nos., Crews, & City Cost Indexes

			CREW	DAILY OUTPUT	LABOR-HOURS	UNIT	2003 BARE COSTS				TOTAL INCL O&P	
08344		**Industrial Doors**					MAT.	LABOR	EQUIP.	TOTAL		
120	2060	Multiples of 5' wide	2 Shee	10	1.600	L.F.	355	59		414	485	**120**
	2250	Maximum wind stoppage, 12' high, multiples of 3' wide		6	2.667		350	98.50		448.50	535	
	2450	Conveyor openings or service windows, unheated, 5' high		5	3.200		268	118		386	475	
	2460	Heated, electric, 5' high, 2'-6" wide		5	3.200		263	118		381	470	

For information about Means Estimating Seminars, see yellow pages 12 and 13 in back of book

DOORS & WINDOWS **8**

For expanded coverage of these items see *Means Interior Cost Data 2003*

Division Notes

		CREW	DAILY OUTPUT	LABOR-HOURS	UNIT	2003 BARE COSTS				TOTAL INCL O&P
						MAT.	LABOR	EQUIP.	TOTAL	

Division 9
Finishes

Estimating Tips
General
- Room Finish Schedule: A complete set of plans should contain a room finish schedule. If one is not available, it would be well worth the time and effort to put one together. A room finish schedule should contain the room number, room name (for clarity), floor materials, base materials, wainscot materials, wainscot height, wall materials (for each wall), ceiling materials, ceiling height and special instructions.
- Surplus Finishes: Review the specifications to determine if there is any requirement to provide certain amounts of extra materials for the owner's maintenance department. In some cases the owner may require a substantial amount of materials, especially when it is a special order item or long lead time item.

09200 Plaster & Gypsum Board
- Lath is estimated by the square yard for both gypsum and metal lath, plus usually 5% allowance for waste. Furring, channels and accessories are measured by the linear foot. An extra foot should be allowed for each accessory miter or stop.
- Plaster is also estimated by the square yard. Deductions for openings vary by preference, from zero deduction to 50% of all openings over 2 feet in width. Some estimators deduct a percentage of the total yardage for openings. The estimator should allow one extra square foot for each linear foot of horizontal interior or exterior angle located below the ceiling level. Also, double the areas of small radius work.
- Each room should be measured, perimeter times maximum wall height. Ceiling areas are equal to length times width.
- Drywall accessories, studs, track, and acoustical caulking are all measured by the linear foot. Drywall taping is figured by the square foot. Gypsum wallboard is estimated by the square foot. No material deductions should be made for door or window openings under 32 S.F. Coreboard can be obtained in a 1″ thickness for solid wall and shaft work. Additions should be made to price out the inside or outside corners.
- Different types of partition construction should be listed separately on the quantity sheets. There may be walls with studs of various widths, double studded, and similar or dissimilar surface materials. Shaft work is usually different construction from surrounding partitions requiring separate quantities and pricing of the work.

09300 Tile
09400 Terrazzo
- Tile and terrazzo areas are taken off on a square foot basis. Trim and base materials are measured by the linear foot. Accent tiles are listed per each. Two basic methods of installation are used. Mud set is approximately 30% more expensive than the thin set. In terrazzo work, be sure to include the linear footage of embedded decorative strips, grounds, machine rubbing and power cleanup.

09600 Flooring
- Wood flooring is available in strip, parquet, or block configuration. The latter two types are set in adhesives with quantities estimated by the square foot. The laying pattern will influence labor costs and material waste. In addition to the material and labor for laying wood floors, the estimator must make allowances for sanding and finishing these areas unless the flooring is prefinished.
- Most of the various types of flooring are all measured on a square foot basis. Base is measured by the linear foot. If adhesive materials are to be quantified, they are estimated at a specified coverage rate by the gallon depending upon the specified type and the manufacturer's recommendations.
- Sheet flooring is measured by the square yard. Roll widths vary, so consideration should be given to use the most economical width, as waste must be figured into the total quantity. Consider also the installation methods available, direct glue down or stretched.

09700 Wall Finishes
- Wall coverings are estimated by the square foot. The area to be covered is measured, length by height of wall above baseboards, to calculate the square footage of each wall. This figure is divided by the number of square feet in the single roll which is being used. Deduct, in full, the areas of openings such as doors and windows. Where a pattern match is required allow 25%-30% waste. One gallon of paste should be sufficient to hang 12 single rolls of light to medium weight paper.

09800 Acoustical Treatment
- Acoustical systems fall into several categories. The takeoff of these materials is by the square foot of area with a 5% allowance for waste. Do not forget about scaffolding, if applicable, when estimating these systems.

09900 Paints & Coatings
- Painting is one area where bids vary to a greater extent than almost any other section of a project. This arises from the many methods of measuring surfaces to be painted. The estimator should check the plans and specifications carefully to be sure of the required number of coats.
- Protection of adjacent surfaces is not included in painting costs. When considering the method of paint application, an important factor is the amount of protection and masking required. These must be estimated separately and may be the determining factor in choosing the method of application.

Reference Numbers
Reference numbers are shown in bold squares at the beginning of some major classifications. These numbers refer to related items in the Reference Section. The reference information may be an estimating procedure, an alternate pricing method or technical information.

Note: Not all subdivisions listed here necessarily appear in this publication.

		09910	**Paints & Coatings**	CREW	DAILY OUTPUT	LABOR-HOURS	UNIT	2003 BARE COSTS				TOTAL INCL O&P	
								MAT.	LABOR	EQUIP.	TOTAL		
630	0010		**MISCELLANEOUS, INTERIOR**										**630**
	3800		Grilles, per side, oil base, primer coat, brushwork	1 Pord	520	.015	S.F.	.09	.43		.52	.75	
	3850		Spray		1,140	.007		.10	.20		.30	.41	
	3880		Paint 1 coat, brushwork		520	.015		.10	.43		.53	.76	
	3900		Spray		1,140	.007		.12	.20		.32	.43	
	3920		Paint 2 coats, brushwork		325	.025		.20	.69		.89	1.27	
	3940		Spray		650	.012		.23	.35		.58	.77	
	3950		Prime & paint 1 coat		325	.025		.20	.69		.89	1.27	
	3960		Prime & paint 2 coats		270	.030		.19	.83		1.02	1.47	
	4250		Paint 1 coat, brushwork	2 Pord	1,300	.012	L.F.	.11	.35		.46	.64	
	4500		Louvers, one side, primer, brushwork	1 Pord	524	.015	S.F.	.06	.43		.49	.71	
	4520		Paint one coat, brushwork		520	.015		.05	.43		.48	.71	
	4530		Spray		1,140	.007		.06	.20		.26	.37	
	4540		Paint two coats, brushwork		325	.025		.11	.69		.80	1.17	
	4550		Spray		650	.012		.12	.35		.47	.65	
	4560		Paint three coats, brushwork		270	.030		.16	.83		.99	1.43	
	4570		Spray		500	.016		.17	.45		.62	.87	
	5000		Pipe, to 4" diameter, primer or sealer coat, oil base, brushwork	2 Pord	1,250	.013	L.F.	.06	.36		.42	.60	
	5100		Spray		2,165	.007		.06	.21		.27	.37	
	5200		Paint 1 coat, brushwork		1,250	.013		.06	.36		.42	.61	
	5300		Spray		2,165	.007		.06	.21		.27	.37	
	5350		Paint 2 coats, brushwork		775	.021		.11	.58		.69	1	
	5400		Spray		1,240	.013		.12	.36		.48	.68	
	5420		Paint 3 coats, brushwork		775	.021		.16	.58		.74	1.06	
	5450		To 8" diameter, primer or sealer coat, brushwork		620	.026		.12	.72		.84	1.23	
	5500		Spray		1,085	.015		.19	.41		.60	.84	
	5550		Paint 1 coat, brushwork		620	.026		.17	.72		.89	1.28	
	5600		Spray		1,085	.015		.19	.41		.60	.83	
	5650		Paint 2 coats, brushwork		385	.042		.22	1.17		1.39	2	
	5700		Spray		620	.026		.24	.72		.96	1.37	
	5720		Paint 3 coats, brushwork		385	.042		.33	1.17		1.50	2.12	
	5750		To 12" diameter, primer or sealer coat, brushwork		415	.039		.18	1.08		1.26	1.83	
	5800		Spray		725	.022		.20	.62		.82	1.16	
	5850		Paint 1 coat, brushwork		415	.039		.17	1.08		1.25	1.83	
	6000		Spray		725	.022		.19	.62		.81	1.15	
	6200		Paint 2 coats, brushwork		260	.062		.33	1.73		2.06	2.97	
	6250		Spray		415	.039		.37	1.08		1.45	2.04	
	6270		Paint 3 coats, brushwork		260	.062		.49	1.73		2.22	3.15	
	6300		To 16" diameter, primer or sealer coat, brushwork		310	.052		.23	1.45		1.68	2.45	
	6350		Spray		540	.030		.26	.83		1.09	1.55	
	6400		Paint 1 coat, brushwork		310	.052		.23	1.45		1.68	2.44	
	6450		Spray		540	.030		.25	.83		1.08	1.54	
	6500		Paint 2 coats, brushwork		195	.082		.44	2.30		2.74	3.97	
	6550		Spray		310	.052		.49	1.45		1.94	2.73	
	6600		Radiators, per side, primer, brushwork	1 Pord	520	.015	S.F.	.06	.43		.49	.71	
	6620		Paint one coat, brushwork		520	.015		.05	.43		.48	.71	
	6640		Paint two coats, brushwork		340	.024		.11	.66		.77	1.12	
	6660		Paint three coats, brushwork		283	.028		.16	.79		.95	1.37	

For information about Means Estimating Seminars, see yellow pages 12 and 13 in back of book

Important: See the Reference Section for critical supporting data - Reference Nos., Crews, & City Cost Indexes

Division 10
Specialties

Estimating Tips

General

- The items in this division are usually priced per square foot or each.
- Many items in Division 10 require some type of support system or special anchors that are not usually furnished with the item. The required anchors must be added to the estimate in the appropriate division.
- Some items in Division 10, such as lockers, may require assembly before installation. Verify the amount of assembly required. Assembly can often exceed installation time.

10150 Compartments & Cubicles

- Support angles and blocking are not included in the installation of toilet compartments, shower/dressing compartments or cubicles. Appropriate line items from Divisions 5 or 6 may need to be added to support the installations.
- Toilet partitions are priced by the stall. A stall consists of a side wall, pilaster and door with hardware. Toilet tissue holders and grab bars are extra.

10600 Partitions

- The required acoustical rating of a folding partition can have a significant impact on costs. Verify the sound transmission coefficient rating of the panel priced to the specification requirements.

10800 Toilet/Bath/Laundry Accessories

- Grab bar installation does not include supplemental blocking or backing to support the required load. When grab bars are installed at an existing facility provisions must be made to attach the grab bars to solid structure.

Reference Numbers

Reference numbers are shown in bold squares at the beginning of some major classifications. These numbers refer to related items in the Reference Section. The reference information may be an estimating procedure, an alternate pricing method or technical information.

Note: Not all subdivisions listed here necessarily appear in this publication.

10210 | Wall Louvers

			CREW	DAILY OUTPUT	LABOR-HOURS	UNIT	2003 BARE COSTS MAT.	LABOR	EQUIP.	TOTAL	TOTAL INCL O&P	
800	0010	**LOUVERS** Aluminum with screen, residential, 8" x 8"	1 Carp	38	.211	Ea.	6.60	6.65		13.25	17.70	800
	0100	12" x 12"		38	.211		8.45	6.65		15.10	19.70	
	0200	12" x 18"		35	.229		12.30	7.20		19.50	25	
	0250	14" x 24"		30	.267		15.95	8.40		24.35	30.50	
	0300	18" x 24"		27	.296		18.55	9.35		27.90	35	
	0500	24" x 30"		24	.333		24	10.50		34.50	42.50	
	0700	Triangle, adjustable, small		20	.400		21	12.60		33.60	42.50	
	0800	Large		15	.533		40	16.85		56.85	70.50	
	1200	Extruded aluminum, see division 15850-600										
	2100	Midget, aluminum, 3/4" deep, 1" diameter	1 Carp	85	.094	Ea.	.67	2.97		3.64	5.40	
	2150	3" diameter		60	.133		1.41	4.21		5.62	8.10	
	2200	4" diameter		50	.160		1.99	5.05		7.04	10.10	
	2250	6" diameter		30	.267		2.58	8.40		10.98	16	
	2300	Ridge vent strip, mill finish	1 Shee	155	.052	L.F.	2.14	1.91		4.05	5.30	
	2400	Under eaves vent, aluminum, mill finish, 16" x 4"	1 Carp	48	.167	Ea.	1.42	5.25		6.67	9.75	
	2500	16" x 8"	"	48	.167	"	1.85	5.25		7.10	10.25	

10275 | Access Flooring

			CREW	DAILY OUTPUT	LABOR-HOURS	UNIT	2003 BARE COSTS MAT.	LABOR	EQUIP.	TOTAL	TOTAL INCL O&P	
150	0010	**PEDESTAL ACCESS FLOORS** Computer room application, metal										150
	0020	Particle board or steel panels, no covering, under 6,000 S.F.	2 Carp	400	.040	S.F.	10.45	1.26		11.71	13.45	
	0300	Metal covered, over 6,000 S.F.		450	.036		7.55	1.12		8.67	10.05	
	0400	Aluminum, 24" panels		500	.032		25.50	1.01		26.51	30	
	0600	For carpet covering, add					5.45			5.45	5.95	
	0700	For vinyl floor covering, add					5.50			5.50	6.05	
	0900	For high pressure laminate covering, add					3.67			3.67	4.04	
	0910	For snap on stringer system, add	2 Carp	1,000	.016		1.25	.50		1.75	2.17	
	0950	Office applications, to 8" high, steel panels,										
	0960	no covering, over 6,000 S.F.	2 Carp	500	.032	S.F.	8	1.01		9.01	10.40	
	1000	Machine cutouts after initial installation	1 Carp	10	.800	Ea.	41	25		66	84.50	
	1050	Pedestals, 6" to 12"	2 Carp	85	.188		7.70	5.95		13.65	17.80	
	1100	Air conditioning grilles, 4" x 12"	1 Carp	17	.471		48.50	14.85		63.35	76	
	1150	4" x 18"	"	14	.571		66.50	18.05		84.55	101	
	1200	Approach ramps, minimum	2 Carp	85	.188	S.F.	24	5.95		29.95	36	
	1300	Maximum	"	60	.267	"	29.50	8.40		37.90	45	
	1500	Handrail, 2 rail, aluminum	1 Carp	15	.533	L.F.	86.50	16.85		103.35	122	

10305 | Manufactured Fireplaces

			CREW	DAILY OUTPUT	LABOR-HOURS	UNIT	2003 BARE COSTS MAT.	LABOR	EQUIP.	TOTAL	TOTAL INCL O&P	
100	0010	**FIREPLACE, PREFABRICATED** Free standing or wall hung										100
	0100	with hood & screen, minimum	1 Carp	1.30	6.154	Ea.	1,025	194		1,219	1,425	

10 SPECIALTIES

10300 | Fireplaces & Stoves

		10305 | **Manufactured Fireplaces**	CREW	DAILY OUTPUT	LABOR-HOURS	UNIT	2003 BARE COSTS				TOTAL INCL O&P	
							MAT.	LABOR	EQUIP.	TOTAL		
100	0150	Average	1 Carp	1	8	Ea.	1,225	252		1,477	1,750	100
	0200	Maximum		.90	8.889	↓	3,000	280		3,280	3,725	
	0500	Chimney dbl. wall, all stainless, over 8'-6", 7" diam., add		33	.242	V.L.F.	47	7.65		54.65	64	
	0600	10" diameter, add		32	.250		49	7.90		56.90	66.50	
	0700	12" diameter, add		31	.258		64	8.15		72.15	83	
	0800	14" diameter, add		30	.267	↓	81	8.40		89.40	102	
	1000	Simulated brick chimney top, 4' high, 16" x 16"		10	.800	Ea.	175	25		200	232	
	1100	24" x 24"		7	1.143	"	325	36		361	415	
	1500	Simulated logs, gas fired, 40,000 BTU, 2' long, minimum		7	1.143	Set	435	36		471	530	
	1600	Maximum		6	1.333		605	42		647	730	
	1700	Electric, 1,500 BTU, 1'-6" long, minimum		7	1.143		121	36		157	190	
	1800	11,500 BTU, maximum	↓	6	1.333	↓	261	42		303	355	

		10320 | **Stoves**										
100	0010	**WOODBURNING STOVES** Cast iron, minimum	2 Carp	1.30	12.308	Ea.	760	390		1,150	1,450	100
	0020	Average		1	16		1,200	505		1,705	2,125	
	0030	Maximum	↓	.80	20		2,400	630		3,030	3,600	
	0050	For gas log lighter, add				↓	38			38	42	

For information about Means Estimating Seminars, see yellow pages 12 and 13 in back of book

SPECIALTIES 10

		CREW	DAILY OUTPUT	LABOR-HOURS	UNIT	2003 BARE COSTS				TOTAL INCL O&P
						MAT.	LABOR	EQUIP.	TOTAL	

Division 11
Equipment

Estimating Tips

General

- The items in this division are usually priced per square foot or each. Many of these items are purchased by the owner for installation by the contractor. Check the specifications for responsibilities, and include time for receiving, storage, installation and mechanical and electrical hook-ups in the appropriate divisions.

- Many items in Division 11 require some type of support system that is not usually furnished with the item. Examples of these systems include blocking for the attachment of casework and support angles for ceiling hung projection screens. The required blocking or supports must be added to the estimate in the appropriate division.

- Some items in Division 11 may require assembly or electrical hook-ups. Verify the amount of assembly required or the need for a hard electrical connection and add the appropriate costs.

Reference Numbers

Reference numbers are shown in bold squares at the beginning of some major classifications. These numbers refer to related items in the Reference Section. The reference information may be an estimating procedure, an alternate pricing method or technical information.

Note: Not all subdivisions listed here necessarily appear in this publication.

11010 | Maintenance Equipment

11013	Floor/Wall Cleaning Equipment	CREW	DAILY OUTPUT	LABOR-HOURS	UNIT	2003 BARE COSTS				TOTAL INCL O&P	
						MAT.	LABOR	EQUIP.	TOTAL		
800	0010	**VACUUM CLEANING**										800
	0020	Central, 3 inlet, residential	1 Skwk	.90	8.889	Total	580	287		867	1,100
	0200	Commercial		.70	11.429		1,075	370		1,445	1,775
	0400	5 inlet system, residential		.50	16		870	515		1,385	1,775
	0600	7 inlet system, commercial		.40	20		980	645		1,625	2,075
	0800	9 inlet system, residential		.30	26.667		1,175	860		2,035	2,650
	4010	Rule of thumb: First 1200 S.F., installed									1,125
	4020	For each additional S.F., add				S.F.					.18

11100 | Mercantile Equipment

11104	Display Cases & Systems	CREW	DAILY OUTPUT	LABOR-HOURS	UNIT	2003 BARE COSTS				TOTAL INCL O&P	
						MAT.	LABOR	EQUIP.	TOTAL		
700	0010	**REFRIGERATED FOOD CASES**										700
	0030	Dairy, multi-deck, 12' long	Q-5	3	5.333	Ea.	7,800	181		7,981	8,850
	0100	For rear sliding doors, add					1,075			1,075	1,175
	0200	Delicatessen case, service deli, 12' long, single deck	Q-5	3.90	4.103		5,250	139		5,389	5,975
	0300	Multi-deck, 18 S.F. shelf display		3	5.333		6,425	181		6,606	7,350
	0400	Freezer, self-contained, chest-type, 30 C.F.		3.90	4.103		3,825	139		3,964	4,400
	0500	Glass door, upright, 78 C.F.		3.30	4.848		7,300	164		7,464	8,275
	0600	Frozen food, chest type, 12' long		3.30	4.848		5,275	164		5,439	6,050
	0700	Glass door, reach-in, 5 door		3	5.333		10,100	181		10,281	11,400
	0800	Island case, 12' long, single deck		3.30	4.848		5,975	164		6,139	6,800
	0900	Multi-deck		3	5.333		12,600	181		12,781	14,200
	1000	Meat case, 12' long, single deck		3.30	4.848		4,325	164		4,489	5,025
	1050	Multi-deck		3.10	5.161		7,450	175		7,625	8,475
	1100	Produce, 12' long, single deck		3.30	4.848		5,725	164		5,889	6,550
	1200	Multi-deck		3.10	5.161		6,375	175		6,550	7,300

11110 | Commercial Laundry & Dry Cleaning Equipment

11119	Laundry Cleaning	CREW	DAILY OUTPUT	LABOR-HOURS	UNIT	2003 BARE COSTS				TOTAL INCL O&P	
						MAT.	LABOR	EQUIP.	TOTAL		
450	0010	**LAUNDRY EQUIPMENT** Not incl. rough-in										450
	0500	Dryers, gas fired residential, 16 lb. capacity, average	1 Plum	3	2.667	Ea.	540	99.50		639.50	740
	1000	Commercial, 30 lb. capacity, coin operated, single		3	2.667		2,475	99.50		2,574.50	2,850
	1100	Double stacked		2	4		5,275	149		5,424	6,025
	1500	Industrial, 30 lb. capacity		2	4		2,100	149		2,249	2,525
	1600	50 lb. capacity		1.70	4.706		2,750	176		2,926	3,300
	5000	Washers, residential, 4 cycle, average		3	2.667		615	99.50		714.50	825
	5300	Commercial, coin operated, average		3	2.667		990	99.50		1,089.50	1,250
	6000	Combination washer/extractor, 20 lb. capacity	L-6	1.50	8		3,550	299		3,849	4,350
	6100	30 lb. capacity		.80	15		6,850	560		7,410	8,375
	6200	50 lb. capacity		.68	17.647		8,525	660		9,185	10,400
	6300	75 lb. capacity		.30	40		17,600	1,500		19,100	21,700

11 EQUIPMENT

11110 | Commercial Laundry & Dry Cleaning Equipment

11119	Laundry Cleaning	CREW	DAILY OUTPUT	LABOR-HOURS	UNIT	2003 BARE COSTS				TOTAL INCL O&P	
						MAT.	LABOR	EQUIP.	TOTAL		
450 6350	125 lb. capacity	L-6	.16	75	Ea.	21,300	2,800		24,100	27,700	**450**

11140 | Vehicle Service Equipment

11141	Service Station Equipment	CREW	DAILY OUTPUT	LABOR-HOURS	UNIT	2003 BARE COSTS				TOTAL INCL O&P	
						MAT.	LABOR	EQUIP.	TOTAL		
150 0010	**AUTOMOTIVE**										**150**
0030	Compressors, electric, 1-1/2 H.P., standard controls	L-4	1.50	16	Ea.	330	470		800	1,100	
0550	Dual controls		1.50	16		520	470		990	1,300	
0600	5 H.P., 115/230 volt, standard controls		1	24		1,575	705		2,280	2,825	
0650	Dual controls		1	24		1,675	705		2,380	2,950	
1100	Product dispenser with vapor recovery for 6 nozzles, installed, not										
1110	including piping to storage tanks				Ea.	15,400			15,400	16,900	
2200	Hoists, single post, 8,000# capacity, swivel arms	L-4	.40	60		3,600	1,775		5,375	6,700	
2400	Two posts, adjustable frames, 11,000# capacity		.25	96		4,625	2,825		7,450	9,500	
2500	24,000# capacity		.15	160		6,150	4,700		10,850	14,200	
2700	7,500# capacity, frame supports		.50	48		5,125	1,400		6,525	7,850	
2800	Four post, roll on ramp		.50	48		4,625	1,400		6,025	7,275	
2810	Hydraulic lifts, above ground, 2 post, clear floor, 6000 lb cap		2.67	8.989		4,825	264		5,089	5,725	
2815	9000 lb capacity		2.29	10.480		11,400	310		11,710	13,100	
2820	15,000 lb capacity		2	12		12,900	355		13,255	14,800	
2825	30,000 lb capacity		1.60	15		28,500	440		28,940	32,100	
2830	4 post, ramp style, 25,000 lb capacity		2	12		11,200	355		11,555	12,900	
2835	35,000 lb capacity		1	24		52,000	705		52,705	58,500	
2840	50,000 lb capacity		1	24		58,000	705		58,705	65,000	
2845	75,000 lb capacity		1	24		67,500	705		68,205	75,500	
2850	For drive thru tracks, add, minimum					720			720	790	
2855	Maximum					1,225			1,225	1,350	
2860	Ramp extensions, 3'(set of 2)					590			590	650	
2865	Rolling jack platform					2,050			2,050	2,250	
2870	Elec/hyd jacking beam					5,475			5,475	6,025	
2880	Scissor lift, portable, 6000 lb capacity					5,375			5,375	5,925	
3000	Lube equipment, 3 reel type, with pumps, not including piping	L-4	.50	48	Set	6,150	1,400		7,550	8,975	
3700	Pump lubrication, pneumatic, not incl. air compressor										
3710	Oil/gear lube	Q-1	9.60	1.667	Ea.	680	56		736	835	
3720	Grease	"	9.60	1.667		630	56		686	775	
4000	Spray painting booth, 26' long, complete	L-4	.40	60		12,700	1,775		14,475	16,700	

11170 | Solid Waste Handling Equipment

11179	Waste Handling Equipment	CREW	DAILY OUTPUT	LABOR-HOURS	UNIT	2003 BARE COSTS				TOTAL INCL O&P	
						MAT.	LABOR	EQUIP.	TOTAL		
150 0010	**WASTE HANDLING**										**150**
0020	Compactors, 115 volt, 250#/hr., chute fed	L-4	1	24	Ea.	8,600	705		9,305	10,600	

EQUIPMENT 11

89

11179	Waste Handling Equipment	CREW	DAILY OUTPUT	LABOR-HOURS	UNIT	2003 BARE COSTS				TOTAL INCL O&P	
						MAT.	LABOR	EQUIP.	TOTAL		
150 0100	Hand fed	L-4	2.40	10	Ea.	6,200	294		6,494	7,275	150
0300	Multi-bag, 230 volt, 600#/hr, chute fed		1	24		7,575	705		8,280	9,425	
0400	Hand fed		1	24		6,500	705		7,205	8,250	
0500	Containerized, hand fed, 2 to 6 C.Y. containers, 250#/hr.		1	24		8,175	705		8,880	10,100	
0550	For chute fed, add per floor		1	24		940	705		1,645	2,125	
1000	Heavy duty industrial compactor, 0.5 C.Y. capacity		1	24		5,525	705		6,230	7,175	
1050	1.0 C.Y. capacity		1	24		8,325	705		9,030	10,300	
1100	3 C.Y. capacity		.50	48		11,300	1,400		12,700	14,600	
1150	5.0 C.Y. capacity		.50	48		14,800	1,400		16,200	18,400	
1200	Combination shredder/compactor (5,000 lbs./hr.)		.50	48		28,100	1,400		29,500	33,100	
1400	For handling hazardous waste materials, 55 gallon drum packer, std.					13,000			13,000	14,300	
1410	55 gallon drum packer w/HEPA filter					16,300			16,300	17,900	
1420	55 gallon drum packer w/charcoal & HEPA filter					21,700			21,700	23,900	
1430	All of the above made explosion proof, add					9,925			9,925	10,900	
1500	Crematory, not including building, 1 place	Q-3	.20	160		46,300	5,700		52,000	59,500	
1750	2 place		.10	320		66,000	11,400		77,400	89,500	
4400	Incinerator, gas, not incl. chimney, elec. or pipe, 50#/hr., minimum		.80	40		18,100	1,425		19,525	22,100	
4420	Maximum		.70	45.714		23,500	1,625		25,125	28,400	
4440	200 lb. per hr., minimum (batch type)		.60	53.333		23,500	1,900		25,400	28,800	
4460	Maximum (with feeder)		.50	64		45,800	2,275		48,075	54,000	
4480	400 lb. per hr., minimum (batch type)		.30	106		27,900	3,800		31,700	36,300	
4500	Maximum (with feeder)		.25	128		52,500	4,550		57,050	65,000	
4520	800 lb. per hr., with feeder, minimum		.20	160		68,500	5,700		74,200	84,000	
4540	Maximum		.17	188		93,500	6,700		100,200	113,000	
4560	1,200 lb. per hr., with feeder, minimum		.15	213		99,000	7,600		106,600	120,500	
4580	Maximum		.11	290		119,000	10,400		129,400	146,500	
4600	2,000 lb. per hr., with feeder, minimum		.10	320		173,500	11,400		184,900	207,500	
4620	Maximum		.05	640		291,500	22,800		314,300	355,000	
4700	For heat recovery system, add, minimum		.25	128		56,500	4,550		61,050	69,000	
4710	Add, maximum		.11	290		180,000	10,400		190,400	213,500	
4720	For automatic ash conveyer, add		.50	64		23,600	2,275		25,875	29,500	
4750	Large municipal incinerators, incl. stack, minimum		.25	128	Ton/day	14,400	4,550		18,950	22,700	
4850	Maximum		.10	320	"	38,200	11,400		49,600	59,000	
5800	Shredder, industrial, minimum				Ea.	16,800			16,800	18,500	
5850	Maximum					90,500			90,500	99,500	
5900	Baler, industrial, minimum					6,750			6,750	7,425	
5950	Maximum					394,000			394,000	433,500	

11310	Sewage & Sludge Pumps	CREW	DAILY OUTPUT	LABOR-HOURS	UNIT	2003 BARE COSTS				TOTAL INCL O&P	
						MAT.	LABOR	EQUIP.	TOTAL		
350 0010	**PUMPS, PNEUMATIC EJECTOR**										350
0020	With steel receiver, level controls, inlet/outlet gate/check valves										
0030	Cross connect. not incl. compressor, fittings or piping										
0040	Duplex										
0050	30 GPM	Q-2	1.70	14.118	Ea.	14,400	490		14,890	16,500	
0060	50 GPM		1.56	15.385		20,300	535		20,835	23,100	
0070	100 GPM		1.30	18.462		24,500	645		25,145	28,000	
0080	150 GPM		1.10	21.818		47,300	760		48,060	53,000	

11300 | Fluid Waste Treatment & Disposal Equipment

11310	Sewage & Sludge Pumps	CREW	DAILY OUTPUT	LABOR-HOURS	UNIT	2003 BARE COSTS				TOTAL INCL O&P	
						MAT.	LABOR	EQUIP.	TOTAL		
350	0090	200 GPM	Q-2	.85	28.235	Ea.	60,000	985		60,985	67,500	350
	0100	250 GPM	Q-3	.91	35.165		77,500	1,250		78,750	87,000	
	0110	300 GPM	"	.57	56.140	↓	94,500	2,000		96,500	107,000	

11400 | Food Service Equipment

11410	Food Prep Equipment	CREW	DAILY OUTPUT	LABOR-HOURS	UNIT	2003 BARE COSTS				TOTAL INCL O&P	
						MAT.	LABOR	EQUIP.	TOTAL		
150	0010	**COMMERCIAL KITCHEN EQUIPMENT**										150
	0020	Bake oven, gas, one section	Q-1	8	2	Ea.	4,150	67.50		4,217.50	4,650	
	0300	Two sections		7	2.286		7,900	77		7,977	8,800	
	0600	Three sections	↓	6	2.667		11,600	89.50		11,689.50	12,900	
	0900	Electric convection, single deck	L-7	4	7		4,150	213		4,363	4,875	
	2350	Cooler, reach-in, beverage, 6' long	Q-1	6	2.667	↓	3,125	89.50		3,214.50	3,550	
	2700	Dishwasher, commercial, rack type										
	2720	10 to 12 racks per hour	Q-1	3.20	5	Ea.	2,900	168		3,068	3,450	
	2750	Semi-automatic 38 to 50 racks per hour	"	1.30	12.308		5,775	415		6,190	7,000	
	2800	Automatic, 190 to 230 racks per hour	L-6	.35	34.286		9,625	1,275		10,900	12,500	
	2820	235 to 275 racks per hour		.25	48		23,000	1,800		24,800	28,000	
	2840	8,750 to 12,500 dishes per hour	↓	.10	120		41,200	4,500		45,700	52,000	
	4300	Freezers, reach-in, 44 C.F.	Q-1	4	4		7,650	135		7,785	8,600	
	4500	68 C.F.		3	5.333		8,875	179		9,054	10,000	
	5800	Ice cube maker, 50 pounds per day		6	2.667		1,425	89.50		1,514.50	1,700	
	6050	500 pounds per day	↓	4	4		2,800	135		2,935	3,300	
	6350	20 Gal kettle w/steam jacket, tilting w/positive lock, S/S	L-7	7	4		5,900	122		6,022	6,675	
	6600	60 gallons	"	6	4.667		6,450	142		6,592	7,300	
	7950	Hood fire protection system, minimum	Q-1	3	5.333		3,050	179		3,229	3,625	
	8050	Maximum		1	16		22,100	540		22,640	25,100	
	8300	Refrigerators, reach-in type, 44 C.F.		5	3.200		4,500	108		4,608	5,125	
	8310	With glass doors, 68 C.F.	↓	4	4	↓	6,725	135		6,860	7,600	

11420	Food Cooking Equipment										
550	0010	**KITCHEN VENTILATION,** Commercial										550
	1010	Heat reclaim unit, air to air										
	1100	Heat pipe exchanger										
	1110	Combined supply/exhaust air volume										
	1120	2.5 to 6.0 MCFM	Q-10	2.80	8.571	MCFM	4,850	296		5,146	5,800	
	1130	6 to 16 MCFM		5	4.800		2,975	166		3,141	3,525	
	1140	16 to 22 MCFM	↓	6	4	↓	2,325	138		2,463	2,775	
	2010	Packaged ventilation										
	2100	Combined supply/exhaust air volume										
	2110	With ambient supply										
	2120	3 to 7 MCFM	Q-10	5	4.800	MCFM	880	166		1,046	1,225	
	2130	7 to 12 MCFM		7.10	3.380		585	117		702	825	
	2140	12 to 22 MCFM	↓	10.70	2.243	↓	415	77.50		492.50	575	
	2400	With tempered supply										
	2410	3 to 7 MCFM	Q-10	5	4.800	MCFM	1,525	166		1,691	1,925	
	2420	7 to 12 MCFM		7.10	3.380		980	117		1,097	1,250	
	2430	12 to 22 MCFM	↓	10.70	2.243	↓	720	77.50		797.50	910	
	4000	Electrostatic precipitators										

EQUIPMENT 11

91

11420	Food Cooking Equipment	CREW	DAILY OUTPUT	LABOR-HOURS	UNIT	2003 BARE COSTS				TOTAL INCL O&P	
						MAT.	LABOR	EQUIP.	TOTAL		
550	4100	Smoke pollution control										**550**
	4110	Exhaust air volume									
	4120	1 to 2.4 MCFM	Q-10	2.15	11.163	MCFM	5,700	385		6,085	6,850
	4130	2.4 to 4.8 MCFM		2.70	8.889		6,475	305		6,780	7,600
	4140	4.8 to 7.2 MCFM		3.60	6.667		6,600	230		6,830	7,600
	4150	7.2 to 12 MCFM		5.40	4.444		6,900	153		7,053	7,825
	4400	Electrostatic precipitators with built-in exhaust fans									
	4410	Smoke pollution control									
	4420	Exhaust air volume									
	4440	1 to 2.4 MCFM	Q-10	2.15	11.163	MCFM	6,850	385		7,235	8,125
	4450	2.4 to 4.8 MCFM		2.70	8.889		5,425	305		5,730	6,450
	4460	4.8 to 7.2 MCFM		3.60	6.667		4,650	230		4,880	5,475
	4470	7.2 to 12 MCFM		5.40	4.444		4,000	153		4,153	4,625
	6000	Exhaust hoods									
	6100	Centrifugal grease extraction									
	6110	Water wash type									
	6120	Per foot of length, minimum	Q-10	7.10	3.380	L.F.	1,150	117		1,267	1,450
	6130	Maximum	"	7.10	3.380	"	1,975	117		2,092	2,350
	6200	Non water wash type									
	6210	Per foot of length	Q-10	13.40	1.791	L.F.	850	62		912	1,025
	6400	Island style ventilator									
	6410	Water wash type									
	6420	Per foot of length, minimum	Q-10	7.10	3.380	L.F.	1,550	117		1,667	1,875
	6430	Maximum	"	7.10	3.380	"	2,575	117		2,692	3,000
	6460	Non water wash type									
	6470	Per foot of length, minimum	Q-10	13.40	1.791	L.F.	670	62		732	830
	6480	Maximum	"	13.40	1.791	"	1,375	62		1,437	1,600

11 EQUIPMENT

11454	Residential Appliances	CREW	DAILY OUTPUT	LABOR-HOURS	UNIT	2003 BARE COSTS				TOTAL INCL O&P	
						MAT.	LABOR	EQUIP.	TOTAL		
500	0010	**RESIDENTIAL APPLIANCES**										**500**
	0020	Cooking range, 30" free standing, 1 oven, minimum	2 Clab	10	1.600	Ea.	249	39.50		288.50	335
	0050	Maximum		4	4		1,250	98.50		1,348.50	1,525
	0150	2 oven, minimum		10	1.600		1,400	39.50		1,439.50	1,600
	0200	Maximum		10	1.600		1,450	39.50		1,489.50	1,650
	1500	Combination range, refrigerator and sink, 30" wide, minimum	L-1	2	8		730	300		1,030	1,250
	1550	Maximum		1	16		1,450	600		2,050	2,500
	1570	60" wide, average		1.40	11.429		2,400	430		2,830	3,300
	1590	72" wide, average		1.20	13.333		2,725	500		3,225	3,750
	1600	Office model, 48" wide		2	8		2,050	300		2,350	2,725
	1620	Refrigerator and sink only		2.40	6.667		2,050	250		2,300	2,650
	1640	Combination range, refrigerator, sink, microwave									
	1660	oven and ice maker	L-1	.80	20	Ea.	4,025	750		4,775	5,550
	2450	Dehumidifier, portable, automatic, 15 pint					149			149	164
	2550	40 pint					175			175	192
	2750	Dishwasher, built-in, 2 cycles, minimum	L-1	4	4		245	150		395	495
	2800	Maximum		2	8		289	300		589	770
	2950	4 or more cycles, minimum		4	4		266	150		416	520

Important: See the Reference Section for critical supporting data - Reference Nos., Crews, & City Cost Indexes

11454	Residential Appliances	CREW	DAILY OUTPUT	LABOR-HOURS	UNIT	2003 BARE COSTS				TOTAL INCL O&P	
						MAT.	LABOR	EQUIP.	TOTAL		
500	2960	Average	L-1	4	4	Ea.	355	150		505	615	500
	3000	Maximum	↓	2	8		540	300		840	1,050	
	3200	Dryer, automatic, minimum	L-2	3	5.333		256	147		403	510	
	3250	Maximum	"	2	8		780	221		1,001	1,200	
	3300	Garbage disposer, sink type, minimum	L-1	10	1.600		41.50	60		101.50	136	
	3350	Maximum	"	10	1.600		144	60		204	248	
	3550	Heater, electric, built-in, 1250 watt, ceiling type, minimum	1 Elec	4	2		70	75		145	189	
	3600	Maximum		3	2.667		114	100		214	276	
	3700	Wall type, minimum		4	2		99.50	75		174.50	221	
	3750	Maximum		3	2.667		132	100		232	295	
	3900	1500 watt wall type, with blower		4	2		123	75		198	247	
	3950	3000 watt	↓	3	2.667		251	100		351	425	
	4150	Hood for range, 2 speed, vented, 30" wide, minimum	L-3	5	3.200		36	110		146	210	
	4200	Maximum		3	5.333		530	184		714	870	
	4300	42" wide, minimum		5	3.200		223	110		333	415	
	4330	Custom		5	3.200		620	110		730	855	
	4350	Maximum	↓	3	5.333		755	184		939	1,125	
	4500	For ventless hood, 2 speed, add					15			15	16.50	
	4650	For vented 1 speed, deduct from maximum					39			39	43	
	4850	Humidifier, portable, 8 gallons per day					149			149	164	
	5000	15 gallons per day					179			179	197	
	5200	Icemaker, automatic, 20 lb. per day	1 Plum	7	1.143		360	42.50		402.50	460	
	5350	51 lb. per day		2	4		980	149		1,129	1,300	
	6400	Sump pump cellar drainer, pedestal, 1/3 H.P., molded PVC base		3	2.667		87	99.50		186.50	246	
	6450	Solid brass	↓	2	4	↓	179	149		328	425	
	6460	Sump pump, see also division 15440-940										
	6650	Washing machine, automatic, minimum	1 Plum	3	2.667	Ea.	271	99.50		370.50	450	
	6700	Maximum	"	1	8		1,100	299		1,399	1,650	
	6900	Water heater, electric, glass lined, 30 gallon, minimum	L-1	5	3.200		243	120		363	445	
	6950	Maximum		3	5.333		335	200		535	670	
	7100	80 gallon, minimum		2	8		460	300		760	955	
	7150	Maximum	↓	1	16		640	600		1,240	1,600	
	7180	Water heater, gas, glass lined, 30 gallon, minimum	2 Plum	5	3.200		286	120		406	495	
	7220	Maximum		3	5.333		395	199		594	735	
	7260	50 gallon, minimum		2.50	6.400		350	239		589	745	
	7300	Maximum	↓	1.50	10.667	↓	490	400		890	1,125	
	7310	Water heater, see also division 15480-200										
	7350	Water softener, automatic, to 30 grains per gallon	2 Plum	5	3.200	Ea.	430	120		550	655	
	7400	To 100 grains per gallon	"	4	4		550	149		699	830	
	7450	Vent kits for dryers	1 Carp	10	.800	↓	12	25		37	52.50	

11471	Darkroom Processing	CREW	DAILY OUTPUT	LABOR-HOURS	UNIT	2003 BARE COSTS				TOTAL INCL O&P	
						MAT.	LABOR	EQUIP.	TOTAL		
700	0010	**DARKROOM EQUIPMENT**										700
	0020	Developing sink, 5" deep, 24" x 48"	Q-1	2	8	Ea.	4,100	269		4,369	4,925	
	0050	48" x 52"		1.70	9.412		4,175	315		4,490	5,050	
	0200	10" deep, 24" x 48"		1.70	9.412		5,125	315		5,440	6,100	
	0250	24" x 108"		1.50	10.667		1,850	360		2,210	2,600	
	3500	Washers, round, minimum sheet 11" x 14"	↓	2	8	↓	2,525	269		2,794	3,175	

EQUIPMENT 11

93

11470 | Darkroom Equipment

		11471	Darkroom Processing	CREW	DAILY OUTPUT	LABOR-HOURS	UNIT	2003 BARE COSTS MAT.	LABOR	EQUIP.	TOTAL	TOTAL INCL O&P	
700	3550		Maximum sheet 20" x 24"	Q-1	1	16	Ea.	2,975	540		3,515	4,075	700
	3800		Square, minimum sheet 20" x 24"		1	16		2,650	540		3,190	3,725	
	3900		Maximum sheet 50" x 56"	↓	.80	20	↓	4,150	675		4,825	5,575	
	4500		Combination tank sink, tray sink, washers, with										
	4510		dry side tables, average	Q-1	.45	35.556	Ea.	7,875	1,200		9,075	10,500	

11500 | Industrial & Process Equipment

		11520	Industrial Equipment	CREW	DAILY OUTPUT	LABOR-HOURS	UNIT	2003 BARE COSTS MAT.	LABOR	EQUIP.	TOTAL	TOTAL INCL O&P	
250	0010	**DUST COLLECTION SYSTEMS** Commercial / industrial											250
	0120		Central vacuum units										
	0130		Includes stand, filters and motorized shaker										
	0200		500 CFM, 10" inlet, 2 HP	Q-20	2.40	8.333	Ea.	3,250	285		3,535	4,000	
	0220		1000 CFM, 10" inlet, 3 HP		2.20	9.091		3,400	310		3,710	4,200	
	0240		1500 CFM, 10" inlet, 5 HP		2	10		3,475	340		3,815	4,325	
	0260		3000 CFM, 13" inlet, 10 HP		1.50	13.333		9,050	455		9,505	10,700	
	0280		5000 CFM, 16" inlet, 2 @ 10 HP	↓	1	20	↓	9,575	685		10,260	11,600	
	1000		Vacuum tubing, galvanized										
	1100		2-1/8" OD, 16 ga.	Q-9	440	.036	L.F.	1.76	1.21		2.97	3.81	
	1110		2-1/2" OD, 16 ga.		420	.038		2.04	1.27		3.31	4.20	
	1120		3" OD, 16 ga.		400	.040		2.59	1.33		3.92	4.91	
	1130		3-1/2" OD, 16 ga.		380	.042		3.68	1.40		5.08	6.20	
	1140		4" OD, 16 ga.		360	.044		3.76	1.48		5.24	6.45	
	1150		5" OD, 14 ga.		320	.050		6.95	1.67		8.62	10.20	
	1160		6" OD, 14 ga.		280	.057		7.80	1.90		9.70	11.55	
	1170		8" OD, 14 ga.		200	.080		11.80	2.66		14.46	17.10	
	1180		10" OD, 12 ga.		160	.100		11.80	3.33		15.13	18.15	
	1190		12" OD, 12 ga.		120	.133		34	4.44		38.44	44	
	1200		14" OD, 12 ga.	↓	80	.200	↓	37.50	6.65		44.15	51.50	
	1940		Hose, flexible wire reinforced rubber										
	1956		3" dia.	Q-9	400	.040	L.F.	5.55	1.33		6.88	8.15	
	1960		4" dia.		360	.044		7.50	1.48		8.98	10.55	
	1970		5" dia.		320	.050		8.65	1.67		10.32	12.05	
	1980		6" dia.	↓	280	.057	↓	10.10	1.90		12	14.05	
	2000		90° Elbow, slip fit										
	2110		2-1/8" dia.	Q-9	70	.229	Ea.	7.35	7.60		14.95	19.85	
	2120		2-1/2" dia.		65	.246		10.30	8.20		18.50	24	
	2130		3" dia.		60	.267		13.95	8.90		22.85	29	
	2140		3-1/2" dia.		55	.291		17.25	9.70		26.95	34	
	2150		4" dia.		50	.320		21.50	10.65		32.15	40	
	2160		5" dia.		45	.356		40.50	11.85		52.35	63	
	2170		6" dia.		40	.400		56	13.30		69.30	82	
	2180		8" dia.	↓	30	.533	↓	106	17.75		123.75	145	
	2400		45° Elbow, slip fit										
	2410		2-1/8" dia.	Q-9	70	.229	Ea.	6.45	7.60		14.05	18.85	
	2420		2-1/2" dia.		65	.246		9.50	8.20		17.70	23	
	2430		3" dia.		60	.267		11.45	8.90		20.35	26.50	
	2440		3-1/2" dia.		55	.291		14.40	9.70		24.10	31	
	2450		4" dia.	↓	50	.320	↓	18.70	10.65		29.35	37	

			DAILY	LABOR-		2003 BARE COSTS				TOTAL	
11520	**Industrial Equipment**	CREW	OUTPUT	HOURS	UNIT	MAT.	LABOR	EQUIP.	TOTAL	INCL O&P	
250	2460	5" dia.	Q-9	45	.356	Ea.	31.50	11.85		43.35	53	**250**
	2470	6" dia.		40	.400		42.50	13.30		55.80	67.50	
	2480	8" dia.	▼	35	.457	▼	83.50	15.20		98.70	116	
	2800	90° TY, slip fit thru 6" dia										
	2810	2-1/8" dia.	Q-9	42	.381	Ea.	14.40	12.70		27.10	35.50	
	2820	2-1/2" dia.		39	.410		18.70	13.65		32.35	41.50	
	2830	3" dia.		36	.444		24.50	14.80		39.30	50	
	2840	3-1/2" dia.		33	.485		31.50	16.15		47.65	59.50	
	2850	4" dia.		30	.533		45.50	17.75		63.25	77.50	
	2860	5" dia.		27	.593		86	19.75		105.75	125	
	2870	6" dia.	▼	24	.667		117	22		139	164	
	2880	8" dia., butt end					249			249	274	
	2890	10" dia., butt end					450			450	495	
	2900	12" dia., butt end					450			450	495	
	2910	14" dia., butt end					815			815	895	
	2920	6 x 4" dia., butt end					71.50			71.50	78.50	
	2930	8" x 4" dia., butt end					135			135	149	
	2940	10" x 4" dia., butt end					177			177	195	
	2950	12" x 4" dia., butt end	▼				204			204	224	
	3100	90° Elbow, butt end segmented										
	3110	8" dia., butt end, segmented				Ea.	161			161	177	
	3120	10" dia., butt end, segmented					264			264	291	
	3130	12" dia., butt end, segmented					335			335	370	
	3140	14" dia., butt end, segmented				▼	420			420	460	
	3200	45° Elbow, butt end segmented										
	3210	8" dia., butt end, segmented				Ea.	137			137	151	
	3220	10" dia., butt end, segmented					191			191	210	
	3230	12" dia., butt end, segmented					195			195	215	
	3240	14" dia., butt end, segmented				▼	240			240	264	
	3400	All butt end fittings require one coupling per joint.										
	3410	Labor for fitting included with couplings.										
	3460	Compression coupling, galvanized, neoprene gasket										
	3470	2-1/8" dia.	Q-9	44	.364	Ea.	11.20	12.10		23.30	31	
	3480	2-1/2" dia.		44	.364		11.20	12.10		23.30	31	
	3490	3" dia.		38	.421		14.70	14		28.70	37.50	
	3500	3-1/2" dia.		35	.457		16.45	15.20		31.65	41.50	
	3510	4" dia.		33	.485		17.90	16.15		34.05	44.50	
	3520	5" dia.		29	.552		20.50	18.35		38.85	51	
	3530	6" dia.		26	.615		24	20.50		44.50	57.50	
	3540	8" dia.		22	.727		43.50	24		67.50	85	
	3550	10" dia.		20	.800		63.50	26.50		90	111	
	3560	12" dia.		18	.889		79	29.50		108.50	133	
	3570	14" dia.	▼	16	1	▼	112	33.50		145.50	175	
	3800	Air gate valves, galvanized										
	3810	2-1/8" dia.	Q-9	30	.533	Ea.	85	17.75		102.75	121	
	3820	2-1/2" dia.		28	.571		89.50	19.05		108.55	128	
	3830	3" dia.		26	.615		104	20.50		124.50	146	
	3840	4" dia.		23	.696		136	23		159	186	
	3850	6" dia.	▼	18	.889	▼	188	29.50		217.50	252	

EQUIPMENT 11

11620 | Laboratory Equipment

		CREW	DAILY OUTPUT	LABOR-HOURS	UNIT	2003 BARE COSTS				TOTAL INCL O&P
						MAT.	LABOR	EQUIP.	TOTAL	
350	0010	**LABORATORY EQUIPMENT**									350
	0020	Cabinets, base, door units, metal	2 Carp	18	.889	L.F.	137	28		165	194
	0300	Drawer units	"	18	.889	"	295	28		323	370
	2000	Fume hood, with countertop & base, not including HVAC									
	2020	Simple, minimum	2 Carp	5.40	2.963	L.F.	520	93.50		613.50	715
	2050	Complex, including fixtures		2.40	6.667		1,225	210		1,435	1,675
	2100	Special, maximum		1.70	9.412		1,275	297		1,572	1,875
	2200	Ductwork, minimum	2 Shee	1	16	Hood	860	590		1,450	1,875
	2250	Maximum	"	.50	32	"	4,350	1,175		5,525	6,625
	2300	Service fixtures, average				Ea.	44			44	48.50
	2500	For sink assembly with hot and cold water, add	1 Plum	1.40	5.714		540	213		753	910
	2550	Glassware washer, undercounter, minimum	L-1	1.80	8.889		5,975	335		6,310	7,075
	2600	Maximum	"	1	16		7,675	600		8,275	9,350
	2800	Sink, one piece plastic, flask wash, hose, free standing	1 Plum	1.60	5		1,375	187		1,562	1,775
	2850	Epoxy resin sink, 25" x 16" x 10"	"	2	4		151	149		300	390
	4200	Alternate pricing method: as percent of lab furniture									
	4400	Installation, not incl. plumbing & duct work				% Furn.					22%
	4800	Plumbing, final connections, simple system									10%
	5000	Moderately complex system									15%
	5200	Complex system									20%
	5400	Electrical, simple system									10%
	5600	Moderately complex system									20%
	5800	Complex system									35%
	6000	Safety equipment, eye wash, hand held				Ea.	294			294	325
	6200	Deluge shower				"	565			565	625
	6300	Rule of thumb: lab furniture including installation & connection									
	6320	High school				S.F.					30.50
	6340	College									45
	6360	Clinical, health care									38.50
	6380	Industrial									62

11710 | Medical Equipment

		CREW	DAILY OUTPUT	LABOR-HOURS	UNIT	2003 BARE COSTS				TOTAL INCL O&P
						MAT.	LABOR	EQUIP.	TOTAL	
500	0010	**MEDICAL EQUIPMENT**									500
	0015	Autopsy table, standard	1 Plum	1	8	Ea.	7,275	299		7,574	8,475
	0020	Deluxe		.60	13.333		10,800	500		11,300	12,700
	0700	Distiller, water, steam heated, 50 gal. capacity		1.40	5.714		15,200	213		15,413	17,000
	1800	Heat therapy unit, humidified, 26" x 78" x 28"					2,800			2,800	3,075
	3200	Mortuary refrigerator, end operated, 2 capacity					11,100			11,100	12,200
	3300	6 capacity					20,400			20,400	22,500
	5600	Sterilizers, floor loading, 26" x 62" x 42", single door, steam					130,500			130,500	143,500
	5650	Double door, steam					167,500			167,500	184,500
	5800	General purpose, 20" x 20" x 38", single door					12,600			12,600	13,800
	6000	Portable, counter top, steam, minimum					3,125			3,125	3,450
	6020	Maximum					4,900			4,900	5,375
	6050	Portable, counter top, gas, 17"x15"x32-1/2"					32,300			32,300	35,500
	6150	Manual washer/sterilizer, 16"x16"x26"	1 Plum	2	4		44,300	149		44,449	48,900
	6200	Steam generators, electric 10 KW to 180 KW, freestanding									
	6250	Minimum	1 Elec	3	2.667	Ea.	6,575	100		6,675	7,400

11 EQUIPMENT

11700 | Medical Equipment

11710 | Medical Equipment

		CREW	DAILY OUTPUT	LABOR-HOURS	UNIT	MAT.	LABOR	EQUIP.	TOTAL	TOTAL INCL O&P	
500	6300	Maximum	1 Elec	.70	11.429	Ea.	23,700	430		24,130	26,700	**500**
	8100	Utensil washer-sanitizer	1 Plum	2	4		8,975	149		9,124	10,100	
	8200	Bed pan washer-sanitizer	"	2	4		5,875	149		6,024	6,675	
	8400	Whirlpool bath, mobile, sst, 18" x 24" x 60"					3,700			3,700	4,075	
	8450	Fixed, incl. mixing valves	1 Plum	2	4	↓	7,575	149		7,724	8,550	

11740 | Dental Equipment

		CREW	DAILY OUTPUT	LABOR-HOURS	UNIT	MAT.	LABOR	EQUIP.	TOTAL	TOTAL INCL O&P	
200	0010	**DENTAL EQUIPMENT**										**200**
	0020	Central suction system, minimum	1 Plum	1.20	6.667	Ea.	2,900	249		3,149	3,550	
	0100	Maximum	"	.90	8.889		22,300	330		22,630	25,000	
	0600	Chair, electric or hydraulic, minimum	1 Skwk	.50	16		3,625	515		4,140	4,775	
	0700	Maximum		.25	32		7,675	1,025		8,700	10,100	
	2000	Light, ceiling mounted, minimum		8	1		1,875	32.50		1,907.50	2,125	
	2100	Maximum	↓	8	1		3,800	32.50		3,832.50	4,225	
	2200	Unit light, minimum	2 Skwk	5.33	3.002		1,400	97		1,497	1,700	
	2210	Maximum		5.33	3.002		4,600	97		4,697	5,225	
	2220	Track light, minimum		3.20	5		2,500	161		2,661	3,000	
	2230	Maximum	↓	3.20	5		5,475	161		5,636	6,275	
	2300	Sterilizers, steam portable, minimum					3,175			3,175	3,500	
	2350	Maximum					11,000			11,000	12,100	
	2600	Steam, institutional					13,700			13,700	15,000	
	2650	Dry heat, electric, portable, 3 trays				↓	1,075			1,075	1,175	

For information about Means Estimating Seminars, see yellow pages 12 and 13 in back of book

EQUIPMENT 11

Division Notes

	CREW	DAILY OUTPUT	LABOR-HOURS	UNIT	2003 BARE COSTS				TOTAL INCL O&P
					MAT.	LABOR	EQUIP.	TOTAL	

Division 13
Special Construction

Estimating Tips

General

- The items and systems in this division are usually estimated, purchased, supplied and installed as a unit by one or more subcontractors. The estimator must ensure that all parties are operating from the same set of specifications and assumptions and that all necessary items are estimated and will be provided. Many times the complex items and systems are covered but the more common ones such as excavation or a crane are overlooked for the very reason that everyone assumes nobody could miss them. The estimator should be the central focus and be able to ensure that all systems are complete.

- Another area where problems can develop in this division is at the interface between systems. The estimator must ensure, for instance, that anchor bolts, nuts and washers are estimated and included for the air-supported structures and pre-engineered buildings to be bolted to their foundations.

Utility supply is a common area where essential items or pieces of equipment can be missed or overlooked due to the fact that each subcontractor may feel it is the others' responsibility. The estimator should also be aware of certain items which may be supplied as part of a package but installed by others, and ensure that the installing contractor's estimate includes the cost of installation. Conversely, the estimator must also ensure that items are not costed by two different subcontractors, resulting in an inflated overall estimate.

13120 Pre-Engineered Structures

- The foundations and floor slab, as well as rough mechanical and electrical, should be estimated, as this work is required for the assembly and erection of the structure. Generally, as noted in the book, the pre-engineered building comes as a shell and additional features must be included by the estimator. Here again, the estimator must have a clear understanding of the scope of each portion of the work and all the necessary interfaces.

13200 Storage Tanks

- The prices in this subdivision for above and below ground storage tanks do not include foundations or hold-down slabs. The estimator should refer to Divisions 2 and 3 for foundation system pricing. In addition to the foundations, required tank accessories such as tank gauges, leak detection devices, and additional manholes and piping must be added to the tank prices.

Reference Numbers

Reference numbers are shown in bold squares at the beginning of some major classifications. These numbers refer to related items in the Reference Section. The reference information may be an estimating procedure, an alternate pricing method or technical information.

Note: Not all subdivisions listed here necessarily appear in this publication.

		13035	Special Purpose Rooms	CREW	DAILY OUTPUT	LABOR-HOURS	UNIT	MAT.	LABOR	EQUIP.	TOTAL	TOTAL INCL O&P	
								2003 BARE COSTS					
200	0010	**CLEAN ROOMS**											**200**
	1100	Clean room, soft wall, 12' x 12', Class 100	1 Carp	.18	44.444	Ea.	12,400	1,400		13,800	15,800	
	1110	Class 1,000		.18	44.444		9,600	1,400		11,000	12,800	
	1120	Class 10,000		.21	38.095		8,175	1,200		9,375	10,900	
	1130	Class 100,000	▼	.21	38.095	▼	7,550	1,200		8,750	10,200	
	2800	Ceiling grid support, slotted channel struts 4'-0" O.C., ea. way				S.F.					6.50	
	3000	Ceiling panel, vinyl coated foil on mineral substrate										
	3020	Sealed, non-perforated				S.F.					1.40	
	4000	Ceiling panel seal, silicone sealant, 150 L.F./gal.	1 Carp	150	.053	L.F.	.22	1.68		1.90	2.87	
	4100	Two sided adhesive tape	"	240	.033	"	.10	1.05		1.15	1.75	
	4200	Clips, one per panel				Ea.	.86			.86	.95	
	6000	HEPA filter,2'x4',99.97% eff.,3" dp beveled frame(silicone seal)					264			264	290	
	6040	6" deep skirted frame (channel seal)					305			305	335	
	6100	99.99% efficient, 3" deep beveled frame (silicone seal)					284			284	315	
	6140	6" deep skirted frame (channel seal)					325			325	355	
	6200	99.999% efficient, 3" deep beveled frame (silicone seal)					325			325	355	
	6240	6" deep skirted frame (channel seal)				▼	365			365	400	
	7000	Wall panel systems, including channel strut framing										
	7020	Polyester coated aluminum, particle board				S.F.					20	
	7100	Porcelain coated aluminum, particle board									35	
	7400	Wall panel support, slotted channel struts, to 12' high				▼					18	
700	0010	**REFRIGERATION** Curbs, 12" high, 4" thick, concrete	2 Carp	58	.276	L.F.	3.02	8.70		11.72	16.90	**700**
	1000	Doors, see division 08341-200										
	2400	Finishes, 2 coat portland cement plaster, 1/2" thick	1 Plas	48	.167	S.F.	.83	4.83		5.66	8.35	
	2500	For galvanized reinforcing mesh, add	1 Lath	335	.024		.56	.71		1.27	1.68	
	2700	3/16" thick latex cement	1 Plas	88	.091		1.42	2.64		4.06	5.60	
	2900	For glass cloth reinforced ceilings, add	"	450	.018		.34	.52		.86	1.16	
	3100	Fiberglass panels, 1/8" thick	1 Carp	149.45	.054		1.88	1.69		3.57	4.71	
	3200	Polystyrene, plastic finish ceiling, 1" thick		274	.029		1.74	.92		2.66	3.35	
	3400	2" thick		274	.029		1.97	.92		2.89	3.61	
	3500	4" thick	▼	219	.037		2.20	1.15		3.35	4.22	
	3800	Floors, concrete, 4" thick	1 Cefi	93	.086		.82	2.60		3.42	4.75	
	3900	6" thick	"	85	.094	▼	1.24	2.84		4.08	5.55	
	4000	Insulation, 1" to 6" thick, cork				B.F.	.80			.80	.88	
	4100	Urethane					.79			.79	.87	
	4300	Polystyrene, regular					.54			.54	.59	
	4400	Bead board				▼	.41			.41	.45	
	4600	Installation of above, add per layer	2 Carp	657.60	.024	S.F.	.25	.77		1.02	1.48	
	4700	Wall and ceiling juncture		298.90	.054	L.F.	1.28	1.69		2.97	4.05	
	4900	Partitions, galvanized sandwich panels, 4" thick, stock		219.20	.073	S.F.	5.40	2.30		7.70	9.55	
	5000	Aluminum or fiberglass	▼	219.20	.073	"	5.95	2.30		8.25	10.10	
	5200	Prefab walk-in, 7'-6" high, aluminum, incl. door & floors,										
	5210	not incl. partitions or refrigeration, 6' x 6' O.D. nominal	2 Carp	54.80	.292	SF Flr.	97	9.20		106.20	121	
	5500	10' x 10' O.D. nominal		82.20	.195		78	6.15		84.15	95.50	
	5700	12' x 14' O.D. nominal		109.60	.146		70	4.61		74.61	84	
	5800	12' x 20' O.D. nominal	▼	109.60	.146		61	4.61		65.61	74	
	6100	For 8'-6" high, add					5%					
	6300	Rule of thumb for complete units, w/o doors & refrigeration, cooler	2 Carp	146	.110		88	3.46		91.46	102	
	6400	Freezer		109.60	.146	▼	104	4.61		108.61	121	
	6600	Shelving, plated or galvanized, steel wire type		360	.044	SF Hor.	7.70	1.40		9.10	10.65	
	6700	Slat shelf type	▼	375	.043		9.50	1.35		10.85	12.55	
	6900	For stainless steel shelving, add				▼	300%					
	7000	Vapor barrier, on wood walls	2 Carp	1,644	.010	S.F.	.11	.31		.42	.60	
	7200	On masonry walls	"	1,315	.012	"	.29	.38		.67	.92	
	7500	For air curtain doors, see division 08344-120										

Important: See the Reference Section for critical supporting data - Reference Nos., Crews, & City Cost Indexes

T3

SPECIAL CONSTRUCTION

		CREW	DAILY OUTPUT	LABOR-HOURS	UNIT	2003 BARE COSTS				TOTAL INCL O&P	
13035	**Special Purpose Rooms**					MAT.	LABOR	EQUIP.	TOTAL		
800	0010	**SAUNA** Prefabricated, incl. heater & controls, 7' high, 6' x 4', C/C	L-7	2.20	12.727	Ea.	3,475	385		3,860	4,425	**800**
	0050	6' x 4', C/P		2	14		3,225	425		3,650	4,200
	0400	6' x 5', C/C		2	14		3,850	425		4,275	4,900
	0450	6' x 5', C/P		2	14		3,600	425		4,025	4,625
	0600	6' x 6', C/C		1.80	15.556		4,100	475		4,575	5,250
	0650	6' x 6', C/P		1.80	15.556		3,850	475		4,325	4,950
	0800	6' x 9', C/C		1.60	17.500		5,150	535		5,685	6,475
	0850	6' x 9', C/P		1.60	17.500		5,050	535		5,585	6,375
	1000	8' x 12', C/C		1.10	25.455		8,025	775		8,800	10,000
	1050	8' x 12', C/P		1.10	25.455		7,400	775		8,175	9,325
	1200	8' x 8', C/C		1.40	20		6,100	610		6,710	7,650
	1250	8' x 8', C/P		1.40	20		5,750	610		6,360	7,275
	1400	8' x 10', C/C		1.20	23.333		6,775	710		7,485	8,550
	1450	8' x 10', C/P		1.20	23.333		6,300	710		7,010	8,025
	1600	10' x 12', C/C		1	28		8,500	850		9,350	10,700
	1650	10' x 12', C/P	↓	1	28		7,725	850		8,575	9,800
	2500	Heaters only (incl. above), wall mounted, to 200 C.F.					485			485	535
	2750	To 300 C.F.					560			560	620
	3000	Floor standing, to 720 C.F., 10,000 watts, w/controls	1 Elec	3	2.667	↓	1,375	100		1,475	1,675
	3250	To 1,000 C.F., 16,000 watts	"	3	2.667		1,650	100		1,750	1,950
940	0010	**STEAM BATH** Heater, timer & head, single, to 140 C.F.	1 Plum	1.20	6.667	Ea.	950	249		1,199	1,425	**940**
	0500	To 300 C.F.		1.10	7.273		1,025	272		1,297	1,525
	1000	Commercial size, with blow-down assembly, to 800 C.F.		.90	8.889		3,975	330		4,305	4,875
	1500	To 2500 C.F.	↓	.80	10		6,250	375		6,625	7,450
	2000	Multiple, motels, apts., 2 baths, w/ blow-down assm., 500 C.F.	Q-1	1.30	12.308		3,475	415		3,890	4,450
	2500	4 baths	"	.70	22.857		3,550	770		4,320	5,050
	2700	Conversion unit for residential tub, including door				↓	3,225			3,225	3,550

		CREW	DAILY OUTPUT	LABOR-HOURS	UNIT	2003 BARE COSTS				TOTAL INCL O&P	
13128	**Pre-Eng. Structures**					MAT.	LABOR	EQUIP.	TOTAL		
200	0010	**COMFORT STATIONS** Prefab., stock, w/doors, windows & fixt.										**200**
	0100	Not incl. interior finish or electrical									
	0300	Mobile, on steel frame, minimum				S.F.	36.50			36.50	40
	0350	Maximum					57.50			57.50	63.50
	0400	Permanent, including concrete slab, minimum	B-12J	50	.320		142	9.90	16.95	168.85	190
	0500	Maximum	"	43	.372	↓	206	11.55	19.70	237.25	266
	0600	Alternate pricing method, mobile, minimum				Fixture	1,775			1,775	1,950
	0650	Maximum					2,650			2,650	2,925
	0700	Permanent, minimum	B-12J	.70	22.857		10,200	710	1,200	12,110	13,600
	0750	Maximum	"	.50	32	↓	17,000	990	1,700	19,690	22,100

13150 | Swimming Pools

13151 | Swimming Pools

200		CREW	DAILY OUTPUT	LABOR-HOURS	UNIT	2003 BARE COSTS MAT.	LABOR	EQUIP.	TOTAL	TOTAL INCL O&P	200
0010	**SWIMMING POOLS** Residential in-ground, vinyl lined, concrete sides										
0020	Sides including equipment, sand bottom	B-52	300	.187	SF Surf	10.60	5.40	1.30	17.30	21.50	
0100	Metal or polystyrene sides	B-14	410	.117		8.85	3.05	.53	12.43	15.10	
0200	Add for vermiculite bottom				↓	.68			.68	.75	
0500	Gunite bottom and sides, white plaster finish										
0600	12' x 30' pool	B-52	145	.386	SF Surf	17.55	11.20	2.68	31.43	40	
0720	16' x 32' pool	↓	155	.361		15.85	10.50	2.51	28.86	36.50	
0750	20' x 40' pool	↓	250	.224	↓	14.15	6.50	1.56	22.21	27.50	
0810	Concrete bottom and sides, tile finish										
0820	12' x 30' pool	B-52	80	.700	SF Surf	17.75	20.50	4.87	43.12	56.50	
0830	16' x 32' pool	↓	95	.589		14.65	17.10	4.10	35.85	47	
0840	20' x 40' pool	↓	130	.431	↓	11.65	12.50	2.99	27.14	35.50	
1100	Motel, gunite with plaster finish, incl. medium										
1150	capacity filtration & chlorination	B-52	115	.487	SF Surf	21.50	14.15	3.38	39.03	49.50	
1200	Municipal, gunite with plaster finish, incl. high										
1250	capacity filtration & chlorination	B-52	100	.560	SF Surf	28	16.25	3.89	48.14	61	
1350	Add for formed gutters				L.F.	48.50			48.50	53.50	
1360	Add for stainless steel gutters				"	144			144	158	
1700	Filtration and deck equipment only, as % of total				Total				20%	20%	
1800	Deck equipment, rule of thumb, 20' x 40' pool				SF Pool					1.30	
1900	5000 S.F. pool				"					1.90	
3000	Painting pools, preparation + 3 coats, 20' x 40' pool, epoxy	2 Pord	.33	48.485	Total	605	1,350		1,955	2,725	
3100	Rubber base paint, 18 gallons	"	.33	48.485		460	1,350		1,810	2,550	
3500	42' x 82' pool, 75 gallons, epoxy paint	3 Pord	.14	171		2,550	4,800		7,350	10,100	
3600	Rubber base paint	"	.14	171	↓	1,975	4,800		6,775	9,450	

700		CREW	DAILY OUTPUT	LABOR-HOURS	UNIT	2003 BARE COSTS MAT.	LABOR	EQUIP.	TOTAL	TOTAL INCL O&P	700
0010	**SWIMMING POOL EQUIPMENT** Diving stand, stainless steel, 3 meter	2 Carp	.40	40	Ea.	4,650	1,250		5,900	7,075	
0300	1 meter	"	2.70	5.926	"	3,775	187		3,962	4,450	
0900	Filter system, sand or diatomite type, incl. pump, 6,000 gal./hr.	2 Plum	1.80	8.889	Total	1,050	330		1,380	1,650	
1020	Add for chlorination system, 800 S.F. pool		3	5.333	Ea.	199	199		398	520	
1040	5,000 S.F. pool	↓	3	5.333	"	1,200	199		1,399	1,600	
1100	Gutter system, stainless steel, with grating, stock,										
1110	contains supply and drainage system	E-1	20	1.200	L.F.	160	42	3.94	205.94	252	
1120	Integral gutter and 5' high wall system, stainless steel	"	10	2.400	"	240	83.50	7.85	331.35	415	
2100	Lights, underwater, 12 volt, with transformer, 300 watt	1 Elec	.40	20	Ea.	116	750		866	1,250	
2200	110 volt, 500 watt, standard		.40	20		126	750		876	1,275	
2400	Low water cutoff type	↓	.40	20	↓	130	750		880	1,275	
2800	Heaters, see division 15510-880										

13175 | Ice Rinks

13176 | Ice Rinks

500		CREW	DAILY OUTPUT	LABOR-HOURS	UNIT	2003 BARE COSTS MAT.	LABOR	EQUIP.	TOTAL	TOTAL INCL O&P	500
0010	**ICE SKATING** Equipment incl. refrigeration, plumbing & cooling										
0020	coils & concrete slab, 85' x 200' rink										
0300	55° system, 5 mos., 100 ton				Total					451,000	
0700	90° system, 12 mos., 135 ton				"					500,500	
1000	Dasher boards, 1/2" H.D. polyethylene faced steel frame, 3' acrylic										
1020	screen at sides, 5' acrylic ends, 85' x 200'	F-5	.06	533	Ea.	115,000	17,100		132,100	153,000	
1100	Fiberglass & aluminum construction, same sides and ends	"	.06	533	↓	160,000	17,100		177,100	202,500	
1200	Subsoil heating system (recycled from compressor), 85' x 200'	Q-7	.27	118		25,000	4,250		29,250	33,900	

T3 SPECIAL CONSTRUCTION

Important: See the Reference Section for critical supporting data - Reference Nos., Crews, & City Cost Indexes

13175 | Ice Rinks

		13176	Ice Rinks	CREW	DAILY OUTPUT	LABOR-HOURS	UNIT	2003 BARE COSTS MAT.	LABOR	EQUIP.	TOTAL	TOTAL INCL O&P	
500	1300		Subsoil insulation, 2 lb. polystyrene with vapor barrier, 85' x 200'	2 Carp	.14	114	Ea.	30,000	3,600		33,600	38,600	500

13200 | Storage Tanks

		13201	Storage Tanks	CREW	DAILY OUTPUT	LABOR-HOURS	UNIT	2003 BARE COSTS MAT.	LABOR	EQUIP.	TOTAL	TOTAL INCL O&P	
200	0010		**ELEVATED STORAGE TANKS**, not incl pipe, pumps or foundation										200
	3000		Elevated water tanks, 100' to bottom capacity line, incl painting										
	3010		50,000 gallons				Ea.					166,000	
	3300		100,000 gallons									229,000	
	3400		250,000 gallons									316,000	
	3600		500,000 gallons									510,000	
	3700		750,000 gallons									703,000	
	3900		1,000,000 gallons									812,000	
300	0010		**GROUND TANKS** Not incl. pipe or pumps, prestress conc., 250,000 gal.				Ea.					280,000	300
	0100		500,000 gallons									380,000	
	0300		1,000,000 gallons									540,000	
	0400		2,000,000 gallons									816,000	
	0600		4,000,000 gallons									1,290,000	
	0700		6,000,000 gallons									1,760,000	
	0750		8,000,000 gallons									2,240,000	
	0800		10,000,000 gallons									2,700,000	
	0900		Steel, ground level, ht/dia less than 1, not incl. fdn, 100,000 gallons									99,500	
	1000		250,000 gallons									137,000	
	1200		500,000 gallons									209,000	
	1250		750,000 gallons									249,000	
	1300		1,000,000 gallons									336,000	
	1500		2,000,000 gallons									615,000	
	1600		4,000,000 gallons									955,000	
	1800		6,000,000 gallons									1,325,000	
	1850		8,000,000 gallons									1,735,000	
	1900		10,000,000 gallons									1,962,000	
	2100		Steel standpipes, hgt/dia more than 1, 100' to overflow, no fdn										
	2200		500,000 gallons				Ea.					275,000	
	2400		750,000 gallons									335,000	
	2500		1,000,000 gallons									408,000	
	2700		1,500,000 gallons									570,000	
	2800		2,000,000 gallons									700,000	
	3000		Steel, storage, above ground, including cradles, coating,										
	3020		fittings, not including fdn, pumps or piping										
	3040		Single wall, interior, 275 gallon	Q-5	5	3.200	Ea.	241	108		349	430	
	3060		550 gallon	"	2.70	5.926		1,150	201		1,351	1,550	
	3080		1,000 gallon	Q-7	5	6.400		1,800	229		2,029	2,325	
	3100		1,500 gallon		4.75	6.737		2,700	242		2,942	3,350	
	3120		2,000 gallon		4.60	6.957		3,025	249		3,274	3,700	
	3140		5,000 gallon		3.20	10		4,200	360		4,560	5,175	
	3150		10,000 gallon		2	16		11,300	575		11,875	13,400	
	3160		15,000 gallon		1.70	18.824		13,700	675		14,375	16,100	
	3170		20,000 gallon		1.45	22.069		17,300	790		18,090	20,200	
	3180		25,000 gallon capacity		1.30	24.615		19,200	880		20,080	22,400	

SPECIAL CONSTRUCTION **13**

13201	Storage Tanks	CREW	DAILY OUTPUT	LABOR-HOURS	UNIT	2003 BARE COSTS				TOTAL INCL O&P
						MAT.	LABOR	EQUIP.	TOTAL	
300	3190	30,000 gallon	Q-7	1.10	29.091	Ea.	23,900	1,050		24,950	27,900	300
3320	Double wall, 500 gallon capacity	Q-5	2.40	6.667		2,725	226		2,951	3,325
3330	2000 gallon capacity	Q-7	4.15	7.711		5,075	276		5,351	6,025
3340	4000 gallon capacity		3.60	8.889		9,025	320		9,345	10,400
3350	6000 gallon capacity		2.40	13.333		10,800	480		11,280	12,500
3360	8000 gallon capacity		2	16		12,300	575		12,875	14,500
3370	10000 gallon capacity		1.80	17.778		13,600	635		14,235	15,900
3380	15000 gallon capacity		1.50	21.333		20,200	765		20,965	23,400
3390	20000 gallon capacity		1.30	24.615		23,000	880		23,880	26,600
3400	25000 gallon capacity		1.15	27.826		28,000	1,000		29,000	32,300
3410	30000 gallon capacity	↓	1	32	↓	30,700	1,150		31,850	35,400
4000	Fixed roof oil storage tanks, steel, (1 BBL=42 GAL w/ fdn 3'D x 1'W)									
4200	5,000 barrels				Ea.					144,000
4300	25,000 barrels									202,000
4500	55,000 barrels									342,000
4600	100,000 barrels									550,000
4800	150,000 barrels									775,000
4900	225,000 barrels									1,168,000
5100	Floating roof gasoline tanks, steel, 5,000 barrels									122,000
5200	25,000 barrels									273,000
5400	55,000 barrels									413,000
5500	100,000 barrels									625,000
5700	150,000 barrels									846,000
5800	225,000 barrels									1,173,000
6000	Wood tanks, ground level, 2" cypress, 3,000 gallons	C-1	.19	168		5,775	5,025		10,800	14,200
6100	2-1/2" cypress, 10,000 gallons		.12	266		15,600	7,950		23,550	29,600
6300	3" redwood or 3" fir, 20,000 gallons		.10	320		23,800	9,550		33,350	41,100
6400	30,000 gallons		.08	400		29,500	11,900		41,400	51,000
6600	45,000 gallons	↓	.07	457	↓	44,600	13,600		58,200	70,500
6700	Larger sizes, minimum				Gal.					.66
6900	Maximum				"					.81
7000	Vinyl coated fabric pillow tanks, freestanding, 5,000 gallons	4 Clab	4	8	Ea.	3,525	197		3,722	4,175
7100	Supporting embankment not included, 25,000 gallons	6 Clab	2	24		6,700	590		7,290	8,300
7200	50,000 gallons	8 Clab	1.50	42.667		12,700	1,050		13,750	15,600
7300	100,000 gallons	9 Clab	.90	80		25,900	1,975		27,875	31,600
7400	150,000 gallons		.50	144		31,600	3,550		35,150	40,300
7500	200,000 gallons		.40	180		37,600	4,425		42,025	48,300
7600	250,000 gallons	↓	.30	240	↓	41,700	5,925		47,625	55,000
800	0010	**UNDERGROUND STORAGE TANKS**										800
0210	Fiberglass, underground, single wall, U.L. listed, not including									
0220	manway or hold-down strap									
0225	550 gallon capacity	Q-5	2.67	5.993	Ea.	1,450	203		1,653	1,900
0230	1,000 gallon capacity	"	2.46	6.504		1,825	220		2,045	2,325
0240	2,000 gallon capacity	Q-7	4.57	7.002		2,375	251		2,626	2,975
0250	4,000 gallon capacity		3.55	9.014		3,425	325		3,750	4,275
0260	6,000 gallon capacity		2.67	11.985		3,825	430		4,255	4,850
0270	8,000 gallon capacity		2.29	13.974		4,800	500		5,300	6,025
0280	10,000 gallon capacity		2	16		5,200	575		5,775	6,600
0282	12,000 gallon capacity		1.88	17.021		7,100	610		7,710	8,725
0284	15,000 gallon capacity		1.68	19.048		7,575	685		8,260	9,375
0290	20,000 gallon capacity		1.45	22.069		10,500	790		11,290	12,700
0300	25,000 gallon capacity		1.28	25		17,400	895		18,295	20,500
0320	30,000 gallon capacity		1.14	28.070		21,700	1,000		22,700	25,400
0340	40,000 gallon capacity		.89	35.955		30,500	1,300		31,800	35,500
0360	48,000 gallon capacity	↓	.81	39.506		50,000	1,425		51,425	57,000
0500	For manway, fittings and hold-downs, add				↓	20%	15%			

T3 SPECIAL CONSTRUCTION

Important: See the Reference Section for critical supporting data - Reference Nos., Crews, & City Cost Indexes

13201	Storage Tanks	CREW	DAILY OUTPUT	LABOR-HOURS	UNIT	2003 BARE COSTS				TOTAL INCL O&P
						MAT.	LABOR	EQUIP.	TOTAL	
0600	For manways, add				Ea.	840			840	925
1000	For helical heating coil, add	Q-5	2.50	6.400	↓	2,125	217		2,342	2,675
1020	Fiberglass, underground, double wall, U.L. listed									
1030	includes manways, not incl. hold-down straps									
1040	600 gallon capacity	Q-5	2.42	6.612	Ea.	3,200	224		3,424	3,850
1050	1,000 gallon capacity	"	2.25	7.111		4,250	241		4,491	5,050
1060	2,500 gallon capacity	Q-7	4.16	7.692		6,050	276		6,326	7,075
1070	3,000 gallon capacity		3.90	8.205		6,500	294		6,794	7,600
1080	4,000 gallon capacity		3.64	8.791		7,100	315		7,415	8,275
1090	6,000 gallon capacity		2.42	13.223		8,225	475		8,700	9,775
1100	8,000 gallon capacity		2.08	15.385		10,400	550		10,950	12,200
1110	10,000 gallon capacity		1.82	17.582		10,800	630		11,430	12,900
1120	12,000 gallon capacity		1.70	18.824		13,200	675		13,875	15,500
1122	15,000 gallon capacity		1.52	21.053		18,600	755		19,355	21,700
1124	20,000 gallon capacity		1.33	24.060		22,600	865		23,465	26,200
1126	25,000 gallon capacity		1.16	27.586		29,600	990		30,590	34,100
1128	30,000 gallon capacity	↓	1.03	31.068		35,400	1,125		36,525	40,700
1140	For hold-down straps, add				↓	2%	10%			
1150	For hold-downs 500-4000 gal, add	Q-7	16	2	Set	162	71.50		233.50	286
1160	For hold-downs 5000-15000 gal, add		8	4		325	143		468	570
1170	For hold-downs 20,000 gal, add		5.33	6.004		485	215		700	860
1180	For hold-downs 25,000 gal, add		4	8		650	287		937	1,150
1190	For hold-downs 30,000 gal, add	↓	2.60	12.308	↓	975	440		1,415	1,750
2210	Fiberglass, underground, single wall, U.L. listed, including									
2220	hold-down straps, no manways									
2225	550 gallon capacity	Q-5	2	8	Ea.	1,625	271		1,896	2,175
2230	1,000 gallon capacity	"	1.88	8.511		1,975	288		2,263	2,600
2240	2,000 gallon capacity	Q-7	3.55	9.014		2,525	325		2,850	3,275
2250	4,000 gallon capacity		2.90	11.034		3,600	395		3,995	4,550
2260	6,000 gallon capacity		2	16		4,150	575		4,725	5,425
2270	8,000 gallon capacity		1.78	17.978		5,125	645		5,770	6,600
2280	10,000 gallon capacity		1.60	20		5,525	715		6,240	7,150
2282	12,000 gallon capacity		1.52	21.053		7,425	755		8,180	9,300
2284	15,000 gallon capacity		1.39	23.022		7,900	825		8,725	9,950
2290	20,000 gallon capacity		1.14	28.070		11,000	1,000		12,000	13,600
2300	25,000 gallon capacity		.96	33.333		18,000	1,200		19,200	21,600
2320	30,000 gallon capacity	↓	.80	40	↓	22,700	1,425		24,125	27,200
3020	Fiberglass, underground, double wall, U.L. listed									
3030	includes manways and hold-down straps									
3040	600 gallon capacity	Q-5	1.86	8.602	Ea.	3,350	291		3,641	4,125
3050	1,000 gallon capacity	"	1.70	9.412		4,425	320		4,745	5,325
3060	2,500 gallon capacity	Q-7	3.29	9.726		6,225	350		6,575	7,350
3070	3,000 gallon capacity		3.13	10.224		6,675	365		7,040	7,900
3080	4,000 gallon capacity		2.93	10.921		7,275	390		7,665	8,600
3090	6,000 gallon capacity		1.86	17.204		8,550	615		9,165	10,300
3100	8,000 gallon capacity		1.65	19.394		10,700	695		11,395	12,900
3110	10,000 gallon capacity		1.48	21.622		11,100	775		11,875	13,400
3120	12,000 gallon capacity		1.40	22.857		13,500	820		14,320	16,200
3122	15,000 gallon capacity		1.28	25		18,900	895		19,795	22,200
3124	20,000 gallon capacity		1.06	30.189		23,100	1,075		24,175	27,000
3126	25,000 gallon capacity		.90	35.556		30,200	1,275		31,475	35,200
3128	30,000 gallon capacity	↓	.74	43.243	↓	36,400	1,550		37,950	42,400
5000	Steel underground, sti-P3, set in place, not incl. hold-down bars.									
5500	Excavation, pad, pumps and piping not included									
5510	Single wall, 500 gallon capacity, 7 gauge shell	Q-5	2.70	5.926	Ea.	820	201		1,021	1,200
5520	1,000 gallon capacity, 7 gauge shell	"	2.50	6.400	↓	1,175	217		1,392	1,600

SPECIAL CONSTRUCTION **13**

		CREW	DAILY OUTPUT	LABOR-HOURS	UNIT	2003 BARE COSTS				TOTAL INCL O&P	
13201	Storage Tanks					MAT.	LABOR	EQUIP.	TOTAL		
800											800
5530	2,000 gallon capacity, 1/4" thick shell	Q-7	4.60	6.957	Ea.	2,425	249		2,674	3,050	
5535	2,500 gallon capacity, 7 gauge shell	Q-5	3	5.333		2,500	181		2,681	3,025	
5540	5,000 gallon capacity, 1/4" thick shell	Q-7	3.20	10		4,675	360		5,035	5,675	
5560	10,000 gallon capacity, 1/4" thick shell		2	16		7,600	575		8,175	9,225	
5580	15,000 gallon capacity, 5/16" thick shell		1.70	18.824		10,900	675		11,575	12,900	
5600	20,000 gallon capacity, 5/16" thick shell		1.50	21.333		13,500	765		14,265	16,100	
5610	25,000 gallon capacity, 3/8" thick shell		1.30	24.615		17,600	880		18,480	20,700	
5620	30,000 gallon capacity, 3/8" thick shell		1.10	29.091		22,300	1,050		23,350	26,100	
5630	40,000 gallon capacity, 3/8" thick shell		.90	35.556		30,700	1,275		31,975	35,600	
5640	50,000 gallon capacity, 3/8" thick shell	▼	.80	40	▼	38,300	1,425		39,725	44,400	
6200	Steel, underground, 360°, double wall, U.L. listed,										
6210	with sti-P3 corrosion protection,										
6220	(dielectric coating, cathodic protection, electrical										
6230	isolation) 30 year warranty,										
6240	not incl. manholes or hold-downs.										
6250	500 gallon capacity	Q-5	2.40	6.667	Ea.	2,425	226		2,651	3,025	
6260	1,000 gallon capactiy	"	2.25	7.111		3,550	241		3,791	4,275	
6270	2,000 gallon capacity	Q-7	4.16	7.692		4,650	276		4,926	5,550	
6280	3,000 gallon capacity		3.90	8.205		5,250	294		5,544	6,225	
6290	4,000 gallon capacity		3.64	8.791		6,625	315		6,940	7,750	
6300	5,000 gallon capacity		2.91	10.997		8,575	395		8,970	10,000	
6310	6,000 gallon capacity		2.42	13.223		10,200	475		10,675	11,900	
6320	8,000 gallon capacity		2.08	15.385		10,600	550		11,150	12,500	
6330	10,000 gallon capacity		1.82	17.582		12,000	630		12,630	14,200	
6340	12,000 gallon capacity		1.70	18.824		14,400	675		15,075	16,900	
6350	15,000 gallon capacity		1.33	24.060		16,300	865		17,165	19,200	
6360	20,000 gallon capacity		1.33	24.060		19,600	865		20,465	22,900	
6370	25,000 gallon capacity		1.16	27.586		32,200	990		33,190	36,900	
6380	30,000 gallon capacity		1.03	31.068		38,800	1,125		39,925	44,400	
6390	40,000 gallon capacity		.80	40		49,900	1,425		51,325	57,000	
6395	50,000 gallon capacity		.73	43.836	▼	60,500	1,575		62,075	69,000	
6400	For hold-downs 500-2000 gal, add		16	2	Set	284	71.50		355.50	420	
6410	For hold-downs 3000-6000 gal, add		12	2.667		370	95.50		465.50	550	
6420	For hold-downs 8000-12,000 gal, add		11	2.909		905	104		1,009	1,150	
6430	For hold-downs 15,000 gal, add		9	3.556		1,125	127		1,252	1,450	
6440	For hold-downs 20,000 gal, add		8	4		1,250	143		1,393	1,600	
6450	For hold-downs 20,000 gal plus, add	▼	6	5.333	▼	2,675	191		2,866	3,225	
6500	For manways, add				Ea.	590			590	650	
6600	In place with hold-downs										
6652	550 gallon capacity	Q-5	1.84	8.696	Ea.	2,725	294		3,019	3,425	

		CREW	DAILY OUTPUT	LABOR-HOURS	UNIT	2003 BARE COSTS				TOTAL INCL O&P	
13281	Hazardous Material Remediation					MAT.	LABOR	EQUIP.	TOTAL		
120	0010	**BULK ASBESTOS REMOVAL**										120
	0020	Includes disposable tools and 2 suits and 1respirator filter/day/worker										
	0200	Boiler insulation	A-9	480	.133	S.F.	.46	4.63		5.09	7.85	
	0210	With metal lath add				%				50%		
	0300	Boiler breeching or flue insulation	A-9	520	.123	S.F.	.37	4.28		4.65	7.20	
	0310	For active boiler, add				%				100%		

Important: See the Reference Section for critical supporting data - Reference Nos., Crews, & City Cost Indexes

		13281	Hazardous Material Remediation	CREW	DAILY OUTPUT	LABOR-HOURS	UNIT	2003 BARE COSTS				TOTAL INCL O&P	
								MAT.	LABOR	EQUIP.	TOTAL		
120	0400		Duct or AHU insulation	A-10B	440	.073	S.F.	.22	2.53		2.75	4.27	120
	0500		Duct vibration isolation joints, up to 24 Sq. In. duct	A-9	56	1.143	Ea.	3.42	39.50		42.92	67	
	0520		25 Sq. In. to 48 Sq. In. duct		48	1.333		4	46.50		50.50	78	
	0530		49 Sq. In. to 76 Sq. In. duct		40	1.600	↓	4.79	55.50		60.29	94	
	0600		Pipe insulation, air cell type, up to 4" diameter pipe		900	.071	L.F.	.21	2.47		2.68	4.16	
	0610		4" to 8" diameter pipe		800	.080		.24	2.78		3.02	4.68	
	0620		10" to 12" diameter pipe		700	.091		.27	3.18		3.45	5.35	
	0630		14" to 16" diameter pipe		550	.116	↓	.35	4.04		4.39	6.85	
	0650		Over 16" diameter pipe		650	.098	S.F.	.30	3.42		3.72	5.75	
	0700		With glove bag up to 3" diameter pipe		100	.640	L.F.	3.15	22.50		25.65	39	
	1000		Pipe fitting insulation up to 4" diameter pipe		320	.200	Ea.	.60	6.95		7.55	11.70	
	1100		6" to 8" diameter pipe		304	.211		.63	7.30		7.93	12.35	
	1110		10" to 12" diameter pipe		192	.333		1	11.60		12.60	19.55	
	1120		14" to 16" diameter pipe		128	.500	↓	1.50	17.40		18.90	29	
	1130		Over 16" diameter pipe		176	.364	S.F.	1.09	12.65		13.74	21	
	1200		With glove bag, up to 8" diameter pipe		40	1.600	Ea.	6.55	55.50		62.05	95.50	
	2000		Scrape foam fireproofing from flat surface		2,400	.027	S.F.	.08	.93		1.01	1.56	
	2100		Irregular surfaces		1,200	.053		.16	1.85		2.01	3.13	
	3000		Remove cementitious material from flat surface		1,800	.036		.11	1.24		1.35	2.09	
	3100		Irregular surface		1,400	.046	↓	.14	1.59		1.73	2.68	
	6000		Remove contaminated soil from crawl space by hand	↓	400	.160	C.F.	.48	5.55		6.03	9.40	
	6100		With large production vacuum loader	A-12	700	.091	"	.27	3.18	.70	4.15	6.10	
	7000		Radiator backing, not including radiator removal	A-9	1,200	.053	S.F.	.16	1.85		2.01	3.13	
	9000		For type C (supplied air) respirator equipment, add				%					10%	
125	0010		**ASBESTOS ABATEMENT WORK AREA** Containment and preparation.										125
	0100		Pre-cleaning, HEPA vacuum and wet wipe, flat surfaces	A-10	12,000	.005	S.F.	.02	.19		.21	.31	
	0200		Protect carpeted area, 2 layers 6 mil poly on 3/4" plywood	"	1,000	.064		1.50	2.22		3.72	5.20	
	0300		Separation barrier, 2" x 4" @ 16", 1/2" plywood ea. side, 8' high	2 Carp	400	.040		1.25	1.26		2.51	3.35	
	0310		12' high		320	.050		1.40	1.58		2.98	4.01	
	0320		16' high		200	.080		1.50	2.52		4.02	5.60	
	0400		Personnel decontam. chamber, 2" x 4" @ 16", 3/4" ply ea. side		280	.057		2.50	1.80		4.30	5.55	
	0450		Waste decontam. chamber, 2" x 4" studs @ 16", 3/4" ply ea. side	↓	360	.044	↓	3	1.40		4.40	5.50	
	0500		Cover surfaces with polyethelene sheeting										
	0501		Including glue and tape										
	0550		Floors, each layer, 6 mil	A-10	8,000	.008	S.F.	.10	.28		.38	.55	
	0551		4 mil		9,000	.007		.06	.25		.31	.46	
	0560		Walls, each layer, 6 mil		6,000	.011		.09	.37		.46	.69	
	0561		4 mil	↓	7,000	.009	↓	.07	.32		.39	.59	
	0570		For heights above 12', add						20%				
	0575		For heights above 20', add						30%				
	0580		For fire retardant poly, add					100%					
	0590		For large open areas, deduct					10%	20%				
	0600		Seal floor penetrations with foam firestop to 36 Sq. In.	2 Carp	200	.080	Ea.	6.25	2.52		8.77	10.85	
	0610		36 Sq. In. to 72 Sq. In.		125	.128		12.50	4.04		16.54	20	
	0615		72 Sq. In. to 144 Sq. In.		80	.200		25	6.30		31.30	37.50	
	0620		Wall penetrations, to 36 square inches		180	.089		6.25	2.80		9.05	11.30	
	0630		36 Sq. In. to 72 Sq. In.		100	.160		12.50	5.05		17.55	21.50	
	0640		72 Sq. In. to 144 Sq. In.	↓	60	.267	↓	25	8.40		33.40	40.50	
	0800		Caulk seams with latex	1 Carp	230	.035	L.F.	.15	1.10		1.25	1.88	
	0900		Set up neg. air machine, 1-2k C.F.M. /25 M.C.F. volume	1 Asbe	4.30	1.860	Ea.		64.50		64.50	103	
130	0010		**DEMOLITION IN ASBESTOS CONTAMINATED AREA**										130
	0200		Ceiling, including suspension system, plaster and lath	A-9	2,100	.030	S.F.	.09	1.06		1.15	1.79	
	0210		Finished plaster, leaving wire lath		585	.109		.33	3.80		4.13	6.40	
	0220		Suspended acoustical tile		3,500	.018		.05	.64		.69	1.07	
	0230		Concealed tile grid system		3,000	.021		.06	.74		.80	1.25	
	0240		Metal pan grid system	↓	1,500	.043	↓	.13	1.48		1.61	2.50	

Note: Row 0310/0320 references R13281-120

SPECIAL CONSTRUCTION 13

13281	Hazardous Material Remediation	CREW	DAILY OUTPUT	LABOR-HOURS	UNIT	2003 BARE COSTS				TOTAL INCL O&P	
						MAT.	LABOR	EQUIP.	TOTAL		
130	0250	Gypsum board	A-9	2,500	.026	S.F.	.08	.89		.97	1.50	**130**
	0260	Lighting fixtures up to 2' x 4'	↓	72	.889	Ea.	2.66	31		33.66	52	
	0400	Partitions, non load bearing										
	0410	Plaster, lath, and studs	A-9	690	.093	S.F.	.83	3.22		4.05	6.05	
	0450	Gypsum board and studs	"	1,390	.046	"	.14	1.60		1.74	2.70	
	9000	For type C (supplied air) respirator equipment, add				%					10%	
135	0010	**ASBESTOS ABATEMENT EQUIPMENT** and supplies, buy [R13281-120]										**135**
	0200	Air filtration device, 2000 C.F.M.				Ea.	2,500			2,500	2,750	
	0250	Large volume air sampling pump, minimum					385			385	420	
	0260	Maximum					700			700	770	
	0300	Airless sprayer unit, 2 gun					2,000			2,000	2,200	
	0350	Light stand, 500 watt				↓	250			250	275	
	0400	Personal respirators										
	0410	Negative pressure, 1/2 face, dual operation, min.				Ea.	22.50			22.50	25	
	0420	Maximum					24			24	26.50	
	0450	P.A.P.R., full face, minimum					400			400	440	
	0460	Maximum					700			700	770	
	0470	Supplied air, full face, incl. air line, minimum					450			450	495	
	0480	Maximum					600			600	660	
	0500	Personnel sampling pump, minimum					450			450	495	
	0510	Maximum					750			750	825	
	1500	Power panel, 20 unit, incl. G.F.I.					1,800			1,800	1,975	
	1600	Shower unit, including pump and filters					1,125			1,125	1,250	
	1700	Supplied air system (type C)					10,000			10,000	11,000	
	1750	Vacuum cleaner, HEPA, 16 gal., stainless steel, wet/dry					1,000			1,000	1,100	
	1760	55 gallon					2,200			2,200	2,425	
	1800	Vacuum loader, 9-18 ton/hr					90,000			90,000	99,000	
	1900	Water atomizer unit, including 55 gal. drum					230			230	253	
	2000	Worker protection, whole body, foot, head cover & gloves, plastic					5.15			5.15	5.65	
	2500	Respirator, single use					10.50			10.50	11.55	
	2550	Cartridge for respirator					13.65			13.65	15.05	
	2570	Glove bag, 7 mil, 50" x 64"					8.50			8.50	9.35	
	2580	10 mil, 44" x 60"					8.40			8.40	9.25	
	3000	HEPA vacuum for work area, minimun					1,000			1,000	1,100	
	3050	Maximum					2,775			2,775	3,050	
	6000	Disposable polyethelene bags, 6 mil, 3 C.F.					1.15			1.15	1.27	
	6300	Disposable fiber drums, 3 C.F.					6.50			6.50	7.15	
	6400	Pressure sensitive caution lables, 3" x 5"					1.64			1.64	1.80	
	6450	11" x 17"					6.50			6.50	7.15	
	6500	Negative air machine, 1800 C.F.M.	↓			↓	775			775	855	
140	0010	**DECONTAMINATION CONTAINMENT AREA DEMOLITION** and clean-up										**140**
	0100	Spray exposed substrate with surfactant (bridging)										
	0200	Flat surfaces	A-9	6,000	.011	S.F.	.35	.37		.72	.98	
	0250	Irregular surfaces		4,000	.016	"	.30	.56		.86	1.21	
	0300	Pipes, beams, and columns		2,000	.032	L.F.	.55	1.11		1.66	2.38	
	1000	Spray encapsulate polyethelene sheeting		8,000	.008	S.F.	.30	.28		.58	.77	
	1100	Roll down polyethelene sheeting		8,000	.008	"		.28		.28	.44	
	1500	Bag polyethelene sheeting		400	.160	Ea.	.75	5.55		6.30	9.70	
	2000	Fine clean exposed substrate, with nylon brush		2,400	.027	S.F.		.93		.93	1.47	
	2500	Wet wipe substrate		4,800	.013			.46		.46	.74	
	2600	Vacuum surfaces, fine brush	↓	6,400	.010	↓		.35		.35	.55	
	3000	Structural demolition										
	3100	Wood stud walls	A-9	2,800	.023	S.F.		.79		.79	1.26	
	3500	Window manifolds, not incl. window replacement		4,200	.015			.53		.53	.84	
	3600	Plywood carpet protection	↓	2,000	.032	↓		1.11		1.11	1.77	
	4000	Remove custom decontamination facility	A-10A	8	3	Ea.	15	105		120	183	

Important: See the Reference Section for critical supporting data - Reference Nos., Crews, & City Cost Indexes

13280 | Hazardous Material Remediation

		13281	Hazardous Material Remediation	CREW	DAILY OUTPUT	LABOR-HOURS	UNIT	2003 BARE COSTS MAT.	LABOR	EQUIP.	TOTAL	TOTAL INCL O&P	
140	4100		Remove portable decontamination facility	3 Asbe	12	2	Ea.	12.50	69.50		82	124	140
	5000		HEPA vacuum, shampoo carpeting	A-9	4,800	.013	S.F.	.05	.46		.51	.80	
	9000		Final cleaning of protected surfaces	A-10A	8,000	.003	"		.10		.10	.17	
145	0010		**OSHA TESTING**										145
	0100		Certified technician, minimum				Day					300	
	0110		Maximum				"					500	
	0200		Personal sampling, PCM analysis, minimum	1 Asbe	8	1	Ea.	2.75	34.50		37.25	58	
	0210		Maximum	"	4	2	"	3	69.50		72.50	113	
	0300		Industrial hygenist, minimum				Day					400	
	0310		Maximum				"					550	
	1000		Cleaned area samples	1 Asbe	8	1	Ea.	2.63	34.50		37.13	58	
	1100		PCM air sample analysis, minimum		8	1		30	34.50		64.50	88	
	1110		Maximum		4	2		3.09	69.50		72.59	113	
	1200		TEM air sample analysis, minimum								100	125	
	1210		Maximum								400	500	
150	0010		**ENCAPSULATION WITH SEALANTS**										150
	0100		Ceilings and walls, minimum	A-9	21,000	.003	S.F.	.26	.11		.37	.46	
	0110		Maximum		10,600	.006	"	.41	.21		.62	.78	
	0300		Pipes to 12" diameter including minor repairs, minimum		800	.080	L.F.	.36	2.78		3.14	4.82	
	0310		Maximum		400	.160	"	1.02	5.55		6.57	9.95	
155	0010		**WASTE PACKAGING, HANDLING, & DISPOSAL**										155
	0100		Collect and bag bulk material, 3 C.F. bags, by hand	A-9	400	.160	Ea.	1.15	5.55		6.70	10.10	
	0200		Large production vacuum loader	A-12	880	.073		.80	2.53	.56	3.89	5.50	
	1000		Double bag and decontaminate	A-9	960	.067		2.30	2.32		4.62	6.20	
	2000		Containerize bagged material in drums, per 3 C.F. drum	"	800	.080		6.50	2.78		9.28	11.55	
	3000		Cart bags 50' to dumpster	2 Asbe	400	.040			1.39		1.39	2.21	
	5000		Disposal charges, not including haul, minimum				C.Y.					50	
	5020		Maximum				"					175	
	5100		Remove refrigerant from system	1 Plum	40	.200	Lb.		7.45		7.45	11.30	
	9000		For type C (supplied air) respirator equipment, add				%					10%	

13600 | Solar and Wind Energy Equipment

		13630	Solar Collector Components	CREW	DAILY OUTPUT	LABOR-HOURS	UNIT	2003 BARE COSTS MAT.	LABOR	EQUIP.	TOTAL	TOTAL INCL O&P	
200	0010		**SOLAR ENERGY**	D3010 715									200
	0020		System/Package prices, not including connecting										
	0030		pipe, insulation, or special heating/plumbing fixtures	D3010 720									
	0150		For solar ultraviolet pipe insulation see Div. 15086										
	0500		Hot water, standard package, low temperature	D3010 730									
	0540		1 collector, circulator, fittings, 65 gal. tank		Q-1	.50	32	Ea.	1,350	1,075		2,425	3,125	
	0580		2 collectors, circulator, fittings, 120 gal. tank			.40	40		1,925	1,350		3,275	4,150	
	0620		3 collectors, circulator, fittings, 120 gal. tank	D3010 735		.34	47.059		2,350	1,575		3,925	4,975	
	0700		Medium temperature package	D3010 745									
	0720		1 collector, circulator, fittings, 80 gal. tank		Q-1	.50	32	Ea.	1,450	1,075		2,525	3,225	
	0740		2 collectors, circulator, fittings, 120 gal. tank	D3010 750		.40	40		2,075	1,350		3,425	4,325	
	0780		3 collectors, circulator, fittings, 120 gal. tank			.30	53.333		3,775	1,800		5,575	6,850	
	0980		For each additional 120 gal. tank, add						660			660	730	
	2250		Controller, liquid temperature	D3010 760	1 Plum	5	1.600		92	60		152	191	

109

		CREW	DAILY OUTPUT	LABOR-HOURS	UNIT	2003 BARE COSTS				TOTAL INCL O&P
13630	**Solar Collector Components**					MAT.	LABOR	EQUIP.	TOTAL	
2300	Circulators, air									
2310	Blowers									
2330	100-300 S.F. system, 1/10 HP	Q-9	16	1	Ea.	128	33.50		161.50	193
2340	300-500 S.F. system, 1/5 HP		15	1.067		177	35.50		212.50	249
2350	Two speed, 100-300 S.F., 1/10 HP		14	1.143		128	38		166	200
2400	Reversible fan, 20" diameter, 2 speed		18	.889		102	29.50		131.50	158
2520	Space & DHW system, less duct work		.50	32		1,400	1,075		2,475	3,200
2550	Booster fan 6" diameter, 120 CFM		16	1		33	33.50		66.50	88
2570	6" diameter, 225 CFM		16	1		40.50	33.50		74	96
2580	8" diameter, 150 CFM		16	1		37	33.50		70.50	92.50
2590	8" diameter, 310 CFM		14	1.143		57	38		95	122
2600	8" diameter, 425 CFM		14	1.143		64	38		102	130
2650	Rheostat		32	.500		13.95	16.65		30.60	41
2660	Shutter/damper		12	1.333		45	44.50		89.50	118
2670	Shutter motor		16	1		97	33.50		130.50	159
2800	Circulators, liquid, 1/25 HP, 5.3 GPM	Q-1	14	1.143		139	38.50		177.50	211
2820	1/20 HP, 17 GPM		12	1.333		146	45		191	229
2850	1/20 HP, 17 GPM, stainless steel		12	1.333		230	45		275	320
2870	1/12 HP, 30 GPM		10	1.600		315	54		369	425
3000	Collector panels, air with aluminum absorber plate									
3010	Wall or roof mount									
3040	Flat black, plastic glazing									
3080	4' x 8'	Q-9	6	2.667	Ea.	600	89		689	795
3100	4' x 10'		5	3.200	"	740	107		847	980
3200	Flush roof mount, 10' to 16' x 22" wide		96	.167	L.F.	146	5.55		151.55	170
3210	Manifold, by L.F. width of collectors		160	.100	"	27	3.33		30.33	34.50
3300	Collector panels, liquid with copper absorber plate									
3320	Black chrome, tempered glass glazing									
3330	Alum. frame, 4' x 8', 5/32" single glazing	Q-1	9.50	1.684	Ea.	545	56.50		601.50	685
3390	Alum. frame, 4' x 10', 5/32" single glazing	"	6	2.667	"	650	89.50		739.50	845
3440	Flat black									
3450	Alum. frame, 3' x 8'	Q-1	9	1.778	Ea.	430	60		490	565
3500	Alum. frame, 4' x 8.5'		5.50	2.909		490	98		588	690
3520	Alum. frame, 4' x 10.5'		10	1.600		595	54		649	735
3540	Alum. frame, 4' x 12.5'		5	3.200		735	108		843	975
3550	Liquid with fin tube absorber plate									
3560	Alum. frame 4' x 8' tempered glass	Q-1	10	1.600	Ea.	495	54		549	625
3580	Liquid with vacuum tubes, 4' x 6'-10"		9	1.778		845	60		905	1,025
3600	Liquid, full wetted, plastic, alum. frame, 3' x 10'		5	3.200		195	108		303	380
3650	Collector panel mounting, flat roof or ground rack		7	2.286		64.50	77		141.50	187
3670	Roof clamps		70	.229	Set	1.61	7.70		9.31	13.35
3700	Roof strap, teflon	1 Plum	205	.039	L.F.	9.50	1.46		10.96	12.65
3900	Differential controller with two sensors									
3930	Thermostat, hard wired	1 Plum	8	1	Ea.	92	37.50		129.50	158
3950	Line cord and receptacle		12	.667		77.50	25		102.50	123
4050	Pool valve system		2.50	3.200		199	120		319	400
4070	With 12 VAC actuator		2	4		256	149		405	510
4080	Pool pump system, 2" pipe size		6	1.333		160	50		210	251
4100	Five station with digital read-out		3	2.667		202	99.50		301.50	370
4150	Sensors									
4200	Brass plug, 1/2" MPT	1 Plum	32	.250	Ea.	16.60	9.35		25.95	32.50
4210	Brass plug, reversed		32	.250		22.50	9.35		31.85	39
4220	Freeze prevention		32	.250		20	9.35		29.35	36
4240	Screw attached		32	.250		10.15	9.35		19.50	25.50
4250	Brass, immersion		32	.250		24.50	9.35		33.85	41
4300	Heat exchanger									

Reference boxes: D3010 775, D3010 780, R13600 -610

SPECIAL CONSTRUCTION 13

			CREW	DAILY OUTPUT	LABOR-HOURS	UNIT	2003 BARE COSTS				TOTAL INCL O&P	
							MAT.	LABOR	EQUIP.	TOTAL		
200		**13630	Solar Collector Components**										**200**
4330	Fluid to air coil, up flow, 45 MBH	D3010 715	Q-1	4	4	Ea.	279	135		414	510	
4380	70 MBH			3.50	4.571		315	154		469	575	
4400	80 MBH	D3010 720	▼	3	5.333	▼	420	179		599	730	
4580	Fluid to fluid package includes two circulating pumps											
4590	expansion tank, check valve, relief valve	D3010 730										
4600	controller, high temperature cutoff and sensors		Q-1	2.50	6.400	Ea.	695	215		910	1,100	
4650	Heat transfer fluid	D3010 735										
4700	Propylene glycol, inhibited anti-freeze		1 Plum	28	.286	Gal.	8.80	10.65		19.45	26	
4800	Solar storage tanks, knocked down	D3010 745										
4810	Air, galvanized steel clad, double wall, 4" fiberglass											
4820	insulation, 20 Mil PVC lining	D3010 750										
4870	4' high, 4' x 4', = 64 C.F./450 gallons		Q-9	2	8	Ea.	1,775	266		2,041	2,350	
4880	4' x 8' = 128 C.F./900 gallons	D3010 760		1.50	10.667		2,250	355		2,605	3,025	
4890	4' x 12' = 190 C.F./1300 gallons			1.30	12.308		3,000	410		3,410	3,925	
4900	8' x 8' = 250 C.F./1700 gallons	D3010 775	▼	1	16		3,350	535		3,885	4,500	
5010	6'-3" high, 7' x 7' = 306 C.F./2000 gallons		Q-10	1.20	20		8,150	690		8,840	10,000	
5020	7' x 10'-6" = 459 C.F./3000 gallons	D3010 780		.80	30		10,300	1,025		11,325	12,900	
5030	7' x 14' = 613 C.F./4000 gallons			.60	40		12,000	1,375		13,375	15,300	
5040	10'-6" x 10'-6" = 689 C.F./4500 gallons	R13600 -610		.50	48		12,400	1,650		14,050	16,200	
5050	10'-6" x 14' = 919 C.F./6000 gallons		▼	.40	60		14,600	2,075		16,675	19,200	
5060	14' x 14' = 1225 C.F./8000 gallons		Q-11	.40	80		17,100	2,825		19,925	23,200	
5070	14' x 17'-6" = 1531 C.F./10,000 gallons			.30	106		19,200	3,775		22,975	26,900	
5080	17'-6" x 17'-6" = 1914 C.F./12,500 gallons			.25	128		22,600	4,525		27,125	31,900	
5090	17'-6" x 21' = 2297 C.F./15,000 gallons			.20	160		25,200	5,650		30,850	36,500	
5100	21' x 21' = 2756 C.F./18,000 gallons		▼	.18	177	▼	28,900	6,275		35,175	41,500	
5120	30 Mil reinforced Chemflex lining,											
5140	4' high, 4' x 4' = 64 C.F./450 gallons		Q-9	2	8	Ea.	1,750	266		2,016	2,325	
5150	4' x 8' = 128 C.F./900 gallons			1.50	10.667		2,650	355		3,005	3,475	
5160	4' x 12' = 190 C.F./1300 gallons			1.30	12.308		3,525	410		3,935	4,500	
5170	8' x 8' = 250 C.F./1700 gallons		▼	1	16		3,925	535		4,460	5,150	
5190	6'-3" high, 7' x 7' = 306 C.F./2000 gallons		Q-10	1.20	20		8,700	690		9,390	10,700	
5200	7' x 10'-6" = 459 C.F./3000 gallons			.80	30		10,600	1,025		11,625	13,300	
5210	7' x 14' = 613 C.F./4000 gallons			.60	40		12,300	1,375		13,675	15,700	
5220	10'-6" x 10'-6" = 689 C.F./4500 gallons			.50	48		12,700	1,650		14,350	16,600	
5230	10'-6" x 14' = 919 C.F./6000 gallons		▼	.40	60		15,000	2,075		17,075	19,700	
5240	14' x 14' = 1225 C.F./8000 gallons		Q-11	.40	80		17,500	2,825		20,325	23,700	
5250	14' x 17'-6" = 1531 C.F./10,000 gallons			.30	106		19,600	3,775		23,375	27,400	
5260	17'-6" x 17'-6" = 1914 C.F./12,500 gallons			.25	128		23,100	4,525		27,625	32,400	
5270	17'-6" x 21' = 2297 C.F./15,000 gallons			.20	160		26,000	5,650		31,650	37,300	
5280	21' x 21' = 2756 C.F./18,000 gallons		▼	.18	177	▼	29,800	6,275		36,075	42,500	
5290	30 Mil reinforced Hypalon lining, add						.02%					
7000	Solar control valves and vents											
7050	Air purger, 1" pipe size		1 Plum	12	.667	Ea.	41	25		66	82.50	
7070	Air eliminator, automatic 3/4" size			32	.250		27	9.35		36.35	43.50	
7090	Air vent, automatic, 1/8" fitting			32	.250		8.80	9.35		18.15	24	
7100	Manual, 1/8" NPT			32	.250		1.94	9.35		11.29	16.25	
7120	Backflow preventer, 1/2" pipe size			16	.500		58.50	18.70		77.20	92	
7130	3/4" pipe size			16	.500		59	18.70		77.70	93	
7150	Balancing valve, 3/4" pipe size			20	.400		22.50	14.95		37.45	47	
7180	Draindown valve, 1/2" copper tube			9	.889		187	33		220	256	
7200	Flow control valve, 1/2" pipe size			22	.364		49.50	13.60		63.10	75	
7220	Expansion tank, up to 5 gal.			32	.250		58	9.35		67.35	78	
7250	Hydronic controller (aquastat)			8	1		41.50	37.50		79	102	
7400	Pressure gauge, 2" dial			32	.250		21	9.35		30.35	37	
7450	Relief valve, temp. and pressure 3/4" pipe size		▼	30	.267	▼	7.90	9.95		17.85	23.50	
7500	Solenoid valve, normally closed											

SPECIAL CONSTRUCTION **13**

13600 | Solar and Wind Energy Equipment

13630	Solar Collector Components		CREW	DAILY OUTPUT	LABOR-HOURS	UNIT	2003 BARE COSTS				TOTAL INCL O&P
							MAT.	LABOR	EQUIP.	TOTAL	
7520	Brass, 3/4" NPT, 24V	D3010 715	1 Plum	9	.889	Ea.	212	33		245	283
7530	1" NPT, 24V			9	.889		212	33		245	283
7750	Vacuum relief valve, 3/4" pipe size	D3010 720		32	.250		25	9.35		34.35	41.50
7800	Thermometers										
7820	Digital temperature monitoring, 4 locations	D3010 730	1 Plum	2.50	3.200	Ea.	135	120		255	330
7900	Upright, 1/2" NPT			8	1		25.50	37.50		63	84.50
7970	Remote probe, 2" dial	D3010 735		8	1		29	37.50		66.50	88.50
7990	Stem, 2" dial, 9" stem			16	.500		19	18.70		37.70	49
8250	Water storage tank with heat exchanger and electric element	D3010 745									
8270	66 gal. with 2" x 2 lb. density insulation		1 Plum	1.60	5	Ea.	650	187		837	995
8300	80 gal. with 2" x 2 lb. density insulation	D3010 750		1.60	5		745	187		932	1,100
8380	120 gal. with 2" x 2 lb. density insulation			1.40	5.714		840	213		1,053	1,250
8400	120 gal. with 2" x 2 lb. density insul., 40 S.F. heat coil	D3010 760		1.40	5.714		965	213		1,178	1,400
8500	Water storage module, plastic										
8600	Tubular, 12" diameter, 4' high	D3010 775	1 Carp	48	.167	Ea.	76	5.25		81.25	91.50
8610	12" diameter, 8' high			40	.200		116	6.30		122.30	138
8620	18" diameter, 5' high	D3010 780		38	.211		127	6.65		133.65	150
8630	18" diameter, 10' high			32	.250		167	7.90		174.90	196
8640	58" diameter, 5' high	R13600 -610	2 Carp	32	.500		460	15.80		475.80	535
8650	Cap, 12" diameter						12			12	13.20
8660	18" diameter						15			15	16.50

13800 | Building Automation & Control

13832	Direct Digital Controls	CREW	DAILY OUTPUT	LABOR-HOURS	UNIT	2003 BARE COSTS				TOTAL INCL O&P
						MAT.	LABOR	EQUIP.	TOTAL	
0010	CONTROL COMPONENTS / DDC SYSTEMS (Sub's quote incl. M & L)									
0100	Analog inputs									
0110	Sensors (avg. 50' run in 1/2" EMT)									
0120	Duct temperature				Ea.					346.80
0130	Space temperature									622.20
0140	Duct humidity, +/- 3%									652.80
0150	Space humidity, +/- 2%									999.60
0160	Duct static pressure									530.40
0170	C.F.M./Transducer									714
0172	Water temp. (see 15120 for well tap add)									612
0174	Water flow (see 15120 for circuit sensor add)									2,244
0176	Water pressure differential (see 15120 for tap add)									918
0177	Steam flow (see 15120 for circuit sensor add)									2,244
0178	Steam pressure (see 15120 for tap add)									958.80
0180	K.W./Transducer									1,275
0182	K.W.H. totalization (not incl. elec. meter pulse xmtr.)									586
0190	Space static pressure									1,000
1000	Analog outputs (avg. 50' run in 1/2" EMT)									
1010	P/I Transducer				Ea.					591.60
1020	Analog output, matl. in MUX									285.60
1030	Pneumatic (not incl. control device)									601.80
1040	Electric (not incl. control device)									357
2000	Status (Alarms)									
2100	Digital inputs (avg. 50' run in 1/2" EMT)									

T3 SPECIAL CONSTRUCTION

112

Important: See the Reference Section for critical supporting data - Reference Nos., Crews, & City Cost Indexes

13832	Direct Digital Controls	CREW	DAILY OUTPUT	LABOR-HOURS	UNIT	2003 BARE COSTS				TOTAL INCL O&P	
						MAT.	LABOR	EQUIP.	TOTAL		
200	2110	Freeze				Ea.					408	**200**
	2120	Fire									367.20	
	2130	Differential pressure, (air)									561	
	2140	Differential pressure, (water)									795.60	
	2150	Current sensor									408	
	2160	Duct high temperature thermostat									535.50	
	2170	Duct smoke detector				↓					663	
	2200	Digital output (avg. 50' run in 1/2" EMT)										
	2210	Start/stop				Ea.					326.40	
	2220	On/off (maintained contact)				"					561	
	3000	Controller M.U.X. panel, incl. function boards										
	3100	48 point				Ea.					4,947	
	3110	128 point				"					6,783	
	3200	D.D.C. controller (avg. 50' run in conduit)										
	3210	Mechanical room										
	3214	16 point controller (incl. 120v/1ph power supply)				Ea.					3,060	
	3229	32 point controller (incl. 120v/1ph power supply)				"					5,100	
	3230	Includes software programming and checkout										
	3260	Space										
	3266	V.A.V. terminal box (incl. space temp. sensor)				Ea.					790.50	
	3280	Host computer (avg. 50' run in conduit)										
	3281	Package complete with PC, keyboard,										
	3282	printer, color CRT, modem, basic software				Ea.					9,180	
	4000	Front end costs										
	4100	Computer (P.C.)/software program				Ea.					6,120	
	4200	Color graphics software									3,672	
	4300	Color graphics slides									459	
	4350	Additional dot matrix printer				↓					918	
	4400	Communications trunk cable				L.F.					3.57	
	4500	Engineering labor, (not incl. dftg.)				Point					77.52	
	4600	Calibration labor									77.52	
	4700	Start-up, checkout labor				↓					117.30	
	4800	Drafting labor, as req'd										
	5000	Communications bus (data transmission cable)										
	5010	#18 twisted shielded pair in 1/2" EMT conduit				C.L.F.					357	
	8000	Applications software										
	8050	Basic maintenance manager software (not incl. data base entry)				Ea.					1,836	
	8100	Time program				Point					6.43	
	8120	Duty cycle									12.80	
	8140	Optimum start/stop									38.76	
	8160	Demand limiting									19.18	
	8180	Enthalpy program				↓					38.76	
	8200	Boiler optimization				Ea.					1,147.50	
	8220	Chiller optimization				"					1,530	
	8240	Custom applications										
	8260	Cost varies with complexity										

13834	Electric/Electronic Control										
200	0010	**CONTROL SYSTEMS, ELECTRONIC**	R15050 -710									**200**
	0020	For electronic costs, add to division 13836-200				Ea.					15%	

13836	Pneumatic Controls										
200	0010	**CONTROL SYSTEMS, PNEUMATIC** (Sub's quote incl. mat. & labor)										**200**
	0011	and nominal 50 Ft. of tubing. Add control panelboard if req'd.										
	0100	Heating and Ventilating, split system	R15050 -710									
	0200	Mixed air bunrol, economizer cycle, panel readout, tubing										

SPECIAL CONSTRUCTION 13

13

	13836	Pneumatic Controls	CREW	DAILY OUTPUT	LABOR-HOURS	UNIT	2003 BARE COSTS				TOTAL INCL O&P	
							MAT.	LABOR	EQUIP.	TOTAL		
200	0220	Up to 10 tons R15050 -710	Q-19	.68	35.294	Ea.	2,825	1,250		4,075	5,000	200
	0240	For 10 to 20 tons		.63	37.915		3,025	1,325		4,350	5,325	
	0260	For over 20 tons		.58	41.096		3,275	1,450		4,725	5,775	
	0270	Enthalpy cycle, up to 10 tons		.50	48.387		3,125	1,700		4,825	6,000	
	0280	For 10 to 20 tons		.46	52.174		3,375	1,825		5,200	6,475	
	0290	For over 20 tons		.42	56.604		3,650	1,975		5,625	7,025	
	0300	Heating coil, hot water, 3 way valve,										
	0320	Freezestat, limit control on discharge, readout	Q-5	.69	23.088	Ea.	2,100	780		2,880	3,475	
	0500	Cooling coil, chilled water, room										
	0520	Thermostat, 3 way valve	Q-5	2	8	Ea.	935	271		1,206	1,425	
	0600	Cooling tower, fan cycle, damper control,										
	0620	Control system including water readout in/out at panel	Q-19	.67	35.821	Ea.	3,725	1,250		4,975	6,000	
	1000	Unit ventilator, day/night operation,										
	1100	freezestat, ASHRAE, cycle 2	Q-19	.91	26.374	Ea.	2,050	925		2,975	3,675	
	2000	Compensated hot water from boiler, valve control,										
	2100	readout and reset at panel, up to 60 GPM	Q-19	.55	43.956	Ea.	3,850	1,550		5,400	6,575	
	2120	For 120 GPM		.51	47.059		4,125	1,650		5,775	7,000	
	2140	For 240 GPM		.49	49.180		4,325	1,725		6,050	7,350	
	3000	Boiler room combustion air, damper to 5 SF, controls		1.36	17.582		1,850	615		2,465	2,975	
	3500	Fan coil, heating and cooling valves, 4 pipe control system		3	8		840	281		1,121	1,350	
	3600	Heat exchanger system controls		.86	27.907		1,800	980		2,780	3,475	
	3900	Multizone control (one per zone), includes thermostat, damper										
	3910	motor and reset of discharge temperature	Q-5	.51	31.373	Ea.	1,850	1,050		2,900	3,650	
	4000	Pneumatic thermostat, including controlling room radiator valve	"	2.43	6.593		560	223		783	950	
	4040	Program energy saving optimizer	Q-19	1.21	19.786		4,650	695		5,345	6,175	
	4060	Pump control system	"	3	8		860	281		1,141	1,375	
	4080	Reheat coil control system, not incl coil	Q-5	2.43	6.593		730	223		953	1,125	
	4500	Air supply for pneumatic control system										
	4600	Tank mounted duplex compressor, starter, alternator,										
	4620	piping, dryer, PRV station and filter										
	4630	1/2 HP	Q-19	.68	35.139	Ea.	6,925	1,225		8,150	9,450	
	4640	3/4 HP		.64	37.383		7,250	1,300		8,550	9,950	
	4650	1 HP		.61	39.539		7,925	1,400		9,325	10,800	
	4660	1-1/2 HP		.57	41.739		8,450	1,475		9,925	11,500	
	4680	3 HP		.55	43.956		11,500	1,550		13,050	15,000	
	4690	5 HP		.42	57.143		20,100	2,000		22,100	25,100	
	4800	Main air supply, includes 3/8" copper main and labor	Q-5	1.82	8.791	C.L.F.	233	298		531	705	
	4810	If poly tubing used, deduct									30%	
	7000	Static pressure control for air handling unit, includes pressure										
	7010	sensor, receiver controller, readout and damper motors	Q-19	.64	37.383	Ea.	5,500	1,300		6,800	8,025	
	7020	If return air fan requires control, add									70%	
	8600	VAV boxes, incl. thermostat, damper motor, reheat coil & tubing	Q-5	1.46	10.989		865	370		1,235	1,500	
	8610	If no reheat coil, deduct									204	
	9400	Sub assemblies for assembly systems										

	13838	Pneumatic/Electric Controls										
200	0010	CONTROL COMPONENTS										200
	0500	Aquastats										
	0508	Immersion type										
	0510	High/low limit, breaks contact w/temp rise	1 Stpi	4	2	Ea.	107	75		182	232	
	0514	Sequencing, break 2 switches w/temp rise		4	2		87.50	75		162.50	211	
	0518	Circulating, makes contact w/temp rise		4	2		79.50	75		154.50	201	
	0600	Carbon monoxide detector system										
	0606	Panel	1 Stpi	4	2	Ea.	2,350	75		2,425	2,725	
	0610	Sensor	"	7.30	1.096	"	320	41		361	415	
	0700	Controller, receiver										

Important: See the Reference Section for critical supporting data - Reference Nos., Crews, & City Cost Indexes

T3 SPECIAL CONSTRUCTION

	13838	**Pneumatic/Electric Controls**	CREW	DAILY OUTPUT	LABOR-HOURS	UNIT	2003 BARE COSTS				TOTAL INCL O&P	
							MAT.	LABOR	EQUIP.	TOTAL		
200	0730	Pneumatic, panel mount, single input	1 Plum	8	1	Ea.	268	37.50		305.50	350	200
	0740	With conversion mounting bracket		8	1		277	37.50		314.50	360	
	0750	Dual input, with control point adjustment	↓	7	1.143		370	42.50		412.50	470	
	0850	Electric, single snap switch	1 Elec	4	2		360	75		435	505	
	0860	Dual snap switches		3	2.667		450	100		550	645	
	0870	Humidity controller		8	1		201	37.50		238.50	277	
	0880	Load limiting controller		8	1		440	37.50		477.50	540	
	0890	Temperature controller	↓	8	1	↓	251	37.50		288.50	330	
	0900	Control panel readout										
	0910	Panel up to 12 indicators	1 Stpi	1.20	6.667	Ea.	134	251		385	525	
	0914	Panel up to 24 indicators		.86	9.302		167	350		517	715	
	0918	Panel up to 48 indicators	↓	.50	16	↓	395	600		995	1,350	
	1000	Enthalpy control, boiler water temperature control										
	1010	governed by outdoor temperature, with timer	1 Elec	3	2.667	Ea.	199	100		299	370	
	1300	Energy control/monitor										
	1320	BTU computer meter and controller	1 Stpi	1	8	Ea.	1,425	300		1,725	2,025	
	1600	Flow meters										
	1610	Gas	1 Stpi	4	2	Ea.	79.50	75		154.50	202	
	1620	Liquid	"	4	2	"	79.50	75		154.50	202	
	1640	Freezestat										
	1644	20' sensing element, adjustable	1 Stpi	3.50	2.286	Ea.	87.50	86		173.50	226	
	2000	Gauges, pressure or vacuum										
	2100	2" diameter dial	1 Stpi	32	.250	Ea.	14.55	9.40		23.95	30	
	2200	2-1/2" diameter dial		32	.250		16.95	9.40		26.35	33	
	2300	3-1/2" diameter dial		32	.250		21	9.40		30.40	37	
	2400	4-1/2" diameter dial	↓	32	.250	↓	31.50	9.40		40.90	48.50	
	2700	Flanged iron case, black ring										
	2800	3-1/2" diameter dial	1 Stpi	32	.250	Ea.	66.50	9.40		75.90	87.50	
	2900	4-1/2" diameter dial		32	.250		91	9.40		100.40	114	
	3000	6" diameter dial	↓	32	.250	↓	72.50	9.40		81.90	93.50	
	3010	Steel case, 0 - 300 psi										
	3012	2" diam. dial	1 Stpi	16	.500	Ea.	21	18.80		39.80	51.50	
	3014	4" diam. dial	"	16	.500	"	35	18.80		53.80	67	
	3020	Aluminum case, 0 - 300 psi										
	3022	3-1/2" diam. dial	1 Stpi	16	.500	Ea.	48	18.80		66.80	81.50	
	3024	4-1/2" diam. dial		16	.500		71	18.80		89.80	107	
	3026	6" diam. dial		16	.500		118	18.80		136.80	159	
	3028	8-1/2" diam. dial	↓	16	.500	↓	281	18.80		299.80	340	
	3030	Brass case, 0 - 300 psi										
	3032	2" diam. dial	1 Stpi	16	.500	Ea.	23.50	18.80		42.30	54.50	
	3034	4-1/2" diam. dial	"	16	.500	"	53.50	18.80		72.30	87.50	
	3040	Steel case, high pressure, 0 -10,000 psi										
	3042	4-1/2" diam. dial	1 Stpi	16	.500	Ea.	134	18.80		152.80	176	
	3044	6-1/2" diam. dial		16	.500		215	18.80		233.80	265	
	3046	8-1/2" diam. dial	↓	16	.500	↓	390	18.80		408.80	460	
	3080	Pressure gauge, differential, magnehelic										
	3084	0 - 2" W.C., with air filter kit	1 Stpi	6	1.333	Ea.	74	50		124	157	
	3300	For compound pressure-vacuum, add					18%					
	3350	Humidistat										
	3360	Pneumatic operation										
	3361	Room humidistat, direct acting	1 Stpi	12	.667	Ea.	158	25		183	211	
	3362	Room humidistat, reverse acting		12	.667		158	25		183	211	
	3363	Room humidity transmitter		17	.471		170	17.70		187.70	214	
	3364	Duct mounted controller		12	.667		208	25		233	267	
	3365	Duct mounted transmitter		12	.667		187	25		212	244	
	3366	Humidity indicator, 3-1/2"	↓	28	.286	↓	75.50	10.75		86.25	99.50	

SPECIAL CONSTRUCTION 13

	13838	Pneumatic/Electric Controls	CREW	DAILY OUTPUT	LABOR-HOURS	UNIT	2003 BARE COSTS				TOTAL INCL O&P	
							MAT.	LABOR	EQUIP.	TOTAL		
200	3390	Electric operated	1 Shee	8	1	Ea.	30	37		67	90	**200**
	3400	Relays										
	3430	Pneumatic/electric	1 Plum	16	.500	Ea.	184	18.70		202.70	230	
	3440	Pneumatic proportioning		8	1		100	37.50		137.50	167	
	3450	Pneumatic switching		12	.667		88.50	25		113.50	135	
	3460	Selector, 3 point		6	1.333		63.50	50		113.50	145	
	3470	Pneumatic time delay	▼	8	1	▼	235	37.50		272.50	315	
	3500	Sensor, air operated										
	3520	Humidity	1 Plum	16	.500	Ea.	208	18.70		226.70	257	
	3540	Pressure		16	.500		47	18.70		65.70	79.50	
	3560	Temperature	▼	12	.667	▼	99.50	25		124.50	147	
	3600	Electric operated										
	3620	Humidity	1 Elec	8	1	Ea.	26	37.50		63.50	85	
	3650	Pressure		8	1		830	37.50		867.50	970	
	3680	Temperature	▼	10	.800	▼	96	30		126	150	
	3700	Switches										
	3710	Minimum position										
	3720	Electrical	1 Stpi	4	2	Ea.	25.50	75		100.50	142	
	3730	Pneumatic	"	4	2	"	36.50	75		111.50	154	
	4000	Thermometers										
	4100	Dial type, 3-1/2" diameter, vapor type, union connection	1 Stpi	32	.250	Ea.	97.50	9.40		106.90	121	
	4120	Liquid type, union connection		32	.250		136	9.40		145.40	164	
	4130	Remote reading, 15' capillary		32	.250		107	9.40		116.40	131	
	4500	Stem type, 6-1/2" case, 2" stem, 1/2" NPT		32	.250		55.50	9.40		64.90	75	
	4520	4" stem, 1/2" NPT		32	.250		76	9.40		85.40	97.50	
	4600	9" case, 3-1/2" stem, 3/4" NPT		28	.286		92.50	10.75		103.25	117	
	4620	6" stem, 3/4" NPT		28	.286		98	10.75		108.75	124	
	4640	8" stem, 3/4" NPT		28	.286		121	10.75		131.75	149	
	4660	12" stem, 1" NPT	▼	26	.308	▼	133	11.55		144.55	165	
	4670	Bi-metal, dial type, steel case brass stem										
	4672	2" dial, 4" - 9" stem	1 Stpi	16	.500	Ea.	30	18.80		48.80	61.50	
	4673	2-1/2" dial, 4" - 9" stem		16	.500		32.50	18.80		51.30	64.50	
	4674	3-1/2" dial, 4" - 9" stem	▼	16	.500	▼	32.50	18.80		51.30	64.50	
	4680	Mercury filled, industrial, union connection type										
	4682	Angle stem, 7" scale	1 Stpi	16	.500	Ea.	76.50	18.80		95.30	113	
	4683	9" scale		16	.500		112	18.80		130.80	152	
	4684	12" scale		16	.500		124	18.80		142.80	165	
	4686	Straight stem, 7" scale		16	.500		98	18.80		116.80	137	
	4687	9" scale		16	.500		105	18.80		123.80	144	
	4688	12" scale	▼	16	.500	▼	125	18.80		143.80	167	
	4690	Mercury filled, industrial, separable socket type										
	4692	Angle stem, with socket, 7" scale	1 Stpi	16	.500	Ea.	64.50	18.80		83.30	99	
	4693	9" scale		16	.500		77	18.80		95.80	113	
	4694	12" scale		16	.500		156	18.80		174.80	200	
	4696	Straight stem, with socket, 7" scale		16	.500		64.50	18.80		83.30	99	
	4697	9" scale		16	.500		77	18.80		95.80	113	
	4698	12" scale	▼	16	.500	▼	90.50	18.80		109.30	129	
	5000	Thermostats										
	5030	Manual	1 Shee	8	1	Ea.	27	37		64	87	
	5040	1 set back, electric, timed		8	1		81	37		118	146	
	5050	2 set back, electric, timed	▼	8	1		111	37		148	179	
	5100	Locking cover					14.05			14.05	15.45	
	5200	24 hour, automatic, clock	1 Shee	8	1		109	37		146	177	
	5220	Electric, low voltage, 2 wire	1 Elec	13	.615		18	23		41	54.50	
	5230	3 wire		10	.800		22	30		52	69.50	
	5236	Heating/cooling, low voltage, with clock	▼	8	1	▼	105	37.50		142.50	172	

Important: See the Reference Section for critical supporting data - Reference Nos., Crews, & City Cost Indexes

13 SPECIAL CONSTRUCTION

13838	Pneumatic/Electric Controls	CREW	DAILY OUTPUT	LABOR-HOURS	UNIT	2003 BARE COSTS				TOTAL INCL O&P
						MAT.	LABOR	EQUIP.	TOTAL	
200 5240	Pneumatic									**200**
5250	Single temp., single pressure	1 Stpi	8	1	Ea.	115	37.50		152.50	184
5251	Dual pressure		8	1		170	37.50		207.50	244
5252	Dual temp., dual pressure		8	1		161	37.50		198.50	234
5253	Reverse acting w/averaging element		8	1		147	37.50		184.50	219
5254	Heating-cooling w/deadband		8	1		160	37.50		197.50	233
5255	Integral w/piston top valve actuator		8	1		157	37.50		194.50	230
5256	Dual temp., dual pressure		8	1		161	37.50		198.50	234
5257	Low limit, 8' averaging element		8	1		112	37.50		149.50	180
5258	Room single temp. proportional		8	1		46	37.50		83.50	108
5260	Dual temp, direct acting for VAV fan		8	1		146	37.50		183.50	218
5262	Capillary tube type, 20', -30 to +100°F	↓	8	1	↓	95.50	37.50		133	162
5300	Transmitter, pneumatic									
5320	Temperature averaging element	Q-1	8	2	Ea.	95.50	67.50		163	207
5350	Pressure differential	1 Plum	7	1.143		560	42.50		602.50	680
5370	Humidity, duct		8	1		187	37.50		224.50	263
5380	Room		12	.667		170	25		195	225
5390	Temperature, with averaging element	↓	6	1.333		107	50		157	193
5420	Electric operated, humidity	1 Elec	8	1		50.50	37.50		88	112
5430	DPST	"	8	1	↓	71	37.50		108.50	134
5500	Timer/time clocks									
5510	7 day, 12 hr. battery	1 Stpi	2.60	3.077	Ea.	226	116		342	425
5600	Recorders									
5610	Hydrograph humidity, wall mtd., aluminum case, grad. chart									
5612	8"	1 Stpi	1.60	5	Ea.	545	188		733	885
5614	10"	"	1.60	5	"	980	188		1,168	1,350
5620	Thermo-hydrograph, wall mtd., grad. chart, 5' SS probe									
5622	8"	1 Stpi	1.60	5	Ea.	475	188		663	810
5624	10"	"	1.60	5	"	790	188		978	1,150
5630	Time of operation, wall mtd., 1 pen cap, 24 hr. clock									
5632	6"	1 Stpi	1.60	5	Ea.	400	188		588	725
5634	8"	"	1.60	5	"	400	188		588	725
5640	Pressure & vacuum, bourdon tube type									
5642	Flush mount, 1 pen									
5644	6"	1 Stpi	1.60	5	Ea.	365	188		553	685
5646	8"	"	1.60	5	"	720	188		908	1,075
6000	Valves, motorized zone									
6100	Sweat connections, 1/2" C x C	1 Stpi	20	.400	Ea.	67.50	15.05		82.55	97
6110	3/4" C x C		20	.400		72.50	15.05		87.55	102
6120	1" C x C		19	.421		84.50	15.85		100.35	117
6140	1/2" C x C, with end switch, 2 wire		20	.400		71.50	15.05		86.55	101
6150	3/4" C x C, with end switch, 2 wire		20	.400		76	15.05		91.05	106
6160	1" C x C, with end switch, 2 wire		19	.421		86	15.85		101.85	119
6180	1-1/4" C x C, w/end switch, 2 wire	↓	15	.533	↓	103	20		123	145
7090	Valves, motor controlled, including actuator									
7100	Electric motor actuated									
7200	Brass, two way, screwed									
7210	1/2" pipe size	L-6	36	.333	Ea.	182	12.50		194.50	219
7220	3/4" pipe size		30	.400		276	14.95		290.95	330
7230	1" pipe size		28	.429		330	16.05		346.05	390
7240	1-1/2" pipe size		19	.632		355	23.50		378.50	425
7250	2" pipe size	↓	16	.750	↓	530	28		558	620
7350	Brass, three way, screwed									
7360	1/2" pipe size	L-6	33	.364	Ea.	212	13.60		225.60	254
7370	3/4" pipe size		27	.444		238	16.65		254.65	286
7380	1" pipe size	↓	25.50	.471	↓	210	17.60		227.60	258

SPECIAL CONSTRUCTION 13

	13838	Pneumatic/Electric Controls	CREW	DAILY OUTPUT	LABOR-HOURS	UNIT	2003 BARE COSTS				TOTAL INCL O&P	
							MAT.	LABOR	EQUIP.	TOTAL		
200	7384	1-1/4" pipe size	L-6	21	.571	Ea.	266	21.50		287.50	325	**200**
	7390	1-1/2" pipe size		17	.706		300	26.50		326.50	370	
	7400	2" pipe size	↓	14	.857	↓	465	32		497	565	
	7550	Iron body, two way, flanged										
	7560	2-1/2" pipe size	L-6	4	3	Ea.	355	112		467	560	
	7570	3" pipe size		3	4		470	150		620	740	
	7580	4" pipe size	↓	2	6	↓	655	225		880	1,075	
	7850	Iron body, three way, flanged										
	7860	2-1/2" pipe size	L-6	3	4	Ea.	770	150		920	1,075	
	7870	3" pipe size		2.50	4.800		930	180		1,110	1,300	
	7880	4" pipe size	↓	2	6	↓	1,200	225		1,425	1,675	
	8000	Pneumatic, air operated										
	8050	Brass, two way, screwed										
	8060	1/2" pipe size, class 250	1 Plum	24	.333	Ea.	100	12.45		112.45	129	
	8070	3/4" pipe size, class 250		20	.400		133	14.95		147.95	169	
	8080	1" pipe size, class 250		19	.421		151	15.75		166.75	190	
	8090	1-1/4" pipe size, class 125		15	.533		217	19.90		236.90	268	
	8100	1-1/2" pipe size, class 125		13	.615		251	23		274	310	
	8110	2" pipe size, class 125	↓	11	.727	↓	285	27		312	355	
	8180	Brass, three way, screwed										
	8190	1/2" pipe size, class 250	1 Plum	22	.364	Ea.	110	13.60		123.60	142	
	8200	3/4" pipe size, class 250		18	.444		134	16.60		150.60	172	
	8210	1" pipe size, class 250		17	.471		153	17.60		170.60	196	
	8214	1-1/4" pipe size, class 250		14	.571		210	21.50		231.50	263	
	8220	1-1/2" pipe size, class 125		11	.727		246	27		273	310	
	8230	2" pipe size, class 125	↓	9	.889	↓	310	33		343	390	
	8250	Iron body, two way, screwed										
	8260	1/2" pipe size, class 300	1 Plum	24	.333	Ea.	277	12.45		289.45	325	
	8270	3/4" pipe size, class 300		20	.400		305	14.95		319.95	360	
	8280	1" pipe size, class 300		19	.421		495	15.75		510.75	570	
	8290	1-1/4" pipe size, class 300		15	.533		525	19.90		544.90	610	
	8300	1-1/2" pipe size, class 300		13	.615		695	23		718	800	
	8310	2" pipe size, class 300		11	.727		765	27		792	885	
	8320	1-1/4" pipe size, class 125		15	.533		415	19.90		434.90	490	
	8330	1-1/2" pipe size, class 125		13	.615		375	23		398	445	
	8340	2" pipe size, class 125	↓	11	.727	↓	435	27		462	520	
	8450	Iron body, two way, flanged										
	8460	2-1/2" pipe size, 250 lb. flanges	Q-1	5	3.200	Ea.	2,775	108		2,883	3,225	
	8470	3" pipe size, 250 lb. flanges		4.50	3.556		2,950	120		3,070	3,425	
	8480	4" pipe size, 250 lb. flanges		3	5.333		3,600	179		3,779	4,225	
	8510	2-1/2" pipe size, class 125		5	3.200		595	108		703	820	
	8520	3" pipe size, class 125		4.50	3.556		710	120		830	965	
	8530	4" pipe size, class 125	↓	3	5.333		915	179		1,094	1,275	
	8540	5" pipe size, class 125	Q-2	3.40	7.059		1,900	246		2,146	2,450	
	8550	6" pipe size, class 125	"	3	8	↓	2,250	279		2,529	2,900	
	8560	Iron body, three way, flanged										
	8570	2-1/2" pipe size, class 125	Q-1	4.50	3.556	Ea.	680	120		800	925	
	8580	3" pipe size, class 125		4	4		845	135		980	1,125	
	8590	4" pipe size, class 125	↓	2.50	6.400		1,700	215		1,915	2,200	
	8600	6" pipe size, class 125	Q-2	3	8	↓	2,325	279		2,604	2,975	
	9005	Pneumatic system misc. components										
	9010	Adding/subtracting repeater	1 Stpi	20	.400	Ea.	98.50	15.05		113.55	131	
	9020	Adjustable ratio network		16	.500		92	18.80		110.80	130	
	9030	Comparator	↓	20	.400	↓	188	15.05		203.05	230	
	9040	Cumulator										
	9041	Air switching	1 Stpi	20	.400	Ea.	88.50	15.05		103.55	120	

Important: See the Reference Section for critical supporting data - Reference Nos., Crews, & City Cost Indexes

	13838	Pneumatic/Electric Controls	CREW	DAILY OUTPUT	LABOR-HOURS	UNIT	MAT.	LABOR	EQUIP.	TOTAL	TOTAL INCL O&P	
							2003 BARE COSTS					
200	9042	Averaging	1 Stpi	20	.400	Ea.	92	15.05		107.05	124	200
	9043	2:1 ratio		20	.400		83	15.05		98.05	114	
	9044	Sequencing		20	.400		56.50	15.05		71.55	84.50	
	9045	Two-position		16	.500		102	18.80		120.80	141	
	9046	Two-position pilot	▼	16	.500	▼	95.50	18.80		114.30	134	
	9050	Damper actuator										
	9051	For smoke control	1 Stpi	8	1	Ea.	128	37.50		165.50	198	
	9052	For ventilation	"	8	1	"	140	37.50		177.50	211	
	9053	Series duplex pedestal mounted										
	9054	For inlet vanes on fans, compressors	1 Stpi	6	1.333	Ea.	640	50		690	780	
	9060	Enthalpy logic center		16	.500		199	18.80		217.80	248	
	9070	High/low pressure selector		22	.364		35.50	13.65		49.15	59.50	
	9080	Signal limiter		20	.400		56.50	15.05		71.55	84.50	
	9081	Transmitter	▼	28	.286	▼	21.50	10.75		32.25	40	
	9090	Optimal start										
	9091	Warmup or cooldown programmer	1 Stpi	10	.800	Ea.	213	30		243	280	
	9092	Mass temperature transmitter	"	18	.444	"	92.50	16.70		109.20	127	
	9100	Step controller with positioner, time delay restrictor										
	9110	recycler air valve, switches and cam settings										
	9112	With cabinet										
	9113	6 points	1 Stpi	4	2	Ea.	435	75		510	595	
	9114	6 points, 6 manual sequences		2	4		460	150		610	735	
	9115	8 points	▼	3	2.667	▼	555	100		655	760	
	9200	Pressure controller and switches										
	9210	High static pressure limit	1 Stpi	8	1	Ea.	129	37.50		166.50	198	
	9220	Pressure transmitter		8	1		139	37.50		176.50	210	
	9230	Differential pressure transmitter		8	1		560	37.50		597.50	670	
	9240	Static pressure transmitter		8	1		325	37.50		362.50	410	
	9250	Proportional-only control, single input		6	1.333		268	50		318	370	
	9260	Proportional plus integral, single input		6	1.333		400	50		450	515	
	9270	Proportional-only, dual input		6	1.333		370	50		420	480	
	9280	Proportional plus integral, dual input		6	1.333		485	50		535	610	
	9281	Time delay for above proportional units		16	.500		106	18.80		124.80	145	
	9290	Proportional/2 position controller units		6	1.333		283	50		333	385	
	9300	Differential press. control dir./rev.	▼	6	1.333	▼	190	50		240	285	
	9310	Air pressure reducing valve										
	9311	1/8" size	1 Stpi	18	.444	Ea.	20.50	16.70		37.20	47.50	
	9315	Precision valve, 1/4" size		17	.471		109	17.70		126.70	147	
	9320	Air flow controller		12	.667		110	25		135	159	
	9330	Booster relay volume amplifier		18	.444		60.50	16.70		77.20	91.50	
	9340	Series restrictor, straight		60	.133		3.24	5		8.24	11.10	
	9341	T-fitting		40	.200		4.75	7.50		12.25	16.60	
	9342	In-line adjustable		48	.167		20.50	6.25		26.75	32	
	9350	Diode tee		40	.200		10.90	7.50		18.40	23.50	
	9351	Restrictor tee		40	.200		10.90	7.50		18.40	23.50	
	9360	Pneumatic gradual switch		8	1		145	37.50		182.50	216	
	9361	Selector switch		8	1		63.50	37.50		101	127	
	9370	Electro-pneumatic motor driven servo		10	.800		685	30		715	795	
	9380	Regulated power supply		10	.800		475	30		505	565	
	9390	Fan control switch and mounting base		16	.500		64	18.80		82.80	99	
	9400	Circulating pump sequencer	▼	14	.571	▼	660	21.50		681.50	760	
	9410	Pneumatic tubing, fittings and accessories										
	9414	Tubing, urethane										
	9415	1/8" OD x 1/16" ID	1 Stpi	120	.067	L.F.	.43	2.51		2.94	4.26	
	9416	1/4" OD x 1/8" ID		115	.070		.85	2.62		3.47	4.89	
	9417	5/32" OD x 3/32" ID	▼	110	.073	▼	.48	2.73		3.21	4.66	

13838	Pneumatic/Electric Controls	CREW	DAILY OUTPUT	LABOR-HOURS	UNIT	2003 BARE COSTS				TOTAL INCL O&P
						MAT.	LABOR	EQUIP.	TOTAL	
9420	Coupling, straight									
9422	Barb x barb									
9423	1/4" x 5/32"	1 Stpi	160	.050	Ea.	.48	1.88		2.36	3.37
9424	1/4" x 1/4"		158	.051		.25	1.90		2.15	3.16
9425	3/8" x 3/8"		154	.052		.33	1.95		2.28	3.31
9426	3/8" x 1/4"		150	.053		.33	2.01		2.34	3.39
9427	1/2" x 1/4"		148	.054		.71	2.03		2.74	3.85
9428	1/2" x 3/8"		144	.056		.65	2.09		2.74	3.88
9429	1/2" x 1/2"		140	.057		.48	2.15		2.63	3.78
9440	Tube x tube									
9441	1/4" x 1/4"	1 Stpi	100	.080	Ea.	1.34	3.01		4.35	6
9442	3/8" x 1/4"		96	.083		1.65	3.13		4.78	6.55
9443	3/8" x 3/8"		92	.087		1.65	3.27		4.92	6.75
9444	1/2" x 3/8"		88	.091		2.33	3.42		5.75	7.70
9445	1/2" x 1/2"		84	.095		3.03	3.58		6.61	8.75
9450	Elbow coupling									
9452	Barb x barb									
9454	1/4" x 1/4"	1 Stpi	158	.051	Ea.	.53	1.90		2.43	3.46
9455	3/8" x 3/8"		154	.052		.71	1.95		2.66	3.73
9456	1/2" x 1/2"		140	.057		.95	2.15		3.10	4.30
9460	Tube x tube									
9462	1/2" x 1/2"	1 Stpi	84	.095	Ea.	3.71	3.58		7.29	9.50
9470	Tee coupling									
9472	Barb x barb x barb									
9474	1/4" x 1/4" x 5/32"	1 Stpi	108	.074	Ea.	.86	2.79		3.65	5.15
9475	1/4" x 1/4" x 1/4"		104	.077		.58	2.89		3.47	5
9476	3/8" x 3/8" x 5/32"		102	.078		1.85	2.95		4.80	6.50
9477	3/8" x 3/8" x 1/4"		99	.081		.86	3.04		3.90	5.55
9478	3/8" x 3/8" x 3/8"		98	.082		.92	3.07		3.99	5.65
9479	1/2" x 1/2" x 1/4"		96	.083		1.78	3.13		4.91	6.70
9480	1/2" x 1/2" x 3/8"		95	.084		1	3.17		4.17	5.90
9481	1/2" x 1/2" x 1/2"		92	.087		1.04	3.27		4.31	6.10
9484	Tube x tube x tube									
9485	1/4" x 1/4" x 1/4"	1 Stpi	66	.121	Ea.	2.57	4.56		7.13	9.75
9486	3/8" x 1/4" x 1/4"		64	.125		4.87	4.70		9.57	12.45
9487	3/8" x 3/8" x 1/4"		61	.131		3.82	4.93		8.75	11.65
9488	3/8" x 3/8" x 3/8"		60	.133		4.01	5		9.01	11.95
9489	1/2" x 1/2" x 3/8"		58	.138		6.20	5.20		11.40	14.65
9490	1/2" x 1/2" x 1/2"		56	.143		5.50	5.35		10.85	14.15
9492	Needle valve									
9494	Tube x tube									
9495	1/4" x 1/4"	1 Stpi	86	.093	Ea.	4.68	3.50		8.18	10.45
9496	3/8" x 3/8"	"	82	.098	"	5.20	3.67		8.87	11.30
9600	Electronic system misc. components									
9610	Electric motor damper actuator	1 Elec	8	1	Ea.	243	37.50		280.50	325
9620	Damper position indicator		10	.800		77	30		107	130
9630	Pneumatic-electronic transducer		12	.667		283	25		308	350
9640	Electro-pneumatic transducer		12	.667		283	25		308	350
9650	Remote reset control with motor drive		6	1.333		325	50		375	430
9660	Manual remote set point control		16	.500		62	18.80		80.80	96.50
9670	Alarm unit with two adjustable points		12	.667		299	25		324	370
9700	Modulating step controller									
9701	2-5 steps	1 Elec	5.30	1.509	Ea.	1,325	57		1,382	1,550
9702	6-10 steps		3.20	2.500		1,650	94		1,744	1,975
9703	11-15 steps		2.70	2.963		2,050	111		2,161	2,425
9704	16-20 steps		1.60	5		2,425	188		2,613	2,925

Important: See the Reference Section for critical supporting data - Reference Nos., Crews, & City Cost Indexes

13800 | Building Automation & Control

13838 | Pneumatic/Electric Controls

			CREW	DAILY OUTPUT	LABOR-HOURS	UNIT	2003 BARE COSTS MAT.	LABOR	EQUIP.	TOTAL	TOTAL INCL O&P	
200	9710	Staging network										200
	9711	Master one stage	1 Elec	10	.800	Ea.	390	30		420	475	
	9712	Two stage		9	.889		515	33.50		548.50	615	
	9720	Sequencing network		8	1		590	37.50		627.50	700	
	9730	Regulated DC power supply		16	.500		360	18.80		378.80	425	
	9740	Accessory unit regulator										
	9741	Three function	1 Elec	9	.889	Ea.	194	33.50		227.50	263	
	9742	Five function		8	1		375	37.50		412.50	465	
	9750	Accessory power supply		18	.444		74.50	16.70		91.20	107	
	9760	Step down transformer		16	.500		126	18.80		144.80	166	

13850 | Detection & Alarm

13851 | Detection & Alarm

			CREW	DAILY OUTPUT	LABOR-HOURS	UNIT	2003 BARE COSTS MAT.	LABOR	EQUIP.	TOTAL	TOTAL INCL O&P	
065	0010	**DETECTION SYSTEMS**, not including wires & conduits										065
	0100	Burglar alarm, battery operated, mechanical trigger	1 Elec	4	2	Ea.	249	75		324	385	
	0200	Electrical trigger		4	2		297	75		372	435	
	0400	For outside key control, add		8	1		70.50	37.50		108	134	
	0600	For remote signaling circuitry, add		8	1		112	37.50		149.50	179	
	0800	Card reader, flush type, standard		2.70	2.963		835	111		946	1,075	
	1000	Multi-code		2.70	2.963		1,075	111		1,186	1,350	
	3594	Fire, alarm control panel										
	3600	4 zone	2 Elec	2	8	Ea.	920	300		1,220	1,450	
	3800	8 zone		1	16		1,400	600		2,000	2,450	
	4000	12 zone		.67	23.988		1,825	900		2,725	3,350	
	4020	Alarm device	1 Elec	8	1		122	37.50		159.50	190	
	4050	Actuating device		8	1		292	37.50		329.50	375	
	4200	Battery and rack		4	2		690	75		765	870	
	4400	Automatic charger		8	1		445	37.50		482.50	545	
	4600	Signal bell		8	1		49.50	37.50		87	111	
	4800	Trouble buzzer or manual station		8	1		37	37.50		74.50	96.50	
	5000	Detector, rate of rise		8	1		33.50	37.50		71	93	
	5100	Fixed temperature		8	1		28	37.50		65.50	87	
	5200	Smoke detector, ceiling type		6.20	1.290		75	48.50		123.50	155	
	5400	Duct type		3.20	2.500		250	94		344	415	
	5600	Strobe and horn		5.30	1.509		95	57		152	190	
	5800	Fire alarm horn		6.70	1.194		36.50	45		81.50	107	
	6000	Door holder, electro-magnetic		4	2		77.50	75		152.50	198	
	6200	Combination holder and closer		3.20	2.500		430	94		524	615	
	6400	Code transmitter		4	2		690	75		765	870	
	6600	Drill switch		8	1		86.50	37.50		124	151	
	6800	Master box		2.70	2.963		3,100	111		3,211	3,575	
	7000	Break glass station		8	1		50	37.50		87.50	111	
	7800	Remote annunciator, 8 zone lamp		1.80	4.444		175	167		342	445	
	8000	12 zone lamp	2 Elec	2.60	6.154		300	231		531	675	
	8200	16 zone lamp	"	2.20	7.273		300	273		573	740	
350	0010	**TANK LEAK DETECTION SYSTEMS** Liquid and vapor										350
	0100	For hydrocarbons and hazardous liquids/vapors										
	0120	Controller, data acquisition, incl. printer, modem, RS232 port										
	0140	24 channel, for use with all probes				Ea.	2,050			2,050	2,250	

13851	Detection & Alarm	CREW	DAILY OUTPUT	LABOR-HOURS	UNIT	2003 BARE COSTS				TOTAL INCL O&P	
						MAT.	LABOR	EQUIP.	TOTAL		
350	0160	9 channel, for external monitoring				Ea.	1,575			1,575	1,750	**350**
	0200	Probes									
	0210	Well monitoring									
	0220	Liquid phase detection				Ea.	710			710	785
	0230	Hydrocarbon vapor, fixed position					655			655	720
	0240	Hydrocarbon vapor, float mounted					580			580	640
	0250	Both liquid and vapor hydrocarbon				↓	880			880	970
	0300	Secondary containment, liquid phase									
	0310	Pipe trench/manway sump				Ea.	320			320	350
	0320	Double wall pipe and manual sump					310			310	340
	0330	Double wall fiberglass annular space					286			286	315
	0340	Double wall steel tank annular space				↓	286			286	315
	0500	Accessories									
	0510	Modem, non-dedicated phone line				Ea.	256			256	282
	0600	Monitoring, internal									
	0610	Automatic tank gauge, incl. overfill				Ea.	1,025			1,025	1,125
	0620	Product line				"	1,100			1,100	1,200
	0700	Monitoring, special									
	0710	Cathodic protection				Ea.	610			610	670
	0720	Annular space chemical monitor				"	830			830	915

For information about Means Estimating Seminars, see yellow pages 12 and 13 in back of book

13

SPECIAL CONSTRUCTION

Important: See the Reference Section for critical supporting data - Reference Nos., Crews, & City Cost Indexes

Division 14
Conveying Systems

Estimating Tips

General

- Many products in Division 14 will require some type of support or blocking for installation not included with the item itself. Examples are supports for conveyors or tube systems, attachment points for lifts, and footings for hoists or cranes. Add these supports in the appropriate division.

14100 Dumbwaiters
14200 Elevators

- Dumbwaiters and elevators are estimated and purchased in a method similar to buying a car. The manufacturer has a base unit with standard features. Added to this base unit price will be whatever options the owner or specifications require. Increased load capacity, additional vertical travel, additional stops, higher speed, and cab finish options are items to be considered. When developing an estimate for dumbwaiters and elevators, remember that some items needed by the installers may have to be included as part of the general contract.

Examples are:
  - shaftway
  - rail support brackets
  - machine room
  - electrical supply
  - sill angles
  - electrical connections
  - pits
  - roof penthouses
  - pit ladders

Check the job specifications and drawings before pricing.

- Installation of elevators and handicapped lifts in historic structures can require significant additional costs. The associated structural requirements may involve cutting into and repairing finishes, mouldings, flooring, etc. The estimator must account for these special conditions.

14300 Escalators & Moving Walks

- Escalators and moving walks are specialty items installed by specialty contractors. There are numerous options associated with these items. For specific options contact a manufacturer or contractor. In a method similar to estimating dumbwaiters and elevators, you should verify the extent of general contract work and add items as necessary.

14400 Lifts
14500 Material Handling
14600 Hoists & Cranes

- Products such as correspondence lifts, conveyors, chutes, pneumatic tube systems, material handling cranes and hoists as well as other items specified in this subdivision may require trained installers. The general contractor might not have any choice as to who will perform the installation or when it will be performed. Long lead times are often required for these products, making early decisions in scheduling necessary.

Reference Numbers

Reference numbers are shown in bold squares at the beginning of some major classifications. These numbers refer to related items in the Reference Section. The reference information may be an estimating procedure, an alternate pricing method or technical information.

Note: Not all subdivisions listed here necessarily appear in this publication.

14560	Chutes	CREW	DAILY OUTPUT	LABOR-HOURS	UNIT	2003 BARE COSTS				TOTAL INCL O&P	
						MAT.	LABOR	EQUIP.	TOTAL		
250	**0010** **CHUTES** Linen or refuse, incl. sprinklers, 12' floor height										**250**
0050	Aluminized steel, 16 ga., 18" diameter	2 Shee	3.50	4.571	Floor	765	169		934	1,100	
0100	24" diameter		3.20	5		835	185		1,020	1,200	
0200	30" diameter		3	5.333		915	197		1,112	1,300	
0300	36" diameter		2.80	5.714		1,050	211		1,261	1,475	
0400	Galvanized steel, 16 ga., 18" diameter		3.50	4.571		700	169		869	1,025	
0500	24" diameter		3.20	5		790	185		975	1,150	
0600	30" diameter		3	5.333		885	197		1,082	1,275	
0700	36" diameter		2.80	5.714		1,050	211		1,261	1,475	
0800	Stainless steel, 18" diameter		3.50	4.571		1,225	169		1,394	1,600	
0900	24" diameter		3.20	5		1,300	185		1,485	1,725	
1000	30" diameter		3	5.333		1,575	197		1,772	2,050	
1005	36" diameter		2.80	5.714	▼	2,025	211		2,236	2,550	
1200	Linen chute bottom collector, aluminized steel		4	4	Ea.	980	148		1,128	1,300	
1300	Stainless steel		4	4		1,250	148		1,398	1,600	
1500	Refuse, bottom hopper, aluminized steel, 18" diameter		3	5.333		640	197		837	1,000	
1600	24" diameter		3	5.333	▼	690	197		887	1,075	
1800	36" diameter		3	5.333		1,175	197		1,372	1,600	
2900	Package chutes, spiral type, minimum		4.50	3.556	Floor	1,700	132		1,832	2,075	
3000	Maximum	▼	1.50	10.667	"	4,425	395		4,820	5,475	

14580	Pneumatic Tube Systems										
800	**0010** **PNEUMATIC TUBE SYSTEM** Single tube, 2 stations, blower										**800**
0020	100' long, stock										
0100	3" diameter	2 Stpi	.12	133	Total	4,800	5,025		9,825	12,900	
0300	4" diameter	"	.09	177	"	5,400	6,675		12,075	16,100	
0400	Twin tube, two stations or more, conventional system										
0600	2-1/2" round	2 Stpi	62.50	.256	L.F.	9.65	9.65		19.30	25	
0700	3" round		46	.348		11.10	13.10		24.20	32	
0900	4" round		49.60	.323		12.30	12.15		24.45	32	
1000	4" x 7" oval		37.60	.426	▼	17.90	16		33.90	43.50	
1050	Add for blower		2	8	System	3,775	300		4,075	4,625	
1110	Plus for each round station, add		7.50	2.133	Ea.	425	80		505	590	
1150	Plus for each oval station, add		7.50	2.133	"	425	80		505	590	
1200	Alternate pricing method: base cost, minimum		.75	21.333	Total	4,650	800		5,450	6,300	
1300	Maximum		.25	64	"	9,200	2,400		11,600	13,700	
1500	Plus total system length, add, minimum		93.40	.171	L.F.	5.95	6.45		12.40	16.30	
1600	Maximum		37.60	.426	"	17.65	16		33.65	43.50	
1800	Completely automatic system, 4" round, 15 to 50 stations		.29	55.172	Station	14,600	2,075		16,675	19,200	
2200	51 to 144 stations		.32	50		11,300	1,875		13,175	15,400	
2400	6" round or 4" x 7" oval, 15 to 50 stations		.24	66.667		18,300	2,500		20,800	23,900	
2800	51 to 144 stations	▼	.23	69.565	▼	15,400	2,625		18,025	20,900	

For information about Means Estimating Seminars, see yellow pages 12 and 13 in back of book

Important: See the Reference Section for critical supporting data - Reference Nos., Crews, & City Cost Indexes

14 CONVEYING SYSTEMS

Division 15
Mechanical

Estimating Tips

15100 Building Services Piping

This subdivision is primarily basic pipe and related materials. The pipe may be used by any of the mechanical disciplines, i.e., plumbing, fire protection, heating, and air conditioning.

- The piping section lists the add to labor for elevated pipe installation. These adds apply to all elevated pipe, fittings, valves, insulation, etc., that are placed above 10' high. CAUTION: the correct percentage may vary for the same pipe. For example, the percentage add for the basic pipe installation should be based on the maximum height that the craftsman must install for that particular section. If the pipe is to be located 14' above the floor but it is suspended on threaded rod from beams, the bottom flange of which is 18' high (4' rods), then the height is actually 18' and the add is 20%. The pipe coverer, however, does not have to go above the 14' and so his add should be 10%.

- Most pipe is priced first as straight pipe with a joint (coupling, weld, etc.) every 10' and a hanger usually every 10'. There are exceptions with hanger spacing such as: for cast iron pipe (5') and plastic pipe (3 per 10'). Following each type of pipe there are several lines listing sizes and the amount to be subtracted to delete couplings and hangers. This is for pipe that is to be buried or supported together on trapeze hangers. The reason that the couplings are deleted is that these runs are usually long and frequently longer lengths of pipe are used. By deleting the couplings the estimator is expected to look up and add back the correct reduced number of couplings.

- When preparing an estimate it may be necessary to approximate the fittings. Fittings usually run between 25% and 50% of the cost of the pipe. The lower percentage is for simpler runs, and the higher number is for complex areas like mechanical rooms.

- For historic restoration projects, the systems must be as invisible as possible, and pathways must be sought for pipes, conduits, and ductwork. While installations in accessible spaces (such as basements and attics) are relatively straightforward to estimate, labor costs may be more difficult to determine when delivery systems must be concealed.

15400 Plumbing Fixtures & Equipment

- Plumbing fixture costs usually require two lines, the fixture itself and its "rough-in, supply and waste".

- In the Assemblies Section (Plumbing D2010) for the desired fixture, the System Components Group in the center of the page shows the fixture itself on the first line while the rest of the list (fittings, pipe, tubing, etc.) will total up to what we refer to in the Unit Price section as "Rough-in, supply, waste and vent". Note that for most fixtures we allow a nominal 5' of tubing to reach from the fixture to a main or riser.

- Remember that gas and oil fired units need venting.

15500 Heat Generation Equipment

- When estimating the cost of an HVAC system, check to see who is responsible for providing and installing the temperature control system. It is possible to overlook controls, assuming that they would be included in the electrical estimate.

- When looking up a boiler be careful on specified capacity. Some manufacturers rate their products on output while others use input.

- Include HVAC insulation for pipe, boiler and duct (wrap and liner).

- Be careful when looking up mechanical items to get the correct pressure rating and connection type (thread, weld, flange).

15700 Heating/Ventilation/ Air Conditioning Equipment

- Combination heating and cooling units are sized by the air conditioning requirements. (See Reference No. R15710-020 for preliminary sizing guide.)

- A ton of air conditioning is nominally 400 CFM.

- Rectangular duct is taken off by the linear foot for each size, but its cost is usually estimated by the pound. Remember that SMACNA standards now base duct on internal pressure.

- Prefabricated duct is estimated and purchased like pipe: straight sections and fittings.

- Note that cranes or other lifting equipment are not included on any lines in Division 15. For example, if a crane is required to lift a heavy piece of pipe into place high above a gym floor, or to put a rooftop unit on the roof of a four-story building, etc., it must be added. Due to the potential for extreme variation—from nothing additional required, to a major crane or helicopter—we feel that including a nominal amount for "lifting contingency" would be useless and detract from the accuracy of the estimate. When using equipment rental from Means do not forget to include the cost of the operator(s).

Reference Numbers

Reference numbers are shown in bold squares at the beginning of some major classifications. These numbers refer to related items in the Reference Section. The reference information may be an estimating procedure, an alternate pricing method or technical information.

Note: Not all subdivisions listed here necessarily appear in this publication.

15051 | Mechanical General

		CREW	DAILY OUTPUT	LABOR-HOURS	UNIT	MAT.	2003 BARE COSTS LABOR	EQUIP.	TOTAL	TOTAL INCL O&P	
100	0010 **BOILERS, GENERAL** Prices do not include flue piping, elec. wiring,										**100**
	0020 gas or oil piping, boiler base, pad, or tankless unless noted										
	0100 Boiler H.P.: 10 KW = 34 lbs/steam/hr = 33,475 BTU/hr. R15500										
	0120 -020										
	0150 To convert SFR to BTU rating: Hot water, 150 x SFR;										
	0160 Forced hot water, 180 x SFR; steam, 240 x SFR										
700	0010 **PIPING** See also divisions 02500 & 02600 for site work R15100										**700**
	1000 Add to labor for elevated installation -070										
	1080 10' to 15' high						10%				
	1100 15' to 20' high						20%				
	1120 20' to 25' high						25%				
	1140 25' to 30' high						35%				
	1160 30' to 35' high						40%				
	1180 35' to 40' high						50%				
	1200 Over 40' high						55%				

15055 | Selective Mech Demolition

		CREW	DAILY OUTPUT	LABOR-HOURS	UNIT	MAT.	2003 BARE COSTS LABOR	EQUIP.	TOTAL	TOTAL INCL O&P	
300	0010 **HVAC DEMOLITION** R15050 -720										**300**
	0100 Air conditioner, split unit, 3 ton	Q-5	2	8	Ea.		271		271	410	
	0150 Package unit, 3 ton	Q-6	3	8			281		281	425	
	0300 Boiler, electric	Q-19	2	12			420		420	635	
	0340 Gas or oil, steel, under 150 MBH	Q-6	3	8			281		281	425	
	0380 Over 150 MBH	"	2	12			420		420	635	
	1000 Ductwork, 4" high, 8" wide	1 Clab	200	.040	L.F.		.99		.99	1.54	
	1020 10" wide		190	.042			1.04		1.04	1.62	
	1040 14" wide		180	.044			1.10		1.10	1.71	
	1100 6" high, 8" wide		165	.048			1.20		1.20	1.87	
	1120 12" wide		150	.053			1.31		1.31	2.05	
	1140 18" wide		135	.059			1.46		1.46	2.28	
	1200 10" high, 12" wide		125	.064			1.58		1.58	2.46	
	1220 18" wide		115	.070			1.71		1.71	2.68	
	1240 24" wide		110	.073			1.79		1.79	2.80	
	1300 12"-14" high, 16"-18" wide		85	.094			2.32		2.32	3.62	
	1320 24" wide		75	.107			2.63		2.63	4.11	
	1340 48" wide		71	.113			2.78		2.78	4.34	
	1400 18" high, 24" wide		67	.119			2.94		2.94	4.60	
	1420 36" wide		63	.127			3.13		3.13	4.89	
	1440 48" wide		59	.136			3.34		3.34	5.20	
	1500 30" high, 36" wide		56	.143			3.52		3.52	5.50	
	1520 48" wide		53	.151			3.72		3.72	5.80	
	1540 72" wide		50	.160			3.94		3.94	6.15	
	1550 Duct heater, electric strip	1 Elec	8	1	Ea.		37.50		37.50	56	
	2200 Furnace, electric	Q-20	2	10			340		340	525	
	2300 Gas or oil, under 120 MBH	Q-9	4	4			133		133	206	
	2340 Over 120 MBH	"	3	5.333			178		178	274	
	2800 Heat pump, package unit, 3 ton	Q-5	2.40	6.667			226		226	340	
	2840 Split unit, 3 ton		2	8			271		271	410	
	3000 Mechanical equipment, light items. Unit is weight, not cooling.		.90	17.778	Ton		600		600	910	
	3600 Heavy items		1.10	14.545	"		490		490	745	
	3700 Deduct for salvage (when applicable), minimum				Job					55	
	3750 Maximum				"					420	
600	0010 **PLUMBING DEMOLITION** R15050 -720										**600**
	1020 Fixtures, including 10' piping										
	1100 Bath tubs, cast iron	1 Plum	4	2	Ea.		74.50		74.50	113	
	1120 Fiberglass		6	1.333			50		50	75	

Important: See the Reference Section for critical supporting data - Reference Nos., Crews, & City Cost Indexes

		15055	Selective Mech Demolition		DAILY	LABOR-		2003 BARE COSTS				TOTAL	
				CREW	OUTPUT	HOURS	UNIT	MAT.	LABOR	EQUIP.	TOTAL	INCL O&P	
600	1140		Steel	1 Plum	5	1.600	Ea.		60		60	90	600
	1200		Lavatory, wall hung		10	.800			30		30	45	
	1220		Counter top		8	1			37.50		37.50	56.50	
	1300		Sink, steel or cast iron, single		8	1			37.50		37.50	56.50	
	1320		Double		7	1.143			42.50		42.50	64.50	
	1400		Water closet, floor mounted		8	1			37.50		37.50	56.50	
	1420		Wall mounted		7	1.143			42.50		42.50	64.50	
	1500		Urinal, floor mounted		4	2			74.50		74.50	113	
	1520		Wall mounted		7	1.143			42.50		42.50	64.50	
	1600		Water fountains, free standing		8	1			37.50		37.50	56.50	
	1620		Recessed		6	1.333			50		50	75	
	2000		Piping, metal, to 1-1/2" diameter		200	.040	L.F.		1.49		1.49	2.26	
	2050		2" to 3-1/2" diameter		150	.053			1.99		1.99	3.01	
	2100		4" to 6" diameter	2 Plum	100	.160			6		6	9	
	2150		8" to 14" diameter		60	.267			9.95		9.95	15.05	
	2153		16" diameter		55	.291			10.85		10.85	16.40	
	2160		Deduct for salvage, aluminum scrap				Ton					525	
	2170		Brass scrap									420	
	2180		Copper scrap									1,025	
	2200		Lead scrap									235	
	2220		Steel scrap									65	
	2250		Water heater, 40 gal.	1 Plum	6	1.333	Ea.		50		50	75	
	6000		Remove and reset fixtures, minimum		6	1.333			50		50	75	
	6100		Maximum		4	2			74.50		74.50	113	

R15050-720

		15060	Hangers & Supports										
300	0010	**PIPE HANGERS AND SUPPORTS**, TYPE numbers per MSS-SP58											300
	0050		Brackets										
	0060		Beam side or wall, malleable iron, TYPE 34										
	0070		3/8" threaded rod size	1 Plum	48	.167	Ea.	2.74	6.25		8.99	12.40	
	0080		1/2" threaded rod size		48	.167		5.30	6.25		11.55	15.20	
	0090		5/8" threaded rod size		48	.167		7.05	6.25		13.30	17.15	
	0100		3/4" threaded rod size		48	.167		8.45	6.25		14.70	18.70	
	0110		7/8" threaded rod size		48	.167		10.15	6.25		16.40	20.50	
	0120		For concrete installation, add						30%				
	0150		Wall, welded steel, medium, TYPE 32										
	0160		0 size, 12" wide, 18" deep	1 Plum	34	.235	Ea.	206	8.80		214.80	239	
	0170		1 size, 18" wide 24" deep		34	.235		244	8.80		252.80	282	
	0180		2 size, 24" wide, 30" deep		34	.235		325	8.80		333.80	370	
	0200		Beam attachment, welded, TYPE 22										
	0202		3/8"	Q-15	80	.200	Ea.	12.40	6.75	.98	20.13	25	
	0203		1/2"		76	.211		13.20	7.10	1.03	21.33	26.50	
	0204		5/8"		72	.222		14	7.45	1.09	22.54	28	
	0205		3/4"		68	.235		19.45	7.90	1.16	28.51	34.50	
	0206		7/8"		64	.250		24.50	8.40	1.23	34.13	41	
	0207		1"		56	.286		37	9.60	1.40	48	56.50	
	0300		Clamps										
	0310		C-clamp, for mounting on steel beam flange, w/locknut, TYPE 23										
	0320		3/8" threaded rod size	1 Plum	160	.050	Ea.	1.35	1.87		3.22	4.31	
	0330		1/2" threaded rod size		160	.050		1.57	1.87		3.44	4.55	
	0340		5/8" threaded rod size		160	.050		2.28	1.87		4.15	5.35	
	0350		3/4" threaded rod size		160	.050		3.29	1.87		5.16	6.45	
	0352		7/8" threaded rod size		140	.057		14	2.13		16.13	18.60	
	0400		High temperature to 1050°F, alloy steel										
	0410		4" pipe size	Q-1	106	.151	Ea.	33	5.10		38.10	44	
	0420		6" pipe size		106	.151		55.50	5.10		60.60	69	

MECHANICAL 15

15060	Hangers & Supports	CREW	DAILY OUTPUT	LABOR-HOURS	UNIT	2003 BARE COSTS				TOTAL INCL O&P
						MAT.	LABOR	EQUIP.	TOTAL	
300 0430	8″ pipe size	Q-1	97	.165	Ea.	63	5.55		68.55	77.50	**300**
0440	10″ pipe size		84	.190		85	6.40		91.40	103
0450	12″ pipe size		72	.222		99	7.45		106.45	120
0460	14″ pipe size		64	.250		219	8.40		227.40	254
0470	16″ pipe size		56	.286		253	9.60		262.60	293
0480	Beam clamp, flange type, TYPE 25									
0482	For 3/8″ bolt	1 Plum	48	.167	Ea.	4.09	6.25		10.34	13.90
0483	For 1/2″ bolt		44	.182		4.30	6.80		11.10	15
0484	For 5/8″ bolt		40	.200		12.10	7.45		19.55	24.50
0485	For 3/4″ bolt		36	.222		14.65	8.30		22.95	28.50
0486	For 1″ bolt		32	.250		71	9.35		80.35	92
0500	I-beam, for mounting on bottom flange, strap iron, TYPE 21									
0510	2″ flange size	1 Plum	96	.083	Ea.	4.09	3.11		7.20	9.20
0520	3″ flange size		95	.084		4.22	3.15		7.37	9.40
0530	4″ flange size		93	.086		4.32	3.21		7.53	9.60
0540	5″ flange size		92	.087		4.48	3.25		7.73	9.85
0550	6″ flange size		90	.089		4.59	3.32		7.91	10.05
0560	7″ flange size		88	.091		4.74	3.40		8.14	10.35
0570	8″ flange size		86	.093		4.90	3.47		8.37	10.65
0600	One hole, vertical mounting, malleable iron									
0610	1/2″ pipe size	1 Plum	160	.050	Ea.	.71	1.87		2.58	3.60
0620	3/4″ pipe size		145	.055		.88	2.06		2.94	4.08
0630	1″ pipe size		136	.059		1.33	2.20		3.53	4.78
0640	1-1/4″ pipe size		128	.063		1.66	2.33		3.99	5.35
0650	1-1/2″ pipe size		120	.067		1.87	2.49		4.36	5.80
0660	2″ pipe size		112	.071		2.82	2.67		5.49	7.15
0670	2-1/2″ pipe size		104	.077		7.70	2.87		10.57	12.80
0680	3″ pipe size		96	.083		10.65	3.11		13.76	16.45
0690	3-1/2″ pipe size		90	.089		10.65	3.32		13.97	16.75
0700	4″ pipe size		84	.095		41.50	3.56		45.06	51
0750	Riser or extension pipe, carbon steel, TYPE 8									
0756	1/2″ pipe size	1 Plum	52	.154	Ea.	3.77	5.75		9.52	12.85
0760	3/4″ pipe size		48	.167		2.61	6.25		8.86	12.25
0770	1″ pipe size		47	.170		2.71	6.35		9.06	12.60
0780	1-1/4″ pipe size		46	.174		3.31	6.50		9.81	13.45
0790	1-1/2″ pipe size		45	.178		3.47	6.65		10.12	13.85
0800	2″ pipe size		43	.186		3.65	6.95		10.60	14.50
0810	2-1/2″ pipe size		41	.195		3.85	7.30		11.15	15.25
0820	3″ pipe size		40	.200		4.20	7.45		11.65	15.90
0830	3-1/2″ pipe size		39	.205		5.15	7.65		12.80	17.20
0840	4″ pipe size		38	.211		5.30	7.85		13.15	17.70
0850	5″ pipe size		37	.216		7.50	8.10		15.60	20.50
0860	6″ pipe size		36	.222		8.80	8.30		17.10	22
0870	8″ pipe size		34	.235		14.25	8.80		23.05	29
0880	10″ pipe size		32	.250		22.50	9.35		31.85	39
0890	12″ pipe size		28	.286		27	10.65		37.65	45.50
0900	For plastic coating 3/4″ to 4″, add					190%				
0910	For copper plating 3/4″ to 4″, add					58%				
0950	Two piece, complete, carbon steel, medium weight, TYPE 4									
0960	1/2″ pipe size	Q-1	137	.117	Ea.	2.81	3.93		6.74	9.05
0970	3/4″ pipe size		134	.119		2.81	4.02		6.83	9.15
0980	1″ pipe size		132	.121		2.96	4.08		7.04	9.40
0990	1-1/4″ pipe size		130	.123		3.69	4.14		7.83	10.30
1000	1-1/2″ pipe size		126	.127		3.69	4.27		7.96	10.50
1010	2″ pipe size		124	.129		4.56	4.34		8.90	11.55
1020	2-1/2″ pipe size		120	.133		4.91	4.48		9.39	12.15

Important: See the Reference Section for critical supporting data - Reference Nos., Crews, & City Cost Indexes

15060 | Hangers & Supports

		CREW	DAILY OUTPUT	LABOR-HOURS	UNIT	2003 BARE COSTS				TOTAL INCL O&P
						MAT.	LABOR	EQUIP.	TOTAL	
300	1030	3" pipe size	Q-1	117	.137	Ea.	5.45	4.60		10.05	12.95	300
1040	3-1/2" pipe size		114	.140		7.75	4.72		12.47	15.70
1050	4" pipe size		110	.145		7.75	4.89		12.64	15.95
1060	5" pipe size		106	.151		11.15	5.10		16.25	19.95
1070	6" pipe size		104	.154		18.90	5.15		24.05	29
1080	8" pipe size		100	.160		22.50	5.40		27.90	33
1090	10" pipe size		96	.167		41	5.60		46.60	53.50
1100	12" pipe size		89	.180		47.50	6.05		53.55	61
1110	14" pipe size		82	.195		82.50	6.55		89.05	100
1120	16" pipe size		68	.235		89	7.90		96.90	110
1130	For galvanized, add					45%				
1150	Insert, concrete									
1160	Wedge type, carbon steel body, malleable iron nut									
1170	1/4" threaded rod size	1 Plum	96	.083	Ea.	4.75	3.11		7.86	9.95
1180	3/8" threaded rod size		96	.083		4.99	3.11		8.10	10.20
1190	1/2" threaded rod size		96	.083		5.15	3.11		8.26	10.40
1200	5/8" threaded rod size		96	.083		5.25	3.11		8.36	10.45
1210	3/4" threaded rod size		96	.083		5.75	3.11		8.86	11.05
1220	7/8" threaded rod size		96	.083		6.70	3.11		9.81	12.05
1230	For galvanized, add					1.07			1.07	1.18
1250	Pipe guide sized for insulation									
1260	No. 1, 1" pipe size, 1" thick insulation	1 Stpi	26	.308	Ea.	138	11.55		149.55	170
1270	No. 2, 1-1/4"-2" pipe size, 1" thick insulation		23	.348		163	13.10		176.10	199
1280	No. 3, 1-1/4"-2" pipe size, 1-1/2" thick insulation		21	.381		163	14.30		177.30	201
1290	No. 4, 2-1/2"-3-1/2" pipe size, 1-1/2" thick insulation		18	.444		163	16.70		179.70	204
1300	No. 5, 4"-5" pipe size, 1-1/2" thick insulation		16	.500		181	18.80		199.80	228
1310	No. 6, 5"-6" pipe size, 2" thick insulation	Q-5	21	.762		201	26		227	260
1320	No. 7, 8" pipe size, 2" thick insulation		16	1		350	34		384	435
1330	No. 8, 10" pipe size, 2" thick insulation		12	1.333		665	45		710	800
1340	No. 9, 12" pipe size, 2" thick insulation	Q-6	17	1.412		665	49.50		714.50	805
1350	No. 10, 12"-14" pipe size, 2-1/2" thick insulation		16	1.500		865	52.50		917.50	1,025
1360	No. 11, 16" pipe size, 2-1/2" thick insulation		10.50	2.286		865	80		945	1,075
1370	No. 12, 16"-18" pipe size, 3" thick insulation		9	2.667		915	93.50		1,008.50	1,150
1380	No. 13, 20" pipe size, 3" thick insulation		7.50	3.200		915	112		1,027	1,175
1390	No. 14, 24" pipe size, 3" thick insulation		7	3.429		1,175	120		1,295	1,450
1400	Bands									
1410	Adjustable band, carbon steel, for non-insulated pipe, TYPE 7									
1420	1/2" pipe size	Q-1	142	.113	Ea.	.81	3.79		4.60	6.60
1430	3/4" pipe size		140	.114		.81	3.84		4.65	6.70
1440	1" pipe size		137	.117		.94	3.93		4.87	7
1450	1-1/4" pipe size		134	.119		.96	4.02		4.98	7.10
1460	1-1/2" pipe size		131	.122		1.09	4.11		5.20	7.40
1470	2" pipe size		129	.124		1.50	4.17		5.67	7.95
1480	2-1/2" pipe size		125	.128		1.57	4.30		5.87	8.25
1490	3" pipe size		122	.131		1.72	4.41		6.13	8.55
1500	3-1/2" pipe size		119	.134		1.96	4.52		6.48	9
1510	4" pipe size		114	.140		2.68	4.72		7.40	10.10
1520	5" pipe size		110	.145		2.84	4.89		7.73	10.50
1530	6" pipe size		108	.148		3.75	4.98		8.73	11.65
1540	8" pipe size		104	.154		6.15	5.15		11.30	14.55
1550	For copper plated, add					50%				
1560	For galvanized, add					30%				
1570	For plastic coating, add					30%				
1600	Adjusting nut malleable iron, steel band, TYPE 9									
1610	1/2" pipe size, galvanized band	Q-1	137	.117	Ea.	3.31	3.93		7.24	9.60
1620	3/4" pipe size, galvanized band		135	.119		3.31	3.99		7.30	9.65

MECHANICAL 15

	15060	Hangers & Supports	CREW	DAILY OUTPUT	LABOR-HOURS	UNIT	2003 BARE COSTS				TOTAL INCL O&P	
							MAT.	LABOR	EQUIP.	TOTAL		
300	1630	1" pipe size, gavanized band	Q-1	132	.121	Ea.	3.31	4.08		7.39	9.80	300
	1640	1-1/4" pipe size, galvanized band		129	.124		3.99	4.17		8.16	10.70	
	1650	1-1/2" pipe size, galvanized band		126	.127		3.99	4.27		8.26	10.85	
	1660	2" pipe size, galvanized band		124	.129		4.33	4.34		8.67	11.30	
	1670	2-1/2" pipe size, galvanized band		120	.133		7.25	4.48		11.73	14.75	
	1680	3" pipe size, galvanized band		117	.137		7.90	4.60		12.50	15.65	
	1690	3-1/2" pipe size, galvanized band		114	.140		10.40	4.72		15.12	18.60	
	1700	4" pipe size, cadmium plated band		110	.145		12.70	4.89		17.59	21.50	
	1740	For plastic coated band, add					35%					
	1750	For completely copper coated, add					45%					
	1800	Clevis, adjustable, carbon steel, for non-insulated pipe, TYPE 1										
	1810	1/2" pipe size	Q-1	137	.117	Ea.	.86	3.93		4.79	6.90	
	1820	3/4" pipe size		135	.119		.86	3.99		4.85	6.95	
	1830	1" pipe size		132	.121		.91	4.08		4.99	7.15	
	1840	1-1/4" pipe size		129	.124		1.04	4.17		5.21	7.45	
	1850	1-1/2" pipe size		126	.127		1.14	4.27		5.41	7.70	
	1860	2" pipe size		124	.129		1.27	4.34		5.61	7.95	
	1870	2-1/2" pipe size		120	.133		2.05	4.48		6.53	9	
	1880	3" pipe size		117	.137		2.59	4.60		7.19	9.80	
	1890	3-1/2" pipe size		114	.140		2.71	4.72		7.43	10.15	
	1900	4" pipe size		110	.145		3.12	4.89		8.01	10.85	
	1910	5" pipe size		106	.151		4.31	5.10		9.41	12.40	
	1920	6" pipe size		104	.154		4.97	5.15		10.12	13.25	
	1930	8" pipe size		100	.160		8.20	5.40		13.60	17.10	
	1940	10" pipe size		96	.167		22	5.60		27.60	32.50	
	1950	12" pipe size		89	.180		27.50	6.05		33.55	39.50	
	1960	14" pipe size		82	.195		40	6.55		46.55	54	
	1970	16" pipe size		68	.235		56.50	7.90		64.40	74.50	
	1971	18" pipe size		54	.296		67.50	9.95		77.45	89.50	
	1972	20" pipe size		38	.421		144	14.15		158.15	181	
	1980	For galvanized, add					66%					
	1990	For copper plated 1/2" to 4", add					77%					
	2000	For light weight 1/2" to 4", deduct					13%					
	2010	Insulated pipe type, 3/4" to 12" pipe, add					180%					
	2020	Insulated pipe type, chrome-moly U-strap, add					530%					
	2250	Split ring, malleable iron, for non-insulated pipe, TYPE 11										
	2260	1/2" pipe size	Q-1	137	.117	Ea.	3.15	3.93		7.08	9.40	
	2270	3/4" pipe size		135	.119		3.17	3.99		7.16	9.50	
	2280	1" pipe size		132	.121		3.35	4.08		7.43	9.85	
	2290	1-1/4" pipe size		129	.124		3.88	4.17		8.05	10.55	
	2300	1-1/2" pipe size		126	.127		5.05	4.27		9.32	12	
	2310	2" pipe size		124	.129		5.75	4.34		10.09	12.85	
	2320	2-1/2" pipe size		120	.133		8.25	4.48		12.73	15.85	
	2330	3" pipe size		117	.137		10.30	4.60		14.90	18.30	
	2340	3-1/2" pipe size		114	.140		14.30	4.72		19.02	23	
	2350	4" pipe size		110	.145		14.30	4.89		19.19	23	
	2360	5" pipe size		106	.151		32	5.10		37.10	42.50	
	2370	6" pipe size		104	.154		32	5.15		37.15	43	
	2380	8" pipe size		100	.160		56.50	5.40		61.90	70	
	2390	For copper plated, add					8%					
	2500	Washer, flat steel										
	2502	3/8"	1 Plum	240	.033	Ea.	.05	1.24		1.29	1.94	
	2503	1/2"		220	.036		.07	1.36		1.43	2.13	
	2504	5/8"		200	.040		.15	1.49		1.64	2.43	
	2505	3/4"		180	.044		.11	1.66		1.77	2.63	
	2506	7/8"		160	.050		.16	1.87		2.03	3	

T5 MECHANICAL

Important: See the Reference Section for critical supporting data - Reference Nos., Crews, & City Cost Indexes

15060	Hangers & Supports	CREW	DAILY OUTPUT	LABOR-HOURS	UNIT	2003 BARE COSTS				TOTAL INCL O&P
						MAT.	LABOR	EQUIP.	TOTAL	
2507	1"	1 Plum	140	.057	Ea.	.16	2.13		2.29	3.40
2508	1-1/8"		120	.067		.48	2.49		2.97	4.29
2520	Nut, steel, hex									
2522	3/8"	1 Plum	200	.040	Ea.	.08	1.49		1.57	2.35
2523	1/2"		180	.044		.14	1.66		1.80	2.66
2524	5/8"		160	.050		.28	1.87		2.15	3.13
2525	3/4"		140	.057		.37	2.13		2.50	3.63
2526	7/8"		120	.067		.63	2.49		3.12	4.45
2527	1"		100	.080		2	2.99		4.99	6.70
2528	1-1/8"		80	.100		10.30	3.74		14.04	17
2532	Turnbuckle, TYPE 13									
2534	3/8"	1 Plum	80	.100	Ea.	13.35	3.74		17.09	20.50
2535	1/2"		72	.111		16.60	4.15		20.75	24.50
2536	5/8"		64	.125		19.30	4.67		23.97	28
2538	7/8"		48	.167		35.50	6.25		41.75	48.50
2539	1"		40	.200		43	7.45		50.45	59
2540	1-1/8"		32	.250		70	9.35		79.35	91
2544	Eye rod, welded									
2546	3/8"	1 Plum	56	.143	Ea.	17.95	5.35		23.30	28
2547	1/2"		48	.167		26.50	6.25		32.75	39
2548	5/8"		40	.200		40	7.45		47.45	56
2549	3/4"		32	.250		51	9.35		60.35	70
2550	7/8"		24	.333		62.50	12.45		74.95	88
2551	1"		16	.500		79	18.70		97.70	115
2556	Eye rod, welded, linked									
2558	3/8"	1 Plum	56	.143	Ea.	55.50	5.35		60.85	69
2559	1/2"		48	.167		75.50	6.25		81.75	92.50
2560	5/8"		40	.200		108	7.45		115.45	130
2561	3/4"		32	.250		140	9.35		149.35	168
2562	7/8"		24	.333		171	12.45		183.45	207
2563	1"		16	.500		220	18.70		238.70	269
2650	Rods, carbon steel									
2660	Continuous thread									
2670	1/4" thread size	1 Plum	144	.056	L.F.	.48	2.08		2.56	3.66
2680	3/8" thread size		144	.056		.53	2.08		2.61	3.71
2690	1/2" thread size		144	.056		.84	2.08		2.92	4.05
2700	5/8" thread size		144	.056		1.19	2.08		3.27	4.44
2710	3/4" thread size		144	.056		2.11	2.08		4.19	5.45
2720	7/8" thread size		144	.056		2.64	2.08		4.72	6.05
2721	1" thread size	Q-1	160	.100		4.49	3.36		7.85	10.05
2722	1-1/8" thread size	"	120	.133		5.60	4.48		10.08	12.90
2725	1/4" thread size, bright finish	1 Plum	144	.056		.68	2.08		2.76	3.88
2726	1/2" thread size, bright finish	"	144	.056		1.17	2.08		3.25	4.42
2730	For galvanized, add					40%				
2750	Both ends machine threaded 18" length									
2760	3/8" thread size	1 Plum	240	.033	Ea.	3.94	1.24		5.18	6.20
2770	1/2" thread size		240	.033		6.30	1.24		7.54	8.85
2780	5/8" thread size		240	.033		9.05	1.24		10.29	11.90
2790	3/4" thread size		240	.033		13.90	1.24		15.14	17.20
2800	7/8" thread size		240	.033		23.50	1.24		24.74	27.50
2810	1" thread size		240	.033		31.50	1.24		32.74	36.50
2820	Rod couplings									
2821	3/8"	1 Plum	60	.133	Ea.	1.29	4.98		6.27	8.90
2822	1/2"		54	.148		1.65	5.55		7.20	10.15
2823	5/8"		48	.167		2.55	6.25		8.80	12.20
2824	3/4"		44	.182		3.70	6.80		10.50	14.30

300

MECHANICAL 15

15060 | Hangers & Supports

		CREW	DAILY OUTPUT	LABOR-HOURS	UNIT	2003 BARE COSTS				TOTAL INCL O&P
						MAT.	LABOR	EQUIP.	TOTAL	
300	2825	7/8"	1 Plum	40	.200	Ea.	5.45	7.45		12.90	17.30	300
2826	1"		34	.235		7.35	8.80		16.15	21.50
2827	1-1/8"		30	.267		16.55	9.95		26.50	33.50
2860	Pipe hanger assy, adj. clevis, saddle, rod, clamp, insul. allowance									
2864	1/2" pipe size	Q-5	35	.457	Ea.	24	15.45		39.45	49.50
2868	1" pipe size		34.60	.462		33	15.65		48.65	59.50
2872	2" pipe size		33.30	.480		36	16.25		52.25	64
2876	3" pipe size		31.20	.513		50.50	17.35		67.85	81.50
2880	4" pipe size		30.70	.521		53.50	17.65		71.15	85.50
2884	6" pipe size		29.80	.537		66.50	18.15		84.65	101
2888	8" pipe size		28	.571		104	19.35		123.35	144
2892	10" pipe size		25.20	.635		172	21.50		193.50	223
2896	12" pipe size		23.20	.690		193	23.50		216.50	248
2900	Rolls									
2910	Adjustable yoke, carbon steel with CI roll, TYPE 43									
2918	2" pipe size	Q-1	140	.114	Ea.	11.75	3.84		15.59	18.75
2920	2-1/2" pipe size		137	.117		11.75	3.93		15.68	18.90
2930	3" pipe size		131	.122		13.10	4.11		17.21	20.50
2940	3-1/2" pipe size		124	.129		17.40	4.34		21.74	25.50
2950	4" pipe size		117	.137		17.65	4.60		22.25	26.50
2960	5" pipe size		110	.145		20.50	4.89		25.39	30
2970	6" pipe size		104	.154		25.50	5.15		30.65	36.50
2980	8" pipe size		96	.167		38.50	5.60		44.10	50.50
2990	10" pipe size		80	.200		45.50	6.75		52.25	60
3000	12" pipe size		68	.235		67	7.90		74.90	86
3010	14" pipe size		56	.286		180	9.60		189.60	213
3020	16" pipe size		48	.333		231	11.20		242.20	271
3050	Chair, carbon steel with CI roll									
3060	2" pipe size	1 Plum	68	.118	Ea.	15.90	4.39		20.29	24
3070	2-1/2" pipe size		65	.123		17.20	4.60		21.80	26
3080	3" pipe size		62	.129		18.60	4.82		23.42	28
3090	3-1/2" pipe size		60	.133		22.50	4.98		27.48	32
3100	4" pipe size		58	.138		25	5.15		30.15	35.50
3110	5" pipe size		56	.143		27.50	5.35		32.85	38
3120	6" pipe size		53	.151		38	5.65		43.65	50.50
3130	8" pipe size		50	.160		51.50	6		57.50	65.50
3140	10" pipe size		48	.167		64.50	6.25		70.75	80.50
3150	12" pipe size		46	.174		94	6.50		100.50	113
3170	Single pipe roll, (see line 2650 for rods), TYPE 41, 1" pipe size	Q-1	137	.117		9.35	3.93		13.28	16.25
3180	1-1/4" pipe size		131	.122		9.35	4.11		13.46	16.50
3190	1-1/2" pipe size		129	.124		9.50	4.17		13.67	16.75
3200	2" pipe size		124	.129		9.95	4.34		14.29	17.50
3210	2-1/2" pipe size		118	.136		10.70	4.56		15.26	18.65
3220	3" pipe size		115	.139		11.30	4.68		15.98	19.45
3230	3-1/2" pipe size		113	.142		13.10	4.76		17.86	21.50
3240	4" pipe size		112	.143		14.35	4.80		19.15	23
3250	5" pipe size		110	.145		15.90	4.89		20.79	25
3260	6" pipe size		101	.158		20.50	5.35		25.85	30.50
3270	8" pipe size		90	.178		32	6		38	44
3280	10" pipe size		80	.200		38	6.75		44.75	51.50
3290	12" pipe size		68	.235		55	7.90		62.90	72.50
3291	14" pipe size		56	.286		90.50	9.60		100.10	114
3292	16" pipe size		48	.333		123	11.20		134.20	152
3293	18" pipe size		40	.400		132	13.45		145.45	167
3294	20" pipe size		35	.457		181	15.35		196.35	222
3296	24" pipe size		30	.533		287	17.95		304.95	340

T5 MECHANICAL

Important: See the Reference Section for critical supporting data - Reference Nos., Crews, & City Cost Indexes

15060	Hangers & Supports	CREW	DAILY OUTPUT	LABOR-HOURS	UNIT	2003 BARE COSTS				TOTAL INCL O&P
						MAT.	LABOR	EQUIP.	TOTAL	
300 3297	30" pipe size	Q-1	25	.640	Ea.	565	21.50		586.50	655
3298	36" pipe size	↓	20	.800	↓	580	27		607	675
3300	Saddles (add vertical pipe riser, usually 3" diameter)									
3310	Pipe support, complete, adjust., CI saddle, TYPE 36									
3320	2-1/2" pipe size	1 Plum	96	.083	Ea.	147	3.11		150.11	167
3330	3" pipe size		88	.091		149	3.40		152.40	169
3340	3-1/2" pipe size		79	.101		150	3.78		153.78	171
3350	4" pipe size		68	.118		195	4.39		199.39	221
3360	5" pipe size		64	.125		198	4.67		202.67	225
3370	6" pipe size		59	.136		201	5.05		206.05	229
3380	8" pipe size		53	.151		213	5.65		218.65	243
3390	10" pipe size		50	.160		235	6		241	268
3400	12" pipe size	↓	48	.167	↓	222	6.25		228.25	253
3450	For standard pipe support, one piece, CI deduct					34%				
3460	For stanchion support, CI with steel yoke, deduct					60%				
3500	Multiple pipe support 2' long strut	1 Plum	30	.267	Ea.	4.40	9.95		14.35	19.90
3510	3' long strut		20	.400		6.60	14.95		21.55	30
3520	4' long strut	↓	17	.471		8.80	17.60		26.40	36
3550	Insulation shield 1" thick, 1/2" pipe size, TYPE 40	1 Asbe	100	.080		5.95	2.78		8.73	10.95
3560	3/4" pipe size		100	.080		5.95	2.78		8.73	10.95
3570	1" pipe size		98	.082		7.40	2.83		10.23	12.65
3580	1-1/4" pipe size		98	.082		7.40	2.83		10.23	12.65
3590	1-1/2" pipe size		96	.083		7.85	2.89		10.74	13.20
3600	2" pipe size		96	.083		8.15	2.89		11.04	13.60
3610	2-1/2" pipe size		94	.085		8.55	2.95		11.50	14.10
3620	3" pipe size		94	.085		9	2.95		11.95	14.60
3630	2" thick, 3-1/2" pipe size		92	.087		11.80	3.02		14.82	17.75
3640	4" pipe size		92	.087		11.80	3.02		14.82	17.75
3650	5" pipe size		90	.089		17.45	3.08		20.53	24
3660	6" pipe size		90	.089		19.90	3.08		22.98	27
3670	8" pipe size		88	.091		30	3.15		33.15	37.50
3680	10" pipe size		88	.091		44	3.15		47.15	53.50
3690	12" pipe size		86	.093		50.50	3.23		53.73	60.50
3700	14" pipe size		86	.093		59	3.23		62.23	70
3710	16" pipe size		84	.095		67.50	3.30		70.80	80
3720	18" pipe size		84	.095		74.50	3.30		77.80	87
3730	20" pipe size		82	.098		77.50	3.39		80.89	91
3732	24" pipe size		80	.100		88	3.47		91.47	102
3733	30" pipe size		78	.103		101	3.56		104.56	117
3735	36" pipe size	↓	75	.107	↓	108	3.70		111.70	125
3750	Covering protection saddle, TYPE 39									
3760	1" covering size									
3770	3/4" pipe size	1 Plum	68	.118	Ea.	10	4.39		14.39	17.65
3780	1" pipe size		68	.118		10	4.39		14.39	17.65
3790	1-1/4" pipe size		68	.118		10	4.39		14.39	17.65
3800	1-1/2" pipe size		66	.121		10.85	4.53		15.38	18.80
3810	2" pipe size		66	.121		10.85	4.53		15.38	18.80
3820	2-1/2" pipe size		64	.125		10.85	4.67		15.52	19
3830	3" pipe size		64	.125		15.65	4.67		20.32	24.50
3840	3-1/2" pipe size		62	.129		17.85	4.82		22.67	27
3850	4" pipe size		62	.129		16.70	4.82		21.52	25.50
3860	5" pipe size		60	.133		17.85	4.98		22.83	27
3870	6" pipe size	↓	60	.133	↓	21.50	4.98		26.48	31
3900	1-1/2" covering size									
3910	3/4" pipe size	1 Plum	68	.118	Ea.	12.50	4.39		16.89	20.50
3920	1" pipe size	↓	68	.118	↓	12.50	4.39		16.89	20.50

MECHANICAL 15

	15060	Hangers & Supports	CREW	DAILY OUTPUT	LABOR-HOURS	UNIT	2003 BARE COSTS				TOTAL INCL O&P	
							MAT.	LABOR	EQUIP.	TOTAL		
300	3930	1-1/4" pipe size	1 Plum	68	.118	Ea.	12.50	4.39		16.89	20.50	300
	3940	1-1/2" pipe size		66	.121		12.50	4.53		17.03	20.50	
	3950	2" pipe size		66	.121		13.55	4.53		18.08	22	
	3960	2-1/2" pipe size		64	.125		16	4.67		20.67	24.50	
	3970	3" pipe size		64	.125		16	4.67		20.67	24.50	
	3980	3-1/2" pipe size		62	.129		16.25	4.82		21.07	25	
	3990	4" pipe size		62	.129		16.25	4.82		21.07	25	
	4000	5" pipe size		60	.133		16.25	4.98		21.23	25.50	
	4010	6" pipe size		60	.133		23.50	4.98		28.48	33	
	4020	8" pipe size		58	.138		29.50	5.15		34.65	40	
	4022	10" pipe size		56	.143		29.50	5.35		34.85	40	
	4024	12" pipe size		54	.148		43	5.55		48.55	55.50	
	4028	2" covering size										
	4029	2-1/2" pipe size	1 Plum	62	.129	Ea.	15.35	4.82		20.17	24	
	4032	3" pipe size		60	.133		18.05	4.98		23.03	27.50	
	4033	4" pipe size		58	.138		18.05	5.15		23.20	27.50	
	4034	6" pipe size		56	.143		26	5.35		31.35	36.50	
	4035	8" pipe size		54	.148		30.50	5.55		36.05	42	
	4080	10" pipe size		58	.138		32.50	5.15		37.65	43.50	
	4090	12" pipe size		56	.143		48	5.35		53.35	61	
	4100	14" pipe size		56	.143		48	5.35		53.35	61	
	4110	16" pipe size		54	.148		67.50	5.55		73.05	82.50	
	4120	18" pipe size		54	.148		67.50	5.55		73.05	82.50	
	4130	20" pipe size		52	.154		75	5.75		80.75	91	
	4150	24" pipe size		50	.160		87.50	6		93.50	105	
	4160	30" pipe size		48	.167		96	6.25		102.25	115	
	4180	36" pipe size		45	.178		106	6.65		112.65	127	
	4186	2-1/2" covering size										
	4187	3" pipe size	1 Plum	58	.138	Ea.	17.60	5.15		22.75	27	
	4188	4" pipe size		56	.143		19.10	5.35		24.45	29	
	4189	6" pipe size		52	.154		29	5.75		34.75	40.50	
	4190	8" pipe size		48	.167		35.50	6.25		41.75	49	
	4191	10" pipe size		44	.182		39	6.80		45.80	53.50	
	4192	12" pipe size		40	.200		52	7.45		59.45	69	
	4193	14" pipe size		36	.222		52	8.30		60.30	70	
	4194	16" pipe size		32	.250		77	9.35		86.35	98.50	
	4195	18" pipe size		28	.286		82	10.65		92.65	107	
	4200	Sockets										
	4210	Rod end, malleable iron, TYPE 16										
	4220	1/4" thread size	1 Plum	240	.033	Ea.	.99	1.24		2.23	2.97	
	4230	3/8" thread size		240	.033		1.59	1.24		2.83	3.63	
	4240	1/2" thread size		230	.035		1.90	1.30		3.20	4.05	
	4250	5/8" thread size		225	.036		3.76	1.33		5.09	6.15	
	4260	3/4" thread size		220	.036		6.10	1.36		7.46	8.80	
	4270	7/8" thread size		210	.038		12	1.42		13.42	15.35	
	4290	Strap, 1/2" pipe size, TYPE 26	Q-1	142	.113		2.39	3.79		6.18	8.35	
	4300	3/4" pipe size		140	.114		2.43	3.84		6.27	8.45	
	4310	1" pipe size		137	.117		2.48	3.93		6.41	8.70	
	4320	1-1/4" pipe size		134	.119		2.79	4.02		6.81	9.10	
	4330	1-1/2" pipe size		131	.122		3.50	4.11		7.61	10.05	
	4340	2" pipe size		129	.124		3.60	4.17		7.77	10.25	
	4350	2-1/2" pipe size		125	.128		6.30	4.30		10.60	13.40	
	4360	3" pipe size		122	.131		6.90	4.41		11.31	14.25	
	4370	3-1/2" pipe size		119	.134		11.55	4.52		16.07	19.55	
	4380	4" pipe size		114	.140		12.90	4.72		17.62	21.50	
	4400	U-bolt, carbon steel										

15 MECHANICAL

15060 | Hangers & Supports

		CREW	DAILY OUTPUT	LABOR-HOURS	UNIT	2003 BARE COSTS MAT.	LABOR	EQUIP.	TOTAL	TOTAL INCL O&P	
300	4410	Standard, with nuts, TYPE 42										**300**
	4420	1/2" pipe size	1 Plum	160	.050	Ea.	1.12	1.87		2.99	4.05	
	4430	3/4" pipe size		158	.051		1.12	1.89		3.01	4.09	
	4450	1" pipe size		152	.053		1.22	1.97		3.19	4.31	
	4460	1-1/4" pipe size		148	.054		1.35	2.02		3.37	4.54	
	4470	1-1/2" pipe size		143	.056		1.40	2.09		3.49	4.70	
	4480	2" pipe size		139	.058		1.44	2.15		3.59	4.83	
	4490	2-1/2" pipe size		134	.060		2	2.23		4.23	5.55	
	4500	3" pipe size		128	.063		3.72	2.33		6.05	7.60	
	4510	3-1/2" pipe size		122	.066		3.84	2.45		6.29	7.90	
	4520	4" pipe size		117	.068		3.89	2.55		6.44	8.15	
	4530	5" pipe size		114	.070		3.96	2.62		6.58	8.30	
	4540	6" pipe size		111	.072		4.23	2.69		6.92	8.70	
	4550	8" pipe size		109	.073		5.05	2.74		7.79	9.75	
	4560	10" pipe size		107	.075		8.85	2.79		11.64	13.95	
	4570	12" pipe size		104	.077		11.35	2.87		14.22	16.85	
	4580	For plastic coating on 1/2" thru 6" size, add					150%					
	4700	U-hook, carbon steel, requires mounting screws or bolts										
	4710	3/4" thru 2" pipe size										
	4720	6" long	1 Plum	96	.083	Ea.	3.20	3.11		6.31	8.20	
	4730	8" long		96	.083		3.42	3.11		6.53	8.45	
	4740	10" long		96	.083		4.32	3.11		7.43	9.45	
	4750	12" long		96	.083		4.79	3.11		7.90	9.95	
	4760	For copper plated, add					50%					
	8000	Pipe clamp, plastic, 1/2" CTS	1 Plum	80	.100		.19	3.74		3.93	5.85	
	8010	3/4" CTS		73	.110		.20	4.09		4.29	6.40	
	8020	1" CTS		68	.118		.46	4.39		4.85	7.15	
	8080	Economy clamp, 1/4" CTS		175	.046		.05	1.71		1.76	2.64	
	8090	3/8" CTS		168	.048		.06	1.78		1.84	2.76	
	8100	1/2" CTS		160	.050		.06	1.87		1.93	2.89	
	8110	3/4" CTS		145	.055		.12	2.06		2.18	3.24	
	8200	Half clamp, 1/2" CTS		80	.100		.06	3.74		3.80	5.70	
	8210	3/4" CTS		73	.110		.09	4.09		4.18	6.30	
	8300	Suspension clamp, 1/2" CTS		80	.100		.19	3.74		3.93	5.85	
	8310	3/4" CTS		73	.110		.20	4.09		4.29	6.40	
	8320	1" CTS		68	.118		.56	4.39		4.95	7.25	
	8400	Insulator, 1/2" CTS		80	.100		.33	3.74		4.07	6	
	8410	3/4" CTS		73	.110		.34	4.09		4.43	6.55	
	8420	1" CTS		68	.118		.36	4.39		4.75	7.05	
	8500	J hook clamp with nail 1/2" CTS		240	.033		.10	1.24		1.34	1.99	
	8501	3/4" CTS		240	.033		.12	1.24		1.36	2.01	

15070 | Mech Sound/Vibration/Seismic Ctrl

		CREW	DAILY OUTPUT	LABOR-HOURS	UNIT	MAT.	LABOR	EQUIP.	TOTAL	TOTAL INCL O&P	
800	0010	**VIBRATION ABSORBERS**										**800**
	0100	Hangers, neoprene flex										
	0200	10 - 120 lb. capacity				Ea.	12.95			12.95	14.25	
	0220	75 - 550 lb. capacity					16.85			16.85	18.55	
	0240	250 - 1100 lb. capacity					34			34	37.50	
	0260	1000 - 4000 lb. capacity					57			57	62.50	
	0500	Spring flex, 60 lb. capacity					18.15			18.15	20	
	0520	450 lb. capacity					22.50			22.50	25	
	0540	900 lb. capacity					30			30	33	
	0560	1100-1300 lb. capacity					30			30	33	
	0600	Rubber in shear										
	0610	45 - 340 lb., up to 1/2" rod size	1 Stpi	22	.364	Ea.	12	13.65		25.65	33.50	

MECHANICAL 15

		15070	Mech Sound/Vibration/Seismic Ctrl	CREW	DAILY OUTPUT	LABOR-HOURS	UNIT	2003 BARE COSTS				TOTAL INCL O&P	
								MAT.	LABOR	EQUIP.	TOTAL		
800	0620		130 - 700 lb., up to 3/4" rod size	1 Stpi	20	.400	Ea.	27	15.05		42.05	52	800
	0630		50 - 1000 lb., up to 3/4" rod size	↓	18	.444		31	16.70		47.70	59	
	1000		Mounts, neoprene, 45 - 380 lb. capacity					10			10	11	
	1020		250 - 1100 lb. capacity					26			26	28.50	
	1040		1000 - 4000 lb. capacity					58.50			58.50	64	
	1100		Spring flex, 60 lb. capacity					35			35	38.50	
	1120		165 lb. capacity					35			35	38.50	
	1140		260 lb. capacity					35			35	38.50	
	1160		450 lb. capacity					46.50			46.50	51	
	1180		600 lb. capacity					46.50			46.50	51	
	1200		750 lb. capacity					46.50			46.50	51	
	1220		900 lb. capacity					48			48	52.50	
	1240		1100 lb. capacity					49.50			49.50	54.50	
	1260		1300 lb. capacity					51			51	56	
	1280		1500 lb. capacity					75			75	82.50	
	1300		1800 lb. capacity					76.50			76.50	84	
	1320		2200 lb. capacity					80.50			80.50	88.50	
	1340		2600 lb. capacity				↓	80.50			80.50	88.50	
	1399		Spring type										
	1400		2 Piece										
	1410		50 - 1000 lb.	1 Stpi	12	.667	Ea.	102	25		127	150	
	1420		1100 - 1600 lb.	"	12	.667	"	110	25		135	159	
	1500		Double spring open										
	1510		150 - 450 lb.	1 Stpi	24	.333	Ea.	74	12.55		86.55	100	
	1520		500 - 1000 lb.		24	.333		74	12.55		86.55	100	
	1530		1100 - 1600 lb.		24	.333		74	12.55		86.55	100	
	1540		1700 - 2400 lb.		24	.333		80	12.55		92.55	107	
	1550		2500 - 3400 lb.	↓	24	.333		136	12.55		148.55	169	
	2000		Pads, cork rib, 18" x 18" x 1", 10-50 psi					96			96	106	
	2020		18" x 36" x 1", 10-50 psi					192			192	211	
	2100		Shear flexible pads, 18" x 18" x 3/8", 20-70 psi					45.50			45.50	50	
	2120		18" x 36" x 3/8", 20-70 psi				↓	86.50			86.50	95	
	2150		Laminated neoprene and cork										
	2160		1" thick	1 Stpi	16	.500	S.F.	39	18.80		57.80	71	
	3000		Note overlap in capacities due to deflections										

		15075	Mech Identification										
400	0010		**PIPING SYSTEM IDENTIFICATION LABELS**										400
	0100		Indicate contents and flow direction										
	0106		Pipe markers										
	0110		Plastic snap around										
	0114		1/2" pipe	1 Plum	80	.100	Ea.	3.15	3.74		6.89	9.10	
	0116		3/4" pipe		80	.100		3.15	3.74		6.89	9.10	
	0118		1" pipe		80	.100		3.15	3.74		6.89	9.10	
	0120		2" pipe		75	.107		4.15	3.98		8.13	10.55	
	0122		3" pipe		70	.114		7.75	4.27		12.02	15	
	0124		4" pipe		60	.133		7.75	4.98		12.73	16.05	
	0126		6" pipe		60	.133		8.40	4.98		13.38	16.75	
	0128		8" pipe		56	.143		11.55	5.35		16.90	21	
	0130		10" pipe		56	.143		11.55	5.35		16.90	21	
	0200		Over 10" pipe size	↓	50	.160	↓	14.20	6		20.20	24.50	
	1110		Self adhesive										
	1114		1" pipe	1 Plum	80	.100	Ea.	2.20	3.74		5.94	8.05	
	1116		2" pipe		75	.107		2.20	3.98		6.18	8.40	
	1118		3" pipe	↓	70	.114	↓	3.55	4.27		7.82	10.35	

Important: See the Reference Section for critical supporting data - Reference Nos., Crews, & City Cost Indexes

15075 | Mech Identification

		CREW	DAILY OUTPUT	LABOR-HOURS	UNIT	2003 BARE COSTS MAT.	LABOR	EQUIP.	TOTAL	TOTAL INCL O&P	
400	1120	4" pipe	1 Plum	60	.133	Ea.	3.55	4.98		8.53	11.40	**400**
	1122	6" pipe		60	.133		3.55	4.98		8.53	11.40	
	1124	8" pipe		56	.143		4.95	5.35		10.30	13.50	
	1126	10" pipe		56	.143		4.70	5.35		10.05	13.20	
	1200	Over 10" pipe size	↓	50	.160	↓	6.90	6		12.90	16.60	
	2000	Valve tags										
	2010	Numbered plus identifying legend										
	2100	Brass, 2" diameter	1 Plum	40	.200	Ea.	2.05	7.45		9.50	13.55	
	2200	Plastic, 1-1/2" Dia.	"	40	.200	"	1.75	7.45		9.20	13.25	

15082 | Mechanical Insulation

		CREW	DAILY OUTPUT	LABOR-HOURS	UNIT	2003 BARE COSTS MAT.	LABOR	EQUIP.	TOTAL	TOTAL INCL O&P	
200	0010	**INSULATION**										**200**
	0100	Rule of thumb, as a percentage of total mechanical costs				Job				10%		
	0110	Insulation req'd is based on the surface size/area to be covered										
	1000	Boiler, 1-1/2" calcium silicate, 1/2" cement finish	Q-14	50	.320	S.F.	3.23	10		13.23	19.45	
	1020	2" fiberglass	"	80	.200	"	1.90	6.25		8.15	12.05	
	2000	Breeching, 2" calcium silicate with 1/2" cement finish, no lath										
	2020	Rectangular	Q-14	42	.381	S.F.	3.65	11.90		15.55	23	
	2040	Round	"	38.70	.413	"	4.09	12.90		16.99	25	
	2300	Calcium silicate block, + 200°F to + 1200°F										
	2310	On irregular surfaces, valves and fittings										
	2340	1" thick	Q-14	30	.533	S.F.	2.55	16.65		19.20	29.50	
	2360	1-1/2" thick		25	.640		2.99	20		22.99	35.50	
	2380	2" thick		22	.727		4.11	22.50		26.61	40.50	
	2400	3" thick	↓	18	.889	↓	6	28		34	50.50	
	2410	On plane surfaces										
	2420	1" thick	Q-14	126	.127	S.F.	2.55	3.97		6.52	9.10	
	2430	1-1/2" thick		120	.133		2.99	4.16		7.15	9.90	
	2440	2" thick		100	.160		4.11	5		9.11	12.45	
	2450	3" thick	↓	70	.229	↓	6	7.15		13.15	17.95	
	2900	Domestic water heater wrap kit										
	2920	1-1/2" with vinyl jacket, 20-60 gal.	1 Plum	8	1	Ea.	16.25	37.50		53.75	74.50	
	2930	Insulated protectors, (ADA)										
	2935	For exposed piping under sinks or lavatories.										
	2940	Vinyl coated foam, velcro tabs										
	2945	P Trap, 1-1/4" or 1-1/2"	1 Plum	32	.250	Ea.	16.80	9.35		26.15	32.50	
	2960	Valve and supply cover										
	2965	1/2", 3/8", and 7/16" pipe size	1 Plum	32	.250	Ea.	16.80	9.35		26.15	32.50	
	2970	Extension drain cover										
	2975	1-1/4", or 1-1/2" pipe size	1 Plum	32	.250	Ea.	17.20	9.35		26.55	33	
	2980	Tailpiece offset (wheelchair)										
	2985	1-1/4" pipe size	1 Plum	32	.250	Ea.	19.60	9.35		28.95	35.50	
	3000	Ductwork										
	3020	Blanket type, fiberglass, flexible										
	3030	Fire resistant liner, black coating one side										
	3050	1/2" thick, 2 lb. density	Q-14	380	.042	S.F.	.34	1.32		1.66	2.46	
	3060	1" thick, 1-1/2 lb. density		350	.046		.46	1.43		1.89	2.78	
	3070	1-1/2" thick, 1-1/2 lb. density		320	.050		.80	1.56		2.36	3.36	
	3080	2" thick, 1-1/2 lb. density	↓	300	.053	↓	.95	1.67		2.62	3.70	
	3140	FRK vapor barrier wrap, .75 lb. density										
	3160	1" thick	Q-14	350	.046	S.F.	.32	1.43		1.75	2.62	
	3170	1-1/2" thick		320	.050		.29	1.56		1.85	2.80	
	3180	2" thick		300	.053		.34	1.67		2.01	3.02	
	3190	3" thick		260	.062		.51	1.92		2.43	3.62	
	3200	4" thick	↓	242	.066	↓	.68	2.06		2.74	4.03	

MECHANICAL 15

15082	Mechanical Insulation	CREW	DAILY OUTPUT	LABOR-HOURS	UNIT	2003 BARE COSTS				TOTAL INCL O&P
						MAT.	LABOR	EQUIP.	TOTAL	
200	3210	Vinyl jacket, same as FRK										200
	3280	Unfaced, 1 lb. density									
	3310	1" thick	Q-14	360	.044	S.F.	.29	1.39		1.68	2.53
	3320	1-1/2" thick		330	.048		.40	1.51		1.91	2.85
	3330	2" thick	▼	310	.052	▼	.46	1.61		2.07	3.07
	3340	Board type fiberglass liner, 1-1/2 lb. density									
	3344	1" thick	Q-14	150	.107	S.F.	.63	3.33		3.96	6
	3345	1-1/2" thick		130	.123		.83	3.84		4.67	7
	3346	2" thick	▼	120	.133	▼	1.22	4.16		5.38	7.95
	3490	Board type, fiberglass liner, 3 lb. density									
	3500	Fire resistant, black pigmented, 1 side									
	3520	1" thick	Q-14	150	.107	S.F.	1.38	3.33		4.71	6.80
	3540	1-1/2" thick		130	.123		1.70	3.84		5.54	7.95
	3560	2" thick	▼	120	.133	▼	2.04	4.16		6.20	8.85
	3600	FRK vapor barrier									
	3620	1" thick	Q-14	150	.107	S.F.	1.51	3.33		4.84	6.95
	3630	1-1/2" thick		130	.123		1.95	3.84		5.79	8.25
	3640	2" thick	▼	120	.133	▼	2.36	4.16		6.52	9.20
	3680	No finish									
	3700	1" thick	Q-14	170	.094	S.F.	.80	2.94		3.74	5.55
	3710	1-1/2" thick		140	.114		1.21	3.57		4.78	7.05
	3720	2" thick	▼	130	.123	▼	1.51	3.84		5.35	7.75
	3730	Sheet insulation									
	3760	Polyethylene foam, closed cell, UV resistant									
	3770	Standard temperature (-90° to +212° F)									
	3771	1/4" thick	Q-14	450	.036	S.F.	1.10	1.11		2.21	2.98
	3772	3/8" thick		440	.036		1.62	1.14		2.76	3.59
	3773	1/2" thick		420	.038		2.02	1.19		3.21	4.11
	3774	3/4" thick		400	.040		2.95	1.25		4.20	5.25
	3775	1" thick	▼	380	.042	▼	3.75	1.32		5.07	6.20
	3779	Adhesive (see line 7878)									
	3780	Foam, rubber									
	3782	1" thick	1 Stpi	50	.160	S.F.	2.66	6		8.66	12.05
	3795	Finishes									
	3800	1/2" cement over 1" wire mesh, incl. corner bead	Q-14	100	.160	S.F.	1.44	5		6.44	9.55
	3820	Glass cloth, pasted on		170	.094		2.33	2.94		5.27	7.25
	3900	Weatherproof, non-metallic, 2 lb. per S.F.	▼	100	.160	▼	3.33	5		8.33	11.60
	4000	Pipe covering (price copper tube one size less than IPS)									
	4040	Air cell, asbestos free corrugated felt, with cover									
	4100	3 ply, 1/2" iron pipe size	Q-14	145	.110	L.F.	.75	3.45		4.20	6.35
	4130	3/4" iron pipe size		145	.110		.85	3.45		4.30	6.45
	4140	1" iron pipe size		145	.110		.90	3.45		4.35	6.50
	4150	1-1/4" iron pipe size		140	.114		1	3.57		4.57	6.80
	4160	1-1/2" iron pipe size		140	.114		1.10	3.57		4.67	6.90
	4170	2" iron pipe size		135	.119		1.20	3.70		4.90	7.20
	4180	2-1/2" iron pipe size		135	.119		1.35	3.70		5.05	7.40
	4190	3" iron pipe size		130	.123		1.50	3.84		5.34	7.75
	4200	3-1/2" iron pipe size		130	.123		1.76	3.84		5.60	8.05
	4210	4" iron pipe size		125	.128		1.95	4		5.95	8.50
	4220	5" iron pipe size		120	.133		2.25	4.16		6.41	9.10
	4230	6" iron pipe size		115	.139		2.40	4.35		6.75	9.55
	4240	8" iron pipe size		105	.152		3.30	4.76		8.06	11.20
	4250	10" iron pipe size		95	.168		3.90	5.25		9.15	12.65
	4260	12" iron pipe size	▼	90	.178	▼	4.50	5.55		10.05	13.80
	4280	Cellular glass, closed cell foam, all service jacket, sealant,									
	4281	working temp. (-450°F to +900°F), 0 water vapor xmission									

Important: See the Reference Section for critical supporting data - Reference Nos., Crews, & City Cost Indexes

15 MECHANICAL

15082	Mechanical Insulation	CREW	DAILY OUTPUT	LABOR-HOURS	UNIT	MAT.	LABOR	EQUIP.	TOTAL	TOTAL INCL O&P
							2003 BARE COSTS			
200 4284	1" wall,									**200**
4286	1/2" iron pipe size	Q-14	120	.133	L.F.	1.60	4.16		5.76	8.35
4300	1-1/2" wall,									
4301	1" iron pipe size	Q-14	105	.152	L.F.	2.57	4.76		7.33	10.40
4304	2-1/2" iron pipe size		90	.178		4.33	5.55		9.88	13.60
4306	3" iron pipe size		85	.188		4.38	5.90		10.28	14.15
4308	4" iron pipe size		70	.229		5.80	7.15		12.95	17.75
4310	5" iron pipe size	↓	65	.246	↓	6.70	7.70		14.40	19.60
4320	2" wall,									
4322	1" iron pipe size	Q-14	100	.160	L.F.	3.82	5		8.82	12.15
4324	2-1/2" iron pipe size		85	.188		5.65	5.90		11.55	15.60
4326	3" iron pipe size		80	.200		5.80	6.25		12.05	16.35
4328	4" iron pipe size		65	.246		6.50	7.70		14.20	19.40
4330	5" iron pipe size		60	.267		7.85	8.35		16.20	22
4332	6" iron pipe size		50	.320		9.25	10		19.25	26
4336	8" iron pipe size		40	.400		11.95	12.50		24.45	33
4338	10" iron pipe size	↓	35	.457	↓	12.35	14.30		26.65	36
4350	2-1/2" wall,									
4360	12" iron pipe size	Q-14	32	.500	L.F.	19.60	15.60		35.20	46.50
4362	14" iron pipe size	"	28	.571	"	21	17.85		38.85	51.50
4370	3" wall,									
4378	6" iron pipe size	Q-14	48	.333	L.F.	12.35	10.40		22.75	30
4380	8" iron pipe size		38	.421		14.40	13.15		27.55	37
4382	10" iron pipe size		33	.485		19.05	15.15		34.20	45
4384	16" iron pipe size		25	.640		24.50	20		44.50	59
4386	18" iron pipe size		22	.727		27	22.50		49.50	65.50
4388	20" iron pipe size	↓	20	.800	↓	31.50	25		56.50	74.50
4400	3-1/2" wall,									
4412	12" iron pipe size	Q-14	27	.593	L.F.	28	18.50		46.50	60
4414	14" iron pipe size	"	25	.640	"	29.50	20		49.50	64.50
4430	4" wall,									
4446	16" iron pipe size	Q-14	22	.727	L.F.	30	22.50		52.50	69
4448	18" iron pipe size		20	.800		38.50	25		63.50	82
4450	20" iron pipe size	↓	18	.889	↓	40	28		68	88
4480	Fittings, average with fabric and mastic									
4484	1" wall,									
4486	1/2" iron pipe size	1 Asbe	40	.200	Ea.	2.84	6.95		9.79	14.15
4500	1-1/2" wall,									
4502	1" iron pipe size	1 Asbe	38	.211	Ea.	4.89	7.30		12.19	17
4504	2-1/2" iron pipe size		32	.250		6.10	8.70		14.80	20.50
4506	3" iron pipe size		30	.267		6.70	9.25		15.95	22
4508	4" iron pipe size		28	.286		10.30	9.90		20.20	27
4510	5" iron pipe size	↓	24	.333	↓	11.35	11.55		22.90	31
4520	2" wall,									
4522	1" iron pipe size	1 Asbe	36	.222	Ea.	5.65	7.70		13.35	18.50
4524	2-1/2" iron pipe size		30	.267		6.70	9.25		15.95	22
4526	3" iron pipe size		28	.286		8.05	9.90		17.95	24.50
4528	4" iron pipe size		24	.333		10.80	11.55		22.35	30.50
4530	5" iron pipe size		22	.364		13.65	12.60		26.25	35
4532	6" iron pipe size		20	.400		15.95	13.90		29.85	39.50
4536	8" iron pipe size		12	.667		24	23		47	63.50
4538	10" iron pipe size	↓	8	1	↓	30	34.50		64.50	88
4550	2-1/2" wall,									
4560	12" iron pipe size	1 Asbe	6	1.333	Ea.	51.50	46.50		98	130
4562	14" iron pipe size	"	4	2	"	60	69.50		129.50	176
4570	3" wall,									

15082	Mechanical Insulation	CREW	DAILY OUTPUT	LABOR-HOURS	UNIT	2003 BARE COSTS				TOTAL INCL O&P
						MAT.	LABOR	EQUIP.	TOTAL	
200 4578	6" iron pipe size	1 Asbe	16	.500	Ea.	18.55	17.35		35.90	48 **200**
4580	8" iron pipe size		10	.800		27.50	28		55.50	74
4582	10" iron pipe size		6	1.333		38.50	46.50		85	116
4584	16" iron pipe size		2	4		92.50	139		231.50	325
4586	18" iron pipe size		2	4		120	139		259	350
4588	20" iron pipe size	▼	2	4	▼	141	139		280	375
4600	3-1/2" wall,									
4612	12" iron pipe size	1 Asbe	4	2	Ea.	67	69.50		136.50	184
4614	14" iron pipe size	"	2	4	"	74	139		213	305
4630	4" wall,									
4646	16" iron pipe size	1 Asbe	2	4	Ea.	111	139		250	345
4648	18" iron pipe size		2	4		136	139		275	370
4650	20" iron pipe size	▼	2	4	▼	159	139		298	395
4900	Calcium silicate, with cover									
5100	1" wall, 1/2" iron pipe size	Q-14	170	.094	L.F.	2.64	2.94		5.58	7.60
5130	3/4" iron pipe size		170	.094		2.65	2.94		5.59	7.60
5140	1" iron pipe size		170	.094		2.50	2.94		5.44	7.45
5150	1-1/4" iron pipe size		165	.097		2.54	3.03		5.57	7.60
5160	1-1/2" iron pipe size		165	.097		2.58	3.03		5.61	7.65
5170	2" iron pipe size		160	.100		3.06	3.12		6.18	8.35
5180	2-1/2" iron pipe size		160	.100		3.27	3.12		6.39	8.55
5190	3" iron pipe size		150	.107		3.46	3.33		6.79	9.10
5200	4" iron pipe size		140	.114		4.03	3.57		7.60	10.15
5210	5" iron pipe size		135	.119		4.43	3.70		8.13	10.75
5220	6" iron pipe size		130	.123		4.82	3.84		8.66	11.40
5280	1-1/2" wall, 1/2" iron pipe size		150	.107		2.88	3.33		6.21	8.45
5310	3/4" iron pipe size		150	.107		2.89	3.33		6.22	8.50
5320	1" iron pipe size		150	.107		3.15	3.33		6.48	8.75
5330	1-1/4" iron pipe size		145	.110		3.40	3.45		6.85	9.25
5340	1-1/2" iron pipe size		145	.110		3.57	3.45		7.02	9.45
5350	2" iron pipe size		140	.114		3.89	3.57		7.46	10
5360	2-1/2" iron pipe size		140	.114		4.27	3.57		7.84	10.40
5370	3" iron pipe size		135	.119		4.48	3.70		8.18	10.85
5380	4" iron pipe size		125	.128		5.20	4		9.20	12.05
5390	5" iron pipe size		120	.133		5.80	4.16		9.96	13
5400	6" iron pipe size		110	.145		6	4.54		10.54	13.85
5402	8" iron pipe size		95	.168		8.05	5.25		13.30	17.25
5404	10" iron pipe size		85	.188		10.65	5.90		16.55	21
5406	12" iron pipe size		80	.200		12.55	6.25		18.80	24
5408	14" iron pipe size		75	.213		14.20	6.65		20.85	26
5410	16" iron pipe size		70	.229		15.90	7.15		23.05	29
5412	18" iron pipe size		65	.246		18.10	7.70		25.80	32
5460	2" wall, 1/2" iron pipe size		135	.119		4.62	3.70		8.32	11
5490	3/4" iron pipe size		135	.119		4.63	3.70		8.33	11
5500	1" iron pipe size		135	.119		4.86	3.70		8.56	11.25
5510	1-1/4" iron pipe size		130	.123		5.40	3.84		9.24	12.05
5520	1-1/2" iron pipe size		130	.123		5.50	3.84		9.34	12.15
5530	2" iron pipe size		125	.128		5.75	4		9.75	12.70
5540	2-1/2" iron pipe size		125	.128		5.95	4		9.95	12.85
5550	3" iron pipe size		120	.133		6.65	4.16		10.81	13.90
5560	4" iron pipe size		115	.139		7.80	4.35		12.15	15.45
5570	5" iron pipe size		110	.145		8.85	4.54		13.39	17
5580	6" iron pipe size		105	.152		9.95	4.76		14.71	18.50
5581	8" iron pipe size		95	.168		11.90	5.25		17.15	21.50
5582	10" iron pipe size		85	.188		14.80	5.90		20.70	25.50
5583	12" iron pipe size	▼	80	.200	▼	16.55	6.25		22.80	28

Important: See the Reference Section for critical supporting data - Reference Nos., Crews, & City Cost Indexes

15082 | Mechanical Insulation

		CREW	DAILY OUTPUT	LABOR-HOURS	UNIT	2003 BARE COSTS				TOTAL INCL O&P
						MAT.	LABOR	EQUIP.	TOTAL	
5584	14" iron pipe size	Q-14	75	.213	L.F.	18.20	6.65		24.85	30.50
5585	16" iron pipe size		70	.229		20.50	7.15		27.65	34.50
5586	18" iron pipe size		65	.246		22.50	7.70		30.20	37.50
5587	3" wall, 1-1/4" iron pipe size		100	.160		7.80	5		12.80	16.50
5588	1-1/2" iron pipe size		100	.160		8.30	5		13.30	17.05
5589	2" iron pipe size		95	.168		8.80	5.25		14.05	18
5590	2-1/2" iron pipe size		95	.168		9.95	5.25		15.20	19.30
5591	3" iron pipe size		90	.178		10.15	5.55		15.70	20
5592	4" iron pipe size		85	.188		13.20	5.90		19.10	24
5593	6" iron pipe size		80	.200		16.15	6.25		22.40	27.50
5594	8" iron pipe size		75	.213		19.10	6.65		25.75	31.50
5595	10" iron pipe size		65	.246		23	7.70		30.70	38
5596	12" iron pipe size		60	.267		25.50	8.35		33.85	41.50
5597	14" iron pipe size		55	.291		28.50	9.10		37.60	46
5598	16" iron pipe size		50	.320		31.50	10		41.50	50.50
5599	18" iron pipe size	▼	45	.356	▼	34.50	11.10		45.60	55.50
5600	Calcium silicate, no cover									
5720	1" wall, 1/2" iron pipe size	Q-14	180	.089	L.F.	2.47	2.78		5.25	7.15
5740	3/4" iron pipe size		180	.089		2.47	2.78		5.25	7.15
5750	1" iron pipe size		180	.089		2.31	2.78		5.09	6.95
5760	1-1/4" iron pipe size		175	.091		2.32	2.86		5.18	7.10
5770	1-1/2" iron pipe size		175	.091		2.37	2.86		5.23	7.15
5780	2" iron pipe size		170	.094		2.81	2.94		5.75	7.75
5790	2-1/2" iron pipe size		170	.094		3	2.94		5.94	8
5800	3" iron pipe size		160	.100		3.16	3.12		6.28	8.45
5810	4" iron pipe size		150	.107		3.67	3.33		7	9.35
5820	5" iron pipe size		145	.110		4.03	3.45		7.48	9.95
5830	6" iron pipe size		140	.114		4.35	3.57		7.92	10.50
5900	1-1/2" wall, 1/2" iron pipe size		160	.100		2.67	3.12		5.79	7.90
5920	3/4" iron pipe size		160	.100		2.67	3.12		5.79	7.90
5930	1" iron pipe size		160	.100		2.91	3.12		6.03	8.15
5940	1-1/4" iron pipe size		155	.103		3.14	3.22		6.36	8.60
5950	1-1/2" iron pipe size		155	.103		3.29	3.22		6.51	8.75
5960	2" iron pipe size		150	.107		3.59	3.33		6.92	9.25
5970	2-1/2" iron pipe size		150	.107		3.96	3.33		7.29	9.65
5980	3" iron pipe size		145	.110		4.12	3.45		7.57	10.05
5990	4" iron pipe size		135	.119		4.79	3.70		8.49	11.15
6000	5" iron pipe size		130	.123		5.35	3.84		9.19	12
6010	6" iron pipe size		120	.133		5.50	4.16		9.66	12.65
6020	7" iron pipe size		115	.139		6.60	4.35		10.95	14.15
6030	8" iron pipe size		105	.152		7.45	4.76		12.21	15.75
6040	9" iron pipe size		100	.160		8.95	5		13.95	17.80
6050	10" iron pipe size		95	.168		9.90	5.25		15.15	19.25
6060	12" iron pipe size		90	.178		11.70	5.55		17.25	21.50
6070	14" iron pipe size		85	.188		13.25	5.90		19.15	24
6080	16" iron pipe size		80	.200		14.90	6.25		21.15	26.50
6090	18" iron pipe size		75	.213		17	6.65		23.65	29.50
6120	2" wall, 1/2" iron pipe size		145	.110		4.35	3.45		7.80	10.30
6140	3/4" iron pipe size		145	.110		4.35	3.45		7.80	10.30
6150	1" iron pipe size		145	.110		4.57	3.45		8.02	10.55
6160	1-1/4" iron pipe size		140	.114		5.10	3.57		8.67	11.30
6170	1-1/2" iron pipe size		140	.114		5.15	3.57		8.72	11.40
6180	2" iron pipe size		135	.119		5.40	3.70		9.10	11.85
6190	2-1/2" iron pipe size		135	.119		6.55	3.70		10.25	13.10
6200	3" iron pipe size		130	.123		6.70	3.84		10.54	13.50
6210	4" iron pipe size	▼	125	.128	▼	7.30	4		11.30	14.40

MECHANICAL 15

	15082	Mechanical Insulation	CREW	DAILY OUTPUT	LABOR-HOURS	UNIT	2003 BARE COSTS				TOTAL INCL O&P	
							MAT.	LABOR	EQUIP.	TOTAL		
200	6220	5" iron pipe size	Q-14	120	.133	L.F.	8.35	4.16		12.51	15.80	**200**
	6230	6" iron pipe size		115	.139		9.40	4.35		13.75	17.20	
	6240	7" iron pipe size		110	.145		10.20	4.54		14.74	18.50	
	6250	8" iron pipe size		105	.152		11.20	4.76		15.96	19.90	
	6260	9" iron pipe size		100	.160		12.65	5		17.65	22	
	6270	10" iron pipe size		95	.168		14	5.25		19.25	24	
	6280	12" iron pipe size		90	.178		15.65	5.55		21.20	26	
	6290	14" iron pipe size		85	.188		17.25	5.90		23.15	28.50	
	6300	16" iron pipe size		80	.200		19.65	6.25		25.90	31.50	
	6310	18" iron pipe size		75	.213		21.50	6.65		28.15	34	
	6320	20" iron pipe size		65	.246		25	7.70		32.70	40	
	6330	22" iron pipe size		60	.267		27	8.35		35.35	43	
	6340	24" iron pipe size		55	.291		28	9.10		37.10	45	
	6360	3" wall, 1/2" iron pipe size		115	.139		7.90	4.35		12.25	15.60	
	6380	3/4" iron pipe size		115	.139		7.90	4.35		12.25	15.60	
	6390	1" iron pipe size		115	.139		7.15	4.35		11.50	14.75	
	6400	1-1/4" iron pipe size		110	.145		7.35	4.54		11.89	15.35	
	6410	1-1/2" iron pipe size		110	.145		7.85	4.54		12.39	15.90	
	6420	2" iron pipe size		105	.152		8.35	4.76		13.11	16.70	
	6430	2-1/2" iron pipe size		105	.152		10	4.76		14.76	18.55	
	6440	3" iron pipe size		100	.160		10.15	5		15.15	19.10	
	6450	4" iron pipe size		95	.168		12.60	5.25		17.85	22.50	
	6460	5" iron pipe size		90	.178		13.70	5.55		19.25	24	
	6470	6" iron pipe size		90	.178		15.45	5.55		21	26	
	6480	7" iron pipe size		85	.188		17.65	5.90		23.55	29	
	6490	8" iron pipe size		85	.188		18.35	5.90		24.25	29.50	
	6500	9" iron pipe size		80	.200		20.50	6.25		26.75	33	
	6510	10" iron pipe size		75	.213		22	6.65		28.65	35	
	6520	12" iron pipe size		70	.229		24.50	7.15		31.65	38.50	
	6530	14" iron pipe size		65	.246		27.50	7.70		35.20	42.50	
	6540	16" iron pipe size		60	.267		30.50	8.35		38.85	47	
	6550	18" iron pipe size		55	.291		33.50	9.10		42.60	51	
	6560	20" iron pipe size		50	.320		37.50	10		47.50	57	
	6570	22" iron pipe size		45	.356		40.50	11.10		51.60	62	
	6580	24" iron pipe size	▼	40	.400	▼	43.50	12.50		56	68	
	6600	Fiberglass, with all service jacket										
	6640	1/2" wall, 1/2" iron pipe size	Q-14	250	.064	L.F.	1.10	2		3.10	4.39	
	6660	3/4" iron pipe size		240	.067		1.22	2.08		3.30	4.65	
	6670	1" iron pipe size		230	.070		1.31	2.17		3.48	4.90	
	6680	1-1/4" iron pipe size		220	.073		1.44	2.27		3.71	5.20	
	6690	1-1/2" iron pipe size		220	.073		1.53	2.27		3.80	5.30	
	6700	2" iron pipe size		210	.076		1.64	2.38		4.02	5.60	
	6710	2-1/2" iron pipe size		200	.080		1.88	2.50		4.38	6.05	
	6720	3" iron pipe size		190	.084		2.10	2.63		4.73	6.50	
	6730	3-1/2" iron pipe size		180	.089		2.25	2.78		5.03	6.90	
	6740	4" iron pipe size		160	.100		2.69	3.12		5.81	7.95	
	6750	5" iron pipe size		150	.107		3.09	3.33		6.42	8.70	
	6760	6" iron pipe size		120	.133		3.26	4.16		7.42	10.20	
	6840	1" wall, 1/2" iron pipe size		240	.067		1.31	2.08		3.39	4.75	
	6860	3/4" iron pipe size		230	.070		1.50	2.17		3.67	5.10	
	6870	1" iron pipe size		220	.073		1.53	2.27		3.80	5.30	
	6880	1-1/4" iron pipe size		210	.076		1.73	2.38		4.11	5.70	
	6890	1-1/2" iron pipe size		210	.076		1.86	2.38		4.24	5.85	
	6900	2" iron pipe size		200	.080		2.03	2.50		4.53	6.20	
	6910	2-1/2" iron pipe size		190	.084		2.24	2.63		4.87	6.65	
	6920	3" iron pipe size	▼	180	.089	▼	2.53	2.78		5.31	7.20	

15 MECHANICAL

Important: See the Reference Section for critical supporting data - Reference Nos., Crews, & City Cost Indexes

15050 | Basic Materials & Methods

15082	Mechanical Insulation	CREW	DAILY OUTPUT	LABOR-HOURS	UNIT	MAT.	LABOR	EQUIP.	TOTAL	TOTAL INCL O&P	
						2003 BARE COSTS					
200 6930	3-1/2" iron pipe size	Q-14	170	.094	L.F.	2.78	2.94		5.72	7.75	**200**
6940	4" iron pipe size		150	.107		3.26	3.33		6.59	8.90	
6950	5" iron pipe size		140	.114		3.75	3.57		7.32	9.85	
6960	6" iron pipe size		120	.133		3.99	4.16		8.15	11	
6970	7" iron pipe size		110	.145		4.90	4.54		9.44	12.65	
6980	8" iron pipe size		100	.160		5.70	5		10.70	14.20	
6990	9" iron pipe size		90	.178		6.20	5.55		11.75	15.65	
7000	10" iron pipe size		90	.178		6.65	5.55		12.20	16.20	
7010	12" iron pipe size		80	.200		7.55	6.25		13.80	18.30	
7020	14" iron pipe size		80	.200		8.75	6.25		15	19.60	
7030	16" iron pipe size		70	.229		10.40	7.15		17.55	23	
7040	18" iron pipe size		70	.229		12.70	7.15		19.85	25.50	
7050	20" iron pipe size		60	.267		14.15	8.35		22.50	29	
7060	24" iron pipe size		60	.267		17.30	8.35		25.65	32.50	
7080	1-1/2" wall, 1/2" iron pipe size		230	.070		2.66	2.17		4.83	6.40	
7100	3/4" iron pipe size		220	.073		2.78	2.27		5.05	6.65	
7110	1" iron pipe size		210	.076		2.92	2.38		5.30	7	
7120	1-1/4" iron pipe size		200	.080		3.14	2.50		5.64	7.40	
7130	1-1/2" iron pipe size		200	.080		3.31	2.50		5.81	7.60	
7140	2" iron pipe size		190	.084		3.61	2.63		6.24	8.15	
7150	2-1/2" iron pipe size		180	.089		3.97	2.78		6.75	8.80	
7160	3" iron pipe size		170	.094		4.09	2.94		7.03	9.20	
7170	3-1/2" iron pipe size		160	.100		4.48	3.12		7.60	9.90	
7180	4" iron pipe size		140	.114		4.69	3.57		8.26	10.85	
7190	5" iron pipe size		130	.123		5.20	3.84		9.04	11.85	
7200	6" iron pipe size		110	.145		5.35	4.54		9.89	13.15	
7210	7" iron pipe size		100	.160		6.20	5		11.20	14.75	
7220	8" iron pipe size		90	.178		6.70	5.55		12.25	16.25	
7230	9" iron pipe size		85	.188		7.65	5.90		13.55	17.80	
7240	10" iron pipe size		80	.200		8.45	6.25		14.70	19.25	
7250	12" iron pipe size		75	.213		9.20	6.65		15.85	21	
7260	14" iron pipe size		70	.229		10.65	7.15		17.80	23	
7270	16" iron pipe size		65	.246		12	7.70		19.70	25.50	
7280	18" iron pipe size		60	.267		15.50	8.35		23.85	30.50	
7290	20" iron pipe size		55	.291		16.40	9.10		25.50	32.50	
7300	24" iron pipe size		50	.320		19.20	10		29.20	37	
7320	2" wall, 1/2" iron pipe size		220	.073		4.17	2.27		6.44	8.20	
7340	3/4" iron pipe size		210	.076		4.31	2.38		6.69	8.55	
7350	1" iron pipe size		200	.080		4.61	2.50		7.11	9	
7360	1-1/4" iron pipe size		190	.084		4.87	2.63		7.50	9.55	
7370	1-1/2" iron pipe size		190	.084		5.15	2.63		7.78	9.85	
7380	2" iron pipe size		180	.089		5.35	2.78		8.13	10.25	
7390	2-1/2" iron pipe size		170	.094		5.75	2.94		8.69	11	
7400	3" iron pipe size		160	.100		6.20	3.12		9.32	11.75	
7410	3-1/2" iron pipe size		150	.107		6.70	3.33		10.03	12.65	
7420	4" iron pipe size		130	.123		7.10	3.84		10.94	13.90	
7430	5" iron pipe size		120	.133		8	4.16		12.16	15.40	
7440	6" iron pipe size		100	.160		8.25	5		13.25	17	
7450	7" iron pipe size		90	.178		9.45	5.55		15	19.25	
7460	8" iron pipe size		80	.200		10.20	6.25		16.45	21	
7470	9" iron pipe size		75	.213		11.25	6.65		17.90	23	
7480	10" iron pipe size		70	.229		12.20	7.15		19.35	25	
7490	12" iron pipe size		65	.246		13.40	7.70		21.10	27	
7500	14" iron pipe size		60	.267		15.15	8.35		23.50	30	
7510	16" iron pipe size		55	.291		17.25	9.10		26.35	33.50	
7520	18" iron pipe size		50	.320		20.50	10		30.50	38.50	

MECHANICAL 15

15082	Mechanical Insulation	CREW	DAILY OUTPUT	LABOR-HOURS	UNIT	2003 BARE COSTS				TOTAL INCL O&P
						MAT.	LABOR	EQUIP.	TOTAL	
200										**200**
7530	20" iron pipe size	Q-14	45	.356	L.F.	23	11.10		34.10	42.50
7540	24" iron pipe size	↓	40	.400	↓	25.50	12.50		38	48
7600	For fiberglass with standard canvas jacket, deduct					5%				
7660	For fittings, add 3 L.F. for each fitting									
7680	plus 4 L.F. for each flange of the fitting									
7700	For equipment or congested areas, add						20%			
7720	Finishes, for .010" aluminum jacket, add	Q-14	200	.080	S.F.	.33	2.50		2.83	4.33
7740	For .016" aluminum jacket, add	↓	200	.080		.42	2.50		2.92	4.43
7760	For .010" stainless steel, add	↓	160	.100	↓	1.23	3.12		4.35	6.30
7780	For single layer of felt, add					10%	10%			
7782	For roofing paper, 45 lb to 55 lb, add					25%	10%			
7784	Polyethylene tubing flexible closed cell foam, UV resistant									
7828	Standard temperature (-90° to +212° F)									
7830	3/8" wall, 1/8" iron pipe size	1 Asbe	130	.062	L.F.	.25	2.14		2.39	3.68
7831	1/4" iron pipe size		130	.062		.28	2.14		2.42	3.71
7832	3/8" iron pipe size		130	.062		.29	2.14		2.43	3.72
7833	1/2" iron pipe size		126	.063		.33	2.20		2.53	3.86
7834	3/4" iron pipe size		122	.066		.36	2.28		2.64	4.02
7835	1" iron pipe size		120	.067		.46	2.31		2.77	4.19
7836	1-1/4" iron pipe size		118	.068		.59	2.35		2.94	4.39
7837	1-1/2" iron pipe size		118	.068		.66	2.35		3.01	4.47
7838	2" iron pipe size		116	.069		.88	2.39		3.27	4.78
7839	2-1/2" iron pipe size		114	.070		1.20	2.44		3.64	5.20
7840	3" iron pipe size		112	.071		1.34	2.48		3.82	5.40
7842	1/2" wall, 1/8" iron pipe size		120	.067		.39	2.31		2.70	4.11
7843	1/4" iron pipe size		120	.067		.41	2.31		2.72	4.13
7844	3/8" iron pipe size		120	.067		.42	2.31		2.73	4.14
7845	1/2" iron pipe size		118	.068		.51	2.35		2.86	4.30
7846	3/4" iron pipe size		116	.069		.53	2.39		2.92	4.39
7847	1" iron pipe size		114	.070		.68	2.44		3.12	4.62
7848	1-1/4" iron pipe size		112	.071		.83	2.48		3.31	4.85
7849	1-1/2" iron pipe size		110	.073		1.01	2.52		3.53	5.10
7850	2" iron pipe size		108	.074		1.25	2.57		3.82	5.45
7851	2-1/2" iron pipe size		106	.075		1.62	2.62		4.24	5.95
7852	3" iron pipe size		104	.077		1.76	2.67		4.43	6.20
7853	3-1/2" iron pipe size		102	.078		2.55	2.72		5.27	7.15
7854	4" iron pipe size		100	.080		2.79	2.78		5.57	7.50
7855	3/4" wall, 1/8" iron pipe size		110	.073		.91	2.52		3.43	5
7856	1/4" iron pipe size		110	.073		.91	2.52		3.43	5
7857	3/8" iron pipe size		108	.074		1.05	2.57		3.62	5.25
7858	1/2" iron pipe size		106	.075		1.26	2.62		3.88	5.55
7859	3/4" iron pipe size		104	.077		1.50	2.67		4.17	5.90
7860	1" iron pipe size		102	.078		1.75	2.72		4.47	6.25
7861	1-1/4" iron pipe size		100	.080		2.37	2.78		5.15	7.05
7862	1-1/2" iron pipe size		100	.080		2.70	2.78		5.48	7.40
7863	2" iron pipe size		98	.082		3.22	2.83		6.05	8.05
7864	2-1/2" iron pipe size		96	.083		3.99	2.89		6.88	9
7865	3" iron pipe size		94	.085		5.10	2.95		8.05	10.35
7866	3-1/2" iron pipe size		92	.087		6.30	3.02		9.32	11.75
7867	4" iron pipe size		90	.089		6.70	3.08		9.78	12.30
7868	1" wall, 1/4" iron pipe size		100	.080		1.96	2.78		4.74	6.60
7869	3/8" iron pipe size		98	.082		1.96	2.83		4.79	6.65
7870	1/2" iron pipe size		96	.083		2.45	2.89		5.34	7.30
7871	3/4" iron pipe size		94	.085		2.48	2.95		5.43	7.45
7872	1" iron pipe size		92	.087		2.87	3.02		5.89	7.95
7873	1-1/4" iron pipe size	↓	90	.089	↓	4.02	3.08		7.10	9.35

Important: See the Reference Section for critical supporting data - Reference Nos., Crews, & City Cost Indexes

15082	Mechanical Insulation	CREW	DAILY OUTPUT	LABOR-HOURS	UNIT	2003 BARE COSTS				TOTAL INCL O&P
						MAT.	LABOR	EQUIP.	TOTAL	
200 7874	1-1/2" iron pipe size	1 Asbe	90	.089	L.F.	4.31	3.08		7.39	9.65 **200**
7875	2" iron pipe size		88	.091		5.65	3.15		8.80	11.25
7876	2-1/2" iron pipe size		86	.093		7.85	3.23		11.08	13.80
7877	3" iron pipe size	▼	84	.095	▼	8.20	3.30		11.50	14.30
7878	Contact cement, quart can				Ea.	18.55			18.55	20.50
7879	Rubber tubing, flexible closed cell foam									
7880	3/8" wall, 1/4" iron pipe size	1 Asbe	120	.067	L.F.	.28	2.31		2.59	3.99
7900	3/8" iron pipe size		120	.067		.31	2.31		2.62	4.02
7910	1/2" iron pipe size		115	.070		.35	2.41		2.76	4.23
7920	3/4" iron pipe size		115	.070		.40	2.41		2.81	4.28
7930	1" iron pipe size		110	.073		.49	2.52		3.01	4.55
7940	1-1/4" iron pipe size		110	.073		.60	2.52		3.12	4.67
7950	1-1/2" iron pipe size		110	.073		.72	2.52		3.24	4.80
8100	1/2" wall, 1/4" iron pipe size		90	.089		.60	3.08		3.68	5.55
8120	3/8" iron pipe size		90	.089		.66	3.08		3.74	5.65
8130	1/2" iron pipe size		89	.090		.51	3.12		3.63	5.50
8140	3/4" iron pipe size		89	.090		.57	3.12		3.69	5.60
8150	1" iron pipe size		88	.091		.63	3.15		3.78	5.70
8160	1-1/4" iron pipe size		87	.092		.77	3.19		3.96	5.95
8170	1-1/2" iron pipe size		87	.092		.93	3.19		4.12	6.10
8180	2" iron pipe size		86	.093		1.52	3.23		4.75	6.80
8190	2-1/2" iron pipe size		86	.093		1.68	3.23		4.91	7
8200	3" iron pipe size		85	.094		1.72	3.27		4.99	7.10
8210	3-1/2" iron pipe size		85	.094		2.30	3.27		5.57	7.75
8220	4" iron pipe size		80	.100		2.73	3.47		6.20	8.50
8230	5" iron pipe size		80	.100		3.75	3.47		7.22	9.65
8300	3/4" wall, 1/4" iron pipe size		90	.089		.67	3.08		3.75	5.65
8320	3/8" iron pipe size		90	.089		.74	3.08		3.82	5.70
8330	1/2" iron pipe size		89	.090		.81	3.12		3.93	5.85
8340	3/4" iron pipe size		89	.090		.97	3.12		4.09	6.05
8350	1" iron pipe size		88	.091		1.18	3.15		4.33	6.30
8360	1-1/4" iron pipe size		87	.092		1.36	3.19		4.55	6.60
8370	1-1/2" iron pipe size		87	.092		2.07	3.19		5.26	7.40
8380	2" iron pipe size		86	.093		2.43	3.23		5.66	7.80
8390	2-1/2" iron pipe size		86	.093		2.85	3.23		6.08	8.30
8400	3" iron pipe size		85	.094		3.19	3.27		6.46	8.70
8410	3-1/2" iron pipe size		85	.094		3.90	3.27		7.17	9.50
8420	4" iron pipe size		80	.100		4.64	3.47		8.11	10.60
8430	5" iron pipe size		80	.100		5.65	3.47		9.12	11.75
8440	6" iron pipe size		80	.100		6.85	3.47		10.32	13.05
8444	1" wall, 1/2" iron pipe size		86	.093		1.63	3.23		4.86	6.95
8445	3/4" iron pipe size		84	.095		1.82	3.30		5.12	7.25
8446	1" iron pipe size		84	.095		2.21	3.30		5.51	7.70
8447	1-1/4" iron pipe size		82	.098		2.57	3.39		5.96	8.25
8448	1-1/2" iron pipe size		82	.098		3.38	3.39		6.77	9.10
8449	2" iron pipe size		80	.100		4.70	3.47		8.17	10.65
8450	2-1/2" iron pipe size	▼	80	.100	▼	6.15	3.47		9.62	12.25
8456	Rubber insulation tape, 1/8" x 2" x 30'				Ea.	10.80			10.80	11.90
8460	Polyolefin tubing, flexible closed cell foam, UV stabilized, work									
8462	temp.-165° F to +210° F, 0 water vapor transmission.									
8464	3/8" wall, 1/8" iron pipe size	1 Asbe	140	.057	L.F.	.26	1.98		2.24	3.44
8466	1/4" iron pipe size		140	.057		.27	1.98		2.25	3.45
8468	3/8" iron pipe size		140	.057		.30	1.98		2.28	3.48
8470	1/2" iron pipe size		136	.059		.33	2.04		2.37	3.61
8472	3/4" iron pipe size		132	.061		.38	2.10		2.48	3.77
8474	1" iron pipe size	▼	130	.062	▼	.42	2.14		2.56	3.86

MECHANICAL 15

	15082	Mechanical Insulation	CREW	DAILY OUTPUT	LABOR-HOURS	UNIT	| 2003 BARE COSTS				TOTAL INCL O&P	
							MAT.	LABOR	EQUIP.	TOTAL		
200	8476	1-1/4" iron pipe size	1 Asbe	128	.063	L.F.	.55	2.17		2.72	4.06	200
	8478	1-1/2" iron pipe size		128	.063		.67	2.17		2.84	4.19	
	8480	2" iron pipe size		126	.063		.75	2.20		2.95	4.33	
	8482	2-1/2" iron pipe size		123	.065		1.08	2.26		3.34	4.78	
	8484	3" iron pipe size		121	.066		1.68	2.29		3.97	5.50	
	8486	4" iron pipe size		118	.068		2.34	2.35		4.69	6.30	
	8500	1/2" wall, 1/8" iron pipe size		130	.062		.38	2.14		2.52	3.82	
	8502	1/4" iron pipe size		130	.062		.41	2.14		2.55	3.85	
	8504	3/8" iron pipe size		130	.062		.44	2.14		2.58	3.88	
	8506	1/2" iron pipe size		128	.063		.48	2.17		2.65	3.98	
	8508	3/4" iron pipe size		126	.063		.55	2.20		2.75	4.11	
	8510	1" iron pipe size		123	.065		.60	2.26		2.86	4.25	
	8512	1-1/4" iron pipe size		121	.066		.75	2.29		3.04	4.48	
	8514	1-1/2" iron pipe size		119	.067		.89	2.33		3.22	4.69	
	8516	2" iron pipe size		117	.068		1.05	2.37		3.42	4.93	
	8518	2-1/2" iron pipe size		114	.070		1.42	2.44		3.86	5.45	
	8520	3" iron pipe size		112	.071		1.99	2.48		4.47	6.15	
	8522	4" iron pipe size		110	.073		2.46	2.52		4.98	6.70	
	8534	3/4" wall, 1/8" iron pipe size		120	.067		.58	2.31		2.89	4.32	
	8536	1/4" iron pipe size		120	.067		.61	2.31		2.92	4.35	
	8538	3/8" iron pipe size		117	.068		.70	2.37		3.07	4.54	
	8540	1/2" iron pipe size		114	.070		.78	2.44		3.22	4.73	
	8542	3/4" iron pipe size		112	.071		1.01	2.48		3.49	5.05	
	8544	1" iron pipe size		110	.073		1.18	2.52		3.70	5.30	
	8546	1-1/4" iron pipe size		108	.074		1.58	2.57		4.15	5.85	
	8548	1-1/2" iron pipe size		108	.074		1.84	2.57		4.41	6.10	
	8550	2" iron pipe size		106	.075		2.16	2.62		4.78	6.55	
	8552	2-1/2" iron pipe size		104	.077		2.71	2.67		5.38	7.25	
	8554	3" iron pipe size		102	.078		3.50	2.72		6.22	8.20	
	8556	4" iron pipe size		100	.080		4.31	2.78		7.09	9.15	
	8570	1" wall, 1/8" iron pipe size		110	.073		1.04	2.52		3.56	5.15	
	8572	1/4" iron pipe size		108	.074		1.07	2.57		3.64	5.25	
	8574	3/8" iron pipe size		106	.075		1.09	2.62		3.71	5.35	
	8576	1/2" iron pipe size		104	.077		1.16	2.67		3.83	5.55	
	8578	3/4" iron pipe size		102	.078		1.52	2.72		4.24	6	
	8580	1" iron pipe size		100	.080		1.76	2.78		4.54	6.35	
	8582	1-1/4" iron pipe size		97	.082		1.99	2.86		4.85	6.75	
	8584	1-1/2" iron pipe size		97	.082		2.28	2.86		5.14	7.05	
	8586	2" iron pipe size		95	.084		3	2.92		5.92	7.95	
	8588	2-1/2" iron pipe size		93	.086		3.90	2.98		6.88	9.05	
	8590	3" iron pipe size	▼	91	.088	▼	4.27	3.05		7.32	9.55	
	8600	Seam Roller				Ea.	6.85			6.85	7.55	
	8602	Fuse Seal applicator (Butane)					176			176	193	
	8604	Fuse Seal 1/2" sticks				▼	12.90			12.90	14.20	
	8606	Leaktite adhesive				Qt.	22			22	24	
	8608	Leaktite adhesive				Gal.	61.50			61.50	68	
	8610	NOTE: Preslit/preglued vs unslit, same price										

15 MECHANICAL

Important: See the Reference Section for critical supporting data - Reference Nos., Crews, & City Cost Indexes

15107 | Metal Pipe & Fittings

		CREW	DAILY OUTPUT	LABOR-HOURS	UNIT	2003 BARE COSTS				TOTAL INCL O&P	
						MAT.	LABOR	EQUIP.	TOTAL		
220	0010	**PIPE, BRASS** Plain end,										220
	0900	Field threaded, coupling & clevis hanger 10' O.C.										
	0920	Regular weight										
	0980	1/8" diameter	1 Plum	62	.129	L.F.	2.58	4.82		7.40	10.15	
	1000	1/4" diameter		57	.140		3.21	5.25		8.46	11.45	
	1100	3/8" diameter		52	.154		3.82	5.75		9.57	12.90	
	1120	1/2" diameter		48	.167		4.55	6.25		10.80	14.40	
	1140	3/4" diameter		46	.174		6.05	6.50		12.55	16.45	
	1160	1" diameter		43	.186		8.60	6.95		15.55	19.95	
	1180	1-1/4" diameter	Q-1	72	.222		12.85	7.45		20.30	25.50	
	1200	1-1/2" diameter		65	.246		15.10	8.30		23.40	29	
	1220	2" diameter		53	.302		20.50	10.15		30.65	38	
	1240	2-1/2" diameter		41	.390		31	13.10		44.10	54	
	1260	3" diameter		31	.516		44	17.35		61.35	74	
	1280	3-1/2" diameter	Q-2	39	.615		72.50	21.50		94	113	
	1300	4" diameter	"	37	.649		70.50	22.50		93	112	
	1930	To delete coupling & hanger, subtract										
	1940	1/8" diam. to 1/2" diam.					14%	46%				
	1950	3/4" diam. to 1-1/2" diam.					8%	47%				
	1960	2" diam. to 4" diam.					10%	37%				
260	0010	**PIPE, BRASS, FITTINGS** Rough bronze, threaded										260
	1000	Standard wt., 90° Elbow										
	1040	1/8"	1 Plum	13	.615	Ea.	7.35	23		30.35	42.50	
	1060	1/4"		13	.615		7.35	23		30.35	42.50	
	1080	3/8"		13	.615		7.35	23		30.35	42.50	
	1100	1/2"		12	.667		7.35	25		32.35	45.50	
	1120	3/4"		11	.727		9.80	27		36.80	52	
	1140	1"		10	.800		15.95	30		45.95	62.50	
	1160	1-1/4"	Q-1	17	.941		26	31.50		57.50	76.50	
	1180	1-1/2"		16	1		32	33.50		65.50	86	
	1200	2"		14	1.143		51.50	38.50		90	115	
	1220	2-1/2"		11	1.455		125	49		174	211	
	1240	3"		8	2		190	67.50		257.50	310	
	1260	4"	Q-2	11	2.182		385	76		461	540	
	1280	5"		8	3		1,075	105		1,180	1,325	
	1300	6"		7	3.429		1,575	120		1,695	1,900	
	1500	45° Elbow, 1/8"	1 Plum	13	.615		9	23		32	44.50	
	1540	1/4"		13	.615		9	23		32	44.50	
	1560	3/8"		13	.615		9	23		32	44.50	
	1580	1/2"		12	.667		9	25		34	47.50	
	1600	3/4"		11	.727		12.80	27		39.80	55	
	1620	1"		10	.800		22	30		52	69	
	1640	1-1/4"	Q-1	17	.941		34.50	31.50		66	86	
	1660	1-1/2"		16	1		43.50	33.50		77	99	
	1680	2"		14	1.143		70.50	38.50		109	136	
	1700	2-1/2"		11	1.455		135	49		184	222	
	1720	3"		8	2		208	67.50		275.50	330	
	1740	4"	Q-2	11	2.182		475	76		551	640	
	1760	5"		8	3		880	105		985	1,125	
	1780	6"		7	3.429		1,225	120		1,345	1,500	
	2000	Tee, 1/8"	1 Plum	9	.889		8.60	33		41.60	59.50	
	2040	1/4"		9	.889		8.60	33		41.60	59.50	
	2060	3/8"		9	.889		8.60	33		41.60	59.50	
	2080	1/2"		8	1		8.60	37.50		46.10	66	
	2100	3/4"		7	1.143		12.25	42.50		54.75	78	
	2120	1"		6	1.333		22	50		72	99.50	

15 MECHANICAL

260	15107	Metal Pipe & Fittings	CREW	DAILY OUTPUT	LABOR-HOURS	UNIT	2003 BARE COSTS				TOTAL INCL O&P	260
							MAT.	LABOR	EQUIP.	TOTAL		
	2140	1-1/4"	Q-1	10	1.600	Ea.	38	54		92	124	
	2160	1-1/2"		9	1.778		43	60		103	138	
	2180	2"		8	2		71	67.50		138.50	181	
	2200	2-1/2"		7	2.286		170	77		247	300	
	2220	3"		5	3.200		259	108		367	450	
	2240	4"	Q-2	7	3.429		645	120		765	885	
	2260	5"		5	4.800		1,350	167		1,517	1,725	
	2280	6"		4	6		2,100	209		2,309	2,625	
	2500	Coupling, 1/8"	1 Plum	26	.308		6.15	11.50		17.65	24	
	2540	1/4"		22	.364		6.15	13.60		19.75	27.50	
	2560	3/8"		18	.444		6.15	16.60		22.75	32	
	2580	1/2"		15	.533		6.15	19.90		26.05	37	
	2600	3/4"		14	.571		8.60	21.50		30.10	41.50	
	2620	1"		13	.615		14.70	23		37.70	50.50	
	2640	1-1/4"	Q-1	22	.727		24.50	24.50		49	64	
	2660	1-1/2"		20	.800		32	27		59	75.50	
	2680	2"		18	.889		53	30		83	103	
	2700	2-1/2"		14	1.143		90	38.50		128.50	157	
	2720	3"		10	1.600		125	54		179	219	
	2740	4"	Q-2	12	2		257	69.50		326.50	385	
	2760	5"		10	2.400		480	83.50		563.50	655	
	2780	6"		9	2.667		685	93		778	890	
	3000	Union, 125 lb										
	3020	1/8"	1 Plum	12	.667	Ea.	21.50	25		46.50	61.50	
	3040	1/4"		12	.667		21.50	25		46.50	61.50	
	3060	3/8"		12	.667		21.50	25		46.50	61.50	
	3080	1/2"		11	.727		21.50	27		48.50	65	
	3100	3/4"		10	.800		30	30		60	78	
	3120	1"		9	.889		39.50	33		72.50	93.50	
	3140	1-1/4"	Q-1	16	1		57	33.50		90.50	114	
	3160	1-1/2"		15	1.067		68	36		104	129	
	3180	2"		13	1.231		105	41.50		146.50	178	
	3200	2-1/2"		10	1.600		274	54		328	380	
	3220	3"		7	2.286		430	77		507	585	
	3240	4"	Q-2	10	2.400		1,075	83.50		1,158.50	1,300	
	3320	For 250 lb. (navy pattern), add					100%					
420	0010	**PIPE, COPPER** Solder joints	R15100 -050									420
	0100	Solder										
	0120	Solder, lead free, roll				Lb.	7.85			7.85	8.60	
	1000	Type K tubing, couplings & clevis hangers 10' O.C.										
	1100	1/4" diameter	1 Plum	84	.095	L.F.	1.35	3.56		4.91	6.85	
	1120	3/8" diameter		82	.098		1.39	3.64		5.03	7.05	
	1140	1/2" diameter		78	.103		1.49	3.83		5.32	7.45	
	1160	5/8" diameter		77	.104		1.72	3.88		5.60	7.75	
	1180	3/4" diameter		74	.108		2.20	4.04		6.24	8.50	
	1200	1" diameter		66	.121		2.80	4.53		7.33	9.95	
	1220	1-1/4" diameter		56	.143		3.36	5.35		8.71	11.75	
	1240	1-1/2" diameter		50	.160		4.27	6		10.27	13.70	
	1260	2" diameter		40	.200		6.30	7.45		13.75	18.25	
	1280	2-1/2" diameter	Q-1	60	.267		9.40	8.95		18.35	24	
	1300	3" diameter		54	.296		12.95	9.95		22.90	29.50	
	1320	3-1/2" diameter		42	.381		17.60	12.80		30.40	39	
	1330	4" diameter		38	.421		21.50	14.15		35.65	45.50	
	1340	5" diameter		32	.500		52	16.80		68.80	82.50	
	1360	6" diameter	Q-2	38	.632		75.50	22		97.50	117	
	1380	8" diameter	"	34	.706		129	24.50		153.50	179	

Important: See the Reference Section for critical supporting data - Reference Nos., Crews, & City Cost Indexes

15107	Metal Pipe & Fittings		CREW	DAILY OUTPUT	LABOR-HOURS	UNIT	2003 BARE COSTS				TOTAL INCL O&P
							MAT.	LABOR	EQUIP.	TOTAL	
420	1390	For other than full hard temper, add					L.F.	13%				420
	1440	For silver solder, add	R15100 -050						15%			
	1800	For medical clean, (oxygen class), add						12%				
	1950	To delete cplgs. & hngrs., 1/4"-1" pipe, subtract						27%	60%			
	1960	1-1/4"-3" pipe, subtract						14%	52%			
	1970	3-1/2"-5" pipe, subtract						10%	60%			
	1980	6"-8" pipe, subtract						19%	53%			
	2000	Type L tubing, couplings & hangers 10' O.C.										
	2100	1/4" diameter		1 Plum	88	.091	L.F.	1.03	3.40		4.43	6.30
	2120	3/8" diameter			84	.095		1.22	3.56		4.78	6.70
	2140	1/2" diameter			81	.099		1.35	3.69		5.04	7.05
	2160	5/8" diameter			79	.101		1.67	3.78		5.45	7.55
	2180	3/4" diameter			76	.105		1.77	3.93		5.70	7.90
	2200	1" diameter			68	.118		2.34	4.39		6.73	9.25
	2220	1-1/4" diameter			58	.138		3	5.15		8.15	11.10
	2240	1-1/2" diameter			52	.154		3.68	5.75		9.43	12.75
	2260	2" diameter			42	.190		5.45	7.10		12.55	16.75
	2280	2-1/2" diameter		Q-1	62	.258		8.25	8.70		16.95	22
	2300	3" diameter			56	.286		11	9.60		20.60	26.50
	2320	3-1/2" diameter			43	.372		15.20	12.50		27.70	35.50
	2340	4" diameter			39	.410		18.40	13.80		32.20	41.50
	2360	5" diameter			34	.471		47.50	15.85		63.35	76
	2380	6" diameter		Q-2	40	.600		61	21		82	98.50
	2400	8" diameter		"	36	.667		101	23.50		124.50	146
	2410	For other than full hard temper, add						21%				
	2590	For silver solder, add							15%			
	2900	For medical clean, (oxygen class), add						12%				
	2940	To delete cplgs. & hngrs., 1/4"-1" pipe, subtract						37%	63%			
	2960	1-1/4"-3" pipe, subtract						12%	53%			
	2970	3-1/2"-5" pipe, subtract						12%	63%			
	2980	6"-8" pipe, subtract						24%	55%			
	3000	Type M tubing, couplings & hangers 10' O.C.										
	3100	1/4" diameter		1 Plum	90	.089	L.F.	1.08	3.32		4.40	6.20
	3120	3/8" diameter			87	.092		1.09	3.43		4.52	6.40
	3140	1/2" diameter			84	.095		1.09	3.56		4.65	6.55
	3160	5/8" diameter			81	.099		1.42	3.69		5.11	7.10
	3180	3/4" diameter			78	.103		1.48	3.83		5.31	7.45
	3200	1" diameter			70	.114		1.93	4.27		6.20	8.55
	3220	1-1/4" diameter			60	.133		2.56	4.98		7.54	10.30
	3240	1-1/2" diameter			54	.148		3.27	5.55		8.82	11.95
	3260	2" diameter			44	.182		4.83	6.80		11.63	15.55
	3280	2-1/2" diameter		Q-1	64	.250		7.15	8.40		15.55	20.50
	3300	3" diameter			58	.276		9.50	9.30		18.80	24.50
	3320	3-1/2" diameter			45	.356		13.55	11.95		25.50	33
	3340	4" diameter			40	.400		18.40	13.45		31.85	41
	3360	5" diameter			36	.444		47.50	14.95		62.45	74.50
	3370	6" diameter		Q-2	42	.571		61	19.95		80.95	97
	3380	8" diameter		"	38	.632		84.50	22		106.50	127
	3440	For silver solder, add							15%			
	3960	To delete cplgs. & hngrs., 1/4"-1" pipe, subtract						35%	65%			
	3970	1-1/4"-3" pipe, subtract						19%	56%			
	3980	3-1/2"-5" pipe, subtract						13%	65%			
	3990	6"-8" pipe, subtract						28%	58%			
	4000	Type DWV tubing, couplings & hangers 10' O.C.										
	4100	1-1/4" diameter		1 Plum	60	.133	L.F.	2.56	4.98		7.54	10.30
	4120	1-1/2" diameter			54	.148		3.06	5.55		8.61	11.70

MECHANICAL 15

		CREW	DAILY OUTPUT	LABOR-HOURS	UNIT	2003 BARE COSTS				TOTAL INCL O&P
						MAT.	LABOR	EQUIP.	TOTAL	
420	4140	2" diameter	1 Plum	44	.182	L.F.	3.95	6.80		10.75	14.60	**420**
	4160	3" diameter	Q-1	58	.276		7.15	9.30		16.45	22
	4180	4" diameter		40	.400		12.15	13.45		25.60	34
	4200	5" diameter		36	.444		34	14.95		48.95	59.50
	4220	6" diameter	Q-2	42	.571		48	19.95		67.95	83
	4240	8" diameter	"	38	.632		114	22		136	159
	4730	To delete cplgs. & hngrs., 1-1/4"-2" pipe, subtract					16%	53%			
	4740	2-1/2"-4" pipe, subtract					13%	60%			
	4750	5"-8" pipe, subtract					23%	58%			
	5200	ACR tubing, type L, hard temper, cleaned and									
	5220	capped, no couplings or hangers									
	5240	3/8" OD				L.F.	.42			.42	.46
	5250	1/2" OD					.63			.63	.69
	5260	5/8" OD					.80			.80	.88
	5270	3/4" OD					1.12			1.12	1.23
	5280	7/8" OD					1.25			1.25	1.38
	5290	1-1/8" OD					1.80			1.80	1.98
	5300	1-3/8" OD					2.44			2.44	2.68
	5310	1-5/8" OD					3.12			3.12	3.43
	5320	2-1/8" OD					4.91			4.91	5.40
	5330	2-5/8" OD					7.30			7.30	8.05
	5340	3-1/8" OD					9.85			9.85	10.80
	5350	3-5/8" OD					16.35			16.35	18
	5360	4-1/8" OD					16.35			16.35	18
	5380	ACR tubing, type L, annealed, cleaned and capped									
	5381	No couplings or hangers									
	5384	3/8"	1 Stpi	160	.050	L.F.	.42	1.88		2.30	3.30
	5385	1/2"		160	.050		.63	1.88		2.51	3.53
	5386	5/8"		160	.050		.80	1.88		2.68	3.72
	5387	3/4"		130	.062		1.12	2.31		3.43	4.73
	5388	7/8"		130	.062		1.25	2.31		3.56	4.88
	5389	1-1/8"		115	.070		1.80	2.62		4.42	5.95
	5390	1-3/8"		100	.080		2.44	3.01		5.45	7.20
	5391	1-5/8"		90	.089		3.12	3.34		6.46	8.50
	5392	2-1/8"		80	.100		4.91	3.76		8.67	11.10
	5393	2-5/8"	Q-5	125	.128		7.30	4.33		11.63	14.60
	5394	3-1/8"		105	.152		9.85	5.15		15	18.60
	5395	4-1/8"		95	.168		16.35	5.70		22.05	26.50
	5800	Refrigeration tubing, dryseal, 50' coils									
	5840	1/8" OD				Coil	10.65			10.65	11.70
	5850	3/16" OD					11.65			11.65	12.80
	5860	1/4" OD					13.90			13.90	15.30
	5870	5/16" OD					17.90			17.90	19.70
	5880	3/8" OD					20			20	22
	5890	1/2" OD					26			26	28.50
	5900	5/8" OD					35			35	38.50
	5910	3/4" OD					41			41	45
	5920	7/8" OD					60.50			60.50	66.50
	5930	1-1/8" OD					88.50			88.50	97.50
	5940	1-3/8" OD					130			130	143
	5950	1-5/8" OD					154			154	170
	9400	Sub assemblies used in assembly systems									
	9410	Chilled water unit, coil connections per unit under 10 ton	Q-5	.80	20	System	505	675		1,180	1,575
	9420	Chilled water unit, coil connections per unit 10 ton and up		1	16		935	540		1,475	1,850
	9430	Chilled water dist. piping per ton, less than 61 ton systems		26	.615		11.40	21		32.40	44
	9440	Chilled water dist. piping per ton, 61 through 120 ton systems	Q-6	31	.774		21.50	27		48.50	64.50

R15100 -050

15 MECHANICAL

15107	Metal Pipe & Fittings		CREW	DAILY OUTPUT	LABOR-HOURS	UNIT	MAT.	LABOR	EQUIP.	TOTAL	TOTAL INCL O&P	
							2003 BARE COSTS					
420	9450	Chilled water dist. piping/ton, 135 ton systems and up	R15100 -050	Q-8	25.40	1.260	System	28	45	3.09	76.09	102	**420**
	9510	Refrigerant piping/ton of cooling for remote condensers		Q-5	2	8		109	271		380	530	
	9520	Refrigerant piping per ton up to 10 ton w/remote condensing unit			2.40	6.667		60	226		286	405	
	9530	Refrigerant piping per ton, 20 ton w/remote condensing unit			2	8		79	271		350	495	
	9540	Refrigerant piping per ton, 40 ton w/remote condensing unit			1.90	8.421		100	285		385	540	
	9550	Refrigerant piping per ton, 75-80 ton w/remote condensing unit		Q-6	2.40	10		142	350		492	685	
	9560	Refrigerant piping per ton, 100 ton w/remote condensing unit		"	2.20	10.909		185	385		570	785	
460	0010	**PIPE, COPPER, FITTINGS** Wrought unless otherwise noted											**460**
	0040	Solder joints, copper x copper											
	0070	90° elbow, 1/4"		1 Plum	22	.364	Ea.	1.38	13.60		14.98	22	
	0090	3/8"			22	.364		1.38	13.60		14.98	22	
	0100	1/2"			20	.400		.48	14.95		15.43	23	
	0110	5/8"			19	.421		2.64	15.75		18.39	27	
	0120	3/4"			19	.421		1.04	15.75		16.79	25	
	0130	1"			16	.500		2.54	18.70		21.24	31	
	0140	1-1/4"			15	.533		3.84	19.90		23.74	34	
	0150	1-1/2"			13	.615		6	23		29	41	
	0160	2"			11	.727		10.90	27		37.90	53	
	0170	2-1/2"		Q-1	13	1.231		21	41.50		62.50	85.50	
	0180	3"			11	1.455		29	49		78	106	
	0190	3-1/2"			10	1.600		96.50	54		150.50	188	
	0200	4"			9	1.778		74.50	60		134.50	173	
	0210	5"			6	2.667		274	89.50		363.50	435	
	0220	6"		Q-2	9	2.667		385	93		478	560	
	0230	8"		"	8	3		1,550	105		1,655	1,850	
	0250	45° elbow, 1/4"		1 Plum	22	.364		2.71	13.60		16.31	23.50	
	0270	3/8"			22	.364		2.21	13.60		15.81	23	
	0280	1/2"			20	.400		.85	14.95		15.80	23.50	
	0290	5/8"			19	.421		4.22	15.75		19.97	28.50	
	0300	3/4"			19	.421		1.48	15.75		17.23	25.50	
	0310	1"			16	.500		3.72	18.70		22.42	32	
	0320	1-1/4"			15	.533		5.05	19.90		24.95	35.50	
	0330	1-1/2"			13	.615		6.05	23		29.05	41	
	0340	2"			11	.727		10.10	27		37.10	52	
	0350	2-1/2"		Q-1	13	1.231		21.50	41.50		63	86	
	0360	3"			13	1.231		32	41.50		73.50	97.50	
	0370	3-1/2"			10	1.600		52.50	54		106.50	139	
	0380	4"			9	1.778		68	60		128	166	
	0390	5"			6	2.667		263	89.50		352.50	425	
	0400	6"		Q-2	9	2.667		395	93		488	575	
	0410	8"		"	8	3		1,400	105		1,505	1,700	
	0450	Tee, 1/4"		1 Plum	14	.571		3.05	21.50		24.55	35.50	
	0470	3/8"			14	.571		2.33	21.50		23.83	34.50	
	0480	1/2"			13	.615		.77	23		23.77	35.50	
	0490	5/8"			12	.667		5.05	25		30.05	43	
	0500	3/4"			12	.667		1.90	25		26.90	39.50	
	0510	1"			10	.800		5.85	30		35.85	51.50	
	0520	1-1/4"			9	.889		8.90	33		41.90	60	
	0530	1-1/2"			8	1		12.35	37.50		49.85	70	
	0540	2"			7	1.143		19.25	42.50		61.75	85.50	
	0550	2-1/2"		Q-1	8	2		39	67.50		106.50	145	
	0560	3"			7	2.286		59	77		136	181	
	0570	3-1/2"			6	2.667		172	89.50		261.50	325	
	0580	4"			5	3.200		143	108		251	320	
	0590	5"			4	4		445	135		580	695	

MECHANICAL 15

			DAILY	LABOR-			2003 BARE COSTS				TOTAL
15107	**Metal Pipe & Fittings**	CREW	OUTPUT	HOURS	UNIT	MAT.	LABOR	EQUIP.	TOTAL		INCL O&P
460	0600	6"	Q-2	6	4	Ea.	610	139		749	880	**460**
	0610	8"	"	5	4.800		2,475	167		2,642	2,975
	0612	Tee, reducing on the outlet, 1/4"	1 Plum	15	.533		4.91	19.90		24.81	35.50
	0613	3/8"		15	.533		4.62	19.90		24.52	35
	0614	1/2"		14	.571		4.05	21.50		25.55	36.50
	0615	5/8"		13	.615		7.80	23		30.80	43
	0616	3/4"		12	.667		1.84	25		26.84	39.50
	0617	1"		11	.727		5.85	27		32.85	47.50
	0618	1-1/4"		10	.800		8.65	30		38.65	54.50
	0619	1-1/2"		9	.889		9.40	33		42.40	60.50
	0620	2"	↓	8	1		14.50	37.50		52	72.50
	0621	2-1/2"	Q-1	9	1.778		45	60		105	141
	0622	3"		8	2		50	67.50		117.50	157
	0623	4"		6	2.667		97	89.50		186.50	241
	0624	5"	↓	5	3.200		445	108		553	655
	0625	6"	Q-2	7	3.429		610	120		730	850
	0626	8"	"	6	4		2,475	139		2,614	2,925
	0630	Tee, reducing on the run, 1/4"	1 Plum	15	.533		6.15	19.90		26.05	37
	0631	3/8"		15	.533		6.95	19.90		26.85	37.50
	0632	1/2"		14	.571		5.55	21.50		27.05	38
	0633	5/8"		13	.615		8.20	23		31.20	43.50
	0634	3/4"		12	.667		2.25	25		27.25	40
	0635	1"		11	.727		7	27		34	48.50
	0636	1-1/4"		10	.800		11.35	30		41.35	57.50
	0637	1-1/2"		9	.889		19.55	33		52.55	71.50
	0638	2"	↓	8	1		25	37.50		62.50	84
	0639	2-1/2"	Q-1	9	1.778		58.50	60		118.50	155
	0640	3"		8	2		86	67.50		153.50	197
	0641	4"		6	2.667		191	89.50		280.50	345
	0642	5"	↓	5	3.200		445	108		553	655
	0643	6"	Q-2	7	3.429		610	120		730	850
	0644	8"	"	6	4		2,475	139		2,614	2,925
	0650	Coupling, 1/4"	1 Plum	24	.333		.35	12.45		12.80	19.20
	0670	3/8"		24	.333		.45	12.45		12.90	19.30
	0680	1/2"		22	.364		.35	13.60		13.95	21
	0690	5/8"		21	.381		1.04	14.25		15.29	22.50
	0700	3/4"		21	.381		.68	14.25		14.93	22.50
	0710	1"		18	.444		1.42	16.60		18.02	26.50
	0715	1-1/4"		17	.471		2.48	17.60		20.08	29
	0716	1-1/2"		15	.533		3.28	19.90		23.18	33.50
	0718	2"	↓	13	.615		5.45	23		28.45	40.50
	0721	2-1/2"	Q-1	15	1.067		11.60	36		47.60	67
	0722	3"		13	1.231		17.45	41.50		58.95	81.50
	0724	3-1/2"		8	2		33.50	67.50		101	139
	0726	4"		7	2.286		35	77		112	155
	0728	5"	↓	6	2.667		81.50	89.50		171	225
	0731	6"	Q-2	8	3		128	105		233	299
	0732	8"	"	7	3.429		440	120		560	660
	0741	Coupling, reducing, concentric									
	0743	1/2"	1 Plum	23	.348	Ea.	.79	13		13.79	20.50
	0745	3/4"		21.50	.372		1.45	13.90		15.35	22.50
	0747	1"		19.50	.410		2.69	15.30		17.99	26
	0749	1-1/2"		16	.500		3.34	18.70		22.04	31.50
	0751	2"	↓	14	.571		8.80	21.50		30.30	41.50
	0753	3"	Q-1	14	1.143		26.50	38.50		65	87
	0755	4"	"	8	2		56	67.50		123.50	164

15107	Metal Pipe & Fittings	CREW	DAILY OUTPUT	LABOR-HOURS	UNIT	MAT.	LABOR	EQUIP.	TOTAL	TOTAL INCL O&P	
						2003 BARE COSTS					
460 0757	5"	Q-2	7.50	3.200	Ea.	176	112		288	360	**460**
0759	6"		7	3.429		305	120		425	515	
0761	8"		6.50	3.692		500	129		629	745	
0771	Cap, sweat										
0773	1/2"	1 Plum	40	.200	Ea.	.34	7.45		7.79	11.65	
0775	3/4"		38	.211		.58	7.85		8.43	12.50	
0777	1"		32	.250		1.41	9.35		10.76	15.65	
0779	1-1/2"		26	.308		2.89	11.50		14.39	20.50	
0781	2"		22	.364		5.30	13.60		18.90	26.50	
0791	Flange, sweat										
0793	3"	Q-1	22	.727	Ea.	48	24.50		72.50	90	
0795	4"		18	.889		68.50	30		98.50	121	
0797	5"		12	1.333		132	45		177	213	
0799	6"	Q-2	18	1.333		138	46.50		184.50	222	
0801	8"	"	16	1.500		440	52.50		492.50	560	
0850	Unions, 1/4"	1 Plum	21	.381		8.70	14.25		22.95	31	
0870	3/8"		21	.381		8.30	14.25		22.55	30.50	
0880	1/2"		19	.421		4.75	15.75		20.50	29.50	
0890	5/8"		18	.444		19.80	16.60		36.40	47	
0900	3/4"		18	.444		5.95	16.60		22.55	31.50	
0910	1"		15	.533		10.60	19.90		30.50	41.50	
0920	1-1/4"		14	.571		18.30	21.50		39.80	52	
0930	1-1/2"		12	.667		24	25		49	64	
0940	2"		10	.800		41	30		71	90.50	
0950	2-1/2"	Q-1	12	1.333		85.50	45		130.50	162	
0960	3"	"	10	1.600		221	54		275	325	
0980	Adapter, copper x male IPS, 1/4"	1 Plum	20	.400		4.80	14.95		19.75	28	
0990	3/8"		20	.400		2.38	14.95		17.33	25	
1000	1/2"		18	.444		.97	16.60		17.57	26	
1010	3/4"		17	.471		1.61	17.60		19.21	28.50	
1020	1"		15	.533		4.18	19.90		24.08	34.50	
1030	1-1/4"		13	.615		6.15	23		29.15	41.50	
1040	1-1/2"		12	.667		7.10	25		32.10	45.50	
1050	2"		11	.727		12	27		39	54	
1060	2-1/2"	Q-1	10.50	1.524		38.50	51.50		90	120	
1070	3"		10	1.600		52	54		106	139	
1080	3-1/2"		9	1.778		87.50	60		147.50	187	
1090	4"		8	2		87.50	67.50		155	199	
1200	5"		6	2.667		410	89.50		499.50	590	
1210	6"	Q-2	8.50	2.824		460	98.50		558.50	655	
1214	Adapter, copper x female IPS										
1216	1/2"	1 Plum	18	.444	Ea.	1.52	16.60		18.12	26.50	
1218	3/4"		17	.471		1.63	17.60		19.23	28.50	
1220	1"		15	.533		4.85	19.90		24.75	35.50	
1222	1-1/2"		12	.667		11.10	25		36.10	49.50	
1224	2"		11	.727		15.05	27		42.05	57.50	
1250	Cross, 1/2"		10	.800		6.30	30		36.30	52	
1260	3/4"		9.50	.842		12.25	31.50		43.75	61	
1270	1"		8	1		20.50	37.50		58	79	
1280	1-1/4"		7.50	1.067		28.50	40		68.50	91	
1290	1-1/2"		6.50	1.231		42.50	46		88.50	117	
1300	2"		5.50	1.455		80.50	54.50		135	171	
1310	2-1/2"	Q-1	6.50	2.462		185	83		268	330	
1320	3"		5.50	2.909		243	98		341	415	
1340	4"		4.70	3.404		340	114		454	550	
1350	5"		3.20	5		485	168		653	790	

15107	Metal Pipe & Fittings	CREW	DAILY OUTPUT	LABOR-HOURS	UNIT	2003 BARE COSTS				TOTAL INCL O&P	
						MAT.	LABOR	EQUIP.	TOTAL		
460	1360	6"	Q-2	4.70	5.106	Ea.	680	178		858	1,025	460
	1500	Tee fitting, mechanically formed, (Type 1).									
	1520	1/2" run size, 3/8" to 1/2" branch size	1 Plum	80	.100	Ea.		3.74		3.74	5.65
	1530	3/4" run size, 3/8" to 3/4" branch size		60	.133			4.98		4.98	7.50
	1540	1" run size, 3/8" to 1" branch size		54	.148			5.55		5.55	8.35
	1550	1-1/4" run size, 3/8" to 1-1/4" branch size		48	.167			6.25		6.25	9.40
	1560	1-1/2" run size, 3/8" to 1-1/2" branch size		40	.200			7.45		7.45	11.30
	1570	2" run size, 3/8" to 2" branch size		35	.229			8.55		8.55	12.90
	1580	2-1/2" run size, 1/2" to 2" branch size		32	.250			9.35		9.35	14.10
	1590	3" run size, 1" to 2" branch size		26	.308			11.50		11.50	17.35
	1600	4" run size, 1" to 2" branch size	▼	24	.333	▼		12.45		12.45	18.80
	1640	Tee fitting, mechanically formed, (Type 2)									
	1650	2-1/2" run size, 2-1/2" branch size	1 Plum	12.50	.640	Ea.		24		24	36
	1660	3" run size, 2-1/2" to 3" branch size		12	.667			25		25	37.50
	1670	3-1/2" run size, 2-1/2" to 3-1/2" branch size		11	.727			27		27	41
	1680	4" run size, 2-1/2" to 4" branch size		10.50	.762			28.50		28.50	43
	1698	5" run size, 2" to 4" branch size		9.50	.842			31.50		31.50	47.50
	1700	6" run size, 2" to 4" branch size		8.50	.941			35		35	53
	1710	8" run size, 2" to 4" branch size	▼	7	1.143	▼		42.50		42.50	64.50
	1800	ACR fittings									
	1802	Tee, straight									
	1808	5/8"	1 Stpi	12	.667	Ea.	5.05	25		30.05	43.50
	1810	3/4"		12	.667		1.90	25		26.90	40
	1812	7/8"		10	.800		13.75	30		43.75	60.50
	1814	1-1/8"		10	.800		5.85	30		35.85	52
	1816	1-3/8"		9	.889		8.90	33.50		42.40	60.50
	1818	1-5/8"		8	1		12.35	37.50		49.85	70.50
	1820	2-1/8"	▼	7	1.143		19.25	43		62.25	86
	1822	2-5/8"	Q-5	8	2		39	67.50		106.50	145
	1824	3-1/8"		7	2.286		59	77.50		136.50	182
	1826	4-1/8"	▼	5	3.200	▼	143	108		251	320
	1830	90° elbow									
	1836	5/8"	1 Stpi	19	.421	Ea.	1.53	15.85		17.38	25.50
	1838	3/4"		19	.421		2.80	15.85		18.65	27
	1840	7/8"		16	.500		2.67	18.80		21.47	31.50
	1842	1-1/8"		16	.500		3.72	18.80		22.52	32.50
	1844	1-3/8"		15	.533		3.84	20		23.84	34.50
	1846	1-5/8"		13	.615		6	23		29	41.50
	1848	2-1/8"	▼	11	.727		10.90	27.50		38.40	53.50
	1850	2-5/8"	Q-5	13	1.231		21	41.50		62.50	86
	1852	3-1/8"		11	1.455		29	49		78	107
	1854	4-1/8"	▼	9	1.778	▼	74.50	60		134.50	173
	1860	Coupling									
	1866	5/8"	1 Stpi	21	.381	Ea.	.35	14.30		14.65	22
	1868	3/4"		21	.381		1.04	14.30		15.34	22.50
	1870	7/8"		18	.444		.68	16.70		17.38	26
	1872	1-1/8"		18	.444		1.42	16.70		18.12	26.50
	1874	1-3/8"		17	.471		2.48	17.70		20.18	29
	1876	1-5/8"		15	.533		3.28	20		23.28	34
	1878	2-1/8"	▼	13	.615		5.45	23		28.45	41
	1880	2-5/8"	Q-5	15	1.067		11.60	36		47.60	67.50
	1882	3-1/8"		13	1.231		17.45	41.50		58.95	82
	1884	4-1/8"	▼	7	2.286	▼	35	77.50		112.50	156
	2000	DWV, solder joints, copper x copper									
	2030	90° Elbow, 1-1/4"	1 Plum	13	.615	Ea.	4.29	23		27.29	39
	2050	1-1/2"	▼	12	.667	▼	6	25		31	44

15107	Metal Pipe & Fittings	CREW	DAILY OUTPUT	LABOR-HOURS	UNIT	2003 BARE COSTS				TOTAL INCL O&P
						MAT.	LABOR	EQUIP.	TOTAL	
2070	2"	1 Plum	10	.800	Ea.	8.70	30		38.70	54.50
2090	3"	Q-1	10	1.600		21.50	54		75.50	105
2100	4"	"	9	1.778		100	60		160	201
2150	45° Elbow, 1-1/4"	1 Plum	13	.615		4.10	23		27.10	39
2170	1-1/2"		12	.667		3.40	25		28.40	41
2180	2"		10	.800		7.80	30		37.80	53.50
2190	3"	Q-1	10	1.600		16.50	54		70.50	99.50
2200	4"	"	9	1.778		49.50	60		109.50	145
2250	Tee, Sanitary, 1-1/4"	1 Plum	9	.889		8.80	33		41.80	59.50
2270	1-1/2"		8	1		10.95	37.50		48.45	68.50
2290	2"		7	1.143		12.75	42.50		55.25	78.50
2310	3"	Q-1	7	2.286		45.50	77		122.50	166
2330	4"	"	6	2.667		116	89.50		205.50	263
2400	Coupling, 1-1/4"	1 Plum	14	.571		2.08	21.50		23.58	34.50
2420	1-1/2"		13	.615		2.56	23		25.56	37.50
2440	2"		11	.727		3.58	27		30.58	45
2460	3"	Q-1	11	1.455		6.95	49		55.95	81.50
2480	4"	"	10	1.600		22	54		76	106
2600	Traps, see division 15155									
3500	Compression joint fittings									
3510	As used for plumbing and oil burner work									
3520	Fitting price includes nuts and sleeves									
3540	Sleeve, 1/8"				Ea.	.09			.09	.10
3550	3/16"					.09			.09	.10
3560	1/4"					.05			.05	.06
3570	5/16"					.10			.10	.11
3580	3/8"					.21			.21	.23
3600	1/2"					.28			.28	.31
3620	Nut, 1/8"					.15			.15	.17
3630	3/16"					.17			.17	.19
3640	1/4"					.16			.16	.18
3650	5/16"					.22			.22	.24
3660	3/8"					.24			.24	.26
3670	1/2"					.44			.44	.48
3710	Union, 1/8"	1 Plum	26	.308		.80	11.50		12.30	18.25
3720	3/16"		24	.333		.88	12.45		13.33	19.75
3730	1/4"		24	.333		.92	12.45		13.37	19.80
3740	5/16"		23	.348		1.17	13		14.17	21
3750	3/8"		22	.364		1.19	13.60		14.79	22
3760	1/2"		22	.364		1.82	13.60		15.42	22.50
3780	5/8"		21	.381		2.02	14.25		16.27	23.50
3820	Union tee, 1/8"		17	.471		3.47	17.60		21.07	30.50
3830	3/16"		16	.500		2.66	18.70		21.36	31
3840	1/4"		15	.533		2.24	19.90		22.14	32.50
3850	5/16"		15	.533		4.50	19.90		24.40	35
3860	3/8"		15	.533		2.61	19.90		22.51	33
3870	1/2"		15	.533		4.39	19.90		24.29	35
3910	Union elbow, 1/4"		24	.333		1.12	12.45		13.57	20
3920	5/16"		23	.348		1.97	13		14.97	22
3930	3/8"		22	.364		2.24	13.60		15.84	23
3940	1/2"		22	.364		3.42	13.60		17.02	24.50
3980	Female connector, 1/8"		26	.308		.74	11.50		12.24	18.15
4000	3/16" x 1/8"		24	.333		.98	12.45		13.43	19.90
4010	1/4" x 1/8"		24	.333		.83	12.45		13.28	19.70
4020	1/4"		24	.333		1.04	12.45		13.49	19.95
4030	3/8" x 1/4"		22	.364		1.16	13.60		14.76	22

MECHANICAL 15

15107	Metal Pipe & Fittings	CREW	DAILY OUTPUT	LABOR-HOURS	UNIT	MAT.	LABOR	EQUIP.	TOTAL	TOTAL INCL O&P	
						2003 BARE COSTS					
460	4040	1/2" x 3/8"	1 Plum	22	.364	Ea.	1.84	13.60		15.44	22.50	**460**
4050	5/8" x 1/2"		21	.381		2.15	14.25		16.40	24
4090	Male connector, 1/8"		26	.308		.57	11.50		12.07	18
4100	3/16" x 1/8"		24	.333		.59	12.45		13.04	19.45
4110	1/4" x 1/8"		24	.333		.61	12.45		13.06	19.45
4120	1/4"		24	.333		.61	12.45		13.06	19.45
4130	5/16" x 1/8"		23	.348		.71	13		13.71	20.50
4140	5/16" x 1/4"		23	.348		.82	13		13.82	20.50
4150	3/8" x 1/8"		22	.364		.85	13.60		14.45	21.50
4160	3/8" x 1/4"		22	.364		.98	13.60		14.58	21.50
4170	3/8"		22	.364		1.19	13.60		14.79	22
4180	3/8" x 1/2"		22	.364		1.34	13.60		14.94	22
4190	1/2" x 3/8"		22	.364		1.30	13.60		14.90	22
4200	1/2"		22	.364		1.91	13.60		15.51	22.50
4210	5/8" x 1/2"		21	.381		1.69	14.25		15.94	23.50
4240	Male elbow, 1/8"		26	.308		1.17	11.50		12.67	18.65
4250	3/16" x 1/8"		24	.333		1.29	12.45		13.74	20
4260	1/4" x 1/8"		24	.333		.98	12.45		13.43	19.90
4270	1/4"		24	.333		1.11	12.45		13.56	20
4280	3/8" x 1/4"		22	.364		1.41	13.60		15.01	22
4290	3/8"		22	.364		2.42	13.60		16.02	23
4300	1/2" x 1/4"		22	.364		2.35	13.60		15.95	23
4310	1/2" x 3/8"		22	.364		3.68	13.60		17.28	24.50
4340	Female elbow, 1/8"		26	.308		2.51	11.50		14.01	20
4350	1/4" x 1/8"		24	.333		1.43	12.45		13.88	20.50
4360	1/4"		24	.333		1.88	12.45		14.33	21
4370	3/8" x 1/4"		22	.364		1.85	13.60		15.45	22.50
4380	1/2" x 3/8"		22	.364		3.16	13.60		16.76	24
4390	1/2"		22	.364		4.25	13.60		17.85	25
4420	Male run tee, 1/4" x 1/8"		15	.533		1.76	19.90		21.66	32
4430	5/16" x 1/8"		15	.533		2.62	19.90		22.52	33
4440	3/8" x 1/4"		15	.533		2.71	19.90		22.61	33
4480	1/4" x 1/8"		15	.533		2.03	19.90		21.93	32
4490	1/4"		15	.533		2.19	19.90		22.09	32.50
4500	3/8" x 1/4"		15	.533		2.54	19.90		22.44	33
4510	1/2" x 3/8"		15	.533		3.95	19.90		23.85	34.50
4520	1/2"	▼	15	.533	▼	6.35	19.90		26.25	37
4800	Flare joint fittings									
4810	Refrigeration fittings									
4820	Flare joint nuts and labor not incl. in price. Add 1 nut per jnt.									
4830	90° Elbow, 1/4"				Ea.	1.96			1.96	2.16
4840	3/8"					2.10			2.10	2.31
4850	1/2"					2.82			2.82	3.10
4860	5/8"					4.13			4.13	4.54
4870	3/4"					11.30			11.30	12.45
5030	Tee, 1/4"					3.15			3.15	3.47
5040	5/16"					3.32			3.32	3.65
5050	3/8"					2.35			2.35	2.59
5060	1/2"					3.58			3.58	3.94
5070	5/8"					5.55			5.55	6.10
5080	3/4"					11.20			11.20	12.35
5140	Union, 3/16"					1.97			1.97	2.17
5150	1/4"					.60			.60	.66
5160	5/16"					.67			.67	.74
5170	3/8"					.71			.71	.78
5180	1/2"			▼		1.07			1.07	1.18

Important: See the Reference Section for critical supporting data - Reference Nos., Crews, & City Cost Indexes

			DAILY	LABOR-		2003 BARE COSTS				TOTAL
15107	**Metal Pipe & Fittings**	CREW	OUTPUT	HOURS	UNIT	MAT.	LABOR	EQUIP.	TOTAL	INCL O&P
5190	5/8"				Ea.	1.73			1.73	1.90
5200	3/4"					5.20			5.20	5.70
5260	Long flare nut, 3/16"	1 Stpi	42	.190		.42	7.15		7.57	11.25
5270	1/4"		41	.195		.48	7.35		7.83	11.65
5280	5/16"		40	.200		.73	7.50		8.23	12.15
5290	3/8"		39	.205		1.13	7.70		8.83	12.90
5300	1/2"		38	.211		1.70	7.90		9.60	13.80
5310	5/8"		37	.216		3.42	8.15		11.57	16.05
5320	3/4"		34	.235		5.80	8.85		14.65	19.70
5380	Short flare nut, 3/16"		42	.190		.29	7.15		7.44	11.10
5390	1/4"		41	.195		.39	7.35		7.74	11.55
5400	5/16"		40	.200		.57	7.50		8.07	12
5410	3/8"		39	.205		.49	7.70		8.19	12.20
5420	1/2"		38	.211		.71	7.90		8.61	12.75
5430	5/8"		36	.222		1.01	8.35		9.36	13.70
5440	3/4"		34	.235		3.36	8.85		12.21	17.05
5500	90° Elbow flare by MIPS, 1/4"					1.38			1.38	1.52
5510	3/8"					1.92			1.92	2.11
5520	1/2"					3.08			3.08	3.39
5530	5/8"					3.91			3.91	4.30
5540	3/4"					5.60			5.60	6.15
5600	Flare by FIPS, 1/4"					4.09			4.09	4.50
5610	3/8"					3.57			3.57	3.93
5620	1/2"					6.75			6.75	7.45
5670	Flare by sweat, 1/4"					1.42			1.42	1.56
5680	3/8"					1.38			1.38	1.52
5690	1/2"					2.51			2.51	2.76
5700	5/8"					3.47			3.47	3.82
5760	Tee flare by IPS, 1/4"					2.19			2.19	2.41
5770	3/8"					2.16			2.16	2.38
5780	1/2"					6.30			6.30	6.95
5790	5/8"					6.40			6.40	7.05
5850	Connector, 1/4"					.43			.43	.47
5860	3/8"					1.13			1.13	1.24
5870	1/2"					1.16			1.16	1.28
5880	5/8"					1.33			1.33	1.46
5890	3/4"					7.55			7.55	8.30
5950	Seal cap, 1/4"					.29			.29	.32
5960	3/8"					.37			.37	.41
5970	1/2"					.59			.59	.65
5980	5/8"					1.79			1.79	1.97
5990	3/4"					6.20			6.20	6.85
6000	Water service fittings									
6010	Flare joints nut and labor are included in the fitting price.									
6020	90° Elbow, cxc, 3/8"	1 Plum	19	.421	Ea.	9.20	15.75		24.95	34
6030	1/2"		18	.444		9.20	16.60		25.80	35
6040	3/4"		16	.500		13.05	18.70		31.75	42.50
6050	1"		15	.533		24.50	19.90		44.40	57
6060	1-1/4"		12	.667		85	25		110	131
6070	1-1/2"		12	.667		85	25		110	131
6080	2"		10	.800		85	30		115	139
6090	C x MPT, 3/8"		19	.421		6.65	15.75		22.40	31.50
6100	1/2"		18	.444		6.65	16.60		23.25	32.50
6110	3/4"		16	.500		7.95	18.70		26.65	36.50
6120	1"		15	.533		22	19.90		41.90	54.50
6130	1-1/4"		13	.615		48	23		71	87.50

460

MECHANICAL 15

15107	Metal Pipe & Fittings	CREW	DAILY OUTPUT	LABOR-HOURS	UNIT	2003 BARE COSTS				TOTAL INCL O&P	
						MAT.	LABOR	EQUIP.	TOTAL		
460	6140	1-1/2"	1 Plum	12	.667	Ea.	48	25		73	90.50	460
	6150	2"		10	.800		85	30		115	139
	6160	C x FPT, 3/8"		19	.421		7.15	15.75		22.90	32
	6170	1/2"		18	.444		7.15	16.60		23.75	33
	6180	3/4"		16	.500		9.80	18.70		28.50	39
	6190	1"		15	.533		21.50	19.90		41.40	53.50
	6200	1-1/4"		13	.615		57	23		80	97.50
	6210	1-1/2"		12	.667		132	25		157	183
	6220	2"		10	.800		201	30		231	266
	6230	Tee, cxcxc, 3/8"		13	.615		14.95	23		37.95	51
	6240	1/2"		12	.667		14.95	25		39.95	54
	6250	3/4"		11	.727		21.50	27		48.50	64.50
	6260	1"		10	.800		31	30		61	79
	6270	1-1/4"		8	1		103	37.50		140.50	171
	6280	1-1/2"		8	1		124	37.50		161.50	193
	6290	2"		7	1.143		222	42.50		264.50	310
	6300	Tee, cxcxFPT, 3/8"		13	.615		11.65	23		34.65	47.50
	6310	1/2"		12	.667		17.30	25		42.30	56.50
	6320	3/4"		11	.727		24.50	27		51.50	68
	6330	Tube nut, cxnut seat, 3/8"		40	.200		1.54	7.45		8.99	13
	6340	1/2"		38	.211		1.54	7.85		9.39	13.55
	6350	3/4"		34	.235		3.04	8.80		11.84	16.60
	6360	1"		32	.250		5.50	9.35		14.85	20
	6370	1-1/4"		25	.320		28	11.95		39.95	48.50
	6380	Coupling, cxc, 3/8"		19	.421		9.20	15.75		24.95	34
	6390	1/2"		18	.444		9.20	16.60		25.80	35
	6400	3/4"		16	.500		10.80	18.70		29.50	40
	6410	1"		15	.533		21	19.90		40.90	53
	6420	1-1/4"		12	.667		33.50	25		58.50	74.50
	6430	1-1/2"		12	.667		52.50	25		77.50	95.50
	6440	2"		10	.800		79	30		109	132
	6450	Adapter, cxFPT, 3/8"		19	.421		6.70	15.75		22.45	31.50
	6460	1/2"		18	.444		6.70	16.60		23.30	32.50
	6470	3/4"		16	.500		7.95	18.70		26.65	37
	6480	1"		15	.533		17.90	19.90		37.80	49.50
	6490	1-1/4"		13	.615		34	23		57	72
	6500	1-1/2"		12	.667		34	25		59	75
	6510	2"		10	.800		48	30		78	98
	6520	Adapter, cxMPT, 3/8"		19	.421		5.75	15.75		21.50	30.50
	6530	1/2"		18	.444		5.75	16.60		22.35	31.50
	6540	3/4"		16	.500		8.15	18.70		26.85	37
	6550	1"		15	.533		14.35	19.90		34.25	46
	6560	1-1/4"		13	.615		37.50	23		60.50	76
	6570	1-1/2"		12	.667		37.50	25		62.50	79
	6580	2"	▼	10	.800	▼	51	30		81	101
	6990	Polybutylene/polyethylene pipe, See 15108 for plastic ftng.									
	7000	Insert type Brass/copper, 100 psi @ 180°F, CTS									
	7010	Adapter MPT 3/8" x 3/8" CTS	1 Plum	29	.276	Ea.	.97	10.30		11.27	16.60
	7020	1/2" x 1/2"		26	.308		.97	11.50		12.47	18.40
	7030	3/4" x 1/2"		26	.308		1.44	11.50		12.94	18.95
	7040	3/4" x 3/4"		25	.320		1.42	11.95		13.37	19.60
	7050	Adapter CTS 1/2" x 1/2" sweat		24	.333		.38	12.45		12.83	19.20
	7060	3/4" x 3/4" sweat		22	.364		.47	13.60		14.07	21
	7070	Coupler center set 3/8" CTS		25	.320		.49	11.95		12.44	18.60
	7080	1/2" CTS		23	.348		.38	13		13.38	20
	7090	3/4" CTS	▼	22	.364	▼	.47	13.60		14.07	21

Important: See the Reference Section for critical supporting data - Reference Nos., Crews, & City Cost Indexes

15107	Metal Pipe & Fittings	CREW	DAILY OUTPUT	LABOR-HOURS	UNIT	2003 BARE COSTS				TOTAL INCL O&P	
						MAT.	LABOR	EQUIP.	TOTAL		
460	7100	Elbow 90°, copper 3/8"	1 Plum	25	.320	Ea.	1.08	11.95		13.03	19.25	**460**
	7110	1/2" CTS		23	.348		.69	13		13.69	20.50	
	7120	3/4" CTS		22	.364		.84	13.60		14.44	21.50	
	7130	Tee copper 3/8" CTS		17	.471		1.28	17.60		18.88	28	
	7140	1/2" CTS		15	.533		.88	19.90		20.78	31	
	7150	3/4" CTS		14	.571		1.35	21.50		22.85	33.50	
	7160	3/8" x 3/8" x 1/2"		16	.500		.88	18.70		19.58	29	
	7170	1/2" x 3/8" x 1/2"		15	.533		.88	19.90		20.78	31	
	7180	3/4" x 1/2" x 3/4"		14	.571		1.35	21.50		22.85	33.50	
480	0010	**PIPE/TUBE, GROOVED-JOINT FOR COPPER**										**480**
	4000	Fittings: coupling and labor required at joints not incl.										
	4001	in fitting price. add 1 per joint for installed price.										
	4010	Coupling, rigid style										
	4018	2" diameter	1 Plum	50	.160	Ea.	6.90	6		12.90	16.55	
	4020	2-1/2" diameter	Q-1	80	.200		7.85	6.75		14.60	18.75	
	4022	3" diameter		67	.239		8.75	8.05		16.80	22	
	4024	4" diameter		50	.320		13.40	10.75		24.15	31	
	4026	5" diameter		40	.400		23	13.45		36.45	46	
	4028	6" diameter	Q-2	50	.480		31	16.75		47.75	59.50	
	4100	Elbow, 90° or 45°										
	4108	2" diameter	1 Plum	25	.320	Ea.	10.50	11.95		22.45	29.50	
	4110	2-1/2" diameter	Q-1	40	.400		11.45	13.45		24.90	33	
	4112	3" diameter		33	.485		16.20	16.30		32.50	42.50	
	4114	4" diameter		25	.640		38	21.50		59.50	74	
	4116	5" diameter		20	.800		108	27		135	160	
	4118	6" diameter	Q-2	25	.960		173	33.50		206.50	242	
	4200	Tee										
	4208	2" diameter	1 Plum	17	.471	Ea.	17.60	17.60		35.20	46	
	4210	2-1/2" diameter	Q-1	27	.593		18.70	19.95		38.65	50.50	
	4212	3" diameter		22	.727		28	24.50		52.50	68	
	4214	4" diameter		17	.941		62	31.50		93.50	116	
	4216	5" diameter		13	1.231		175	41.50		216.50	255	
	4218	6" diameter	Q-2	17	1.412		216	49		265	315	
	4300	Reducer, concentric										
	4310	3" x 2-1/2" diameter	Q-1	35	.457	Ea.	15.35	15.35		30.70	40	
	4312	4" x 2-1/2" diameter		32	.500		32	16.80		48.80	60.50	
	4314	4" x 3" diameter		29	.552		32	18.55		50.55	63	
	4316	5" x 3" diameter		25	.640		90.50	21.50		112	132	
	4318	5" x 4" diameter		22	.727		90.50	24.50		115	137	
	4320	6" x 3" diameter	Q-2	28	.857		98	30		128	153	
	4322	6" x 4" diameter		26	.923		98	32		130	157	
	4324	6" x 5" diameter		24	1		98	35		133	161	
	4350	Flange, w/groove gasket										
	4351	ANSI class 125 and 150										
	4355	2" diameter	1 Plum	23	.348	Ea.	56	13		69	81	
	4356	2-1/2" diameter	Q-1	37	.432		58.50	14.55		73.05	86	
	4358	3" diameter		31	.516		61	17.35		78.35	93	
	4360	4" diameter		23	.696		67	23.50		90.50	109	
	4362	5" diameter		19	.842		90	28.50		118.50	142	
	4364	6" diameter	Q-2	23	1.043		98.50	36.50		135	163	
620	0010	**PIPE, STEEL**	R15100 -050									**620**
	0020	All pipe sizes are to Spec. A-53 unless noted otherwise										
	0030	Schedule 10, see 15107-690-0500										
	0050	Schedule 40, threaded, with couplings, and clevis type										

MECHANICAL 15

15107 | Metal Pipe & Fittings

		CREW	DAILY OUTPUT	LABOR-HOURS	UNIT	2003 BARE COSTS				TOTAL INCL O&P
						MAT.	LABOR	EQUIP.	TOTAL	
620	0060	hangers sized for covering, 10' O.C.	R15100 -050								620
	0540	Black, 1/4" diameter	1 Plum	66	.121	L.F.	1.40	4.53		5.93	8.40
	0550	3/8" diameter		65	.123		1.53	4.60		6.13	8.65
	0560	1/2" diameter		63	.127		1.32	4.74		6.06	8.60
	0570	3/4" diameter		61	.131		1.47	4.90		6.37	9
	0580	1" diameter		53	.151		1.88	5.65		7.53	10.55
	0590	1-1/4" diameter	Q-1	89	.180		2.23	6.05		8.28	11.60
	0600	1-1/2" diameter		80	.200		2.51	6.75		9.26	12.90
	0610	2" diameter		64	.250		3.18	8.40		11.58	16.20
	0620	2-1/2" diameter		50	.320		5.05	10.75		15.80	22
	0630	3" diameter		43	.372		6.85	12.50		19.35	26.50
	0640	3-1/2" diameter		40	.400		8.95	13.45		22.40	30.50
	0650	4" diameter		36	.444		10.05	14.95		25	33.50
	0660	5" diameter		26	.615		17.60	20.50		38.10	51
	0670	6" diameter	Q-2	31	.774		21.50	27		48.50	65
	0680	8" diameter		27	.889		32	31		63	82
	0690	10" diameter		23	1.043		48.50	36.50		85	108
	0700	12" diameter		18	1.333		61	46.50		107.50	137
	1220	To delete coupling & hanger, subtract									
	1230	1/4" diam. to 3/4" diam.					31%	56%			
	1240	1" diam. to 1-1/2" diam.					23%	51%			
	1250	2" diam. to 4" diam.					23%	41%			
	1260	5" diam. to 12" diam.					21%	45%			
	1290	Galvanized, 1/4" diameter	1 Plum	66	.121	L.F.	1.73	4.53		6.26	8.75
	1300	3/8" diameter		65	.123		1.86	4.60		6.46	9
	1310	1/2" diameter		63	.127		1.73	4.74		6.47	9.05
	1320	3/4" diameter		61	.131		1.93	4.90		6.83	9.50
	1330	1" diameter		53	.151		2.29	5.65		7.94	11
	1340	1-1/4" diameter	Q-1	89	.180		2.84	6.05		8.89	12.25
	1350	1-1/2" diameter		80	.200		3.06	6.75		9.81	13.50
	1360	2" diameter		64	.250		4.11	8.40		12.51	17.20
	1370	2-1/2" diameter		50	.320		5.50	10.75		16.25	22.50
	1380	3" diameter		43	.372		6.70	12.50		19.20	26.50
	1390	3-1/2" diameter		40	.400		8.60	13.45		22.05	30
	1400	4" diameter		36	.444		9.75	14.95		24.70	33
	1410	5" diameter		26	.615		20	20.50		40.50	53.50
	1420	6" diameter	Q-2	31	.774		22	27		49	65.50
	1430	8" diameter		27	.889		35	31		66	85.50
	1440	10" diameter		23	1.043		53.50	36.50		90	114
	1450	12" diameter		18	1.333		67.50	46.50		114	144
	1750	To delete coupling & hanger, subtract									
	1760	1/4" diam. to 3/4" diam.					31%	56%			
	1770	1" diam. to 1-1/2" diam.					23%	51%			
	1780	2" diam. to 4" diam.					23%	41%			
	1790	5" diam. to 12" diam.					21%	45%			
	2000	Welded, sch. 40, on yoke & roll hangers, sized for covering,									
	2010	10' O.C. (no hangers incl. for 14" diam. and up)									
	2040	Black, 1" diameter	Q-15	93	.172	L.F.	2.99	5.80	.84	9.63	12.95
	2050	1-1/4" diameter		84	.190		3.82	6.40	.94	11.16	14.90
	2060	1-1/2" diameter		76	.211		4.13	7.10	1.03	12.26	16.40
	2070	2" diameter		61	.262		3.91	8.80	1.29	14	19
	2080	2-1/2" diameter		47	.340		6.15	11.45	1.67	19.27	26
	2090	3" diameter		43	.372		7.40	12.50	1.83	21.73	29
	2100	3-1/2" diameter		39	.410		9.30	13.80	2.01	25.11	33.50
	2110	4" diameter		37	.432		9	14.55	2.12	25.67	34
	2120	5" diameter		32	.500		13.35	16.80	2.46	32.61	43

Important: See the Reference Section for critical supporting data - Reference Nos., Crews, & City Cost Indexes

15107	Metal Pipe & Fittings		CREW	DAILY OUTPUT	LABOR-HOURS	UNIT	2003 BARE COSTS				TOTAL INCL O&P
							MAT.	LABOR	EQUIP.	TOTAL	
620	2130	6" diameter	R15100 -050	Q-16	36	.667	L.F.	15.70	23.50	2.18	41.38	54.50	620
	2140	8" diameter			29	.828		23.50	29	2.71	55.21	72.50
	2150	10" diameter			24	1		43	35	3.27	81.27	104
	2160	12" diameter			19	1.263		48	44	4.13	96.13	124
	2170	14" diameter			15	1.600		40	56	5.25	101.25	134
	2180	16" diameter			13	1.846		41.50	64.50	6.05	112.05	149
	2190	18" diameter			11	2.182		55.50	76	7.15	138.65	184
	2200	20" diameter			9	2.667		61	93	8.70	162.70	217
	2220	24" diameter			8	3		76	105	9.80	190.80	252
	2230	26" diameter			7.50	3.200		145	112	10.45	267.45	340
	2240	30" diameter			6	4		158	139	13.10	310.10	400
	2260	36" diameter			4.50	5.333		181	186	17.45	384.45	500
	2323	.375" wall, A-53, gr. A/B, seamless, welded w/hngrs.										
	2324	12" diameter		Q-16	20	1.200	L.F.	26	42	3.92	71.92	96
	2325	14" diameter			17	1.412		31.50	49	4.62	85.12	115
	2326	16" diameter			14	1.714		34.50	60	5.60	100.10	135
	2334	.50" wall, 8" diameter			25	.960		17.10	33.50	3.14	53.74	73
	2335	10" diameter			20	1.200		27	42	3.92	72.92	97
	2336	12" diameter			16	1.500		32.50	52.50	4.91	89.91	120
	2337	14" diameter			13	1.846		48	64.50	6.05	118.55	157
	2345	Sch. 40, A-53, gr. A/B, ERW, welded w/hngrs.										
	2346	2" diameter		Q-15	61	.262	L.F.	2.64	8.80	1.29	12.73	17.60
	2347	2-1/2" diameter			47	.340		3.84	11.45	1.67	16.96	23.50
	2348	3" diameter			43	.372		4.82	12.50	1.83	19.15	26
	2349	4" diameter			38	.421		6.60	14.15	2.07	22.82	31
	2350	6" diameter		Q-16	37	.649		8.25	22.50	2.12	32.87	45.50
	2351	8" diameter			29	.828		12.40	29	2.71	44.11	60
	2352	10" diameter			24	1		22	35	3.27	60.27	80.50
	2363	.375" wall, A-53, gr. A/B, ERW, welded w/hngrs.										
	2364	12" diameter		Q-16	20	1.200	L.F.	26	42	3.92	71.92	96
	2365	14" diameter			17	1.412		31.50	49	4.62	85.12	115
	2366	16" diameter			14	1.714		34.50	60	5.60	100.10	135
	2560	To delete hanger, subtract										
	2570	1" diam. to 1-1/2" diam.						15%	34%			
	2580	2" diam. to 3-1/2" diam.						9%	21%			
	2590	4" diam. to 12" diam.						5%	12%			
	2640	Galvanized, 1" diameter		Q-15	93	.172	L.F.	2.90	5.80	.84	9.54	12.85
	2650	1-1/4" diameter			84	.190		3.65	6.40	.94	10.99	14.70
	2660	1-1/2" diameter			76	.211		3.91	7.10	1.03	12.04	16.15
	2670	2" diameter			61	.262		4.52	8.80	1.29	14.61	19.70
	2680	2-1/2" diameter			47	.340		5.75	11.45	1.67	18.87	25.50
	2690	3" diameter			43	.372		6.75	12.50	1.83	21.08	28.50
	2700	3-1/2" diameter			39	.410		8.40	13.80	2.01	24.21	32.50
	2710	4" diameter			37	.432		8.55	14.55	2.12	25.22	34
	2720	5" diameter			32	.500		16.40	16.80	2.46	35.66	46
	2730	6" diameter		Q-16	36	.667		19.30	23.50	2.18	44.98	58.50
	2740	8" diameter			29	.828		29	29	2.71	60.71	78.50
	2750	10" diameter			24	1		51.50	35	3.27	89.77	113
	2760	12" diameter			19	1.263		58.50	44	4.13	106.63	136
	3160	To delete hanger, subtract										
	3170	1" diam. to 1-1/2" diam.						30%	34%			
	3180	2" diam. to 3-1/2" diam.						19%	21%			
	3190	4" diam. to 12" diam.						10%	12%			
	3250	Flanged, 150 lb. weld neck, on yoke & roll hangers										
	3260	sized for covering, 10' O.C.										
	3290	Black, 1" diameter		Q-15	70	.229	L.F.	6.55	7.70	1.12	15.37	20

MECHANICAL 15

15107 | Metal Pipe & Fittings

		CREW	DAILY OUTPUT	LABOR-HOURS	UNIT	2003 BARE COSTS				TOTAL INCL O&P
						MAT.	LABOR	EQUIP.	TOTAL	
620 3300	1-1/4" diameter	Q-15	64	.250	L.F.	7.40	8.40	1.23	17.03	22	**620**
3310	1-1/2" diameter		58	.276		7.70	9.30	1.35	18.35	24
3320	2" diameter		45	.356		7.95	11.95	1.75	21.65	28.50
3330	2-1/2" diameter		36	.444		10.80	14.95	2.18	27.93	37
3340	3" diameter		32	.500		12.50	16.80	2.46	31.76	42
3350	3-1/2" diameter		29	.552		15.30	18.55	2.71	36.56	48
3360	4" diameter		26	.615		14.85	20.50	3.02	38.37	51
3370	5" diameter		21	.762		22.50	25.50	3.74	51.74	67.50
3380	6" diameter	Q-16	25	.960		25	33.50	3.14	61.64	81.50
3390	8" diameter		19	1.263		38	44	4.13	86.13	113
3400	10" diameter		16	1.500		68.50	52.50	4.91	125.91	160
3410	12" diameter		14	1.714		89.50	60	5.60	155.10	195
3470	For 300 lb. flanges, add					63%				
3480	For 600 lb. flanges, add					310%				
3960	To delete flanges & hanger, subtract									
3970	1" diam. to 2" diam.					76%	65%			
3980	2-1/2" diam. to 4" diam.					62%	59%			
3990	5" diam. to 12" diam.					60%	46%			
4040	Galvanized, 1" diameter	Q-15	70	.229	L.F.	6.45	7.70	1.12	15.27	19.95
4050	1-1/4" diameter		64	.250		7.20	8.40	1.23	16.83	22
4060	1-1/2" diameter		58	.276		7.50	9.30	1.35	18.15	23.50
4070	2" diameter		45	.356		8.20	11.95	1.75	21.90	29
4080	2-1/2" diameter		36	.444		10.15	14.95	2.18	27.28	36
4090	3" diameter		32	.500		11.40	16.80	2.46	30.66	41
4100	3-1/2" diameter		29	.552		13.55	18.55	2.71	34.81	46
4110	4" diameter		26	.615		14.45	20.50	3.02	37.97	50.50
4120	5" diameter		21	.762		25.50	25.50	3.74	54.74	71
4130	6" diameter	Q-16	25	.960		29	33.50	3.14	65.64	86
4140	8" diameter		19	1.263		38	44	4.13	86.13	113
4150	10" diameter		16	1.500		67.50	52.50	4.91	124.91	159
4160	12" diameter		14	1.714		80	60	5.60	145.60	185
4220	For 300 lb. flanges, add					65%				
4240	For 600 lb. flanges, add					350%				
4660	To delete flanges & hanger, subtract									
4670	1" diam. to 2" diam.					73%	65%			
4680	2-1/2" diam. to 4" diam.					58%	59%			
4690	5" diam. to 12" diam.					43%	46%			
4750	Schedule 80, threaded, with couplings, and clevis type hangers									
4760	sized for covering, 10' O.C.									
4790	Black, 1/4" diameter	1 Plum	54	.148	L.F.	1.91	5.55		7.46	10.45
4800	3/8" diameter		53	.151		1.69	5.65		7.34	10.35
4810	1/2" diameter		52	.154		1.83	5.75		7.58	10.70
4820	3/4" diameter		50	.160		2.08	6		8.08	11.30
4830	1" diameter		45	.178		2.52	6.65		9.17	12.80
4840	1-1/4" diameter	Q-1	75	.213		3.31	7.15		10.46	14.50
4850	1-1/2" diameter		69	.232		3.64	7.80		11.44	15.75
4860	2" diameter		56	.286		4.88	9.60		14.48	19.85
4870	2-1/2" diameter		44	.364		7.10	12.25		19.35	26.50
4880	3" diameter		38	.421		8.80	14.15		22.95	31
4890	3-1/2" diameter		35	.457		10.90	15.35		26.25	35
4900	4" diameter		32	.500		12.85	16.80		29.65	39.50
4910	5" diameter		23	.696		22.50	23.50		46	60.50
4920	6" diameter	Q-2	28	.857		31.50	30		61.50	79.50
4930	8" diameter		23	1.043		53.50	36.50		90	114
4940	10" diameter		20	1.200		82.50	42		124.50	154
4950	12" diameter		15	1.600		95	56		151	188

R15100 -050

Important: See the Reference Section for critical supporting data - Reference Nos., Crews, & City Cost Indexes

15107	Metal Pipe & Fittings		CREW	DAILY OUTPUT	LABOR-HOURS	UNIT	2003 BARE COSTS				TOTAL INCL O&P
							MAT.	LABOR	EQUIP.	TOTAL	
620	5061	A-106, gr. A/B seamless with cplgs. & hgrs, 1/4" dia. R15100 -050	1 Plum	63	.127	L.F.	2.03	4.74		6.77	9.40	**620**
	5062	3/8" diameter		62	.129		2.18	4.82		7	9.70
	5063	1/2" diameter		61	.131		2.74	4.90		7.64	10.40
	5064	3/4" diameter		57	.140		2.53	5.25		7.78	10.70
	5065	1" diameter	↓	51	.157		3.45	5.85		9.30	12.65
	5066	1-1/4" diameter	Q-1	85	.188		5	6.35		11.35	15.05
	5067	1-1/2" diameter		77	.208		5.45	7		12.45	16.55
	5071	A-53 2" diameter		61	.262		6.10	8.80		14.90	20
	5072	2-1/2" diameter		48	.333		6.30	11.20		17.50	24
	5073	3" diameter		41	.390		7.80	13.10		20.90	28.50
	5074	4" diameter	↓	35	.457	↓	16.45	15.35		31.80	41
	5430	To delete coupling & hanger, subtract									
	5440	1/4" diam. to 1/2" diam.					31%	54%			
	5450	3/4" diam. to 1-1/2" diam.					28%	49%			
	5460	2" diam. to 4" diam.					21%	40%			
	5470	5" diam. to 12" diam.					17%	43%			
	5510	Galvanized, 1/4" diameter	1 Plum	54	.148	L.F.	1.91	5.55		7.46	10.45
	5520	3/8" diameter		53	.151		1.94	5.65		7.59	10.65
	5530	1/2" diameter		52	.154		2.08	5.75		7.83	11
	5540	3/4" diameter		50	.160		2.39	6		8.39	11.60
	5550	1" diameter	↓	45	.178		2.75	6.65		9.40	13.10
	5560	1-1/4" diameter	Q-1	75	.213		3.63	7.15		10.78	14.85
	5570	1-1/2" diameter		69	.232		4.03	7.80		11.83	16.20
	5580	2" diameter		56	.286		5.40	9.60		15	20.50
	5590	2-1/2" diameter		44	.364		7.90	12.25		20.15	27
	5600	3" diameter		38	.421		9.90	14.15		24.05	32.50
	5610	3-1/2" diameter		35	.457		12.25	15.35		27.60	36.50
	5620	4" diameter		32	.500		14.45	16.80		31.25	41.50
	5630	5" diameter	↓	23	.696		23	23.50		46.50	61
	5640	6" diameter	Q-2	28	.857		30.50	30		60.50	78.50
	5650	8" diameter		23	1.043		52.50	36.50		89	113
	5660	10" diameter		20	1.200		69	42		111	139
	5670	12" diameter	↓	15	1.600	↓	93.50	56		149.50	187
	5930	To delete coupling & hanger, subtract									
	5940	1/4" diam. to 1/2" diam.					31%	54%			
	5950	3/4" diam. to 1-1/2" diam.					28%	49%			
	5960	2" diam. to 4" diam.					21%	40%			
	5970	5" diam. to 12" diam.					17%	43%			
	6000	Welded, on yoke & roller hangers									
	6010	sized for covering, 10' O.C.									
	6040	Black, 1" diameter	Q-15	85	.188	L.F.	3	6.35	.92	10.27	13.85
	6050	1-1/4" diameter		79	.203		3.86	6.80	.99	11.65	15.65
	6060	1-1/2" diameter		72	.222		4.19	7.45	1.09	12.73	17.10
	6070	2" diameter		57	.281		4.93	9.45	1.38	15.76	21
	6080	2-1/2" diameter		44	.364		6.65	12.25	1.79	20.69	27.50
	6090	3" diameter		40	.400		8	13.45	1.96	23.41	31.50
	6100	3-1/2" diameter		34	.471		9.65	15.85	2.31	27.81	37
	6110	4" diameter		33	.485		11	16.30	2.38	29.68	39
	6120	5" diameter	↓	26	.615		20	20.50	3.02	43.52	57
	6130	6" diameter	Q-16	30	.800		29	28	2.62	59.62	76.50
	6140	8" diameter		25	.960		51	33.50	3.14	87.64	110
	6150	10" diameter		20	1.200		83	42	3.92	128.92	159
	6160	12" diameter	↓	15	1.600	↓	94.50	56	5.25	155.75	194
	6540	To delete hanger, subtract									
	6550	1" diam. to 1-1/2" diam.					30%	14%			
	6560	2" diam. to 3" diam.					23%	9%			

MECHANICAL 15

	15107	**Metal Pipe & Fittings**		CREW	DAILY OUTPUT	LABOR-HOURS	UNIT	MAT.	LABOR	EQUIP.	TOTAL	TOTAL INCL O&P	
620	6570	3-1/2″ diam. to 5″ diam.	R15100					12%	6%				620
	6580	6″ diam. to 12″ diam.	-050					10%	4%				
	6610	Galvanized, 1″ diameter		Q-15	85	.188	L.F.	3.23	6.35	.92	10.50	14.10	
	6650	1-1/4″ diameter			79	.203		4.18	6.80	.99	11.97	16	
	6660	1-1/2″ diameter			72	.222		4.58	7.45	1.09	13.12	17.55	
	6670	2″ diameter			57	.281		5.45	9.45	1.38	16.28	22	
	6680	2-1/2″ diameter			44	.364		7.50	12.25	1.79	21.54	28.50	
	6690	3″ diameter			40	.400		9.10	13.45	1.96	24.51	32.50	
	6700	3-1/2″ diameter			34	.471		11	15.85	2.31	29.16	38.50	
	6710	4″ diameter			33	.485		12.65	16.30	2.38	31.33	41	
	6720	5″ diameter		▼	26	.615		20.50	20.50	3.02	44.02	57.50	
	6730	6″ diameter		Q-16	30	.800		28	28	2.62	58.62	75.50	
	6740	8″ diameter			25	.960		50	33.50	3.14	86.64	109	
	6750	10″ diameter			20	1.200		70	42	3.92	115.92	144	
	6760	12″ diameter		▼	15	1.600	▼	93	56	5.25	154.25	192	
	7150	To delete hanger, subtract											
	7160	1″ diam. to 1-1/2″ diam.						26%	14%				
	7170	2″ diam. to 3″ diam.						16%	9%				
	7180	3-1/2″ diam. to 5″ diam.						10%	6%				
	7190	6″ diam. to 12″ diam.						7%	4%				
	7250	Flanged, 300 lb. weld neck, on yoke & roll hangers											
	7260	sized for covering, 10′ O.C.											
	7290	Black, 1″ diameter		Q-15	66	.242	L.F.	7.50	8.15	1.19	16.84	22	
	7300	1-1/4″ diameter			61	.262		8.55	8.80	1.29	18.64	24	
	7310	1-1/2″ diameter			54	.296		8.70	9.95	1.45	20.10	26.50	
	7320	2″ diameter			42	.381		10.30	12.80	1.87	24.97	32.50	
	7330	2-1/2″ diameter			33	.485		12.70	16.30	2.38	31.38	41	
	7340	3″ diameter			29	.552		14.05	18.55	2.71	35.31	46.50	
	7350	3-1/2″ diameter			24	.667		19.60	22.50	3.27	45.37	59	
	7360	4″ diameter			23	.696		21	23.50	3.42	47.92	62.50	
	7370	5″ diameter		▼	19	.842		35.50	28.50	4.13	68.13	86.50	
	7380	6″ diameter		Q-16	23	1.043		41.50	36.50	3.41	81.41	105	
	7390	8″ diameter			17	1.412		73.50	49	4.62	127.12	160	
	7400	10″ diameter			14	1.714		130	60	5.60	195.60	240	
	7410	12″ diameter		▼	12	2		155	69.50	6.55	231.05	282	
	7470	For 600 lb. flanges, add					▼	100%					
	7940	To delete flanges & hanger, subtract											
	7950	1″ diam. to 1-1/2″ diam.						75%	66%				
	7960	2″ diam. to 3″ diam.						62%	60%				
	7970	3-1/2″ diam. to 5″ diam.						54%	66%				
	7980	6″ diam. to 12″ diam.						55%	62%				
	8040	Galvanized, 1″ diameter		Q-15	66	.242	L.F.	7.75	8.15	1.19	17.09	22	
	8050	1-1/4″ diameter			61	.262		8.90	8.80	1.29	18.99	24.50	
	8060	1-1/2″ diameter			54	.296		9.10	9.95	1.45	20.50	26.50	
	8070	2″ diameter			42	.381		10.85	12.80	1.87	25.52	33.50	
	8080	2-1/2″ diameter			33	.485		13.50	16.30	2.38	32.18	42	
	8090	3″ diameter			29	.552		15.15	18.55	2.71	36.41	47.50	
	8100	3-1/2″ diameter			24	.667		21	22.50	3.27	46.77	60.50	
	8110	4″ diameter			23	.696		22.50	23.50	3.42	49.42	64.50	
	8120	5″ diameter		▼	19	.842		36	28.50	4.13	68.63	87	
	8130	6″ diameter		Q-16	23	1.043		40.50	36.50	3.41	80.41	103	
	8140	8″ diameter			17	1.412		72.50	49	4.62	126.12	159	
	8150	10″ diameter			14	1.714		117	60	5.60	182.60	226	
	8160	12″ diameter		▼	12	2		153	69.50	6.55	229.05	280	
	8240	For 600 lb. flanges, add					▼	100%					
	8900	To delete flanges & hangers, subtract		▼									

T5 MECHANICAL

Important: See the Reference Section for critical supporting data - Reference Nos., Crews, & City Cost Indexes

	15107	Metal Pipe & Fittings	CREW	DAILY OUTPUT	LABOR-HOURS	UNIT	2003 BARE COSTS				TOTAL INCL O&P
							MAT.	LABOR	EQUIP.	TOTAL	
620	8910	1" diam. to 1-1/2" diam.	R15100 -050				72%	66%			620
	8920	2" diam. to 3" diam.					59%	60%			
	8930	3-1/2" diam. to 5" diam.					51%	66%			
	8940	6" diam. to 12" diam.					49%	62%			
	9000	Threading pipe labor, one end, all schedules through 80									
	9010	1/4" through 3/4" pipe size	1 Plum	80	.100	Ea.		3.74		3.74	5.65
	9020	1" through 2" pipe size		73	.110			4.09		4.09	6.20
	9030	2-1/2" pipe size		53	.151			5.65		5.65	8.50
	9040	3" pipe size	↓	50	.160			6		6	9
	9050	3-1/2" pipe size	Q-1	89	.180			6.05		6.05	9.15
	9060	4" pipe size		73	.219			7.35		7.35	11.15
	9070	5" pipe size		53	.302			10.15		10.15	15.35
	9080	6" pipe size		46	.348			11.70		11.70	17.65
	9090	8" pipe size		29	.552			18.55		18.55	28
	9100	10" pipe size		21	.762			25.50		25.50	38.50
	9110	12" pipe size	↓	13	1.231	↓		41.50		41.50	62.50
	9200	Welding labor per joint									
	9210	Schedule 40,									
	9230	1/2" pipe size	Q-15	32	.500	Ea.		16.80	2.46	19.26	28
	9240	3/4" pipe size		27	.593			19.95	2.91	22.86	33
	9250	1" pipe size		23	.696			23.50	3.42	26.92	39.50
	9260	1-1/4" pipe size		20	.800			27	3.93	30.93	45
	9270	1-1/2" pipe size		19	.842			28.50	4.13	32.63	47.50
	9280	2" pipe size		16	1			33.50	4.91	38.41	56.50
	9290	2-1/2" pipe size		13	1.231			41.50	6.05	47.55	69
	9300	3" pipe size		12	1.333			45	6.55	51.55	74.50
	9310	4" pipe size		10	1.600			54	7.85	61.85	90
	9320	5" pipe size		9	1.778			60	8.75	68.75	100
	9330	6" pipe size		8	2			67.50	9.80	77.30	113
	9340	8" pipe size		5	3.200			108	15.70	123.70	180
	9350	10" pipe size		4	4			135	19.65	154.65	225
	9360	12" pipe size		3	5.333			179	26	205	300
	9370	14" pipe size		2.60	6.154			207	30	237	345
	9380	16" pipe size		2.20	7.273			245	35.50	280.50	410
	9390	18" pipe size		2	8			269	39.50	308.50	450
	9400	20" pipe size		1.80	8.889			299	43.50	342.50	500
	9410	22" pipe size		1.70	9.412			315	46	361	530
	9420	24" pipe size	↓	1.50	10.667	↓		360	52.50	412.50	600
	9450	Schedule 80,	Q-15	27	.593	Ea.					
	9460	1/2" pipe size		27	.593			19.95	2.91	22.86	33
	9470	3/4" pipe size		23	.696			23.50	3.42	26.92	39.50
	9480	1" pipe size		20	.800			27	3.93	30.93	45
	9490	1-1/4" pipe size		19	.842			28.50	4.13	32.63	47.50
	9500	1-1/2" pipe size		18	.889			30	4.36	34.36	50
	9510	2" pipe size		15	1.067			36	5.25	41.25	60
	9520	2-1/2" pipe size		12	1.333			45	6.55	51.55	74.50
	9530	3" pipe size		11	1.455			49	7.15	56.15	82
	9540	4" pipe size		8	2			67.50	9.80	77.30	113
	9550	5" pipe size		6	2.667			89.50	13.10	102.60	149
	9560	6" pipe size		5	3.200			108	15.70	123.70	180
	9570	8" pipe size		4	4			135	19.65	154.65	225
	9580	10" pipe size		3	5.333			179	26	205	300
	9590	12" pipe size	↓	2	8			269	39.50	308.50	450
	9600	14" pipe size	Q-16	2.60	9.231			320	30	350	520
	9610	16" pipe size		2.30	10.435			365	34	399	590
	9620	18" pipe size	↓	2	12	↓		420	39	459	675

15107	Metal Pipe & Fittings	CREW	DAILY OUTPUT	LABOR-HOURS	UNIT	2003 BARE COSTS				TOTAL INCL O&P	
						MAT.	LABOR	EQUIP.	TOTAL		
620	9630	20" pipe size	Q-16	1.80	13.333	Ea.		465	43.50	508.50	750	**620**
	9640	22" pipe size		1.60	15			525	49	574	845	
	9650	24" pipe size		1.50	16			560	52.50	612.50	900	
640	0010	**PIPE, STEEL, FITTINGS** Threaded										**640**
	0020	Cast Iron										
	0040	Standard weight, black										
	0060	90° Elbow, straight										
	0070	1/4"	1 Plum	16	.500	Ea.	3.17	18.70		21.87	31.50	
	0080	3/8"		16	.500		4.75	18.70		23.45	33.50	
	0090	1/2"		15	.533		2.13	19.90		22.03	32.50	
	0100	3/4"		14	.571		2.23	21.50		23.73	34.50	
	0110	1"		13	.615		2.63	23		25.63	37.50	
	0120	1-1/4"	Q-1	22	.727		3.75	24.50		28.25	41	
	0130	1-1/2"		20	.800		5.20	27		32.20	46	
	0140	2"		18	.889		8.10	30		38.10	54	
	0150	2-1/2"		14	1.143		19.40	38.50		57.90	79.50	
	0160	3"		10	1.600		32	54		86	117	
	0170	3-1/2"		8	2		65.50	67.50		133	174	
	0180	4"		6	2.667		59	89.50		148.50	200	
	0190	5"		5	3.200		114	108		222	289	
	0200	6"	Q-2	7	3.429		145	120		265	340	
	0210	8"	"	6	4		300	139		439	540	
	0250	45° Elbow, straight										
	0260	1/4"	1 Plum	16	.500	Ea.	3.60	18.70		22.30	32	
	0270	3/8"		16	.500		3.87	18.70		22.57	32.50	
	0280	1/2"		15	.533		3.29	19.90		23.19	33.50	
	0300	3/4"		14	.571		3.29	21.50		24.79	35.50	
	0320	1"		13	.615		3.85	23		26.85	38.50	
	0330	1-1/4"	Q-1	22	.727		5.20	24.50		29.70	42.50	
	0340	1-1/2"		20	.800		8.60	27		35.60	50	
	0350	2"		18	.889		9.90	30		39.90	56	
	0360	2-1/2"		14	1.143		25	38.50		63.50	85.50	
	0370	3"		10	1.600		41	54		95	127	
	0380	3-1/2"		8	2		69	67.50		136.50	178	
	0400	4"		6	2.667		83	89.50		172.50	227	
	0420	5"		5	3.200		142	108		250	320	
	0440	6"	Q-2	7	3.429		182	120		302	380	
	0460	8"	"	6	4		400	139		539	650	
	0500	Tee, straight										
	0510	1/4"	1 Plum	10	.800	Ea.	4.70	30		34.70	50	
	0520	3/8"		10	.800		5	30		35	50.50	
	0530	1/2"		9	.889		3.33	33		36.33	53.50	
	0540	3/4"		9	.889		3.89	33		36.89	54.50	
	0550	1"		8	1		3.45	37.50		40.95	60.50	
	0560	1-1/4"	Q-1	14	1.143		6.30	38.50		44.80	65	
	0570	1-1/2"		13	1.231		8.20	41.50		49.70	71.50	
	0580	2"		11	1.455		11.40	49		60.40	86.50	
	0590	2-1/2"		9	1.778		29.50	60		89.50	123	
	0600	3"		6	2.667		45.50	89.50		135	185	
	0610	3-1/2"		5	3.200		89.50	108		197.50	262	
	0620	4"		4	4		89	135		224	300	
	0630	5"		3	5.333		182	179		361	470	
	0640	6"	Q-2	4	6		204	209		413	540	
	0650	8"	"	3	8		460	279		739	925	
	0660	Tee, reducing, run or outlet										

R15100 -050

15107	Metal Pipe & Fittings	CREW	DAILY OUTPUT	LABOR-HOURS	UNIT	2003 BARE COSTS				TOTAL INCL O&P	
						MAT.	LABOR	EQUIP.	TOTAL		
0661	1/2"	1 Plum	9	.889	Ea.	7.25	33		40.25	58	640
0662	3/4"		9	.889		6.10	33		39.10	57	
0663	1"	↓	8	1		6.95	37.50		44.45	64	
0664	1-1/4"	Q-1	14	1.143		11	38.50		49.50	70	
0665	1-1/2"		13	1.231		13.40	41.50		54.90	77	
0666	2"		11	1.455		18.05	49		67.05	94	
0667	2-1/2"		9	1.778		36.50	60		96.50	131	
0668	3"		6	2.667		58.50	89.50		148	200	
0669	3-1/2"		5	3.200		102	108		210	275	
0670	4"		4	4		117	135		252	330	
0671	5"	↓	3	5.333		230	179		409	525	
0672	6"	Q-2	4	6		295	209		504	640	
0673	8"	"	3	8	↓	350	279		629	805	
0674	Reducer, concentric										
0675	3/4"	1 Plum	18	.444	Ea.	5.45	16.60		22.05	31	
0676	1"	"	15	.533		4.04	19.90		23.94	34.50	
0677	1-1/4"	Q-1	26	.615		10.50	20.50		31	43	
0678	1-1/2"		24	.667		14.85	22.50		37.35	50.50	
0679	2"		21	.762		20.50	25.50		46	61	
0680	2-1/2"		18	.889		27.50	30		57.50	75.50	
0681	3"		14	1.143		38	38.50		76.50	100	
0682	3-1/2"		12	1.333		66.50	45		111.50	141	
0683	4"		10	1.600		66.50	54		120.50	155	
0684	5"	↓	6	2.667		91	89.50		180.50	235	
0685	6"	Q-2	8	3		120	105		225	290	
0686	8"	"	7	3.429	↓	157	120		277	355	
0687	Reducer, eccentric										
0688	3/4"	1 Plum	16	.500	Ea.	12.40	18.70		31.10	41.50	
0689	1"	"	14	.571		12.40	21.50		33.90	45.50	
0690	1-1/4"	Q-1	25	.640		18.45	21.50		39.95	53	
0691	1-1/2"		22	.727		26	24.50		50.50	66	
0692	2"		20	.800		33	27		60	77	
0693	2-1/2"		16	1		49.50	33.50		83	105	
0694	3"		12	1.333		61	45		106	135	
0695	3-1/2"		10	1.600		92	54		146	183	
0696	4"		9	1.778		120	60		180	223	
0697	5"	↓	5	3.200		160	108		268	340	
0698	6"	Q-2	8	3		219	105		324	400	
0699	8"	"	7	3.429	↓	825	120		945	1,100	
0700	Standard weight, galvanized cast iron										
0720	90° Elbow, straight										
0730	1/4"	1 Plum	16	.500	Ea.	6.10	18.70		24.80	34.50	
0740	3/8"		16	.500		6.10	18.70		24.80	34.50	
0750	1/2"		15	.533		6.10	19.90		26	37	
0760	3/4"		14	.571		6.85	21.50		28.35	39.50	
0770	1"	↓	13	.615		7.90	23		30.90	43	
0780	1-1/4"	Q-1	22	.727		12.30	24.50		36.80	50.50	
0790	1-1/2"		20	.800		16.85	27		43.85	59	
0800	2"		18	.889		25	30		55	72.50	
0810	2-1/2"		14	1.143		51	38.50		89.50	115	
0820	3"		10	1.600		78	54		132	167	
0830	3-1/2"		8	2		124	67.50		191.50	239	
0840	4"		6	2.667		143	89.50		232.50	292	
0850	5"	↓	5	3.200		144	108		252	320	
0860	6"	Q-2	7	3.429		320	120		440	530	
0870	8"	"	6	4	↓	650	139		789	925	

15107 | Metal Pipe & Fittings

		CREW	DAILY OUTPUT	LABOR-HOURS	UNIT	2003 BARE COSTS				TOTAL INCL O&P
						MAT.	LABOR	EQUIP.	TOTAL	
640	0900	45° Elbow, straight									**640**
	0910	1/4"	1 Plum	16	.500	Ea.	6.95	18.70		25.65	35.50
	0920	3/8"		16	.500		6.95	18.70		25.65	35.50
	0930	1/2"		15	.533		6.95	19.90		26.85	37.50
	0940	3/4"		14	.571		8	21.50		29.50	41
	0950	1"		13	.615		10.10	23		33.10	45.50
	0960	1-1/4"	Q-1	22	.727		13.95	24.50		38.45	52.50
	0970	1-1/2"		20	.800		20.50	27		47.50	63.50
	0980	2"		18	.889		29	30		59	77
	0990	2-1/2"		14	1.143		57	38.50		95.50	121
	1000	3"		10	1.600		91.50	54		145.50	182
	1010	3-1/2"		8	2		168	67.50		235.50	286
	1020	4"		6	2.667		168	89.50		257.50	320
	1030	5"		5	3.200		207	108		315	390
	1040	6"	Q-2	7	3.429		266	120		386	475
	1050	8"	"	6	4		605	139		744	875
	1100	Tee, straight									
	1110	1/4"	1 Plum	10	.800	Ea.	7.80	30		37.80	53.50
	1120	3/8"		10	.800		7.80	30		37.80	53.50
	1130	1/2"		9	.889		6.75	33		39.75	57.50
	1140	3/4"		9	.889		9.85	33		42.85	61
	1150	1"		8	1		10.65	37.50		48.15	68.50
	1160	1-1/4"	Q-1	14	1.143		18.65	38.50		57.15	78.50
	1170	1-1/2"		13	1.231		24.50	41.50		66	89.50
	1180	2"		11	1.455		31	49		80	108
	1190	2-1/2"		9	1.778		66.50	60		126.50	164
	1200	3"		6	2.667		172	89.50		261.50	325
	1210	3-1/2"		5	3.200		176	108		284	355
	1220	4"		4	4		198	135		333	420
	1230	5"		3	5.333		249	179		428	545
	1240	6"	Q-2	4	6		299	209		508	645
	1250	8"	"	3	8		680	279		959	1,175
	1300	Extra heavy weight, black									
	1310	Couplings, steel straight									
	1320	1/4"	1 Plum	19	.421	Ea.	2.86	15.75		18.61	27
	1330	3/8"		19	.421		3.12	15.75		18.87	27.50
	1340	1/2"		19	.421		4.22	15.75		19.97	28.50
	1350	3/4"		18	.444		4.50	16.60		21.10	30
	1360	1"		15	.533		5.75	19.90		25.65	36.50
	1370	1-1/4"	Q-1	26	.615		9.20	20.50		29.70	41.50
	1380	1-1/2"		24	.667		9.20	22.50		31.70	44
	1390	2"		21	.762		14	25.50		39.50	54
	1400	2-1/2"		18	.889		21	30		51	68
	1410	3"		14	1.143		24.50	38.50		63	85
	1420	3-1/2"		12	1.333		33	45		78	104
	1430	4"		10	1.600		39	54		93	125
	1440	5"		6	2.667		55.50	89.50		145	197
	1450	6"	Q-2	8	3		68.50	105		173.50	233
	1460	8"		7	3.429		83.50	120		203.50	273
	1470	10"		6	4		132	139		271	355
	1480	12"		4	6		177	209		386	510
	1510	90° Elbow, straight									
	1520	1/2"	1 Plum	15	.533	Ea.	10.60	19.90		30.50	41.50
	1530	3/4"		14	.571		11.20	21.50		32.70	44.50
	1540	1"		13	.615		13.55	23		36.55	49.50
	1550	1-1/4"	Q-1	22	.727		20	24.50		44.50	59

Important: See the Reference Section for critical supporting data - Reference Nos., Crews, & City Cost Indexes

15107	Metal Pipe & Fittings	CREW	DAILY OUTPUT	LABOR-HOURS	UNIT	2003 BARE COSTS				TOTAL INCL O&P
						MAT.	LABOR	EQUIP.	TOTAL	
1560	1-1/2"	Q-1	20	.800	Ea.	25	27		52	68
1580	2"		18	.889		31	30		61	79
1590	2-1/2"		14	1.143		60	38.50		98.50	124
1600	3"		10	1.600		100	54		154	192
1610	4"	↓	6	2.667		167	89.50		256.50	320
1620	6"	Q-2	7	3.429	↓	435	120		555	655
1650	45° Elbow, straight									
1660	1/2"	1 Plum	15	.533	Ea.	12.60	19.90		32.50	44
1670	3/4"		14	.571		14.25	21.50		35.75	47.50
1680	1"	↓	13	.615		17.10	23		40.10	53.50
1690	1-1/4"	Q-1	22	.727		24	24.50		48.50	63.50
1700	1-1/2"		20	.800		31	27		58	74.50
1710	2"		18	.889		44	30		74	93.50
1720	2-1/2"	↓	14	1.143	↓	66.50	38.50		105	131
1800	Tee, straight									
1810	1/2"	1 Plum	9	.889	Ea.	16.35	33		49.35	68
1820	3/4"		9	.889		16.70	33		49.70	68.50
1830	1"	↓	8	1		20	37.50		57.50	78.50
1840	1-1/4"	Q-1	14	1.143		30	38.50		68.50	91
1850	1-1/2"		13	1.231		38.50	41.50		80	105
1860	2"		11	1.455		48	49		97	127
1870	2-1/2"		9	1.778		85.50	60		145.50	185
1880	3"		6	2.667		148	89.50		237.50	298
1890	4"	↓	4	4		233	135		368	460
1900	6"	Q-2	4	6	↓	515	209		724	880
4000	Standard weight, black									
4010	Couplings, steel straight, merchants									
4030	1/4"	1 Plum	19	.421	Ea.	.70	15.75		16.45	25
4040	3/8"		19	.421		.84	15.75		16.59	25
4050	1/2"		19	.421		.90	15.75		16.65	25
4060	3/4"		18	.444		1.14	16.60		17.74	26.50
4070	1"	↓	15	.533		1.60	19.90		21.50	32
4080	1-1/4"	Q-1	26	.615		2.09	20.50		22.59	34
4090	1-1/2"		24	.667		2.59	22.50		25.09	37
4100	2"		21	.762		3.70	25.50		29.20	42.50
4110	2-1/2"		18	.889		10.55	30		40.55	56.50
4120	3"		14	1.143		14.80	38.50		53.30	74.50
4130	3-1/2"		12	1.333		26	45		71	96.50
4140	4"		10	1.600		26	54		80	111
4150	5"	↓	6	2.667		48	89.50		137.50	188
4160	6"	Q-2	8	3		57.50	105		162.50	222
4166	Plug, 1/4"	1 Plum	38	.211		1.15	7.85		9	13.10
4167	3/8"		38	.211		1.07	7.85		8.92	13.05
4168	1/2"		38	.211		.93	7.85		8.78	12.85
4169	3/4"		32	.250		1.18	9.35		10.53	15.40
4170	1"	↓	30	.267		1.18	9.95		11.13	16.35
4171	1-1/4"	Q-1	52	.308		1.88	10.35		12.23	17.65
4172	1-1/2"		48	.333		2.66	11.20		13.86	19.90
4173	2"		42	.381		3.10	12.80		15.90	23
4176	2-1/2"	↓	36	.444	↓	7.55	14.95		22.50	31
4200	Standard weight, galvanized									
4210	Couplings, steel straight, merchants									
4230	1/4"	1 Plum	19	.421	Ea.	1.46	15.75		17.21	25.50
4240	3/8"		19	.421		1.61	15.75		17.36	26
4250	1/2"		19	.421		3.21	15.75		18.96	27.50
4260	3/4"	↓	18	.444	↓	3.44	16.60		20.04	29

640

		CREW	DAILY OUTPUT	LABOR-HOURS	UNIT	2003 BARE COSTS				TOTAL INCL O&P
15107	**Metal Pipe & Fittings**					MAT.	LABOR	EQUIP.	TOTAL	
640	4270	1"	1 Plum	15	.533	Ea.	4.43	19.90		24.33	35	640
	4280	1-1/4"	Q-1	26	.615		6.15	20.50		26.65	38.50
	4290	1-1/2"		24	.667		6.15	22.50		28.65	41
	4300	2"		21	.762		10.45	25.50		35.95	50
	4310	2-1/2"		18	.889		11.20	30		41.20	57.50
	4320	3"		14	1.143		12.90	38.50		51.40	72
	4330	3-1/2"		12	1.333		22.50	45		67.50	92.50
	4340	4"		10	1.600		22.50	54		76.50	107
	4350	5"		6	2.667		42.50	89.50		132	182
	4360	6"	Q-2	8	3		52	105		157	216
	4370	Plug, galvanized, square head									
	4374	1/2"	1 Plum	38	.211	Ea.	2	7.85		9.85	14.05
	4375	3/4"		32	.250		2.17	9.35		11.52	16.50
	4376	1"		30	.267		2.17	9.95		12.12	17.45
	4377	1-1/4"	Q-1	52	.308		3.62	10.35		13.97	19.60
	4378	1-1/2"		48	.333		4.89	11.20		16.09	22.50
	4379	2"		42	.381		6.05	12.80		18.85	26
	4380	2-1/2"		36	.444		12.85	14.95		27.80	36.50
	4381	3"		28	.571		16.45	19.20		35.65	47
	4382	4"		20	.800		40	27		67	84.50
	4384	6"		8	2		75.50	67.50		143	185
	5000	Malleable iron, 150 lb.									
	5020	Black									
	5040	90° elbow, straight									
	5060	1/4"	1 Plum	16	.500	Ea.	1.79	18.70		20.49	30
	5070	3/8"		16	.500		1.79	18.70		20.49	30
	5080	1/2"		15	.533		1.16	19.90		21.06	31.50
	5090	3/4"		14	.571		1.38	21.50		22.88	33.50
	5100	1"		13	.615		2.45	23		25.45	37
	5110	1-1/4"	Q-1	22	.727		4.02	24.50		28.52	41.50
	5120	1-1/2"		20	.800		5.30	27		32.30	46.50
	5130	2"		18	.889		9.15	30		39.15	55
	5140	2-1/2"		14	1.143		20.50	38.50		59	80.50
	5150	3"		10	1.600		30	54		84	115
	5160	3-1/2"		8	2		64	67.50		131.50	173
	5170	4"		6	2.667		64	89.50		153.50	206
	5180	5"		5	3.200		197	108		305	380
	5190	6"	Q-2	7	3.429		197	120		317	400
	5250	45° elbow, straight									
	5270	1/4"	1 Plum	16	.500	Ea.	2.53	18.70		21.23	31
	5280	3/8"		16	.500		2.53	18.70		21.23	31
	5290	1/2"		15	.533		1.92	19.90		21.82	32
	5300	3/4"		14	.571		2.37	21.50		23.87	34.50
	5310	1"		13	.615		3	23		26	38
	5320	1-1/4"	Q-1	22	.727		5.30	24.50		29.80	43
	5330	1-1/2"		20	.800		6.55	27		33.55	47.50
	5340	2"		18	.889		9.90	30		39.90	56
	5350	2-1/2"		14	1.143		28.50	38.50		67	89.50
	5360	3"		10	1.600		37.50	54		91.50	123
	5370	3-1/2"		8	2		73	67.50		140.50	183
	5380	4"		6	2.667		73	89.50		162.50	216
	5390	5"		5	3.200		262	108		370	450
	5400	6"	Q-2	7	3.429		262	120		382	470
	5450	Tee, straight									
	5470	1/4"	1 Plum	10	.800	Ea.	2.45	30		32.45	47.50
	5480	3/8"		10	.800		2.45	30		32.45	47.50

Important: See the Reference Section for critical supporting data - Reference Nos., Crews, & City Cost Indexes

		CREW	DAILY OUTPUT	LABOR-HOURS	UNIT	2003 BARE COSTS				TOTAL INCL O&P	
15107	**Metal Pipe & Fittings**					MAT.	LABOR	EQUIP.	TOTAL		
640 5490	1/2"	1 Plum	9	.889	Ea.	1.56	33		34.56	51.50	640
5500	3/4"		9	.889		2.24	33		35.24	52.50	
5510	1"	▼	8	1		3.82	37.50		41.32	60.50	
5520	1-1/4"	Q-1	14	1.143		6.20	38.50		44.70	65	
5530	1-1/2"		13	1.231		7.70	41.50		49.20	71	
5540	2"		11	1.455		13.15	49		62.15	88.50	
5550	2-1/2"		9	1.778		28.50	60		88.50	122	
5560	3"		6	2.667		41.50	89.50		131	181	
5570	3-1/2"		5	3.200		101	108		209	274	
5580	4"		4	4		101	135		236	315	
5590	5"	▼	3	5.333		225	179		404	520	
5600	6"	Q-2	4	6	▼	291	209		500	635	
5601	Tee, reducing, on outlet										
5602	1/2"	1 Plum	9	.889	Ea.	3.40	33		36.40	53.50	
5603	3/4"		9	.889		4.25	33		37.25	54.50	
5604	1"	▼	8	1		4.36	37.50		41.86	61.50	
5605	1-1/4"	Q-1	14	1.143		7.55	38.50		46.05	66.50	
5606	1-1/2"		13	1.231		10.85	41.50		52.35	74.50	
5607	2"		11	1.455		15.80	49		64.80	91.50	
5608	2-1/2"		9	1.778		31.50	60		91.50	126	
5609	3"		6	2.667		62	89.50		151.50	204	
5610	3-1/2"		5	3.200		113	108		221	287	
5611	4"		4	4		110	135		245	325	
5612	5"	▼	3	5.333	▼	137	179		316	420	
5650	Coupling										
5670	1/4"	1 Plum	19	.421	Ea.	2.08	15.75		17.83	26.50	
5680	3/8"		19	.421		2.08	15.75		17.83	26.50	
5690	1/2"		19	.421		1.61	15.75		17.36	26	
5700	3/4"		18	.444		2	16.60		18.60	27	
5710	1"	▼	15	.533		2.82	19.90		22.72	33	
5720	1-1/4"	Q-1	26	.615		3.66	20.50		24.16	35.50	
5730	1-1/2"		24	.667		4.94	22.50		27.44	39.50	
5740	2"		21	.762		7.30	25.50		32.80	46.50	
5750	2-1/2"		18	.889		20	30		50	67	
5760	3"		14	1.143		27.50	38.50		66	88	
5770	3-1/2"		12	1.333		55	45		100	128	
5780	4"		10	1.600		55	54		109	142	
5790	5"	▼	6	2.667		114	89.50		203.50	260	
5800	6"	Q-2	8	3		114	105		219	283	
5840	Reducer, concentric, 1/4"	1 Plum	19	.421		2.20	15.75		17.95	26.50	
5850	3/8"		19	.421		2.20	15.75		17.95	26.50	
5860	1/2"		19	.421		2	15.75		17.75	26	
5870	3/4"		16	.500		2.12	18.70		20.82	30.50	
5880	1"	▼	15	.533		3.37	19.90		23.27	33.50	
5890	1-1/4"	Q-1	26	.615		4.97	20.50		25.47	37	
5900	1-1/2"		24	.667		6.20	22.50		28.70	41	
5910	2"		21	.762		8.90	25.50		34.40	48.50	
5911	2-1/2"		18	.889		21	30		51	68	
5912	3"		14	1.143		28	38.50		66.50	88.50	
5913	3-1/2"		12	1.333		71	45		116	146	
5914	4"		10	1.600		56	54		110	143	
5915	5"	▼	6	2.667		112	89.50		201.50	258	
5916	6"	Q-2	8	3		143	105		248	315	
5981	Bushing, 1/4"	1 Plum	19	.421		2.09	15.75		17.84	26.50	
5982	3/8"		19	.421		2.09	15.75		17.84	26.50	
5983	1/2"	▼	19	.421	▼	2.09	15.75		17.84	26.50	

MECHANICAL 15

640

		CREW	DAILY OUTPUT	LABOR-HOURS	UNIT	MAT.	LABOR	EQUIP.	TOTAL	TOTAL INCL O&P	
	**15107	Metal Pipe & Fittings**						2003 BARE COSTS				
5984	3/4"	1 Plum	16	.500	Ea.	2.09	18.70		20.79	30.50	640
5985	1"	↓	15	.533		2.73	19.90		22.63	33	
5986	1-1/4"	Q-1	26	.615		3.81	20.50		24.31	35.50	
5987	1-1/2"		24	.667		4.56	22.50		27.06	39	
5988	2"	↓	21	.762		4.95	25.50		30.45	44	
5989	Cap, 1/4"	1 Plum	38	.211		1.54	7.85		9.39	13.55	
5991	3/8"		38	.211		1.30	7.85		9.15	13.30	
5992	1/2"		38	.211		1.17	7.85		9.02	13.15	
5993	3/4"		32	.250		1.59	9.35		10.94	15.85	
5994	1"	↓	30	.267		1.93	9.95		11.88	17.15	
5995	1-1/4"	Q-1	52	.308		2.54	10.35		12.89	18.40	
5996	1-1/2"		48	.333		3.49	11.20		14.69	21	
5997	2"	↓	42	.381		5.10	12.80		17.90	25	
6000	For galvanized elbows, tees, and couplings add				↓	20%					
6058	For galvanized reducers, caps and plugs add					20%					
6100	90° elbow, galvanized, 150 lb., reducing										
6110	3/4" x 1/2"	1 Plum	15.40	.519	Ea.	3.37	19.40		22.77	33	
6112	1" x 3/4"	↓	14	.571		4.32	21.50		25.82	37	
6114	1" x 1/2"	↓	14.50	.552		4.58	20.50		25.08	36	
6116	1-1/4" x 1"	Q-1	24.20	.661		7.25	22		29.25	41.50	
6118	1-1/4" x 3/4"		25.40	.630		8.70	21		29.70	41.50	
6120	1-1/4" x 1/2"		26.20	.611		9.25	20.50		29.75	41	
6122	1-1/2" x 1-1/4"		21.60	.741		11.60	25		36.60	50.50	
6124	1-1/2" x 1"		23.50	.681		11.60	23		34.60	47.50	
6126	1-1/2" x 3/4"		24.60	.650		11.60	22		33.60	46	
6128	2" x 1-1/2"		20.50	.780		12.10	26.50		38.60	53	
6130	2" x 1-1/4"		21	.762		15.50	25.50		41	55.50	
6132	2" x 1"		22.80	.702		16.10	23.50		39.60	53	
6134	2" x 3/4"		23.90	.669		16.40	22.50		38.90	52	
6136	2-1/2" x 2"		12.30	1.301		46	44		90	117	
6138	2-1/2" x 1-1/2"		12.50	1.280		46	43		89	116	
6140	3" x 2-1/2"		8.60	1.860		74	62.50		136.50	176	
6142	3" x 2"		11.80	1.356		72.50	45.50		118	149	
6144	4" x 3"	↓	8.20	1.951	↓	143	65.50		208.50	256	
6160	90° elbow, black, 150 lb., reducing										
6170	1" x 3/4"	1 Plum	14	.571	Ea.	3.12	21.50		24.62	35.50	
6174	1-1/2" x 1"	Q-1	23.50	.681		7.05	23		30.05	42.50	
6178	1-1/2" x 3/4"		24.60	.650		8.45	22		30.45	42.50	
6182	2" x 1-1/2"		20.50	.780		10.70	26.50		37.20	51.50	
6186	2" x 1"		22.80	.702		12.20	23.50		35.70	49	
6190	2" x 3/4"		23.90	.669		12.90	22.50		35.40	48	
6194	2-1/2" x 2"	↓	12.30	1.301	↓	29.50	44		73.50	98.50	
7000	Union, with brass seat										
7010	1/4"	1 Plum	15	.533	Ea.	9.60	19.90		29.50	40.50	
7020	3/8"		15	.533		9.60	19.90		29.50	40.50	
7030	1/2"		14	.571		6	21.50		27.50	38.50	
7040	3/4"		13	.615		6.30	23		29.30	41.50	
7050	1"	↓	12	.667		9	25		34	47.50	
7060	1-1/4"	Q-1	21	.762		12	25.50		37.50	51.50	
7070	1-1/2"		19	.842		16.10	28.50		44.60	60.50	
7080	2"		17	.941		18.70	31.50		50.20	68.50	
7090	2-1/2"		13	1.231		55.50	41.50		97	124	
7100	3"	↓	9	1.778	↓	67	60		127	165	
7120	Union, galvanized										
7124	1/2"	1 Plum	14	.571	Ea.	6.85	21.50		28.35	39.50	
7125	3/4"	↓	13	.615	↓	7.90	23		30.90	43	

Important: See the Reference Section for critical supporting data - Reference Nos., Crews, & City Cost Indexes

15 MECHANICAL

15107	Metal Pipe & Fittings	CREW	DAILY OUTPUT	LABOR-HOURS	UNIT	2003 BARE COSTS				TOTAL INCL O&P
						MAT.	LABOR	EQUIP.	TOTAL	
640	7126	1"	1 Plum	12	.667	Ea.	10.35	25		35.35	49	640
7127	1-1/4"	Q-1	21	.762		14.95	25.50		40.45	55
7128	1-1/2"		19	.842		18.10	28.50		46.60	63
7129	2"		17	.941		21	31.50		52.50	71
7130	2-1/2"		13	1.231		72.50	41.50		114	142
7131	3"		9	1.778		101	60		161	202
7500	Malleable iron, 300 lb									
7520	Black									
7540	90° Elbow, straight, 1/4"	1 Plum	16	.500	Ea.	6.90	18.70		25.60	35.50
7560	3/8"		16	.500		6.10	18.70		24.80	34.50
7570	1/2"		15	.533		7.90	19.90		27.80	38.50
7580	3/4"		14	.571		8.95	21.50		30.45	42
7590	1"		13	.615		11.45	23		34.45	47
7600	1-1/4"	Q-1	22	.727		16.70	24.50		41.20	55.50
7610	1-1/2"		20	.800		19.70	27		46.70	62
7620	2"		18	.889		28.50	30		58.50	76
7630	2-1/2"		14	1.143		67	38.50		105.50	132
7640	3"		10	1.600		84.50	54		138.50	174
7650	4"		6	2.667		176	89.50		265.50	330
7700	45° Elbow, straight, 1/4"	1 Plum	16	.500		10.25	18.70		28.95	39.50
7720	3/8"		16	.500		10.25	18.70		28.95	39.50
7730	1/2"		15	.533		11.40	19.90		31.30	42.50
7740	3/4"		14	.571		12.70	21.50		34.20	46
7750	1"		13	.615		14.05	23		37.05	50
7760	1-1/4"	Q-1	22	.727		22.50	24.50		47	61.50
7770	1-1/2"		20	.800		29.50	27		56.50	72.50
7780	2"		18	.889		44.50	30		74.50	93.50
7790	2-1/2"		14	1.143		88	38.50		126.50	155
7800	3"		10	1.600		115	54		169	209
7810	4"		6	2.667		223	89.50		312.50	380
7850	Tee, straight, 1/4"	1 Plum	10	.800		8.65	30		38.65	54.50
7870	3/8"		10	.800		9.20	30		39.20	55
7880	1/2"		9	.889		11.70	33		44.70	63
7890	3/4"		9	.889		12.60	33		45.60	64
7900	1"		8	1		15.15	37.50		52.65	73
7910	1-1/4"	Q-1	14	1.143		21.50	38.50		60	82
7920	1-1/2"		13	1.231		25.50	41.50		67	90.50
7930	2"		11	1.455		37.50	49		86.50	115
7940	2-1/2"		9	1.778		81	60		141	180
7950	3"		6	2.667		130	89.50		219.50	278
7960	4"		4	4		279	135		414	510
7980	6"	Q-2	4	6		296	209		505	640
7990	8"	"	3	8		465	279		744	930
8050	Couplings, straight, 1/4"	1 Plum	19	.421		6.20	15.75		21.95	31
8070	3/8"		19	.421		6.20	15.75		21.95	31
8080	1/2"		19	.421		6.95	15.75		22.70	31.50
8090	3/4"		18	.444		7.95	16.60		24.55	34
8100	1"		15	.533		9.10	19.90		29	40
8110	1-1/4"	Q-1	26	.615		10.85	20.50		31.35	43.50
8120	1-1/2"		24	.667		16.20	22.50		38.70	52
8130	2"		21	.762		23.50	25.50		49	64
8140	2-1/2"		18	.889		38	30		68	87
8150	3"		14	1.143		55	38.50		93.50	119
8160	4"		10	1.600		69	54		123	157
8200	Galvanized									
8220	90° Elbow, straight, 1/4"	1 Plum	16	.500	Ea.	11.65	18.70		30.35	41

		CREW	DAILY OUTPUT	LABOR-HOURS	UNIT	2003 BARE COSTS				TOTAL INCL O&P	
15107	**Metal Pipe & Fittings**					MAT.	LABOR	EQUIP.	TOTAL		
640	8222	3/8"	1 Plum	16	.500	Ea.	11.90	18.70		30.60	41	640
	8224	1/2"		15	.533		14.15	19.90		34.05	45.50
	8226	3/4"		14	.571		15.90	21.50		37.40	49.50
	8228	1"		13	.615		20.50	23		43.50	57
	8230	1-1/4"	Q-1	22	.727		32	24.50		56.50	72
	8232	1-1/2"		20	.800		34.50	27		61.50	78.50
	8234	2"		18	.889		58.50	30		88.50	109
	8236	2-1/2"		14	1.143		114	38.50		152.50	183
	8238	3"		10	1.600		153	54		207	250
	8240	4"		6	2.667		345	89.50		434.50	515
	8280	45° Elbow, straight									
	8282	1/2"	1 Plum	15	.533	Ea.	21	19.90		40.90	53
	8284	3/4"		14	.571		24.50	21.50		46	58.50
	8286	1"		13	.615		26.50	23		49.50	63.50
	8288	1-1/4"	Q-1	22	.727		41	24.50		65.50	82.50
	8290	1-1/2"		20	.800		53.50	27		80.50	99
	8292	2"		18	.889		74	30		104	126
	8310	Tee, straight, 1/4"	1 Plum	10	.800		16.55	30		46.55	63
	8312	3/8"		10	.800		16.90	30		46.90	63.50
	8314	1/2"		9	.889		21	33		54	73
	8316	3/4"		9	.889		23.50	33		56.50	75.50
	8318	1"		8	1		28.50	37.50		66	88
	8320	1-1/4"	Q-1	14	1.143		41	38.50		79.50	103
	8322	1-1/2"		13	1.231		42.50	41.50		84	110
	8324	2"		11	1.455		68	49		117	149
	8326	2-1/2"		9	1.778		163	60		223	270
	8328	3"		6	2.667		244	89.50		333.50	405
	8330	4"		4	4		475	135		610	725
	8380	Couplings, straight, 1/4"	1 Plum	19	.421		8.90	15.75		24.65	34
	8382	3/8"		19	.421		8.90	15.75		24.65	34
	8384	1/2"		19	.421		9.15	15.75		24.90	34
	8386	3/4"		18	.444		10.35	16.60		26.95	36.50
	8388	1"		15	.533		13.65	19.90		33.55	45
	8390	1-1/4"	Q-1	26	.615		17.80	20.50		38.30	51
	8392	1-1/2"		24	.667		28.50	22.50		51	65.50
	8394	2"		21	.762		34.50	25.50		60	76.50
	8396	2-1/2"		18	.889		68.50	30		98.50	121
	8398	3"		14	1.143		96	38.50		134.50	163
	8399	4"		10	1.600		102	54		156	194
	8529	Black									
	8530	Reducer, concentric, 1/4"	1 Plum	19	.421	Ea.	7.55	15.75		23.30	32.50
	8531	3/8"		19	.421		7.55	15.75		23.30	32.50
	8532	1/2"		17	.471		9.10	17.60		26.70	36.50
	8533	3/4"		16	.500		9.30	18.70		28	38
	8534	1"		15	.533		11.10	19.90		31	42
	8535	1-1/4"	Q-1	26	.615		15.15	20.50		35.65	48
	8536	1-1/2"		24	.667		21	22.50		43.50	57
	8537	2"		21	.762		29	25.50		54.50	70.50
	8550	Cap, 1/4"	1 Plum	38	.211		5.85	7.85		13.70	18.25
	8551	3/8"		38	.211		5.85	7.85		13.70	18.25
	8552	1/2"		34	.235		5.80	8.80		14.60	19.65
	8553	3/4"		32	.250		7.30	9.35		16.65	22
	8554	1"		30	.267		9.60	9.95		19.55	25.50
	8555	1-1/4"	Q-1	52	.308		10.65	10.35		21	27.50
	8556	1-1/2"		48	.333		15.55	11.20		26.75	34
	8557	2"		42	.381		21.50	12.80		34.30	43.50

Important: See the Reference Section for critical supporting data - Reference Nos., Crews, & City Cost Indexes

15107	Metal Pipe & Fittings	CREW	DAILY OUTPUT	LABOR-HOURS	UNIT	2003 BARE COSTS				TOTAL INCL O&P
						MAT.	LABOR	EQUIP.	TOTAL	
640 8570	Plug, 1/4"	1 Plum	38	.211	Ea.	1.94	7.85		9.79	14	**640**
8571	3/8"		38	.211		2.39	7.85		10.24	14.50
8572	1/2"		34	.235		2.39	8.80		11.19	15.90
8573	3/4"		32	.250		3.11	9.35		12.46	17.50
8574	1"	↓	30	.267		4.76	9.95		14.71	20.50
8575	1-1/4"	Q-1	52	.308		8.70	10.35		19.05	25
8576	1-1/2"		48	.333		10.40	11.20		21.60	28.50
8577	2"	↓	42	.381		17.35	12.80		30.15	38.50
9500	Union with brass seat, 1/4"	1 Plum	15	.533		12.80	19.90		32.70	44
9530	3/8"		15	.533		13	19.90		32.90	44.50
9540	1/2"		14	.571		10.50	21.50		32	43.50
9550	3/4"		13	.615		11.60	23		34.60	47.50
9560	1"	↓	12	.667		15.20	25		40.20	54
9570	1-1/4"	Q-1	21	.762		25	25.50		50.50	66
9580	1-1/2"		19	.842		26	28.50		54.50	71.50
9590	2"		17	.941		32	31.50		63.50	83.50
9600	2-1/2"		13	1.231		104	41.50		145.50	177
9610	3"		9	1.778		134	60		194	238
9620	4"	↓	5	3.200		330	108		438	530
9630	Union, all iron, 1/4"	1 Plum	15	.533		22	19.90		41.90	54
9650	3/8"		15	.533		23	19.90		42.90	55
9660	1/2"		14	.571		23	21.50		44.50	57
9670	3/4"		13	.615		24.50	23		47.50	61.50
9680	1"	↓	12	.667		30.50	25		55.50	71
9690	1-1/4"	Q-1	21	.762		46.50	25.50		72	89.50
9700	1-1/2"		19	.842		56.50	28.50		85	105
9710	2"		17	.941		71.50	31.50		103	127
9720	2-1/2"		13	1.231		261	41.50		302.50	350
9730	3"	↓	9	1.778		261	60		321	380
9750	For galvanized unions, add				↓	15%				
9757	Forged steel, 3000 lb.									
9758	Black									
9760	90° Elbow, 1/4"	1 Plum	16	.500	Ea.	8.25	18.70		26.95	37
9761	3/8"		16	.500		11.30	18.70		30	40.50
9762	1/2"		15	.533		8.40	19.90		28.30	39
9763	3/4"		14	.571		10.05	21.50		31.55	43
9764	1"	↓	13	.615		14.55	23		37.55	50.50
9765	1-1/4"	Q-1	22	.727		28.50	24.50		53	68.50
9766	1-1/2"		20	.800		36.50	27		63.50	80.50
9767	2"	↓	18	.889		44.50	30		74.50	94
9780	45° Elbow 1/4"	1 Plum	16	.500		12.40	18.70		31.10	41.50
9781	3/8"		16	.500		12.40	18.70		31.10	41.50
9782	1/2"		15	.533		12.40	19.90		32.30	43.50
9783	3/4"		14	.571		14.30	21.50		35.80	47.50
9784	1"	↓	13	.615		19.45	23		42.45	56
9785	1-1/4"	Q-1	22	.727		27	24.50		51.50	66.50
9786	1-1/2"		20	.800		40.50	27		67.50	85
9787	2"	↓	18	.889		55.50	30		85.50	107
9800	Tee, 1/4"	1 Plum	10	.800		12.80	30		42.80	59
9801	3/8"		10	.800		12.80	30		42.80	59
9802	1/2"		9	.889		11.05	33		44.05	62
9803	3/4"		9	.889		15.70	33		48.70	67.50
9804	1"	↓	8	1		19.95	37.50		57.45	78.50
9805	1-1/4"	Q-1	14	1.143		38.50	38.50		77	101
9806	1-1/2"		13	1.231		45.50	41.50		87	113
9807	2"	↓	11	1.455	↓	60.50	49		109.50	141

			DAILY	LABOR-			2003 BARE COSTS				TOTAL	
	15107	**Metal Pipe & Fittings**									INCL O&P	
			CREW	OUTPUT	HOURS	UNIT	MAT.	LABOR	EQUIP.	TOTAL		
640	9820	Reducer, concentric, 1/4"	1 Plum	19	.421	Ea.	6.60	15.75		22.35	31.50	**640**
	9821	3/8"		19	.421		6.60	15.75		22.35	31.50	
	9822	1/2"		17	.471		6.60	17.60		24.20	34	
	9823	3/4"		16	.500		7.40	18.70		26.10	36	
	9824	1"		15	.533		10.15	19.90		30.05	41	
	9825	1-1/4"	Q-1	26	.615		15.90	20.50		36.40	49	
	9826	1-1/2"		24	.667		18.50	22.50		41	54.50	
	9827	2"		21	.762		26.50	25.50		52	68	
	9840	Cap, 1/4"	1 Plum	38	.211		4.07	7.85		11.92	16.35	
	9841	3/8"		38	.211		4.07	7.85		11.92	16.35	
	9842	1/2"		34	.235		4.07	8.80		12.87	17.75	
	9843	3/4"		32	.250		5.65	9.35		15	20.50	
	9844	1"		30	.267		8.75	9.95		18.70	24.50	
	9845	1-1/4"	Q-1	52	.308		13.10	10.35		23.45	30	
	9846	1-1/2"		48	.333		15.15	11.20		26.35	33.50	
	9847	2"		42	.381		23.50	12.80		36.30	45	
	9860	Plug, 1/4"	1 Plum	38	.211		1.28	7.85		9.13	13.25	
	9861	3/8"		38	.211		1.60	7.85		9.45	13.60	
	9862	1/2"		34	.235		1.60	8.80		10.40	15	
	9863	3/4"		32	.250		2.79	9.35		12.14	17.15	
	9864	1"		30	.267		3.66	9.95		13.61	19.10	
	9865	1-1/4"	Q-1	52	.308		6.35	10.35		16.70	22.50	
	9866	1-1/2"		48	.333		8.20	11.20		19.40	26	
	9867	2"		42	.381		15.40	12.80		28.20	36.50	
	9880	Union, bronze seat, 1/4"	1 Plum	15	.533		14.80	19.90		34.70	46.50	
	9881	3/8"		15	.533		14.80	19.90		34.70	46.50	
	9882	1/2"		14	.571		14.80	21.50		36.30	48.50	
	9883	3/4"		13	.615		17.80	23		40.80	54	
	9884	1"		12	.667		23	25		48	63	
	9885	1-1/4"	Q-1	21	.762		39	25.50		64.50	81.50	
	9886	1-1/2"		19	.842		40.50	28.50		69	87.50	
	9887	2"		17	.941		52	31.50		83.50	106	
	9900	Coupling, 1/4"	1 Plum	19	.421		3.07	15.75		18.82	27.50	
	9901	3/8"		19	.421		3.41	15.75		19.16	28	
	9902	1/2"		17	.471		2.91	17.60		20.51	29.50	
	9903	3/4"		16	.500		4.13	18.70		22.83	32.50	
	9904	1"		15	.533		6.05	19.90		25.95	36.50	
	9905	1-1/4"	Q-1	26	.615		11.20	20.50		31.70	44	
	9906	1-1/2"		24	.667		15.40	22.50		37.90	51	
	9907	2"		21	.762		19.15	25.50		44.65	59.50	
660	0010	**PIPE, STEEL, FITTINGS** Flanged, welded and special type										**660**
	0020	flanged joints, C.I., standard weight, black. One gasket & bolt										
	0040	set, mat'l only, required at each joint, not included (see line 0620)										
	0060	90° Elbow, straight, 1-1/2" pipe size	Q-1	14	1.143	Ea.	143	38.50		181.50	216	
	0080	2" pipe size		13	1.231		112	41.50		153.50	187	
	0090	2-1/2" pipe size		12	1.333		122	45		167	202	
	0100	3" pipe size		11	1.455		101	49		150	186	
	0110	4" pipe size		8	2		126	67.50		193.50	240	
	0120	5" pipe size		7	2.286		231	77		308	370	
	0130	6" pipe size	Q-2	9	2.667		197	93		290	355	
	0140	8" pipe size		8	3		310	105		415	500	
	0150	10" pipe size		7	3.429		680	120		800	925	
	0160	12" pipe size		6	4		1,175	139		1,314	1,475	
	0171	90° Elbow, reducing										
	0172	2-1/2" by 2" pipe size	Q-1	12	1.333	Ea.	278	45		323	375	
	0173	3" by 2-1/2" pipe size		11	1.455		292	49		341	395	

Important: See the Reference Section for critical supporting data - Reference Nos., Crews, & City Cost Indexes

15107	Metal Pipe & Fittings	CREW	DAILY OUTPUT	LABOR-HOURS	UNIT	2003 BARE COSTS				TOTAL INCL O&P
						MAT.	LABOR	EQUIP.	TOTAL	
660 0174	4" by 3" pipe size	Q-1	8	2	Ea.	330	67.50		397.50	465 **660**
0175	5" by 3" pipe size	↓	7	2.286		425	77		502	580
0176	6" by 4" pipe size	Q-2	9	2.667		475	93		568	665
0177	8" by 6" pipe size		8	3		700	105		805	930
0178	10" by 8" pipe size		7	3.429		1,100	120		1,220	1,375
0179	12" by 10" pipe size	↓	6	4		1,900	139		2,039	2,275
0200	45° Elbow, straight, 1-1/2" pipe size	Q-1	14	1.143		173	38.50		211.50	248
0220	2" pipe size		13	1.231		172	41.50		213.50	252
0230	2-1/2" pipe size		12	1.333		173	45		218	259
0240	3" pipe size		11	1.455		169	49		218	260
0250	4" pipe size		8	2		190	67.50		257.50	310
0260	5" pipe size	↓	7	2.286		355	77		432	505
0270	6" pipe size	Q-2	9	2.667		310	93		403	480
0280	8" pipe size		8	3		455	105		560	660
0290	10" pipe size		7	3.429		960	120		1,080	1,225
0300	12" pipe size	↓	6	4	↓	1,475	139		1,614	1,825
0310	Cross, straight									
0311	2-1/2" pipe size	Q-1	6	2.667	Ea.	375	89.50		464.50	550
0312	3" pipe size		5	3.200		375	108		483	580
0313	4" pipe size		4	4		490	135		625	745
0314	5" pipe size	↓	3	5.333		800	179		979	1,150
0315	6" pipe size	Q-2	5	4.800		800	167		967	1,125
0316	8" pipe size		4	6		1,275	209		1,484	1,750
0317	10" pipe size		3	8		1,975	279		2,254	2,575
0318	12" pipe size	↓	2	12		2,975	420		3,395	3,900
0350	Tee, straight, 1-1/2" pipe size	Q-1	10	1.600		123	54		177	217
0370	2" pipe size		9	1.778		123	60		183	226
0380	2-1/2" pipe size		8	2		180	67.50		247.50	300
0390	3" pipe size		7	2.286		126	77		203	254
0400	4" pipe size		5	3.200		192	108		300	375
0410	5" pipe size	↓	4	4		400	135		535	645
0420	6" pipe size	Q-2	6	4		277	139		416	515
0430	8" pipe size		5	4.800		435	167		602	730
0440	10" pipe size		4	6		990	209		1,199	1,425
0450	12" pipe size	↓	3	8	↓	1,550	279		1,829	2,150
0459	Tee, reducing on outlet									
0460	2-1/2" by 2" pipe size	Q-1	8	2	Ea.	265	67.50		332.50	395
0461	3" by 2-1/2" pipe size		7	2.286		305	77		382	450
0462	4" by 3" pipe size		5	3.200		745	108		853	985
0463	5" by 4" pipe size	↓	4	4		940	135		1,075	1,225
0464	6" by 4" pipe size	Q-2	6	4		940	139		1,079	1,225
0465	8" by 6" pipe size		5	4.800		940	167		1,107	1,275
0466	10" by 8" pipe size		4	6		1,025	209		1,234	1,450
0467	12" by 10" pipe size	↓	3	8	↓	1,600	279		1,879	2,200
0476	Reducer, concentric									
0477	3" by 2-1/2"	Q-1	12	1.333	Ea.	281	45		326	380
0478	4" by 3"		9	1.778		300	60		360	420
0479	5" by 4"	↓	8	2		475	67.50		542.50	620
0480	6" by 4"	Q-2	10	2.400		530	83.50		613.50	710
0481	8" by 6"		9	2.667		600	93		693	800
0482	10" by 8"		8	3		960	105		1,065	1,200
0483	12" by 10"	↓	7	3.429		1,750	120		1,870	2,100
0492	Reducer, eccentric									
0493	4" by 3"	Q-1	8	2	Ea.	370	67.50		437.50	505
0494	5" by 4"	"	7	2.286		475	77		552	635
0495	6" by 4"	Q-2	9	2.667	↓	505	93		598	695

15107 | Metal Pipe & Fittings

		CREW	DAILY OUTPUT	LABOR-HOURS	UNIT	2003 BARE COSTS				TOTAL INCL O&P
						MAT.	LABOR	EQUIP.	TOTAL	
660 0496	8" by 6"	Q-2	8	3	Ea.	665	105		770	895	**660**
0497	10" by 8"		7	3.429		1,150	120		1,270	1,425
0498	12" by 10"	↓	6	4		1,750	139		1,889	2,125
0500	For galvanized elbows and tees, add					100%				
0520	For extra heavy weight elbows and tees, add					140%				
0620	Gasket and bolt set, 150#, 1/2" pipe size	1 Plum	20	.400		2.36	14.95		17.31	25
0622	3/4" pipe size		19	.421		2.36	15.75		18.11	26.50
0624	1" pipe size		18	.444		2.42	16.60		19.02	27.50
0626	1-1/4" pipe size		17	.471		2.54	17.60		20.14	29.50
0628	1-1/2" pipe size		15	.533		2.71	19.90		22.61	33
0630	2" pipe size		13	.615		4.33	23		27.33	39.50
0640	2-1/2" pipe size		12	.667		4.58	25		29.58	42.50
0650	3" pipe size		11	.727		4.80	27		31.80	46.50
0660	3-1/2" pipe size		9	.889		8.70	33		41.70	59.50
0670	4" pipe size		8	1		8.70	37.50		46.20	66
0680	5" pipe size		7	1.143		12.90	42.50		55.40	78.50
0690	6" pipe size		6	1.333		13.55	50		63.55	90
0700	8" pipe size		5	1.600		14.45	60		74.45	106
0710	10" pipe size		4.50	1.778		30	66.50		96.50	133
0720	12" pipe size		4.20	1.905		32	71		103	143
0730	14" pipe size		4	2		45.50	74.50		120	163
0740	16" pipe size		3	2.667		59.50	99.50		159	216
0750	18" pipe size		2.70	2.963		83	111		194	259
0760	20" pipe size		2.30	3.478		111	130		241	320
0780	24" pipe size		1.90	4.211		149	157		306	400
0790	26" pipe size		1.60	5		168	187		355	465
0810	30" pipe size		1.40	5.714		208	213		421	550
0830	36" pipe size	↓	1.10	7.273	↓	256	272		528	690
0850	For 300 lb gasket set, add					40%				
2000	Flanged unions, 125 lb., black, 1/2" pipe size	1 Plum	17	.471	Ea.	28	17.60		45.60	57.50
2040	3/4" pipe size		17	.471		37	17.60		54.60	67
2050	1" pipe size	↓	16	.500		36	18.70		54.70	67.50
2060	1-1/4" pipe size	Q-1	28	.571		42.50	19.20		61.70	75.50
2070	1-1/2" pipe size		27	.593		39	19.95		58.95	73
2080	2" pipe size		26	.615		47	20.50		67.50	83.50
2090	2-1/2" pipe size		24	.667		63.50	22.50		86	104
2100	3" pipe size		22	.727		72	24.50		96.50	117
2110	3-1/2" pipe size		18	.889		150	30		180	210
2120	4" pipe size		16	1		98	33.50		131.50	159
2130	5" pipe size	↓	14	1.143		174	38.50		212.50	249
2140	6" pipe size	Q-2	19	1.263		210	44		254	297
2150	8" pipe size	"	16	1.500		360	52.50		412.50	480
2200	For galvanized unions, add				↓	150%				
2290	Threaded flange									
2300	Cast iron									
2310	Black, 125 lb., per flange									
2320	1" pipe size	1 Plum	27	.296	Ea.	16.40	11.05		27.45	35
2330	1-1/4" pipe size	Q-1	44	.364		19.60	12.25		31.85	40
2340	1-1/2" pipe size		40	.400		18.15	13.45		31.60	40.50
2350	2" pipe size		36	.444		18.10	14.95		33.05	42.50
2360	2-1/2" pipe size		28	.571		21	19.20		40.20	52.50
2370	3" pipe size		20	.800		27.50	27		54.50	70.50
2380	3-1/2" pipe size		16	1		36.50	33.50		70	91.50
2390	4" pipe size		12	1.333		36.50	45		81.50	108
2400	5" pipe size	↓	10	1.600		52	54		106	139
2410	6" pipe size	Q-2	14	1.714	↓	58.50	60		118.50	155

Important: See the Reference Section for critical supporting data - Reference Nos., Crews, & City Cost Indexes

		DAILY	LABOR-		2003 BARE COSTS				TOTAL
15107	**Metal Pipe & Fittings**	CREW	OUTPUT	HOURS	UNIT	MAT.	LABOR	EQUIP.	TOTAL	INCL O&P

		CREW	DAILY OUTPUT	LABOR-HOURS	UNIT	MAT.	LABOR	EQUIP.	TOTAL	INCL O&P
2420	8" pipe size	Q-2	12	2	Ea.	92.50	69.50		162	207
2430	10" pipe size		10	2.400		161	83.50		244.50	305
2440	12" pipe size	↓	8	3	↓	267	105		372	450
2460	For galvanized flanges, add					95%				
2580	Forged steel,									
2590	Black 150 lb., per flange									
2600	1/2" pipe size	1 Plum	30	.267	Ea.	17.30	9.95		27.25	34
2610	3/4" pipe size		28	.286		17.30	10.65		27.95	35
2620	1" pipe size	↓	27	.296		17.30	11.05		28.35	35.50
2630	1-1/4" pipe size	Q-1	44	.364		17.30	12.25		29.55	37.50
2640	1-1/2" pipe size		40	.400		17.30	13.45		30.75	39.50
2650	2" pipe size		36	.444		17.30	14.95		32.25	41.50
2660	2-1/2" pipe size		28	.571		21	19.20		40.20	52
2670	3" pipe size		20	.800		21	27		48	63.50
2690	4" pipe size		12	1.333		24.50	45		69.50	94.50
2700	5" pipe size	↓	10	1.600		36	54		90	122
2710	6" pipe size	Q-2	14	1.714		40.50	60		100.50	135
2720	8" pipe size		12	2		69	69.50		138.50	181
2730	10" pipe size	↓	10	2.400	↓	125	83.50		208.50	263
2860	Black 300 lb., per flange									
2870	1/2" pipe size	1 Plum	30	.267	Ea.	19.20	9.95		29.15	36
2880	3/4" pipe size		28	.286		19.20	10.65		29.85	37
2890	1" pipe size	↓	27	.296		19.20	11.05		30.25	37.50
2900	1-1/4" pipe size	Q-1	44	.364		19.20	12.25		31.45	39.50
2910	1-1/2" pipe size		40	.400		19.20	13.45		32.65	41.50
2920	2" pipe size		36	.444		20	14.95		34.95	44.50
2930	2-1/2" pipe size		28	.571		28	19.20		47.20	59.50
2940	3" pipe size		20	.800		30	27		57	73.50
2960	4" pipe size	↓	12	1.333		42.50	45		87.50	115
2970	6" pipe size	Q-2	14	1.714	↓	75	60		135	173
3000	Weld joint, butt, carbon steel, standard weight									
3040	90° elbow, long radius									
3050	1/2" pipe size	Q-15	16	1	Ea.	14.10	33.50	4.91	52.51	72
3060	3/4" pipe size		16	1		14.10	33.50	4.91	52.51	72
3070	1" pipe size		16	1		7.05	33.50	4.91	45.46	64
3080	1-1/4" pipe size		14	1.143		7.05	38.50	5.60	51.15	72
3090	1-1/2" pipe size		13	1.231		7.05	41.50	6.05	54.60	77
3100	2" pipe size		10	1.600		7.05	54	7.85	68.90	98
3110	2-1/2" pipe size		8	2		9.30	67.50	9.80	86.60	123
3120	3" pipe size		7	2.286		10.90	77	11.20	99.10	140
3130	4" pipe size		5	3.200		18.55	108	15.70	142.25	201
3136	5" pipe size	↓	4	4		39.50	135	19.65	194.15	268
3140	6" pipe size	Q-16	5	4.800		44	167	15.70	226.70	320
3150	8" pipe size		3.75	6.400		82.50	223	21	326.50	450
3160	10" pipe size		3	8		151	279	26	456	615
3170	12" pipe size		2.50	9.600		217	335	31.50	583.50	780
3180	14" pipe size		2	12		299	420	39	758	1,000
3190	16" pipe size		1.50	16		525	560	52.50	1,137.50	1,475
3191	18" pipe size		1.25	19.200		525	670	63	1,258	1,650
3192	20" pipe size		1.15	20.870		775	730	68	1,573	2,025
3194	24" pipe size	↓	1.02	23.529	↓	1,025	820	77	1,922	2,450
3200	45° Elbow, long									
3210	1/2" pipe size	Q-15	16	1	Ea.	13.75	33.50	4.91	52.16	71.50
3220	3/4" pipe size		16	1		13.75	33.50	4.91	52.16	71.50
3230	1" pipe size		16	1		6.70	33.50	4.91	45.11	64
3240	1-1/4" pipe size	↓	14	1.143	↓	6.70	38.50	5.60	50.80	71.50

660

MECHANICAL 15

15107	Metal Pipe & Fittings	CREW	DAILY OUTPUT	LABOR-HOURS	UNIT	MAT.	LABOR	EQUIP.	TOTAL	TOTAL INCL O&P	
						\multicolumn{4}{}{2003 BARE COSTS}			
3250	1-1/2" pipe size	Q-15	13	1.231	Ea.	6.70	41.50	6.05	54.25	76.50	660
3260	2" pipe size		10	1.600		6.70	54	7.85	68.55	97.50	
3270	2-1/2" pipe size		8	2		8	67.50	9.80	85.30	122	
3280	3" pipe size		7	2.286		9.30	77	11.20	97.50	139	
3290	4" pipe size		5	3.200		16.30	108	15.70	140	198	
3296	5" pipe size		4	4		25	135	19.65	179.65	252	
3300	6" pipe size	Q-16	5	4.800		35	167	15.70	217.70	310	
3310	8" pipe size		3.75	6.400		59	223	21	303	425	
3320	10" pipe size		3	8		107	279	26	412	565	
3330	12" pipe size		2.50	9.600		151	335	31.50	517.50	705	
3340	14" pipe size		2	12		205	420	39	664	900	
3341	16" pipe size		1.50	16		281	560	52.50	893.50	1,200	
3342	18" pipe size		1.25	19.200		375	670	63	1,108	1,475	
3343	20" pipe size		1.15	20.870		495	730	68	1,293	1,725	
3345	24" pipe size		1.05	22.857		715	795	74.50	1,584.50	2,075	
3346	26" pipe size		.85	28.235		1,475	985	92.50	2,552.50	3,200	
3347	30" pipe size		.45	53.333		1,625	1,850	174	3,649	4,800	
3349	36" pipe size		.38	63.158		2,700	2,200	207	5,107	6,525	
3350	Tee, straight										
3352	For reducing tees and concentrics see starting line 4600										
3360	1/2" pipe size	Q-15	10	1.600	Ea.	33	54	7.85	94.85	127	
3370	3/4" pipe size		10	1.600		33	54	7.85	94.85	127	
3380	1" pipe size		10	1.600		18.55	54	7.85	80.40	111	
3390	1-1/4" pipe size		9	1.778		21	60	8.75	89.75	123	
3400	1-1/2" pipe size		8	2		21	67.50	9.80	98.30	136	
3410	2" pipe size		6	2.667		18.55	89.50	13.10	121.15	170	
3420	2-1/2" pipe size		5	3.200		23.50	108	15.70	147.20	206	
3430	3" pipe size		4	4		25.50	135	19.65	180.15	253	
3440	4" pipe size		3	5.333		36.50	179	26	241.50	340	
3446	5" pipe size		2.50	6.400		66	215	31.50	312.50	435	
3450	6" pipe size	Q-16	3	8		66	279	26	371	520	
3460	8" pipe size		2.50	9.600		121	335	31.50	487.50	675	
3470	10" pipe size		2	12		202	420	39	661	895	
3480	12" pipe size		1.60	15		305	525	49	879	1,175	
3481	14" pipe size		1.30	18.462		510	645	60.50	1,215.50	1,600	
3482	16" pipe size		1	24		615	835	78.50	1,528.50	2,025	
3483	18" pipe size		.80	30		975	1,050	98	2,123	2,750	
3484	20" pipe size		.75	32		1,500	1,125	105	2,730	3,450	
3486	24" pipe size		.70	34.286		1,875	1,200	112	3,187	3,975	
3487	26" pipe size		.55	43.636		2,600	1,525	143	4,268	5,300	
3488	30" pipe size		.30	80		2,825	2,800	262	5,887	7,600	
3490	36" pipe size		.20	120		4,550	4,175	390	9,115	11,800	
3491	Eccentric reducer, 1-1/2" pipe size	Q-15	14	1.143		25.50	38.50	5.60	69.60	92	
3492	2" pipe size		11	1.455		19.50	49	7.15	75.65	103	
3493	2-1/2" pipe size		9	1.778		16.65	60	8.75	85.40	118	
3494	3" pipe size		8	2		18.90	67.50	9.80	96.20	134	
3495	4" pipe size		6	2.667		25	89.50	13.10	127.60	177	
3496	6" pipe size	Q-16	5	4.800		71	167	15.70	253.70	350	
3497	8" pipe size		4	6		99.50	209	19.60	328.10	445	
3498	10" pipe size		3	8		121	279	26	426	580	
3499	12" pipe size		2.50	9.600		180	335	31.50	546.50	740	
3501	Cap, 1-1/2" pipe size	Q-15	28	.571		7.70	19.20	2.81	29.71	40.50	
3502	2" pipe size		22	.727		7.05	24.50	3.57	35.12	48.50	
3503	2-1/2" pipe size		18	.889		7.05	30	4.36	41.41	57.50	
3504	3" pipe size		16	1		7.05	33.50	4.91	45.46	64	
3505	4" pipe size		12	1.333		8.95	45	6.55	60.50	84.50	

MECHANICAL

Important: See the Reference Section for critical supporting data - Reference Nos., Crews, & City Cost Indexes

		CREW	DAILY OUTPUT	LABOR-HOURS	UNIT	2003 BARE COSTS				TOTAL INCL O&P
						MAT.	LABOR	EQUIP.	TOTAL	
3506	6" pipe size	Q-16	10	2.400	Ea.	16.95	83.50	7.85	108.30	153
3507	8" pipe size		8	3		26	105	9.80	140.80	198
3508	10" pipe size		6	4		46	139	13.10	198.10	276
3509	12" pipe size		5	4.800		59	167	15.70	241.70	335
3511	14" pipe size		4	6		105	209	19.60	333.60	450
3512	16" pipe size		4	6		123	209	19.60	351.60	475
3513	18" pipe size	▼	3	8	▼	141	279	26	446	605
3517	Weld joint, butt, carbon steel, extra strong									
3519	90° elbow, long									
3520	1/2" pipe size	Q-15	13	1.231	Ea.	16.95	41.50	6.05	64.50	88
3530	3/4" pipe size		12	1.333		16.95	45	6.55	68.50	93.50
3540	1" pipe size		11	1.455		8.95	49	7.15	65.10	91.50
3550	1-1/4" pipe size		10	1.600		8.95	54	7.85	70.80	100
3560	1-1/2" pipe size		9	1.778		8.95	60	8.75	77.70	110
3570	2" pipe size		8	2		8.95	67.50	9.80	86.25	123
3580	2-1/2" pipe size		7	2.286		13.10	77	11.20	101.30	143
3590	3" pipe size		6	2.667		16.30	89.50	13.10	118.90	167
3600	4" pipe size		4	4		27.50	135	19.65	182.15	255
3606	5" pipe size	▼	3.50	4.571		59.50	154	22.50	236	320
3610	6" pipe size	Q-16	4.50	5.333		66	186	17.45	269.45	375
3620	8" pipe size		3.50	6.857		125	239	22.50	386.50	520
3630	10" pipe size		2.50	9.600		224	335	31.50	590.50	785
3640	12" pipe size	▼	2.25	10.667	▼	315	370	35	720	945
3650	45° Elbow, long									
3660	1/2" pipe size	Q-15	13	1.231	Ea.	19.50	41.50	6.05	67.05	90.50
3670	3/4" pipe size		12	1.333		19.50	45	6.55	71.05	96
3680	1" pipe size		11	1.455		9.30	49	7.15	65.45	92
3690	1-1/4" pipe size		10	1.600		9.30	54	7.85	71.15	100
3700	1-1/2" pipe size		9	1.778		9.30	60	8.75	78.05	110
3710	2" pipe size		8	2		8.30	67.50	9.80	85.60	122
3720	2-1/2" pipe size		7	2.286		11.50	77	11.20	99.70	141
3730	3" pipe size		6	2.667		12.80	89.50	13.10	115.40	164
3740	4" pipe size		4	4		19.50	135	19.65	174.15	246
3746	5" pipe size	▼	3.50	4.571		47.50	154	22.50	224	310
3750	6" pipe size	Q-16	4.50	5.333		51.50	186	17.45	254.95	355
3760	8" pipe size		3.50	6.857		88	239	22.50	349.50	480
3770	10" pipe size		2.50	9.600		142	335	31.50	508.50	695
3780	12" pipe size	▼	2.25	10.667	▼	211	370	35	616	830
3800	Tee, straight									
3810	1/2" pipe size	Q-15	9	1.778	Ea.	41.50	60	8.75	110.25	146
3820	3/4" pipe size		8.50	1.882		41.50	63.50	9.25	114.25	151
3830	1" pipe size		8	2		22	67.50	9.80	99.30	137
3840	1-1/4" pipe size		7	2.286		22	77	11.20	110.20	152
3850	1-1/2" pipe size		6	2.667		25.50	89.50	13.10	128.10	177
3860	2" pipe size		5	3.200		22	108	15.70	145.70	204
3870	2-1/2" pipe size		4	4		38.50	135	19.65	193.15	267
3880	3" pipe size		3.50	4.571		36	154	22.50	212.50	296
3890	4" pipe size		2.50	6.400		58.50	215	31.50	305	425
3896	5" pipe size	▼	2.25	7.111		119	239	35	393	530
3900	6" pipe size	Q-16	2.25	10.667		92	370	35	497	700
3910	8" pipe size		2	12		187	420	39	646	880
3920	10" pipe size		1.75	13.714		300	480	45	825	1,100
3930	12" pipe size	▼	1.50	16		440	560	52.50	1,052.50	1,375
4000	Eccentric reducer, 1-1/2" pipe size	Q-15	10	1.600		38	54	7.85	99.85	132
4010	2" pipe size		9	1.778		29	60	8.75	97.75	132
4020	2-1/2" pipe size	▼	8	2	▼	30.50	67.50	9.80	107.80	147

660

MECHANICAL 15

		CREW	DAILY OUTPUT	LABOR-HOURS	UNIT	2003 BARE COSTS MAT.	LABOR	EQUIP.	TOTAL	TOTAL INCL O&P	
660	4030	3" pipe size	Q-15	7	2.286	Ea.	30.50	77	11.20	118.70	162	**660**
	4040	4" pipe size		5	3.200		30	108	15.70	153.70	213	
	4046	5" pipe size	▼	4.70	3.404		85	114	16.70	215.70	285	
	4050	6" pipe size	Q-16	4.50	5.333		74.50	186	17.45	277.95	380	
	4060	8" pipe size		3.50	6.857		214	239	22.50	475.50	620	
	4070	10" pipe size		2.50	9.600		223	335	31.50	589.50	785	
	4080	12" pipe size		2.25	10.667		221	370	35	626	840	
	4090	14" pipe size		2.10	11.429		500	400	37.50	937.50	1,175	
	4100	16" pipe size	▼	1.90	12.632		470	440	41.50	951.50	1,225	
	4151	Cap, 1-1/2" pipe size	Q-15	24	.667		8	22.50	3.27	33.77	46.50	
	4152	2" pipe size		18	.889		7.05	30	4.36	41.41	57.50	
	4153	2-1/2" pipe size		16	1		8	33.50	4.91	46.41	65	
	4154	3" pipe size		14	1.143		8.95	38.50	5.60	53.05	74	
	4155	4" pipe size	▼	10	1.600		12.50	54	7.85	74.35	104	
	4156	6" pipe size	Q-16	9	2.667		31.50	93	8.70	133.20	184	
	4157	8" pipe size		7	3.429		42	120	11.20	173.20	240	
	4158	10" pipe size		5	4.800		66	167	15.70	248.70	345	
	4159	12" pipe size	▼	4	6	▼	93.50	209	19.60	322.10	440	
	4190	Weld fittings, reducing, standard weight										
	4200	Welding ring w/spacer pins, 2" pipe size				Ea.	1.95			1.95	2.15	
	4210	2-1/2" pipe size					2.35			2.35	2.59	
	4220	3" pipe size					2.44			2.44	2.68	
	4230	4" pipe size					2.54			2.54	2.79	
	4236	5" pipe size					2.99			2.99	3.29	
	4240	6" pipe size					2.99			2.99	3.29	
	4250	8" pipe size					3.53			3.53	3.88	
	4260	10" pipe size					4.32			4.32	4.75	
	4270	12" pipe size					4.75			4.75	5.25	
	4280	14" pipe size					5.75			5.75	6.35	
	4290	16" pipe size					6.50			6.50	7.15	
	4300	18" pipe size					7.50			7.50	8.25	
	4310	20" pipe size					7.95			7.95	8.75	
	4330	24" pipe size					9.25			9.25	10.20	
	4340	26" pipe size					12.35			12.35	13.60	
	4350	30" pipe size					12.35			12.35	13.60	
	4370	36" pipe size				▼	17.75			17.75	19.55	
	4600	Tee, reducing on outlet										
	4601	2-1/2" x 2" pipe size	Q-15	5	3.200	Ea.	33	108	15.70	156.70	217	
	4602	3" x 2-1/2" pipe size		4	4		33.50	135	19.65	188.15	261	
	4603	3-1/2" x 3" pipe size		3.50	4.571		42	154	22.50	218.50	305	
	4604	4" x 3" pipe size		3	5.333		42	179	26	247	345	
	4605	5" x 4" pipe size	▼	2.50	6.400		91.50	215	31.50	338	460	
	4606	6" x 5" pipe size	Q-16	3	8		82.50	279	26	387.50	540	
	4607	8" x 6" pipe size		2.50	9.600		170	335	31.50	536.50	725	
	4608	10" x 8" pipe size		2	12		266	420	39	725	965	
	4609	12" x 10" pipe size		1.60	15		395	525	49	969	1,275	
	4610	16" x 12" pipe size		1.50	16		645	560	52.50	1,257.50	1,600	
	4611	14" x 12" pipe size	▼	1.52	15.789	▼	420	550	51.50	1,021.50	1,350	
	4618	Reducer, concentric										
	4619	2-1/2" by 2" pipe size	Q-15	10	1.600	Ea.	11.85	54	7.85	73.70	103	
	4620	3" by 2-1/2" pipe size		9	1.778		15.05	60	8.75	83.80	117	
	4621	3-1/2" by 3" pipe size		8	2		25.50	67.50	9.80	102.80	141	
	4622	4" by 2-1/2" pipe size		7	2.286		14.70	77	11.20	102.90	145	
	4623	5" by 3" pipe size	▼	7	2.286		35	77	11.20	123.20	167	
	4624	6" by 4" pipe size	Q-16	6	4		28.50	139	13.10	180.60	257	
	4625	8" by 6" pipe size	▼	5	4.800	▼	38	167	15.70	220.70	310	

Important: See the Reference Section for critical supporting data - Reference Nos., Crews, & City Cost Indexes

15107	Metal Pipe & Fittings	CREW	DAILY OUTPUT	LABOR-HOURS	UNIT	2003 BARE COSTS				TOTAL INCL O&P
						MAT.	LABOR	EQUIP.	TOTAL	
4626	10" by 8" pipe size	Q-16	4	6	Ea.	57.50	209	19.60	286.10	400
4627	12" by 10" pipe size	↓	3	8	↓	89	279	26	394	545
4660	Reducer, eccentric									
4662	3" x 2"	Q-15	8	2	Ea.	13.35	67.50	9.80	90.65	127
4664	4" x 3"		6	2.667		18.90	89.50	13.10	121.50	170
4666	4" x 2"	↓	6	2.667		25	89.50	13.10	127.60	177
4670	6" x 4"	Q-16	5	4.800		38	167	15.70	220.70	310
4672	6" x 3"		5	4.800		71	167	15.70	253.70	350
4676	8" x 6"		4	6		38	209	19.60	266.60	380
4678	8" x 4"		4	6		99.50	209	19.60	328.10	445
4682	10" x 8"		3	8		68	279	26	373	525
4684	10" x 6"		3	8		121	279	26	426	580
4688	12" x 10"		2.50	9.600		114	335	31.50	480.50	665
4690	12" x 8"		2.50	9.600		168	335	31.50	534.50	725
4691	14" x 12"		2.20	10.909		167	380	35.50	582.50	800
4693	16" x 14"		1.80	13.333		223	465	43.50	731.50	995
4694	16" x 12"		2	12		340	420	39	799	1,050
4696	18" x 16"	↓	1.60	15	↓	360	525	49	934	1,250
5000	Weld joint, socket, forged steel, 3000 lb., schedule 40 pipe									
5010	90° elbow, straight									
5020	1/4" pipe size	Q-15	22	.727	Ea.	7.45	24.50	3.57	35.52	49
5030	3/8" pipe size		22	.727		9.20	24.50	3.57	37.27	51
5040	1/2" pipe size		20	.800		7.45	27	3.93	38.38	53
5050	3/4" pipe size		20	.800		7.65	27	3.93	38.58	53.50
5060	1" pipe size		20	.800		9.90	27	3.93	40.83	55.50
5070	1-1/4" pipe size		18	.889		18.90	30	4.36	53.26	71
5080	1-1/2" pipe size		16	1		21	33.50	4.91	59.41	79.50
5090	2" pipe size		12	1.333		31.50	45	6.55	83.05	110
5100	2-1/2" pipe size		10	1.600		93	54	7.85	154.85	192
5110	3" pipe size		8	2		157	67.50	9.80	234.30	286
5120	4" pipe size	↓	6	2.667	↓	425	89.50	13.10	527.60	620
5130	45° Elbow, straight									
5134	1/4" pipe size	Q-15	22	.727	Ea.	12.10	24.50	3.57	40.17	54
5135	3/8" pipe size		22	.727		12.10	24.50	3.57	40.17	54
5136	1/2" pipe size		20	.800		9.40	27	3.93	40.33	55
5137	3/4" pipe size		20	.800		10.95	27	3.93	41.88	57
5140	1" pipe size		20	.800		14.50	27	3.93	45.43	61
5150	1-1/4" pipe size		18	.889		19.30	30	4.36	53.66	71
5160	1-1/2" pipe size		16	1		24	33.50	4.91	62.41	83
5170	2" pipe size		12	1.333		41	45	6.55	92.55	120
5180	2-1/2" pipe size		10	1.600		106	54	7.85	167.85	206
5190	3" pipe size		8	2		178	67.50	9.80	255.30	310
5200	4" pipe size	↓	6	2.667	↓	350	89.50	13.10	452.60	535
5250	Tee, straight									
5254	1/4" pipe size	Q-15	15	1.067	Ea.	8.55	36	5.25	49.80	69
5255	3/8" pipe size		15	1.067		10.55	36	5.25	51.80	71.50
5256	1/2" pipe size		13	1.231		8.55	41.50	6.05	56.10	78.50
5257	3/4" pipe size		13	1.231		10.80	41.50	6.05	58.35	81
5260	1" pipe size		13	1.231		14.85	41.50	6.05	62.40	85.50
5270	1-1/4" pipe size		12	1.333		23.50	45	6.55	75.05	100
5280	1-1/2" pipe size		11	1.455		31.50	49	7.15	87.65	116
5290	2" pipe size		8	2		44	67.50	9.80	121.30	161
5300	2-1/2" pipe size		6	2.667		130	89.50	13.10	232.60	292
5310	3" pipe size		5	3.200		315	108	15.70	438.70	525
5320	4" pipe size	↓	4	4	↓	500	135	19.65	654.65	775
5350	For reducing sizes, add					60%				

MECHANICAL 15

			DAILY	LABOR-			2003 BARE COSTS				TOTAL
15107	**Metal Pipe & Fittings**	CREW	OUTPUT	HOURS	UNIT	MAT.	LABOR	EQUIP.	TOTAL		INCL O&P
5450	Couplings										
5451	1/4″ pipe size	Q-15	23	.696	Ea.	6.55	23.50	3.42	33.47		46.50
5452	3/8″ pipe size		23	.696		6.55	23.50	3.42	33.47		46.50
5453	1/2″ pipe size		21	.762		3.31	25.50	3.74	32.55		46.50
5454	3/4″ pipe size		21	.762		4.41	25.50	3.74	33.65		47.50
5460	1″ pipe size		20	.800		5.15	27	3.93	36.08		50.50
5470	1-1/4″ pipe size		20	.800		10.10	27	3.93	41.03		56
5480	1-1/2″ pipe size		18	.889		11.15	30	4.36	45.51		62
5490	2″ pipe size		14	1.143		17.55	38.50	5.60	61.65		83.50
5500	2-1/2″ pipe size		12	1.333		37	45	6.55	88.55		116
5510	3″ pipe size		9	1.778		86.50	60	8.75	155.25		195
5520	4″ pipe size		7	2.286		130	77	11.20	218.20		271
5570	Union, 1/4″ pipe size		21	.762		14.35	25.50	3.74	43.59		58.50
5571	3/8″ pipe size		21	.762		14.35	25.50	3.74	43.59		58.50
5572	1/2″ pipe size		19	.842		14.35	28.50	4.13	46.98		63.50
5573	3/4″ pipe size		19	.842		16.30	28.50	4.13	48.93		65.50
5574	1″ pipe size		19	.842		21	28.50	4.13	53.63		70.50
5575	1-1/4″ pipe size		17	.941		34	31.50	4.62	70.12		90.50
5576	1-1/2″ pipe size		15	1.067		36	36	5.25	77.25		99.50
5577	2″ pipe size		11	1.455		50	49	7.15	106.15		137
5600	Reducer, 1/4″ pipe size		23	.696		17.65	23.50	3.42	44.57		58.50
5601	3/8″ pipe size		23	.696		17.65	23.50	3.42	44.57		58.50
5602	1/2″ pipe size		21	.762		12	25.50	3.74	41.24		56
5603	3/4″ pipe size		21	.762		13.95	25.50	3.74	43.19		58
5604	1″ pipe size		21	.762		15.30	25.50	3.74	44.54		59.50
5605	1-1/4″ pipe size		19	.842		27.50	28.50	4.13	60.13		78
5607	1-1/2″ pipe size		17	.941		29.50	31.50	4.62	65.62		85.50
5608	2″ pipe size		13	1.231		32.50	41.50	6.05	80.05		105
5612	Cap, 1/4″ pipe size		46	.348		6.95	11.70	1.71	20.36		27
5613	3/8″ pipe size		46	.348		6.95	11.70	1.71	20.36		27
5614	1/2″ pipe size		42	.381		4.91	12.80	1.87	19.58		27
5615	3/4″ pipe size		42	.381		5.70	12.80	1.87	20.37		27.50
5616	1″ pipe size		42	.381		8.80	12.80	1.87	23.47		31
5617	1-1/4″ pipe size		38	.421		10.25	14.15	2.07	26.47		35
5618	1-1/2″ pipe size		34	.471		13.95	15.85	2.31	32.11		42
5619	2″ pipe size		26	.615		22	20.50	3.02	45.52		59
5630	T-O-L, 1/4″ pipe size, nozzle		23	.696		4.31	23.50	3.42	31.23		44
5631	3/8″ pipe size, nozzle		23	.696		4.31	23.50	3.42	31.23		44
5632	1/2″ pipe size, nozzle		22	.727		4.31	24.50	3.57	32.38		45.50
5633	3/4″ pipe size, nozzle		21	.762		5.05	25.50	3.74	34.29		48
5634	1″ pipe size, nozzle		20	.800		5.70	27	3.93	36.63		51
5635	1-1/4″ pipe size, nozzle		18	.889		8.05	30	4.36	42.41		58.50
5636	1-1/2″ pipe size, nozzle		16	1		8.75	33.50	4.91	47.16		66
5637	2″ pipe size, nozzle		12	1.333		10.05	45	6.55	61.60		86
5638	2-1/2″ pipe size, nozzle		10	1.600		31.50	54	7.85	93.35		125
5639	4″ pipe size, nozzle		6	2.667		72	89.50	13.10	174.60		228
5640	W-O-L, 1/4″ pipe size, nozzle		23	.696		8.25	23.50	3.42	35.17		48.50
5641	3/8″ pipe size, nozzle		23	.696		8.25	23.50	3.42	35.17		48.50
5642	1/2″ pipe size, nozzle		22	.727		8.25	24.50	3.57	36.32		50
5643	3/4″ pipe size, nozzle		21	.762		8.45	25.50	3.74	37.69		52
5644	1″ pipe size, nozzle		20	.800		8.85	27	3.93	39.78		54.50
5645	1-1/4″ pipe size, nozzle		18	.889		10.75	30	4.36	45.11		61.50
5646	1-1/2″ pipe size, nozzle		16	1		10.75	33.50	4.91	49.16		68
5647	2″ pipe size, nozzle		12	1.333		11.20	45	6.55	62.75		87
5648	2-1/2″ pipe size, nozzle		10	1.600		25.50	54	7.85	87.35		118
5649	3″ pipe size, nozzle		8	2		29.50	67.50	9.80	106.80		145

Important: See the Reference Section for critical supporting data - Reference Nos., Crews, & City Cost Indexes

		CREW	DAILY OUTPUT	LABOR-HOURS	UNIT	2003 BARE COSTS				TOTAL INCL O&P
15107	**Metal Pipe & Fittings**					MAT.	LABOR	EQUIP.	TOTAL	
660 5650	4" pipe size, nozzle	Q-15	6	2.667	Ea.	37.50	89.50	13.10	140.10	191 660
5651	5" pipe size, nozzle		5	3.200		93.50	108	15.70	217.20	283
5652	6" pipe size, nozzle		4	4		119	135	19.65	273.65	355
5653	8" pipe size, nozzle		3	5.333		215	179	26	420	535
5654	10" pipe size, nozzle		2.60	6.154		310	207	30	547	690
5655	12" pipe size, nozzle		2.20	7.273		490	245	35.50	770.50	945
5656	14" pipe size, nozzle		2	8		905	269	39.50	1,213.50	1,450
5660	L-O-L, 1/4" pipe size, outlet		23	.696		47	23.50	3.42	73.92	91
5661	3/8" pipe size, outlet		23	.696		47	23.50	3.42	73.92	91
5662	1/2" pipe size, outlet		22	.727		47	24.50	3.57	75.07	92.50
5663	3/4" pipe size, outlet		21	.762		51.50	25.50	3.74	80.74	99.50
5664	1" pipe size, outlet		20	.800		69.50	27	3.93	100.43	121
5665	1-1/4" pipe size, outlet		18	.889		108	30	4.36	142.36	169
5666	1-1/2" pipe size, outlet		16	1		108	33.50	4.91	146.41	175
5667	2" pipe size, outlet		12	1.333		171	45	6.55	222.55	264
5674	S-O-L, 1/4" pipe size, outlet		23	.696		4.81	23.50	3.42	31.73	44.50
5675	3/8" pipe size, outlet		23	.696		4.81	23.50	3.42	31.73	44.50
5676	1/2" pipe size, outlet		22	.727		4.81	24.50	3.57	32.88	46
5677	3/4" pipe size, outlet		21	.762		4.88	25.50	3.74	34.12	48
5678	1" pipe size, outlet		20	.800		5.30	27	3.93	36.23	50.50
5679	1-1/4" pipe size, outlet		18	.889		8.20	30	4.36	42.56	59
5680	1-1/2" pipe size, outlet		16	1		8.60	33.50	4.91	47.01	66
5681	2" pipe size, outlet	▼	12	1.333	▼	9.65	45	6.55	61.20	85.50
6000	Weld-on flange, forged steel									
6020	Slip-on, 150 lb. flange (welded front and back)									
6050	1/2" pipe size	Q-15	18	.889	Ea.	11.85	30	4.36	46.21	63
6060	3/4" pipe size		18	.889		11.85	30	4.36	46.21	63
6070	1" pipe size		17	.941		11.85	31.50	4.62	47.97	66
6080	1-1/4" pipe size		16	1		11.85	33.50	4.91	50.26	69.50
6090	1-1/2" pipe size		15	1.067		11.85	36	5.25	53.10	73
6100	2" pipe size		12	1.333		11.85	45	6.55	63.40	87.50
6110	2-1/2" pipe size		10	1.600		14.70	54	7.85	76.55	106
6120	3" pipe size		9	1.778		16.65	60	8.75	85.40	118
6130	3-1/2" pipe size		7	2.286		19.50	77	11.20	107.70	150
6140	4" pipe size		6	2.667		19.50	89.50	13.10	122.10	171
6150	5" pipe size	▼	5	3.200		38	108	15.70	161.70	222
6160	6" pipe size	Q-16	6	4		34	139	13.10	186.10	263
6170	8" pipe size		5	4.800		51	167	15.70	233.70	325
6180	10" pipe size		4	6		89.50	209	19.60	318.10	435
6190	12" pipe size		3	8		132	279	26	437	595
6191	14" pipe size		2.50	9.600		175	335	31.50	541.50	735
6192	16" pipe size	▼	1.80	13.333	▼	290	465	43.50	798.50	1,075
6200	300 lb. flange									
6210	1/2" pipe size	Q-15	17	.941	Ea.	17.90	31.50	4.62	54.02	73
6220	3/4" pipe size		17	.941		17.90	31.50	4.62	54.02	73
6230	1" pipe size		16	1		17.90	33.50	4.91	56.31	76
6240	1-1/4" pipe size		13	1.231		17.90	41.50	6.05	65.45	89
6250	1-1/2" pipe size		12	1.333		17.90	45	6.55	69.45	94.50
6260	2" pipe size		11	1.455		22	49	7.15	78.15	106
6270	2-1/2" pipe size		9	1.778		24.50	60	8.75	93.25	127
6280	3" pipe size		7	2.286		29	77	11.20	117.20	160
6290	4" pipe size		6	2.667		41	89.50	13.10	143.60	194
6300	5" pipe size	▼	4	4		67.50	135	19.65	222.15	299
6310	6" pipe size	Q-16	5	4.800		67	167	15.70	249.70	345
6320	8" pipe size		4	6		115	209	19.60	343.60	465
6330	10" pipe size	▼	3.40	7.059	▼	204	246	23	473	620

MECHANICAL 15

15107 | Metal Pipe & Fittings

		CREW	DAILY OUTPUT	LABOR-HOURS	UNIT	MAT.	2003 BARE COSTS LABOR	EQUIP.	TOTAL	TOTAL INCL O&P
660	6340	12" pipe size	Q-16	2.80	8.571	Ea.	222	299	28	549	725	660
6400	Welding neck, 150 lb. flange									
6410	1/2" pipe size	Q-15	40	.400	Ea.	16.95	13.45	1.96	32.36	41.50
6420	3/4" pipe size		36	.444		16.95	14.95	2.18	34.08	43.50
6430	1" pipe size		32	.500		16.95	16.80	2.46	36.21	47
6440	1-1/4" pipe size		29	.552		16.95	18.55	2.71	38.21	49.50
6450	1-1/2" pipe size		26	.615		16.95	20.50	3.02	40.47	53.50
6460	2" pipe size		20	.800		18.55	27	3.93	49.48	65.50
6470	2-1/2" pipe size		16	1		19.85	33.50	4.91	58.26	78.50
6480	3" pipe size		14	1.143		22	38.50	5.60	66.10	88
6500	4" pipe size		10	1.600		26	54	7.85	87.85	119
6510	5" pipe size		8	2		41.50	67.50	9.80	118.80	158
6520	6" pipe size	Q-16	10	2.400		40	83.50	7.85	131.35	179
6530	8" pipe size		7	3.429		65.50	120	11.20	196.70	265
6540	10" pipe size		6	4		113	139	13.10	265.10	350
6550	12" pipe size		5	4.800		164	167	15.70	346.70	450
6551	14" pipe size		4.50	5.333		236	186	17.45	439.45	560
6552	16" pipe size		3	8		400	279	26	705	890
6553	18" pipe size		2.50	9.600		570	335	31.50	936.50	1,175
6554	20" pipe size		2.30	10.435		690	365	34	1,089	1,350
6556	24" pipe size		2	12		925	420	39	1,384	1,700
6557	26" pipe size		1.70	14.118		1,500	490	46	2,036	2,450
6558	30" pipe size		.90	26.667		2,000	930	87	3,017	3,700
6559	36" pipe size		.75	32		2,300	1,125	105	3,530	4,325
6560	300 lb. flange									
6570	1/2" pipe size	Q-15	36	.444	Ea.	21.50	14.95	2.18	38.63	48.50
6580	3/4" pipe size		34	.471		21.50	15.85	2.31	39.66	50
6590	1" pipe size		30	.533		21.50	17.95	2.62	42.07	53.50
6600	1-1/4" pipe size		28	.571		22.50	19.20	2.81	44.51	56.50
6610	1-1/2" pipe size		24	.667		21.50	22.50	3.27	47.27	61
6620	2" pipe size		18	.889		24.50	30	4.36	58.86	77
6630	2-1/2" pipe size		14	1.143		28	38.50	5.60	72.10	94.50
6640	3" pipe size		12	1.333		28	45	6.55	79.55	105
6650	4" pipe size		8	2		45.50	67.50	9.80	122.80	163
6660	5" pipe size		7	2.286		70.50	77	11.20	158.70	206
6670	6" pipe size	Q-16	9	2.667		66	93	8.70	167.70	223
6680	8" pipe size		6	4		114	139	13.10	266.10	350
6690	10" pipe size		5	4.800		221	167	15.70	403.70	515
6700	12" pipe size		4	6		284	209	19.60	512.60	645
6710	14" pipe size		3.50	6.857		630	239	22.50	891.50	1,075
6720	16" pipe size		2	12		745	420	39	1,204	1,500
7740	Plain ends for plain end pipe, mechanically coupled									
7750	Cplg & labor required at joints not included add 1 per									
7760	joint for installed price, see line 9180									
7770	Malleable iron, painted, unless noted otherwise									
7800	90° Elbow 1"				Ea.	17.85			17.85	19.60
7810	1-1/2"					34			34	37.50
7820	2"					34			34	37.50
7830	2-1/2"					34			34	37.50
7840	3"					41			41	45
7860	4"					59.50			59.50	65.50
7870	5" welded steel					108			108	119
7880	6"					86			86	94.50
7890	8" welded steel					181			181	199
7900	10" welded steel					228			228	251
7910	12" welded steel					290			290	320

Important: See the Reference Section for critical supporting data - Reference Nos., Crews, & City Cost Indexes

		DAILY	LABOR-			2003 BARE COSTS			TOTAL
		CREW	OUTPUT	HOURS	UNIT	MAT.	LABOR	EQUIP.	TOTAL	INCL O&P

		CREW	OUTPUT	HOURS	UNIT	MAT.	LABOR	EQUIP.	TOTAL	INCL O&P
660 7970	45° Elbow 1"				Ea.	17.85			17.85	19.60 **660**
7980	1-1/2"					34			34	37.50
7990	2"					34			34	37.50
8000	2-1/2"					34			34	37.50
8010	3"					41			41	45
8030	4"					59.50			59.50	65.50
8040	5" welded steel					108			108	119
8050	6"					86			86	94.50
8060	8"					99.50			99.50	110
8070	10" welded steel					139			139	153
8080	12" welded steel					180			180	197
8140	Tee, straight 1"					24			24	26.50
8150	1-1/2"					46			46	50.50
8160	2"					46			46	50.50
8170	2-1/2"					46			46	50.50
8180	3"					66			66	72.50
8200	4"					92			92	101
8210	5" welded steel					105			105	116
8220	6"					130			130	143
8230	8" welded steel					180			180	198
8240	10" welded steel					275			275	305
8250	12" welded steel				↓	370			370	410
8340	Segmentally welded steel, painted									
8390	Wye 2"				Ea.	70			70	77
8400	2-1/2"					70			70	77
8410	3"					79			79	87
8430	4"					117			117	129
8440	5"					143			143	157
8450	6"					196			196	216
8460	8"					254			254	280
8470	10"					375			375	410
8480	12"					580			580	635
8540	Wye, lateral 2"					75			75	82.50
8550	2-1/2"					75			75	82.50
8560	3"					84			84	92.50
8580	4"					127			127	139
8590	5"					165			165	181
8600	6"					211			211	232
8610	8"					305			305	340
8620	10"					435			435	475
8630	12"					585			585	645
8690	Cross, 2"					80			80	88
8700	2-1/2"					80			80	88
8710	3"					94.50			94.50	104
8730	4"					130			130	143
8740	5"					185			185	203
8750	6"					244			244	268
8760	8"					315			315	345
8770	10"					455			455	500
8780	12"					660			660	725
8800	Tees, reducing 2" x 1"					47.50			47.50	52
8810	2" x 1-1/2"					47.50			47.50	52
8820	3" x 1"					47.50			47.50	52
8830	3" x 1-1/2"					47.50			47.50	52
8840	3" x 2"					47.50			47.50	52
8850	4" x 1"				↓	69			69	76

MECHANICAL 15

15107 | Metal Pipe & Fittings

		CREW	DAILY OUTPUT	LABOR-HOURS	UNIT	2003 BARE COSTS				TOTAL INCL O&P
						MAT.	LABOR	EQUIP.	TOTAL	
660 8860	4" x 1-1/2"				Ea.	69			69	76 **660**
8870	4" x 2"					69			69	76
8880	4" x 2-1/2"					71			71	78
8890	4" x 3"					71			71	78
8900	6" x 2"					115			115	126
8910	6" x 3"					121			121	133
8920	6" x 4"					121			121	133
8930	8" x 2"					147			147	162
8940	8" x 3"					155			155	170
8950	8" x 4"					157			157	173
8960	8" x 5"					162			162	178
8970	8" x 6"					162			162	178
8980	10" x 4"					232			232	255
8990	10" x 6"					246			246	270
9000	10" x 8"					246			246	270
9010	12" x 6"					355			355	395
9020	12" x 8"					360			360	395
9030	12" x 10"					370			370	410
9080	Adapter nipples 3" long									
9090	1"				Ea.	6.20			6.20	6.80
9100	1-1/2"					6.20			6.20	6.80
9110	2"					6.20			6.20	6.80
9120	2-1/2"					7.25			7.25	8
9130	3"					8.95			8.95	9.85
9140	4"					14.95			14.95	16.40
9150	6"					40.50			40.50	44.50
9180	Coupling, mechanical, plain end pipe to plain end pipe or fitting									
9190	1"	Q-1	29	.552	Ea.	23	18.55		41.55	53
9200	1-1/2"		28	.571		23	19.20		42.20	54
9210	2"		27	.593		23	19.95		42.95	55
9220	2-1/2"		26	.615		23	20.50		43.50	56.50
9230	3"		25	.640		34	21.50		55.50	70
9240	3-1/2"		24	.667		39	22.50		61.50	77
9250	4"		22	.727		39	24.50		63.50	80
9260	5"	Q-2	28	.857		56	30		86	107
9270	6"		24	1		68.50	35		103.50	128
9280	8"		19	1.263		121	44		165	200
9290	10"		16	1.500		158	52.50		210.50	252
9300	12"		12	2		198	69.50		267.50	320
9310	Outlets for precut holes through pipe wall									
9331	Strapless type, with gasket									
9332	4" to 8" pipe x 1/2"	1 Plum	13	.615	Ea.	22.50	23		45.50	59
9333	4" to 8" pipe x 3/4"		13	.615		24	23		47	61
9334	10" pipe and larger x 1/2"		11	.727		22.50	27		49.50	65.50
9335	10" pipe and larger x 3/4"		11	.727		24	27		51	67.50
9341	Thermometer wells with gasket									
9342	4" to 8" pipe, 6" stem	1 Plum	14	.571	Ea.	34.50	21.50		56	69.50
9343	8" pipe and larger, 6" stem	"	13	.615	"	34.50	23		57.50	72
9400	Mechanical joint ends for plain end pipe									
9410	Malleable iron, black									
9420	90° Elbows, 1-1/4"	Q-1	29	.552	Ea.	7.95	18.55		26.50	37
9430	1-1/2"		27	.593		8.75	19.95		28.70	39.50
9440	2"		24	.667		10.30	22.50		32.80	45.50
9490	Tee, reducing outlet									
9510	1-1/4" x 1/2"	Q-1	18	.889	Ea.	6.40	30		36.40	52
9520	1-1/4" x 3/4"		18	.889		6.40	30		36.40	52

15 MECHANICAL

Important: See the Reference Section for critical supporting data - Reference Nos., Crews, & City Cost Indexes

15107 | Metal Pipe & Fittings

		CREW	DAILY OUTPUT	LABOR-HOURS	UNIT	2003 BARE COSTS MAT.	LABOR	EQUIP.	TOTAL	TOTAL INCL O&P	
660	9530	1-1/4" x 1"	Q-1	18	.889	Ea.	6.40	30		36.40	52	660
	9540	1-1/2" x 1/2"		17	.941		6.75	31.50		38.25	55.50	
	9550	1-1/2" x 3/4"		17	.941		6.75	31.50		38.25	55.50	
	9560	1-1/2" x 1"		17	.941		6.75	31.50		38.25	55.50	
	9570	2" x 1/2"		15	1.067		8.25	36		44.25	63	
	9580	2" x 3/4"		15	1.067		8.25	36		44.25	63	
	9590	2" x 1"		15	1.067		8.25	36		44.25	63	
	9640	Tee, reducing run and outlet										
	9660	1-1/4" x 1" x 1/2"	Q-1	18	.889	Ea.	6.85	30		36.85	52.50	
	9670	1-1/4" x 1" x 3/4"		18	.889		6.85	30		36.85	52.50	
	9680	1-1/4" x 1" x 1"		18	.889		6.85	30		36.85	52.50	
	9690	1-1/2" x 1-1/4" x 1/2"		17	.941		7.25	31.50		38.75	56	
	9700	1-1/2" x 1-1/4" x 3/4"		17	.941		7.25	31.50		38.75	56	
	9710	1-1/2" x 1-1/4" x 1"		17	.941		7.25	31.50		38.75	56	
	9720	2" x 1-1/2" x 1/2"		15	1.067		8.75	36		44.75	63.50	
	9730	2" x 1-1/2" x 3/4"		15	1.067		8.75	36		44.75	63.50	
	9740	2" x 1-1/2" x 1"		15	1.067		8.75	36		44.75	63.50	
	9790	Tee, outlet										
	9810	3" x 1-1/4"	Q-1	29	.552	Ea.	15.45	18.55		34	45	
	9820	3" x 1-1/2"		28	.571		15.45	19.20		34.65	46	
	9830	3" x 2"		26	.615		17.25	20.50		37.75	50.50	
	9840	4" x 1-1/4"		28	.571		19.20	19.20		38.40	50	
	9850	4" x 1-1/2"		26	.615		20	20.50		40.50	53.50	
	9860	4" x 2"		24	.667		20	22.50		42.50	56	
	9940	For galvanized fittings for plain end pipe, add					20%					
690	0010	**PIPE, GROOVED-JOINT STEEL FITTINGS & VALVES**										690
	0020	Pipe includes coupling & clevis type hanger 10' O.C.										
	0500	Schedule 10, black										
	0550	2" diameter	1 Plum	43	.186	L.F.	2.94	6.95		9.89	13.75	
	0560	2-1/2" diameter	Q-1	61	.262		3.78	8.80		12.58	17.45	
	0570	3" diameter		55	.291		4.29	9.80		14.09	19.45	
	0580	3-1/2" diameter		53	.302		5.30	10.15		15.45	21	
	0590	4" diameter		49	.327		5.40	11		16.40	22.50	
	0600	5" diameter		40	.400		9	13.45		22.45	30.50	
	0610	6" diameter	Q-2	46	.522		9.60	18.20		27.80	38	
	0620	8" diameter	"	41	.585		17.40	20.50		37.90	50	
	0700	To delete couplings & hangers, subtract										
	0710	2" diam. to 5" diam.					25%	20%				
	0720	6" diam. to 8" diam.					27%	15%				
	1000	Schedule 40, black										
	1040	3/4" diameter	1 Plum	71	.113	L.F.	1.88	4.21		6.09	8.40	
	1050	1" diameter		63	.127		2.28	4.74		7.02	9.65	
	1060	1-1/4" diameter		58	.138		2.88	5.15		8.03	10.95	
	1070	1-1/2" diameter		51	.157		3.08	5.85		8.93	12.25	
	1080	2" diameter		40	.200		3.63	7.45		11.08	15.30	
	1090	2-1/2" diameter	Q-1	57	.281		5.55	9.45		15	20.50	
	1100	3" diameter		50	.320		7	10.75		17.75	24	
	1110	4" diameter		45	.356		8.25	11.95		20.20	27	
	1120	5" diameter		37	.432		12.35	14.55		26.90	35.50	
	1130	6" diameter	Q-2	42	.571		14.70	19.95		34.65	46	
	1140	8" diameter		37	.649		22.50	22.50		45	58.50	
	1150	10" diameter		31	.774		35.50	27		62.50	80	
	1160	12" diameter		27	.889		42.50	31		73.50	93.50	
	1170	14" diameter		20	1.200		56	42		98	125	
	1180	16" diameter		17	1.412		63	49		112	144	

MECHANICAL 15

15107	Metal Pipe & Fittings	CREW	DAILY OUTPUT	LABOR-HOURS	UNIT	MAT.	LABOR	EQUIP.	TOTAL	TOTAL INCL O&P
							2003 BARE COSTS			
690	1190	18" diameter	Q-2	14	1.714	L.F.	81	60		141	180	690
1200	20" diameter		12	2		104	69.50		173.50	219
1210	24" diameter	↓	10	2.400	↓	122	83.50		205.50	260
1740	To delete coupling & hanger, subtract									
1750	3/4" diam. to 2" diam.					65%	27%			
1760	2-1/2" diam. to 5" diam.					41%	18%			
1770	6" diam. to 12" diam.					31%	13%			
1780	14" diam. to 24" diam.					35%	10%			
1800	Galvanized									
1840	3/4" diameter	1 Plum	71	.113	L.F.	2.11	4.21		6.32	8.70
1850	1" diameter		63	.127		2.19	4.74		6.93	9.55
1860	1-1/4" diameter		58	.138		2.71	5.15		7.86	10.80
1870	1-1/2" diameter		51	.157		3.05	5.85		8.90	12.20
1880	2" diameter	↓	40	.200		3.63	7.45		11.08	15.30
1890	2-1/2" diameter	Q-1	57	.281		5.15	9.45		14.60	19.90
1900	3" diameter		50	.320		6.30	10.75		17.05	23
1910	4" diameter		45	.356		7.80	11.95		19.75	26.50
1920	5" diameter	↓	37	.432		15.40	14.55		29.95	39
1930	6" diameter	Q-2	42	.571		18.30	19.95		38.25	50
1940	8" diameter		37	.649		28	22.50		50.50	64.50
1950	10" diameter		31	.774		44	27		71	89.50
1960	12" diameter	↓	27	.889	↓	52.50	31		83.50	105
2540	To delete coupling & hanger, subtract									
2550	3/4" diam. to 2" diam.					36%	27%			
2560	2-1/2" diam. to 5" diam.					19%	18%			
2570	6" diam. to 12" diam.					14%	13%			
2600	Schedule 80, black									
2610	3/4" diameter	1 Plum	65	.123	L.F.	2.15	4.60		6.75	9.30
2650	1" diameter		61	.131		2.35	4.90		7.25	10
2660	1-1/4" diameter		55	.145		2.98	5.45		8.43	11.50
2670	1-1/2" diameter		49	.163		3.39	6.10		9.49	12.95
2680	2" diameter	↓	38	.211		4.17	7.85		12.02	16.45
2690	2-1/2" diameter	Q-1	54	.296		5.85	9.95		15.80	21.50
2700	3" diameter		48	.333		7.35	11.20		18.55	25
2710	4" diameter		44	.364		10.45	12.25		22.70	30
2720	5" diameter	↓	35	.457		19.30	15.35		34.65	44
2730	6" diameter	Q-2	40	.600		26	21		47	60
2740	8" diameter		35	.686		48	24		72	89
2750	10" diameter		29	.828		75.50	29		104.50	127
2760	12" diameter	↓	24	1	↓	84	35		119	145
3240	To delete coupling & hanger, subtract									
3250	3/4" diam. to 2" diam.					30%	25%			
3260	2-1/2" diam. to 5" diam.					14%	17%			
3270	6" diam. to 12" diam.					12%	12%			
3300	Galvanized									
3310	3/4" diameter	1 Plum	65	.123	L.F.	2.46	4.60		7.06	9.65
3350	1" diameter		61	.131		2.58	4.90		7.48	10.25
3360	1-1/4" diameter		55	.145		3.30	5.45		8.75	11.85
3370	1-1/2" diameter		46	.174		3.78	6.50		10.28	13.95
3380	2" diameter	↓	38	.211		4.65	7.85		12.50	16.95
3390	2-1/2" diameter	Q-1	54	.296		6.55	9.95		16.50	22.50
3400	3" diameter		48	.333		8.35	11.20		19.55	26
3410	4" diameter		44	.364		11.95	12.25		24.20	31.50
3420	5" diameter	↓	35	.457		19.25	15.35		34.60	44
3430	6" diameter	Q-2	40	.600		24.50	21		45.50	58.50
3440	8" diameter	↓	35	.686	↓	46.50	24		70.50	87

Important: See the Reference Section for critical supporting data - Reference Nos., Crews, & City Cost Indexes

15107	Metal Pipe & Fittings	CREW	DAILY OUTPUT	LABOR-HOURS	UNIT	2003 BARE COSTS				TOTAL INCL O&P
						MAT.	LABOR	EQUIP.	TOTAL	
690	3450	10" diameter	Q-2	29	.828	L.F.	61.50	29		90.50	112	690
3460	12" diameter	↓	24	1	↓	81.50	35		116.50	142
3920	To delete coupling & hanger, subtract									
3930	3/4" diam. to 2" diam.					30%	25%			
3940	2-1/2" diam. to 5" diam.					15%	17%			
3950	6" diam. to 12" diam.					11%	12%			
3990	Fittings: cplg. & labor required at joints not incl. in fitting									
3994	price. Add 1 per joint for installed price.									
4000	Elbow, 90° or 45°, painted									
4030	3/4" diameter	1 Plum	50	.160	Ea.	10.75	6		16.75	21
4040	1" diameter		50	.160		10.75	6		16.75	21
4050	1-1/4" diameter		40	.200		10.75	7.45		18.20	23
4060	1-1/2" diameter		33	.242		10.75	9.05		19.80	25.50
4070	2" diameter	↓	25	.320		10.75	11.95		22.70	30
4080	2-1/2" diameter	Q-1	40	.400		10.75	13.45		24.20	32.50
4090	3" diameter		33	.485		19.15	16.30		35.45	45.50
4100	4" diameter		25	.640		21	21.50		42.50	55.50
4110	5" diameter	↓	20	.800		50.50	27		77.50	96
4120	6" diameter	Q-2	25	.960		59	33.50		92.50	116
4130	8" diameter		21	1.143		124	40		164	196
4140	10" diameter		18	1.333		225	46.50		271.50	315
4150	12" diameter		15	1.600		360	56		416	480
4170	14" diameter		12	2		645	69.50		714.50	815
4180	16" diameter	↓	11	2.182		835	76		911	1,025
4190	18" diameter	Q-3	15	2.133		1,075	76		1,151	1,300
4200	20" diameter		13	2.462		1,400	87.50		1,487.50	1,675
4210	24" diameter	↓	11	2.909		2,025	104		2,129	2,375
4250	For galvanized elbows, add				↓	26%				
4690	Tee, painted									
4700	3/4" diameter	1 Plum	38	.211	Ea.	22	7.85		29.85	36
4740	1" diameter		33	.242		16.60	9.05		25.65	32
4750	1-1/4" diameter		27	.296		16.60	11.05		27.65	35
4760	1-1/2" diameter		22	.364		16.60	13.60		30.20	39
4770	2" diameter	↓	17	.471		16.60	17.60		34.20	45
4780	2-1/2" diameter	Q-1	27	.593		16.60	19.95		36.55	48.50
4790	3" diameter		22	.727		23	24.50		47.50	62.50
4800	4" diameter		17	.941		35.50	31.50		67	87
4810	5" diameter	↓	13	1.231		83	41.50		124.50	154
4820	6" diameter	Q-2	17	1.412		96	49		145	181
4830	8" diameter		14	1.714		211	60		271	325
4840	10" diameter		12	2		440	69.50		509.50	585
4850	12" diameter		10	2.400		610	83.50		693.50	795
4851	14" diameter		9	2.667		695	93		788	905
4852	16" diameter	↓	8	3		730	105		835	965
4853	18" diameter	Q-3	11	2.909		915	104		1,019	1,150
4854	20" diameter		10	3.200		1,300	114		1,414	1,625
4855	24" diameter	↓	8	4		2,000	142		2,142	2,425
4900	For galvanized tees, add				↓	24%				
4906	Couplings, rigid style, painted									
4908	1" diameter	1 Plum	100	.080	Ea.	7.95	2.99		10.94	13.20
4909	1-1/4" diameter		100	.080		7.95	2.99		10.94	13.20
4910	1-1/2" diameter		67	.119		7.95	4.46		12.41	15.45
4912	2" diameter	↓	50	.160		8.15	6		14.15	17.95
4914	2-1/2" diameter	Q-1	80	.200		9.40	6.75		16.15	20.50
4916	3" diameter		67	.239		10.95	8.05		19	24
4918	4" diameter	↓	50	.320	↓	15.50	10.75		26.25	33.50

15107	Metal Pipe & Fittings	CREW	DAILY OUTPUT	LABOR-HOURS	UNIT	2003 BARE COSTS				TOTAL INCL O&P	
						MAT.	LABOR	EQUIP.	TOTAL		
690	4920	5" diameter	Q-1	40	.400	Ea.	20	13.45		33.45	42.50	**690**
	4922	6" diameter	Q-2	50	.480		27	16.75		43.75	55	
	4924	8" diameter		42	.571		42.50	19.95		62.45	76.50	
	4926	10" diameter		35	.686		76	24		100	120	
	4928	12" diameter		32	.750		85	26		111	133	
	4930	14" diameter		24	1		111	35		146	175	
	4931	16" diameter		20	1.200		144	42		186	222	
	4932	18" diameter		18	1.333		167	46.50		213.50	254	
	4933	20" diameter		16	1.500		199	52.50		251.50	298	
	4934	24" diameter	Q-9	13	1.231		293	41		334	385	
	4940	Flexible, standard, painted										
	4950	3/4" diameter	1 Plum	100	.080	Ea.	5.70	2.99		8.69	10.75	
	4960	1" diameter		100	.080		5.70	2.99		8.69	10.75	
	4970	1-1/4" diameter		80	.100		7.60	3.74		11.34	14	
	4980	1-1/2" diameter		67	.119		8.30	4.46		12.76	15.90	
	4990	2" diameter		50	.160		8.80	6		14.80	18.65	
	5000	2-1/2" diameter	Q-1	80	.200		10.45	6.75		17.20	21.50	
	5010	3" diameter		67	.239		11.60	8.05		19.65	25	
	5020	3-1/2" diameter		57	.281		16.85	9.45		26.30	33	
	5030	4" diameter		50	.320		16.85	10.75		27.60	35	
	5040	5" diameter		40	.400		26	13.45		39.45	49	
	5050	6" diameter	Q-2	50	.480		30.50	16.75		47.25	59	
	5070	8" diameter		42	.571		49.50	19.95		69.45	84.50	
	5090	10" diameter		35	.686		82.50	24		106.50	127	
	5110	12" diameter		32	.750		94	26		120	143	
	5120	14" diameter		24	1		114	35		149	179	
	5130	16" diameter		20	1.200		150	42		192	228	
	5140	18" diameter		18	1.333		175	46.50		221.50	263	
	5150	20" diameter		16	1.500		276	52.50		328.50	385	
	5160	24" diameter		13	1.846		305	64.50		369.50	430	
	5176	Lightweight style, painted										
	5178	1-1/2" diameter	1 Plum	67	.119	Ea.	7.25	4.46		11.71	14.70	
	5180	2" diameter	"	50	.160		7.40	6		13.40	17.10	
	5182	2-1/2" diameter	Q-1	80	.200		8.60	6.75		15.35	19.60	
	5184	3" diameter		67	.239		9.95	8.05		18	23	
	5186	3-1/2" diameter		57	.281		14.10	9.45		23.55	30	
	5188	4" diameter		50	.320		14.10	10.75		24.85	32	
	5190	5" diameter		40	.400		20.50	13.45		33.95	43	
	5192	6" diameter	Q-2	50	.480		24.50	16.75		41.25	52.50	
	5194	8" diameter		42	.571		38.50	19.95		58.45	72.50	
	5196	10" diameter		35	.686		103	24		127	150	
	5198	12" diameter		32	.750		115	26		141	167	
	5200	For galvanized couplings, add					33%					
	5220	Tee, reducing, painted										
	5225	2" x 1-1/2" diameter	Q-1	38	.421	Ea.	36	14.15		50.15	61	
	5226	2-1/2" x 2" diameter		28	.571		36	19.20		55.20	68.50	
	5227	3" x 2-1/2" diameter		23	.696		31.50	23.50		55	70	
	5228	4" x 3" diameter		18	.889		42.50	30		72.50	92	
	5229	5" x 4" diameter		15	1.067		91.50	36		127.50	155	
	5230	6" x 4" diameter	Q-2	18	1.333		101	46.50		147.50	181	
	5231	8" x 6" diameter		15	1.600		211	56		267	315	
	5232	10" x 8" diameter		13	1.846		277	64.50		341.50	400	
	5233	12" x 10" diameter		11	2.182		420	76		496	580	
	5234	14" x 12" diameter		10	2.400		470	83.50		553.50	645	
	5235	16" x 12" diameter		9	2.667		585	93		678	780	
	5236	18" x 12" diameter	Q-3	12	2.667		700	95		795	915	

15 MECHANICAL

Important: See the Reference Section for critical supporting data - Reference Nos., Crews, & City Cost Indexes

			CREW	DAILY OUTPUT	LABOR-HOURS	UNIT	2003 BARE COSTS				TOTAL INCL O&P
	15107	**Metal Pipe & Fittings**					MAT.	LABOR	EQUIP.	TOTAL	
690	5237	18" x 16" diameter	Q-3	11	2.909	Ea.	885	104		989	1,125
	5238	20" x 16" diameter		10	3.200		1,150	114		1,264	1,450
	5239	24" x 20" diameter	▼	9	3.556	▼	1,825	127		1,952	2,200
	5240	Reducer, concentric, painted									
	5241	2-1/2" x 2" diameter	Q-1	43	.372	Ea.	12.55	12.50		25.05	32.50
	5242	3" x 2-1/2" diameter		35	.457		15.15	15.35		30.50	39.50
	5243	4" x 3" diameter		29	.552		18.35	18.55		36.90	48
	5244	5" x 4" diameter	▼	22	.727		25.50	24.50		50	65
	5245	6" x 4" diameter	Q-2	26	.923		29.50	32		61.50	81
	5246	8" x 6" diameter		23	1.043		77.50	36.50		114	140
	5247	10" x 8" diameter		20	1.200		158	42		200	237
	5248	12" x 10" diameter	▼	16	1.500	▼	283	52.50		335.50	390
	5255	Eccentric, painted									
	5256	2-1/2" x 2" diameter	Q-1	42	.381	Ea.	26.50	12.80		39.30	48.50
	5257	3" x 2-1/2" diameter		34	.471		30.50	15.85		46.35	57.50
	5258	4" x 3" diameter		28	.571		37	19.20		56.20	70
	5259	5" x 4" diameter	▼	21	.762		50.50	25.50		76	94
	5260	6" x 4" diameter	Q-2	25	.960		59	33.50		92.50	116
	5261	8" x 6" diameter		22	1.091		120	38		158	189
	5262	10" x 8" diameter		19	1.263		330	44		374	425
	5263	12" x 10" diameter	▼	15	1.600	▼	450	56		506	580
	5270	Coupling, reducing, painted									
	5272	2" x 1-1/2" diameter	1 Plum	52	.154	Ea.	12.50	5.75		18.25	22.50
	5274	2-1/2" x 2" diameter	Q-1	82	.195		16.35	6.55		22.90	28
	5276	3" x 2" diameter		69	.232		18.70	7.80		26.50	32.50
	5278	4" x 2" diameter		52	.308		30	10.35		40.35	48.50
	5280	5" x 4" diameter	▼	42	.381		33.50	12.80		46.30	56.50
	5282	6" x 4" diameter	Q-2	52	.462		51	16.10		67.10	80.50
	5284	8" x 6" diameter	"	44	.545	▼	76.50	19		95.50	113
	5290	Outlet coupling, painted									
	5294	1-1/2" x 1" pipe size	1 Plum	65	.123	Ea.	18	4.60		22.60	27
	5296	2" x 1" pipe size	"	48	.167		19.40	6.25		25.65	31
	5298	2-1/2" x 1" pipe size	Q-1	78	.205		30	6.90		36.90	43.50
	5300	2-1/2" x 1" pipe size	1 Plum	70	.114		34.50	4.27		38.77	44.50
	5302	3" x 1" pipe size	Q-1	65	.246		28.50	8.30		36.80	44
	5304	4" x 3/4" pipe size		48	.333		43	11.20		54.20	64
	5306	4" x 1-1/2" pipe size	▼	46	.348		62.50	11.70		74.20	86
	5308	6" x 1-1/2" pipe size	Q-2	44	.545	▼	90	19		109	128
	5750	Flange, w/groove gasket, black steel									
	5752	See 15107-660-0620 for gasket & bolt set									
	5760	ANSI class 125 and 150, painted									
	5780	2" pipe size	1 Plum	23	.348	Ea.	36	13		49	59.50
	5790	2-1/2" pipe size	Q-1	37	.432		45	14.55		59.55	71.50
	5800	3" pipe size		31	.516		48.50	17.35		65.85	79.50
	5820	4" pipe size		23	.696		65	23.50		88.50	108
	5830	5" pipe size	▼	19	.842		75.50	28.50		104	127
	5840	6" pipe size	Q-2	23	1.043		82.50	36.50		119	146
	5850	8" pipe size		17	1.412		93.50	49		142.50	178
	5860	10" pipe size		14	1.714		148	60		208	254
	5870	12" pipe size		12	2		194	69.50		263.50	320
	5880	14" pipe size		10	2.400		445	83.50		528.50	615
	5890	16" pipe size		9	2.667		515	93		608	705
	5900	18" pipe size		6	4		635	139		774	910
	5910	20" pipe size		5	4.800		765	167		932	1,100
	5920	24" pipe size	▼	4.50	5.333	▼	980	186		1,166	1,350
	5940	ANSI class 350, painted									

690

15107 | Metal Pipe & Fittings

		CREW	DAILY OUTPUT	LABOR-HOURS	UNIT	2003 BARE COSTS				TOTAL INCL O&P
						MAT.	LABOR	EQUIP.	TOTAL	
5946	2" pipe size	1 Plum	23	.348	Ea.	45.50	13		58.50	69.50
5948	2-1/2" pipe size	Q-1	37	.432		52.50	14.55		67.05	80
5950	3" pipe size		31	.516		72	17.35		89.35	105
5952	4" pipe size		23	.696		96	23.50		119.50	142
5954	5" pipe size		19	.842		109	28.50		137.50	163
5956	6" pipe size	Q-2	23	1.043		127	36.50		163.50	195
5958	8" pipe size		17	1.412		146	49		195	236
5960	10" pipe size		14	1.714		233	60		293	345
5962	12" pipe size	1 Plum	12	.667		249	25		274	310
6100	Coupling, for PVC plastic pipe									
6110	2" diameter	1 Plum	50	.160	Ea.	9.75	6		15.75	19.75
6112	2-1/2" diameter	Q-1	80	.200		13.70	6.75		20.45	25
6114	3" diameter		67	.239		16.65	8.05		24.70	30.50
6116	4" diameter		50	.320		21.50	10.75		32.25	40.50
6118	6" diameter	Q-2	50	.480		36	16.75		52.75	65
6120	8" diameter		42	.571		59	19.95		78.95	95
6122	10" diameter		35	.686		97	24		121	142
6124	12" diameter		32	.750		122	26		148	175
7400	Suction diffuser									
7402	Grooved end inlet x flanged outlet									
7410	3" x 3"	Q-1	50	.320	Ea.	375	10.75		385.75	430
7412	4" x 4"		38	.421		510	14.15		524.15	580
7414	5" x 5"		30	.533		600	17.95		617.95	685
7416	6" x 6"	Q-2	38	.632		755	22		777	865
7418	8" x 8"		27	.889		1,400	31		1,431	1,600
7420	10" x 10"		20	1.200		1,900	42		1,942	2,175
7422	12" x 12"		16	1.500		3,150	52.50		3,202.50	3,525
7424	14" x 14"		15	1.600		3,975	56		4,031	4,450
7426	16" x 14"		14	1.714		4,075	60		4,135	4,600
7500	Strainer, tee type, painted									
7506	2" pipe size	1 Plum	38	.211	Ea.	235	7.85		242.85	270
7508	2-1/2" pipe size	Q-1	62	.258		247	8.70		255.70	284
7510	3" pipe size		50	.320		277	10.75		287.75	320
7512	4" pipe size		38	.421		315	14.15		329.15	365
7514	5" pipe size		30	.533		455	17.95		472.95	525
7516	6" pipe size	Q-2	38	.632		495	22		517	580
7518	8" pipe size		27	.889		755	31		786	875
7520	10" pipe size		20	1.200		1,100	42		1,142	1,300
7522	12" pipe size		16	1.500		1,425	52.50		1,477.50	1,650
7524	14" pipe size		15	1.600		5,100	56		5,156	5,675
7526	16" pipe size		14	1.714		6,325	60		6,385	7,075
7570	Expansion joint, max. 3" travel									
7572	2" diameter	1 Plum	38	.211	Ea.	193	7.85		200.85	224
7574	3" diameter	Q-1	50	.320		226	10.75		236.75	265
7576	4" diameter	"	38	.421		284	14.15		298.15	330
7578	6" diameter	Q-2	38	.632		495	22		517	580
7800	Ball valve w/handle, carbon steel trim									
7810	1-1/2" pipe size	1 Plum	50	.160	Ea.	59.50	6		65.50	74.50
7812	2" pipe size	"	38	.211		66	7.85		73.85	84.50
7814	2-1/2" pipe size	Q-1	62	.258		144	8.70		152.70	172
7816	3" pipe size		50	.320		233	10.75		243.75	272
7818	4" pipe size		38	.421		360	14.15		374.15	415
7820	6" pipe size	Q-2	30	.800		1,100	28		1,128	1,275
7830	With gear operator									
7834	2-1/2" pipe size	Q-1	62	.258	Ea.	405	8.70		413.70	460
7836	3" pipe size		50	.320		495	10.75		505.75	560

690

Important: See the Reference Section for critical supporting data - Reference Nos., Crews, & City Cost Indexes

15107	Metal Pipe & Fittings	CREW	DAILY OUTPUT	LABOR-HOURS	UNIT	2003 BARE COSTS				TOTAL INCL O&P	
						MAT.	LABOR	EQUIP.	TOTAL		
690	7838	4" pipe size	Q-1	38	.421	Ea.	620	14.15		634.15	700	690
7840	6" pipe size	Q-2	30	.800	↓	1,400	28		1,428	1,600
7870	Check valve									
7874	2-1/2" pipe size	Q-1	62	.258	Ea.	103	8.70		111.70	126
7876	3" pipe size		50	.320		122	10.75		132.75	150
7878	4" pipe size		38	.421		128	14.15		142.15	163
7880	5" pipe size	↓	30	.533		214	17.95		231.95	262
7882	6" pipe size	Q-2	38	.632		254	22		276	315
7884	8" pipe size		27	.889		345	31		376	425
7886	10" pipe size		20	1.200		1,000	42		1,042	1,175
7888	12" pipe size	↓	16	1.500	↓	1,200	52.50		1,252.50	1,375
7900	Plug valve, balancing, w/lever operator									
7906	3" pipe size	Q-1	50	.320	Ea.	232	10.75		242.75	272
7908	4" pipe size	"	38	.421		286	14.15		300.15	335
7909	6" pipe size	Q-2	30	.800	↓	440	28		468	520
7916	With gear operator									
7920	3" pipe size	Q-1	50	.320	Ea.	420	10.75		430.75	475
7922	4" pipe size	"	38	.421		435	14.15		449.15	500
7924	6" pipe size	Q-2	38	.632		585	22		607	680
7926	8" pipe size		27	.889		795	31		826	920
7928	10" pipe size		20	1.200		1,225	42		1,267	1,425
7930	12" pipe size	↓	16	1.500	↓	1,875	52.50		1,927.50	2,150
8000	Butterfly valve, 2 position handle, with standard trim									
8010	1-1/2" pipe size	1 Plum	50	.160	Ea.	91	6		97	109
8020	2" pipe size	"	38	.211		91	7.85		98.85	112
8030	3" pipe size	Q-1	50	.320		130	10.75		140.75	160
8050	4" pipe size	"	38	.421		144	14.15		158.15	180
8070	6" pipe size	Q-2	38	.632		290	22		312	355
8080	8" pipe size		27	.889		410	31		441	495
8090	10" pipe size	↓	20	1.200	↓	590	42		632	715
8200	With stainless steel trim									
8240	1-1/2" pipe size	1 Plum	50	.160	Ea.	115	6		121	136
8250	2" pipe size	"	38	.211		115	7.85		122.85	139
8270	3" pipe size	Q-1	50	.320		155	10.75		165.75	187
8280	4" pipe size	"	38	.421		168	14.15		182.15	207
8300	6" pipe size	Q-2	38	.632		315	22		337	380
8310	8" pipe size		27	.889		580	31		611	680
8320	10" pipe size		20	1.200		965	42		1,007	1,150
8322	12" pipe size		16	1.500		1,075	52.50		1,127.50	1,250
8324	14" pipe size		15	1.600		1,425	56		1,481	1,650
8326	16" pipe size	↓	14	1.714		1,950	60		2,010	2,250
8328	18" pipe size	Q-3	12	2.667		2,400	95		2,495	2,800
8330	20" pipe size		11	2.909		3,250	104		3,354	3,725
8332	24" pipe size	↓	10	3.200	↓	4,125	114		4,239	4,725
8336	Note: sizes 8" up w/manual gear operator									
9000	Cut one groove, labor									
9010	3/4" pipe size	Q-1	152	.105	Ea.		3.54		3.54	5.35
9020	1" pipe size		140	.114			3.84		3.84	5.80
9030	1-1/4" pipe size		124	.129			4.34		4.34	6.55
9040	1-1/2" pipe size		114	.140			4.72		4.72	7.15
9050	2" pipe size		104	.154			5.15		5.15	7.80
9060	2-1/2" pipe size		96	.167			5.60		5.60	8.45
9070	3" pipe size		88	.182			6.10		6.10	9.25
9080	3-1/2" pipe size		83	.193			6.50		6.50	9.80
9090	4" pipe size		78	.205			6.90		6.90	10.40
9100	5" pipe size	↓	72	.222	↓		7.45		7.45	11.30

MECHANICAL **15**

195

15107	Metal Pipe & Fittings	CREW	DAILY OUTPUT	LABOR-HOURS	UNIT	2003 BARE COSTS				TOTAL INCL O&P
						MAT.	LABOR	EQUIP.	TOTAL	
690 9110	6" pipe size	Q-1	70	.229	Ea.		7.70		7.70	11.60 **690**
9120	8" pipe size		54	.296			9.95		9.95	15.05
9130	10" pipe size		38	.421			14.15		14.15	21.50
9140	12" pipe size		30	.533			17.95		17.95	27
9150	14" pipe size		20	.800			27		27	40.50
9160	16" pipe size		19	.842			28.50		28.50	43
9170	18" pipe size		18	.889			30		30	45
9180	20" pipe size		17	.941			31.50		31.50	48
9190	24" pipe size	▼	15	1.067	▼		36		36	54
9210	Roll one groove									
9220	3/4" pipe size	Q-1	266	.060	Ea.		2.02		2.02	3.05
9230	1" pipe size		228	.070			2.36		2.36	3.56
9240	1-1/4" pipe size		200	.080			2.69		2.69	4.06
9250	1-1/2" pipe size		178	.090			3.02		3.02	4.56
9260	2" pipe size		116	.138			4.64		4.64	7
9270	2-1/2" pipe size		110	.145			4.89		4.89	7.40
9280	3" pipe size		100	.160			5.40		5.40	8.10
9290	3-1/2" pipe size		94	.170			5.70		5.70	8.65
9300	4" pipe size		86	.186			6.25		6.25	9.45
9310	5" pipe size		84	.190			6.40		6.40	9.65
9320	6" pipe size		80	.200			6.75		6.75	10.15
9330	8" pipe size		66	.242			8.15		8.15	12.30
9340	10" pipe size		58	.276			9.30		9.30	14
9350	12" pipe size		46	.348			11.70		11.70	17.65
9360	14" pipe size		30	.533			17.95		17.95	27
9370	16" pipe size		28	.571			19.20		19.20	29
9380	18" pipe size		27	.593			19.95		19.95	30
9390	20" pipe size		25	.640			21.50		21.50	32.50
9400	24" pipe size	▼	23	.696	▼		23.50		23.50	35.50
920 0010	**PIPE, STAINLESS STEEL**									**920**
0020	Welded, with clevis type hangers 10' O.C.									
0500	Schedule 5, type 304									
0540	1/2" diameter	Q-15	128	.125	L.F.	4.86	4.20	.61	9.67	12.40
0550	3/4" diameter		116	.138		5.50	4.64	.68	10.82	13.80
0560	1" diameter		103	.155		6.70	5.20	.76	12.66	16.15
0570	1-1/4" diameter		93	.172		7.65	5.80	.84	14.29	18.15
0580	1-1/2" diameter		85	.188		8.80	6.35	.92	16.07	20
0590	2" diameter		69	.232		10.55	7.80	1.14	19.49	24.50
0600	2-1/2" diameter		53	.302		14.70	10.15	1.48	26.33	33
0610	3" diameter		48	.333		17.70	11.20	1.64	30.54	38
0620	4" diameter		44	.364		22.50	12.25	1.79	36.54	45
0630	5" diameter	▼	36	.444		45	14.95	2.18	62.13	75
0640	6" diameter	Q-16	42	.571		42	19.95	1.87	63.82	78
0650	8" diameter		34	.706		64	24.50	2.31	90.81	110
0660	10" diameter		26	.923		89.50	32	3.02	124.52	150
0670	12" diameter	▼	21	1.143	▼	118	40	3.74	161.74	194
0700	To delete hangers, subtract									
0710	1/2" diam. to 1-1/2" diam.					8%	19%			
0720	2" diam. to 5" diam.					4%	9%			
0730	6" diam. to 12" diam.					3%	4%			
0750	For small quantities, add				L.F.	10%				
1250	Schedule 5, type 316									
1290	1/2" diameter	Q-15	128	.125	L.F.	6.10	4.20	.61	10.91	13.75
1300	3/4" diameter		116	.138		7.05	4.64	.68	12.37	15.50
1310	1" diameter	▼	103	.155	▼	8	5.20	.76	13.96	17.55

15107	Metal Pipe & Fittings	CREW	DAILY OUTPUT	LABOR-HOURS	UNIT	2003 BARE COSTS				TOTAL INCL O&P
						MAT.	LABOR	EQUIP.	TOTAL	
920 1320	1-1/4" diameter	Q-15	93	.172	L.F.	9.60	5.80	.84	16.24	20
1330	1-1/2" diameter		85	.188		10.90	6.35	.92	18.17	22.50
1340	2" diameter		69	.232		13.20	7.80	1.14	22.14	27.50
1350	2-1/2" diameter		53	.302		18.95	10.15	1.48	30.58	38
1360	3" diameter		48	.333		24	11.20	1.64	36.84	45.50
1370	4" diameter		44	.364		29.50	12.25	1.79	43.54	53
1380	5" diameter		36	.444		56.50	14.95	2.18	73.63	87
1390	6" diameter	Q-16	42	.571		55	19.95	1.87	76.82	92.50
1400	8" diameter		34	.706		82.50	24.50	2.31	109.31	131
1410	10" diameter		26	.923		119	32	3.02	154.02	183
1420	12" diameter		21	1.143		153	40	3.74	196.74	232
1490	For small quantities, add					10%				
1940	To delete hanger, subtract									
1950	1/2" diam. to 1-1/2" diam.					5%	19%			
1960	2" diam. to 5" diam.					3%	9%			
1970	6" diam. to 12" diam.					2%	4%			
2000	Schedule 10, type 304									
2040	1/4" diameter	Q-15	131	.122	L.F.	3.83	4.11	.60	8.54	11.10
2050	3/8" diameter		128	.125		4.04	4.20	.61	8.85	11.50
2060	1/2" diameter		125	.128		4.97	4.30	.63	9.90	12.65
2070	3/4" diameter		113	.142		5.75	4.76	.70	11.21	14.30
2080	1" diameter		100	.160		8.25	5.40	.79	14.44	18
2090	1-1/4" diameter		91	.176		9.90	5.90	.86	16.66	21
2100	1-1/2" diameter		83	.193		10.90	6.50	.95	18.35	23
2110	2" diameter		67	.239		13.50	8.05	1.17	22.72	28
2120	2-1/2" diameter		51	.314		16.75	10.55	1.54	28.84	36
2130	3" diameter		46	.348		20.50	11.70	1.71	33.91	42
2140	4" diameter		42	.381		26.50	12.80	1.87	41.17	51
2150	5" diameter		35	.457		36	15.35	2.24	53.59	65
2160	6" diameter	Q-16	40	.600		42	21	1.96	64.96	79.50
2170	8" diameter		33	.727		64	25.50	2.38	91.88	112
2180	10" diameter		25	.960		89.50	33.50	3.14	126.14	152
2190	12" diameter		21	1.143		112	40	3.74	155.74	187
2250	For small quantities, add					10%				
2650	To delete hanger, subtract									
2660	1/4" diam. to 3/4" diam.					9%	22%			
2670	1" diam. to 2" diam.					4%	15%			
2680	2-1/2" diam. to 5" diam.					3%	8%			
2690	6" diam. to 12" diam.					3%	4%			
2750	Schedule 10, type 316									
2790	1/4" diameter	Q-15	131	.122	L.F.	4.31	4.11	.60	9.02	11.60
2800	3/8" diameter		128	.125		4.78	4.20	.61	9.59	12.30
2810	1/2" diameter		125	.128		6.10	4.30	.63	11.03	13.90
2820	3/4" diameter		113	.142		7.05	4.76	.70	12.51	15.75
2830	1" diameter		100	.160		9.30	5.40	.79	15.49	19.15
2840	1-1/4" diameter		91	.176		12.45	5.90	.86	19.21	23.50
2850	1-1/2" diameter		83	.193		13.90	6.50	.95	21.35	26
2860	2" diameter		67	.239		17.15	8.05	1.17	26.37	32.50
2870	2-1/2" diameter		51	.314		22.50	10.55	1.54	34.59	42
2880	3" diameter		46	.348		28	11.70	1.71	41.41	50.50
2890	4" diameter		42	.381		34	12.80	1.87	48.67	59
2900	5" diameter		35	.457		50.50	15.35	2.24	68.09	81.50
2910	6" diameter	Q-16	40	.600		54.50	21	1.96	77.46	93.50
2920	8" diameter		33	.727		90	25.50	2.38	117.88	140
2930	10" diameter		25	.960		119	33.50	3.14	155.64	185
2940	12" diameter		21	1.143		149	40	3.74	192.74	228

MECHANICAL 15

15107	Metal Pipe & Fittings	CREW	DAILY OUTPUT	LABOR-HOURS	UNIT	2003 BARE COSTS				TOTAL INCL O&P
						MAT.	LABOR	EQUIP.	TOTAL	
920	2990	For small quantities, add				L.F.	10%				920
	3430	To delete hanger, subtract									
	3440	1/4″ diam. to 3/4″ diam.					6%	22%			
	3450	1″ diam. to 2″ diam.					3%	15%			
	3460	2-1/2″ diam. to 5″ diam.					2%	8%			
	3470	6″ diam. to 12″ diam.					2%	4%			
	3500	Threaded, couplings and hangers 10′ O.C.									
	3520	Schedule 40, type 304									
	3540	1/4″ diameter	1 Plum	54	.148	L.F.	5.05	5.55		10.60	13.90
	3550	3/8″ diameter		53	.151		5.75	5.65		11.40	14.80
	3560	1/2″ diameter		52	.154		6.60	5.75		12.35	15.95
	3570	3/4″ diameter		51	.157		8.90	5.85		14.75	18.65
	3580	1″ diameter		45	.178		11.20	6.65		17.85	22.50
	3590	1-1/4″ diameter	Q-1	76	.211		13.20	7.10		20.30	25
	3600	1-1/2″ diameter		69	.232		15.40	7.80		23.20	28.50
	3610	2″ diameter		57	.281		20.50	9.45		29.95	37
	3620	2-1/2″ diameter		44	.364		32.50	12.25		44.75	54.50
	3630	3″ diameter		38	.421		42	14.15		56.15	68
	3640	4″ diameter	Q-2	51	.471		58	16.40		74.40	89
	3740	For small quantities, add					10%				
	4200	To delete couplings & hangers, subtract									
	4210	1/4″ diam. to 3/4″ diam.					15%	56%			
	4220	1″ diam. to 2″ diam.					18%	49%			
	4230	2-1/2″ diam. to 4″ diam.					34%	40%			
	4250	Schedule 40, type 316									
	4290	1/4″ diameter	1 Plum	54	.148	L.F.	5.80	5.55		11.35	14.75
	4300	3/8″ diameter		53	.151		6.85	5.65		12.50	16.05
	4310	1/2″ diameter		52	.154		8.60	5.75		14.35	18.20
	4320	3/4″ diameter		51	.157		10.85	5.85		16.70	21
	4330	1″ diameter		45	.178		14.25	6.65		20.90	25.50
	4340	1-1/4″ diameter	Q-1	76	.211		16.80	7.10		23.90	29
	4350	1-1/2″ diameter		69	.232		19.65	7.80		27.45	33.50
	4360	2″ diameter		57	.281		26	9.45		35.45	43
	4370	2-1/2″ diameter		44	.364		42.50	12.25		54.75	65
	4380	3″ diameter		38	.421		57	14.15		71.15	84.50
	4390	4″ diameter	Q-2	51	.471		77	16.40		93.40	110
	4490	For small quantities, add					10%				
	4900	To delete couplings & hangers, subtract									
	4910	1/4″ diam. to 3/4″ diam.					12%	56%			
	4920	1″ diam. to 2″ diam.					14%	49%			
	4930	2-1/2″ diam. to 4″ diam.					27%	40%			
	5000	Schedule 80, type 304									
	5040	1/4″ diameter	1 Plum	53	.151	L.F.	6.65	5.65		12.30	15.85
	5050	3/8″ diameter		52	.154		7.55	5.75		13.30	17
	5060	1/2″ diameter		51	.157		9.45	5.85		15.30	19.25
	5070	3/4″ diameter		48	.167		12.15	6.25		18.40	23
	5080	1″ diameter		43	.186		16.60	6.95		23.55	29
	5090	1-1/4″ diameter	Q-1	73	.219		23.50	7.35		30.85	37
	5100	1-1/2″ diameter		67	.239		27.50	8.05		35.55	42
	5110	2″ diameter		54	.296		34.50	9.95		44.45	53
	5190	For small quantities, add					10%				
	5700	To delete couplings & hangers, subtract									
	5710	1/4″ diam. to 3/4″ diam.					10%	53%			
	5720	1″ diam. to 2″ diam.					14%	47%			
	5750	Schedule 80, type 316									
	5790	1/4″ diameter	1 Plum	53	.151	L.F.	8.20	5.65		13.85	17.50

Important: See the Reference Section for critical supporting data - Reference Nos., Crews, & City Cost Indexes

15107	Metal Pipe & Fittings	CREW	DAILY OUTPUT	LABOR-HOURS	UNIT	2003 BARE COSTS				TOTAL INCL O&P
						MAT.	LABOR	EQUIP.	TOTAL	
920 5800	3/8" diameter	1 Plum	52	.154	L.F.	9.30	5.75		15.05	18.95 **920**
5810	1/2" diameter		51	.157		13.30	5.85		19.15	23.50
5820	3/4" diameter		48	.167		15.05	6.25		21.30	26
5830	1" diameter		43	.186		20.50	6.95		27.45	33.50
5840	1-1/4" diameter	Q-1	73	.219		30	7.35		37.35	44
5850	1-1/2" diameter		67	.239		34.50	8.05		42.55	50
5860	2" diameter		54	.296		44.50	9.95		54.45	63.50
5950	For small quantities, add					10%				
7000	To delete couplings & hangers, subtract									
7010	1/4" diam. to 3/4" diam.					9%	53%			
7020	1" diam. to 2" diam.					14%	47%			
8000	Weld joints with clevis type hangers 10' O.C.									
8010	Schedule 40, type 304									
8050	1/8" pipe size	Q-15	126	.127	L.F.	3.57	4.27	.62	8.46	11.05
8060	1/4" pipe size		125	.128		4.45	4.30	.63	9.38	12.10
8070	3/8" pipe size		122	.131		5.05	4.41	.64	10.10	12.90
8080	1/2" pipe size		118	.136		6.10	4.56	.67	11.33	14.35
8090	3/4" pipe size		109	.147		7.60	4.94	.72	13.26	16.60
8100	1" pipe size		95	.168		9.50	5.65	.83	15.98	19.90
8110	1-1/4" pipe size		86	.186		11.70	6.25	.91	18.86	23.50
8120	1-1/2" pipe size		78	.205		13.60	6.90	1.01	21.51	26.50
8130	2" pipe size		62	.258		17.45	8.70	1.27	27.42	33.50
8140	2-1/2" pipe size		49	.327		26	11	1.60	38.60	47.50
8150	3" pipe size		44	.364		33.50	12.25	1.79	47.54	57.50
8160	3-1/2" pipe size		44	.364		39.50	12.25	1.79	53.54	64
8170	4" pipe size		39	.410		45	13.80	2.01	60.81	72.50
8180	5" pipe size		32	.500		65.50	16.80	2.46	84.76	100
8190	6" pipe size	Q-16	37	.649		78.50	22.50	2.12	103.12	123
8200	8" pipe size		29	.828		127	29	2.71	158.71	185
8210	10" pipe size		24	1		215	35	3.27	253.27	292
8220	12" pipe size		20	1.200		243	42	3.92	288.92	335
8300	Schedule 40, type 316									
8310	1/8" pipe size	Q-15	126	.127	L.F.	4.32	4.27	.62	9.21	11.90
8320	1/4" pipe size		125	.128		5.10	4.30	.63	10.03	12.80
8330	3/8" pipe size		122	.131		6.05	4.41	.64	11.10	14
8340	1/2" pipe size		118	.136		8	4.56	.67	13.23	16.45
8350	3/4" pipe size		109	.147		9.25	4.94	.72	14.91	18.45
8360	1" pipe size		95	.168		12.15	5.65	.83	18.63	23
8370	1-1/4" pipe size		86	.186		15	6.25	.91	22.16	27
8380	1-1/2" pipe size		78	.205		17.50	6.90	1.01	25.41	31
8390	2" pipe size		62	.258		22.50	8.70	1.27	32.47	39.50
8400	2-1/2" pipe size		49	.327		35	11	1.60	47.60	56.50
8410	3" pipe size		44	.364		47	12.25	1.79	61.04	72
8420	3-1/2" pipe size		44	.364		54.50	12.25	1.79	68.54	80
8430	4" pipe size		39	.410		61.50	13.80	2.01	77.31	91
8440	5" pipe size		32	.500		93	16.80	2.46	112.26	130
8450	6" pipe size	Q-16	37	.649		104	22.50	2.12	128.62	151
8460	8" pipe size		29	.828		195	29	2.71	226.71	260
8470	10" pipe size		24	1		261	35	3.27	299.27	345
8480	12" pipe size		20	1.200		355	42	3.92	400.92	455
8500	Schedule 80, type 304									
8510	1/4" pipe size	Q-15	110	.145	L.F.	6	4.89	.71	11.60	14.80
8520	3/8" pipe size		109	.147		6.75	4.94	.72	12.41	15.70
8530	1/2" pipe size		106	.151		8.50	5.10	.74	14.34	17.80
8540	3/4" pipe size		96	.167		10.85	5.60	.82	17.27	21.50
8550	1" pipe size		87	.184		14.45	6.20	.90	21.55	26

15107 | Metal Pipe & Fittings

		CREW	DAILY OUTPUT	LABOR-HOURS	UNIT	2003 BARE COSTS				TOTAL INCL O&P
						MAT.	LABOR	EQUIP.	TOTAL	
8560	1-1/4" pipe size	Q-15	81	.198	L.F.	18.60	6.65	.97	26.22	31.50
8570	1-1/2" pipe size		74	.216		21	7.25	1.06	29.31	35.50
8580	2" pipe size		58	.276		26.50	9.30	1.35	37.15	44.50
8590	2-1/2" pipe size		46	.348		54	11.70	1.71	67.41	79
8600	3" pipe size		41	.390		72.50	13.10	1.92	87.52	102
8610	4" pipe size	↓	33	.485		113	16.30	2.38	131.68	152
8630	6" pipe size	Q-16	30	.800	↓	249	28	2.62	279.62	320
8640	Schedule 80, type 316									
8650	1/4" pipe size	Q-15	110	.145	L.F.	7.45	4.89	.71	13.05	16.40
8660	3/8" pipe size		109	.147		8.50	4.94	.72	14.16	17.60
8670	1/2" pipe size		106	.151		12.15	5.10	.74	17.99	22
8680	3/4" pipe size		96	.167		13.40	5.60	.82	19.82	24
8690	1" pipe size		87	.184		17.95	6.20	.90	25.05	30
8700	1-1/4" pipe size		81	.198		24.50	6.65	.97	32.12	37.50
8710	1-1/2" pipe size		74	.216		27.50	7.25	1.06	35.81	42
8720	2" pipe size		58	.276		34.50	9.30	1.35	45.15	53
8730	2-1/2" pipe size		46	.348		89	11.70	1.71	102.41	118
8740	3" pipe size		41	.390		104	13.10	1.92	119.02	136
8760	4" pipe size	↓	33	.485		144	16.30	2.38	162.68	186
8770	6" pipe size	Q-16	30	.800	↓	390	28	2.62	420.62	470
8790	Schedule 160, type 304									
8800	1/2" pipe size	Q-15	96	.167	L.F.	11.95	5.60	.82	18.37	22.50
8810	3/4" pipe size		88	.182		14	6.10	.89	20.99	25.50
8820	1" pipe size		80	.200		23.50	6.75	.98	31.23	37
8840	1-1/2" pipe size		67	.239		32.50	8.05	1.17	41.72	49
8850	2" pipe size		53	.302		46.50	10.15	1.48	58.13	68.50
8870	3" pipe size	↓	37	.432	↓	76	14.55	2.12	92.67	108
8900	Schedule 160, type 316									
8910	1/2" pipe size	Q-15	96	.167	L.F.	17.65	5.60	.82	24.07	29
8920	3/4" pipe size		88	.182		18.55	6.10	.89	25.54	30.50
8930	1" pipe size		80	.200		26.50	6.75	.98	34.23	40
8940	1-1/4" pipe size		73	.219		37.50	7.35	1.08	45.93	53.50
8950	1-1/2" pipe size		67	.239		43	8.05	1.17	52.22	61
8960	2" pipe size		53	.302		59.50	10.15	1.48	71.13	82.50
8990	4" pipe size	↓	30	.533	↓	150	17.95	2.62	170.57	195
9100	Threading pipe labor, sst, one end, schedules 40 & 80									
9110	1/4" through 3/4" pipe size	1 Plum	61.50	.130	Ea.		4.86		4.86	7.35
9120	1" through 2" pipe size		55.90	.143			5.35		5.35	8.05
9130	2-1/2" pipe size		41.50	.193			7.20		7.20	10.85
9140	3" pipe size	↓	38.50	.208			7.75		7.75	11.70
9150	3-1/2" pipe size	Q-1	68.40	.234			7.85		7.85	11.90
9160	4" pipe size		73	.219			7.35		7.35	11.15
9170	5" pipe size		40.70	.393			13.20		13.20	19.95
9180	6" pipe size		35.40	.452			15.20		15.20	23
9190	8" pipe size		22.30	.717			24		24	36.50
9200	10" pipe size		16.10	.994			33.50		33.50	50.50
9210	12" pipe size	↓	12.30	1.301	↓		44		44	66
9250	Welding labor per joint for stainless steel									
9260	Schedule 5 and 10									
9270	1/4" pipe size	Q-15	36	.444	Ea.		14.95	2.18	17.13	25
9280	3/8" pipe size		35	.457			15.35	2.24	17.59	25.50
9290	1/2" pipe size		35	.457			15.35	2.24	17.59	25.50
9300	3/4" pipe size		28	.571			19.20	2.81	22.01	32
9310	1" pipe size		25	.640			21.50	3.14	24.64	36
9320	1-1/4" pipe size		22	.727			24.50	3.57	28.07	41
9330	1-1/2" pipe size	↓	21	.762	↓		25.50	3.74	29.24	42.50

920 (left margin) **920** (right margin)

Important: See the Reference Section for critical supporting data - Reference Nos., Crews, & City Cost Indexes

15 MECHANICAL (side tab)

15107	Metal Pipe & Fittings	CREW	DAILY OUTPUT	LABOR-HOURS	UNIT	2003 BARE COSTS				TOTAL INCL O&P
						MAT.	LABOR	EQUIP.	TOTAL	
920 9340	2" pipe size	Q-15	18	.889	Ea.		30	4.36	34.36	50	**920**
9350	2-1/2" pipe size		12	1.333			45	6.55	51.55	74.50
9360	3" pipe size		9.73	1.644			55.50	8.05	63.55	92.50
9370	4" pipe size		7.37	2.171			73	10.65	83.65	122
9380	5" pipe size		6.15	2.602			87.50	12.75	100.25	146
9390	6" pipe size		5.71	2.802			94	13.75	107.75	157
9400	8" pipe size		3.69	4.336			146	21.50	167.50	244
9410	10" pipe size		2.91	5.498			185	27	212	310
9420	12" pipe size	↓	2.31	6.926	↓		233	34	267	390
9500	Schedule 40									
9510	1/4" pipe size	Q-15	28	.571	Ea.		19.20	2.81	22.01	32
9520	3/8" pipe size		27	.593			19.95	2.91	22.86	33
9530	1/2" pipe size		25.40	.630			21	3.09	24.09	35.50
9540	3/4" pipe size		22.22	.720			24	3.54	27.54	40.50
9550	1" pipe size		20.25	.790			26.50	3.88	30.38	44.50
9560	1-1/4" pipe size		18.82	.850			28.50	4.17	32.67	47.50
9570	1-1/2" pipe size		17.78	.900			30.50	4.42	34.92	50.50
9580	2" pipe size		15.09	1.060			35.50	5.20	40.70	60
9590	2-1/2" pipe size		7.96	2.010			67.50	9.85	77.35	113
9600	3" pipe size		6.43	2.488			83.50	12.20	95.70	139
9610	4" pipe size		4.88	3.279			110	16.10	126.10	184
9620	5" pipe size		4.26	3.756			126	18.45	144.45	212
9630	6" pipe size		3.77	4.244			143	21	164	239
9640	8" pipe size		2.44	6.557			221	32	253	370
9650	10" pipe size		1.92	8.333			280	41	321	470
9660	12" pipe size	↓	1.52	10.526	↓		355	51.50	406.50	590
9750	Schedule 80									
9760	1/4" pipe size	Q-15	21.55	.742	Ea.		25	3.65	28.65	41.50
9770	3/8" pipe size		20.75	.771			26	3.79	29.79	43
9780	1/2" pipe size		19.54	.819			27.50	4.02	31.52	46
9790	3/4" pipe size		17.09	.936			31.50	4.60	36.10	52.50
9800	1" pipe size		15.58	1.027			34.50	5.05	39.55	57.50
9810	1-1/4" pipe size		14.48	1.105			37	5.45	42.45	62
9820	1-1/2" pipe size		13.68	1.170			39.50	5.75	45.25	66
9830	2" pipe size		11.61	1.378			46.50	6.75	53.25	77.50
9840	2-1/2" pipe size		6.12	2.614			88	12.85	100.85	147
9850	3" pipe size		4.94	3.239			109	15.90	124.90	182
9860	4" pipe size		3.75	4.267			143	21	164	240
9870	5" pipe size		3.27	4.893			165	24	189	275
9880	6" pipe size		2.90	5.517			186	27	213	310
9890	8" pipe size		1.87	8.556			288	42	330	480
9900	10" pipe size		1.48	10.811			365	53	418	610
9910	12" pipe size		1.17	13.675			460	67	527	770
9920	Schedule 160, 1/2" pipe size		17	.941			31.50	4.62	36.12	53
9930	3/4" pipe size		14.81	1.080			36.50	5.30	41.80	61
9940	1" pipe size		13.50	1.185			40	5.80	45.80	66.50
9950	1-1/4" pipe size		12.55	1.275			43	6.25	49.25	71.50
9960	1-1/2" pipe size		11.85	1.350			45.50	6.65	52.15	76
9970	2" pipe size		10	1.600			54	7.85	61.85	90
9980	3" pipe size		4.28	3.738			126	18.35	144.35	210
9990	4" pipe size	↓	3.25	4.923	↓		166	24	190	277
960 0010	**PIPE, STAINLESS STEEL, FITTINGS**										**960**
0100	Butt weld joint, schedule 5, type 304									
0120	90° Elbow, long									
0140	1/2"	Q-15	17.50	.914	Ea.	10.05	31	4.49	45.54	62.50

MECHANICAL 15

	15107	Metal Pipe & Fittings	CREW	DAILY OUTPUT	LABOR-HOURS	UNIT	2003 BARE COSTS				TOTAL INCL O&P	
							MAT.	LABOR	EQUIP.	TOTAL		
960	0150	3/4"	Q-15	14	1.143	Ea.	15.75	38.50	5.60	59.85	81.50	960
	0160	1"		12.50	1.280		11.50	43	6.30	60.80	84.50	
	0170	1-1/4"		11	1.455		18.40	49	7.15	74.55	102	
	0180	1-1/2"		10.50	1.524		14.45	51.50	7.50	73.45	102	
	0190	2"		9	1.778		19.05	60	8.75	87.80	121	
	0200	2-1/2"		6	2.667		27.50	89.50	13.10	130.10	180	
	0210	3"		4.86	3.292		35	111	16.15	162.15	223	
	0220	3-1/2"		4.27	3.747		184	126	18.40	328.40	410	
	0230	4"		3.69	4.336		64.50	146	21.50	232	315	
	0240	5"		3.08	5.195		460	175	25.50	660.50	795	
	0250	6"	Q-16	4.29	5.594		184	195	18.30	397.30	515	
	0260	8"		2.76	8.696		310	305	28.50	643.50	835	
	0270	10"		2.18	11.009		645	385	36	1,066	1,325	
	0280	12"		1.73	13.873		1,025	485	45.50	1,555.50	1,900	
	0320	For schedule 5, type 316, add					30%					
	0600	45° Elbow, long										
	0620	1/2"	Q-15	17.50	.914	Ea.	14.45	31	4.49	49.94	67.50	
	0630	3/4"		14	1.143		14.45	38.50	5.60	58.55	80	
	0640	1"		12.50	1.280		16.30	43	6.30	65.60	90	
	0650	1-1/4"		11	1.455		27.50	49	7.15	83.65	112	
	0660	1-1/2"		10.50	1.524		14.80	51.50	7.50	73.80	102	
	0670	2"		9	1.778		17	60	8.75	85.75	119	
	0680	2-1/2"		6	2.667		28	89.50	13.10	130.60	180	
	0690	3"		4.86	3.292		29	111	16.15	156.15	217	
	0700	3-1/2"		4.27	3.747		184	126	18.40	328.40	410	
	0710	4"		3.69	4.336		51.50	146	21.50	219	300	
	0720	5"		3.08	5.195		320	175	25.50	520.50	645	
	0730	6"	Q-16	4.29	5.594		110	195	18.30	323.30	435	
	0740	8"		2.76	8.696		193	305	28.50	526.50	705	
	0750	10"		2.18	11.009		430	385	36	851	1,100	
	0760	12"		1.73	13.873		690	485	45.50	1,220.50	1,550	
	0800	For schedule 5, type 316, add					25%					
	1100	Tee, straight										
	1130	1/2"	Q-15	11.66	1.372	Ea.	40.50	46	6.75	93.25	121	
	1140	3/4"		9.33	1.715		40.50	57.50	8.40	106.40	141	
	1150	1"		8.33	1.921		38.50	64.50	9.45	112.45	150	
	1160	1-1/4"		7.33	2.183		66	73.50	10.70	150.20	196	
	1170	1-1/2"		7	2.286		38.50	77	11.20	126.70	171	
	1180	2"		6	2.667		40.50	89.50	13.10	143.10	194	
	1190	2-1/2"		4	4		116	135	19.65	270.65	350	
	1200	3"		3.24	4.938		75.50	166	24.50	266	360	
	1210	3-1/2"		2.85	5.614		202	189	27.50	418.50	540	
	1220	4"		2.46	6.504		107	219	32	358	485	
	1230	5"		2	8		430	269	39.50	738.50	925	
	1240	6"	Q-16	2.85	8.421		320	294	27.50	641.50	830	
	1250	8"		1.84	13.043		735	455	42.50	1,232.50	1,550	
	1260	10"		1.45	16.552		1,150	575	54	1,779	2,200	
	1270	12"		1.15	20.870		1,450	730	68	2,248	2,775	
	1320	For schedule 5, type 316, add					25%					
	2000	Butt weld joint, schedule 10, type 304										
	2020	90° elbow, long										
	2040	1/2"	Q-15	17	.941	Ea.	9.45	31.50	4.62	45.57	63.50	
	2050	3/4"		14	1.143		9.45	38.50	5.60	53.55	74.50	
	2060	1"		12.50	1.280		9.45	43	6.30	58.75	82.50	
	2070	1-1/4"		11	1.455		17.65	49	7.15	73.80	101	
	2080	1-1/2"		10.50	1.524		13.75	51.50	7.50	72.75	101	

15 MECHANICAL

Important: See the Reference Section for critical supporting data - Reference Nos., Crews, & City Cost Indexes

		CREW	DAILY OUTPUT	LABOR-HOURS	UNIT	2003 BARE COSTS				TOTAL INCL O&P
15107	**Metal Pipe & Fittings**					MAT.	LABOR	EQUIP.	TOTAL	
960 2090	2"	Q-15	9	1.778	Ea.	18.40	60	8.75	87.15	120 **960**
2100	2-1/2"		6	2.667		40	89.50	13.10	142.60	193
2110	3"		4.86	3.292		47.50	111	16.15	174.65	237
2120	3-1/2"		4.27	3.747		202	126	18.40	346.40	430
2130	4"		3.69	4.336		77.50	146	21.50	245	330
2140	5"		3.08	5.195		415	175	25.50	615.50	745
2150	6"	Q-16	4.29	5.594		176	195	18.30	389.30	510
2160	8"		2.76	8.696		350	305	28.50	683.50	875
2170	10"		2.18	11.009		755	385	36	1,176	1,450
2180	12"		1.73	13.873		1,175	485	45.50	1,705.50	2,050
2500	45° elbow, long									
2520	1/2"	Q-15	17.50	.914	Ea.	16.50	31	4.49	51.99	69.50
2530	3/4"		14	1.143		16.50	38.50	5.60	60.60	82.50
2540	1"		12.50	1.280		15.45	43	6.30	64.75	89
2550	1-1/4"		11	1.455		31.50	49	7.15	87.65	116
2560	1-1/2"		10.50	1.524		13.75	51.50	7.50	72.75	101
2570	2"		9	1.778		15.80	60	8.75	84.55	118
2580	2-1/2"		6	2.667		36	89.50	13.10	138.60	189
2590	3"		4.86	3.292		40.50	111	16.15	167.65	229
2600	3-1/2"		4.27	3.747		202	126	18.40	346.40	430
2610	4"		3.69	4.336		48	146	21.50	215.50	296
2620	5"		3.08	5.195		315	175	25.50	515.50	635
2630	6"	Q-16	4.29	5.594		103	195	18.30	316.30	430
2640	8"		2.76	8.696		233	305	28.50	566.50	750
2650	10"		2.18	11.009		540	385	36	961	1,225
2660	12"		1.73	13.873		825	485	45.50	1,355.50	1,700
3000	Tee, straight									
3030	1/2"	Q-15	11.66	1.372	Ea.	41.50	46	6.75	94.25	122
3040	3/4"		9.33	1.715		41.50	57.50	8.40	107.40	142
3050	1"		8.33	1.921		33	64.50	9.45	106.95	144
3060	1-1/4"		7.33	2.183		63	73.50	10.70	147.20	192
3070	1-1/2"		7	2.286		36	77	11.20	124.20	168
3080	2"		6	2.667		37	89.50	13.10	139.60	190
3090	2-1/2"		4	4		116	135	19.65	270.65	350
3100	3"		3.24	4.938		81	166	24.50	271.50	365
3110	3-1/2"		2.85	5.614		122	189	27.50	338.50	450
3120	4"		2.46	6.504		101	219	32	352	475
3130	5"		2	8		390	269	39.50	698.50	875
3140	6"	Q-16	2.85	8.421		305	294	27.50	626.50	810
3150	8"		1.84	13.043		705	455	42.50	1,202.50	1,500
3151	10"		1.45	16.552		1,175	575	54	1,804	2,200
3152	12"		1.15	20.870		1,625	730	68	2,423	2,950
3154	For schedule 10, type 316, add					25%				
3281	Butt weld joint, schedule 40, type 304									
3284	90° Elbow, long, 1/2"	Q-15	12.70	1.260	Ea.	15.75	42.50	6.20	64.45	88
3288	3/4"		11.10	1.441		15.75	48.50	7.10	71.35	98
3289	1"		10.13	1.579		17.25	53	7.75	78	108
3290	1-1/4"		9.40	1.702		25.50	57	8.35	90.85	124
3300	1-1/2"		8.89	1.800		20.50	60.50	8.85	89.85	124
3310	2"		7.55	2.119		28.50	71.50	10.40	110.40	151
3320	2-1/2"		3.98	4.020		59.50	135	19.75	214.25	291
3330	3"		3.21	4.984		76	168	24.50	268.50	365
3340	3-1/2"		2.83	5.654		315	190	28	533	670
3350	4"		2.44	6.557		140	221	32	393	525
3360	5"		2.13	7.512		495	253	37	785	965
3370	6"	Q-16	2.83	8.481		440	296	27.50	763.50	960

MECHANICAL **15**

			DAILY	LABOR-		2003 BARE COSTS				TOTAL	
960	**15107**	**Metal Pipe & Fittings**	CREW	OUTPUT	HOURS	UNIT	MAT.	LABOR	EQUIP.	TOTAL	INCL O&P
	3380	8"	Q-16	1.83	13.115	Ea.	740	455	43	1,238	1,550	960
	3390	10"		1.44	16.667		1,550	580	54.50	2,184.50	2,650
	3400	12"		1.14	21.053		2,175	735	69	2,979	3,575
	3410	For schedule 40, type 316, add					25%				
	3460	45° Elbow, long, 1/2"	Q-15	12.70	1.260	Ea.	21	42.50	6.20	69.70	94
	3470	3/4"		11.10	1.441		21	48.50	7.10	76.60	104
	3480	1"		10.13	1.579		21	53	7.75	81.75	112
	3490	1-1/4"		9.40	1.702		31	57	8.35	96.35	130
	3500	1-1/2"		8.89	1.800		24	60.50	8.85	93.35	128
	3510	2"		7.55	2.119		29.50	71.50	10.40	111.40	151
	3520	2-1/2"		3.98	4.020		59.50	135	19.75	214.25	291
	3530	3"		3.21	4.984		55	168	24.50	247.50	340
	3540	3-1/2"		2.83	5.654		320	190	28	538	675
	3550	4"		2.44	6.557		99	221	32	352	480
	3560	5"		2.13	7.512		345	253	37	635	800
	3570	6"	Q-16	2.83	8.481		258	296	27.50	581.50	760
	3580	8"		1.83	13.115		490	455	43	988	1,275
	3590	10"		1.44	16.667		920	580	54.50	1,554.50	1,950
	3600	12"		1.14	21.053		1,275	735	69	2,079	2,575
	3610	For schedule 40, type 316, add					25%				
	3660	Tee, straight 1/2"	Q-15	8.46	1.891	Ea.	44.50	63.50	9.30	117.30	155
	3670	3/4"		7.40	2.162		44.50	72.50	10.60	127.60	170
	3680	1"		6.74	2.374		44.50	80	11.65	136.15	182
	3690	1-1/4"		6.27	2.552		73.50	86	12.55	172.05	225
	3700	1-1/2"		5.92	2.703		44.50	91	13.25	148.75	200
	3710	2"		5.03	3.181		48	107	15.60	170.60	232
	3720	2-1/2"		2.65	6.038		120	203	29.50	352.50	470
	3730	3"		2.14	7.477		97	251	36.50	384.50	525
	3740	3-1/2"		1.88	8.511		139	286	42	467	630
	3750	4"		1.62	9.877		139	330	48.50	517.50	705
	3760	5"		1.42	11.268		515	380	55.50	950.50	1,200
	3770	6"	Q-16	1.88	12.766		445	445	41.50	931.50	1,200
	3780	8"		1.22	19.672		1,025	685	64.50	1,774.50	2,225
	3790	10"		.96	25		2,175	870	82	3,127	3,800
	3800	12"		.76	31.579		2,300	1,100	103	3,503	4,325
	3810	For schedule 40, type 316, add					25%				
	3820	Tee, reducing on outlet, 3/4" x 1/2"	Q-15	7.73	2.070	Ea.	50.50	69.50	10.15	130.15	172
	3822	1" x 1/2"		7.24	2.210		70	74.50	10.85	155.35	201
	3824	1" x 3/4"		6.96	2.299		58.50	77.50	11.30	147.30	193
	3826	1-1/4" x 1"		6.43	2.488		185	83.50	12.20	280.70	340
	3828	1-1/2" x 1/2"		6.58	2.432		97	82	11.95	190.95	243
	3830	1-1/2" x 3/4"		6.35	2.520		97	84.50	12.35	193.85	249
	3832	1-1/2" x 1"		6.18	2.589		70	87	12.70	169.70	222
	3834	2" x 1"		5.50	2.909		68	98	14.30	180.30	239
	3836	2" x 1-1/2"		5.30	3.019		64	102	14.80	180.80	240
	3838	2-1/2" x 2"		3.15	5.079		165	171	25	361	465
	3840	3" x 1-1/2"		2.72	5.882		141	198	29	368	485
	3842	3" x 2"		2.65	6.038		119	203	29.50	351.50	470
	3844	4" x 2"		2.10	7.619		214	256	37.50	507.50	660
	3846	4" x 3"		1.77	9.040		175	305	44.50	524.50	700
	3848	5" x 4"		1.48	10.811		550	365	53	968	1,225
	3850	6" x 3"	Q-16	2.19	10.959		440	380	36	856	1,100
	3852	6" x 4"		2.04	11.765		400	410	38.50	848.50	1,100
	3854	8" x 4"		1.46	16.438		735	575	54	1,364	1,725
	3856	10" x 8"		.69	34.783		2,025	1,225	114	3,364	4,200
	3858	12" x 10"		.55	43.636		2,525	1,525	143	4,193	5,225

Important: See the Reference Section for critical supporting data - Reference Nos., Crews, & City Cost Indexes

			DAILY	LABOR-		2003 BARE COSTS				TOTAL	
15107	**Metal Pipe & Fittings**	CREW	OUTPUT	HOURS	UNIT	MAT.	LABOR	EQUIP.	TOTAL	INCL O&P	
960	3950	Reducer, concentric, 3/4" x 1/2"	Q-15	11.85	1.350	Ea.	23.50	45.50	6.65	75.65	101	**960**
	3952	1" x 3/4"		10.60	1.509		27	51	7.40	85.40	115	
	3954	1-1/4" x 3/4"		10.19	1.570		77.50	53	7.70	138.20	174	
	3956	1-1/4" x 1"		9.76	1.639		42	55	8.05	105.05	138	
	3958	1-1/2" x 3/4"		9.88	1.619		48.50	54.50	7.95	110.95	144	
	3960	1-1/2" x 1"		9.47	1.690		16.20	57	8.30	81.50	113	
	3962	2" x 1"		8.65	1.850		19.45	62	9.10	90.55	126	
	3964	2" x 1-1/2"		8.16	1.961		16.20	66	9.65	91.85	128	
	3966	2-1/2" x 1"		5.71	2.802		117	94	13.75	224.75	285	
	3968	2-1/2" x 2"		5.21	3.071		58.50	103	15.10	176.60	237	
	3970	3" x 1"		4.88	3.279		77.50	110	16.10	203.60	269	
	3972	3" x 1-1/2"		4.72	3.390		36	114	16.65	166.65	230	
	3974	3" x 2"		4.51	3.548		29	119	17.40	165.40	231	
	3976	4" x 2"		3.69	4.336		40	146	21.50	207.50	288	
	3978	4" x 3"		2.77	5.776		35	194	28.50	257.50	365	
	3980	5" x 3"		2.56	6.250		291	210	30.50	531.50	670	
	3982	5" x 4"	▼	2.27	7.048		233	237	34.50	504.50	655	
	3984	6" x 3"	Q-16	3.57	6.723		111	234	22	367	500	
	3986	6" x 4"		3.19	7.524		92.50	262	24.50	379	525	
	3988	8" x 4"		2.44	9.836		350	345	32	727	940	
	3990	8" x 6"		2.22	10.811		224	375	35.50	634.50	855	
	3992	10" x 6"		1.91	12.565		465	440	41	946	1,225	
	3994	10" x 8"		1.61	14.907		276	520	49	845	1,150	
	3995	12" x 6"		1.63	14.724		700	515	48	1,263	1,600	
	3996	12" x 8"		1.41	17.021		640	595	55.50	1,290.50	1,650	
	3997	12" x 10"	▼	1.27	18.898	▼	385	660	62	1,107	1,500	
	4000	Socket weld joint, 3000 lb., type 304										
	4100	90° Elbow										
	4140	1/4"	Q-15	13.47	1.188	Ea.	21.50	40	5.85	67.35	90.50	
	4150	3/8"		12.97	1.234		27.50	41.50	6.05	75.05	99.50	
	4160	1/2"		12.21	1.310		29	44	6.45	79.45	106	
	4170	3/4"		10.68	1.498		35	50.50	7.35	92.85	123	
	4180	1"		9.74	1.643		53	55	8.05	116.05	150	
	4190	1-1/4"		9.05	1.768		92	59.50	8.70	160.20	201	
	4200	1-1/2"		8.55	1.871		114	63	9.20	186.20	230	
	4210	2"	▼	7.26	2.204	▼	184	74	10.80	268.80	325	
	4300	45° Elbow										
	4340	1/4"	Q-15	13.47	1.188	Ea.	41	40	5.85	86.85	112	
	4350	3/8"		12.97	1.234		41	41.50	6.05	88.55	114	
	4360	1/2"		12.21	1.310		41	44	6.45	91.45	119	
	4370	3/4"		10.68	1.498		47	50.50	7.35	104.85	136	
	4380	1"		9.74	1.643		68	55	8.05	131.05	167	
	4390	1-1/4"		9.05	1.768		110	59.50	8.70	178.20	221	
	4400	1-1/2"		8.55	1.871		111	63	9.20	183.20	227	
	4410	2"	▼	7.26	2.204	▼	201	74	10.80	285.80	345	
	4500	Tee										
	4540	1/4"	Q-15	8.97	1.784	Ea.	26	60	8.75	94.75	129	
	4550	3/8"		8.64	1.852		31	62.50	9.10	102.60	139	
	4560	1/2"		8.13	1.968		41.50	66	9.65	117.15	156	
	4570	3/4"		7.12	2.247		48	75.50	11.05	134.55	179	
	4580	1"		6.48	2.469		64.50	83	12.10	159.60	209	
	4590	1-1/4"		6.03	2.653		114	89	13.05	216.05	275	
	4600	1-1/2"		5.69	2.812		167	94.50	13.80	275.30	340	
	4610	2"	▼	4.83	3.313	▼	253	111	16.25	380.25	465	
	5000	Socket weld joint, 3000 lb., type 316										
	5100	90° Elbow										

MECHANICAL 15

15107	Metal Pipe & Fittings	CREW	DAILY OUTPUT	LABOR-HOURS	UNIT	2003 BARE COSTS				TOTAL INCL O&P	
						MAT.	LABOR	EQUIP.	TOTAL		
960 5140	1/4"	Q-15	13.47	1.188	Ea.	25	40	5.85	70.85	94.50	**960**
5150	3/8"		12.97	1.234		29	41.50	6.05	76.55	101	
5160	1/2"		12.21	1.310		35	44	6.45	85.45	112	
5170	3/4"		10.68	1.498		46	50.50	7.35	103.85	135	
5180	1"		9.74	1.643		68.50	55	8.05	131.55	168	
5190	1-1/4"		9.05	1.768		122	59.50	8.70	190.20	235	
5200	1-1/2"		8.55	1.871		141	63	9.20	213.20	260	
5210	2"		7.26	2.204		240	74	10.80	324.80	390	
5300	45° Elbow										
5340	1/4"	Q-15	13.47	1.188	Ea.	52.50	40	5.85	98.35	125	
5350	3/8"		12.97	1.234		52.50	41.50	6.05	100.05	127	
5360	1/2"		12.21	1.310		52.50	44	6.45	102.95	132	
5370	3/4"		10.68	1.498		59.50	50.50	7.35	117.35	150	
5380	1"		9.74	1.643		90	55	8.05	153.05	191	
5390	1-1/4"		9.05	1.768		126	59.50	8.70	194.20	239	
5400	1-1/2"		8.55	1.871		141	63	9.20	213.20	261	
5410	2"		7.26	2.204		211	74	10.80	295.80	355	
5500	Tee										
5540	1/4"	Q-15	8.97	1.784	Ea.	31.50	60	8.75	100.25	135	
5550	3/8"		8.64	1.852		38.50	62.50	9.10	110.10	147	
5560	1/2"		8.13	1.968		48	66	9.65	123.65	164	
5570	3/4"		7.12	2.247		59.50	75.50	11.05	146.05	192	
5580	1"		6.48	2.469		91	83	12.10	186.10	238	
5590	1-1/4"		6.03	2.653		145	89	13.05	247.05	310	
5600	1-1/2"		5.69	2.812		208	94.50	13.80	316.30	385	
5610	2"		4.83	3.313		330	111	16.25	457.25	550	
5700	For socket weld joint, 6000 lb., type 304 and 316, add					100%					
6000	Threaded companion flange										
6010	Stainless steel, 150 lb., type 304										
6020	1/2" diam.	1 Plum	30	.267	Ea.	45.50	9.95		55.45	65	
6030	3/4" diam.		28	.286		50.50	10.65		61.15	72	
6040	1" diam.		27	.296		56.50	11.05		67.55	79	
6050	1-1/4" diam.	Q-1	44	.364		77	12.25		89.25	103	
6060	1-1/2" diam.		40	.400		79	13.45		92.45	108	
6070	2" diam.		36	.444		88	14.95		102.95	120	
6080	2-1/2" diam.		28	.571		135	19.20		154.20	177	
6090	3" diam.		20	.800		147	27		174	202	
6110	4" diam.		12	1.333		231	45		276	320	
6130	6" diam.	Q-2	14	1.714		440	60		500	575	
6140	8" diam.	"	12	2		825	69.50		894.50	1,000	
6150	For type 316 add					40%					
6260	Weld flanges, stainless steel, type 304										
6270	Slip on, 150 lb. (welded, front and back)										
6280	1/2" diam.	Q-15	12.70	1.260	Ea.	45.50	42.50	6.20	94.20	121	
6290	3/4" diam.		11.11	1.440		50.50	48.50	7.05	106.05	137	
6300	1" diam.		10.13	1.579		36	53	7.75	96.75	129	
6310	1-1/4" diam.		9.41	1.700		77	57	8.35	142.35	181	
6320	1-1/2" diam.		8.89	1.800		53.50	60.50	8.85	122.85	160	
6330	2" diam.		7.55	2.119		70	71.50	10.40	151.90	196	
6340	2-1/2" diam.		3.98	4.020		135	135	19.75	289.75	375	
6350	3" diam.		3.21	4.984		147	168	24.50	339.50	440	
6370	4" diam.		2.44	6.557		147	221	32	400	535	
6390	6" diam.	Q-16	1.89	12.698		231	445	41.50	717.50	970	
6400	8" diam.	"	1.22	19.672		825	685	64.50	1,574.50	2,000	
6410	For type 316, add					40%					
6530	Weld neck 150 lb.										

Important: See the Reference Section for critical supporting data - Reference Nos., Crews, & City Cost Indexes

15107	Metal Pipe & Fittings	CREW	DAILY OUTPUT	LABOR-HOURS	UNIT	2003 BARE COSTS				TOTAL INCL O&P
						MAT.	LABOR	EQUIP.	TOTAL	
6540	1/2" diam.	Q-15	25.40	.630	Ea.	34	21	3.09	58.09	73
6550	3/4" diam.		22.22	.720		38.50	24	3.54	66.04	83
6560	1" diam.		20.25	.790		46	26.50	3.88	76.38	95
6570	1-1/4" diam.		18.82	.850		55	28.50	4.17	87.67	108
6580	1-1/2" diam.		17.78	.900		64.50	30.50	4.42	99.42	121
6590	2" diam.		15.09	1.060		82.50	35.50	5.20	123.20	151
6600	2-1/2" diam.		7.96	2.010		138	67.50	9.85	215.35	264
6610	3" diam.		6.43	2.488		156	83.50	12.20	251.70	310
6630	4" diam.		4.88	3.279		202	110	16.10	328.10	405
6640	5" diam.		4.26	3.756		268	126	18.45	412.45	505
6650	6" diam.	Q-16	5.66	4.240		315	148	13.85	476.85	585
6652	8" diam.		3.65	6.575		440	229	21.50	690.50	855
6654	10" diam.		2.88	8.333		600	291	27.50	918.50	1,125
6656	12" diam.		2.28	10.526		955	365	34.50	1,354.50	1,650
6670	For type 316 add					23%				
7000	Threaded joint, 150 lb., type 304									
7030	90° elbow									
7040	1/8"	1 Plum	13	.615	Ea.	12.80	23		35.80	48.50
7050	1/4"		13	.615		12.80	23		35.80	48.50
7070	3/8"		13	.615		15.45	23		38.45	51.50
7080	1/2"		12	.667		8.30	25		33.30	46.50
7090	3/4"		11	.727		9.40	27		36.40	51.50
7100	1"		10	.800		12.50	30		42.50	59
7110	1-1/4"	Q-1	17	.941		33	31.50		64.50	84
7120	1-1/2"		16	1		21	33.50		54.50	74.50
7130	2"		14	1.143		30.50	38.50		69	91.50
7140	2-1/2"		11	1.455		144	49		193	233
7150	3"		8	2		182	67.50		249.50	305
7160	4"	Q-2	11	2.182		211	76		287	345
7180	45° elbow									
7190	1/8"	1 Plum	13	.615	Ea.	17.60	23		40.60	54
7200	1/4"		13	.615		17.60	23		40.60	54
7210	3/8"		13	.615		19.60	23		42.60	56
7220	1/2"		12	.667		10	25		35	48.50
7230	3/4"		11	.727		12.25	27		39.25	54.50
7240	1"		10	.800		13.50	30		43.50	60
7250	1-1/4"	Q-1	17	.941		31	31.50		62.50	82
7260	1-1/2"		16	1		20.50	33.50		54	73.50
7270	2"		14	1.143		32	38.50		70.50	93
7280	2-1/2"		11	1.455		204	49		253	298
7290	3"		8	2		298	67.50		365.50	430
7300	4"	Q-2	11	2.182		267	76		343	410
7320	Tee, straight									
7330	1/8"	1 Plum	9	.889	Ea.	19.70	33		52.70	71.50
7340	1/4"		9	.889		19.70	33		52.70	71.50
7350	3/8"		9	.889		22.50	33		55.50	75
7360	1/2"		8	1		12.75	37.50		50.25	70.50
7370	3/4"		7	1.143		13.40	42.50		55.90	79.50
7380	1"		6.50	1.231		16	46		62	87
7390	1-1/4"	Q-1	11	1.455		45	49		94	123
7400	1-1/2"		10	1.600		32.50	54		86.50	118
7410	2"		9	1.778		38.50	60		98.50	133
7420	2-1/2"		7	2.286		208	77		285	345
7430	3"		5	3.200		315	108		423	515
7440	4"	Q-2	7	3.429		315	120		435	525
7460	Coupling, straight									

MECHANICAL 15

				DAILY	LABOR-			2003 BARE COSTS				TOTAL	
		### 15107 | Metal Pipe & Fittings	CREW	OUTPUT	HOURS	UNIT	MAT.	LABOR	EQUIP.	TOTAL	INCL O&P	
960	7470	1/8"	1 Plum	19	.421	Ea.	4.86	15.75		20.61	29.50	960
	7480	1/4"		19	.421		5.85	15.75		21.60	30.50	
	7490	3/8"		19	.421		6.90	15.75		22.65	31.50	
	7500	1/2"		19	.421		5.05	15.75		20.80	29.50	
	7510	3/4"		18	.444		13.05	16.60		29.65	39.50	
	7520	1"		15	.533		17.35	19.90		37.25	49	
	7530	1-1/4"	Q-1	26	.615		14.85	20.50		35.35	48	
	7540	1-1/2"		24	.667		17.90	22.50		40.40	53.50	
	7550	2"		21	.762		28.50	25.50		54	69.50	
	7560	2-1/2"		18	.889		64.50	30		94.50	116	
	7570	3"		14	1.143		86.50	38.50		125	153	
	7580	4"	Q-2	16	1.500		129	52.50		181.50	220	
	7600	Reducer, concentric, 1/2"	1 Plum	12	.667		4.44	25		29.44	42.50	
	7610	3/4"		11	.727		5.80	27		32.80	47.50	
	7612	1"		10	.800		10.10	30		40.10	56	
	7614	1-1/4"	Q-1	17	.941		20.50	31.50		52	70.50	
	7616	1-1/2"		16	1		22.50	33.50		56	76	
	7618	2"		14	1.143		35.50	38.50		74	97	
	7620	2-1/2"		11	1.455		74.50	49		123.50	156	
	7622	3"		8	2		108	67.50		175.50	221	
	7624	4"	Q-2	11	2.182		155	76		231	285	
	7710	Union										
	7720	1/8"	1 Plum	12	.667	Ea.	23.50	25		48.50	63.50	
	7730	1/4"		12	.667		23.50	25		48.50	63.50	
	7740	3/8"		12	.667		27.50	25		52.50	67.50	
	7750	1/2"		11	.727		17.15	27		44.15	60	
	7760	3/4"		10	.800		24	30		54	71.50	
	7770	1"		9	.889		34.50	33		67.50	88	
	7780	1-1/4"	Q-1	16	1		151	33.50		184.50	217	
	7790	1-1/2"		15	1.067		168	36		204	239	
	7800	2"		13	1.231		212	41.50		253.50	297	
	7810	2-1/2"		10	1.600		435	54		489	555	
	7820	3"		7	2.286		575	77		652	750	
	7830	4"	Q-2	10	2.400		795	83.50		878.50	1,000	
	7838	Caps										
	7840	1/2"	1 Plum	24	.333	Ea.	4.61	12.45		17.06	24	
	7841	3/4"		22	.364		6.70	13.60		20.30	28	
	7842	1"		20	.400		10.70	14.95		25.65	34.50	
	7843	1-1/2"	Q-1	32	.500		28	16.80		44.80	56.50	
	7844	2"	"	28	.571		36	19.20		55.20	68.50	
	7845	4"	Q-2	22	1.091		141	38		179	213	
	7850	For 150 lb., type 316, add					25%					
	8750	Threaded joint, 2000 lb., type 304										
	8770	90° Elbow										
	8780	1/8"	1 Plum	13	.615	Ea.	15.40	23		38.40	51.50	
	8790	1/4"		13	.615		15.40	23		38.40	51.50	
	8800	3/8"		13	.615		19.05	23		42.05	55.50	
	8810	1/2"		12	.667		26	25		51	66	
	8820	3/4"		11	.727		30	27		57	74	
	8830	1"		10	.800		39.50	30		69.50	88.50	
	8840	1-1/4"	Q-1	17	.941		63.50	31.50		95	118	
	8850	1-1/2"		16	1		102	33.50		135.50	163	
	8860	2"		14	1.143		143	38.50		181.50	215	
	8880	45° Elbow										
	8890	1/8"	1 Plum	13	.615	Ea.	32	23		55	70	
	8900	1/4"		13	.615		32	23		55	70	

Important: See the Reference Section for critical supporting data - Reference Nos., Crews, & City Cost Indexes

15 MECHANICAL

15107	Metal Pipe & Fittings	CREW	DAILY OUTPUT	LABOR-HOURS	UNIT	2003 BARE COSTS				TOTAL INCL O&P
						MAT.	LABOR	EQUIP.	TOTAL	
960	8910	3/8"	1 Plum	13	.615	Ea.	40	23		63	78.50	960
8920	1/2"		12	.667		40	25		65	81.50
8930	3/4"		11	.727		44.50	27		71.50	89.50
8940	1"		10	.800		52.50	30		82.50	103
8950	1-1/4"	Q-1	17	.941		90.50	31.50		122	148
8960	1-1/2"		16	1		121	33.50		154.50	184
8970	2"		14	1.143		140	38.50		178.50	212
8990	Tee, straight									
9000	1/8"	1 Plum	9	.889	Ea.	19.90	33		52.90	72
9010	1/4"		9	.889		19.90	33		52.90	72
9020	3/8"		9	.889		25.50	33		58.50	78
9030	1/2"		8	1		33	37.50		70.50	92.50
9040	3/4"		7	1.143		37	42.50		79.50	106
9050	1"		6.50	1.231		51.50	46		97.50	126
9060	1-1/4"	Q-1	11	1.455		85	49		134	168
9070	1-1/2"		10	1.600		141	54		195	238
9080	2"		9	1.778		197	60		257	305
9100	For couplings and unions use 3000 lb., type 304									
9120	2000 lb., type 316									
9130	90° Elbow									
9140	1/8"	1 Plum	13	.615	Ea.	18.70	23		41.70	55
9150	1/4"		13	.615		18.70	23		41.70	55
9160	3/8"		13	.615		22	23		45	58.50
9170	1/2"		12	.667		26.50	25		51.50	66.50
9180	3/4"		11	.727		34.50	27		61.50	79
9190	1"		10	.800		51.50	30		81.50	102
9200	1-1/4"	Q-1	17	.941		92	31.50		123.50	149
9210	1-1/2"		16	1		106	33.50		139.50	167
9220	2"		14	1.143		180	38.50		218.50	256
9240	45° Elbow									
9250	1/8"	1 Plum	13	.615	Ea.	39.50	23		62.50	78
9260	1/4"		13	.615		39.50	23		62.50	78
9270	3/8"		13	.615		39.50	23		62.50	78
9280	1/2"		12	.667		39.50	25		64.50	81
9300	3/4"		11	.727		44.50	27		71.50	90
9310	1"		10	.800		67.50	30		97.50	119
9320	1-1/4"	Q-1	17	.941		94.50	31.50		126	152
9330	1-1/2"		16	1		129	33.50		162.50	193
9340	2"		14	1.143		159	38.50		197.50	232
9360	Tee, straight									
9370	1/8"	1 Plum	9	.889	Ea.	24	33		57	76
9380	1/4"		9	.889		24	33		57	76
9390	3/8"		9	.889		29	33		62	82
9400	1/2"		8	1		36	37.50		73.50	96
9410	3/4"		7	1.143		44.50	42.50		87	114
9420	1"		6.50	1.231		68.50	46		114.50	145
9430	1-1/4"	Q-1	11	1.455		109	49		158	193
9440	1-1/2"		10	1.600		156	54		210	253
9450	2"		9	1.778		249	60		309	365
9470	For couplings and unions use 3000 lb., type 316									
9490	3000 lb., type 304									
9510	Coupling									
9520	1/8"	1 Plum	19	.421	Ea.	6.20	15.75		21.95	31
9530	1/4"		19	.421		6.80	15.75		22.55	31.50
9540	3/8"		19	.421		7.90	15.75		23.65	32.50
9550	1/2"		19	.421		9.55	15.75		25.30	34.50

15107	Metal Pipe & Fittings	CREW	DAILY OUTPUT	LABOR-HOURS	UNIT	MAT.	LABOR	EQUIP.	TOTAL	TOTAL INCL O&P	
						2003 BARE COSTS					
960 9560	3/4"	1 Plum	18	.444	Ea.	12.70	16.60		29.30	39	**960**
9570	1"	↓	15	.533		21.50	19.90		41.40	53.50	
9580	1-1/4"	Q-1	26	.615		48.50	20.50		69	85	
9590	1-1/2"		24	.667		61	22.50		83.50	102	
9600	2"	↓	21	.762	↓	80.50	25.50		106	127	
9620	Union										
9630	1/8"	1 Plum	12	.667	Ea.	48	25		73	90	
9640	1/4"		12	.667		48	25		73	90	
9650	3/8"		12	.667		51	25		76	94	
9660	1/2"		11	.727		51	27		78	97.50	
9670	3/4"		10	.800		62.50	30		92.50	114	
9680	1"	↓	9	.889		96.50	33		129.50	156	
9690	1-1/4"	Q-1	16	1		162	33.50		195.50	229	
9700	1-1/2"		15	1.067		182	36		218	255	
9710	2"	↓	13	1.231	↓	244	41.50		285.50	330	
9730	3000 lb., type 316										
9750	Coupling										
9770	1/8"	1 Plum	19	.421	Ea.	6.55	15.75		22.30	31	
9780	1/4"		19	.421		7.40	15.75		23.15	32	
9790	3/8"		19	.421		7.90	15.75		23.65	32.50	
9800	1/2"		19	.421		11.20	15.75		26.95	36.50	
9810	3/4"		18	.444		16.50	16.60		33.10	43	
9820	1"	↓	15	.533		27.50	19.90		47.40	60.50	
9830	1-1/4"	Q-1	26	.615		57.50	20.50		78	95	
9840	1-1/2"		24	.667		72	22.50		94.50	114	
9850	2"	↓	21	.762	↓	100	25.50		125.50	149	
9870	Union										
9880	1/8"	1 Plum	12	.667	Ea.	49	25		74	91.50	
9890	1/4"		12	.667		49	25		74	91.50	
9900	3/8"		12	.667		56.50	25		81.50	100	
9910	1/2"		11	.727		56.50	27		83.50	104	
9920	3/4"		10	.800		74	30		104	127	
9930	1"	↓	9	.889		117	33		150	178	
9940	1-1/4"	Q-1	16	1		185	33.50		218.50	254	
9950	1-1/2"		15	1.067		228	36		264	305	
9960	2"	↓	13	1.231	↓	290	41.50		331.50	385	

15108	Plastic Pipe & Fittings										
520 0010	**PIPE, PLASTIC**										**520**
0020	Fiberglass reinforced, couplings 10' O.C., hangers 3 per 10'										
0080	General service										
0120	2" diameter	Q-1	59	.271	L.F.	10.20	9.10		19.30	25	
0140	3" diameter		52	.308		12.90	10.35		23.25	30	
0150	4" diameter		48	.333		14.65	11.20		25.85	33	
0160	6" diameter	↓	39	.410		22	13.80		35.80	45.50	
0170	8" diameter	Q-2	49	.490		34	17.10		51.10	63.50	
0180	10" diameter		41	.585		51.50	20.50		72	87.50	
0190	12" diameter		36	.667	↓	65.50	23.50		89	107	
0200	High strength										
0240	2" diameter	Q-1	58	.276	L.F.	10.50	9.30		19.80	25.50	
0260	3" diameter		51	.314		14.75	10.55		25.30	32	
0280	4" diameter		47	.340		17.45	11.45		28.90	36.50	
0300	6" diameter	↓	38	.421		26.50	14.15		40.65	50.50	
0320	8" diameter	Q-2	48	.500		41.50	17.45		58.95	72	
0340	10" diameter		40	.600		60.50	21		81.50	98.50	
0360	12" diameter	↓	36	.667	↓	74.50	23.50		98	117	

Important: See the Reference Section for critical supporting data - Reference Nos., Crews, & City Cost Indexes

15108 | Plastic Pipe & Fittings

		DAILY	LABOR-		2003 BARE COSTS				TOTAL
		CREW	OUTPUT	HOURS	UNIT	MAT.	LABOR	EQUIP.	TOTAL	INCL O&P

520											520
0550	To delete coupling & hangers, subtract									
0560	2" diam. to 6" diam.					33%	56%			
0570	8" diam. to 12" diam.					31%	52%			
0600	PVC, high impact/pressure, cplgs. 10' O.C., hangers 3 per 10'									
1020	Schedule 80									
1040	1/4" diameter	1 Plum	58	.138	L.F.	2.57	5.15		7.72	10.60
1060	3/8" diameter		55	.145		2.76	5.45		8.21	11.25
1070	1/2" diameter		50	.160		2.45	6		8.45	11.70
1080	3/4" diameter		47	.170		2.63	6.35		8.98	12.50
1090	1" diameter		43	.186		3.01	6.95		9.96	13.80
1100	1-1/4" diameter		39	.205		3.46	7.65		11.11	15.35
1110	1-1/2" diameter		34	.235		3.74	8.80		12.54	17.35
1120	2" diameter	Q-1	55	.291		4.34	9.80		14.14	19.50
1130	2-1/2" diameter		52	.308		5.80	10.35		16.15	22
1140	3" diameter		50	.320		7.45	10.75		18.20	24.50
1150	4" diameter		46	.348		9.80	11.70		21.50	28.50
1160	5" diameter		42	.381		13.15	12.80		25.95	34
1170	6" diameter		38	.421		17.35	14.15		31.50	40.50
1180	8" diameter	Q-2	47	.511		29	17.80		46.80	58.50
1190	10" diameter		42	.571		39	19.95		58.95	73
1200	12" diameter		38	.632		50.50	22		72.50	89
1730	To delete coupling & hangers, subtract									
1740	1/4" diam. to 1/2" diam.					62%	80%			
1750	3/4" diam. to 1-1/4" diam.					58%	73%			
1760	1-1/2" diam. to 6" diam.					40%	57%			
1770	8" diam. to 12" diam.					34%	50%			
1800	PVC, couplings 10' O.C., hangers 3 per 10'									
1820	Schedule 40									
1860	1/2" diameter	1 Plum	54	.148	L.F.	2.15	5.55		7.70	10.70
1870	3/4" diameter		51	.157		2.23	5.85		8.08	11.30
1880	1" diameter		46	.174		2.42	6.50		8.92	12.45
1890	1-1/4" diameter		42	.190		2.63	7.10		9.73	13.65
1900	1-1/2" diameter		36	.222		2.75	8.30		11.05	15.55
1910	2" diameter	Q-1	59	.271		3.08	9.10		12.18	17.15
1920	2-1/2" diameter		56	.286		3.82	9.60		13.42	18.70
1930	3" diameter		53	.302		4.63	10.15		14.78	20.50
1940	4" diameter		48	.333		5.65	11.20		16.85	23
1950	5" diameter		43	.372		7.10	12.50		19.60	26.50
1960	6" diameter		39	.410		9.15	13.80		22.95	31
1970	8" diameter	Q-2	48	.500		16.30	17.45		33.75	44.50
1980	10" diameter		43	.558		26	19.45		45.45	58.50
1990	12" diameter		42	.571		34	19.95		53.95	67.50
2000	14" diameter		31	.774		52.50	27		79.50	99
2010	16" diameter		23	1.043		83	36.50		119.50	147
2340	To delete coupling & hangers, subtract									
2360	1/2" diam. to 1-1/4" diam.					65%	74%			
2370	1-1/2" diam. to 6" diam.					44%	57%			
2380	8" diam. to 12" diam.					41%	53%			
2390	14" diam. to 16" diam.					48%	45%			
2420	Schedule 80									
2440	1/4" diameter	1 Plum	58	.138	L.F.	2.02	5.15		7.17	10
2450	3/8" diameter		55	.145		2.02	5.45		7.47	10.40
2460	1/2" diameter		50	.160		2.19	6		8.19	11.40
2470	3/4" diameter		47	.170		2.29	6.35		8.64	12.10
2480	1" diameter		43	.186		2.51	6.95		9.46	13.25
2490	1-1/4" diameter		39	.205		2.76	7.65		10.41	14.60

MECHANICAL

15108	Plastic Pipe & Fittings	CREW	DAILY OUTPUT	LABOR-HOURS	UNIT	2003 BARE COSTS				TOTAL INCL O&P	
						MAT.	LABOR	EQUIP.	TOTAL		
520	2500	1-1/2" diameter	1 Plum	34	.235	L.F.	2.91	8.80		11.71	16.45	520
	2510	2" diameter	Q-1	55	.291		3.20	9.80		13	18.25
	2520	2-1/2" diameter		52	.308		3.94	10.35		14.29	19.95
	2530	3" diameter		50	.320		5.10	10.75		15.85	22
	2540	4" diameter		46	.348		6.40	11.70		18.10	24.50
	2550	5" diameter		42	.381		9.35	12.80		22.15	29.50
	2560	6" diameter		38	.421		10.80	14.15		24.95	33.50
	2570	8" diameter	Q-2	47	.511		18.85	17.80		36.65	48
	2580	10" diameter		42	.571		32	19.95		51.95	65
	2590	12" diameter		38	.632		40.50	22		62.50	78
	2830	To delete coupling & hangers, subtract									
	2840	1/4" diam. to 1/2" diam.					66%	80%			
	2850	3/4" diam. to 1-1/4" diam.					61%	73%			
	2860	1-1/2" diam. to 6" diam.					41%	57%			
	2870	8" diam. to 12" diam.					31%	50%			
	2900	Schedule 120									
	2910	1/2" diameter	1 Plum	50	.160	L.F.	2.58	6		8.58	11.85
	2950	3/4" diameter		47	.170		2.79	6.35		9.14	12.65
	2960	1" diameter		43	.186		3.25	6.95		10.20	14.10
	2970	1-1/4" diameter		39	.205		3.80	7.65		11.45	15.75
	2980	1-1/2" diameter		33	.242		4.16	9.05		13.21	18.25
	2990	2" diameter	Q-1	54	.296		4.97	9.95		14.92	20.50
	3000	2-1/2" diameter		52	.308		7	10.35		17.35	23.50
	3010	3" diameter		49	.327		8.75	11		19.75	26
	3020	4" diameter		45	.356		12.45	11.95		24.40	31.50
	3030	6" diameter		37	.432		22	14.55		36.55	46.50
	3240	To delete coupling & hangers, subtract									
	3250	1/2" diam. to 1-1/4" diam.					52%	74%			
	3260	1-1/2" diam. to 4" diam.					30%	57%			
	3270	6" diam.					17%	50%			
	3300	PVC, pressure, couplings 10' O.C., hangers 3 per 10'									
	3310	SDR 26, 160 psi									
	3350	1-1/4" diameter	1 Plum	42	.190	L.F.	2.46	7.10		9.56	13.45
	3360	1-1/2" diameter	"	36	.222		2.56	8.30		10.86	15.35
	3370	2" diameter	Q-1	59	.271		2.73	9.10		11.83	16.75
	3380	2-1/2" diameter		56	.286		3.45	9.60		13.05	18.30
	3390	3" diameter		53	.302		4.23	10.15		14.38	20
	3400	4" diameter		48	.333		5.25	11.20		16.45	22.50
	3420	6" diameter		39	.410		8.90	13.80		22.70	31
	3430	8" diameter	Q-2	48	.500		16.30	17.45		33.75	44.50
	3660	To delete coupling & hangers, subtract									
	3670	1-1/4" diam.					63%	68%			
	3680	1-1/2" diam. to 4" diam.					48%	57%			
	3690	6" diam. to 8" diam.					60%	54%			
	3720	SDR 21, 200 psi, 1/2" diameter	1 Plum	54	.148	L.F.	2.11	5.55		7.66	10.65
	3740	3/4" diameter		51	.157		2.15	5.85		8	11.20
	3750	1" diameter		46	.174		2.30	6.50		8.80	12.35
	3760	1-1/4" diameter		42	.190		2.49	7.10		9.59	13.50
	3770	1-1/2" diameter		36	.222		2.60	8.30		10.90	15.40
	3780	2" diameter	Q-1	59	.271		2.81	9.10		11.91	16.85
	3790	2-1/2" diameter		56	.286		3.94	9.60		13.54	18.85
	3800	3" diameter		53	.302		4.39	10.15		14.54	20
	3810	4" diameter		48	.333		5.50	11.20		16.70	23
	3830	6" diameter		39	.410		9.45	13.80		23.25	31.50
	3840	8" diameter	Q-2	48	.500		17.45	17.45		34.90	45.50
	4000	To delete coupling & hangers, subtract									

15 MECHANICAL

Important: See the Reference Section for critical supporting data - Reference Nos., Crews, & City Cost Indexes

15108	Plastic Pipe & Fittings	CREW	DAILY OUTPUT	LABOR-HOURS	UNIT	2003 BARE COSTS				TOTAL INCL O&P
						MAT.	LABOR	EQUIP.	TOTAL	
4010	1/2" diam. to 3/4" diam.					71%	77%			
4020	1" diam. to 1-1/4" diam.					63%	70%			
4030	1-1/2" diam. to 6" diam.					44%	57%			
4040	8" diam.					46%	54%			
4100	DWV type, schedule 40, couplings 10' O.C., hangers 3 per 10'									
4120	ABS									
4140	1-1/4" diameter	1 Plum	42	.190	L.F.	2.32	7.10		9.42	13.30
4150	1-1/2" diameter	"	36	.222		2.35	8.30		10.65	15.15
4160	2" diameter	Q-1	59	.271		2.46	9.10		11.56	16.45
4170	3" diameter		53	.302		3.57	10.15		13.72	19.25
4180	4" diameter		48	.333		4.33	11.20		15.53	21.50
4190	6" diameter	↓	39	.410	↓	8.25	13.80		22.05	30
4360	To delete coupling & hangers, subtract									
4370	1-1/4" diam.					64%	68%			
4380	1-1/2" diam. to 6" diam.					54%	57%			
4400	PVC									
4410	1-1/4" diameter	1 Plum	42	.190	L.F.	2.45	7.10		9.55	13.45
4420	1-1/2" diameter	"	36	.222		2.48	8.30		10.78	15.30
4460	2" diameter	Q-1	59	.271		2.65	9.10		11.75	16.65
4470	3" diameter		53	.302		3.96	10.15		14.11	19.70
4480	4" diameter		48	.333		4.83	11.20		16.03	22.50
4490	6" diameter	↓	39	.410		8.15	13.80		21.95	30
4500	8" diameter	Q-2	48	.500	↓	18.95	17.45		36.40	47.50
4750	To delete coupling & hangers, subtract									
4760	1-1/4" diam. to 1-1/2" diam.					71%	64%			
4770	2" diam. to 8" diam.					60%	57%			
4800	PVC, clear pipe, cplgs. 10' O.C., hangers 3 per 10', Sched. 40									
4840	1/4" diameter	1 Plum	59	.136	L.F.	2.50	5.05		7.55	10.40
4850	3/8" diameter		56	.143		2.70	5.35		8.05	11
4860	1/2" diameter		54	.148		3.08	5.55		8.63	11.75
4870	3/4" diameter		51	.157		3.49	5.85		9.34	12.70
4880	1" diameter		46	.174		4.35	6.50		10.85	14.60
4890	1-1/4" diameter		42	.190		5.25	7.10		12.35	16.50
4900	1-1/2" diameter	↓	36	.222		5.85	8.30		14.15	19
4910	2" diameter	Q-1	59	.271		7.25	9.10		16.35	21.50
4920	2-1/2" diameter		56	.286		10.70	9.60		20.30	26.50
4930	3" diameter		53	.302		13.40	10.15		23.55	30
4940	3-1/2" diameter		50	.320		17.60	10.75		28.35	35.50
4950	4" diameter	↓	48	.333	↓	19.15	11.20		30.35	38
5250	To delete coupling & hangers, subtract									
5260	1/4" diam. to 3/8" diam.					60%	81%			
5270	1/2" diam. to 3/4" diam.					41%	77%			
5280	1" diam. to 1-1/2" diam.					26%	67%			
5290	2" diam. to 4" diam.					16%	58%			
5360	CPVC, couplings 10' O.C., hangers 3 per 10'									
5380	Schedule 40									
5460	1/2" diameter	1 Plum	54	.148	L.F.	2.88	5.55		8.43	11.50
5470	3/4" diameter		51	.157		3.24	5.85		9.09	12.40
5480	1" diameter		46	.174		3.81	6.50		10.31	14
5490	1-1/4" diameter		42	.190		4.42	7.10		11.52	15.60
5500	1-1/2" diameter	↓	36	.222		5.50	8.30		13.80	18.60
5510	2" diameter	Q-1	59	.271		6.35	9.10		15.45	20.50
5520	2-1/2" diameter		56	.286		10.25	9.60		19.85	26
5530	3" diameter		53	.302		12.35	10.15		22.50	29
5540	4" diameter		48	.333		17.90	11.20		29.10	36.50
5550	6" diameter	↓	43	.372	↓	25	12.50		37.50	46.50

520

MECHANICAL 15

15108	Plastic Pipe & Fittings	CREW	DAILY OUTPUT	LABOR-HOURS	UNIT	2003 BARE COSTS MAT.	LABOR	EQUIP.	TOTAL	TOTAL INCL O&P
520	5730	To delete coupling & hangers, subtract										520
	5740	1/2" diam. to 3/4" diam.					37%	77%			
	5750	1" diam. to 1-1/4" diam.					27%	70%			
	5760	1-1/2" diam. to 3" diam.					21%	57%			
	5770	4" diam. to 6" diam.					16%	57%			
	5800	Schedule 80									
	5860	1/2" diameter	1 Plum	50	.160	L.F.	3.27	6		9.27	12.60
	5870	3/4" diameter		47	.170		3.42	6.35		9.77	13.35
	5880	1" diameter		43	.186		4.08	6.95		11.03	15
	5890	1-1/4" diameter		39	.205		4.83	7.65		12.48	16.85
	5900	1-1/2" diameter		34	.235		6	8.80		14.80	19.85
	5910	2" diameter	Q-1	55	.291		7.15	9.80		16.95	22.50
	5920	2-1/2" diameter		52	.308		11.35	10.35		21.70	28
	5930	3" diameter		50	.320		14	10.75		24.75	31.50
	5940	4" diameter		46	.348		20.50	11.70		32.20	40
	5950	6" diameter		38	.421		31	14.15		45.15	55.50
	5960	8" diameter	Q-2	47	.511		66	17.80		83.80	99.50
	6060	To delete couplings & hangers, subtract									
	6070	1/2" diam. to 3/4" diam.					44%	77%			
	6080	1" diam. to 1-1/4" diam.					32%	71%			
	6090	1-1/2" diam. to 4" diam.					25%	58%			
	6100	6" diam. to 8" diam.					20%	53%			
	6240	CTS, 1/2" diameter	1 Plum	54	.148	L.F.	2.14	5.55		7.69	10.70
	6250	3/4" diameter		51	.157		3.02	5.85		8.87	12.15
	6260	1" diameter		46	.174		3.09	6.50		9.59	13.20
	6270	1 1/4"		42	.190		3.68	7.10		10.78	14.80
	6280	1 1/2" diameter		36	.222		4.28	8.30		12.58	17.25
	6290	2" diameter	Q-1	59	.271		5.90	9.10		15	20.50
	6370	To delete coupling & hangers, subtract									
	6380	1/2" diam.					51%	79%			
	6390	3/4" diam.					40%	76%			
	6392	1" thru 2" diam.					72%	68%			
	7280	Polyethylene, flexible, no couplings or hangers									
	7282	Note: For labor costs add 25% to the couplings and fittings labor total.									
	7300	SDR 15, 100 psi									
	7310	3/4" diameter				L.F.	.22			.22	.24
	7350	1" diameter					.29			.29	.32
	7360	1-1/4" diameter					.53			.53	.58
	7370	1-1/2" diameter					.64			.64	.70
	7380	2" diameter					1.05			1.05	1.16
	7700	SIDR 9, 160 psi									
	7710	1/2" diameter				L.F.	.29			.29	.32
	7750	3/4" diameter					.29			.29	.32
	7760	1" diameter					.45			.45	.50
	7770	1-1/4" diameter					.79			.79	.87
	7780	1-1/2" diameter					1.10			1.10	1.21
	7790	2" diameter					1.82			1.82	2
	8120	SDR 9, 200 psi									
	8150	3/4" diameter				L.F.	.26			.26	.29
	8160	1" diameter					.45			.45	.50
	8170	1-1/4" diameter					.64			.64	.70
	8420	SDR 7, 250 psi									
	8440	3/4" diameter				L.F.	.37			.37	.41
	8450	1" diameter					.60			.60	.66
	8460	1-1/4" diameter					1.05			1.05	1.16
	8800	PVC, type PSP, drain & sewer, belled end gasket jnt., no hngr.									

Important: See the Reference Section for critical supporting data - Reference Nos., Crews, & City Cost Indexes

15 MECHANICAL

15108 | Plastic Pipe & Fittings

		CREW	DAILY OUTPUT	LABOR-HOURS	UNIT	MAT.	LABOR	EQUIP.	TOTAL	TOTAL INCL O&P	
520	8840	3" diameter				L.F.	1.25			1.25	1.38	520
	8850	4" diameter				↓	1.55			1.55	1.71	
	8860	6" diameter				↓	2.99			2.99	3.29	
	9000	Perforated										
	9040	4" diameter				L.F.	1.55			1.55	1.71	
560	0010	**PIPE, PLASTIC, FITTINGS**										560
	0030	Epoxy resin, fiberglass reinforced, general service										
	0090	Elbow, 90°, 2"	Q-1	33.10	.483	Ea.	45.50	16.25		61.75	74.50	
	0100	3"		20.80	.769		64.50	26		90.50	110	
	0110	4"		16.50	.970		88	32.50		120.50	146	
	0120	6"	↓	10.10	1.584		134	53.50		187.50	229	
	0130	8"	Q-2	9.30	2.581		248	90		338	410	
	0140	10"		8.50	2.824		310	98.50		408.50	490	
	0150	12"	↓	7.60	3.158	↓	445	110		555	655	
	0160	45° Elbow, same as 90°										
	0170	Elbow, 90°, flanged										
	0172	2"	Q-1	23	.696	Ea.	91	23.50		114.50	136	
	0173	3"		16	1		105	33.50		138.50	167	
	0174	4"		13	1.231		138	41.50		179.50	214	
	0176	6"	↓	8	2		249	67.50		316.50	375	
	0177	8"	Q-2	9	2.667		450	93		543	630	
	0178	10"		7	3.429		615	120		735	855	
	0179	12"	↓	5	4.800	↓	835	167		1,002	1,175	
	0186	Elbow, 45°, flanged										
	0188	2"	Q-1	23	.696	Ea.	91.50	23.50		115	137	
	0189	3"		16	1		105	33.50		138.50	167	
	0190	4"		13	1.231		137	41.50		178.50	214	
	0192	6"	↓	8	2		251	67.50		318.50	380	
	0193	8"	Q-2	9	2.667		450	93		543	635	
	0194	10"		7	3.429		615	120		735	855	
	0195	12"	↓	5	4.800		835	167		1,002	1,175	
	0290	Tee, 2"	Q-1	20	.800		55	27		82	101	
	0300	3"		13.90	1.151		64.50	38.50		103	129	
	0310	4"		11	1.455		76.50	49		125.50	158	
	0320	6"	↓	6.70	2.388		205	80.50		285.50	345	
	0330	8"	Q-2	6.20	3.871		237	135		372	465	
	0340	10"		5.70	4.211		380	147		527	635	
	0350	12"	↓	5.10	4.706	↓	460	164		624	755	
	0352	Tee, flanged										
	0354	2"	Q-1	17	.941	Ea.	124	31.50		155.50	184	
	0355	3"		10	1.600		166	54		220	265	
	0356	4"		8	2		185	67.50		252.50	305	
	0358	6"	↓	5	3.200		365	108		473	565	
	0359	8"	Q-2	6	4		625	139		764	895	
	0360	10"		5	4.800		905	167		1,072	1,250	
	0361	12"	↓	4	6	↓	1,150	209		1,359	1,600	
	0365	Wye, flanged										
	0367	2"	Q-1	17	.941	Ea.	320	31.50		351.50	400	
	0368	3"		10	1.600		355	54		409	470	
	0369	4"		8	2		440	67.50		507.50	585	
	0371	6"	↓	5	3.200		530	108		638	745	
	0372	8"	Q-2	6	4		690	139		829	965	
	0373	10"		5	4.800		1,175	167		1,342	1,525	
	0374	12"	↓	4	6	↓	1,425	209		1,634	1,875	
	0380	Couplings										

			DAILY	LABOR-			2003 BARE COSTS				TOTAL	
15108	**Plastic Pipe & Fittings**	CREW	OUTPUT	HOURS	UNIT	MAT.	LABOR	EQUIP.	TOTAL		INCL O&P	
560	0410	2"	Q-1	33.10	.483	Ea.	9.85	16.25		26.10	35.50	560
	0420	3"		20.80	.769		16.95	26		42.95	57.50	
	0430	4"		16.50	.970		20	32.50		52.50	71	
	0440	6"		10.10	1.584		38	53.50		91.50	123	
	0450	8"	Q-2	9.30	2.581		58	90		148	200	
	0460	10"		8.50	2.824		87.50	98.50		186	245	
	0470	12"		7.60	3.158		117	110		227	295	
	0473	High corrosion resistant couplings, add					30%					
	0474	Reducer, concentric, flanged										
	0475	2" x 1-1/2"	Q-1	30	.533	Ea.	132	17.95		149.95	172	
	0476	3" x 2"		24	.667		150	22.50		172.50	199	
	0477	4" x 3"		19	.842		158	28.50		186.50	217	
	0479	6" x 4"		15	1.067		206	36		242	281	
	0480	8" x 6"	Q-2	16	1.500		289	52.50		341.50	400	
	0481	10" x 8"		13	1.846		385	64.50		449.50	520	
	0482	12" x 10"		11	2.182		555	76		631	725	
	0486	Adapter, bell x male or female										
	0488	2"	Q-1	28	.571	Ea.	17.70	19.20		36.90	48.50	
	0489	3"		20	.800		22.50	27		49.50	65.50	
	0491	4"		17	.941		31	31.50		62.50	82	
	0492	6"		12	1.333		61.50	45		106.50	135	
	0493	8"	Q-2	15	1.600		86	56		142	179	
	0494	10"	"	11	2.182		151	76		227	281	
	0528	Flange										
	0532	2"	Q-1	46	.348	Ea.	17.40	11.70		29.10	37	
	0533	3"		32	.500		21	16.80		37.80	49	
	0534	4"		26	.615		29.50	20.50		50	64	
	0536	6"		16	1		50	33.50		83.50	106	
	0537	8"	Q-2	18	1.333		82	46.50		128.50	160	
	0538	10"		14	1.714		114	60		174	217	
	0539	12"		10	2.400		161	83.50		244.50	305	
	2100	PVC schedule 80										
	2110	90° elbow, 1/2"	1 Plum	30.30	.264	Ea.	.66	9.85		10.51	15.65	
	2130	3/4"		26	.308		.84	11.50		12.34	18.25	
	2140	1"		22.70	.352		1.36	13.15		14.51	21.50	
	2150	1-1/4"		20.20	.396		1.81	14.80		16.61	24.50	
	2160	1-1/2"		18.20	.440		1.94	16.40		18.34	27	
	2170	2"	Q-1	33.10	.483		2.34	16.25		18.59	27	
	2180	3"		20.80	.769		6.15	26		32.15	46	
	2190	4"		16.50	.970		7.15	32.50		39.65	57	
	2200	6"		10.10	1.584		57.50	53.50		111	144	
	2210	8"	Q-2	9.30	2.581		158	90		248	310	
	2250	45° elbow, 1/2"	1 Plum	30.30	.264		1.24	9.85		11.09	16.25	
	2270	3/4"		26	.308		1.89	11.50		13.39	19.45	
	2280	1"		22.70	.352		2.83	13.15		15.98	23	
	2290	1-1/4"		20.20	.396		3.16	14.80		17.96	26	
	2300	1-1/2"		18.20	.440		4.26	16.40		20.66	29.50	
	2310	2"	Q-1	33.10	.483		5.50	16.25		21.75	30.50	
	2320	3"		20.80	.769		14.10	26		40.10	54.50	
	2330	4"		16.50	.970		25.50	32.50		58	77	
	2340	6"		10.10	1.584		32	53.50		85.50	116	
	2350	8"	Q-2	9.30	2.581		149	90		239	300	
	2400	Tee, 1/2"	1 Plum	20.20	.396		1.85	14.80		16.65	24.50	
	2420	3/4"		17.30	.462		1.94	17.25		19.19	28	
	2430	1"		15.20	.526		2.43	19.65		22.08	32	
	2440	1-1/4"		13.50	.593		6.65	22		28.65	41	

Important: See the Reference Section for critical supporting data - Reference Nos., Crews, & City Cost Indexes

15 MECHANICAL

15108	Plastic Pipe & Fittings	CREW	DAILY OUTPUT	LABOR-HOURS	UNIT	2003 BARE COSTS				TOTAL INCL O&P
						MAT.	LABOR	EQUIP.	TOTAL	
2450	1-1/2″	1 Plum	12.10	.661	Ea.	6.65	24.50		31.15	45
2460	2″	Q-1	20	.800		8.35	27		35.35	49.50
2470	3″		13.90	1.151		11.25	38.50		49.75	71
2480	4″		11	1.455		13.10	49		62.10	88.50
2490	6″		6.70	2.388		45	80.50		125.50	170
2500	8″	Q-2	6.20	3.871		213	135		348	440
2510	Flange, socket, 150 lb., 1/2″	1 Plum	55.60	.144		3.59	5.35		8.94	12.05
2514	3/4″		47.60	.168		3.84	6.30		10.14	13.70
2518	1″		41.70	.192		4.28	7.15		11.43	15.50
2522	1-1/2″		33.30	.240		4.50	8.95		13.45	18.50
2526	2″	Q-1	60.60	.264		6	8.90		14.90	20
2530	4″		30.30	.528		12.90	17.75		30.65	41
2534	6″		18.50	.865		20.50	29		49.50	66.50
2538	8″	Q-2	17.10	1.404		78.50	49		127.50	160
2550	Coupling, 1/2″	1 Plum	30.30	.264		1.19	9.85		11.04	16.20
2570	3/4″		26	.308		1.61	11.50		13.11	19.10
2580	1″		22.70	.352		1.65	13.15		14.80	21.50
2590	1-1/4″		20.20	.396		2.52	14.80		17.32	25.50
2600	1-1/2″		18.20	.440		2.72	16.40		19.12	28
2610	2″	Q-1	33.10	.483		2.91	16.25		19.16	27.50
2620	3″		20.80	.769		8.25	26		34.25	48
2630	4″		16.50	.970		10.30	32.50		42.80	60.50
2640	6″		10.10	1.584		22	53.50		75.50	105
2650	8″	Q-2	9.30	2.581		35	90		125	175
2660	10″		8.50	2.824		41	98.50		139.50	194
2670	12″		7.60	3.158		44	110		154	215
2700	PVC (white), schedule 40, socket joints									
2760	90° elbow, 1/2″	1 Plum	33.30	.240	Ea.	.29	8.95		9.24	13.85
2770	3/4″		28.60	.280		.32	10.45		10.77	16.15
2780	1″		25	.320		.57	11.95		12.52	18.70
2790	1-1/4″		22.20	.360		1	13.45		14.45	21.50
2800	1-1/2″		20	.400		1	14.95		15.95	23.50
2810	2″	Q-1	36.40	.440		1.68	14.80		16.48	24.50
2820	2-1/2″		26.70	.599		5.10	20		25.10	36
2830	3″		22.90	.699		6.10	23.50		29.60	42
2840	4″		18.20	.879		10.95	29.50		40.45	56.50
2850	5″		12.10	1.322		28.50	44.50		73	98
2860	6″		11.10	1.441		35	48.50		83.50	112
2870	8″	Q-2	10.30	2.330		89	81.50		170.50	221
2980	45° elbow, 1/2″	1 Plum	33.30	.240		.46	8.95		9.41	14.05
2990	3/4″		28.60	.280		.71	10.45		11.16	16.60
3000	1″		25	.320		.86	11.95		12.81	19
3010	1-1/4″		22.20	.360		1.20	13.45		14.65	22
3020	1-1/2″		20	.400		1.51	14.95		16.46	24
3030	2″	Q-1	36.40	.440		1.97	14.80		16.77	24.50
3040	2-1/2″		26.70	.599		5.10	20		25.10	36
3050	3″		22.90	.699		7.95	23.50		31.45	44.50
3060	4″		18.20	.879		14.25	29.50		43.75	60
3070	5″		12.10	1.322		28.50	44.50		73	98
3080	6″		11.10	1.441		35	48.50		83.50	112
3090	8″	Q-2	10.30	2.330		100	81.50		181.50	233
3180	Tee, 1/2″	1 Plum	22.20	.360		.35	13.45		13.80	21
3190	3/4″		19	.421		.40	15.75		16.15	24.50
3200	1″		16.70	.479		.75	17.90		18.65	28
3210	1-1/4″		14.80	.541		1.18	20		21.18	32
3220	1-1/2″		13.30	.601		1.43	22.50		23.93	35.50

560

560

15108	Plastic Pipe & Fittings	CREW	DAILY OUTPUT	LABOR-HOURS	UNIT	2003 BARE COSTS				TOTAL INCL O&P
						MAT.	LABOR	EQUIP.	TOTAL	
3230	2"	Q-1	24.20	.661	Ea.	2.07	22		24.07	36
3240	2-1/2"		17.80	.899		6.85	30		36.85	53
3250	3"		15.20	1.053		9	35.50		44.50	63.50
3260	4"		12.10	1.322		16.25	44.50		60.75	85
3270	5"		8.10	1.975		39	66.50		105.50	143
3280	6"		7.40	2.162		54.50	72.50		127	170
3290	8"	Q-2	6.80	3.529		56.50	123		179.50	248
3380	Coupling, 1/2"	1 Plum	33.30	.240		.19	8.95		9.14	13.75
3390	3/4"		28.60	.280		.26	10.45		10.71	16.10
3400	1"		25	.320		.44	11.95		12.39	18.55
3410	1-1/4"		22.20	.360		.61	13.45		14.06	21
3420	1-1/2"		20	.400		.65	14.95		15.60	23
3430	2"	Q-1	36.40	.440		1.01	14.80		15.81	23.50
3440	2-1/2"		26.70	.599		2.22	20		22.22	33
3450	3"		22.90	.699		3.49	23.50		26.99	39.50
3460	4"		18.20	.879		5	29.50		34.50	50
3470	5"		12.10	1.322		12.25	44.50		56.75	80.50
3480	6"		11.10	1.441		15.90	48.50		64.40	90.50
3490	8"	Q-2	10.30	2.330		39.50	81.50		121	167
3710	Reducing insert, schedule 40, socket weld									
3712	3/4"	1 Plum	31.50	.254	Ea.	.29	9.50		9.79	14.60
3713	1"		27.50	.291		.53	10.85		11.38	17
3715	1-1/2"		22	.364		.74	13.60		14.34	21.50
3716	2"	Q-1	40	.400		1.24	13.45		14.69	22
3717	4"		20	.800		6.60	27		33.60	48
3718	6"		12.20	1.311		16.30	44		60.30	84.50
3719	8"	Q-2	11.30	2.124		57	74		131	175
3730	Reducing insert, socket weld x female/male thread									
3732	1/2"	1 Plum	38.30	.209	Ea.	1.32	7.80		9.12	13.25
3733	3/4"		32.90	.243		.82	9.10		9.92	14.60
3734	1"		28.80	.278		1.14	10.40		11.54	16.90
3736	1-1/2"		23	.348		2.06	13		15.06	22
3737	2"	Q-1	41.90	.382		1.10	12.85		13.95	20.50
3738	4"	"	20.90	.766		6.15	26		32.15	46
3742	Male adapter, socket weld x male thread									
3744	1/2"	1 Plum	38.30	.209	Ea.	.26	7.80		8.06	12.10
3745	3/4"		32.90	.243		.38	9.10		9.48	14.10
3746	1"		28.80	.278		.50	10.40		10.90	16.20
3748	1-1/2"		23	.348		.82	13		13.82	20.50
3749	2"	Q-1	41.90	.382		1.07	12.85		13.92	20.50
3750	4"	"	20.90	.766		5.90	26		31.90	45.50
3754	Female adapter, socket weld x female thread									
3756	1/2"	1 Plum	38.30	.209	Ea.	.41	7.80		8.21	12.25
3757	3/4"		32.90	.243		.53	9.10		9.63	14.30
3758	1"		28.80	.278		.61	10.40		11.01	16.30
3760	1-1/2"		23	.348		.82	13		13.82	20.50
3761	2"	Q-1	41.90	.382		1.10	12.85		13.95	20.50
3762	4"	"	20.90	.766		6.15	26		32.15	46
3800	PVC, schedule 80, socket joints									
3810	Reducing insert									
3812	3/4"	1 Plum	28.60	.280	Ea.	1.59	10.45		12.04	17.55
3813	1"		25	.320		2.40	11.95		14.35	20.50
3815	1-1/2"		20	.400		5.15	14.95		20.10	28
3816	2"	Q-1	36.40	.440		7.35	14.80		22.15	30.50
3817	4"		18.20	.879		28	29.50		57.50	75.50
3818	6"		11.10	1.441		39	48.50		87.50	116

560

Important: See the Reference Section for critical supporting data - Reference Nos., Crews, & City Cost Indexes

15108	Plastic Pipe & Fittings	CREW	DAILY OUTPUT	LABOR-HOURS	UNIT	2003 BARE COSTS				TOTAL INCL O&P
						MAT.	LABOR	EQUIP.	TOTAL	
3819	8"	Q-2	10.20	2.353	Ea.	55.50	82		137.50	186
3830	Reducing insert, socket weld x female/male thread									
3832	1/2"	1 Plum	34.80	.230	Ea.	2.40	8.60		11	15.60
3833	3/4"		29.90	.268		2.96	10		12.96	18.35
3834	1"		26.10	.307		5.15	11.45		16.60	23
3836	1-1/2"		20.90	.383		8.60	14.30		22.90	31
3837	2"	Q-1	38	.421		12.45	14.15		26.60	35
3838	4"	"	19	.842		44.50	28.50		73	92
3844	Adapter, male socket x male thread									
3846	1/2"	1 Plum	34.80	.230	Ea.	2.68	8.60		11.28	15.90
3847	3/4"		29.90	.268		2.96	10		12.96	18.35
3848	1"		26.10	.307		3.43	11.45		14.88	21
3850	1-1/2"		20.90	.383		8.60	14.30		22.90	31
3851	2"	Q-1	38	.421		12.45	14.15		26.60	35
3852	4"	"	19	.842		28	28.50		56.50	74
3860	Adapter, female socket x female thread									
3862	1/2"	1 Plum	34.80	.230	Ea.	2.19	8.60		10.79	15.35
3863	3/4"		29.90	.268		3.62	10		13.62	19.10
3864	1"		26.10	.307		5.35	11.45		16.80	23
3866	1-1/2"		20.90	.383		10.60	14.30		24.90	33
3867	2"	Q-1	38	.421		18.55	14.15		32.70	42
3868	4"	"	19	.842		56.50	28.50		85	106
3872	Union, socket joints									
3874	1/2"	1 Plum	25.80	.310	Ea.	5.35	11.60		16.95	23.50
3875	3/4"		22.10	.362		6.80	13.50		20.30	28
3876	1"		19.30	.415		7.75	15.50		23.25	32
3878	1-1/2"		15.50	.516		17.45	19.30		36.75	48
3879	2"	Q-1	28.10	.569		23.50	19.15		42.65	55
3888	Cap									
3890	1/2"	1 Plum	54.50	.147	Ea.	2.56	5.50		8.06	11.10
3891	3/4"		46.70	.171		2.69	6.40		9.09	12.60
3892	1"		41	.195		4.78	7.30		12.08	16.25
3894	1-1/2"		32.80	.244		5.75	9.10		14.85	20
3895	2"	Q-1	59.50	.269		15.15	9.05		24.20	30.50
3896	4"		30	.533		46	17.95		63.95	77.50
3897	6"		18.20	.879		112	29.50		141.50	168
3898	8"	Q-2	16.70	1.437		143	50		193	234
4500	DWV, ABS, non pressure, socket joints									
4540	1/4 Bend, 1-1/4"	1 Plum	20.20	.396	Ea.	1.72	14.80		16.52	24.50
4560	1-1/2"	"	18.20	.440		1.03	16.40		17.43	26
4570	2"	Q-1	33.10	.483		1.73	16.25		17.98	26.50
4580	3"		20.80	.769		3.82	26		29.82	43
4590	4"		16.50	.970		6.20	32.50		38.70	56
4600	6"		10.10	1.584		58	53.50		111.50	144
4650	1/8 Bend, same as 1/4 Bend									
4800	Tee, sanitary									
4820	1-1/4"	1 Plum	13.50	.593	Ea.	2.17	22		24.17	36
4830	1-1/2"	"	12.10	.661		1.52	24.50		26.02	39
4840	2"	Q-1	20	.800		2.26	27		29.26	43
4850	3"		13.90	1.151		4.85	38.50		43.35	64
4860	4"		11	1.455		12	49		61	87
4862	Tee, sanitary, reducing, 2" x 1-1/2"		22	.727		2.03	24.50		26.53	39
4864	3" x 2"		15.30	1.046		4.09	35		39.09	57.50
4868	4" x 3"		12.10	1.322		10.90	44.50		55.40	79
4870	Combination Y and 1/8 bend									
4872	1-1/2"	1 Plum	12.10	.661	Ea.	4.06	24.50		28.56	42

560

MECHANICAL 15

15108	Plastic Pipe & Fittings	CREW	DAILY OUTPUT	LABOR-HOURS	UNIT	2003 BARE COSTS				TOTAL INCL O&P
						MAT.	LABOR	EQUIP.	TOTAL	
4874	2"	Q-1	20	.800	Ea.	8.15	27		35.15	49.50
4876	3"		13.90	1.151		8.70	38.50		47.20	68
4878	4"		11	1.455		17.10	49		66.10	93
4880	3" x 1-1/2"		15.50	1.032		10.15	34.50		44.65	63.50
4882	4" x 3"	▼	12.10	1.322		15.25	44.50		59.75	84
4900	Wye, 1-1/4"	1 Plum	13.50	.593		4.14	22		26.14	38
4902	1-1/2"	"	12.10	.661		2.55	24.50		27.05	40.50
4904	2"	Q-1	20	.800		2.83	27		29.83	43.50
4906	3"		13.90	1.151		6.75	38.50		45.25	66
4908	4"		11	1.455		14.25	49		63.25	89.50
4910	6"		6.70	2.388		54	80.50		134.50	181
4918	3" x 1-1/2"		15.50	1.032		5.95	34.50		40.45	59
4920	4" x 3"		12.10	1.322		9.85	44.50		54.35	78
4922	6" x 4"	▼	6.90	2.319		44.50	78		122.50	167
4930	Double Wye, 1-1/2"	1 Plum	9.10	.879		5.40	33		38.40	55.50
4932	2"	Q-1	16.60	.964		6.95	32.50		39.45	56.50
4934	3"		10.40	1.538		18	51.50		69.50	98
4936	4"		8.25	1.939		36.50	65		101.50	139
4940	2" x 1-1/2"		16.80	.952		6.95	32		38.95	56
4942	3" x 2"		10.60	1.509		13.35	51		64.35	91
4944	4" x 3"		8.45	1.893		29	63.50		92.50	128
4946	6" x 4"		7.25	2.207		100	74		174	222
4950	Reducer bushing, 2" x 1-1/2"		36.40	.440		.89	14.80		15.69	23.50
4952	3" x 1-1/2"		27.30	.586		3.22	19.70		22.92	33.50
4954	4" x 2"		18.20	.879		7.40	29.50		36.90	52.50
4956	6" x 4"	▼	11.10	1.441		19.75	48.50		68.25	94.50
4960	Couplings, 1-1/2"	1 Plum	18.20	.440		.62	16.40		17.02	25.50
4962	2"	Q-1	33.10	.483		.74	16.25		16.99	25.50
4963	3"		20.80	.769		2.15	26		28.15	41.50
4964	4"		16.50	.970		3.07	32.50		35.57	52.50
4966	6"		10.10	1.584		16.20	53.50		69.70	98.50
4970	2" x 1-1/2"		33.30	.480		2.62	16.15		18.77	27.50
4972	3" x 1-1/2"		21	.762		7.50	25.50		33	47
4974	4" x 3"	▼	16.70	.958		13	32		45	63
4978	Closet flange, 4"	1 Plum	32	.250		3.88	9.35		13.23	18.35
4980	4" x 3"	"	34	.235	▼	6.05	8.80		14.85	19.90
5000	DWV, PVC, schedule 40, socket joints									
5040	1/4 bend, 1-1/4"	1 Plum	20.20	.396	Ea.	1.43	14.80		16.23	24
5060	1-1/2"	"	18.20	.440		.81	16.40		17.21	26
5070	2"	Q-1	33.10	.483		1.25	16.25		17.50	26
5080	3"		20.80	.769		3.10	26		29.10	42.50
5090	4"		16.50	.970		5.55	32.50		38.05	55
5100	6"	▼	10.10	1.584		27.50	53.50		81	111
5105	8"	Q-2	9.30	2.581		55.50	90		145.50	197
5106	10"	"	8.50	2.824		65.50	98.50		164	221
5110	1/4 bend, long sweep, 1-1/2"	1 Plum	18.20	.440		2.27	16.40		18.67	27.50
5112	2"	Q-1	33.10	.483		2.06	16.25		18.31	27
5114	3"		20.80	.769		5.10	26		31.10	44.50
5116	4"	▼	16.50	.970		9.55	32.50		42.05	59.50
5150	1/8 bend, 1-1/4"	1 Plum	20.20	.396		2.31	14.80		17.11	25
5170	1-1/2"	"	18.20	.440		1.60	16.40		18	27
5180	2"	Q-1	33.10	.483		1.19	16.25		17.44	26
5190	3"		20.80	.769		2.88	26		28.88	42
5200	4"		16.50	.970		5	32.50		37.50	54.50
5210	6"	▼	10.10	1.584		33.50	53.50		87	118
5215	8"	Q-2	9.30	2.581		38.50	90		128.50	178

560

15108	Plastic Pipe & Fittings	CREW	DAILY OUTPUT	LABOR-HOURS	UNIT	2003 BARE COSTS				TOTAL INCL O&P
						MAT.	LABOR	EQUIP.	TOTAL	
5216	10"	Q-2	8.50	2.824	Ea.	69	98.50		167.50	225
5217	12"	↓	7.60	3.158		95	110		205	271
5250	Tee, sanitary 1-1/4"	1 Plum	13.50	.593		2.57	22		24.57	36.50
5254	1-1/2"	"	12.10	.661		1.21	24.50		25.71	39
5255	2"	Q-1	20	.800		1.77	27		28.77	42.50
5256	3"		13.90	1.151		4.54	38.50		43.04	63.50
5257	4"		11	1.455		8.65	49		57.65	83.50
5259	6"	↓	6.70	2.388		74.50	80.50		155	203
5261	8"	Q-2	6.20	3.871		223	135		358	450
5264	2" x 1-1/2"	Q-1	22	.727		3.71	24.50		28.21	41
5266	3" x 1-1/2"		15.50	1.032		5.15	34.50		39.65	58
5268	4" x 3"		12.10	1.322		23.50	44.50		68	93
5271	6" x 4"	↓	6.90	2.319	↓	80	78		158	206
5276	Tee, sanitary, reducing									
5281	2" x 1-1/2" x 1-1/2"	Q-1	23	.696	Ea.	3.59	23.50		27.09	39.50
5282	2" x 1-1/2" x 2"		22	.727		2.45	24.50		26.95	39.50
5283	2" x 2" x 1-1/2"		22	.727		4.90	24.50		29.40	42.50
5284	3" x 3" x 1-1/2"		15.50	1.032		5.70	34.50		40.20	59
5285	3" x 3" x 2"		15.30	1.046		6.80	35		41.80	60.50
5286	4" x 4" x 1-1/2"		12.30	1.301		19.65	44		63.65	87.50
5287	4" x 4" x 2"		12.20	1.311		19.30	44		63.30	87.50
5288	4" x 4" x 3"		12.10	1.322		26	44.50		70.50	96
5291	6" x 6" x 4"	↓	6.90	2.319	↓	80	78		158	206
5294	Tee, double sanitary									
5295	1-1/2"	1 Plum	9.10	.879	Ea.	3.21	33		36.21	53
5296	2"	Q-1	16.60	.964		4.71	32.50		37.21	54
5297	3"		10.40	1.538		13.20	51.50		64.70	92.50
5298	4"	↓	8.25	1.939	↓	21	65		86	122
5303	Wye, reducing									
5304	2" x 1-1/2" x 1-1/2"	Q-1	23	.696	Ea.	3.98	23.50		27.48	40
5305	2" x 2" x 1-1/2"		22	.727		3.86	24.50		28.36	41.50
5306	3" x 3" x 2"		15.30	1.046		14.40	35		49.40	69
5307	4" x 4" x 2"		12.20	1.311		7.25	44		51.25	74.50
5309	4" x 4" x 3"	↓	12.10	1.322		8.45	44.50		52.95	76.50
5314	Combination Y & 1/8 bend, 1-1/2"	1 Plum	12.10	.661		5.35	24.50		29.85	43.50
5315	2"	Q-1	20	.800		3.98	27		30.98	45
5317	3"		13.90	1.151		7.30	38.50		45.80	66.50
5318	4"		11	1.455		14.35	49		63.35	90
5319	6"	↓	6.70	2.388		66.50	80.50		147	195
5320	8"	Q-2	6.20	3.871	↓	106	135		241	320
5324	Combination Y & 1/8 bend, reducing									
5325	2" x 2" x 1-1/2"	Q-1	22	.727	Ea.	4.89	24.50		29.39	42.50
5327	3" x 3" x 1-1/2"		15.50	1.032		8.50	34.50		43	62
5328	3" x 3" x 2"		15.30	1.046		5.30	35		40.30	59
5329	4" x 4" x 2"	↓	12.20	1.311		11.85	44		55.85	79.50
5331	Wye, 1-1/4"	1 Plum	13.50	.593		4.07	22		26.07	38
5332	1-1/2"	"	12.10	.661		2.45	24.50		26.95	40
5333	2"	Q-1	20	.800		2.14	27		29.14	43
5334	3"		13.90	1.151		5.45	38.50		43.95	64.50
5335	4"		11	1.455		14.25	49		63.25	89.50
5336	6"	↓	6.70	2.388		35.50	80.50		116	160
5337	8"	Q-2	6.20	3.871		64	135		199	275
5338	10"		5.70	4.211		124	147		271	360
5339	12"	↓	5.10	4.706		202	164		366	470
5341	2" x 1-1/2"	Q-1	22	.727		5.90	24.50		30.40	43.50
5342	3" x 1-1/2"	↓	15.50	1.032	↓	7.85	34.50		42.35	61

560

MECHANICAL 15

15108 | Plastic Pipe & Fittings

		CREW	DAILY OUTPUT	LABOR-HOURS	UNIT	2003 BARE COSTS MAT.	LABOR	EQUIP.	TOTAL	TOTAL INCL O&P
560 5343	4" x 3"	Q-1	12.10	1.322	Ea.	13.30	44.50		57.80	81.50 **560**
5344	6" x 4"	↓	6.90	2.319		52.50	78		130.50	176
5345	8" x 6"	Q-2	6.40	3.750		48.50	131		179.50	251
5347	Double wye, 1-1/2"	1 Plum	9.10	.879		7.15	33		40.15	57.50
5348	2"	Q-1	16.60	.964		9.15	32.50		41.65	59
5349	3"		10.40	1.538		23.50	51.50		75	104
5350	4"		8.25	1.939		48	65		113	152
5354	2" x 1-1/2"		16.80	.952		8.40	32		40.40	58
5355	3" x 2"		10.60	1.509		17.65	51		68.65	96
5356	4" x 3"		8.45	1.893		38	63.50		101.50	138
5357	6" x 4"	↓	7.25	2.207		79	74		153	199
5374	Coupling, 1-1/4"	1 Plum	20.20	.396		1.38	14.80		16.18	24
5376	1-1/2"	"	18.20	.440		.47	16.40		16.87	25.50
5378	2"	Q-1	33.10	.483		.54	16.25		16.79	25
5380	3"		20.80	.769		1.53	26		27.53	40.50
5390	4"		16.50	.970		2.54	32.50		35.04	52
5400	6"	↓	10.10	1.584		12.15	53.50		65.65	94
5402	8"	Q-2	9.30	2.581		45.50	90		135.50	186
5404	2" x 1-1/2"	Q-1	33.30	.480		1.20	16.15		17.35	26
5406	3" x 1-1/2"		21	.762		3.82	25.50		29.32	42.50
5408	4" x 3"		16.70	.958		6.60	32		38.60	56
5410	Reducer bushing, 2" x 1-1/4"		36.50	.438		.81	14.75		15.56	23.50
5411	2" x 1-1/2"		36.40	.440		.81	14.80		15.61	23.50
5412	3" x 1-1/2"		27.30	.586		3.56	19.70		23.26	34
5413	3" x 2"		27.10	.590		5.85	19.85		25.70	36.50
5414	4" x 2"		18.20	.879		11.40	29.50		40.90	57
5415	4" x 3"		16.70	.958		12.35	32		44.35	62
5416	6" x 4"	↓	11.10	1.441		33	48.50		81.50	109
5418	8" x 6"	Q-2	10.20	2.353		65.50	82		147.50	196
5425	Closet flange 4"	Q-1	32	.500		6.30	16.80		23.10	32.50
5426	4" x 3"	"	34	.471	↓	9.90	15.85		25.75	35
5450	Solvent cement for PVC, industrial grade, per quart				Qt.	11.30			11.30	12.45
5500	CPVC, Schedule 80, threaded joints									
5540	90° Elbow, 1/4"	1 Plum	32	.250	Ea.	7.05	9.35		16.40	22
5560	1/2"		30.30	.264		4.10	9.85		13.95	19.40
5570	3/4"		26	.308		6.15	11.50		17.65	24
5580	1"		22.70	.352		8.60	13.15		21.75	29.50
5590	1-1/4"		20.20	.396		16.55	14.80		31.35	41
5600	1-1/2"	↓	18.20	.440		17.85	16.40		34.25	44.50
5610	2"	Q-1	33.10	.483		24	16.25		40.25	51
5620	2-1/2"		24.20	.661		74	22		96	115
5630	3"		20.80	.769		79.50	26		105.50	127
5640	4"		16.50	.970		124	32.50		156.50	186
5650	6"	↓	10.10	1.584	↓	147	53.50		200.50	242
5700	45° Elbow same as 90° Elbow									
5850	Tee, 1/4"	1 Plum	22	.364	Ea.	13.70	13.60		27.30	35.50
5870	1/2"		20.20	.396		13.70	14.80		28.50	37.50
5880	3/4"		17.30	.462		19.75	17.25		37	47.50
5890	1"		15.20	.526		21.50	19.65		41.15	53
5900	1-1/4"		13.50	.593		21.50	22		43.50	57
5910	1-1/2"	↓	12.10	.661		22.50	24.50		47	62
5920	2"	Q-1	20	.800		25	27		52	68
5930	2-1/2"		16.20	.988		123	33		156	185
5940	3"		13.90	1.151		142	38.50		180.50	216
5950	4"		11	1.455		335	49		384	445
5960	6"	↓	6.70	2.388	↓	183	80.50		263.50	320

Important: See the Reference Section for critical supporting data - Reference Nos., Crews, & City Cost Indexes

15 MECHANICAL

15108	Plastic Pipe & Fittings	CREW	DAILY OUTPUT	LABOR-HOURS	UNIT	2003 BARE COSTS				TOTAL INCL O&P
						MAT.	LABOR	EQUIP.	TOTAL	
560	6000	Coupling, 1/4"	1 Plum	32	.250	Ea.	3.14	9.35		12.49	17.55	**560**
	6020	1/2"		30.30	.264		2.92	9.85		12.77	18.10
	6030	3/4"		26	.308		4.16	11.50		15.66	22
	6040	1"		22.70	.352		4.47	13.15		17.62	25
	6050	1-1/4"		20.20	.396		5.55	14.80		20.35	28.50
	6060	1-1/2"		18.20	.440		12.30	16.40		28.70	38.50
	6070	2"	Q-1	33.10	.483		12.90	16.25		29.15	38.50
	6080	2-1/2"		24.20	.661		31	22		53	67.50
	6090	3"		20.80	.769		36	26		62	78.50
	6100	4"		16.50	.970		62.50	32.50		95	118
	6110	6"	Q-2	10.10	1.584		48.50	53.50		102	134
	6120	8"		9.30	2.581		227	90		317	385
	6200	CTS, 100 psi at 180°F, hot and cold water									
	6230	90° Elbow, 1/2"	1 Plum	20	.400	Ea.	.08	14.95		15.03	22.50
	6250	3/4"		19	.421		.18	15.75		15.93	24
	6251	1"		16	.500		.58	18.70		19.28	28.50
	6252	1-1/4"		15	.533		.99	19.90		20.89	31
	6253	1-1/2"		14	.571		1.59	21.50		23.09	34
	6254	2"	Q-1	23	.696		3.43	23.50		26.93	39.50
	6260	45° Elbow, 1/2"	1 Plum	20	.400		.13	14.95		15.08	22.50
	6280	3/4"		19	.421		.20	15.75		15.95	24
	6281	1"		16	.500		.53	18.70		19.23	28.50
	6282	1-1/4"		15	.533		1.04	19.90		20.94	31
	6283	1-1/2"		14	.571		1.65	21.50		23.15	34
	6284	2"	Q-1	23	.696		3.48	23.50		26.98	39.50
	6290	Tee, 1/2"	1 Plum	13	.615		.12	23		23.12	34.50
	6310	3/4"		12	.667		.26	25		25.26	38
	6311	1"		11	.727		1.31	27		28.31	42.50
	6312	1-1/4"		10	.800		2	30		32	47
	6313	1-1/2"		10	.800		2.60	30		32.60	48
	6314	2"	Q-1	17	.941		4.22	31.50		35.72	52.50
	6320	Coupling, 1/2"	1 Plum	22	.364		.10	13.60		13.70	20.50
	6340	3/4"		21	.381		.12	14.25		14.37	21.50
	6341	1"		18	.444		.50	16.60		17.10	25.50
	6342	1-1/4"		17	.471		.62	17.60		18.22	27
	6343	1-1/2"		16	.500		.88	18.70		19.58	29
	6344	2"	Q-1	28	.571		1.75	19.20		20.95	31
	6360	Solvent cement for CPVC, commercial grade, per quart				Qt.	17.85			17.85	19.65
	7990	Polybutyl/polyethyl pipe, for copper fittings see 15107-460-7000									
	8000	Compression type, PVC, 160 psi cold water									
	8010	Coupling, 3/4" CTS	1 Plum	21	.381	Ea.	2.36	14.25		16.61	24
	8020	1" CTS		18	.444		2.92	16.60		19.52	28
	8030	1-1/4" CTS		17	.471		4.08	17.60		21.68	31
	8040	1-1/2" CTS		16	.500		5.60	18.70		24.30	34
	8050	2" CTS		15	.533		7.85	19.90		27.75	38.50
	8060	Female adapter, 3/4" FPT x 3/4" CTS		23	.348		3.76	13		16.76	23.50
	8070	3/4" FPT x 1" CTS		21	.381		5.60	14.25		19.85	27.50
	8080	1" FPT x 1" CTS		20	.400		5.65	14.95		20.60	28.50
	8090	1-1/4" FPT x 1-1/4" CTS		18	.444		7.40	16.60		24	33
	8100	1-1/2" FPT x 1-1/2" CTS		16	.500		8.65	18.70		27.35	37.50
	8110	2" FPT x 2" CTS		13	.615		12.80	23		35.80	48.50
	8130	Male adapter, 3/4" MPT x 3/4" CTS		23	.348		3.39	13		16.39	23.50
	8140	3/4" MPT x 1" CTS		21	.381		4.17	14.25		18.42	26
	8150	1" MPT x 1" CTS		20	.400		4.26	14.95		19.21	27
	8160	1-1/4" MPT x 1-1/4" CTS		18	.444		6.55	16.60		23.15	32
	8170	1-1/2" MPT x 1-1/2" CTS		16	.500		7.55	18.70		26.25	36.50

15108 | Plastic Pipe & Fittings

		DAILY	LABOR-		2003 BARE COSTS				TOTAL
					MAT.	**LABOR**	**EQUIP.**	**TOTAL**	**INCL O&P**
		CREW	OUTPUT	HOURS	UNIT				

		CREW	DAILY OUTPUT	LABOR HOURS	UNIT	MAT.	LABOR	EQUIP.	TOTAL	TOTAL INCL O&P
560	8180	2" MPT x 2" CTS	1 Plum	13	.615	Ea.	11.75	23		34.75	47.50
	8200	Spigot adapter, 3/4" IPS x 3/4" CTS		23	.348		3.39	13		16.39	23.50
	8210	3/4" IPS x 1" CTS		21	.381		3.39	14.25		17.64	25
	8220	1" IPS x 1" CTS		20	.400		3.66	14.95		18.61	26.50
	8230	1-1/4" IPS x 1-1/4" CTS		18	.444		6.55	16.60		23.15	32
	8240	1-1/2" IPS x 1-1/2" CTS		16	.500		7.55	18.70		26.25	36.50
	8250	2" IPS x 2" CTS		13	.615		12.10	23		35.10	48
	8270	Price includes insert stiffeners									
	8280	250 psi is same price as 160 psi									
	8300	Insert type, nylon, 160 & 250 psi, cold water									
	8310	Clamp ring stainless steel, 3/4" IPS	1 Plum	115	.070	Ea.	.92	2.60		3.52	4.93
	8320	1" IPS		107	.075		.94	2.79		3.73	5.25
	8330	1-1/4" IPS		101	.079		1.25	2.96		4.21	5.85
	8340	1-1/2" IPS		95	.084		1.25	3.15		4.40	6.15
	8350	2" IPS		85	.094		1.43	3.52		4.95	6.85
	8370	Coupling, 3/4" IPS		22	.364		.62	13.60		14.22	21
	8390	1-1/4" IPS		18	.444		1.56	16.60		18.16	26.50
	8400	1-1/2" IPS		17	.471		2.04	17.60		19.64	28.50
	8410	2" IPS		16	.500		2.45	18.70		21.15	30.50
	8430	Elbow, 90°, 3/4" IPS		22	.364		.95	13.60		14.55	21.50
	8440	1" IPS		19	.421		1.05	15.75		16.80	25
	8450	1-1/4" IPS		18	.444		1.18	16.60		17.78	26.50
	8460	1-1/2" IPS		17	.471		1.40	17.60		19	28
	8470	2" IPS		16	.500		1.95	18.70		20.65	30
	8490	Male adapter, 3/4" IPS x 3/4" MPT		25	.320		.66	11.95		12.61	18.80
	8500	1" IPS x 1" MPT		21	.381		.87	14.25		15.12	22.50
	8510	1-1/4" IPS x 1-1/4" MPT		20	.400		1.44	14.95		16.39	24
	8520	1-1/2" IPS x 1-1/2" MPT		18	.444		1.91	16.60		18.51	27
	8530	2" IPS x 2" MPT		15	.533		2.45	19.90		22.35	32.50
	8550	Tee, 3/4" IPS		14	.571		1.25	21.50		22.75	33.50
	8560	1" IPS		13	.615		2.05	23		25.05	37
	8570	1-1/4" IPS		12	.667		3.68	25		28.68	41.50
	8580	1-1/2" IPS		11	.727		4.63	27		31.63	46
	8590	2" IPS		10	.800		6.25	30		36.25	52
	8610	Insert type, PVC, 100 psi @ 180°F, hot & cold water									
	8620	Coupler, male, 3/8" CTS x 3/8" MPT	1 Plum	29	.276	Ea.	.40	10.30		10.70	16
	8630	3/8" CTS x 1/2" MPT		28	.286		.47	10.65		11.12	16.60
	8640	1/2" CTS x 1/2" MPT		27	.296		.40	11.05		11.45	17.15
	8650	1/2" CTS x 3/4" MPT		26	.308		1.12	11.50		12.62	18.60
	8660	3/4" CTS x 1/2" MPT		25	.320		1.12	11.95		13.07	19.30
	8670	3/4" CTS x 3/4" MPT		25	.320		1.12	11.95		13.07	19.30
	8700	Coupling, 3/8" CTS x 1/2" CTS		25	.320		2.15	11.95		14.10	20.50
	8710	1/2" CTS		23	.348		2.50	13		15.50	22.50
	8720	1/2" CTS x stub		23	.348		2.52	13		15.52	22.50
	8730	3/4" CTS		22	.364		3.96	13.60		17.56	25
	8750	Elbow 90°, 3/8" CTS		25	.320		2.55	11.95		14.50	21
	8760	1/2" CTS		23	.348		3.01	13		16.01	23
	8770	3/4" CTS		22	.364		4.61	13.60		18.21	25.50
	8800	Rings, crimp, copper, 3/8" CTS		120	.067		.11	2.49		2.60	3.88
	8810	1/2" CTS		117	.068		.12	2.55		2.67	3.99
	8820	3/4" CTS		115	.070		.15	2.60		2.75	4.09
	8850	Reducer tee, 3/8" x 3/8" x 1/2" CTS		17	.471		1.46	17.60		19.06	28
	8860	1/2" x 3/8" x 1/2" CTS		15	.533		1.25	19.90		21.15	31.50
	8870	3/4" x 1/2" x 1/2" CTS		14	.571		1.31	21.50		22.81	33.50
	8890	3/4" x 3/4" x 1/2" CTS		14	.571		1.24	21.50		22.74	33.50
	8900	3/4" x 1/2" x 3/8" CTS		14	.571		1.09	21.50		22.59	33

Important: See the Reference Section for critical supporting data - Reference Nos., Crews, & City Cost Indexes

15108	Plastic Pipe & Fittings	CREW	DAILY OUTPUT	LABOR-HOURS	UNIT	2003 BARE COSTS				TOTAL INCL O&P
						MAT.	LABOR	EQUIP.	TOTAL	
560 8930	Tee, 3/8" CTS	1 Plum	17	.471	Ea.	1.05	17.60		18.65	27.50	**560**
8940	1/2" CTS		15	.533		1.19	19.90		21.09	31.50
8950	3/4" CTS		14	.571		1.29	21.50		22.79	33.50
8960	Copper rings included in fitting price									
9000	Flare type, assembled, acetal, hot & cold water									
9010	Coupling, 1/4" & 3/8" CTS	1 Plum	24	.333	Ea.	2.14	12.45		14.59	21
9020	1/2" CTS		22	.364		2.45	13.60		16.05	23
9030	3/4" CTS		21	.381		3.96	14.25		18.21	26
9040	1" CTS		18	.444		3.96	16.60		20.56	29.50
9050	Elbow 90°, 1/4" CTS		26	.308		2.38	11.50		13.88	19.95
9060	3/8" CTS		24	.333		2.55	12.45		15	21.50
9070	1/2" CTS		22	.364		3.01	13.60		16.61	24
9080	3/4" CTS		21	.381		4.61	14.25		18.86	26.50
9090	1" CTS		18	.444		5.80	16.60		22.40	31.50
9110	Tee ,1/4" & 3/8" CTS		15	.533		2.62	19.90		22.52	33
9120	1/2" CTS		14	.571		3.44	21.50		24.94	36
9130	3/4" CTS		13	.615		5.25	23		28.25	40.50
9140	1" CTS		12	.667		7	25		32	45
9550	For plastic hangers see 15060-300-8000									
9560	For copper/brass fittings see 15107-460-7000									
590 0010	**PIPE, HIGH DENSITY POLYETHYLENE PLASTIC (HDPE)**										**590**
0020	Not incl. hangers, trenching, backfill, hoisting or digging equipment.									
0030	Standard length is 40', add a weld for each joint.									
0040	Single wall									
0050	Straight									
0054	1" diameter DR 11				L.F.	.28			.28	.31
0058	1-1/2" diameter DR 11					.62			.62	.68
0062	2" diameter DR 11					.79			.79	.87
0066	3" diameter DR 11					1.72			1.72	1.89
0070	3" diameter DR 17					1.15			1.15	1.27
0074	4" diameter DR 11					2.84			2.84	3.12
0078	4" diameter DR 17					1.91			1.91	2.10
0082	6" diameter DR 11					6.15			6.15	6.75
0086	6" diameter DR 17					4.13			4.13	4.54
0090	8" diameter DR 11					10.40			10.40	11.45
0094	8" diameter DR 26					4.69			4.69	5.15
0098	10" diameter DR 11					16.20			16.20	17.85
0102	10" diameter DR 26					7.25			7.25	8
0106	12" diameter DR 11					23			23	25
0110	12" diameter DR 26					10.25			10.25	11.25
0114	16" diameter DR 11					34			34	37
0118	16" diameter DR 26					15.20			15.20	16.70
0122	18" diameter DR 11					43			43	47
0126	18" diameter DR 26					19.20			19.20	21
0130	20" diameter DR 11					53			53	58
0134	20" diameter DR 26					23.50			23.50	26
0138	22" diameter DR 11					64			64	70.50
0142	22" diameter DR 26					28.50			28.50	31.50
0146	24" diameter DR 11					76			76	84
0150	24" diameter DR 26					34			34	37.50
0154	28" diameter DR 17					69.50			69.50	76.50
0158	28" diameter DR 26					46.50			46.50	51
0162	30" diameter DR 21					65.50			65.50	72
0166	30" diameter DR 26					43			43	47.50
0170	36" diameter DR 26					43			43	47.50
0174	42" diameter DR 26					105			105	115

15108	Plastic Pipe & Fittings	CREW	DAILY OUTPUT	LABOR-HOURS	UNIT	2003 BARE COSTS				TOTAL INCL O&P	
						MAT.	LABOR	EQUIP.	TOTAL		
590	0178	48" diameter DR 26				L.F.	132			132	146	590
0182	54" diameter DR 26				↓	173			173	190
0300	90° Elbow									
0304	1" diameter DR 11				Ea.	6.45			6.45	7.10
0308	1-1/2" diameter DR 11					10.75			10.75	11.85
0312	2" diameter DR 11					12			12	13.20
0316	3" diameter DR 11					23			23	25.50
0320	3" diameter DR 17					23			23	25.50
0324	4" diameter DR 11					27			27	29.50
0328	4" diameter DR 17					27			27	29.50
0332	6" diameter DR 11					79			79	87
0336	6" diameter DR 17					79			79	87
0340	8" diameter DR 11					260			260	286
0344	8" diameter DR 26					177			177	195
0348	10" diameter DR 11					645			645	710
0352	10" diameter DR 26					325			325	360
0356	12" diameter DR 11					645			645	710
0360	12" diameter DR 26					355			355	390
0364	16" diameter DR 11					720			720	795
0368	16" diameter DR 26					580			580	640
0372	18" diameter DR 11					755			755	830
0376	18" diameter DR 26					755			755	830
0380	20" diameter DR 11					1,125			1,125	1,225
0384	20" diameter DR 26					1,125			1,125	1,225
0388	22" diameter DR 11					1,400			1,400	1,550
0392	22" diameter DR 26					1,400			1,400	1,550
0396	24" diameter DR 11					1,625			1,625	1,775
0400	24" diameter DR 26					1,625			1,625	1,775
0404	28" diameter DR 17					2,450			2,450	2,700
0408	28" diameter DR 26					2,450			2,450	2,700
0412	30" diameter DR 17					2,250			2,250	2,475
0416	30" diameter DR 26					2,250			2,250	2,475
0420	36" diameter DR 26					3,450			3,450	3,775
0424	42" diameter DR 26					3,875			3,875	4,250
0428	48" diameter DR 26					4,075			4,075	4,500
0432	54" diameter DR 26				↓	9,475			9,475	10,400
0500	45° Elbow									
0512	2" diameter DR 11				Ea.	24			24	26.50
0516	3" diameter DR 11					21			21	23
0520	3" diameter DR 17					34			34	37.50
0524	4" diameter DR 11					35			35	38.50
0528	4" diameter DR 17					42			42	46
0532	6" diameter DR 11					74			74	81.50
0536	6" diameter DR 17					70			70	77
0540	8" diameter DR 11					175			175	193
0544	8" diameter DR 26					105			105	116
0548	10" diameter DR 11					490			490	540
0552	10" diameter DR 26					142			142	156
0556	12" diameter DR 11					590			590	650
0560	12" diameter DR 26					224			224	246
0564	16" diameter DR 11					282			282	310
0568	16" diameter DR 26					282			282	310
0572	18" diameter DR 11					415			415	455
0576	18" diameter DR 26					415			415	455
0580	20" diameter DR 11					645			645	710
0584	20" diameter DR 26				↓	645			645	710

Important: See the Reference Section for critical supporting data - Reference Nos., Crews, & City Cost Indexes

			DAILY	LABOR-		2003 BARE COSTS				TOTAL	
15108	**Plastic Pipe & Fittings**	CREW	OUTPUT	HOURS	UNIT	MAT.	LABOR	EQUIP.	TOTAL	INCL O&P	
590	0588	22" diameter DR 11				Ea.	755			755	830	590
	0592	22" diameter DR 26					755			755	830	
	0596	24" diameter DR 11					925			925	1,025	
	0600	24" diameter DR 26					925			925	1,025	
	0604	28" diameter DR 17					1,350			1,350	1,500	
	0608	28" diameter DR 26					1,350			1,350	1,500	
	0612	30" diameter DR 17					1,300			1,300	1,425	
	0616	30" diameter DR 26					1,300			1,300	1,425	
	0620	36" diameter DR 26					2,125			2,125	2,350	
	0624	42" diameter DR 26					2,600			2,600	2,850	
	0628	48" diameter DR 26					2,675			2,675	2,925	
	0632	54" diameter DR 26					4,350			4,350	4,775	
	0700	Tee										
	0704	1" diameter DR 11				Ea.	6.45			6.45	7.10	
	0708	1-1/2" diameter DR 11					10.75			10.75	11.85	
	0712	2" diameter DR 11					15			15	16.50	
	0716	3" diameter DR 11					24			24	26.50	
	0720	3" diameter DR 17					24			24	26.50	
	0724	4" diameter DR 11					38			38	42	
	0728	4" diameter DR 17					38			38	42	
	0732	6" diameter DR 11					94			94	103	
	0736	6" diameter DR 17					94			94	103	
	0740	8" diameter DR 11					217			217	239	
	0744	8" diameter DR 17					197			197	217	
	0748	10" diameter DR 11					675			675	745	
	0752	10" diameter DR 17					242			242	266	
	0756	12" diameter DR 11					975			975	1,075	
	0760	12" diameter DR 17					276			276	305	
	0764	16" diameter DR 11					645			645	710	
	0768	16" diameter DR 17					645			645	710	
	0772	18" diameter DR 11					685			685	755	
	0776	18" diameter DR 17					685			685	755	
	0780	20" diameter DR 11					850			850	935	
	0784	20" diameter DR 17					850			850	935	
	0788	22" diameter DR 11					1,175			1,175	1,300	
	0792	22" diameter DR 17					1,175			1,175	1,300	
	0796	24" diameter DR 11					1,175			1,175	1,300	
	0800	24" diameter DR 17					1,175			1,175	1,300	
	0804	28" diameter DR 17					2,100			2,100	2,300	
	0812	30" diameter DR 17					2,350			2,350	2,575	
	0820	36" diameter DR 17					4,675			4,675	5,125	
	0824	42" diameter DR 26					5,125			5,125	5,650	
	0828	48" diameter DR 26					6,000			6,000	6,600	
	1000	Flange adptr, w/back-up ring and 1/2 cost of plated bolt set										
	1004	1" diameter DR 11				Ea.	59			59	65	
	1008	1-1/2" diameter DR 11					60.50			60.50	66.50	
	1012	2" diameter DR 11					33.50			33.50	36.50	
	1016	3" diameter DR 11					41			41	45	
	1020	3" diameter DR 17					41			41	45	
	1024	4" diameter DR 11					54			54	59	
	1028	4" diameter DR 17					54			54	59	
	1032	6" diameter DR 11					78.50			78.50	86	
	1036	6" diameter DR 17					78.50			78.50	86	
	1040	8" diameter DR 11					112			112	123	
	1044	8" diameter DR 26					112			112	123	
	1048	10" diameter DR 11					168			168	185	

MECHANICAL 15

15108	Plastic Pipe & Fittings	CREW	DAILY OUTPUT	LABOR-HOURS	UNIT	2003 BARE COSTS				TOTAL INCL O&P
						MAT.	LABOR	EQUIP.	TOTAL	
1052	10" diameter DR 26				Ea.	168			168	185
1056	12" diameter DR 11					243			243	267
1060	12" diameter DR 26					243			243	267
1064	16" diameter DR 11					520			520	570
1068	16" diameter DR 26					520			520	570
1072	18" diameter DR 11					665			665	730
1076	18" diameter DR 26					665			665	730
1080	20" diameter DR 11					950			950	1,050
1084	20" diameter DR 26					950			950	1,050
1088	22" diameter DR 11					1,125			1,125	1,225
1092	22" diameter DR 26					1,125			1,125	1,225
1096	24" diameter DR 17					1,175			1,175	1,300
1100	24" diameter DR 32.5					1,175			1,175	1,300
1104	28" diameter DR 15.5					2,025			2,025	2,225
1108	28" diameter DR 32.5					1,975			1,975	2,175
1112	30" diameter DR 11					2,175			2,175	2,375
1116	30" diameter DR 21					1,875			1,875	2,075
1120	36" diameter DR 26					2,800			2,800	3,075
1124	42" diameter DR 26					3,150			3,150	3,475
1128	48" diameter DR 26					4,450			4,450	4,900
1132	54" diameter DR 26				▼	4,975			4,975	5,475
1200	Reducer									
1204	1-1/2" x 1" diameter DR 11				Ea.	6.90			6.90	7.55
1208	2" x 1-1/2" diameter DR 11					10.75			10.75	11.85
1212	3" x 2" diameter DR 11					14			14	15.40
1216	4" x 2" diameter DR 11					21			21	23
1220	4" x 3" diameter DR 11					21			21	23
1224	6" x 4" diameter DR 11					49			49	54
1228	8" x 6" diameter DR 11					86			86	94.50
1232	10" x 8" diameter DR 11					115			115	127
1236	12" x 8" diameter DR 11					215			215	237
1240	12" x 10" diameter DR 11					154			154	169
1244	14" x 12" diameter DR 11					173			173	190
1248	16" x 14" diameter DR 11					208			208	229
1252	18" x 16" diameter DR 11					415			415	460
1256	20" x 18" diameter DR 11					565			565	620
1260	22" x 20" diameter DR 11					1,150			1,150	1,275
1264	24" x 22" diameter DR 11					740			740	810
1268	26" x 24" diameter DR 11					865			865	950
1272	28" x 24" diameter DR 11					995			995	1,100
1276	32" x 28" diameter DR 17					1,200			1,200	1,325
1280	36" x 32" diameter DR 17				▼	1,425			1,425	1,550
4000	Welding labor per joint, not including welding machine									
4010	Pipe joint size (cost based on thickest wall for each dia.)									
4030	1" pipe size	4 Skwk	273	.117	Ea.		3.78		3.78	5.90
4040	1-1/2" pipe size		175	.183			5.90		5.90	9.25
4050	2" pipe size		128	.250			8.05		8.05	12.65
4060	3" pipe size		100	.320			10.30		10.30	16.15
4070	4" pipe size	▼	77	.416			13.40		13.40	21
4080	6" pipe size	5 Skwk	63	.635			20.50		20.50	32
4090	8" pipe size		48	.833			27		27	42
4100	10" pipe size	▼	40	1			32.50		32.50	50.50
4110	12" pipe size	6 Skwk	41	1.171			38		38	59
4120	16" pipe size		34	1.412			45.50		45.50	71.50
4130	18" pipe size	▼	32	1.500			48.50		48.50	76
4140	20" pipe size	8 Skwk	37	1.730	▼		56		56	87.50

590

Important: See the Reference Section for critical supporting data - Reference Nos., Crews, & City Cost Indexes

15108	Plastic Pipe & Fittings	CREW	DAILY OUTPUT	LABOR-HOURS	UNIT	2003 BARE COSTS				TOTAL INCL O&P
						MAT.	LABOR	EQUIP.	TOTAL	
590 4150	22" pipe size	8 Skwk	35	1.829	Ea.		59		59	92.50 **590**
4160	24" pipe size		34	1.882			60.50		60.50	95
4170	28" pipe size		33	1.939			62.50		62.50	98
4180	30" pipe size		32	2			64.50		64.50	101
4190	36" pipe size		31	2.065			66.50		66.50	104
4200	42" pipe size		30	2.133			69		69	108
4210	48" pipe size	9 Skwk	33	2.182			70.50		70.50	110
4220	54" pipe size	"	31	2.323			75		75	117
4300	Note: Cost for set up each time welder is moved.									
4301	Add 50% of a weld cost									
4310	Welder usually remains stationary with pipe moved through it.									
4340	Weld machine, rental per day based on dia. capacity									
4350	1" thru 2" diameter				Ea.			35	35	38.50
4360	3" thru 4" diameter							40	40	44
4370	6" thru 8" diameter							90	90	99
4380	10" thru 12" diameter							155	155	170.50
4390	16" thru 18" diameter							225	225	247.50
4400	20" thru 24" diameter							435	435	478.50
4410	28" thru 32" diameter							475	475	525
4420	36" diameter							495	495	545
4430	42" thru 54" diameter							775	775	855
5000	Dual wall contained pipe									
5040	Straight									
5054	1" DR 11 x 3" DR 11				L.F.	5.05			5.05	5.55
5058	1" DR 11 x 4" DR 11					5.55			5.55	6.10
5062	1-1/2" DR 11 x 4" DR 17					5.85			5.85	6.45
5066	2" DR 11 x 4" DR 17					5.90			5.90	6.50
5070	2" DR 11 x 6" DR 17					9.25			9.25	10.15
5074	3" DR 11 x 6" DR 17					10.15			10.15	11.20
5078	3" DR 11 x 6" DR 26					8.15			8.15	8.95
5082	3" DR 17 x 8" DR 11					16.80			16.80	18.45
5086	3" DR 17 x 8" DR 17					13.25			13.25	14.60
5090	4" DR 11 x 8" DR 17					15			15	16.50
5094	4" DR 17 x 8" DR 26					11.85			11.85	13.05
5098	6" DR 11 x 10" DR 17					24			24	26.50
5102	6" DR 17 x 10" DR 26					18.25			18.25	20
5106	6" DR 26 x 10" DR 26					16.85			16.85	18.50
5110	8" DR 17 x 12" DR 26					26.50			26.50	29
5114	8" DR 26 x 12" DR 32.5					22			22	24
5118	10" DR 17 x 14" DR 26					34.50			34.50	38
5122	10" DR 17 x 16" DR 26					39			39	43
5126	10" DR 26 x 16" DR 26					35.50			35.50	39
5130	12" DR 26 x 16" DR 26					40			40	44
5134	12" DR 17 x 18" DR 26					51.50			51.50	56.50
5138	12" DR 26 x 18" DR 26					46			46	50.50
5142	14" DR 26 x 20" DR 32.5					50.50			50.50	55.50
5146	16" DR 26 x 22" DR 32.5					76			76	83.50
5150	18" DR 26 x 24" DR 32.5					70			70	77
5154	20" DR 32.5 x 28" DR 32.5					90.50			90.50	99.50
5158	22" DR 32.5 x 30" DR 32.5					90.50			90.50	99.50
5162	24" DR 32.5 x 32" DR 32.5					115			115	127
5166	36" DR 32.5 x 42" DR 32.5					195			195	215
5300	Force transfer coupling									
5354	1" DR 11 x 3" DR 11				Ea.	182			182	200
5358	1" DR 11 x 4" DR 17					191			191	210
5362	1-1/2" DR 11 x 4" DR 17					200			200	220

MECHANICAL 15

15108	Plastic Pipe & Fittings	CREW	DAILY OUTPUT	LABOR-HOURS	UNIT	2003 BARE COSTS				TOTAL INCL O&P	
						MAT.	LABOR	EQUIP.	TOTAL		
590	5366	2" DR 11 x 4" DR 17				Ea.	209			209	230	**590**
	5370	2" DR 11 x 6" DR 17					305			305	335	
	5374	3" DR 11 x 6" DR 17					305			305	335	
	5378	3" DR 11 x 6" DR 26					305			305	335	
	5382	3" DR 11 x 8" DR 11					320			320	355	
	5386	3" DR 11 x 8" DR 17					320			320	355	
	5390	4" DR 11 x 8" DR 17					320			320	355	
	5394	4" DR 17 x 8" DR 26					282			282	310	
	5398	6" DR 11 x 10" DR 17					410			410	450	
	5402	6" DR 17 x 10" DR 26					335			335	370	
	5406	6" DR 26 x 10" DR 26					335			335	370	
	5410	8" DR 17 x 12" DR 26					445			445	490	
	5414	8" DR 26 x 12" DR 32.5					360			360	395	
	5418	10" DR 17 x 14" DR 26					585			585	645	
	5422	10" DR 17 x 16" DR 26					585			585	645	
	5426	10" DR 26 x 16" DR 26					585			585	645	
	5430	12" DR 26 x 16" DR 26					735			735	805	
	5434	12" DR 17 x 18" DR 26					750			750	825	
	5438	12" DR 26 x 18" DR 26					750			750	825	
	5442	14" DR 26 x 20" DR 32.5					795			795	870	
	5446	16" DR 26 x 22" DR 32.5					815			815	900	
	5450	18" DR 26 x 24" DR 32.5					885			885	975	
	5454	20" DR 32.5 x 28" DR 32.5					1,075			1,075	1,175	
	5458	22" DR 32.5 x 30" DR 32.5					1,200			1,200	1,300	
	5462	24" DR 32.5 x 32" DR 32.5					1,300			1,300	1,425	
	5466	36" DR 32.5 x 42" DR 32.5				↓	2,150			2,150	2,350	
	5600	90° Elbow										
	5654	1" DR 11 x 3" DR 11				Ea.	150			150	165	
	5658	1" DR 11 x 4" DR 17					144			144	158	
	5662	1-1/2" DR 11 x 4" DR 17					157			157	173	
	5666	2" DR 11 x 4" DR 17					167			167	184	
	5670	2" DR 11 x 6" DR 17					212			212	233	
	5674	3" DR 11 x 6" DR 17					243			243	267	
	5678	3" DR 17 x 6" DR 26					194			194	213	
	5682	3" DR 17 x 8" DR 11					390			390	430	
	5686	3" DR 17 x 8" DR 17					305			305	335	
	5690	4" DR 11 x 8" DR 17					360			360	395	
	5694	4" DR 17 x 8" DR 26					285			285	315	
	5698	6" DR 11 x 10" DR 17					480			480	525	
	5702	6" DR 17 x 10" DR 26					365			365	400	
	5706	6" DR 26 x 10" DR 26					340			340	375	
	5710	8" DR 17 x 12" DR 26					545			545	600	
	5714	8" DR 26 x 12" DR 32.5					505			505	555	
	5718	10" DR 17 x 14" DR 26					855			855	940	
	5722	10" DR 17 x 16" DR 26					820			820	900	
	5726	10" DR 26 x 16" DR 26					760			760	835	
	5730	12" DR 26 x 16" DR 26					840			840	925	
	5734	12" DR 17 x 18" DR 26					1,000			1,000	1,100	
	5738	12" DR 26 x 18" DR 26					920			920	1,025	
	5742	14" DR 26 x 20" DR 32.5					1,225			1,225	1,325	
	5746	16" DR 26 x 22" DR 32.5					1,225			1,225	1,350	
	5750	18" DR 26 x 24" DR 32.5					1,500			1,500	1,650	
	5754	20" DR 32.5 x 28" DR 32.5					1,725			1,725	1,900	
	5758	22" DR 32.5 x 30" DR 32.5					2,350			2,350	2,600	
	5762	24" DR 32.5 x 32" DR 32.5					2,300			2,300	2,525	
	5766	36" DR 32.5 x 42" DR 32.5				↓	3,725			3,725	4,100	

15 MECHANICAL

Important: See the Reference Section for critical supporting data - Reference Nos., Crews, & City Cost Indexes

		CREW	DAILY OUTPUT	LABOR-HOURS	UNIT	2003 BARE COSTS				TOTAL INCL O&P
15108	**Plastic Pipe & Fittings**					MAT.	LABOR	EQUIP.	TOTAL	
5800	45° Elbow									
5804	1" DR 11 x 3" DR 11				Ea.	90			90	99
5808	1" DR 11 x 4" DR 17					89.50			89.50	98.50
5812	1-1/2" DR 11 x 4" DR 17					96			96	106
5816	2" DR 11 x 4" DR 17					105			105	115
5820	2" DR 11 x 6" DR 17					132			132	145
5824	3" DR 11 x 6" DR 17					146			146	161
5828	3" DR 17 x 6" DR 26					121			121	133
5832	3" DR 17 x 8" DR 11					224			224	246
5836	3" DR 17 x 8" DR 17					179			179	197
5840	4" DR 11 x 8" DR 17					209			209	230
5844	4" DR 17 x 8" DR 26					170			170	187
5848	6" DR 11 x 10" DR 17					276			276	305
5852	6" DR 17 x 10" DR 26					216			216	238
5856	6" DR 26 x 10" DR 26					204			204	224
5860	8" DR 17 x 12" DR 26					330			330	360
5864	8" DR 26 x 12" DR 32.5					305			305	340
5868	10" DR 17 x 14" DR 26					515			515	565
5872	10" DR 17 x 16" DR 26					505			505	555
5876	10" DR 26 x 16" DR 26					470			470	520
5880	12" DR 26 x 16" DR 26					530			530	585
5884	12" DR 17 x 18" DR 26					625			625	690
5888	12" DR 26 x 18" DR 26					585			585	645
5892	14" DR 26 x 20" DR 32.5					755			755	830
5896	16" DR 26 x 22" DR 32.5					790			790	870
5900	18" DR 26 x 24" DR 32.5					935			935	1,025
5904	20" DR 32.5 x 28" DR 32.5					1,150			1,150	1,250
5908	22" DR 32.5 x 30" DR 32.5					1,450			1,450	1,600
5912	24" DR 32.5 x 32" DR 32.5					1,450			1,450	1,575
5916	36" DR 32.5 x 42" DR 32.5					2,250			2,250	2,475
6000	Access port with 4" riser									
6050	1" DR 11 x 4" DR 17				Ea.	197			197	216
6054	1-1/2" DR 11 x 4" DR 17					201			201	221
6058	2" DR 11 x 6" DR 17					245			245	269
6062	3" DR 11 x 6" DR 17					246			246	271
6066	3" DR 17 x 6" DR 26					241			241	265
6070	3" DR 17 x 8" DR 11					276			276	305
6074	3" DR 17 x 8" DR 17					264			264	290
6078	4" DR 11 x 8" DR 17					273			273	300
6082	4" DR 17 x 8" DR 26					261			261	287
6086	6" DR 11 x 10" DR 17					340			340	375
6090	6" DR 17 x 10" DR 26					315			315	345
6094	6" DR 26 x 10" DR 26					310			310	340
6098	8" DR 17 x 12" DR 26					330			330	365
6102	8" DR 26 x 12" DR 32.5					315			315	345
6200	End termination with vent plug									
6204	1" DR 11 x 3" DR 11				Ea.	163			163	179
6208	1" DR 11 x 4" DR 17					174			174	192
6212	1-1/2" DR 11 x 4" DR 17					174			174	192
6216	2" DR 11 x 4" DR 17					189			189	208
6220	2" DR 11 x 6" DR 17					216			216	237
6224	3" DR 11 x 6" DR 17					216			216	237
6228	3" DR 17 x 6" DR 26					216			216	237
6232	3" DR 17 x 8" DR 11					298			298	330
6236	3" DR 17 x 8" DR 17					298			298	330
6240	4" DR 11 x 8" DR 17					298			298	330

MECHANICAL 15

15108 | Plastic Pipe & Fittings

			DAILY	LABOR-		2003 BARE COSTS				TOTAL	
			CREW	OUTPUT	HOURS	UNIT	MAT.	LABOR	EQUIP.	TOTAL	INCL O&P
590	6244	4" DR 17 x 8" DR 26				Ea.	259			259	285	**590**
	6248	6" DR 11 x 10" DR 17					355			355	390	
	6252	6" DR 17 x 10" DR 26					291			291	320	
	6256	6" DR 26 x 10" DR 26					291			291	320	
	6260	8" DR 17 x 12" DR 26					430			430	475	
	6264	8" DR 26 x 12" DR 32.5					345			345	380	
	6268	10" DR 17 x 14" DR 26					450			450	495	
	6272	10" DR 17 x 16" DR 26					450			450	495	
	6276	10" DR 26 x 16" DR 26					450			450	495	
	6280	12" DR 26 x 16" DR 26					450			450	495	
	6284	12" DR 17 x 18" DR 26					590			590	650	
	6288	12" DR 26 x 18" DR 26					590			590	650	
	6292	14" DR 26 x 20" DR 32.5					600			600	660	
	6296	16" DR 26 x 22" DR 32.5					695			695	765	
	6300	18" DR 26 x 24" DR 32.5					720			720	790	
	6304	20" DR 32.5 x 28" DR 32.5					920			920	1,000	
	6308	22" DR 32.5 x 30" DR 32.5					1,000			1,000	1,100	
	6312	24" DR 32.5 x 32" DR 32.5					1,125			1,125	1,225	
	6316	36" DR 32.5 x 42" DR 32.5				▼	1,625			1,625	1,800	
	6600	Tee										
	6604	1" DR 11 x 3" DR 11				Ea.	166			166	182	
	6608	1" DR 11 x 4" DR 17					206			206	226	
	6612	1-1/2" DR 11 x 4" DR 17					229			229	252	
	6616	2" DR 11 x 4" DR 17					250			250	275	
	6620	2" DR 11 x 6" DR 17					305			305	335	
	6624	3" DR 11 x 6" DR 17					350			350	385	
	6628	3" DR 17 x 6" DR 26					340			340	375	
	6632	3" DR 17 x 8" DR 11					410			410	455	
	6636	3" DR 17 x 8" DR 17					390			390	430	
	6640	4" DR 11 x 8" DR 17					425			425	470	
	6644	4" DR 17 x 8" DR 26					410			410	450	
	6648	6" DR 11 x 10" DR 17					535			535	585	
	6652	6" DR 17 x 10" DR 26					505			505	555	
	6656	6" DR 26 x 10" DR 26					500			500	550	
	6660	8" DR 17 x 12" DR 26					500			500	550	
	6664	8" DR 26 x 12" DR 32.5					655			655	720	
	6668	10" DR 17 x 14" DR 26					755			755	830	
	6672	10" DR 17 x 16" DR 26					795			795	870	
	6676	10" DR 26 x 16" DR 26					795			795	870	
	6680	12" DR 26 x 16" DR 26					855			855	940	
	6684	12" DR 17 x 18" DR 26					855			855	940	
	6688	12" DR 26 x 18" DR 26					950			950	1,050	
	6692	14" DR 26 x 20" DR 32.5					1,250			1,250	1,375	
	6696	16" DR 26 x 22" DR 32.5					1,525			1,525	1,675	
	6700	18" DR 26 x 24" DR 32.5					1,675			1,675	1,825	
	6704	20" DR 32.5 x 28" DR 32.5					2,100			2,100	2,325	
	6708	22" DR 32.5 x 30" DR 32.5					2,500			2,500	2,750	
	6712	24" DR 32.5 x 32" DR 32.5					3,000			3,000	3,300	
	6716	36" DR 32.5 x 42" DR 32.5				▼	6,100			6,100	6,725	
	6800	Wye										
	6816	2" DR 11 x 4" DR 17				Ea.	315			315	345	
	6820	2" DR 11 x 6" DR 17					360			360	395	
	6824	3" DR 11 x 6" DR 17					360			360	400	
	6828	3" DR 17 x 6" DR 26					350			350	385	
	6832	3" DR 17 x 8" DR 11					440			440	485	
	6836	3" DR 17 x 8" DR 17				▼	410			410	450	

Important: See the Reference Section for critical supporting data - Reference Nos., Crews, & City Cost Indexes

15108	Plastic Pipe & Fittings	CREW	DAILY OUTPUT	LABOR-HOURS	UNIT	2003 BARE COSTS				TOTAL INCL O&P
						MAT.	LABOR	EQUIP.	TOTAL	
590	6840	4" DR 11 x 8" DR 17				Ea.	450			450	495	590
6844	4" DR 17 x 8" DR 26					425			425	470
6848	6" DR 11 x 10" DR 17					555			555	610
6852	6" DR 17 x 10" DR 26					515			515	565
6856	6" DR 26 x 10" DR 26					510			510	565
6860	8" DR 17 x 12" DR 26					735			735	805
6864	8" DR 26 x 12" DR 32.5					710			710	780
6868	10" DR 17 x 14" DR 26					885			885	975
6872	10" DR 17 x 16" DR 26					940			940	1,025
6876	10" DR 26 x 16" DR 26					925			925	1,025
6880	12" DR 26 x 16" DR 26					1,050			1,050	1,150
6884	12" DR 17 x 18" DR 26					1,175			1,175	1,275
6888	12" DR 26 x 18" DR 26					1,175			1,175	1,300
6892	14" DR 26 x 20" DR 32.5					1,450			1,450	1,575
6896	16" DR 26 x 22" DR 32.5					2,075			2,075	2,300
6900	18" DR 26 x 24" DR 32.5					2,450			2,450	2,700
6904	20" DR 32.5 x 28" DR 32.5					3,075			3,075	3,375
6908	22" DR 32.5 x 30" DR 32.5					3,500			3,500	3,850
6912	24" DR 32.5 x 32" DR 32.5				▼	3,975			3,975	4,375
9000	Welding labor per joint, not including welding machine									
9010	Pipe joint size, outer pipe (cost based on the thickest walls)									
9020	Straight pipe									
9050	3" pipe size	4 Skwk	96	.333	Ea.		10.75		10.75	16.85
9060	4" pipe size	"	77	.416			13.40		13.40	21
9070	6" pipe size	5 Skwk	60	.667			21.50		21.50	33.50
9080	8" pipe size	"	40	1			32.50		32.50	50.50
9090	10" pipe size	6 Skwk	41	1.171			38		38	59
9100	12" pipe size		39	1.231			39.50		39.50	62
9110	14" pipe size		38	1.263			40.50		40.50	64
9120	16" pipe size	▼	35	1.371			44		44	69.50
9130	18" pipe size	8 Skwk	45	1.422			46		46	72
9140	20" pipe size		42	1.524			49		49	77
9150	22" pipe size		40	1.600			51.50		51.50	81
9160	24" pipe size		38	1.684			54.50		54.50	85
9170	28" pipe size		37	1.730			56		56	87.50
9180	30" pipe size		36	1.778			57.50		57.50	90
9190	32" pipe size		35	1.829			59		59	92.50
9200	42" pipe size	▼	32	2	▼		64.50		64.50	101
9300	Note: Cost for set up each time welder is moved.									
9301	Add 100% of weld labor cost									
9310	For handling between fitting welds add 50% of weld labor cost									
9320	Welder usually remains stationary with pipe moved through it									
9360	Weld machine, rental per day based on dia. capacity									
9380	3" thru 4" diameter				Ea.			55	55	60.50
9390	6" thru 8" diameter							180	180	198
9400	10" thru 12" diameter							270	270	297
9410	14" thru 18" diameter							365	365	402
9420	20" thru 24" diameter							650	650	715
9430	28" thru 32" diameter							765	765	840
9440	42" thru 58" diameter				▼			790	790	870

15110 | Valves

		CREW	DAILY OUTPUT	LABOR-HOURS	UNIT	MAT.	LABOR	EQUIP.	TOTAL	TOTAL INCL O&P
100	0010	**VALVES, BRASS**										100
0030	For motorized valves, see Division 13838-200									
0500	Gas cocks, threaded									
0510	1/4"	1 Plum	26	.308	Ea.	6.15	11.50		17.65	24

		CREW	DAILY OUTPUT	LABOR-HOURS	UNIT	2003 BARE COSTS				TOTAL INCL O&P	
15110	**Valves**					MAT.	LABOR	EQUIP.	TOTAL		
100	0520	3/8"	1 Plum	24	.333	Ea.	6.15	12.45		18.60	25.50	**100**
	0530	1/2"		24	.333		6	12.45		18.45	25.50	
	0540	3/4"		22	.364		10.30	13.60		23.90	32	
	0550	1"		19	.421		12.60	15.75		28.35	38	
	0560	1-1/4"		15	.533		36	19.90		55.90	69.50	
	0570	1-1/2"		13	.615		61.50	23		84.50	102	
	0580	2"	▼	11	.727	▼	65.50	27		92.50	113	
	0670	For larger sizes use lubricated plug valve, Section 15110-600										
160	0010	**VALVES, BRONZE** [R15100-090]										**160**
	1020	Angle, 150 lb., rising stem, threaded										
	1030	1/8"	1 Plum	24	.333	Ea.	46.50	12.45		58.95	70	
	1040	1/4"		24	.333		46.50	12.45		58.95	70	
	1050	3/8"		24	.333		46.50	12.45		58.95	70	
	1060	1/2"		22	.364		46.50	13.60		60.10	71.50	
	1070	3/4"		20	.400		62.50	14.95		77.45	91	
	1080	1"		19	.421		89.50	15.75		105.25	122	
	1090	1-1/4"		15	.533		116	19.90		135.90	157	
	1100	1-1/2"		13	.615		149	23		172	199	
	1110	2"	▼	11	.727	▼	242	27		269	305	
	1380	Ball, 150 psi, threaded										
	1400	1/4"	1 Plum	24	.333	Ea.	6.75	12.45		19.20	26	
	1430	3/8"		24	.333		6.75	12.45		19.20	26	
	1450	1/2"		22	.364		6.75	13.60		20.35	28	
	1460	3/4"		20	.400		11.10	14.95		26.05	34.50	
	1470	1"		19	.421		13.95	15.75		29.70	39.50	
	1480	1-1/4"		15	.533		23.50	19.90		43.40	56	
	1490	1-1/2"		13	.615		30	23		53	67.50	
	1500	2"		11	.727		37.50	27		64.50	82.50	
	1510	2-1/2"		9	.889		177	33		210	245	
	1520	3"	▼	8	1	▼	289	37.50		326.50	375	
	1600	Butterfly, 175 psi, full port, solder or threaded ends										
	1610	Stainless steel disc and stem										
	1620	1/4"	1 Plum	24	.333	Ea.	5.75	12.45		18.20	25	
	1630	3/8"		24	.333		5.75	12.45		18.20	25	
	1640	1/2"		22	.364		6.10	13.60		19.70	27	
	1650	3/4"		20	.400		9.85	14.95		24.80	33.50	
	1660	1"		19	.421		12.25	15.75		28	37.50	
	1670	1-1/4"		15	.533		19.45	19.90		39.35	51.50	
	1680	1-1/2"		13	.615		25	23		48	62	
	1690	2"	▼	11	.727	▼	31.50	27		58.50	75.50	
	1750	Check, swing, class 150, regrinding disc, threaded										
	1800	1/8"	1 Plum	24	.333	Ea.	23.50	12.45		35.95	44.50	
	1830	1/4"		24	.333		23.50	12.45		35.95	44.50	
	1840	3/8"		24	.333		23.50	12.45		35.95	44.50	
	1850	1/2"		24	.333		24	12.45		36.45	45.50	
	1860	3/4"		20	.400		32.50	14.95		47.45	58	
	1870	1"		19	.421		49	15.75		64.75	77.50	
	1880	1-1/4"		15	.533		67.50	19.90		87.40	105	
	1890	1-1/2"		13	.615		79.50	23		102.50	122	
	1900	2"	▼	11	.727		117	27		144	170	
	1910	2-1/2"	Q-1	15	1.067		194	36		230	268	
	1920	3"	"	13	1.231	▼	286	41.50		327.50	380	
	2000	For 200 lb, add					5%	10%				
	2040	For 300 lb, add					15%	15%				
	2060	Check swing, 300#, sweat, 3/8" size	1 Plum	24	.333	Ea.	13.55	12.45		26	33.50	
	2070	1/2"	▼	24	.333	▼	13.55	12.45		26	33.50	

Important: See the Reference Section for critical supporting data - Reference Nos., Crews, & City Cost Indexes

15110	Valves		CREW	DAILY OUTPUT	LABOR-HOURS	UNIT	2003 BARE COSTS				TOTAL INCL O&P	
							MAT.	LABOR	EQUIP.	TOTAL		
160	2080	3/4"	1 Plum	20	.400	Ea.	17.30	14.95		32.25	41.50	**160**
	2090	1" R15100 -090		19	.421		24	15.75		39.75	50.50	
	2100	1-1/4"		15	.533		31	19.90		50.90	64	
	2110	1-1/2 "		13	.615		41.50	23		64.50	80	
	2120	2"		11	.727		62	27		89	110	
	2130	2-1/2"	Q-1	15	1.067		148	36		184	217	
	2140	3"	"	13	1.231		197	41.50		238.50	280	
	2350	Check, lift, class 150, horizontal composition disc, threaded										
	2430	1/4"	1 Plum	24	.333	Ea.	46.50	12.45		58.95	70	
	2440	3/8"		24	.333		46.50	12.45		58.95	70	
	2450	1/2"		24	.333		51.50	12.45		63.95	75.50	
	2460	3/4"		20	.400		66.50	14.95		81.45	95.50	
	2470	1"		19	.421		89	15.75		104.75	122	
	2480	1-1/4"		15	.533		159	19.90		178.90	205	
	2490	1-1/2"		13	.615		190	23		213	244	
	2500	2"		11	.727		315	27		342	390	
	2850	Gate, N.R.S., soldered, 125 psi										
	2900	3/8"	1 Plum	24	.333	Ea.	18.30	12.45		30.75	39	
	2920	1/2"		24	.333		16	12.45		28.45	36.50	
	2940	3/4"		20	.400		18.30	14.95		33.25	42.50	
	2950	1"		19	.421		26	15.75		41.75	52.50	
	2960	1-1/4"		15	.533		40	19.90		59.90	73.50	
	2970	1-1/2"		13	.615		45	23		68	84	
	2980	2"		11	.727		62.50	27		89.50	110	
	2990	2-1/2"	Q-1	15	1.067		157	36		193	227	
	3000	3"	"	13	1.231		205	41.50		246.50	289	
	3350	Threaded, class 150										
	3410	1/4"	1 Plum	24	.333	Ea.	27	12.45		39.45	48.50	
	3420	3/8"		24	.333		27	12.45		39.45	48.50	
	3430	1/2"		24	.333		25	12.45		37.45	46.50	
	3440	3/4"		20	.400		27.50	14.95		42.45	53	
	3450	1"		19	.421		36	15.75		51.75	63.50	
	3460	1-1/4" size		15	.533		48	19.90		67.90	83	
	3470	1-1/2"		13	.615		61	23		84	102	
	3480	2"		11	.727		83	27		110	132	
	3490	2-1/2"	Q-1	15	1.067		192	36		228	265	
	3500	3" size	"	13	1.231		271	41.50		312.50	360	
	3600	Gate, flanged, 150 lb.										
	3610	1"	1 Plum	7	1.143	Ea.	510	42.50		552.50	625	
	3620	1-1/2"		6	1.333		635	50		685	770	
	3630	2"		5	1.600		935	60		995	1,125	
	3640	3"	Q-1	4.50	3.556		1,700	120		1,820	2,025	
	3850	Rising stem, soldered, 300 psi										
	3900	3/8"	1 Plum	24	.333	Ea.	43	12.45		55.45	66	
	3920	1/2"		24	.333		38.50	12.45		50.95	61	
	3940	3/4"		20	.400		49	14.95		63.95	76	
	3950	1"		19	.421		59.50	15.75		75.25	89	
	3960	1-1/4"		15	.533		88	19.90		107.90	127	
	3970	1-1/2"		13	.615		107	23		130	153	
	3980	2"		11	.727		161	27		188	218	
	3990	2-1/2"	Q-1	15	1.067		390	36		426	480	
	4000	3"	"	13	1.231		455	41.50		496.50	565	
	4250	Threaded, class 150										
	4310	1/4"	1 Plum	24	.333	Ea.	23.50	12.45		35.95	45	
	4320	3/8"		24	.333		23.50	12.45		35.95	45	
	4330	1/2"		24	.333		22	12.45		34.45	43	

MECHANICAL 15

				DAILY	LABOR-			2003 BARE COSTS				TOTAL	
	15110	**Valves**		CREW	OUTPUT	HOURS	UNIT	MAT.	LABOR	EQUIP.	TOTAL	INCL O&P	
160	4340	3/4"	R15100 -090	1 Plum	20	.400	Ea.	26	14.95		40.95	51	**160**
	4350	1"			19	.421		34.50	15.75		50.25	62	
	4360	1-1/4"			15	.533		47	19.90		66.90	81.50	
	4370	1-1/2"			13	.615		58.50	23		81.50	98.50	
	4380	2"			11	.727		79	27		106	128	
	4390	2-1/2"		Q-1	15	1.067		185	36		221	258	
	4400	3"		"	13	1.231		259	41.50		300.50	350	
	4500	For 300 psi, threaded, add						100%	15%				
	4540	For chain operated type, add						15%					
	4850	Globe, class 150, rising stem, threaded											
	4920	1/4"		1 Plum	24	.333	Ea.	34	12.45		46.45	56.50	
	4940	3/8" size			24	.333		34	12.45		46.45	56.50	
	4950	1/2"			24	.333		34	12.45		46.45	56.50	
	4960	3/4"			20	.400		46.50	14.95		61.45	73.50	
	4970	1"			19	.421		72	15.75		87.75	103	
	4980	1-1/4"			15	.533		114	19.90		133.90	155	
	4990	1-1/2"			13	.615		138	23		161	187	
	5000	2"			11	.727		207	27		234	269	
	5010	2-1/2"		Q-1	15	1.067		415	36		451	515	
	5020	3"		"	13	1.231		590	41.50		631.50	715	
	5120	For 300 lb threaded, add						50%	15%				
	5130	Globe, 300 lb., sweat, 3/8" size		1 Plum	24	.333	Ea.	36	12.45		48.45	58.50	
	5140	1/2"			24	.333		35.50	12.45		47.95	58	
	5150	3/4"			20	.400		51	14.95		65.95	78.50	
	5160	1"			19	.421		77	15.75		92.75	109	
	5170	1-1/4"			15	.533		120	19.90		139.90	161	
	5180	1-1/2"			13	.615		142	23		165	191	
	5190	2"			11	.727		212	27		239	275	
	5200	2-1/2"		Q-1	15	1.067		455	36		491	555	
	5210	3"		"	13	1.231		665	41.50		706.50	795	
	5600	Relief, pressure & temperature, self-closing, ASME, threaded											
	5640	3/4"		1 Plum	28	.286	Ea.	71.50	10.65		82.15	95	
	5650	1"			24	.333		140	12.45		152.45	173	
	5660	1-1/4"			20	.400		267	14.95		281.95	315	
	5670	1-1/2"			18	.444		510	16.60		526.60	585	
	5680	2"			16	.500		560	18.70		578.70	645	
	5950	Pressure, poppet type, threaded											
	6000	1/2"		1 Plum	30	.267	Ea.	10.80	9.95		20.75	27	
	6040	3/4"		"	28	.286	"	45	10.65		55.65	65.50	
	6400	Pressure, water, ASME, threaded											
	6440	3/4"		1 Plum	28	.286	Ea.	44.50	10.65		55.15	65	
	6450	1"			24	.333		153	12.45		165.45	187	
	6460	1-1/4"			20	.400		153	14.95		167.95	191	
	6470	1-1/2"			18	.444		211	16.60		227.60	257	
	6480	2"			16	.500		305	18.70		323.70	365	
	6490	2-1/2"			15	.533		965	19.90		984.90	1,075	
	6900	Reducing, water pressure											
	6920	300 psi to 25-75 psi, threaded or sweat											
	6940	1/2"		1 Plum	24	.333	Ea.	119	12.45		131.45	150	
	6950	3/4"			20	.400		119	14.95		133.95	154	
	6960	1"			19	.421		185	15.75		200.75	227	
	6970	1-1/4"			15	.533		330	19.90		349.90	395	
	6980	1-1/2"			13	.615		500	23		523	585	
	6990	2"			11	.727		720	27		747	830	
	7100	For built-in by-pass or 10-35 psi, add							7			7	7.70	
	7700	High capacity, 250 psi to 25-75 psi, threaded												

Important: See the Reference Section for critical supporting data - Reference Nos., Crews, & City Cost Indexes

					DAILY	LABOR-		2003 BARE COSTS				TOTAL
	15110	**Valves**		CREW	OUTPUT	HOURS	UNIT	MAT.	LABOR	EQUIP.	TOTAL	INCL O&P
160	7740	1/2"	R15100	1 Plum	24	.333	Ea.	119	12.45		131.45	150
	7780	3/4"	-090		20	.400		119	14.95		133.95	154
	7790	1"			19	.421		185	15.75		200.75	227
	7800	1-1/4"			15	.533		320	19.90		339.90	385
	7810	1-1/2"			13	.615		500	23		523	585
	7820	2"			11	.727		720	27		747	830
	7830	2-1/2"			9	.889		1,050	33		1,083	1,200
	7840	3"		↓	8	1		1,100	37.50		1,137.50	1,275
	7850	3" flanged (iron body)		Q-1	10	1.600		1,175	54		1,229	1,350
	7860	4" flanged (iron body)		"	8	2	↓	1,600	67.50		1,667.50	1,875
	7920	For higher pressure, add						25%				
	8000	Silent check, bronze trim										
	8010	Compact wafer type, for 125 or 150 lb. flanges										
	8020	1-1/2"		1 Plum	11	.727	Ea.	255	27		282	320
	8021	2"		"	9	.889		274	33		307	350
	8022	2-1/2"		Q-1	9	1.778		288	60		348	405
	8023	3"			8	2		320	67.50		387.50	455
	8024	4"		↓	5	3.200		550	108		658	770
	8025	5"		Q-2	6	4		760	139		899	1,050
	8026	6"			5	4.800		865	167		1,032	1,200
	8027	8"			4.50	5.333		1,850	186		2,036	2,300
	8028	10"			4	6		3,250	209		3,459	3,900
	8029	12"		↓	3	8	↓	7,600	279		7,879	8,775
	8050	For 250 or 300 lb flanges, thru 6" no change										
	8051	For 8" and 10", add					Ea.	40%	10%			
	8060	Full flange wafer type, 150 lb.										
	8061	1"		1 Plum	14	.571	Ea.	272	21.50		293.50	330
	8062	1-1/4"			12	.667		284	25		309	350
	8063	1-1/2"			11	.727		340	27		367	415
	8064	2"		↓	9	.889		460	33		493	555
	8065	2-1/2"		Q-1	9	1.778		520	60		580	660
	8066	3"			8	2		575	67.50		642.50	730
	8067	4"		↓	5	3.200		850	108		958	1,100
	8068	5"		Q-2	6	4		1,100	139		1,239	1,400
	8069	6"		"	5	4.800		1,450	167		1,617	1,825
	8080	For 300 lb., add					↓	40%	10%			
	8100	Globe type, 150 lb.										
	8110	2"		1 Plum	9	.889	Ea.	535	33		568	640
	8111	2-1/2"		Q-1	9	1.778		660	60		720	820
	8112	3"			8	2		790	67.50		857.50	970
	8113	4"		↓	5	3.200		1,100	108		1,208	1,375
	8114	5"		Q-2	6	4		1,350	139		1,489	1,675
	8115	6"		"	5	4.800	↓	1,875	167		2,042	2,300
	8130	For 300 lb., add						20%	10%			
	8140	Screwed end type, 250 lb.										
	8141	1/2"		1 Plum	24	.333	Ea.	48	12.45		60.45	72
	8142	3/4"			20	.400		48	14.95		62.95	75.50
	8143	1"			19	.421		55	15.75		70.75	84.50
	8144	1-1/4"			15	.533		76	19.90		95.90	114
	8145	1-1/2"			13	.615		85	23		108	128
	8146	2"		↓	11	.727	↓	114	27		141	167
	8350	Tempering, water, sweat connections										
	8400	1/2"		1 Plum	24	.333	Ea.	44	12.45		56.45	67.50
	8440	3/4"		"	20	.400	"	54	14.95		68.95	82
	8650	Threaded connections										
	8700	1/2"		1 Plum	24	.333	Ea.	54	12.45		66.45	78.50

160

			DAILY	LABOR-			2003 BARE COSTS			TOTAL
15110	**Valves**	**CREW**	**OUTPUT**	**HOURS**	**UNIT**	**MAT.**	**LABOR**	**EQUIP.**	**TOTAL**	**INCL O&P**

160	8740	3/4"	1 Plum	20	.400	Ea.	207	14.95		221.95	250	**160**
	8750	1"	R15100 -090	19	.421		228	15.75		243.75	274	
	8760	1-1/4"		15	.533		360	19.90		379.90	425	
	8770	1-1/2"		13	.615		395	23		418	470	
	8780	2"		11	.727		595	27		622	690	
200	0010	**VALVES, IRON BODY**										**200**
	0020	For grooved joint, see Division 15107-690	R15100 -090									
	0100	Angle, 125 lb.										
	0110	Flanged										
	0116	2"	1 Plum	5	1.600	Ea.	650	60		710	805	
	0118	4"	Q-1	3	5.333		1,075	179		1,254	1,450	
	0120	6"	Q-2	3	8		2,100	279		2,379	2,725	
	0122	8"	"	2.50	9.600		3,750	335		4,085	4,625	
	0560	Butterfly, lug type, pneumatic operator, 2" size	1 Stpi	14	.571		221	21.50		242.50	276	
	0570	3"	Q-1	8	2		229	67.50		296.50	355	
	0580	4"	"	5	3.200		250	108		358	440	
	0590	6"	Q-2	5	4.800		335	167		502	625	
	0600	8"		4.50	5.333		405	186		591	725	
	0610	10"		4	6		510	209		719	875	
	0620	12"		3	8		685	279		964	1,175	
	0630	14"		2.30	10.435		985	365		1,350	1,625	
	0640	18"		1.50	16		1,700	560		2,260	2,725	
	0650	20"		1	24		2,050	835		2,885	3,525	
	0800	Butterfly, lug type, gear operated, 2-1/2" size, 200 lb. except noted	Q-1	9	1.778		182	60		242	291	
	0810	3"		8	2		191	67.50		258.50	310	
	0820	4"		5	3.200		216	108		324	400	
	0830	5"	Q-2	5	4.800		269	167		436	550	
	0840	6"		5	4.800		298	167		465	585	
	0850	8"		4.50	5.333		385	186		571	705	
	0860	10"		4	6		510	209		719	875	
	0870	12"		3	8		740	279		1,019	1,225	
	0880	14", 150 lb.		2.30	10.435		815	365		1,180	1,450	
	0890	16", 150 lb.		1.75	13.714		1,150	480		1,630	2,000	
	0900	18", 150 lb.		1.50	16		1,500	560		2,060	2,500	
	0910	20", 150 lb.		1	24		1,775	835		2,610	3,225	
	0930	24", 150 lb.		.75	32		2,475	1,125		3,600	4,400	
	1020	Butterfly, wafer type, gear actuator, 200 lb.										
	1030	2"	1 Plum	14	.571	Ea.	109	21.50		130.50	151	
	1040	2-1/2"	Q-1	9	1.778		112	60		172	214	
	1050	3"		8	2		116	67.50		183.50	230	
	1060	4"		5	3.200		145	108		253	325	
	1070	5"	Q-2	5	4.800		175	167		342	445	
	1080	6"		5	4.800		198	167		365	470	
	1090	8"		4.50	5.333		264	186		450	570	
	1100	10"		4	6		330	209		539	680	
	1110	12"		3	8		460	279		739	925	
	1200	Wafer type, lever actuator, 200 lb.										
	1220	2"	1 Plum	14	.571	Ea.	79.50	21.50		101	120	
	1230	2-1/2"	Q-1	9	1.778		82.50	60		142.50	182	
	1240	3"		8	2		87	67.50		154.50	198	
	1250	4"		5	3.200		107	108		215	280	
	1260	5"	Q-2	5	4.800		148	167		315	415	
	1270	6"		5	4.800		178	167		345	450	
	1280	8"		4.50	5.333		263	186		449	570	
	1290	10"		4	6		475	209		684	835	

Important: See the Reference Section for critical supporting data - Reference Nos., Crews, & City Cost Indexes

				DAILY	LABOR-		2003 BARE COSTS				TOTAL	
15110	**Valves**		CREW	OUTPUT	HOURS	UNIT	MAT.	LABOR	EQUIP.	TOTAL	INCL O&P	
200	1300	12"	R15100	Q-2	3	8	Ea.	670	279		949	1,150	**200**
	1400	Diverter, 150 lb. flanged, bronze or iron plugs	-090										
	1440	2"		Q-1	2	8	Ea.	2,725	269		2,994	3,400	
	1450	3"		"	1.50	10.667	"	3,900	360		4,260	4,850	
	1600	Gate, threaded, 125 lb.											
	1604	1-1/2"		1 Plum	13	.615	Ea.	218	23		241	274	
	1605	2"			11	.727		605	27		632	705	
	1606	3"			8	1		705	37.50		742.50	830	
	1607	4"			5	1.600		1,100	60		1,160	1,300	
	1650	Gate, 125 lb., N.R.S.											
	2150	Flanged											
	2200	2"		1 Plum	5	1.600	Ea.	296	60		356	415	
	2240	2-1/2"		Q-1	5	3.200		305	108		413	500	
	2260	3"			4.50	3.556		340	120		460	555	
	2280	4"			3	5.333		485	179		664	805	
	2290	5"		Q-2	3.40	7.059		830	246		1,076	1,275	
	2300	6"			3	8		830	279		1,109	1,325	
	2320	8"			2.50	9.600		1,425	335		1,760	2,075	
	2340	10"			2.20	10.909		2,525	380		2,905	3,350	
	2360	12"			1.70	14.118		3,450	490		3,940	4,525	
	2370	14"			1.30	18.462		5,825	645		6,470	7,400	
	2380	16"			1	24		8,150	835		8,985	10,200	
	2420	For 250 lb flanged, add						200%	10%				
	3500	OS&Y, 125 lb., (225# noted), threaded											
	3504	3/4" (225 lb.)		1 Plum	18	.444	Ea.	385	16.60		401.60	450	
	3505	1" (225 lb.)			16	.500		440	18.70		458.70	510	
	3506	1-1/2" (225 lb.)			13	.615		640	23		663	735	
	3507	2"			11	.727		540	27		567	635	
	3508	2-1/2"			9	.889		560	33		593	665	
	3510	4"			7	1.143		900	42.50		942.50	1,050	
	3550	OS&Y, 125 lb., flanged											
	3600	2"		1 Plum	5	1.600	Ea.	221	60		281	335	
	3640	2-1/2"		Q-1	5	3.200		230	108		338	415	
	3660	3"			4.50	3.556		257	120		377	465	
	3670	3-1/2"			3	5.333		410	179		589	720	
	3680	4"			3	5.333		365	179		544	675	
	3690	5"		Q-2	3.40	7.059		605	246		851	1,025	
	3700	6"			3	8		605	279		884	1,075	
	3720	8"			2.50	9.600		1,075	335		1,410	1,675	
	3740	10"			2.20	10.909		1,975	380		2,355	2,725	
	3760	12"			1.70	14.118		2,625	490		3,115	3,650	
	3770	14"			1.30	18.462		5,000	645		5,645	6,475	
	3780	16"			1	24		7,500	835		8,335	9,525	
	3790	18"			.80	30		10,100	1,050		11,150	12,700	
	3800	20"			.60	40		14,000	1,400		15,400	17,500	
	3830	24"			.50	48		20,900	1,675		22,575	25,400	
	3900	For 175 lb, flanged, add						200%	10%				
	4350	Globe, OS&Y											
	4540	Class 125, flanged											
	4550	2"		1 Plum	5	1.600	Ea.	455	60		515	590	
	4560	2-1/2"		Q-1	5	3.200		480	108		588	695	
	4570	3"			4.50	3.556		555	120		675	790	
	4580	4"			3	5.333		795	179		974	1,150	
	4590	5"		Q-2	3.40	7.059		1,450	246		1,696	1,975	
	4600	6"			3	8		1,450	279		1,729	2,025	
	4610	8"			2.50	9.600		2,850	335		3,185	3,625	

15110	Valves		CREW	DAILY OUTPUT	LABOR-HOURS	UNIT	2003 BARE COSTS				TOTAL INCL O&P
							MAT.	LABOR	EQUIP.	TOTAL	
200 4612	10"	R15100 -090	Q-2	2.20	10.909	Ea.	4,450	380		4,830	5,450 **200**
4614	12"		↓	1.70	14.118	↓	4,850	490		5,340	6,075
5040	Class 250, flanged										
5050	2"		1 Plum	4.50	1.778	Ea.	630	66.50		696.50	795
5060	2-1/2"		Q-1	4.50	3.556		820	120		940	1,075
5070	3"			4	4		850	135		985	1,150
5080	4"		↓	2.70	5.926		1,250	199		1,449	1,675
5090	5"		Q-2	3	8		2,225	279		2,504	2,875
5100	6"			2.70	8.889		2,225	310		2,535	2,925
5110	8"			2.20	10.909		3,775	380		4,155	4,725
5120	10"			2	12		5,550	420		5,970	6,725
5130	12"		↓	1.60	15		9,900	525		10,425	11,700
5240	Valve sprocket rim w/chain, for 2" valve		1 Stpi	30	.267		71.50	10.05		81.55	93.50
5250	2-1/2" valve			27	.296		71.50	11.15		82.65	95.50
5260	3-1/2" valve			25	.320		71.50	12.05		83.55	96.50
5270	6" valve			20	.400		71.50	15.05		86.55	101
5280	8" valve			18	.444		108	16.70		124.70	144
5290	12" valve			16	.500		108	18.80		126.80	148
5300	16" valve			12	.667		147	25		172	199
5310	20" valve			10	.800		147	30		177	207
5320	36" valve		↓	8	1	↓	272	37.50		309.50	355
5450	Swing check, 125 lb., threaded										
5500	2"		1 Plum	11	.727	Ea.	385	27		412	465
5540	2-1/2"		Q-1	15	1.067		455	36		491	555
5550	3"			13	1.231		510	41.50		551.50	625
5560	4"		↓	10	1.600	↓	815	54		869	975
5950	Flanged										
5994	1"		1 Plum	7	1.143	Ea.	82.50	42.50		125	156
5998	1-1/2"			6	1.333		120	50		170	206
6000	2"			5	1.600		153	60		213	258
6040	2-1/2"		Q-1	5	3.200		185	108		293	365
6050	3"			4.50	3.556		198	120		318	400
6060	4"		↓	3	5.333		310	179		489	615
6070	6"		Q-2	3	8		535	279		814	1,000
6080	8"			2.50	9.600		1,000	335		1,335	1,600
6090	10"			2.20	10.909		1,700	380		2,080	2,450
6100	12"			1.70	14.118		2,675	490		3,165	3,675
6110	18"			1.30	18.462		18,000	645		18,645	20,800
6114	24"		↓	.75	32	↓	33,400	1,125		34,525	38,400
6160	For 250 lb flanged, add						200%	20%			
6600	Silent check, bronze trim										
6610	Compact wafer type, for 125 or 150 lb. flanges										
6630	1-1/2"		1 Plum	11	.727	Ea.	118	27		145	170
6640	2"		"	9	.889		140	33		173	204
6650	2-1/2"		Q-1	9	1.778		154	60		214	260
6660	3"			8	2		165	67.50		232.50	283
6670	4"		↓	5	3.200		210	108		318	395
6680	5"		Q-2	6	4		275	139		414	515
6690	6"			6	4		370	139		509	615
6700	8"			4.50	5.333		660	186		846	1,000
6710	10"			4	6		1,150	209		1,359	1,575
6720	12"		↓	3	8		2,175	279		2,454	2,825
6740	For 250 or 300 lb. flanges, thru 6" no change										
6741	For 8" and 10", add					Ea.	11%	10%			
6750	Twin disc										
6752	2"		1 Plum	9	.889	Ea.	168	33		201	235

Important: See the Reference Section for critical supporting data - Reference Nos., Crews, & City Cost Indexes

15110 | Valves

			CREW	DAILY OUTPUT	LABOR-HOURS	UNIT	MAT.	LABOR	EQUIP.	TOTAL	TOTAL INCL O&P	
200	6754	4"	Q-1	5	3.200	Ea.	282	108		390	475	**200**
	6756	6"	Q-2	5	4.800		415	167		582	710	
	6758	8"		4.50	5.333		645	186		831	985	
	6760	10"		4	6		1,025	209		1,234	1,450	
	6762	12"		3	8		1,250	279		1,529	1,775	
	6764	18"		1.50	16		5,050	560		5,610	6,400	
	6766	24"		.75	32		7,475	1,125		8,600	9,900	
	6800	Full flange type, 150 lb.										
	6810	1"	1 Plum	14	.571	Ea.	138	21.50		159.50	183	
	6811	1-1/4"		12	.667		145	25		170	197	
	6812	1-1/2"		11	.727		168	27		195	226	
	6813	2"		9	.889		193	33		226	262	
	6814	2-1/2"	Q-1	9	1.778		234	60		294	350	
	6815	3"		8	2		242	67.50		309.50	370	
	6816	4"		5	3.200		299	108		407	495	
	6817	5"	Q-2	6	4		470	139		609	730	
	6818	6"		5	4.800		620	167		787	935	
	6819	8"		4.50	5.333		1,050	186		1,236	1,425	
	6820	10"		4	6		1,325	209		1,534	1,775	
	6840	For 250 lb., add					30%	10%				
	6900	Globe type, 125 lb.										
	6910	2"	1 Plum	9	.889	Ea.	267	33		300	345	
	6911	2-1/2"	Q-1	9	1.778		294	60		354	415	
	6912	3"		8	2		315	67.50		382.50	450	
	6913	4"		5	3.200		425	108		533	635	
	6914	5"	Q-2	6	4		550	139		689	815	
	6915	6"		5	4.800		685	167		852	1,000	
	6916	8"		4.50	5.333		1,250	186		1,436	1,650	
	6917	10"		4	6		1,550	209		1,759	2,025	
	6918	12"		3	8		2,625	279		2,904	3,325	
	6919	14"		2.30	10.435		3,600	365		3,965	4,500	
	6920	16"		1.75	13.714		5,175	480		5,655	6,425	
	6921	18"		1.50	16		7,475	560		8,035	9,075	
	6922	20"		1	24		9,075	835		9,910	11,300	
	6923	24"		.75	32		13,200	1,125		14,325	16,300	
	6940	For 250 lb., add					40%	10%				
	6980	Screwed end type, 125 lb.										
	6981	1"	1 Plum	19	.421	Ea.	50.50	15.75		66.25	79.50	
	6982	1-1/4"		15	.533		65.50	19.90		85.40	102	
	6983	1-1/2"		13	.615		81	23		104	124	
	6984	2"		11	.727		110	27		137	162	
400	0010	**MULTIPURPOSE VALVES**										**400**
	0100	Functions as a shut off, balancing, check & metering valve										
	1000	Cast iron body										
	1010	Threaded										
	1020	1-1/2" size	1 Stpi	11	.727	Ea.	288	27.50		315.50	355	
	1030	2" size	"	8	1		350	37.50		387.50	440	
	1040	2-1/2" size	Q-5	5	3.200		380	108		488	580	
	1050	3" size	"	4.50	3.556		400	120		520	620	
	1200	Flanged										
	1210	3" size	Q-5	4.20	3.810	Ea.	435	129		564	675	
	1220	4" size	"	3	5.333		880	181		1,061	1,250	
	1230	5" size	Q-6	3.80	6.316		1,050	222		1,272	1,475	
	1240	6" size		3	8		1,300	281		1,581	1,850	
	1250	8" size		2.50	9.600		1,900	335		2,235	2,600	

R15100 -090

MECHANICAL **15**

			DAILY	LABOR-		2003 BARE COSTS				TOTAL	
15110	**Valves**	CREW	OUTPUT	HOURS	UNIT	MAT.	LABOR	EQUIP.	TOTAL	INCL O&P	
400	1260	10" size	Q-6	2.20	10.909	Ea.	2,725	385		3,110	3,575	**400**
	1270	12" size		2.10	11.429		7,450	400		7,850	8,775	
	1280	14" size	↓	2	12	↓	10,500	420		10,920	12,200	
500	0010	**VALVES, PLASTIC** R15100-090										**500**
	1100	Angle, PVC, threaded										
	1110	1/4"	1 Plum	26	.308	Ea.	49.50	11.50		61	72	
	1120	1/2"		26	.308		49.50	11.50		61	72	
	1130	3/4"		25	.320		58.50	11.95		70.45	82.50	
	1140	1"	↓	23	.348	↓	70	13		83	96.50	
	1150	Ball, PVC, socket or threaded, single union										
	1200	1/4"	1 Plum	26	.308	Ea.	33.50	11.50		45	54	
	1220	3/8"		26	.308		33.50	11.50		45	54	
	1230	1/2"		26	.308		33.50	11.50		45	54	
	1240	3/4"		25	.320		40	11.95		51.95	61.50	
	1250	1"		23	.348		48	13		61	72.50	
	1260	1-1/4"		21	.381		79.50	14.25		93.75	109	
	1270	1-1/2"		20	.400		79.50	14.95		94.45	110	
	1280	2"	↓	17	.471		114	17.60		131.60	152	
	1290	2-1/2"	Q-1	26	.615		285	20.50		305.50	345	
	1300	3"		24	.667		285	22.50		307.50	350	
	1310	4"	↓	20	.800		385	27		412	465	
	1360	For PVC, flanged, add					100%	15%				
	1450	Double union 1/2"	1 Plum	26	.308		13.80	11.50		25.30	32.50	
	1460	3/4"		25	.320		16.20	11.95		28.15	36	
	1470	1"		23	.348		17.75	13		30.75	39	
	1480	1-1/4"		21	.381		27	14.25		41.25	51	
	1490	1-1/2"		20	.400		30	14.95		44.95	55.50	
	1500	2"	↓	17	.471	↓	42.50	17.60		60.10	73	
	1650	CPVC, socket or threaded, single union										
	1700	1/2"	1 Plum	26	.308	Ea.	33.50	11.50		45	54	
	1720	3/4"		25	.320		42	11.95		53.95	64	
	1730	1"		23	.348		50	13		63	74.50	
	1750	1-1/4"		21	.381		83.50	14.25		97.75	114	
	1760	1-1/2"		20	.400		83.50	14.95		98.45	115	
	1770	2"	↓	17	.471		116	17.60		133.60	155	
	1780	3"	Q-1	24	.667		345	22.50		367.50	415	
	1840	For CPVC, flanged, add					65%	15%				
	1880	For true union, socket or threaded, add				↓	50%	5%				
	2050	Polypropylene, threaded										
	2100	1/4"	1 Plum	26	.308	Ea.	33	11.50		44.50	54	
	2120	3/8"		26	.308		33	11.50		44.50	54	
	2130	1/2"		26	.308		33	11.50		44.50	54	
	2140	3/4"		25	.320		41	11.95		52.95	63.50	
	2150	1"		23	.348		49	13		62	73.50	
	2160	1-1/4"		21	.381		71	14.25		85.25	99.50	
	2170	1-1/2"		20	.400		82	14.95		96.95	113	
	2180	2"	↓	17	.471		111	17.60		128.60	149	
	2190	3"	Q-1	24	.667		294	22.50		316.50	360	
	2200	4"	"	20	.800	↓	490	27		517	580	
	2550	PVC, three way, socket or threaded										
	2600	1/2"	1 Plum	26	.308	Ea.	44.50	11.50		56	66.50	
	2640	3/4"		25	.320		50.50	11.95		62.45	73.50	
	2650	1"		23	.348		54.50	13		67.50	79.50	
	2660	1-1/2"		20	.400		111	14.95		125.95	145	
	2670	2"	↓	17	.471		148	17.60		165.60	190	

Important: See the Reference Section for critical supporting data - Reference Nos., Crews, & City Cost Indexes

15110	Valves		CREW	DAILY OUTPUT	LABOR-HOURS	UNIT	2003 BARE COSTS				TOTAL INCL O&P
							MAT.	LABOR	EQUIP.	TOTAL	
500 2680	3"	R15100 -090	Q-1	24	.667	Ea.	320	22.50		342.50	390 **500**
2740	For flanged, add					↓	60%	15%			
3150	Ball check, PVC, socket or threaded										
3200	1/4"		1 Plum	26	.308	Ea.	29	11.50		40.50	49.50
3220	3/8"			26	.308		29	11.50		40.50	49.50
3240	1/2"			26	.308		29	11.50		40.50	49.50
3250	3/4"			25	.320		32.50	11.95		44.45	53.50
3260	1"			23	.348		40.50	13		53.50	64.50
3270	1-1/4"			21	.381		68	14.25		82.25	96.50
3280	1-1/2"			20	.400		68	14.95		82.95	97.50
3290	2"		↓	17	.471		93	17.60		110.60	129
3310	3"		Q-1	24	.667		257	22.50		279.50	315
3320	4"		"	20	.800		365	27		392	440
3360	For PVC, flanged, add					↓	50%	15%			
3750	CPVC, socket or threaded										
3800	1/2"		1 Plum	26	.308	Ea.	42.50	11.50		54	64
3840	3/4"			25	.320		50.50	11.95		62.45	73.50
3850	1"			23	.348		60	13		73	85
3860	1-1/2"			20	.400		103	14.95		117.95	136
3870	2"		↓	17	.471		138	17.60		155.60	179
3880	3"		Q-1	24	.667		375	22.50		397.50	445
3920	4"		"	20	.800		505.	27		532	595
3930	For CPVC, flanged, add					↓	40%	15%			
4340	Polypropylene, threaded										
4360	1/2"		1 Plum	26	.308	Ea.	28	11.50		39.50	48
4400	3/4"			25	.320		32	11.95		43.95	53.50
4440	1"			23	.348		41	13		54	65
4450	1-1/2"			20	.400		80	14.95		94.95	111
4460	2"		↓	17	.471		101	17.60		118.60	138
4500	For polypropylene flanged, add					↓	200%	15%			
4850	Foot valve, PVC, socket or threaded										
4900	1/2"		1 Plum	34	.235	Ea.	45	8.80		53.80	63
4930	3/4"			32	.250		51	9.35		60.35	70
4940	1"			28	.286		66.50	10.65		77.15	89
4950	1-1/4"			27	.296		127	11.05		138.05	156
4960	1-1/2"			26	.308		127	11.50		138.50	156
4970	2"			24	.333		147	12.45		159.45	181
4980	3"			20	.400		350	14.95		364.95	410
4990	4"		↓	18	.444		615	16.60		631.60	705
5000	For flanged, add					↓	25%	10%			
5050	CPVC, socket or threaded										
5060	1/2"		1 Plum	34	.235	Ea.	93.50	8.80		102.30	116
5070	3/4"			32	.250		114	9.35		123.35	139
5080	1"			28	.286		134	10.65		144.65	163
5090	1-1/4"			27	.296		218	11.05		229.05	257
5100	1-1/2"			26	.308		218	11.50		229.50	257
5110	2"			24	.333		280	12.45		292.45	330
5120	3"			20	.400		670	14.95		684.95	765
5130	4"		↓	18	.444		1,150	16.60		1,166.60	1,275
5140	For flanged, add					↓	25%	10%			
5280	Needle valve, PVC, threaded										
5300	1/4"		1 Plum	26	.308	Ea.	35	11.50		46.50	56
5340	3/8"			26	.308		41	11.50		52.50	62.50
5360	1/2"		↓	26	.308		41	11.50		52.50	62.50
5380	For polypropylene, add					↓	10%				
5800	Y check, PVC, socket or threaded		↓								

MECHANICAL **15**

			CREW	DAILY OUTPUT	LABOR-HOURS	UNIT	2003 BARE COSTS				TOTAL INCL O&P	
	15110	**Valves**					MAT.	LABOR	EQUIP.	TOTAL		
500	5820	1/2"	1 Plum	26	.308	Ea.	55	11.50		66.50	78	**500**
	5840	3/4"		25	.320		59.50	11.95		71.45	83.50	
	5850	1"		23	.348		65	13		78	91	
	5860	1-1/4"		21	.381		102	14.25		116.25	134	
	5870	1-1/2"		20	.400		111	14.95		125.95	145	
	5880	2"		17	.471		138	17.60		155.60	178	
	5890	2-1/2"		15	.533		291	19.90		310.90	350	
	5900	3"	Q-1	24	.667		273	22.50		295.50	335	
	5910	4"	"	20	.800		475	27		502	565	
	5960	For PVC flanged, add					45%	15%				
	6350	Y sediment strainer, PVC, socket or threaded										
	6400	1/2"	1 Plum	26	.308	Ea.	32.50	11.50		44	53	
	6440	3/4"		24	.333		35.50	12.45		47.95	58	
	6450	1"		23	.348		43	13		56	67	
	6460	1-1/4"		21	.381		71	14.25		85.25	99.50	
	6470	1-1/2"		20	.400		71	14.95		85.95	101	
	6480	2"		17	.471		86.50	17.60		104.10	122	
	6490	2-1/2"		15	.533		212	19.90		231.90	263	
	6500	3"	Q-1	24	.667		212	22.50		234.50	267	
	6510	4"	"	20	.800		355	27		382	430	
	6560	For PVC, flanged, add					55%	15%				
600	0010	**VALVES, SEMI-STEEL**										**600**
	1020	Lubricated plug valve, threaded, 200 psi										
	1030	1/2"	1 Plum	18	.444	Ea.	61.50	16.60		78.10	92.50	
	1040	3/4"		16	.500		61.50	18.70		80.20	95.50	
	1050	1"		14	.571		79	21.50		100.50	119	
	1060	1-1/4"		12	.667		94.50	25		119.50	142	
	1070	1-1/2"		11	.727		102	27		129	153	
	1080	2"		8	1		120	37.50		157.50	189	
	1090	2-1/2"	Q-1	5	3.200		186	108		294	370	
	1100	3"	"	4.50	3.556		228	120		348	430	
	6990	Flanged, 200 psi										
	7000	2"	1 Plum	8	1	Ea.	145	37.50		182.50	216	
	7010	2-1/2"	Q-1	5	3.200		218	108		326	400	
	7020	3"		4.50	3.556		263	120		383	470	
	7030	4"		3	5.333		335	179		514	635	
	7036	5"		2.50	6.400		500	215		715	875	
	7040	6"	Q-2	3	8		655	279		934	1,150	
	7050	8"		2.50	9.600		1,175	335		1,510	1,800	
	7060	10"		2.20	10.909		1,725	380		2,105	2,475	
	7070	12"		1.70	14.118		2,975	490		3,465	4,000	
700	0010	**VALVES, STEEL**										**700**
	0800	Cast										
	1350	Check valve, swing type, 150 lb., flanged										
	1400	2"	1 Plum	8	1	Ea.	465	37.50		502.50	570	
	1440	2-1/2"	Q-1	5	3.200		505	108		613	720	
	1450	3"		4.50	3.556		595	120		715	835	
	1460	4"		3	5.333		875	179		1,054	1,225	
	1470	6"	Q-2	3	8		1,350	279		1,629	1,925	
	1480	8"		2.50	9.600		2,225	335		2,560	2,950	
	1490	10"		2.20	10.909		3,300	380		3,680	4,200	
	1500	12"		1.70	14.118		4,725	490		5,215	5,950	
	1510	14"		1.30	18.462		6,750	645		7,395	8,400	
	1520	16"		1	24		8,225	835		9,060	10,300	
	1548	For 600 lb., flanged, add					110%	20%				

R15100-090

Important: See the Reference Section for critical supporting data - Reference Nos., Crews, & City Cost Indexes

15110 | Valves

		CREW	DAILY OUTPUT	LABOR-HOURS	UNIT	2003 BARE COSTS MAT.	LABOR	EQUIP.	TOTAL	TOTAL INCL O&P	
700	**1571**	300 lb., 2"	1 Plum	7.40	1.081	Ea.	545	40.50		585.50	660	**700**
	1572	2-1/2"	Q-1	4.20	3.810		825	128		953	1,100	
	1573	3"		4	4		825	135		960	1,100	
	1574	4"		2.80	5.714		1,175	192		1,367	1,575	
	1575	6"	Q-2	2.90	8.276		2,100	289		2,389	2,750	
	1576	8"		2.40	10		3,475	350		3,825	4,325	
	1577	10"		2.10	11.429		4,400	400		4,800	5,450	
	1578	12"		1.60	15		5,975	525		6,500	7,375	
	1579	14"		1.20	20		9,725	695		10,420	11,800	
	1581	16"		.90	26.667		13,500	930		14,430	16,200	
	1700	Diverter, 150 lb., flanged, bronze or iron plugs										
	1720	1-1/2" pipe size	Q-1	3	5.333	Ea.	2,450	179		2,629	2,975	
	1730	2" pipe size		2	8		2,775	269		3,044	3,450	
	1740	3" pipe size		1.50	10.667		3,950	360		4,310	4,900	
	1750	4" pipe size		1	16		6,025	540		6,565	7,425	
	1760	6" pipe size	Q-2	1	24		11,500	835		12,335	14,000	
	1790	For 300 lb., add					20%	5%				
	1810	Diverter, 150 lb., flanged, stainless steel plugs										
	1830	1-1/2" pipe size	Q-1	3	5.333	Ea.	2,400	179		2,579	2,925	
	1840	2" pipe size		2	8		3,200	269		3,469	3,925	
	1850	3" pipe size		1.50	10.667		4,325	360		4,685	5,300	
	1860	4" pipe size		1	16		6,525	540		7,065	7,975	
	1870	6" pipe size	Q-2	1	24		13,200	835		14,035	15,800	
	1890	For 300 lb., add					10%	5%				
	1950	Gate valve, 150 lb., flanged										
	2000	2"	1 Plum	8	1	Ea.	540	37.50		577.50	650	
	2040	2-1/2"	Q-1	5	3.200		765	108		873	1,000	
	2050	3"		4.50	3.556		765	120		885	1,025	
	2060	4"		3	5.333		930	179		1,109	1,300	
	2070	6"	Q-2	3	8		1,475	279		1,754	2,050	
	2080	8"		2.50	9.600		2,350	335		2,685	3,100	
	2090	10"		2.20	10.909		3,700	380		4,080	4,650	
	2100	12"		1.70	14.118		5,050	490		5,540	6,300	
	2110	14"		1.30	18.462		8,400	645		9,045	10,200	
	2120	16"		1	24		10,200	835		11,035	12,500	
	2130	18"		.80	30		11,300	1,050		12,350	14,000	
	2140	20"		.60	40		13,100	1,400		14,500	16,500	
	2650	300 lb., flanged										
	2700	2"	1 Plum	7.40	1.081	Ea.	740	40.50		780.50	870	
	2740	2-1/2"	Q-1	4.20	3.810		995	128		1,123	1,300	
	2750	3"		4	4		995	135		1,130	1,300	
	2760	4"		2.80	5.714		1,375	192		1,567	1,825	
	2770	6"	Q-2	2.90	8.276		2,350	289		2,639	3,000	
	2780	8"		2.40	10		3,675	350		4,025	4,550	
	2790	10"		2.10	11.429		5,425	400		5,825	6,575	
	2800	12"		1.60	15		7,200	525		7,725	8,725	
	2810	14"		1.20	20		11,300	695		11,995	13,600	
	2820	16"		.90	26.667		14,200	930		15,130	17,000	
	2830	18"		.70	34.286		17,800	1,200		19,000	21,300	
	2840	20"		.50	48		21,500	1,675		23,175	26,200	
	3650	Globe valve, 150 lb., flanged										
	3700	2"	1 Plum	8	1	Ea.	675	37.50		712.50	800	
	3740	2-1/2"	Q-1	5	3.200		850	108		958	1,100	
	3750	3"		4.50	3.556		850	120		970	1,125	
	3760	4"		3	5.333		1,250	179		1,429	1,650	
	3770	6"	Q-2	3	8		1,950	279		2,229	2,575	

R15100 -090

MECHANICAL 15

15110 | Valves

		CREW	DAILY OUTPUT	LABOR-HOURS	UNIT	MAT.	LABOR	EQUIP.	TOTAL	TOTAL INCL O&P	
							2003 BARE COSTS				
700	3780	8"	Q-2	2.50	9.600	Ea.	3,625	335		3,960	4,475	**700**
	3790	10"		2.20	10.909		8,925	380		9,305	10,400	
	3800	12"		1.70	14.118		11,400	490		11,890	13,200	
	4080	300 lb., flanged										
	4100	2"	1 Plum	7.40	1.081	Ea.	935	40.50		975.50	1,075	
	4140	2-1/2"	Q-1	4.20	3.810		1,250	128		1,378	1,600	
	4150	3"		4	4		1,250	135		1,385	1,600	
	4160	4"		2.80	5.714		1,750	192		1,942	2,225	
	4170	6"	Q-2	2.90	8.276		3,150	289		3,439	3,875	
	4180	8"		2.40	10		5,175	350		5,525	6,225	
	4190	10"		2.10	11.429		12,700	400		13,100	14,500	
	4200	12"		1.60	15		14,900	525		15,425	17,200	
	4680	600 lb., flanged										
	4700	2"	1 Plum	7	1.143	Ea.	1,325	42.50		1,367.50	1,525	
	4740	2-1/2"	Q-1	4	4		1,975	135		2,110	2,375	
	4750	3"		3.60	4.444		1,975	149		2,124	2,400	
	4760	4"		2.50	6.400		3,200	215		3,415	3,825	
	4770	6"	Q-2	2.60	9.231		6,600	320		6,920	7,750	
	4780	8"	"	2.10	11.429		10,600	400		11,000	12,300	
	4800	Silent check, 316 S.S. trim										
	4810	Full flange type, 150 lb.										
	4811	1"	1 Plum	14	.571	Ea.	305	21.50		326.50	365	
	4812	1-1/4"		12	.667		305	25		330	375	
	4813	1-1/2"		11	.727		335	27		362	405	
	4814	2"		9	.889		400	33		433	495	
	4815	2-1/2"	Q-1	9	1.778		485	60		545	625	
	4816	3"		8	2		565	67.50		632.50	720	
	4817	4"		5	3.200		770	108		878	1,000	
	4818	5"	Q-2	6	4		965	139		1,104	1,250	
	4819	6"		5	4.800		1,250	167		1,417	1,625	
	4820	8"		4.50	5.333		1,800	186		1,986	2,250	
	4821	10"		4	6		2,550	209		2,759	3,150	
	4840	For 300 lb., add					20%	10%				
	4860	For 600 lb., add					50%	15%				
	4900	Globe type, 150 lb.										
	4910	2"	1 Plum	9	.889	Ea.	465	33		498	565	
	4911	2-1/2"	Q-1	9	1.778		575	60		635	720	
	4912	3"		8	2		640	67.50		707.50	805	
	4913	4"		5	3.200		850	108		958	1,100	
	4914	5"	Q-2	6	4		1,075	139		1,214	1,375	
	4915	6"		5	4.800		1,325	167		1,492	1,700	
	4916	8"		4.50	5.333		1,850	186		2,036	2,300	
	4917	10"		4	6		3,200	209		3,409	3,825	
	4918	12"		3	8		3,925	279		4,204	4,725	
	4919	14"		2.30	10.435		5,925	365		6,290	7,075	
	4920	16"		1.75	13.714		8,275	480		8,755	9,850	
	4921	18"		1.50	16		8,700	560		9,260	10,400	
	4940	For 300 lb., add					30%	10%				
	4960	For 600 lb., add					60%	15%				
	5150	Forged										
	5340	Ball valve, 800 psi, threaded, 1/4" size	1 Plum	24	.333	Ea.	35	12.45		47.45	57.50	
	5350	3/8"		24	.333		35	12.45		47.45	57.50	
	5360	1/2"		24	.333		43.50	12.45		55.95	67	
	5370	3/4"		20	.400		58	14.95		72.95	86	
	5380	1"		19	.421		72	15.75		87.75	103	
	5390	1-1/4"		15	.533		96	19.90		115.90	136	

Note at row 3780: R15100 -090

Important: See the Reference Section for critical supporting data - Reference Nos., Crews, & City Cost Indexes

15110 | Valves

		CREW	DAILY OUTPUT	LABOR-HOURS	UNIT	2003 BARE COSTS				TOTAL INCL O&P
						MAT.	LABOR	EQUIP.	TOTAL	
700	5400	1-1/2"	1 Plum	13	.615	Ea.	125	23		148	173	700
	5410	2"		11	.727		166	27		193	223
	5460	Ball valve, 800 lb., socket weld, 1/4" size	Q-15	19	.842		36.50	28.50	4.13	69.13	87.50
	5470	3/8"		19	.842		36.50	28.50	4.13	69.13	87.50
	5480	1/2"		19	.842		43.50	28.50	4.13	76.13	95.50
	5490	3/4"		19	.842		58	28.50	4.13	90.63	111
	5500	1"		15	1.067		72	36	5.25	113.25	139
	5510	1-1/4"		13	1.231		96	41.50	6.05	143.55	175
	5520	1-1/2"		11	1.455		125	49	7.15	181.15	220
	5530	2"		8.50	1.882		166	63.50	9.25	238.75	288
	5550	Ball valve, 150 lb., flanged									
	5560	4"	Q-1	3	5.333	Ea.	415	179		594	730
	5570	6"	Q-2	3	8		1,000	279		1,279	1,525
	5580	8"	"	2.50	9.600		1,850	335		2,185	2,550
	5650	Check valve, class 800, horizontal, socket									
	5652	1/4"	Q-15	19	.842	Ea.	41	28.50	4.13	73.63	92.50
	5654	3/8"		19	.842		41	28.50	4.13	73.63	92.50
	5656	1/2"		19	.842		41	28.50	4.13	73.63	92.50
	5658	3/4"		19	.842		44.50	28.50	4.13	77.13	96.50
	5660	1"		15	1.067		52	36	5.25	93.25	117
	5662	1-1/4"		13	1.231		99.50	41.50	6.05	147.05	178
	5664	1-1/2"		11	1.455		99.50	49	7.15	155.65	191
	5666	2"		8.50	1.882		139	63.50	9.25	211.75	259
	5698	Threaded									
	5700	1/4"	1 Plum	24	.333	Ea.	41	12.45		53.45	64
	5720	3/8"		24	.333		41	12.45		53.45	64
	5730	1/2"		24	.333		41	12.45		53.45	64
	5740	3/4"		20	.400		44.50	14.95		59.45	71.50
	5750	1"		19	.421		52	15.75		67.75	81
	5760	1-1/4"		15	.533		99.50	19.90		119.40	139
	5770	1-1/2"		13	.615		99.50	23		122.50	144
	5780	2"		11	.727		139	27		166	194
	5840	For class 150, flanged, add					100%	15%			
	5860	For class 300, flanged, add					120%	20%			
	6100	Gate, class 800, OS&Y, socket									
	6102	3/8"	Q-15	19	.842	Ea.	29.50	28.50	4.13	62.13	80
	6103	1/2"		19	.842		29.50	28.50	4.13	62.13	80
	6104	3/4"		19	.842		32.50	28.50	4.13	65.13	83
	6105	1"		15	1.067		39.50	36	5.25	80.75	103
	6106	1-1/4"		13	1.231		76.50	41.50	6.05	124.05	153
	6107	1-1/2"		11	1.455		76.50	49	7.15	132.65	166
	6108	2"		8.50	1.882		96	63.50	9.25	168.75	211
	6118	Threaded									
	6120	3/8"	1 Plum	24	.333	Ea.	29.50	12.45		41.95	51.50
	6130	1/2"		24	.333		29.50	12.45		41.95	51.50
	6140	3/4"		20	.400		32.50	14.95		47.45	58
	6150	1"		19	.421		39.50	15.75		55.25	67.50
	6160	1-1/4"		15	.533		76.50	19.90		96.40	114
	6170	1-1/2"		13	.615		76.50	23		99.50	119
	6180	2"		11	.727		96	27		123	146
	6260	For OS&Y, flanged, add					100%	20%			
	6700	Globe, OS&Y, class 800, socket									
	6710	1/4"	Q-15	19	.842	Ea.	49.50	28.50	4.13	82.13	102
	6720	3/8"		19	.842		49.50	28.50	4.13	82.13	102
	6730	1/2"		19	.842		49.50	28.50	4.13	82.13	102
	6740	3/4"		19	.842		52.50	28.50	4.13	85.13	106

R15100 -090

MECHANICAL 15

			DAILY	LABOR-			2003 BARE COSTS				TOTAL	
15110	**Valves**	CREW	OUTPUT	HOURS	UNIT	MAT.	LABOR	EQUIP.	TOTAL	INCL O&P	

700	6750	1"	R15100 -090	Q-15	15	1.067	Ea.	66	36	5.25	107.25	133	**700**
	6760	1-1/4"			13	1.231		126	41.50	6.05	173.55	208	
	6770	1-1/2"			11	1.455		126	49	7.15	182.15	221	
	6780	2"			8.50	1.882		162	63.50	9.25	234.75	284	
	6860	For OS&Y, flanged, add						300%	20%				
	6880	Threaded											
	6882	1/4"		1 Plum	24	.333	Ea.	49.50	12.45		61.95	73.50	
	6884	3/8"			24	.333		49.50	12.45		61.95	73.50	
	6886	1/2"			24	.333		49.50	12.45		61.95	73.50	
	6888	3/4"			20	.400		52.50	14.95		67.45	80.50	
	6890	1"			19	.421		66	15.75		81.75	97	
	6892	1-1/4"			15	.533		126	19.90		145.90	169	
	6894	1-1/2"			13	.615		126	23		149	174	
	6896	2"			11	.727		162	27		189	219	
800	0010	**VALVES, STAINLESS STEEL**	R15100 -090										**800**
	1610	Ball, threaded 1/4"		1 Stpi	24	.333	Ea.	29.50	12.55		42.05	51	
	1620	3/8"			24	.333		29.50	12.55		42.05	51	
	1630	1/2"			22	.364		29.50	13.65		43.15	52.50	
	1640	3/4"			20	.400		48.50	15.05		63.55	76	
	1650	1"			19	.421		59	15.85		74.85	89	
	1660	1-1/4			15	.533		113	20		133	155	
	1670	1-1/2"			13	.615		118	23		141	165	
	1680	2"			11	.727		153	27.50		180.50	210	
	1700	Check, 200 lb., threaded											
	1710	1/4"		1 Plum	24	.333	Ea.	72	12.45		84.45	98	
	1720	1/2"			22	.364		72	13.60		85.60	99.50	
	1730	3/4"			20	.400		79	14.95		93.95	110	
	1750	1"			19	.421		101	15.75		116.75	135	
	1760	1-1/2"			13	.615		191	23		214	245	
	1770	2"			11	.727		320	27		347	390	
	1800	150 lb., flanged											
	1810	2-1/2"		Q-1	5	3.200	Ea.	765	108		873	1,000	
	1820	3"			4.50	3.556		765	120		885	1,025	
	1830	4"			3	5.333		1,150	179		1,329	1,550	
	1840	6"		Q-2	3	8		2,050	279		2,329	2,675	
	1850	8"		"	2.50	9.600		3,425	335		3,760	4,250	
	1900	Diverter, 150 lb. flanged, stainless steel plugs											
	1910	1-1/2" pipe size		Q-1	3	5.333	Ea.	4,750	179		4,929	5,500	
	1920	2" pipe size			2	8		6,175	269		6,444	7,175	
	1940	3" pipe size			1.50	10.667		7,850	360		8,210	9,175	
	1950	4" pipe size			1	16		13,200	540		13,740	15,300	
	2100	Gate, OS&Y, 150 lb., flanged											
	2120	1/2"		1 Plum	18	.444	Ea.	310	16.60		326.60	365	
	2140	3/4"			16	.500		335	18.70		353.70	400	
	2150	1"			14	.571		375	21.50		396.50	440	
	2155	1-1/4"			12	.667		470	25		495	555	
	2160	1-1/2"			11	.727		470	27		497	555	
	2170	2"			8	1		520	37.50		557.50	630	
	2180	2-1/2"		Q-1	5	3.200		810	108		918	1,050	
	2190	3"			4.50	3.556		885	120		1,005	1,150	
	2200	4"			3	5.333		1,225	179		1,404	1,625	
	2205	5"			2.80	5.714		2,200	192		2,392	2,725	
	2210	6"		Q-2	3	8		2,200	279		2,479	2,850	
	2220	8"			2.50	9.600		3,400	335		3,735	4,250	
	2230	10"			2.30	10.435		4,950	365		5,315	6,000	
	2240	12"			1.90	12.632		6,900	440		7,340	8,275	

15
MECHANICAL

Important: See the Reference Section for critical supporting data - Reference Nos., Crews, & City Cost Indexes

			CREW	DAILY OUTPUT	LABOR-HOURS	UNIT	2003 BARE COSTS				TOTAL INCL O&P	
15110		**Valves**					MAT.	LABOR	EQUIP.	TOTAL		
800	2260	For 300 lb., flanged, add	R15100 -090			Ea.	120%	15%				800
	2600	600 lb., flanged										
	2620	1/2"	1 Plum	16	.500	Ea.	850	18.70		868.70	965	
	2640	3/4"		14	.571		925	21.50		946.50	1,050	
	2650	1"		12	.667		1,175	25		1,200	1,325	
	2660	1-1/2"		10	.800		1,650	30		1,680	1,850	
	2670	2"		7	1.143		2,350	42.50		2,392.50	2,675	
	2680	2-1/2"	Q-1	4	4		3,650	135		3,785	4,225	
	2690	3"	"	3.60	4.444		3,900	149		4,049	4,500	
	3100	Globe, OS&Y, 150 lb., flanged										
	3120	1/2"	1 Plum	18	.444	Ea.	370	16.60		386.60	430	
	3140	3/4"		16	.500		410	18.70		428.70	480	
	3150	1"		14	.571		505	21.50		526.50	585	
	3160	1-1/2"		11	.727		750	27		777	865	
	3170	2"		8	1		910	37.50		947.50	1,050	
	3180	2-1/2"	Q-1	5	3.200		1,325	108		1,433	1,625	
	3190	3"		4.50	3.556		1,625	120		1,745	1,950	
	3200	4"		3	5.333		2,475	179		2,654	3,000	
	3210	6"	Q-2	3	8		4,075	279		4,354	4,900	
	5000	Silent check, 316 S.S. body and trim										
	5010	Compact wafer type, 300 lb.										
	5020	1"	1 Plum	14	.571	Ea.	555	21.50		576.50	640	
	5021	1-1/4"		12	.667		600	25		625	700	
	5022	1-1/2"		11	.727		620	27		647	725	
	5023	2"		9	.889		685	33		718	800	
	5024	2-1/2"	Q-1	9	1.778		975	60		1,035	1,175	
	5025	3"	"	8	2		1,075	67.50		1,142.50	1,275	
	5100	Full flange wafer type, 150 lb.										
	5110	1"	1 Plum	14	.571	Ea.	390	21.50		411.50	460	
	5111	1-1/4"		12	.667		420	25		445	500	
	5112	1-1/2"		11	.727		480	27		507	570	
	5113	2"		9	.889		625	33		658	740	
	5114	2-1/2"	Q-1	9	1.778		750	60		810	910	
	5115	3"		8	2		845	67.50		912.50	1,025	
	5116	4"		5	3.200		1,250	108		1,358	1,550	
	5117	5"	Q-2	6	4		1,475	139		1,614	1,800	
	5118	6"		5	4.800		1,825	167		1,992	2,275	
	5119	8"		4.50	5.333		3,050	186		3,236	3,625	
	5120	10"		4	6		4,225	209		4,434	4,975	
	5200	Globe type, 300 lb.										
	5210	2"	1 Plum	9	.889	Ea.	715	33		748	835	
	5211	2-1/2"	Q-1	9	1.778		880	60		940	1,050	
	5212	3"		8	2		1,100	67.50		1,167.50	1,300	
	5213	4"		5	3.200		1,500	108		1,608	1,825	
	5214	5"	Q-2	6	4		2,125	139		2,264	2,525	
	5215	6"		5	4.800		2,400	167		2,567	2,875	
	5216	8"		4.50	5.333		3,375	186		3,561	4,000	
	5217	10"		4	6		4,775	209		4,984	5,575	
	5300	Screwed end type, 300 lb.										
	5310	1/2"	1 Plum	24	.333	Ea.	60.50	12.45		72.95	85.50	
	5311	3/4"		20	.400		72	14.95		86.95	102	
	5312	1"		19	.421		84	15.75		99.75	116	
	5313	1-1/4"		15	.533		111	19.90		130.90	152	
	5314	1-1/2"		13	.615		120	23		143	167	
	5315	2"		11	.727		149	27		176	205	

MECHANICAL 15

15120	Piping Specialties	CREW	DAILY OUTPUT	LABOR-HOURS	UNIT	2003 BARE COSTS				TOTAL INCL O&P
						MAT.	LABOR	EQUIP.	TOTAL	
0010	**AIR CONTROL**									
0030	Air separator, with strainer									
0040	2" diameter	Q-5	6	2.667	Ea.	550	90.50		640.50	740
0080	2-1/2" diameter		5	3.200		620	108		728	850
0100	3" diameter		4	4		945	135		1,080	1,250
0120	4" diameter		3	5.333		1,375	181		1,556	1,775
0130	5" diameter	Q-6	3.60	6.667		1,750	234		1,984	2,275
0140	6" diameter		3.40	7.059		2,075	248		2,323	2,675
0160	8" diameter		3	8		3,125	281		3,406	3,850
0180	10" diameter		2.20	10.909		4,850	385		5,235	5,900
0200	12" diameter		1.70	14.118		8,125	495		8,620	9,675
0210	14" diameter		1.30	18.462		9,675	650		10,325	11,600
0220	16" diameter		1	24		14,600	840		15,440	17,300
0230	18" diameter		.80	30		19,100	1,050		20,150	22,600
0240	20" diameter		.60	40		21,400	1,400		22,800	25,700
0300	Without strainer									
0310	2" diameter	Q-5	6	2.667	Ea.	425	90.50		515.50	600
0320	2-1/2" diameter		5	3.200		505	108		613	720
0330	3" diameter		4	4		715	135		850	990
0340	4" diameter		3	5.333		1,125	181		1,306	1,500
0350	5" diameter		2.40	6.667		1,525	226		1,751	2,025
0360	6" diameter	Q-6	3.40	7.059		1,750	248		1,998	2,300
0370	8" diameter		3	8		2,400	281		2,681	3,075
0380	10" diameter		2.20	10.909		3,525	385		3,910	4,450
0390	12" diameter		1.70	14.118		5,425	495		5,920	6,700
0400	14" diameter		1.30	18.462		8,125	650		8,775	9,900
0410	16" diameter		1	24		11,300	840		12,140	13,800
0420	18" diameter		.80	30		14,800	1,050		15,850	17,900
0430	20" diameter		.60	40		17,700	1,400		19,100	21,600
1000	Micro-bubble separator for total air removal, closed loop system									
1010	Requires bladder type tank in system.									
1020	Water (hot or chilled) or glycol system									
1030	Threaded									
1040	3/4" diameter	1 Stpi	20	.400	Ea.	67	15.05		82.05	96
1050	1" diameter		19	.421		75	15.85		90.85	107
1060	1-1/4" diameter		16	.500		103	18.80		121.80	143
1070	1-1/2" diameter		13	.615		135	23		158	183
1080	2" diameter		11	.727		655	27.50		682.50	760
1090	2-1/2" diameter	Q-5	15	1.067		725	36		761	855
1100	3" diameter		13	1.231		1,050	41.50		1,091.50	1,225
1110	4" diameter		10	1.600		1,125	54		1,179	1,300
1230	Flanged									
1250	2" diameter	1 Stpi	8	1	Ea.	835	37.50		872.50	970
1260	2-1/2" diameter	Q-5	5	3.200		905	108		1,013	1,150
1270	3" diameter		4.50	3.556		1,225	120		1,345	1,525
1280	4" diameter		3	5.333		1,350	181		1,531	1,750
1290	5" diameter	Q-6	3.40	7.059		2,525	248		2,773	3,150
1300	6" diameter	"	3	8		2,625	281		2,906	3,300
1400	Larger sizes available									
1590	With extended tank dirt catcher									
1600	Flanged									
1620	2" diameter	1 Stpi	7.40	1.081	Ea.	1,100	40.50		1,140.50	1,275
1630	2-1/2" diameter	Q-5	4.20	3.810		1,150	129		1,279	1,475
1640	3" diameter		4	4		1,525	135		1,660	1,875
1650	4" diameter		2.80	5.714		1,650	193		1,843	2,100
1660	5" diameter	Q-6	3.10	7.742		2,775	272		3,047	3,450

120 ... 120

T5 MECHANICAL

Important: See the Reference Section for critical supporting data - Reference Nos., Crews, & City Cost Indexes

15120		Piping Specialties	CREW	DAILY OUTPUT	LABOR-HOURS	UNIT	2003 BARE COSTS				TOTAL INCL O&P	
							MAT.	LABOR	EQUIP.	TOTAL		
120	1670	6" diameter	Q-6	2.90	8.276	Ea.	2,950	290		3,240	3,700	120
	1800	With drain/dismantle/cleaning access flange										
	1810	Flanged										
	1820	2" diameter	1 Stpi	7	1.143	Ea.	2,100	43		2,143	2,400	
	1830	2-1/2" diameter	Q-5	4	4		2,175	135		2,310	2,600	
	1840	3" diameter		3.60	4.444		2,850	150		3,000	3,350	
	1850	4" diameter	▼	2.50	6.400		3,000	217		3,217	3,625	
	1860	5" diameter	Q-6	2.80	8.571		5,550	300		5,850	6,575	
	1870	6" diameter	"	2.60	9.231	▼	5,925	325		6,250	7,025	
	2000	Boiler air fitting, separator										
	2010	1-1/4"	Q-5	22	.727	Ea.	119	24.50		143.50	168	
	2020	1-1/2"		20	.800		238	27		265	305	
	2021	2"	▼	18	.889	▼	415	30		445	500	
	2400	Compression tank air fitting										
	2410	For tanks 9" to 24" diameter	1 Stpi	15	.533	Ea.	34.50	20		54.50	68.50	
	2420	For tanks 100 gallon or larger	"	12	.667	"	144	25		169	196	
140	0010	**AIR PURGING SCOOP** with tappings										140
	0020	for air vent and expansion tank connection										
	0100	1" pipe size, threaded	1 Stpi	19	.421	Ea.	13.60	15.85		29.45	39	
	0110	1-1/4" pipe size, threaded		15	.533		13.70	20		33.70	45.50	
	0120	1-1/2" pipe size, threaded		13	.615		30	23		53	68	
	0130	2" pipe size, threaded	▼	11	.727		34	27.50		61.50	79	
	0140	2-1/2" pipe size, threaded	Q-5	15	1.067		76.50	36		112.50	139	
	0150	3" pipe size, threaded		13	1.231		95	41.50		136.50	168	
	0160	4" 150 lb. flanges	▼	3	5.333	▼	214	181		395	510	
160	0010	**AUTOMATIC AIR VENT**										160
	0020	Cast iron body, stainless steel internals, float type										
	0060	1/2" NPT inlet, 300 psi	1 Stpi	12	.667	Ea.	66.50	25		91.50	111	
	0140	3/4" NPT inlet, 300 psi		12	.667		66.50	25		91.50	111	
	0180	1/2" NPT inlet, 250 psi		10	.800		212	30		242	279	
	0220	3/4" NPT inlet, 250 psi		10	.800		212	30		242	279	
	0260	1" NPT inlet, 250 psi	▼	10	.800		310	30		340	385	
	0340	1-1/2" NPT inlet, 250 psi	Q-5	12	1.333		660	45		705	795	
	0380	2" NPT inlet, 250 psi	"	12	1.333	▼	660	45		705	795	
	0600	Forged steel body, stainless steel internals, float type										
	0640	1/2" NPT inlet, 750 psi	1 Stpi	12	.667	Ea.	655	25		680	765	
	0680	3/4" NPT inlet, 750 psi		12	.667		655	25		680	765	
	0760	3/4" NPT inlet, 1000 psi	▼	10	.800		985	30		1,015	1,125	
	0800	1" NPT inlet, 1000 psi	Q-5	12	1.333		985	45		1,030	1,150	
	0880	1-1/2" NPT inlet, 1000 psi		10	1.600		2,775	54		2,829	3,125	
	0920	2" NPT inlet, 1000 psi	▼	10	1.600	▼	2,775	54		2,829	3,125	
	1100	Formed steel body, non corrosive										
	1110	1/8" NPT inlet 150 psi	1 Stpi	32	.250	Ea.	6.90	9.40		16.30	22	
	1120	1/4" NPT inlet 150 psi		32	.250		23	9.40		32.40	39.50	
	1130	3/4" NPT inlet 150 psi	▼	32	.250	▼	23	9.40		32.40	39.50	
	1300	Chrome plated brass, automatic/manual, for radiators										
	1310	1/8" NPT inlet, nickel plated brass	1 Stpi	32	.250	Ea.	4.03	9.40		13.43	18.65	
180	0010	**CIRCUIT SENSOR** Flow meter										180
	0020	Metering stations										
	0040	Wafer orifice insert type										
	0060	2-1/2" pipe size	Q-5	12	1.333	Ea.	126	45		171	207	
	0100	3" pipe size		11	1.455		140	49		189	229	
	0140	4" pipe size		8	2		158	67.50		225.50	275	
	0180	5" pipe size		7.30	2.192		203	74		277	335	
	0220	6" pipe size	▼	6.40	2.500	▼	238	84.50		322.50	390	

15120	Piping Specialties	CREW	DAILY OUTPUT	LABOR-HOURS	UNIT	2003 BARE COSTS				TOTAL INCL O&P
						MAT.	LABOR	EQUIP.	TOTAL	
180 0260	8" pipe size	Q-6	5.30	4.528	Ea.	335	159		494	610	180
0280	10" pipe size		4.60	5.217		390	183		573	700
0360	12" pipe size	▼	4.20	5.714	▼	645	201		846	1,025
2000	In-line probe type									
2200	Copper, soldered									
2210	1/2" size	1 Stpi	24	.333	Ea.	107	12.55		119.55	136
2220	3/4" size		20	.400		111	15.05		126.05	145
2230	1" size		19	.421		126	15.85		141.85	162
2240	1-1/4" size		15	.533		132	20		152	177
2250	1-1/2" size		13	.615		155	23		178	206
2260	2" size	▼	11	.727		174	27.50		201.50	233
2270	2-1/2" size	Q-5	15	1.067		184	36		220	258
2280	3" size	"	13	1.231	▼	200	41.50		241.50	283
2500	Brass, threaded									
2510	1" size	1 Stpi	19	.421	Ea.	132	15.85		147.85	170
2520	1-1/4" size		15	.533		146	20		166	192
2530	1-1/2" size		13	.615		175	23		198	228
2540	2" size	▼	11	.727		184	27.50		211.50	245
2550	2-1/2" size	Q-5	15	1.067		201	36		237	276
2560	3" size	"	13	1.231	▼	205	41.50		246.50	288
2700	Steel, threaded									
2710	1" size	1 Stpi	19	.421	Ea.	127	15.85		142.85	163
2720	1-1/4" size		15	.533		138	20		158	183
2730	1-1/2" size		13	.615		156	23		179	207
2740	2" size	▼	11	.727		174	27.50		201.50	233
2750	2-1/2" size	Q-5	15	1.067		184	36		220	258
2760	3" size	"	13	1.231	▼	204	41.50		245.50	288
3000	Pitot tube probe type									
3100	Weld-on mounting with probe									
3110	2" size	Q-17	18	.889	Ea.	143	30	4.36	177.36	207
3120	2-1/2" size		18	.889		146	30	4.36	180.36	211
3130	3" size		18	.889		155	30	4.36	189.36	221
3140	4" size		18	.889		205	30	4.36	239.36	275
3150	5" size		17	.941		230	32	4.62	266.62	305
3160	6" size		17	.941		297	32	4.62	333.62	380
3170	8" size		16	1		425	34	4.91	463.91	520
3180	10" size		15	1.067		460	36	5.25	501.25	565
3190	12" size		15	1.067		490	36	5.25	531.25	600
3200	14" size		14	1.143		560	38.50	5.60	604.10	685
3210	16" size		14	1.143		665	38.50	5.60	709.10	795
3220	18" size		13	1.231		710	41.50	6.05	757.55	850
3230	20" size		13	1.231		765	41.50	6.05	812.55	915
3240	24" size	▼	12	1.333	▼	860	45	6.55	911.55	1,025
3400	Weld-on wet tap with probe									
3410	2" size, 5/16" probe dia.	Q-17	14	1.143	Ea.	345	38.50	5.60	389.10	445
3420	2-1/2" size, 5/16" probe dia.		14	1.143		355	38.50	5.60	399.10	455
3430	3" size, 5/16" probe dia.		14	1.143		375	38.50	5.60	419.10	475
3440	4" size, 5/16" probe dia.		14	1.143		385	38.50	5.60	429.10	485
3450	5" size, 3/8" probe dia.		13	1.231		400	41.50	6.05	447.55	510
3460	6" size, 3/8" probe dia.		13	1.231		425	41.50	6.05	472.55	540
3470	8" size, 1/2" probe dia.		12	1.333		445	45	6.55	496.55	560
3480	10" size, 3/4" probe dia.		11	1.455		470	49	7.15	526.15	600
3490	12" size, 3/4" probe dia.		11	1.455		515	49	7.15	571.15	645
3500	14" size, 1" probe dia.		10	1.600		580	54	7.85	641.85	730
3510	16" size, 1" probe dia.		10	1.600		695	54	7.85	756.85	855
3520	18" size, 1" probe dia.	▼	9	1.778	▼	745	60	8.75	813.75	920

Important: See the Reference Section for critical supporting data - Reference Nos., Crews, & City Cost Indexes

15120 | Piping Specialties

		CREW	DAILY OUTPUT	LABOR-HOURS	UNIT	2003 BARE COSTS MAT.	LABOR	EQUIP.	TOTAL	TOTAL INCL O&P	
180	3530	20" size, 1" probe dia.	Q-17	8	2	Ea.	800	67.50	9.80	877.30	995	**180**
	3540	24" size, 1" probe dia.	↓	8	2	↓	895	67.50	9.80	972.30	1,100	
	3700	Clamp-on wet tap with probe										
	3710	2" size, 5/16" probe dia.	Q-5	16	1	Ea.	405	34		439	495	
	3720	2-1/2" size, 5/16" probe dia.		16	1		420	34		454	515	
	3730	3" size, 5/16" probe dia.		15	1.067		435	36		471	530	
	3740	4" size, 5/16" probe dia.		15	1.067		500	36		536	605	
	3750	5" size, 3/8" probe dia.		14	1.143		525	38.50		563.50	640	
	3760	6" size, 3/8" probe dia.		14	1.143		560	38.50		598.50	675	
	3770	8" size, 1/2" probe dia.		13	1.231		615	41.50		656.50	745	
	3780	10" size, 3/4" probe dia.		12	1.333		645	45		690	780	
	3790	12" size, 3/4" probe dia.		12	1.333		720	45		765	865	
	3800	14" size, 1" probe dia.		11	1.455		975	49		1,024	1,150	
	3810	16" size, 1" probe dia.		11	1.455		1,125	49		1,174	1,325	
	3820	18" size, 1" probe dia.		10	1.600		1,200	54		1,254	1,400	
	3830	20" size, 1" probe dia		9	1.778		1,325	60		1,385	1,550	
	3840	24" size, 1" probe dia.	↓	9	1.778	↓	1,425	60		1,485	1,675	
	4000	Wet tap drills										
	4010	3/8" dia. for 5/16" probe				Ea.	83			83	91.50	
	4020	1/2" dia. for 3/8" probe					86.50			86.50	95	
	4030	5/8" dia for 1/2" probe					103			103	113	
	4040	7/8" dia for 3/4" probe					134			134	147	
	4050	1-3/16" dia for 1" probe				↓	143			143	157	
	4100	Wet tap punch										
	4110	5/16" dia for 5/16" probe				Ea.	86.50			86.50	95	
	4120	3/8" dia for 3/8" probe					93			93	102	
	4130	1/2" dia for 1/2" probe				↓	111			111	123	
	4150	Note: labor for wet tap drill or punch										
	4160	is included with the mounting and probe assembly										
	9000	Readout instruments										
	9200	Gauges										
	9220	GPM gauge				Ea.	1,400			1,400	1,525	
	9230	Dual gauge kit				"	1,450			1,450	1,600	
200	0010	**CIRCUIT SETTER** Balance valve										**200**
	0012	Bronze body, soldered										
	0013	1/2" size	1 Stpi	24	.333	Ea.	43	12.55		55.55	66	
	0014	3/4" size	"	20	.400	"	45	15.05		60.05	72	
	0018	Threaded										
	0019	1/2" pipe size	1 Stpi	22	.364	Ea.	44.50	13.65		58.15	69.50	
	0020	3/4" pipe size		20	.400		47.50	15.05		62.55	74.50	
	0040	1" pipe size		18	.444		61	16.70		77.70	92.50	
	0050	1-1/4" pipe size		15	.533		90.50	20		110.50	130	
	0060	1-1/2" pipe size		12	.667		106	25		131	155	
	0080	2" pipe size	↓	10	.800		152	30		182	213	
	0100	2-1/2" pipe size	Q-5	15	1.067		345	36		381	435	
	0120	3" pipe size	"	10	1.600	↓	485	54		539	615	
	0130	Cast iron body, flanged										
	0136	3" pipe size, flanged	Q-5	4	4	Ea.	515	135		650	770	
	0140	4" pipe size	"	3	5.333		725	181		906	1,075	
	0200	For differential meter, accurate to 1%, add					590			590	650	
	0300	For differential meter 400 psi/200° F continuous, add				↓	1,450			1,450	1,600	
220	0010	**COCKS, DRAINS & SPECIALTIES**										**220**
	1000	Boiler drain										
	1010	Pipe thread to hose										
	1020	Bronze										

MECHANICAL 15

15120	Piping Specialties	CREW	DAILY OUTPUT	LABOR-HOURS	UNIT	2003 BARE COSTS				TOTAL INCL O&P
						MAT.	LABOR	EQUIP.	TOTAL	
220 1030	1/2" size	1 Stpi	36	.222	Ea.	4.10	8.35		12.45	17.10 **220**
1040	3/4" size	"	34	.235	"	4.44	8.85		13.29	18.25
1100	Solder to hose									
1110	Bronze									
1120	1/2" size	1 Stpi	46	.174	Ea.	3.87	6.55		10.42	14.15
1130	3/4" size	"	44	.182	"	3.68	6.85		10.53	14.40
1600	With built-in vacuum breaker									
1610	1/2" IP or solder	1 Stpi	36	.222	Ea.	13.90	8.35		22.25	28
1630	With tamper proof vacuum breaker									
1640	1/2" IP or solder	1 Stpi	36	.222	Ea.	16.45	8.35		24.80	30.50
1650	3/4" IP or solder	"	34	.235	"	16.75	8.85		25.60	32
3000	Cocks									
3010	Air, lever or tee handle									
3020	Bronze, single thread									
3030	1/8" size	1 Stpi	52	.154	Ea.	6.05	5.80		11.85	15.40
3040	1/4" size		46	.174		6.55	6.55		13.10	17.10
3050	3/8" size		40	.200		7.35	7.50		14.85	19.45
3060	1/2" size		36	.222		7.40	8.35		15.75	21
3100	Bronze, double thread									
3110	1/8" size	1 Stpi	26	.308	Ea.	8.10	11.55		19.65	26.50
3120	1/4" size		22	.364		7.80	13.65		21.45	29
3130	3/8" size		18	.444		8.60	16.70		25.30	34.50
3140	1/2" size		15	.533		10.25	20		30.25	42
4000	Steam									
4010	Bronze									
4020	1/8" size	1 Stpi	26	.308	Ea.	14.80	11.55		26.35	34
4030	1/4" size		22	.364		14.50	13.65		28.15	36.50
4040	3/8" size		18	.444		14.10	16.70		30.80	40.50
4050	1/2" size		15	.533		13.25	20		33.25	45
4060	3/4" size		14	.571		14.60	21.50		36.10	48.50
4070	1" size		13	.615		27	23		50	64.50
4080	1-1/4" size		11	.727		64	27.50		91.50	112
4090	1-1/2" size		10	.800		83.50	30		113.50	137
4100	2" size		9	.889		128	33.50		161.50	191
4300	Bronze, 3 way									
4320	1/4" size	1 Stpi	15	.533	Ea.	16.85	20		36.85	49
4330	3/8" size		12	.667		21.50	25		46.50	62
4340	1/2" size		10	.800		16.85	30		46.85	64
4350	3/4" size		9.50	.842		29	31.50		60.50	80
4360	1" size		8.50	.941		35	35.50		70.50	92
4370	1-1/4" size		7.50	1.067		52.50	40		92.50	118
4380	1-1/2" size		6.50	1.231		61	46.50		107.50	137
4390	2" size		6	1.333		90	50		140	175
4500	Gauge cock, brass									
4510	1/4" FPT	1 Stpi	24	.333	Ea.	6.85	12.55		19.40	26.50
4512	1/4" MPT	"	24	.333	"	7.90	12.55		20.45	27.50
4600	Pigtail, steam syphon									
4604	1/4"	1 Stpi	24	.333	Ea.	9.80	12.55		22.35	29.50
4650	Snubber valve									
4654	1/4"	1 Stpi	22	.364	Ea.	9.75	13.65		23.40	31.50
4660	Nipple, black steel									
4664	1/4" x 3"	1 Stpi	37	.216	Ea.	.44	8.15		8.59	12.80
250 0010	**DIELECTRIC UNIONS** Standard gaskets for water and air									**250**
0020	250 psi maximum pressure									
0280	Female IPT to sweat, straight									
0300	1/2" pipe size	1 Plum	24	.333	Ea.	3.15	12.45		15.60	22.50

Important: See the Reference Section for critical supporting data - Reference Nos., Crews, & City Cost Indexes

15120	Piping Specialties	CREW	DAILY OUTPUT	LABOR-HOURS	UNIT	2003 BARE COSTS				TOTAL INCL O&P
						MAT.	LABOR	EQUIP.	TOTAL	
250 0340	3/4" pipe size	1 Plum	20	.400	Ea.	3.15	14.95		18.10	26	250
0360	1" pipe size		19	.421		6.10	15.75		21.85	30.50
0380	1-1/4" pipe size		15	.533		10	19.90		29.90	41
0400	1-1/2" pipe size		13	.615		15	23		38	51
0420	2" pipe size	▼	11	.727	▼	21	27		48	64
0580	Female IPT to brass pipe thread, straight									
0600	1/2" pipe size	1 Plum	24	.333	Ea.	5.95	12.45		18.40	25.50
0640	3/4" pipe size		20	.400		7.60	14.95		22.55	31
0660	1" pipe size		19	.421		11.40	15.75		27.15	36.50
0680	1-1/4" pipe size		15	.533		17.05	19.90		36.95	49
0700	1-1/2" pipe size		13	.615		23.50	23		46.50	60.50
0720	2" pipe size	▼	11	.727	▼	30.50	27		57.50	75
0780	Female IPT to female IPT, straight									
0800	1/2" pipe size	1 Plum	24	.333	Ea.	4.75	12.45		17.20	24
0840	3/4" pipe size		20	.400		5.30	14.95		20.25	28.50
0860	1" pipe size		19	.421		7	15.75		22.75	31.50
0880	1-1/4" pipe size		15	.533		9.75	19.90		29.65	41
0900	1-1/2" pipe size		13	.615		14.95	23		37.95	51
0920	2" pipe size	▼	11	.727	▼	21.50	27		48.50	64.50
2000	175 psi maximum pressure									
2180	Female IPT to sweat									
2200	1-1/2" pipe size	1 Plum	11	.727	Ea.	38.50	27		65.50	83
2240	2" pipe size	"	9	.889		44	33		77	98.50
2260	2-1/2" pipe size	Q-1	15	1.067		48	36		84	107
2280	3" pipe size		14	1.143		65.50	38.50		104	130
2300	4" pipe size	▼	11	1.455		174	49		223	266
2320	5" pipe size	Q-2	14	1.714		284	60		344	400
2340	6" pipe size		12	2		330	69.50		399.50	465
2360	8" pipe size	▼	10	2.400	▼	910	83.50		993.50	1,125
2480	Female IPT to brass pipe									
2500	1-1/2" pipe size	1 Plum	11	.727	Ea.	58	27		85	105
2540	2" pipe size	"	9	.889		69	33		102	126
2560	2-1/2" pipe size	Q-1	15	1.067		100	36		136	164
2580	3" pipe size		14	1.143		116	38.50		154.50	186
2600	4" pipe size	▼	11	1.455		199	49		248	293
2620	5" pipe size	Q-2	14	1.714		224	60		284	335
2640	6" pipe size		12	2		267	69.50		336.50	400
2680	8" pipe size	▼	10	2.400	▼	435	83.50		518.50	605
2880	Female IPT to female IPT									
2900	1-1/2" pipe size	1 Plum	11	.727	Ea.	36	27		63	80.50
2940	2" pipe size	"	9	.889		38.50	33		71.50	92.50
2960	2-1/2" pipe size	Q-1	15	1.067		48	36		84	107
2980	3" pipe size		14	1.143		60	38.50		98.50	125
3000	4" pipe size	▼	11	1.455		91	49		140	174
3020	5" pipe size	Q-2	14	1.714		105	60		165	207
3040	6" pipe size		12	2		141	69.50		210.50	260
3060	8" pipe size	▼	10	2.400	▼	192	83.50		275.50	335
3380	Copper to copper									
3400	1-1/2" pipe size	1 Plum	11	.727	Ea.	47.50	27		74.50	93.50
3440	2" pipe size	"	9	.889		54	33		87	110
3460	2-1/2" pipe size	Q-1	15	1.067		58	36		94	118
3480	3" pipe size		14	1.143		72.50	38.50		111	138
3500	4" pipe size	▼	11	1.455		230	49		279	325
3520	5" pipe size	Q-2	14	1.714		370	60		430	495
3540	6" pipe size		12	2		450	69.50		519.50	600
3560	8" pipe size	▼	10	2.400		1,200	83.50		1,283.50	1,450

MECHANICAL **15**

15120	Piping Specialties	CREW	DAILY OUTPUT	LABOR-HOURS	UNIT	2003 BARE COSTS				TOTAL INCL O&P
						MAT.	LABOR	EQUIP.	TOTAL	
280	**0010 EXPANSION COUPLINGS** Hydronic									**280**
	0100 Copper to copper, sweat									
	0120 3/4" diameter	1 Stpi	20	.400	Ea.	28	15.05		43.05	53.50
	0140 1" diameter		19	.421		35.50	15.85		51.35	63
	0160 1-1/4" diameter		15	.533		48.50	20		68.50	84
	1000 Baseboard riser fitting, 5" stub by coupling 12" long									
	1020 1/2" diameter	1 Stpi	24	.333	Ea.	10.50	12.55		23.05	30.50
	1040 3/4" diameter		20	.400		14.90	15.05		29.95	39
	1060 1" diameter		19	.421		20.50	15.85		36.35	47
	1080 1-1/4" diameter		15	.533		27	20		47	60.50
	1180 9" Stub by tubing 8" long									
	1200 1/2" diameter	1 Stpi	24	.333	Ea.	9.40	12.55		21.95	29.50
	1220 3/4" diameter		20	.400		12.45	15.05		27.50	36
	1240 1" diameter		19	.421		17.30	15.85		33.15	43
	1260 1-1/4" diameter		15	.533		24	20		44	57
300	**0010 EXPANSION JOINTS**									**300**
	0100 Bellows type, neoprene cover, flanged spool									
	0140 6" face to face, 1-1/4" diameter	1 Stpi	11	.727	Ea.	161	27.50		188.50	219
	0160 1-1/2" diameter	"	10.60	.755		161	28.50		189.50	220
	0180 2" diameter	Q-5	13.30	1.203		163	40.50		203.50	241
	0190 2-1/2" diameter		12.40	1.290		170	43.50		213.50	253
	0200 3" diameter		11.40	1.404		193	47.50		240.50	284
	0210 4" diameter		8.40	1.905		206	64.50		270.50	325
	0220 5" diameter		7.60	2.105		254	71.50		325.50	385
	0230 6" diameter		6.80	2.353		261	79.50		340.50	405
	0240 8" diameter		5.40	2.963		299	100		399	480
	0250 10" diameter		5	3.200		400	108		508	605
	0260 12" diameter		4.60	3.478		455	118		573	680
	0480 10" face to face, 2" diameter		13	1.231		240	41.50		281.50	325
	0500 2-1/2" diameter		12	1.333		253	45		298	345
	0520 3" diameter		11	1.455		257	49		306	360
	0540 4" diameter		8	2		287	67.50		354.50	415
	0560 5" diameter		7	2.286		390	77.50		467.50	545
	0580 6" diameter		6	2.667		355	90.50		445.50	530
	0600 8" diameter		5	3.200		425	108		533	630
	0620 10" diameter		4.60	3.478		455	118		573	680
	0640 12" diameter		4	4		520	135		655	775
	0660 14" diameter		3.80	4.211		650	143		793	930
	0680 16" diameter		2.90	5.517		745	187		932	1,100
	0700 18" diameter		2.50	6.400		840	217		1,057	1,250
	0720 20" diameter		2.10	7.619		880	258		1,138	1,350
	0740 24" diameter		1.80	8.889		1,000	300		1,300	1,575
	0760 26" diameter		1.40	11.429		1,125	385		1,510	1,800
	0780 30" diameter		1.20	13.333		1,225	450		1,675	2,025
	0800 36" diameter		1	16		1,525	540		2,065	2,500
	1000 Bellows with internal sleeves and external covers									
	1010 Stainless steel, 150 lb.									
	1020 With male threads									
	1030 1/2" diameter	1 Stpi	22	.364	Ea.	71.50	13.65		85.15	99
	1040 3/4" diameter		20	.400		71.50	15.05		86.55	101
	1050 1" diameter		19	.421		76	15.85		91.85	108
	1060 1-1/4" diameter		16	.500		79	18.80		97.80	115
	1070 1-1/2" diameter		13	.615		96	23		119	141
	1080 2" diameter		11	.727		122	27.50		149.50	176
	1110 With flanged ends									

15 MECHANICAL

Important: See the Reference Section for critical supporting data - Reference Nos., Crews, & City Cost Indexes

			CREW	DAILY OUTPUT	LABOR-HOURS	UNIT	2003 BARE COSTS				TOTAL INCL O&P	
	15120	**Piping Specialties**					MAT.	LABOR	EQUIP.	TOTAL		
300	1120	1-1/4" diameter	1 Stpi	12	.667	Ea.	320	25		345	390	300
	1130	1-1/2" diameter		11	.727		320	27.50		347.50	390	
	1140	2" diameter	↓	9	.889		510	33.50		543.50	610	
	1150	3" diameter	Q-5	8	2		515	67.50		582.50	665	
	1160	4" diameter	"	5	3.200		550	108		658	770	
	1170	5" diameter	Q-6	6	4		585	140		725	855	
	1180	6" diameter		5	4.800		645	168		813	965	
	1190	8" diameter		4.50	5.333		770	187		957	1,125	
	1200	10" diameter		4	6		940	211		1,151	1,350	
	1210	12" diameter	↓	3.60	6.667	↓	1,200	234		1,434	1,675	
320	0010	**EXPANSION TANKS**										320
	1400	Plastic, corrosion resistant, see Plumbing Cost Data										
	1410	Div. 15220										
	1505	Fiberglass and steel single / double wall storage, see Div 13201										
	1510	Tank leak detection systems, see Div 13851-350										
	2000	Steel, liquid expansion, ASME, painted, 15 gallon capacity	Q-5	17	.941	Ea.	355	32		387	440	
	2020	24 gallon capacity		14	1.143		360	38.50		398.50	455	
	2040	30 gallon capacity		12	1.333		395	45		440	505	
	2060	40 gallon capacity		10	1.600		450	54		504	575	
	2080	60 gallon capacity		8	2		540	67.50		607.50	695	
	2100	80 gallon capacity		7	2.286		555	77.50		632.50	725	
	2120	100 gallon capacity		6	2.667		715	90.50		805.50	925	
	2130	120 gallon capacity		5	3.200		795	108		903	1,050	
	2140	135 gallon capacity		4.50	3.556		830	120		950	1,100	
	2150	175 gallon capacity		4	4		1,150	135		1,285	1,475	
	2160	220 gallon capacity		3.60	4.444		1,325	150		1,475	1,675	
	2170	240 gallon capacity		3.30	4.848		1,475	164		1,639	1,875	
	2180	305 gallon capacity		3	5.333		1,975	181		2,156	2,450	
	2190	400 gallon capacity	↓	2.80	5.714	↓	2,225	193		2,418	2,750	
	2360	Galvanized										
	2370	15 gallon capacity	Q-5	17	.941	Ea.	570	32		602	675	
	2380	24 gallon capacity		14	1.143		640	38.50		678.50	765	
	2390	30 gallon capacity		12	1.333		710	45		755	850	
	2400	40 gallon capacity		10	1.600		810	54		864	970	
	2410	60 gallon capacity		8	2		930	67.50		997.50	1,125	
	2420	80 gallon capacity		7	2.286		1,150	77.50		1,227.50	1,375	
	2430	100 gallon capacity		6	2.667		1,400	90.50		1,490.50	1,650	
	2440	120 gallon capacity		5	3.200		1,625	108		1,733	1,975	
	2450	135 gallon capacity		4.50	3.556		1,725	120		1,845	2,075	
	2460	175 gallon capacity		4	4		2,125	135		2,260	2,550	
	2470	220 gallon capacity		3.60	4.444		2,450	150		2,600	2,925	
	2480	240 gallon capacity		3.30	4.848		2,550	164		2,714	3,050	
	2490	305 gallon capacity		3	5.333		3,450	181		3,631	4,075	
	2500	400 gallon capacity		2.80	5.714		4,200	193		4,393	4,925	
	3000	Steel ASME expansion, rubber diaphragm, 19 gal. cap. accept.		12	1.333		1,375	45		1,420	1,600	
	3020	31 gallon capacity		8	2		1,525	67.50		1,592.50	1,775	
	3040	61 gallon capacity		6	2.667		2,150	90.50		2,240.50	2,500	
	3060	79 gallon capacity		5	3.200		2,250	108		2,358	2,650	
	3080	119 gallon capacity		4	4		2,375	135		2,510	2,825	
	3100	158 gallon capacity		3.80	4.211		3,300	143		3,443	3,875	
	3120	211 gallon capacity		3.30	4.848		3,825	164		3,989	4,450	
	3140	317 gallon capacity		2.80	5.714		5,000	193		5,193	5,800	
	3160	422 gallon capacity		2.60	6.154		7,375	208		7,583	8,450	
	3180	528 gallon capacity	↓	2.40	6.667	↓	8,100	226		8,326	9,250	
350	0010	**FLEXIBLE CONNECTORS**, Corrugated, 7/8" O.D., 1/2" I.D.										350
	0050	Gas, seamless brass, steel fittings										

			DAILY	LABOR-			2003 BARE COSTS				TOTAL	
			CREW	OUTPUT	HOURS	UNIT	MAT.	LABOR	EQUIP.	TOTAL	INCL O&P	

15120 | Piping Specialties

350	0200	12" long	1 Plum	36	.222	Ea.	10.45	8.30		18.75	24	350
	0220	18" long		36	.222		12.95	8.30		21.25	27	
	0240	24" long		34	.235		15.35	8.80		24.15	30	
	0260	30" long		34	.235		16.55	8.80		25.35	31.50	
	0280	36" long		32	.250		18.30	9.35		27.65	34	
	0320	48" long		30	.267		23	9.95		32.95	40.50	
	0340	60" long		30	.267		27.50	9.95		37.45	45.50	
	0360	72" long	▼	30	.267	▼	32	9.95		41.95	50	
	2000	Water, copper tubing, dielectric separators										
	2100	12" long	1 Plum	36	.222	Ea.	8.95	8.30		17.25	22.50	
	2220	15" long		36	.222		9.95	8.30		18.25	23.50	
	2240	18" long		36	.222		10.80	8.30		19.10	24.50	
	2260	24" long	▼	34	.235	▼	13.25	8.80		22.05	28	
370	0010	**FLEXIBLE METAL HOSE** Connectors, standard lengths										370
	0100	Bronze braided, bronze ends										
	0120	3/8" diameter x 12"	1 Stpi	26	.308	Ea.	15.40	11.55		26.95	34.50	
	0140	1/2" diameter x 12"		24	.333		14.65	12.55		27.20	35	
	0160	3/4" diameter x 12"		20	.400		21.50	15.05		36.55	46	
	0180	1" diameter x 18"		19	.421		27.50	15.85		43.35	54.50	
	0200	1-1/2" diameter x 18"		13	.615		44	23		67	83.50	
	0220	2" diameter x 18"	▼	11	.727	▼	53	27.50		80.50	99.50	
	1000	Carbon steel ends										
	1020	1/4" diameter x 12"	1 Stpi	28	.286	Ea.	12.65	10.75		23.40	30	
	1040	3/8" diameter x 12"		26	.308		13	11.55		24.55	32	
	1060	1/2" diameter x 12"		24	.333		12.65	12.55		25.20	33	
	1080	1/2" diameter x 24"		24	.333		16.20	12.55		28.75	37	
	1100	1/2" diameter x 36"		24	.333		19.80	12.55		32.35	41	
	1120	3/4" diameter x 12"		20	.400		16.80	15.05		31.85	41	
	1140	3/4" diameter x 24"		20	.400		20.50	15.05		35.55	45	
	1160	3/4" diameter x 36"		20	.400		24.50	15.05		39.55	49.50	
	1180	1" diameter x 18"		19	.421		22	15.85		37.85	48.50	
	1200	1" diameter x 30"		19	.421		26	15.85		41.85	52.50	
	1220	1" diameter x 36"		19	.421		28	15.85		43.85	55	
	1240	1-1/4" diameter x 18"		15	.533		22.50	20		42.50	55.50	
	1260	1-1/4" diameter x 36"		15	.533		30	20		50	63.50	
	1280	1-1/2" diameter x 18"		13	.615		24.50	23		47.50	61.50	
	1300	1-1/2" diameter x 36"		13	.615		32.50	23		55.50	70.50	
	1320	2" diameter x 24"		11	.727		32	27.50		59.50	76.50	
	1340	2" diameter x 36"		11	.727		39.50	27.50		67	85	
	1360	2-1/2" diameter x 24"		9	.889		45	33.50		78.50	100	
	1380	2-1/2" diameter x 36"		9	.889		53	33.50		86.50	109	
	1400	3" diameter x 24"		7	1.143		53.50	43		96.50	124	
	1420	3" diameter x 36"	▼	7	1.143	▼	62	43		105	134	
	2000	Carbon steel braid, carbon steel solid ends										
	2100	1/2" diameter x 12"	1 Stpi	24	.333	Ea.	12.65	12.55		25.20	33	
	2120	3/4" diameter x 12"		20	.400		16.80	15.05		31.85	41	
	2140	1" diameter x 12"		19	.421		20	15.85		35.85	46	
	2160	1-1/4" diameter x 12"		15	.533		20	20		40	52.50	
	2180	1-1/2" diameter x 12"	▼	13	.615	▼	21.50	23		44.50	59	
	3000	Stainless steel braid, welded on carbon steel ends										
	3100	1/2" diameter x 12"	1 Stpi	24	.333	Ea.	18.50	12.55		31.05	39.50	
	3120	3/4" diameter x 12"		20	.400		26.50	15.05		41.55	51.50	
	3140	3/4" diameter x 24"		20	.400		32	15.05		47.05	57.50	
	3160	3/4" diameter x 36"		20	.400		37.50	15.05		52.55	63.50	
	3180	1" diameter x 12"	▼	19	.421	▼	30	15.85		45.85	57	

15 MECHANICAL

Important: See the Reference Section for critical supporting data - Reference Nos., Crews, & City Cost Indexes

15120	Piping Specialties	CREW	DAILY OUTPUT	LABOR-HOURS	UNIT	2003 BARE COSTS				TOTAL INCL O&P	
						MAT.	LABOR	EQUIP.	TOTAL		
370	3200	1" diameter x 24"	1 Stpi	19	.421	Ea.	36	15.85		51.85	64	**370**
	3220	1" diameter x 36"		19	.421		42.50	15.85		58.35	71	
	3240	1-1/4" diameter x 12"		15	.533		32.50	20		52.50	66.50	
	3260	1-1/4" diameter x 24"		15	.533		41.50	20		61.50	76	
	3280	1-1/4" diameter x 36"		15	.533		50	20		70	85.50	
	3300	1-1/2" diameter x 12"		13	.615		36	23		59	74.50	
	3320	1-1/2" diameter x 24"		13	.615		46	23		69	85.50	
	3340	1-1/2" diameter x 36"		13	.615		56	23		79	96.50	
	3400	Metal sst braid, over corrugated stainless steel, flanged ends										
	3410	150 PSI										
	3420	1/2" diameter x 12"	1 Stpi	24	.333	Ea.	20	12.55		32.55	41	
	3430	1" diameter x 12"		20	.400		32	15.05		47.05	57.50	
	3440	1-1/2" diameter x 12"		15	.533		37	20		57	71	
	3450	2-1/2" diameter x 9"		12	.667		54	25		79	97.50	
	3460	3" diameter x 9"		9	.889		65	33.50		98.50	122	
	3470	4" diameter x 9"		7	1.143		80	43		123	153	
	3480	4" diameter x 30"		5	1.600		170	60		230	278	
	3490	4" diameter x 36"		4.80	1.667		190	62.50		252.50	305	
	3500	6" diameter x 11"		5	1.600		132	60		192	236	
	3510	6" diameter x 36"		3.80	2.105		251	79		330	395	
	3520	8" diameter x 12"		4	2		264	75		339	405	
	3530	10" diameter x 13"		3	2.667		335	100		435	520	
	3540	12" diameter x 14"	Q-5	4	4		655	135		790	925	
	6000	Molded rubber with helical wire reinforcement										
	6010	150 PSI										
	6020	1-1/2" diameter x 12"	1 Stpi	15	.533	Ea.	39	20		59	73	
	6030	2" diameter x 12"		12	.667		133	25		158	184	
	6040	3" diameter x 12"		8	1		170	37.50		207.50	244	
	6050	4" diameter x 12"		6	1.333		198	50		248	294	
	6060	6" diameter x 18"		4	2		250	75		325	390	
	6070	8" diameter x 24"		3	2.667		325	100		425	510	
	6080	10" diameter x 24"		2	4		385	150		535	650	
	6090	12" diameter x 24"	Q-5	3	5.333		415	181		596	730	
	7000	Molded teflon with stainless steel flanges										
	7010	150 PSI										
	7020	2-1/2" diameter x 3-3/16"	Q-1	7.80	2.051	Ea.	256	69		325	385	
	7030	3" diameter x 3-5/8"		6.50	2.462		330	83		413	490	
	7040	4" diameter x 3-5/8"		5	3.200		445	108		553	655	
	7050	6" diameter x 4"		4.30	3.721		650	125		775	905	
	7060	8" diameter x 6"		3.80	4.211		860	142		1,002	1,150	
400	0010	**FLOAT VALVES**										**400**
	0020	With ball and bracket										
	0030	Single seat, threaded										
	0040	Brass body										
	0050	1/2"	1 Stpi	11	.727	Ea.	46	27.50		73.50	92	
	0060	3/4"		9	.889		51	33.50		84.50	107	
	0070	1"		7	1.143		64.50	43		107.50	136	
	0080	1-1/2"		4.50	1.778		93	67		160	203	
	0090	2"		3.60	2.222		94.50	83.50		178	230	
	0300	For condensate receivers, CI, in-line mount										
	0320	1" inlet	1 Stpi	7	1.143	Ea.	100	43		143	175	
	0360	For condensate receiver, CI, external float, flanged tank mount										
	0370	3/4" inlet	1 Stpi	5	1.600	Ea.	85	60		145	185	
420	0010	**FLOW CHECK CONTROL**										**420**
	0100	Bronze body, soldered										

			DAILY	LABOR-		2003 BARE COSTS				TOTAL	
15120		**Piping Specialties**	CREW	OUTPUT	HOURS	UNIT	MAT.	LABOR	EQUIP.	TOTAL	INCL O&P	
420	0110	3/4" size	1 Stpi	20	.400	Ea.	35.50	15.05		50.55	61.50	**420**
	0120	1" size	"	19	.421	"	42.50	15.85		58.35	71	
	0200	Cast iron body, threaded										
	0210	3/4" size	1 Stpi	20	.400	Ea.	28.50	15.05		43.55	53.50	
	0220	1" size		19	.421		32	15.85		47.85	59	
	0230	1-1/4" size		15	.533		39	20		59	73	
	0240	1-1/2" size		13	.615		59.50	23		82.50	100	
	0250	2" size	↓	11	.727	↓	85	27.50		112.50	135	
	0300	Flanged inlet, threaded outlet										
	0310	2-1/2" size	1 Stpi	8	1	Ea.	189	37.50		226.50	265	
	0320	3" size	Q-5	9	1.778		225	60		285	340	
	0330	4" size	"	7	2.286	↓	440	77.50		517.50	595	
520	0010	**HYDRONIC HEATING CONTROL VALVES**										**520**
	0050	Hot water, nonelectric, thermostatic										
	0100	Radiator supply, 1/2" diameter	1 Stpi	24	.333	Ea.	38.50	12.55		51.05	61	
	0120	3/4" diameter		20	.400		39	15.05		54.05	65.50	
	0140	1" diameter		19	.421		48.50	15.85		64.35	77.50	
	0160	1-1/4" diameter	↓	15	.533	↓	68	20		88	106	
	0500	For low pressure steam, add					25%					
	1000	Manual, radiator supply										
	1010	1/2" pipe size, angle union	1 Stpi	24	.333	Ea.	28.50	12.55		41.05	50	
	1020	3/4" pipe size, angle union		20	.400		26.50	15.05		41.55	52	
	1030	1" pipe size, angle union	↓	19	.421	↓	34.50	15.85		50.35	62	
	1100	Radiator, balancing, straight, sweat connections										
	1110	1/2" pipe size	1 Stpi	24	.333	Ea.	8.10	12.55		20.65	28	
	1120	3/4" pipe size		20	.400		11.30	15.05		26.35	35	
	1130	1" pipe size	↓	19	.421	↓	18.40	15.85		34.25	44.50	
	1200	Steam, radiator, supply										
	1210	1/2" pipe size, angle union	1 Stpi	24	.333	Ea.	20.50	12.55		33.05	42	
	1220	3/4" pipe size, angle union		20	.400		22.50	15.05		37.55	47.50	
	1230	1" pipe size, angle union		19	.421		26	15.85		41.85	52.50	
	1240	1-1/4" pipe size, angle union	↓	15	.533	↓	33.50	20		53.50	67	
	8000	System balancing and shut-off										
	8020	Butterfly, quarter turn, calibrated, threaded or solder										
	8040	Bronze, -30° F to +350° F, pressure to 175 psi										
	8060	1/2" size	1 Stpi	22	.364	Ea.	5.50	13.65		19.15	26.50	
	8070	3/4" size		20	.400		8.90	15.05		23.95	32.50	
	8080	1" size		19	.421		10.90	15.85		26.75	36	
	8090	1-1/4" size		15	.533		17.50	20		37.50	50	
	8100	1-1/2" size		13	.615		22.50	23		45.50	60	
	8110	2" size	↓	11	.727	↓	28.50	27.50		56	72.50	
550	0010	**LIQUID DRAINERS**										**550**
	0100	Guided lever type										
	0110	Cast iron body, threaded										
	0120	1/2" pipe size	1 Stpi	22	.364	Ea.	68	13.65		81.65	95.50	
	0130	3/4" pipe size		20	.400		181	15.05		196.05	222	
	0140	1" pipe size		19	.421		289	15.85		304.85	345	
	0150	1-1/2" pipe size		11	.727		630	27.50		657.50	735	
	0160	2" pipe size	↓	8	1	↓	630	37.50		667.50	750	
	0200	Forged steel body, threaded										
	0210	1/2" pipe size	1 Stpi	18	.444	Ea.	635	16.70		651.70	725	
	0220	3/4" pipe size		16	.500		635	18.80		653.80	730	
	0230	1" pipe size	↓	14	.571		955	21.50		976.50	1,075	
	0240	1-1/2" pipe size	Q-5	10	1.600		2,775	54		2,829	3,125	
	0250	2" pipe size	"	8	2	↓	2,775	67.50		2,842.50	3,150	

Important: See the Reference Section for critical supporting data - Reference Nos., Crews, & City Cost Indexes

15 MECHANICAL

			DAILY	LABOR-		2003 BARE COSTS				TOTAL
	15120 | Piping Specialties	CREW	OUTPUT	HOURS	UNIT	MAT.	LABOR	EQUIP.	TOTAL	INCL O&P
550	0300	Socket weld									**550**
	0310	1/2" pipe size	Q-17	16	1	Ea.	730	34	4.91	768.91	860
	0320	3/4" pipe size		14	1.143		730	38.50	5.60	774.10	870
	0330	1" pipe size		12	1.333		1,075	45	6.55	1,126.55	1,250
	0340	1-1/2" pipe size		9	1.778		2,875	60	8.75	2,943.75	3,275
	0350	2" pipe size	↓	7	2.286	↓	2,875	77.50	11.20	2,963.70	3,300
	0400	Flanged									
	0410	1/2" pipe size	1 Stpi	12	.667	Ea.	1,350	25		1,375	1,550
	0420	3/4" pipe size		10	.800		1,350	30		1,380	1,550
	0430	1" pipe size	↓	7	1.143		1,775	43		1,818	2,025
	0440	1-1/2" pipe size	Q-5	9	1.778		3,875	60		3,935	4,375
	0450	2" pipe size	"	7	2.286	↓	3,875	77.50		3,952.50	4,400
	1000	Fixed lever and snap action type									
	1100	Cast iron body, threaded									
	1110	1/2" pipe size	1 Stpi	22	.364	Ea.	107	13.65		120.65	138
	1120	3/4" pipe size		20	.400		281	15.05		296.05	335
	1130	1" pipe size	↓	19	.421	↓	281	15.85		296.85	335
	1200	Forged steel body, threaded									
	1210	1/2" pipe size	1 Stpi	18	.444	Ea.	690	16.70		706.70	785
	1220	3/4" pipe size		16	.500		690	18.80		708.80	790
	1230	1" pipe size		14	.571		1,700	21.50		1,721.50	1,900
	1240	1-1/4" pipe size	↓	12	.667	↓	1,700	25		1,725	1,925
	2000	High pressure high leverage type									
	2100	Forged steel body, threaded									
	2110	1/2" pipe size	1 Stpi	18	.444	Ea.	1,350	16.70		1,366.70	1,525
	2120	3/4" pipe size		16	.500		1,350	18.80		1,368.80	1,525
	2130	1" pipe size		14	.571		1,600	21.50		1,621.50	1,775
	2140	1-1/4" pipe size	↓	12	.667		2,250	25		2,275	2,525
	2150	1-1/2" pipe size	Q-5	10	1.600		3,425	54		3,479	3,850
	2160	2" pipe size	"	8	2	↓	3,425	67.50		3,492.50	3,875
	2200	Socket weld									
	2210	1/2" pipe size	Q-17	16	1	Ea.	1,475	34	4.91	1,513.91	1,675
	2220	3/4" pipe size		14	1.143		1,475	38.50	5.60	1,519.10	1,700
	2230	1" pipe size		12	1.333		2,375	45	6.55	2,426.55	2,675
	2240	1-1/4" pipe size		11	1.455		2,375	49	7.15	2,431.15	2,675
	2250	1-1/2" pipe size		9	1.778		3,525	60	8.75	3,593.75	3,975
	2260	2" pipe size	↓	7	2.286	↓	3,525	77.50	11.20	3,613.70	4,000
	2300	Flanged									
	2310	1/2" pipe size	1 Stpi	12	.667	Ea.	2,175	25		2,200	2,450
	2320	3/4" pipe size		10	.800		2,175	30		2,205	2,450
	2330	1" pipe size		7	1.143		3,350	43		3,393	3,750
	2340	1-1/4" pipe size	↓	6	1.333		3,350	50		3,400	3,750
	2350	1-1/2" pipe size	Q-5	9	1.778		4,525	60		4,585	5,075
	2360	2" pipe size	"	7	2.286	↓	4,525	77.50		4,602.50	5,100
	3000	Guided lever dual gravity type									
	3100	Cast iron body, threaded									
	3110	1/2" pipe size	1 Stpi	22	.364	Ea.	261	13.65		274.65	310
	3120	3/4" pipe size		20	.400		261	15.05		276.05	310
	3130	1" pipe size		19	.421		380	15.85		395.85	445
	3140	1-1/2" pipe size		11	.727		735	27.50		762.50	850
	3150	2" pipe size	↓	8	1	↓	735	37.50		772.50	865
	3200	Forged steel body, threaded									
	3210	1/2" pipe size	1 Stpi	18	.444	Ea.	740	16.70		756.70	840
	3220	3/4" pipe size		16	.500		735	18.80		753.80	840
	3230	1" pipe size	↓	14	.571		1,100	21.50		1,121.50	1,225
	3240	1-1/2" pipe size	Q-5	10	1.600	↓	2,925	54		2,979	3,275

MECHANICAL 15

			DAILY	LABOR-		2003 BARE COSTS				TOTAL	
15120		**Piping Specialties**	CREW	OUTPUT	HOURS	UNIT	MAT.	LABOR	EQUIP.	TOTAL	INCL O&P
550	3250	2" pipe size	Q-5	8	2	Ea.	2,925	67.50		2,992.50	3,300	**550**
	3300	Socket weld									
	3310	1/2" pipe size	Q-17	16	1	Ea.	835	34	4.91	873.91	975
	3320	3/4" pipe size		14	1.143		835	38.50	5.60	879.10	985
	3330	1" pipe size		12	1.333		1,200	45	6.55	1,251.55	1,400
	3340	1-1/2" pipe size		9	1.778		3,025	60	8.75	3,093.75	3,425
	3350	2" pipe size	▼	7	2.286	▼	3,025	77.50	11.20	3,113.70	3,450
	3400	Flanged									
	3410	1/2" pipe size	1 Stpi	12	.667	Ea.	1,475	25		1,500	1,650
	3420	3/4" pipe size		10	.800		1,475	30		1,505	1,650
	3430	1" pipe size	▼	7	1.143		1,900	43		1,943	2,175
	3440	1-1/2" pipe size	Q-5	9	1.778		4,025	60		4,085	4,525
	3450	2" pipe size	"	7	2.286	▼	4,025	77.50		4,102.50	4,550
580	0010	**MIXING VALVE** Automatic water tempering										**580**
	0040	1/2" size	1 Stpi	19	.421	Ea.	375	15.85		390.85	440
	0050	3/4" size		18	.444		375	16.70		391.70	440
	0100	1" size		16	.500		540	18.80		558.80	625
	0120	1-1/4" size		13	.615		780	23		803	895
	0140	1-1/2" size		10	.800		930	30		960	1,075
	0160	2" size		8	1		1,175	37.50		1,212.50	1,350
	0170	2-1/2" size		6	1.333		1,175	50		1,225	1,375
	0180	3" size		4	2		2,700	75		2,775	3,100
	0190	4" size	▼	3	2.667	▼	2,700	100		2,800	3,125
610	0010	**MONOFLOW TEE FITTING**										**610**
	1100	For one pipe hydronic, supply and return									
	1110	Copper, soldered									
	1120	3/4" x 1/2" size	1 Stpi	13	.615	Ea.	12.75	23		35.75	49
	1130	1" x 1/2" size		12	.667		12.75	25		37.75	52
	1140	1" x 3/4" size		11	.727		12.75	27.50		40.25	55.50
	1150	1-1/4" x 1/2" size		11	.727		13.50	27.50		41	56.50
	1160	1-1/4" x 3/4" size		10	.800		13.50	30		43.50	60.50
	1170	1-1/2" x 3/4" size		10	.800		18.95	30		48.95	66.50
	1180	1-1/2" x 1" size		9	.889		18.95	33.50		52.45	71.50
	1190	2" x 3/4" size		9	.889		28.50	33.50		62	82
	1200	2" x 1" size	▼	8	1	▼	28.50	37.50		66	88.50
640	0010	**PRESSURE REDUCING VALVE** Steam, pilot operated										**640**
	0100	Threaded, iron body									
	0200	1-1/2" size	1 Stpi	8	1	Ea.	640	37.50		677.50	760
	0220	2" size	"	5	1.600	"	835	60		895	1,000
	1000	Flanged, iron body, 125 lb. flanges									
	1020	2" size	1 Stpi	8	1	Ea.	880	37.50		917.50	1,025
	1040	2-1/2" size	"	4	2		1,100	75		1,175	1,350
	1060	3" size	Q-5	4.50	3.556		1,125	120		1,245	1,425
	1080	4" size	"	3	5.333		1,425	181		1,606	1,850
	1500	For 250 lb. flanges, add				▼	5%				
670	0010	**PRESSURE REGULATOR**										**670**
	0100	Gas appliance regulators									
	0106	Main burner and pilot applications									
	0108	Rubber seat poppet type									
	0109	1/8" pipe size	1 Stpi	24	.333	Ea.	10.60	12.55		23.15	30.50
	0110	1/4" pipe size		24	.333		12.15	12.55		24.70	32.50
	0112	3/8" pipe size		24	.333		13.10	12.55		25.65	33.50
	0113	1/2" pipe size		24	.333		14.55	12.55		27.10	35
	0114	3/4" pipe size	▼	20	.400	▼	17.45	15.05		32.50	41.50
	0122	Lever action type									

Important: See the Reference Section for critical supporting data - Reference Nos., Crews, & City Cost Indexes

		DAILY	LABOR-		2003 BARE COSTS				TOTAL
15120 | **Piping Specialties**	CREW	OUTPUT	HOURS	UNIT	MAT.	LABOR	EQUIP.	TOTAL	INCL O&P
670	0123	3/8" pipe size	1 Stpi	24	.333	Ea.	21.50	12.55		34.05	42.50	670
	0124	1/2" pipe size		24	.333		21.50	12.55		34.05	42.50
	0125	3/4" pipe size		20	.400		43	15.05		58.05	70
	0126	1" pipe size		19	.421		43	15.85		58.85	71.50
	0132	Double diaphragm type									
	0133	3/8" pipe size	1 Stpi	24	.333	Ea.	24.50	12.55		37.05	46
	0134	1/2" pipe size		24	.333		37.50	12.55		50.05	60
	0135	3/4" pipe size		20	.400		64	15.05		79.05	93
	0136	1" pipe size		19	.421		78.50	15.85		94.35	110
	0137	1-1/4" pipe size		15	.533		255	20		275	310
	0138	1-1/2" pipe size		13	.615		475	23		498	555
	0139	2" pipe size		11	.727		475	27.50		502.50	560
	0140	2-1/2" pipe size	Q-5	15	1.067		925	36		961	1,075
	0141	3" pipe size		13	1.231		925	41.50		966.50	1,100
	0142	4" pipe size (flanged)		8	2		1,700	67.50		1,767.50	1,975
	0160	Main burner only									
	0162	Straight-thru-flow design									
	0163	1/2" pipe size	1 Stpi	24	.333	Ea.	27.50	12.55		40.05	49.50
	0164	3/4" pipe size		20	.400		37.50	15.05		52.55	64
	0165	1" pipe size		19	.421		56	15.85		71.85	85.50
	0166	1-1/4" pipe size		15	.533		56	20		76	92
	0200	Oil, light, hot water, ordinary steam, threaded									
	0220	Bronze body, 1/4" size	1 Stpi	24	.333	Ea.	100	12.55		112.55	129
	0230	3/8" size		24	.333		108	12.55		120.55	138
	0240	1/2" size		24	.333		134	12.55		146.55	166
	0250	3/4" size		20	.400		160	15.05		175.05	199
	0260	1" size		19	.421		243	15.85		258.85	292
	0270	1-1/4" size		15	.533		325	20		345	390
	0280	1-1/2" size		13	.615		350	23		373	425
	0290	2" size		11	.727		640	27.50		667.50	745
	0320	Iron body, 1/4" size		24	.333		78	12.55		90.55	105
	0330	3/8" size		24	.333		92	12.55		104.55	120
	0340	1/2" size		24	.333		97.50	12.55		110.05	126
	0350	3/4" size		20	.400		118	15.05		133.05	153
	0360	1" size		19	.421		154	15.85		169.85	194
	0370	1-1/4" size		15	.533		214	20		234	266
	0380	1-1/2" size		13	.615		238	23		261	297
	0390	2" size		11	.727		345	27.50		372.50	420
	0500	Oil, heavy, viscous fluids, threaded									
	0520	Bronze body, 3/8" size	1 Stpi	24	.333	Ea.	172	12.55		184.55	208
	0530	1/2" size		24	.333		221	12.55		233.55	262
	0540	3/4" size		20	.400		245	15.05		260.05	293
	0550	1" size		19	.421		305	15.85		320.85	365
	0560	1-1/4" size		15	.533		435	20		455	505
	0570	1-1/2" size		13	.615		495	23		518	580
	0600	Iron body, 3/8" size		24	.333		137	12.55		149.55	169
	0620	1/2" size		24	.333		166	12.55		178.55	201
	0630	3/4" size		20	.400		188	15.05		203.05	229
	0640	1" size		19	.421		222	15.85		237.85	268
	0650	1-1/4" size		15	.533		320	20		340	385
	0660	1-1/2" size		13	.615		345	23		368	415
	0800	Process steam, wet or super heated, monel trim, threaded									
	0820	Bronze body, 1/4" size	1 Stpi	24	.333	Ea.	287	12.55		299.55	335
	0830	3/8" size		24	.333		287	12.55		299.55	335
	0840	1/2" size		24	.333		315	12.55		327.55	365
	0850	3/4" size		20	.400		350	15.05		365.05	405

MECHANICAL 15

15120	Piping Specialties	CREW	DAILY OUTPUT	LABOR-HOURS	UNIT	2003 BARE COSTS				TOTAL INCL O&P
						MAT.	LABOR	EQUIP.	TOTAL	
670	0860	1" size	1 Stpi	19	.421	Ea.	445	15.85		460.85	515	670
	0870	1-1/4" size		15	.533		540	20		560	620
	0880	1-1/2" size		13	.615		645	23		668	745
	0890	2" size		11	.727		935	27.50		962.50	1,075
	0920	Iron body, max 125 PSIG press out, 1/4" size		24	.333		98	12.55		110.55	127
	0930	3/8" size		24	.333		116	12.55		128.55	146
	0940	1/2" size		24	.333		123	12.55		135.55	154
	0950	3/4" size		20	.400		148	15.05		163.05	186
	0960	1" size		19	.421		194	15.85		209.85	237
	0970	1-1/4" size		15	.533		269	20		289	325
	0980	1-1/2" size		13	.615		299	23		322	365
	0990	2" size		11	.727		435	27.50		462.50	515
	1000	Flanged, Class 125									
	1006	2-1/2" size	Q-5	10	1.600	Ea.	1,100	54		1,154	1,300
	1008	3" size		9	1.778		1,125	60		1,185	1,350
	1010	4" size		8	2		1,425	67.50		1,492.50	1,675
	1020	6" size		7	2.286		4,575	77.50		4,652.50	5,150
	3000	Steam, high capacity, bronze body, stainless steel trim									
	3020	Threaded, 1/2" diameter	1 Stpi	24	.333	Ea.	795	12.55		807.55	890
	3030	3/4" diameter		24	.333		795	12.55		807.55	890
	3040	1" diameter		19	.421		885	15.85		900.85	1,000
	3060	1-1/4" diameter		15	.533		980	20		1,000	1,100
	3080	1-1/2" diameter		13	.615		1,125	23		1,148	1,250
	3100	2" diameter		11	.727		1,375	27.50		1,402.50	1,550
	3120	2-1/2" diameter	Q-5	12	1.333		1,725	45		1,770	1,975
	3140	3" diameter	"	11	1.455		1,950	49		1,999	2,225
	3500	Flanged connection, iron body, 125 lb. W.S.P.									
	3520	3" diameter	Q-5	11	1.455	Ea.	2,150	49		2,199	2,425
	3540	4" diameter	"	5	3.200	"	2,700	108		2,808	3,150
	9000	For water pressure regulators, see Div. 15110-160									
700	0010	**PRESSURE & TEMPERATURE SAFETY PLUG**									700
	1000	3/4" external thread, 3/8" diam. element									
	1020	Carbon steel									
	1050	7-1/2" insertion	1 Stpi	32	.250	Ea.	59.50	9.40		68.90	79
	1120	304 stainless steel									
	1150	7-1/2" insertion	1 Stpi	32	.250	Ea.	54	9.40		63.40	73.50
	1220	316 Stainless steel									
	1250	7-1/2" insertion	1 Stpi	32	.250	Ea.	57	9.40		66.40	77
730	0010	**SLEEVES AND ESCUTCHEONS**									730
	0100	Pipe sleeve									
	0110	Steel, w/water stop, 12" long, with link seal									
	0120	2" diam. for 1/2" carrier pipe	1 Plum	8.40	.952	Ea.	25	35.50		60.50	81
	0130	2-1/2" diam. for 3/4" carrier pipe		8	1		30	37.50		67.50	89.50
	0140	2-1/2" diam. for 1" carrier pipe		8	1		27.50	37.50		65	87
	0150	3" diam. for 1-1/4" carrier pipe		7.20	1.111		39	41.50		80.50	106
	0160	3-1/2" diam. for 1-1/2" carrier pipe		6.80	1.176		40.50	44		84.50	111
	0170	4" diam. for 2" carrier pipe		6	1.333		45	50		95	125
	0180	4" diam. for 2-1/2" carrier pipe		6	1.333		44.50	50		94.50	124
	0190	5" diam. for 3" carrier pipe		5.40	1.481		56	55.50		111.50	145
	0200	6" diam. for 4" carrier pipe		4.80	1.667		65	62.50		127.50	166
	0210	10" diam. for 6" carrier pipe	Q-1	8	2		121	67.50		188.50	235
	0220	12" diam. for 8" carrier pipe		7.20	2.222		174	74.50		248.50	305
	0230	14" diam. for 10" carrier pipe		6.40	2.500		183	84		267	330
	0240	16" diam. for 12" carrier pipe		5.80	2.759		214	93		307	375
	0250	18" diam. for 14" carrier pipe		5.20	3.077		241	103		344	420
	0260	24" diam. for 18" carrier pipe		4	4		400	135		535	645

15 MECHANICAL

Important: See the Reference Section for critical supporting data - Reference Nos., Crews, & City Cost Indexes

15120 | Piping Specialties

		CREW	DAILY OUTPUT	LABOR-HOURS	UNIT	2003 BARE COSTS				TOTAL INCL O&P	
						MAT.	LABOR	EQUIP.	TOTAL		
730	0270	24" diam. for 20" carrier pipe	Q-1	4	4	Ea.	325	135		460	560	730
	0280	30" diam. for 24" carrier pipe	↓	3.20	5	↓	735	168		903	1,075	
	0500	Wall sleeve										
	0510	Ductile iron with rubber gasket seal										
	0520	3"	1 Plum	8.40	.952	Ea.	96	35.50		131.50	160	
	0530	4"		7.20	1.111		120	41.50		161.50	195	
	0540	6"		6	1.333		161	50		211	252	
	0550	8"		4	2		206	74.50		280.50	340	
	0560	10"		3	2.667		231	99.50		330.50	405	
	0570	12"	↓	2.40	3.333	↓	274	125		399	490	
	5000	Escutcheon										
	5100	Split ring, pipe										
	5110	Chrome plated										
	5120	1/2"	1 Plum	160	.050	Ea.	.95	1.87		2.82	3.87	
	5130	3/4"		160	.050		.98	1.87		2.85	3.90	
	5140	1"		135	.059		1.06	2.21		3.27	4.51	
	5150	1-1/2"		115	.070		1.60	2.60		4.20	5.70	
	5160	2"		100	.080		1.66	2.99		4.65	6.35	
	5170	4"		80	.100		3.69	3.74		7.43	9.70	
	5180	6"	↓	68	.118	↓	6.10	4.39		10.49	13.40	
760	0010	**STEAM TRAP**										760
	0030	Cast iron body, threaded										
	0040	Inverted bucket										
	0050	1/2" pipe size	1 Stpi	12	.667	Ea.	86.50	25		111.50	134	
	0070	3/4" pipe size		10	.800		143	30		173	204	
	0100	1" pipe size		9	.889		230	33.50		263.50	305	
	0120	1-1/4" pipe size		8	1		345	37.50		382.50	435	
	0130	1-1/2" pipe size		7	1.143		385	43		428	490	
	0140	2" pipe size	↓	6	1.333	↓	575	50		625	710	
	0200	With thermic vent & check valve										
	0202	1/2" pipe size	1 Stpi	12	.667	Ea.	141	25		166	193	
	0204	3/4" pipe size		10	.800		209	30		239	276	
	0206	1" pipe size		9	.889		305	33.50		338.50	385	
	0208	1-1/4" pipe size		8	1		460	37.50		497.50	560	
	0210	1-1/2" pipe size		7	1.143		620	43		663	750	
	0212	2" pipe size	↓	6	1.333		945	50		995	1,100	
	1000	Float & thermostatic, 15 psi										
	1010	3/4" pipe size	1 Stpi	16	.500	Ea.	100	18.80		118.80	139	
	1020	1" pipe size		15	.533		100	20		120	141	
	1030	1-1/4" pipe size		13	.615		123	23		146	170	
	1040	1-1/2" pipe size		9	.889		179	33.50		212.50	248	
	1060	2" pipe size	↓	6	1.333		330	50		380	435	
	1100	For vacuum breaker, add				↓	44.50			44.50	49	
	1110	Differential condensate controller										
	1120	With union and manual metering valve										
	1130	1/2" pipe size	1 Stpi	24	.333	Ea.	360	12.55		372.55	415	
	1140	3/4" pipe size		20	.400		415	15.05		430.05	485	
	1150	1" pipe size		19	.421		560	15.85		575.85	645	
	1160	1-1/4" pipe size		15	.533		755	20		775	865	
	1170	1-1/2" pipe size		13	.615		915	23		938	1,025	
	1180	2" pipe size	↓	11	.727	↓	1,200	27.50		1,227.50	1,375	
	1290	Brass body, threaded										
	1300	Thermostatic, angle union, 25 psi										
	1310	1/2" pipe size	1 Stpi	24	.333	Ea.	40	12.55		52.55	62.50	
	1320	3/4" pipe size		20	.400		70.50	15.05		85.55	100	
	1330	1" pipe size	↓	19	.421	↓	104	15.85		119.85	138	

MECHANICAL 15

15120	Piping Specialties	CREW	DAILY OUTPUT	LABOR-HOURS	UNIT	2003 BARE COSTS				TOTAL INCL O&P
						MAT.	LABOR	EQUIP.	TOTAL	
1990	Carbon steel body, threaded									
2000	Controlled disc, integral strainer & blow down									
2010	Threaded									
2020	3/8" pipe size	1 Stpi	24	.333	Ea.	158	12.55		170.55	193
2030	1/2" pipe size		24	.333		158	12.55		170.55	193
2040	3/4" pipe size		20	.400		212	15.05		227.05	256
2050	1" pipe size	▼	19	.421	▼	272	15.85		287.85	325
2100	Socket weld									
2110	3/8" pipe size	Q-17	17	.941	Ea.	189	32	4.62	225.62	260
2120	1/2" pipe size		16	1		189	34	4.91	227.91	263
2130	3/4" pipe size		14	1.143		245	38.50	5.60	289.10	335
2140	1" pipe size	▼	12	1.333	▼	305	45	6.55	356.55	415
2170	Flanged, 600 lb. ASA									
2180	3/8" pipe size	1 Stpi	12	.667	Ea.	465	25		490	550
2190	1/2" pipe size		12	.667		465	25		490	550
2200	3/4" pipe size		10	.800		555	30		585	655
2210	1" pipe size	▼	7	1.143	▼	655	43		698	785
5000	Forged steel body									
5010	Inverted bucket									
5020	Threaded									
5030	1/2" pipe size	1 Stpi	12	.667	Ea.	281	25		306	350
5040	3/4" pipe size		10	.800		485	30		515	580
5050	1" pipe size		9	.889		730	33.50		763.50	850
5060	1-1/4" pipe size		8	1		1,050	37.50		1,087.50	1,225
5070	1-1/2" pipe size		7	1.143		1,500	43		1,543	1,725
5080	2" pipe size	▼	6	1.333	▼	2,625	50		2,675	2,975
5100	Socket weld									
5110	1/2" pipe size	Q-17	16	1	Ea.	340	34	4.91	378.91	425
5120	3/4" pipe size		14	1.143		545	38.50	5.60	589.10	665
5130	1" pipe size		12	1.333		790	45	6.55	841.55	945
5140	1-1/4" pipe size		10	1.600		1,125	54	7.85	1,186.85	1,325
5150	1-1/2" pipe size		9	1.778		1,550	60	8.75	1,618.75	1,800
5160	2" pipe size	▼	7	2.286	▼	2,700	77.50	11.20	2,788.70	3,100
5200	Flanged									
5210	1/2" pipe size	1 Stpi	12	.667	Ea.	570	25		595	665
5220	3/4" pipe size		10	.800		920	30		950	1,050
5230	1" pipe size		7	1.143		1,300	43		1,343	1,500
5240	1-1/4" pipe size	▼	6	1.333		1,800	50		1,850	2,075
5250	1-1/2" pipe size	Q-5	9	1.778		2,250	60		2,310	2,575
5260	2" pipe size	"	7	2.286	▼	3,450	77.50		3,527.50	3,925
6000	Stainless steel body									
6010	Inverted bucket									
6020	Threaded									
6030	1/2" pipe size	1 Stpi	12	.667	Ea.	93.50	25		118.50	141
6040	3/4" pipe size		10	.800		122	30		152	180
6050	1" pipe size	▼	9	.889	▼	435	33.50		468.50	530
6100	Threaded with thermic vent									
6110	1/2" pipe size	1 Stpi	12	.667	Ea.	107	25		132	155
6120	3/4" pipe size		10	.800		137	30		167	197
6130	1" pipe size	▼	9	.889	▼	450	33.50		483.50	550
6200	Threaded with check valve									
6210	1/2" pipe size	1 Stpi	12	.667	Ea.	133	25		158	184
6220	3/4" pipe size	"	10	.800	"	181	30		211	245
6300	Socket weld									
6310	1/2" pipe size	Q-17	16	1	Ea.	108	34	4.91	146.91	175
6320	3/4" pipe size	"	14	1.143	"	139	38.50	5.60	183.10	218

Important: See the Reference Section for critical supporting data - Reference Nos., Crews, & City Cost Indexes

15120	Piping Specialties	CREW	DAILY OUTPUT	LABOR-HOURS	UNIT	2003 BARE COSTS				TOTAL INCL O&P	
						MAT.	LABOR	EQUIP.	TOTAL		
790	**0010**	**STRAINERS, BASKET TYPE** Perforated stainless steel basket										**790**
	0100	Brass or monel available									
	2000	Simplex style									
	2300	Bronze body									
	2320	Screwed, 3/8" pipe size	1 Stpi	22	.364	Ea.	98	13.65		111.65	129
	2340	1/2" pipe size		20	.400		100	15.05		115.05	133
	2360	3/4" pipe size		17	.471		103	17.70		120.70	140
	2380	1" pipe size		15	.533		134	20		154	178
	2400	1-1/4" pipe size		13	.615		201	23		224	256
	2420	1-1/2" pipe size		12	.667		204	25		229	262
	2440	2" pipe size	▼	10	.800		310	30		340	385
	2460	2-1/2" pipe size	Q-5	15	1.067		440	36		476	540
	2480	3" pipe size	"	14	1.143		635	38.50		673.50	760
	2600	Flanged, 2" pipe size	1 Stpi	6	1.333		465	50		515	590
	2620	2-1/2" pipe size	Q-5	4.50	3.556		730	120		850	985
	2640	3" pipe size		3.50	4.571		840	155		995	1,150
	2660	4" pipe size	▼	3	5.333		1,425	181		1,606	1,850
	2680	5" pipe size	Q-6	3.40	7.059		2,150	248		2,398	2,725
	2700	6" pipe size		3	8		2,725	281		3,006	3,400
	2710	8" pipe size	▼	2.50	9.600	▼	4,500	335		4,835	5,450
	3600	Iron body									
	3700	Screwed, 3/8" pipe size	1 Stpi	22	.364	Ea.	112	13.65		125.65	144
	3720	1/2" pipe size		20	.400		112	15.05		127.05	146
	3740	3/4" pipe size		17	.471		147	17.70		164.70	189
	3760	1" pipe size		15	.533		147	20		167	193
	3780	1-1/4" pipe size		13	.615		197	23		220	251
	3800	1-1/2" pipe size		12	.667		216	25		241	275
	3820	2" pipe size	▼	10	.800		258	30		288	330
	3840	2-1/2" pipe size	Q-5	15	1.067		370	36		406	460
	3860	3" pipe size	"	14	1.143		420	38.50		458.50	520
	4000	Flanged, 2" pipe size	1 Stpi	6	1.333		350	50		400	460
	4020	2-1/2" pipe size	Q-5	4.50	3.556		475	120		595	705
	4040	3" pipe size		3.50	4.571		505	155		660	790
	4060	4" pipe size	▼	3	5.333		760	181		941	1,100
	4080	5" pipe size	Q-6	3.40	7.059		1,200	248		1,448	1,700
	4100	6" pipe size		3	8		1,475	281		1,756	2,050
	4120	8" pipe size		2.50	9.600		2,725	335		3,060	3,500
	4140	10" pipe size		2.20	10.909		4,100	385		4,485	5,075
	4160	12" pipe size		1.70	14.118		4,875	495		5,370	6,125
	4180	14" pipe size		1.40	17.143		6,150	600		6,750	7,650
	4200	16" pipe size	▼	1	24	▼	6,425	840		7,265	8,350
	6000	Cast steel body									
	6400	Screwed, 1" pipe size	1 Stpi	15	.533	Ea.	241	20		261	296
	6410	1-1/4" pipe size		13	.615		365	23		388	435
	6420	1-1/2" pipe size		12	.667		365	25		390	440
	6440	2" pipe size	▼	10	.800		515	30		545	610
	6460	2-1/2" pipe size	Q-5	15	1.067		720	36		756	850
	6480	3" pipe size	"	14	1.143		945	38.50		983.50	1,100
	6560	Flanged, 2" pipe size	1 Stpi	6	1.333		750	50		800	900
	6580	2-1/2" pipe size	Q-5	4.50	3.556		1,100	120		1,220	1,375
	6600	3" pipe size		3.50	4.571		1,200	155		1,355	1,525
	6620	4" pipe size	▼	3	5.333		1,675	181		1,856	2,125
	6640	6" pipe size	Q-6	3	8		3,050	281		3,331	3,775
	6660	8" pipe size	"	2.50	9.600	▼	4,950	335		5,285	5,950
	7000	Stainless steel body									
	7200	Screwed, 1" pipe size	1 Stpi	15	.533	Ea.	395	20		415	465

MECHANICAL 15

15120	Piping Specialties	CREW	DAILY OUTPUT	LABOR-HOURS	UNIT	2003 BARE COSTS				TOTAL INCL O&P	
						MAT.	LABOR	EQUIP.	TOTAL		
790	7210	1-1/4" pipe size	1 Stpi	13	.615	Ea.	575	23		598	670	790
	7220	1-1/2" pipe size		12	.667		575	25		600	675
	7240	2" pipe size		10	.800		845	30		875	975
	7260	2-1/2" pipe size	Q-5	15	1.067		1,400	36		1,436	1,600
	7280	3" pipe size	"	14	1.143		1,850	38.50		1,888.50	2,100
	7400	Flanged, 2" pipe size	1 Stpi	6	1.333		1,150	50		1,200	1,350
	7420	2-1/2" pipe size	Q-5	4.50	3.556		2,100	120		2,220	2,500
	7440	3" pipe size		3.50	4.571		2,150	155		2,305	2,600
	7460	4" pipe size		3	5.333		3,375	181		3,556	4,000
	7480	6" pipe size	Q-6	3	8		5,850	281		6,131	6,875
	7500	8" pipe size	"	2.50	9.600		9,925	335		10,260	11,400
	8100	Duplex style									
	8200	Bronze body									
	8240	Screwed, 3/4" pipe size	1 Stpi	16	.500	Ea.	1,050	18.80		1,068.80	1,175
	8260	1" pipe size		14	.571		1,050	21.50		1,071.50	1,175
	8280	1-1/4" pipe size		12	.667		2,050	25		2,075	2,300
	8300	1-1/2" pipe size		11	.727		2,050	27.50		2,077.50	2,300
	8320	2" pipe size		9	.889		3,250	33.50		3,283.50	3,625
	8340	2-1/2" pipe size	Q-5	14	1.143		3,250	38.50		3,288.50	3,625
	8420	Flanged, 2" pipe size	1 Stpi	6	1.333		3,675	50		3,725	4,125
	8440	2-1/2" pipe size	Q-5	4.50	3.556		3,775	120		3,895	4,325
	8460	3" pipe size		3.50	4.571		5,200	155		5,355	5,925
	8480	4" pipe size		3	5.333		8,200	181		8,381	9,300
	8500	5" pipe size	Q-6	3.40	7.059		14,300	248		14,548	16,100
	8520	6" pipe size	"	3	8		18,000	281		18,281	20,200
	8700	Iron body									
	8740	Screwed, 3/4" pipe size	1 Stpi	16	.500	Ea.	545	18.80		563.80	630
	8760	1" pipe size		14	.571		545	21.50		566.50	635
	8780	1-1/4" pipe size		12	.667		950	25		975	1,100
	8800	1-1/2" pipe size		11	.727		950	27.50		977.50	1,100
	8820	2" pipe size		9	.889		1,625	33.50		1,658.50	1,825
	8840	2-1/2" pipe size	Q-5	14	1.143		1,625	38.50		1,663.50	1,825
	9000	Flanged, 2" pipe size	1 Stpi	6	1.333		1,700	50		1,750	1,950
	9020	2-1/2" pipe size	Q-5	4.50	3.556		1,775	120		1,895	2,125
	9040	3" pipe size		3.50	4.571		2,250	155		2,405	2,700
	9060	4" pipe size		3	5.333		3,550	181		3,731	4,200
	9080	5" pipe size	Q-6	3.40	7.059		7,150	248		7,398	8,250
	9100	6" pipe size		3	8		7,150	281		7,431	8,300
	9120	8" pipe size		2.50	9.600		13,700	335		14,035	15,600
	9140	10" pipe size		2.20	10.909		19,100	385		19,485	21,600
	9160	12" pipe size		1.70	14.118		21,200	495		21,695	24,100
	9170	14" pipe size		1.40	17.143		25,700	600		26,300	29,200
	9180	16" pipe size		1	24		32,200	840		33,040	36,800
	9300	Cast steel body									
	9340	Screwed, 1" pipe size	1 Stpi	14	.571	Ea.	1,525	21.50		1,546.50	1,700
	9360	1-1/2" pipe size		11	.727		2,250	27.50		2,277.50	2,525
	9380	2" pipe size		9	.889		3,825	33.50		3,858.50	4,250
	9460	Flanged, 2" pipe size		6	1.333		3,825	50		3,875	4,300
	9480	2-1/2" pipe size	Q-5	4.50	3.556		4,225	120		4,345	4,825
	9500	3" pipe size		3.50	4.571		5,575	155		5,730	6,375
	9520	4" pipe size		3	5.333		8,175	181		8,356	9,275
	9540	6" pipe size	Q-6	3	8		14,000	281		14,281	15,800
	9560	8" pipe size	"	2.50	9.600		27,700	335		28,035	31,000
	9700	Stainless steel body									
	9740	Screwed, 1" pipe size	1 Stpi	14	.571	Ea.	2,775	21.50		2,796.50	3,075
	9760	1-1/2" pipe size		11	.727		4,200	27.50		4,227.50	4,675

Important: See the Reference Section for critical supporting data - Reference Nos., Crews, & City Cost Indexes

15120	Piping Specialties		CREW	DAILY OUTPUT	LABOR-HOURS	UNIT	2003 BARE COSTS				TOTAL INCL O&P	
							MAT.	LABOR	EQUIP.	TOTAL		
790	9780	2" pipe size	1 Stpi	9	.889	Ea.	6,775	33.50		6,808.50	7,500	**790**
	9860	Flanged, 2" pipe size	↓	6	1.333		7,050	50		7,100	7,825	
	9880	2-1/2" pipe size	Q-5	4.50	3.556		7,400	120		7,520	8,325	
	9900	3" pipe size		3.50	4.571		11,600	155		11,755	13,000	
	9920	4" pipe size	↓	3	5.333		14,200	181		14,381	15,900	
	9940	6" pipe size	Q-6	3	8		25,000	281		25,281	27,900	
	9960	8" pipe size	"	2.50	9.600	↓	46,300	335		46,635	51,500	
820	0010	**STRAINERS, Y TYPE** Bronze body										**820**
	0050	Screwed, 150 lb., 1/4" pipe size	1 Stpi	24	.333	Ea.	11.40	12.55		23.95	31.50	
	0070	3/8" pipe size		24	.333		15.90	12.55		28.45	36.50	
	0100	1/2" pipe size		20	.400		15.90	15.05		30.95	40	
	0120	3/4" pipe size		19	.421		17.40	15.85		33.25	43	
	0140	1" pipe size		17	.471		18.70	17.70		36.40	47	
	0150	1-1/4" pipe size		15	.533		42	20		62	76.50	
	0160	1-1/2" pipe size		14	.571		40.50	21.50		62	77	
	0180	2" pipe size		13	.615		53.50	23		76.50	94	
	0182	3" pipe size		12	.667		310	25		335	380	
	0200	300 lb., 2-1/2" pipe size	Q-5	17	.941		286	32		318	365	
	0220	3" pipe size		16	1		565	34		599	675	
	0240	4" pipe size	↓	15	1.067	↓	1,300	36		1,336	1,475	
	0500	For 300 lb rating 1/4" thru 2", add					15%					
	1000	Flanged, 150 lb., 1-1/2" pipe size	1 Stpi	11	.727	Ea.	305	27.50		332.50	375	
	1020	2" pipe size	"	8	1		390	37.50		427.50	485	
	1030	2-1/2" pipe size	Q-5	5	3.200		505	108		613	720	
	1040	3" pipe size		4.50	3.556		620	120		740	865	
	1060	4" pipe size	↓	3	5.333		945	181		1,126	1,300	
	1080	5" pipe size	Q-6	3.40	7.059		1,925	248		2,173	2,500	
	1100	6" pipe size		3	8		2,500	281		2,781	3,175	
	1106	8" pipe size	↓	2.60	9.231	↓	3,900	325		4,225	4,800	
	1500	For 300 lb rating, add					40%					
840	0010	**STRAINERS, Y TYPE** Iron body										**840**
	0050	Screwed, 250 lb., 1/4" pipe size	1 Stpi	20	.400	Ea.	6.65	15.05		21.70	30	
	0070	3/8" pipe size		20	.400		6.65	15.05		21.70	30	
	0100	1/2" pipe size		20	.400		6.65	15.05		21.70	30	
	0120	3/4" pipe size		18	.444		7.95	16.70		24.65	34	
	0140	1" pipe size		16	.500		10.75	18.80		29.55	40.50	
	0150	1-1/4" pipe size		15	.533		14.75	20		34.75	46.50	
	0160	1-1/2" pipe size		12	.667		17.75	25		42.75	57.50	
	0180	2" pipe size	↓	8	1		27.50	37.50		65	87	
	0200	2-1/2" pipe size	Q-5	12	1.333		140	45		185	222	
	0220	3" pipe size		11	1.455		151	49		200	241	
	0240	4" pipe size	↓	5	3.200	↓	255	108		363	445	
	0500	For galvanized body, add					50%					
	1000	Flanged, 125 lb., 1-1/2" pipe size	1 Stpi	11	.727	Ea.	86	27.50		113.50	136	
	1020	2" pipe size	"	8	1		64	37.50		101.50	128	
	1030	2-1/2" pipe size	Q-5	5	3.200		72.50	108		180.50	244	
	1040	3" pipe size		4.50	3.556		84.50	120		204.50	275	
	1060	4" pipe size	↓	3	5.333		155	181		336	445	
	1080	5" pipe size	Q-6	3.40	7.059		242	248		490	640	
	1100	6" pipe size		3	8		295	281		576	750	
	1120	8" pipe size		2.50	9.600		495	335		830	1,050	
	1140	10" pipe size		2	12		950	420		1,370	1,675	
	1160	12" pipe size		1.70	14.118		1,425	495		1,920	2,325	
	1170	14" pipe size	↓	1.30	18.462	↓	2,625	650		3,275	3,875	

MECHANICAL 15

		15120	Piping Specialties	CREW	DAILY OUTPUT	LABOR-HOURS	UNIT	MAT.	LABOR	EQUIP.	TOTAL	TOTAL INCL O&P	
								2003 BARE COSTS					
840	1180		16" pipe size	Q-6	1	24	Ea.	3,725	840		4,565	5,375	840
	1500		For 250 lb rating, add					20%					
	2000		For galvanized body, add					50%					
	2500		For steel body, add					40%					
870	0010	**SUCTION DIFFUSERS**											870
	0100	Cast iron body with integral straightening vanes, strainer											
	1000	Flanged											
	1010	2" inlet, 1-1/2" pump side		1 Stpi	6	1.333	Ea.	219	50		269	315	
	1020	2" pump side		"	5	1.600		227	60		287	340	
	1030	3" inlet, 2" pump side		Q-5	6.50	2.462		278	83.50		361.50	430	
	1040	2-1/2" pump side			5.30	3.019		405	102		507	605	
	1050	3" pump side			4.50	3.556		410	120		530	630	
	1060	4" inlet, 3" pump side			3.50	4.571		480	155		635	765	
	1070	4" pump side			3	5.333		560	181		741	890	
	1080	5" inlet, 4" pump side		Q-6	3.80	6.316		635	222		857	1,025	
	1090	5" pump side			3.40	7.059		760	248		1,008	1,200	
	1100	6" inlet, 4" pump side			3.30	7.273		660	255		915	1,100	
	1110	5" pump side			3.10	7.742		795	272		1,067	1,275	
	1120	6" pump side			3	8		835	281		1,116	1,350	
	1130	8" inlet, 6" pump side			2.70	8.889		900	310		1,210	1,450	
	1140	8" pump side			2.50	9.600		1,550	335		1,885	2,225	
	1150	10" inlet, 6" pump side			2.40	10		2,175	350		2,525	2,900	
	1160	8" pump side			2.30	10.435		1,825	365		2,190	2,550	
	1170	10" pump side			2.20	10.909		2,600	385		2,985	3,450	
	1180	12" inlet, 8" pump side			2	12		2,750	420		3,170	3,650	
	1190	10" pump side			1.80	13.333		3,050	470		3,520	4,050	
	1200	12" pump side			1.70	14.118		3,425	495		3,920	4,525	
	1210	14" inlet, 12" pump side			1.50	16		4,450	560		5,010	5,750	
	1220	14" pump side			1.30	18.462		4,625	650		5,275	6,075	
900	0010	**THERMOFLO INDICATOR** For balancing											900
	1000	Sweat connections, 1-1/4" pipe size		1 Stpi	12	.667	Ea.	365	25		390	440	
	1020	1-1/2" pipe size			10	.800		370	30		400	455	
	1040	2" pipe size			8	1		390	37.50		427.50	485	
	1060	2-1/2" pipe size			7	1.143		595	43		638	720	
	2000	Flange connections, 3" pipe size		Q-5	5	3.200		710	108		818	945	
	2020	4" pipe size			4	4		855	135		990	1,150	
	2030	5" pipe size			3.50	4.571		1,075	155		1,230	1,400	
	2040	6" pipe size			3	5.333		1,150	181		1,331	1,550	
	2060	8" pipe size			2	8		1,400	271		1,671	1,925	
920	0010	**VENTURI FLOW** Measuring device											920
	0050	1/2" diameter		1 Stpi	24	.333	Ea.	175	12.55		187.55	212	
	0100	3/4" diameter			20	.400		163	15.05		178.05	202	
	0120	1" diameter			19	.421		169	15.85		184.85	210	
	0140	1-1/4" diameter			15	.533		203	20		223	254	
	0160	1-1/2" diameter			13	.615		206	23		229	262	
	0180	2" diameter			11	.727		222	27.50		249.50	286	
	0200	2-1/2" diameter		Q-5	16	1		330	34		364	415	
	0220	3" diameter			14	1.143		380	38.50		418.50	480	
	0240	4" diameter			11	1.455		595	49		644	730	
	0260	5" diameter		Q-6	4	6		720	211		931	1,100	
	0280	6" diameter			3.50	6.857		780	241		1,021	1,225	
	0300	8" diameter			3	8		1,075	281		1,356	1,600	
	0320	10" diameter			2	12		2,050	420		2,470	2,900	
	0330	12" diameter			1.80	13.333		3,400	470		3,870	4,425	
	0340	14" diameter			1.60	15		3,650	525		4,175	4,825	

Important: See the Reference Section for critical supporting data - Reference Nos., Crews, & City Cost Indexes

			DAILY	LABOR-		2003 BARE COSTS				TOTAL	
15120		**Piping Specialties**	CREW	OUTPUT	HOURS	UNIT	MAT.	LABOR	EQUIP.	TOTAL	INCL O&P	
920	0350	16" diameter	Q-6	1.40	17.143	Ea.	3,950	600		4,550	5,250	**920**
	0500	For meter, add				↓	1,175			1,175	1,300	
940	0010	**WATER SUPPLY METERS**										**940**
	1000	Detector, serves dual systems such as fire and domestic or										
	1020	process water, wide range cap., UL and FM approved										
	1100	3" mainline x 2" by-pass, 400 GPM	Q-1	3.60	4.444	Ea.	5,150	149		5,299	5,875	
	1140	4" mainline x 2" by-pass, 700 GPM	"	2.50	6.400		5,150	215		5,365	5,975	
	1180	6" mainline x 3" by-pass, 1600 GPM	Q-2	2.60	9.231		7,775	320		8,095	9,025	
	1220	8" mainline x 4" by-pass, 2800 GPM		2.10	11.429		11,500	400		11,900	13,200	
	1260	10" mainline x 6" by-pass, 4400 GPM		2	12		15,700	420		16,120	17,900	
	1300	10"x12" mainlines x 6" by-pass,5400 GPM	↓	1.70	14.118	↓	20,700	490		21,190	23,500	
	2000	Domestic/commercial, bronze										
	2020	Threaded										
	2060	5/8" diameter, to 20 GPM	1 Plum	16	.500	Ea.	40	18.70		58.70	72	
	2080	3/4" diameter, to 30 GPM		14	.571		67.50	21.50		89	107	
	2100	1" diameter, to 50 GPM	↓	12	.667	↓	94	25		119	141	
	2300	Threaded/flanged										
	2340	1-1/2" diameter, to 100 GPM	1 Plum	8	1	Ea.	330	37.50		367.50	420	
	2360	2" diameter, to 160 GPM	"	6	1.333	"	415	50		465	530	
	2600	Flanged, compound										
	2640	3" diameter, 320 GPM	Q-1	3	5.333	Ea.	1,950	179		2,129	2,400	
	2660	4" diameter, to 500 GPM		1.50	10.667		3,025	360		3,385	3,875	
	2680	6" diameter, to 1,000 GPM		1	16		4,350	540		4,890	5,575	
	2700	8" diameter, to 1,800 GPM	↓	.80	20	↓	8,600	675		9,275	10,500	
	7000	Turbine										
	7260	Flanged										
	7300	2" diameter, to 160 GPM	1 Plum	7	1.143	Ea.	455	42.50		497.50	565	
	7320	3" diameter, to 450 GPM	Q-1	3.60	4.444		745	149		894	1,050	
	7340	4" diameter, to 650 GPM	"	2.50	6.400		1,350	215		1,565	1,800	
	7360	6" diameter, to 1800 GPM	Q-2	2.60	9.231		2,550	320		2,870	3,275	
	7380	8" diameter, to 2500 GPM		2.10	11.429		4,050	400		4,450	5,050	
	7400	10" diameter, to 5500 GPM	↓	1.70	14.118	↓	5,400	490		5,890	6,700	
960	0010	**WELD END BALL JOINTS** Steel										**960**
	0050	2-1/2" diameter	Q-17	13	1.231	Ea.	455	41.50	6.05	502.55	570	
	0100	3" diameter		12	1.333		475	45	6.55	526.55	600	
	0120	4" diameter		11	1.455		765	49	7.15	821.15	925	
	0140	5" diameter	Q-18	14	1.714		1,100	60	5.60	1,165.60	1,325	
	0160	6" diameter		12	2		1,250	70	6.55	1,326.55	1,500	
	0180	8" diameter		9	2.667		1,775	93.50	8.70	1,877.20	2,100	
	0200	10" diameter		8	3		2,625	105	9.80	2,739.80	3,075	
	0220	12" diameter		6	4		3,475	140	13.10	3,628.10	4,050	
	0240	14" diameter	↓	4	6	↓	4,175	211	19.60	4,405.60	4,925	
980	0010	**ZONE VALVES**										**980**
	1000	Bronze body										
	1010	2 way, 125 psi										
	1020	Standard, 65' pump head										
	1030	1/2" soldered	1 Stpi	24	.333	Ea.	71.50	12.55		84.05	97.50	
	1040	3/4" soldered		20	.400		71.50	15.05		86.55	101	
	1050	1" soldered		19	.421		76.50	15.85		92.35	109	
	1060	1-1/4" soldered	↓	15	.533	↓	89.50	20		109.50	129	
	1200	High head, 125' pump head										
	1210	1/2" soldered	1 Stpi	24	.333	Ea.	96.50	12.55		109.05	125	
	1220	3/4" soldered		20	.400		100	15.05		115.05	133	
	1230	1" soldered	↓	19	.421	↓	110	15.85		125.85	145	
	1300	Geothermal, 150' pump head										
	1310	3/4" soldered	1 Stpi	20	.400	Ea.	115	15.05		130.05	150	

MECHANICAL 15

		15120	Piping Specialties		CREW	DAILY OUTPUT	LABOR-HOURS	UNIT	2003 BARE COSTS				TOTAL INCL O&P	
									MAT.	LABOR	EQUIP.	TOTAL		
980	1320		1" soldered		1 Stpi	19	.421	Ea.	123	15.85		138.85	159	980
	1330		3/4" threaded		↓	20	.400	↓	115	15.05		130.05	150	
	3000		3 way, 125 psi											
	3200		By-pass, 65' pump head											
	3210		1/2" soldered		1 Stpi	24	.333	Ea.	101	12.55		113.55	130	
	3220		3/4" soldered			20	.400		105	15.05		120.05	138	
	3230		1" soldered			19	.421		119	15.85		134.85	155	
	9000		Transformer, for up to 3 zone valves		↓	20	.400	↓	14.70	15.05		29.75	38.50	

		15140	Domestic Water Piping											
100	0010	BACKFLOW PREVENTER Includes valves												100
	0020	and four test cocks, corrosion resistant, automatic operation												
	1000	Double check principle												
	1010	Threaded, with ball valves												
	1020	3/4" pipe size			1 Plum	16	.500	Ea.	191	18.70		209.70	238	
	1030	1" pipe size				14	.571		194	21.50		215.50	245	
	1040	1-1/2" pipe size				10	.800		340	30		370	415	
	1050	2" pipe size			↓	7	1.143		405	42.50		447.50	510	
	1080	Threaded, with gate valves												
	1100	3/4" pipe size			1 Plum	16	.500	Ea.	620	18.70		638.70	710	
	1120	1" pipe size				14	.571		645	21.50		666.50	735	
	1140	1-1/2" pipe size				10	.800		735	30		765	855	
	1160	2" pipe size			↓	7	1.143		905	42.50		947.50	1,050	
	1200	Flanged, valves are gate												
	1210	3" pipe size			Q-1	4.50	3.556	Ea.	1,425	120		1,545	1,750	
	1220	4" pipe size			"	3	5.333		1,800	179		1,979	2,250	
	1230	6" pipe size			Q-2	3	8		3,025	279		3,304	3,750	
	1240	8" pipe size				2	12		5,625	420		6,045	6,825	
	1250	10" pipe size			↓	1	24	↓	8,000	835		8,835	10,100	
	1300	Flanged, valves are OS&Y												
	1370	1" pipe size			1 Plum	5	1.600	Ea.	675	60		735	835	
	1374	1-1/2" pipe size				5	1.600		770	60		830	940	
	1378	2" pipe size			↓	4.80	1.667		940	62.50		1,002.50	1,125	
	1380	3" pipe size			Q-1	4.50	3.556		1,525	120		1,645	1,850	
	1400	4" pipe size			"	3	5.333		2,175	179		2,354	2,675	
	1420	6" pipe size			Q-2	3	8		3,425	279		3,704	4,200	
	1430	8" pipe size			"	2	12	↓	6,350	420		6,770	7,625	
	4000	Reduced pressure principle												
	4100	Threaded, valves are ball												
	4120	3/4" pipe size			1 Plum	16	.500	Ea.	715	18.70		733.70	815	
	4140	1" pipe size				14	.571		760	21.50		781.50	865	
	4150	1-1/4" pipe size				12	.667		1,050	25		1,075	1,200	
	4160	1-1/2" pipe size				10	.800		1,025	30		1,055	1,175	
	4180	2" pipe size			↓	7	1.143	↓	1,100	42.50		1,142.50	1,275	
	5000	Flanged, bronze, valves are OS&Y												
	5060	2-1/2" pipe size			Q-1	5	3.200	Ea.	2,750	108		2,858	3,200	
	5080	3" pipe size				4.50	3.556		3,425	120		3,545	3,950	
	5100	4" pipe size			↓	3	5.333		4,850	179		5,029	5,625	
	5120	6" pipe size			Q-2	3	8	↓	10,300	279		10,579	11,700	
	5200	Flanged, iron, valves are gate												
	5210	2-1/2" pipe size			Q-1	5	3.200	Ea.	1,775	108		1,883	2,125	
	5220	3" pipe size				4.50	3.556		1,975	120		2,095	2,350	
	5230	4" pipe size			↓	3	5.333		2,525	179		2,704	3,050	
	5240	6" pipe size			Q-2	3	8		3,875	279		4,154	4,700	
	5250	8" pipe size				2	12		8,350	420		8,770	9,800	
	5260	10" pipe size			↓	1	24		11,000	835		11,835	13,400	

15 MECHANICAL

Important: See the Reference Section for critical supporting data - Reference Nos., Crews, & City Cost Indexes

15140	Domestic Water Piping	CREW	DAILY OUTPUT	LABOR-HOURS	UNIT	2003 BARE COSTS				TOTAL INCL O&P	
						MAT.	LABOR	EQUIP.	TOTAL		
100	5600	Flanged, iron, valves are OS&Y									**100**
	5660	2-1/2" pipe size	Q-1	5	3.200	Ea.	1,900	108		2,008	2,275
	5680	3" pipe size	↓	4.50	3.556		2,025	120		2,145	2,400
	5700	4" pipe size	↓	3	5.333		2,675	179		2,854	3,225
	5720	6" pipe size	Q-2	3	8		4,200	279		4,479	5,050
	5740	8" pipe size		2	12		8,500	420		8,920	9,975
	5760	10" pipe size	↓	1	24	↓	11,700	835		12,535	14,100
600	0010	**VACUUM BREAKERS** Hot or cold water									**600**
	1030	Anti-siphon, brass									
	1040	1/4" size	1 Plum	24	.333	Ea.	19.75	12.45		32.20	40.50
	1050	3/8" size		24	.333		19.75	12.45		32.20	40.50
	1060	1/2" size		24	.333		22.50	12.45		34.95	43.50
	1080	3/4" size		20	.400		26.50	14.95		41.45	51.50
	1100	1" size		19	.421		41.50	15.75		57.25	69.50
	1120	1-1/4" size		15	.533		72.50	19.90		92.40	110
	1140	1-1/2" size		13	.615		83	23		106	126
	1160	2" size		11	.727		134	27		161	188
	1180	2-1/2" size		9	.889		380	33		413	470
	1200	3" size	↓	7	1.143	↓	510	42.50		552.50	625
	1300	For polished chrome, (1/4" thru 1"), add					50%				
	1900	Vacuum relief, water service, bronze									
	2000	1/2" size	1 Plum	30	.267	Ea.	19.50	9.95		29.45	36.50
	2040	3/4" size	"	28	.286	"	24.50	10.65		35.15	43
700	0010	**VACUUM BREAKERS**									**700**
	0011	See also backflow preventers 15140-100									
	1000	Anti-siphon continuous pressure type									
	1010	Max. 150 PSI - 210°F									
	1020	Bronze body									
	1030	1/2" size	1 Stpi	24	.333	Ea.	84	12.55		96.55	111
	1040	3/4" size		20	.400		84	15.05		99.05	115
	1050	1" size		19	.421		86.50	15.85		102.35	120
	1060	1-1/4" size		15	.533		171	20		191	219
	1070	1-1/2" size		13	.615		210	23		233	267
	1080	2" size	↓	11	.727	↓	217	27.50		244.50	281
	1200	Max. 125 PSI with atmospheric vent									
	1210	Brass, in-line construction									
	1220	1/4" size	1 Stpi	24	.333	Ea.	32	12.55		44.55	54.50
	1230	3/8" size	"	24	.333		32	12.55		44.55	54.50
	1260	For polished chrome finish, add				↓	13%				
	2000	Anti-siphon, non-continuous pressure type									
	2010	Hot or cold water 125 PSI - 210° F									
	2020	Bronze body									
	2030	1/4" size	1 Stpi	24	.333	Ea.	19.75	12.55		32.30	40.50
	2040	3/8" size		24	.333		19.75	12.55		32.30	40.50
	2050	1/2" size		24	.333		22.50	12.55		35.05	43.50
	2060	3/4" size		20	.400		26.50	15.05		41.55	51.50
	2070	1" size		19	.421		41.50	15.85		57.35	69.50
	2080	1-1/4" size		15	.533		72.50	20		92.50	111
	2090	1-1/2" size		13	.615		85.50	23		108.50	129
	2100	2" size		11	.727		133	27.50		160.50	188
	2110	2-1/2" size		8	1		380	37.50		417.50	475
	2120	3" size	↓	6	1.333	↓	510	50		560	635
	2150	For polished chrome finish, add					50%				
800	0010	**WATER HAMMER ARRESTORS / SHOCK ABSORBERS**									**800**
	0490	Copper									

			DAILY	LABOR-			2003 BARE COSTS			TOTAL	
15140	**Domestic Water Piping**	CREW	OUTPUT	HOURS	UNIT	MAT.	LABOR	EQUIP.	TOTAL	INCL O&P	
800	0500	3/4" male I.P.S. For 1 to 11 fixtures	1 Plum	12	.667	Ea.	14.50	25		39.50	53.50	**800**
	0600	1" male I.P.S., For 12 to 32 fixtures		8	1		37	37.50		74.50	97.50	
	0700	1-1/4" male I.P.S. For 33 to 60 fixtures		8	1		43	37.50		80.50	104	
	0800	1-1/2" male I.P.S. For 61 to 113 fixtures		8	1		58.50	37.50		96	121	
	0900	2" male I.P.S.For 114 to 154 fixtures		8	1		94	37.50		131.50	160	
	1000	2-1/2" male I.P.S. For 155 to 330 fixtures	▼	4	2	▼	315	74.50		389.50	460	
	4000	Bellows type										
	4010	3/4" FNPT, to 11 fixture units	1 Plum	10	.800	Ea.	68.50	30		98.50	120	
	4020	1" FNPT, to 32 fixture units		8.80	.909		138	34		172	204	
	4030	1" FNPT, to 60 fixture units		8.80	.909		206	34		240	279	
	4040	1" FNPT, to 113 fixture units		8.80	.909		515	34		549	620	
	4050	1" FNPT, to 154 fixture units		6.60	1.212		585	45.50		630.50	710	
	4060	1-1/2" FNPT, to 300 fixture units	▼	5	1.600	▼	720	60		780	880	

15155	**Drainage Specialties**										
680	0010	**SEPARATORS** Entrainment eliminator, steel body, 150 PSIG										**680**
	0100	1/4" size	1 Stpi	24	.333	Ea.	154	12.55		166.55	188	
	0120	1/2" size		24	.333		166	12.55		178.55	202	
	0140	3/4" size		20	.400		178	15.05		193.05	219	
	0160	1" size		19	.421		237	15.85		252.85	285	
	0180	1-1/4" size		15	.533		272	20		292	330	
	0200	1-1/2" size		13	.615		310	23		333	375	
	0220	2" size	▼	11	.727		345	27.50		372.50	415	
	0240	2-1/2" size	Q-5	15	1.067		1,650	36		1,686	1,875	
	0260	3" size		13	1.231		1,600	41.50		1,641.50	1,850	
	0280	4" size		10	1.600		1,875	54		1,929	2,150	
	0300	5" size		6	2.667		2,125	90.50		2,215.50	2,475	
	0320	6" size	▼	3	5.333		2,425	181		2,606	2,950	
	0340	8" size	Q-6	4.40	5.455		2,925	191		3,116	3,500	
	0360	10" size	"	4	6		4,375	211		4,586	5,150	
	1000	For 300 PSIG, add				▼	15%					
780	0010	**TRAPS**										**780**
	4700	Copper, drainage, drum trap										
	4800	3" x 5" solid, 1-1/2" pipe size	1 Plum	16	.500	Ea.	33.50	18.70		52.20	65	
	4840	3" x 6" swivel, 1-1/2" pipe size	"	16	.500	"	40.50	18.70		59.20	72.50	
	5100	P trap, standard pattern										
	5200	1-1/4" pipe size	1 Plum	18	.444	Ea.	24	16.60		40.60	51.50	
	5240	1-1/2" pipe size		17	.471		24	17.60		41.60	52.50	
	5260	2" pipe size		15	.533		37	19.90		56.90	70.50	
	5280	3" pipe size	▼	11	.727	▼	88.50	27		115.50	138	
	5340	With cleanout and slip joint										
	5360	1-1/4" pipe size	1 Plum	18	.444	Ea.	21	16.60		37.60	48	
	5400	1-1/2" pipe size		17	.471		34.50	17.60		52.10	64.50	
	5420	2" pipe size	▼	15	.533		57	19.90		76.90	92.50	
	5460	For swivel, add					3.63			3.63	3.99	
	6710	ABS DWV P trap, solvent weld joint										
	6720	1-1/2" pipe size	1 Plum	18	.444	Ea.	4.98	16.60		21.58	30.50	
	6722	2" pipe size		17	.471		8.05	17.60		25.65	35.50	
	6724	3" pipe size		15	.533		34	19.90		53.90	67.50	
	6726	4" pipe size	▼	14	.571	▼	70	21.50		91.50	109	
	6732	PVC DWV P trap, solvent weld joint										
	6733	1-1/2" pipe size	1 Plum	18	.444	Ea.	3.22	16.60		19.82	28.50	
	6734	2" pipe size		17	.471		4.48	17.60		22.08	31.50	
	6735	3" pipe size		15	.533		18.95	19.90		38.85	51	
	6736	4" pipe size	▼	14	.571		39	21.50		60.50	74.50	

15 MECHANICAL

Important: See the Reference Section for critical supporting data - Reference Nos., Crews, & City Cost Indexes

			DAILY	LABOR-		2003 BARE COSTS				TOTAL	
15155	**Drainage Specialties**	CREW	OUTPUT	HOURS	UNIT	MAT.	LABOR	EQUIP.	TOTAL	INCL O&P	
780	6760	PP DWV, dilution trap, 1-1/2" pipe size	1 Plum	16	.500	Ea.	108	18.70		126.70	147	**780**
	6770	P trap, 1-1/2" pipe size		17	.471		29	17.60		46.60	58.50	
	6780	2" pipe size		16	.500		39.50	18.70		58.20	71.50	
	6790	3" pipe size		14	.571		91.50	21.50		113	133	
	6800	4" pipe size		13	.615		117	23		140	163	
	6810	Running trap, 1-1/2" pipe size		16	.500		28.50	18.70		47.20	59.50	
	6820	2" pipe size		15	.533		40	19.90		59.90	74	
	6830	S trap, 1-1/2" pipe size		16	.500		29	18.70		47.70	60	
	6840	2" pipe size		15	.533		36.50	19.90		56.40	70	
	6850	Universal trap, 1-1/2" pipe size		14	.571		82.50	21.50		104	123	
	6860	PVC DWV hub x hub, basin trap, 1-1/4" pipe size		18	.444		5.30	16.60		21.90	31	
	6870	Sink P trap, 1-1/2" pipe size		18	.444		5.30	16.60		21.90	31	
	6880	Tubular S trap, 1-1/2" pipe size	↓	17	.471	↓	9.80	17.60		27.40	37.50	
	6890	PVC sch. 40 DWV, drum trap										
	6900	1-1/2" pipe size	1 Plum	16	.500	Ea.	11.60	18.70		30.30	41	
	6910	P trap, 1-1/2" pipe size		18	.444		3.29	16.60		19.89	28.50	
	6920	2" pipe size		17	.471		4.48	17.60		22.08	31.50	
	6930	3" pipe size		15	.533		18.95	19.90		38.85	51	
	6940	4" pipe size		14	.571		39	21.50		60.50	74.50	
	6950	P trap w/clean out, 1-1/2" pipe size		18	.444		5.50	16.60		22.10	31	
	6960	2" pipe size		17	.471		8.50	17.60		26.10	36	
	6970	P trap adjustable, 1-1/2" pipe size		17	.471		2.99	17.60		20.59	30	
	6980	P trap adj. w/union & cleanout, 1-1/2" pipe size	↓	16	.500	↓	11.25	18.70		29.95	40.50	
940	0010	**VENT FLASHING, CAPS**										**940**
	0120	Vent caps										
	0140	Cast iron										
	0180	2-1/2" - 3-5/8" pipe	1 Plum	21	.381	Ea.	33	14.25		47.25	58	
	0190	4" - 4-1/8" pipe	"	19	.421	"	39.50	15.75		55.25	67.50	
	0900	Vent flashing										
	1000	Aluminum with lead ring										
	1020	1-1/4" pipe	1 Plum	20	.400	Ea.	6.65	14.95		21.60	30	
	1030	1-1/2" pipe		20	.400		7.15	14.95		22.10	30.50	
	1040	2" pipe		18	.444		7.15	16.60		23.75	33	
	1050	3" pipe		17	.471		7.90	17.60		25.50	35	
	1060	4" pipe	↓	16	.500	↓	9.55	18.70		28.25	38.50	
	1350	Copper with neoprene ring										
	1400	1-1/4" pipe	1 Plum	20	.400	Ea.	14.05	14.95		29	38	
	1430	1-1/2" pipe		20	.400		14.05	14.95		29	38	
	1440	2" pipe		18	.444		14.85	16.60		31.45	41.50	
	1450	3" pipe		17	.471		17.40	17.60		35	45.50	
	1460	4" pipe	↓	16	.500	↓	19.20	18.70		37.90	49	
	2000	Galvanized with neoprene ring										
	2020	1-1/4" pipe	1 Plum	20	.400	Ea.	5.30	14.95		20.25	28.50	
	2030	1-1/2" pipe		20	.400		5.30	14.95		20.25	28.50	
	2040	2" pipe		18	.444		5.60	16.60		22.20	31	
	2050	3" pipe		17	.471		6.10	17.60		23.70	33	
	2060	4" pipe	↓	16	.500	↓	7.90	18.70		26.60	36.50	
	2980	Neoprene, one piece										
	3000	1-1/4" pipe	1 Plum	24	.333	Ea.	5.15	12.45		17.60	24.50	
	3030	1-1/2" pipe		24	.333		5.40	12.45		17.85	25	
	3040	2" pipe		23	.348		5.40	13		18.40	25.50	
	3050	3" pipe		21	.381		6.55	14.25		20.80	29	
	3060	4" pipe	↓	20	.400	↓	8.90	14.95		23.85	32.50	

	15180	Heating and Cooling Piping	CREW	DAILY OUTPUT	LABOR-HOURS	UNIT	2003 BARE COSTS				TOTAL INCL O&P	
							MAT.	LABOR	EQUIP.	TOTAL		
100	0010	**ANTI-FREEZE** Inhibited										**100**
	0900	Ethylene glycol concentrated										
	1000	55 gallon drums, small quantities				Gal.	7.20			7.20	7.95	
	1200	Large quantities					6.65			6.65	7.30	
	2000	Propylene glycol, for solar heat, small quantities					7.95			7.95	8.75	
	2100	Large quantities				↓	7.75			7.75	8.55	
200	0010	**PUMPS, CIRCULATING** Heated or chilled water application										**200**
	0600	Bronze, sweat connections, 1/40 HP, in line										
	0640	3/4" size	Q-1	16	1	Ea.	117	33.50		150.50	180	
	1000	Flange connection, 3/4" to 1-1/2" size										
	1040	1/12 HP	Q-1	6	2.667	Ea.	320	89.50		409.50	485	
	1060	1/8 HP		6	2.667		550	89.50		639.50	740	
	1100	1/3 HP		6	2.667		615	89.50		704.50	810	
	1140	2" size, 1/6 HP		5	3.200		790	108		898	1,025	
	1180	2-1/2" size, 1/4 HP		5	3.200		1,025	108		1,133	1,300	
	1220	3" size, 1/4 HP		4	4		1,075	135		1,210	1,375	
	1260	1/3 HP		4	4		1,300	135		1,435	1,650	
	1300	1/2 HP		4	4		1,350	135		1,485	1,675	
	1340	3/4 HP		4	4		1,450	135		1,585	1,800	
	1380	1 HP	↓	4	4	↓	2,350	135		2,485	2,800	
	2000	Cast iron, flange connection										
	2040	3/4" to 1-1/2" size, in line, 1/12 HP	Q-1	6	2.667	Ea.	209	89.50		298.50	365	
	2060	1/8 HP		6	2.667		350	89.50		439.50	520	
	2100	1/3 HP		6	2.667		390	89.50		479.50	565	
	2140	2" size, 1/6 HP		5	3.200		425	108		533	635	
	2180	2-1/2" size, 1/4 HP		5	3.200		560	108		668	780	
	2220	3" size, 1/4 HP		4	4		565	135		700	830	
	2260	1/3 HP		4	4		765	135		900	1,050	
	2300	1/2 HP		4	4		795	135		930	1,075	
	2340	3/4 HP		4	4		920	135		1,055	1,200	
	2380	1 HP	↓	4	4	↓	1,325	135		1,460	1,650	
	2600	For non-ferrous impeller, add					3%					
	3000	High head, bronze impeller										
	3030	1-1/2" size 1/2 HP	Q-1	5	3.200	Ea.	680	108		788	910	
	3040	1-1/2" size 3/4 HP		5	3.200		725	108		833	965	
	3050	2" size 1 HP		4	4		870	135		1,005	1,175	
	3090	2" size 1-1/2 HP	↓	4	4	↓	1,050	135		1,185	1,350	
	4000	Close coupled, end suction, bronze impeller										
	4040	1-1/2" size, 1-1/2 HP, to 40 GPM	Q-1	3	5.333	Ea.	1,100	179		1,279	1,500	
	4090	2" size, 2 HP, to 50 GPM		3	5.333		1,125	179		1,304	1,500	
	4100	2" size, 3 HP, to 90 GPM		2.30	6.957		1,225	234		1,459	1,700	
	4190	2-1/2" size, 3 HP, to 150 GPM		2	8		1,325	269		1,594	1,850	
	4300	3" size, 5 HP, to 225 GPM		1.80	8.889		1,425	299		1,724	2,025	
	4410	3" size, 10 HP, to 350 GPM		1.60	10		2,075	335		2,410	2,775	
	4420	4" size, 7-1/2 HP, to 350 GPM	↓	1.60	10		2,000	335		2,335	2,700	
	4520	4" size, 10 HP, to 600 GPM	Q-2	1.70	14.118		2,475	490		2,965	3,450	
	4530	5" size, 15 HP, to 1000 GPM		1.70	14.118		2,975	490		3,465	4,025	
	4610	5" size, 20 HP, to 1350 GPM		1.50	16		3,275	560		3,835	4,450	
	4620	5" size, 25 HP, to 1550 GPM	↓	1.50	16	↓	3,900	560		4,460	5,150	
	5000	Base mounted, bronze impeller, coupling guard										
	5040	1-1/2" size, 1-1/2 HP, to 40 GPM	Q-1	2.30	6.957	Ea.	1,575	234		1,809	2,075	
	5090	2" size, 2 HP, to 50 GPM		2.30	6.957		1,775	234		2,009	2,325	
	5100	2" size, 3 HP, to 90 GPM		2	8		1,800	269		2,069	2,375	
	5190	2-1/2" size, 3 HP, to 150 GPM		1.80	8.889		1,900	299		2,199	2,525	
	5300	3" size, 5 HP, to 225 GPM		1.60	10		1,950	335		2,285	2,650	
	5410	4" size, 5 HP, to 350 GPM	↓	1.50	10.667		2,125	360		2,485	2,900	

15 MECHANICAL

Important: See the Reference Section for critical supporting data - Reference Nos., Crews, & City Cost Indexes

15180	Heating and Cooling Piping	CREW	DAILY OUTPUT	LABOR-HOURS	UNIT	2003 BARE COSTS				TOTAL INCL O&P
						MAT.	LABOR	EQUIP.	TOTAL	
200										**200**
5420	4" size, 7-1/2 HP, to 350 GPM	Q-1	1.50	10.667	Ea.	2,450	360		2,810	3,250
5520	5" size, 10 HP, to 600 GPM	Q-2	1.60	15		2,450	525		2,975	3,475
5530	5" size, 15 HP, to 1000 GPM		1.60	15		3,375	525		3,900	4,500
5610	6" size, 20 HP, to 1350 GPM		1.40	17.143		3,725	600		4,325	4,975
5620	6" size, 25 HP, to 1550 GPM		1.40	17.143		4,200	600		4,800	5,525
5800	The above pump capacities are based on 1800 RPM,									
5810	at a 60 foot head. Increasing the RPM									
5820	or decreasing the head will increase the GPM.									
300	0010	**PUMPS, CONDENSATE RETURN SYSTEM**									**300**
2000	Simplex									
2010	With pump, motor, CI receiver, float switch									
2020	3/4 HP, 15 GPM	Q-1	1.80	8.889	Ea.	3,525	299		3,824	4,325
2100	Duplex									
2110	With 2 pumps and motors, CI receiver, float switch, alternator									
2120	3/4 HP, 15 GPM, 15 Gal CI rcvr	Q-1	1.40	11.429	Ea.	3,850	385		4,235	4,800
2130	1 HP, 25 GPM		1.20	13.333		4,200	450		4,650	5,300
2140	1-1/2 HP, 45 GPM		1	16		5,050	540		5,590	6,350
2150	1-1/2 HP, 60 GPM		1	16		4,025	540		4,565	5,225
700	0010	**REFRIGERATION SPECIALTIES**									**700**
0600	Accumulator									
0610	3/4"	1 Stpi	8.80	.909	Ea.	56	34		90	113
0614	7/8"		6.40	1.250		66	47		113	144
0618	1-1/8"		4.80	1.667		81	62.50		143.50	184
0622	1-3/8"		4	2		110	75		185	235
0626	1-5/8"		3.20	2.500		134	94		228	289
0630	2-1/8"		2.40	3.333		445	125		570	680
1000	Filter dryer									
1010	Replaceable core type, solder									
1020	1/2"	1 Stpi	20	.400	Ea.	102	15.05		117.05	136
1030	5/8"		19	.421		115	15.85		130.85	150
1040	7/8"		18	.444		115	16.70		131.70	151
1050	1-1/8"		15	.533		120	20		140	163
1060	1-3/8"		14	.571		144	21.50		165.50	191
1070	1-5/8"		12	.667		144	25		169	196
1080	2-1/8"		10	.800		315	30		345	390
1090	2-5/8"		9	.889		365	33.50		398.50	450
1100	3-1/8"		8	1		375	37.50		412.50	470
1200	Sealed in-line, solder									
1210	1/4", 3 cubic inches	1 Stpi	22	.364	Ea.	8.10	13.65		21.75	29.50
1220	3/8", 5 cubic inches		21	.381		7.40	14.30		21.70	29.50
1230	1/2", 9 cubic inches		20	.400		12	15.05		27.05	35.50
1240	1/2", 16 cubic inches		20	.400		15.60	15.05		30.65	39.50
1250	5/8", 16 cubic inches		19	.421		16.35	15.85		32.20	42
1260	5/8", 30 cubic inches		19	.421		26.50	15.85		42.35	53
1270	7/8", 30 cubic inches		18	.444		28	16.70		44.70	56
1280	7/8", 41 cubic inches		17	.471		37.50	17.70		55.20	68
1290	1-1/8", 60 cubic inches		15	.533		52	20		72	88
4000	P-Trap, suction line, solder									
4010	5/8"	1 Stpi	19	.421	Ea.	19.85	15.85		35.70	46
4020	3/4"		19	.421		13.80	15.85		29.65	39
4030	7/8"		18	.444		7.35	16.70		24.05	33
4040	1-1/8"		15	.533		35	20		55	69
4050	1-3/8"		14	.571		35	21.50		56.50	71
4060	1-5/8"		12	.667		53.50	25		78.50	96.50
4070	2-1/8"		10	.800		109	30		139	166
4440	Refrigerant, R-507, 25 lb. disposable cylinder				Lb.	256			256	281

MECHANICAL 15

15180 | Heating and Cooling Piping

		CREW	DAILY OUTPUT	LABOR-HOURS	UNIT	2003 BARE COSTS MAT.	LABOR	EQUIP.	TOTAL	TOTAL INCL O&P
700	5000 Sightglass									700
	5010 Moisture and liquid indicator, solder									
	5020 1/4"	1 Stpi	22	.364	Ea.	11.35	13.65		25	33
	5030 3/8"		21	.381		11.55	14.30		25.85	34
	5040 1/2"		20	.400		14.55	15.05		29.60	38.50
	5050 5/8"		19	.421		14.80	15.85		30.65	40.50
	5060 7/8"		18	.444		20	16.70		36.70	47
	5070 1-1/8"		15	.533		21	20		41	53.50
	5080 1-3/8"		14	.571		26	21.50		47.50	61
	5090 1-5/8"		12	.667		30	25		55	71
	5100 2-1/8"		10	.800		37.50	30		67.50	86.50
	7400 Vacuum pump set									
	7410 Two stage, high vacuum continuous duty	1 Stpi	8	1	Ea.	2,325	37.50		2,362.50	2,600
	8000 Valves									
	8100 Check valve, soldered									
	8110 5/8"	1 Stpi	36	.222	Ea.	22	8.35		30.35	37
	8114 7/8"		26	.308		46	11.55		57.55	68
	8118 1-1/8"		18	.444		105	16.70		121.70	141
	8122 1-3/8"		14	.571		144	21.50		165.50	191
	8126 1-5/8"		13	.615		195	23		218	250
	8130 2-1/8"		12	.667		296	25		321	365
	8134 2-5/8"	Q-5	22	.727		730	24.50		754.50	840
	8138 3-1/8"	"	20	.800		900	27		927	1,025
	8500 Refrigeration valve, packless, soldered									
	8510 1/2"	1 Stpi	38	.211	Ea.	25.50	7.90		33.40	40
	8514 5/8"		36	.222		30.50	8.35		38.85	46.50
	8518 7/8"		26	.308		76.50	11.55		88.05	102
	8520 Packed, soldered									
	8522 1-1/8"	1 Stpi	18	.444	Ea.	108	16.70		124.70	144
	8526 1-3/8"		14	.571		246	21.50		267.50	305
	8530 1-5/8"		13	.615		305	23		328	370
	8534 2-1/8"		12	.667		450	25		475	535
	8538 2-5/8"	Q-5	22	.727		670	24.50		694.50	770
	8542 3-1/8"		20	.800		890	27		917	1,025
	8546 4-1/8"		18	.889		1,250	30		1,280	1,425
	8600 Solenoid valve, flange/solder									
	8610 1/2"	1 Stpi	38	.211	Ea.	126	7.90		133.90	150
	8614 5/8"		36	.222		156	8.35		164.35	185
	8618 3/4"		30	.267		196	10.05		206.05	231
	8622 7/8"		26	.308		291	11.55		302.55	340
	8626 1-1/8"		18	.444		325	16.70		341.70	380
	8630 1-3/8"		14	.571		360	21.50		381.50	430
	8634 1-5/8"		13	.615		550	23		573	640
	8638 2-1/8"		12	.667		600	25		625	700
	8800 Thermostatic valve, flange/solder									
	8810 1/2 - 3 ton, 3/8" x 5/8"	1 Stpi	9	.889	Ea.	111	33.50		144.50	173
	8814 4 - 5 ton, 1/2" x 7/8"		7	1.143		111	43		154	187
	8818 6 - 8 ton, 5/8" x 7/8"		5	1.600		111	60		171	213
	8822 7 - 12 ton, 7/8" x 1-1/8"		4	2		129	75		204	256
	8826 15 - 20 ton, 7/8" x 1-3/8"		3.20	2.500		129	94		223	284
800	0010 **STEAM CONDENSATE METER**									800
	0100 500 lb. per hour	1 Stpi	14	.571	Ea.	2,300	21.50		2,321.50	2,550
	0140 1500 lb. per hour		7	1.143		2,475	43		2,518	2,800
	0160 3000 lb. per hour		5	1.600		3,325	60		3,385	3,750
	0200 12,000 lb. per hour	Q-5	3.50	4.571		4,450	155		4,605	5,100

15 MECHANICAL

Important: See the Reference Section for critical supporting data - Reference Nos., Crews, & City Cost Indexes

15195	Fuel Systems		CREW	DAILY OUTPUT	LABOR-HOURS	UNIT	2003 BARE COSTS				TOTAL INCL O&P
							MAT.	LABOR	EQUIP.	TOTAL	
490	0010	**FUEL OIL SPECIALTIES**										**490**
	0020	Foot valve, single poppet, metal to metal construction									
	0040	Bevel seat, 1/2" diameter	1 Stpi	20	.400	Ea.	36	15.05		51.05	62
	0060	3/4" diameter		18	.444		37.50	16.70		54.20	66.50
	0080	1" diameter		16	.500		27	18.80		45.80	58
	0100	1-1/4" diameter		15	.533		53	20		73	88.50
	0120	1-1/2" diameter		13	.615		47	23		70	86.50
	0140	2" diameter		11	.727		52	27.50		79.50	98.50
	0160	Foot valve, double poppet, metal to metal construction									
	0164	1" diameter	1 Stpi	15	.533	Ea.	83	20		103	122
	0166	1-1/2" diameter	"	12	.667	"	111	25		136	160
	0400	Fuel fill box, flush type									
	0408	Nonlocking, watertight									
	0410	1-1/2" diameter	1 Stpi	12	.667	Ea.	13.60	25		38.60	53
	0440	2" diameter		10	.800		14.65	30		44.65	61.50
	0450	2-1/2" diameter		9	.889		49	33.50		82.50	105
	0460	3" diameter		7	1.143		47.50	43		90.50	118
	0470	4" diameter		5	1.600		65	60		125	163
	0500	Locking inner cover									
	0510	2" diameter	1 Stpi	8	1	Ea.	71	37.50		108.50	136
	0520	2-1/2" diameter		7	1.143		95.50	43		138.50	170
	0530	3" diameter		5	1.600		102	60		162	203
	0540	4" diameter		4	2		131	75		206	258
	0600	Fuel system components									
	0620	Spill container	1 Stpi	4	2	Ea.	485	75		560	650
	0640	Fill adapter, 4", straight drop		8	1		30	37.50		67.50	90
	0680	Fill cap, 4"		30	.267		25.50	10.05		35.55	43
	0700	Extractor fitting, 4" x 1-1/2"		8	1		340	37.50		377.50	430
	0740	Vapor hose adapter, 4"		8	1		64.50	37.50		102	128
	0760	Wood gage stick, 10'					9.20			9.20	10.10
	1000	Oil filters, 3/8" IPT., 20 gal. per hour	1 Stpi	20	.400		19.25	15.05		34.30	43.50
	1020	32 gal. per hour		18	.444		32.50	16.70		49.20	61
	1040	40 gal. per hour		16	.500		37	18.80		55.80	69
	1060	50 gal. per hour		14	.571		38	21.50		59.50	74.50
	1800	Pump and motor sets									
	1810	Light fuel and diesel oils, 100 PSI									
	1820	25 GPH 1/4 HP	Q-5	6	2.667	Ea.	795	90.50		885.50	1,000
	1830	45 GPH, 1/4 HP		6	2.667		800	90.50		890.50	1,025
	1840	90 GPH, 1/4 HP		5	3.200		860	108		968	1,100
	1850	160 GPH, 1/3 HP		4	4		885	135		1,020	1,175
	1860	325 GPH, 3/4 HP		4	4		970	135		1,105	1,275
	1870	700 GPH, 1-1/2 HP		3	5.333		1,750	181		1,931	2,200
	1880	1000 GPH, 2 HP		3	5.333		1,800	181		1,981	2,250
	1890	1800 GPH, 5 HP		1.80	8.889		2,750	300		3,050	3,500
	2000	Remote tank gauging system, self contained									
	2100	Gage and one probe	1 Stpi	2.50	3.200	Ea.	2,700	120		2,820	3,150
	2120	Gage with two probes		2	4		3,700	150		3,850	4,300
	3000	Valve, ball check, globe type, 3/8" diameter		24	.333		8.40	12.55		20.95	28
	3500	Fusible, 3/8" diameter		24	.333		6.85	12.55		19.40	26.50
	3600	1/2" diameter		24	.333		17.85	12.55		30.40	38.50
	3610	3/4" diameter		20	.400		35.50	15.05		50.55	61.50
	3620	1" diameter		19	.421		108	15.85		123.85	143
	4000	Nonfusible, 3/8" diameter		24	.333		12.75	12.55		25.30	33
	4500	Shutoff, gate type, lever handle, spring-fusible kit									
	4520	1/4" diameter	1 Stpi	14	.571	Ea.	20	21.50		41.50	54.50
	4540	3/8" diameter		12	.667		18.75	25		43.75	58.50

MECHANICAL 15

			DAILY	LABOR-		2003 BARE COSTS				TOTAL	
15195		**Fuel Systems**	CREW	OUTPUT	HOURS	UNIT	MAT.	LABOR	EQUIP.	TOTAL	INCL O&P
490	4560	1/2" diameter	1 Stpi	10	.800	Ea.	22.50	30		52.50	70.50	**490**
	4570	3/4" diameter	↓	8	1	↓	42.50	37.50		80	104
	4580	Lever handle, requires weight and fusible kit									
	4600	1" diameter	1 Stpi	9	.889	Ea.	284	33.50		317.50	360
	4620	1-1/4" diameter		8	1		294	37.50		331.50	380
	4640	1-1/2" diameter		7	1.143		400	43		443	505
	4660	2" diameter	↓	6	1.333		420	50		470	535
	4680	For fusible link, weight and braided wire, add					5%				
	5000	Vent alarm, whistling signal					26			26	28.50
	5500	Vent protector/breather, 1-1/4" diameter	1 Stpi	32	.250		7.20	9.40		16.60	22
	5520	1-1/2" diameter		32	.250		8	9.40		17.40	23
	5540	2" diameter		32	.250		12.80	9.40		22.20	28.50
	5560	3" diameter		28	.286		40	10.75		50.75	60.50
	5580	4" diameter		24	.333		47	12.55		59.55	71
	5600	Dust cap, breather, 2"	↓	40	.200	↓	10.20	7.50		17.70	22.50
	8000	Fuel oil and tank heaters									
	8020	Electric, capacity rated at 230 volts									
	8040	Immersion element in steel manifold									
	8060	96 GPH at 50°F rise	Q-5	6.40	2.500	Ea.	630	84.50		714.50	820
	8070	128 GPH at 50°F rise		6.20	2.581		665	87.50		752.50	860
	8080	160 GPH at 50°F rise		5.90	2.712		695	92		787	905
	8090	192 GPH at 50°F rise		5.50	2.909		740	98.50		838.50	965
	8100	240 GPH at 50°F rise		5.10	3.137		890	106		996	1,150
	8110	288 GPH at 50°F rise		4.60	3.478		1,050	118		1,168	1,350
	8120	384 GPH at 50°F rise		3.10	5.161		1,200	175		1,375	1,600
	8130	480 GPH at 50°F rise		2.30	6.957		1,500	235		1,735	2,000
	8140	576 GPH at 50°F rise	↓	2.10	7.619	↓	1,625	258		1,883	2,200
	8300	Suction stub, immersion type									
	8320	75" long, 750 watts	1 Stpi	14	.571	Ea.	530	21.50		551.50	620
	8330	99" long, 2000 watts		12	.667		745	25		770	860
	8340	123" long, 3000 watts	↓	10	.800	↓	805	30		835	930
	8660	Steam, cross flow, rated at 5 PSIG									
	8680	42 GPH	Q-5	7	2.286	Ea.	1,225	77.50		1,302.50	1,475
	8690	73 GPH		6.70	2.388		1,425	81		1,506	1,700
	8700	112 GPH		6.20	2.581		1,475	87.50		1,562.50	1,750
	8710	158 GPH		5.80	2.759		1,625	93.50		1,718.50	1,925
	8720	187 GPH		4	4		2,225	135		2,360	2,650
	8730	270 GPH	↓	3.60	4.444		2,400	150		2,550	2,850
	8740	365 GPH	Q-6	4.90	4.898		2,800	172		2,972	3,350
	8750	635 GPH		3.70	6.486		3,550	228		3,778	4,250
	8760	845 GPH		2.50	9.600		5,200	335		5,535	6,225
	8770	1420 GPH		1.60	15		8,475	525		9,000	10,100
	8780	2100 GPH	↓	1.10	21.818	↓	11,600	765		12,365	14,000
605	0010	**METERS**										**605**
	4000	Residential									
	4010	Gas meter, residential, 3/4" pipe size	1 Plum	14	.571	Ea.	112	21.50		133.50	155
	4020	Gas meter, residential, 1" pipe size		12	.667		136	25		161	188
	4030	Gas meter, residential, 1-1/4" pipe size	↓	10	.800	↓	136	30		166	195

Important: See the Reference Section for critical supporting data - Reference Nos., Crews, & City Cost Indexes

15210	Process Air/Gas Piping	CREW	DAILY OUTPUT	LABOR-HOURS	UNIT	2003 BARE COSTS				TOTAL INCL O&P
						MAT.	LABOR	EQUIP.	TOTAL	
100	0010 **COMPRESSORS**									100
0100	Refrigeration, hermetic, switches and protective devices									
0200	High back pressure types for R-22									
0210	1.25 ton	1 Stpi	3	2.667	Ea.	330	100		430	515
0220	1.42 ton		3	2.667		340	100		440	525
0230	1.68 ton		2.80	2.857		385	107		492	580
0240	2.00 ton		2.60	3.077		475	116		591	695
0250	2.37 ton		2.50	3.200		575	120		695	810
0260	2.67 ton		2.40	3.333		365	125		490	590
0270	3.53 ton		2.30	3.478		425	131		556	670
0280	4.43 ton		2.20	3.636		605	137		742	870
0290	5.08 ton		2.10	3.810		745	143		888	1,025
0300	5.22 ton	▼	2	4		755	150		905	1,050
0310	7.31 ton	Q-5	3	5.333		1,400	181		1,581	1,800
0320	9.95 ton	"	2.60	6.154	▼	1,525	208		1,733	2,000
0990	Refrigeration, recip. hermetic, switches & protective devices									
1000	(Ratings are ARI standard 515 group IV using R-22) 10 ton	Q-5	1	16	Ea.	12,100	540		12,640	14,100
1100	20 ton	Q-6	.72	33.333		18,500	1,175		19,675	22,200
1200	30 ton		.64	37.500		22,300	1,325		23,625	26,500
1300	40 ton		.44	54.545		24,000	1,925		25,925	29,300
1400	50 ton	▼	.20	120		25,200	4,200		29,400	34,100
1500	75 ton	Q-7	.27	118		27,900	4,250		32,150	37,100
1600	130 ton	"	.21	152	▼	31,800	5,475		37,275	43,300
1800	Refrigeration, scroll type									
1810	1.9 ton	1 Stpi	2.70	2.963	Ea.	410	111		521	620
1820	2.35 ton		2.50	3.200		470	120		590	700
1830	2.82 ton		2.35	3.404		505	128		633	750
1840	3.3 ton		2.30	3.478		530	131		661	785
1850	3.83 ton		2.25	3.556		510	134		644	765
1860	4.1 ton		2.20	3.636		690	137		827	960
1870	4.8 ton		2.10	3.810		720	143		863	1,000
1880	5 ton	▼	2.07	3.865	▼	735	145		880	1,025
5250	Air, reciprocating air cooled, splash lubricated, tank mounted									
5300	Single stage, 1 phase, 140 psi									
5303	1/2 HP, 30 gal tank	1 Stpi	3	2.667	Ea.	940	100		1,040	1,175
5305	3/4 HP, 30 gal tank		2.60	3.077		945	116		1,061	1,225
5307	1 HP, 30 gal tank	▼	2.20	3.636		1,200	137		1,337	1,525
5309	2 HP, 30 gal tank	Q-5	4	4		1,625	135		1,760	2,000
5310	3 HP, 30 gal tank		3.60	4.444		2,125	150		2,275	2,575
5314	3 HP, 60 gal tank		3.50	4.571		2,450	155		2,605	2,900
5320	5 HP, 60 gal tank		3.20	5		2,825	169		2,994	3,350
5330	5 HP, 80 gal tank		3	5.333		2,800	181		2,981	3,375
5340	7.5 HP, 80 gal tank	▼	2.60	6.154	▼	4,550	208		4,758	5,325
5600	2 stage pkg., 3 phase									
5650	6 CFM at 125 psi 1-1/2 HP, 60 gal tank	Q-5	3	5.333	Ea.	2,100	181		2,281	2,600
5670	10.9 CFM at 125 psi, 3 HP, 80 gal tank		1.50	10.667		2,500	360		2,860	3,300
5680	38.7 CFM at 125 psi, 10 HP, 120 gal tank	▼	.60	26.667		4,925	905		5,830	6,800
5690	105 CFM at 125 psi, 25 HP, 250 gal tank	Q-6	.60	40	▼	9,200	1,400		10,600	12,200
5800	With single stage pump									
5850	8.3 CFM at 125 psi, 2 HP, 80 gal tank	Q-6	3.50	6.857	Ea.	2,575	241		2,816	3,225
5860	38.7 CFM at 125 psi, 10 HP, 120 gal tank	"	.90	26.667	"	5,450	935		6,385	7,425
6000	Reciprocating, 2 stage, tank mtd, 3 Ph., Cap. rated @175 PSIG									
6050	Pressure lubcatd, hvy duty, 9.7 CFM, 3 HP, 120 Gal. tank	Q-5	1.30	12.308	Ea.	4,050	415		4,465	5,075
6054	5 CFM, 1-1/2 HP, 80 gal tank		2.80	5.714		2,450	193		2,643	3,000
6056	6.4 CFM, 2 HP, 80 gal tank		2	8		2,575	271		2,846	3,250
6058	8.1 CFM, 3 HP, 80 gal tank	▼	1.70	9.412	▼	2,750	320		3,070	3,500

	15210	Process Air/Gas Piping	CREW	DAILY OUTPUT	LABOR-HOURS	UNIT	2003 BARE COSTS				TOTAL INCL O&P	
							MAT.	LABOR	EQUIP.	TOTAL		
100	6059	14.8 CFM, 5 HP, 80 gal tank	Q-5	1	16	Ea.	2,850	540		3,390	3,950	**100**
	6060	16.5 CFM, 5 HP, 120 gal. tank		1	16		4,100	540		4,640	5,325	
	6063	13 CFM, 6 HP, 80 gal tank		.90	17.778		4,325	600		4,925	5,650	
	6066	19.8 CFM, 7.5 HP, 80 gal tank		.80	20		4,550	675		5,225	6,025	
	6070	25.8 CFM, 7-1/2 HP, 120 gal. tank		.80	20		6,100	675		6,775	7,750	
	6078	34.8 CFM, 10 HP, 80 gal tank		.70	22.857		5,775	775		6,550	7,525	
	6080	34.8 CFM, 10 HP, 120 gal. tank	▼	.60	26.667		6,075	905		6,980	8,050	
	6090	53.7 CFM, 15 HP, 120 gal. tank	Q-6	.80	30		6,300	1,050		7,350	8,525	
	6100	76.7 CFM, 20 HP, 120 gal. tank		.70	34.286		9,400	1,200		10,600	12,100	
	6104	76.7 CFM, 20 HP, 240 gal tank		.68	35.294		9,325	1,250		10,575	12,200	
	6110	90.1 CFM, 25 HP, 120 gal. tank		.63	38.095		9,375	1,325		10,700	12,300	
	6120	101 CFM, 30 HP, 120 gal. tank		.57	42.105		9,350	1,475		10,825	12,500	
	6130	101 CFM, 30 HP, 250 gal. tank	▼	.52	46.154		11,500	1,625		13,125	15,100	
	6200	Oil-less, 13.6 CFM, 5 HP, 120 gal. tank	Q-5	.88	18.182		12,600	615		13,215	14,800	
	6210	13.6 CFM, 5 HP, 250 gal. tank		.80	20		13,500	675		14,175	15,900	
	6220	18.2 CFM, 7.5 HP, 120 gal. tank		.73	21.918		12,800	740		13,540	15,200	
	6230	18.2 CFM, 7.5 HP, 250 gal. tank		.67	23.881		13,700	810		14,510	16,300	
	6250	30.5 CFM, 10 HP, 120 gal. tank		.57	28.070		15,000	950		15,950	17,800	
	6260	30.5 CFM, 10 HP, 250 gal. tank	▼	.53	30.189		15,800	1,025		16,825	19,000	
	6270	41.3 CFM, 15 HP, 120 gal. tank	Q-6	.70	34.286		16,200	1,200		17,400	19,600	
	6280	41.3 CFM, 15 HP, 250 gal. tank	"	.67	35.821	▼	17,100	1,250		18,350	20,700	
200	0010	**COMPRESSOR ACCESSORIES**										**200**
	4000	Couplers, air line, sleeve type										
	4010	Female, connection size NPT										
	4020	1/4"	1 Stpi	38	.211	Ea.	4.06	7.90		11.96	16.40	
	4030	3/8"		36	.222		6.05	8.35		14.40	19.25	
	4040	1/2"		35	.229		10.60	8.60		19.20	24.50	
	4050	3/4"	▼	34	.235	▼	12.45	8.85		21.30	27	
	4100	Male										
	4110	1/4"	1 Stpi	38	.211	Ea.	4.01	7.90		11.91	16.35	
	4120	3/8"		36	.222		5.05	8.35		13.40	18.15	
	4130	1/2"		35	.229		10	8.60		18.60	24	
	4140	3/4"	▼	34	.235	▼	11.85	8.85		20.70	26.50	
	4150	Coupler, combined male and female halves										
	4160	1/2"	1 Stpi	17	.471	Ea.	20.50	17.70		38.20	49	
	4170	3/4"	"	15	.533	"	24.50	20		44.50	57	
900	0010	**VACUUM PUMPS**										**900**
	0100	Medical, with receiver										
	0110	Duplex										
	0120	20 SCFM	Q-1	1.14	14.035	Ea.	14,200	470		14,670	16,300	
	0130	60 SCFM, 10 HP	Q-2	1.20	20	"	25,500	695		26,195	29,200	
	0200	Triplex										
	0220	180 SCFM	Q-2	.86	27.907	Ea.	46,400	975		47,375	52,500	
	0300	Dental oral										
	0310	Duplex										
	0330	165 SCFM with 77 Gal separator	Q-2	1.30	18.462	Ea.	42,000	645		42,645	47,200	

	15230	Industrial Process Piping										
500	0010	**PUMPS, GENERAL UTILITY** With motor										**500**
	0200	Multi-stage, horizontal split, for boiler feed applications										
	0300	Two stage, 3" discharge x 4" suction, 75 HP	Q-7	.30	106	Ea.	14,900	3,825		18,725	22,200	
	0340	Four stage, 3" discharge x 4" suction, 150 HP	"	.18	177	"	24,500	6,375		30,875	36,600	
	2000	Single stage										
	2060	End suction, 1"D. x 2"S., 3 HP	Q-1	.50	32	Ea.	3,650	1,075		4,725	5,625	

T5

MECHANICAL

Important: See the Reference Section for critical supporting data - Reference Nos., Crews, & City Cost Indexes

15230	Industrial Process Piping	CREW	DAILY OUTPUT	LABOR-HOURS	UNIT	2003 BARE COSTS				TOTAL INCL O&P	
						MAT.	LABOR	EQUIP.	TOTAL		
500 2100	1-1/2"D. x 3"S., 10 HP	Q-1	.40	40	Ea.	3,925	1,350		5,275	6,350	**500**
2140	2"D. x 3"S., 15 HP	Q-2	.60	40		4,275	1,400		5,675	6,800	
2180	3"D. x 4"S., 20 HP		.50	48		4,550	1,675		6,225	7,525	
2220	4"D. x 6"S., 30 HP		.40	60		4,925	2,100		7,025	8,575	
3000	Double suction, 2"D. x 2-1/2"S., 10 HP	Q-1	.30	53.333		5,050	1,800		6,850	8,250	
3060	3"D. x 4"S., 15 HP	Q-2	.46	52.174		5,200	1,825		7,025	8,475	
3100	4"D. x 5"S., 30 HP		.40	60		7,050	2,100		9,150	10,900	
3140	5"D. x 6"S., 50 HP		.33	72.727		7,575	2,525		10,100	12,200	
3180	6"D. x 8"S., 60 HP	Q-3	.30	106		8,475	3,800		12,275	15,100	
3190	75 HP, to 2500 GPM		.28	114		10,800	4,075		14,875	18,100	
3220	100 HP, to 3000 GPM		.26	123		13,700	4,375		18,075	21,700	
3240	150 HP, to 4000 GPM		.24	133		18,500	4,750		23,250	27,500	
4000	Centrifugal, end suction, mounted on base										
4010	Horizontal mounted, with drip proof motor, rated @ 100' head										
4020	Vertical split case, single stage										
4040	100 GPM, 5 HP, 1-1/2" discharge	Q-1	1.70	9.412	Ea.	1,975	315		2,290	2,650	
4050	200 GPM, 10 HP, 2" discharge		1.30	12.308		2,400	415		2,815	3,250	
4060	250 GPM, 10 HP, 3" discharge		1.28	12.500		2,500	420		2,920	3,375	
4070	300 GPM, 15 HP, 2" discharge	Q-2	1.56	15.385		2,750	535		3,285	3,825	
4080	500 GPM, 20 HP, 4" discharge		1.44	16.667		3,200	580		3,780	4,375	
4090	750 GPM, 30 HP, 4" discharge		1.20	20		4,275	695		4,970	5,750	
4100	1050 GPM, 40 HP, 5" discharge		1	24		4,625	835		5,460	6,350	
4110	1500 GPM, 60 HP, 6" discharge		.60	40		6,725	1,400		8,125	9,500	
4120	2000 GPM, 75 HP, 6" discharge		.50	48		7,725	1,675		9,400	11,000	
4130	3000 GPM, 100 HP, 8" discharge		.40	60		11,600	2,100		13,700	16,000	
4200	Horizontal split case, single stage										
4210	100 GPM, 7.5 HP, 1-1/2" discharge	Q-1	1.70	9.412	Ea.	3,275	315		3,590	4,100	
4220	250 GPM, 15 HP, 2-1/2" discharge	"	1.30	12.308		4,550	415		4,965	5,625	
4230	500 GPM, 20 HP, 4" discharge	Q-2	1.60	15		5,200	525		5,725	6,525	
4240	750 GPM, 25 HP, 5" discharge		1.54	15.584		5,375	545		5,920	6,750	
4250	1000 GPM, 40 HP, 5" discharge		1.20	20		7,050	695		7,745	8,800	
4260	1500 GPM, 50 HP, 6" discharge	Q-3	1.42	22.535		8,850	800		9,650	11,000	
4270	2000 GPM, 75 HP, 8" discharge		1.14	28.070		12,400	1,000		13,400	15,200	
4280	3000 GPM, 100 HP, 10" discharge		.96	33.333		13,400	1,175		14,575	16,500	
4290	3500 GPM, 150 HP, 10" discharge		.86	37.209		16,700	1,325		18,025	20,400	
4300	4000 GPM, 200 HP, 10" discharge		.66	48.485		18,500	1,725		20,225	22,900	
4330	Horizontal split case, two stage, 500' head										
4340	100 GPM, 40 HP, 1-1/2" discharge	Q-2	1.70	14.118	Ea.	4,300	490		4,790	5,475	
4350	200 GPM, 50 HP, 1-1/2" discharge	"	1.44	16.667		4,800	580		5,380	6,150	
4360	300 GPM, 75 HP, 2" discharge	Q-3	1.57	20.382		5,100	725		5,825	6,700	
4370	400 GPM, 100 HP, 3" discharge		1.14	28.070		6,550	1,000		7,550	8,700	
4380	800 GPM, 200 HP, 4" discharge		.86	37.209		15,300	1,325		16,625	18,900	
5000	Centrifugal, in-line										
5006	Vertical mount, iron body, 125 lb. flgd, 3550 RPM TEFC mtr										
5010	Single stage										
5020	50 GPM, 3 HP, 1-1/2" discharge	Q-1	2.30	6.957	Ea.	550	234		784	960	
5030	75 GPM, 5 HP, 1-1/2" discharge		1.60	10		930	335		1,265	1,525	
5040	100 GPM, 7.5 HP, 1-1/2" discharge		1.30	12.308		1,000	415		1,415	1,725	
5050	125 GPM, 10 HP, 1-1/2" discharge	Q-2	1.70	14.118		1,250	490		1,740	2,125	
5060	150 GPM, 15 HP, 1-1/2" discharge		1.60	15		1,600	525		2,125	2,550	
5070	200 GPM, 30 HP, 1-1/2" discharge		1.50	16		2,600	560		3,160	3,725	
5080	250 GPM, 40 HP, 2" discharge		1.40	17.143		3,450	600		4,050	4,675	
5090	300 GPM, 50 HP, 3" discharge		1.30	18.462		4,175	645		4,820	5,575	
5100	400 GPM, 75 HP, 3" discharge	Q-3	.60	53.333		6,800	1,900		8,700	10,400	
5110	600 GPM, 100 HP, 3" discharge	"	.50	64		9,125	2,275		11,400	13,500	
8000	Vertical submerged, with non-submerged motor										

MECHANICAL 15

15230 | Industrial Process Piping

		CREW	DAILY OUTPUT	LABOR-HOURS	UNIT	2003 BARE COSTS				TOTAL INCL O&P	
						MAT.	LABOR	EQUIP.	TOTAL		
500	8060	1"D., 3 HP	Q-1	.50	32	Ea.	5,275	1,075		6,350	7,425	500
	8100	1-1/2"D., 10 HP	"	.30	53.333		5,525	1,800		7,325	8,775	
	8140	2"D., 15 HP	Q-2	.40	60		5,675	2,100		7,775	9,400	
	8180	3"D., 25 HP		.30	80		5,800	2,800		8,600	10,600	
	8220	4"D., 30 HP	▼	.20	120	▼	6,500	4,175		10,675	13,500	
600	0010	**PUMPS MISCELLANEOUS**										600
	0020	Water pump, portable, gasoline powered										
	0100	6000 GPH, 2" discharge	Q-1	11	1.455	Ea.	1,000	49		1,049	1,175	
	0110	8000 GPH, 2" discharge		10.50	1.524		1,175	51.50		1,226.50	1,375	
	0120	10,000 GPH, 2" discharge	▼	10	1.600	▼	1,800	54		1,854	2,050	
	0500	Pump, propylene body, housing and impeller										
	0510	22 GPM, 1/3 HP, 40' HD	1 Plum	5	1.600	Ea.	485	60		545	625	
	0520	33 GPM, 1/2 HP, 40' HD	Q-1	5	3.200		495	108		603	705	
	0530	53 GPM, 3/4 HP, 40' HD	"	4	4	▼	530	135		665	785	
	0600	Rotary pump, CI										
	0610	10 GPM, 1/2 HP, 1" discharge	Q-1	5	3.200	Ea.	945	108		1,053	1,225	
	0614	10 GPM, 3/4 HP, 1" discharge		4.50	3.556		945	120		1,065	1,225	
	0618	10 GPM 1 HP, 1" discharge		4	4		980	135		1,115	1,275	
	0620	25 GPM, 1 HP, 1-1/4" discharge		3.80	4.211		980	142		1,122	1,300	
	0624	25 GPM, 1.5 HP, 1-1/4" discharge		3.60	4.444		1,075	149		1,224	1,400	
	0628	25 GPM, 2 HP, 1-1/4" discharge	▼	3.20	5	▼	1,075	168		1,243	1,425	
	0700	Fuel oil pump, 2 stage, 3450 RPM										
	0710	1/4 HP, 100 PSI	Q-1	8	2	Ea.	1,600	67.50		1,667.50	1,875	
	0800	Hydraulic oil pump										
	0810	75 HP, 3 PH, 480 V	Q-3	1.20	26.667	Ea.	5,175	950		6,125	7,100	
	1000	Turbine pump, CI										
	1010	50 GPM, 2 HP, 3" discharge	Q-1	.80	20	Ea.	1,075	675		1,750	2,225	
	1020	100 GPM, 3 HP, 4" discharge	Q-2	.96	25		1,725	870		2,595	3,225	
	1030	250 GPM, 15 HP, 6" discharge	"	.94	25.532		3,450	890		4,340	5,125	
	1040	500 GPM, 25 HP, 6" discharge	Q-3	1.30	24.615		5,725	875		6,600	7,625	
	1050	1000 GPM, 50 HP, 8" discharge		1.14	28.070		9,175	1,000		10,175	11,600	
	1060	2000 GPM, 100 HP, 10" discharge		1	32		10,300	1,150		11,450	13,000	
	1070	3000 GPM, 150 HP, 10" discharge		.80	40		10,800	1,425		12,225	14,100	
	1080	4000 GPM, 200 HP, 12" discharge		.70	45.714		17,200	1,625		18,825	21,400	
	1090	6000 GPM, 300 HP, 14" discharge		.60	53.333		17,900	1,900		19,800	22,600	
	1100	10,000 GPM, 300 HP, 18" discharge	▼	.58	55.172	▼	40,100	1,975		42,075	47,100	

15400 | Plumbing Fixtures & Equipment

15410 | Plumbing Fixtures

		CREW	DAILY OUTPUT	LABOR-HOURS	UNIT	2003 BARE COSTS				TOTAL INCL O&P	
						MAT.	LABOR	EQUIP.	TOTAL		
300	0010	**FAUCETS/FITTINGS**										300
	5000	Sillcock, compact, brass, IPS or copper to hose	1 Plum	24	.333	Ea.	4.74	12.45		17.19	24	
	6000	Stop and waste valves, bronze										
	6100	Angle, solder end 1/2"	1 Plum	24	.333	Ea.	4.18	12.45		16.63	23.50	
	6110	3/4"		20	.400		4.59	14.95		19.54	27.50	
	6300	Straightway, solder end 3/8"		24	.333		3.59	12.45		16.04	23	
	6310	1/2"		24	.333		2.32	12.45		14.77	21.50	
	6320	3/4"	▼	20	.400		2.68	14.95		17.63	25.50	

15410 | Plumbing Fixtures

		CREW	DAILY OUTPUT	LABOR-HOURS	UNIT	2003 BARE COSTS				TOTAL INCL O&P	
						MAT.	LABOR	EQUIP.	TOTAL		
300	6330	1"	1 Plum	19	.421	Ea.	4	15.75		19.75	28.50	300
	6400	Straightway, threaded 3/8"		24	.333		4.07	12.45		16.52	23.50	
	6410	1/2"		24	.333		3.13	12.45		15.58	22	
	6420	3/4"		20	.400		3.73	14.95		18.68	26.50	
	6430	1"	▼	19	.421	▼	10.50	15.75		26.25	35.50	

15411 | Commercial/Indust Fixtures

		CREW	DAILY OUTPUT	LABOR-HOURS	UNIT	2003 BARE COSTS				TOTAL INCL O&P	
						MAT.	LABOR	EQUIP.	TOTAL		
500	0010	**HYDRANTS**										500
	0050	Wall type, moderate climate, bronze, encased										
	0200	3/4" IPS connection	1 Plum	16	.500	Ea.	226	18.70		244.70	277	
	0300	1" IPS connection	"	14	.571		226	21.50		247.50	281	
	0500	Anti-siphon type				▼	30.50			30.50	33.50	
	1000	Non-freeze, bronze, exposed										
	1100	3/4" IPS connection, 4" to 9" thick wall	1 Plum	14	.571	Ea.	153	21.50		174.50	201	
	1120	10" to 14" thick wall		12	.667		166	25		191	221	
	1140	15" to 19" thick wall		12	.667		185	25		210	242	
	1160	20" to 24" thick wall	▼	10	.800	▼	201	30		231	266	
	1200	For 1" IPS connection, add					15%	10%				
	1240	For 3/4" adapter type vacuum breaker, add				Ea.	19.25			19.25	21	
	1280	For anti-siphon type				"	40.50			40.50	44.50	
	2000	Non-freeze bronze, encased										
	2100	3/4" IPS connection, 5" to 9" thick wall	1 Plum	14	.571	Ea.	264	21.50		285.50	325	
	2120	10" to 14" thick wall		12	.667		277	25		302	345	
	2140	15" to 19" thick wall		12	.667		297	25		322	365	
	2160	20" to 24" thick wall	▼	10	.800	▼	310	30		340	385	
	2200	For 1" IPS connection, add					10%	10%				
	2280	Anti-siphon type				Ea.	124			124	136	
	3000	Ground box type, bronze frame, 3/4" IPS connection										
	3080	Non-freeze, all bronze, polished face, set flush										
	3100	2 feet depth of bury	1 Plum	8	1	Ea.	282	37.50		319.50	365	
	3120	3 feet depth of bury		8	1		305	37.50		342.50	390	
	3140	4 feet depth of bury		8	1		325	37.50		362.50	410	
	3160	5 feet depth of bury		7	1.143		350	42.50		392.50	445	
	3180	6 feet depth of bury		7	1.143		370	42.50		412.50	470	
	3200	7 feet depth of bury		6	1.333		390	50		440	505	
	3220	8 feet depth of bury		5	1.600		415	60		475	545	
	3240	9 feet depth of bury		4	2		435	74.50		509.50	595	
	3260	10 feet depth of bury	▼	4	2		455	74.50		529.50	615	
	3400	For 1" IPS connection, add					15%	10%				
	3450	For 1-1/4" IPS connection, add					325%	14%				
	3500	For 1-1/2" connection, add					370%	18%				
	3550	For 2" connection, add					445%	24%				
	3600	For tapped drain port in box, add				▼	25.50			25.50	28.50	
	4000	Non-freeze, CI body, bronze frame & scoriated cover										
	4010	with hose storage										
	4100	2 feet depth of bury	1 Plum	7	1.143	Ea.	535	42.50		577.50	650	
	4120	3 feet depth of bury		7	1.143		555	42.50		597.50	675	
	4140	4 feet depth of bury		7	1.143		575	42.50		617.50	695	
	4160	5 feet depth of bury		6.50	1.231		585	46		631	715	
	4180	6 feet depth of bury		6	1.333		595	50		645	730	
	4200	7 feet depth of bury		5.50	1.455		615	54.50		669.50	760	
	4220	8 feet depth of bury		5	1.600		640	60		700	790	
	4240	9 feet depth of bury		4.50	1.778		660	66.50		726.50	825	
	4260	10 feet depth of bury	▼	4	2		680	74.50		754.50	865	
	4280	For 1" IPS connection, add				▼	117			117	129	

MECHANICAL 15

15411	Commercial/Indust Fixtures	CREW	DAILY OUTPUT	LABOR-HOURS	UNIT	2003 BARE COSTS				TOTAL INCL O&P	
						MAT.	LABOR	EQUIP.	TOTAL		
500	4300	For tapped drain port in box, add				Ea.	25.50			25.50	28.50	**500**
	5000	Moderate climate, all bronze, polished face										
	5020	and scoriated cover, set flush										
	5100	3/4" IPS connection	1 Plum	16	.500	Ea.	201	18.70		219.70	249	
	5120	1" IPS connection	"	14	.571		200	21.50		221.50	252	
	5200	For tapped drain port in box, add					25.50			25.50	28.50	
	6000	Ground post type, all non-freeze, all bronze, aluminum casing										
	6010	guard, exposed head, 3/4" IPS connection										
	6100	2 feet depth of bury	1 Plum	8	1	Ea.	283	37.50		320.50	365	
	6120	3 feet depth of bury		8	1		305	37.50		342.50	390	
	6140	4 feet depth of bury		8	1		330	37.50		367.50	420	
	6160	5 feet depth of bury		7	1.143		355	42.50		397.50	455	
	6180	6 feet depth of bury		7	1.143		375	42.50		417.50	480	
	6200	7 feet depth of bury		6	1.333		405	50		455	520	
	6220	8 feet depth of bury		5	1.600		430	60		490	560	
	6240	9 feet depth of bury		4	2		455	74.50		529.50	615	
	6260	10 feet depth of bury		4	2		480	74.50		554.50	645	
	6300	For 1" IPS connection, add					40%	10%				
	6350	For 1-1/4" IPS connection, add					140%	14%				
	6400	For 1-1/2" IPS connection, add					225%	18%				
	6450	For 2" IPS connection, add					315%	24%				

15440 | Plumbing Pumps

			CREW	DAILY OUTPUT	LABOR-HOURS	UNIT	MAT.	LABOR	EQUIP.	TOTAL	INCL O&P	
240	0010	**PUMPS, PRESSURE BOOSTER SYSTEM**										**240**
	0200	Pump system, with diaphragm tank, control, press. switch										
	0300	1 HP pump	Q-1	1.30	12.308	Ea.	3,275	415		3,690	4,225	
	0400	1-1/2 HP pump		1.25	12.800		3,325	430		3,755	4,300	
	0420	2 HP pump		1.20	13.333		3,400	450		3,850	4,400	
	0440	3 HP pump		1.10	14.545		3,450	490		3,940	4,550	
	0460	5 HP pump	Q-2	1.50	16		3,800	560		4,360	5,050	
	0480	7-1/2 HP pump		1.42	16.901		4,250	590		4,840	5,575	
	0500	10 HP pump		1.34	17.910		4,450	625		5,075	5,825	
	1000	Pump/ energy storage system, diaphragm tank, 3 HP pump										
	1100	motor, PRV, switch, gauge, control center, flow switch										
	1200	125 lb. working pressure	Q-2	.70	34.286	Ea.	9,550	1,200		10,750	12,300	
	1300	250 lb. working pressure	"	.64	37.500	"	10,400	1,300		11,700	13,500	
400	0010	**PUMPS, GRINDER SYSTEM** Complete, incl. check valve, tank, std.										**400**
	0020	controls incl. alarm/disconnect panel w/wire. Excavation not included										
	0260	Simplex, 9 GPM at 60 PSIG, 70 gal. tank				Ea.	2,225			2,225	2,450	
	0300	For manway, 26" I.D., 18" high, add					420			420	460	
	0340	26" I.D., 36" high, add					490			490	540	
	0380	43" I.D., 4' high, add					540			540	595	
	0600	Simplex, 9 GPM at 60 PSIG, 150 gal. tank					2,525			2,525	2,775	
	0660	For manway, add										
	0700	26" I.D., 36" high, add				Ea.	505			505	555	
	0740	26" I.D., 4' high, add					585			585	645	
	2000	Duplex, 18 GPM at 60 PSIG, 150 gal. tank					4,175			4,175	4,600	
	2060	For manway 43" I.D., 4' high, add					1,325			1,325	1,475	
	2400	For core only					1,550			1,550	1,700	
800	0010	**PUMPS, SEWAGE EJECTOR** With operating and level controls										**800**
	0100	Simplex system incl. tank, cover, pump 15' head										
	0500	37 gal PE tank, 12 GPM, 1/2 HP, 2" discharge	Q-1	3.20	5	Ea.	370	168		538	665	
	0510	3" discharge		3.10	5.161		400	174		574	700	

Important: See the Reference Section for critical supporting data - Reference Nos., Crews, & City Cost Indexes

15440	Plumbing Pumps	CREW	DAILY OUTPUT	LABOR-HOURS	UNIT	2003 BARE COSTS				TOTAL INCL O&P	
						MAT.	LABOR	EQUIP.	TOTAL		
800	0530	87 GPM, .7 HP, 2" discharge	Q-1	3.20	5	Ea.	570	168		738	880	**800**
	0540	3" discharge		3.10	5.161		615	174		789	940	
	0600	45 gal. coated stl tank, 12 GPM, 1/2 HP, 2" discharge		3	5.333		665	179		844	1,000	
	0610	3" discharge		2.90	5.517		695	186		881	1,050	
	0630	87 GPM, .7 HP, 2" discharge		3	5.333		855	179		1,034	1,200	
	0640	3" discharge		2.90	5.517		905	186		1,091	1,275	
	0660	134 GPM, 1 HP, 2" discharge		2.80	5.714		925	192		1,117	1,325	
	0680	3" discharge		2.70	5.926		975	199		1,174	1,375	
	0700	70 gal. PE tank, 12 GPM, 1/2 HP, 2" discharge		2.60	6.154		730	207		937	1,125	
	0710	3" discharge		2.40	6.667		780	224		1,004	1,200	
	0730	87 GPM, 0.7 HP, 2" discharge		2.50	6.400		935	215		1,150	1,350	
	0740	3" discharge		2.30	6.957		1,000	234		1,234	1,450	
	0760	134 GPM, 1 HP, 2" discharge		2.20	7.273		1,025	245		1,270	1,500	
	0770	3" discharge		2	8		1,100	269		1,369	1,600	
	0800	75 gal. coated stl. tank, 12 GPM, 1/2 HP, 2 " discharge		2.40	6.667		755	224		979	1,175	
	0810	3" discharge		2.20	7.273		785	245		1,030	1,225	
	0830	87 GPM, .7 HP, 2" discharge		2.30	6.957		950	234		1,184	1,400	
	0840	3" discharge		2.10	7.619		1,000	256		1,256	1,475	
	0860	134 GPM, 1 HP, 2" discharge		2	8		1,025	269		1,294	1,525	
	0880	3" discharge		1.80	8.889		1,075	299		1,374	1,625	
	1040	Duplex system incl. tank, covers, pumps										
	1060	110 gal. fiberglass tank, 24 GPM, 1/2 HP, 2" discharge	Q-1	1.60	10	Ea.	1,425	335		1,760	2,075	
	1080	3" discharge		1.40	11.429		1,475	385		1,860	2,200	
	1100	174 GPM, .7 HP, 2" discharge		1.50	10.667		1,875	360		2,235	2,600	
	1120	3" discharge		1.30	12.308		1,950	415		2,365	2,750	
	1140	268 GPM, 1 HP, 2" discharge		1.20	13.333		2,025	450		2,475	2,900	
	1160	3" discharge		1	16		2,100	540		2,640	3,100	
	1260	135 gal. coated stl. tank, 24 GPM, 1/2 HP, 2" discharge	Q-2	1.70	14.118		1,500	490		1,990	2,375	
	2000	3" discharge		1.60	15		1,575	525		2,100	2,525	
	2640	174 GPM, .7 HP, 2" discharge		1.60	15		1,950	525		2,475	2,950	
	2660	3" discharge		1.50	16		2,050	560		2,610	3,100	
	2700	268 GPM, 1 HP, 2" discharge		1.30	18.462		2,125	645		2,770	3,300	
	3040	3" discharge		1.10	21.818		2,225	760		2,985	3,600	
	3060	275 gal. coated stl. tank, 24 GPM, 1/2 HP, 2" discharge		1.50	16		1,850	560		2,410	2,900	
	3080	3" discharge		1.40	17.143		1,875	600		2,475	2,975	
	3100	174 GPM, .7 HP, 2" discharge		1.40	17.143		2,400	600		3,000	3,525	
	3120	3" discharge		1.30	18.462		2,425	645		3,070	3,625	
	3140	268 GPM, 1 HP, 2" discharge		1.10	21.818		2,625	760		3,385	4,050	
	3160	3" discharge		.90	26.667		2,725	930		3,655	4,400	
	3260	Pump system accessories, add										
	3300	Alarm horn and lights, 115V mercury switch	Q-1	8	2	Ea.	82.50	67.50		150	193	
	3340	Switch, mag. contactor, alarm bell, light, 3 level control		5	3.200		390	108		498	595	
	3380	Alternator, mercury switch activated		4	4		750	135		885	1,025	
900	0010	**PUMPS, PEDESTAL SUMP** With float control										**900**
	0400	Molded PVC base, 21 GPM at 15' head, 1/3 HP	1 Plum	5	1.600	Ea.	87	60		147	186	
	0800	Iron base, 21 GPM at 15' head, 1/3 HP		5	1.600		110	60		170	211	
	1200	Solid brass, 21 GPM at 15' head, 1/3 HP		5	1.600		179	60		239	287	
940	0010	**PUMPS, SUBMERSIBLE** Sump										**940**
	7000	Sump pump, automatic										
	7100	Plastic, 1-1/4" discharge, 1/4 HP	1 Plum	6	1.333	Ea.	105	50		155	191	
	7140	1/3 HP		5	1.600		127	60		187	230	
	7160	1/2 HP		5	1.600		153	60		213	258	
	7180	1-1/2" discharge, 1/2 HP		4	2		171	74.50		245.50	300	
	7500	Cast iron, 1-1/4" discharge, 1/4 HP		6	1.333		125	50		175	212	
	7540	1/3 HP		6	1.333		140	50		190	229	

MECHANICAL 15

		15440	Plumbing Pumps	CREW	DAILY OUTPUT	LABOR-HOURS	UNIT	2003 BARE COSTS				TOTAL INCL O&P	
								MAT.	LABOR	EQUIP.	TOTAL		
940	7560		1/2 HP	1 Plum	5	1.600	Ea.	176	60		236	283	940

		15450	Potable Water Tanks	CREW	DAILY OUTPUT	LABOR-HOURS	UNIT	MAT.	LABOR	EQUIP.	TOTAL	TOTAL INCL O&P	
900	0010	**WATER HEATER STORAGE TANKS** 125 psi ASME											900
	2000	Galvanized steel, 15 gal., 14" diam., 26" LOA		1 Plum	12	.667	Ea.	530	25		555	620	
	2060	30 gal., 14" diam. x 49" LOA			11	.727		630	27		657	730	
	2080	80 gal., 20" diam. x 64" LOA			9	.889		975	33		1,008	1,125	
	2100	135 gal., 24" diam. x 75" LOA			6	1.333		1,450	50		1,500	1,650	
	2120	240 gal., 30" diam. x 86" LOA			4	2		2,500	74.50		2,574.50	2,875	
	2140	300 gal., 36" diam. x 76" LOA		↓	3	2.667		4,050	99.50		4,149.50	4,600	
	2160	400 gal., 36" diam. x 100" LOA		Q-1	4	4		4,975	135		5,110	5,675	
	2180	500 gal., 36" diam., x 126" LOA		"	3	5.333		6,150	179		6,329	7,050	
	3000	Glass lined, P.E., 80 gal., 20" diam. x 60" LOA		1 Plum	9	.889		1,600	33		1,633	1,800	
	3060	140 gal., 24" diam. x 80" LOA			6	1.333		2,275	50		2,325	2,575	
	3080	225 gal., 30" diam. x 78" LOA			4	2		2,650	74.50		2,724.50	3,050	
	3100	325 gal., 36" diam. x 81" LOA		↓	3	2.667		3,400	99.50		3,499.50	3,900	
	3120	460 gal., 42" diam. x 84" LOA		Q-1	4	4		3,575	135		3,710	4,150	
	3140	605 gal., 48" diam. x 87" LOA			3	5.333		6,850	179		7,029	7,800	
	3160	740 gal., 54" diam. x 91" LOA			3	5.333		7,700	179		7,879	8,750	
	3180	940 gal., 60" diam. x 93" LOA			2.50	6.400		8,900	215		9,115	10,100	
	3200	1330 gal., 66" diam. x 107" LOA			2	8		12,200	269		12,469	13,800	
	3220	1615 gal., 72" diam. x 110" LOA			1.50	10.667		13,900	360		14,260	15,700	
	3240	2285 gal., 84" diam. x 128" LOA		↓	1	16		17,100	540		17,640	19,600	
	3260	3440 gal., 96" diam. x 157" LOA		Q-2	1.50	16	↓	25,400	560		25,960	28,700	

		15480	Domestic Water Heaters	CREW	DAILY OUTPUT	LABOR-HOURS	UNIT	MAT.	LABOR	EQUIP.	TOTAL	TOTAL INCL O&P	
900	0010	**HEAT TRANSFER PACKAGES** Complete, controls,											900
	0020	expansion tank, converter, air separator											
	1000	Hot water, 180°F enter, 200°F leaving, 15# steam											
	1010	One pump system, 28 GPM		Q-6	.75	32	Ea.	11,300	1,125		12,425	14,200	
	1020	35 GPM			.70	34.286		12,500	1,200		13,700	15,500	
	1040	55 GPM			.65	36.923		14,200	1,300		15,500	17,700	
	1060	130 GPM			.55	43.636		17,600	1,525		19,125	21,700	
	1080	255 GPM			.40	60		24,200	2,100		26,300	29,800	
	1100	550 GPM			.30	80		30,800	2,800		33,600	38,200	
	1120	800 GPM			.25	96		37,800	3,375		41,175	46,600	
	1220	Two pump system, 28 GPM			.70	34.286		15,000	1,200		16,200	18,300	
	1240	35 GPM			.65	36.923		18,000	1,300		19,300	21,800	
	1260	55 GPM			.60	40		18,500	1,400		19,900	22,400	
	1280	130 GPM			.50	48		23,700	1,675		25,375	28,600	
	1300	255 GPM			.35	68.571		32,400	2,400		34,800	39,300	
	1320	550 GPM			.25	96		41,000	3,375		44,375	50,000	
	1340	800 GPM		↓	.20	120	↓	50,000	4,200		54,200	61,500	

Important: See the Reference Section for critical supporting data - Reference Nos., Crews, & City Cost Indexes

15510	Heating Boilers and Accessories	CREW	DAILY OUTPUT	LABOR-HOURS	UNIT	2003 BARE COSTS				TOTAL INCL O&P
						MAT.	LABOR	EQUIP.	TOTAL	
110	**0010**	**BOILER BLOWDOWN SYSTEMS**										110
	1010	Boiler blowdown, auto/manual to 2000 MBH	Q-5	3.75	4.267	Ea.	2,625	144		2,769	3,100
	1020	7300 MBH	"	3	5.333	"	3,000	181		3,181	3,575
120	**0010**	**BURNERS**										120
	0990	Residential, conversion, gas fired, LP or natural									
	1000	Gun type, atmospheric input 72 to 200 MBH	Q-1	2.50	6.400	Ea.	655	215		870	1,050
	1020	120 to 360 MBH		2	8		725	269		994	1,200
	1040	280 to 800 MBH	↓	1.70	9.412	↓	1,400	315		1,715	2,025
	2000	Commercial and industrial, gas/oil, input									
	2050	400 MBH	Q-1	1.50	10.667	Ea.	1,950	360		2,310	2,700
	2080	For mounting plate, add		35	.457		141	15.35		156.35	178
	2090	670 MBH		1.40	11.429		2,400	385		2,785	3,225
	2110	For mounting plate, add		35	.457		141	15.35		156.35	178
	2140	1155 MBH		1.30	12.308		3,125	415		3,540	4,050
	2170	For mounting plate, add		35	.457		141	15.35		156.35	178
	2200	1800 MBH		1.20	13.333		3,400	450		3,850	4,400
	2230	For mounting plate, add		35	.457		141	15.35		156.35	178
	2260	3000 MBH		1.10	14.545		4,675	490		5,165	5,875
	2290	For mounting plate, add		35	.457		141	15.35		156.35	178
	2320	4100 MBH		1	16		4,825	540		5,365	6,100
	2350	For mounting plate, add	↓	35	.457	↓	141	15.35		156.35	178
	2500	Impinged jet, rectangular, burner only, 3-1/2" WC,									
	2520	Gas fired, input, 420 to 640 MBH	Q-1	2	8	Ea.	570	269		839	1,025
	2540	560 to 860 MBH		2	8		820	269		1,089	1,300
	2560	630 to 950 MBH		1.90	8.421		855	283		1,138	1,375
	2580	950 to 1200 MBH		1.80	8.889		1,225	299		1,524	1,800
	2600	1120 to 1700 MBH		1.80	8.889		1,625	299		1,924	2,250
	2620	1700 to 1900 MBH		1.70	9.412		1,850	315		2,165	2,500
	2640	1400 to 2100 MBH		1.60	10		2,050	335		2,385	2,750
	2660	1600 to 2400 MBH		1.60	10		2,150	335		2,485	2,850
	2680	1700 to 2500 MBH		1.50	10.667		2,450	360		2,810	3,250
	2700	1880 to 2880 MBH		1.40	11.429		2,575	385		2,960	3,400
	2720	2260 to 3680 MBH		1.40	11.429		3,000	385		3,385	3,875
	2740	2590 to 4320 MBH		1.30	12.308		3,650	415		4,065	4,650
	2760	3020 to 5040 MBH		1.20	13.333		4,275	450		4,725	5,375
	2780	3450 to 5760 MBH		1.20	13.333		4,875	450		5,325	6,025
	2850	Burner pilot		50	.320		15.25	10.75		26	33
	2860	Thermocouple		50	.320		15.70	10.75		26.45	33.50
	2870	Thermocouple & pilot bracket	↓	45	.356	↓	41.50	11.95		53.45	64
	2900	For 7" W.C. pressure, increase MBH 40%									
	3000	Flame retention oil fired assembly, input									
	3020	.50 to 2.25 GPH	Q-1	2.40	6.667	Ea.	291	224		515	660
	3040	2.0 to 5.0 GPH		2	8		525	269		794	985
	3060	3.0 to 7.0 GPH		1.80	8.889		535	299		834	1,050
	3080	6.0 to 12.0 GPH	↓	1.60	10		835	335		1,170	1,425
	4600	Gas safety, shut off valve, 3/4" threaded	1 Stpi	20	.400		283	15.05		298.05	335
	4610	1" threaded		19	.421		285	15.85		300.85	340
	4620	1-1/4" threaded		15	.533		310	20		330	370
	4630	1-1/2" threaded		13	.615		335	23		358	405
	4640	2" threaded	↓	11	.727		375	27.50		402.50	450
	4650	2-1/2" threaded	Q-1	15	1.067		430	36		466	530
	4660	3" threaded		13	1.231		590	41.50		631.50	715
	4670	4" flanged	↓	3	5.333		2,375	179		2,554	2,875
	4680	6" flanged	Q-2	3	8	↓	5,275	279		5,554	6,225

MECHANICAL 15

15510	Heating Boilers and Accessories	CREW	DAILY OUTPUT	LABOR-HOURS	UNIT	2003 BARE COSTS				TOTAL INCL O&P
						MAT.	LABOR	EQUIP.	TOTAL	
300										300
0010	**BOILERS, ELECTRIC, ASME** Standard controls and trim R15500-030									
1000	Steam, 6 KW, 20.5 MBH	Q-19	1.20	20	Ea.	2,550	700		3,250	3,850
1040	9 KW, 30.7 MBH		1.20	20		2,600	700		3,300	3,900
1060	18 KW, 61.4 MBH D3020 102		1.20	20		2,675	700		3,375	4,000
1080	24 KW, 81.8 MBH		1.10	21.818		3,100	765		3,865	4,550
1120	36 KW, 123 MBH D3020 104		1.10	21.818		3,450	765		4,215	4,925
1160	60 KW, 205 MBH		1	24		4,450	840		5,290	6,175
1220	112 KW, 382 MBH		.75	32		5,900	1,125		7,025	8,175
1240	148 KW, 505 MBH		.65	36.923		6,075	1,300		7,375	8,625
1260	168 KW, 573 MBH		.60	40		8,575	1,400		9,975	11,600
1280	222 KW, 758 MBH		.55	43.636		9,825	1,525		11,350	13,100
1300	296 KW, 1010 MBH		.45	53.333		9,900	1,875		11,775	13,700
1320	300 KW, 1023 MBH		.40	60		12,700	2,100		14,800	17,200
1340	370 KW, 1263 MBH		.35	68.571		13,700	2,400		16,100	18,700
1360	444 KW, 1515 MBH		.30	80		14,600	2,800		17,400	20,300
1380	518 KW, 1768 MBH	Q-21	.36	88.889		16,600	3,175		19,775	23,000
1400	592 KW, 2020 MBH		.34	94.118		18,700	3,375		22,075	25,700
1420	666 KW, 2273 MBH		.32	100		19,700	3,575		23,275	27,100
1460	740 KW, 2526 MBH		.28	114		20,700	4,075		24,775	29,000
1480	814 KW, 2778 MBH		.25	128		23,200	4,575		27,775	32,400
1500	962 KW, 3283 MBH		.22	145		23,800	5,200		29,000	34,000
1520	1036 KW, 3536 MBH		.20	160		25,000	5,725		30,725	36,100
1540	1110 KW, 3788 MBH		.19	168		26,200	6,025		32,225	37,900
1560	2070 KW, 7063 MBH		.18	177		37,800	6,350		44,150	51,000
1580	2250 KW, 7677 MBH		.17	188		44,100	6,725		50,825	58,500
1600	2,340 KW, 7984 MBH		.16	200		48,900	7,150		56,050	65,000
2000	Hot water, 6 KW, 20.5 MBH	Q-19	1.30	18.462		2,375	650		3,025	3,600
2020	9 KW, 30.7 MBH		1.30	18.462		2,400	650		3,050	3,625
2040	24 KW, 82 MBH		1.20	20		2,850	700		3,550	4,175
2060	30 KW, 103 MBH		1.20	20		3,100	700		3,800	4,450
2070	36 KW, 123 MBH		1.20	20		3,325	700		4,025	4,700
2080	48 KW, 164 MBH		1.10	21.818		4,625	765		5,390	6,225
2100	74 KW, 253 MBH		1.10	21.818		4,800	765		5,565	6,425
2120	93 KW, 317 MBH		1	24		5,800	840		6,640	7,675
2140	112 KW, 382 MBH		.90	26.667		5,875	935		6,810	7,850
2160	148 KW, 505 MBH		.75	32		6,900	1,125		8,025	9,300
2180	168 KW, 573 MBH		.65	36.923		7,100	1,300		8,400	9,750
2200	222 KW, 758 MBH		.60	40		7,850	1,400		9,250	10,700
2220	296 KW, 1010 MBH		.55	43.636		8,675	1,525		10,200	11,800
2280	370 KW, 1263 MBH		.40	60		9,600	2,100		11,700	13,800
2300	444 KW, 1515 MBH		.35	68.571		11,600	2,400		14,000	16,400
2340	518 KW, 1768 MBH	Q-21	.44	72.727		12,900	2,600		15,500	18,000
2360	592 KW, 2020 MBH		.43	74.419		14,200	2,650		16,850	19,600
2400	666 KW, 2273 MBH		.40	80		15,200	2,850		18,050	21,000
2420	740 KW, 2526 MBH		.39	82.051		15,600	2,925		18,525	21,500
2440	814 KW, 2778 MBH		.38	84.211		16,100	3,000		19,100	22,200
2460	888 KW, 3031 MBH		.37	86.486		16,900	3,100		20,000	23,300
2480	962 KW, 3283 MBH		.36	88.889		19,000	3,175		22,175	25,700
2500	1,036 KW, 3536 MBH		.34	94.118		20,100	3,375		23,475	27,300
2520	1110 KW, 3788 MBH		.33	96.970		22,300	3,475		25,775	29,700
2540	1440 KW, 4915 MBH		.32	100		25,600	3,575		29,175	33,500
2560	1560 KW, 5323 MBH		.31	103		28,700	3,700		32,400	37,200
2580	1680 KW, 5733 MBH		.30	106		30,700	3,800		34,500	39,600
2600	1800 KW, 6143 MBH		.29	110		32,300	3,950		36,250	41,400
2620	1980 KW, 6757 MBH		.28	114		34,500	4,075		38,575	44,200
2640	2100 KW, 7167 MBH		.27	118		38,500	4,225		42,725	48,800

Important: See the Reference Section for critical supporting data - Reference Nos., Crews, & City Cost Indexes

		15510 | **Heating Boilers and Accessories**	CREW	DAILY OUTPUT	LABOR-HOURS	UNIT	2003 BARE COSTS MAT.	LABOR	EQUIP.	TOTAL	TOTAL INCL O&P	
300	2660	2220 KW, 7576 MBH R15500-030	Q-21	.26	123	Ea.	39,800	4,400		44,200	50,500	**300**
	2680	2,400 KW, 8191 MBH		.25	128		42,000	4,575		46,575	53,000	
	2700	2610 KW, 8905 MBH		.24	133		45,100	4,775		49,875	57,000	
	2720	2790 KW, 9519 MBH		.23	139		47,200	4,975		52,175	59,500	
	2740	2970 KW, 10133 MBH		.21	152		48,300	5,450		53,750	61,000	
	2760	3150 KW, 10748 MBH		.19	168		53,000	6,025		59,025	67,000	
	2780	3240 KW, 11055 MBH		.18	177		54,000	6,350		60,350	69,000	
	2800	3420 KW, 11669 MBH		.17	188		57,000	6,725		63,725	73,000	
	2820	3,600 KW, 12,283 MBH		.16	200		59,000	7,150		66,150	76,000	
400	0010	**BOILERS, GAS FIRED** Natural or propane, standard controls										**400**
	1000	Cast iron, with insulated jacket										
	2000	Steam, gross output, 81 MBH	Q-7	1.40	22.857	Ea.	1,150	820		1,970	2,525	
	2020	102 MBH		1.30	24.615		1,325	880		2,205	2,775	
	2040	122 MBH		1	32		1,450	1,150		2,600	3,325	
	2060	163 MBH		.90	35.556		1,775	1,275		3,050	3,875	
	2080	203 MBH		.90	35.556		1,975	1,275		3,250	4,075	
	2100	240 MBH		.85	37.647		1,975	1,350		3,325	4,200	
	2120	280 MBH		.80	40		2,500	1,425		3,925	4,925	
	2140	320 MBH		.70	45.714		2,725	1,650		4,375	5,475	
	2160	360 MBH		.63	51.200		3,075	1,825		4,900	6,150	
	2180	400 MBH		.56	56.838		3,275	2,050		5,325	6,675	
	2200	440 MBH		.51	62.500		3,500	2,250		5,750	7,225	
	2220	544 MBH		.45	71.588		4,175	2,575		6,750	8,475	
	2240	765 MBH		.43	74.419		5,325	2,675		8,000	9,900	
	2260	892 MBH		.38	84.211		6,400	3,025		9,425	11,600	
	2280	1275 MBH		.34	94.118		8,125	3,375		11,500	14,000	
	2300	1530 MBH		.32	100		9,525	3,575		13,100	15,900	
	2320	1,875 MBH		.30	106		11,100	3,825		14,925	18,000	
	2340	2170 MBH		.26	122		12,700	4,375		17,075	20,500	
	2360	2675 MBH		.20	163		14,500	5,850		20,350	24,800	
	2380	3060 MBH		.19	172		15,700	6,175		21,875	26,500	
	2400	3570 MBH		.18	181		17,500	6,525		24,025	29,100	
	2420	4207 MBH		.16	205		19,300	7,350		26,650	32,300	
	2440	4,720 MBH		.15	207		21,200	7,450		28,650	34,600	
	2460	5660 MBH		.15	220		45,000	7,900		52,900	61,500	
	2480	6,100 MBH		.13	246		51,500	8,825		60,325	70,000	
	2500	6390 MBH		.12	266		54,500	9,550		64,050	74,500	
	2520	6680 MBH		.11	290		55,000	10,400		65,400	76,500	
	2540	6,970 MBH		.10	320		58,000	11,500		69,500	81,000	
	3000	Hot water, gross output, 80 MBH		1.46	21.918		1,050	785		1,835	2,325	
	3020	100 MBH		1.35	23.704		1,200	850		2,050	2,600	
	3040	122 MBH		1.10	29.091		1,325	1,050		2,375	3,025	
	3060	163 MBH		1	32		1,525	1,150		2,675	3,400	
	3080	203 MBH		1	32		1,850	1,150		3,000	3,750	
	3100	240 MBH		.95	33.684		1,850	1,200		3,050	3,875	
	3120	280 MBH		.90	35.556		2,325	1,275		3,600	4,475	
	3140	320 MBH		.80	40		2,550	1,425		3,975	4,975	
	3160	360 MBH		.71	45.070		2,900	1,625		4,525	5,625	
	3180	400 MBH		.64	50		3,100	1,800		4,900	6,100	
	3200	440 MBH		.58	54.983		3,325	1,975		5,300	6,625	
	3220	544 MBH		.51	62.992		3,975	2,250		6,225	7,775	
	3240	765 MBH		.46	70.022		5,125	2,500		7,625	9,425	
	3260	1,088 MBH		.40	80		6,700	2,875		9,575	11,700	
	3280	1,275 MBH		.36	89.888		7,875	3,225		11,100	13,600	
	3300	1,530 MBH		.31	104		9,325	3,750		13,075	16,000	

MECHANICAL 15

15510	Heating Boilers and Accessories	CREW	DAILY OUTPUT	LABOR- HOURS	UNIT	2003 BARE COSTS				TOTAL INCL O&P
						MAT.	LABOR	EQUIP.	TOTAL	
400	3320	2,000 MBH	Q-7	.26	125	Ea.	11,500	4,475		15,975	19,500	**400**
3340	2,312 MBH		.22	148		13,400	5,300		18,700	22,700
3360	2,856 MBH		.20	160		15,700	5,725		21,425	25,900
3380	3,264 MBH		.18	179		16,600	6,450		23,050	28,000
3400	3,808 MBH		.16	195		18,300	7,000		25,300	30,800
3420	4,488 MBH		.15	210		27,900	7,550		35,450	42,000
3440	4,720 MBH		.15	220		46,100	7,900		54,000	62,500
3460	5,520 MBH		.14	228		52,000	8,200		60,200	69,500
3480	6,100 MBH		.13	250		52,500	8,975		61,475	71,500
3500	6,390 MBH		.11	285		53,500	10,200		63,700	74,500
3520	6,680 MBH		.10	310		57,000	11,100		68,100	79,500
3540	6,970 MBH		.09	359		59,000	12,900		71,900	84,500
4000	Steel, insulating jacket									
4500	Steam, not including burner, gross output									
5500	1,440 MBH	Q-6	.35	68.571	Ea.	16,600	2,400		19,000	21,800
5520	1630 MBH		.30	80		17,200	2,800		20,000	23,200
5540	1800 MBH		.26	92.308		17,800	3,250		21,050	24,500
5560	2520 MBH		.20	120		22,100	4,200		26,300	30,700
5580	3,065 MBH		.18	133		23,600	4,675		28,275	33,100
5600	3600 MBH		.16	150		25,100	5,275		30,375	35,600
5620	5400 MBH		.14	171		35,000	6,025		41,025	47,600
5640	7,200 MBH	Q-7	.18	177		43,300	6,375		49,675	57,500
5660	10,800 MBH		.17	188		54,000	6,750		60,750	69,500
5680	12,600 MBH		.15	213		60,000	7,650		67,650	77,500
5700	16,190 MBH		.12	266		76,000	9,550		85,550	98,000
5720	17,990 MBH		.10	320		81,000	11,500		92,500	106,500
6000	Hot water, including burner & one zone valve, gross output									
6010	51.2 MBH	Q-6	2	12	Ea.	1,600	420		2,020	2,400
6020	72 MBH		2	12		1,800	420		2,220	2,600
6040	89 MBH		1.90	12.632		1,850	445		2,295	2,700
6060	105 MBH		1.80	13.333		2,050	470		2,520	2,975
6080	132 MBH		1.70	14.118		2,350	495		2,845	3,325
6100	155 MBH		1.50	16		2,725	560		3,285	3,850
6110	186 MBH		1.40	17.143		3,275	600		3,875	4,525
6120	227 MBH		1.30	18.462		4,175	650		4,825	5,575
6140	292 MBH		1.20	20		4,700	700		5,400	6,225
6160	400 MBH		.80	30		5,950	1,050		7,000	8,150
6180	480 MBH		.70	34.286		6,800	1,200		8,000	9,300
6200	640 MBH		.60	40		8,075	1,400		9,475	11,000
6220	800 MBH		.50	48		9,425	1,675		11,100	13,000
6240	960 MBH		.45	53.333		11,700	1,875		13,575	15,700
6260	1200 MBH		.40	60		14,400	2,100		16,500	19,000
6280	1440 MBH		.35	68.571		16,300	2,400		18,700	21,500
6300	1960 MBH		.30	80		21,600	2,800		24,400	28,000
6320	2400 MBH		.20	120		26,300	4,200		30,500	35,300
6340	3,000 MBH		.15	160		32,500	5,625		38,125	44,300
7000	For tankless water heater on smaller gas units, add					10%				
7050	For additional zone valves up to 312 MBH add				Ea.	104			104	114
7990	Special feature gas fired boilers									
8000	Pulse combustion, standard controls / trim, 44,000 BTU	Q-5	1.50	10.667	Ea.	2,350	360		2,710	3,125
8050	88,000 BTU		1.40	11.429		2,775	385		3,160	3,625
8080	134,000 BTU		1.20	13.333		3,600	450		4,050	4,625
460	0010	**BOILERS, GAS/OIL** Combination with burners and controls										**460**
1000	Cast iron with insulated jacket									
2000	Steam, gross output, 720 MBH	Q-7	.43	74.074	Ea.	7,000	2,650		9,650	11,700
2020	810 MBH		.38	83.989		7,000	3,000		10,000	12,300

Important: See the Reference Section for critical supporting data - Reference Nos., Crews, & City Cost Indexes

15510	Heating Boilers and Accessories	CREW	DAILY OUTPUT	LABOR-HOURS	UNIT	2003 BARE COSTS				TOTAL INCL O&P
						MAT.	LABOR	EQUIP.	TOTAL	
2040	1,084 MBH	Q-7	.34	93.023	Ea.	8,025	3,325		11,350	13,900
2060	1,360 MBH		.33	98.160		9,500	3,525		13,025	15,800
2080	1,600 MBH		.30	107		10,700	3,825		14,525	17,600
2100	2,040 MBH		.25	130		13,200	4,675		17,875	21,600
2120	2,450 MBH		.20	156		14,400	5,600		20,000	24,400
2140	2,700 MBH		.19	165		15,300	5,950		21,250	25,800
2160	3,000 MBH		.18	175		16,500	6,300		22,800	27,600
2180	3,270 MBH		.17	183		17,700	6,600		24,300	29,500
2200	3,770 MBH		.17	191		40,700	6,875		47,575	55,000
2220	4,070 MBH		.16	200		42,800	7,175		49,975	58,000
2240	4,650 MBH		.15	210		45,100	7,550		52,650	61,000
2260	5,230 MBH		.14	223		50,000	8,025		58,025	67,000
2280	5,520 MBH		.14	235		54,500	8,425		62,925	72,500
2300	5,810 MBH		.13	248		55,000	8,900		63,900	74,000
2320	6,100 MBH		.12	260		55,500	9,325		64,825	75,000
2340	6,390 MBH		.11	296		58,500	10,600		69,100	80,000
2360	6,680 MBH		.10	320		61,500	11,500		73,000	85,000
2380	6,970 MBH		.09	372		62,000	13,300		75,300	88,000
2900	Hot water, gross output									
2910	200 MBH	Q-6	.61	39.024	Ea.	4,800	1,375		6,175	7,350
2920	300 MBH		.49	49.080		4,800	1,725		6,525	7,875
2930	400 MBH		.41	57.971		5,625	2,025		7,650	9,250
2940	500 MBH		.36	67.039		6,050	2,350		8,400	10,200
3000	584 MBH	Q-7	.44	72.072		6,000	2,575		8,575	10,500
3020	876 MBH		.40	79.012		8,850	2,825		11,675	14,000
3040	1,168 MBH		.31	103		10,100	3,725		13,825	16,700
3060	1,460 MBH		.28	113		11,800	4,050		15,850	19,000
3080	2,044 MBH		.26	122		14,300	4,375		18,675	22,300
3100	2,628 MBH		.21	150		17,900	5,375		23,275	27,800
3120	3,210 MBH		.18	174		22,100	6,275		28,375	33,800
3140	3,796 MBH		.17	186		24,700	6,675		31,375	37,300
3160	4,088 MBH		.16	195		28,100	7,000		35,100	41,500
3180	4,672 MBH		.16	203		30,700	7,300		38,000	44,700
3200	5,256 MBH		.15	217		35,300	7,800		43,100	50,500
3220	6,000 MBH, 179 BHP		.13	256		49,200	9,175		58,375	68,000
3240	7,130 MBH, 213 BHP		.08	385		51,500	13,800		65,300	77,500
3260	9,800 MBH, 286 BHP		.06	533		59,500	19,100		78,600	94,000
3280	10,900 MBH, 325.6 BHP		.05	592		72,000	21,200		93,200	111,000
3290	12,200 MBH, 364.5 BHP		.05	666		78,500	23,900		102,400	122,500
3300	13,500 MBH, 403.3 BHP		.04	727		87,000	26,100		113,100	135,000
4000	Steel, insulated jacket, skid base, tubeless									
4100	Steam, 15 psi, gross output 517 MBH, 15 BHP	Q-6	.41	57.971	Ea.	12,900	2,025		14,925	17,300
4110	690 MBH, 20 BHP		.38	62.992		13,200	2,200		15,400	17,900
4120	862 MBH, 25 BHP		.36	67.039		14,100	2,350		16,450	19,100
4130	1,004 MBH, 30 BHP		.32	74.074		16,600	2,600		19,200	22,100
4140	1,340 MBH, 40 BHP		.31	78.431		18,200	2,750		20,950	24,200
4150	1,675 MBH, 50 BHP		.28	85.714		18,900	3,000		21,900	25,400
4160	2,175 MBH, 65 BHP		.23	104		19,400	3,675		23,075	26,900
4170	2,510 MBH, 75 BHP		.19	125		20,900	4,400		25,300	29,600
4180	3,015 MBH, 90 BHP		.17	140		23,700	4,925		28,625	33,600
4190	3,350 MBH, 100 BHP	Q-7	.22	146		27,100	5,250		32,350	37,800
4200	3,880 MBH, 116 BHP		.21	153		29,200	5,525		34,725	40,500
4210	4,685 MBH, 140 BHP		.19	168		31,100	6,050		37,150	43,400
4220	5,025 MBH, 150 BHP		.18	177		32,100	6,375		38,475	45,000
4230	5,550 MBH, 166 BHP		.17	189		41,700	6,800		48,500	56,000
4240	6,700 MBH, 200 BHP		.13	256		45,500	9,175		54,675	64,000

460

MECHANICAL 15

15510	Heating Boilers and Accessories	CREW	DAILY OUTPUT	LABOR-HOURS	UNIT	2003 BARE COSTS				TOTAL INCL O&P
						MAT.	LABOR	EQUIP.	TOTAL	
460	4250	7,760 MBH, 232 BHP	Q-7	.11	296	Ea.	51,000	10,600		61,600	72,000	460
	4260	8,870 MBH, 265 BHP		.09	336		55,500	12,100		67,600	79,000
	4270	11,100 MBH, 322 BHP		.07	426		59,000	15,300		74,300	88,000
	4280	13,400 MBH, 400 BHP		.06	516		67,500	18,500		86,000	102,500
	4290	17,750 MBH, 530 BHP		.05	680		85,000	24,400		109,400	130,500
	4300	20,750 MBH, 620 BHP		.04	800		95,000	28,700		123,700	147,500
	4500	Steam, 150 psi gross output, 335 MBH, 10 BHP	Q-6	.54	44.037		10,400	1,550		11,950	13,700
	4520	502 MBH, 15 BHP		.40	60		11,800	2,100		13,900	16,200
	4560	670 MBH, 20 BHP		.36	65.934		12,900	2,325		15,225	17,700
	4600	1,005 MBH, 30 BHP		.31	77.922		16,900	2,725		19,625	22,600
	4640	1,339 MBH, 40 BHP		.29	81.911		20,400	2,875		23,275	26,800
	4660	1,674 MBH, 50 BHP		.24	100		22,800	3,500		26,300	30,300
	4680	2,009 MBH, 60 BHP		.21	114		27,000	4,000		31,000	35,800
	4720	2,511 MBH, 75 BHP		.17	141		27,800	4,950		32,750	38,100
	5000	Hot water, gross output, 525 MBH		.45	52.980		10,100	1,850		11,950	14,000
	5020	630 MBH		.43	55.944		10,200	1,975		12,175	14,300
	5040	735 MBH		.41	58.968		12,600	2,075		14,675	17,000
	5060	840 MBH		.38	64		14,000	2,250		16,250	18,800
	5080	1,050 MBH		.35	67.989		16,300	2,375		18,675	21,600
	5100	1,365 MBH		.33	72.727		20,400	2,550		22,950	26,300
	5120	1,680 MBH		.29	82.759		23,300	2,900		26,200	30,100
	5140	2,310 MBH		.20	120		30,400	4,200		34,600	39,800
	5160	2,835 MBH		.18	129		36,600	4,550		41,150	47,200
	5180	3,150 MBH		.16	146		40,500	5,125		45,625	52,500
500	0010	**BOILERS, OIL FIRED** Standard controls, flame retention burner										500
	1000	Cast iron, with insulated flush jacket									
	2000	Steam, gross output, 109 MBH	Q-7	1.20	26.667	Ea.	1,450	955		2,405	3,050
	2020	144 MBH		1.10	29.091		1,675	1,050		2,725	3,425
	2040	173 MBH		1	32		1,825	1,150		2,975	3,725
	2060	207 MBH		.90	35.556		1,975	1,275		3,250	4,100
	2080	236 MBH		.85	37.647		2,125	1,350		3,475	4,400
	2100	300 MBH		.70	45.714		2,750	1,650		4,400	5,500
	2120	480 MBH		.50	64		3,525	2,300		5,825	7,375
	2140	665 MBH		.45	71.111		4,925	2,550		7,475	9,275
	2160	794 MBH		.41	78.049		5,750	2,800		8,550	10,600
	2180	1,084 MBH		.38	85.106		6,800	3,050		9,850	12,100
	2200	1,360 MBH		.33	98.160		7,950	3,525		11,475	14,100
	2220	1,600 MBH		.26	122		9,025	4,375		13,400	16,600
	2240	2,175 MBH		.24	133		10,900	4,800		15,700	19,300
	2260	2,480 MBH		.20	156		12,800	5,600		18,400	22,500
	2280	3,000 MBH		.19	170		14,600	6,100		20,700	25,200
	2300	3,550 MBH		.17	187		17,000	6,700		23,700	28,800
	2320	3,820 MBH		.16	200		17,900	7,175		25,075	30,500
	2340	4,360 MBH		.15	214		19,600	7,700		27,300	33,200
	2360	4,940 MBH		.14	225		38,900	8,075		46,975	55,000
	2380	5,520 MBH		.14	235		45,600	8,425		54,025	62,500
	2400	6,100 MBH		.13	256		52,500	9,175		61,675	71,500
	2420	6,390 MBH		.11	290		55,000	10,400		65,400	77,000
	2440	6,680 MBH		.10	313		56,500	11,200		67,700	79,500
	2460	6,970 MBH		.09	363		59,000	13,000		72,000	84,000
	3000	Hot water, same price as steam									
	4000	For tankless coil in smaller sizes, add					15%				
	5000	Steel, insulated jacket, burner									
	6000	Steam, full water leg construction, gross output									
	6020	144 MBH	Q-6	1.33	18.045	Ea.	3,500	635		4,135	4,825
	6040	198 MBH		1.20	20		3,825	700		4,525	5,250

Important: See the Reference Section for critical supporting data - Reference Nos., Crews, & City Cost Indexes

15510	Heating Boilers and Accessories	CREW	DAILY OUTPUT	LABOR-HOURS	UNIT	2003 BARE COSTS				TOTAL INCL O&P	
						MAT.	LABOR	EQUIP.	TOTAL		
500	6060	252 MBH	Q-6	1.13	21.295	Ea.	4,775	745		5,520	6,375	**500**
	6080	324 MBH		1.04	23.011		5,450	810		6,260	7,200	
	6100	396 MBH		.78	30.691		5,900	1,075		6,975	8,125	
	6120	468 MBH		.70	34.483		6,475	1,200		7,675	8,950	
	6140	648 MBH		.52	45.977		6,900	1,625		8,525	10,000	
	6160	792 MBH		.44	55.046		8,100	1,925		10,025	11,900	
	6180	1,008 MBH		.39	61.381		10,000	2,150		12,150	14,400	
	6200	1,260 MBH		.35	68.966		11,100	2,425		13,525	15,900	
	6220	1,512 MBH		.31	78.689		13,300	2,750		16,050	18,800	
	6240	1,800 MBH		.29	83.624		14,600	2,925		17,525	20,500	
	6260	2,100 MBH		.23	106		17,200	3,725		20,925	24,500	
	6280	2,400 MBH	▼	.19	125	▼	19,200	4,400		23,600	27,700	
	6400	Larger sizes are same as steel, gas fired										
	7000	Hot water, gross output, 103 MBH	Q-6	1.60	15	Ea.	1,300	525		1,825	2,225	
	7020	122 MBH		1.45	16.506		2,200	580		2,780	3,300	
	7040	137 MBH		1.36	17.595		2,525	620		3,145	3,700	
	7060	168 MBH		1.30	18.405		2,775	645		3,420	4,025	
	7080	225 MBH		1.22	19.704		2,850	690		3,540	4,175	
	7100	315 MBH		.96	25.105		3,700	880		4,580	5,400	
	7120	420 MBH		.70	34.483		4,375	1,200		5,575	6,650	
	7140	525 MBH		.57	42.403		5,000	1,500		6,500	7,750	
	7160	630 MBH		.52	45.977		5,725	1,625		7,350	8,750	
	7180	735 MBH		.48	50.104		5,850	1,750		7,600	9,075	
	7200	840 MBH		.43	55.172		9,150	1,925		11,075	13,000	
	7220	1,050 MBH		.36	65.753		9,450	2,300		11,750	13,900	
	7240	1,365 MBH		.32	74.534		10,800	2,625		13,425	15,900	
	7260	1,680 MBH		.29	83.624		12,200	2,925		15,125	17,800	
	7280	2,310 MBH		.21	114		13,200	4,025		17,225	20,600	
	7320	3,150 MBH	▼	.13	184	▼	18,200	6,475		24,675	29,800	
	7340	For tankless coil in steam or hot water, add					7%					
700	0010	**BOILERS, PACKAGED SCOTCH MARINE** Steam or hot water										**700**
	1000	Packaged fire tube, #2 oil, gross output										
	1006	15 PSI steam										
	1010	1005 MBH, 30 HP	Q-6	.33	72.072	Ea.	23,200	2,525		25,725	29,300	
	1014	1675 MBH, 50 HP	"	.26	93.023		29,000	3,275		32,275	36,800	
	1020	3348 MBH, 100 HP	Q-7	.21	152		32,500	5,475		37,975	44,100	
	1030	5025 MBH, 150 HP		.17	192		41,100	6,900		48,000	55,500	
	1040	6696 MBH, 200 HP		.14	223		50,500	8,025		58,525	68,000	
	1050	8375 MBH, 250 HP		.14	231		52,500	8,325		60,825	70,000	
	1060	10,044 MBH, 300 HP		.13	251		68,500	9,025		77,525	89,000	
	1080	16,740 MBH, 500 HP		.08	380		99,500	13,700		113,200	130,000	
	1100	23,435 MBH, 700 HP		.07	484		121,500	17,400		138,900	160,500	
	1140	To fire #6, and gas, add	▼	.42	76.190	▼	10,800	2,725		13,525	16,000	
	1180	For duplex package feed system										
	1200	To 3348 MBH boiler, add	Q-7	.54	59.259	Ea.	5,825	2,125		7,950	9,600	
	1220	To 6696 MBH boiler, add		.41	78.049		6,900	2,800		9,700	11,800	
	1240	To 10,044 MBH boiler, add		.38	84.211		7,925	3,025		10,950	13,300	
	1260	To 16,740 MBH boiler, add		.28	114		9,025	4,100		13,125	16,100	
	1280	To 23,435 MBH boiler, add	▼	.25	128	▼	10,700	4,600		15,300	18,600	
	1300	150 PSI steam										
	1310	1005 MBH, 30 HP	Q-6	.33	72.072	Ea.	23,700	2,525		26,225	29,800	
	1320	1675 MBH, 50 HP	"	.26	93.023		29,600	3,275		32,875	37,500	
	1330	3350 MBH, 100 HP	Q-7	.21	152		40,400	5,475		45,875	52,500	
	1340	5025 MBH, 150 HP		.17	192		51,500	6,900		58,400	67,000	
	1350	6700 MBH, 200 HP		.14	223		57,000	8,025		65,025	75,000	
	1360	10,050 MBH, 300 HP	▼	.13	251		84,500	9,025		93,525	106,500	

15510	Heating Boilers and Accessories	CREW	DAILY OUTPUT	LABOR-HOURS	UNIT	2003 BARE COSTS				TOTAL INCL O&P	
						MAT.	LABOR	EQUIP.	TOTAL		
700	1400	Packaged scotch marine, #6 oil										**700**
	1410	15 PSI steam									
	1420	1675 MBH, 50 HP	Q-6	.25	97.166	Ea.	35,600	3,400		39,000	44,400
	1430	3350 MBH, 100 HP	Q-7	.20	159		40,200	5,700		45,900	53,000
	1440	5025 MBH, 150 HP		.16	202		50,500	7,250		57,750	66,500
	1450	6700 MBH, 200 HP		.14	233		55,000	8,375		63,375	73,000
	1460	8375 MBH, 250 HP		.13	244		70,500	8,750		79,250	90,500
	1470	10,050 MBH, 300 HP		.12	262		83,000	9,400		92,400	105,000
	1480	13,400 MBH, 400 HP	↓	.09	351	↓	95,000	12,600		107,600	123,500
	1500	150 PSI steam									
	1510	1675 MBH, 50 HP	Q-6	.25	97.166	Ea.	42,200	3,400		45,600	51,500
	1520	3350 MBH, 100 HP	Q-7	.20	159		46,800	5,700		52,500	60,000
	1530	5025 MBH, 150 HP		.16	202		58,500	7,250		65,750	75,000
	1540	6700 MBH, 200 HP		.14	233		62,500	8,375		70,875	81,000
	1550	8375 MBH, 250 HP		.13	244		79,500	8,750		88,250	100,500
	1560	10,050 MBH, 300 HP		.12	262		90,000	9,400		99,400	113,000
	1570	13,400 MBH, 400 HP	↓	.09	351	↓	104,500	12,600		117,100	134,000
	2000	Packaged water tube, #2 oil, steam or hot water, gross output									
	2010	200 MBH	Q-6	1.50	16	Ea.	7,425	560		7,985	9,025
	2014	275 MBH		1.40	17.143		8,050	600		8,650	9,775
	2018	360 MBH		1.10	21.818		9,025	765		9,790	11,100
	2022	520 MBH		.65	36.923		9,250	1,300		10,550	12,200
	2026	600 MBH		.60	40		9,400	1,400		10,800	12,400
	2030	720 MBH		.55	43.636		9,800	1,525		11,325	13,100
	2034	960 MBH	↓	.48	50		10,800	1,750		12,550	14,600
	2040	1200 MBH	Q-7	.50	64		12,300	2,300		14,600	17,000
	2044	1440 MBH		.45	71.111		15,800	2,550		18,350	21,300
	2060	1600 MBH		.40	80		16,200	2,875		19,075	22,100
	2068	1920 MBH		.35	91.429		17,500	3,275		20,775	24,300
	2072	2160 MBH		.33	96.970		19,400	3,475		22,875	26,700
	2080	2400 MBH		.30	106		21,800	3,825		25,625	29,700
	2100	3200 MBH		.25	128		24,900	4,600		29,500	34,300
	2120	4800 MBH	↓	.20	160	↓	32,500	5,725		38,225	44,400
	2200	Gas fired									
	2204	200 MBH	Q-6	1.50	16	Ea.	5,350	560		5,910	6,750
	2208	275 MBH		1.40	17.143		5,575	600		6,175	7,025
	2212	360 MBH		1.10	21.818		5,575	765		6,340	7,275
	2216	520 MBH		.65	36.923		6,575	1,300		7,875	9,175
	2220	600 MBH		.60	40		6,575	1,400		7,975	9,350
	2224	720 MBH		.55	43.636		7,250	1,525		8,775	10,300
	2228	960 MBH	↓	.48	50		8,550	1,750		10,300	12,100
	2232	1220 MBH	Q-7	.50	64		9,200	2,300		11,500	13,600
	2236	1440 MBH		.45	71.111		9,800	2,550		12,350	14,700
	2240	1680 MBH		.40	80		10,900	2,875		13,775	16,300
	2244	1920 MBH		.35	91.429		11,600	3,275		14,875	17,700
	2248	2160 MBH		.33	96.970		12,500	3,475		15,975	19,000
	2252	2400 MBH	↓	.30	106	↓	14,000	3,825		17,825	21,200
760	0010	**BOILERS, SOLID FUEL**										**760**
	0020	Wood, coal, wood/paper waste, peat. Steel, 3" diam. horiz.									
	0040	fire tubes, low pressure steam or hot water, natural or									
	0060	mechanical draft. Not including dry base.									
	1000	Natural draft, gross output, 1445 MBH	Q-6	.45	53.333	Ea.	12,600	1,875		14,475	16,700
	1050	1635 MBH	↓	.43	55.814	↓	13,200	1,950		15,150	17,500
	1100	1805 MBH		.41	58.537		13,800	2,050		15,850	18,200
	1150	2160 MBH	↓	.38	63.158	↓	14,800	2,225		17,025	19,700

15510	Heating Boilers and Accessories	CREW	DAILY OUTPUT	LABOR-HOURS	UNIT	2003 BARE COSTS				TOTAL INCL O&P
						MAT.	LABOR	EQUIP.	TOTAL	
760 1200	2520 MBH	Q-6	.32	75	Ea.	17,900	2,625		20,525	23,700	**760**
1250	3065 MBH	Q-7	.40	80		19,200	2,875		22,075	25,400
1300	3600 MBH		.38	84.211		20,600	3,025		23,625	27,200
1350	4500 MBH		.29	110		26,600	3,950		30,550	35,300
1400	5400 MBH		.26	123		28,600	4,400		33,000	38,200
1450	6295 MBH		.23	139		33,200	5,000		38,200	44,000
1500	7200 MBH		.21	152		37,600	5,475		43,075	49,700
1550	9000 MBH		.19	168		40,800	6,050		46,850	54,000
1600	10,800 MBH		.17	188		46,700	6,750		53,450	61,500
1650	12,600 MBH		.15	213		51,000	7,650		58,650	68,000
1700	14,390 MBH		.13	246		61,500	8,825		70,325	81,500
1750	16,190 MBH		.12	266		67,000	9,550		76,550	88,000
1800	17,990 MBH	↓	.10	320	↓	74,500	11,500		86,000	99,500
3000	Stoker fired (coal) cast iron with flush jacket and									
3080	insulation, steam or water, gross output, 148 MBH	Q-6	1.50	16	Ea.	3,800	560		4,360	5,050
3100	200 MBH		1.30	18.462		4,350	650		5,000	5,775
3120	247 MBH		1.20	20		4,875	700		5,575	6,400
3140	300 MBH		1	24		5,450	840		6,290	7,275
3160	410 MBH		.96	25		6,250	880		7,130	8,200
3180	495 MBH		.88	27.273		6,975	955		7,930	9,125
3200	580 MBH		.80	30		7,725	1,050		8,775	10,100
3400	1280 MBH		.36	66.667		19,000	2,350		21,350	24,400
3420	1460 MBH		.30	80		20,900	2,800		23,700	27,300
3440	1640 MBH		.28	85.714		22,300	3,000		25,300	29,100
3460	1820 MBH		.26	92.308		23,900	3,250		27,150	31,200
3480	2000 MBH		.25	96		25,400	3,375		28,775	33,100
3500	2360 MBH		.23	104		28,800	3,675		32,475	37,200
3540	2725 MBH	↓	.20	120		32,000	4,200		36,200	41,600
3800	2950 MBH	Q-7	.16	200		53,500	7,175		60,675	70,000
3820	3210 MBH		.15	213		56,000	7,650		63,650	73,500
3840	3480 MBH		.14	228		58,500	8,200		66,700	77,000
3860	3745 MBH		.14	228		61,500	8,200		69,700	80,000
3880	4000 MBH		.13	246		64,000	8,825		72,825	84,000
3900	4200 MBH		.13	246		67,000	8,825		75,825	87,000
3920	4400 MBH		.12	266		69,500	9,550		79,050	91,000
3940	4600 MBH	↓	.12	266	↓	72,000	9,550		81,550	94,000
880 0010	**SWIMMING POOL HEATERS** Not including wiring, external										**880**
0020	piping, base or pad,									
0060	Gas fired, input, 115 MBH	Q-6	3	8	Ea.	1,600	281		1,881	2,200
0100	135 MBH		2	12		1,800	420		2,220	2,625
0160	155 MBH		1.50	16		1,925	560		2,485	2,950
0180	175 MBH		1.30	18.462		2,125	650		2,775	3,325
0200	190 MBH		1	24		2,500	840		3,340	4,000
0220	235 MBH		.70	34.286		3,050	1,200		4,250	5,175
0240	285 MBH		.60	40		3,900	1,400		5,300	6,425
0260	365 MBH		.50	48		4,400	1,675		6,075	7,375
0280	500 MBH		.40	60		6,025	2,100		8,125	9,800
0300	600 MBH		.35	68.571		6,925	2,400		9,325	11,300
0320	800 MBH		.33	72.727		8,250	2,550		10,800	12,900
0360	1000 MBH		.22	109		9,650	3,825		13,475	16,400
0370	1,200 MBH		.21	114		12,000	4,000		16,000	19,300
0380	1,500 MBH		.19	126		14,800	4,425		19,225	23,000
0400	1,800 MBH		.14	171		16,800	6,025		22,825	27,600
0410	2,450 MBH		.13	184		22,300	6,475		28,775	34,400
0420	3,000 MBH		.11	218		27,200	7,650		34,850	41,600
0440	3,750 MBH	↓	.09	266	↓	33,800	9,350		43,150	51,500

15510	Heating Boilers and Accessories	CREW	DAILY OUTPUT	LABOR-HOURS	UNIT	2003 BARE COSTS				TOTAL INCL O&P	
						MAT.	LABOR	EQUIP.	TOTAL		
880	1000	Oil fired, gross output, 97 MBH	Q-6	1.70	14.118	Ea.	1,850	495		2,345	2,775	880
1020	118 MBH		1.70	14.118		1,875	495		2,370	2,825
1040	134 MBH		1.60	15		2,000	525		2,525	3,000
1060	161 MBH		1.30	18.462		2,425	650		3,075	3,650
1080	187 MBH		1.20	20		2,575	700		3,275	3,875
1100	214 MBH		1.10	21.818		2,925	765		3,690	4,375
1120	262 MBH		.89	26.966		3,300	945		4,245	5,075
1140	340 MBH		.68	35.294		4,375	1,250		5,625	6,675
1160	420 MBH		.58	41.379		5,000	1,450		6,450	7,725
1180	525 MBH		.46	52.174		6,425	1,825		8,250	9,825
1200	630 MBH		.41	58.537		7,175	2,050		9,225	11,000
1220	735 MBH		.35	68.571		8,825	2,400		11,225	13,300
1240	840 MBH		.33	72.727		10,100	2,550		12,650	15,000
1260	1,050 MBH		.27	88.889		12,300	3,125		15,425	18,200
1280	1,365 MBH		.21	114		15,800	4,000		19,800	23,500
1300	1,680 MBH		.18	133		18,600	4,675		23,275	27,500
1320	2,310 MBH		.13	184		23,900	6,475		30,375	36,100
1340	2,835 MBH		.11	218		29,800	7,650		37,450	44,400
1360	3,150 MBH	▼	.09	266		33,500	9,350		42,850	51,000
2000	Electric, 12 KW, 4,800 gallon pool	Q-19	3	8		1,550	281		1,831	2,150
2020	15 KW, 7,200 gallon pool		2.80	8.571		1,550	300		1,850	2,175
2040	24 KW, 9,600 gallon pool		2.40	10		2,100	350		2,450	2,850
2060	30 KW, 12,000 gallon pool		2	12		2,175	420		2,595	3,025
2080	35 KW, 14,400 gallon pool		1.60	15		2,175	525		2,700	3,200
2100	55 KW, 24,000 gallon pool	▼	1.20	20	▼	3,000	700		3,700	4,350
9000	To select pool heater: 12 BTUH x S.F. pool area									
9010	X temperature differential =required output									
9050	For electric, KW = gallons x 2.5 divided by 1000									
9100	For family home type pool, double the									
9110	Rated gallon capacity = 1/2°F rise per hour									

15520	Feedwater Equipment										
050	0010	**SHOT CHEMICAL FEEDER**										050
0200	Shot chem feeder, by pass, in line mount, 125 PSIG, 1.7 gallon	2 Stpi	1.80	8.889	Ea.	315	335		650	850
0220	Floor mount, 175 PSIG, 5 gallon		1.24	12.903		400	485		885	1,175
0230	12 gallon		.93	17.204		635	645		1,280	1,675
0300	150 lb., ASME, 5 gallon		1.24	12.903		1,200	485		1,685	2,050
0320	10 gallon		.93	17.204		1,300	645		1,945	2,400
0400	300 lb., ASME, 5 gallon		1.24	12.903		3,225	485		3,710	4,250
0420	10 gallon	▼	.93	17.204	▼	4,000	645		4,645	5,375

15530	Furnaces										
200	0010	**FURNACE COMPONENTS AND COMBINATIONS**										200
0080	Coils, A/C evaporator, for gas or oil furnaces									
0090	Add-on, with holding charge									
0100	Upflow									
0120	1-1/2 ton cooling	Q-5	4	4	Ea.	127	135		262	345
0130	2 ton cooling		3.70	4.324		153	146		299	390
0140	3 ton cooling		3.30	4.848		191	164		355	460
0150	4 ton cooling		3	5.333		270	181		451	570
0160	5 ton cooling	▼	2.70	5.926	▼	345	201		546	685
0300	Downflow									
0330	2-1/2 ton cooling	Q-5	3	5.333	Ea.	178	181		359	470
0340	3-1/2 ton cooling		2.60	6.154		239	208		447	580
0350	5 ton cooling	▼	2.20	7.273	▼	345	246		591	750
0600	Horizontal									

15530	Furnaces	CREW	DAILY OUTPUT	LABOR-HOURS	UNIT	2003 BARE COSTS				TOTAL INCL O&P	
						MAT.	LABOR	EQUIP.	TOTAL		
200											**200**
0630	2 ton cooling	Q-5	3.90	4.103	Ea.	180	139		319	410	
0640	3 ton cooling		3.50	4.571		206	155		361	460	
0650	4 ton cooling		3.20	5		259	169		428	540	
0660	5 ton cooling		2.90	5.517		345	187		532	660	
2000	Cased evaporator coils for air handlers										
2100	1-1/2 ton cooling	Q-5	4.40	3.636	Ea.	191	123		314	395	
2110	2 ton cooling		4.10	3.902		194	132		326	415	
2120	2-1/2 ton cooling		3.90	4.103		211	139		350	440	
2130	3 ton cooling		3.70	4.324		249	146		395	495	
2140	3-1/2 ton cooling		3.50	4.571		234	155		389	490	
2150	4 ton cooling		3.20	5		280	169		449	565	
2160	5 ton cooling		2.90	5.517		325	187		512	635	
3010	Air handler, modular										
3100	With cased evaporator cooling coil										
3120	1-1/2 ton cooling	Q-5	3.80	4.211	Ea.	530	143		673	800	
3130	2 ton cooling		3.50	4.571		555	155		710	845	
3140	2-1/2 ton cooling		3.30	4.848		605	164		769	915	
3150	3 ton cooling		3.10	5.161		660	175		835	990	
3160	3-1/2 ton cooling		2.90	5.517		790	187		977	1,150	
3170	4 ton cooling		2.50	6.400		900	217		1,117	1,325	
3180	5 ton cooling		2.10	7.619		950	258		1,208	1,450	
3500	With no cooling coil										
3520	1-1/2 ton coil size	Q-5	12	1.333	Ea.	340	45		385	445	
3530	2 ton coil size		10	1.600		365	54		419	480	
3540	2-1/2 ton coil size		10	1.600		405	54		459	525	
3554	3 ton coil size		9	1.778		450	60		510	585	
3560	3-1/2 ton coil size		9	1.778		450	60		510	585	
3570	4 ton coil size		8.50	1.882		640	63.50		703.50	800	
3580	5 ton coil size		8	2		690	67.50		757.50	860	
4000	With heater										
4120	5 kW, 17.1 MBH	Q-5	16	1	Ea.	264	34		298	340	
4130	7.5 kW, 25.6 MBH		15.60	1.026		274	34.50		308.50	355	
4140	10 kW, 34.2 MBH		15.20	1.053		330	35.50		365.50	420	
4150	12.5 KW, 42.7 MBH		14.80	1.081		375	36.50		411.50	470	
4160	15 KW, 51.2 MBH		14.40	1.111		460	37.50		497.50	560	
4170	25 KW, 85.4 MBH		14	1.143		505	38.50		543.50	615	
4180	30 KW, 102 MBH		13	1.231		615	41.50		656.50	745	
400	**FURNACES** Hot air heating, blowers, standard controls										**400**
0010											
0020	not including gas, oil or flue piping										
1000	Electric, UL listed										
1020	10.2 MBH	Q-20	5	4	Ea.	315	137		452	555	
1040	17.1 MBH		4.80	4.167		330	142		472	580	
1060	27.3 MBH		4.60	4.348		395	149		544	665	
1100	34.1 MBH		4.40	4.545		405	155		560	685	
1120	51.6 MBH		4.20	4.762		505	163		668	805	
1140	68.3 MBH		4	5		610	171		781	930	
1160	85.3 MBH		3.80	5.263		670	180		850	1,000	
3000	Gas, AGA certified, upflow, direct drive models										
3020	45 MBH input	Q-9	4	4	Ea.	435	133		568	685	
3040	60 MBH input		3.80	4.211		580	140		720	855	
3060	75 MBH input		3.60	4.444		615	148		763	910	
3100	100 MBH input		3.20	5		655	167		822	975	
3120	125 MBH input		3	5.333		755	178		933	1,100	
3130	150 MBH input		2.80	5.714		875	190		1,065	1,250	
3140	200 MBH input		2.60	6.154		1,675	205		1,880	2,150	

MECHANICAL 15

15530	Furnaces	CREW	DAILY OUTPUT	LABOR-HOURS	UNIT	2003 BARE COSTS				TOTAL INCL O&P
						MAT.	LABOR	EQUIP.	TOTAL	
400 3160	300 MBH input	Q-9	2.30	6.957	Ea.	1,750	232		1,982	2,275 **400**
3180	400 MBH input	↓	2	8	↓	2,275	266		2,541	2,900
3200	Gas fired wall furnace									
3204	Horizontal flow									
3208	7.7 MBH	Q-9	7	2.286	Ea.	400	76		476	560
3212	14 MBH		6.50	2.462		430	82		512	600
3216	24 MBH		5	3.200		520	107		627	735
3220	49 MBH		4	4		515	133		648	770
3224	65 MBH	↓	3.60	4.444	↓	695	148		843	990
3240	Up-flow									
3244	7.7 MBH	Q-9	7	2.286	Ea.	254	76		330	395
3248	14 MBH		6.50	2.462		294	82		376	450
3252	24 MBH		5	3.200		360	107		467	565
3256	49 MBH		4	4		705	133		838	980
3260	65 MBH	↓	3.60	4.444	↓	730	148		878	1,025
3500	Gas furnace									
3510	Up-flow									
3514	1200 CFM, 45 MBH	Q-9	4	4	Ea.	685	133		818	960
3518	75 MBH		3.60	4.444		755	148		903	1,050
3522	100 MBH		3.20	5		865	167		1,032	1,200
3526	125 MBH		3	5.333		1,000	178		1,178	1,375
3530	150 MBH		2.80	5.714		1,225	190		1,415	1,650
3540	2000 CFM, 75 MBH		3.60	4.444		830	148		978	1,150
3544	100 MBH		3.20	5		940	167		1,107	1,275
3548	120 MBH		3	5.333		1,075	178		1,253	1,475
3552	150 MBH	↓	2.80	5.714	↓	1,300	190		1,490	1,725
3560	Horizontal flow									
3564	1200 CFM, 45 MBH	Q-9	4	4	Ea.	690	133		823	960
3568	60 MBH		3.80	4.211		790	140		930	1,075
3572	70 MBH		3.60	4.444		800	148		948	1,100
3576	2000 CFM, 90 MBH		3.30	4.848		865	161		1,026	1,200
3580	110 MBH	↓	3.10	5.161	↓	905	172		1,077	1,275
6000	Oil, UL listed, atomizing gun type burner									
6020	56 MBH output	Q-9	3.60	4.444	Ea.	745	148		893	1,050
6030	84 MBH output		3.50	4.571		775	152		927	1,075
6040	95 MBH output		3.40	4.706		790	157		947	1,100
6060	134 MBH output		3.20	5		1,100	167		1,267	1,450
6080	151 MBH output		3	5.333		1,225	178		1,403	1,600
6100	200 MBH input		2.60	6.154		1,725	205		1,930	2,225
6120	300 MBH input		2.30	6.957		2,000	232		2,232	2,550
6140	400 MBH input	↓	2	8	↓	2,350	266		2,616	3,000
6200	Up flow									
6210	105 MBH	Q-9	3.40	4.706	Ea.	1,125	157		1,282	1,500
6220	140 MBH		3.30	4.848		1,300	161		1,461	1,675
6230	168 MBH	↓	2.90	5.517	↓	1,400	184		1,584	1,825
6260	Lo-boy style									
6270	105 MBH	Q-9	3.40	4.706	Ea.	1,300	157		1,457	1,700
6280	140 MBH		3.30	4.848		1,425	161		1,586	1,800
6290	189 MBH	↓	2.80	5.714	↓	1,575	190		1,765	2,050
6320	Down flow									
6330	105 MBH	Q-9	3.40	4.706	Ea.	1,250	157		1,407	1,625
6340	140 MBH		3.30	4.848		1,425	161		1,586	1,825
6350	168 MBH	↓	2.90	5.517	↓	1,475	184		1,659	1,900
6380	Horizontal									
6390	105 MBH	Q-9	3.40	4.706	Ea.	1,225	157		1,382	1,600
6400	140 MBH	↓	3.30	4.848	↓	1,350	161		1,511	1,750

15 MECHANICAL

Important: See the Reference Section for critical supporting data - Reference Nos., Crews, & City Cost Indexes

15530	Furnaces	CREW	DAILY OUTPUT	LABOR-HOURS	UNIT	2003 BARE COSTS				TOTAL INCL O&P
						MAT.	LABOR	EQUIP.	TOTAL	
400 6410	189 MBH	Q-9	2.80	5.714	Ea.	1,475	190		1,665	1,925 **400**
440 0010	**FURNACES, COMBINATION SYSTEMS** Heating, cooling,									**440**
0020	electric air cleaner, humidification, dehumidification.									
2000	Gas fired, 80 MBH heat output, 24 MBH cooling	Q-9	1.20	13.333	Ea.	2,950	445		3,395	3,925
2020	80 MBH heat output, 36 MBH cooling		1.20	13.333		3,225	445		3,670	4,225
2040	100 MBH heat output, 29 MBH cooling		1	16		3,300	535		3,835	4,450
2060	100 MBH heat output, 36 MBH cooling		1	16		3,450	535		3,985	4,625
2080	100 MBH heat output, 47 MBH cooling		.90	17.778		4,100	590		4,690	5,450
2100	120 MBH heat output, 29 MBH cooling	Q-10	1.30	18.462		3,500	635		4,135	4,825
2120	120 MBH heat output, 42 MBH cooling		1.30	18.462		3,700	635		4,335	5,050
2140	120 MBH heat output, 47 MBH cooling		1.20	20		4,125	690		4,815	5,600
2160	120 MBH heat output, 55 MBH cooling		1.10	21.818		4,425	755		5,180	6,050
2180	144 MBH heat output, 42 MBH cooling		1.20	20		3,850	690		4,540	5,300
2200	144 MBH heat output, 47 MBH cooling		1.20	20		4,275	690		4,965	5,775
2220	144 MBH heat output, 58 MBH cooling		1	24		4,475	830		5,305	6,200
2500	For precharged tubing with connection									
2520	15 feet				Ea.	123			123	135
2540	25 feet					164			164	180
2560	35 feet					211			211	232
3000	Oil fired, 84 MBH heat output, 24 MBH cooling	Q-9	1.20	13.333		3,300	445		3,745	4,300
3020	84 MBH heat output, 36 MBH cooling		1.20	13.333		3,625	445		4,070	4,675
3040	95.2 MBH heat output, 29 MBH cooling		1	16		3,550	535		4,085	4,725
3060	95.2 MBH heat output, 36 MBH cooling		1	16		3,700	535		4,235	4,900
3080	95.2 MBH heat output, 47 MBH cooling		.90	17.778		4,275	590		4,865	5,625
3100	112 MBH heat output, 29 MBH cooling	Q-10	1.30	18.462		3,800	635		4,435	5,150
3120	112 MBH heat output, 42 MBH cooling		1.30	18.462		4,125	635		4,760	5,500
3140	112 MBH heat output, 47 MBH cooling		1.20	20		4,550	690		5,240	6,075
3160	112 MBH heat output, 58 MBH cooling		1.10	21.818		4,975	755		5,730	6,650
3180	151.2 MBH heat output, 42 MBH cooling		1.20	20		4,425	690		5,115	5,950
3200	151.2 MBH heat output, 47 MBH cooling		1.20	20		4,850	690		5,540	6,425
3240	151.2 MBH heat output, 58 MBH cooling		1	24		5,075	830		5,905	6,850
3500	For precharged tubing with connection, add									
3520	15 feet				Ea.	123			123	135
3540	25 feet					164			164	180
3560	35 feet					196			196	215

15540	Fuel-Fired Heaters	CREW	DAILY OUTPUT	LABOR-HOURS	UNIT	MAT.	LABOR	EQUIP.	TOTAL	TOTAL INCL O&P
300 0010	**DUCT FURNACES** Includes burner, controls, stainless steel									**300**
0020	heat exchanger. Gas fired, electric ignition									
0030	Indoor installation									
0080	100 MBH output	Q-5	5	3.200	Ea.	1,150	108		1,258	1,450
0100	120 MBH output		4	4		1,325	135		1,460	1,650
0130	200 MBH output		2.70	5.926		1,800	201		2,001	2,300
0140	240 MBH output		2.30	6.957		1,875	235		2,110	2,425
0160	280 MBH output		2	8		2,100	271		2,371	2,700
0180	320 MBH output		1.60	10		2,275	340		2,615	3,000
0300	For powered venter and adapter, add					293			293	325
0500	For required flue pipe, see 15550									
1000	Outdoor installation, with vent cap									
1020	75 MBH output	Q-5	4	4	Ea.	1,650	135		1,785	2,025
1040	94 MBH output		4	4		1,875	135		2,010	2,250
1060	120 MBH output		4	4		2,150	135		2,285	2,550
1080	157 MBH output		3.50	4.571		2,500	155		2,655	2,975
1100	187 MBH output		3	5.333		3,000	181		3,181	3,575
1120	225 MBH output		2.50	6.400		3,200	217		3,417	3,850

	15540	**Fuel-Fired Heaters**	CREW	DAILY OUTPUT	LABOR-HOURS	UNIT	|2003 BARE COSTS| MAT.	LABOR	EQUIP.	TOTAL	TOTAL INCL O&P	
300	1140	300 MBH output	Q-5	1.80	8.889	Ea.	4,050	300		4,350	4,900	300
	1160	375 MBH output		1.60	10		4,150	340		4,490	5,075	
	1180	450 MBH output		1.40	11.429		4,350	385		4,735	5,375	
	1200	600 MBH output		1	16		5,300	540		5,840	6,650	
	1300	Aluminized exchanger, subtract					15%					
	1400	For powered venter, add					10%					
	1500	For two stage gas valve, add					160			160	176	
900	0010	**SPACE HEATERS** Cabinet, grilles, fan, controls, burner,										900
	0020	thermostat, no piping. For flue see 15550										
	1000	Gas fired, floor mounted										
	1100	60 MBH output	Q-5	10	1.600	Ea.	545	54		599	680	
	1120	80 MBH output		9	1.778		555	60		615	700	
	1140	100 MBH output		8	2		595	67.50		662.50	755	
	1160	120 MBH output		7	2.286		670	77.50		747.50	855	
	1180	180 MBH output		6	2.667		815	90.50		905.50	1,025	
	1500	Rooftop mounted, gravity vent, stainless steel exchanger										
	1520	75 MBH output	Q-6	4	6	Ea.	3,300	211		3,511	3,975	
	1540	95 MBH output		3.60	6.667		3,375	234		3,609	4,050	
	1560	120 MBH output		3.30	7.273		3,600	255		3,855	4,350	
	1580	159 MBH output		3	8		3,725	281		4,006	4,525	
	1600	190 MBH output		2.60	9.231		4,200	325		4,525	5,100	
	1620	225 MBH output		2.30	10.435		4,250	365		4,615	5,225	
	1640	300 MBH output		1.90	12.632		5,075	445		5,520	6,250	
	1660	375 MBH output		1.40	17.143		5,750	600		6,350	7,225	
	1680	450 MBH output		1.20	20		5,875	700		6,575	7,500	
	1700	600 MBH output		1	24		7,075	840		7,915	9,075	
	1720	750 MBH output		.80	30		12,100	1,050		13,150	14,900	
	1740	900 MBH output		.60	40		13,600	1,400		15,000	17,000	
	1760	1200 MBH output		.30	80		16,500	2,800		19,300	22,400	
	1800	For power vent, add					10%					
	1900	For aluminized steel exchanger, subtract					10%					
	2000	Suspension mounted, propeller fan, 20 MBH output	Q-5	8.50	1.882		425	63.50		488.50	560	
	2020	40 MBH output		7.50	2.133		465	72		537	620	
	2040	60 MBH output		7	2.286		525	77.50		602.50	690	
	2060	80 MBH output		6	2.667		585	90.50		675.50	775	
	2080	100 MBH output		5.50	2.909		695	98.50		793.50	910	
	2100	130 MBH output		5	3.200		775	108		883	1,025	
	2120	140 MBH output		4.50	3.556		860	120		980	1,125	
	2140	160 MBH output		4	4		860	135		995	1,150	
	2160	180 MBH output		3.50	4.571		975	155		1,130	1,300	
	2180	200 MBH output		3	5.333		1,000	181		1,181	1,375	
	2200	240 MBH output		2.70	5.926		1,200	201		1,401	1,600	
	2220	280 MBH output		2.30	6.957		1,450	235		1,685	1,950	
	2240	320 MBH output		2	8		1,600	271		1,871	2,175	
	2500	For powered venter and adapter, add					258			258	284	
	3000	Suspension mounted, blower type, 40 MBH output	Q-5	6.80	2.353		730	79.50		809.50	925	
	3020	60 MBH output		6.60	2.424		810	82		892	1,025	
	3040	84 MBH output		5.80	2.759		865	93.50		958.50	1,100	
	3060	104 MBH output		5.20	3.077		1,000	104		1,104	1,250	
	3080	140 MBH output		4.30	3.721		1,200	126		1,326	1,525	
	3100	180 MBH output		3.30	4.848		1,250	164		1,414	1,625	
	3120	240 MBH output		2.50	6.400		1,825	217		2,042	2,350	
	3140	280 MBH output		2	8		2,050	271		2,321	2,650	
	4000	Suspension mounted, sealed combustion system,										
	4020	Aluminized steel exchanger, powered vent										

Important: See the Reference Section for critical supporting data - Reference Nos., Crews, & City Cost Indexes

15540	Fuel-Fired Heaters	CREW	DAILY OUTPUT	LABOR-HOURS	UNIT	2003 BARE COSTS				TOTAL INCL O&P
						MAT.	LABOR	EQUIP.	TOTAL	
900 4040	100 MBH output	Q-5	5	3.200	Ea.	1,725	108		1,833	2,075
4060	120 MBH output		4.70	3.404		1,800	115		1,915	2,150
4080	160 MBH output		3.70	4.324		2,000	146		2,146	2,425
4100	200 MBH output		2.90	5.517		2,325	187		2,512	2,825
4120	240 MBH output		2.50	6.400		2,500	217		2,717	3,100
4140	320 MBH output		1.70	9.412		2,900	320		3,220	3,650
5000	Wall furnace, 17.5 MBH output		6	2.667		545	90.50		635.50	735
5020	24 MBH output		5	3.200		555	108		663	775
5040	35 MBH output		4	4		755	135		890	1,050
6000	Oil fired, suspension mounted, 94 MBH output		4	4		2,000	135		2,135	2,400
6040	140 MBH output		3	5.333		2,125	181		2,306	2,625
6060	184 MBH output		3	5.333		2,300	181		2,481	2,800

15550	Breechings, Chimneys & Stacks									
200 0010	**DRAFT CONTROLS**									**200**
1000	Barometric, gas fired system only, 6" size for 5" and 6" pipes	1 Shee	20	.400	Ea.	39.50	14.80		54.30	66.50
1020	7" size, for 6" and 7" pipes		19	.421		47.50	15.60		63.10	76
1040	8" size, for 7" and 8" pipes		18	.444		54.50	16.45		70.95	85.50
1060	9" size, for 8" and 9" pipes		16	.500		64	18.50		82.50	99
2000	All fuel, oil, oil/gas, coal									
2020	10" for 9" and 10" pipes	1 Shee	15	.533	Ea.	93.50	19.75		113.25	134
2040	12" for 11" and 12" pipes		15	.533		124	19.75		143.75	168
2060	14" for 13" and 14" pipes		14	.571		155	21		176	204
2080	16" for 15" and 16" pipes		13	.615		217	23		240	274
2100	18" for 17" and 18" pipes		12	.667		284	24.50		308.50	350
2120	20" for 19" and 21" pipes		10	.800		340	29.50		369.50	420
2140	24" for 22" and 25" pipes	Q-9	12	1.333		425	44.50		469.50	535
2160	28" for 26" and 30" pipes		10	1.600		510	53.50		563.50	645
2180	32" for 31" and 34" pipes		8	2		665	66.50		731.50	835
3260	For thermal switch for above, add	1 Shee	24	.333		39.50	12.35		51.85	62
5000	Vent damper, bi-metal, gas, 3" diameter	Q-9	24	.667		22	22		44	58.50
5010	4" diameter		24	.667		22	22		44	58.50
5020	5" diameter		23	.696		22	23		45	60
5030	6" diameter		22	.727		22	24		46	61.50
5040	7" diameter		21	.762		23.50	25.50		49	65
5050	8" diameter		20	.800		25.50	26.50		52	69
5101	Electric, automatic, gas, 4" diameter		24	.667		137	22		159	185
5110	5" diameter		23	.696		139	23		162	189
5121	6" diameter		22	.727		142	24		166	194
5130	7" diameter		21	.762		144	25.50		169.50	198
5140	8" diameter		20	.800		147	26.50		173.50	203
5150	9" diameter		20	.800		150	26.50		176.50	206
5160	10" diameter		19	.842		155	28		183	214
5170	12" diameter		19	.842		157	28		185	217
5250	Automatic, oil, 4" diameter		24	.667		153	22		175	203
5260	5" diameter		23	.696		156	23		179	207
5270	6" diameter		22	.727		159	24		183	213
5280	7" diameter		21	.762		162	25.50		187.50	217
5290	8" diameter		20	.800		169	26.50		195.50	227
5300	9" diameter		20	.800		171	26.50		197.50	230
5310	10" diameter		19	.842		172	28		200	233
5320	12" diameter		19	.842		175	28		203	236
5330	14" diameter		18	.889		234	29.50		263.50	305
5340	16" diameter		17	.941		271	31.50		302.50	345
5350	18" diameter		16	1		305	33.50		338.50	385

15550	Breechings, Chimneys & Stacks	CREW	DAILY OUTPUT	LABOR-HOURS	UNIT	2003 BARE COSTS				TOTAL INCL O&P
						MAT.	LABOR	EQUIP.	TOTAL	
440	**0010 VENT CHIMNEY** Prefab metal, U.L. listed									**440**
0020	Gas, double wall, galvanized steel									
0080	3" diameter	Q-9	72	.222	V.L.F.	3.48	7.40		10.88	15.30
0100	4" diameter		68	.235		4.35	7.85		12.20	16.90
0120	5" diameter		64	.250		5.05	8.35		13.40	18.45
0140	6" diameter		60	.267		6.25	8.90		15.15	20.50
0160	7" diameter		56	.286		8.65	9.50		18.15	24.50
0180	8" diameter		52	.308		9.65	10.25		19.90	26.50
0200	10" diameter		48	.333		20.50	11.10		31.60	39.50
0220	12" diameter		44	.364		27	12.10		39.10	48.50
0240	14" diameter		42	.381		46	12.70		58.70	70
0260	16" diameter		40	.400		62	13.30		75.30	88.50
0280	18" diameter	▼	38	.421		80	14		94	110
0300	20" diameter	Q-10	36	.667		94.50	23		117.50	140
0320	22" diameter		34	.706		120	24.50		144.50	170
0340	24" diameter		32	.750		147	26		173	202
0360	26" diameter		31	.774		177	26.50		203.50	237
0380	28" diameter		30	.800		187	27.50		214.50	248
0400	30" diameter		28	.857		197	29.50		226.50	263
0420	32" diameter		27	.889		229	30.50		259.50	300
0440	34" diameter		26	.923		261	32		293	335
0460	36" diameter		25	.960		275	33		308	355
0480	38" diameter		24	1		300	34.50		334.50	385
0500	40" diameter		23	1.043		335	36		371	425
0520	42" diameter		22	1.091		350	37.50		387.50	445
0540	44" diameter		21	1.143		390	39.50		429.50	490
0560	46" diameter		20	1.200		430	41.50		471.50	535
0580	48" diameter	▼	19	1.263	▼	475	43.50		518.50	590
0600	For 4", 5" and 6" oval, add					50%				
0650	Gas, double wall, galvanized steel, fittings									
0660	Elbow 45°, 3" diameter	Q-9	36	.444	Ea.	8.40	14.80		23.20	32.50
0670	4" diameter		34	.471		10.20	15.65		25.85	35.50
0680	5" diameter		32	.500		12	16.65		28.65	38.50
0690	6" diameter		30	.533		14.35	17.75		32.10	43.50
0700	7" diameter		28	.571		20.50	19.05		39.55	52.50
0710	8" diameter		26	.615		24	20.50		44.50	57.50
0720	10" diameter		24	.667		59	22		81	99.50
0730	12" diameter		22	.727		75	24		99	120
0740	14" diameter		21	.762		105	25.50		130.50	154
0750	16" diameter		20	.800		131	26.50		157.50	186
0760	18" diameter	▼	19	.842		177	28		205	238
0770	20" diameter	Q-10	18	1.333		194	46		240	284
0780	22" diameter		17	1.412		292	49		341	395
0790	24" diameter		16	1.500		380	52		432	500
0800	26" diameter		16	1.500		750	52		802	905
0810	28" diameter		15	1.600		770	55.50		825.50	935
0820	30" diameter		14	1.714		815	59		874	985
0830	32" diameter		14	1.714		915	59		974	1,100
0840	34" diameter		13	1.846		1,000	64		1,064	1,200
0850	36" diameter		12	2		1,025	69		1,094	1,225
0860	38" diameter		12	2		1,125	69		1,194	1,325
0870	40" diameter		11	2.182		1,175	75.50		1,250.50	1,425
0880	42" diameter		11	2.182		1,225	75.50		1,300.50	1,475
0890	44" diameter		10	2.400		1,325	83		1,408	1,575
0900	46" diameter		10	2.400		1,400	83		1,483	1,675
0910	48" diameter	▼	10	2.400	▼	1,475	83		1,558	1,750

Important: See the Reference Section for critical supporting data - Reference Nos., Crews, & City Cost Indexes

15550	Breechings, Chimneys & Stacks	CREW	DAILY OUTPUT	LABOR-HOURS	UNIT	2003 BARE COSTS				TOTAL INCL O&P
						MAT.	LABOR	EQUIP.	TOTAL	
0916	Adjustable length									
0918	3" diameter, to 12"	Q-9	36	.444	Ea.	7.05	14.80		21.85	31
0920	4" diameter, to 12"		34	.471		8.35	15.65		24	33
0924	6" diameter, to 12"		30	.533		10.80	17.75		28.55	39.50
0928	8" diameter, to 12"		26	.615		15.85	20.50		36.35	49
0930	10" diameter, to 18"		24	.667		59.50	22		81.50	100
0932	12" diameter, to 18"		22	.727		72.50	24		96.50	118
0936	16" diameter, to 18"		20	.800		140	26.50		166.50	195
0938	18" diameter, to 18"	▼	19	.842		167	28		195	228
0944	24" diameter, to 18"	Q-10	16	1.500		299	52		351	410
0950	Elbow 90°, adjustable, 3" diameter	Q-9	36	.444		14.40	14.80		29.20	39
0960	4" diameter		34	.471		17.50	15.65		33.15	43.50
0970	5" diameter		32	.500		20.50	16.65		37.15	48
0980	6" diameter		30	.533		24.50	17.75		42.25	54.50
0990	7" diameter		28	.571		38.50	19.05		57.55	72
1010	8" diameter		26	.615		41	20.50		61.50	76.50
1020	Wall thimble, 4 to 7" adjustable, 3" diameter		36	.444		9.10	14.80		23.90	33
1022	4" diameter		34	.471		10.15	15.65		25.80	35
1024	5" diameter		32	.500		11.40	16.65		28.05	38
1026	6" diameter		30	.533		14	17.75		31.75	43
1028	7" diameter		28	.571		16.25	19.05		35.30	47.50
1030	8" diameter		26	.615		16.85	20.50		37.35	50
1040	Roof flashing, 3" diameter		36	.444		5.20	14.80		20	28.50
1050	4" diameter		34	.471		5.85	15.65		21.50	30.50
1060	5" diameter		32	.500		9.10	16.65		25.75	35.50
1070	6" diameter		30	.533		12.75	17.75		30.50	41.50
1080	7" diameter		28	.571		16.10	19.05		35.15	47
1090	8" diameter		26	.615		17.50	20.50		38	51
1100	10" diameter		24	.667		47.50	22		69.50	87
1110	12" diameter		22	.727		61	24		85	105
1120	14" diameter		20	.800		75	26.50		101.50	124
1130	16" diameter		18	.889		92.50	29.50		122	148
1140	18" diameter	▼	16	1		125	33.50		158.50	190
1150	20" diameter	Q-10	18	1.333		163	46		209	250
1160	22" diameter		14	1.714		206	59		265	320
1170	24" diameter	▼	12	2		247	69		316	380
1200	Tee, 3" diameter	Q-9	27	.593		21.50	19.75		41.25	54
1210	4" diameter		26	.615		22	20.50		42.50	56
1220	5" diameter		25	.640		24	21.50		45.50	59.50
1230	6" diameter		24	.667		26	22		48	63
1240	7" diameter		23	.696		28.50	23		51.50	67.50
1250	8" diameter		22	.727		39	24		63	80.50
1260	10" diameter		21	.762		106	25.50		131.50	156
1270	12" diameter		20	.800		128	26.50		154.50	181
1280	14" diameter		18	.889		208	29.50		237.50	275
1290	16" diameter		16	1		295	33.50		328.50	375
1300	18" diameter	▼	14	1.143		370	38		408	465
1310	20" diameter	Q-10	17	1.412		500	49		549	625
1320	22" diameter		13	1.846		620	64		684	785
1330	24" diameter		12	2		705	69		774	885
1340	26" diameter		12	2		1,025	69		1,094	1,225
1350	28" diameter		11	2.182		1,150	75.50		1,225.50	1,400
1360	30" diameter		10	2.400		1,250	83		1,333	1,500
1370	32" diameter		10	2.400		1,475	83		1,558	1,750
1380	34" diameter		9	2.667		1,475	92		1,567	1,775
1390	36" diameter	▼	9	2.667	▼	1,775	92		1,867	2,100

440

MECHANICAL 15

	15550	Breechings, Chimneys & Stacks	CREW	DAILY OUTPUT	LABOR-HOURS	UNIT	2003 BARE COSTS				TOTAL INCL O&P	
							MAT.	LABOR	EQUIP.	TOTAL		
440	1400	38" diameter	Q-10	8	3	Ea.	1,775	104		1,879	2,100	**440**
	1410	40" diameter		7	3.429		1,925	118		2,043	2,300	
	1420	42" diameter		6	4		2,100	138		2,238	2,550	
	1430	44" diameter		5	4.800		2,225	166		2,391	2,700	
	1440	46" diameter		5	4.800		2,450	166		2,616	2,925	
	1450	48" diameter	▼	5	4.800		2,600	166		2,766	3,125	
	1460	Tee cap, 3" diameter	Q-9	45	.356		1.53	11.85		13.38	20	
	1470	4" diameter		42	.381		1.69	12.70		14.39	21.50	
	1480	5" diameter		40	.400		2.28	13.30		15.58	23	
	1490	6" diameter		37	.432		2.63	14.40		17.03	25	
	1500	7" diameter		35	.457		4.29	15.20		19.49	28	
	1510	8" diameter		34	.471		5	15.65		20.65	29.50	
	1520	10" diameter		32	.500		9.60	16.65		26.25	36	
	1530	12" diameter		30	.533		12.80	17.75		30.55	41.50	
	1540	14" diameter		28	.571		26.50	19.05		45.55	58.50	
	1550	16" diameter		25	.640		30.50	21.50		52	66.50	
	1560	18" diameter	▼	24	.667		34.50	22		56.50	72.50	
	1570	20" diameter	Q-10	27	.889		37	30.50		67.50	88	
	1580	22" diameter		22	1.091		42.50	37.50		80	105	
	1590	24" diameter		21	1.143		48	39.50		87.50	114	
	1600	26" diameter		20	1.200		83	41.50		124.50	155	
	1610	28" diameter		19	1.263		103	43.50		146.50	181	
	1620	30" diameter		18	1.333		113	46		159	195	
	1630	32" diameter		17	1.412		131	49		180	220	
	1640	34" diameter		16	1.500		146	52		198	241	
	1650	36" diameter		15	1.600		163	55.50		218.50	265	
	1660	38" diameter		14	1.714		177	59		236	286	
	1670	40" diameter		14	1.714		193	59		252	305	
	1680	42" diameter		13	1.846		205	64		269	325	
	1690	44" diameter		12	2		222	69		291	350	
	1700	46" diameter		12	2		235	69		304	365	
	1710	48" diameter	▼	11	2.182		265	75.50		340.50	405	
	1750	Top, 3" diameter	Q-9	46	.348		8	11.60		19.60	26.50	
	1760	4" diameter		44	.364		8.40	12.10		20.50	28	
	1770	5" diameter		42	.381		11.55	12.70		24.25	32.50	
	1780	6" diameter		40	.400		14.65	13.30		27.95	36.50	
	1790	7" diameter		38	.421		22	14		36	45.50	
	1800	8" diameter		36	.444		27.50	14.80		42.30	53	
	1810	10" diameter		34	.471		66	15.65		81.65	96.50	
	1820	12" diameter		32	.500		99.50	16.65		116.15	135	
	1830	14" diameter		30	.533		113	17.75		130.75	152	
	1840	16" diameter		28	.571		204	19.05		223.05	254	
	1850	18" diameter	▼	26	.615		370	20.50		390.50	435	
	1860	20" diameter	Q-10	28	.857		400	29.50		429.50	485	
	1870	22" diameter		22	1.091		545	37.50		582.50	660	
	1880	24" diameter		20	1.200		680	41.50		721.50	810	
	1882	26" diameter		19	1.263		1,025	43.50		1,068.50	1,200	
	1884	28" diameter		18	1.333		1,175	46		1,221	1,350	
	1886	30" diameter		17	1.412		1,300	49		1,349	1,525	
	1888	32" diameter		16	1.500		1,375	52		1,427	1,600	
	1890	34" diameter		15	1.600		1,475	55.50		1,530.50	1,700	
	1892	36" diameter	▼	14	1.714	▼	1,575	59		1,634	1,850	
	1900	Gas, double wall, galvanized steel, oval										
	1904	4" x 1'	Q-9	68	.235	V.L.F.	11.20	7.85		19.05	24.50	
	1906	5"		64	.250		23	8.35		31.35	38	
	1908	5"/6" x 1'	▼	60	.267	▼	23	8.90		31.90	39	

Important: See the Reference Section for critical supporting data - Reference Nos., Crews, & City Cost Indexes

15550	Breechings, Chimneys & Stacks	CREW	DAILY OUTPUT	LABOR-HOURS	UNIT	2003 BARE COSTS				TOTAL INCL O&P
						MAT.	LABOR	EQUIP.	TOTAL	
440 1910	Oval fittings									**440**
1912	Adjustable length									
1914	4" diameter to 12" lg	Q-9	34	.471	Ea.	11.85	15.65		27.50	37
1916	5" diameter to 12" lg		32	.500		33	16.65		49.65	61.50
1918	5"/6" diameter to 12"	▼	30	.533	▼	30	17.75		47.75	60.50
1920	Elbow 45°									
1922	4"	Q-9	34	.471	Ea.	21.50	15.65		37.15	48
1924	5"		32	.500		41	16.65		57.65	70.50
1926	5"/6"	▼	30	.533	▼	39	17.75		56.75	70
1930	Elbow 45°, flat									
1932	4"	Q-9	34	.471	Ea.	21.50	15.65		37.15	48
1934	5"		32	.500		41	16.65		57.65	70.50
1936	5"/6"	▼	30	.533	▼	41	17.75		58.75	72.50
1940	Top									
1942	4"	Q-9	44	.364	Ea.	16.10	12.10		28.20	36.50
1944	5"		42	.381		24.50	12.70		37.20	46.50
1946	5"/6"	▼	40	.400	▼	24.50	13.30		37.80	47.50
1950	Adjustable flashing									
1952	4"	Q-9	34	.471	Ea.	9.70	15.65		25.35	34.50
1954	5"		32	.500		28.50	16.65		45.15	57
1956	5"/6"	▼	30	.533	▼	19.20	17.75		36.95	48.50
1960	Tee									
1962	4"	Q-9	26	.615	Ea.	32	20.50		52.50	67
1964	5"		25	.640		61.50	21.50		83	101
1966	5"/6"	▼	24	.667	▼	62	22		84	103
1970	Tee with short snout									
1972	4"	Q-9	26	.615	Ea.	32	20.50		52.50	67
3200	All fuel, pressure tight, double wall, U.L. listed, 1400°F.									
3210	304 stainless steel liner, aluminized steel outer jacket									
3220	6" diameter	Q-9	60	.267	L.F.	35.50	8.90		44.40	52.50
3221	8" diameter		52	.308		43.50	10.25		53.75	64
3222	10" diameter		48	.333		48.50	11.10		59.60	70.50
3223	12" diameter		44	.364		55.50	12.10		67.60	79.50
3224	14" diameter		42	.381		63.50	12.70		76.20	89.50
3225	16" diameter		40	.400		67.50	13.30		80.80	94.50
3226	18" diameter	▼	38	.421		76	14		90	105
3227	20" diameter	Q-10	36	.667		87	23		110	131
3228	24" diameter		32	.750		112	26		138	163
3229	28" diameter		30	.800		128	27.50		155.50	184
3230	32" diameter		27	.889		144	30.50		174.50	206
3231	36" diameter		25	.960		161	33		194	228
3232	42" diameter		22	1.091		184	37.50		221.50	261
3233	48" diameter	▼	19	1.263		217	43.50		260.50	305
3260	For 316 stainless steel liner add				▼	6%				
3280	All fuel, pressure tight, double wall fittings									
3284	304 stainless steel inner, aluminized steel jacket									
3288	Adjustable 18"/30" section									
3291	5" diameter	Q-9	32	.500	Ea.	91	16.65		107.65	126
3292	6" diameter		30	.533		115	17.75		132.75	155
3293	8" diameter		26	.615		133	20.50		153.50	178
3294	10" diameter		24	.667		149	22		171	199
3295	12" diameter		22	.727		170	24		194	225
3296	14" diameter		21	.762		196	25.50		221.50	255
3297	16" diameter		20	.800		204	26.50		230.50	265
3298	18" diameter	▼	19	.842		251	28		279	320
3299	20" diameter	Q-10	18	1.333	▼	290	46		336	390

MECHANICAL 15

15550	Breechings, Chimneys & Stacks	CREW	DAILY OUTPUT	LABOR-HOURS	UNIT	2003 BARE COSTS				TOTAL INCL O&P	
						MAT.	LABOR	EQUIP.	TOTAL		
440	3300	24" diameter	Q-10	16	1.500	Ea.	375	52		427	490	**440**
3301	28" diameter		15	1.600		415	55.50		470.50	545
3302	32" diameter		14	1.714		430	59		489	565
3303	36" diameter		12	2		450	69		519	600
3304	42" diameter		11	2.182		515	75.50		590.50	685
3305	48" diameter		10	2.400		605	83		688	795
3326	For 316 stainless steel liner, add					15%				
3350	Elbow 90° fixed									
3353	5" diameter	Q-9	32	.500	Ea.	170	16.65		186.65	213
3354	6" diameter		30	.533		200	17.75		217.75	248
3355	8" diameter		26	.615		221	20.50		241.50	276
3356	10" diameter		24	.667		254	22		276	315
3357	12" diameter		22	.727		287	24		311	355
3358	14" diameter		21	.762		335	25.50		360.50	410
3359	16" diameter		20	.800		375	26.50		401.50	455
3360	18" diameter		19	.842		440	28		468	525
3361	20" diameter	Q-10	18	1.333		490	46		536	605
3362	24" diameter		16	1.500		605	52		657	745
3363	28" diameter		15	1.600		755	55.50		810.50	915
3364	32" diameter		14	1.714		1,000	59		1,059	1,200
3365	36" diameter		12	2		1,125	69		1,194	1,350
3366	42" diameter		11	2.182		1,725	75.50		1,800.50	2,025
3367	48" diameter		10	2.400		2,175	83		2,258	2,500
3380	For 316 stainless steel liner, add					20%				
3400	Elbow 45°									
3404	6" diameter	Q-9	30	.533	Ea.	127	17.75		144.75	167
3405	8" diameter		26	.615		143	20.50		163.50	190
3406	10" diameter		24	.667		161	22		183	212
3407	12" diameter		22	.727		184	24		208	240
3408	14" diameter		21	.762		211	25.50		236.50	271
3409	16" diameter		20	.800		238	26.50		264.50	300
3410	18" diameter		19	.842		269	28		297	340
3411	20" diameter	Q-10	18	1.333		310	46		356	410
3412	24" diameter		16	1.500		395	52		447	515
3413	28" diameter		15	1.600		445	55.50		500.50	575
3414	32" diameter		14	1.714		600	59		659	750
3415	36" diameter		12	2		805	69		874	990
3416	42" diameter		11	2.182		1,100	75.50		1,175.50	1,325
3417	48" diameter		10	2.400		1,475	83		1,558	1,750
3430	For 316 stainless steel liner, add					20%				
3450	Tee 90°									
3454	6" diameter	Q-9	24	.667	Ea.	132	22		154	180
3455	8" diameter		22	.727		152	24		176	205
3456	10" diameter		21	.762		178	25.50		203.50	234
3457	12" diameter		20	.800		208	26.50		234.50	269
3458	14" diameter		18	.889		245	29.50		274.50	315
3459	16" diameter		16	1		275	33.50		308.50	355
3460	18" diameter		14	1.143		325	38		363	415
3461	20" diameter	Q-10	17	1.412		365	49		414	475
3462	24" diameter		12	2		450	69		519	600
3463	28" diameter		11	2.182		565	75.50		640.50	735
3464	32" diameter		10	2.400		770	83		853	980
3465	36" diameter		9	2.667		860	92		952	1,075
3466	42" diameter		6	4		1,375	138		1,513	1,725
3467	48" diameter		5	4.800		1,750	166		1,916	2,175
3480	For Tee Cap, add					35%	20%			

Important: See the Reference Section for critical supporting data - Reference Nos., Crews, & City Cost Indexes

		CREW	DAILY OUTPUT	LABOR-HOURS	UNIT	2003 BARE COSTS				TOTAL INCL O&P
						MAT.	LABOR	EQUIP.	TOTAL	
15550	**Breechings, Chimneys & Stacks**									
440										440
3500	For 316 stainless steel liner, add				Ea.	22%				
3520	Plate support, galvanized									
3524	6" diameter	Q-9	26	.615	Ea.	83	20.50		103.50	123
3525	8" diameter		22	.727		93.50	24		117.50	141
3526	10" diameter		20	.800		102	26.50		128.50	153
3527	12" diameter		18	.889		106	29.50		135.50	163
3528	14" diameter		17	.941		128	31.50		159.50	190
3529	16" diameter		16	1		136	33.50		169.50	201
3530	18" diameter		15	1.067		143	35.50		178.50	212
3531	20" diameter	Q-10	16	1.500		151	52		203	246
3532	24" diameter		14	1.714		159	59		218	267
3533	28" diameter		13	1.846		196	64		260	315
3534	32" diameter		12	2		254	69		323	385
3535	36" diameter		10	2.400		315	83		398	480
3536	42" diameter		9	2.667		560	92		652	755
3537	48" diameter		8	3		675	104		779	905
3570	Bellows, lined, 316 stainless steel only									
3574	6" diameter	Q-9	30	.533	Ea.	284	17.75		301.75	345
3575	8" diameter		26	.615		310	20.50		330.50	370
3576	10" diameter		24	.667		340	22		362	410
3577	12" diameter		22	.727		375	24		399	455
3578	14" diameter		21	.762		425	25.50		450.50	505
3579	16" diameter		20	.800		470	26.50		496.50	555
3580	18" diameter		19	.842		525	28		553	620
3581	20" diameter	Q-10	18	1.333		590	46		636	720
3582	24" diameter		16	1.500		740	52		792	895
3583	28" diameter		15	1.600		830	55.50		885.50	1,000
3584	32" diameter		14	1.714		955	59		1,014	1,150
3585	36" diameter		12	2		1,100	69		1,169	1,300
3586	42" diameter		11	2.182		1,375	75.50		1,450.50	1,650
3587	48" diameter		10	2.400		1,700	83		1,783	2,000
3600	Ventilated roof thimble, 304 stainless steel									
3620	6" diameter	Q-9	26	.615	Ea.	191	20.50		211.50	242
3624	8" diameter		22	.727		205	24		229	263
3625	10" diameter		20	.800		206	26.50		232.50	268
3626	12" diameter		18	.889		222	29.50		251.50	290
3627	14" diameter		17	.941		239	31.50		270.50	310
3628	16" diameter		16	1		254	33.50		287.50	330
3629	18" diameter		15	1.067		274	35.50		309.50	355
3630	20" diameter	Q-10	16	1.500		293	52		345	405
3631	24" diameter		14	1.714		330	59		389	455
3632	28" diameter		13	1.846		380	64		444	520
3633	32" diameter		12	2		425	69		494	575
3634	36" diameter		10	2.400		465	83		548	640
3635	42" diameter		9	2.667		620	92		712	820
3636	48" diameter		8	3		780	104		884	1,025
3650	For 316 stainless steel, add					10%				
3670	Exit cone, 316 stainless steel only									
3674	6" diameter	Q-9	46	.348	Ea.	125	11.60		136.60	156
3675	8" diameter		42	.381		131	12.70		143.70	165
3676	10" diameter		40	.400		142	13.30		155.30	177
3677	12" diameter		38	.421		149	14		163	186
3678	14" diameter		37	.432		166	14.40		180.40	204
3679	16" diameter		36	.444		182	14.80		196.80	224
3680	18" diameter		35	.457		224	15.20		239.20	271
3681	20" diameter	Q-10	28	.857		267	29.50		296.50	340

15550	Breechings, Chimneys & Stacks	CREW	DAILY OUTPUT	LABOR-HOURS	UNIT	2003 BARE COSTS				TOTAL INCL O&P
						MAT.	LABOR	EQUIP.	TOTAL	
3682	24" diameter	Q-10	26	.923	Ea.	345	32		377	430
3683	28" diameter		25	.960		415	33		448	510
3684	32" diameter		24	1		520	34.50		554.50	625
3685	36" diameter		22	1.091		595	37.50		632.50	715
3686	42" diameter		21	1.143		760	39.50		799.50	895
3687	48" diameter	▼	20	1.200	▼	980	41.50		1,021.50	1,150
3720	Roof support assy., incl. 30" pipe sect, 304 st. st.									
3724	6" diameter	Q-9	25	.640	Ea.	218	21.50		239.50	273
3725	8" diameter		21	.762		238	25.50		263.50	300
3726	10" diameter		19	.842		261	28		289	330
3727	12" diameter		17	.941		278	31.50		309.50	355
3728	14" diameter		16	1		310	33.50		343.50	390
3729	16" diameter		15	1.067		330	35.50		365.50	420
3730	18" diameter	▼	14	1.143		355	38		393	450
3731	20" diameter	Q-10	15	1.600		380	55.50		435.50	505
3732	24" diameter		13	1.846		425	64		489	570
3733	28" diameter		12	2		500	69		569	655
3734	32" diameter		11	2.182		600	75.50		675.50	775
3735	36" diameter		9	2.667		690	92		782	900
3736	42" diameter		8	3		1,075	104		1,179	1,325
3737	48" diameter	▼	7	3.429		1,325	118		1,443	1,625
3750	For 316 stainless steel, add				▼	20%				
3770	Stack cap, 316 stainless steel only									
3773	5" diameter	Q-9	48	.333	Ea.	119	11.10		130.10	148
3774	6" diameter		46	.348		140	11.60		151.60	172
3775	8" diameter		42	.381		162	12.70		174.70	198
3776	10" diameter		40	.400		190	13.30		203.30	230
3777	12" diameter		38	.421		221	14		235	265
3778	14" diameter		37	.432		260	14.40		274.40	310
3779	16" diameter		36	.444		297	14.80		311.80	350
3780	18" diameter	▼	35	.457		335	15.20		350.20	395
3781	20" diameter	Q-10	28	.857		385	29.50		414.50	465
3782	24" diameter		26	.923		460	32		492	560
3783	28" diameter		25	.960		505	33		538	605
3784	32" diameter		24	1		580	34.50		614.50	690
3785	36" diameter		22	1.091		675	37.50		712.50	805
3786	42" diameter		21	1.143		805	39.50		844.50	945
3787	48" diameter	▼	20	1.200	▼	975	41.50		1,016.50	1,150
3800	Round storm collar									
3806	5" diameter	Q-9	32	.500	Ea.	25	16.65		41.65	53
7800	All fuel, double wall, stainless steel, 6" diameter		60	.267	V.L.F.	29	8.90		37.90	45.50
7802	7" diameter		56	.286		37.50	9.50		47	56
7804	8" diameter		52	.308		44	10.25		54.25	64.50
7806	10" diameter		48	.333		64	11.10		75.10	87.50
7808	12" diameter		44	.364		86	12.10		98.10	113
7810	14" diameter	▼	42	.381	▼	113	12.70		125.70	144
8000	All fuel, double wall, stainless steel fittings									
8010	Roof support 6" diameter	Q-9	30	.533	Ea.	76	17.75		93.75	111
8020	7" diameter		28	.571		86	19.05		105.05	124
8030	8" diameter		26	.615		95.50	20.50		116	137
8040	10" diameter		24	.667		110	22		132	156
8050	12" diameter		22	.727		133	24		157	184
8060	14" diameter		21	.762		168	25.50		193.50	224
8100	Elbow 15°, 6" diameter		30	.533		61	17.75		78.75	94.50
8120	7" diameter		28	.571		68.50	19.05		87.55	105
8140	8" diameter	▼	26	.615	▼	78	20.50		98.50	117

440

Important: See the Reference Section for critical supporting data - Reference Nos., Crews, & City Cost Indexes

		CREW	DAILY OUTPUT	LABOR-HOURS	UNIT	2003 BARE COSTS				TOTAL INCL O&P
15550	**Breechings, Chimneys & Stacks**					MAT.	LABOR	EQUIP.	TOTAL	
8160	10" diameter	Q-9	24	.667	Ea.	102	22		124	147
8180	12" diameter		22	.727		124	24		148	174
8200	14" diameter		21	.762		148	25.50		173.50	202
8300	Insulated tee with insulated tee cap, 6" diameter		30	.533		115	17.75		132.75	154
8340	7" diameter		28	.571		150	19.05		169.05	196
8360	8" diameter		26	.615		170	20.50		190.50	219
8380	10" diameter		24	.667		239	22		261	298
8400	12" diameter		22	.727		335	24		359	410
8420	14" diameter		21	.762		440	25.50		465.50	525
8500	Joist shield, 6" diameter		30	.533		35	17.75		52.75	66
8510	7" diameter		28	.571		38	19.05		57.05	71.50
8520	8" diameter		26	.615		47	20.50		67.50	83
8530	10" diameter		24	.667		63	22		85	104
8540	12" diameter		22	.727		78.50	24		102.50	124
8550	14" diameter		21	.762		97.50	25.50		123	146
8600	Round top, 6" diameter		30	.533		39	17.75		56.75	70.50
8620	7" diameter		28	.571		53	19.05		72.05	88
8640	8" diameter		26	.615		68.50	20.50		89	107
8660	10" diameter		24	.667		128	22		150	176
8680	12" diameter		22	.727		183	24		207	239
8700	14" diameter		21	.762		244	25.50		269.50	310
8800	Adjustable roof flashing, 6" diameter		30	.533		46	17.75		63.75	78.50
8820	7" diameter		28	.571		53	19.05		72.05	87.50
8840	8" diameter		26	.615		57.50	20.50		78	94.50
8860	10" diameter		24	.667		74	22		96	116
8880	12" diameter		22	.727		95	24		119	143
8900	14" diameter	▼	21	.762	▼	119	25.50		144.50	170
9000	High temp. (2000° F), steel jacket, acid resist. refractory lining									
9010	11 ga. galvanized jacket, U.L. listed									
9020	Straight section, 48" long, 10" diameter	Q-10	13.30	1.805	Ea.	335	62.50		397.50	460
9030	12" diameter		11.20	2.143		360	74		434	515
9040	18" diameter		7.40	3.243		510	112		622	735
9050	24" diameter		4.60	5.217		695	180		875	1,050
9060	30" diameter		3.70	6.486		885	224		1,109	1,325
9070	36" diameter	▼	2.70	8.889		1,075	305		1,380	1,675
9080	42" diameter	Q-11	3.10	10.323		1,350	365		1,715	2,025
9090	48" diameter		2.70	11.852		1,575	420		1,995	2,400
9100	54" diameter		2.20	14.545		1,925	515		2,440	2,925
9110	60" diameter	▼	2	16		2,525	565		3,090	3,650
9120	Tee section, 10" diameter	Q-10	4.40	5.455		705	188		893	1,075
9130	12" diameter		3.70	6.486		745	224		969	1,150
9140	18" diameter		2.40	10		1,000	345		1,345	1,625
9150	24" diameter		1.50	16		1,400	550		1,950	2,375
9160	30" diameter		1.20	20		1,950	690		2,640	3,225
9170	36" diameter	▼	.80	30		3,300	1,025		4,325	5,225
9180	42" diameter	Q-11	1	32		4,325	1,125		5,450	6,525
9190	48" diameter		.90	35.556		5,600	1,250		6,850	8,100
9200	54" diameter		.70	45.714		8,175	1,625		9,800	11,500
9210	60" diameter	▼	.60	53.333		9,875	1,875		11,750	13,800
9220	Cleanout pier section, 10" diameter	Q-10	3.50	6.857		545	237		782	965
9230	12" diameter		2.50	9.600		570	330		900	1,125
9240	18" diameter		1.90	12.632		730	435		1,165	1,475
9250	24" diameter		1.30	18.462		925	635		1,560	2,000
9260	30" diameter		1	24		1,125	830		1,955	2,525
9270	36" diameter	▼	.75	32		1,350	1,100		2,450	3,175
9280	42" diameter	Q-11	.90	35.556	▼	1,700	1,250		2,950	3,800

15550	Breechings, Chimneys & Stacks	CREW	DAILY OUTPUT	LABOR-HOURS	UNIT	2003 BARE COSTS				TOTAL INCL O&P
						MAT.	LABOR	EQUIP.	TOTAL	
9290	48" diameter	Q-11	.75	42.667	Ea.	2,050	1,500		3,550	4,575
9300	54" diameter		.60	53.333		2,450	1,875		4,325	5,600
9310	60" diameter	▼	.50	64		3,125	2,250		5,375	6,900
9320	For drain, add					53%				
9330	Elbow, 30° and 45°, 10" diameter	Q-10	6.60	3.636		460	126		586	700
9340	12" diameter		5.60	4.286		495	148		643	775
9350	18" diameter		3.70	6.486		745	224		969	1,150
9360	24" diameter		2.30	10.435		1,100	360		1,460	1,775
9370	30" diameter		1.85	12.973		1,525	450		1,975	2,375
9380	36" diameter	▼	1.30	18.462		1,850	635		2,485	3,000
9390	42" diameter	Q-11	1.60	20		2,525	705		3,230	3,875
9400	48" diameter		1.35	23.704		3,225	835		4,060	4,850
9410	54" diameter		1.10	29.091		4,175	1,025		5,200	6,175
9420	60" diameter	▼	1	32		5,075	1,125		6,200	7,350
9430	For 60° and 90° elbow, add					112%				
9440	End cap, 10" diameter	Q-10	26	.923		790	32		822	915
9450	12" diameter		22	1.091		900	37.50		937.50	1,050
9460	18" diameter		15	1.600		1,125	55.50		1,180.50	1,300
9470	24" diameter		9	2.667		1,375	92		1,467	1,650
9480	30" diameter		7.30	3.288		1,700	114		1,814	2,025
9490	36" diameter	▼	5	4.800		1,950	166		2,116	2,400
9500	42" diameter	Q-11	6.30	5.079		2,425	179		2,604	2,950
9510	48" diameter		5.30	6.038		2,875	213		3,088	3,475
9520	54" diameter		4.40	7.273		3,275	257		3,532	4,000
9530	60" diameter	▼	3.90	8.205		3,275	289		3,564	4,050
9540	Increaser (1 diameter), 10" diameter	Q-10	6.60	3.636		390	126		516	625
9550	12" diameter		5.60	4.286		420	148		568	690
9560	18" diameter		3.70	6.486		610	224		834	1,025
9570	24" diameter		2.30	10.435		880	360		1,240	1,525
9580	30" diameter		1.85	12.973		1,075	450		1,525	1,900
9590	36" diameter	▼	1.30	18.462		1,200	635		1,835	2,300
9592	42" diameter	Q-11	1.60	20		1,325	705		2,030	2,550
9594	48" diameter		1.35	23.704		1,575	835		2,410	3,050
9596	54" diameter	▼	1.10	29.091	▼	1,925	1,025		2,950	3,675
9600	For expansion joints, add to straight section					7%				
9610	For 1/4" hot rolled steel jacket, add					157%				
9620	For 2950°F very high temperature, add					89%				
9630	26 ga. aluminized jacket, straight section, 48" long									
9640	10" diameter	Q-10	15.30	1.569	V.L.F.	197	54		251	300
9650	12" diameter		12.90	1.860		208	64		272	330
9660	18" diameter		8.50	2.824		300	97.50		397.50	480
9670	24" diameter		5.30	4.528		430	156		586	715
9680	30" diameter		4.30	5.581		550	193		743	905
9690	36" diameter	▼	3.10	7.742	▼	695	267		962	1,175
9700	Accessories (all models)									
9710	Guy band, 10" diameter	Q-10	32	.750	Ea.	61	26		87	107
9720	12" diameter		30	.800		64.50	27.50		92	114
9730	18" diameter		26	.923		83.50	32		115.50	141
9740	24" diameter		24	1		124	34.50		158.50	190
9750	30" diameter		20	1.200		143	41.50		184.50	221
9760	36" diameter	▼	18	1.333		158	46		204	244
9770	42" diameter	Q-11	22	1.455		287	51.50		338.50	395
9780	48" diameter		20	1.600		315	56.50		371.50	430
9790	54" diameter		16	2		350	70.50		420.50	490
9800	60" diameter	▼	12	2.667		375	94		469	560
9810	Draw band, galv. stl., 11 gauge, 10" diameter	Q-10	32	.750	▼	53	26		79	98.50

440

440

15
MECHANICAL

Estimating...
Budgeting...
Job Costing...

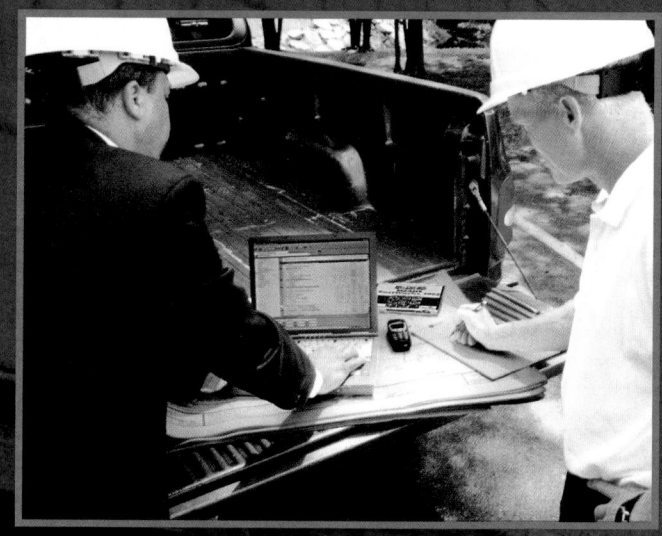

Means
CostWorks 2003

All Means Data is included on one CD – Means CostWorks 2003 will help simplify your estimating process

Means CostWorks 2003 provides 18 detailed cost titles that cover every job costing requirement you can imagine... it turns your desktop or laptop PC into a comprehensive cost library. Each title addresses a specific area of the commercial or residential building and maintenance industry.

Plus you get the productivity-enhancing tools to really put this valuable information to work.

Bonus! Get the Means Construction Dictionary with over 17,000 construction terms — a $99.95 value — FREE on the Means CostWorks 2003 CD!

Save hundreds on specially priced packages! Choose the one that's right for you...

Builder's Package
Includes:
- Residential Cost Data
- Light Commercial Cost Data
- Site Work & Landscape Cost Data
- Concrete & Masonry Cost Data
- Open Shop Building Construction Cost Data

Just $345.99

Save over $300 off the price of buying each book separately!

Building Professional's Package
Includes:
- Building Construction Cost Data
- Mechanical Cost Data
- Plumbing Cost Data
- Electrical Cost Data
- Square Foot Cost Data

Just $404.99

Save over $250 off the price of buying each book separately!

Facility Manager's Package
Includes:
- Facilities Construction Cost Data
- Facilities Maintenance & Repair
- Repair & Remodeling Cost Data
- Building Construction Cost Data

Just $519.99

Save over $300 off the price of buying each book separately!

Design Professional's Package
Includes:
- Building Construction Cost Data
- Square Foot Cost Data
- Interior Cost Data
- Assemblies Cost Data

Just $404.99

Save over $200 off the price of buying each book separately!

Your 2003 Annual Means CostWorks license includes four free quarterly updates from the Means CostWorks Web site

15550	Breechings, Chimneys & Stacks	CREW	DAILY OUTPUT	LABOR-HOURS	UNIT	2003 BARE COSTS				TOTAL INCL O&P	
						MAT.	LABOR	EQUIP.	TOTAL		
440	9820	12" diameter	Q-10	30	.800	Ea.	57	27.50		84.50	105	**440**
	9830	18" diameter		26	.923		71.50	32		103.50	128	
	9840	24" diameter		24	1		90.50	34.50		125	153	
	9850	30" diameter		20	1.200		105	41.50		146.50	180	
	9860	36" diameter	▼	18	1.333		121	46		167	204	
	9870	42" diameter	Q-11	22	1.455		162	51.50		213.50	257	
	9880	48" diameter		20	1.600		207	56.50		263.50	315	
	9890	54" diameter		16	2		241	70.50		311.50	375	
	9900	60" diameter	▼	12	2.667		305	94		399	480	
	9910	Draw band, aluminized stl., 26 gauge, 10" diameter	Q-10	32	.750		15.20	26		41.20	56.50	
	9920	12" diameter		30	.800		15.20	27.50		42.70	59	
	9930	18" diameter		26	.923		19	32		51	70	
	9940	24" diameter		24	1		23	34.50		57.50	78.50	
	9950	30" diameter		20	1.200		26.50	41.50		68	93.50	
	9960	36" diameter	▼	18	1.333		30.50	46		76.50	105	
	9962	42" diameter	Q-11	22	1.455		34	51.50		85.50	117	
	9964	48" diameter		20	1.600		42	56.50		98.50	133	
	9966	54" diameter		16	2		42	70.50		112.50	155	
	9968	60" diameter	▼	12	2.667	▼	45.50	94		139.50	195	
600	0010	**INDUCED DRAFT FANS**										**600**
	1000	Breeching installation										
	1800	Hot gas, 600°F, variable pitch pulley and motor										
	1840	6" diam. inlet, 1/4 H.P., 1 phase, 400 CFM	Q-9	6	2.667	Ea.	1,025	89		1,114	1,250	
	1850	7" diam. inlet, 1/4 H.P., 1 phase, 800 CFM		5	3.200		1,100	107		1,207	1,375	
	1860	8" diam. inlet, 1/4 H.P., 1phase, 1120 CFM		4	4		1,600	133		1,733	1,950	
	1870	9" diam. inlet, 3/4 H.P., 1 phase, 1440 CFM		3.60	4.444		1,875	148		2,023	2,300	
	1880	10" diam. inlet, 3/4 H.P., 1 phase, 2000 CFM		3.30	4.848		2,125	161		2,286	2,575	
	1900	12" diam. inlet, 3/4 H.P., 3 phase, 2960 CFM		3	5.333		2,175	178		2,353	2,675	
	1910	14" diam. inlet, 1 H.P., 3 phase, 4160 CFM		2.60	6.154		2,575	205		2,780	3,150	
	1920	16" diam. inlet, 2 H.P., 3 phase, 6720 CFM		2.30	6.957		2,800	232		3,032	3,425	
	1940	18" diam. inlet, 3 H.P., 3 phase, 9120 CFM		2	8		3,375	266		3,641	4,125	
	1950	20" diam. inlet, 3 H.P., 3 phase, 9760 CFM		1.50	10.667		3,675	355		4,030	4,600	
	1960	22" diam. inlet, 5 H.P., 3 phase, 13,360 CFM		1	16		4,325	535		4,860	5,575	
	1980	24" diam. inlet, 7-1/2 H.P., 3 phase, 17,760 CFM	▼	.80	20	▼	5,300	665		5,965	6,875	
	2300	For multi-blade damper at fan inlet, add					20%					
	3600	Chimneytop installation										
	3700	6" size	1 Shee	8	1	Ea.	485	37		522	585	
	3740	8" size		7	1.143		485	42.50		527.50	600	
	3780	13" size	▼	6	1.333		665	49.50		714.50	805	
	3880	For speed control switch, add					39.50			39.50	43.50	
	3920	For thermal fan control, add				▼	39.50			39.50	43.50	
	5500	Flue shutter damper for draft control,										
	5510	parallel blades										
	5550	8" size	Q-9	8	2	Ea.	360	66.50		426.50	500	
	5560	9" size		7.50	2.133		390	71		461	535	
	5570	10" size		7	2.286		400	76		476	560	
	5580	12" size		6.50	2.462		440	82		522	605	
	5590	14" size		6	2.667		500	89		589	685	
	5600	16" size		5.50	2.909		540	97		637	745	
	5610	18" size		5	3.200		595	107		702	820	
	5620	20" size		4.50	3.556		655	118		773	910	
	5630	22" size		4	4		725	133		858	1,000	
	5640	24" size		3.50	4.571		790	152		942	1,100	
	5650	27" size		3	5.333		790	178		968	1,150	
	5660	30" size	▼	2.50	6.400	▼	840	213		1,053	1,250	

MECHANICAL **15**

		15550	Breechings, Chimneys & Stacks	CREW	DAILY OUTPUT	LABOR-HOURS	UNIT	2003 BARE COSTS				TOTAL INCL O&P	
								MAT.	LABOR	EQUIP.	TOTAL		
600	5670		32" size	Q-9	2	8	Ea.	880	266		1,146	1,375	600
	5680		36" size	↓	1.50	10.667	↓	950	355		1,305	1,600	

		15620	Packaged Water Chillers	CREW	DAILY OUTPUT	LABOR-HOURS	UNIT	2003 BARE COSTS				TOTAL INCL O&P	
								MAT.	LABOR	EQUIP.	TOTAL		
100	0010	**ABSORPTION WATER CHILLERS**											100
	0020		Steam or hot water, water cooled										
	0050		100 ton	Q-7	.13	240	Ea.	115,500	8,600		124,100	140,500	
	0100		148 ton		.12	258		117,000	9,250		126,250	143,000	
	0200		200 ton		.12	275		130,500	9,900		140,400	158,500	
	0240		250 ton		.11	290		143,500	10,400		153,900	174,000	
	0300		354 ton		.10	304		179,500	10,900		190,400	213,500	
	0400		420 ton		.10	323		196,000	11,600		207,600	233,000	
	0500		665 ton		.09	340		275,000	12,200		287,200	321,000	
	0600		750 ton		.09	363		321,500	13,000		334,500	373,000	
	0700		850 ton		.08	385		343,000	13,800		356,800	398,500	
	0800		955 ton		.08	410		367,000	14,700		381,700	425,500	
	0900		1125 ton		.08	421		423,500	15,100		438,600	489,000	
	1000		1250 ton		.07	444		455,500	15,900		471,400	525,000	
	1100		1465 ton		.07	463		540,000	16,600		556,600	619,000	
	1200		1660 ton	↓	.07	477	↓	635,500	17,100		652,600	725,000	
	2000		For two stage unit, add					80%	25%				
	3000		Gas fired, air cooled										
	3180		3 ton	Q-5	1.30	12.308	Ea.	6,025	415		6,440	7,275	
	3220		5 ton	↓	.60	26.667		7,050	905		7,955	9,125	
	3270		10 ton	↓	.40	40	↓	16,900	1,350		18,250	20,700	
	4000		Water cooled, duplex										
	4130		100 ton	Q-7	.13	246	Ea.	122,000	8,825		130,825	147,500	
	4140		200 ton		.11	283		168,500	10,200		178,700	201,000	
	4150		300 ton		.11	299		195,000	10,700		205,700	230,500	
	4160		400 ton		.10	316		253,500	11,400		264,900	296,000	
	4170		500 ton		.10	329		319,000	11,800		330,800	369,000	
	4180		600 ton		.09	340		377,000	12,200		389,200	433,000	
	4190		700 ton		.09	359		418,000	12,900		430,900	479,500	
	4200		800 ton		.08	380		504,500	13,700		518,200	575,500	
	4210		900 ton		.08	400		568,000	14,300		582,300	646,000	
	4220		1000 ton	↓	.08	421	↓	631,000	15,100		646,100	717,000	
600	0010	**CENTRIFUGAL/SCREW/RECIP. WATER CHILLERS,** W/ standard controls											600
	0020		Centrifugal liquid chiller, water cooled										
	0030		not including water tower										
	0100		Open drive, 2000 ton	Q-7	.07	477	Ea.	521,500	17,100		538,600	599,500	
	0110		Screw, liquid chiller, air cooled, insulated evaporator										
	0120		130 ton	Q-7	.14	228	Ea.	58,500	8,200		66,700	77,000	
	0124		160 ton		.13	246		72,000	8,825		80,825	92,500	
	0128		180 ton		.13	250		81,000	8,975		89,975	102,500	
	0132		210 ton		.12	258		89,500	9,250		98,750	112,000	
	0136		270 ton		.12	266		102,000	9,550		111,550	126,500	
	0140		320 ton	↓	.12	275	↓	128,000	9,900		137,900	156,000	
	0200		Packaged unit, water cooled, not incl. tower										

Important: See the Reference Section for critical supporting data - Reference Nos., Crews, & City Cost Indexes

15620	Packaged Water Chillers	CREW	DAILY OUTPUT	LABOR-HOURS	UNIT	2003 BARE COSTS				TOTAL INCL O&P	
						MAT.	LABOR	EQUIP.	TOTAL		
600 0210	80 ton	Q-7	.14	223	Ea.	33,000	8,025		41,025	48,400	**600**
0220	100 ton		.14	230		38,500	8,250		46,750	55,000	
0230	150 ton		.13	240		55,000	8,625		63,625	73,500	
0240	200 ton		.13	251		64,000	9,025		73,025	84,000	
0250	250 ton		.12	260		66,500	9,325		75,825	87,000	
0260	300 ton		.12	266		76,000	9,550		85,550	98,000	
0270	350 ton		.12	275		106,500	9,900		116,400	132,000	
0274	Centrifugal, packaged unit, water cooled, not incl. tower										
0280	400 ton	Q-7	.11	283	Ea.	118,500	10,200		128,700	146,000	
0282	450 ton		.11	290		129,500	10,400		139,900	158,500	
0290	500 ton		.11	296		150,000	10,600		160,600	181,000	
0292	550 ton		.10	304		165,000	10,900		175,900	198,000	
0300	600 ton		.10	310		177,500	11,100		188,600	212,500	
0302	650 ton		.10	320		192,500	11,500		204,000	229,500	
0304	700 ton		.10	326		207,500	11,700		219,200	245,500	
0306	750 ton		.10	333		222,000	12,000		234,000	262,500	
0310	800 ton		.09	340		232,500	12,200		244,700	274,000	
0312	850 ton		.09	351		251,000	12,600		263,600	295,000	
0316	900 ton		.09	359		276,000	12,900		288,900	323,500	
0318	950 ton		.09	363		291,500	13,000		304,500	340,000	
0320	1000 ton		.09	372		312,000	13,300		325,300	363,000	
0324	1100 ton		.08	385		337,500	13,800		351,300	392,500	
0330	1200 ton		.08	395		368,000	14,200		382,200	426,500	
0340	1300 ton		.08	410		400,500	14,700		415,200	463,000	
0350	1400 ton		.08	421		429,500	15,100		444,600	495,500	
0360	1500 ton		.07	426		460,000	15,300		475,300	529,000	
0370	1600 ton		.07	438		491,000	15,700		506,700	563,500	
0380	1700 ton		.07	450		521,500	16,200		537,700	598,500	
0390	1800 ton		.07	457		552,500	16,400		568,900	632,500	
0400	1900 ton		.07	470		583,000	16,900		599,900	667,000	
0490	Reciprocating, packaged w/integral air cooled condenser, 15 ton cool		.37	86.486		21,500	3,100		24,600	28,300	
0500	20 ton cooling		.34	94.955		27,300	3,400		30,700	35,200	
0510	25 ton cooling		.33	97.859		31,000	3,500		34,500	39,400	
0515	30 ton cooling		.31	101		33,400	3,650		37,050	42,200	
0517	35 ton cooling		.31	104		38,000	3,750		41,750	47,500	
0520	40 ton cooling		.30	108		45,900	3,875		49,775	56,500	
0528	45 ton cooling		.29	109		52,000	3,950		55,950	63,000	
0536	50 ton cooling		.28	113		58,500	4,050		62,550	70,500	
0538	60 ton cooling		.28	115		67,000	4,125		71,125	80,000	
0540	65 ton cooling		.27	118		67,500	4,225		71,725	80,500	
0546	70 ton cooling		.27	119		77,000	4,300		81,300	91,000	
0552	80 ton cooling		.26	123		82,500	4,400		86,900	97,000	
0554	90 ton cooling		.25	125		89,500	4,525		94,025	105,500	
0600	100 ton cooling		.25	129		99,500	4,625		104,125	116,500	
0620	110 ton cooling		.24	132	Ea.	109,000	4,750		113,750	126,500	
0630	130 ton cooling		.24	135		120,500	4,850		125,350	140,000	
0640	150 ton cooling		.23	137		139,000	4,950		143,950	160,500	
0650	175 ton cooling		.23	140		148,000	5,050		153,050	170,500	
0654	190 ton cooling		.22	144		168,000	5,175		173,175	192,500	
0660	210 ton cooling		.22	148		183,500	5,300		188,800	209,500	
0662	240 ton cooling		.21	151		220,000	5,425		225,425	250,500	
0664	270 ton cooling		.20	156		248,000	5,600		253,600	281,000	
0666	300 ton cooling		.20	160		268,000	5,725		273,725	303,000	
0668	330 ton cooling		.19	164		296,500	5,875		302,375	335,500	
0670	360 ton cooling		.19	168		335,500	6,050		341,550	378,500	
0672	390 ton cooling		.18	172		352,000	6,200		358,200	396,500	

MECHANICAL **15**

15620	Packaged Water Chillers	CREW	DAILY OUTPUT	LABOR-HOURS	UNIT	2003 BARE COSTS				TOTAL INCL O&P	
						MAT.	LABOR	EQUIP.	TOTAL		
600											**600**
0674	420 ton cooling	Q-7	.18	177	Ea.	368,500	6,375		374,875	414,500	
0680	Water cooled, single compressor, semi-hermetic, tower not incl.										
0700	2 to 5 ton cooling	Q-5	.57	28.021	Ea.	6,200	950		7,150	8,250	
0720	6 ton cooling		.42	38.005		9,150	1,275		10,425	12,100	
0740	8 ton cooling	▼	.31	52.117		9,825	1,775		11,600	13,500	
0760	10 ton cooling	Q-6	.36	67.039		11,000	2,350		13,350	15,700	
0780	15 ton cooling	"	.33	72.727		11,000	2,550		13,550	15,900	
0800	20 ton cooling	Q-7	.38	83.989		12,800	3,000		15,800	18,600	
0820	30 ton cooling		.33	96.096		15,400	3,450		18,850	22,100	
0840	35 ton cooling	▼	.33	97.859	▼	16,100	3,500		19,600	23,000	
0980	Water cooled, multiple compress., semi-hermetic, tower not incl.										
1000	15 ton cooling	Q-6	.36	65.934	Ea.	21,300	2,325		23,625	26,900	
1020	20 ton cooling	Q-7	.41	78.049		23,200	2,800		26,000	29,700	
1040	25 ton cooling		.36	89.888		26,200	3,225		29,425	33,700	
1060	30 ton cooling		.31	101		27,600	3,650		31,250	35,800	
1080	40 ton cooling		.30	108		32,000	3,875		35,875	41,100	
1100	50 ton cooling		.28	113		49,400	4,075		53,475	60,500	
1120	60 ton cooling		.25	125		54,500	4,525		59,025	67,000	
1130	70 ton cooling		.23	139		73,500	5,000		78,500	88,500	
1140	80 ton cooling		.21	151		76,500	5,425		81,925	92,500	
1150	90 ton cooling		.19	164		82,500	5,925		88,425	99,500	
1160	100 ton cooling		.18	179		88,000	6,450		94,450	106,500	
1170	110 ton cooling		.17	190		92,500	6,825		99,325	112,500	
1180	120 ton cooling		.16	196		100,000	7,050		107,050	120,500	
1200	140 ton cooling		.16	202		117,500	7,250		124,750	140,000	
1210	160 ton cooling	▼	.15	210	▼	125,000	7,550		132,550	149,000	
1300	Water cooled, single compressor, direct drive, tower not incl.										
1320	40 ton cooling,	Q-7	.30	108	Ea.	24,000	3,875		27,875	32,300	
1340	50 ton cooling		.28	113		28,000	4,075		32,075	37,000	
1360	60 ton cooling		.25	125		30,400	4,525		34,925	40,300	
1380	80 ton cooling	▼	.21	151	▼	40,200	5,425		45,625	52,500	
1450	Water cooled, dual compressors, direct drive, tower not incl.										
1500	80 ton cooling	Q-7	.14	222	Ea.	46,400	7,975		54,375	63,000	
1520	100 ton cooling		.14	228		54,000	8,200		62,200	72,000	
1540	120 ton cooling		.14	231		59,500	8,325		67,825	78,000	
1560	135 ton cooling		.14	235		68,000	8,425		76,425	87,000	
1580	150 ton cooling		.13	240		72,500	8,625		81,125	92,500	
1600	175 ton cooling		.13	244		83,000	8,750		91,750	104,000	
1620	200 ton cooling		.13	250		88,500	8,975		97,475	110,500	
1640	225 ton cooling		.13	256		96,000	9,175		105,175	120,000	
1660	250 ton cooling	▼	.12	260	▼	102,000	9,325		111,325	126,000	
4000	Packaged chiller, remote air cooled condensers not incl.										
4020	15 ton cooling	Q-7	.30	108	Ea.	21,300	3,875		25,175	29,300	
4030	20 ton cooling		.28	115		23,200	4,150		27,350	31,800	
4040	25 ton cooling		.25	125		26,200	4,525		30,725	35,600	
4050	30 ton cooling		.24	133		27,600	4,800		32,400	37,600	
4060	40 ton cooling		.22	144		32,000	5,175		37,175	43,000	
4070	50 ton cooling		.21	153		39,200	5,525		44,725	51,500	
4080	60 ton cooling		.19	164		45,700	5,875		51,575	59,000	
4090	70 ton cooling		.18	173		73,500	6,225		79,725	90,500	
4100	80 ton cooling		.17	183		76,500	6,600		83,100	94,500	
4110	90 ton cooling		.16	193		82,500	6,950		89,450	101,000	
4120	100 ton cooling		.16	203		88,000	7,300		95,300	108,000	
4130	110 ton cooling		.15	213		92,500	7,650		100,150	113,500	
4140	120 ton cooling		.14	223		100,000	8,025		108,025	122,000	
4150	140 ton cooling	▼	.14	233	▼	118,500	8,375		126,875	142,500	

Important: See the Reference Section for critical supporting data - Reference Nos., Crews, & City Cost Indexes

15620 | Packaged Water Chillers

		CREW	DAILY OUTPUT	LABOR-HOURS	UNIT	2003 BARE COSTS				TOTAL INCL O&P
						MAT.	LABOR	EQUIP.	TOTAL	
600	8000	Direct expansion, shell and tube type, for built up systems									600
	8020	1 ton	Q-5	2	8	Ea.	4,775	271		5,046	5,650
	8030	5 ton		1.90	8.421		7,925	285		8,210	9,125
	8040	10 ton		1.70	9.412		9,825	320		10,145	11,300
	8050	20 ton		1.50	10.667		11,000	360		11,360	12,600
	8060	30 ton		1	16		14,000	540		14,540	16,200
	8070	50 ton	↓	.90	17.778		22,500	600		23,100	25,700
	8080	100 ton	Q-6	.90	26.667	↓	40,000	935		40,935	45,400

15640 | Packaged Cooling Towers

			CREW	DAILY OUTPUT	LABOR-HOURS	UNIT	MAT.	LABOR	EQUIP.	TOTAL	TOTAL INCL O&P
400	0010	**COOLING TOWERS** Packaged units	D3030 115									400
	0070	Galvanized steel									
	0080	Induced draft, crossflow	D3030 210								
	0100	Vertical, belt drive, 61 tons	Q-6	90	.267	TonAC	79	9.35		88.35	101
	0150	100 ton		100	.240		62.50	8.40		70.90	81.50
	0200	115 ton		109	.220		61	7.75		68.75	78.50
	0250	131 ton		120	.200		54	7		61	69.50
	0260	162 ton	↓	132	.182	↓	45.50	6.40		51.90	59.50
	1000	For higher capacities, use multiples									
	1500	Induced air, double flow									
	1900	Vertical, gear drive, 167 ton	Q-6	126	.190	TonAC	64	6.70		70.70	80.50
	2000	297 ton		129	.186		60	6.55		66.55	76
	2100	582 ton		132	.182		49	6.40		55.40	63
	2150	849 ton		142	.169		49	5.95		54.95	63
	2200	1016 ton	↓	150	.160	↓	47	5.60		52.60	60
	3000	For higher capacities, use multiples									
	3500	For pumps and piping, add	Q-6	38	.632	TonAC	39.50	22		61.50	77
	4000	For absorption systems, add				"	75%	75%			
	4100	Cooling water chemical feeder	Q-5	3	5.333	Ea.	286	181		467	590
	4500	For rigging, see division 01559									
	5000	Fiberglass									
	5010	Draw thru									
	5100	60 ton	Q-6	1.50	16	Ea.	2,775	560		3,335	3,900
	5120	125 ton		.99	24.242		6,250	850		7,100	8,150
	5140	300 ton		.43	55.814		14,600	1,950		16,550	19,100
	5160	600 ton		.22	109		26,600	3,825		30,425	35,000
	5180	1000 ton	↓	.15	160	↓	45,600	5,625		51,225	58,500
	6000	Stainless steel									
	6010	Induced draft, crossflow, horizontal, belt drive									
	6100	57 ton	Q-6	1.50	16	Ea.	7,975	560		8,535	9,600
	6120	91 ton		.99	24.242		11,500	850		12,350	13,900
	6140	111 ton		.43	55.814		13,500	1,950		15,450	17,800
	6160	126 ton	↓	.22	109	↓	14,600	3,825		18,425	21,900
	6170	Induced draft, crossflow, vertical, gear drive									
	6172	167 ton	Q-6	126	.190	Ea.	23,500	6.70		23,506.70	25,800
	6174	297 ton		129	.186		30,200	6.55		30,206.55	33,200
	6176	582 ton		132	.182		47,500	6.40		47,506.40	52,500
	6178	849 ton		142	.169		76,500	5.95		76,505.95	84,000
	6180	1016 ton	↓	150	.160	↓	83,000	5.60		83,005.60	91,500

15660 | Liquid Coolers/Evap Condensers

			CREW	DAILY OUTPUT	LABOR-HOURS	UNIT	MAT.	LABOR	EQUIP.	TOTAL	TOTAL INCL O&P
100	0010	**CONDENSERS** Ratings are for 30°F TD, R-22	D3030 110									100
	0080	Air cooled, belt drive, propeller fan									
	0220	45 ton	Q-6	.70	34.286	Ea.	10,600	1,200		11,800	13,400
	0240	50 ton	↓	.69	34.985	↓	10,900	1,225		12,125	13,900

MECHANICAL 15

		CREW	DAILY OUTPUT	LABOR-HOURS	UNIT	2003 BARE COSTS				TOTAL INCL O&P
15660	**Liquid Coolers/Evap Condensers**					MAT.	LABOR	EQUIP.	TOTAL	
0260	54 ton D3030 110	Q-6	.63	37.795	Ea.	11,200	1,325		12,525	14,400
0280	59 ton		.58	41.308		12,000	1,450		13,450	15,400
0300	65 ton		.53	45.541		12,500	1,600		14,100	16,100
0320	73 ton		.47	51.173		14,000	1,800		15,800	18,100
0340	81 ton		.42	56.738		16,800	2,000		18,800	21,500
0360	86 ton		.40	60.302		17,500	2,125		19,625	22,400
0380	88 ton		.39	61.697		18,600	2,175		20,775	23,800
0400	101 ton	Q-7	.45	70.640		21,500	2,525		24,025	27,500
0500	159 ton		.31	102		34,600	3,700		38,300	43,700
0600	228 ton		.22	148		50,000	5,300		55,300	63,000
0700	314 ton		.16	203		62,500	7,300		69,800	79,500
0800	471 ton		.11	283		72,500	10,200		82,700	95,500
0900	630 ton		.08	376		116,500	13,500		130,000	148,500
1000	840 ton		.06	507		156,500	18,200		174,700	199,500
1500	May be specified single or multi-circuit									
1550	Air cooled, direct drive, propeller fan									
1590	1 ton	Q-5	3.80	4.211	Ea.	515	143		658	780
1600	1-1/2 ton		3.60	4.444		610	150		760	895
1620	2 ton		3.20	5		675	169		844	995
1630	3 ton		2.40	6.667		750	226		976	1,175
1640	5 ton		2	8		1,375	271		1,646	1,900
1650	8 ton		1.80	8.889		1,725	300		2,025	2,350
1660	10 ton		1.40	11.429		2,125	385		2,510	2,900
1670	12 ton		1.30	12.308		2,375	415		2,790	3,250
1680	14 ton		1.20	13.333		2,725	450		3,175	3,675
1690	16 ton		1.10	14.545		2,975	490		3,465	4,025
1700	21 ton		1	16		3,450	540		3,990	4,600
1720	26 ton		.84	19.002		3,850	645		4,495	5,225
1740	30 ton		.70	22.792		6,425	770		7,195	8,250
1760	41 ton	Q-6	.77	31.008		7,075	1,100		8,175	9,425
1780	52 ton		.66	36.419		10,100	1,275		11,375	13,000
1800	63 ton		.54	44.037		11,200	1,550		12,750	14,600
1820	76 ton		.45	52.980		12,700	1,850		14,550	16,700
1840	86 ton		.40	60		15,700	2,100		17,800	20,400
1860	97 ton		.35	67.989		17,800	2,375		20,175	23,200
1880	105 ton	Q-7	.43	73.563		20,500	2,625		23,125	26,600
1890	118 ton		.39	82.687		22,100	2,975		25,075	28,800
1900	126 ton		.36	88.154		25,700	3,150		28,850	33,100
1910	136 ton		.34	95.238		28,800	3,425		32,225	36,800
1920	142 ton		.32	99.379		29,500	3,575		33,075	37,900
3400	Evaporative, copper coil, pump, fan motor									
3440	10 ton	Q-5	.54	29.630	Ea.	4,225	1,000		5,225	6,175
3460	15 ton		.50	32		4,375	1,075		5,450	6,450
3480	20 ton		.47	34.043		4,575	1,150		5,725	6,775
3500	25 ton		.45	35.556		4,875	1,200		6,075	7,200
3520	30 ton		.42	38.095		5,200	1,300		6,500	7,675
3540	40 ton	Q-6	.49	48.980		6,875	1,725		8,600	10,200
3560	50 ton		.39	61.538		8,625	2,150		10,775	12,800
3580	65 ton		.35	68.571		9,575	2,400		11,975	14,100
3600	80 ton		.33	72.727		10,600	2,550		13,150	15,600
3620	90 ton		.29	82.759		11,400	2,900		14,300	16,900
3640	100 ton	Q-7	.36	88.889		12,400	3,175		15,575	18,500
3660	110 ton		.33	96.970		13,800	3,475		17,275	20,400
3680	125 ton		.30	106		14,900	3,825		18,725	22,200
3700	135 ton		.28	114		16,000	4,100		20,100	23,800
3720	150 ton		.25	128		17,600	4,600		22,200	26,300

15 MECHANICAL

Important: See the Reference Section for critical supporting data - Reference Nos., Crews, & City Cost Indexes

15660	Liquid Coolers/Evap Condensers		CREW	DAILY OUTPUT	LABOR-HOURS	UNIT	2003 BARE COSTS				TOTAL INCL O&P
							MAT.	LABOR	EQUIP.	TOTAL	
100	3740	165 ton		Q-7	.23	139	Ea.	19,400	5,000		24,400	28,800	100
	3760	185 ton	D3030 110	↓	.22	145		20,500	5,225		25,725	30,400
	3860	For fan damper control, add		Q-5	2	8	↓	440	271		711	895
	8000	Oils										
	8100	Lubricating										
	8120	Oil, lubricating					Oz.	.17			.17	.19
	8500	Refrigeration										
	8520	Oil, refrigeration		↓			Gal.	10			10	11

15670	Refrig Condensing Units		CREW	DAILY OUTPUT	LABOR-HOURS	UNIT	MAT.	LABOR	EQUIP.	TOTAL	TOTAL INCL O&P
300	0010	**CONDENSING UNITS**										300
	0030	Air cooled, compressor, standard controls									
	0050	1.5 ton	Q-5	2.50	6.400	Ea.	790	217		1,007	1,200
	0100	2 ton		2.10	7.619		835	258		1,093	1,300
	0200	2.5 ton		1.70	9.412		835	320		1,155	1,400
	0300	3 ton		1.30	12.308		990	415		1,405	1,700
	0350	3.5 ton		1.10	14.545		1,100	490		1,590	1,950
	0400	4 ton		.90	17.778		1,275	600		1,875	2,300
	0500	5 ton		.60	26.667		1,550	905		2,455	3,075
	0550	7.5 ton		.55	29.091		3,175	985		4,160	4,975
	0560	8.5 ton		.53	30.189		4,525	1,025		5,550	6,550
	0600	10 ton		.50	32		5,400	1,075		6,475	7,575
	0620	11 ton		.48	33.333		9,475	1,125		10,600	12,100
	0650	15 ton	↓	.40	40		7,475	1,350		8,825	10,300
	0700	20 ton	Q-6	.40	60		12,100	2,100		14,200	16,500
	0720	25 ton		.35	68.571		21,300	2,400		23,700	27,000
	0750	30 ton		.30	80		29,500	2,800		32,300	36,800
	0800	40 ton		.20	120		31,000	4,200		35,200	40,500
	0840	50 ton		.18	133		37,000	4,675		41,675	47,800
	0860	60 ton		.16	150		42,600	5,275		47,875	55,000
	0900	75 ton		.14	171		48,600	6,025		54,625	62,500
	1000	80 ton		.12	200		54,000	7,025		61,025	70,000
	1100	100 ton	↓	.09	266	↓	66,000	9,350		75,350	87,000
	2000	Water cooled, compressor, heat exchanger, controls									
	2100	5 ton	Q-5	.70	22.857	Ea.	16,500	775		17,275	19,400
	2110	10 ton		.60	26.667		19,100	905		20,005	22,400
	2200	15 ton	↓	.50	32		21,600	1,075		22,675	25,400
	2300	20 ton	Q-6	.40	60		29,100	2,100		31,200	35,200
	2310	30 ton		.30	80		34,600	2,800		37,400	42,400
	2400	40 ton		.20	120		39,400	4,200		43,600	49,800
	2410	60 ton		.18	133		59,500	4,675		64,175	72,000
	2420	80 ton		.14	171		68,500	6,025		74,525	84,500
	2500	100 ton		.11	218		84,000	7,650		91,650	104,000
	2510	120 ton	↓	.10	240	↓	135,500	8,425		143,925	161,500

15680	Evaporators, DX Coils		CREW	DAILY OUTPUT	LABOR-HOURS	UNIT	MAT.	LABOR	EQUIP.	TOTAL	TOTAL INCL O&P
300	0010	**EVAPORATORS** DX coils, remote compressors not included										300
	1000	Coolers, reach-in type, above freezing temperatures									
	1300	Shallow depth, wall mount, 7 fins per inch, air defrost									
	1310	600 BTUH, 8" fan, rust-proof core	Q-5	3.80	4.211	Ea.	276	143		419	520
	1320	900 BTUH, 8" fan, rust-proof core		3.30	4.848		340	164		504	625
	1330	1200 BTUH, 8" fan, rust-proof core		3	5.333		400	181		581	720
	1340	1800 BTUH, 8" fan, rust-proof core		2.40	6.667		470	226		696	855
	1350	2500 BTUH, 10" fan, rust-proof core		2.20	7.273		505	246		751	925
	1360	3500 BTUH, 10" fan		2	8		595	271		866	1,075
	1370	4500 BTUH, 10" fan	↓	1.90	8.421	↓	625	285		910	1,125

MECHANICAL 15

	15680	Evaporators, DX Coils	CREW	DAILY OUTPUT	LABOR-HOURS	UNIT	2003 BARE COSTS				TOTAL INCL O&P	
							MAT.	LABOR	EQUIP.	TOTAL		
300	1390	For pan drain, add				Ea.	38			38	41.50	300
	1600	Undercounter refrigerators, ceiling or wall mount,										
	1610	8 fins per inch, air defrost, rust-proof core										
	1630	800 BTUH, one 6" fan	Q-5	5	3.200	Ea.	173	108		281	355	
	1640	1300 BTUH, two 6" fans	↓	4	4		211	135		346	435	
	1650	1700 BTUH, two 6" fans	↓	3.60	4.444	↓	236	150		386	485	
	2000	Coolers, reach-in and walk-in types, above freezing temperatures										
	2600	Two-way discharge, ceiling mount, 150-4100 CFM,										
	2610	above 34°F applications, air defrost										
	2630	900 BTUH, 7 fins per inch, 8" fan	Q-5	4	4	Ea.	197	135		332	420	
	2660	2500 BTUH, 7 fins per inch, 10" fan		2.60	6.154		370	208		578	720	
	2690	5500 BTUH, 7 fins per inch, 12" fan		1.70	9.412		735	320		1,055	1,300	
	2720	8500 BTUH, 8 fins per inch, 16" fan		1.20	13.333		955	450		1,405	1,725	
	2750	15,000 BTUH, 7 fins per inch, 18" fan		1.10	14.545		1,325	490		1,815	2,200	
	2770	24,000 BTUH, 7 fins per inch, two 16" fans		1	16		2,000	540		2,540	3,025	
	2790	30,000 BTUH, 7 fins per inch, two 18" fans	↓	.90	17.778	↓	2,475	600		3,075	3,625	
	2850	Two-way discharge, low profile, ceiling mount,										
	2860	8 fins per inch, 200-570 CFM, air defrost										
	2880	800 BTUH, one 6" fan	Q-5	4	4	Ea.	224	135		359	450	
	2890	1300 BTUH, two 6" fans		3.50	4.571		266	155		421	525	
	2900	1800 BTUH, two 6" fans		3.30	4.848		281	164		445	560	
	2910	2700 BTUH, three 6" fans	↓	2.70	5.926	↓	330	201		531	670	
	3000	Coolers, walk-in type, above freezing temperatures										
	3300	General use, ceiling mount, 108-2080 CFM										
	3320	600 BTUH, 7 fins per inch, 6" fan	Q-5	5.20	3.077	Ea.	176	104		280	350	
	3340	1200 BTUH, 7 fins per inch, 8" fan		3.70	4.324		243	146		389	490	
	3360	1800 BTUH, 7 fins per inch, 10" fan		2.70	5.926		340	201		541	680	
	3380	3500 BTUH, 7 fins per inch, 12" fan		2.40	6.667		500	226		726	890	
	3400	5500 BTUH, 8 fins per inch, 12" fan		2	8		560	271		831	1,025	
	3430	8500 BTUH, 8 fins per inch, 16" fan		1.40	11.429		830	385		1,215	1,500	
	3460	15,000 BTUH, 7 fins per inch, two 16" fans	↓	1.10	14.545	↓	1,325	490		1,815	2,225	
	3640	Low velocity, high latent load, ceiling mount,										
	3650	1050-2420 CFM, air defrost, 6 fins per inch										
	3670	6700 BTUH, two 10" fans	Q-5	1.30	12.308	Ea.	1,100	415		1,515	1,825	
	3680	10,000 BTUH, three 10" fans		1	16		1,275	540		1,815	2,225	
	3690	13,500 BTUH, three 10" fans		1	16		1,525	540		2,065	2,525	
	3700	18,000 BTUH, four 10" fans		1	16		2,025	540		2,565	3,075	
	3710	26,500 BTUH, four 10" fans	↓	.80	20		2,525	675		3,200	3,800	
	3730	For electric defrost, add				↓	17%					
	5000	Freezers and coolers, reach-in type, above 34°F										
	5030	to sub-freezing temperature range, low latent load,										
	5050	air defrost, 7 fins per inch										
	5070	1200 BTUH, 8" fan, rustproof core	Q-5	4.40	3.636	Ea.	229	123		352	440	
	5080	1500 BTUH, 8" fan, rustproof core		4.20	3.810		236	129		365	455	
	5090	1800 BTUH, 8" fan, rustproof core		3.70	4.324		247	146		393	495	
	5100	2500 BTUH, 10" fan, rustproof core		2.90	5.517		282	187		469	590	
	5110	3500 BTUH, 10" fan, rustproof core		2.50	6.400		390	217		607	750	
	5120	4500 BTUH, 12" fan, rustproof core	↓	2.10	7.619	↓	515	258		773	955	
	6000	Freezers and coolers, walk-in type										
	6050	1960-18,000 CFM, medium profile										
	6060	Standard motor, 6 fins per inch, to -30°F, all aluminum										
	6080	10,500 BTUH, air defrost, one 18" fan	Q-5	1.40	11.429	Ea.	985	385		1,370	1,650	
	6090	12,500 BTUH, air defrost, one 18" fan		1.40	11.429		1,075	385		1,460	1,750	
	6100	16,400 BTUH, air defrost, one 18" fan		1.30	12.308		1,100	415		1,515	1,825	
	6110	20,900 BTUH, air defrost, one 18" fans		1	16		1,275	540		1,815	2,225	
	6120	27,000 BTUH, air defrost, two 18" fans	↓	1	16	↓	1,900	540		2,440	2,900	

Important: See the Reference Section for critical supporting data - Reference Nos., Crews, & City Cost Indexes

15680	Evaporators, DX Coils	CREW	DAILY OUTPUT	LABOR-HOURS	UNIT	2003 BARE COSTS				TOTAL INCL O&P
						MAT.	LABOR	EQUIP.	TOTAL	
300 6130	32,900 BTUH, air defrost, two 20" fans	Q-6	1.30	18.462	Ea.	2,125	650		2,775	3,300	**300**
6140	39,000 BTUH, air defrost, two 20" fans		1.25	19.200		2,450	675		3,125	3,725
6150	44,100 BTUH, air defrost, three 20" fans		1	24		2,800	840		3,640	4,350
6155	53,000 BTUH, air defrost, three 20" fans		1	24		2,800	840		3,640	4,350
6160	66,200 BTUH, air defrost, four 20" fans		1	24		3,175	840		4,015	4,775
6170	78,000 BTUH, air defrost, four 20" fans		.80	30		3,950	1,050		5,000	5,950
6180	88,200 BTUH, air defrost, four 24" fans		.60	40		4,200	1,400		5,600	6,750
6190	110,000 BTUH, air defrost, four 24" fans		.50	48		4,775	1,675		6,450	7,800
6330	Hot gas defrost, standard motor units, add					35%				
6370	Electric defrost, 230V, standard motor, add					16%				
6410	460V, standard or high capacity motor, add					25%				
6800	Eight fins per inch, increases BTUH 75%									
6810	Standard capacity, add				Ea.	5%				
6830	Hot gas & electric defrost not recommended									
7000	800-4150 CFM, 12" fans									
7010	Four fins per inch									
7030	3400 BTUH, 1 fan	Q-5	2.80	5.714	Ea.	345	193		538	670
7040	4200 BTUH, 1 fan		2.40	6.667		350	226		576	730
7050	5300 BTUH, 1 fan		1.90	8.421		410	285		695	880
7060	6800 BTUH, 1 fan		1.70	9.412		460	320		780	985
7070	8400 BTUH, 2 fans		1.60	10		610	340		950	1,175
7080	10,500 BTUH, 2 fans		1.40	11.429		665	385		1,050	1,325
7090	13,000 BTUH, 3 fans		1.30	12.308		860	415		1,275	1,575
7100	17,000 BTUH, 4 fans		1.20	13.333		1,075	450		1,525	1,875
7110	21,500 BTUH, 5 fans		1	16		1,350	540		1,890	2,300
7160	Six fins per inch increases BTUH 35%, add					5%				
7180	Eight fins per inch increases BTUH 53%,									
7190	not recommended for below 35°F, add				Ea.	8%				
7210	For 12°F temperature differential, add					20%				
7230	For adjustable thermostat control, add					73			73	80.50
7240	For hot gas defrost, except 8 fin models									
7250	on applications below 35°F, add				Ea.	58%				
7260	For electric defrost, except 8 fin models									
7270	on applications below 35°F, add				Ea.	36%				
8000	Freezers, pass-thru door uprights, walk-in storage									
8300	Low temperature, thin profile, electric defrost									
8310	138-1800 CFM, 5 fins per inch									
8320	900 BTUH, one 6" fan	Q-5	3.30	4.848	Ea.	260	164		424	535
8340	1500 BTUH, two 6" fans		2.20	7.273		390	246		636	800
8360	2600 BTUH, three 6" fans		1.90	8.421		560	285		845	1,050
8370	3300 BTUH, four 6" fans		1.70	9.412		625	320		945	1,175
8380	4400 BTUH, five 6" fans		1.40	11.429		880	385		1,265	1,550
8400	7000 BTUH, three 10" fans		1.30	12.308		1,050	415		1,465	1,800
8410	8700 BTUH, four 10" fans		1.10	14.545		1,200	490		1,690	2,075
8450	For air defrost, deduct					130			130	143

15705 | Curbs/Pads/Stands Prefab

		CREW	DAILY OUTPUT	LABOR-HOURS	UNIT	2003 BARE COSTS				TOTAL INCL O&P
						MAT.	LABOR	EQUIP.	TOTAL	
600	**0010**	**CURBS/PADS PREFABRICATED**									600
	6000	Pad, fiberglass reinforced concrete with polystyrene foam core								
	6050	Condenser, 2" thick, 20" x 38"	1 Shee	8	1	Ea.	8.55	37		45.55	66.50
	6070	24" x 24"		16	.500		7.60	18.50		26.10	37
	6090	24" x 36"		12	.667		11.80	24.50		36.30	51
	6110	24" x 42"	↓	8	1		13.70	37		50.70	72
	6150	26" x 36"	Q-9	8	2		12.90	66.50		79.40	117
	6170	28" x 38"	1 Shee	8	1		14.15	37		51.15	72.50
	6190	30" x 30"		12	.667		12.20	24.50		36.70	51.50
	6220	30" x 36"	↓	8	1		14.60	37		51.60	73
	6240	30" x 40"	Q-9	8	2		15.90	66.50		82.40	121
	6260	32" x 32"		7	2.286		14.20	76		90.20	134
	6280	36" x 36"		8	2		17	66.50		83.50	122
	6300	36" x 40"		7	2.286		19.30	76		95.30	139
	6320	36" x 48"		7	2.286		22.50	76		98.50	143
	6340	36" x 54"		6	2.667		27	89		116	167
	6360	36" x 60"		7	2.286		29	76		105	150
	6400	41" x 47"		6	2.667		27.50	89		116.50	168
	6490	26" round		10	1.600		7.30	53.50		60.80	90.50
	6550	30" round		9	1.778		13.70	59		72.70	107
	6600	36" round	↓	8	2		22.50	66.50		89	128
	6700	For 3" thick pad, add				↓	20%	5%			
	8800	Stand, corrosion-resistant plastic									
	8850	Heat pump, 6" high	1 Shee	32	.250	Ea.	6.80	9.25		16.05	22
	8870	12" high	"	32	.250	"	9.20	9.25		18.45	24.50

15710 | Heat Exchangers

		CREW	DAILY OUTPUT	LABOR-HOURS	UNIT	MAT.	LABOR	EQUIP.	TOTAL	TOTAL INCL O&P
900	**0010**	**HEAT EXCHANGERS**									900
	0016	Shell & tube type, 2 or 4 pass, 3/4" O.D. copper tubes,								
	0020	C.I. heads, C.I. tube sheet, steel shell								
	0100	Hot water 40°F to 180°F, by steam at 10 PSI								
	0120	8 GPM	Q-5	6	2.667	Ea.	945	90.50		1,035.50	1,175
	0140	10 GPM		5	3.200		1,425	108		1,533	1,750
	0160	40 GPM		4	4		2,200	135		2,335	2,625
	0180	64 GPM		2	8		3,400	271		3,671	4,125
	0200	96 GPM	↓	1	16		4,525	540		5,065	5,800
	0220	120 GPM	Q-6	1.50	16		5,950	560		6,510	7,400
	0240	168 GPM		1	24		7,300	840		8,140	9,300
	0260	240 GPM		.80	30		11,400	1,050		12,450	14,200
	0300	600 GPM	↓	.70	34.286	↓	24,500	1,200		25,700	28,700
	0500	For bronze head and tube sheet, add					50%				
	1000	Hot water 40°F to 140°F, by water at 200°F								
	1020	7 GPM	Q-5	6	2.667	Ea.	1,175	90.50		1,265.50	1,400
	1040	16 GPM		5	3.200		1,650	108		1,758	2,000
	1060	34 GPM		4	4		2,500	135		2,635	2,950
	1080	55 GPM		3	5.333		3,625	181		3,806	4,250
	1100	74 GPM		1.50	10.667		4,525	360		4,885	5,525
	1120	86 GPM	↓	1.40	11.429		6,050	385		6,435	7,225
	1140	112 GPM	Q-6	2	12		7,525	420		7,945	8,900
	1160	126 GPM		1.80	13.333		9,400	470		9,870	11,000
	1180	152 GPM	↓	1	24	↓	12,000	840		12,840	14,500
	3000	Plate type,								
	3100	400 GPM	Q-6	.80	30	Ea.	16,000	1,050		17,050	19,200
	3120	800 GPM	"	.50	48		27,500	1,675		29,175	32,900
	3140	1200 GPM	Q-7	.34	94.118	↓	33,500	3,375		36,875	42,000

Important: See the Reference Section for critical supporting data - Reference Nos., Crews, & City Cost Indexes

		CREW	DAILY OUTPUT	LABOR-HOURS	UNIT	2003 BARE COSTS				TOTAL INCL O&P	
						MAT.	LABOR	EQUIP.	TOTAL		
15710	**Heat Exchangers**										
900	3160	1800 GPM	Q-7	.24	133	Ea.	50,000	4,775		54,775	62,000	**900**
	8000	Heat pipe type, glycol, 50% efficient										
	8010	100 MBH, 1700 CFM	1 Stpi	.80	10	Ea.	17,800	375		18,175	20,200	
	8020	160 MBH, 2700 CFM	↓	.60	13.333	↓	22,400	500		22,900	25,400	
	8030	620 MBH, 4000 CFM	↓	.40	20	↓	30,400	750		31,150	34,600	
15720	**Air Handling Units**										
100	0010	**AIR HANDLING UNIT** Built-up										**100**
	0100	With cooling/heating coil section, filters, mixing box										
	0880	Single zone, horizontal / vertical										
	0890	Constant volume										
	0900	1600 CFM	Q-5	1.20	13.333	Ea.	2,400	450		2,850	3,325	
	0906	2000 CFM		1.10	14.545		3,000	490		3,490	4,050	
	0910	3000 CFM		1	16		3,750	540		4,290	4,950	
	0916	4000 CFM	↓	.95	16.842		5,000	570		5,570	6,350	
	0920	5000 CFM	Q-6	1.40	17.143		6,250	600		6,850	7,775	
	0926	6500 CFM		1.30	18.462		8,125	650		8,775	9,925	
	0930	7500 CFM		1.20	20		9,375	700		10,075	11,400	
	0936	9200 CFM		1.10	21.818		11,500	765		12,265	13,900	
	0940	11,500 CFM		1	24		12,700	840		13,540	15,200	
	0946	13,200 CFM		.90	26.667		14,500	935		15,435	17,400	
	0950	16,500 CFM		.80	30		18,200	1,050		19,250	21,600	
	0960	19,500 CFM		.70	34.286		21,500	1,200		22,700	25,400	
	0970	22,000 CFM		.60	40		24,200	1,400		25,600	28,700	
	0980	27,000 CFM		.50	48		27,000	1,675		28,675	32,300	
	0990	34,000 CFM		.40	60		34,000	2,100		36,100	40,600	
	1000	40,000 CFM		.30	80		38,500	2,800		41,300	46,700	
	1010	47,000 CFM	↓	.20	120		47,000	4,200		51,200	58,000	
	1140	60,000 CFM	Q-7	.18	177		50,500	6,375		56,875	65,000	
	1160	75,000 CFM	"	.16	200	↓	63,000	7,175		70,175	80,500	
	2000	Variable air volume										
	2100	75,000 CFM	Q-7	.12	266	Ea.	64,500	9,550		74,050	85,500	
	2140	100,000 CFM		.11	290		92,000	10,400		102,400	117,000	
	2160	150,000 CFM	↓	.10	320	↓	138,500	11,500		150,000	170,000	
	2300	Multi-zone, horizontal or vertical, blow-thru fan										
	2310	3000 CFM	Q-5	1	16	Ea.	1,625	540		2,165	2,600	
	2314	4000 CFM	"	.95	16.842		2,175	570		2,745	3,225	
	2318	5000 CFM	Q-6	1.40	17.143		2,700	600		3,300	3,875	
	2322	6500 CFM		1.30	18.462		3,525	650		4,175	4,850	
	2326	7500 CFM		1.20	20		4,050	700		4,750	5,525	
	2330	9200 CFM		1.10	21.818		4,975	765		5,740	6,625	
	2334	11,500 CFM		1	24		6,225	840		7,065	8,125	
	2336	13,200 CFM		.90	26.667		7,150	935		8,085	9,275	
	2340	16,500 CFM		.80	30		8,925	1,050		9,975	11,400	
	2344	19,500 CFM		.70	34.286		10,600	1,200		11,800	13,400	
	2348	22,000 CFM		.60	40		11,900	1,400		13,300	15,200	
	2352	27,000 CFM		.50	48		14,600	1,675		16,275	18,700	
	2356	34,000 CFM		.40	60		18,400	2,100		20,500	23,400	
	2360	40,000 CFM		.30	80		21,600	2,800		24,400	28,100	
	2364	47,000 CFM	↓	.20	120	↓	25,400	4,200		29,600	34,400	
	3000	Rooftop										
	3010	Constant volume										
	3100	2000 CFM	Q-5	1	16	Ea.	3,800	540		4,340	5,000	
	3140	5000 CFM	Q-6	.79	30.380		6,800	1,075		7,875	9,075	
	3150	10,000 CFM	↓	.52	46.154	↓	8,600	1,625		10,225	11,900	

MECHANICAL 15

		15720	**Air Handling Units**	CREW	DAILY OUTPUT	LABOR-HOURS	UNIT	2003 BARE COSTS				TOTAL INCL O&P	
								MAT.	LABOR	EQUIP.	TOTAL		
100	3160		15,000 CFM	Q-6	.39	61.538	Ea.	13,000	2,150		15,150	17,600	100
	3170		20,000 CFM	↓	.30	80	↓	16,200	2,800		19,000	22,100	
	3200		Variable air volume										
	3240		5000 CFM	Q-6	.70	34.286	Ea.	8,375	1,200		9,575	11,100	
	3250		10,000 CFM		.50	48		10,500	1,675		12,175	14,100	
	3260		15,000 CFM		.40	60		14,900	2,100		17,000	19,600	
	3270		20,000 CFM		.29	82.759		19,400	2,900		22,300	25,800	
	3280		30,000 CFM	↓	.20	120	↓	32,700	4,200		36,900	42,400	
200	0010		**CENTRAL STATION AIR-HANDLING UNIT** Chilled water										200
	2300		Variable air volume, also incl. heating coil										
	2340		5000 CFM	Q-6	.70	34.286	Ea.	7,850	1,200		9,050	10,500	
	2350		10,000 CFM		.50	48		9,825	1,675		11,500	13,400	
	2360		15,000 CFM		.40	60		13,900	2,100		16,000	18,500	
	2370		20,000 CFM		.29	82.759		18,200	2,900		21,100	24,400	
	2380		30,000 CFM		.20	120		30,500	4,200		34,700	40,000	
	3000		Packaged, 3000 CFM, 7.5 ton		1.70	14.118		3,350	495		3,845	4,425	
	3050		3200 CFM, 8 ton		1.60	15		3,750	525		4,275	4,925	
	3100		4000 CFM, 10 ton		1.40	17.143		4,025	600		4,625	5,325	
	3150		4400 CFM, 11 ton		1.35	17.778		4,800	625		5,425	6,225	
	3200		6000 CFM, 15 ton		1	24		5,450	840		6,290	7,250	
	3250		7000 CFM, 17.5 ton		.90	26.667		6,475	935		7,410	8,550	
	3300		10,000 CFM, 25 ton		.65	36.923		9,875	1,300		11,175	12,900	
	3400		12,000 CFM, 30 ton	↓	.50	48		10,500	1,675		12,175	14,100	
	3500		For hot water heat coil, add				↓	30%	5%				
300	0010		**EVAPORATIVE COOLERS** Ducted, not incl. duct.										300
	0100		Side discharge style, capacities at .25" S.P.										
	0120		1785 CFM, 1/3 HP, 115 V	Q-9	5	3.200	Ea.	350	107		457	550	
	0140		2740 CFM, 1/3 HP, 115 V		4.50	3.556		425	118		543	655	
	0160		3235 CFM, 1/2 HP, 115 V		4	4		440	133		573	685	
	0180		3615 CFM, 1/2 HP, 230 V		3.60	4.444		445	148		593	720	
	0200		4215 CFM, 3/4 HP, 230 V		3.20	5		570	167		737	880	
	0220		5255 CFM. 1 HP, 115/230 V		3	5.333		935	178		1,113	1,300	
	0240		6090 CFM, 1 HP, 230/460 V		2.80	5.714		1,300	190		1,490	1,750	
	0260		8300 CFM, 1-1/2 HP, 230/460 V		2.60	6.154		1,300	205		1,505	1,775	
	0280		8360 CFM, 1-1/2 HP, 230/460 V		2.20	7.273		1,750	242		1,992	2,300	
	0300		9725 CFM, 2 HP, 230/460 V		1.80	8.889		1,775	296		2,071	2,400	
	0320		11,715 CFM, 3 HP, 230/460 V		1.40	11.429		1,775	380		2,155	2,575	
	0340		14,410 CFM, 5 HP, 230/460 V	↓	1	16	↓	1,925	535		2,460	2,950	
	0400		For two-speed motor, add					5%					
	0500		For down discharge style, add					10%					
	1000		High capacities at .4" S.P.										
	1040		15,000 CFM, 5 HP, 230/460 V	Q-9	.54	29.630	Ea.	5,575	985		6,560	7,650	
	1080		20,000 CFM, 10HP, 230/460 V	"	.48	33.333		7,150	1,100		8,250	9,575	
	1100		25,000 CFM, 20 HP, 230/460 V	Q-10	.54	44.444		7,875	1,525		9,400	11,000	
	1140		30,000 CFM, 20 HP, 230/460 V		.46	52.174		9,075	1,800		10,875	12,800	
	1160		35,000 CFM, 20 HP, 230/460 V		.44	54.545		9,925	1,875		11,800	13,800	
	1180		40,000 CFM, 30 HP, 230/460 V	↓	.42	57.143	↓	10,400	1,975		12,375	14,500	
500	0010		**MAKE-UP AIR UNIT**										500
	0020		Indoor suspension, natural/LP gas, direct fired,										
	0030		standard control. For flue see division 15550-440										
	0040		70°F temperature rise, MBH is input										
	0100		2000 CFM, 168 MBH	Q-6	3	8	Ea.	5,725	281		6,006	6,725	
	0120		3000 CFM, 252 MBH		2	12		5,725	420		6,145	6,925	
	0140		4000 CFM, 336 MBH		1.80	13.333		5,975	470		6,445	7,275	
	0160		6000 CFM, 502 MBH	↓	1.50	16	↓	7,200	560		7,760	8,775	

Important: See the Reference Section for critical supporting data - Reference Nos., Crews, & City Cost Indexes

15 MECHANICAL

15720	Air Handling Units	CREW	DAILY OUTPUT	LABOR-HOURS	UNIT	2003 BARE COSTS				TOTAL INCL O&P	
						MAT.	LABOR	EQUIP.	TOTAL		
500	0180	8000 CFM, 670 MBH	Q-6	1.40	17.143	Ea.	7,900	600		8,500	9,575	500
	0200	10,000 CFM, 838 MBH		1.20	20		8,900	700		9,600	10,800	
	0220	12,000 CFM, 1005 MBH		1	24		8,975	840		9,815	11,200	
	0240	14,000 CFM, 1180 MBH		.94	25.532		9,100	895		9,995	11,400	
	0260	18,000 CFM, 1340 MBH		.86	27.907		9,175	980		10,155	11,600	
	0280	20,000 CFM, 1675 MBH		.76	31.579		11,300	1,100		12,400	14,100	
	0300	24,000 CFM, 2007 MBH	Q-7	1	32		11,700	1,150		12,850	14,600	
	0320	30,000 CFM, 2510 MBH		.96	33.333		12,000	1,200		13,200	15,000	
	0340	35,000 CFM, 2930 MBH		.92	34.783		15,100	1,250		16,350	18,500	
	0360	40,000 CFM, 3350 MBH		.88	36.364		15,500	1,300		16,800	19,000	
	0380	45,000 CFM, 3770 MBH		.84	38.095		15,900	1,375		17,275	19,600	
	0400	50,000 CFM, 4180 MBH		.80	40		16,100	1,425		17,525	19,900	
	0420	55,000 CFM, 4600 MBH		.75	42.667		20,500	1,525		22,025	24,800	
	0440	60,000 CFM, 5020 MBH		.70	45.714		21,300	1,650		22,950	25,900	
	0460	65,000 CFM, 5435 MBH		.60	53.333		21,400	1,900		23,300	26,500	
	0480	75,000 CFM, 6275 MBH		.50	64		22,600	2,300		24,900	28,300	
	0600	For discharge louver assembly, add					10%					
	0700	For filters, add					20%					
	0800	For air shut-off damper section, add					10%					
	0900	For vertical unit, add					10%					
	1000	Rooftop unit, natural gas, gravity vent, S.S. exchanger										
	1010	70°F temperature rise, MBH is input										
	1020	995 CFM, 75 MBH	Q-6	4	6	Ea.	3,850	211		4,061	4,550	
	1040	1245 CFM, 95 MBH		3.60	6.667		3,875	234		4,109	4,625	
	1060	1590 CFM, 120 MBH		3.30	7.273		4,175	255		4,430	4,975	
	1080	2090 CFM, 159 MBH		3	8		4,300	281		4,581	5,175	
	1100	2490 CFM, 190 MBH		2.60	9.231		4,925	325		5,250	5,925	
	1120	2985 CFM, 225 MBH		2.30	10.435		5,000	365		5,365	6,050	
	1140	3975 CFM, 300 MBH		1.90	12.632		5,925	445		6,370	7,200	
	1160	4970 CFM, 375 MBH		1.40	17.143		6,875	600		7,475	8,450	
	1180	5950 CFM, 450 MBH		1.20	20		7,250	700		7,950	9,025	
	1200	7940 CFM, 600 MBH		1	24		8,425	840		9,265	10,600	
	1220	9930 CFM, 750 MBH		.80	30		14,400	1,050		15,450	17,400	
	1240	11,900 CFM, 900 MBH		.60	40		16,100	1,400		17,500	19,800	
	1260	15,880 CFM, 1200 MBH		.30	80		19,300	2,800		22,100	25,500	
	1500	For power vent, add					8%					
	1600	For cleanable filters, add				Ea.	112			112	123	
	1700	For electric modulating gas control, add				"	1,200			1,200	1,325	
	1800	For aluminized steel exchanger, deduct					10%					
	8000	Rooftop unit, electric, with curb, controls										
	8010	33 kW	Q-5	4	4	Ea.	3,175	135		3,310	3,675	
	8020	100 kW		3.60	4.444		5,250	150		5,400	6,000	
	8030	150 kW		3.20	5		6,100	169		6,269	6,950	
	8040	250 kW		2.80	5.714		7,925	193		8,118	9,000	
	8050	400 kW		2.20	7.273		11,600	246		11,846	13,100	

15730	Unitary Air Conditioning Equip										
200	0010	**COMPUTER ROOM UNITS** [D3030 222]										200
	1000	Air cooled, includes remote condenser but not										
	1020	interconnecting tubing or refrigerant										
	1080	3 ton	Q-5	.50	32	Ea.	8,050	1,075		9,125	10,500	
	1120	5 ton		.45	35.556		10,800	1,200		12,000	13,600	
	1160	6 ton		.30	53.333		19,800	1,800		21,600	24,500	
	1200	8 ton		.27	59.259		22,400	2,000		24,400	27,700	
	1240	10 ton		.25	64		23,400	2,175		25,575	29,000	

MECHANICAL 15

15730	Unitary Air Conditioning Equip		CREW	DAILY OUTPUT	LABOR-HOURS	UNIT	2003 BARE COSTS				TOTAL INCL O&P	
							MAT.	LABOR	EQUIP.	TOTAL		
200	1280	15 ton		Q-5	.22	72.727	Ea.	26,100	2,450		28,550	32,400	**200**
	1290	18 ton		↓	.20	80		29,700	2,700		32,400	36,800	
	1320	20 ton		Q-6	.29	82.759		31,100	2,900		34,000	38,600	
	1360	23 ton		"	.28	85.714	↓	32,700	3,000		35,700	40,600	
	2200	Chilled water, for connection to											
	2220	existing chiller system of adequate capacity											
	2260	5 ton		Q-5	.74	21.622	Ea.	9,075	730		9,805	11,100	
	2280	6 ton			.52	30.769		12,100	1,050		13,150	14,900	
	2300	8 ton			.50	32		13,000	1,075		14,075	15,900	
	2320	10 ton			.49	32.653		14,400	1,100		15,500	17,500	
	2330	12 ton			.48	32.990		14,900	1,125		16,025	18,100	
	2360	15 ton			.48	33.333		15,400	1,125		16,525	18,700	
	2400	20 ton			.46	34.783		15,800	1,175		16,975	19,200	
	2440	23 ton		↓	.42	38.095	↓	16,600	1,300		17,900	20,300	
	4000	Glycol system, complete except for interconnecting tubing											
	4060	3 ton		Q-5	.40	40	Ea.	9,850	1,350		11,200	12,900	
	4100	5 ton			.38	42.105		12,800	1,425		14,225	16,300	
	4120	6 ton			.25	64		25,700	2,175		27,875	31,600	
	4140	8 ton			.23	69.565		27,800	2,350		30,150	34,100	
	4160	10 ton		↓	.21	76.190		30,100	2,575		32,675	37,000	
	4200	15 ton		Q-6	.26	92.308		36,500	3,250		39,750	45,000	
	4240	20 ton			.24	100		39,000	3,500		42,500	48,200	
	4280	23 ton		↓	.22	109	↓	42,500	3,825		46,325	52,500	
	8000	Water cooled system, not including condenser,											
	8020	water supply or cooling tower											
	8060	3 ton		Q-5	.62	25.806	Ea.	7,275	875		8,150	9,325	
	8100	5 ton			.54	29.630		10,200	1,000		11,200	12,800	
	8120	6 ton			.35	45.714		19,200	1,550		20,750	23,400	
	8140	8 ton			.33	48.485		21,100	1,650		22,750	25,700	
	8160	10 ton			.31	51.613		22,400	1,750		24,150	27,400	
	8200	15 ton		↓	.27	59.259		25,800	2,000		27,800	31,400	
	8240	20 ton		Q-6	.38	63.158		30,500	2,225		32,725	36,900	
	8280	23 ton		"	.34	70.588	↓	32,300	2,475		34,775	39,300	
500	0010	**PACKAGED TERMINAL AIR CONDITIONER** Cabinet, wall sleeve,											**500**
	0100	louver, electric heat, thermostat, manual changeover, 208 V											
	0200	6,000 BTUH cooling, 8800 BTU heat		Q-5	6	2.667	Ea.	955	90.50		1,045.50	1,175	
	0220	9,000 BTUH cooling, 13,900 BTU heat			5	3.200		980	108		1,088	1,250	
	0240	12,000 BTUH cooling, 13,900 BTU heat			4	4		1,025	135		1,160	1,325	
	0260	15,000 BTUH cooling, 13,900 BTU heat			3	5.333		1,100	181		1,281	1,475	
	0280	18,000 BTUH cooling, 10 KW heat			2.40	6.667		1,350	226		1,576	1,850	
	0300	24,000 BTUH cooling, 10 KW heat			1.90	8.421		1,375	285		1,660	1,925	
	0320	30,000 BTUH cooling, 10 KW heat			1.40	11.429		1,550	385		1,935	2,275	
	0340	36,000 BTUH cooling, 10 KW heat			1.25	12.800		1,625	435		2,060	2,450	
	0360	42,000 BTUH cooling, 10 KW heat			1	16		2,050	540		2,590	3,100	
	0380	48,000 BTUH cooling, 10 KW heat		↓	.90	17.778		2,225	600		2,825	3,350	
	0500	For hot water coil, increase heat by 10%, add						5%	10%				
	1000	For steam, increase heat output by 30%, add					↓	8%	10%				
600	0010	**ROOF TOP AIR CONDITIONERS** Standard controls, curb, economizer											**600**
	1000	Single zone, electric cool, gas heat											
	1090	2 ton cooling, 55 MBH heating	D3030 202	Q-5	.93	17.204	Ea.	3,800	580		4,380	5,075	
	1100	3 ton cooling, 60 MBH heating		↓	.70	22.857		3,825	775		4,600	5,375	
	1120	4 ton cooling, 95 MBH heating	D3030 206		.61	26.403		4,550	895		5,445	6,350	
	1140	5 ton cooling, 112 MBH heating			.56	28.520		5,175	965		6,140	7,125	
	1145	6 ton cooling, 140 MBH heating	R15700 -010		.52	30.769		6,325	1,050		7,375	8,525	
	1150	7.5 ton cooling, 170 MBH heating		↓	.50	32.258	↓	8,400	1,100		9,500	10,900	

15 MECHANICAL

Important: See the Reference Section for critical supporting data - Reference Nos., Crews, & City Cost Indexes

15730	Unitary Air Conditioning Equip		CREW	DAILY OUTPUT	LABOR-HOURS	UNIT	2003 BARE COSTS				TOTAL INCL O&P
							MAT.	LABOR	EQUIP.	TOTAL	
600	1160	10 ton cooling, 200 MBH heating	D3030 202	Q-6	.67	35.982	Ea.	11,500	1,275		12,775	14,500
	1170	12.5 ton cooling, 230 MBH heating			.63	37.975		13,400	1,325		14,725	16,700
	1180	15 ton cooling, 270 MBH heating	D3030 206		.57	42.032		13,600	1,475		15,075	17,200
	1190	18 ton cooling, 330 MBH heating			.52	45.889		22,700	1,600		24,300	27,400
	1200	20 ton cooling, 360 MBH heating	R15700 -010	Q-7	.67	47.976		28,300	1,725		30,025	33,800
	1210	25 ton cooling, 450 MBH heating			.56	57.554		34,600	2,075		36,675	41,100
	1220	30 ton cooling, 540 MBH heating	R15700 -020		.47	68.376		44,900	2,450		47,350	53,000
	1240	40 ton cooling, 675 MBH heating			.35	91.168		58,000	3,275		61,275	69,000
	1300	Electric cool, electric heat										
	1310	5 ton		Q-5	.63	25.600	Ea.	2,750	865		3,615	4,325
	1314	10 ton		Q-6	.74	32.432		5,000	1,150		6,150	7,225
	1318	15 ton		"	.63	37.795		7,500	1,325		8,825	10,300
	1322	20 ton		Q-7	.74	43.185		10,000	1,550		11,550	13,400
	1326	25 ton			.62	51.696		12,500	1,850		14,350	16,600
	1330	30 ton			.52	61.657		15,000	2,200		17,200	19,900
	1334	40 ton			.39	82.051		20,000	2,950		22,950	26,500
	1338	50 ton			.31	102		23,800	3,675		27,475	31,700
	1342	60 ton			.26	123		28,500	4,400		32,900	38,100
	1346	75 ton			.21	153		35,600	5,525		41,125	47,500
	1700	Gas cool, gas heat										
	1720	3 ton cooling, 90 MBH heating		Q-5	.93	17.204	Ea.	7,400	580		7,980	9,025
	1730	4 ton cooling, 112.5 MBH heating			.61	26.403		8,700	895		9,595	10,900
	1750	5 ton cooling, 135 MBH heating			.56	28.571		10,400	965		11,365	13,000
	2000	Multizone, electric cool, gas heat, economizer										
	2100	15 ton cooling, 360 MBH heating		Q-7	.61	52.545	Ea.	70,500	1,875		72,375	81,000
	2120	20 ton cooling, 360 MBH heating			.53	60.038		81,000	2,150		83,150	93,000
	2140	25 ton cooling, 450 MBH heating			.44	71.910		92,500	2,575		95,075	105,500
	2160	28 ton cooling, 450 MBH heating			.40	79.012		96,500	2,825		99,325	110,500
	2180	30 ton cooling, 540 MBH heating			.37	85.562		105,000	3,075		108,075	120,000
	2200	40 ton cooling, 540 MBH heating			.28	113		109,500	4,075		113,575	126,500
	2210	50 ton cooling, 540 MBH heating			.22	142		112,000	5,100		117,100	130,500
	2220	70 ton cooling, 1500 MBH heating			.16	198		118,000	7,125		125,125	140,500
	2240	80 ton cooling, 1500 MBH heating			.14	228		123,500	8,200		131,700	148,500
	2260	90 ton cooling, 1500 MBH heating			.13	256		129,500	9,175		138,675	156,500
	2280	105 ton cooling, 1500 MBH heating			.11	290		151,000	10,400		161,400	182,500
	2400	For hot water heat coil, deduct						5%				
	2500	For steam heat coil, deduct						2%				
	2600	For electric heat, deduct						3%	5%			
	4000	Single zone, electric cool, gas heat, variable air volume										
	4100	12.5 ton cooling, 230 MBH heating		Q-6	.55	43.716	Ea.	15,300	1,525		16,825	19,100
	4120	18 ton cooling, 330 MBH heating		"	.45	52.747		24,900	1,850		26,750	30,200
	4140	25 ton cooling, 450 MBH heating		Q-7	.48	66.116		37,800	2,375		40,175	45,200
	4160	40 ton cooling, 675 MBH heating			.31	104		65,000	3,750		68,750	77,000
	4180	60 ton cooling, 900 MBH heating			.20	157		91,000	5,650		96,650	108,500
	4200	80 ton cooling, 1000 MBH heating			.15	209		142,500	7,500		150,000	168,000
	5000	Single zone, electric cool only										
	5050	3 ton cooling		Q-5	.88	18.203	Ea.	3,125	615		3,740	4,375
	5060	4 ton cooling			.76	21.108		3,700	715		4,415	5,150
	5070	5 ton cooling			.70	22.792		4,325	770		5,095	5,925
	5080	6 ton cooling			.65	24.502		5,300	830		6,130	7,100
	5090	7.5 ton cooling			.62	25.806		7,275	875		8,150	9,325
	5100	8.5 ton cooling			.59	26.981		8,225	915		9,140	10,400
	5110	10 ton cooling		Q-6	.83	28.812		9,525	1,000		10,525	12,000
	5120	12 ton cooling			.80	30		12,300	1,050		13,350	15,100
	5130	15 ton cooling			.71	33.613		16,900	1,175		18,075	20,400
	5140	18 ton cooling			.65	36.697		20,700	1,300		22,000	24,800

MECHANICAL 15

	15730	Unitary Air Conditioning Equip	CREW	DAILY OUTPUT	LABOR-HOURS	UNIT	2003 BARE COSTS				TOTAL INCL O&P	
							MAT.	LABOR	EQUIP.	TOTAL		
600	5150	20 ton cooling **D3030 202**	Q-7	.83	38.415	Ea.	23,900	1,375		25,275	28,400	600
	5160	25 ton cooling		.70	45.977		29,700	1,650		31,350	35,100	
	5170	30 ton cooling **D3030 206**		.58	54.701		38,600	1,950		40,550	45,500	
	5180	40 ton cooling		.44	73.059		51,000	2,625		53,625	60,000	
	5400	For low heat, add					7%					
	5410	For high heat, add **R15700 -010**					10%					
	6000	Single zone electric cooling w/variable vol. distribution **R15700 -020**										
	6020	20 ton cooling	Q-7	.72	44.199	Ea.	27,300	1,575		28,875	32,400	
	6030	25 ton cooling		.60	52.893		33,000	1,900		34,900	39,200	
	6040	30 ton cooling		.51	62.868		42,000	2,250		44,250	49,600	
	6050	40 ton cooling		.38	83.989		54,500	3,000		57,500	64,000	
	6060	50 ton cooling		.31	104		59,000	3,750		62,750	70,500	
	6070	60 ton cooling		.25	125		66,000	4,525		70,525	80,000	
	6200	For low heat, add					7%					
	6210	For high heat, add					10%					
	7000	Multizone, cool/heat, variable volume distribution										
	7100	50 ton cooling	Q-7	.19	164	Ea.	62,000	5,875		67,875	77,000	
	7110	70 ton cooling		.14	228		87,000	8,200		95,200	108,000	
	7120	90 ton cooling		.11	296		94,500	10,600		105,100	120,000	
	7130	105 ton cooling		.10	333		110,500	12,000		122,500	139,500	
	7140	120 ton cooling		.08	380		126,000	13,700		139,700	159,000	
	7150	140 ton cooling		.07	444		145,500	15,900		161,400	184,000	
	7300	Penthouse unit, cool/heat, variable volume distribution										
	7400	150 ton cooling	Q-7	.10	310	Ea.	225,000	11,100		236,100	264,500	
	7410	170 ton cooling		.10	320		245,000	11,500		256,500	287,000	
	7420	200 ton cooling		.10	326		286,000	11,700		297,700	332,000	
	7430	225 ton cooling		.09	340		322,000	12,200		334,200	372,500	
	7440	250 ton cooling		.09	359		322,500	12,900		335,400	374,500	
	7450	270 ton cooling		.09	363		348,500	13,000		361,500	402,500	
	7460	300 ton cooling		.09	372		387,000	13,300		400,300	445,500	
	9400	Sub assemblies for assembly systems										
	9410	Duct work per ton roof top 1 zone units	Q-9	.96	16.667	Ton	233	555		788	1,100	
	9420	Ductwork per ton rooftop multizone units		.50	32		310	1,075		1,385	2,000	
	9430	Ductwork, VAV cooling/ton rooftop multizone, 1 zone/ton		.32	50		420	1,675		2,095	3,050	
	9440	Duct work per ton packaged water & air cooled units		1.26	12.698		55	425		480	715	
	9450	Duct work per ton split system remote condensing units		1.32	12.121		52	405		457	680	
800	0010	**WINDOW UNIT AIR CONDITIONERS**										800
	4000	Portable/window, 15 amp 125V grounded receptacle required										
	4060	5000 BTUH	1 Carp	8	1	Ea.	296	31.50		327.50	375	
	4340	6000 BTUH		8	1		345	31.50		376.50	430	
	4480	8000 BTUH		6	1.333		400	42		442	505	
	4500	10,000 BTUH		6	1.333		470	42		512	585	
	4520	12,000 BTUH	L-2	8	2		510	55.50		565.50	650	
	4600	Window/thru-the-wall, 15 amp 230V grounded receptacle required										
	4780	17,000 BTUH	L-2	6	2.667	Ea.	740	73.50		813.50	925	
	4940	25,000 BTUH		4	4		1,050	111		1,161	1,350	
	4960	29,000 BTUH		4	4		1,250	111		1,361	1,550	
840	0010	**SELF-CONTAINED SINGLE PACKAGE** **D3030 210**										840
	0100	Air cooled, for free blow or duct, not incl. remote condenser										
	0110	Constant volume										
	0200	3 ton cooling **D3030 214**	Q-5	1	16	Ea.	4,825	540		5,365	6,125	
	0210	4 ton cooling	"	.80	20		5,250	675		5,925	6,800	
	0220	5 ton cooling	Q-6	1.20	20		5,675	700		6,375	7,300	
	0230	7.5 ton cooling	"	.90	26.667		7,700	935		8,635	9,900	
	0240	10 ton cooling	Q-7	1	32		10,200	1,150		11,350	12,900	

Important: See the Reference Section for critical supporting data - Reference Nos., Crews, & City Cost Indexes

			DAILY	LABOR-		2003 BARE COSTS				TOTAL	
	15730 | Unitary Air Conditioning Equip	CREW	OUTPUT	HOURS	UNIT	MAT.	LABOR	EQUIP.	TOTAL	INCL O&P	
840	0250	15 ton cooling	D3030 210	Q-7	.95	33.684	Ea.	17,000	1,200		18,200	20,500	**840**
	0260	20 ton cooling			.90	35.556		23,600	1,275		24,875	27,900	
	0270	25 ton cooling	D3030 214		.85	37.647		23,900	1,350		25,250	28,400	
	0280	30 ton cooling			.80	40		24,600	1,425		26,025	29,300	
	0300	40 ton cooling			.60	53.333		34,100	1,900		36,000	40,500	
	0320	50 ton cooling			.50	64		41,900	2,300		44,200	49,600	
	0340	60 ton cooling		Q-8	.40	80		49,000	2,875	196	52,071	58,500	
	0380	With hot water heat, variable air volume											
	0390	10 ton cooling		Q-7	.96	33.333	Ea.	13,100	1,200		14,300	16,200	
	0400	20 ton cooling			.86	37.209		29,200	1,325		30,525	34,100	
	0410	30 ton cooling			.76	42.105		31,600	1,500		33,100	37,100	
	0420	40 ton cooling			.58	55.172		44,300	1,975		46,275	51,500	
	0430	50 ton cooling			.48	66.667		53,500	2,400		55,900	62,500	
	0440	60 ton cooling		Q-8	.38	84.211		65,000	3,025	206	68,231	76,500	
	0490	For duct mounting no price change											
	0500	For steam heating coils, add					Ea.	10%	10%				
	0550	Packaged, with electric heat											
	0552	Constant volume											
	0560	5 ton cooling		Q-6	1.20	20	Ea.	8,075	700		8,775	9,950	
	0570	10 ton cooling		Q-7	1	32		13,900	1,150		15,050	17,000	
	0580	20 ton cooling			.90	35.556		28,900	1,275		30,175	33,700	
	0590	30 ton cooling			.80	40		33,500	1,425		34,925	39,000	
	0600	40 ton cooling			.60	53.333		43,700	1,900		45,600	51,000	
	0610	50 ton cooling			.50	64		55,000	2,300		57,300	64,000	
	0700	Variable air volume											
	0710	10 ton cooling		Q-7	1	32	Ea.	15,800	1,150		16,950	19,000	
	0720	20 ton cooling			.90	35.556		32,100	1,275		33,375	37,200	
	0730	30 ton cooling			.80	40		38,000	1,425		39,425	44,000	
	0740	40 ton cooling			.60	53.333		50,500	1,900		52,400	58,500	
	0750	50 ton cooling			.50	64		62,500	2,300		64,800	72,500	
	0760	60 ton cooling		Q-8	.40	80		72,000	2,875	196	75,071	83,500	
	1000	Water cooled for free blow or duct, not including tower											
	1010	Constant volume											
	1100	3 ton cooling		Q-6	1	24	Ea.	4,675	840		5,515	6,425	
	1120	5 ton cooling			1	24		6,025	840		6,865	7,900	
	1130	7.5 ton cooling			.80	30		8,500	1,050		9,550	11,000	
	1140	10 ton cooling		Q-7	.90	35.556		12,200	1,275		13,475	15,300	
	1150	15 ton cooling			.85	37.647		17,000	1,350		18,350	20,800	
	1160	20 ton cooling			.80	40		20,300	1,425		21,725	24,500	
	1170	25 ton cooling			.75	42.667		23,900	1,525		25,425	28,600	
	1180	30 ton cooling			.70	45.714		26,900	1,650		28,550	32,100	
	1200	40 ton cooling			.40	80		34,100	2,875		36,975	41,800	
	1220	50 ton cooling			.30	106		42,400	3,825		46,225	52,500	
	1240	60 ton cooling		Q-8	.30	106		49,500	3,825	261	53,586	60,500	
	1300	For hot water or steam heat coils, add						12%	10%				
	2000	Water cooled with electric heat, not including tower											
	2010	Constant volume											
	2100	5 ton cooling		Q-6	1.20	20	Ea.	8,425	700		9,125	10,300	
	2120	10 ton cooling		Q-7	1	32		15,900	1,150		17,050	19,200	
	2130	15 ton cooling		"	.90	35.556		22,200	1,275		23,475	26,300	
	2400	Variable air volume											
	2440	10 ton cooling		Q-7	1	32	Ea.	17,800	1,150		18,950	21,200	
	2450	15 ton cooling		"	.90	35.556	"	25,500	1,275		26,775	29,900	
	2600	Water cooled, hot water coils, not including tower											
	2610	Variable air volume											
	2640	10 ton cooling		Q-7	1	32	Ea.	15,300	1,150		16,450	18,500	

MECHANICAL 15

		15730	Unitary Air Conditioning Equip	CREW	DAILY OUTPUT	LABOR-HOURS	UNIT	MAT.	2003 BARE COSTS LABOR	EQUIP.	TOTAL	TOTAL INCL O&P	
840	2650		15 ton cooling	Q-7	.80	40	Ea.	21,900	1,425		23,325	26,300	840
900	0010	**SPLIT DUCTLESS SYSTEM**											900
	0100		Cooling only, single zone										
	0110		Wall mount										
	0120		3/4 ton cooling	Q-5	2	8	Ea.	1,425	271		1,696	1,950	
	0130		1 ton cooling		1.80	8.889		1,675	300		1,975	2,300	
	0140		1-1/2 ton cooling		1.60	10		2,300	340		2,640	3,025	
	0150		2 ton cooling	▼	1.40	11.429	▼	3,175	385		3,560	4,075	
	1000		Ceiling mount										
	1020		2 ton cooling	Q-5	1.40	11.429	Ea.	3,150	385		3,535	4,025	
	1030		3 ton cooling	"	1.20	13.333	"	4,975	450		5,425	6,150	
	2000		T-Bar mount										
	2010		2 ton cooling	Q-5	1.40	11.429	Ea.	3,700	385		4,085	4,650	
	2020		3 ton cooling		1.20	13.333		4,450	450		4,900	5,550	
	2030		3-1/2 ton cooling	▼	1.10	14.545	▼	5,350	490		5,840	6,625	
	3000		Multizone										
	3010		Wall mount										
	3020		2 @ 3/4 ton cooling	Q-5	1.80	8.889	Ea.	4,450	300		4,750	5,325	
	5000		Cooling / Heating										
	5010		Wall mount										
	5110		1 ton cooling	Q-5	1.70	9.412	Ea.	2,175	320		2,495	2,875	
	5120		1-1/2 ton cooling	"	1.50	10.667	"	2,925	360		3,285	3,775	
	5300		Ceiling mount										
	5310		3 ton cooling	Q-5	1	16	Ea.	5,750	540		6,290	7,150	
	7000		Accessories for all split ductless systems										
	7010		Add for ambient frost control	Q-5	8	2	Ea.	182	67.50		249.50	300	
	7020		Add for tube / wiring kit										
	7030		15' kit	Q-5	32	.500	Ea.	67.50	16.95		84.45	100	
	7040		35' kit	"	24	.667	"	134	22.50		156.50	181	

		15740	**Heat Pumps**										
100	0010	**HEAT PUMPS** (Not including interconnecting tubing)											100
	1000		Air to air, split system, not including curbs, pads, or ductwork										
	1010		For curbs/pads see division 15705										
	1015		1.5 ton cooling, 7 MBH heat @ 0°F	Q-5	1.22	13.115	Ea.	1,225	445		1,670	2,025	
	1020		2 ton cooling, 8.5 MBH heat @ 0°F		1.20	13.333		1,275	450		1,725	2,075	
	1030		2.5 ton cooling, 10 MBH heat @ 0°F		1	16		1,425	540		1,965	2,375	
	1040		3 ton cooling, 13 MBH heat @ 0°F		.80	20		1,575	675		2,250	2,750	
	1050		3.5 ton cooling, 18 MBH heat @ 0°F		.75	21.333		1,775	720		2,495	3,050	
	1054		4 ton cooling, 24 MBH heat @ 0°F		.60	26.667		2,000	905		2,905	3,550	
	1060		5 ton cooling, 27 MBH heat @ 0°F		.50	32		2,300	1,075		3,375	4,150	
	1080		7.5 ton cooling, 33 MBH heat @ 0°F	▼	.30	53.333		7,775	1,800		9,575	11,300	
	1100		10 ton cooling, 50 MBH heat @ 0°F	Q-6	.38	63.158		10,800	2,225		13,025	15,300	
	1120		15 ton cooling, 64 MBH heat @ 0°F		.26	92.308		14,100	3,250		17,350	20,400	
	1130		20 ton cooling, 85 MBH heat @ 0°F		.20	120		19,700	4,200		23,900	28,100	
	1140		25 ton cooling, 119 MBH heat @ 0°F	▼	.20	120	▼	23,400	4,200		27,600	32,100	
	1300		Supplementary electric heat coil, included										
	1500		Single package, not including curbs, pads, or plenums										
	1502		1/2 ton cooling, supplementary heat not incl.	Q-5	8	2	Ea.	1,200	67.50		1,267.50	1,425	
	1504		3/4 ton cooling, supplementary heat not incl.		6	2.667		1,300	90.50		1,390.50	1,550	
	1506		1 ton cooling, supplementary heat not incl.		4	4		1,375	135		1,510	1,725	
	1510		1.5 ton cooling, 5 MBH heat @ 0°F		1.55	10.323		3,100	350		3,450	3,925	
	1520		2 ton cooling, 6.5 MBH heat @ 0°F		1.50	10.667		3,175	360		3,535	4,025	
	1540		2.5 ton cooling, 8 MBH heat @ 0°F		1.40	11.429		3,325	385		3,710	4,225	
	1560		3 ton cooling, 10 MBH heat @ 0°F	▼	1.20	13.333	▼	3,625	450		4,075	4,650	

15 MECHANICAL

15740	Heat Pumps	CREW	DAILY OUTPUT	LABOR-HOURS	UNIT	MAT.	LABOR	EQUIP.	TOTAL	TOTAL INCL O&P	
100	1570	3.5 ton cooling, 11 MBH heat @ 0°F	Q-5	1	16	Ea.	3,950	540		4,490	5,150	**100**
	1580	4 ton cooling, 13 MBH heat @ 0°F		.96	16.667		4,275	565		4,840	5,550	
	1620	5 ton cooling, 27 MBH heat @ 0°F		.65	24.615		4,875	835		5,710	6,600	
	1640	7.5 ton cooling, 35 MBH heat @ 0°F	↓	.40	40		8,025	1,350		9,375	10,900	
	1648	10 ton cooling, 45 MBH heat @ 0°F	Q-6	.40	60		10,800	2,100		12,900	15,100	
	1652	12 ton cooling, 50 MBH heat @ 0°F	"	.36	66.667	↓	13,200	2,350		15,550	18,000	
	1696	Supplementary electric heat coil incl., except as noted										
	1800	Rooftop, with curb, economizer and supplementary electric heat										
	1820	10 ton cooling	Q-6	.40	60	Ea.	36,100	2,100		38,200	43,000	
	1830	20 ton cooling	"	.20	120		51,500	4,200		55,700	63,500	
	1840	30 ton cooling	Q-7	.19	168		69,500	6,050		75,550	85,500	
	1850	40 ton cooling		.17	188		77,500	6,750		84,250	95,000	
	1860	50 ton cooling		.15	213		90,500	7,650		98,150	111,000	
	1870	60 ton cooling	↓	.13	246	↓	95,500	8,825		104,325	118,500	
	2000	Water source to air, single package										
	2100	1 ton cooling, 13 MBH heat @ 75°F	Q-5	2	8	Ea.	1,725	271		1,996	2,300	
	2120	1.5 ton cooling, 17 MBH heat @ 75°F		1.80	8.889		2,050	300		2,350	2,700	
	2140	2 ton cooling, 19 MBH heat @ 75°F		1.70	9.412		2,100	320		2,420	2,800	
	2160	2.5 ton cooling, 25 MBH heat @ 75°F		1.60	10		2,175	340		2,515	2,900	
	2180	3 ton cooling, 27 MBH heat @ 75°F		1.40	11.429		2,375	385		2,760	3,175	
	2190	3.5 ton cooling, 29 MBH heat @ 75°F		1.30	12.308		2,450	415		2,865	3,325	
	2200	4 ton cooling, 31 MBH heat @ 75°F		1.20	13.333		2,850	450		3,300	3,825	
	2220	5 ton cooling, 29 MBH heat @ 75°F		.90	17.778		3,450	600		4,050	4,700	
	2240	7.5 ton cooling, 35 MBH heat @ 75°F		.60	26.667		3,900	905		4,805	5,675	
	2250	8.5 ton cooling, 40 MBH heat @ 75°F		.58	27.586		11,200	935		12,135	13,700	
	2260	10 ton cooling, 50 MBH heat @ 75°F	↓	.53	30.189		12,300	1,025		13,325	15,100	
	2280	15 ton cooling, 64 MBH heat @ 75°F	Q-6	.47	51.064		13,500	1,800		15,300	17,600	
	2300	20 ton cooling, 100 MBH heat @ 75°F		.41	58.537		15,100	2,050		17,150	19,700	
	2320	30 ton cooling, (twin 15 ton units)		.23	102		27,000	3,575		30,575	35,100	
	2340	40 ton cooling, (twin 20 ton units)		.20	117		30,200	4,100		34,300	39,400	
	2360	50 ton cooling, (twin 15 + 20 ton unit)	↓	.15	160		42,100	5,625		47,725	55,000	
	3960	For supplementary heat coil, add				↓	10%					
	4000	For increase in capacity thru use										
	4020	of solar collector, size boiler at 60%										

15750	Humidity Control Equipment										
300	0010	**DEHUMIDIFIERS**										**300**
	6000	Selfcontained with filters and standard controls										
	6040	1.5 lb/hr, 50 cfm	1 Plum	8	1	Ea.	2,950	37.50		2,987.50	3,300	
	6060	3 lb/hr, 150 cfm	Q-1	12	1.333		3,350	45		3,395	3,775	
	6065	6 lb/hr, 150 cfm		9	1.778		6,200	60		6,260	6,925	
	6070	16 to 20 lb/hr, 600 cfm		5	3.200		12,400	108		12,508	13,800	
	6080	30 to 40 lb/hr, 1125 cfm		4	4		20,400	135		20,535	22,700	
	6090	60 to 75 lb/hr, 2250 cfm		3	5.333		26,800	179		26,979	29,800	
	6100	120 to 155 lb/hr, 4500 cfm		2	8		49,000	269		49,269	54,500	
	6110	240 to 310 lb/hr, 9000 cfm	↓	1.50	10.667		69,000	360		69,360	76,500	
	6120	400 to 515 lb/hr, 15,000 cfm	Q-2	1.60	15		94,000	525		94,525	104,000	
	6130	530 to 690 lb/hr, 20,000 cfm		1.40	17.143		101,500	600		102,100	113,000	
	6140	800 to 1030 lb/hr, 30,000 cfm		1.20	20		116,000	695		116,695	128,500	
	6150	1060 to 1375 lb/hr, 40,000 cfm	↓	1	24	↓	161,500	835		162,335	179,500	
500	0010	**HUMIDIFIERS**										**500**
	0520	Steam, room or duct, filter, regulators, auto. controls, 220 V										
	0540	11 lb. per hour	Q-5	6	2.667	Ea.	2,050	90.50		2,140.50	2,375	
	0560	22 lb. per hour		5	3.200		2,250	108		2,358	2,650	
	0580	33 lb. per hour		4	4		2,325	135		2,460	2,750	
	0600	50 lb. per hour	↓	4	4	↓	2,850	135		2,985	3,325	

MECHANICAL 15

	15750	Humidity Control Equipment	CREW	DAILY OUTPUT	LABOR-HOURS	UNIT	2003 BARE COSTS				TOTAL INCL O&P	
							MAT.	LABOR	EQUIP.	TOTAL		
500	0620	100 lb. per hour	Q-5	3	5.333	Ea.	3,400	181		3,581	4,025	500
	0640	150 lb. per hour		2.50	6.400		4,500	217		4,717	5,275	
	0660	200 lb. per hour	▼	2	8	▼	5,575	271		5,846	6,550	
	0700	With blower										
	0720	11 lb. per hour	Q-5	5.50	2.909	Ea.	2,850	98.50		2,948.50	3,300	
	0740	22 lb. per hour		4.75	3.368		3,075	114		3,189	3,550	
	0760	33 lb. per hour		3.75	4.267		3,125	144		3,269	3,675	
	0780	50 lb. per hour		3.50	4.571		3,775	155		3,930	4,375	
	0800	100 lb. per hour		2.75	5.818		4,250	197		4,447	4,975	
	0820	150 lb. per hour		2	8		6,300	271		6,571	7,325	
	0840	200 lb. per hour	▼	1.50	10.667	▼	7,375	360		7,735	8,675	
	1000	Steam for duct installation, manifold with pneum. controls										
	1010	3 - 191 lb/hr, 24" x 24"	Q-5	5	3.200	Ea.	1,775	108		1,883	2,125	
	1020	3 - 191 lb/hr, 48" x 48"		4.60	3.478		2,275	118		2,393	2,675	
	1030	65 - 334 lb/hr, 36" x 36"		4.80	3.333		1,800	113		1,913	2,175	
	1040	65 - 334 lb/hr, 72" x 72"		4	4		3,075	135		3,210	3,575	
	1050	100 - 690 lb/hr, 48" x 48"		4	4		3,150	135		3,285	3,650	
	1060	100 - 690 lb/hr, 84" x 84"		3.60	4.444		5,650	150		5,800	6,425	
	1070	220 - 2000 lb/hr, 72" x 72"		3.80	4.211		4,175	143		4,318	4,825	
	1080	220 - 2000 lb/hr, 144" x 144"	▼	3.40	4.706		8,700	159		8,859	9,800	
	1090	For electrically controlled unit, add				▼	700			700	770	
	1092	For additional manifold, add					25%	15%				
	1200	Electric steam, self generating, with manifold										
	1210	1 - 96 lb/hr, 12" x 72"	Q-5	3.60	4.444	Ea.	5,050	150		5,200	5,775	
	1220	20 - 168 lb/hr, 12" x 72"		3.60	4.444		5,050	150		5,200	5,775	
	1290	For room mounting, add	▼	16	1	▼	146	34		180	212	
	5000	Furnace type, wheel bypass.										
	5020	10 GPD	1 Stpi	4	2	Ea.	129	75		204	255	
	5040	14 GPD		3.80	2.105		171	79		250	310	
	5060	19 GPD	▼	3.60	2.222	▼	198	83.50		281.50	345	

	15761	Air Coils										
200	0010	**DUCT HEATERS** Electric, 480 V, 3 Ph										200
	0020	Finned tubular insert, 500°F										
	0100	8" wide x 6" high, 4.0 kW	Q-20	16	1.250	Ea.	525	42.50		567.50	645	
	0120	12" high, 8.0 kW		15	1.333		875	45.50		920.50	1,025	
	0140	18" high, 12.0 kW		14	1.429		1,225	49		1,274	1,425	
	0160	24" high, 16.0 kW		13	1.538		1,575	52.50		1,627.50	1,800	
	0180	30" high, 20.0 kW		12	1.667		1,925	57		1,982	2,225	
	0300	12" wide x 6" high, 6.7 kW		15	1.333		560	45.50		605.50	685	
	0320	12" high, 13.3 kW		14	1.429		905	49		954	1,075	
	0340	18" high, 20.0 kW		13	1.538		1,275	52.50		1,327.50	1,475	
	0360	24" high, 26.7 kW		12	1.667		1,625	57		1,682	1,900	
	0380	30" high, 33.3 kW		11	1.818		2,000	62		2,062	2,300	
	0500	18" wide x 6" high, 13.3 kW		14	1.429		600	49		649	735	
	0520	12" high, 26.7 kW		13	1.538		1,025	52.50		1,077.50	1,225	
	0540	18" high, 40.0 kW		12	1.667		1,375	57		1,432	1,625	
	0560	24" high, 53.3 kW		11	1.818		1,825	62		1,887	2,125	
	0580	30" high, 66.7 kW		10	2		2,300	68.50		2,368.50	2,625	
	0700	24" wide x 6" high, 17.8 kW		13	1.538		660	52.50		712.50	805	
	0720	12" high, 35.6 kW		12	1.667		1,125	57		1,182	1,325	
	0740	18" high, 53.3 kW		11	1.818		1,575	62		1,637	1,825	
	0760	24" high, 71.1 kW		10	2		2,025	68.50		2,093.50	2,325	
	0780	30" high, 88.9 kW		9	2.222		2,500	76		2,576	2,875	
	0900	30" wide x 6" high, 22.2 kW		12	1.667		700	57		757	860	
	0920	12" high, 44.4 kW	▼	11	1.818	▼	1,175	62		1,237	1,400	

Important: See the Reference Section for critical supporting data - Reference Nos., Crews, & City Cost Indexes

			DAILY	LABOR-		2003 BARE COSTS				TOTAL	
	15761	Air Coils	CREW	OUTPUT	HOURS	UNIT	MAT.	LABOR	EQUIP.	TOTAL	INCL O&P	
200	0940	18" high, 66.7 kW	Q-20	10	2	Ea.	1,675	68.50		1,743.50	1,925	200
	0960	24" high, 88.9 kW		9	2.222		2,150	76		2,226	2,500	
	0980	30" high, 111.0 kW		8	2.500		2,625	85.50		2,710.50	3,025	
	1400	Note decreased kW available for										
	1410	each duct size at same cost										
	1420	See line 5000 for modifications and accessories										
	2000	Finned tubular flange with insulated										
	2020	terminal box, 500°F										
	2100	12" wide x 36" high, 54 kW	Q-20	10	2	Ea.	2,600	68.50		2,668.50	2,950	
	2120	40" high, 60 kW		9	2.222		2,975	76		3,051	3,400	
	2200	24" wide x 36" high, 118.8 kW		9	2.222		2,875	76		2,951	3,300	
	2220	40" high, 132 kW		8	2.500		3,300	85.50		3,385.50	3,750	
	2400	36" wide x 8" high, 40 kW		11	1.818		1,325	62		1,387	1,550	
	2420	16" high, 80 kW		10	2		1,775	68.50		1,843.50	2,050	
	2440	24" high, 120 kW		9	2.222		2,300	76		2,376	2,675	
	2460	32" high, 160 kW		8	2.500		2,950	85.50		3,035.50	3,375	
	2480	36" high, 180 kW		7	2.857		3,650	97.50		3,747.50	4,150	
	2500	40" high, 200 kW		6	3.333		4,025	114		4,139	4,625	
	2600	40" wide x 8" high, 45 kW		11	1.818		1,375	62		1,437	1,625	
	2620	16" high, 90 kW		10	2		1,900	68.50		1,968.50	2,200	
	2640	24" high, 135 kW		9	2.222		2,400	76		2,476	2,775	
	2660	32" high, 180 kW		8	2.500		3,200	85.50		3,285.50	3,650	
	2680	36" high, 202.5 kW		7	2.857		3,700	97.50		3,797.50	4,225	
	2700	40" high, 225 kW		6	3.333		4,200	114		4,314	4,800	
	2800	48" wide x 8" high, 54.8 kW		10	2		1,450	68.50		1,518.50	1,700	
	2820	16" high, 109.8 kW		9	2.222		2,000	76		2,076	2,325	
	2840	24" high, 164.4 kW		8	2.500		2,550	85.50		2,635.50	2,925	
	2860	32" high, 219.2 kW		7	2.857		3,400	97.50		3,497.50	3,900	
	2880	36" high, 246.6 kW		6	3.333		3,925	114		4,039	4,475	
	2900	40" high, 274 kW		5	4		4,425	137		4,562	5,075	
	3000	56" wide x 8" high, 64 kW		9	2.222		1,650	76		1,726	1,925	
	3020	16" high, 128 kW		8	2.500		2,275	85.50		2,360.50	2,625	
	3040	24" high, 192 kW		7	2.857		2,775	97.50		2,872.50	3,200	
	3060	32" high, 256 kW		6	3.333		3,875	114		3,989	4,425	
	3080	36" high, 288kW		5	4		4,425	137		4,562	5,075	
	3100	40" high, 320kW		4	5		4,900	171		5,071	5,650	
	3200	64" wide x 8" high, 74kW		8	2.500		1,700	85.50		1,785.50	2,000	
	3220	16" high, 148kW		7	2.857		2,350	97.50		2,447.50	2,725	
	3240	24" high, 222kW		6	3.333		3,000	114		3,114	3,450	
	3260	32" high, 296kW		5	4		4,000	137		4,137	4,600	
	3280	36" high, 333kW		4	5		4,775	171		4,946	5,500	
	3300	40" high, 370kW		3	6.667		5,275	228		5,503	6,150	
	3800	Note decreased kW available for										
	3820	each duct size at same cost										
	5000	Duct heater modifications and accessories										
	5120	T.C.O. limit auto or manual reset	Q-20	42	.476	Ea.	86.50	16.25		102.75	120	
	5140	Thermostat		28	.714		370	24.50		394.50	445	
	5160	Overheat thermocouple (removable)		7	2.857		520	97.50		617.50	725	
	5180	Fan interlock relay		18	1.111		126	38		164	197	
	5200	Air flow switch		20	1		109	34		143	173	
	5220	Split terminal box cover		100	.200		36	6.85		42.85	50	
	8000	To obtain BTU multiply kW by 3413										
700	0010	**COILS, FLANGED**										700
	0100	Basic water, DX, or condenser coils										
	0110	Copper tubes, alum. fins, galv. end sheets										
	0112	H is finned height, L is finned length										

MECHANICAL 15

	15761 | **Air Coils**	CREW	DAILY OUTPUT	LABOR-HOURS	UNIT	2003 BARE COSTS				TOTAL INCL O&P	
						MAT.	LABOR	EQUIP.	TOTAL		
700	0120	3/8" x .016 tube, .0065 aluminum fins										**700**
	0130	2 row, 8 fins per inch									
	0140	4" H x 12" L	Q-5	48	.333	Ea.	278	11.30		289.30	320
	0150	4" H x 24" L		38.38	.417		285	14.10		299.10	335
	0160	4" H x 48" L		19.25	.831		300	28		328	375
	0170	4" H x 72" L		12.80	1.250		415	42.50		457.50	520
	0180	6" H x 12" L		48	.333		286	11.30		297.30	330
	0190	6" H x 24" L		25.60	.625		296	21		317	355
	0200	6" H x 48" L		12.80	1.250		315	42.50		357.50	410
	0210	6" H x 72" L		8.53	1.876		435	63.50		498.50	575
	0220	10" H x 12" L		30.73	.521		305	17.60		322.60	360
	0230	10" H x 24" L		15.33	1.044		315	35.50		350.50	405
	0240	10" H x 48" L		7.69	2.081		340	70.50		410.50	480
	0250	10" H x 72" L		5.12	3.125		480	106		586	685
	0260	12" H x 12" L		25.60	.625		310	21		331	375
	0270	12" H x 24" L		12.80	1.250		325	42.50		367.50	425
	0280	12" H x 48" L		6.40	2.500		355	84.50		439.50	520
	0290	12" H x 72" L		4.27	3.747		500	127		627	740
	0300	16" H x 16" L		14.38	1.113		335	37.50		372.50	425
	0310	16" H x 24" L		9.59	1.668		350	56.50		406.50	465
	0320	16" H x 48" L		4.80	3.333		435	113		548	650
	0330	16" H x 72" L		3.20	5		540	169		709	850
	0340	20" H x 24" L		7.69	2.081		370	70.50		440.50	510
	0350	20" H x 48" L		3.84	4.167		470	141		611	730
	0360	20" H x 72" L		2.56	6.250		585	212		797	965
	0370	24" H x 24" L		6.40	2.500		390	84.50		474.50	560
	0380	24" H x 48" L		3.20	5		500	169		669	805
	0390	24" H x 72" L		2.13	7.512		625	254		879	1,075
	0400	30" H x 28" L		4.39	3.645		475	123		598	705
	0410	30" H x 48" L		2.56	6.250		550	212		762	925
	0420	30" H x 72" L		1.71	9.357		690	315		1,005	1,250
	0430	30" H x 84" L		1.46	10.959		765	370		1,135	1,400
	0440	34" H x 32" L		3.39	4.720		515	160		675	810
	0450	34" H x 48" L		2.26	7.080		585	240		825	1,000
	0460	34" H x 72" L		1.51	10.596		735	360		1,095	1,350
	0470	34" H x 84" L	▼	1.29	12.403	▼	805	420		1,225	1,525
	0600	For 10 fins per inch, add					1%				
	0610	For 12 fins per inch, add					3%				
	0620	For 4 row, add				Ea.	30%	50%			
	0630	For 6 row, add					60%	100%			
	0640	For 8 row, add				▼	80%	150%			
	1100	1/2" x .017 tube, .0065 aluminum fins									
	1110	2 row, 8 fins per inch									
	1120	5" H x 5" L	Q-5	40	.400	Ea.	277	13.55		290.55	325
	1130	5" H x 20" L		37.10	.431		288	14.60		302.60	335
	1140	5" H x 45" L		16.41	.975		305	33		338	385
	1150	5" H x 90" L		8.21	1.949		490	66		556	640
	1160	7.5" H x 5" L		38.50	.416		286	14.05		300.05	335
	1170	7.5" H x 20" L		24.62	.650		300	22		322	365
	1180	7.5" H x 45" L		10.94	1.463		325	49.50		374.50	430
	1190	7.5" H x 90" L		5.46	2.930		525	99		624	725
	1200	12.5" H x 10" L		29.49	.543		310	18.35		328.35	370
	1210	12.5" H x 20" L		14.75	1.085		325	36.50		361.50	410
	1220	12.5" H x 45" L		6.55	2.443		360	82.50		442.50	520
	1230	12.5" H x 90" L		3.28	4.878		585	165		750	895
	1240	15" H x 15" L	▼	16.41	.975	▼	330	33		363	410

Important: See the Reference Section for critical supporting data - Reference Nos., Crews, & City Cost Indexes

	15761	Air Coils	CREW	DAILY OUTPUT	LABOR-HOURS	UNIT	2003 BARE COSTS				TOTAL INCL O&P	
							MAT.	LABOR	EQUIP.	TOTAL		
700	1250	15" H x 30" L	Q-5	8.21	1.949	Ea.	355	66		421	490	700
	1260	15" H x 45" L		5.46	2.930		375	99		474	565	
	1270	15" H x 90" L		2.73	5.861		620	198		818	980	
	1280	20" H x 20" L		9.22	1.735		360	58.50		418.50	490	
	1290	20" H x 45" L		4.10	3.902		465	132		597	715	
	1300	20" H x 90" L		2.05	7.805		685	264		949	1,150	
	1310	25" H x 25" L		5.90	2.712		400	92		492	580	
	1320	25" H x 45" L		3.28	4.878		505	165		670	805	
	1330	25" H x 90" L		1.64	9.756		750	330		1,080	1,325	
	1340	30" H x 30" L		4.10	3.902		485	132		617	735	
	1350	30" H x 45" L		2.73	5.861		545	198		743	900	
	1360	30" H x 90" L		1.37	11.679		815	395		1,210	1,500	
	1370	35" H x 35" L		3.01	5.316		540	180		720	865	
	1380	35" H x 45" L		2.34	6.838		590	231		821	995	
	1390	35" H x 90" L		1.17	13.675		880	465		1,345	1,675	
	1400	42.5" H x 45" L		1.93	8.290		650	281		931	1,150	
	1410	42.5" H x 75" L		1.16	13.793		855	465		1,320	1,650	
	1420	42.5" H x 105" L	▼	.83	19.277		1,050	655		1,705	2,150	
	1600	For 10 fins per inch, add					2%					
	1610	For 12 fins per inch, add					5%					
	1620	For 4 row, add					35%	50%				
	1630	For 6 row, add					65%	100%				
	1640	For 8 row, add				▼	90%	150%				
	2010	5/8" x .020" tube, .0065 aluminum fins										
	2020	2 row, 8 fins per inch										
	2030	6" H x 6" L	Q-5	38	.421	Ea.	340	14.25		354.25	395	
	2040	6" H x 24" L		24.96	.641		370	21.50		391.50	445	
	2050	6" H x 48" L		12.48	1.282		410	43.50		453.50	520	
	2060	6" H x 96" L		6.24	2.564		795	87		882	1,000	
	2070	12" H x 12" L		24.96	.641		305	21.50		326.50	370	
	2080	12" H x 24" L		12.48	1.282		330	43.50		373.50	425	
	2090	12" H x 48" L		6.24	2.564		570	87		657	760	
	2100	12" H x 96" L		3.12	5.128		1,025	174		1,199	1,375	
	2110	18" H x 24" L		8.32	1.923		520	65		585	675	
	2120	18" H x 48" L		4.16	3.846		685	130		815	950	
	2130	18" H x 96" L		2.08	7.692		1,250	260		1,510	1,775	
	2140	24" H x 24" L		6.24	2.564		635	87		722	830	
	2150	24" H x 48" L		3.12	5.128		525	174		699	835	
	2160	24" H x 96" L		1.56	10.256		1,475	345		1,820	2,150	
	2170	30" H x 30" L		3.99	4.010		770	136		906	1,050	
	2180	30" H x 48" L		2.50	6.400		585	217		802	970	
	2190	30" H x 96" L		1.25	12.800		1,700	435		2,135	2,525	
	2200	30" H x 120" L		1	16		2,125	540		2,665	3,175	
	2210	36" H x 36" L		2.77	5.776		915	196		1,111	1,300	
	2220	36" H x 48" L		2.08	7.692		1,025	260		1,285	1,525	
	2230	36" H x 90" L		1.11	14.414		1,850	490		2,340	2,775	
	2240	36" H x 120" L		.83	19.277		2,425	655		3,080	3,650	
	2250	42" H x 48" L		1.78	8.989		1,150	305		1,455	1,700	
	2260	42" H x 90" L		.95	16.842		2,075	570		2,645	3,125	
	2270	42" H x 120" L		.71	22.535		2,750	765		3,515	4,175	
	2280	51" H x 48" L		1.47	10.884		1,325	370		1,695	2,000	
	2290	51" H x 96" L		.73	21.918		2,500	740		3,240	3,875	
	2300	51" H x 120" L	▼	.59	27.119	▼	3,200	920		4,120	4,900	
	2500	For 10 fins per inch, add					3%					
	2510	For 12 fins per inch, add					6%					
	2520	For 4 row, add				Ea.	40%	50%				

MECHANICAL 15

			DAILY	LABOR-			2003 BARE COSTS			TOTAL	
15761	**Air Coils**	CREW	OUTPUT	HOURS	UNIT	MAT.	LABOR	EQUIP.	TOTAL	INCL O&P	
700 2530	For 6 row, add				Ea.	80%	100%				**700**
2540	For 8 row, add			↓		110%	150%				
3000	Hot water booster coils										
3010	Copper tubes, alum. fins, galv. end sheets										
3012	H is finned height, L is finned length										
3020	1/2" x .017" tube, .0065 Al fins										
3030	1 row, 10 fins per inch										
3040	5"H x 10"L	Q-5	40	.400	Ea.	365	13.55		378.55	425	
3050	5"H x 25"L		33.18	.482		375	16.30		391.30	435	
3060	10"H x 10"L		39	.410		375	13.90		388.90	430	
3070	10"H x 20"L		20.73	.772		380	26		406	460	
3080	10"H x 30"L		13.83	1.157		385	39		424	485	
3090	15"H x 15"L		18.44	.868		390	29.50		419.50	470	
3110	15"H x 20"L		13.83	1.157		395	39		434	495	
3120	15"H x 30"L		9.22	1.735		405	58.50		463.50	535	
3130	20"H x 20"L		10.37	1.543		405	52		457	525	
3140	20"H x 30"L		6.91	2.315		420	78.50		498.50	580	
3150	20"H x 40"L		5.18	3.089		480	105		585	690	
3160	25"H x 25"L		6.64	2.410		465	81.50		546.50	635	
3170	25"H x 35"L		4.74	3.376		490	114		604	715	
3180	25"H x 45"L		3.69	4.336		515	147		662	785	
3190	30"H x 30"L		4.61	3.471		495	117		612	720	
3200	30"H x 40"L		3.46	4.624		525	157		682	810	
3210	30"H x 50"L	↓	2.76	5.797	↓	560	196		756	910	
3300	For 2 row, add				↓	20%	50%				
3400	1/2" x .020" tube, .008 Al fins										
3410	2 row, 10 fins per inch										
3420	20" H x 27" L	Q-5	7.27	2.201	Ea.	465	74.50		539.50	625	
3430	20" H x 36" L		5.44	2.941		535	99.50		634.50	740	
3440	20" H x 45" L		4.35	3.678		565	125		690	810	
3450	25" H x 45" L		3.49	4.585		605	155		760	900	
3460	25" H x 54" L		2.90	5.517		685	187		872	1,025	
3470	30" H x 54" L		2.42	6.612		750	224		974	1,175	
3480	35" H x 54" L		2.07	7.729		800	262		1,062	1,275	
3490	35" H x 63" L		1.78	8.989		810	305		1,115	1,350	
3500	40" H x 72" L		1.36	11.765		960	400		1,360	1,650	
3510	55" H x 63" L		1.13	14.159		1,075	480		1,555	1,925	
3520	55" H x 72" L		.99	16.162		1,150	545		1,695	2,100	
3530	55" H x 81" L		.88	18.182		1,275	615		1,890	2,325	
3540	75" H x 81" L		.64	25		1,550	845		2,395	2,975	
3550	75" H x 90" L		.58	27.586		1,650	935		2,585	3,200	
3560	75" H x 108" L	↓	.48	33.333	↓	1,900	1,125		3,025	3,775	
3800	5/8" x .020" tubing, .0065 aluminum fins										
3810	1 row, 10 fins per inch										
3820	6" H x 12" L	Q-5	38	.421	Ea.	370	14.25		384.25	430	
3830	6" H x 24" L		28.80	.556		380	18.80		398.80	445	
3840	6" H x 30" L		23.04	.694		380	23.50		403.50	455	
3850	12" H x 12" L		28.80	.556		380	18.80		398.80	450	
3860	12" H x 24" L		14.40	1.111		395	37.50		432.50	490	
3870	12" H x 30" L		11.52	1.389		400	47		447	510	
3880	18" H x 18" L		12.80	1.250		400	42.50		442.50	505	
3890	18" H x 24" L		9.60	1.667		410	56.50		466.50	535	
3900	18" H x 36" L		6.40	2.500		425	84.50		509.50	600	
3910	24" H x 24" L		7.20	2.222		425	75		500	585	
3920	24" H x 30" L		4.80	3.333		465	113		578	685	
3930	24" H x 36" L	↓	5.05	3.168	↓	480	107		587	685	

Important: See the Reference Section for critical supporting data - Reference Nos., Crews, & City Cost Indexes

15761	Air Coils	CREW	DAILY OUTPUT	LABOR-HOURS	UNIT	MAT.	2003 BARE COSTS LABOR	EQUIP.	TOTAL	TOTAL INCL O&P	
700	3940	30″ H x 30″ L	Q-5	4.61	3.471	Ea.	485	117		602	710	700
	3950	30″ H x 36″ L		3.85	4.159		500	141		641	765	
	3960	30″ H x 42″ L		3.29	4.863		515	165		680	820	
	3970	36″ H x 36″ L		3.20	5		525	169		694	835	
	3980	36″ H x 42″ L	▼	2.74	5.839		540	198		738	895	
	4100	For 2 row, add				▼	30%	50%				
	4400	Steam and hot water heating coils										
	4410	Copper tubes, alum. fins. galv. end sheets										
	4412	H is finned height, L is finned length										
	4420	5/8″ tubing, to 175 psig working pressure										
	4430	1 row, 8 fins per inch										
	4440	6″H x 6″L	Q-5	38	.421	Ea.	305	14.25		319.25	355	
	4450	6″H x 12″L		37	.432		310	14.65		324.65	360	
	4460	6″H x 24″L		24.48	.654		320	22		342	385	
	4470	6″H x 36″L		16.32	.980		330	33		363	415	
	4480	6″H x 48″L		12.24	1.307		345	44.50		389.50	440	
	4490	6″H x 84″L		6.99	2.289		640	77.50		717.50	820	
	4500	6″ H x 108″ L		5.44	2.941		730	99.50		829.50	955	
	4510	12″ H x 12″ L		24.48	.654		345	22		367	415	
	4520	12″ H x 24″ L		12.24	1.307		365	44.50		409.50	465	
	4530	12″ H x 36″ L		8.16	1.961		380	66.50		446.50	520	
	4540	12″ H x 48″ L		6.12	2.614		450	88.50		538.50	630	
	4550	12″ H x 84″ L		3.50	4.571		755	155		910	1,075	
	4560	12″ H x 108″ L		2.72	5.882		890	199		1,089	1,275	
	4570	18″ H x18″ L		10.88	1.471		390	50		440	505	
	4580	18″ H x 24″ L		8.16	1.961		405	66.50		471.50	545	
	4590	18″ H x 48″ L		4.08	3.922		515	133		648	770	
	4600	18″ H x 84″ L		2.33	6.867		875	232		1,107	1,300	
	4610	18″ H x 108″ L		1.81	8.840		1,050	299		1,349	1,600	
	4620	24″ H x 24″ L		6.12	2.614		490	88.50		578.50	670	
	4630	24″ H x 48″ L		3.06	5.229		580	177		757	905	
	4640	24″ H x 84″ L		1.75	9.143		995	310		1,305	1,575	
	4650	24″ H x 108″ L		1.36	11.765		1,200	400		1,600	1,925	
	4660	36″ H x 36″ L		2.72	5.882		650	199		849	1,025	
	4670	36″ H x 48″ L		2.04	7.843		710	265		975	1,175	
	4680	36″ H x 84″ L		1.17	13.675		1,225	465		1,690	2,050	
	4690	36″ H x 108″ L		.91	17.582		1,525	595		2,120	2,575	
	4700	48″H x 48″L		1.53	10.458		850	355		1,205	1,475	
	4710	48″H x 84″L		.87	18.391		1,475	625		2,100	2,575	
	4720	48″H x 108″L		.68	23.529		1,850	795		2,645	3,250	
	4730	48″H x 126″L	▼	.58	27.586	▼	2,000	935		2,935	3,575	
	4910	For 10 fins per inch, add					2%					
	4920	For 12 fins per inch, add					5%					
	4930	For 2 row, add					30%	50%				
	4940	For 3 row, add					50%	100%				
	4950	For 4 row, add					80%	123%				
	4960	For 230 PSIG heavy duty, add					20%	30%				
	6000	Steam, heavy duty, 200 PSIG										
	6010	Copper tubes, alum. fins, galv. end sheets										
	6012	H is finned height, L is finned length										
	6020	1″ x .035″ tubing, .010 aluminum fins										
	6030	1 row, 8 fins per inch										
	6040	12″H x 12″L	Q-5	24.48	.654	Ea.	330	22		352	400	
	6050	12″H x 24″L		12.24	1.307		355	44.50		399.50	455	
	6060	12″H x 36″L		8.16	1.961		380	66.50		446.50	520	
	6070	12″H x 48″L	▼	6.12	2.614	▼	455	88.50		543.50	635	

MECHANICAL 15

			DAILY	LABOR-		2003 BARE COSTS				TOTAL	
15761	**Air Coils**	CREW	OUTPUT	HOURS	UNIT	MAT.	LABOR	EQUIP.	TOTAL	INCL O&P	
700	6080	12"H x 84"L	Q-5	3.50	4.571	Ea.	645	155		800	945	**700**
	6090	12"H x 108"L		2.72	5.882		760	199		959	1,125	
	6100	18"H x 24"L		8.16	1.961		400	66.50		466.50	540	
	6110	18"H x 36"L		5.44	2.941		485	99.50		584.50	685	
	6120	18"H x 48"L		4.08	3.922		530	133		663	785	
	6130	18"H x 84"L		2.33	6.867		760	232		992	1,175	
	6140	18"H x 108"L		1.81	8.840		905	299		1,204	1,450	
	6150	24"H x 24"L		6.12	2.614		485	88.50		573.50	670	
	6160	24"H x 36"L		4.08	3.922		545	133		678	800	
	6170	24"H x 48"L		3.06	5.229		605	177		782	930	
	6180	24"H x 84"L		1.75	9.143		870	310		1,180	1,425	
	6190	24"H x 108"L		1.36	11.765		1,050	400		1,450	1,750	
	6200	30"H x 36"L		3.26	4.908		610	166		776	920	
	6210	30"H x 48"L		2.45	6.531		675	221		896	1,075	
	6220	30"H x 84"L		1.40	11.429		980	385		1,365	1,650	
	6230	30"H x 108"L		1.09	14.679		1,200	495		1,695	2,050	
	6240	36"H x 36"L		2.72	5.882		670	199		869	1,025	
	6250	36"H x 48"L		2.04	7.843		750	265		1,015	1,225	
	6260	36"H x 84"L		1.17	13.675		1,100	465		1,565	1,900	
	6270	36"H x 108"L		.91	17.582		1,325	595		1,920	2,375	
	6280	36"H x 120"L		.82	19.512		1,425	660		2,085	2,550	
	6290	48"H x 48"L		1.53	10.458		905	355		1,260	1,525	
	6300	48"H x 84"L		.87	18.391		1,325	625		1,950	2,400	
	6310	48"H x 108"L		.68	23.529		1,625	795		2,420	2,975	
	6320	48"H x 120"L	↓	.61	26.230		1,725	890		2,615	3,250	
	6400	For 10 fins per inch, add					4%					
	6410	For 12 fins per inch, add				↓	10%					

			DAILY	LABOR-		2003 BARE COSTS				TOTAL	
15765	**Fan Coil Unit/Unit Ventilators**	CREW	OUTPUT	HOURS	UNIT	MAT.	LABOR	EQUIP.	TOTAL	INCL O&P	
200	0010	**FAN COIL AIR CONDITIONING** Cabinet mounted, filters, controls										**200**
	0100	Chilled water, 1/2 ton cooling	Q-5	8	2	Ea.	765	67.50		832.50	940	
	0110	3/4 ton cooling		7	2.286		805	77.50		882.50	1,000	
	0120	1 ton cooling		6	2.667		920	90.50		1,010.50	1,125	
	0140	1.5 ton cooling		5.50	2.909		1,050	98.50		1,148.50	1,300	
	0150	2 ton cooling		5.25	3.048		1,350	103		1,453	1,625	
	0160	2.5 ton cooling		5	3.200		1,975	108		2,083	2,350	
	0180	3 ton cooling		4	4		2,175	135		2,310	2,600	
	0190	7.5 ton cooling	↓	2.70	5.926	↓	2,500	201		2,701	3,050	
	0262	For hot water coil, add					40%	10%				
	0300	Console, 2 pipe with electric heat										
	0310	1/2 ton cooling	Q-5	8	2	Ea.	925	67.50		992.50	1,125	
	0315	3/4 ton cooling		7	2.286		1,000	77.50		1,077.50	1,225	
	0320	1 ton cooling		6	2.667		1,100	90.50		1,190.50	1,325	
	0330	1.5 ton cooling		5.50	2.909		1,350	98.50		1,448.50	1,625	
	0340	2 ton cooling		4.70	3.404		1,775	115		1,890	2,125	
	0345	2-1/2 ton cooling		4.70	3.404		2,575	115		2,690	3,000	
	0350	3 ton cooling		4	4		3,450	135		3,585	3,975	
	0360	4 ton cooling		4	4		3,575	135		3,710	4,150	
	0940	Direct expansion, for use w/air cooled condensing, 1.5 ton cooling		5	3.200		575	108		683	795	
	0950	2 ton cooling		4.80	3.333		610	113		723	840	
	0960	2.5 ton cooling		4.40	3.636		715	123		838	970	
	0970	3 ton cooling		3.80	4.211		745	143		888	1,025	
	0980	3.5 ton cooling		3.60	4.444		910	150		1,060	1,225	
	0990	4 ton cooling		3.40	4.706		965	159		1,124	1,300	
	1000	5 ton cooling	↓	3	5.333	↓	1,025	181		1,206	1,400	

15 MECHANICAL

Important: See the Reference Section for critical supporting data - Reference Nos., Crews, & City Cost Indexes

			CREW	DAILY OUTPUT	LABOR-HOURS	UNIT	2003 BARE COSTS				TOTAL INCL O&P	
	15765	**Fan Coil Unit/Unit Ventilators**					MAT.	LABOR	EQUIP.	TOTAL		
200	1020	7.5 ton cooling	Q-5	3	5.333	Ea.	3,100	181		3,281	3,700	**200**
	1040	10 ton cooling	Q-6	2.60	9.231		3,775	325		4,100	4,650	
	1042	11 ton cooling		2.30	10.435		4,950	365		5,315	6,000	
	1050	15 ton cooling		1.60	15		5,300	525		5,825	6,650	
	1060	20 ton cooling		.70	34.286		6,925	1,200		8,125	9,450	
	1070	25 ton cooling		.65	36.923		8,675	1,300		9,975	11,500	
	1080	30 ton cooling	▼	.60	40		9,475	1,400		10,875	12,500	
	1500	For hot water coil, add				▼	40%	10%				
	1510	For condensing unit add see division 15670.										
	3000	Chilled water, horizontal unit, housing, 2 pipe, controls										
	3100	1/2 ton cooling	Q-5	8	2	Ea.	860	67.50		927.50	1,050	
	3110	1 ton cooling		6	2.667		1,050	90.50		1,140.50	1,275	
	3120	1.5 ton cooling		5.50	2.909		1,250	98.50		1,348.50	1,525	
	3130	2 ton cooling		5.25	3.048		1,500	103		1,603	1,800	
	3140	3 ton cooling		4	4		1,725	135		1,860	2,100	
	3150	3.5 ton cooling		3.80	4.211		1,900	143		2,043	2,325	
	3160	4 ton cooling		3.80	4.211		1,900	143		2,043	2,325	
	3170	5 ton cooling		3.40	4.706		2,200	159		2,359	2,675	
	3180	6 ton cooling		3.40	4.706		2,200	159		2,359	2,675	
	3190	7 ton cooling		2.90	5.517		2,400	187		2,587	2,925	
	3200	8 ton cooling	▼	2.90	5.517		2,400	187		2,587	2,925	
	3210	10 ton cooling	Q-6	2.80	8.571	▼	2,650	300		2,950	3,375	
	3240	For hot water coil, add					40%	10%				
	4000	With electric heat, 2 pipe										
	4100	1/2 ton cooling	Q-5	8	2	Ea.	1,050	67.50		1,117.50	1,250	
	4105	3/4 ton cooling		7	2.286		1,150	77.50		1,227.50	1,400	
	4110	1 ton cooling		6	2.667		1,275	90.50		1,365.50	1,525	
	4120	1.5 ton cooling		5.50	2.909		1,425	98.50		1,523.50	1,700	
	4130	2 ton cooling		5.25	3.048		1,775	103		1,878	2,100	
	4135	2.5 ton cooling		4.80	3.333		2,450	113		2,563	2,875	
	4140	3 ton cooling		4	4		3,300	135		3,435	3,825	
	4150	3.5 ton cooling		3.80	4.211		3,925	143		4,068	4,550	
	4160	4 ton cooling		3.60	4.444		4,550	150		4,700	5,225	
	4170	5 ton cooling		3.40	4.706		5,100	159		5,259	5,850	
	4180	6 ton cooling		3.20	5		5,100	169		5,269	5,850	
	4190	7 ton cooling		2.90	5.517		5,350	187		5,537	6,175	
	4200	8 ton cooling	▼	2.20	7.273		6,050	246		6,296	7,025	
	4210	10 ton cooling	Q-6	2.80	8.571	▼	6,400	300		6,700	7,500	
600	0010	**HEATING & VENTILATING UNITS** Classroom										**600**
	0020	Includes filter, heating/cooling coils, standard controls										
	0080	750 CFM, 2 tons cooling	Q-6	2	12	Ea.	3,050	420		3,470	3,975	
	0100	1000 CFM, 2-1/2 tons cooling		1.60	15		3,600	525		4,125	4,775	
	0120	1250 CFM, 3 tons cooling		1.40	17.143		3,725	600		4,325	5,000	
	0140	1500 CFM, 4 tons cooling		.80	30		3,975	1,050		5,025	5,975	
	0160	2000 CFM 5 tons cooling	▼	.50	48		4,675	1,675		6,350	7,700	
	0500	For electric heat, add					35%					
	1000	For no cooling, deduct				▼	25%	10%				

			CREW	DAILY OUTPUT	LABOR-HOURS	UNIT	MAT.	LABOR	EQUIP.	TOTAL	TOTAL INCL O&P	
	15766	**Fin Tube Radiation**										
190	0010	**HYDRONIC HEATING** Terminal units, not incl. main supply pipe										**190**
	1000	Radiation										
	1100	Panel, baseboard, C.I., including supports, no covers	Q-5	46	.348	L.F.	21.50	11.75		33.25	41.50	
	1150	Fin tube, wall hung, 14" slope top cover, with damper										
	1200	1-1/4" copper tube, 4-1/4" alum. fin	Q-5	38	.421	L.F.	27	14.25		41.25	51.50	
	1250	1-1/4" steel tube, 4-1/4" steel fin	▼	36	.444	▼	24.50	15.05		39.55	49.50	

MECHANICAL 15

	15766	Fin Tube Radiation	CREW	DAILY OUTPUT	LABOR-HOURS	UNIT	2003 BARE COSTS				TOTAL INCL O&P	
							MAT.	LABOR	EQUIP.	TOTAL		
190	1255	2" steel tube, 4-1/4" steel fin	Q-5	32	.500	L.F.	26.50	16.95		43.45	54.50	190
	1260	21", two tier, slope top, w/damper										
	1262	1-1/4" copper tube, 4-1/4" alum. fin	Q-5	30	.533	L.F.	32	18.05		50.05	62.50	
	1264	1-1/4" steel tube, 4-1/4" steel fin		28	.571		29.50	19.35		48.85	61	
	1266	2" steel tube, 4-1/4" steel fin	↓	25	.640	↓	31	21.50		52.50	67	
	1270	10", flat top, wtih damper										
	1272	1-1/4" copper tube, 4-1/4" alum. fin	Q-5	38	.421	L.F.	26	14.25		40.25	50	
	1274	1-1/4" steel tube, 4-1/4" steel fin		36	.444		24.50	15.05		39.55	49.50	
	1276	2" steel tube, 4-1/4" steel fin	↓	32	.500	↓	26.50	16.95		43.45	54.50	
	1280	17", two tier, flat top, w/damper										
	1282	1-1/4" copper tube, 4-1/4" alum. fin	Q-5	30	.533	L.F.	30.50	18.05		48.55	61	
	1284	1-1/4" steel tube, 4-1/4" steel fin		28	.571		29.50	19.35		48.85	61	
	1286	2" steel tube, 4-1/4" steel fin	↓	25	.640	↓	31	21.50		52.50	67	
	1301	Rough in wall hung, steel fin w/supply & balance valves	1 Stpi	.91	8.791	Ea.	147	330		477	660	
	1310	Baseboard, pkgd, 1/2" copper tube, alum. fin, 7" high	Q-5	60	.267	L.F.	5.35	9.05		14.40	19.50	
	1320	3/4" copper tube, alum. fin, 7" high		58	.276		5.65	9.35		15	20.50	
	1340	1" copper tube, alum. fin, 8-7/8" high		56	.286		11.65	9.65		21.30	27.50	
	1360	1-1/4" copper tube, alum. fin, 8-7/8" high		54	.296		17.25	10.05		27.30	34	
	1380	1-1/4" IPS steel tube with steel fins	↓	52	.308	↓	16.60	10.40		27	34	
	1381	Rough in baseboard panel & fin tube, supply & balance valves	1 Stpi	1.06	7.547	Ea.	106	284		390	545	
	1500	Note: fin tube may also require corners, caps, etc.										
	1990	Convector unit, floor recessed, flush, with trim										
	2000	for under large glass wall areas, no damper	Q-5	20	.800	L.F.	23	27		50	66.50	
	2100	For unit with damper				"	9.95			9.95	10.95	
	3000	Radiators, cast iron										
	3100	Free standing or wall hung, 6 tube, 25" high	Q-5	96	.167	Section	21.50	5.65		27.15	32	
	3150	4 tube 25" high		96	.167		15.90	5.65		21.55	26	
	3200	4 tube, 19" high	↓	96	.167	↓	15.85	5.65		21.50	26	
	3250	Adj. brackets, 2 per wall radiator up to 30 sections	1 Stpi	32	.250	Ea.	32	9.40		41.40	49	
	3500	Recessed, 20" high x 5" deep, without grille	Q-5	60	.267	Section	17.60	9.05		26.65	33	
	3600	For inlet grille, add				"	1.25			1.25	1.38	
	3950	Unit heaters, propeller, 115 V 2 psi steam, 60°F entering air										
	4000	Horizontal, 12 MBH	Q-5	12	1.333	Ea.	247	45		292	340	
	4020	28.2 MBH		10	1.600		340	54		394	455	
	4040	36.5 MBH		8	2		400	67.50		467.50	540	
	4060	43.9 MBH		8	2		405	67.50		472.50	545	
	4080	56.5 MBH		7.50	2.133		425	72		497	580	
	4100	65.6 MBH		7	2.286		445	77.50		522.50	605	
	4120	87.6 MBH		6.50	2.462		470	83.50		553.50	640	
	4140	96.8 MBH		6	2.667		570	90.50		660.50	765	
	4160	133.3 MBH		5	3.200		590	108		698	815	
	4180	157.6 MBH		4	4		740	135		875	1,025	
	4200	197.7 MBH		3	5.333		875	181		1,056	1,250	
	4220	257.2 MBH		2.50	6.400		1,150	217		1,367	1,575	
	4240	286.9 MBH		2	8		1,225	271		1,496	1,750	
	4250	326.0 MBH		1.90	8.421		1,375	285		1,660	1,950	
	4260	364 MBH		1.80	8.889		1,450	300		1,750	2,050	
	4270	404 MBH	↓	1.60	10	↓	2,000	340		2,340	2,700	
	4280	Rough in unit heater, 72.7 MBH, piping, supports & controls	1 Stpi	.47	17.021	Set	330	640		970	1,325	
	4300	For vertical diffuser, add				Ea.	140			140	154	
	4310	Vertical flow, 40 MBH	Q-5	11	1.455		395	49		444	505	
	4314	58.5 MBH		8	2		445	67.50		512.50	590	
	4318	92 MBH		7	2.286		535	77.50		612.50	705	
	4322	109.7 MBH		6	2.667		625	90.50		715.50	820	
	4326	131.0 MBH		4	4		625	135		760	890	
	4330	160.0 MBH	↓	3	5.333	↓	745	181		926	1,100	

Important: See the Reference Section for critical supporting data - Reference Nos., Crews, & City Cost Indexes

15
MECHANICAL

15766	Fin Tube Radiation	CREW	DAILY OUTPUT	LABOR-HOURS	UNIT	2003 BARE COSTS				TOTAL INCL O&P
						MAT.	LABOR	EQUIP.	TOTAL	
190 4334	194.0 MBH	Q-5	2.20	7.273	Ea.	805	246		1,051	1,250	190
4338	212.0 MBH		2.10	7.619		1,275	258		1,533	1,800
4342	247.0 MBH		1.96	8.163		1,325	276		1,601	1,875
4346	297.0 MBH		1.80	8.889		1,425	300		1,725	2,025
4350	333.0 MBH	Q-6	1.90	12.632		1,450	445		1,895	2,275
4354	420 MBH, (460 V)		1.80	13.333		1,450	470		1,920	2,300
4358	500 MBH, (460 V)		1.71	14.035		2,075	495		2,570	3,025
4362	570 MBH, (460 V)		1.40	17.143		2,975	600		3,575	4,175
4366	620 MBH, (460 V)		1.30	18.462		3,400	650		4,050	4,725
4370	960 MBH, (460 V)		1.10	21.818		5,975	765		6,740	7,725
4400	Unit heaters, propeller, hot water, 115 V									
4404	Horizontal									
4408	6 MBH	Q-5	11	1.455	Ea.	247	49		296	345
4412	12 MBH		11	1.455		247	49		296	345
4416	16 MBH		10	1.600		267	54		321	375
4420	24 MBH		9	1.778		340	60		400	465
4424	29 MBH		8	2		400	67.50		467.50	540
4428	47 MBH		7	2.286		420	77.50		497.50	575
4432	63 MBH		6	2.667		470	90.50		560.50	650
4436	81 MBH		5.50	2.909		570	98.50		668.50	780
4440	90 MBH		5	3.200		590	108		698	815
4460	133 MBH		4	4		740	135		875	1,025
4464	139 MBH		3.50	4.571		875	155		1,030	1,200
4468	198 MBH		2.50	6.400		1,200	217		1,417	1,650
4472	224 MBH		2	8		1,250	271		1,521	1,775
4476	273 MBH		1.50	10.667		1,325	360		1,685	2,000
4500	Vertical									
4504	8 MBH	Q-5	13	1.231	Ea.	395	41.50		436.50	495
4508	12 MBH		13	1.231		395	41.50		436.50	495
4512	16 MBH		13	1.231		395	41.50		436.50	495
4516	30 MBH		12	1.333		395	45		440	500
4520	43 MBH		11	1.455		445	49		494	565
4524	57 MBH		8	2		500	67.50		567.50	650
4528	68 MBH		7	2.286		535	77.50		612.50	705
4540	105 MBH		6	2.667		625	90.50		715.50	820
4544	123 MBH		5	3.200		740	108		848	980
4550	140 MBH		5	3.200		805	108		913	1,050
4554	156 MBH		4	4		910	135		1,045	1,200
4560	210 MBH		4	4		995	135		1,130	1,300
4564	223 MBH		3	5.333		1,475	181		1,656	1,900
4568	257 MBH		2	8		1,475	271		1,746	2,025
5400	Cabinet, horizontal, hot water, blower type									
5420	20 MBH	Q-5	10	1.600	Ea.	675	54		729	825
5430	60 MBH		7	2.286		1,175	77.50		1,252.50	1,425
5440	100 MBH		5	3.200		1,225	108		1,333	1,525
5450	120 MBH		4	4		1,275	135		1,410	1,625
6000	Valance units, complete with 1/2" cooling coil, enclosure									
6020	2 tube	Q-5	18	.889	L.F.	29	30		59	77.50
6040	3 tube		16	1		34	34		68	88.50
6060	4 tube		16	1		39	34		73	94
6080	5 tube		15	1.067		45	36		81	104
6100	6 tube		15	1.067		50	36		86	110
6120	8 tube		14	1.143		75	38.50		113.50	141
6200	For 3/4" cooling coil, add					10%				
9500	To convert SFR to BTU rating: Hot water, 150 x SFR									
9510	Forced hot water, 180 x SFR; steam, 240 x SFR									

MECHANICAL 15

15768	Infrared Heaters	CREW	DAILY OUTPUT	LABOR-HOURS	UNIT	2003 BARE COSTS				TOTAL INCL O&P
						MAT.	LABOR	EQUIP.	TOTAL	
600	0010	**INFRA-RED UNIT**									**600**
	0020	Gas fired, unvented, electric ignition, 100% shutoff.									
	0030	Piping and wiring not included									
	0060	Input, 15 MBH	Q-5	7	2.286	Ea.	350	77.50		427.50	500
	0100	30 MBH		6	2.667		365	90.50		455.50	535
	0120	45 MBH		5	3.200		430	108		538	640
	0140	50 MBH		4.50	3.556		450	120		570	675
	0160	60 MBH		4	4		465	135		600	715
	0180	75 MBH		3	5.333		540	181		721	870
	0200	90 MBH		2.50	6.400		545	217		762	925
	0220	105 MBH		2	8		680	271		951	1,150
	0240	120 MBH	▼	2	8	▼	895	271		1,166	1,400
	2000	Electric, single or three phase									
	2050	6 kW, 20,478 BTU	1 Elec	2.30	3.478	Ea.	360	131		491	595
	2100	13.5 KW, 40,956 BTU		2.20	3.636		565	137		702	830
	2150	24 KW, 81,912 BTU	▼	2	4	▼	1,225	150		1,375	1,575
	3000	Oil fired, pump, controls, fusible valve, oil supply tank									
	3050	91,000 BTU	Q-5	2.50	6.400	Ea.	5,225	217		5,442	6,075
	3080	105.000 BTU		2.25	7.111		5,475	241		5,716	6,400
	3110	119,000 BTU	▼	2	8	▼	5,775	271		6,046	6,750

15770	Floor-Heating & Snow-Melting Eq.	CREW	DAILY OUTPUT	LABOR-HOURS	UNIT	MAT.	LABOR	EQUIP.	TOTAL	TOTAL INCL O&P
200	0010	**ELECTRIC HEATING**, not incl. conduit or feed wiring									**200**
	1100	Rule of thumb: Baseboard units, including control	1 Elec	4.40	1.818	kW	77	68.50		145.50	187
	1300	Baseboard heaters, 2' long, 375 watt		8	1	Ea.	37.50	37.50		75	97
	1400	3' long, 500 watt		8	1		42	37.50		79.50	102
	1600	4' long, 750 watt		6.70	1.194		47	45		92	119
	1800	5' long, 935 watt		5.70	1.404		53	53		106	138
	2000	6' long, 1125 watt		5	1.600		57.50	60		117.50	153
	2200	7' long, 1310 watt		4.40	1.818		68.50	68.50		137	178
	2400	8' long, 1500 watt		4	2		78.50	75		153.50	198
	2600	9' long, 1680 watt		3.60	2.222		88	83.50		171.50	222
	2800	10' long, 1875 watt	▼	3.30	2.424	▼	98	91		189	244
	2950	Wall heaters with fan, 120 to 277 volt									
	3160	Recessed, residential, 750 watt	1 Elec	6	1.333	Ea.	80.50	50		130.50	164
	3170	1000 watt		6	1.333		82.50	50		132.50	166
	3180	1250 watt		5	1.600		90.50	60		150.50	190
	3190	1500 watt		4	2		90.50	75		165.50	212
	3210	2000 watt		4	2		95.50	75		170.50	217
	3230	2500 watt		3.50	2.286		152	86		238	295
	3240	3000 watt		3	2.667		310	100		410	490
	3250	4000 watt		2.70	2.963		310	111		421	510
	3260	Commercial, 750 watt		6	1.333		140	50		190	229
	3270	1000 watt		6	1.333		140	50		190	229
	3280	1250 watt		5	1.600		140	60		200	244
	3290	1500 watt		4	2		140	75		215	266
	3300	2000 watt		4	2		140	75		215	266
	3310	2500 watt		3.50	2.286		140	86		226	282
	3320	3000 watt		3	2.667		247	100		347	420
	3330	4000 watt		2.70	2.963		247	111		358	440
	3600	Thermostats, integral		16	.500		16.65	18.80		35.45	46.50
	3800	Line voltage, 1 pole		8	1		20	37.50		57.50	78
	3810	2 pole		8	1		26	37.50		63.50	84.50
	3820	Low voltage, 1 pole	▼	8	1	▼	22	37.50		59.50	80
	4000	Heat trace system, 400 degree									

T5 MECHANICAL

Important: See the Reference Section for critical supporting data - Reference Nos., Crews, & City Cost Indexes

15770	Floor-Heating & Snow-Melting Eq.	CREW	DAILY OUTPUT	LABOR-HOURS	UNIT	2003 BARE COSTS				TOTAL INCL O&P	
						MAT.	LABOR	EQUIP.	TOTAL		
200 4020	115V, 2.5 watts per L.F.	1 Elec	530	.015	L.F.	6.70	.57		7.27	8.20	**200**
4030	5 watts per L.F.		530	.015		6.70	.57		7.27	8.20	
4050	10 watts per L.F.		530	.015		6.70	.57		7.27	8.20	
4060	208V, 5 watts per L.F.		530	.015		6.70	.57		7.27	8.20	
4080	480V, 8 watts per L.F.		530	.015		6.70	.57		7.27	8.20	
4200	Heater raceway										
4260	Heat transfer cement										
4280	1 gallon				Ea.	52.50			52.50	57.50	
4300	5 gallon				"	214			214	236	
4320	Snap band, clamp										
4340	3/4" pipe size	1 Elec	470	.017	Ea.		.64		.64	.96	
4360	1" pipe size		444	.018			.68		.68	1.01	
4380	1-1/4" pipe size		400	.020			.75		.75	1.12	
4400	1-1/2" pipe size		355	.023			.85		.85	1.27	
4420	2" pipe size		320	.025			.94		.94	1.40	
4440	3" pipe size		160	.050			1.88		1.88	2.81	
4460	4" pipe size		100	.080			3.01		3.01	4.49	
4480	Thermostat NEMA 3R, 22 amp, 0-150 Deg, 10' cap.		8	1		179	37.50		216.50	252	
4500	Thermostat NEMA 4X, 25 amp, 40 Deg, 5-1/2' cap.		7	1.143		179	43		222	260	
4520	Thermostat NEMA 4X, 22 amp, 25-325 Deg, 10' cap.		7	1.143		480	43		523	595	
4540	Thermostat NEMA 4X, 22 amp, 15-140 Deg,		6	1.333		405	50		455	520	
4580	Thermostat NEMA 4,7,9, 22 amp, 25-325 Deg, 10' cap.		3.60	2.222		595	83.50		678.50	780	
4600	Thermostat NEMA 4,7,9, 22 amp, 15-140 Deg,		3	2.667		590	100		690	795	
4720	Fiberglass application tape, 36 yard roll		11	.727		54	27.50		81.50	101	
5000	Radiant heating ceiling panels, 2' x 4', 500 watt		16	.500		195	18.80		213.80	242	
5050	750 watt		16	.500		215	18.80		233.80	264	
5200	For recessed plaster frame, add		32	.250		39	9.40		48.40	57	
5300	Infra-red quartz heaters, 120 volts, 1000 watts		6.70	1.194		119	45		164	198	
5350	1500 watt		5	1.600		119	60		179	221	
5400	240 volts, 1500 watt		5	1.600		119	60		179	221	
5450	2000 watt		4	2		119	75		194	243	
5500	3000 watt		3	2.667		139	100		239	300	
5550	4000 watt		2.60	3.077		139	116		255	325	
5570	Modulating control		.80	10		72.50	375		447.50	640	
5600	Unit heaters, heavy duty, with fan & mounting bracket										
5650	Single phase, 208-240-277 volt, 3 kW	1 Elec	3.20	2.500	Ea.	320	94		414	490	
5750	5 kW		2.40	3.333		330	125		455	545	
5800	7 kW		1.90	4.211		510	158		668	795	
5850	10 kW		1.30	6.154		580	231		811	985	
5950	15 kW		.90	8.889		940	335		1,275	1,525	
6000	480 volt, 3 kW		3.30	2.424		400	91		491	575	
6020	4 kW		3	2.667		360	100		460	545	
6040	5 kW		2.60	3.077		365	116		481	580	
6060	7 kW		2	4		545	150		695	825	
6080	10 kW		1.40	5.714		595	215		810	975	
6100	13 kW		1.10	7.273		940	273		1,213	1,425	
6120	15 kW		1	8		985	300		1,285	1,525	
6140	20 kW		.90	8.889		1,225	335		1,560	1,825	
6300	3 phase, 208-240 volt, 5 kW		2.40	3.333		310	125		435	525	
6320	7 kW		1.90	4.211		480	158		638	765	
6340	10 kW		1.30	6.154		520	231		751	915	
6360	15 kW		.90	8.889		870	335		1,205	1,450	
6380	20 kW		.70	11.429		1,225	430		1,655	2,000	
6400	25 kW		.50	16		1,450	600		2,050	2,500	
6500	480 volt, 5 kW		2.60	3.077		435	116		551	655	
6520	7 kW		2	4		550	150		700	830	

MECHANICAL 15

15770	Floor-Heating & Snow-Melting Eq.	CREW	DAILY OUTPUT	LABOR-HOURS	UNIT	2003 BARE COSTS				TOTAL INCL O&P	
						MAT.	LABOR	EQUIP.	TOTAL		
200 6540	10 kW	1 Elec	1.40	5.714	Ea.	580	215		795	960	**200**
6560	13 kW		1.10	7.273		940	273		1,213	1,425	
6580	15 kW		1	8		940	300		1,240	1,475	
6600	20 kW		.90	8.889		1,225	335		1,560	1,825	
6620	25 kW		.60	13.333		1,450	500		1,950	2,350	
6630	30 kW		.70	11.429		1,700	430		2,130	2,500	
6640	40 kW		.60	13.333		2,150	500		2,650	3,125	
6650	50 kW	▼	.50	16	▼	2,600	600		3,200	3,750	
6800	Vertical discharge heaters, with fan										
6820	Single phase, 208-240-277 volt, 10 kW	1 Elec	1.30	6.154	Ea.	535	231		766	935	
6840	15 kW		.90	8.889		900	335		1,235	1,500	
6900	3 phase, 208-240 volt, 10 kW		1.30	6.154		520	231		751	915	
6920	15 kW		.90	8.889		870	335		1,205	1,450	
6940	20 kW		.70	11.429		1,225	430		1,655	2,000	
6960	25 kW		.50	16		1,450	600		2,050	2,500	
6980	30 kW		.40	20		1,700	750		2,450	2,975	
7000	40 kW		.36	22.222		2,150	835		2,985	3,625	
7020	50 kW		.32	25		2,600	940		3,540	4,250	
7100	480 volt, 10 kW		1.40	5.714		580	215		795	960	
7120	15 kW		1	8		940	300		1,240	1,475	
7140	20 kW		.90	8.889		1,225	335		1,560	1,825	
7160	25 kW		.60	13.333		1,450	500		1,950	2,350	
7180	30 kW		.50	16		1,700	600		2,300	2,750	
7200	40 kW		.40	20		2,150	750		2,900	3,500	
7220	50 kW	▼	.35	22.857	▼	2,600	860		3,460	4,125	
7900	Cabinet convector heaters, 240 volt										
7920	3' long, 2000 watt	1 Elec	5.30	1.509	Ea.	1,375	57		1,432	1,600	
7940	3000 watt		5.30	1.509		1,450	57		1,507	1,675	
7960	4000 watt		5.30	1.509		1,500	57		1,557	1,725	
7980	6000 watt		4.60	1.739		1,550	65.50		1,615.50	1,800	
8000	8000 watt		4.60	1.739		1,625	65.50		1,690.50	1,875	
8020	4' long, 4000 watt		4.60	1.739		1,500	65.50		1,565.50	1,775	
8040	6000 watt		4	2		1,575	75		1,650	1,825	
8060	8000 watt		4	2		1,650	75		1,725	1,900	
8080	10,000 watt	▼	4	2	▼	1,650	75		1,725	1,925	
8100	Available also in 208 or 277 volt										
8200	Cabinet unit heaters, 120 to 277 volt, 1 pole,										
8220	wall mounted, 2 kW	1 Elec	4.60	1.739	Ea.	1,375	65.50		1,440.50	1,625	
8230	3 kW		4.60	1.739		1,450	65.50		1,515.50	1,700	
8240	4 kW		4.40	1.818		1,500	68.50		1,568.50	1,750	
8250	5 kW		4.40	1.818		1,550	68.50		1,618.50	1,800	
8260	6 kW		4.20	1.905		1,550	71.50		1,621.50	1,800	
8270	8 kW		4	2		1,625	75		1,700	1,875	
8280	10 kW		3.80	2.105		1,650	79		1,729	1,950	
8290	12 kW		3.50	2.286		1,700	86		1,786	1,975	
8300	13.5 kW		2.90	2.759		1,675	104		1,779	1,975	
8310	16 kW		2.70	2.963		1,750	111		1,861	2,100	
8320	20 kW		2.30	3.478		2,650	131		2,781	3,125	
8330	24 kW		1.90	4.211		2,700	158		2,858	3,175	
8350	Recessed, 2 kW		4.40	1.818		1,375	68.50		1,443.50	1,625	
8370	3 kW		4.40	1.818		1,450	68.50		1,518.50	1,700	
8380	4 kW		4.20	1.905		1,500	71.50		1,571.50	1,750	
8390	5 kW		4.20	1.905		1,550	71.50		1,621.50	1,800	
8400	6 kW		4	2		1,575	75		1,650	1,825	
8410	8 kW		3.80	2.105		1,650	79		1,729	1,925	
8420	10 kW	▼	3.50	2.286	▼	2,025	86		2,111	2,350	

Important: See the Reference Section for critical supporting data - Reference Nos., Crews, & City Cost Indexes

15770	Floor-Heating & Snow-Melting Eq.	CREW	DAILY OUTPUT	LABOR-HOURS	UNIT	2003 BARE COSTS				TOTAL INCL O&P
						MAT.	LABOR	EQUIP.	TOTAL	
200											200
8430	12 kW	1 Elec	2.90	2.759	Ea.	2,050	104		2,154	2,400
8440	13.5 kW		2.70	2.963		2,075	111		2,186	2,450
8450	16 kW		2.30	3.478		2,075	131		2,206	2,475
8460	20 kW		1.90	4.211		2,500	158		2,658	2,975
8470	24 kW		1.60	5		2,575	188		2,763	3,100
8490	Ceiling mounted, 2 kW		3.20	2.500		1,375	94		1,469	1,675
8510	3 kW		3.20	2.500		1,450	94		1,544	1,750
8520	4 kW		3	2.667		1,500	100		1,600	1,800
8530	5 kW		3	2.667		1,550	100		1,650	1,850
8540	6 kW		2.80	2.857		1,550	107		1,657	1,850
8550	8 kW		2.40	3.333		1,625	125		1,750	1,950
8560	10 kW		2.20	3.636		1,650	137		1,787	2,025
8570	12 kW		2	4		1,700	150		1,850	2,075
8580	13.5 kW		1.50	5.333		1,725	201		1,926	2,200
8590	16 kW		1.30	6.154		1,750	231		1,981	2,275
8600	20 kW		.90	8.889		2,500	335		2,835	3,250
8610	24 kW	▼	.60	13.333	▼	2,550	500		3,050	3,550
8630	208 to 480 V, 3 pole									
8650	Wall mounted, 2 kW	1 Elec	4.60	1.739	Ea.	1,375	65.50		1,440.50	1,625
8670	3 kW		4.60	1.739		1,450	65.50		1,515.50	1,700
8680	4 kW		4.40	1.818		1,500	68.50		1,568.50	1,750
8690	5 kW		4.40	1.818		1,550	68.50		1,618.50	1,800
8700	6 kW		4.20	1.905		1,550	71.50		1,621.50	1,800
8710	8 kW		4	2		1,625	75		1,700	1,875
8720	10 kW		3.80	2.105		1,650	79		1,729	1,950
8730	12 kW		3.50	2.286		1,700	86		1,786	1,975
8740	13.5 kW		2.90	2.759		1,725	104		1,829	2,050
8750	16 kW		2.70	2.963		1,750	111		1,861	2,100
8760	20 kW		2.30	3.478		2,500	131		2,631	2,950
8770	24 kW		1.90	4.211		2,550	158		2,708	3,025
8790	Recessed, 2 kW		4.40	1.818		1,375	68.50		1,443.50	1,625
8810	3 kW		4.40	1.818		1,450	68.50		1,518.50	1,700
8820	4 kW		4.20	1.905		1,500	71.50		1,571.50	1,750
8830	5 kW		4.20	1.905		1,550	71.50		1,621.50	1,800
8840	6 kW		4	2		1,550	75		1,625	1,800
8850	8 kW		3.80	2.105		1,625	79		1,704	1,900
8860	10 kW		3.50	2.286		1,650	86		1,736	1,950
8870	12 kW		2.90	2.759		1,700	104		1,804	2,000
8880	13.5 kW		2.70	2.963		1,725	111		1,836	2,075
8890	16 kW		2.30	3.478		1,750	131		1,881	2,125
8900	20 kW		1.90	4.211		2,500	158		2,658	2,975
8920	24 kW		1.60	5		2,550	188		2,738	3,075
8940	Ceiling mount, 2 kW		3.20	2.500		1,375	94		1,469	1,675
8950	3 kW		3.20	2.500		1,450	94		1,544	1,750
8960	4 kW		3	2.667		1,500	100		1,600	1,800
8970	5 kW		3	2.667		1,550	100		1,650	1,850
8980	6 kW		2.80	2.857		1,550	107		1,657	1,850
8990	8 kW		2.40	3.333		1,625	125		1,750	1,950
9000	10 kW		2.20	3.636		1,650	137		1,787	2,025
9020	13.5 kW		1.50	5.333		1,725	201		1,926	2,200
9030	16 kW		1.30	6.154		1,750	231		1,981	2,275
9040	20 kW		.90	8.889		2,500	335		2,835	3,250
9060	24 kW	▼	.60	13.333	▼	2,550	500		3,050	3,550
500											500
0010	**RADIANT FLOOR HEATING**									
0100	Tubing, PEX (cross-linked polyethylene)									
0110	Oxygen barrier type for systems with ferrous materials									
0120	1/2"	Q-5	800	.020	L.F.	.49	.68		1.17	1.56

15770	Floor-Heating & Snow-Melting Eq.	CREW	DAILY OUTPUT	LABOR-HOURS	UNIT	2003 BARE COSTS				TOTAL INCL O&P	
						MAT.	LABOR	EQUIP.	TOTAL		
500	0130	3/4"	Q-5	535	.030	L.F.	.62	1.01		1.63	2.21	500
	0140	1"	↓	400	.040	↓	1.25	1.35		2.60	3.43	
	0200	Non barrier type for ferrous free systems										
	0210	1/2"	Q-5	800	.020	L.F.	.27	.68		.95	1.32	
	0220	3/4"		535	.030		.44	1.01		1.45	2.01	
	0230	1"	↓	400	.040	↓	.77	1.35		2.12	2.90	
	1000	Manifolds										
	1110	Brass										
	1120	Valved										
	1130	1", 2 circuit	Q-5	13	1.231	Ea.	37.50	41.50		79	105	
	1140	1", 3 circuit		11	1.455		51	49		100	131	
	1150	1", 4 circuit		9	1.778		69	60		129	167	
	1160	1-1/4", 3 circuit		10	1.600		58	54		112	146	
	1170	1-1/4", 4 circuit	↓	8	2	↓	70.50	67.50		138	180	
	1300	Valveless										
	1310	1", 2 circuit	Q-5	14	1.143	Ea.	18.75	38.50		57.25	79	
	1320	1", 3 circuit		12	1.333		25.50	45		70.50	96	
	1330	1", 4 circuit		10	1.600		33	54		87	119	
	1340	1-1/4", 3 circuit		11	1.455		31.50	49		80.50	109	
	1350	1-1/4", 4 circuit	↓	9	1.778	↓	36	60		96	131	
	1400	Manifold connector kit										
	1410	Mounting brackets, couplings, end cap, air vent, valve, plug										
	1420	1"	Q-5	8	2	Ea.	64.50	67.50		132	173	
	1430	1-1/4"	"	7	2.286	"	77.50	77.50		155	202	
	1610	Copper, (cut to size)										
	1620	1" x 1/2" x 72" Lg, 2" sweat drops, 24 circuit	Q-5	3.33	4.805	Ea.	65.50	163		228.50	320	
	1630	1-1/4" x 1/2" x 72" Lg, 2" sweat drops, 24 circuit		3.20	5		77.50	169		246.50	340	
	1640	1-1/2" x 1/2" x 72" Lg, 2" sweat drops, 24 circuit		3	5.333		94	181		275	375	
	1650	1-1/4" x 3/4" x 72" Lg, 2" sweat drops, 24 circuit		3.10	5.161		82.50	175		257.50	355	
	1660	1-1/2" x 3/4" x 72" Lg, 2" sweat drops, 24 circuit	↓	2.90	5.517	↓	99	187		286	390	
	1710	Copper, (modular)										
	1720	1" x 1/2", 2" sweat drops, 2 circuit	Q-5	17.50	.914	Ea.	6.15	31		37.15	54	
	1730	1" x 1/2", 2" sweat drops, 3 circuit		14.50	1.103		9.25	37.50		46.75	66.50	
	1740	1" x 1/2", 2" sweat drops, 4 circuit		12.30	1.301		10.70	44		54.70	78.50	
	1750	1-1/4" x 1/2", 2" sweat drops, 2 circuit		17	.941		7.15	32		39.15	56	
	1760	1-1/4" x 1/2", 2" sweat drops, 3 circuit		14	1.143		10.70	38.50		49.20	70.50	
	1770	1-1/4" x 1/2", 2" sweat drops, 4 circuit	↓	11.80	1.356	↓	14.25	46		60.25	85	
	3000	Valves										
	3110	Motorized zone valve operator, 24V	Q-5	30	.533	Ea.	40	18.05		58.05	71	
	3120	Motorized zone valve with operator complete, 24V										
	3130	3/4"	Q-5	35	.457	Ea.	68.50	15.45		83.95	99	
	3140	1"		32	.500		77.50	16.95		94.45	111	
	3150	1-1/4"		29.60	.541		95.50	18.30		113.80	133	
	3200	Flow control/isolation valve, 1/2"	↓	32	.500	↓	22	16.95		38.95	49.50	
	3300	Thermostatic mixing valves										
	3310	1/2"	Q-5	26.60	.601	Ea.	71.50	20.50		92	110	
	3320	3/4"		25	.640		87	21.50		108.50	128	
	3330	1"	↓	21.40	.748	↓	93	25.50		118.50	140	
	3400	3 Way mixing / diverting valve, manual, brass										
	3410	1/2"	Q-5	21.30	.751	Ea.	91	25.50		116.50	139	
	3420	3/4"		20	.800		94	27		121	144	
	3430	1"	↓	17.80	.899	↓	110	30.50		140.50	166	
	3500	4 Way mixing / diverting valve, manual, brass										
	3510	1/2"	Q-5	16	1	Ea.	95.50	34		129.50	156	
	3520	3/4"		15	1.067		107	36		143	172	
	3530	1"	↓	13.30	1.203	↓	119	40.50		159.50	193	

T5 MECHANICAL

Important: See the Reference Section for critical supporting data - Reference Nos., Crews, & City Cost Indexes

15770 | Floor-Heating & Snow-Melting Eq.

		CREW	DAILY OUTPUT	LABOR-HOURS	UNIT	MAT.	LABOR	EQUIP.	TOTAL	TOTAL INCL O&P	
500	3540	1-1/4"	Q-5	11.40	1.404	Ea.	131	47.50		178.50	216	**500**
	3550	1-1/2"		10.60	1.509		176	51		227	270	
	3560	2"		10	1.600		189	54		243	290	
	3570	2-1/2"		8	2		520	67.50		587.50	675	
	3800	Motor control, 3 or 4 way, for valves up to 1-1/2"	↓	30	.533	↓	131	18.05		149.05	172	
	5000	Radiant floor heating, zone control panel										
	5110	6 Zone control										
	5120	For zone valves	Q-5	16	1	Ea.	125	34		159	188	
	5130	For circulators	"	16	1	"	176	34		210	245	
	7000	PEX tubing fittings, compression type										
	7110	PEX tubing to brass valve conn., 1/2"	Q-5	54	.296	Ea.	2.12	10.05		12.17	17.50	
	7200	PEX x male NPT										
	7210	1/2" x 1/2"	Q-5	54	.296	Ea.	4.13	10.05		14.18	19.70	
	7220	3/4" x 3/4"		46	.348		6.95	11.75		18.70	25.50	
	7230	1" x 1"	↓	40	.400	↓	16.90	13.55		30.45	39	
	7300	PEX coupling										
	7310	1/2" x 1/2"	Q-5	50	.320	Ea.	5	10.85		15.85	22	
	7320	3/4" x 3/4"		42	.381		14.45	12.90		27.35	35.50	
	7330	1" x 1"	↓	36	.444	↓	19.50	15.05		34.55	44	
	7400	PEX x copper sweat										
	7410	1/2" x 1/2"	Q-5	38	.421	Ea.	3.62	14.25		17.87	25.50	
	7420	3/4" x 3/4"		34	.471		7.70	15.95		23.65	32.50	
	7430	1" x 1"	↓	29	.552	↓	16.90	18.70		35.60	46.50	

15780 | Energy Recovery Equipment

		CREW	DAILY OUTPUT	LABOR-HOURS	UNIT	MAT.	LABOR	EQUIP.	TOTAL	TOTAL INCL O&P	
100	0010	**HEAT RECOVERY PACKAGES**										**100**
	0100	Air to air										
	4000	Enthalpy recovery wheel										
	4010	1000 max CFM	Q-9	1.20	13.333	Ea.	3,825	445		4,270	4,875	
	4020	2000 max CFM		1	16		4,475	535		5,010	5,750	
	4030	4000 max CFM		.80	20		5,175	665		5,840	6,725	
	4040	6000 max CFM	↓	.70	22.857		6,050	760		6,810	7,825	
	4050	8000 max CFM	Q-10	1	24		6,700	830		7,530	8,625	
	4060	10,000 max CFM		.90	26.667		8,000	920		8,920	10,200	
	4070	20,000 max CFM		.80	30		14,500	1,025		15,525	17,600	
	4080	25,000 max CFM		.70	34.286		17,700	1,175		18,875	21,300	
	4090	30,000 max CFM		.50	48		19,700	1,650		21,350	24,200	
	4100	40,000 max CFM		.45	53.333		27,100	1,850		28,950	32,700	
	4110	50,000 max CFM	↓	.40	60	↓	31,500	2,075		33,575	37,900	

15800 | Air Distribution

15810 | Ducts

			CREW	DAILY OUTPUT	LABOR-HOURS	UNIT	MAT.	LABOR	EQUIP.	TOTAL	TOTAL INCL O&P	
600	0010	**DUCTWORK**	R15810 -050									**600**
	0020	Fabricated rectangular, includes fittings, joints, supports,										
	0030	allowance for flexible connections, no insulation	R15810 -070									
	0031	NOTE: Fabrication and installation are combined										
	0040	as LABOR cost. Approx. 25% fittings assumed.	R15810 -100									
	0042	Fabrication/Inst. is to commercial quality standards										

	15810	Ducts		CREW	DAILY OUTPUT	LABOR-HOURS	UNIT	2003 BARE COSTS				TOTAL INCL O&P	
								MAT.	LABOR	EQUIP.	TOTAL		
600	0043	(SMACNA or equiv.) for struct., seal., leak test., etc.	R15810 -050										600
	0050	Add to labor for elevated installation											
	0051	of fabricated ductwork	R15810 -070										
	0052	10' to 15' high							6%				
	0053	15' to 20' high	R15810 -100						12%				
	0054	20' to 25' high							15%				
	0055	25' to 30' high							21%				
	0056	30' to 35' high							24%				
	0057	35' to 40' high							30%				
	0058	Over 40' high							33%				
	0070	For duct insulation and lining see 15082-200-3000											
	0100	Aluminum, alloy 3003-H14, under 100 lb.		Q-10	75	.320	Lb.	2.04	11.05		13.09	19.30	
	0110	100 to 500 lb.			80	.300		1.74	10.35		12.09	17.90	
	0120	500 to 1,000 lb.			95	.253		1.54	8.70		10.24	15.15	
	0140	1,000 to 2,000 lb.			120	.200		1.44	6.90		8.34	12.25	
	0150	2,000 to 5,000 lb.			130	.185		1.41	6.35		7.76	11.40	
	0160	Over 5,000 lb.			145	.166		1.36	5.70		7.06	10.35	
	0500	Galvanized steel, under 200 lb.			235	.102		.81	3.53		4.34	6.35	
	0520	200 to 500 lb.			245	.098		.51	3.38		3.89	5.75	
	0540	500 to 1,000 lb.			255	.094		.49	3.25		3.74	5.55	
	0560	1,000 to 2,000 lb.			265	.091		.43	3.13		3.56	5.30	
	0570	2,000 to 5,000 lb.			275	.087		.39	3.01		3.40	5.10	
	0580	Over 5,000 lb.		↓	285	.084	↓	.36	2.91		3.27	4.89	
	0600	For large quantities special prices available from supplier											
	1000	Stainless steel, type 304, under 100 lb.		Q-10	165	.145	Lb.	1.34	5		6.34	9.20	
	1020	100 to 500 lb.			175	.137		1.24	4.74		5.98	8.65	
	1030	500 to 1,000 lb.			190	.126		1.18	4.36		5.54	8.05	
	1040	1,000 to 2,000 lb.			200	.120		1.12	4.14		5.26	7.65	
	1050	2,000 to 5,000 lb.			225	.107		1.04	3.68		4.72	6.85	
	1060	Over 5,000 lb.		↓	235	.102	↓	1.02	3.53		4.55	6.55	
	1080	Note: Minimum order cost exceeds per lb. cost for min. wt.											
	1100	For medium pressure ductwork, add							15%				
	1200	For high pressure ductwork, add							40%				
	1276	Duct tape, 2" x 60 yd roll					Ea.	5.20			5.20	5.70	
	1280	Add to labor for elevated installation											
	1282	of prefabricated (purchased) ductwork											
	1283	10' to 15' high							10%				
	1284	15' to 20' high							20%				
	1285	20' to 25' high							25%				
	1286	25' to 30' high							35%				
	1287	30' to 35' high							40%				
	1288	35' to 40' high							50%				
	1289	Over 40' high							55%				
	1300	Flexible, coated fiberglass fabric on corr. resist. metal helix											
	1400	pressure to 12" (WG) UL-181											
	1500	Non-insulated, 3" diameter		Q-9	400	.040	L.F.	1.03	1.33		2.36	3.19	
	1520	4" diameter			360	.044		1.03	1.48		2.51	3.42	
	1540	5" diameter			320	.050		1.20	1.67		2.87	3.89	
	1560	6" diameter			280	.057		1.43	1.90		3.33	4.51	
	1580	7" diameter			240	.067		1.69	2.22		3.91	5.30	
	1600	8" diameter			200	.080		1.95	2.66		4.61	6.25	
	1620	9" diameter			180	.089		2.28	2.96		5.24	7.10	
	1640	10" diameter			160	.100		2.47	3.33		5.80	7.85	
	1660	12" diameter			120	.133		2.98	4.44		7.42	10.15	
	1680	14" diameter			80	.200		3.61	6.65		10.26	14.25	
	1700	16" diameter		↓	60	.267	↓	5.35	8.90		14.25	19.60	

Important: See the Reference Section for critical supporting data - Reference Nos., Crews, & City Cost Indexes

15810	Ducts	CREW	DAILY OUTPUT	LABOR-HOURS	UNIT	2003 BARE COSTS				TOTAL INCL O&P
						MAT.	LABOR	EQUIP.	TOTAL	
600 1800	For adjustable clamps, add				Ea.	.74			.74	.81 **600**
1900	Insulated, 1" thick, PE jacket, 3" diameter R15810-050	Q-9	380	.042	L.F.	1.07	1.40		2.47	3.35
1910	4" diameter		340	.047		1.16	1.57		2.73	3.70
1920	5" diameter R15810-070		300	.053		1.27	1.78		3.05	4.14
1940	6" diameter		260	.062		1.47	2.05		3.52	4.79
1960	7" diameter R15810-100		220	.073		1.66	2.42		4.08	5.55
1980	8" diameter		180	.089		1.84	2.96		4.80	6.60
2000	9" diameter		160	.100		2.03	3.33		5.36	7.40
2020	10" diameter		140	.114		2.26	3.81		6.07	8.40
2040	12" diameter		100	.160		2.78	5.35		8.13	11.30
2060	14" diameter		80	.200		3.16	6.65		9.81	13.80
2080	16" diameter		60	.267		3.49	8.90		12.39	17.55
2100	18" diameter		45	.356		3.96	11.85		15.81	22.50
2120	20" diameter	Q-10	65	.369		8.20	12.75		20.95	29
2500	Insulated, heavy duty, coated fiberglass fabric									
2520	4" diameter	Q-9	340	.047	L.F.	2.63	1.57		4.20	5.30
2540	5" diameter		300	.053		3.22	1.78		5	6.30
2560	6" diameter		260	.062		3.39	2.05		5.44	6.90
2580	7" diameter		220	.073		4.10	2.42		6.52	8.25
2600	8" diameter		180	.089		4.21	2.96		7.17	9.20
2620	9" diameter		160	.100		5.05	3.33		8.38	10.70
2640	10" diameter		140	.114		5.40	3.81		9.21	11.85
2660	12" diameter		100	.160		6.25	5.35		11.60	15.10
2680	14" diameter		80	.200		7.50	6.65		14.15	18.55
2700	16" diameter		60	.267		8.40	8.90		17.30	23
2720	18" diameter		45	.356		11.80	11.85		23.65	31.50
2800	Flexible, alum., pressure to 12" (WG) UL-181									
2820	Non-insulated									
2830	3" diameter	Q-9	400	.040	L.F.	.97	1.33		2.30	3.13
2831	4" diameter		360	.044		1.18	1.48		2.66	3.59
2832	5" diameter		320	.050		1.49	1.67		3.16	4.21
2833	6" diameter		280	.057		1.82	1.90		3.72	4.94
2834	7" diameter		240	.067		2.61	2.22		4.83	6.30
2835	8" diameter		200	.080		2.67	2.66		5.33	7.05
2836	9" diameter		180	.089		3.12	2.96		6.08	8
2837	10" diameter		160	.100		3.28	3.33		6.61	8.75
2838	12" diameter		120	.133		4.03	4.44		8.47	11.30
2839	14" diameter		80	.200		4.84	6.65		11.49	15.60
2840	15" diameter		70	.229		5.30	7.60		12.90	17.60
2841	16" diameter		60	.267		6.50	8.90		15.40	21
2842	18" diameter		45	.356		7.05	11.85		18.90	26
2843	20" diameter	Q-10	65	.369		7.55	12.75		20.30	28
2880	Insulated, 1" thick with 3/4 lb., PE jacket									
2890	3" diameter	Q-9	380	.042	L.F.	1.78	1.40		3.18	4.13
2891	4" diameter		340	.047		2.01	1.57		3.58	4.63
2892	5" diameter		300	.053		2.54	1.78		4.32	5.55
2893	6" diameter		260	.062		2.89	2.05		4.94	6.35
2894	7" diameter		220	.073		3.28	2.42		5.70	7.35
2895	8" diameter		180	.089		3.86	2.96		6.82	8.80
2896	9" diameter		160	.100		4.36	3.33		7.69	9.95
2897	10" diameter		140	.114		4.93	3.81		8.74	11.30
2898	12" diameter		100	.160		5.50	5.35		10.85	14.30
2899	14" diameter		80	.200		6.40	6.65		13.05	17.35
2900	15" diameter		70	.229		7.10	7.60		14.70	19.55
2901	16" diameter		60	.267		8.40	8.90		17.30	23
2902	18" diameter		45	.356		9.45	11.85		21.30	28.50

MECHANICAL 15

				DAILY	LABOR-		2003 BARE COSTS				TOTAL
15810	**Ducts**		CREW	OUTPUT	HOURS	UNIT	MAT.	LABOR	EQUIP.	TOTAL	INCL O&P
600	3000	Transitions, 3" deep, square to round	R15810 -050									600
	3010	6" x 6" to round		Q-9	44	.364	Ea.	7.15	12.10		19.25	26.50
	3014	6" x 12" to round	R15810 -070		42	.381		13	12.70		25.70	34
	3018	8" x 8" to round			40	.400		7.80	13.30		21.10	29
	3022	9" x 9" to round	R15810 -100		38	.421		9.10	14		23.10	31.50
	3026	9" x 12" to round			37	.432		13.65	14.40		28.05	37
	3030	9" x 15" to round			36	.444		16.90	14.80		31.70	41.50
	3034	10" x 10" to round			34	.471		10.40	15.65		26.05	35.50
	3038	10" x 22" to round			32	.500		19.50	16.65		36.15	47
	3042	12" x 12" to round			30	.533		11.05	17.75		28.80	39.50
	3046	12" x 18" to round			28	.571		21	19.05		40.05	52.50
	3050	12" x 24" to round			26	.615		23.50	20.50		44	57
	3054	15" x 15" to round			25	.640		14.30	21.50		35.80	49
	3058	15" x 18" to round			24	.667		21	22		43	57.50
	3062	18" x 18" to round			22	.727		16.90	24		40.90	56
	3066	21" x 21" to round			21	.762		30	25.50		55.50	72
	3070	22" x 22" to round			20	.800		30	26.50		56.50	74
	3074	24" x 24" to round		18	.889		30	29.50		59.50	78.50
	3490	Rigid fiberglass duct board, foil reinf. kraft facing										
	3500	Rectangular, 1" thick, alum. faced, (FRK), std. weight		Q-10	350	.069	SF Surf	1.09	2.37		3.46	4.86
	3550	Rigid fiberglass reinforced plastic, FM approved										
	3552	for acid fume and smoke exhaust system, nonflammable										
	3554	Straight, 4" diameter		Q-9	200	.080	L.F.	10.75	2.66		13.41	15.95
	3555	6" diameter			146	.110		14.10	3.65		17.75	21
	3556	8" diameter			106	.151		17.10	5.05		22.15	26.50
	3557	10" diameter			85	.188		20.50	6.25		26.75	32
	3558	12" diameter			68	.235		24	7.85		31.85	38.50
	3559	14" diameter			50	.320		35.50	10.65		46.15	55.50
	3560	16" diameter			40	.400		39.50	13.30		52.80	64
	3561	18" diameter			31	.516		44	17.20		61.20	75
	3562	20" diameter		Q-10	44	.545		48.50	18.85		67.35	82.50
	3563	22" diameter			40	.600		52.50	20.50		73	90
	3564	24" diameter			37	.649		56.50	22.50		79	96.50
	3565	26" diameter			34	.706		61	24.50		85.50	105
	3566	28" diameter			28	.857		65.50	29.50		95	118
	3567	30" diameter			24	1		69.50	34.50		104	130
	3568	32" diameter			22.60	1.062		74	36.50		110.50	138
	3569	34" diameter			21.60	1.111		78	38.50		116.50	146
	3570	36" diameter			20.60	1.165		82.50	40		122.50	153
	3571	38" diameter			19.80	1.212		127	42		169	205
	3572	42" diameter			18.30	1.311		142	45.50		187.50	226
	3573	46" diameter			17	1.412		157	49		206	249
	3574	50" diameter			15.90	1.509		171	52		223	269
	3575	54" diameter			14.90	1.611		188	55.50		243.50	293
	3576	60" diameter			13.60	1.765		215	61		276	330
	3577	64" diameter			12.80	1.875		315	64.50		379.50	445
	3578	68" diameter			12.20	1.967		335	68		403	475
	3579	72" diameter			11.90	2.017		360	69.50		429.50	505
	3584	Note: joints are cemented with										
	3586	fiberglass resin, included in material cost.										
	3590	Elbow, 90°, 4" diameter		Q-9	20.30	.788	Ea.	52	26.50		78.50	97.50
	3591	6" diameter			13.80	1.159		72.50	38.50		111	140
	3592	8" diameter			10	1.600		80	53.50		133.50	171
	3593	10" diameter			7.80	2.051		105	68.50		173.50	222
	3594	12" diameter			6.50	2.462		133	82		215	273
	3595	14" diameter			5.50	2.909		163	97		260	330

Important: See the Reference Section for critical supporting data - Reference Nos., Crews, & City Cost Indexes

			CREW	DAILY OUTPUT	LABOR-HOURS	UNIT	2003 BARE COSTS				TOTAL INCL O&P
	15810	**Ducts**					MAT.	LABOR	EQUIP.	TOTAL	
600	3596	16″ diameter	Q-9	4.80	3.333	Ea.	194	111		305	385	**600**
	3597	18″ diameter		4.40	3.636		278	121		399	490
	3598	20″ diameter	Q-10	5.90	4.068		293	140		433	535
	3599	22″ diameter		5.40	4.444		360	153		513	630
	3600	24″ diameter		4.90	4.898		415	169		584	715
	3601	26″ diameter		4.60	5.217		420	180		600	740
	3602	28″ diameter		4.20	5.714		525	197		722	885
	3603	30″ diameter		3.90	6.154		585	212		797	975
	3604	32″ diameter		3.70	6.486		650	224		874	1,050
	3605	34″ diameter		3.45	6.957		765	240		1,005	1,200
	3606	36″ diameter		3.20	7.500		880	259		1,139	1,375
	3607	38″ diameter		3.07	7.818		1,250	270		1,520	1,800
	3608	42″ diameter		2.70	8.889		1,500	305		1,805	2,125
	3609	46″ diameter		2.54	9.449		1,750	325		2,075	2,425
	3610	50″ diameter		2.33	10.300		2,025	355		2,380	2,775
	3611	54″ diameter		2.14	11.215		2,325	385		2,710	3,150
	3612	60″ diameter		1.94	12.371		3,100	425		3,525	4,050
	3613	64″ diameter		1.82	13.187		4,500	455		4,955	5,650
	3614	68″ diameter		1.71	14.035		5,000	485		5,485	6,250
	3615	72″ diameter		1.62	14.815		5,525	510		6,035	6,875
	3626	Elbow, 45°, 4″ diameter	Q-9	22.30	.717		50	24		74	92
	3627	6″ diameter		15.20	1.053		69	35		104	130
	3628	8″ diameter		11	1.455		73.50	48.50		122	156
	3629	10″ diameter		8.60	1.860		77.50	62		139.50	181
	3630	12″ diameter		7.20	2.222		85	74		159	208
	3631	14″ diameter		6.10	2.623		104	87.50		191.50	249
	3632	16″ diameter		5.28	3.030		124	101		225	292
	3633	18″ diameter		4.84	3.306		145	110		255	330
	3634	20″ diameter	Q-10	6.50	3.692		168	128		296	380
	3635	22″ diameter		5.94	4.040		229	140		369	465
	3636	24″ diameter		5.40	4.444		261	153		414	525
	3637	26″ diameter		5.06	4.743		296	164		460	580
	3638	28″ diameter		4.62	5.195		320	179		499	625
	3639	30″ diameter		4.30	5.581		370	193		563	705
	3640	32″ diameter		4.07	5.897		410	204		614	765
	3641	34″ diameter		3.80	6.316		485	218		703	870
	3642	36″ diameter		3.52	6.818		560	235		795	980
	3643	38″ diameter		3.38	7.101		785	245		1,030	1,250
	3644	42″ diameter		2.97	8.081		940	279		1,219	1,450
	3645	46″ diameter		2.80	8.571		1,100	296		1,396	1,650
	3646	50″ diameter		2.56	9.375		1,275	325		1,600	1,900
	3647	54″ diameter		2.35	10.213		1,450	355		1,805	2,150
	3648	60″ diameter		2.13	11.268		1,950	390		2,340	2,750
	3649	64″ diameter		2	12		2,825	415		3,240	3,750
	3650	68″ diameter		1.88	12.766		3,125	440		3,565	4,125
	3651	72″ diameter		1.78	13.483		3,450	465		3,915	4,525
	3660	Tee, 90°, 4″ diameter	Q-9	14.06	1.138		28.50	38		66.50	90
	3661	6″ diameter		9.52	1.681		43.50	56		99.50	135
	3662	8″ diameter		6.98	2.292		59.50	76.50		136	184
	3663	10″ diameter		5.45	2.936		77.50	98		175.50	237
	3664	12″ diameter		4.60	3.478		97	116		213	286
	3665	14″ diameter		3.86	4.145		141	138		279	370
	3666	16″ diameter		3.40	4.706		168	157		325	425
	3667	18″ diameter		3.05	5.246		199	175		374	490
	3668	20″ diameter	Q-10	4.12	5.825		233	201		434	565
	3669	22″ diameter		3.78	6.349		266	219		485	635

Reference numbers: R15810-050, R15810-070, R15810-100

MECHANICAL 15

			CREW	DAILY OUTPUT	LABOR-HOURS	UNIT	2003 BARE COSTS				TOTAL INCL O&P
							MAT.	LABOR	EQUIP.	TOTAL	
600	3670	24" diameter	Q-10	3.46	6.936	Ea.	300	240		540	700	600
	3671	26" diameter	R15810 -050	3.20	7.500		340	259		599	775
	3672	28" diameter		2.96	8.108		380	280		660	850
	3673	30" diameter	R15810 -070	2.75	8.727		425	300		725	935
	3674	32" diameter		2.54	9.449		470	325		795	1,025
	3675	34" diameter	R15810 -100	2.40	10		515	345		860	1,100
	3676	36" diameter		2.27	10.573		565	365		930	1,175
	3677	38" diameter		2.15	11.163		850	385		1,235	1,525
	3678	42" diameter		1.91	12.565		1,025	435		1,460	1,800
	3679	46" diameter		1.78	13.483		1,200	465		1,665	2,050
	3680	50" diameter		1.63	14.724		1,400	510		1,910	2,325
	3681	54" diameter		1.50	16		1,625	550		2,175	2,650
	3682	60" diameter		1.35	17.778		2,025	615		2,640	3,175
	3683	64" diameter		1.27	18.898		3,050	655		3,705	4,350
	3684	68" diameter		1.20	20		3,400	690		4,090	4,825
	3685	72" diameter		1.13	21.239		3,800	735		4,535	5,300
	3690	For Y @ 45°, add					22%				
	3700	Blast gate, 4" diameter	Q-9	22.30	.717		138	24		162	189
	3701	6" diameter		15.20	1.053		156	35		191	226
	3702	8" diameter		11	1.455		210	48.50		258.50	305
	3703	10" diameter		8.60	1.860		240	62		302	360
	3704	12" diameter		7.20	2.222		293	74		367	435
	3705	14" diameter		6.10	2.623		325	87.50		412.50	495
	3706	16" diameter		5.28	3.030		360	101		461	550
	3707	18" diameter		4.84	3.306		395	110		505	605
	3708	20" diameter	Q-10	6.50	3.692		425	128		553	665
	3709	22" diameter		5.94	4.040		460	140		600	720
	3710	24" diameter		5.40	4.444		495	153		648	780
	3711	26" diameter		5.06	4.743		530	164		694	840
	3712	28" diameter		4.62	5.195		570	179		749	900
	3713	30" diameter		4.30	5.581		605	193		798	965
	3714	32" diameter		4.07	5.897		660	204		864	1,050
	3715	34" diameter		3.80	6.316		720	218		938	1,125
	3716	36" diameter		3.52	6.818		775	235		1,010	1,225
	3717	38" diameter		3.38	7.101		1,050	245		1,295	1,525
	3718	42" diameter		2.97	8.081		1,200	279		1,479	1,750
	3719	46" diameter		2.80	8.571		1,375	296		1,671	1,975
	3720	50" diameter		2.56	9.375		1,575	325		1,900	2,225
	3721	54" diameter		2.35	10.213		1,775	355		2,130	2,500
	3722	60" diameter		2.13	11.268		2,075	390		2,465	2,875
	3723	64" diameter		2	12		2,300	415		2,715	3,175
	3724	68" diameter		1.88	12.766		2,550	440		2,990	3,475
	3725	72" diameter		1.78	13.483		2,800	465		3,265	3,800
	3740	Reducers, 4" diameter	Q-9	24.50	.653		34.50	22		56.50	71.50
	3741	6" diameter		16.70	.958		51.50	32		83.50	106
	3742	8" diameter		12	1.333		69.50	44.50		114	145
	3743	10" diameter		9.50	1.684		86	56		142	181
	3744	12" diameter		7.90	2.025		103	67.50		170.50	217
	3745	14" diameter		6.70	2.388		120	79.50		199.50	255
	3746	16" diameter		5.80	2.759		137	92		229	293
	3747	18" diameter		5.30	3.019		155	101		256	325
	3748	20" diameter	Q-10	7.20	3.333		172	115		287	365
	3749	22" diameter		6.50	3.692		277	128		405	500
	3750	24" diameter		5.90	4.068		300	140		440	545
	3751	26" diameter		5.60	4.286		325	148		473	590
	3752	28" diameter		5.10	4.706		355	162		517	640

	15810	Ducts		CREW	DAILY OUTPUT	LABOR-HOURS	UNIT	2003 BARE COSTS				TOTAL INCL O&P	
								MAT.	LABOR	EQUIP.	TOTAL		
600	3753	30" diameter	R15810 -050	Q-10	4.70	5.106	Ea.	380	176		556	690	**600**
	3754	32" diameter			4.50	5.333		405	184		589	730	
	3755	34" diameter	R15810 -070		4.20	5.714		430	197		627	780	
	3756	36" diameter			3.90	6.154		455	212		667	830	
	3757	38" diameter	R15810 -100		3.70	6.486		690	224		914	1,100	
	3758	42" diameter			3.30	7.273		760	251		1,011	1,225	
	3759	46" diameter			3.10	7.742		835	267		1,102	1,325	
	3760	50" diameter			2.80	8.571		905	296		1,201	1,450	
	3761	54" diameter			2.60	9.231		980	320		1,300	1,575	
	3762	60" diameter			2.34	10.256		1,100	355		1,455	1,750	
	3763	64" diameter			2.20	10.909		1,700	375		2,075	2,450	
	3764	68" diameter			2.06	11.650		1,800	400		2,200	2,600	
	3765	72" diameter		▼	1.96	12.245		1,925	425		2,350	2,775	
	3780	Flange, per each, 4" diameter		Q-9	40.60	.394		83.50	13.10		96.60	113	
	3781	6" diameter			27.70	.578		97.50	19.25		116.75	137	
	3782	8" diameter			20.50	.780		122	26		148	174	
	3783	10" diameter			16.30	.982		146	32.50		178.50	212	
	3784	12" diameter			13.90	1.151		168	38.50		206.50	244	
	3785	14" diameter			11.30	1.416		194	47		241	286	
	3786	16" diameter			9.70	1.649		218	55		273	325	
	3787	18" diameter		▼	8.30	1.928		242	64		306	365	
	3788	20" diameter		Q-10	11.30	2.124		266	73.50		339.50	405	
	3789	22" diameter			10.40	2.308		290	79.50		369.50	445	
	3790	24" diameter			9.50	2.526		315	87		402	480	
	3791	26" diameter			8.80	2.727		340	94		434	520	
	3792	28" diameter			8.20	2.927		365	101		466	555	
	3793	30" diameter			7.60	3.158		385	109		494	595	
	3794	32" diameter			7.20	3.333		410	115		525	630	
	3795	34" diameter			6.80	3.529		435	122		557	670	
	3796	36" diameter			6.40	3.750		460	129		589	705	
	3797	38" diameter			6.10	3.934		485	136		621	745	
	3798	42" diameter			5.50	4.364		530	151		681	820	
	3799	46" diameter			5	4.800		580	166		746	895	
	3800	50" diameter			4.60	5.217		630	180		810	975	
	3801	54" diameter			4.20	5.714		675	197		872	1,050	
	3802	60" diameter			3.80	6.316		750	218		968	1,150	
	3803	64" diameter			3.60	6.667		795	230		1,025	1,225	
	3804	68" diameter			3.40	7.059		845	244		1,089	1,300	
	3805	72" diameter		▼	3.20	7.500	▼	895	259		1,154	1,375	
	4000	Rigid plastic, corrosive fume resistant PVC											
	4020	Straight, 6" diameter		Q-9	220	.073	L.F.	6.50	2.42		8.92	10.90	
	4030	7" diameter			160	.100		9.60	3.33		12.93	15.70	
	4040	8" diameter			160	.100		9.60	3.33		12.93	15.70	
	4050	9" diameter			140	.114		10.85	3.81		14.66	17.85	
	4060	10" diameter			120	.133		12.10	4.44		16.54	20	
	4070	12" diameter			100	.160		14.15	5.35		19.50	24	
	4080	14" diameter			70	.229		18.35	7.60		25.95	32	
	4090	16" diameter			62	.258		21	8.60		29.60	37	
	4100	18" diameter		▼	58	.276		33	9.20		42.20	50	
	4110	20" diameter		Q-10	75	.320		42.50	11.05		53.55	63.50	
	4120	22" diameter			60	.400		48.50	13.80		62.30	75	
	4130	24" diameter			55	.436		54.50	15.05		69.55	83.50	
	4140	26" diameter			42	.571		62	19.75		81.75	98.50	
	4150	28" diameter			35	.686		66.50	23.50		90	110	
	4160	30" diameter			28	.857		72	29.50		101.50	125	
	4170	36" diameter		▼	23	1.043	▼	86	36		122	150	

MECHANICAL **15**

				DAILY	LABOR-			2003 BARE COSTS				TOTAL	
15810	**Ducts**		CREW	OUTPUT	HOURS	UNIT	MAT.	LABOR	EQUIP.	TOTAL		INCL O&P	
600	4180	42″ diameter	R15810 -050	Q-10	20	1.200	L.F.	100	41.50		141.50	174	**600**
	4200	48″ diameter		↓	18	1.333	↓	114	46		160	197	
	4250	Coupling, 6″ diameter	R15810 -070	Q-9	88	.182	Ea.	6.75	6.05		12.80	16.80	
	4260	7″ diameter			82	.195		7.25	6.50		13.75	18	
	4270	8″ diameter	R15810 -100		78	.205		7.55	6.85		14.40	18.85	
	4280	9″ diameter			74	.216		8.25	7.20		15.45	20	
	4290	10″ diameter			70	.229		8.25	7.60		15.85	21	
	4300	12″ diameter			55	.291		8.80	9.70		18.50	24.50	
	4310	14″ diameter			44	.364		10.75	12.10		22.85	30.50	
	4320	16″ diameter			40	.400		12.70	13.30		26	34.50	
	4330	18″ diameter		↓	38	.421		14.75	14		28.75	37.50	
	4340	20″ diameter		Q-10	55	.436		25.50	15.05		40.55	51.50	
	4350	22″ diameter			50	.480		29.50	16.55		46.05	58	
	4360	24″ diameter			45	.533		33	18.40		51.40	65	
	4370	26″ diameter			39	.615		35	21.50		56.50	71.50	
	4380	28″ diameter			33	.727		36.50	25		61.50	79.50	
	4390	30″ diameter			27	.889		40	30.50		70.50	91.50	
	4400	36″ diameter			25	.960		63.50	33		96.50	121	
	4410	42″ diameter			23	1.043		71	36		107	134	
	4420	48″ diameter		↓	21	1.143		83	39.50		122.50	152	
	4470	Elbow, 90°, 6″ diameter		Q-9	44	.364		63.50	12.10		75.60	88	
	4480	7″ diameter			36	.444		68	14.80		82.80	97.50	
	4490	8″ diameter			28	.571		68	19.05		87.05	104	
	4500	9″ diameter			22	.727		93.50	24		117.50	141	
	4510	10″ diameter			18	.889		110	29.50		139.50	167	
	4520	12″ diameter			15	1.067		150	35.50		185.50	220	
	4530	14″ diameter			11	1.455		205	48.50		253.50	300	
	4540	16″ diameter		↓	10	1.600		247	53.50		300.50	355	
	4550	18″ diameter		Q-10	15	1.600		268	55.50		323.50	380	
	4560	20″ diameter			14	1.714		284	59		343	405	
	4570	22″ diameter			13	1.846		360	64		424	500	
	4580	24″ diameter			12	2		410	69		479	555	
	4590	26″ diameter			11	2.182		465	75.50		540.50	630	
	4600	28″ diameter			10	2.400		555	83		638	740	
	4610	30″ diameter			9	2.667		650	92		742	855	
	4620	36″ diameter			7	3.429		895	118		1,013	1,175	
	4630	42″ diameter			5	4.800		1,225	166		1,391	1,600	
	4640	48″ diameter		↓	4	6	↓	1,625	207		1,832	2,100	
	4750	Elbow 45°, use 90° and deduct						45%					
	5400	Spiral preformed, steel, galv., straight lengths, Max 10″ spwg.											
	5410	4″ diameter, 26 ga.		Q-9	360	.044	L.F.	.85	1.48		2.33	3.23	
	5416	5″ diameter, 26 ga.			320	.050		1.08	1.67		2.75	3.76	
	5420	6″ diameter, 26 ga.			280	.057		1.23	1.90		3.13	4.29	
	5425	7″ diameter, 26 ga.			240	.067		1.46	2.22		3.68	5.05	
	5430	8″ diameter, 26 ga.			200	.080		1.62	2.66		4.28	5.90	
	5440	10″ diameter, 26 ga.			160	.100		2	3.33		5.33	7.35	
	5450	12″ diameter, 26 ga.			120	.133		2.46	4.44		6.90	9.55	
	5460	14″ diameter, 26 ga.			80	.200		2.85	6.65		9.50	13.45	
	5480	16″ diameter, 24 ga.			60	.267		4.24	8.90		13.14	18.35	
	5490	18″ diameter, 24 ga.		↓	50	.320		4.70	10.65		15.35	21.50	
	5500	20″ diameter, 24 ga.		Q-10	65	.369		5.25	12.75		18	25.50	
	5510	22″ diameter, 24 ga.			60	.400		5.80	13.80		19.60	28	
	5520	24″ diameter, 24 ga.			55	.436		6.30	15.05		21.35	30.50	
	5540	30″ diameter, 22 ga.			45	.533		9.55	18.40		27.95	39	
	5600	36″ diameter, 22 ga.		↓	40	.600	↓	11.45	20.50		31.95	44.50	
	5800	Connector, 4″ diameter		Q-9	100	.160	Ea.	1.58	5.35		6.93	10	

Important: See the Reference Section for critical supporting data - Reference Nos., Crews, & City Cost Indexes

15810	Ducts		CREW	DAILY OUTPUT	LABOR-HOURS	UNIT	2003 BARE COSTS				TOTAL INCL O&P
							MAT.	LABOR	EQUIP.	TOTAL	
5810	5" diameter	R15810 -050	Q-9	94	.170	Ea.	1.69	5.65		7.34	10.60
5820	6" diameter			88	.182		1.82	6.05		7.87	11.35
5840	8" diameter	R15810 -070		78	.205		2.35	6.85		9.20	13.15
5860	10" diameter			70	.229		3.71	7.60		11.31	15.85
5880	12" diameter	R15810 -100		50	.320		4.47	10.65		15.12	21.50
5900	14" diameter			44	.364		5.20	12.10		17.30	24.50
5920	16" diameter			40	.400		5.80	13.30		19.10	27
5930	18" diameter			37	.432		6.50	14.40		20.90	29
5940	20" diameter			34	.471		7.05	15.65		22.70	32
5950	22" diameter			31	.516		7.70	17.20		24.90	35
5960	24" diameter			28	.571		8.25	19.05		27.30	38.50
5980	30" diameter			22	.727		10.75	24		34.75	49.50
6000	36" diameter			18	.889		16.10	29.50		45.60	63.50
6300	Elbow, 45°, 4" diameter			60	.267		1.79	8.90		10.69	15.65
6310	5" diameter			52	.308		2.30	10.25		12.55	18.35
6320	6" diameter			44	.364		2.51	12.10		14.61	21.50
6340	8" diameter			28	.571		3.31	19.05		22.36	33
6360	10" diameter			18	.889		5.05	29.50		34.55	51
6380	12" diameter			13	1.231		6.50	41		47.50	70.50
6400	14" diameter			11	1.455		12.20	48.50		60.70	88.50
6420	16" diameter			10	1.600		20	53.50		73.50	105
6430	18" diameter		▼	9.60	1.667		24.50	55.50		80	113
6440	20" diameter		Q-10	14	1.714		31	59		90	126
6450	22" diameter			13	1.846		31	64		95	133
6460	24" diameter			12	2		41.50	69		110.50	153
6480	30" diameter			9	2.667		64	92		156	213
6500	36" diameter		▼	7	3.429		78	118		196	269
6600	Elbow, 90°, 4" diameter		Q-9	60	.267		3.05	8.90		11.95	17.05
6610	5" diameter			52	.308		3.51	10.25		13.76	19.65
6620	6" diameter			44	.364		4.02	12.10		16.12	23
6625	7" diameter			36	.444		4.25	14.80		19.05	27.50
6630	8" diameter			28	.571		4.93	19.05		23.98	35
6640	10" diameter			18	.889		7.05	29.50		36.55	53.50
6650	12" diameter			13	1.231		9.30	41		50.30	74
6660	14" diameter			11	1.455		17.45	48.50		65.95	94
6670	16" diameter			10	1.600		27.50	53.50		81	113
6676	18" diameter		▼	9.60	1.667		36	55.50		91.50	125
6680	20" diameter		Q-10	14	1.714		48	59		107	144
6684	22" diameter			13	1.846		74.50	64		138.50	181
6690	24" diameter			12	2		83.50	69		152.50	199
6700	30" diameter			9	2.667		128	92		220	283
6710	36" diameter		▼	7	3.429		156	118		274	355
6720	End cap										
6722	4"		Q-9	110	.145	Ea.	1.58	4.84		6.42	9.25
6724	6"			94	.170		2.10	5.65		7.75	11.05
6725	8"			88	.182		3.01	6.05		9.06	12.65
6726	10"			78	.205		4.37	6.85		11.22	15.35
6727	12"			70	.229		5.80	7.60		13.40	18.15
6728	14"			50	.320		8.05	10.65		18.70	25.50
6729	16"			44	.364		9.65	12.10		21.75	29.50
6730	18"			40	.400		11.30	13.30		24.60	33
6731	20"			37	.432		12.30	14.40		26.70	35.50
6732	22"			34	.471		13.70	15.65		29.35	39
6733	24"			30	.533		16.10	17.75		33.85	45
6800	Reducing coupling, 6" x 4"			46	.348		4.55	11.60		16.15	23
6820	8" x 6"		▼	40	.400	▼	5.60	13.30		18.90	26.50

600

15810	Ducts	CREW	DAILY OUTPUT	LABOR-HOURS	UNIT	2003 BARE COSTS				TOTAL INCL O&P	
						MAT.	LABOR	EQUIP.	TOTAL		
600 6840	10" x 8" R15810-050	Q-9	32	.500	Ea.	7.05	16.65		23.70	33.50	600
6860	12" x 10"		24	.667		10.10	22		32.10	45.50	
6880	14" x 12" R15810-070		20	.800		16.65	26.50		43.15	59.50	
6900	16" x 14"		18	.889		24.50	29.50		54	72	
6920	18" x 16" R15810-100		16	1		30	33.50		63.50	84.50	
6940	20" x 18"	Q-10	24	1		36.50	34.50		71	94	
6950	22" x 20"		23	1.043		43	36		79	103	
6960	24" x 22"		22	1.091		49	37.50		86.50	112	
6980	30" x 28"		18	1.333		73	46		119	152	
7000	36" x 34"		16	1.500		98	52		150	188	
7100	Tee, 90°, 4" diameter	Q-9	40	.400		3.99	13.30		17.29	25	
7110	5" diameter		34.60	.462		5.30	15.40		20.70	30	
7120	6" diameter		30	.533		5.75	17.75		23.50	34	
7130	8" diameter		19	.842		7.50	28		35.50	52	
7140	10" diameter		12	1.333		10.70	44.50		55.20	80.50	
7150	12" diameter		9	1.778		13.65	59		72.65	107	
7160	14" diameter		7	2.286		26	76		102	147	
7170	16" diameter		6.70	2.388		36.50	79.50		116	163	
7176	18" diameter		6	2.667		50.50	89		139.50	193	
7180	20" diameter	Q-10	8.70	2.759		56	95.50		151.50	209	
7190	24" diameter		7	3.429		66.50	118		184.50	256	
7200	30" diameter		6	4		82	138		220	305	
7210	36" diameter		5	4.800		104	166		270	370	
7350	Lateral 45°, reducing										
7354	6" x 6" x 4"	Q-9	30	.533	Ea.	7.35	17.75		25.10	35.50	
7358	8" x 8" x 4"		19	.842		9.05	28		37.05	53.50	
7362	10" x 10" x 4"		12	1.333		12.70	44.50		57.20	82.50	
7366	12" x 12" x 4"		9	1.778		14.35	59		73.35	107	
7370	14" x 14" x 4"		7	2.286		20.50	76		96.50	141	
7374	16" x 16" x 4"		6.70	2.388		32.50	79.50		112	159	
7378	18" x 18" x 4"		6	2.667		41	89		130	182	
7382	20" x 20" x 4"	Q-10	8.70	2.759		44.50	95.50		140	196	
7386	22" x 22" x 4"		8	3		50	104		154	215	
7390	24" x 24" x 4"		7	3.429		53.50	118		171.50	242	
7400	Steel, PVC coated both sides, straight lengths										
7410	4" diameter, 26 ga.	Q-9	360	.044	L.F.	1.19	1.48		2.67	3.60	
7416	5" diameter, 26 ga.		240	.067		1.48	2.22		3.70	5.05	
7420	6" diameter, 26 ga.		280	.057		1.65	1.90		3.55	4.76	
7440	8" diameter, 26 ga.		200	.080		2.28	2.66		4.94	6.60	
7460	10" diameter, 26 ga.		160	.100		3.07	3.33		6.40	8.55	
7480	12" diameter, 26 ga.		120	.133		3.80	4.44		8.24	11.05	
7500	14" diameter, 26 ga.		80	.200		5.20	6.65		11.85	16	
7520	16" diameter, 24 ga.		60	.267		7.35	8.90		16.25	22	
7540	18" diameter, 24 ga.		45	.356		7.95	11.85		19.80	27	
7560	20" diameter, 24 ga.	Q-10	65	.369		9.35	12.75		22.10	30	
7580	24" diameter, 24 ga.		55	.436		11.90	15.05		26.95	36.50	
7600	30" diameter, 22 ga.		45	.533		17.25	18.40		35.65	47.50	
7620	36" diameter, 22 ga.		40	.600		21	20.50		41.50	55	
7890	Connector, 4" diameter	Q-9	100	.160	Ea.	2.49	5.35		7.84	11	
7896	5" diameter		83	.193		2.66	6.40		9.06	12.85	
7900	6" diameter		88	.182		2.80	6.05		8.85	12.45	
7920	8" diameter		78	.205		3.54	6.85		10.39	14.45	
7940	10" diameter		70	.229		4.87	7.60		12.47	17.10	
7960	12" diameter		50	.320		6.25	10.65		16.90	23.50	
7980	14" diameter		44	.364		8	12.10		20.10	27.50	
8000	16" diameter		40	.400		8.90	13.30		22.20	30.50	

Important: See the Reference Section for critical supporting data - Reference Nos., Crews, & City Cost Indexes

	15810	Ducts		CREW	DAILY OUTPUT	LABOR-HOURS	UNIT	MAT.	LABOR	EQUIP.	TOTAL	TOTAL INCL O&P	
600	8020	18″ diameter	R15810-050	Q-9	37	.432	Ea.	10	14.40		24.40	33	600
	8040	20″ diameter			34	.471		10.75	15.65		26.40	36	
	8060	24″ diameter	R15810-070		28	.571		12.05	19.05		31.10	43	
	8080	30″ diameter			22	.727		15.75	24		39.75	55	
	8100	36″ diameter	R15810-100		18	.889		21.50	29.50		51	69	
	8390	Elbow, 45°, 4″ diameter			60	.267		3.14	8.90		12.04	17.15	
	8396	5″ diameter			36	.444		3.79	14.80		18.59	27	
	8400	6″ diameter			44	.364		4.12	12.10		16.22	23	
	8420	8″ diameter			28	.571		5.45	19.05		24.50	35.50	
	8440	10″ diameter			18	.889		7.90	29.50		37.40	54	
	8460	12″ diameter			13	1.231		11.65	41		52.65	76.50	
	8480	14″ diameter			11	1.455		21	48.50		69.50	98	
	8500	16″ diameter			10	1.600		30	53.50		83.50	116	
	8520	18″ diameter			9	1.778		34.50	59		93.50	130	
	8540	20″ diameter		Q-10	13	1.846		46.50	64		110.50	150	
	8560	24″ diameter			11	2.182		59	75.50		134.50	181	
	8580	30″ diameter			9	2.667		90.50	92		182.50	242	
	8600	36″ diameter			7	3.429		110	118		228	305	
	8800	Elbow, 90°, 4″ diameter		Q-9	60	.267		5.20	8.90		14.10	19.45	
	8804	5″ diameter			52	.308		5.95	10.25		16.20	22.50	
	8806	6″ diameter			44	.364		6.50	12.10		18.60	26	
	8808	8″ diameter			28	.571		8.10	19.05		27.15	38.50	
	8810	10″ diameter			18	.889		10.95	29.50		40.45	57.50	
	8812	12″ diameter			13	1.231		15.10	41		56.10	80	
	8814	14″ diameter			11	1.455		27	48.50		75.50	105	
	8816	16″ diameter			10	1.600		42	53.50		95.50	129	
	8818	18″ diameter			9.60	1.667		51.50	55.50		107	142	
	8820	20″ diameter		Q-10	14	1.714		65.50	59		124.50	164	
	8822	24″ diameter			12	2		118	69		187	236	
	8824	30″ diameter			9	2.667		181	92		273	340	
	8826	36″ diameter			7	3.429		220	118		338	425	
	8850	Tee, 90°, 4″ diameter		Q-9	40	.400		5.85	13.30		19.15	27	
	8854	5″ diameter			34.60	.462		7.70	15.40		23.10	32.50	
	8856	6″ diameter			30	.533		8.55	17.75		26.30	37	
	8858	8″ diameter			19	.842		10.95	28		38.95	55.50	
	8860	10″ diameter			12	1.333		14.30	44.50		58.80	84.50	
	8862	12″ diameter			9	1.778		20.50	59		79.50	114	
	8864	14″ diameter			7	2.286		35.50	76		111.50	157	
	8866	16″ diameter			6.70	2.388		51.50	79.50		131	180	
	8868	18″ diameter			6	2.667		74.50	89		163.50	219	
	8870	20″ diameter		Q-10	8.70	2.759		80.50	95.50		176	236	
	8872	24″ diameter			7	3.429		111	118		229	305	
	8874	30″ diameter			6	4		138	138		276	365	
	8876	36″ diameter			5	4.800		175	166		341	450	
	9000	Reducing coupling, tapered offset											
	9020	Diameter x any smaller diameter											
	9040	6″ x		Q-9	46	.348	Ea.	6.50	11.60		18.10	25	
	9060	8″ x			40	.400		8	13.30		21.30	29.50	
	9080	10″ x			32	.500		9.65	16.65		26.30	36	
	9100	12″ x			24	.667		13.45	22		35.45	49.50	
	9120	14″ x			20	.800		21.50	26.50		48	65	
	9140	16″ x			18	.889		33	29.50		62.50	81.50	
	9160	18″ x			16	1		40	33.50		73.50	95.50	
	9180	20″ x		Q-10	24	1		49	34.50		83.50	108	
	9200	24″ x			22	1.091		66.50	37.50		104	131	
	9220	30″ x			18	1.333		94	46		140	174	

MECHANICAL 15

	15810	Ducts		CREW	DAILY OUTPUT	LABOR-HOURS	UNIT	2003 BARE COSTS				TOTAL INCL O&P	
								MAT.	LABOR	EQUIP.	TOTAL		
600	9240	36" x		Q-10	16	1.500	Ea.	122	52		174	214	600
	9400	Lateral 45°, straight barrel	R15810 -050										
	9410	4" size		Q-9	40	.400	Ea.	6.75	13.30		20.05	28	
	9416	5" size	R15810 -070		34.60	.462		8.85	15.40		24.25	34	
	9420	6" size			30	.533		9.80	17.75		27.55	38.50	
	9440	8" size	R15810 -100		19	.842		12.60	28		40.60	57.50	
	9460	10" size			12	1.333		16.45	44.50		60.95	86.50	
	9480	12" size			9	1.778		20.50	59		79.50	114	
	9500	14" size			7	2.286		27	76		103	148	
	9520	16" size			6.70	2.388		51	79.50		130.50	179	
	9540	18" size		▼	6	2.667		72	89		161	216	
	9560	20" size		Q-10	8.70	2.759		85.50	95.50		181	241	
	9580	24" size			7	3.429		119	118		237	315	
	9600	30" size			6	4		139	138		277	365	
	9620	36" size		▼	5	4.800		181	166		347	455	
	9640	For reducing lateral, add					▼	40%					
	9650	Lateral 45°, reducing											
	9654	6" x 6" x 4"		Q-9	30	.533	Ea.	14.55	17.75		32.30	43.50	
	9658	8" x 8" x 4"			19	.842		18.40	28		46.40	64	
	9662	10" x 10" x 4"			12	1.333		24	44.50		68.50	95	
	9666	12" x 12" x 4"			9	1.778		28	59		87	123	
	9670	14" x 14" x 4"			7	2.286		46.50	76		122.50	169	
	9674	16" x 16" x 4"			6.70	2.388		58	79.50		137.50	187	
	9678	18" x 18" x 4"		▼	6	2.667		91	89		180	237	
	9682	20" x 20" x 4"		Q-10	8.70	2.759		116	95.50		211.50	275	
	9686	22" x 22" x 4"			8	3		132	104		236	305	
	9690	24" x 24" x 4"		▼	7	3.429	▼	142	118		260	340	
	9700	Register boot, elbow or perimeter											
	9710	6" duct diameter		Q-9	60	.267	Ea.	24	8.90		32.90	40	
	9720	8" duct diameter			56	.286		29	9.50		38.50	46.50	
	9730	10" duct diameter			45	.356		40	11.85		51.85	62	
	9740	12" duct diameter		▼	36	.444	▼	40	14.80		54.80	66.50	
	9800	Steel, stainless, straight lengths											
	9810	3" diameter, 24 Ga.		Q-9	400	.040	L.F.	4.28	1.33		5.61	6.75	
	9820	4" diameter, 24 Ga.			360	.044		5.40	1.48		6.88	8.25	
	9830	5" diameter, 24 Ga.			320	.050		6.65	1.67		8.32	9.85	
	9840	6" diameter, 24 Ga.			280	.057		7.80	1.90		9.70	11.55	
	9850	7" diameter, 24 Ga.			240	.067		8.95	2.22		11.17	13.30	
	9860	8" diameter, 24 Ga.			200	.080		10.15	2.66		12.81	15.25	
	9870	9" diameter, 24 Ga.			180	.089		11.30	2.96		14.26	17	
	9880	10" diameter, 24 Ga.			160	.100		12.45	3.33		15.78	18.85	
	9890	12" diameter, 24 Ga.			120	.133		14.85	4.44		19.29	23	
	9900	14" diameter, 24 Ga.			80	.200		17.15	6.65		23.80	29	
	9910	16" diameter, 24 Ga.			60	.267		19.50	8.90		28.40	35	
	9920	18" diameter, 24 Ga.		▼	45	.356		22	11.85		33.85	42.50	
	9930	20" diameter, 24 Ga.		Q-10	65	.369		24	12.75		36.75	46	
	9940	24" diameter, 24 Ga.			55	.436		29	15.05		44.05	55.50	
	9950	30" diameter, 22 Ga.			45	.533		44	18.40		62.40	77	
	9960	36" diameter, 22 Ga.			40	.600		53.50	20.50		74	91	
	9970	40" diameter, 20 Ga.		▼	35	.686	▼	71	23.50		94.50	115	

	15820	Duct Accessories											
300	0010	**DUCT ACCESSORIES**											300
	0050	Air extractors, 12" x 4"		1 Shee	24	.333	Ea.	15.55	12.35		27.90	36	
	0100	8" x 6"			22	.364		15.55	13.45		29	38	
	0120	12" x 6"		▼	21	.381	▼	21.50	14.10		35.60	45.50	

Important: See the Reference Section for critical supporting data - Reference Nos., Crews, & City Cost Indexes

		CREW	DAILY OUTPUT	LABOR-HOURS	UNIT	2003 BARE COSTS				TOTAL INCL O&P
15820	**Duct Accessories**					MAT.	LABOR	EQUIP.	TOTAL	
0140	16" x 6"	1 Shee	20	.400	Ea.	27	14.80		41.80	53
0160	24" x 6"		18	.444		45	16.45		61.45	75
0180	12" x 8"		20	.400		24	14.80		38.80	49.50
0200	20" x 8"		16	.500		35	18.50		53.50	67.50
0240	18" x 10"		14	.571		34.50	21		55.50	70
0260	24" x 10"		12	.667		44.50	24.50		69	86.50
0280	24" x 12"		10	.800		48	29.50		77.50	98
0300	30" x 12"		8	1		59.50	37		96.50	123
0350	Duct collar, spin-in type									
0352	Without damper									
0354	Round									
0356	4"	1 Shee	32	.250	Ea.	1.73	9.25		10.98	16.20
0357	5"		30	.267		1.73	9.85		11.58	17.15
0358	6"		28	.286		1.85	10.55		12.40	18.40
0359	7"		26	.308		1.98	11.40		13.38	19.80
0360	8"		24	.333		2.11	12.35		14.46	21.50
0361	9"		22	.364		2.37	13.45		15.82	23.50
0362	10"		20	.400		2.47	14.80		17.27	25.50
0363	12"		17	.471		2.90	17.40		20.30	30
1000	Duct access door, insulated, 6" x 6"		14	.571		11.35	21		32.35	45
1020	10" x 10"		11	.727		16.25	27		43.25	59.50
1040	12" x 12"		10	.800		17.55	29.50		47.05	65
1050	12" x 18"		9	.889		21.50	33		54.50	75
1060	16" x 12"		9	.889		24.50	33		57.50	78
1070	18" x 18"		8	1		25	37		62	84.50
1074	24" x 18"		8	1		31	37		68	91
1080	24" x 24"		8	1		33	37		70	93
1100	Exhaust vent damper, auto., OB with elect. actuator									
1110	6" x 6"	1 Shee	24	.333	Ea.	67	12.35		79.35	93
1114	8" x 6"		23	.348		69	12.85		81.85	96
1116	8" x 8"		22	.364		72	13.45		85.45	100
1118	10" x 6"		23	.348		72	12.85		84.85	99
1120	10" x 10"		22	.364		76	13.45		89.45	105
1122	12" x 6"		22	.364		73	13.45		86.45	102
1124	12" x 10"		21	.381		80	14.10		94.10	110
1126	12" x 12"		20	.400		84	14.80		98.80	116
1128	14" x 6"		21	.381		74	14.10		88.10	104
1130	14" x 10"		19	.421		84	15.60		99.60	117
1132	16" x 6"		21	.381		74	14.10		88.10	104
1134	16" x 8"		20	.400		83	14.80		97.80	115
1136	18" x 6"		20	.400		79	14.80		93.80	110
1138	18" x 8"		19	.421		84	15.60		99.60	117
2000	Fabrics for flexible connections, with metal edge		100	.080	L.F.	1.54	2.96		4.50	6.25
2100	Without metal edge		160	.050	"	1.02	1.85		2.87	3.98
3000	Fire damper, curtain type, 1-1/2 hr rated, vertical, 6" x 6"		24	.333	Ea.	12.80	12.35		25.15	33
3020	8" x 6"		22	.364		19.70	13.45		33.15	42.50
3021	8" x 8"		22	.364		12.90	13.45		26.35	35
3030	10" x 10"		22	.364		13.45	13.45		26.90	36
3040	12" x 6"		22	.364		19.70	13.45		33.15	42.50
3044	12" x 12"		20	.400		14.10	14.80		28.90	38.50
3060	20" x 6"		18	.444		23.50	16.45		39.95	51
3080	12" x 8"		22	.364		19.50	13.45		32.95	42.50
3100	24" x 8"		16	.500		28.50	18.50		47	60
3120	12" x 10"		21	.381		20.50	14.10		34.60	44.50
3140	24" x 10"		15	.533		31.50	19.75		51.25	65
3160	36" x 10"		12	.667		44.50	24.50		69	87

MECHANICAL 15

			DAILY	LABOR-			2003 BARE COSTS			TOTAL	
15820	**Duct Accessories**	CREW	OUTPUT	HOURS	UNIT	MAT.	LABOR	EQUIP.	TOTAL	INCL O&P	
300	3180	16" x 12"	1 Shee	20	.400	Ea.	26	14.80		40.80	51.50	**300**
	3200	24" x 12"		13	.615		22	23		45	59	
	3220	48" x 12"		10	.800		48	29.50		77.50	98.50	
	3238	14" x 14"		19	.421		16.90	15.60		32.50	42.50	
	3240	16" x 14"		18	.444		28.50	16.45		44.95	57	
	3260	24" x 14"		12	.667		34	24.50		58.50	75	
	3280	30" x 14"		11	.727		27	27		54	71	
	3298	16" x 16"		18	.444		21	16.45		37.45	48.50	
	3300	18" x 16"		17	.471		22	17.40		39.40	51	
	3320	24" x 16"		11	.727		36.50	27		63.50	81.50	
	3340	36" x 16"		10	.800		33	29.50		62.50	82	
	3356	18" x 18"		16	.500		22.50	18.50		41	53.50	
	3360	24" x 18"		10	.800		37.50	29.50		67	87	
	3380	48" x 18"		8	1		61	37		98	124	
	3398	20" x 20"		10	.800		25	29.50		54.50	73	
	3400	24" x 20"		8	1		39.50	37		76.50	101	
	3420	36" x 20"		7	1.143		52	42.50		94.50	123	
	3440	24" x 22"		7	1.143		41	42.50		83.50	111	
	3460	30" x 22"		6	1.333		47	49.50		96.50	128	
	3478	24" x 24"		8	1		29	37		66	89	
	3480	26" x 24"		7	1.143		43.50	42.50		86	114	
	3484	32" x 24"	Q-9	13	1.231		36	41		77	103	
	3500	48" x 24"		12	1.333		67.50	44.50		112	143	
	3520	28" x 26"		13	1.231		49.50	41		90.50	118	
	3540	30" x 28"		12	1.333		55.50	44.50		100	130	
	3560	48" x 30"		11	1.455		80	48.50		128.50	163	
	3580	48" x 36"		10	1.600		88	53.50		141.50	179	
	3590	32" x 40"		10	1.600		49.50	53.50		103	137	
	3600	44" x 44"		10	1.600		91.50	53.50		145	184	
	3620	48" x 48"		8	2		85	66.50		151.50	197	
	3700	U.L. label included in above										
	3800	For horizontal operation, add				Ea.	20%					
	3900	For cap for blades out of air stream, add					20%					
	4000	For 10" 22 ga., U.L. approved sleeve, add					35%					
	4100	For 10" 22 ga., U.L. sleeve, 100% free area, add					65%					
	4200	For oversize openings group dampers										
	4210	Fire damper, round, Type A, 2 Hr										
	4214	4" diam.	1 Shee	20	.400	Ea.	39	14.80		53.80	66	
	4216	5" diam.		16	.500		36.50	18.50		55	68.50	
	4218	6" diam.		14.30	.559		36.50	20.50		57	72	
	4220	7" diam.		13.30	.601		36.50	22.50		59	74.50	
	4222	8" diam.		11.40	.702		36.50	26		62.50	80	
	4224	9" diam.		10.80	.741		41.50	27.50		69	88	
	4226	10" diam.		10	.800		42.50	29.50		72	92	
	4228	12" diam.		9.50	.842		45	31		76	97.50	
	4230	14" diam.		8.20	.976		69.50	36		105.50	133	
	4232	16" diam.		7.60	1.053		75	39		114	142	
	4234	18" diam.		7.20	1.111		89	41		130	162	
	4236	20" diam.		6.40	1.250		89	46.50		135.50	170	
	4238	22" diam.		6.10	1.311		106	48.50		154.50	192	
	4240	24" diam.		5.70	1.404		106	52		158	197	
	4242	26" diam.		5.50	1.455		125	54		179	221	
	4244	28" diam.		5.40	1.481		125	55		180	223	
	4246	30" diam.		5.30	1.509		125	56		181	225	
	4500	Fire/smoke combination damper, louver type, UL										
	4506	6" x 6"	1 Shee	24	.333	Ea.	55	12.35		67.35	79.50	

Important: See the Reference Section for critical supporting data - Reference Nos., Crews, & City Cost Indexes

						2003 BARE COSTS				TOTAL
15820	**Duct Accessories**	CREW	DAILY OUTPUT	LABOR-HOURS	UNIT	MAT.	LABOR	EQUIP.	TOTAL	INCL O&P
300										
4510	8" x 8"	1 Shee	22	.364	Ea.	126	13.45		139.45	160
4520	16" x 8"		20	.400		138	14.80		152.80	175
4540	18" x 8"		18	.444		138	16.45		154.45	178
4560	20" x 8"		16	.500		148	18.50		166.50	191
4580	10" x 10"		21	.381		143	14.10		157.10	179
4600	24" x 10"		15	.533		168	19.75		187.75	215
4620	30" x 10"		12	.667		173	24.50		197.50	229
4640	12" x 12"		20	.400		149	14.80		163.80	187
4660	18" x 12"		18	.444		152	16.45		168.45	193
4680	24" x 12"		13	.615		160	23		183	211
4700	30" x 12"		11	.727		166	27		193	225
4720	14" x 14"		17	.471		160	17.40		177.40	203
4740	16" x 14"		18	.444		165	16.45		181.45	207
4760	20" x 14"		14	.571		168	21		189	217
4780	24" x 14"		12	.667		172	24.50		196.50	227
4800	30" x 14"		10	.800		182	29.50		211.50	246
4820	16" x 16"		16	.500		177	18.50		195.50	224
4840	20" x 16"		14	.571		182	21		203	233
4860	24" x 16"		11	.727		188	27		215	249
4880	30" x 16"		8	1		196	37		233	273
4900	18" x 18"		15	.533		176	19.75		195.75	224
5000	24" x 18"		10	.800		188	29.50		217.50	252
5020	36" x 18"		7	1.143		205	42.50		247.50	292
5040	20" x 20"		13	.615		196	23		219	251
5060	24" x 20"		8	1		201	37		238	278
5080	30" x 20"		8	1		211	37		248	289
5100	36" x 20"		7	1.143		222	42.50		264.50	310
5120	24" x 24"		8	1		217	37		254	296
5130	30" x 24"	↓	7.60	1.053		103	39		142	174
5140	36" x 24"	Q-9	12	1.333		237	44.50		281.50	330
5141	36" x 30"		10	1.600		126	53.50		179.50	221
5142	40" x 36"		8	2		445	66.50		511.50	595
5143	48" x 48"	↓	4	4		870	133		1,003	1,150
5150	Damper operator motor, 24 or 120 volt	1 Shee	16	.500	↓	263	18.50		281.50	320
5154	Induction air terminal unit, cooling and heating									
5156	Vertical, cooling w/ electric heating coils									
5158	60 CFM	Q-20	7.60	2.632	Ea.	910	90		1,000	1,150
5160	90 CFM		7.20	2.778		1,125	95		1,220	1,375
5162	120 CFM		6.80	2.941		1,125	100		1,225	1,375
5164	150 CFM		6.40	3.125		1,275	107		1,382	1,575
5166	170 CFM	↓	6.20	3.226	↓	1,275	110		1,385	1,575
5170	Horizontal, cooling w/ electric heating coils									
5172	60 CFM	Q-20	7.60	2.632	Ea.	745	90		835	955
5174	90 CFM		7.20	2.778		910	95		1,005	1,150
5176	120 CFM	↓	6.80	2.941	↓	910	100		1,010	1,150
5180	Mixing box, includes electric or pneumatic motor									
5190	Recommend use with attenuator, see 15820-300-9000									
5200	Constant volume, 150 to 270 CFM	Q-9	12	1.333	Ea.	495	44.50		539.50	610
5210	270 to 600 CFM		11	1.455		505	48.50		553.50	630
5230	550 to 1000 CFM		9	1.778		505	59		564	645
5240	1000 to 1600 CFM		8	2		520	66.50		586.50	675
5250	1300 to 1900 CFM		6	2.667		530	89		619	720
5260	550 to 2640 CFM		5.60	2.857		585	95		680	790
5270	650 to 3120 CFM		5.20	3.077		620	102		722	845
5500	VAV Cool only, pneum. press indep. 300 to 600 CFM		11	1.455		276	48.50		324.50	380
5510	500 to 1000 CFM	↓	9	1.778	↓	288	59		347	405

MECHANICAL **15**

15820	Duct Accessories	CREW	DAILY OUTPUT	LABOR-HOURS	UNIT	2003 BARE COSTS				TOTAL INCL O&P
						MAT.	LABOR	EQUIP.	TOTAL	
5520	800 to 1600 CFM	Q-9	9	1.778	Ea.	300	59		359	420
5530	1100 to 2000 CFM		8	2		315	66.50		381.50	450
5540	1500 to 3000 CFM		7	2.286		330	76		406	485
5550	2000 to 4000 CFM		6	2.667		350	89		439	515
5560	For electric, w/thermostat, press. dependent, add					20			20	22
5600	VAV Cool & HW coils, damper, actuator and t'stat									
5610	200 CFM	Q-9	11	1.455	Ea.	565	48.50		613.50	695
5620	400 CFM		10	1.600		570	53.50		623.50	715
5630	600 CFM		10	1.600		570	53.50		623.50	715
5640	800 CFM		8	2		590	66.50		656.50	750
5650	1000 CFM		8	2		590	66.50		656.50	750
5660	1250 CFM		6	2.667		645	89		734	845
5670	1500 CFM		6	2.667		645	89		734	845
5680	2000 CFM		4	4		700	133		833	975
5684	3000 CFM		3.80	4.211		760	140		900	1,050
5700	VAV Cool only, fan powr'd, damper, actuator, t'stat									
5710	200 CFM	Q-9	10	1.600	Ea.	775	53.50		828.50	940
5720	400 CFM		9	1.778		805	59		864	975
5730	600 CFM		9	1.778		805	59		864	975
5740	800 CFM		7	2.286		860	76		936	1,075
5750	1000 CFM		7	2.286		860	76		936	1,075
5760	1250 CFM		5	3.200		930	107		1,037	1,200
5770	1500 CFM		5	3.200		930	107		1,037	1,200
5780	2000 CFM		4	4		1,025	133		1,158	1,325
5800	VAV Fan powr'd, cool/HW coils, dampers, actuators, t'stat									
5810	200 CFM	Q-9	10	1.600	Ea.	935	53.50		988.50	1,100
5820	400 CFM		9	1.778		960	59		1,019	1,150
5830	600 CFM		9	1.778		960	59		1,019	1,150
5840	800 CFM		7	2.286		1,025	76		1,101	1,250
5850	1000 CFM		7	2.286		1,025	76		1,101	1,250
5860	1250 CFM		5	3.200		1,150	107		1,257	1,450
5870	1500 CFM		5	3.200		1,150	107		1,257	1,450
5880	2000 CFM		3.50	4.571		1,250	152		1,402	1,600
5990	Multi-blade dampers, opposed blade, 8" x 6"	1 Shee	24	.333		18.40	12.35		30.75	39
5994	8" x 8"		22	.364		19.20	13.45		32.65	42
5996	10" x 10"		21	.381		21.50	14.10		35.60	46
6000	12" x 12"		21	.381		25	14.10		39.10	49.50
6020	12" x 18"		18	.444		33.50	16.45		49.95	62.50
6030	14" x 10"		20	.400		21.50	14.80		36.30	47
6031	14" x 14"		17	.471		29.50	17.40		46.90	59.50
6032	16" x 10"		18	.444		21.50	16.45		37.95	49.50
6033	16" x 12"		17	.471		29.50	17.40		46.90	59.50
6035	16" x 16"		16	.500		37	18.50		55.50	69
6036	18" x 14"		16	.500		37	18.50		55.50	69
6037	18" x 16"		15	.533		41	19.75		60.75	75.50
6038	18" x 18"		15	.533		44	19.75		63.75	79
6040	18" x 24"		12	.667		57	24.50		81.50	101
6060	18" x 28"		10	.800		65.50	29.50		95	118
6068	20" x 6"		18	.444		22.50	16.45		38.95	50
6069	20" x 8"		16	.500		25.50	18.50		44	56.50
6070	20" x 16"		14	.571		44	21		65	81
6071	20" x 18"		13	.615		48	23		71	88
6072	20" x 20"		13	.615		53	23		76	93
6073	22" x 6"		20	.400		22.50	14.80		37.30	47.50
6074	22" x 18"		14	.571		53	21		74	90.50
6075	22" x 20"		12	.667		57.50	24.50		82	102

300

15 MECHANICAL

Important: See the Reference Section for critical supporting data - Reference Nos., Crews, & City Cost Indexes

15820	Duct Accessories	CREW	DAILY OUTPUT	LABOR-HOURS	UNIT	2003 BARE COSTS				TOTAL INCL O&P
						MAT.	LABOR	EQUIP.	TOTAL	
300 6076	24" x 16"	1 Shee	11	.727	Ea.	52	27		79	98.50
6077	22" x 22"		10	.800		61.50	29.50		91	114
6078	24" x 20"		8	1		61.50	37		98.50	125
6080	24" x 24"		8	1		72	37		109	136
6100	24" x 28"		6	1.333		83	49.50		132.50	168
6110	26" x 26"		6	1.333		84.50	49.50		134	169
6120	28" x 28"	Q-9	11	1.455		98	48.50		146.50	183
6130	30" x 18"		10	1.600		69.50	53.50		123	159
6132	30" x 24"		7	2.286		88	76		164	215
6133	30" x 30"		6.60	2.424		112	80.50		192.50	249
6135	32" x 32"		6.40	2.500		128	83.50		211.50	270
6150	34" x 34"		6	2.667		144	89		233	295
6151	36" x 12"		10	1.600		59	53.50		112.50	148
6152	36" x 16"		8	2		74.50	66.50		141	185
6156	36" x 32"		6.30	2.540		144	84.50		228.50	289
6157	36" x 34"		6.20	2.581		153	86		239	300
6158	36" x 36"		6	2.667		161	89		250	315
6160	44" x 28"		5.80	2.759		154	92		246	310
6180	48" x 36"		5.60	2.857		215	95		310	385
6200	56" x 36"		5.40	2.963		250	98.50		348.50	425
6220	60" x 36"		5.20	3.077		268	102		370	455
6240	60" x 44"		5	3.200		330	107		437	525
7000	Splitter damper assembly, self-locking, 1' rod	1 Shee	24	.333		16.55	12.35		28.90	37.50
7020	3' rod		22	.364		21.50	13.45		34.95	44.50
7040	4' rod		20	.400		24	14.80		38.80	49.50
7060	6' rod		18	.444		29	16.45		45.45	57.50
7500	Variable volume modulating motorized damper, incl. elect. mtr.									
7504	8" x 6"	1 Shee	15	.533	Ea.	101	19.75		120.75	142
7506	10" x 6"		14	.571		101	21		122	144
7510	10" x 10"		13	.615		104	23		127	149
7520	12" x 12"		12	.667		106	24.50		130.50	155
7522	12" x 16"		11	.727		110	27		137	163
7524	16" x 10"		12	.667		107	24.50		131.50	156
7526	16" x 14"		10	.800		112	29.50		141.50	169
7528	16" x 18"		9	.889		119	33		152	182
7540	18" x 12"		10	.800		114	29.50		143.50	171
7542	18" x 18"		8	1		123	37		160	193
7544	20" x 14"		8	1		126	37		163	195
7546	20" x 18"		7	1.143		134	42.50		176.50	213
7560	24" x 12"		8	1		127	37		164	197
7562	24" x 18"		7	1.143		142	42.50		184.50	222
7568	28" x 10"		7	1.143		211	42.50		253.50	298
7580	28" x 16"		6	1.333		224	49.50		273.50	320
7590	30" x 14"		5	1.600		224	59		283	340
7600	30" x 18"		4	2		238	74		312	375
7610	30" x 24"		3.80	2.105		258	78		336	405
7690	48" x 48"	Q-9	6	2.667		665	89		754	870
7694	6' x 14' w/2 motors	"	3	5.333		2,600	178		2,778	3,125
7700	For thermostat, add	1 Shee	8	1		25.50	37		62.50	85
7800	For transformer 40 VA capacity, add	"	16	.500		20	18.50		38.50	50.50
8000	Multi-blade dampers, parallel blade									
8100	8" x 8"	1 Shee	24	.333	Ea.	51	12.35		63.35	75
8120	12" x 8"		22	.364		51	13.45		64.45	77
8140	16" x 10"		20	.400		67.50	14.80		82.30	97.50
8160	18" x 12"		18	.444		73	16.45		89.45	106
8180	22" x 12"		15	.533		74.50	19.75		94.25	113

	15820	Duct Accessories	CREW	DAILY OUTPUT	LABOR-HOURS	UNIT	2003 BARE COSTS				TOTAL INCL O&P	
							MAT.	LABOR	EQUIP.	TOTAL		
300	8200	24" x 16"	1 Shee	11	.727	Ea.	87.50	27		114.50	138	**300**
	8220	28" x 16"		10	.800		95	29.50		124.50	151	
	8240	30" x 16"		8	1		99.50	37		136.50	166	
	8260	30" x 18"	▼	7	1.143	▼	121	42.50		163.50	199	
	8300	Relief damper, electronic bypass with tight seal										
	8310	8" x 6"	1 Shee	22	.364	Ea.	101	13.45		114.45	132	
	8314	10" x 6"		22	.364		114	13.45		127.45	146	
	8318	10" x 10"		21	.381		147	14.10		161.10	184	
	8322	12" x 12"		20	.400		158	14.80		172.80	196	
	8326	12" x 16"		19	.421		164	15.60		179.60	204	
	8330	16" x 10"		20	.400		158	14.80		172.80	197	
	8334	16" x 14"		18	.444		165	16.45		181.45	208	
	8338	16" x 18"		17	.471		182	17.40		199.40	227	
	8342	18" x 12"		18	.444		164	16.45		180.45	207	
	8346	18" x 18"		15	.533		171	19.75		190.75	219	
	8350	20" x 14"		14	.571		171	21		192	221	
	8354	20" x 18"		13	.615		175	23		198	227	
	8358	24" x 12"		13	.615		182	23		205	236	
	8362	24" x 18"		10	.800		196	29.50		225.50	262	
	8363	24" x 24"		10	.800		204	29.50		233.50	270	
	8364	24" x 36"		9	.889		236	33		269	310	
	8365	24" x 48"		6	1.333		264	49.50		313.50	365	
	8366	28" x 10"		13	.615		171	23		194	223	
	8370	28" x 16"		10	.800		201	29.50		230.50	267	
	8374	30" x 14"		11	.727		197	27		224	259	
	8378	30" x 18"		8	1		199	37		236	276	
	8382	30" x 24"		6	1.333		197	49.50		246.50	293	
	8390	46" x 36"		4	2		277	74		351	420	
	8394	48" x 48"		3	2.667		320	98.50		418.50	500	
	8396	54" x 36"	▼	2	4	▼	340	148		488	605	
	8400	Round damper, butterfly, vol. control w/lever lock reg.										
	8410	6" diam.	1 Shee	22	.364	Ea.	21.50	13.45		34.95	45	
	8412	7" diam.		21	.381		22.50	14.10		36.60	46.50	
	8414	8" diam.		20	.400		23	14.80		37.80	48.50	
	8416	9" diam.		19	.421		24	15.60		39.60	50.50	
	8418	10" diam.		18	.444		25.50	16.45		41.95	53.50	
	8420	12" diam.		16	.500		26.50	18.50		45	57.50	
	8422	14" diam.		14	.571		33.50	21		54.50	69.50	
	8424	16" diam.		13	.615		36	23		59	74.50	
	8426	18" diam.		12	.667		40	24.50		64.50	82	
	8428	20" diam.		11	.727		43	27		70	89	
	8430	24" diam.		10	.800		55	29.50		84.50	106	
	8432	30" diam.		9	.889		76	33		109	135	
	8434	36" diam.	▼	8	1	▼	102	37		139	169	
	8500	Round motor operated damper										
	8510	6" dia.	1 Shee	22	.364	Ea.	51	13.45		64.45	77.50	
	8512	7" dia.		21	.381		53	14.10		67.10	80.50	
	8514	8" dia.		20	.400		56	14.80		70.80	84.50	
	8516	10" dia.		18	.444		61.50	16.45		77.95	93.50	
	8518	12" dia.		16	.500		66.50	18.50		85	102	
	8520	14" dia.	▼	14	.571	▼	75	21		96	115	
	9000	Silencers, noise control for air flow, duct					MCFM	45.50			45.50	50.50	
	9004	Duct sound trap, packaged										
	9010	12" x 18" x 36", 2000 CFM	1 Shee	12	.667	Ea.	275	24.50		299.50	340	
	9014	24" x 36" x 36", 9000 CFM	Q-9	40	.400		620	13.30		633.30	700	
	9018	36" x 18" x 36", 6300 CFM	▼	42	.381	▼	620	12.70		632.70	700	

Important: See the Reference Section for critical supporting data - Reference Nos., Crews, & City Cost Indexes

15820 | Duct Accessories

		CREW	DAILY OUTPUT	LABOR-HOURS	UNIT	2003 BARE COSTS				TOTAL INCL O&P
						MAT.	LABOR	EQUIP.	TOTAL	
300	9022	24" x 24" x 60", 6000 CFM	Q-9	3.60	4.444	Ea.	715	148		863	1,025	300
	9026	24" x 30" x 60", 7000 CFM		3.50	4.571		805	152		957	1,125
	9030	24" x 36" x 60", 8000 CFM		3.40	4.706		885	157		1,042	1,225
	9034	24" x 48" x 60", 11,000 CFM		3.30	4.848		775	161		936	1,100
	9038	36" x 18" x 60", 6300 CFM		3.50	4.571		600	152		752	895
	9042	36" x 36" x 60", 12,000 CFM		3.30	4.848		985	161		1,146	1,325
	9046	36" x 48" x 60", 17,000 CFM		3.10	5.161		1,250	172		1,422	1,675
	9050	36" x 60" x 60", 20,000 CFM		2.80	5.714		1,525	190		1,715	1,975
	9054	48" x 48" x 60", 23,000 CFM		2.60	6.154		815	205		1,020	1,200
	9400	Turning vane components									
	9410	Turning vane rail	1 Shee	160	.050	L.F.	.82	1.85		2.67	3.76
	9420	Double thick, factory fab. vane		300	.027		1.01	.99		2	2.63
	9428	12" high set		170	.047		5.70	1.74		7.44	8.95
	9432	14" high set		160	.050		6.40	1.85		8.25	9.85
	9434	16" high set		150	.053		7.05	1.97		9.02	10.80
	9436	18" high set		144	.056		7.75	2.06		9.81	11.70
	9438	20" high set		138	.058		8.40	2.14		10.54	12.55
	9440	22" high set		130	.062		9.10	2.28		11.38	13.50
	9442	24" high set		124	.065		9.75	2.39		12.14	14.45
	9444	26" high set		116	.069		10.45	2.55		13	15.40
	9446	30" high set		112	.071		11.80	2.64		14.44	17.05

15830 | Fans

		CREW	DAILY OUTPUT	LABOR-HOURS	UNIT	MAT.	LABOR	EQUIP.	TOTAL	TOTAL INCL O&P
100	0010	**FANS** R15700-040										100
	0020	Air conditioning and process air handling									
	0030	Axial flow, compact, low sound, 2.5" S.P. R15800-080									
	0050	3,800 CFM, 5 HP	Q-20	3.40	5.882	Ea.	3,600	201		3,801	4,275
	0080	6,400 CFM, 5 HP		2.80	7.143		4,025	244		4,269	4,800
	0100	10,500 CFM, 7-1/2 HP		2.40	8.333		5,025	285		5,310	5,950
	0120	15,600 CFM, 10 HP		1.60	12.500		6,325	425		6,750	7,600
	0140	23,000 CFM, 15 HP		.70	28.571		9,700	975		10,675	12,200
	0160	28,000 CFM, 20 HP		.40	50		10,800	1,700		12,500	14,500
	0200	In-line centrifugal, supply/exhaust booster									
	0220	aluminum wheel/hub, disconnect switch, 1/4" S.P.									
	0240	500 CFM, 10" diameter connection	Q-20	3	6.667	Ea.	715	228		943	1,125
	0260	1,380 CFM, 12" diameter connection		2	10		815	340		1,155	1,425
	0280	1,520 CFM, 16" diameter connection		2	10		930	340		1,270	1,550
	0300	2,560 CFM, 18" diameter connection		1	20		1,025	685		1,710	2,175
	0320	3,480 CFM, 20" diameter connection		.80	25		1,225	855		2,080	2,650
	0326	5,080 CFM, 20" diameter connection		.75	26.667		1,350	910		2,260	2,900
	0340	7,500 CFM, 22" diameter connection		.70	28.571		1,725	975		2,700	3,400
	0350	10,000 CFM, 27" diameter connection		.65	30.769		2,025	1,050		3,075	3,825
	0500	Axial flow, constant speed									
	0505	Direct drive, 1/8" S.P.									
	0510	12", 1060 CFM, 1/6 HP	Q-20	3	6.667	Ea.	279	228		507	655
	0514	12", 2095 CFM, 1/2 HP		3	6.667		305	228		533	685
	0518	16", 2490 CFM, 1/3 HP		2.80	7.143		330	244		574	735
	0522	20", 4130 CFM, 3/4 HP		2.60	7.692		430	263		693	880
	0526	22", 4700 CFM, 3/4 HP		2.60	7.692		480	263		743	930
	0530	24", 5850 CFM, 1 HP		2.50	8		545	273		818	1,025
	0534	24", 7925 CFM, 1-1/2 HP		2.40	8.333		550	285		835	1,050
	0538	30", 10640 CFM, 2 HP		2.20	9.091		1,000	310		1,310	1,575
	0542	30", 14765 CFM, 2-1/2 HP		2.10	9.524		1,050	325		1,375	1,650
	0546	36", 16780 CFM, 2 HP		2.10	9.524		1,125	325		1,450	1,725
	0550	36", 22920 CFM, 5 HP		1.80	11.111		1,400	380		1,780	2,125

			DAILY OUTPUT	LABOR-HOURS	UNIT	2003 BARE COSTS				TOTAL INCL O&P
	15830 | Fans	CREW				MAT.	LABOR	EQUIP.	TOTAL	
100	0560	Belt drive, 1/8" S.P.									
	0562	15", 2800 CFM, 1/3 HP	Q-20	3.20	6.250	Ea.	505	214		719	880
	0564	15", 3400 CFM, 1/2 HP		3	6.667		515	228		743	915
	0568	18", 3280 CFM, 1/3 HP		2.80	7.143		545	244		789	975
	0572	18", 3900 CFM, 1/2 HP		2.80	7.143		640	244		884	1,075
	0576	18", 5250 CFM, 1 HP		2.70	7.407		685	253		938	1,150
	0584	24", 6430 CFM, 1 HP		2.50	8		790	273		1,063	1,300
	0588	24", 8860 CFM, 2 HP		2.40	8.333		795	285		1,080	1,300
	0592	30", 9250 CFM, 1 HP		2.20	9.091		960	310		1,270	1,525
	0596	30", 16900 CFM, 5 HP		2	10		1,075	340		1,415	1,700
	0604	36", 14475 CFM, 2 HP		2.30	8.696		1,125	297		1,422	1,675
	0608	36", 20080 CFM, 5 HP		1.80	11.111		1,200	380		1,580	1,900
	0612	36", 14475 CFM, 7-1/2 HP		1.60	12.500		1,375	425		1,800	2,175
	0616	42", 29000 CFM, 7-1/2 HP		1.40	14.286		1,600	490		2,090	2,525
	1000	Return air fan									
	1010	9200 CFM	Q-20	1.40	14.286	Ea.	2,950	490		3,440	3,975
	1020	13,200 CFM		1.30	15.385		4,225	525		4,750	5,450
	1030	16,500 CFM		1.20	16.667		5,275	570		5,845	6,675
	1040	19,500 CFM		1	20		6,250	685		6,935	7,900
	1500	Vaneaxial, low pressure, 2000 CFM, 1/2 HP		3.60	5.556		1,575	190		1,765	2,050
	1520	4,000 CFM, 1 HP		3.20	6.250		1,625	214		1,839	2,100
	1540	8,000 CFM, 2 HP		2.80	7.143		1,900	244		2,144	2,475
	1560	16,000 CFM, 5 HP		2.40	8.333		2,525	285		2,810	3,200
	2000	Blowers, direct drive with motor, complete									
	2020	1045 CFM @ .5" S.P., 1/5 HP	Q-20	18	1.111	Ea.	139	38		177	211
	2040	1385 CFM @ .5" S.P., 1/4 HP		18	1.111		142	38		180	214
	2060	1640 CFM @ .5" S.P., 1/3 HP		18	1.111		169	38		207	244
	2080	1760 CFM @ .5" S.P., 1/2 HP		18	1.111		183	38		221	260
	2090	4 speed									
	2100	1164 to 1739 CFM @ .5" S.P., 1/3 HP	Q-20	16	1.250	Ea.	244	42.50		286.50	335
	2120	1467 to 2218 CFM @ 1.0" S.P., 3/4 HP	"	14	1.429	"	315	49		364	425
	2500	Ceiling fan, right angle, extra quiet, 0.10" S.P.									
	2520	95 CFM	Q-20	20	1	Ea.	121	34		155	186
	2540	210 CFM		19	1.053		143	36		179	213
	2560	385 CFM		18	1.111		182	38		220	258
	2580	885 CFM		16	1.250		360	42.50		402.50	460
	2600	1,650 CFM		13	1.538		495	52.50		547.50	625
	2620	2,960 CFM		11	1.818		660	62		722	825
	2640	For wall or roof cap, add	1 Shee	16	.500		121	18.50		139.50	162
	2660	For straight thru fan, add					10%				
	2680	For speed control switch, add	1 Elec	16	.500		66	18.80		84.80	101
	3000	Paddle blade air circulator, 3 speed switch									
	3020	42", 5,000 CFM high, 3000 CFM low	1 Elec	2.40	3.333	Ea.	73	125		198	267
	3040	52", 6,500 CFM high, 4000 CFM low	"	2.20	3.636	"	74	137		211	286
	3100	For antique white motor, same cost									
	3200	For brass plated motor, same cost									
	3300	For light adaptor kit, add				Ea.	27			27	30
	3310	Industrial grade, reversable, 4 blade									
	3312	5500 CFM	Q-20	5	4	Ea.	106	137		243	325
	3314	7600 CFM		5	4		115	137		252	335
	3316	21,015 CFM		4	5		147	171		318	425
	3500	Centrifugal, airfoil, motor and drive, complete									
	3520	1000 CFM, 1/2 HP	Q-20	2.50	8	Ea.	870	273		1,143	1,375
	3540	2,000 CFM, 1 HP		2	10		935	340		1,275	1,550
	3560	4,000 CFM, 3 HP		1.80	11.111		1,200	380		1,580	1,900
	3580	8,000 CFM, 7-1/2 HP		1.40	14.286		1,775	490		2,265	2,700

R15700-040

R15800-080

T5 MECHANICAL

Important: See the Reference Section for critical supporting data - Reference Nos., Crews, & City Cost Indexes

15830	Fans		CREW	DAILY OUTPUT	LABOR-HOURS	UNIT	2003 BARE COSTS				TOTAL INCL O&P
							MAT.	LABOR	EQUIP.	TOTAL	
100 3600	12,000 CFM, 10 HP	R15700 -040	Q-20	1	20	Ea.	2,600	685		3,285	3,900 **100**
4000	Single width, belt drive, not incl. motor, capacities										
4020	at 2000 FPM, 2.5" S.P. for indicated motor	R15800 -080									
4040	6900 CFM, 5 HP		Q-9	2.40	6.667	Ea.	2,625	222		2,847	3,250
4060	10,340 CFM, 7-1/2" HP			2.20	7.273		3,675	242		3,917	4,400
4080	15,320 CFM, 10 HP			2	8		4,325	266		4,591	5,150
4100	22,780 CFM, 15 HP			1.80	8.889		6,425	296		6,721	7,525
4120	33,840 CFM, 20 HP			1.60	10		8,525	335		8,860	9,900
4140	41,400 CFM, 25 HP			1.40	11.429		11,300	380		11,680	13,000
4160	50,100 CFM, 30 HP			.80	20		14,400	665		15,065	16,800
4200	Double width wheel, 12,420 CFM, 7.5 HP			2.20	7.273		3,775	242		4,017	4,525
4220	18,620 CFM, 15 HP			2	8		5,400	266		5,666	6,325
4240	27,580 CFM, 20 HP			1.80	8.889		6,575	296		6,871	7,700
4260	40,980 CFM, 25 HP			1.50	10.667		10,300	355		10,655	11,900
4280	60,920 CFM, 40 HP			1	16		13,900	535		14,435	16,100
4300	74,520 CFM, 50 HP			.80	20		16,500	665		17,165	19,100
4320	90,160 CFM, 50 HP			.70	22.857		22,500	760		23,260	26,000
4340	110,300 CFM, 60 HP			.50	32		30,500	1,075		31,575	35,200
4360	134,960 CFM, 75 HP			.40	40		40,500	1,325		41,825	46,700
4500	Corrosive fume resistant, plastic										
4600	roof ventilators, centrifugal, V belt drive, motor										
4620	1/4" S.P., 250 CFM, 1/4 HP		Q-20	6	3.333	Ea.	2,425	114		2,539	2,825
4640	895 CFM, 1/3 HP			5	4		2,625	137		2,762	3,075
4660	1630 CFM, 1/2 HP			4	5		3,100	171		3,271	3,650
4680	2240 CFM, 1 HP			3	6.667		3,225	228		3,453	3,900
4700	3810 CFM, 2 HP			2	10		3,600	340		3,940	4,475
4710	5000 CFM, 2 HP			1.80	11.111		7,550	380		7,930	8,875
4715	8000 CFM, 5 HP			1.40	14.286		8,400	490		8,890	10,000
4720	11760 CFM, 5 HP			1	20		8,525	685		9,210	10,500
4740	18810 CFM, 10 HP			.70	28.571		8,850	975		9,825	11,200
4800	For intermediate capacity, motors may be varied										
4810	For explosion proof motor, add					Ea.	15%				
5000	Utility set, centrifugal, V belt drive, motor										
5020	1/4" S.P., 1200 CFM, 1/4 HP		Q-20	6	3.333	Ea.	2,550	114		2,664	2,975
5040	1520 CFM, 1/3 HP			5	4		2,550	137		2,687	3,000
5060	1850 CFM, 1/2 HP			4	5		2,550	171		2,721	3,075
5080	2180 CFM, 3/4 HP			3	6.667		2,600	228		2,828	3,200
5100	1/2" S.P., 3600 CFM, 1 HP			2	10		3,750	340		4,090	4,650
5120	4250 CFM, 1-1/2 HP			1.60	12.500		3,800	425		4,225	4,825
5140	4800 CFM, 2 HP			1.40	14.286		3,850	490		4,340	4,975
5160	6920 CFM, 5 HP			1.30	15.385		3,975	525		4,500	5,175
5180	7700 CFM, 7-1/2 HP			1.20	16.667		4,100	570		4,670	5,375
5200	For explosion proof motor, add						15%				
5300	Fume exhauster without hose and intake nozzle										
5310	630 CFM, 1-1/2 HP		Q-20	2	10	Ea.	690	340		1,030	1,275
5320	Hose extension kit 5'		"	20	1	"	51	34		85	109
5500	Industrial exhauster, for air which may contain granular material										
5520	1000 CFM, 1-1/2 HP		Q-20	2.50	8	Ea.	2,000	273		2,273	2,625
5540	2000 CFM, 3 HP			2	10		2,425	340		2,765	3,200
5560	4000 CFM, 7-1/2 HP			1.80	11.111		3,925	380		4,305	4,875
5580	8000 CFM, 15 HP			1.40	14.286		6,125	490		6,615	7,475
5600	12,000 CFM, 30 HP			1	20		8,500	685		9,185	10,400
6000	Propeller exhaust, wall shutter, 1/4" S.P.										
6020	Direct drive, two speed										
6100	375 CFM, 1/10 HP		Q-20	10	2	Ea.	218	68.50		286.50	345
6120	730 CFM, 1/7 HP			9	2.222		243	76		319	385

	15830	Fans	CREW	DAILY OUTPUT	LABOR-HOURS	UNIT	2003 BARE COSTS				TOTAL INCL O&P	
							MAT.	LABOR	EQUIP.	TOTAL		
100	6140	1000 CFM, 1/8 HP	Q-20	8	2.500	Ea.	335	85.50		420.50	500	100
	6160	1890 CFM, 1/4 HP		7	2.857		300	97.50		397.50	480	
	6180	3275 CFM, 1/2 HP		6	3.333		350	114		464	560	
	6200	4720 CFM, 1 HP		5	4		540	137		677	805	
	6300	V-belt drive, 3 phase										
	6320	6175 CFM, 3/4 HP	Q-20	5	4	Ea.	455	137		592	710	
	6340	7500 CFM, 3/4 HP		5	4		475	137		612	735	
	6360	10,100 CFM, 1 HP		4.50	4.444		585	152		737	875	
	6380	14,300 CFM, 1-1/2 HP		4	5		685	171		856	1,025	
	6400	19,800 CFM, 2 HP		3	6.667		785	228		1,013	1,225	
	6420	26,250 CFM, 3 HP		2.60	7.692		940	263		1,203	1,425	
	6440	38,500 CFM, 5 HP		2.20	9.091		1,125	310		1,435	1,700	
	6460	46,000 CFM, 7-1/2 HP		2	10		1,200	340		1,540	1,850	
	6480	51,500 CFM, 10 HP		1.80	11.111		1,225	380		1,605	1,925	
	6490	V-belt drive, 115V., residential, whole house										
	6500	Ceiling-wall, 5200 CFM, 1/4 HP, 30" x 30"	1 Shee	6	1.333	Ea.	237	49.50		286.50	335	
	6510	7500 CFM, 1/3 HP, 36"x36"		5	1.600		248	59		307	365	
	6520	10,500 CFM, 1/3 HP, 42"x42"		5	1.600		259	59		318	375	
	6530	13,200 CFM, 1/3 HP, 48" x 48"		4	2		287	74		361	430	
	6540	15,445 CFM, 1/2 HP, 48" x 48"		4	2		298	74		372	440	
	6550	17,025 CFM, 1/2 HP, 54" x 54"		4	2		575	74		649	745	
	6560	For two speed motor, add					20%					
	6570	Shutter, automatic, ceiling/wall										
	6580	30" x 30"	1 Shee	8	1	Ea.	77	37		114	142	
	6590	36" x 36"		8	1		88	37		125	154	
	6600	42" x 42"		8	1		110	37		147	178	
	6610	48" x 48"		7	1.143		110	42.50		152.50	187	
	6620	54" x 54"		6	1.333		158	49.50		207.50	249	
	6630	Timer, shut off, to 12 Hr.		20	.400		25.50	14.80		40.30	51	
	6650	Residential, bath exhaust, grille, back draft damper										
	6660	50 CFM	Q-20	24	.833	Ea.	29.50	28.50		58	76	
	6670	110 CFM		22	.909		49	31		80	101	
	6672	180 CFM		22	.909		85.50	31		116.50	142	
	6673	210 CFM		22	.909		115	31		146	175	
	6674	260 CFM		22	.909		129	31		160	190	
	6675	300 CFM		22	.909		176	31		207	241	
	6680	Light combination, squirrel cage, 100 watt, 70 CFM		24	.833		60	28.50		88.50	110	
	6700	Light/heater combination, ceiling mounted										
	6710	70 CFM, 1450 watt	Q-20	24	.833	Ea.	73	28.50		101.50	124	
	6800	Heater combination, recessed, 70 CFM		24	.833		33.50	28.50		62	80.50	
	6820	With 2 infrared bulbs		23	.870		52.50	29.50		82	103	
	6840	Wall mount, 170 CFM		22	.909		115	31		146	175	
	6846	Ceiling mount, 180 CFM		22	.909		116	31		147	176	
	6900	Kitchen exhaust, grille, complete, 160 CFM		22	.909		61	31		92	115	
	6910	180 CFM		20	1		51.50	34		85.50	109	
	6920	270 CFM		18	1.111		93	38		131	160	
	6930	350 CFM		16	1.250		72.50	42.50		115	146	
	6940	Residential roof jacks and wall caps										
	6944	Wall cap with back draft damper										
	6946	3" & 4" dia. round duct	1 Shee	11	.727	Ea.	12.50	27		39.50	55.50	
	6948	6" dia. round duct	"	11	.727	"	30	27		57	74.50	
	6958	Roof jack with bird screen and back draft damper										
	6960	3" & 4" dia. round duct	1 Shee	11	.727	Ea.	12	27		39	54.50	
	6962	3-1/4" x 10" rectangular duct	"	10	.800	"	22	29.50		51.50	70	
	6980	Transition										
	6982	3-1/4" x 10" to 6" dia. round	1 Shee	20	.400	Ea.	13.50	14.80		28.30	38	

R15700-040

R15800-080

Important: See the Reference Section for critical supporting data - Reference Nos., Crews, & City Cost Indexes

15 MECHANICAL

15830	Fans		DAILY	LABOR-			2003 BARE COSTS				TOTAL
		CREW	OUTPUT	HOURS	UNIT	MAT.	LABOR	EQUIP.	TOTAL	INCL O&P	
100	7000	Roof exhauster, centrifugal, aluminum housing, 12" galvanized	R15700 -040										100
	7020	curb, bird screen, back draft damper, 1/4" S.P.										
	7100	Direct drive, 320 CFM, 11" sq. damper	R15800 -080	Q-20	7	2.857	Ea.	320	97.50		417.50	500
	7120	600 CFM, 11" sq. damper			6	3.333		325	114		439	530
	7140	815 CFM, 13" sq. damper			5	4		325	137		462	565
	7160	1450 CFM, 13" sq. damper			4.20	4.762		415	163		578	705
	7180	2050 CFM, 16" sq. damper			4	5		415	171		586	715
	7200	V-belt drive, 1650 CFM, 12" sq. damper			6	3.333		700	114		814	945
	7220	2750 CFM, 21" sq. damper			5	4		810	137		947	1,100
	7230	3500 CFM, 21" sq. damper			4.50	4.444		905	152		1,057	1,225
	7240	4910 CFM, 23" sq. damper			4	5		1,100	171		1,271	1,475
	7260	8525 CFM, 28" sq. damper			3	6.667		1,375	228		1,603	1,875
	7280	13,760 CFM, 35" sq. damper			2	10		1,900	340		2,240	2,625
	7300	20,558 CFM, 43" sq. damper			1	20		4,050	685		4,735	5,500
	7320	For 2 speed winding, add						15%				
	7340	For explosionproof motor, add					Ea.	320			320	350
	7360	For belt driven, top discharge, add						15%				
	7400	Roof mounted kitchen exhaust, aluminum, centrifugal										
	7410	Direct drive, 2 speed, temp to 200°F										
	7412	1/12 HP, 9-3/4"		Q-20	8	2.500	Ea.	320	85.50		405.50	485
	7414	1/3 HP, 12-5/8"			7	2.857		395	97.50		492.50	585
	7416	1/2 HP, 13-1/2"			6	3.333		440	114		554	655
	7418	3/4 HP, 15"			5	4		625	137		762	900
	7424	Belt drive, temp to 250°F										
	7426	3/4 HP, 20"		Q-20	5	4	Ea.	1,375	137		1,512	1,725
	7428	1-1/2 HP, 24-1/2"			4	5		1,525	171		1,696	1,925
	7430	1-1/2 HP, 30"			4	5		2,000	171		2,171	2,450
	7450	Upblast, propeller, w/BDD, BS, + C										
	7454	30,300 CFM @ 3/8" S.P., 5 HP		Q-20	1.60	12.500	Ea.	1,975	425		2,400	2,825
	7458	36,000 CFM @ 1/2" S.P., 15 HP		"	1.60	12.500	"	2,800	425		3,225	3,750
	7500	Utility set, steel construction, pedestal, 1/4" S.P.										
	7520	Direct drive, 150 CFM, 1/8 HP		Q-20	6.40	3.125	Ea.	605	107		712	830
	7540	485 CFM, 1/6 HP			5.80	3.448		760	118		878	1,025
	7560	1950 CFM, 1/2 HP			4.80	4.167		895	142		1,037	1,200
	7580	2410 CFM, 3/4 HP			4.40	4.545		1,650	155		1,805	2,075
	7600	3328 CFM, 1-1/2 HP			3	6.667		1,825	228		2,053	2,375
	7680	V-belt drive, drive cover, 3 phase										
	7700	800 CFM, 1/4 HP		Q-20	6	3.333	Ea.	460	114		574	685
	7720	1,300 CFM, 1/3 HP			5	4		485	137		622	745
	7740	2,000 CFM, 1 HP			4.60	4.348		575	149		724	860
	7760	2,900 CFM, 3/4 HP			4.20	4.762		775	163		938	1,100
	7780	3,600 CFM, 3/4 HP			4	5		950	171		1,121	1,300
	7800	4,800 CFM, 1 HP			3.50	5.714		1,125	195		1,320	1,525
	7820	6,700 CFM, 1-1/2 HP			3	6.667		1,375	228		1,603	1,875
	7830	7,500 CFM, 2 HP			2.50	8		1,375	273		1,648	1,950
	7840	11,000 CFM, 3 HP			2	10		2,525	340		2,865	3,300
	7860	13,000 CFM, 3 HP			1.60	12.500		2,550	425		2,975	3,450
	7880	15,000 CFM, 5 HP			1	20		2,650	685		3,335	3,950
	7900	17,000 CFM, 7-1/2 HP			.80	25		2,825	855		3,680	4,400
	7920	20,000 CFM, 7-1/2 HP			.80	25		3,375	855		4,230	5,000
	8000	Ventilation, residential										
	8020	Attic, roof type										
	8030	Aluminum dome, damper & curb										
	8040	6" diameter, 300 CFM		1 Elec	16	.500	Ea.	238	18.80		256.80	289
	8050	7" diameter, 450 CFM			15	.533		259	20		279	315
	8060	9" diameter, 900 CFM			14	.571		415	21.50		436.50	490

MECHANICAL 15

15830 | Fans

			CREW	DAILY OUTPUT	LABOR-HOURS	UNIT	MAT.	LABOR	EQUIP.	TOTAL	TOTAL INCL O&P	
100	8080	12" diameter, 1000 CFM (gravity)	R15700 -040	1 Elec	10	.800	Ea.	250	30		280	320	**100**
	8090	16" diameter, 1500 CFM (gravity)			9	.889		300	33.50		333.50	380	
	8100	20" diameter, 2500 CFM (gravity)	R15800 -080		8	1		370	37.50		407.50	460	
	8110	26" diameter, 4000 CFM (gravity)			7	1.143		450	43		493	560	
	8120	32" diameter, 6500 CFM (gravity)			6	1.333		615	50		665	750	
	8130	38" diameter, 8000 CFM (gravity)			5	1.600		915	60		975	1,100	
	8140	50" diameter, 13,000 CFM (gravity)			4	2		1,325	75		1,400	1,550	
	8160	Plastic, ABS dome											
	8180	1050 CFM		1 Elec	14	.571	Ea.	73.50	21.50		95	113	
	8200	1600 CFM		"	12	.667	"	110	25		135	159	
	8240	Attic, wall type, with shutter, one speed											
	8250	12" diameter, 1000 CFM		1 Elec	14	.571	Ea.	187	21.50		208.50	238	
	8260	14" diameter, 1500 CFM			12	.667		203	25		228	261	
	8270	16" diameter, 2000 CFM			9	.889		229	33.50		262.50	300	
	8290	Whole house, wall type, with shutter, one speed											
	8300	30" diameter, 4800 CFM		1 Elec	7	1.143	Ea.	490	43		533	605	
	8310	36" diameter, 7000 CFM			6	1.333		535	50		585	665	
	8320	42" diameter, 10,000 CFM			5	1.600		600	60		660	750	
	8330	48" diameter, 16,000 CFM			4	2		745	75		820	930	
	8340	For two speed, add						44.50			44.50	49	
	8350	Whole house, lay-down type, with shutter, one speed											
	8360	30" diameter, 4500 CFM		1 Elec	8	1	Ea.	525	37.50		562.50	630	
	8370	36" diameter, 6500 CFM			7	1.143		560	43		603	685	
	8380	42" diameter, 9000 CFM			6	1.333		620	50		670	755	
	8390	48" diameter, 12,000 CFM			5	1.600		700	60		760	860	
	8440	For two speed, add						33.50			33.50	36.50	
	8450	For 12 hour timer switch, add		1 Elec	32	.250		33.50	9.40		42.90	50.50	
	8500	Wall exhausters, centrifugal, auto damper, 1/8" S.P.											
	8520	Direct drive, 610 CFM, 1/20 HP		Q-20	14	1.429	Ea.	176	49		225	269	
	8540	796 CFM, 1/12 HP			13	1.538		182	52.50		234.50	281	
	8560	822 CFM, 1/6 HP			12	1.667		287	57		344	405	
	8580	1,320 CFM, 1/4 HP			12	1.667		290	57		347	410	
	8600	1756 CFM, 1/4 HP			11	1.818		340	62		402	470	
	8620	1983 CFM, 1/4 HP			10	2		405	68.50		473.50	550	
	8640	2900 CFM, 1/2 HP			9	2.222		490	76		566	655	
	8660	3307 CFM, 3/4 HP			8	2.500		540	85.50		625.50	725	
	8670	5940 CFM, 1 HP			7	2.857		875	97.50		972.50	1,125	
	9500	V-belt drive, 3 phase											
	9520	2,800 CFM, 1/4 HP		Q-20	9	2.222	Ea.	800	76		876	995	
	9540	3,740 CFM, 1/2 HP			8	2.500		830	85.50		915.50	1,050	
	9560	4400 CFM, 3/4 HP			7	2.857		840	97.50		937.50	1,075	
	9580	5700 CFM, 1-1/2 HP			6	3.333		875	114		989	1,125	

15850 | Air Outlets & Inlets

			CREW	DAILY OUTPUT	LABOR-HOURS	UNIT	MAT.	LABOR	EQUIP.	TOTAL	TOTAL INCL O&P	
300	0010	**DIFFUSERS** Aluminum, opposed blade damper unless noted	R15850 -060										**300**
	0100	Ceiling, linear, also for sidewall											
	0120	2" wide	R15810 -070	1 Shee	32	.250	L.F.	26.50	9.25		35.75	44	
	0140	3" wide			30	.267		31	9.85		40.85	49.50	
	0160	4" wide	R15800 -080		26	.308		34.50	11.40		45.90	55.50	
	0180	6" wide			24	.333		42	12.35		54.35	65	
	0200	8" wide			22	.364		47.50	13.45		60.95	73.50	
	0220	10" wide			20	.400		53	14.80		67.80	81	
	0240	12" wide			18	.444		67.50	16.45		83.95	99.50	
	0260	For floor or sill application, add						15%					
	0500	Perforated, 24" x 24" lay-in panel size, 6" x 6"		1 Shee	16	.500	Ea.	68	18.50		86.50	104	
	0520	8" x 8"			15	.533		72	19.75		91.75	110	

15 MECHANICAL

Important: See the Reference Section for critical supporting data - Reference Nos., Crews, & City Cost Indexes

15850 | Air Outlets & Inlets

			CREW	DAILY OUTPUT	LABOR-HOURS	UNIT	2003 BARE COSTS				TOTAL INCL O&P
							MAT.	LABOR	EQUIP.	TOTAL	
300	0530	9" x 9"	R15850 -060	1 Shee	14	.571	Ea.	78	21		99	119	300
	0540	10" x 10"			14	.571		78.50	21		99.50	119
	0560	12" x 12"	R15810 -070		12	.667		85.50	24.50		110	132
	0580	15" x 15"			11	.727		108	27		135	161
	0590	16" x 16"	R15800 -080		11	.727		137	27		164	193
	0600	18" x 18"			10	.800		159	29.50		188.50	221
	0610	20" x 20"			10	.800		185	29.50		214.50	249
	0620	24" x 24"			9	.889		226	33		259	299
	1000	Rectangular, 1 to 4 way blow, 6" x 6"			16	.500		37	18.50		55.50	69
	1010	8" x 8"			15	.533		43.50	19.75		63.25	78
	1014	9" x 9"			15	.533		47	19.75		66.75	82.50
	1016	10" x 10"			15	.533		53	19.75		72.75	88.50
	1020	12" x 6"			15	.533		58	19.75		77.75	94
	1040	12" x 9"			14	.571		63	21		84	102
	1060	12" x 12"			12	.667		65	24.50		89.50	110
	1070	14" x 6"			13	.615		62.50	23		85.50	104
	1074	14" x 14"			12	.667		81	24.50		105.50	127
	1080	18" x 12"			11	.727		111	27		138	164
	1120	15" x 15"			10	.800		116	29.50		145.50	174
	1140	18" x 15"			9	.889		130	33		163	194
	1150	18" x 18"			9	.889		119	33		152	182
	1160	21" x 21"			8	1		151	37		188	223
	1170	24" x 12"			10	.800		107	29.50		136.50	164
	1180	24" x 24"			7	1.143		242	42.50		284.50	330
	1500	Round, butterfly damper, 6" diameter			18	.444		15.65	16.45		32.10	42.50
	1520	8" diameter			16	.500		16.80	18.50		35.30	47
	1540	10" diameter			14	.571		21	21		42	55.50
	1560	12" diameter			12	.667		27.50	24.50		52	68
	1580	14" diameter			10	.800		34.50	29.50		64	83.50
	1600	18" diameter			9	.889		81	33		114	140
	1610	20" diameter			9	.889		101	33		134	162
	1620	22" diameter			8	1		112	37		149	181
	1630	24" diameter			8	1		124	37		161	193
	1640	28" diameter			7	1.143		138	42.50		180.50	218
	1650	32" diameter	▼		6	1.333	▼	152	49.50		201.50	243
	1700	Round, stl, adjustable core, annular segmented damper										
	1704	6" diameter		1 Shee	18	.444	Ea.	67	16.45		83.45	99
	1708	8" diameter			16	.500		72	18.50		90.50	108
	1712	10" diameter			14	.571		81	21		102	122
	1716	12" diameter			12	.667		90	24.50		114.50	137
	1720	14" diameter			10	.800		100	29.50		129.50	156
	1724	18" diameter			9	.889		153	33		186	219
	1728	20" diameter			9	.889		177	33		210	246
	1732	24" diameter			8	1		276	37		313	360
	1736	30" diameter			7	1.143		480	42.50		522.50	595
	1740	36" diameter			6	1.333		585	49.50		634.50	720
	2000	T bar mounting, 24" x 24" lay-in frame, 6" x 6"			16	.500		76.50	18.50		95	113
	2020	9" x 9"			14	.571		86.50	21		107.50	128
	2040	12" x 12"			12	.667		113	24.50		137.50	162
	2060	15" x 15"			11	.727		146	27		173	202
	2080	18" x 18"	▼		10	.800	▼	152	29.50		181.50	213
	2500	Combination supply and return										
	2520	21" x 21" supply, 15" x 15" return		Q-9	10	1.600	Ea.	187	53.50		240.50	288
	2540	24" x 24" supply, 18" x 18" return			9.50	1.684		235	56		291	345
	2560	27" x 27" supply, 18" x 18" return			9	1.778		287	59		346	405
	2580	30" x 30" supply, 21" x 21" return			8.50	1.882		365	62.50		427.50	495

MECHANICAL 15

15850 | Air Outlets & Inlets

		CREW	DAILY OUTPUT	LABOR-HOURS	UNIT	2003 BARE COSTS MAT.	LABOR	EQUIP.	TOTAL	TOTAL INCL O&P	
300											**300**
2600	33" x 33" supply, 24" x 24" return R15850-060	Q-9	8	2	Ea.	425	66.50		491.50	570	
2620	36" x 36" supply, 24" x 24" return		7.50	2.133		500	71		571	660	
3000	Baseboard, white enameled steel R15810-070										
3100	18" long	1 Shee	20	.400	Ea.	8.20	14.80		23	32	
3120	24" long R15800-080		18	.444		16.05	16.45		32.50	43	
3140	48" long		16	.500		29.50	18.50		48	61	
3400	For matching return, deduct					10%					
4000	Floor, steel, adjustable pattern										
4100	2" x 10"	1 Shee	34	.235	Ea.	11	8.70		19.70	25.50	
4120	2" x 12"		32	.250		12.10	9.25		21.35	27.50	
4140	2" x 14"		30	.267		13.75	9.85		23.60	30.50	
4200	4" x 10"		28	.286		12.75	10.55		23.30	30.50	
4220	4" x 12"		26	.308		14.10	11.40		25.50	33	
4240	4" x 14"		25	.320		15.95	11.85		27.80	36	
4260	6" x 10"		26	.308		19.30	11.40		30.70	38.50	
4280	6" x 12"		24	.333		20	12.35		32.35	41	
4300	6" x 14"		22	.364		21	13.45		34.45	44	
4500	Linear bar diffuser in wall, sill or ceiling										
4510	2" wide	1 Shee	32	.250	L.F.	26	9.25		35.25	43	
4512	3" wide		30	.267		31	9.85		40.85	49.50	
4514	4" wide		26	.308		33	11.40		44.40	54	
4516	6" wide		24	.333		38.50	12.35		50.85	61.50	
4520	10" wide		20	.400		47	14.80		61.80	75	
4522	12" wide		18	.444		62	16.45		78.45	93.50	
5000	Sidewall, aluminum, 3 way dispersion										
5100	8" x 4"	1 Shee	24	.333	Ea.	6.05	12.35		18.40	25.50	
5110	8" x 6"		24	.333		6.45	12.35		18.80	26	
5120	10" x 4"		22	.364		6.35	13.45		19.80	28	
5130	10" x 6"		22	.364		6.80	13.45		20.25	28.50	
5140	10" x 8"		20	.400		9.30	14.80		24.10	33	
5160	12" x 4"		18	.444		6.85	16.45		23.30	33	
5170	12" x 6"		17	.471		7.55	17.40		24.95	35.50	
5180	12" x 8"		16	.500		9.80	18.50		28.30	39.50	
5200	14" x 4"		15	.533		7.60	19.75		27.35	39	
5220	14" x 6"		14	.571		8.40	21		29.40	42	
5240	14" x 8"		13	.615		10.90	23		33.90	47	
5260	16" x 6"		12.50	.640		12.05	23.50		35.55	50	
6000	For steel diffusers instead of aluminum, deduct					10%					
500	**GRILLES**										**500**
0010											
0020	Aluminum										
0100	Air supply, single deflection, adjustable										
0120	8" x 4"	1 Shee	30	.267	Ea.	11.30	9.85		21.15	27.50	
0140	8" x 8"		28	.286		13.70	10.55		24.25	31.50	
0160	10" x 4"		24	.333		12.45	12.35		24.80	33	
0180	10" x 10"		23	.348		15.85	12.85		28.70	37.50	
0200	12" x 6"		23	.348		14.75	12.85		27.60	36	
0220	12" x 12"		22	.364		19.35	13.45		32.80	42.50	
0230	14" x 6"		23	.348		16.45	12.85		29.30	38	
0240	14" x 8"		23	.348		17.55	12.85		30.40	39	
0260	14" x 14"		22	.364		24	13.45		37.45	47	
0270	18" x 8"		23	.348		20.50	12.85		33.35	42.50	
0280	18" x 10"		23	.348		23	12.85		35.85	45.50	
0300	18" x 18"		21	.381		36	14.10		50.10	62	
0320	20" x 12"		22	.364		26.50	13.45		39.95	50	
0340	20" x 20"		21	.381		44	14.10		58.10	70.50	
0350	24" x 8"		20	.400		25.50	14.80		40.30	51	

15 MECHANICAL

Important: See the Reference Section for critical supporting data - Reference Nos., Crews, & City Cost Indexes

		CREW	DAILY OUTPUT	LABOR-HOURS	UNIT	2003 BARE COSTS				TOTAL INCL O&P	
15850	**Air Outlets & Inlets**					MAT.	LABOR	EQUIP.	TOTAL		
500	0360	24" x 14"	1 Shee	18	.444	Ea.	35.50	16.45		51.95	64.50	**500**
	0380	24" x 24"		15	.533		62	19.75		81.75	98.50
	0400	30" x 8"		20	.400		32.50	14.80		47.30	59
	0420	30" x 10"		19	.421		36	15.60		51.60	63.50
	0440	30" x 12"		18	.444		39.50	16.45		55.95	69
	0460	30" x 16"		17	.471		51.50	17.40		68.90	83.50
	0480	30" x 18"		17	.471		57.50	17.40		74.90	90.50
	0500	30" x 30"		14	.571		103	21		124	146
	0520	36" x 12"		17	.471		49	17.40		66.40	81
	0540	36" x 16"		16	.500		62.50	18.50		81	97.50
	0560	36" x 18"		15	.533		69.50	19.75		89.25	107
	0570	36" x 20"		15	.533		76	19.75		95.75	114
	0580	36" x 24"		14	.571		98.50	21		119.50	141
	0600	36" x 28"		13	.615		121	23		144	168
	0620	36" x 30"		12	.667		133	24.50		157.50	184
	0640	36" x 32"		12	.667		146	24.50		170.50	199
	0660	36" x 34"		11	.727		159	27		186	217
	0680	36" x 36"		11	.727		172	27		199	232
	0700	For double deflecting, add					70%				
	1000	Air return, 6" x 6"	1 Shee	26	.308		12.35	11.40		23.75	31
	1020	10" x 6"		24	.333		14.95	12.35		27.30	35.50
	1040	14" x 6"		23	.348		17.55	12.85		30.40	39
	1060	10" x 8"		23	.348		14.95	12.85		27.80	36.50
	1080	16" x 8"		22	.364		22	13.45		35.45	45.50
	1100	12" x 12"		22	.364		22	13.45		35.45	45.50
	1120	24" x 12"		18	.444		39.50	16.45		55.95	69
	1140	30" x 12"		16	.500		47	18.50		65.50	80
	1160	14" x 14"		22	.364		27.50	13.45		40.95	51
	1180	16" x 16"		22	.364		32	13.45		45.45	56
	1200	18" x 18"		21	.381		37	14.10		51.10	63
	1220	24" x 18"		16	.500		47	18.50		65.50	80
	1240	36" x 18"		15	.533		78.50	19.75		98.25	117
	1260	24" x 24"		15	.533		62	19.75		81.75	98.50
	1280	36" x 24"		14	.571		93.50	21		114.50	136
	1300	48" x 24"		12	.667		153	24.50		177.50	207
	1320	48" x 30"		11	.727		198	27		225	259
	1340	36" x 36"		13	.615		158	23		181	209
	1360	48" x 36"		11	.727		222	27		249	287
	1380	48" x 48"		8	1		296	37		333	380
	2000	Door grilles, 12" x 12"		22	.364		31	13.45		44.45	55.50
	2020	18" x 12"		22	.364		39	13.45		52.45	64
	2040	24" x 12"		18	.444		46	16.45		62.45	76.50
	2060	18" x 18"		18	.444		58.50	16.45		74.95	90
	2080	24" x 18"		16	.500		73	18.50		91.50	109
	2100	24" x 24"		15	.533		97	19.75		116.75	138
	3000	Filter grille with filter, 12" x 12"		24	.333		36.50	12.35		48.85	59
	3020	18" x 12"		20	.400		53.50	14.80		68.30	81.50
	3040	24" x 18"		18	.444		77.50	16.45		93.95	111
	3060	24" x 24"		16	.500		101	18.50		119.50	141
	3080	30" x 24"		14	.571		130	21		151	176
	3100	30" x 30"		13	.615		164	23		187	215
	3950	Eggcrate, framed, 6" x 6" opening		26	.308		13.95	11.40		25.35	33
	3954	8" x 8" opening		24	.333		14.45	12.35		26.80	35
	3960	10" x 10" opening		23	.348		22	12.85		34.85	44.50
	3970	12" x 12" opening		22	.364		19.55	13.45		33	42.50
	3980	14" x 14" opening		22	.364		24	13.45		37.45	47.50

MECHANICAL 15

15850	Air Outlets & Inlets	CREW	DAILY OUTPUT	LABOR-HOURS	UNIT	2003 BARE COSTS				TOTAL INCL O&P	
						MAT.	LABOR	EQUIP.	TOTAL		
500	3984	16" x 16" opening	1 Shee	21	.381	Ea.	30.50	14.10		44.60	55.50	**500**
	3990	18" x 18" opening		21	.381		36	14.10		50.10	61.50	
	4020	22" x 22" opening		17	.471		55.50	17.40		72.90	88	
	4040	24" x 24" opening		15	.533		60	19.75		79.75	96.50	
	4044	28" x 28" opening		14	.571		81.50	21		102.50	123	
	4048	36" x 36" opening		12	.667		98	24.50		122.50	146	
	4050	48" x 24" opening		12	.667		174	24.50		198.50	229	
	4060	Eggcrate, lay-in, T-bar system										
	4070	48" x 24" sheet	1 Shee	40	.200	Ea.	76.50	7.40		83.90	96	
	5000	Transfer grille, vision proof, 8" x 4"		30	.267		21	9.85		30.85	38.50	
	5020	8" x 6"		28	.286		21	10.55		31.55	39.50	
	5040	8" x 8"		26	.308		23.50	11.40		34.90	43	
	5060	10" x 6"		24	.333		21.50	12.35		33.85	43	
	5080	10" x 10"		23	.348		27.50	12.85		40.35	50	
	5090	12" x 6"		23	.348		28.50	12.85		41.35	51	
	5100	12" x 10"		22	.364		30	13.45		43.45	54	
	5110	12" x 12"		22	.364		32.50	13.45		45.95	56.50	
	5120	14" x 10"		22	.364		32.50	13.45		45.95	56.50	
	5140	16" x 8"		21	.381		32.50	14.10		46.60	57.50	
	5160	18" x 12"		20	.400		42.50	14.80		57.30	69.50	
	5170	18" x 18"		21	.381		60	14.10		74.10	88	
	5180	20" x 12"		20	.400		46	14.80		60.80	73.50	
	5200	20" x 20"		19	.421		72.50	15.60		88.10	104	
	5220	24" x 12"		17	.471		51.50	17.40		68.90	83.50	
	5240	24" x 24"		16	.500		99	18.50		117.50	138	
	5260	30" x 6"		15	.533		45.50	19.75		65.25	81	
	5280	30" x 8"		15	.533		53	19.75		72.75	89	
	5300	30" x 12"		14	.571		66.50	21		87.50	106	
	5320	30" x 16"		13	.615		79	23		102	122	
	5340	30" x 20"		12	.667		100	24.50		124.50	148	
	5360	30" x 24"		11	.727		123	27		150	177	
	5380	30" x 30"		10	.800		162	29.50		191.50	224	
	6000	For steel grilles instead of aluminum in above, deduct					10%					
	6200	Plastic, eggcrate, lay-in, T-bar system										
	6210	48" x 24" sheet	1 Shee	50	.160	Ea.	17.55	5.90		23.45	28.50	
	6250	Steel door louver										
	6270	With fire link, steel only										
	6350	12" x 18"	1 Shee	22	.364	Ea.	135	13.45		148.45	170	
	6410	12" x 24"		21	.381		150	14.10		164.10	187	
	6470	18" x 18"		20	.400		150	14.80		164.80	188	
	6530	18" x 24"		19	.421		172	15.60		187.60	213	
	6600	18" x 36"		17	.471		195	17.40		212.40	242	
	6660	24" x 24"		15	.533		181	19.75		200.75	230	
	6720	24" x 30"		15	.533		195	19.75		214.75	246	
	6780	30" x 30"		13	.615		229	23		252	287	
	6840	30" x 36"		12	.667		265	24.50		289.50	330	
	6900	36" x 36"		10	.800		305	29.50		334.50	380	
600	0010	**LOUVERS**										**600**
	0100	Aluminum, extruded, with screen, mill finish										
	1000	Brick vent, (see also division 04090)										
	1100	Standard, 4" deep, 8" wide, 5" high	1 Shee	24	.333	Ea.	23	12.35		35.35	44.50	
	1200	Modular, 4" deep, 7-3/4" wide, 5" high		24	.333		24	12.35		36.35	45.50	
	1300	Speed brick, 4" deep, 11-5/8" wide, 3-7/8" high		24	.333		24	12.35		36.35	45.50	
	1400	Fuel oil brick, 4" deep, 8" wide, 5" high		24	.333		42	12.35		54.35	65	
	2000	Cooling tower and mechanical equip., screens, light weight		40	.200	S.F.	9.95	7.40		17.35	22.50	

Important: See the Reference Section for critical supporting data - Reference Nos., Crews, & City Cost Indexes

15850	Air Outlets & Inlets	CREW	DAILY OUTPUT	LABOR-HOURS	UNIT	2003 BARE COSTS				TOTAL INCL O&P
						MAT.	LABOR	EQUIP.	TOTAL	
600	2020	Standard weight	1 Shee	35	.229	S.F.	27.50	8.45		35.95	43.50	600
2500	Dual combination, automatic, intake or exhaust		20	.400		38	14.80		52.80	65
2520	Manual operation		20	.400		28	14.80		42.80	54
2540	Electric or pneumatic operation		20	.400		28	14.80		42.80	54
2560	Motor, for electric or pneumatic	↓	14	.571	Ea.	330	21		351	400
3000	Fixed blade, continuous line									
3100	Mullion type, stormproof	1 Shee	28	.286	S.F.	28	10.55		38.55	47.50
3200	Stormproof		28	.286		28	10.55		38.55	47.50
3300	Vertical line	↓	28	.286	↓	37	10.55		47.55	57.50
3500	For damper to use with above, add					50%	30%			
3520	Motor, for damper, electric or pneumatic	1 Shee	14	.571	Ea.	330	21		351	400
4000	Operating, 45°, manual, electric or pneumatic		24	.333	S.F.	28	12.35		40.35	50
4100	Motor, for electric or pneumatic		14	.571	Ea.	330	21		351	400
4200	Penthouse, roof		56	.143	S.F.	16.30	5.30		21.60	26
4300	Walls		40	.200	↓	39.50	7.40		46.90	54.50
5000	Thinline, under 4" thick, fixed blade	↓	40	.200		15.80	7.40		23.20	29
5010	Finishes, applied by mfr. at additional cost, available in colors									
5020	Prime coat only, add				S.F.	2.30			2.30	2.53
5040	Baked enamel finish coating, add					4.23			4.23	4.65
5060	Anodized finish, add					4.59			4.59	5.05
5080	Duranodic finish, add					8.30			8.30	9.15
5100	Fluoropolymer finish coating, add					13			13	14.30
9980	For small orders (under 10 pieces), add				↓	25%				
700	0010	**REGISTERS**										700
0980	Air supply									
1000	Ceiling/wall, O.B. damper, anodized aluminum									
1010	One or two way deflection, adj. curved face bars									
1014	6" x 6"	1 Shee	24	.333	Ea.	19.50	12.35		31.85	40.50
1020	8" x 4"		26	.308		19.50	11.40		30.90	39
1040	8" x 8"		24	.333		24.50	12.35		36.85	46
1060	10" x 6"		20	.400		22.50	14.80		37.30	47.50
1080	10" x 10"		19	.421		30	15.60		45.60	57
1100	12" x 6"		19	.421		21.50	15.60		37.10	47.50
1120	12" x 12"		18	.444		31.50	16.45		47.95	60
1140	14" x 8"		17	.471		31	17.40		48.40	61
1160	14" x 14"		18	.444		36.50	16.45		52.95	65.50
1170	16" x 16"		17	.471		40.50	17.40		57.90	71.50
1180	18" x 8"		18	.444		50	16.45		66.45	80.50
1200	18" x 18"		17	.471		63.50	17.40		80.90	96.50
1220	20" x 4"		19	.421		32	15.60		47.60	59
1240	20" x 6"		18	.444		32	16.45		48.45	60.50
1260	20" x 8"		18	.444		35	16.45		51.45	64
1280	20" x 20"		17	.471		77	17.40		94.40	112
1290	22" x 22"		15	.533		88.50	19.75		108.25	128
1300	24" x 4"		17	.471		39.50	17.40		56.90	70
1320	24" x 6"		16	.500		39.50	18.50		58	71.50
1340	24" x 8"		13	.615		43	23		66	82.50
1350	24" x 18"		12	.667		80	24.50		104.50	126
1360	24" x 24"		11	.727		109	27		136	161
1380	30" x 4"		16	.500		49	18.50		67.50	82.50
1400	30" x 6"		15	.533		49	19.75		68.75	84.50
1420	30" x 8"		14	.571		56.50	21		77.50	94.50
1440	30" x 24"		12	.667		140	24.50		164.50	192
1460	30" x 30"	↓	10	.800	↓	180	29.50		209.50	244
1504	4 way deflection, adjustable curved face bars									

		CREW	DAILY OUTPUT	LABOR-HOURS	UNIT	2003 BARE COSTS				TOTAL INCL O&P
	15850 | Air Outlets & Inlets					MAT.	LABOR	EQUIP.	TOTAL	
700 1510	6" x 6"	1 Shee	26	.308	Ea.	23.50	11.40		34.90	43
1514	8" x 8"		24	.333		29	12.35		41.35	51
1518	10" x 10"		19	.421		36	15.60		51.60	64
1522	12" x 6"		19	.421		26	15.60		41.60	52.50
1526	12" x 12"		18	.444		37.50	16.45		53.95	67
1530	14" x 14"		18	.444		44	16.45		60.45	73.50
1534	16" x 16"		17	.471		48.50	17.40		65.90	80.50
1538	18" x 18"		17	.471		76	17.40		93.40	111
1542	22" x 22"	▼	16	.500	▼	106	18.50		124.50	146
1980	One way deflection, adj. vert. or horiz. face bars									
1990	6" x 6"	1 Shee	26	.308	Ea.	27.50	11.40		38.90	48
2000	8" x 4"		26	.308		29	11.40		40.40	49.50
2020	8" x 8"		24	.333		31.50	12.35		43.85	53.50
2040	10" x 6"		20	.400		31.50	14.80		46.30	57.50
2060	10" x 10"		19	.421		36.50	15.60		52.10	64.50
2080	12" x 6"		19	.421		34	15.60		49.60	61.50
2100	12" x 12"		18	.444		44.50	16.45		60.95	74.50
2120	14" x 8"		17	.471		43.50	17.40		60.90	74.50
2140	14" x 12"		18	.444		48.50	16.45		64.95	79
2160	14" x 14"		18	.444		55	16.45		71.45	86
2180	16" x 6"		18	.444		40.50	16.45		56.95	70
2200	16" x 12"		18	.444		53.50	16.45		69.95	84.50
2220	16" x 16"		17	.471		68	17.40		85.40	102
2240	18" x 8"		18	.444		48	16.45		64.45	78
2260	18" x 12"		17	.471		57.50	17.40		74.90	90
2280	18" x 18"		16	.500		84	18.50		102.50	121
2300	20" x 10"		18	.444		56.50	16.45		72.95	87.50
2320	20" x 16"		16	.500		81.50	18.50		100	118
2340	20" x 20"		16	.500		102	18.50		120.50	141
2360	24" x 12"		15	.533		70.50	19.75		90.25	108
2380	24" x 16"		14	.571		94	21		115	136
2400	24" x 20"		12	.667		117	24.50		141.50	167
2420	24" x 24"		10	.800		144	29.50		173.50	204
2440	30" x 12"		13	.615		91	23		114	135
2460	30" x 16"		13	.615		119	23		142	166
2480	30" x 24"		11	.727		185	27		212	246
2500	30" x 30"		9	.889		238	33		271	315
2520	36" x 12"		11	.727		111	27		138	164
2540	36" x 24"		10	.800		229	29.50		258.50	298
2560	36" x 36"	▼	8	1		400	37		437	495
2600	For 2 way deflect., adj. vert. or horiz. face bars, add					40%				
2700	Above registers in steel instead of aluminum, deduct				▼	10%				
3000	Baseboard, hand adj. damper, enameled steel									
3012	8" x 6"	1 Shee	26	.308	Ea.	11.70	11.40		23.10	30.50
3020	10" x 6"		24	.333		13.20	12.35		25.55	33.50
3040	12" x 5"		23	.348		15.10	12.85		27.95	36.50
3060	12" x 6"		23	.348		13.90	12.85		26.75	35
3080	12" x 8"		22	.364		20	13.45		33.45	43
3100	14" x 6"	▼	20	.400	▼	15.10	14.80		29.90	39.50
4000	Floor, toe operated damper, enameled steel									
4020	4" x 8"	1 Shee	32	.250	Ea.	16.65	9.25		25.90	32.50
4040	4" x 12"		26	.308		19.55	11.40		30.95	39
4060	6" x 8"		28	.286		19.50	10.55		30.05	38
4080	6" x 14"		22	.364		22.50	13.45		35.95	46
4100	8" x 10"		22	.364		20.50	13.45		33.95	43.50
4120	8" x 16"	▼	20	.400	▼	34.50	14.80		49.30	60.50

Important: See the Reference Section for critical supporting data - Reference Nos., Crews, & City Cost Indexes

	15850	Air Outlets & Inlets	CREW	DAILY OUTPUT	LABOR-HOURS	UNIT	2003 BARE COSTS				TOTAL INCL O&P	
							MAT.	LABOR	EQUIP.	TOTAL		
700	4140	10" x 10"	1 Shee	20	.400	Ea.	24	14.80		38.80	49.50	**700**
	4160	10" x 16"		18	.444		37	16.45		53.45	66.50	
	4180	12" x 12"		18	.444		30	16.45		46.45	58.50	
	4200	12" x 24"		16	.500		104	18.50		122.50	143	
	4220	14" x 14"		16	.500		85.50	18.50		104	123	
	4240	14" x 20"		15	.533		114	19.75		133.75	156	
	4980	Air return										
	5000	Ceiling or wall, fixed 45° face blades										
	5010	Adjustable O.B. damper, anodized aluminum										
	5020	4" x 8"	1 Shee	26	.308	Ea.	22	11.40		33.40	42	
	5040	6" x 8"		24	.333		22	12.35		34.35	43.50	
	5060	6" x 10"		19	.421		22	15.60		37.60	48.50	
	5070	6" x 12"		19	.421		27.50	15.60		43.10	54	
	5080	6" x 16"		18	.444		30	16.45		46.45	58.50	
	5100	8" x 10"		19	.421		27.50	15.60		43.10	54	
	5120	8" x 12"		16	.500		30	18.50		48.50	61.50	
	5140	10" x 10"		18	.444		30	16.45		46.45	58.50	
	5160	10" x 16"		17	.471		39	17.40		56.40	70	
	5170	12" x 12"		18	.444		37	16.45		53.45	66.50	
	5180	12" x 18"		18	.444		52	16.45		68.45	82.50	
	5184	12" x 24"		14	.571		62	21		83	101	
	5200	12" x 30"		12	.667		76.50	24.50		101	123	
	5210	12" x 36"		10	.800		96	29.50		125.50	152	
	5220	16" x 16"		17	.471		54.50	17.40		71.90	87	
	5240	18" x 18"		16	.500		64.50	18.50		83	99.50	
	5250	18" x 24"		13	.615		81.50	23		104.50	125	
	5254	18" x 30"		11	.727		101	27		128	154	
	5260	18" x 36"		10	.800		129	29.50		158.50	188	
	5270	20" x 20"		10	.800		76.50	29.50		106	130	
	5274	20" x 24"		12	.667		89.50	24.50		114	137	
	5280	24" x 24"		11	.727		106	27		133	159	
	5281	24" x 24"		11	.727		106	27		133	159	
	5290	24" x 30"		10	.800		136	29.50		165.50	195	
	5300	24" x 36"		8	1		166	37		203	239	
	5320	24" x 48"		6	1.333		242	49.50		291.50	340	
	5340	30" x 30"		6.50	1.231		172	45.50		217.50	260	
	5344	30" x 120"	Q-9	6	2.667		925	89		1,014	1,150	
	5360	36" x 36"	1 Shee	6	1.333		278	49.50		327.50	380	
	5370	40" x 32"	"	6	1.333		261	49.50		310.50	365	
	5400	Ceiling or wall, removable-reversible core, single deflection										
	5402	Adjustable O.B. damper, radiused frame with border, aluminum										
	5410	8" x 4"	1 Shee	26	.308	Ea.	32	11.40		43.40	52.50	
	5414	8" x 6"		24	.333		36	12.35		48.35	58.50	
	5418	10" x 6"		19	.421		39	15.60		54.60	67	
	5422	10" x 10"		18	.444		49	16.45		65.45	79.50	
	5426	12" x 6"		19	.421		42	15.60		57.60	70	
	5430	12" x 12"		18	.444		63	16.45		79.45	95	
	5434	16" x 16"		16	.500		94	18.50		112.50	132	
	5438	18" x 18"		17	.471		109	17.40		126.40	147	
	5442	20" x 20"		16	.500		134	18.50		152.50	176	
	5446	24" x 12"		15	.533		92	19.75		111.75	132	
	5450	24" x 18"		13	.615		135	23		158	184	
	5454	24" x 24"		11	.727		175	27		202	235	
	5458	30" x 12"		13	.615		110	23		133	156	
	5462	30" x 18"		12.50	.640		158	23.50		181.50	211	
	5466	30" x 24"		11	.727		210	27		237	273	

MECHANICAL 15

15850 | Air Outlets & Inlets

			DAILY	LABOR-			2003 BARE COSTS				TOTAL	
		CREW	OUTPUT	HOURS	UNIT	MAT.	LABOR	EQUIP.	TOTAL		INCL O&P	
700	5470	30" x 30"	1 Shee	9	.889	Ea.	262	33		295		340	**700**
	5474	36" x 12"		11	.727		133	27		160		188	
	5478	36" x 18"		10.50	.762		187	28		215		250	
	5482	36" x 24"		10	.800		259	29.50		288.50		330	
	5486	36" x 30"		9	.889		340	33		373		420	
	5490	36" x 36"	▼	8	1	▼	425	37		462		520	
	6000	For steel construction instead of aluminum, deduct					10%						

15854 | Ventilators

			DAILY	LABOR-			2003 BARE COSTS			TOTAL	
300	0010	**VENTILATORS** Base & damper										**300**
	1280	Rotary ventilators, wind driven, galvanized										
	1300	4" neck diameter	Q-9	20	.800	Ea.	52	26.50		78.50	98.50	
	1320	5" neck diameter		18	.889		58	29.50		87.50	110	
	1340	6" neck diameter		16	1		58	33.50		91.50	116	
	1360	8" neck diameter		14	1.143		60.50	38		98.50	126	
	1380	10" neck diameter		12	1.333		77.50	44.50		122	154	
	1400	12" neck diameter		10	1.600		84.50	53.50		138	175	
	1420	14" neck diameter		10	1.600		117	53.50		170.50	212	
	1440	16" neck diameter		9	1.778		141	59		200	248	
	1460	18" neck diameter, 1700 CFM		9	1.778		164	59		223	273	
	1480	20" neck diameter, 2100 CFM		8	2		182	66.50		248.50	305	
	1500	24" neck diameter, 3,100 CFM		8	2		245	66.50		311.50	375	
	1520	30" neck diameter, 4500 CFM		7	2.286		510	76		586	680	
	1540	36" neck diameter, 5,500 CFM	▼	6	2.667		720	89		809	925	
	1600	For aluminum, add					300%					
	1620	For stainless steel, add					600%					
	1630	For copper, add				▼	600%					
	2000	Stationary, gravity, syphon, galvanized										
	2100	3" neck diameter, 40 CFM	Q-9	24	.667	Ea.	47	22		69	86	
	2120	4" neck diameter, 50 CFM		20	.800		51.50	26.50		78	98	
	2140	5" neck diameter, 58 CFM		18	.889		52	29.50		81.50	103	
	2160	6" neck diameter, 66 CFM		16	1		45	33.50		78.50	101	
	2180	7" neck diameter, 86 CFM		15	1.067		56.50	35.50		92	118	
	2200	8" neck diameter, 110 CFM		14	1.143		57	38		95	122	
	2220	10" neck diameter, 140 CFM		12	1.333		73.50	44.50		118	150	
	2240	12" neck diameter, 160 CFM		10	1.600		94.50	53.50		148	187	
	2260	14" neck diameter, 250 CFM		10	1.600		119	53.50		172.50	213	
	2280	16" neck diameter, 380 CFM		9	1.778		158	59		217	266	
	2300	18" neck diameter, 500 CFM		9	1.778		146	59		205	252	
	2320	20" neck diameter, 625 CFM		8	2		223	66.50		289.50	350	
	2340	24" neck diameter, 900 CFM		8	2		257	66.50		323.50	385	
	2360	30" neck diameter, 1375 CFM		7	2.286		585	76		661	765	
	2380	36" neck diameter, 2,000 CFM		6	2.667		805	89		894	1,025	
	2400	42" neck diameter, 3000 CFM		4	4		1,125	133		1,258	1,450	
	2410	48" neck diameter, 4000 CFM	▼	3	5.333		1,525	178		1,703	1,950	
	2500	For aluminum, add					300%					
	2520	For stainless steel, add					600%					
	2530	For copper, add					600%					
	3000	Rotating chimney cap, galvanized, 4" neck diameter	Q-9	20	.800		40.50	26.50		67	85.50	
	3020	5" neck diameter		18	.889		41.50	29.50		71	91	
	3040	6" neck diameter		16	1		42	33.50		75.50	97.50	
	3060	7" neck diameter		15	1.067		45.50	35.50		81	105	
	3080	8" neck diameter		14	1.143		47	38		85	111	
	3100	10" neck diameter		12	1.333		61	44.50		105.50	136	
	3600	Stationary chimney rain cap, galvanized, 3" neck diameter	▼	24	.667	▼	35.50	22		57.50	73.50	

Important: See the Reference Section for critical supporting data - Reference Nos., Crews, & City Cost Indexes

15800 | Air Distribution

15854 | Ventilators

			CREW	DAILY OUTPUT	LABOR-HOURS	UNIT	MAT.	LABOR	EQUIP.	TOTAL	TOTAL INCL O&P	
							2003 BARE COSTS					
300	3620	4" neck diameter	Q-9	20	.800	Ea.	36	26.50		62.50	80.50	**300**
	3640	6" neck diameter		16	1		37.50	33.50		71	93	
	3680	8" neck diameter		14	1.143		40.50	38		78.50	104	
	3700	10" neck diameter		12	1.333		51.50	44.50		96	125	
	3720	12" neck diameter		10	1.600		85.50	53.50		139	177	
	3740	14" neck diameter		10	1.600		91.50	53.50		145	184	
	3760	16" neck diameter		9	1.778		112	59		171	215	
	3770	18" neck diameter		9	1.778		162	59		221	270	
	3780	20" neck diameter		8	2		216	66.50		282.50	340	
	3790	24" neck diameter		8	2		252	66.50		318.50	380	
	4200	Stationary mushroom, aluminum, 16" orifice diameter		10	1.600		271	53.50		324.50	380	
	4220	26" orifice diameter		6.15	2.602		400	86.50		486.50	575	
	4230	30" orifice diameter		5.71	2.802		585	93.50		678.50	790	
	4240	38" orifice diameter		5	3.200		840	107		947	1,100	
	4250	42" orifice diameter		4.70	3.404		1,100	113		1,213	1,400	
	4260	50" orifice diameter		4.44	3.604		1,325	120		1,445	1,625	
	5000	Relief vent										
	5500	Rectangular, aluminum, galvanized curb										
	5510	intake/exhaust, 0.05" SP										
	5600	500 CFM, 12" x 16"	Q-9	8	2	Ea.	390	66.50		456.50	535	
	5620	750 CFM, 12" x 20"		7.20	2.222		430	74		504	585	
	5640	1000 CFM, 12" x 24"		6.60	2.424		465	80.50		545.50	635	
	5660	1500 CFM, 12" x 36"		5.80	2.759		560	92		652	755	
	5680	3000 CFM, 20" x 42"		4	4		805	133		938	1,100	
	5700	6000 CFM, 20" x 84"		2.60	6.154		1,225	205		1,430	1,675	
	5720	8000 CFM, 24" x 96"		2.30	6.957		1,400	232		1,632	1,900	
	5740	10,000 CFM, 48" x 60"		1.80	8.889		1,725	296		2,021	2,350	
	5760	12,500 CFM, 48" x 72"		1.60	10		2,025	335		2,360	2,750	
	5780	15,000 CFM, 48" x 96"		1.30	12.308		2,325	410		2,735	3,200	
	5800	20,000 CFM, 48" x 120"		1.20	13.333		2,700	445		3,145	3,650	
	5820	25,000 CFM, 60" x 120"		.90	17.778		3,575	590		4,165	4,850	
	5840	30,000 CFM, 72" x 120"		.70	22.857		4,700	760		5,460	6,325	
	5860	40,000 CFM, 96" x 120"		.60	26.667		6,025	890		6,915	8,000	
	5880	50,000 CFM, 96" x 144"		.50	32		6,400	1,075		7,475	8,675	
	7000	Note: sizes based on exhaust. Intake, with 0.125" SP										
	7100	loss, approximately twice listed capacity.										
	7500	For power open, spring return damper										
	7600	motor for each S.F. of throat area	Q-9	100	.160	S.F.	43.50	5.35		48.85	56	

15860 | Air Cleaning Devices

			CREW	DAILY OUTPUT	LABOR-HOURS	UNIT	MAT.	LABOR	EQUIP.	TOTAL	TOTAL INCL O&P	
100	0010	**AIR FILTERS**										**100**
	0050	Activated charcoal type, full flow				MCFM	600			600	660	
	0060	Activated charcoal type, full flow, impregnated media 12" deep					175			175	193	
	0070	Activated charcoal type, HEPA filter & frame for field erection					175			175	193	
	0080	Activated charcoal type, HEPA filter-diffuser, ceiling install.					165			165	182	
	0500	Chemical media filtration type										
	1100	Industrial air fume & odor scrubber unit w/pump & motor										
	1110	corrosion resistant PVC construction										
	1120	Single pack filter, horizontal type										
	1130	500 CFM	Q-9	14	1.143	Ea.	3,800	38		3,838	4,225	
	1140	1000 CFM		11	1.455		4,450	48.50		4,498.50	4,975	
	1150	2000 CFM		8	2		5,400	66.50		5,466.50	6,050	
	1160	3000 CFM		7	2.286		6,250	76		6,326	7,025	
	1170	5000 CFM		5	3.200		8,150	107		8,257	9,150	
	1180	8000 CFM		4	4		11,400	133		11,533	12,700	
	1190	12,000 CFM		3	5.333		15,100	178		15,278	16,900	

15860	Air Cleaning Devices	CREW	DAILY OUTPUT	LABOR-HOURS	UNIT	2003 BARE COSTS				TOTAL INCL O&P
						MAT.	LABOR	EQUIP.	TOTAL	
100 1200	16,000 CFM	Q-9	2.50	6.400	Ea.	18,700	213		18,913	20,800 **100**
1210	20,000 CFM		2	8		23,400	266		23,666	26,200
1220	26,000 CFM	↓	1.50	10.667		27,000	355		27,355	30,300
1230	30,000 CFM	Q-10	2	12		30,700	415		31,115	34,300
1240	40,000 CFM	"	1.50	16		40,100	550		40,650	45,000
1250	50,000 CFM	Q-11	2	16		48,800	565		49,365	54,500
1260	55,000 CFM		1.80	17.778		53,500	625		54,125	59,500
1270	60,000 CFM	↓	1.50	21.333	↓	57,500	755		58,255	64,000
1300	Double pack filter, horizontal type									
1310	500 CFM	Q-9	10	1.600	Ea.	4,950	53.50		5,003.50	5,525
1320	1000 CFM		8	2		5,925	66.50		5,991.50	6,625
1330	2000 CFM		6	2.667		7,150	89		7,239	7,975
1340	3000 CFM		5	3.200		8,550	107		8,657	9,575
1350	5000 CFM		4	4		10,900	133		11,033	12,100
1360	8000 CFM		3	5.333		14,300	178		14,478	16,100
1370	12,000 CFM		2.50	6.400		19,500	213		19,713	21,700
1380	16,000 CFM		2	8		25,400	266		25,666	28,300
1390	20,000 CFM		1.50	10.667		30,800	355		31,155	34,500
1400	26,000 CFM	↓	1	16		35,700	535		36,235	40,000
1410	30,000 CFM	Q-10	1.50	16		41,800	550		42,350	46,900
1420	40,000 CFM	"	1.30	18.462		53,500	635		54,135	60,000
1430	50,000 CFM	Q-11	1.50	21.333		67,000	755		67,755	74,500
1440	55,000 CFM		1.30	24.615		73,000	870		73,870	82,000
1450	60,000 CFM	↓	1	32	↓	79,500	1,125		80,625	89,000
1500	Single pack filter, vertical type									
1510	500 CFM	Q-9	24	.667	Ea.	3,625	22		3,647	4,025
1520	1000 CFM		18	.889		4,300	29.50		4,329.50	4,775
1530	2000 CFM		12	1.333		5,150	44.50		5,194.50	5,725
1540	3000 CFM		9	1.778		6,000	59		6,059	6,725
1550	5000 CFM		6	2.667		8,000	89		8,089	8,950
1560	8000 CFM		4	4		11,400	133		11,533	12,700
1570	12,000 CFM		3	5.333		15,500	178		15,678	17,400
1580	16,000 CFM		2	8		18,200	266		18,466	20,400
1590	20,000 CFM		1.80	8.889		22,900	296		23,196	25,700
1600	24,000 CFM	↓	1.60	10	↓	25,600	335		25,935	28,600
1650	Double pack filter, vertical type									
1660	500 CFM	Q-9	22	.727	Ea.	4,700	24		4,724	5,225
1670	1000 CFM		16	1		5,625	33.50		5,658.50	6,225
1680	2000 CFM		10	1.600		6,750	53.50		6,803.50	7,500
1690	3000 CFM		7	2.286		7,925	76		8,001	8,850
1700	5000 CFM		5	3.200		10,400	107		10,507	11,600
1710	8000 CFM		3	5.333		14,200	178		14,378	15,900
1720	12,000 CFM		2.50	6.400		19,200	213		19,413	21,400
1730	16,000 CFM		2	8		24,100	266		24,366	27,000
1740	20,000 CFM		1.50	10.667		30,100	355		30,455	33,700
1750	24,000 CFM	↓	1	16	↓	33,600	535		34,135	37,800
1800	Inlet or outlet transition, horizontal									
1810	Single pack to 12,000 CFM, add					4%				
1820	Single pack to 30,000 CFM, add					6%				
1830	Single pack to 60,000 CFM, add					8%				
1840	Double pack to 12,000 CFM, add					3%				
1850	Double pack to 30,000 CFM, add					5%				
1860	Double pack to 60,000 CFM, add					6%				
1870	Inlet or outlet transition, vertical									
1880	Single pack to 5000 CFM, add					2%				
1890	Single pack to 24,000 CFM, add					3%				

15 **MECHANICAL**

Important: See the Reference Section for critical supporting data - Reference Nos., Crews, & City Cost Indexes

15860	Air Cleaning Devices	CREW	DAILY OUTPUT	LABOR-HOURS	UNIT	2003 BARE COSTS				TOTAL INCL O&P
						MAT.	LABOR	EQUIP.	TOTAL	
1900	Double pack to 24,000 CFM, add					2%				
2000	Electronic air cleaner, duct mounted									
2150	400 - 1000 CFM	1 Shee	2.30	3.478	Ea.	660	129		789	925
2200	1000 - 1400 CFM		2.20	3.636		685	135		820	965
2250	1400 - 2000 CFM		2.10	3.810		760	141		901	1,050
2260	2000 - 2500 CFM	↓	2	4	↓	780	148		928	1,100
2950	Mechanical media filtration units									
3000	High efficiency type, with frame, non-supported				MCFM	45			45	49.50
3100	Supported type					55			55	60.50
4000	Medium efficiency, extended surface					5			5	5.50
4500	Permanent washable					20			20	22
5000	Renewable disposable roll				↓	120			120	132
5500	Throwaway glass or paper media type				Ea.	4.60			4.60	5.05
5800	Filter, bag type									
5810	90-95% efficiency									
5820	24" x 12" x 29", .75-1.25 MCFM	1 Shee	1.90	4.211	Ea.	32.50	156		188.50	277
5830	24" x 24" x 29", 1.5-2.5 MCFM	"	1.90	4.211	"	63	156		219	310
5850	80-85% efficiency									
5860	24" x 12" x 29", .75-1.25 MCFM	1 Shee	1.90	4.211	Ea.	29	156		185	273
5870	24" x 24" x 29", 1.5-2.5 MCFM	"	1.90	4.211	"	51.50	156		207.50	298
6000	HEPA filter complete w/particle board,									
6010	kraft paper frame, separator material									
6020	95% DOP efficiency									
6030	12" x 12" x 6", 150 CFM	1 Shee	3.70	2.162	Ea.	29.50	80		109.50	157
6034	24" x 12" x 6", 375 CFM		1.90	4.211		31.50	156		187.50	276
6038	24" x 18" x 6", 450 CFM		1.90	4.211		40	156		196	285
6042	24" x 24" x 6", 700 CFM		1.90	4.211		47	156		203	293
6046	12" x 12" x 12", 250 CFM		3.70	2.162		40	80		120	168
6050	24" x 12" x 12", 500 CFM		1.90	4.211		47	156		203	293
6054	24" x 18" x 12", 875 CFM		1.90	4.211		63.50	156		219.50	310
6058	24" x 24" x 12", 1000 CFM	↓	1.90	4.211	↓	66	156		222	315
6100	99% DOP efficiency									
6110	12" x 12" x 6", 150 CFM	1 Shee	3.70	2.162	Ea.	50	80		130	179
6114	24" x 12" x 6", 325 CFM		1.90	4.211		55	156		211	300
6118	24" x 18" x 6", 550 CFM		1.90	4.211		77	156		233	325
6122	24" x 24" x 6", 775 CFM		1.90	4.211		79	156		235	330
6126	12" x 12" x 12", 250 CFM		3.60	2.222		56.50	82		138.50	189
6130	24" x 12" x 12", 500 CFM		1.90	4.211		68	156		224	315
6134	24" x 18" x 12", 775 CFM		1.90	4.211		109	156		265	360
6138	24" x 24" x 12", 1100 CFM	↓	1.90	4.211	↓	114	156		270	365
6500	HEPA filter housing, 14 ga. galv. sheet metal									
6510	12" x 12" x 6"	1 Shee	2.50	3.200	Ea.	450	118		568	680
6514	24" x 12" x 6"		2	4		500	148		648	780
6518	12" x 12" x 12"		2.40	3.333		450	123		573	685
6522	14" x 12" x 12"		2.30	3.478		500	129		629	750
6526	24" x 18" x 6"		1.90	4.211		585	156		741	880
6530	24" x 24" x 6"		1.80	4.444		585	164		749	895
6534	24" x 48" x 6"		1.70	4.706		780	174		954	1,125
6538	24" x 72" x 6"		1.60	5		1,000	185		1,185	1,375
6542	24" x 18" x 12"		1.80	4.444		585	164		749	895
6546	24" x 24" x 12"		1.70	4.706		585	174		759	910
6550	24" x 48" x 12"		1.60	5		780	185		965	1,150
6554	24" x 72" x 12"	↓	1.50	5.333		1,000	197		1,197	1,400
6558	48" x 48" x 6"	Q-9	2.80	5.714		1,125	190		1,315	1,525
6562	48" x 72" x 6"		2.60	6.154		1,400	205		1,605	1,875
6566	48" x 96" x 6"	↓	2.40	6.667	↓	1,575	222		1,797	2,075

MECHANICAL 15

			15860	**Air Cleaning Devices**	CREW	DAILY OUTPUT	LABOR-HOURS	UNIT	2003 BARE COSTS				TOTAL INCL O&P	
									MAT.	LABOR	EQUIP.	TOTAL		
100	6570			48" x 48" x 12"	Q-9	2.70	5.926	Ea.	1,125	197		1,322	1,525	**100**
	6574			48" x 72" x 12"		2.50	6.400		1,400	213		1,613	1,875	
	6578			48" x 96" x 12"		2.30	6.957		1,575	232		1,807	2,075	
	6582			114" x 72" x 12"		2	8		2,400	266		2,666	3,050	
500	0010		**EXHAUST SYSTEMS**											**500**
	0500		Engine exhaust, garage, in-floor system	D3090 320										
	0510		Single tube outlet assemblies											
	0520		For transite pipe ducting, self-storing tube											
	0530		3" tubing adapter plate	1 Shee	16	.500	Ea.	153	18.50		171.50	197	
	0540		4" tubing adapter plate		16	.500		153	18.50		171.50	197	
	0550		5" tubing adapter plate		16	.500		153	18.50		171.50	197	
	0600		For vitrified tile ducting										
	0610		3" tubing adapter plate, self-storing tube	1 Shee	16	.500	Ea.	153	18.50		171.50	197	
	0620		4" tubing adapter plate, self-storing tube		16	.500		153	18.50		171.50	197	
	0660		5" tubing adapter plate, self-storing tube		16	.500		153	18.50		171.50	197	
	0800		Two tube outlet assemblies										
	0810		For transite pipe ducting, self-storing tube										
	0820		3" tubing, dual exhaust adapter plate	1 Shee	16	.500	Ea.	153	18.50		171.50	197	
	0850		For vitrified tile ducting										
	0860		3" tubing, dual exhaust, self-storing tube	1 Shee	16	.500	Ea.	153	18.50		171.50	197	
	0870		3" tubing, double outlet, non-storing tubes	"	16	.500	"	153	18.50		171.50	197	
	0900		Accessories for metal tubing, (overhead systems also)										
	0910		Adapters, for metal tubing end										
	0920		3" tail pipe type				Ea.	34			34	37.50	
	0930		4" tail pipe type					34			34	37.50	
	0940		5" tail pipe type					34			34	37.50	
	0990		5" diesel stack type					238			238	262	
	1000		6" diesel stack type					238			238	262	
	1100		Bullnose (guide) required for in-floor assemblies										
	1110		3" tubing size				Ea.	17			17	18.70	
	1120		4" tubing size					17			17	18.70	
	1130		5" tubing size					17			17	18.70	
	1150		Plain rings, for tubing end										
	1160		3" tubing size				Ea.	8.10			8.10	8.90	
	1170		4" tubing size				"	10.20			10.20	11.20	
	1200		Tubing, galvanized, flexible, (for overhead systems also)										
	1210		3" ID				L.F.	4.55			4.55	5	
	1220		4" ID					5.50			5.50	6.05	
	1230		5" ID					6.40			6.40	7	
	1240		6" ID					7.30			7.30	8.05	
	1250		Stainless steel, flexible, (for overhead sys, also)										
	1260		3" ID				L.F.	9.45			9.45	10.40	
	1270		4" ID					12.30			12.30	13.50	
	1280		5" ID					14.85			14.85	16.30	
	1290		6" ID					17.70			17.70	19.45	
	1500		Engine exhaust, garage, ovrhd. compon., for neoprene tubing										
	1510		Alternate metal tubing & accessories see above										
	1550		Adapters, for neoprene tubing end										
	1560		3" tail pipe, adjustable, neoprene				Ea.	30			30	32.50	
	1570		3" tail pipe, heavy wall neoprene					38.50			38.50	42	
	1580		4" tail pipe, heavy wall neoprene					41			41	45	
	1590		5" tail pipe, heavy wall neoprene					55.50			55.50	61	
	1650		Connectors, tubing										
	1660		3" interior, aluminum				Ea.	8.10			8.10	8.90	
	1670		4" interior, aluminum					10.20			10.20	11.20	
	1710		5" interior, neoprene					25.50			25.50	28	

Important: See the Reference Section for critical supporting data - Reference Nos., Crews, & City Cost Indexes

15860	Air Cleaning Devices	CREW	DAILY OUTPUT	LABOR-HOURS	UNIT	2003 BARE COSTS				TOTAL INCL O&P
						MAT.	LABOR	EQUIP.	TOTAL	
500										**500**
1750	3" spiralock, neoprene	D3090 320			Ea.	8.10			8.10	8.90
1760	4" spiralock, neoprene					10.20			10.20	11.20
1780	Y for 3" ID tubing, neoprene, dual exhaust					21.50			21.50	23.50
1790	Y for 4" ID tubing, aluminum, dual exhaust					38.50			38.50	42
1850	Elbows, aluminum, splice into tubing for strap									
1860	3" neoprene tubing size				Ea.	25.50			25.50	28
1870	4" neoprene tubing size					30			30	32.50
1900	Flange assemblies, connect tubing to overhead duct					28			28	31
2000	Hardware and accessories									
2020	Cable, galvanized, 1/8" diameter				L.F.	.30			.30	.33
2040	Cleat, tie down cable or rope				Ea.	3.40			3.40	3.74
2060	Pulley					4.68			4.68	5.15
2080	Pulley hook, universal					3.40			3.40	3.74
2100	Rope, nylon, 1/4" diameter				L.F.	.26			.26	.29
2120	Winch, 1" diameter				Ea.	72.50			72.50	79.50
2150	Lifting strap, mounts on neoprene									
2160	3" tubing size				Ea.	17			17	18.70
2170	4" tubing size					17			17	18.70
2180	5" tubing size					17			17	18.70
2190	6" tubing size					17			17	18.70
2200	Tubing, neoprene, 11' lengths									
2210	3" ID				L.F.	5.10			5.10	5.60
2220	4" ID					6.40			6.40	7
2230	5" ID					11.50			11.50	12.65
2500	Engine exhaust, thru-door outlet									
2510	3" tube size	1 Carp	16	.500	Ea.	30	15.80		45.80	57
2530	4" tube size	"	16	.500	"	30	15.80		45.80	57
3000	Tubing, exhaust, flex hose, with									
3010	coupler, damper and tail pipe adapter									
3020	Neoprene									
3040	3" x 20'	1 Shee	6	1.333	Ea.	162	49.50		211.50	254
3050	4" x 15'		5.40	1.481		162	55		217	263
3060	4" x 20'		5	1.600		194	59		253	305
3070	5" x 15'		4.40	1.818		273	67.50		340.50	405
3100	Galvanized									
3110	3" x 20'	1 Shee	6	1.333	Ea.	147	49.50		196.50	237
3120	4" x 17'		5.60	1.429		153	53		206	250
3130	4" x 20'		5	1.600		169	59		228	278
3140	5" x 17'		4.60	1.739		188	64.50		252.50	305
7500	Welding fume elimination accessories for garage exhaust systems									
7600	Cut off (blast gate)									
7610	3" tubing size, 3" x 6" opening	1 Shee	24	.333	Ea.	13.60	12.35		25.95	34
7620	4" tubing size, 4" x 8" opening		24	.333		15.30	12.35		27.65	36
7630	5" tubing size, 5" x 10" opening		24	.333		20.50	12.35		32.85	41.50
7640	6" tubing size		24	.333		24	12.35		36.35	45
7650	8" tubing size		24	.333		34	12.35		46.35	56.50
7700	Hoods, magnetic, with handle & screen									
7710	3" tubing size, 3" x 6" opening	1 Shee	24	.333	Ea.	64	12.35		76.35	89
7720	4" tubing size, 4" x 8" opening		24	.333		64	12.35		76.35	89
7730	5" tubing size, 5" x 10" opening		24	.333		64	12.35		76.35	89
8000	Blower, for tailpipe exhaust system									
8010	Direct drive									
8012	495 CFM, 1/3 HP	Q-9	5	3.200	Ea.	1,100	107		1,207	1,375
8014	1445 CFM, 3/4 HP		4	4		1,250	133		1,383	1,575
8016	1840 CFM, 1-1/2 HP		3	5.333		1,700	178		1,878	2,150
8030	Beltdrive									

MECHANICAL **15**

15860 | Air Cleaning Devices

			CREW	DAILY OUTPUT	LABOR-HOURS	UNIT	2003 BARE COSTS				TOTAL INCL O&P	
							MAT.	LABOR	EQUIP.	TOTAL		
500	8032	1400 CFM, 1 HP	D3090 320	Q-9	4	4	Ea.	1,400	133		1,533	1,750	500
	8034	2023 CFM, 1-1/2 HP			3	5.333		1,975	178		2,153	2,450	
	8036	2750 CFM, 2 HP			2	8		2,350	266		2,616	2,975	
	8038	4400 CFM, 3 HP			1.80	8.889		3,700	296		3,996	4,525	
	8040	7060 CFM, 5 HP			1.40	11.429		6,500	380		6,880	7,750	

15905 | HVAC Instrumentation

			CREW	DAILY OUTPUT	LABOR-HOURS	UNIT	2003 BARE COSTS				TOTAL INCL O&P	
							MAT.	LABOR	EQUIP.	TOTAL		
960	0010	**WATER LEVEL CONTROLS**										960
	1000	Electric water feeder	1 Stpi	12	.667	Ea.	199	25		224	257	
	2000	Feeder cut-off combination										
	2100	Steam system up to 5000 sq. ft.	1 Stpi	12	.667	Ea.	385	25		410	460	
	2200	Steam system above 5000 sq. ft.		12	.667		470	25		495	560	
	2300	Steam and hot water, high pressure		10	.800		555	30		585	655	
	3000	Low water cut-off for hot water boiler, 50 psi maximum										
	3100	1" top & bottom equalizing pipes, manual reset	1 Stpi	14	.571	Ea.	211	21.50		232.50	265	
	3200	1" top & bottom equalizing pipes		14	.571		215	21.50		236.50	269	
	3300	2-1/2" side connection for nipple-to-boiler		14	.571		194	21.50		215.50	246	
	4000	Low water cut-off for low pressure steam with quick hook-up ftgs.										
	4100	For installation in gauge glass tappings	1 Stpi	16	.500	Ea.	151	18.80		169.80	195	
	4200	Built-in type, 2-1/2" tap - 3-1/8" insertion		16	.500		130	18.80		148.80	172	
	4300	Built-in type, 2-1/2" tap - 1-3/4" insertion		16	.500		137	18.80		155.80	180	
	4400	Side connection to 2-1/2" tapping		16	.500		147	18.80		165.80	191	
	5000	Pump control, low water cut-off and alarm switch		14	.571		385	21.50		406.50	455	
	9000	Water gauges, complete										
	9010	Rough brass, wheel type										
	9020	125 PSI at 350° F										
	9030	3/8" pipe size	1 Stpi	11	.727	Ea.	31.50	27.50		59	76	
	9040	1/2" pipe size	"	10	.800	"	34	30		64	83	
	9060	200 PSI at 400° F										
	9070	3/8" pipe size	1 Stpi	11	.727	Ea.	39	27.50		66.50	84.50	
	9080	1/2" pipe size		10	.800		39	30		69	88.50	
	9090	3/4" pipe size		9	.889		49.50	33.50		83	105	
	9130	Rough brass, chain lever type										
	9140	250 PSI at 400° F										
	9200	Polished brass, wheel type										
	9210	200 PSI at 400° F										
	9220	3/8" pipe size	1 Stpi	11	.727	Ea.	74.50	27.50		102	124	
	9230	1/2" pipe size		10	.800		75.50	30		105.50	129	
	9240	3/4" pipe size		9	.889		80	33.50		113.50	139	
	9260	Polished brass, chain lever type										
	9270	250 PSI at 400° F										
	9280	1/2" pipe size	1 Stpi	10	.800	Ea.	127	30		157	185	
	9290	3/4" pipe size	"	9	.889	"	138	33.50		171.50	202	
	9400	Bronze, high pressure, ASME										
	9410	1/2" pipe size	1 Stpi	10	.800	Ea.	166	30		196	228	
	9420	3/4" pipe size	"	9	.889	"	166	33.50		199.50	233	
	9460	Chain lever type										

Important: See the Reference Section for critical supporting data - Reference Nos., Crews, & City Cost Indexes

15 MECHANICAL

		15905	HVAC Instrumentation	CREW	DAILY OUTPUT	LABOR-HOURS	UNIT	2003 BARE COSTS				TOTAL INCL O&P	
								MAT.	LABOR	EQUIP.	TOTAL		
960	9470		1/2" pipe size	1 Stpi	10	.800	Ea.	206	30		236	272	960
	9480		3/4" pipe size	"	9	.889	"	217	33.50		250.50	289	
	9500		316 stainless steel, high pressure, ASME										
	9510		500 PSI at 450° F										
	9520		1/2" pipe size	1 Stpi	10	.800	Ea.	515	30		545	615	
	9530		3/4" pipe size	"	9	.889	"	500	33.50		533.50	600	

		15955	HVAC Test/Adjust/Balance	CREW	DAILY OUTPUT	LABOR-HOURS	UNIT	2003 BARE COSTS				TOTAL INCL O&P	
								MAT.	LABOR	EQUIP.	TOTAL		
100	0010	**BALANCING, AIR** (Subcontractor's quote incl. material & labor)											100
	0900	Heating and ventilating equipment	R15050 -710										
	1000	Centrifugal fans, utility sets					Ea.					274	
	1100	Heating and ventilating unit										411	
	1200	In-line fan										410	
	1300	Propeller and wall fan										78	
	1400	Roof exhaust fan										183	
	2000	Air conditioning equipment, central station										595	
	2100	Built-up low pressure unit										550	
	2200	Built-up high pressure unit										640	
	2300	Built-up high pressure dual duct										1,000	
	2400	Built-up variable volume										1,200	
	2500	Multi-zone A.C. and heating unit										410	
	2600	For each zone over one, add										91	
	2700	Package A.C. unit										229	
	2800	Rooftop heating and cooling unit										320	
	3000	Supply, return, exhaust, registers & diffusers, avg. height ceiling										55	
	3100	High ceiling										82	
	3200	Floor height										45.50	
	3300	Off mixing box										36.50	
	3500	Induction unit										59.50	
	3600	Lab fume hood										274	
	3700	Linear supply										137	
	3800	Linear supply high										160	
	4000	Linear return										45.50	
	4100	Light troffers										55	
	4200	Moduline - master										55	
	4300	Moduline - slaves										27.50	
	4400	Regenerators										365	
	4500	Taps into ceiling plenums										69	
	4600	Variable volume boxes										55	
700	0010	**PIPING, TESTING**											700
	0100	Nondestructive testing											
	0110	Nondestructive hydraulic pressure test, isolate & 1 hr. hold											
	0120	1" - 4" pipe											
	0140	0 - 250 L.F.	1 Stpi	1.33	6.015	Ea.		226		226	340	
	0160	250 - 500 L.F.	"	.80	10			375		375	570	
	0180	500 - 1000 L.F.	Q-5	1.14	14.035			475		475	720	
	0200	1000 - 2000 L.F.	"	.80	20			675		675	1,025	

MECHANICAL 15

15955	HVAC Test/Adjust/Balance	CREW	DAILY OUTPUT	LABOR-HOURS	UNIT	2003 BARE COSTS				TOTAL INCL O&P	
						MAT.	LABOR	EQUIP.	TOTAL		
700	0300	6" - 10" pipe									**700**
	0320	0 - 250 L.F.	Q-5	1	16	Ea.		540		540	820
	0340	250 - 500 L.F.		.73	21.918			740		740	1,125
	0360	500 - 1000 L.F.		.53	30.189			1,025		1,025	1,550
	0380	1000 - 2000 L.F.	↓	.38	42.105	↓		1,425		1,425	2,150
	1000	Pneumatic pressure test, includes soaping joints									
	1120	1" - 4" pipe									
	1140	0 - 250 L.F.	Q-5	2.67	5.993	Ea.	7	203		210	315
	1160	250 - 500 L.F.		1.33	12.030		14.05	405		419.05	630
	1180	500 - 1000 L.F.		.80	20		21	675		696	1,050
	1200	1000 - 2000 L.F.	↓	.50	32	↓	28	1,075		1,103	1,650
	1300	6" - 10" pipe									
	1320	0 - 250 L.F.	Q-5	1.33	12.030	Ea.	7	405		412	625
	1340	250 - 500 L.F.		.67	23.881		14.05	810		824.05	1,250
	1360	500 - 1000 L.F.		.40	40		28	1,350		1,378	2,075
	1380	1000 - 2000 L.F.	↓	.25	64	↓	35	2,175		2,210	3,325
	2000	X-Ray of welds									
	2110	2" diam.	1 Stpi	8	1	Ea.	4.20	37.50		41.70	61.50
	2120	3" diam.		8	1		4.20	37.50		41.70	61.50
	2130	4" diam.		8	1		6.30	37.50		43.80	64
	2140	6" diam.		8	1		6.30	37.50		43.80	64
	2150	8" diam.		6.60	1.212		6.30	45.50		51.80	76
	2160	10" diam.	↓	6	1.333	↓	8.40	50		58.40	85
	3000	Liquid penetration of welds									
	3110	2" diam.	1 Stpi	14	.571	Ea.	1.30	21.50		22.80	34
	3120	3" diam.		13.60	.588		1.30	22		23.30	35
	3130	4" diam.		13.40	.597		1.30	22.50		23.80	35.50
	3140	6" diam.		13.20	.606		1.30	23		24.30	36
	3150	8" diam.		13	.615		1.95	23		24.95	37
	3160	10" diam.	↓	12.80	.625	↓	1.95	23.50		25.45	37.50
900	0010	**BALANCING, WATER** (Subcontractor's quote incl. material & labor)									**900**
	0050	Air cooled condenser				Ea.					159
	0100	Cabinet unit heater									54.50
	0200	Chiller	R15050 -710								385
	0300	Convector									45.50
	0400	Converter									227
	0500	Cooling tower									295
	0600	Fan coil unit, unit ventilator									82
	0700	Fin tube and radiant panels									91
	0800	Main and duct re-heat coils									84
	0810	Heat exchanger									84
	0900	Main balancing cocks									68
	1000	Pumps									200
	1100	Unit heater	↓			↓					63.50

For information about Means Estimating Seminars, see yellow pages 12 and 13 in back of book

Important: See the Reference Section for critical supporting data - Reference Nos., Crews, & City Cost Indexes

Division 16
Estimating Tips

Estimating Tips

16060 Grounding & Bonding

- When taking off grounding system, identify separately the type and size of wire and list each unique type of ground connection.

16100 Wiring Methods

- Conduit should be taken off in three main categories: power distribution, branch power, and branch lighting, so the estimator can concentrate on systems and components, therefore making it easier to ensure all items have been accounted for.

- For cost modifications for elevated conduit installation, add the percentages to labor according to the height of installation and only the quantities exceeding the different height levels, not to the total conduit quantities.

- Remember that aluminum wiring of equal ampacity is larger in diameter than copper and may require larger conduit.

- If more than three wires at a time are being pulled, deduct percentages from the labor hours of that grouping of wires.

- The estimator should take the weights of materials into consideration when completing a takeoff. Topics to consider include: How will the materials be supported? What methods of support are available? How high will the support structure have to reach? Will the final support structure be able to withstand the total burden? Is the support material included or separate from the fixture, equipment and material specified?

16200 Electrical Power

- Do not overlook the costs for equipment used in the installation. If scaffolding or highlifts are available in the field, contractors may use them in lieu of the proposed ladders and rolling staging.

16400 Low-Voltage Distribution

- Supports and concrete pads may be shown on drawings for the larger equipment, or the support system may be just a piece of plywood for the back of a panelboard. In either case, it must be included in the costs.

16500 Lighting

- Fixtures should be taken off room by room, using the fixture schedule, specifications, and the ceiling plan. For large concentrations of lighting fixtures in the same area deduct the percentages from labor hours.

16700 Communications
16800 Sound & Video

- When estimating material costs for special systems, it is always prudent to obtain manufacturers' quotations for equipment prices and special installation requirements which will affect the total costs.

Reference Numbers

Reference numbers are shown in bold squares at the beginning of some major classifications. These numbers refer to related items in the Reference Section. The reference information may be an estimating procedure, an alternate pricing method or technical information.

Note: Not all subdivisions listed here necessarily appear in this publication.

16132	Conduit & Tubing	CREW	DAILY OUTPUT	LABOR-HOURS	UNIT	2003 BARE COSTS				TOTAL INCL O&P	
						MAT.	LABOR	EQUIP.	TOTAL		
260	0010	**CUTTING AND DRILLING**									260
	0100	Hole drilling to 10' high, concrete wall									
	0110	8" thick, 1/2" pipe size	1 Elec	12	.667	Ea.		25		25	37.50
	0120	3/4" pipe size		12	.667			25		25	37.50
	0130	1" pipe size		9.50	.842			31.50		31.50	47.50
	0140	1-1/4" pipe size		9.50	.842			31.50		31.50	47.50
	0150	1-1/2" pipe size		9.50	.842			31.50		31.50	47.50
	0160	2" pipe size		4.40	1.818			68.50		68.50	102
	0170	2-1/2" pipe size		4.40	1.818			68.50		68.50	102
	0180	3" pipe size		4.40	1.818			68.50		68.50	102
	0190	3-1/2" pipe size		3.30	2.424			91		91	136
	0200	4" pipe size		3.30	2.424			91		91	136
	0500	12" thick, 1/2" pipe size		9.40	.851			32		32	48
	0520	3/4" pipe size		9.40	.851			32		32	48
	0540	1" pipe size		7.30	1.096			41		41	61.50
	0560	1-1/4" pipe size		7.30	1.096			41		41	61.50
	0570	1-1/2" pipe size		7.30	1.096			41		41	61.50
	0580	2" pipe size		3.60	2.222			83.50		83.50	125
	0590	2-1/2" pipe size		3.60	2.222			83.50		83.50	125
	0600	3" pipe size		3.60	2.222			83.50		83.50	125
	0610	3-1/2" pipe size		2.80	2.857			107		107	160
	0630	4" pipe size		2.50	3.200			120		120	180
	0650	16" thick, 1/2" pipe size		7.60	1.053			39.50		39.50	59
	0670	3/4" pipe size		7	1.143			43		43	64
	0690	1" pipe size		6	1.333			50		50	75
	0710	1-1/4" pipe size		5.50	1.455			54.50		54.50	81.50
	0730	1-1/2" pipe size		5.50	1.455			54.50		54.50	81.50
	0750	2" pipe size		3	2.667			100		100	150
	0770	2-1/2" pipe size		2.70	2.963			111		111	166
	0790	3" pipe size		2.50	3.200			120		120	180
	0810	3-1/2" pipe size		2.30	3.478			131		131	195
	0830	4" pipe size		2	4			150		150	225
	0850	20" thick, 1/2" pipe size		6.40	1.250			47		47	70
	0870	3/4" pipe size		6	1.333			50		50	75
	0890	1" pipe size		5	1.600			60		60	90
	0910	1-1/4" pipe size		4.80	1.667			62.50		62.50	93.50
	0930	1-1/2" pipe size		4.60	1.739			65.50		65.50	97.50
	0950	2" pipe size		2.70	2.963			111		111	166
	0970	2-1/2" pipe size		2.40	3.333			125		125	187
	0990	3" pipe size		2.20	3.636			137		137	204
	1010	3-1/2" pipe size		2	4			150		150	225
	1030	4" pipe size		1.70	4.706			177		177	264
	1050	24" thick, 1/2" pipe size		5.50	1.455			54.50		54.50	81.50
	1070	3/4" pipe size		5.10	1.569			59		59	88
	1090	1" pipe size		4.30	1.860			70		70	104
	1110	1-1/4" pipe size		4	2			75		75	112
	1130	1-1/2" pipe size		4	2			75		75	112
	1150	2" pipe size		2.40	3.333			125		125	187
	1170	2-1/2" pipe size		2.20	3.636			137		137	204
	1190	3" pipe size		2	4			150		150	225
	1210	3-1/2" pipe size		1.80	4.444			167		167	250
	1230	4" pipe size		1.50	5.333			201		201	299
	1500	Brick wall, 8" thick, 1/2" pipe size		18	.444			16.70		16.70	25
	1520	3/4" pipe size		18	.444			16.70		16.70	25
	1540	1" pipe size		13.30	.601			22.50		22.50	34
	1560	1-1/4" pipe size		13.30	.601			22.50		22.50	34

Important: See the Reference Section for critical supporting data - Reference Nos., Crews, & City Cost Indexes

16132	Conduit & Tubing	CREW	DAILY OUTPUT	LABOR-HOURS	UNIT	2003 BARE COSTS				TOTAL INCL O&P	
						MAT.	LABOR	EQUIP.	TOTAL		
260	1580	1-1/2" pipe size	1 Elec	13.30	.601	Ea.		22.50		22.50	34	260
1600	2" pipe size		5.70	1.404			53		53	79
1620	2-1/2" pipe size		5.70	1.404			53		53	79
1640	3" pipe size		5.70	1.404			53		53	79
1660	3-1/2" pipe size		4.40	1.818			68.50		68.50	102
1680	4" pipe size		4	2			75		75	112
1700	12" thick, 1/2" pipe size		14.50	.552			20.50		20.50	31
1720	3/4" pipe size		14.50	.552			20.50		20.50	31
1740	1" pipe size		11	.727			27.50		27.50	41
1760	1-1/4" pipe size		11	.727			27.50		27.50	41
1780	1-1/2" pipe size		11	.727			27.50		27.50	41
1800	2" pipe size		5	1.600			60		60	90
1820	2-1/2" pipe size		5	1.600			60		60	90
1840	3" pipe size		5	1.600			60		60	90
1860	3-1/2" pipe size		3.80	2.105			79		79	118
1880	4" pipe size		3.30	2.424			91		91	136
1900	16" thick, 1/2" pipe size		12.30	.650			24.50		24.50	36.50
1920	3/4" pipe size		12.30	.650			24.50		24.50	36.50
1940	1" pipe size		9.30	.860			32.50		32.50	48.50
1960	1-1/4" pipe size		9.30	.860			32.50		32.50	48.50
1980	1-1/2" pipe size		9.30	.860			32.50		32.50	48.50
2000	2" pipe size		4.40	1.818			68.50		68.50	102
2010	2-1/2" pipe size		4.40	1.818			68.50		68.50	102
2030	3" pipe size		4.40	1.818			68.50		68.50	102
2050	3-1/2" pipe size		3.30	2.424			91		91	136
2070	4" pipe size		3	2.667			100		100	150
2090	20" thick, 1/2" pipe size		10.70	.748			28		28	42
2110	3/4" pipe size		10.70	.748			28		28	42
2130	1" pipe size		8	1			37.50		37.50	56
2150	1-1/4" pipe size		8	1			37.50		37.50	56
2170	1-1/2" pipe size		8	1			37.50		37.50	56
2190	2" pipe size		4	2			75		75	112
2210	2-1/2" pipe size		4	2			75		75	112
2230	3" pipe size		4	2			75		75	112
2250	3-1/2" pipe size		3	2.667			100		100	150
2270	4" pipe size		2.70	2.963			111		111	166
2290	24" thick, 1/2" pipe size		9.40	.851			32		32	48
2310	3/4" pipe size		9.40	.851			32		32	48
2330	1" pipe size		7.10	1.127			42.50		42.50	63.50
2350	1-1/4" pipe size		7.10	1.127			42.50		42.50	63.50
2370	1-1/2" pipe size		7.10	1.127			42.50		42.50	63.50
2390	2" pipe size		3.60	2.222			83.50		83.50	125
2410	2-1/2" pipe size		3.60	2.222			83.50		83.50	125
2430	3" pipe size		3.60	2.222			83.50		83.50	125
2450	3-1/2" pipe size		2.80	2.857			107		107	160
2470	4" pipe size		2.50	3.200			120		120	180
3000	Knockouts to 8' high, metal boxes & enclosures									
3020	With hole saw, 1/2" pipe size	1 Elec	53	.151	Ea.		5.70		5.70	8.50
3040	3/4" pipe size		47	.170			6.40		6.40	9.55
3050	1" pipe size		40	.200			7.50		7.50	11.25
3060	1-1/4" pipe size		36	.222			8.35		8.35	12.50
3070	1-1/2" pipe size		32	.250			9.40		9.40	14.05
3080	2" pipe size		27	.296			11.15		11.15	16.65
3090	2-1/2" pipe size		20	.400			15.05		15.05	22.50
4010	3" pipe size		16	.500			18.80		18.80	28
4030	3-1/2" pipe size		13	.615			23		23	34.50

ELECTRICAL 16

16132 | Conduit & Tubing

			Description	CREW	DAILY OUTPUT	LABOR-HOURS	UNIT	2003 BARE COSTS				TOTAL INCL O&P	
								MAT.	LABOR	EQUIP.	TOTAL		
260	4050		4" pipe size	1 Elec	11	.727	Ea.		27.50		27.50	41	260
	4070		With hand punch set, 1/2" pipe size		40	.200			7.50		7.50	11.25	
	4090		3/4" pipe size		32	.250			9.40		9.40	14.05	
	4110		1" pipe size		30	.267			10.05		10.05	14.95	
	4130		1-1/4" pipe size		28	.286			10.75		10.75	16.05	
	4150		1-1/2" pipe size		26	.308			11.55		11.55	17.30	
	4170		2" pipe size		20	.400			15.05		15.05	22.50	
	4190		2-1/2" pipe size		17	.471			17.70		17.70	26.50	
	4200		3" pipe size		15	.533			20		20	30	
	4220		3-1/2" pipe size		12	.667			25		25	37.50	
	4240		4" pipe size		10	.800			30		30	45	
	4260		With hydraulic punch, 1/2" pipe size		44	.182			6.85		6.85	10.20	
	4280		3/4" pipe size		38	.211			7.90		7.90	11.80	
	4300		1" pipe size		38	.211			7.90		7.90	11.80	
	4320		1-1/4" pipe size		38	.211			7.90		7.90	11.80	
	4340		1-1/2" pipe size		38	.211			7.90		7.90	11.80	
	4360		2" pipe size		32	.250			9.40		9.40	14.05	
	4380		2-1/2" pipe size		27	.296			11.15		11.15	16.65	
	4400		3" pipe size		23	.348			13.10		13.10	19.55	
	4420		3-1/2" pipe size		20	.400			15.05		15.05	22.50	
	4440		4" pipe size	▼	18	.444	▼		16.70		16.70	25	

16139 | Residential Wiring

			Description	CREW	DAILY OUTPUT	LABOR-HOURS	UNIT	2003 BARE COSTS				TOTAL INCL O&P	
								MAT.	LABOR	EQUIP.	TOTAL		
700	0010		**RESIDENTIAL WIRING**										700
	0020		20' avg. runs and #14/2 wiring incl. unless otherwise noted										
	1000		Service & panel, includes 24' SE-AL cable, service eye, meter,										
	1010		Socket, panel board, main bkr., ground rod, 15 or 20 amp										
	1020		1-pole circuit breakers, and misc. hardware										
	1100		100 amp, with 10 branch breakers	1 Elec	1.19	6.723	Ea.	410	253		663	825	
	1110		With PVC conduit and wire		.92	8.696		445	325		770	980	
	1120		With RGS conduit and wire		.73	10.959		575	410		985	1,250	
	1150		150 amp, with 14 branch breakers		1.03	7.767		635	292		927	1,125	
	1170		With PVC conduit and wire		.82	9.756		705	365		1,070	1,325	
	1180		With RGS conduit and wire	▼	.67	11.940		930	450		1,380	1,700	
	1200		200 amp, with 18 branch breakers	2 Elec	1.80	8.889		835	335		1,170	1,425	
	1220		With PVC conduit and wire		1.46	10.959		905	410		1,315	1,600	
	1230		With RGS conduit and wire	▼	1.24	12.903		1,200	485		1,685	2,050	
	1800		Lightning surge suppressor for above services, add	1 Elec	32	.250	▼	39.50	9.40		48.90	57.50	
	2000		Switch devices										
	2100		Single pole, 15 amp, Ivory, with a 1-gang box, cover plate,										
	2110		Type NM (Romex) cable	1 Elec	17.10	.468	Ea.	6.90	17.60		24.50	34	
	2120		Type MC (BX) cable		14.30	.559		17.10	21		38.10	50.50	
	2130		EMT & wire		5.71	1.401		17.90	52.50		70.40	98	
	2150		3-way, #14/3, type NM cable		14.55	.550		10.35	20.50		30.85	42.50	
	2170		Type MC cable		12.31	.650		21.50	24.50		46	60.50	
	2180		EMT & wire		5	1.600		20	60		80	112	
	2200		4-way, #14/3, type NM cable		14.55	.550		23.50	20.50		44	56.50	
	2220		Type MC cable		12.31	.650		34.50	24.50		59	74.50	
	2230		EMT & wire		5	1.600		33	60		93	126	
	2250		S.P., 20 amp, #12/2, type NM cable		13.33	.600		12.15	22.50		34.65	47	
	2270		Type MC cable		11.43	.700		21.50	26.50		48	63	
	2280		EMT & wire		4.85	1.649		23.50	62		85.50	119	
	2290		S.P. rotary dimmer, 600W, no wiring		17	.471		16	17.70		33.70	44	
	2300		S.P. rotary dimmer, 600W, type NM cable		14.55	.550		18.55	20.50		39.05	51.50	
	2320		Type MC cable	▼	12.31	.650	▼	29	24.50		53.50	68	

Important: See the Reference Section for critical supporting data - Reference Nos., Crews, & City Cost Indexes

16100 | Wiring Methods

16139 | Residential Wiring

		CREW	DAILY OUTPUT	LABOR-HOURS	UNIT	2003 BARE COSTS				TOTAL INCL O&P
						MAT.	LABOR	EQUIP.	TOTAL	
2330	EMT & wire	1 Elec	5	1.600	Ea.	30	60		90	123
2350	3-way rotary dimmer, type NM cable		13.33	.600		16.25	22.50		38.75	51.50
2370	Type MC cable		11.43	.700		26.50	26.50		53	68.50
2380	EMT & wire		4.85	1.649		28	62		90	123
2400	Interval timer wall switch, 20 amp, 1-30 min., #12/2									
2410	Type NM cable	1 Elec	14.55	.550	Ea.	29.50	20.50		50	63.50
2420	Type MC cable		12.31	.650		36.50	24.50		61	76.50
2430	EMT & wire		5	1.600		40.50	60		100.50	135
2500	Decorator style									
2510	S.P., 15 amp, type NM cable	1 Elec	17.10	.468	Ea.	10.45	17.60		28.05	38
2520	Type MC cable		14.30	.559		20.50	21		41.50	54
2530	EMT & wire		5.71	1.401		21.50	52.50		74	102
2550	3-way, #14/3, type NM cable		14.55	.550		13.90	20.50		34.40	46.50
2570	Type MC cable		12.31	.650		25	24.50		49.50	64
2580	EMT & wire		5	1.600		23.50	60		83.50	116
2600	4-way, #14/3, type NM cable		14.55	.550		27	20.50		47.50	60.50
2620	Type MC cable		12.31	.650		38	24.50		62.50	78.50
2630	EMT & wire		5	1.600		36.50	60		96.50	130
2650	S.P., 20 amp, #12/2, type NM cable		13.33	.600		15.70	22.50		38.20	51
2670	Type MC cable		11.43	.700		25	26.50		51.50	67
2680	EMT & wire		4.85	1.649		27	62		89	122
2700	S.P., slide dimmer, type NM cable		17.10	.468		25.50	17.60		43.10	54.50
2720	Type MC cable		14.30	.559		36	21		57	71
2730	EMT & wire		5.71	1.401		37.50	52.50		90	120
2750	S.P., touch dimmer, type NM cable		17.10	.468		22	17.60		39.60	50.50
2770	Type MC cable		14.30	.559		32.50	21		53.50	67
2780	EMT & wire		5.71	1.401		33.50	52.50		86	116
2800	3-way touch dimmer, type NM cable		13.33	.600		40	22.50		62.50	77.50
2820	Type MC cable		11.43	.700		50.50	26.50		77	95
2830	EMT & wire		4.85	1.649		51.50	62		113.50	150
3000	Combination devices									
3100	S.P. switch/15 amp recpt., Ivory, 1-gang box, plate									
3110	Type NM cable	1 Elec	11.43	.700	Ea.	14.95	26.50		41.45	56
3120	Type MC cable		10	.800		25	30		55	72.50
3130	EMT & wire		4.40	1.818		26.50	68.50		95	131
3150	S.P. switch/pilot light, type NM cable		11.43	.700		15.55	26.50		42.05	56.50
3170	Type MC cable		10	.800		26	30		56	73.50
3180	EMT & wire		4.43	1.806		27	68		95	131
3190	2-S.P. switches, 2-#14/2, no wiring		14	.571		5.70	21.50		27.20	38.50
3200	2-S.P. switches, 2-#14/2, type NM cables		10	.800		16.75	30		46.75	63.50
3220	Type MC cable		8.89	.900		34	34		68	88
3230	EMT & wire		4.10	1.951		27.50	73.50		101	141
3250	3-way switch/15 amp recpt., #14/3, type NM cable		10	.800		21.50	30		51.50	68.50
3270	Type MC cable		8.89	.900		32.50	34		66.50	86.50
3280	EMT & wire		4.10	1.951		31	73.50		104.50	144
3300	2-3 way switches, 2-#14/3, type NM cables		8.89	.900		28	34		62	81
3320	Type MC cable		8	1		47	37.50		84.50	108
3330	EMT & wire		4	2		35	75		110	151
3350	S.P. switch/20 amp recpt., #12/2, type NM cable		10	.800		25.50	30		55.50	73
3370	Type MC cable		8.89	.900		32.50	34		66.50	86.50
3380	EMT & wire		4.10	1.951		37	73.50		110.50	151
3400	Decorator style									
3410	S.P. switch/15 amp recpt., type NM cable	1 Elec	11.43	.700	Ea.	18.50	26.50		45	60
3420	Type MC cable		10	.800		28.50	30		58.50	76.50
3430	EMT & wire		4.40	1.818		30	68.50		98.50	135
3450	S.P. switch/pilot light, type NM cable		11.43	.700		19.10	26.50		45.60	60.50

ELECTRICAL 16

For expanded coverage of these items see *Means Electrical Cost Data 2003*

16139	Residential Wiring	CREW	DAILY OUTPUT	LABOR-HOURS	UNIT	2003 BARE COSTS				TOTAL INCL O&P
						MAT.	LABOR	EQUIP.	TOTAL	
3470	Type MC cable	1 Elec	10	.800	Ea.	29.50	30		59.50	77.50	700
3480	EMT & wire		4.40	1.818		30.50	68.50		99	136
3500	2-S.P. switches, 2-#14/2, type NM cables		10	.800		20.50	30		50.50	67.50
3520	Type MC cable		8.89	.900		37.50	34		71.50	91.50
3530	EMT & wire		4.10	1.951		31	73.50		104.50	145
3550	3-way/15 amp recpt., #14/3, type NM cable		10	.800		25	30		55	72.50
3570	Type MC cable		8.89	.900		36	34		70	90
3580	EMT & wire		4.10	1.951		34.50	73.50		108	148
3650	2-3 way switches, 2-#14/3, type NM cables		8.89	.900		31.50	34		65.50	85
3670	Type MC cable		8	1		50.50	37.50		88	112
3680	EMT & wire		4	2		38.50	75		113.50	155
3700	S.P. switch/20 amp recpt., #12/2, type NM cable		10	.800		29	30		59	77
3720	Type MC cable		8.89	.900		36	34		70	90
3730	EMT & wire		4.10	1.951		40.50	73.50		114	155
4000	Receptacle devices									
4010	Duplex outlet, 15 amp recpt., Ivory, 1-gang box, plate									
4015	Type NM cable	1 Elec	14.55	.550	Ea.	5.35	20.50		25.85	37
4020	Type MC cable		12.31	.650		15.60	24.50		40.10	53.50
4030	EMT & wire		5.33	1.501		16.35	56.50		72.85	103
4050	With #12/2, type NM cable		12.31	.650		6.50	24.50		31	43.50
4070	Type MC cable		10.67	.750		15.90	28		43.90	59.50
4080	EMT & wire		4.71	1.699		17.80	64		81.80	115
4100	20 amp recpt., #12/2, type NM cable		12.31	.650		12.25	24.50		36.75	50
4120	Type MC cable		10.67	.750		21.50	28		49.50	66
4130	EMT & wire		4.71	1.699		23.50	64		87.50	122
4140	For GFI see line 4300 below									
4150	Decorator style, 15 amp recpt., type NM cable	1 Elec	14.55	.550	Ea.	8.90	20.50		29.40	41
4170	Type MC cable		12.31	.650		19.15	24.50		43.65	57.50
4180	EMT & wire		5.33	1.501		19.90	56.50		76.40	107
4200	With #12/2, type NM cable		12.31	.650		10.05	24.50		34.55	47.50
4220	Type MC cable		10.67	.750		19.45	28		47.45	63.50
4230	EMT & wire		4.71	1.699		21.50	64		85.50	119
4250	20 amp recpt. #12/2, type NM cable		12.31	.650		15.80	24.50		40.30	54
4270	Type MC cable		10.67	.750		25	28		53	69.50
4280	EMT & wire		4.71	1.699		27	64		91	126
4300	GFI, 15 amp recpt., type NM cable		12.31	.650		33	24.50		57.50	73
4320	Type MC cable		10.67	.750		43.50	28		71.50	89.50
4330	EMT & wire		4.71	1.699		44	64		108	144
4350	GFI with #12/2, type NM cable		10.67	.750		34	28		62	79.50
4370	Type MC cable		9.20	.870		43.50	32.50		76	97
4380	EMT & wire		4.21	1.900		45.50	71.50		117	157
4400	20 amp recpt., #12/2 type NM cable		10.67	.750		36	28		64	81.50
4420	Type MC cable		9.20	.870		45	32.50		77.50	98.50
4430	EMT & wire		4.21	1.900		47	71.50		118.50	159
4500	Weather-proof cover for above receptacles, add		32	.250		4.40	9.40		13.80	18.90
4550	Air conditioner outlet, 20 amp-240 volt recpt.									
4560	30' of #12/2, 2 pole circuit breaker									
4570	Type NM cable	1 Elec	10	.800	Ea.	41	30		71	90
4580	Type MC cable		9	.889		53.50	33.50		87	109
4590	EMT & wire		4	2		51.50	75		126.50	169
4600	Decorator style, type NM cable		10	.800		45	30		75	94.50
4620	Type MC cable		9	.889		57	33.50		90.50	113
4630	EMT & wire		4	2		55.50	75		130.50	173
4650	Dryer outlet, 30 amp-240 volt recpt., 20' of #10/3									
4660	2 pole circuit breaker									
4670	Type NM cable	1 Elec	6.41	1.248	Ea.	48.50	47		95.50	124

Important: See the Reference Section for critical supporting data - Reference Nos., Crews, & City Cost Indexes

	16139	Residential Wiring	CREW	DAILY OUTPUT	LABOR-HOURS	UNIT	2003 BARE COSTS				TOTAL INCL O&P	
							MAT.	LABOR	EQUIP.	TOTAL		
700	4680	Type MC cable	1 Elec	5.71	1.401	Ea.	57.50	52.50		110	142	**700**
	4690	EMT & wire	↓	3.48	2.299	↓	52.50	86.50		139	187	
	4700	Range outlet, 50 amp-240 volt recpt., 30' of #8/3										
	4710	Type NM cable	1 Elec	4.21	1.900	Ea.	70	71.50		141.50	184	
	4720	Type MC cable		4	2		102	75		177	224	
	4730	EMT & wire		2.96	2.703		67.50	102		169.50	226	
	4750	Central vacuum outlet, Type NM cable		6.40	1.250		42.50	47		89.50	117	
	4770	Type MC cable		5.71	1.401		60.50	52.50		113	145	
	4780	EMT & wire	↓	3.48	2.299	↓	54	86.50		140.50	189	
	4800	30 amp-110 volt locking recpt., #10/2 circ. bkr.										
	4810	Type NM cable	1 Elec	6.20	1.290	Ea.	50.50	48.50		99	128	
	4830	EMT & wire	"	3.20	2.500	"	61.50	94		155.50	208	
	4900	Low voltage outlets										
	4910	Telephone recpt., 20' of 4/C phone wire	1 Elec	26	.308	Ea.	6.70	11.55		18.25	24.50	
	4920	TV recpt., 20' of RG59U coax wire, F type connector	"	16	.500	"	11.15	18.80		29.95	40.50	
	4950	Door bell chime, transformer, 2 buttons, 60' of bellwire										
	4970	Economy model	1 Elec	11.50	.696	Ea.	51.50	26		77.50	95.50	
	4980	Custom model	↓	11.50	.696		82.50	26		108.50	130	
	4990	Luxury model, 3 buttons	↓	9.50	.842	↓	224	31.50		255.50	294	
	6000	Lighting outlets										
	6050	Wire only (for fixture), type NM cable	1 Elec	32	.250	Ea.	3.40	9.40		12.80	17.80	
	6070	Type MC cable		24	.333		11.60	12.55		24.15	31.50	
	6080	EMT & wire		10	.800		11.45	30		41.45	57.50	
	6100	Box (4"), and wire (for fixture), type NM cable		25	.320		7.45	12.05		19.50	26	
	6120	Type MC cable		20	.400		15.65	15.05		30.70	39.50	
	6130	EMT & wire	↓	11	.727	↓	15.50	27.50		43	58	
	6200	Fixtures (use with lines 6050 or 6100 above)										
	6210	Canopy style, economy grade	1 Elec	40	.200	Ea.	25.50	7.50		33	39.50	
	6220	Custom grade		40	.200		46	7.50		53.50	62	
	6250	Dining room chandelier, economy grade		19	.421		76	15.85		91.85	107	
	6260	Custom grade		19	.421		225	15.85		240.85	272	
	6270	Luxury grade		15	.533		495	20		515	575	
	6310	Kitchen fixture (fluorescent), economy grade		30	.267		51.50	10.05		61.55	71.50	
	6320	Custom grade		25	.320		161	12.05		173.05	195	
	6350	Outdoor, wall mounted, economy grade		30	.267		27	10.05		37.05	44.50	
	6360	Custom grade		30	.267		101	10.05		111.05	126	
	6370	Luxury grade		25	.320		227	12.05		239.05	268	
	6410	Outdoor PAR floodlights, 1 lamp, 150 watt		20	.400		20	15.05		35.05	44.50	
	6420	2 lamp, 150 watt each		20	.400		33.50	15.05		48.55	59.50	
	6430	For infrared security sensor, add		32	.250		87	9.40		96.40	110	
	6450	Outdoor, quartz-halogen, 300 watt flood		20	.400		38	15.05		53.05	64.50	
	6600	Recessed downlight, round, pre-wired, 50 or 75 watt trim		30	.267		35	10.05		45.05	53.50	
	6610	With shower light trim		30	.267		43	10.05		53.05	62.50	
	6620	With wall washer trim		28	.286		52.50	10.75		63.25	74	
	6630	With eye-ball trim	↓	28	.286		50	10.75		60.75	71	
	6640	For direct contact with insulation, add					1.60			1.60	1.76	
	6700	Porcelain lamp holder	1 Elec	40	.200		3.50	7.50		11	15.10	
	6710	With pull switch		40	.200		3.75	7.50		11.25	15.40	
	6750	Fluorescent strip, 1-20 watt tube, wrap around diffuser, 24"		24	.333		51.50	12.55		64.05	75	
	6770	2-40 watt tubes, 48"		20	.400		79	15.05		94.05	110	
	6780	With residential ballast		20	.400		89.50	15.05		104.55	121	
	6800	Bathroom heat lamp, 1-250 watt		28	.286		34.50	10.75		45.25	54	
	6810	2-250 watt lamps	↓	28	.286	↓	55	10.75		65.75	76.50	
	6820	For timer switch, see line 2400										
	6900	Outdoor post lamp, incl. post, fixture, 35' of #14/2										
	6910	Type NMC cable	1 Elec	3.50	2.286	Ea.	181	86		267	325	

		CREW	DAILY OUTPUT	LABOR-HOURS	UNIT	2003 BARE COSTS				TOTAL INCL O&P	
16139	**Residential Wiring**					MAT.	LABOR	EQUIP.	TOTAL		
700	6920	Photo-eye, add	1 Elec	27	.296	Ea.	29	11.15		40.15	48.50	**700**
	6950	Clock dial time switch, 24 hr., w/enclosure, type NM cable		11.43	.700		50.50	26.50		77	95.50	
	6970	Type MC cable		11	.727		61	27.50		88.50	108	
	6980	EMT & wire		4.85	1.649		61.50	62		123.50	161	
	7000	Alarm systems										
	7050	Smoke detectors, box, #14/3, type NM cable	1 Elec	14.55	.550	Ea.	28	20.50		48.50	62	
	7070	Type MC cable		12.31	.650		37	24.50		61.50	77	
	7080	EMT & wire		5	1.600		35.50	60		95.50	129	
	7090	For relay output to security system, add					11.75			11.75	12.95	
	8000	Residential equipment										
	8050	Disposal hook-up, incl. switch, outlet box, 3' of flex										
	8060	20 amp-1 pole circ. bkr., and 25' of #12/2										
	8070	Type NM cable	1 Elec	10	.800	Ea.	19.90	30		49.90	67	
	8080	Type MC cable		8	1		31	37.50		68.50	90	
	8090	EMT & wire		5	1.600		32.50	60		92.50	126	
	8100	Trash compactor or dishwasher hook-up, incl. outlet box,										
	8110	3' of flex, 15 amp-1 pole circ. bkr., and 25' of #14/2										
	8120	Type NM cable	1 Elec	10	.800	Ea.	14.05	30		44.05	60.50	
	8130	Type MC cable		8	1		26.50	37.50		64	85	
	8140	EMT & wire		5	1.600		27	60		87	120	
	8150	Hot water sink dispensor hook-up, use line 8100										
	8200	Vent/exhaust fan hook-up, type NM cable	1 Elec	32	.250	Ea.	3.40	9.40		12.80	17.80	
	8220	Type MC cable		24	.333		11.60	12.55		24.15	31.50	
	8230	EMT & wire		10	.800		11.45	30		41.45	57.50	
	8250	Bathroom vent fan, 50 CFM (use with above hook-up)										
	8260	Economy model	1 Elec	15	.533	Ea.	21.50	20		41.50	53.50	
	8270	Low noise model		15	.533		30	20		50	63	
	8280	Custom model		12	.667		111	25		136	160	
	8300	Bathroom or kitchen vent fan, 110 CFM										
	8310	Economy model	1 Elec	15	.533	Ea.	57	20		77	92.50	
	8320	Low noise model	"	15	.533	"	75	20		95	113	
	8350	Paddle fan, variable speed (w/o lights)										
	8360	Economy model (AC motor)	1 Elec	10	.800	Ea.	100	30		130	155	
	8370	Custom model (AC motor)		10	.800		173	30		203	235	
	8380	Luxury model (DC motor)		8	1		340	37.50		377.50	430	
	8390	Remote speed switch for above, add		12	.667		24	25		49	64	
	8500	Whole house exhaust fan, ceiling mount, 36", variable speed										
	8510	Remote switch, incl. shutters, 20 amp-1 pole circ. bkr.										
	8520	30' of #12/2, type NM cable	1 Elec	4	2	Ea.	610	75		685	780	
	8530	Type MC cable		3.50	2.286		625	86		711	815	
	8540	EMT & wire		3	2.667		625	100		725	840	
	8600	Whirlpool tub hook-up, incl. timer switch, outlet box										
	8610	3' of flex, 20 amp-1 pole GFI circ. bkr.										
	8620	30' of #12/2, type NM cable	1 Elec	5	1.600	Ea.	78.50	60		138.50	177	
	8630	Type MC cable		4.20	1.905		86	71.50		157.50	202	
	8640	EMT & wire		3.40	2.353		88	88.50		176.50	229	
	8650	Hot water heater hook-up, incl. 1-2 pole circ. bkr., box;										
	8660	3' of flex, 20' of #10/2, type NM cable	1 Elec	5	1.600	Ea.	20.50	60		80.50	113	
	8670	Type MC cable		4.20	1.905		34.50	71.50		106	145	
	8680	EMT & wire		3.40	2.353		29	88.50		117.50	164	
	9000	Heating/air conditioning										
	9050	Furnace/boiler hook-up, incl. firestat, local on-off switch										
	9060	Emergency switch, and 40' of type NM cable	1 Elec	4	2	Ea.	42.50	75		117.50	159	
	9070	Type MC cable		3.50	2.286		60	86		146	194	
	9080	EMT & wire		1.50	5.333		61	201		262	365	
	9100	Air conditioner hook-up, incl. local 60 amp disc. switch										

Important: See the Reference Section for critical supporting data - Reference Nos., Crews, & City Cost Indexes

16 ELECTRICAL

16100 | Wiring Methods

16139	Residential Wiring	CREW	DAILY OUTPUT	LABOR-HOURS	UNIT	2003 BARE COSTS				TOTAL INCL O&P	
						MAT.	LABOR	EQUIP.	TOTAL		
700 9110	3' sealtite, 40 amp, 2 pole circuit breaker										700
9130	40' of #8/2, type NM cable	1 Elec	3.50	2.286	Ea.	139	86		225	281	
9140	Type MC cable		3	2.667		189	100		289	360	
9150	EMT & wire	▼	1.30	6.154	▼	148	231		379	510	
9200	Heat pump hook-up, 1-40 & 1-100 amp 2 pole circ. bkr.										
9210	Local disconnect switch, 3' sealtite										
9220	40' of #8/2 & 30' of #3/2										
9230	Type NM cable	1 Elec	1.30	6.154	Ea.	325	231		556	705	
9240	Type MC cable		1.08	7.407		465	279		744	925	
9250	EMT & wire	▼	.94	8.511	▼	335	320		655	850	
9500	Thermostat hook-up, using low voltage wire										
9520	Heating only	1 Elec	24	.333	Ea.	5.15	12.55		17.70	24.50	
9530	Heating/cooling	"	20	.400	"	6.25	15.05		21.30	29.50	

16150	Wiring Connections	CREW	DAILY OUTPUT	LABOR-HOURS	UNIT	MAT.	LABOR	EQUIP.	TOTAL	TOTAL INCL O&P	
275 0010	**MOTOR CONNECTIONS**										275
0020	Flexible conduit and fittings, 115 volt, 1 phase, up to 1 HP motor	1 Elec	8	1	Ea.	4.37	37.50		41.87	61	
0050	2 HP motor		6.50	1.231		4.45	46.50		50.95	74	
0100	3 HP motor		5.50	1.455		7.45	54.50		61.95	89.50	
0120	230 volt, 10 HP motor, 3 phase		4.20	1.905		8.05	71.50		79.55	116	
0150	15 HP motor		3.30	2.424		15.55	91		106.55	153	
0200	25 HP motor		2.70	2.963		17.45	111		128.45	185	
0400	50 HP motor		2.20	3.636		48	137		185	257	
0600	100 HP motor		1.50	5.333		122	201		323	435	
1500	460 volt, 5 HP motor, 3 phase		8	1		4.87	37.50		42.37	61.50	
1520	10 HP motor		8	1		4.87	37.50		42.37	61.50	
1530	25 HP motor		6	1.333		9.25	50		59.25	85	
1540	30 HP motor		6	1.333		9.25	50		59.25	85	
1550	40 HP motor		5	1.600		14.60	60		74.60	106	
1560	50 HP motor		5	1.600		16.30	60		76.30	108	
1570	60 HP motor		3.80	2.105		22	79		101	142	
1580	75 HP motor		3.50	2.286		29	86		115	160	
1590	100 HP motor		2.50	3.200		40.50	120		160.50	225	
1600	125 HP motor		2	4		53.50	150		203.50	284	
1610	150 HP motor		1.80	4.444		55.50	167		222.50	310	
1620	200 HP motor		1.50	5.333		96	201		297	405	
2005	460 Volt, 5 HP motor, 3 Phase, w/sealtite		8	1		9	37.50		46.50	66	
2010	10 HP motor		8	1		9	37.50		46.50	66	
2015	25 HP motor		6	1.333		16.55	50		66.55	93	
2020	30 HP motor		6	1.333		16.55	50		66.55	93	
2025	40 HP motor		5	1.600		30	60		90	123	
2030	50 HP motor		5	1.600		28.50	60		88.50	121	
2035	60 HP motor		3.80	2.105		42.50	79		121.50	165	
2040	75 HP motor		3.50	2.286		46	86		132	179	
2045	100 HP motor		2.50	3.200		62	120		182	249	
2055	150 HP motor		1.80	4.444		67	167		234	325	
2060	200 HP motor	▼	1.50	5.333	▼	405	201		606	745	

ELECTRICAL 16

16220	Motors & Generators	CREW	DAILY OUTPUT	LABOR-HOURS	UNIT	2003 BARE COSTS				TOTAL INCL O&P	
						MAT.	LABOR	EQUIP.	TOTAL		
900	**0010**	**VARIABLE FREQUENCY DRIVES/ADJUSTABLE FREQUENCY DRIVES**										**900**
0100	Enclosed (NEMA 1), 460 volt, for 3 HP motor size	1 Elec	.80	10	Ea.	1,475	375		1,850	2,150	
0110	5 HP motor size		.80	10		1,600	375		1,975	2,300	
0120	7.5 HP motor size		.67	11.940		1,875	450		2,325	2,750	
0130	10 HP motor size		.67	11.940		1,875	450		2,325	2,750	
0140	15 HP motor size	2 Elec	.89	17.978		2,175	675		2,850	3,400	
0150	20 HP motor size		.89	17.978		3,225	675		3,900	4,550	
0160	25 HP motor size		.67	23.881		3,750	900		4,650	5,475	
0170	30 HP motor size		.67	23.881		4,575	900		5,475	6,375	
0180	40 HP motor size		.67	23.881		6,700	900		7,600	8,725	
0190	50 HP motor size		.53	30.189		7,300	1,125		8,425	9,725	
0200	60 HP motor size	R-3	.56	35.714		8,000	1,325	291	9,616	11,100	
0210	75 HP motor size		.56	35.714		11,000	1,325	291	12,616	14,400	
0220	100 HP motor size		.50	40		11,200	1,475	325	13,000	14,900	
0230	125 HP motor size		.50	40		12,400	1,475	325	14,200	16,200	
0240	150 HP motor size		.50	40		15,800	1,475	325	17,600	20,000	
0250	200 HP motor size		.42	47.619		19,600	1,775	390	21,765	24,600	
1100	Custom-engineered, 460 volt, for 3 HP motor size	1 Elec	.56	14.286		2,425	535		2,960	3,450	
1110	5 HP motor size		.56	14.286		2,550	535		3,085	3,600	
1120	7.5 HP motor size		.47	17.021		3,050	640		3,690	4,300	
1130	10 HP motor size		.47	17.021		3,050	640		3,690	4,300	
1140	15 HP motor size	2 Elec	.62	25.806		3,350	970		4,320	5,125	
1150	20 HP motor size		.62	25.806		4,575	970		5,545	6,475	
1160	25 HP motor size		.47	34.043		4,950	1,275		6,225	7,350	
1170	30 HP motor size		.47	34.043		6,750	1,275		8,025	9,325	
1180	40 HP motor size		.47	34.043		7,550	1,275		8,825	10,200	
1190	50 HP motor size		.37	43.243		7,550	1,625		9,175	10,700	
1200	60 HP motor size	R-3	.39	51.282		11,300	1,900	420	13,620	15,700	
1210	75 HP motor size		.39	51.282		13,000	1,900	420	15,320	17,600	
1220	100 HP motor size		.35	57.143		13,600	2,125	465	16,190	18,600	
1230	125 HP motor size		.35	57.143		14,300	2,125	465	16,890	19,400	
1240	150 HP motor size		.35	57.143		16,100	2,125	465	18,690	21,400	
1250	200 HP motor size		.29	68.966		22,200	2,550	565	25,315	28,800	
2000	For complex & special design systems to meet specific										
2010	requirements, obtain quote from vendor.										

16420	Enclosed Controllers	CREW	DAILY OUTPUT	LABOR-HOURS	UNIT	2003 BARE COSTS				TOTAL INCL O&P	
						MAT.	LABOR	EQUIP.	TOTAL		
220	**0010**	**CONTROL STATIONS**										**220**
0050	NEMA 1, heavy duty, stop/start	1 Elec	8	1	Ea.	119	37.50		156.50	187	
0100	Stop/start, pilot light		6.20	1.290		160	48.50		208.50	249	
0200	Hand/off/automatic		6.20	1.290		87.50	48.50		136	169	
0400	Stop/start/reverse		5.30	1.509		160	57		217	261	
0500	NEMA 7, heavy duty, stop/start		6	1.333		188	50		238	282	
0600	Stop/start, pilot light		4	2		230	75		305	365	

16 ELECTRICAL

Important: See the Reference Section for critical supporting data - Reference Nos., Crews, & City Cost Indexes

16440	Swbds, Panels & Cont Centers	CREW	DAILY OUTPUT	LABOR-HOURS	UNIT	2003 BARE COSTS				TOTAL INCL O&P
						MAT.	LABOR	EQUIP.	TOTAL	
660	0010	**MOTOR STARTERS & CONTROLS**									660
0050	Magnetic, FVNR, with enclosure and heaters, 480 volt									
0080	2 HP, size 00	1 Elec	3.50	2.286	Ea.	173	86		259	320
0100	5 HP, size 0		2.30	3.478		209	131		340	425
0200	10 HP, size 1	↓	1.60	5		235	188		423	540
0300	25 HP, size 2	2 Elec	2.20	7.273		440	273		713	895
0400	50 HP, size 3		1.80	8.889		720	335		1,055	1,300
0500	100 HP, size 4		1.20	13.333		1,600	500		2,100	2,500
0600	200 HP, size 5		.90	17.778		3,725	670		4,395	5,100
0610	400 HP, size 6	↓	.80	20		10,500	750		11,250	12,600
0620	NEMA 7, 5 HP, size 0	1 Elec	1.60	5		865	188		1,053	1,225
0630	10 HP, size 1	"	1.10	7.273		905	273		1,178	1,400
0640	25 HP, size 2	2 Elec	1.80	8.889		1,450	335		1,785	2,100
0650	50 HP, size 3		1.20	13.333		2,150	500		2,650	3,125
0660	100 HP, size 4		.90	17.778		3,525	670		4,195	4,875
0670	200 HP, size 5	↓	.50	32		8,250	1,200		9,450	10,900
0700	Combination, with motor circuit protectors, 5 HP, size 0	1 Elec	1.80	4.444		680	167		847	995
0800	10 HP, size 1	"	1.30	6.154		705	231		936	1,125
0900	25 HP, size 2	2 Elec	2	8		990	300		1,290	1,525
1000	50 HP, size 3		1.32	12.121		1,425	455		1,880	2,250
1200	100 HP, size 4	↓	.80	20		3,100	750		3,850	4,550
1220	NEMA 7, 5 HP, size 0	1 Elec	1.30	6.154		1,525	231		1,756	2,025
1230	10 HP, size 1	"	1	8		1,575	300		1,875	2,175
1240	25 HP, size 2	2 Elec	1.32	12.121		2,075	455		2,530	2,975
1250	50 HP, size 3		.80	20		3,425	750		4,175	4,900
1260	100 HP, size 4		.60	26.667		5,325	1,000		6,325	7,350
1270	200 HP, size 5	↓	.40	40		11,500	1,500		13,000	15,000
1400	Combination, with fused switch, 5 HP, size 0	1 Elec	1.80	4.444		520	167		687	820
1600	10 HP, size 1	"	1.30	6.154		555	231		786	955
1800	25 HP, size 2	2 Elec	2	8		900	300		1,200	1,450
2000	50 HP, size 3		1.32	12.121		1,525	455		1,980	2,350
2200	100 HP, size 4	↓	.80	20	↓	2,650	750		3,400	4,050
3500	Magnetic FVNR with NEMA 12, enclosure & heaters, 480 volt									
3600	5 HP, size 0	1 Elec	2.20	3.636	Ea.	197	137		334	420
3700	10 HP, size 1	"	1.50	5.333		297	201		498	625
3800	25 HP, size 2	2 Elec	2	8		555	300		855	1,050
3900	50 HP, size 3		1.60	10		855	375		1,230	1,500
4000	100 HP, size 4		1	16		2,050	600		2,650	3,150
4100	200 HP, size 5	↓	.80	20		4,900	750		5,650	6,525
4200	Combination, with motor circuit protectors, 5 HP, size 0	1 Elec	1.70	4.706		655	177		832	985
4300	10 HP, size 1	"	1.20	6.667		680	251		931	1,125
4400	25 HP, size 2	2 Elec	1.80	8.889		1,025	335		1,360	1,625
4500	50 HP, size 3		1.20	13.333		1,650	500		2,150	2,575
4600	100 HP, size 4	↓	.74	21.622		3,725	815		4,540	5,325
4700	Combination, with fused switch, 5 HP, size 0	1 Elec	1.70	4.706		630	177		807	960
4800	10 HP, size 1	"	1.20	6.667		660	251		911	1,100
4900	25 HP, size 2	2 Elec	1.80	8.889		1,000	335		1,335	1,600
5000	50 HP, size 3		1.20	13.333		1,600	500		2,100	2,525
5100	100 HP, size 4	↓	.74	21.622	↓	3,250	815		4,065	4,800
5200	Factory installed controls, adders to size 0 thru 5									
5300	Start-stop push button	1 Elec	32	.250	Ea.	41.50	9.40		50.90	59.50
5400	Hand-off-auto-selector switch		32	.250		41.50	9.40		50.90	59.50
5500	Pilot light		32	.250		77.50	9.40		86.90	99
5600	Start-stop-pilot		32	.250		119	9.40		128.40	145
5700	Auxiliary contact, NO or NC		32	.250		57	9.40		66.40	76.50
5800	NO-NC	↓	32	.250	↓	114	9.40		123.40	139

ELECTRICAL 16

For information about Means Estimating Seminars, see yellow pages 12 and 13 in back of book

For expanded coverage of these items see *Means Electrical Cost Data 2003*

		CREW	DAILY OUTPUT	LABOR-HOURS	UNIT	2003 BARE COSTS				TOTAL INCL O&P
						MAT.	LABOR	EQUIP.	TOTAL	

Division 17
Square Foot & Cubic Foot Costs

Estimating Tips

- The cost figures in Division 17 were derived from approximately 10,400 projects contained in the Means database of completed construction projects, and include the contractor's overhead and profit, but do not generally include architectural fees or land costs. The figures have been adjusted to January of the current year. New projects are added to our files each year, and outdated projects are discarded. For this reason, certain costs may not show a uniform annual progression. In no case are all subdivisions of a project listed.

- These projects were located throughout the U.S. and reflect a tremendous variation in square foot (S.F.) and cubic foot (C.F.) costs. This is due to differences, not only in labor and material costs, but also in individual owners' requirements. For instance, a bank in a large city would have different features than one in a rural area. This is true of all the different types of buildings analyzed. Therefore, caution should be exercised when using Division 17 costs. For example, for court houses, costs in the database are local court house costs and will not apply to the larger, more elaborate federal court houses. As a general rule, the projects in the 1/4 column do not include any site work or equipment, while the projects in the 3/4 column may include both equipment and site work. The median figures do not generally include site work.

- None of the figures "go with" any others. All individual cost items were computed and tabulated separately. Thus the sum of the median figures for Plumbing, HVAC and Electrical will not normally total up to the total Mechanical and Electrical costs arrived at by separate analysis and tabulation of the projects.

- Each building was analyzed as to total and component costs and percentages. The figures were arranged in ascending order with the results tabulated as shown. The 1/4 column shows that 25% of the projects had lower costs, 75% higher. The 3/4 column shows that 75% of the projects had lower costs, 25% had higher. The median column shows that 50% of the projects had lower costs, 50% had higher.

- There are two times when square foot costs are useful. The first is in the conceptual stage when no details are available. Then square foot costs make a useful starting point. The second is after the bids are in and the costs can be worked back into their appropriate units for information purposes. As soon as details become available in the project design, the square foot approach should be discontinued and the project priced as to its particular components. When more precision is required or for estimating the replacement cost of specific buildings, the current edition of *Means Square Foot Costs* should be used.

- In using the figures in Division 17, it is recommended that the median column be used for preliminary figures if no additional information is available. The median figures, when multiplied by the total city construction cost index figures (see City Cost Indexes) and then multiplied by the project size modifier in Reference Number R17100-100, should present a fairly accurate base figure, which would then have to be adjusted in view of the estimator's experience, local economic conditions, code requirements and the owner's particular requirements. There is no need to factor the percentage figures, as these should remain constant from city to city. All tabulations mentioning air conditioning had at least partial air conditioning.

- The editors of this book would greatly appreciate receiving cost figures on one or more of your recent projects which would then be included in the averages for next year. All cost figures received will be kept confidential except that they will be averaged with other similar projects to arrive at S.F. and C.F. cost figures for next year's book. See the last page of the book for details and the discount available for submitting one or more of your projects.

17100 | S.F. & C.F. Costs			UNIT	UNIT COSTS			% OF TOTAL		
				1/4	MEDIAN	3/4	1/4	MEDIAN	3/4
010	0010	**APARTMENTS** Low Rise (1 to 3 story) R17100 -100	S.F.	46.95	58.80	78.70				**010**
	0020	Total project cost	C.F.	4.21	5.60	6.90			
	2720	Plumbing	S.F.	3.62	4.78	5.95	6.70%	9%	10.10%
	2770	Heating, ventilating, air conditioning		2.33	2.87	4.22	4.20%	5.80%	7.70%
	2900	Electrical		2.71	3.54	4.84	5.20%	6.70%	8.40%
	3100	Total: Mechanical & Electrical	↓	9.35	11.95	14.95	16%	18.20%	23%
	9000	Per apartment unit, total cost	Apt.	43,300	66,000	98,500			
	9500	Total: Mechanical & Electrical	"	8,300	13,000	17,000			
020	0010	**APARTMENTS** Mid Rise (4 to 7 story) R17100 -100	S.F.	61.10	74.30	90.90				**020**
	0020	Total project costs	C.F.	4.87	6.70	9.15			
	2720	Plumbing	S.F.	3.65	5.85	6.50	6.30%	7.40%	10%
	2900	Electrical		4.30	5.85	6.85	6.70%	7.50%	8.90%
	3100	Total: Mechanical & Electrical	↓	12.40	16.80	20.60	18.80%	21.60%	25.10%
	9000	Per apartment unit, total cost	Apt.	70,500	83,100	137,600			
	9500	Total: Mechanical & Electrical	"	13,000	15,600	19,700			
030	0010	**APARTMENTS** High Rise (8 to 24 story) R17100 -100	S.F.	70.60	85.25	104				**030**
	0020	Total project costs	C.F.	6	8.40	10.20			
	2720	Plumbing	S.F.	5.20	6.15	7.70	7%	9.10%	10.60%
	2900	Electrical		4.85	6.15	8.30	6.40%	7.70%	8.90%
	3100	Total: Mechanical & Electrical	↓	14.55	18.55	22.35	18%	22.30%	24.40%
	9000	Per apartment unit, total cost	Apt.	73,100	79,300	112,100			
	9500	Total: Mechanical & Electrical	"	16,200	18,100	19,200			
040	0010	**AUDITORIUMS** R17100 -100	S.F.	72.65	98.40	133				**040**
	0020	Total project costs	C.F.	4.60	6.40	9.60			
	2720	Plumbing	S.F.	4.65	6.45	8.15	6.30%	7.20%	9.05%
	2900	Electrical		5.90	8.15	10.25	6.80%	9%	10.60%
	3100	Total: Mechanical & Electrical	↓	11.35	15.15	26.40	14.40%	18.50%	23.60%
050	0010	**AUTOMOTIVE SALES** R17100 -100	S.F.	52.90	60.85	89.60				**050**
	0020	Total project costs	C.F.	3.52	4.29	5.55			
	2720	Plumbing	S.F.	2.65	4.30	4.70	4.70%	6.50%	7%
	2900	Electrical		4.22	6.55	7.40	7.30%	9.90%	12.20%
	3100	Total: Mechanical & Electrical	↓	9.55	13.70	17.15	17%	20.40%	26%
060	0010	**BANKS** R17100 -100	S.F.	105	132	167				**060**
	0020	Total project costs	C.F.	7.60	10.35	13.55			
	2720	Plumbing	S.F.	3.35	4.80	6.95	2.80%	3.90%	4.90%
	2770	Heating, ventilating, air conditioning		6.20	8.35	11	4.80%	7.20%	8.50%
	2900	Electrical		10.25	13.50	17.55	8.30%	10.30%	12.20%
	3100	Total: Mechanical & Electrical	↓	24.20	32.60	39.55	16.60%	20.40%	24.40%
130	0010	**CHURCHES** R17100 -100	S.F.	70.80	88.85	115				**130**
	0020	Total project costs	C.F.	4.45	5.60	7.40			
	2720	Plumbing	S.F.	2.79	3.90	5.75	3.60%	5%	6.30%
	2770	Heating, ventilating, air conditioning		6.55	8.50	12.50	7.60%	10%	12.20%
	2900	Electrical		5.90	8.10	10.50	7.30%	8.80%	10.90%
	3100	Total: Mechanical & Electrical	↓	15.25	22.90	33.30	19.60%	22.90%	25.40%
150	0010	**CLUBS, COUNTRY** R17100 -100	S.F.	72.25	88.60	117				**150**
	0020	Total project costs	C.F.	6.20	7.55	10.40			
	2720	Plumbing	S.F.	4.94	6.90	15.70	6%	9.60%	10%
	2900	Electrical		6.05	8.65	11.40	7%	9.30%	11.40%
	3100	Total: Mechanical & Electrical	↓	13.05	16.80	32	16.10%	22.30%	30.90%
170	0010	**CLUBS, SOCIAL** Fraternal R17100 -100	S.F.	60.35	86.20	113				**170**
	0020	Total project costs	C.F.	3.63	5.80	6.75			
	2720	Plumbing	S.F.	3.47	4.55	5.85	5.20%	6.80%	8.30%
	2770	Heating, ventilating, air conditioning	↓	5.55	6.75	8.65	8.70%	11%	14.40%

17 SQUARE FOOT

See R17100-100 and City Cost Indexes in the Reference Section

17100 | S.F. & C.F. Costs

			UNIT	UNIT COSTS			% OF TOTAL		
				1/4	MEDIAN	3/4	1/4	MEDIAN	3/4
170	2900	Electrical	S.F.	4.88	7.60	9.20	7.30%	9.50%	11.40%
	3100	Total: Mechanical & Electrical	↓	13.05	17.70	26.15	19.30%	23%	33.10%
180	0010	**CLUBS, Y.M.C.A.** R17100 -100	S.F.	70.30	86.40	119			
	0020	Total project costs	C.F.	3.55	5.95	8.85			
	2720	Plumbing	S.F.	5.10	9.70	10.90	6%	9.70%	16%
	2900	Electrical		5.85	7.70	11.35	6.40%	9.20%	10.80%
	3100	Total: Mechanical & Electrical	↓	12.75	18.25	28.40	16.20%	21.90%	28.50%
190	0010	**COLLEGES** Classrooms & Administration R17100 -100	S.F.	87.95	116	154			
	0020	Total project costs	C.F.	6.40	8.95	14.45			
	2720	Plumbing	S.F.	4.31	8.45	15.70	4.80%	8.90%	11.80%
	2900	Electrical		7.20	11.10	13.35	8.40%	10%	12.10%
	3100	Total: Mechanical & Electrical	↓	15.30	28.35	43.90	14.20%	24.70%	32.50%
210	0010	**COLLEGES** Science, Engineering, Laboratories R17100 -100	S.F.	140	169	206			
	0020	Total project costs	C.F.	8.30	12.10	13.75			
	2900	Electrical	S.F.	11.95	17.25	26.05	7.10%	9.60%	15.40%
	3100	Total: Mechanical & Electrical	"	47.55	53.35	81.45	30.30%	34%	39.80%
230	0010	**COLLEGES** Student Unions R17100 -100	S.F.	92.30	129	152			
	0020	Total project costs	C.F.	5.15	6.75	8.70			
	3100	Total: Mechanical & Electrical	S.F.	24.30	34.65	37.55	17.40%	23.90%	28.90%
250	0010	**COMMUNITY CENTERS** R17100 -100	"	69	92.60	123			
	0020	Total project costs	C.F.	4.98	7.10	9.20			
	2720	Plumbing	S.F.	3.70	6.30	9.20	4.90%	7.10%	9.50%
	2770	Heating, ventilating, air conditioning		6.10	8.80	12.05	7.70%	10.80%	13%
	2900	Electrical		6.10	8.05	12.50	7.40%	9.40%	9.30%
	3100	Total: Mechanical & Electrical	↓	22.73	26.85	38.75	25.10%	27.90%	34.40%
280	0010	**COURT HOUSES** R17100 -100	S.F.	109	126	145			
	0020	Total project costs	C.F.	8.35	10.10	14			
	2720	Plumbing	S.F.	5.25	7.30	10.50	5.90%	7.40%	9.60%
	2900	Electrical		9.60	11.70	15.35	8.40%	9.40%	11.60%
	3100	Total: Mechanical & Electrical	↓	21.85	29.50	38.25	20.10%	25.70%	28.70%
300	0010	**DEPARTMENT STORES** R17100 -100	S.F.	40.65	55	69.45			
	0020	Total project costs	C.F.	2.14	2.82	3.84			
	2720	Plumbing	S.F.	1.32	1.60	2.43	3.10%	4.20%	5.90%
	2770	Heating, ventilating, air conditioning		3.70	5.70	8.60	8.30%	12.40%	14.80%
	2900	Electrical		4.67	6.40	7.55	9.10%	12.20%	14.90%
	3100	Total: Mechanical & Electrical	↓	8.20	10.65	17.20	20%	19%	26.50%
310	0010	**DORMITORIES** Low Rise (1 to 3 story) R17100 -100	S.F.	66.50	94.90	114			
	0020	Total project costs	C.F.	4.61	7.05	10.55			
	2720	Plumbing	S.F.	4.64	6.20	7.85	8%	9%	9.60%
	2770	Heating, ventilating, air conditioning		4.90	5.90	7.85	4.60%	8%	10%
	2900	Electrical		4.90	7.45	9.35	6.60%	8.90%	9.60%
	3100	Total: Mechanical & Electrical	↓	25.75	27	28.15	22.50%	24.10%	29.10%
	9000	Per bed, total cost	Bed	32,700	36,300	77,800			
320	0010	**DORMITORIES** Mid Rise (4 to 8 story) R17100 -100	S.F.	94.70	124	150			
	0020	Total project costs	C.F.	10.45	11.45	13.70			
	2900	Electrical	S.F.	7	11	11.45	7.40%	8.90%	10.40%
	3100	Total: Mechanical & Electrical	"	21	28.15	36.35	18.10%	19.50%	30.60%
	9000	Per bed, total cost	Bed	13,500	30,800	63,600			
340	0010	**FACTORIES** R17100 -100	S.F.	36.10	53.30	82.50			
	0020	Total project costs	C.F.	2.30	3.41	5.70			
	2720	Plumbing	S.F.	2.08	3.60	6.10	4.10%	6.20%	8.70%
	2770	Heating, ventilating, air conditioning	↓	3.76	5.40	7.30	5.20%	8.50%	11.40%

SQUARE FOOT 17

17100 | S.F. & C.F. Costs

			UNIT	UNIT COSTS			% OF TOTAL			
				1/4	MEDIAN	3/4	1/4	MEDIAN	3/4	
340	2900	Electrical	S.F.	4.39	7	10.75	8.10%	10.50%	14.20%	340
	3100	Total: Mechanical & Electrical	↓	10.60	17	25.85	20.40%	28.10%	34.50%	
360	0010	**FIRE STATIONS** R17100-100	S.F.	69.25	93	120				360
	0020	Total project costs	C.F.	4.05	5.65	7.55				
	2720	Plumbing	S.F.	4.50	6.65	9.75	6.20%	7.70%	9.70%	
	2770	Heating, ventilating, air conditioning		3.88	6.40	9.75	4.90%	7.50%	9.30%	
	2900	Electrical		4.92	8.50	11.25	6.80%	9.20%	10.90%	
	3100	Total: Mechanical & Electrical	↓	21.90	27	32	18.80%	23.80%	27.30%	
370	0010	**FRATERNITY HOUSES** and Sorority Houses R17100-100	S.F.	71.15	91.55	129				370
	0020	Total project costs	C.F.	6.80	7.40	9.45				
	2720	Plumbing	S.F.	5.35	6.15	11.25	5.20%	8%	10.90%	
	2900	Electrical		4.69	10.15	12.40	6.60%	9.90%	10.70%	
	3100	Total: Mechanical & Electrical	↓	13.05	18.45	22.02	14.60%	19.90%	20.70%	
380	0010	**FUNERAL HOMES** R17100-100	S.F.	74.65	102	186				380
	0020	Total project costs	C.F.	7.65	8.55	16.35				
	2900	Electrical	S.F.	3.31	6.05	7.15	3.60%	4.40%	11%	
	3100	Total: Mechanical & Electrical	"	11.90	17.70	23.15	12.90%	17.80%	18.80%	
390	0010	**GARAGES, COMMERCIAL** (Service) R17100-100	S.F.	40.95	65.05	89.35				390
	0020	Total project costs	C.F.	2.60	3.89	5.60				
	2720	Plumbing	S.F.	2.76	4.15	8.15	4.70%	7.40%	10.50%	
	2730	Heating & ventilating		3.86	5.25	7.30	5.30%	6.90%	9.60%	
	2900	Electrical		3.91	6	8.30	7.20%	9%	11.20%	
	3100	Total: Mechanical & Electrical	↓	9.15	17.05	25.15	14.10%	22.50%	26.90%	
400	0010	**GARAGES, MUNICIPAL** (Repair) R17100-100	S.F.	62.45	83.15	119				400
	0020	Total project costs	C.F.	3.90	4.95	8.05				
	2720	Plumbing	S.F.	2.80	5.35	10.15	3.60%	6.70%	8%	
	2730	Heating & ventilating		4.79	6.95	13.35	6.20%	7.40%	13.50%	
	2900	Electrical		4.61	6.80	10.45	6.30%	7.60%	11.70%	
	3100	Total: Mechanical & Electrical	↓	11.65	21.10	34.70	15.20%	25.50%	32.70%	
410	0010	**GARAGES, PARKING** R17100-100	S.F.	23.50	34.05	60.85				410
	0020	Total project costs	C.F.	2.28	3.10	4.51				
	2720	Plumbing	S.F.	.63	1.07	1.65	2.60%	3.40%	3.90%	
	2900	Electrical		.99	1.50	2.15	4.30%	5.20%	6.30%	
	3100	Total: Mechanical & Electrical	↓	1.32	3.89	4.99	6.50%	9.40%	12.80%	
	9000	Per car, total cost	Car	10,200	12,900	16,600				
430	0010	**GYMNASIUMS** R17100-100	S.F.	67.35	86	110				430
	0020	Total project costs	C.F.	3.37	4.30	5.60				
	2720	Plumbing	S.F.	4.27	5.25	6.30	5.40%	7.30%	7.90%	
	2770	Heating, ventilating, air conditioning	"	4.60	7	14.05	9%	11.10%	22.60%	
	2900	Electrical	S.F.	5.10	6.35	8.40	6.60%	8.30%	10.50%	
	3100	Total: Mechanical & Electrical		15.85	21.80	29.25	20.60%	26.20%	29.40%	
460	0010	**HOSPITALS** R17100-100	↓	136	159	241				460
	0020	Total project costs	C.F.	9.85	12.15	17.50				
	2720	Plumbing	S.F.	11.60	15.60	20.10	7.80%	9.40%	10.90%	
	2770	Heating, ventilating, air conditioning		16.40	21.75	29.30	8.40%	14.60%	17%	
	2900	Electrical		13.95	18.10	28.50	9.90%	12%	14.50%	
	3100	Total: Mechanical & Electrical	↓	39.70	52.65	86.20	26.60%	33.10%	39%	
	9000	Per bed or person, total cost	Bed	50,000	106,000	170,000				
480	0010	**HOUSING** For the Elderly R17100-100	S.F.	63.75	80.50	98.85				480
	0020	Total project costs	C.F.	4.52	6.30	8.05				
	2720	Plumbing	S.F.	4.73	6.05	7.90	8.10%	9.60%	10.70%	
	2730	Heating, ventilating, air conditioning		2.43	3.44	5.15	3.30%	5.60%	7.20%	
	2900	Electrical		4.75	6.40	8.20	7.30%	8.50%	10.20%	
	3100	Total: Mechanical & Electrical	↓	16.35	19.20	25.35	18.10%	22%	29.10%	
	9000	Per rental unit, total cost	Unit	59,300	69,000	77,100				
	9500	Total: Mechanical & Electrical	"	13,100	15,200	17,700				

See R17100-100 and City Cost Indexes in the Reference Section

17 SQUARE FOOT

17100	S.F. & C.F. Costs		UNIT	UNIT COSTS			% OF TOTAL		
				1/4	MEDIAN	3/4	1/4	MEDIAN	3/4
500 0010	HOUSING Public (Low Rise)	R17100 -100	S.F.	53.65	74.50	96.90			
0020	Total project costs		C.F.	4.25	5.95	7.40			
2720	Plumbing		S.F.	3.64	5.10	6.45	6.80%	9%	11.60%
2730	Heating, ventilating, air conditioning			1.94	3.75	4.15	4.20%	6%	6.40%
2900	Electrical			3.24	4.83	6.70	5.10%	6.60%	8.30%
3100	Total: Mechanical & Electrical			15.40	19.80	22.20	14.50%	17.60%	26.50%
9000	Per apartment, total cost		Apt.	58,900	66,900	84,000			
9500	Total: Mechanical & Electrical		"	12,600	15,500	17,100			
510 0010	ICE SKATING RINKS	R17100 -100	S.F.	47.65	82.85	118			
0020	Total project costs		C.F.	3.37	3.45	3.97			
2720	Plumbing		S.F.	1.71	3.21	3.28	3.10%	5.60%	6.70%
2900	Electrical			4.91	7.50	7.95	6.80%	15%	15.80%
3100	Total: Mechanical & Electrical			8.25	11.85	14.80	9.90%	25.90%	29.80%
520 0010	JAILS	R17100 -100	S.F.	141	180	232			
0020	Total project costs		C.F.	13.15	16.15	20.55			
2720	Plumbing		S.F.	14.20	18.05	23.80	7%	8.90%	13.80%
2770	Heating, ventilating, air conditioning			12.60	16.80	32.45	7.50%	9.40%	17.70%
2900	Electrical			15.35	19.70	24.40	8.10%	11.40%	12.40%
3100	Total: Mechanical & Electrical			39.15	69.50	82.20	29.20%	31.10%	34.10%
530 0010	LIBRARIES	R17100 -100	S.F.	83.20	106	135			
0020	Total project costs		C.F.	5.85	7.30	9.45			
2720	Plumbing		S.F.	3.26	4.75	6.40	3.40%	4.90%	5.70%
2770	Heating, ventilating, air conditioning			7.20	12.20	15.90	8%	11%	14.60%
2900	Electrical			8.55	10.60	13.75	8.30%	10.70%	12%
3100	Total: Mechanical & Electrical			24.65	33.70	42.20	18.90%	25.30%	27.60%
540 0010	LIVING, ASSISTED	R17100 -100	S.F.	84.80	96.90	114			
0020	Total project costs		C.F.	7.05	8.05	9.50			
2720	Plumbing		S.F.	6.85	9.15	9.70	6%	8.10%	10.60%
2770	Heating, ventilating, air conditioning			8.10	8.50	9.35	7.90%	8.80%	9.70%
2900	Electrical			8.05	9.15	10.20	9.30%	10.40%	10.70%
3100	Total: Mechanical & Electrical			23.10	27.70	30.15	27.90%	30.10%	32%
550 0010	MEDICAL CLINICS	R17100 -100	S.F.	81.85	101	127			
0020	Total project costs		C.F.	6.10	7.90	10.55			
2720	Plumbing		S.F.	5.50	7.75	10.40	6.10%	8.40%	10%
2770	Heating, ventilating, air conditioning			6.55	8.60	12.65	6.70%	9%	11.30%
2900	Electrical			6.95	9.95	13.15	8.10%	10%	12.20%
3100	Total: Mechanical & Electrical			21.85	30.75	43	22%	27.60%	34.30%
570 0010	MEDICAL OFFICES	R17100 -100	S.F.	76.85	95.25	117			
0020	Total project costs		C.F.	5.70	7.85	10.75			
2720	Plumbing		S.F.	4.31	6.65	9	5.70%	6.80%	8.60%
2770	Heating, ventilating, air conditioning			5.20	7.65	9.95	6.20%	8%	9.70%
2900	Electrical			6.10	8.90	12.40	7.60%	9.80%	11.40%
3100	Total: Mechanical & Electrical			15.15	21.75	32.35	18.50%	22%	24.90%
590 0010	MOTELS	R17100 -100	S.F.	49.25	72.95	94.15			
0020	Total project costs		C.F.	4.30	5.90	9.90			
2720	Plumbing		S.F.	4.99	6.35	7.60	9.40%	10.60%	12.50%
2770	Heating, ventilating, air conditioning			3.04	4.54	8.15	5.60%	5.60%	10%
2900	Electrical			4.65	5.95	7.75	7.10%	8.20%	10.40%
3100	Total: Mechanical & Electrical			15.85	19.80	34	18.50%	21%	24.40%
9000	Per rental unit, total cost		Unit	25,100	47,700	51,500			
9500	Total: Mechanical & Electrical		"	4,900	7,400	8,600			
600 0010	NURSING HOMES	R17100 -100	S.F.	74.05	95.05	118			
0020	Total project costs		C.F.	5.95	7.55	10.35			
2720	Plumbing		S.F.	7	8.90	12.30	9.40%	10.70%	14.20%
2770	Heating, ventilating, air conditioning			6.90	10.50	12.30	9.30%	11.80%	11.80%
2900	Electrical			7.65	9.60	12.85	9.70%	11%	13%
3100	Total: Mechanical & Electrical			18.25	25.50	37.40	26%	29.90%	30.50%

SQUARE FOOT 17

For expanded coverage of these items see *Means Square Foot Costs 2003*

17100 | S.F. & C.F. Costs

				UNIT	UNIT COSTS			% OF TOTAL			
					1/4	MEDIAN	3/4	1/4	MEDIAN	3/4	
600	9000	Per bed or person, total cost		Bed	32,000	39,400	52,500				600
610	0010	**OFFICES** Low Rise (1 to 4 story)	R17100 -100	S.F.	62.35	79.45	106				610
	0020	Total project costs		C.F.	4.52	6.30	8.55				
	2720	Plumbing		S.F.	2.38	3.59	5.10	3.70%	4.50%	6.10%	
	2770	Heating, ventilating, air conditioning			5.15	7.10	10.50	7.20%	10.50%	11.90%	
	2900	Electrical			5.30	7.30	10.20	7.50%	9.60%	11.20%	
	3100	Total: Mechanical & Electrical		↓	12.45	17.25	25.20	18%	21.80%	26.50%	
620	0010	**OFFICES** Mid Rise (5 to 10 story)	R17100 -100	S.F.	68.80	83.45	113				620
	0020	Total project costs		C.F.	4.82	6.10	8.85				
	2720	Plumbing		S.F.	2.08	3.23	4.64	2.80%	3.70%	4.50%	
	2770	Heating, ventilating, air conditioning			5.25	7.50	11.95	7.60%	9.40%	11%	
	3100	Total: Mechanical & Electrical		↓	13.05	16.70	33.05	17.90%	22.30%	29.90%	
630	0010	**OFFICES** High Rise (11 to 20 story)	R17100 -100	S.F.	83.25	107	131				630
	0020	Total project costs		C.F.	5.30	7.35	10.60				
	2900	Electrical		S.F.	4.25	6.15	9.30	5.80%	7%	10.50%	
	3100	Total: Mechanical & Electrical		"	15.55	19.90	36.45	15.40%	17%	34.10%	
640	0010	**POLICE STATIONS**	R17100 -100	S.F.	99.75	133	168				640
	0020	Total project costs		C.F.	8.10	9.75	13.55				
	2720	Plumbing		S.F.	5.65	9.15	12.45	5.60%	6.90%	10.70%	
	2770	Heating, ventilating, air conditioning			8.85	11.80	16	7%	10.70%	12%	
	2900	Electrical			11.05	16.55	21	9.70%	11.90%	14.90%	
	3100	Total: Mechanical & Electrical		↓	35.25	42.05	59.20	25.10%	31.30%	36.40%	
650	0010	**POST OFFICES**	R17100 -100	S.F.	77.70	98.40	123				650
	0020	Total project costs		C.F.	4.75	6.10	7.25				
	2720	Plumbing		S.F.	3.61	4.63	5.65	4.40%	5.60%	5.60%	
	2770	Heating, ventilating, air conditioning			5.65	6.95	7.75	7%	8.40%	10.20%	
	2900	Electrical			6.60	9.15	10.40	7.20%	9.60%	11%	
	3100	Total: Mechanical & Electrical		↓	11.60	22.40	28.35	16.40%	20.40%	22.30%	
660	0010	**POWER PLANTS**	R17100 -100	S.F.	570	733	1,350				660
	0020	Total project costs		C.F.	15.30	30.95	71.40				
	2900	Electrical		S.F.	39.25	83.30	124	9.50%	12.80%	18.40%	
	8100	Total: Mechanical & Electrical		"	97.80	318	713	32.50%	32.60%	52.60%	
670	0010	**RELIGIOUS EDUCATION**	R17100 -100	S.F.	62.80	81.50	92.35				670
	0020	Total project costs		C.F.	3.59	5.15	6.65				
	2720	Plumbing		S.F.	2.69	3.81	5.40	4.30%	4.90%	8.40%	
	2770	Heating, ventilating, air conditioning			6.30	7.10	7.85	9.80%	10%	11.80%	
	2900	Electrical			5.10	6.75	8.95	7.90%	9.10%	10.30%	
	3100	Total: Mechanical & Electrical		↓	13.25	23	27.35	17.60%	20.50%	22%	
690	0010	**RESEARCH** Laboratories and facilities	R17100 -100	S.F.	96.40	139	202				690
	0020	Total project costs		C.F.	7.20	12.05	16.25				
	2720	Plumbing		S.F.	9.70	12.40	19.90	7.30%	8.60%	13.30%	
	2770	Heating, ventilating, air conditioning			8.70	29.20	34.55	7.20%	16.50%	17.50%	
	2900	Electrical			11.10	18.65	31.75	9.60%	12%	15.80%	
	3100	Total: Mechanical & Electrical		↓	25.65	48.65	92.75	23.90%	35.60%	41.70%	
700	0010	**RESTAURANTS**	R17100 -100	S.F.	92.90	119	155				700
	0020	Total project costs		C.F.	7.80	10.25	13.50				
	2720	Plumbing		S.F.	7.35	9	12.80	6.10%	8.10%	9.10%	
	2770	Heating, ventilating, air conditioning			9.75	13	17	9.40%	12.20%	13%	
	2900	Electrical			9.90	12.20	15.90	8.50%	10.60%	11.60%	
	3100	Total: Mechanical & Electrical		↓	30.97	32.30	40.85	21.10%	24%	29.50%	
	9000	Per seat unit, total cost		Seat	3,400	4,600	5,800				
	9500	Total: Mechanical & Electrical		"	722	1,000	1,350				
720	0010	**RETAIL STORES**	R17100 -100	S.F.	43.35	58.45	76.80				720
	0020	Total project costs		C.F.	2.95	4.21	5.80				
	2720	Plumbing		S.F.	1.59	2.65	4.52	3.20%	4.60%	6.80%	
	2770	Heating, ventilating, air conditioning		↓	3.43	4.69	7.05	6.80%	8.80%	10.20%	

17 SQUARE FOOT

See R17100-100 and City Cost Indexes in the Reference Section

		17100	S.F. & C.F. Costs		UNIT	UNIT COSTS			% OF TOTAL			
						1/4	MEDIAN	3/4	1/4	MEDIAN	3/4	
720	2900	Electrical			S.F.	3.91	5.30	7.70	7.30%	9.90%	11.60%	720
	3100	Total: Mechanical & Electrical			↓	10.45	13.40	17.40	17.10%	21.40%	23.80%	
740	0010	**SCHOOLS** Elementary		R17100 -100	S.F.	69	85.10	103				740
	0020	Total project costs			C.F.	4.67	6	7.75				
	2720	Plumbing			S.F.	4	5.80	7.75	5.70%	7.10%	9.40%	
	2730	Heating, ventilating, air conditioning				6.15	9.80	13.70	8.10%	10.80%	15.20%	
	2900	Electrical				6.45	8.20	10.50	8.40%	10%	11.60%	
	3100	Total: Mechanical & Electrical			↓	22.25	24.65	34.80	25.40%	28%	30.40%	
	9000	Per pupil, total cost			Ea.	6,900	11,400	37,100				
	9500	Total: Mechanical & Electrical			"	2,300	2,925	11,700				
760	0010	**SCHOOLS** Junior High & Middle		R17100 -100	S.F.	70.10	86.70	102				760
	0020	Total project costs			C.F.	4.61	6	6.80				
	2720	Plumbing			S.F.	4.67	5.15	6.85	5.60%	6.80%	8.10%	
	2770	Heating, ventilating, air conditioning				5.35	10.35	14.45	8.70%	12.70%	17.40%	
	2900	Electrical				6.80	8.50	10.25	7.80%	9.40%	10.60%	
	3100	Total: Mechanical & Electrical			↓	19.55	24.60	32.65	23.30%	25.70%	27.30%	
	9000	Per pupil, total cost			Ea.	8,800	10,100	12,600				
780	0010	**SCHOOLS** Senior High		R17100 -100	S.F.	75.70	86.70	122				780
	0020	Total project costs			C.F.	4.85	6.30	8.90				
	2720	Plumbing			S.F.	3.91	7.15	11.80	5%	6.90%	12%	
	2770	Heating, ventilating, air conditioning				8.75	10.05	19.15	8.90%	11.60%	15%	
	2900	Electrical				7.45	9.80	16.30	8.30%	10.10%	12.60%	
	3100	Total: Mechanical & Electrical			↓	20.35	26.80	50.05	19.80%	23.10%	27.80%	
	9000	Per pupil, total cost			Ea.	7,200	12,300	18,400				
800	0010	**SCHOOLS** Vocational		R17100 -100	S.F.	61.75	86.35	110				800
	0020	Total project costs			C.F.	3.79	5.50	7.65				
	2720	Plumbing			S.F.	3.96	5.90	8.70	5.40%	7%	8.50%	
	2770	Heating, ventilating, air conditioning				5.55	10.30	17.25	8.60%	11.90%	14.60%	
	2900	Electrical				6.35	8.85	12.45	8.40%	11.20%	13.80%	
	3100	Total: Mechanical & Electrical			↓	16.75	22.40	42.55	21.70%	27.30%	33.10%	
	9000	Per pupil, total cost			Ea.	8,600	23,100	34,500				
830	0010	**SPORTS ARENAS**		R17100 -100	S.F.	52.65	72.40	106				830
	0020	Total project costs			C.F.	2.94	5.20	6.80				
	2720	Plumbing			S.F.	2.80	4.66	8.75	4.50%	6.30%	9.40%	
	2770	Heating, ventilating, air conditioning				5.75	8	10.40	5.80%	10.20%	13.50%	
	2900	Electrical				4.65	7.35	9.20	7.70%	9.80%	12.30%	
	3100	Total: Mechanical & Electrical			↓	13.70	24.30	31.40	13.40%	22.50%	30.80%	
850	0010	**SUPERMARKETS**		R17100 -100	S.F.	49.95	58.65	68				850
	0020	Total project costs			C.F.	2.79	3.37	5.10				
	2720	Plumbing			S.F.	2.79	3.52	4.10	5.40%	6%	7.50%	
	2770	Heating, ventilating, air conditioning				4.11	5.05	6	8.60%	8.60%	9.60%	
	2900	Electrical				5.85	7.20	8.55	10.40%	12.40%	13.60%	
	3100	Total: Mechanical & Electrical			↓	16.35	17.65	24.40	17.40%	20.40%	28.40%	
860	0010	**SWIMMING POOLS**		R17100 -100	S.F.	88.30	136	189				860
	0020	Total project costs			C.F.	6.50	8.10	8.85				
	2720	Plumbing			S.F.	7.50	8.55	11.90	9.70%	12.30%	12.30%	
	2900	Electrical				6.10	9.45	12.30	7.50%	8%	8%	
	3100	Total: Mechanical & Electrical			↓	12.90	27.30	50.50	17.70%	24.90%	31.60%	
870	0010	**TELEPHONE EXCHANGES**		R17100 -100	S.F.	108	158	200				870
	0020	Total project costs			C.F.	6.65	10.20	14.50				
	2720	Plumbing			S.F.	4.54	7	10.25	4.50%	5.80%	6.90%	
	2770	Heating, ventilating, air conditioning				9.25	20.75	26.35	11.80%	11.80%	18.40%	
	2900	Electrical				10.95	16.95	30.70	10.90%	14%	17.80%	
	3100	Total: Mechanical & Electrical			↓	23.55	32.35	80.20	27.30%	33.40%	44.40%	

SQUARE FOOT 17

17100 | S.F. & C.F. Costs			UNIT	UNIT COSTS			% OF TOTAL		
				1/4	MEDIAN	3/4	1/4	MEDIAN	3/4
910	0010	**THEATERS**	R17100 -100	S.F.	67.70	86.80	128				**910**
	0020	Total project costs		C.F.	3.13	4.60	6.80			
	2720	Plumbing		S.F.	2.11	2.44	7.35	2.90%	4.70%	6.80%
	2770	Heating, ventilating, air conditioning			6.55	7.95	9.85	8%	12.20%	13.40%
	2900	Electrical			5.90	8	16.25	8%	10%	12.40%
	3100	Total: Mechanical & Electrical			15.25	22.75	46.70	22.90%	26.60%	27.50%
940	0010	**TOWN HALLS** City Halls & Municipal Buildings	R17100 -100	S.F.	75.50	95.45	125				**940**
	0020	Total project costs		C.F.	5.65	7.90	10.55			
	2720	Plumbing		S.F.	2.68	5.55	9.20	4.20%	6.10%	7.90%
	2770	Heating, ventilating, air conditioning			5.70	11.30	16.55	7%	9%	13.50%
	2900	Electrical			6.60	9.15	12.95	8.10%	9.50%	11.70%
	3100	Total: Mechanical & Electrical			23.95	24.90	31.55	22%	25.10%	31.60%
970	0010	**WAREHOUSES** And Storage Buildings	R17100 -100	S.F.	28.15	39.35	60.60				**970**
	0020	Total project costs		C.F.	1.42	2.31	3.83			
	2720	Plumbing		S.F.	.91	1.69	3.15	2.90%	4.80%	6.60%
	2730	Heating, ventilating, air conditioning			1.07	3.02	4.06	2.40%	5%	8.90%
	2900	Electrical			1.67	3.14	5.20	5%	7.30%	10%
	3100	Total: Mechanical & Electrical			4.65	6.85	15.60	12.80%	18.40%	26.20%
990	0010	**WAREHOUSE & OFFICES** Combination	R17100 -100	S.F.	33.90	45.45	60.30				**990**
	0020	Total project costs		C.F.	1.76	2.57	3.80			
	2720	Plumbing		S.F.	1.31	2.26	3.56	3.60%	4.70%	6.30%
	2770	Heating, ventilating, air conditioning			2.08	3.29	4.63	5%	5.60%	9.60%
	2900	Electrical		S.F.	2.29	3.39	5.30	5.70%	7.70%	10%
	3100	Total: Mechanical & Electrical			5.40	9	11.20	13.80%	18.60%	23%

For information about Means Estimating Seminars, see yellow pages 12 and 13 in back of book

17
SQUARE FOOT

See R17100-100 and City Cost Indexes in the Reference Section

Assemblies Section

Table of Contents

How to Use the Assemblies Cost Tables

The following is a detailed explanation of a sample Assemblies Cost Table. Most Assembly Tables are separated into three parts: 1) an illustration of the system to be estimated; 2) the components and related costs of a typical system; and 3) the costs for similar systems with dimensional and/or size variations. For costs of the components that comprise these systems or "assemblies" refer to the Unit Price Section. Next to each bold number below is the item being described with the appropriate component of the sample entry following in parenthesis. In most cases, if the work is to be subcontracted, the general contractor will need to add an additional markup (R.S. Means suggests using 10%) to the "Total" figures.

1 System/Line Numbers (D3010 510 1760)

Each Assemblies Cost Line has been assigned a unique identification number based on the UniFormat classification sytem.

UNIFORMAT II Major Group

D3010   **510**   **1760**

UNIFORMAT II Level 3

Means Major Classification

Means Individual Line Number

D30 HVAC
D3010 Energy Supply

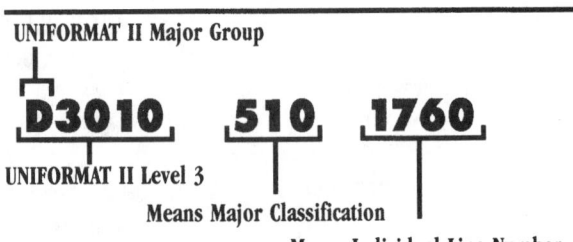

Basis for Heat Loss Estimate, Apartment Type Structures:

1. Masonry walls and flat roof are insulated. U factor is assumed at .08.
2. Window glass area taken as BOCA minimum, 1/10th of floor area. Double insulating glass with 1/4" air space, U = .65.
3. Infiltration = 0.3 C.F. per hour per S.F. of net wall.
4. Concrete floor loss is 2 BTUH per S.F.
5. Temperature difference taken as 70° F.
6. Ventilating or makeup air has not been included and must be added if desired. Air shafts are not used.

System Components	QUANTITY	UNIT	COST EACH MAT.	COST EACH INST.	COST EACH TOTAL
SYSTEM D3010 510 1760					
HEATING SYSTEM, FIN TUBE RADIATION, FORCED HOT WATER					
1,000 S.F. AREA, 10,000 C.F. VOLUME					
Boiler, oil fired, CI, burner, ctrls/insul/breech/pipe/fttng/valves, 109 MBH	1.000		2,800	2,537.50	5,337.50
Circulating pump, CI flange connection, 1/12 HP	1.000	Ea.	229	135	364
Expansion tank, painted steel, ASME 18 Gal capacity	1.000	Ea.	395	58.50	453.50
Storage tank, steel, above ground, 275 Gal capacity w/supports	1.000	Ea.	265		
Copper tubing type L, solder joint, hanger 10' OC, 3/4" diam	100.000	L.F.			
Radiation, 3/4" copper tube w/alum fin baseboard pkg, 7" high	30.000	L.F.			
Pipe covering, calcium silicate w/cover, 1" wall, 3/4" diam	100.000	L.F.	.90		
TOTAL			4,360	4,381	8,741
COST PER S.F.			4.36	4.38	8.74

D3010 510	Apartment Building Heating - Fin Tube Radiation		COST PER S.F. MAT.	COST PER S.F. INST.	COST PER S.F. TOTAL
1740	Heating systems, fin tube radiation, forced hot water				
1760	1,000 S.F. area, 10,000 C.F. volume		4.35	4.38	8.73
1800	10,000 S.F. area, 100,000 C.F. volume	R15500 -010	1.44	2.63	4.07
1840	20,000 S.F. area, 200,000 C.F. volume		1.48	2.99	4.47
1880	30,000 S.F. area, 300,000 C.F. volume	R15500 -020	1.45	2.88	4.33
1890					

2 Illustration

At the top of most assembly pages is an illustration, a brief description, and the design criteria used to develop the cost.

3 System Components

The components of a typical system are listed separately to show what has been included in the development of the total system price. The table below contains prices for other similar systems with dimensional and/or size variations.

4 Quantity

This is the number of line item units required for one system unit. For example, we assume that it will take 30 linear feet of radiation for the 1000 S.F. area shown.

5 Unit of Measure for Each Item

The abbreviated designation indicates the unit of measure, as defined by industry standards, upon which the price of the component is based. For example, baseboard radiation is priced by the linear foot. For a complete listing of abbreviations, see the Reference Section.

6 Unit of Measure for Each System (Each)

Costs shown in the three right hand columns have been adjusted by the component quantity and unit of measure for the entire system. In this example, "Cost Each" is the unit of measure for this system or "assembly."

7 Materials (4,360)

This column contains the Materials Cost of each component. These cost figures are bare costs plus 10% for profit.

8 Installation (4,381)

Installation includes labor and equipment plus the installing contractor's overhead and profit. Equipment costs are the bare rental costs plus 10% for profit. The labor overhead and profit is defined on the inside back cover of this book.

9 Total (8,741)

The figure in this column is the sum of the material and installation costs.

Material Cost	+	Installation Cost	=	Total
$4,360	+	$4,381	=	$8,741

A SUBSTRUCTURE

A2020 180	Subdrainage Piping	COST PER L.F.		
		MAT.	INST.	TOTAL
2000	Piping, excavation & backfill excluded, PVC, perforated			
2110	3" diameter	1.71	2.79	4.50
2130	4" diameter	1.71	2.79	4.50
2140	5" diameter	2.87	2.99	5.86
2150	6" diameter	2.87	2.99	5.86
3000	Metal alum. or steel, perforated asphalt coated			
3150	6" diameter	2.93	5.75	8.68
3160	8" diameter	4.27	5.90	10.17
3170	10" diameter	5.35	6.05	11.40
3180	12" diameter	6	7.65	13.65
3220	18" diameter	8.95	10.60	19.55
4000	Porous wall concrete			
4130	4" diameter	2.05	3.12	5.17
4150	6" diameter	2.66	3.32	5.98
4160	8" diameter	3.28	4.61	7.89
4180	12" diameter	6.95	5	11.95
4200	15" diameter	8	6.25	14.25
4220	18" diameter	10.55	8.65	19.20
5000	Vitrified clay C-211, perforated			
5130	4" diameter	2.20	5.45	7.65
5150	6" diameter	3.64	6.90	10.54
5160	8" diameter	5.20	7.55	12.75
5180	12" diameter	10.90	7.90	18.80

For information about Means Estimating Seminars, see yellow pages 12 and 13 in back of book

SUBSTRUCTURE A

D SERVICES

Basis for Heat Loss Estimate, Apartment Type Structures:

1. Masonry walls and flat roof are insulated. U factor is assumed at .08.
2. Window glass area taken as BOCA minimum, 1/10th of floor area. Double insulating glass with 1/4" air space, U = .65.
3. Infiltration = 0.3 C.F. per hour per S.F. of net wall.
4. Concrete floor loss is 2 BTUH per S.F.
5. Temperature difference taken as 70° F.
6. Ventilating or makeup air has not been included and must be added if desired. Air shafts are not used.

System Components	QUANTITY	UNIT	COST EACH		
			MAT.	INST.	TOTAL
SYSTEM D3010 510 1760					
HEATING SYSTEM, FIN TUBE RADIATION, FORCED HOT WATER					
1,000 S.F. AREA, 10,000 C.F. VOLUME					
Boiler, oil fired, CI, burner, ctrls/insul/breech/pipe/ftng/valves, 109 MBH	1.000	Ea.	2,800	2,537.50	5,337.50
Circulating pump, CI flange connection, 1/12 HP	1.000	Ea.	229	135	364
Expansion tank, painted steel, ASME 18 Gal capacity	1.000	Ea.	395	58.50	453.50
Storage tank, steel, above ground, 275 Gal capacity w/supports	1.000	Ea.	265	164	429
Copper tubing type L, solder joint, hanger 10' OC, 3/4" diam	100.000	L.F.	195	595	790
Radiation, 3/4" copper tube w/alum fin baseboard pkg, 7" high	30.000	L.F.	186	423	609
Pipe covering, calcium silicate w/cover, 1" wall, 3/4" diam	100.000	L.F.	290	468	758
TOTAL			4,360	4,381	8,741
COST PER S.F.			4.36	4.38	8.74

D3010 510	Apartment Building Heating - Fin Tube Radiation		COST PER S.F.		
			MAT.	INST.	TOTAL
1740	Heating systems, fin tube radiation, forced hot water				
1760	1,000 S.F. area, 10,000 C.F. volume		4.35	4.38	8.73
1800	10,000 S.F. area, 100,000 C.F. volume	R15500 -010	1.44	2.63	4.07
1840	20,000 S.F. area, 200,000 C.F. volume		1.48	2.99	4.47
1880	30,000 S.F. area, 300,000 C.F. volume	R15500 -020	1.45	2.88	4.33
1890					

Important: See the Reference Section for critical supporting data - Reference Numbers and City Cost Indexes

D3010 Energy Supply

Fin Tube Radiator

Basis for Heat Loss Estimate, Factory or Commercial Type Structures:

1. Walls and flat roof are of lightly insulated concrete block or metal. U factor is assumed at .17.
2. Windows and doors are figured at 1-1/2 S.F. per linear foot of building perimeter. The U factor for flat single glass is 1.13.
3. Infiltration is approximately 5 C.F. per hour per S.F. of wall.
4. Concrete floor loss is 2 BTUH per S.F. of floor.
5. Temperature difference is assumed at 70° F.
6. Ventilation or makeup air has not been included and must be added if desired.

System Components	QUANTITY	UNIT	COST EACH		
			MAT.	INST.	TOTAL
SYSTEM D3010 520 1960					
HEATING SYSTEM, FIN TUBE RADIATION, FORCED HOT WATER					
1,000 S.F. BLDG., ONE FLOOR					
Boiler, oil fired, CI, burner/ctrls/insul/breech/pipe/ftngs/valves, 109 MBH	1.000	Ea.	2,800	2,537.50	5,337.50
Expansion tank, painted steel, ASME 18 Gal capacity	1.000	Ea.	1,525	68	1,593
Storage tank, steel, above ground, 550 Gal capacity w/supports	1.000	Ea.	1,250	305	1,555
Circulating pump, CI flanged, 1/8 HP	1.000	Ea.	385	135	520
Pipe, steel, black, sch. 40, threaded, cplg & hngr 10'OC, 1-1/2" diam.	260.000	L.F.	717.60	2,639	3,356.60
Pipe covering, calcium silicate w/cover , 1" wall, 1-1/2" diam	260.000	L.F.	738.40	1,253.20	1,991.60
Radiation, steel 1-1/4" tube & 4-1/4" fin w/cover & damper, wall hung	42.000	L.F.	1,134	945	2,079
Rough in, steel fin tube radiation w/supply & balance valves	4.000	Set	644	2,000	2,644
TOTAL			9,194	9,882.70	19,076.70
COST PER S.F.			9.19	9.88	19.07

D3010 520	Commercial Building Heating - Fin Tube Radiation		COST PER S.F.		
			MAT.	INST.	TOTAL
1940	Heating systems, fin tube radiation, forced hot water				
1960	1,000 S.F. bldg, one floor		9.20	9.88	19.08
2000	10,000 S.F., 100,000 C.F., total two floors	R15500 -010	2.50	3.77	6.27
2040	100,000 S.F., 1,000,000 C.F., total three floors		1.13	1.71	2.84
2080	1,000,000 S.F., 10,000,000 C.F., total five floors	R15500 -020	.56	.91	1.47
2090					

SERVICES

D

D3010 Energy Supply

Unit Heater

Basis for Heat Loss Estimate, Factory or Commercial Type Structures:

1. Walls and flat roof are of lightly insulated concrete block or metal. U factor is assumed at .17.
2. Windows and doors are figured at 1-1/2 S.F. per linear foot of building perimeter. The U factor for flat single glass is 1.13.
3. Infiltration is approximately 5 C.F. per hour per S.F. of wall.
4. Concrete floor loss is 2 BTUH per S.F. of floor.
5. Temperature difference is assumed at 70° F.
6. Ventilation or makeup air has not been included and must be added if desired.

System Components	QUANTITY	UNIT	MAT.	INST.	TOTAL
SYSTEM D3010 530 1880					
HEATING SYSTEM, TERMINAL UNIT HEATERS, FORCED HOT WATER					
1,000 S.F. BLDG., ONE FLOOR					
Boiler oil fired, CI, burner/ctrls/insul/breech/pipe/ftngs/valves, 109 MBH	1.000	Ea.	2,800	2,537.50	5,337.50
Expansion tank, painted steel, ASME 18 Gal capacity	1.000	Ea.	1,525	68	1,593
Storage tank, steel, above ground, 550 Gal capacity w/supports	1.000	Ea.	1,250	305	1,555
Circulating pump, CI, flanged, 1/8 HP	1.000	Ea.	385	135	520
Pipe, steel, black, schedule 40, threaded, cplg & hngr 10'OC 1-1/2" diam	260.000	L.F.	717.60	2,639	3,356.60
Pipe covering, calcium silicate w/cover, 1" wall, 1-1/2" diam	260.000	L.F.	738.40	1,253.20	1,991.60
Unit heater, 1 speed propeller, horizontal, 200° EWT, 26.9 MBH	2.000	Ea.	880	204	1,084
Unit heater piping hookup with controls	2.000	Set	730	1,930	2,660
TOTAL			9,026	9,071.70	18,097.70
COST PER S.F.			9.03	9.07	18.10

D3010 530	Commercial Bldg. Heating - Terminal Unit Heaters		MAT.	INST.	TOTAL
1860	Heating systems, terminal unit heaters, forced hot water				
1880	1,000 S.F. bldg., one floor		9.03	9.08	18.11
1920	10,000 S.F. bldg., 100,000 C.F. total two floors	R15500-010	2.25	3.13	5.38
1960	100,000 S.F. bldg., 1,000,000 C.F. total three floors		1.16	1.57	2.73
2000	1,000,000 S.F. bldg., 10,000,000 C.F. total five floors	R15500-020	.74	1.01	1.75
2010					

Many styles of active solar energy systems exist. Those shown on the following page represent the majority of systems now being installed in different regions of the country.

The five active domestic hot water (DHW) systems which follow are typically specified as two or three panel systems with additional variations being the type of glazing and size of the storage tanks. Various combinations have been costed for the user's evaluation and comparison. The basic specifications from which the following systems were developed satisfy the construction detail requirements specified in the HUD Intermediate Minimum Property Standards (IMPS). If these standards are not complied with, the renewable energy system's costs could be significantly lower than shown.

To develop the system's specifications and costs it was necessary to make a number of assumptions about the systems. Certain systems are more appropriate to one climatic region than another or the systems may require modifications to be usable in particular locations. Specific instances in which the systems are not appropriate throughout the country as specified, or in which modification will be needed include the following:

- The freeze protection mechanisms provided in the DHW systems vary greatly. In harsh climates, where freeze is a major concern, a closed-loop indirect collection system may be more appropriate than a direct collection system.

- The thermosyphon water heater system described cannot be used when temperatures drop below 32° F.

- In warm climates it may be necessary to modify the systems installed to prevent overheating.

For each renewable resource (solar) system a schematic diagram and descriptive summary of the system is provided along with a list of all the components priced as part of the system. The costs were developed based on these specifications.

Considerations affecting costs which may increase or decrease beyond the estimates presented here include the following:

- Special structural qualities (allowance for earthquake, future expansion, high winds, and unusual spans or shapes);

- Isolated building site or rough terrain that would affect the transportation of personnel, material, or equipment;

- Unusual climatic conditions during the construction process;

- Substitution of other materials or system components for those used in the system specifications.

SERVICES

D

417

In this closed-loop indirect collection system, fluid with a low freezing temperature, such as propylene glycol, transports heat from the collectors to water storage. The transfer fluid is contained in a closed-loop consisting of collectors, supply and return piping, and a remote heat exchanger. The heat exchanger transfers heat energy from the

fluid in the collector loop to potable water circulated in a storage loop. A typical two-or-three panel system contains 5 to 6 gallons of heat transfer fluid.

When the collectors become approximately 20° F warmer than the storage temperature, a controller activates the circulator on the collector and storage

loops. The circulators will move the fluid and potable water through the heat exchanger until heat collection no longer occurs. At that point, the system shuts down. Since the heat transfer medium is a fluid with a very low freezing temperature, there is no need for it to be drained from the system between periods of collection.

D3010 Energy Supply

System Components	QUANTITY	UNIT	COST EACH		
			MAT.	INST.	TOTAL
SYSTEM D3010 715 2760					
SOLAR, CLOSED LOOP, ADD-ON HOT WATER SYS., EXTERNAL HEAT EXCHANGER					
3/4" TUBING, TWO 3'X7' BLACK CHROME COLLECTORS					
A,B,G,L,K,M Heat exchanger fluid-fluid pkg incl 2 circulators, expansion tank,					
Check valve, relief valve, controller, hi temp cutoff, & 2 sensors	1.000	Ea.	765	325	1,090
C Thermometer, 2" dial	3.000	Ea.	63	84	147
D, T Fill & drain valve, brass, 3/4" connection	1.000	Ea.	5.20	18.80	24
E Air vent, manual, 1/8" fitting	2.000	Ea.	4.26	28.20	32.46
F Air purger	1.000	Ea.	45	37.50	82.50
H Strainer, Y type, bronze body, 3/4" IPS	1.000	Ea.	19.15	24	43.15
I Valve, gate, bronze, NRS, soldered 3/4" diam	6.000	Ea.	120	135	255
J Neoprene vent flashing	2.000	Ea.	19.60	45	64.60
N-1, N Relief valve temp & press, 150 psi 210°F self-closing 3/4" IPS	1.000	Ea.	8.65	15.05	23.70
O Pipe covering, urethane, ultraviolet cover, 1" wall 3/4" diam	20.000	L.F.	33.40	86.60	120
P Pipe covering, fiberglass, all service jacket, 1" wall, 3/4" diam	50.000	L.F.	82.50	173	255.50
Q Collector panel solar energy blk chrome on copper, 1/8" temp glass 3'x7'	2.000	Ea.	1,200	171	1,371
Roof clamps for solar energy collector panels	2.000	Set	3.54	23.20	26.74
R Valve, swing check, bronze, regrinding disc, 3/4" diam	2.000	Ea.	71	45	116
S Pressure gauge, 60 psi, 2" dial	1.000	Ea.	23	14.10	37.10
U Valve, water tempering, bronze, sweat connections, 3/4" diam	1.000	Ea.	59.50	22.50	82
W-2, V Tank water storage w/heating element, drain, relief valve, existing	1.000	Ea.			
Copper tubing type L, solder joint, hanger 10' OC 3/4" diam	20.000	L.F.	39	119	158
Copper tubing, type M, solder joint, hanger 10' OC 3/4" diam	70.000	L.F.	114.10	406	520.10
Sensor wire, #22-2 conductor multistranded	.500	C.L.F.	4.70	22.50	27.20
Solar energy heat transfer fluid, propylene glycol anti-freeze	6.000	Gal.	58.20	96.60	154.80
Wrought copper fittings & solder, 3/4" diam	76.000	Ea.	86.64	1,824	1,910.64
Total			2,825.44	3,716.05	6,541.49

D3010 715	Solar, Closed Loop, Add-On Hot Water Systems		COST EACH		
			MAT.	INST.	TOTAL
2550	Solar, closed loop, add-on hot water system, external heat exchanger				
2570	3/8" tubing, 3 ea. 4' x 4'-4" vacuum tube collectors		4,350	3,450	7,800
2580	1/2" tubing, 4 ea 4 x 4'-4" vacuum tube collectors	R13600	5,200	3,725	8,925
2600	2 ea 3'x7' black chrome collectors	-610	2,675	3,525	6,200
2620	3 ea 3'x7' black chrome collectors		3,275	3,625	6,900
2640	2 ea 3'x7' flat black collectors		2,425	3,525	5,950
2660	3 ea 3'x7' flat black collectors		2,900	3,625	6,525
2700	3/4" tubing, 3 ea 3'x7' black chrome collectors		3,425	3,825	7,250
2720	3 ea 3'x7' flat black absorber plate collectors		3,050	3,825	6,875
2740	2 ea 4'x9' flat black w/plastic glazing collectors		2,700	3,850	6,550
2760	2 ea 3'x7' black chrome collectors		2,825	3,725	6,550
2780	1" tubing, 4 ea 2'x9' plastic absorber & glazing collectors		4,075	4,275	8,350
2800	4 ea 3'x7' black chrome absorber collectors		4,300	4,300	8,600
2820	4 ea 3'x7' flat black absorber collectors		3,800	4,325	8,125

SERVICES

D

SERVICES D

In the drainback indirect-collection system, the heat transfer fluid is distilled water contained in a loop consisting of collectors, supply and return piping, and an unpressurized holding tank. A large heat exchanger containing incoming potable water is immersed in the holding tank. When a controller activates solar collection, the distilled water is pumped through the

collectors and heated and pumped back down to the holding tank. When the temperature differential between the water in the collectors and water in storage is such that collection no longer occurs, the pump turns off and gravity causes the distilled water in the collector loop to drain back to the holding tank. All the loop piping is pitched so that the water can drain out

of the collectors and piping and not freeze there. As hot water is needed in the home, incoming water first flows through the holding tank with the immersed heat exchanger and is warmed and then flows through a conventional heater for any supplemental heating that is necessary.

D3010 Energy Supply

System Components	QUANTITY	UNIT	COST EACH MAT.	INST.	TOTAL
SYSTEM D3010 720 2760					
SOLAR, DRAINBACK, ADD ON, HOT WATER, IMMERSED HEAT EXCHANGER					
3/4" TUBING, THREE EA 3'X7' BLACK CHROME COLLECTOR					
A, B Differential controller 2 sensors, thermostat, solar energy system	1.000	Ea.	85.50	37.50	123
C Thermometer 2" dial	3.000	Ea.	63	84	147
D, T Fill & drain valve, brass, 3/4" connection	1.000	Ea.	5.20	18.80	24
E-1 Automatic air vent 1/8" fitting	1.000	Ea.	9.70	14.10	23.80
H Strainer, Y type, bronze body, 3/4" IPS	1.000	Ea.	19.15	24	43.15
I Valve, gate, bronze, NRS, soldered 3/4" diam	2.000	Ea.	40	45	85
J Neoprene vent flashing	2.000	Ea.	19.60	45	64.60
L Circulator, solar heated liquid, 1/20 HP	1.000	Ea.	161	67.50	228.50
N Relief valve temp. & press. 150 psi 210°F self-closing 3/4" IPS	1.000	Ea.	8.65	15.05	23.70
O Pipe covering, urethane, ultraviolet cover, 1" wall, 3/4" diam	20.000	L.F.	33.40	86.60	120
P Pipe covering, fiberglass, all service jacket, 1" wall, 3/4" diam	50.000	L.F.	82.50	173	255.50
Q Collector panel solar energy blk chrome on copper, 1/8" temp glas 3'x7'	3.000	Ea.	1,800	256.50	2,056.50
Roof clamps for solar energy collector panels	3.000	Set	5.31	34.80	40.11
R Valve, swing check, bronze, regrinding disc, 3/4" diam	1.000	Ea.	35.50	22.50	58
U Valve, water tempering, bronze sweat connections, 3/4" diam	1.000	Ea.	59.50	22.50	82
W Tank, water storage immersed heat exchr elec 2"x1/2# insul 120 gal	1.000	Ea.	925	320	1,245
V Tank, water storage w/heating element, drain, relief valve, existing	1.000	Ea.			
X Valve, globe, bronze, rising stem, 3/4" diam, soldered	3.000	Ea.	153	67.50	220.50
Y Flow control valve	1.000	Ea.	54.50	20.50	75
Z Valve, ball, bronze, solder 3/4" diam, solar loop flow control	1.000	Ea.	12.20	22.50	34.70
Copper tubing, type L, solder joint, hanger 10' OC 3/4" diam	20.000	L.F.	39	119	158
Copper tubing, type M, solder joint, hanger 10' OC 3/4" diam	70.000	L.F.	114.10	406	520.10
Sensor wire, #22-2 conductor, multistranded	.500	C.L.F.	4.70	22.50	27.20
Wrought copper fittings & solder, 3/4" diam	76.000	Ea.	86.64	1,824	1,910.64
Total			3,817.15	3,748.85	7,566

D3010 720	Solar, Drainback, Hot Water Systems		COST EACH MAT.	INST.	TOTAL
2550	Solar, drainback, hot water, immersed heat exchanger	R13600-610			
2560	3/8" tubing, 3 ea. 4' x 4'-4" vacuum tube collectors		4,600	3,350	7,950
2580	1/2" tubing, 4 ea 4' x 4'-4" vacuum tube collectors, 80 gal tank		5,450	3,625	9,075
2600	120 gal tank		5,575	3,675	9,250
2640	2 ea. 3'x7' blk chrome collectors, 80 gal tank		2,925	3,425	6,350
2660	3 ea. 3'x7' blk chrome collectors, 120 gal tank		3,650	3,550	7,200
2700	2 ea. 3'x7' flat blk collectors, 120 gal tank		2,800	3,475	6,275
2720	3 ea. 3'x7' flat blk collectors, 120 gal tank		3,275	3,575	6,850
2760	3/4" tubing, 3 ea 3'x7' black chrome collectors, 120 gal tank		3,825	3,750	7,575
2780	3 ea. 3'x7' flat black absorber collectors, 120 gal tank		3,450	3,775	7,225
2800	2 ea. 4'x9' flat blk w/plastic glazing collectors 120 gal tank		3,100	3,775	6,875
2840	1" tubing, 4 ea. 2'x9' plastic absorber & glazing collectors, 120 gal tank		4,500	4,225	8,725
2860	4 ea. 3'x7' black chrome absorber collectors, 120 gal tank		4,725	4,225	8,950
2880	4 ea. 3'x7' flat black absorber collectors, 120 gal tank		4,225	4,250	8,475

SERVICES

SERVICES D

In the draindown direct-collection system, incoming domestic water is heated in the collectors. When the controller activates solar collection, domestic water is first heated as it flows through the collectors and is then pumped to storage. When conditions are no longer suitable for heat collection, the pump shuts off and the water in the loop drains down and out of the system by means of solenoid valves and properly pitched piping.

Important: See the Reference Section for critical supporting data - Reference Numbers and City Cost Indexes

D3010 Energy Supply

System Components	QUANTITY	UNIT	COST EACH		
			MAT.	INST.	TOTAL
SYSTEM D3010 730 2760					
SOLAR, DRAINDOWN, HOT WATER, DIRECT COLLECTION					
3/4″ TUBING, THREE 3′X7′ BLACK CHROME COLLECTORS					
A, B Differential controller, 2 sensors, thermostat, solar energy system	1.000	Ea.	85.50	37.50	123
A-1 Solenoid valve, solar heating loop, brass, 3/4″ diam, 24 volts	3.000	Ea.	699	150	849
B-1 Solar energy sensor, freeze prevention	1.000	Ea.	22	14.10	36.10
C Thermometer, 2″ dial	3.000	Ea.	63	84	147
E-1 Vacuum relief valve, 3/4″ diam	1.000	Ea.	27.50	14.10	41.60
F-1 Air vent, automatic, 1/8″ fitting	1.000	Ea.	9.70	14.10	23.80
H Strainer, Y type, bronze body, 3/4″ IPS	1.000	Ea.	19.15	24	43.15
I Valve, gate, bronze, NRS, soldered, 3/4″ diam	2.000	Ea.	40	45	85
J Vent flashing neoprene	2.000	Ea.	19.60	45	64.60
K Circulator, solar heated liquid, 1/25 HP	1.000	Ea.	153	58	211
N Relief valve temp & press 150 psi 210°F self-closing 3/4″ IPS	1.000	Ea.	8.65	15.05	23.70
O Pipe covering, urethane, ultraviolet cover, 1″ wall, 3/4″ diam	20.000	L.F.	33.40	86.60	120
P Pipe covering, fiberglass, all service jacket, 1″ wall, 3/4″ diam	50.000	L.F.	82.50	173	255.50
Roof clamps for solar energy collector panels	3.000	Set	5.31	34.80	40.11
Q Collector panel solar energy blk chrome on copper, 1/8″ temp glass 3′x7′	3.000	Ea.	1,800	256.50	2,056.50
R Valve, swing check, bronze, regrinding disc, 3/4″ diam, soldered	2.000	Ea.	71	45	116
T Drain valve, brass, 3/4″ connection	2.000	Ea.	10.40	37.60	48
U Valve, water tempering, bronze, sweat connections, 3/4″ diam	1.000	Ea.	59.50	22.50	82
W-2, W Tank, water storage elec elem 2″x1/2# insul 120 gal	1.000	Ea.	925	320	1,245
X Valve, globe, bronze, rising stem, 3/4″ diam, soldered	1.000	Ea.	51	22.50	73.50
Copper tubing, type L, solder joints, hangers 10′ OC 3/4″ diam	20.000	L.F.	39	119	158
Copper tubing, type M, solder joints, hangers 10′ OC 3/4″ diam	70.000	L.F.	114.10	406	520.10
Sensor wire, #22-2 conductor, multistranded	.500	C.L.F.	4.70	22.50	27.20
Wrought copper fittings & solder, 3/4″ diam	76.000	Ea.	86.64	1,824	1,910.64
Total			4,429.65	3,870.85	8,300.50

D3010 730	Solar, Draindown, Hot Water Systems		COST EACH		
			MAT.	INST.	TOTAL
2550	Solar, draindown, hot water				
2560	3/8″ tubing, 3 ea. 4′ x 4′-4″ vacuum tube collectors, 80 gal tank		5,150	3,650	8,800
2580	1/2″ tubing, 4 ea. 4′ x 4′-4″ vacuum tube collectors, 80 gal tank	R13600 -610	6,100	3,775	9,875
2600	120 gal tank		6,200	3,800	10,000
2640	2 ea 3′x7′ black chrome collectors, 80 gal tank		3,575	3,550	7,125
2660	3 ea 3′x7′ black chrome collectors, 120 gal tank		4,275	3,675	7,950
2700	2 ea 3′x7′ flat black collectors, 120 gal tank		3,425	3,600	7,025
2720	3 ea 3′x7′ flat black collectors, 120 gal tank		3,875	3,675	7,550
2760	3/4″ tubing, 3 ea 3′x7′ black chrome collectors, 120 gal tank		4,425	3,875	8,300
2780	3 ea 3′x7′ flat collectors, 120 gal tank		4,050	3,875	7,925
2800	2 ea. 4′x9′ flat black & plastic glazing collectors, 120 gal tank		3,700	3,900	7,600
2840	1″ tubing, 4 ea. 2′x9′ plastic absorber & glazing collectors, 120 gal tank		5,075	4,325	9,400
2860	4 ea 3′x7′ black chrome absorber collectors, 120 gal tank		5,300	4,350	9,650
2880	4 ea 3′x7′ flat black absorber collectors, 120 gal tank		4,800	4,375	9,175

SERVICES

D

423

SERVICES D

In the recirculation system, (a direct-collection system), incoming domestic water is heated in the collectors. When the controller activates solar collection, domestic water is heated as it flows through the collectors and then it flows back to storage. When conditions are not suitable for heat collection, the pump shuts off and the flow of the water stops. In this type of system, water remains in the collector loop at all times. A "frost sensor" at the collector activates the circulation of warm water from storage through the collectors when protection from freezing is required.

Important: See the Reference Section for critical supporting data - Reference Numbers and City Cost Indexes

D30 HVAC

D3010 Energy Supply

System Components	QUANTITY	UNIT	COST EACH		
			MAT.	INST.	TOTAL
SYSTEM D3010 735 2820					
SOLAR, RECIRCULATION, HOT WATER					
3/4" TUBING, TWO 3'X7' BLACK CHROME COLLECTORS					
A, B Differential controller 2 sensors, thermostat, for solar energy system	1.000	Ea.	85.50	37.50	123
A-1 Solenoid valve, solar heating loop, brass, 3/4" IPS, 24 volts	2.000	Ea.	466	100	566
B-1 Solar energy sensor freeze prevention	1.000	Ea.	22	14.10	36.10
C Thermometer, 2" dial	3.000	Ea.	63	84	147
D Drain valve, brass, 3/4" connection	1.000	Ea.	5.20	18.80	24
F-1 Air vent, automatic, 1/8" fitting	1.000	Ea.	9.70	14.10	23.80
H Strainer, Y type, bronze body, 3/4" IPS	1.000	Ea.	19.15	24	43.15
I Valve, gate, bronze, 125 lb, soldered 3/4" diam	3.000	Ea.	60	67.50	127.50
J Vent flashing, neoprene	2.000	Ea.	19.60	45	64.60
L Circulator, solar heated liquid, 1/20 HP	1.000	Ea.	161	67.50	228.50
N Relief valve, temp & press 150 psi 210°F self-closing 3/4" IPS	2.000	Ea.	17.30	30.10	47.40
O Pipe covering, urethane, ultraviolet cover, 1" wall, 3/4" diam	20.000	L.F.	33.40	86.60	120
P Pipe covering, fiberglass, all service jacket, 1" wall 3/4" diam	50.000	L.F.	82.50	173	255.50
Q Collector panel solar energy blk chrome on copper, 1/8" temp glass 3'x7'	2.000	Ea.	1,200	171	1,371
Roof clamps for solar energy collector panels	2.000	Set	3.54	23.20	26.74
R Valve, swing check, bronze, 125 lb, regrinding disc, soldered 3/4" diam	2.000	Ea.	71	45	116
U Valve, water tempering, bronze, sweat connections, 3/4" diam	1.000	Ea.	59.50	22.50	82
V Tank, water storage, w/heating element, drain, relief valve, existing	1.000	Ea.			
X Valve, globe, bronze, 125 lb, 3/4" diam	2.000	Ea.	102	45	147
Copper tubing, type L, solder joints, hangers 10' OC 3/4" diam	20.000	L.F.	39	119	158
Copper tubing, type M, solder joints, hangers 10' OC 3/4" diam	70.000	L.F.	114.10	406	520.10
Wrought copper fittings & solder, 3/4" diam	76.000	Ea.	86.64	1,824	1,910.64
Sensor wire, #22-2 conductor, multistranded	.500	C.L.F.	4.70	22.50	27.20
Total			2,724.83	3,440.40	6,165.23

D3010 735	Solar Recirculation, Domestic Hot Water Systems		COST EACH		
			MAT.	INST.	TOTAL
2550	Solar recirculation, hot water				
2560	3/8" tubing, 3 ea. 4' x 4'-4" vacuum tube collectors		4,225	3,200	7,425
2580	1/2" tubing, 4 ea. 4' x 4'-4" vacuum tube collectors	R13600 -610	5,075	3,450	8,525
2640	2 ea. 3'x7' black chrome collectors		2,550	3,250	5,800
2660	3 ea. 3'x7' black chrome collectors		3,150	3,350	6,500
2700	2 ea. 3'x7' flat black collectors		2,300	3,250	5,550
2720	3 ea. 3'x7' flat black collectors		2,775	3,350	6,125
2760	3/4" tubing, 3 ea. 3'x7' black chrome collectors		3,325	3,550	6,875
2780	3 ea. 3'x7' flat black absorber plate collectors		2,950	3,550	6,500
2800	2 ea. 4'x9' flat black w/plastic glazing collectors		2,600	3,575	6,175
2820	2 ea. 3'x7' black chrome collectors		2,725	3,450	6,175
2840	1" tubing, 4 ea. 2'x9' black plastic absorber & glazing collectors		4,000	4,000	8,000
2860	4 ea. 3'x7' black chrome absorber collectors		4,225	4,025	8,250
2880	4 ea. 3'x7' flat black absorber collectors		3,725	4,050	7,775

SERVICES

D

SERVICES D

The thermosyphon domestic hot water system, a direct collection system, operates under city water pressure and does not require pumps for system operation. An insulated water storage tank is located above the collectors. As the sun heats the collectors, warm water in them rises by means of natural convection; the colder water in the storage tank flows into the collectors by means of gravity. As long as the sun is shining the water continues to flow through the collectors and to become warmer.

To prevent freezing, the system must be drained or the collectors covered with an insulated lid when the temperature drops below 32°F.

D3010 Energy Supply

System Components	QUANTITY	UNIT	COST EACH		
			MAT.	INST.	TOTAL
SYSTEM D3010 745 0960					
SOLAR, THERMOSYPHON, WATER HEATER					
3/4" TUBING, TWO 3'X7' BLACK CHROME COLLECTORS					
D-1 Framing lumber, fir, 2" x 6" x 8', tank cradle	.008	M.B.F.	4.60	5.04	9.64
F-1 Framing lumber, fir, 2" x 4" x 24', sleepers	.016	M.B.F.	8.64	15.20	23.84
I Valve, gate, bronze, 125 lb, soldered 1/2" diam	2.000	Ea.	35.20	37.60	72.80
J Vent flashing, neoprene	4.000	Ea.	39.20	90	129.20
O Pipe covering, urethane, ultraviolet cover, 1" wall, 1/2" diam	40.000	L.F.	51.20	170	221.20
P Pipe covering fiberglass all service jacket 1" wall 1/2" diam	160.000	L.F.	230.40	529.60	760
Q Collector panel solar, blk chrome on copper, 3/16" temp glass 3'-8" x 6'	2.000	Ea.	950	181	1,131
Y Flow control valve, globe, bronze, 125#, soldered, 1/2" diam	1.000	Ea.	37.50	18.80	56.30
U Valve, water tempering, bronze, sweat connections, 1/2" diam	1.000	Ea.	48.50	18.80	67.30
W Tank, water storage, solar energy system, 80 Gal, 2" x 1/2 lb insul	1.000	Ea.	820	282	1,102
Copper tubing type L, solder joints, hangers 10' OC 1/2" diam	150.000	L.F.	222	832.50	1,054.50
Copper tubing type M, solder joints, hangers 10' OC 1/2" diam	50.000	L.F.	60	267.50	327.50
Sensor wire, #22-2 gauge multistranded	.500	C.L.F.	4.70	22.50	27.20
Wrought copper fittings & solder, 1/2" diam	75.000	Ea.	39.75	1,687.50	1,727.25
TotaL			2,551.69	4,158.04	6,709.73

D3010 745	Thermosyphon, Hot Water		MAT.	INST.	TOTAL
0960 0970	Solar, thermosyphon, hot water, two collector system	R13600 -610	2,550	4,150	6,700

SERVICES

D

SERVICES D

In this closed-loop indirect collection system, fluid with a low freezing temperature, propylene glycol, transports heat from the collectors to water storage. The transfer fluid is contained in a closed loop consisting of collectors, supply and return piping, and a heat exchanger immersed in the storage tank.

When the collectors become approximately 20° F warmer than the storage temperature, the controller activates the circulator. The circulator moves the fluid continuously until the temperature difference between fluid in the collectors and storage is such that the collection will no longer occur and then the circulator turns off. Since the heat transfer fluid has a very low freezing temperature, there is no need for it to be drained from the collectors between periods of collection.

D3010 Energy Supply

System Components	QUANTITY	UNIT	COST EACH		
			MAT.	INST.	TOTAL
SYSTEM D3010 750 2750					
SOLAR, CLOSED LOOP, SPACE/HOT WATER					
1″ TUBING, TEN 3′X7′ BLK CHROME ON COPPER ABSORBER COLLECTORS					
A, B Differential controller 2 sensors, thermostat, solar energy system	2.000	Ea.	171	75	246
C Thermometer, 2″ dial	10.000	Ea.	210	280	490
C-1 Heat exchanger, solar energy system, fluid to air, up flow, 80 MBH	1.000	Ea.	460	271	731
D, T Fill & drain valves, brass, 3/4″ connection	5.000	Ea.	26	94	120
D-1 Fan center	1.000	Ea.	119	100	219
E Air vent, manual, 1/8″ fitting	1.000	Ea.	2.13	14.10	16.23
E-2 Thermostat, 2 stage for sensing room temperature	1.000	Ea.	122	57	179
F Air purger	2.000	Ea.	90	75	165
F-1 Controller, liquid temperature, solar energy system	1.000	Ea.	101	90	191
G Expansion tank	2.000	Ea.	128	28.20	156.20
I Valve, gate, bronze, 125 lb, soldered, 1″ diam	5.000	Ea.	142.50	120	262.50
J Vent flashing, neoprene	2.000	Ea.	19.60	45	64.60
K Circulator, solar heated liquid, 1/25 HP	2.000	Ea.	306	116	422
L Circulator, solar heated liquid, 1/20 HP	1.000	Ea.	161	67.50	228.50
N Relief valve, temp & pressure 150 psi 210°F self-closing	4.000	Ea.	34.60	60.20	94.80
N-1 Relief valve, pressure poppet, bronze, 30 psi, 3/4″ IPS	3.000	Ea.	148.50	48.30	196.80
O Pipe covering, urethane, ultraviolet cover, 1″ wall, 1″ diam	50.000	L.F.	83.50	216.50	300
P Pipe covering, fiberglass, all service jacket, 1″ wall, 1″ diam	60.000	L.F.	99	207.60	306.60
Q Collector panel solar energy blk chrome on copper 1/8″ temp glass 3′x7′	10.000	Ea.	6,000	855	6,855
Roof clamps for solar energy collector panel	10.000	Set	17.70	116	133.70
R Valve, swing check, bronze, 125 lb, regrinding disc, 3/4″ & 1″ diam	4.000	Ea.	214	96	310
S Pressure gage, 0-60 psi, for solar energy system	2.000	Ea.	46	28.20	74.20
U Valve, water tempering, bronze, sweat connections, 3/4″ diam	1.000	Ea.	59.50	22.50	82
W-2, W-1, W Tank, water storage immersed heat xchr elec elem 2″x1/2# ins 120 gal	4.000	Ea.	3,700	1,280	4,980
X Valve, globe, bronze, 125 lb, rising stem, 1″ diam	3.000	Ea.	237	72	309
Y Valve, flow control	1.000	Ea.	54.50	20.50	75
Copper tubing type M, solder joint, hanger 10′ OC, 1″ diam	110.000	L.F.	233.20	709.50	942.70
Copper tubing, type L, solder joint, hanger 10′ OC 3/4″ diam	20.000	L.F.	39	119	158
Wrought copper fittings & solder, 3/4″ & 1″ diam	121.000	Ea.	337.59	3,388	3,725.59
Sensor, wire, #22-2 conductor, multistranded	.700	C.L.F.	6.58	31.50	38.08
Ductwork, galvanized steel, for heat exchanger	8.000	Lb.	7.12	43.60	50.72
Solar energy heat transfer fluid propylene glycol, anti-freeze	25.000	Gal.	242.50	402.50	645
TOTAL			13,618.52	9,149.70	22,768.22

D3010 750	Solar, Closed Loop, Space/Hot Water Systems		COST EACH		
			MAT.	INST.	TOTAL
2540	Solar, closed loop, space/hot water				
2550	1/2″ tubing, 12 ea. 4′x4′4″ vacuum tube collectors		18,000	8,550	26,550
2600	3/4″ tubing, 12 ea. 4′x4′4″ vacuum tube collectors		18,400	8,850	27,250
2650	10 ea. 3′ x 7′ black chrome absorber collectors	R13600 -610	13,300	8,600	21,900
2700	10 ea. 3′ x 7′ flat black absorber collectors		12,000	8,650	20,650
2750	1″ tubing, 10 ea. 3′ x 7′ black chrome absorber collectors		13,600	9,150	22,750
2800	10 ea. 3′ x 7′ flat black absorber collectors		12,400	9,225	21,625
2850	6 ea. 4′ x 9′ flat black w/plastic glazing collectors		10,900	9,150	20,050
2900	12 ea. 2′ x 9′ plastic absorber and glazing collectors		14,200	9,300	23,500

This draindown pool system uses a differential thermostat similar to those used in solar domestic hot water and space heating applications. To heat the pool, the pool water passes through the conventional pump-filter loop and then flows through the collectors. When collection is not possible, or when the pool temperature is reached, all water drains from the solar loop back to the pool through the existing piping. The modes are controlled by solenoid valves or other automatic valves in conjunction with a vacuum breaker relief valve, which facilitates draindown.

SERVICES D

D3010 Energy Supply

System Components	QUANTITY	UNIT	COST EACH		
			MAT.	INST.	TOTAL
SYSTEM D3010 760 2640					
SOLAR SWIMMING POOL HEATER, ROOF MOUNTED COLLECTORS					
TEN 4' X 10' FULLY WETTED UNGLAZED PLASTIC ABSORBERS					
A Differential thermostat/controller, 110V, adj pool pump system	1.000	Ea.	282	226	508
A-1 Solenoid valve, PVC, normally 1 open 1 closed (included)	2.000	Ea.			
B Sensor, thermistor type (included)	2.000	Ea.			
E-1 Valve, vacuum relief	1.000	Ea.	27.50	14.10	41.60
Q Collector panel, solar energy, plastic, liquid full wetted, 4' x 10'	10.000	Ea.	2,150	1,630	3,780
R Valve, ball check, PVC, socket, 1-1/2" diam	1.000	Ea.	75	22.50	97.50
Z Valve, ball, PVC, socket, 1-1/2" diam	3.000	Ea.	262.50	67.50	330
Pipe, PVC, sch 40, 1-1/2" diam	80.000	L.F.	241.60	1,004	1,245.60
Pipe fittings, PVC sch 40, socket joint, 1-1/2" diam	10.000	Ea.	11	225	236
Sensor wire, #22-2 conductor, multistranded	.500	C.L.F.	4.70	22.50	27.20
Roof clamps for solar energy collector panels	10.000	Set	17.70	116	133.70
Roof strap, teflon for solar energy collector panels	26.000	L.F.	271.70	57.20	328.90
TOTAL			3,343.70	3,384.80	6,728.50

D3010 760	Solar Swimming Pool Heater Systems		COST EACH		
			MAT.	INST.	TOTAL
2530	Solar swimming pool heater systems, roof mounted collectors				
2540	10 ea. 3'x7' black chrome absorber, 1/8" temp. glass		7,200	2,600	9,800
2560	10 ea. 4'x8' black chrome absorber, 3/16" temp. glass	R13600 -610	8,300	3,100	11,400
2580	10 ea. 3'8"x6' flat black absorber, 3/16" temp. glass		5,950	2,650	8,600
2600	10 ea. 4'x9' flat black absorber, plastic glazing		6,600	3,225	9,825
2620	10 ea. 2'x9' rubber absorber, plastic glazing		12,100	3,500	15,600
2640	10 ea. 4'x10' fully wetted unglazed plastic absorber		3,350	3,375	6,725
2660	Ground mounted collectors				
2680	10 ea. 3'x7' black chrome absorber, 1/8" temp. glass		7,350	3,000	10,350
2700	10 ea. 4'x8' black chrome absorber, 3/16" temp glass		8,450	3,500	11,950
2720	10 ea. 3'8"x6' flat blk absorber, 3/16" temp. glass		6,100	3,050	9,150
2740	10 ea. 4'x9' flat blk absorber, plastic glazing		6,750	3,625	10,375
2760	10 ea. 2'x9' rubber absorber, plastic glazing		12,300	3,775	16,075
2780	10 ea. 4'x10' fully wetted unglazed plastic absorber		3,500	3,775	7,275

D

SERVICES

SERVICES D

ATTIC FLOOR

CWS

HWS

This domestic hot water pre-heat system
includes heat exchanger with a
circulating pump, blower, air-to-water,
coil and controls, mounted in the upper
collector manifold. Heat from the hot air
coming out of the collectors is transferred
through the heat exchanger. For each
degree of DHW preheating gained, one
degree less heating is needed from the
fuel fired water heater. The system is
simple, inexpensive to operate and can
provide a substantial portion of DHW
requirements for modest additional cost.

D3010 Energy Supply

System Components	QUANTITY	UNIT	COST EACH MAT.	COST EACH INST.	COST EACH TOTAL
SYSTEM D3010 765 2560					
SOLAR, HOT WATER, AIR TO WATER HEAT EXCHANGE					
THREE COLLECTORS, OPTICAL BLACK ON ALUMINUM, 10'X2', 80GAL TANK					
A, B Differential controller, 2 sensors, thermostat, solar energy system	1.000	Ea.	85.50	37.50	123
C Thermometer, 2" dial	2.000	Ea.	42	56	98
C-1 Heat exchanger, air to fluid, up flow 70 MBH	1.000	Ea.	345	232	577
E Air vent, manual, for solar energy system 1/8" fitting	1.000	Ea.	2.13	14.10	16.23
F Air purger	1.000	Ea.	45	37.50	82.50
G Expansion tank	1.000	Ea.	64	14.10	78.10
H Strainer, Y type, bronze body, 1/2" IPS	1.000	Ea.	17.50	22.50	40
I Valve, gate, bronze, 125 lb, NRS, soldered 1/2" diam	5.000	Ea.	88	94	182
N Relief valve, temp & pressure solar 150 psi 210°F self-closing	1.000	Ea.	8.65	15.05	23.70
P Pipe covering, fiberglass, all service jacket, 1" wall, 1/2" diam	60.000	L.F.	86.40	198.60	285
Q Collector panel solar energy, air, black on alum plate, flush mount, 10'x2'	30.000	L.F.	4,830	256.50	5,086.50
B-1, R-1 Shutter damper wfor solar heater circulator	1.000	Ea.	49.50	68.50	118
B-1, R-1 Shutter motor for solar heater circulator	1.000	Ea.	107	51.50	158.50
R Backflow preventer for solar energy system, 1/2" diam	2.000	Ea.	128	56	184
T Drain valve, brass, 3/4" connection	1.000	Ea.	5.20	18.80	24
U Valve, water tempering, bronze, sweat connections, 1/2" diam	1.000	Ea.	48.50	18.80	67.30
W Tank, water storage, solar energy system, 80 Gal, 2" x 2 lb insul	1.000	Ea.	820	282	1,102
V Tank, water storage, w/heating element, drain, relief valve, existing	1.000	System			
X Valve, globe, bronze, 125 lb, rising stem, 1/2" diam	1.000	Ea.	37.50	18.80	56.30
Copper tubing type M, solder joints, hangers 10' OC 1/2" diam	50.000	L.F.	60	267.50	327.50
Copper tubing type L, solder joints, hangers 10' OC 1/2" diam	10.000	L.F.	14.80	55.50	70.30
Wrought copper fittings & solder, 1/2" diam	10.000	Ea.	5.30	225	230.30
Sensor wire, #22-2 conductor multistranded	.500	C.L.F.	4.70	22.50	27.20
Q-1, Q-2 Ductwork, fiberglass, aluminzed jacket, 1-1/2" thick, 8" diam	32.000	S.F.	148.16	146.24	294.40
Q-3 Manifold for flush mount solar energy collector panels, air	6.000	L.F.	177	30.90	207.90
TOTAL			7,219.84	2,239.89	9,459.73

D3010 765	Solar Hot Water, Air To Water Heat Exchange		COST EACH MAT.	COST EACH INST.	COST EACH TOTAL
2550	Solar hot water, air to water heat exchange				
2560	Three collectors, optical black on aluminum, 10' x 2', 80 Gal tank		7,225	2,250	9,475
2580	Four collectors, optical black on aluminum, 10' x 2', 80 Gal tank	R13600 -610	8,925	2,375	11,300
2600	Four collectors, optical black on aluminum, 10' x 2', 120 Gal tank		9,025	2,400	11,425

SERVICES

D

SERVICES D

The complete Solar Air Heating System provides maximum savings of conventional fuel with both space heating and year-round domestic hot water heating. It allows for the home air conditioning to operate simultaneously and independently from the solar domestic water heating in summer. The system's modes of operation are:

Mode 1: The building is heated directly from the collectors with air circulated by the Solar Air Mover.

Mode 2: When heat is not needed in the building, the dampers change within the air mover to circulate the air from the collectors to the rock storage bin.

Mode 3: When heat is not available from the collector array and is available in rock storage, the air mover draws heated air from rock storage and directs it into the building. When heat is not available from the collectors or the rock storage bin, the auxiliary heating unit will provide heat for the building. The size of the collector array is typically 25% the size of the main floor area.

Important: See the Reference Section for critical supporting data - Reference Numbers and City Cost Indexes

D30 HVAC

D3010 Energy Supply

System Components	QUANTITY	UNIT	COST EACH MAT.	COST EACH INST.	COST EACH TOTAL
SYSTEM D3010 775 1210					
SOLAR, SPACE/HOT WATER, AIR TO WATER HEAT EXCHANGE					
A, B Differential controller 2 sensors thermos., solar energy sys liquid loop	1.000	Ea.	85.50	37.50	123
A-1, B Differential controller 2 sensors 6 station solar energy sys air loop	1.000	Ea.	222	150	372
B-1 Solar energy sensor, freeze prevention	1.000	Ea.	22	14.10	36.10
C Thermometer for solar energy system, 2" dial	2.000	Ea.	42	56	98
C-1 Heat exchanger, solar energy system, air to fluid, up flow, 70 MBH	1.000	Ea.	345	232	577
D Drain valve, brass, 3/4" connection	2.000	Ea.	10.40	37.60	48
E Air vent, manual, for solar energy system 1/8" fitting	1.000	Ea.	2.13	14.10	16.23
E-1 Thermostat, 2 stage for sensing room temperature	1.000	Ea.	122	57	179
F Air purger	1.000	Ea.	45	37.50	82.50
G Expansion tank, for solar energy system	1.000	Ea.	64	14.10	78.10
I Valve, gate, bronze, 125 lb, NRS, soldered 3/4" diam	2.000	Ea.	40	45	85
K Circulator, solar heated liquid, 1/25 HP	1.000	Ea.	153	58	211
N Relief valve temp & press 150 psi 210°F self-closing, 3/4" IPS	1.000	Ea.	8.65	15.05	23.70
N-1 Relief valve, pressure, poppet, bronze, 30 psi, 3/4" IPS	1.000	Ea.	49.50	16.10	65.60
P Pipe covering, fiberglass, all service jacket, 1" wall, 3/4" diam	60.000	L.F.	99	207.60	306.60
Q-3 Manifold for flush mount solar energy collector panels	20.000	L.F.	590	103	693
Q Collector panel solar energy, air, black on alum. plate, flush mount 10'x2'	100.000	L.F.	16,100	855	16,955
R Valve, swing check, bronze, 125 lb, regrinding disc, 3/4" diam	2.000	Ea.	71	45	116
S Pressure gage, 2" dial, for solar energy system	1.000	Ea.	23	14.10	37.10
U Valve, water tempering, bronze, sweat connections, 3/4" diam	1.000	Ea.	59.50	22.50	82
W-2, W Tank, water storage, solar, elec element 2"x1/2# insul, 80 Gal	1.000	Ea.	820	282	1,102
X Valve, globe, bronze, 125 lb, soldered, 3/4" diam	1.000	Ea.	51	22.50	73.50
Copper tubing type L, solder joints, hangers 10' OC 3/4" diam	10.000	L.F.	19.50	59.50	79
Copper tubing type M, solder joints, hangers 10' OC 3/4" diam	60.000	L.F.	97.80	348	445.80
Wrought copper fittings & solder, 3/4" diam	26.000	Ea.	29.64	624	653.64
Sensor wire, #22-2 conductor multistranded	1.200	C.L.F.	11.28	54	65.28
Q-1 Duct work, rigid fiberglass, rectangular	400.000	S.F.	480	1,464	1,944
Ductwork, transition pieces	8.000	Ea.	118.40	276	394.40
R-2 Shutter/damper for solar heater circulator	9.000	Ea.	445.50	616.50	1,062
K-1 Shutter motor for solar heater circulator blower	9.000	Ea.	963	463.50	1,426.50
K-1 Fan, solar energy heated air circulator, space & DHW system	1.000	Ea.	1,550	1,650	3,200
Z-1 Tank, solar energy air storage, 6'-3"H 7'x7' = 306 CF/2000 Gal	1.000	Ea.	8,950	1,075	10,025
Z-2 Crushed stone 1-1/2"	11.000	C.Y.	188.65	57.42	246.07
C-2 Thermometer, remote probe, 2" dial	1.000	Ea.	32	56.50	88.50
R-1 Solenoid valve	1.000	Ea.	233	50	283
V, R-4, R-3 Furnace, supply diffusers, return grilles, existing	1.000	Ea.	8,950	1,075	10,025
Total			41,093.45	10,205.17	51,298.62

D3010 775	Air To Water Heat Exchange		MAT.	INST.	TOTAL
1210 1220	Solar, air to water heat exchange, for space / hot water heating	R13600 -610	41,100	10,200	51,300

435

In this closed-loop indirect collection system, fluid with a low freezing temperature, such as propylene glycol, transports heat from the collectors to water storage. The transfer fluid is contained in a closed-loop consisting of collectors, supply and return piping, and a heat exchanger immersed in the storage tank. A typical two-or-three panel system contains 5 to 6 gallons of heat transfer fluid.

When the collectors become approximately 20° F warmer than the storage temperature, a controller activates the circulator. The circulator moves the fluid continuously through the collectors until the temperature difference between the collectors and storage is such that heat collection no longer occurs; at that point, the circulator shuts off. Since the heat transfer fluid has a very low freezing temperature, there is no need for it to be drained from the collectors between periods of collection.

D3010 Energy Supply

System Components	QUANTITY	UNIT	COST EACH		
			MAT.	INST.	TOTAL
SYSTEM D3010 780 2760					
SOLAR, CLOSED LOOP, HOT WATER SYSTEM, IMMERSED HEAT EXCHANGER					
3/4" TUBING, THREE 3' X 7' BLACK CHROME COLLECTORS					
A, B Differential controller, 2 sensors, thermostat, solar energy system	1.000	Ea.	85.50	37.50	123
C Thermometer 2" dial	3.000	Ea.	63	84	147
D, T Fill & drain valves, brass, 3/4" connection	3.000	Ea.	15.60	56.40	72
E Air vent, manual, 1/8" fitting	1.000	Ea.	2.13	14.10	16.23
F Air purger	1.000	Ea.	45	37.50	82.50
G Expansion tank	1.000	Ea.	64	14.10	78.10
I Valve, gate, bronze, NRS, soldered 3/4" diam	3.000	Ea.	60	67.50	127.50
J Neoprene vent flashing	2.000	Ea.	19.60	45	64.60
K Circulator, solar heated liquid, 1/25 HP	1.000	Ea.	153	58	211
N-1, N Relief valve, temp & press 150 psi 210°F self-closing 3/4" IPS	2.000	Ea.	17.30	30.10	47.40
O Pipe covering, urethane ultraviolet cover, 1" wall, 3/4" diam	20.000	L.F.	33.40	86.60	120
P Pipe covering, fiberglass, all service jacket, 1" wall, 3/4" diam	50.000	L.F.	82.50	173	255.50
Roof clamps for solar energy collector panel	3.000	Set	5.31	34.80	40.11
Q Collector panel solar blk chrome on copper, 1/8" temp glass, 3'x7'	3.000	Ea.	1,800	256.50	2,056.50
R-1 Valve, swing check, bronze, regrinding disc, 3/4" diam, soldered	1.000	Ea.	35.50	22.50	58
S Pressure gauge, 60 psi, 2-1/2" dial	1.000	Ea.	23	14.10	37.10
U Valve, water tempering, bronze, sweat connections, 3/4" diam	1.000	Ea.	59.50	22.50	82
W-2, W Tank, water storage immersed heat exchr elec elem 2"x2# insul 120 Gal	1.000	Ea.	925	320	1,245
X Valve, globe, bronze, rising stem, 3/4" diam, soldered	1.000	Ea.	51	22.50	73.50
Copper tubing type L, solder joint, hanger 10' OC 3/4" diam	20.000	L.F.	39	119	158
Copper tubing, type M, solder joint, hanger 10' OC 3/4" diam	70.000	L.F.	114.10	406	520.10
Sensor wire, #22-2 conductor multistranded	.500	C.L.F.	4.70	22.50	27.20
Solar energy heat transfer fluid, propylene glycol, anti-freeze	6.000	Gal.	58.20	96.60	154.80
Wrought copper fittings & solder, 3/4" diam	76.000	Ea.	86.64	1,824	1,910.64
TOTAL			3,842.98	3,864.80	7,707.78

D3010 780	Solar, Closed Loop, Hot Water Systems		COST EACH		
			MAT.	INST.	TOTAL
2550	Solar, closed loop, hot water system, immersed heat exchanger				
2560	3/8" tubing, 3 ea. 4' x 4'-4" vacuum tube collectors, 80 gal. tank		4,650	3,500	8,150
2580	1/2" tubing, 4 ea. 4' x 4'-4" vacuum tube collectors, 80 gal. tank		5,525	3,750	9,275
2600	120 gal. tank		5,625	3,800	9,425
2640	2 ea. 3'x7' black chrome collectors, 80 gal. tank	R13600 -610	2,975	3,600	6,575
2660	120 gal. tank		3,100	3,575	6,675
2700	2 ea. 3'x7' flat black collectors, 120 gal. tank		2,850	3,600	6,450
2720	3 ea. 3'x7' flat black collectors, 120 gal. tank		3,325	3,700	7,025
2760	3/4" tubing, 3 ea. 3'x7' black chrome collectors, 120 gal. tank		3,850	3,875	7,725
2780	3 ea. 3'x7' flat black collectors, 120 gal. tank		3,475	3,875	7,350
2800	2 ea. 4'x9' flat black w/plastic glazing collectors 120 gal. tank		3,125	3,900	7,025
2840	1" tubing, 4 ea. 2'x9' plastic absorber & glazing collectors 120 gal. tank		4,475	4,300	8,775
2860	4 ea. 3'x7' black chrome collectors, 120 gal. tank		4,675	4,325	9,000
2880	4 ea. 3'x7' flat black absorber collectors, 120 gal. tank		4,175	4,350	8,525

SERVICES

D

Boiler

Baseboard Radiation

Small Electric Boiler
System Considerations:
1. Terminal units are fin tube baseboard radiation rated at 720 BTU/hr with 200° water temperature or 820 BTU/hr steam.
2. Primary use being for residential or smaller supplementary areas, the floor levels are based on 7-1/2' ceiling heights.
3. All distribution piping is copper for boilers through 205 MBH. All piping for larger systems is steel pipe.

System Components	QUANTITY	UNIT	COST EACH		
			MAT.	INST.	TOTAL
SYSTEM D3020 102 1120					
SMALL HEATING SYSTEM, HYDRONIC, ELECTRIC BOILER					
1,480 S.F., 61 MBH, STEAM, 1 FLOOR					
Boiler, electric steam, std cntrls, trim, ftngs and valves, 18 KW, 61.4 MBH	1.000	Ea.	3,245	1,155	4,400
Copper tubing type L, solder joint, hanger 10'OC, 1-1/4" diam	160.000	L.F.	528	1,248	1,776
Radiation, 3/4" copper tube w/alum fin baseboard pkg 7" high	60.000	L.F.	372	846	1,218
Rough in baseboard panel or fin tube with valves & traps	10.000	Set	1,170	4,300	5,470
Pipe covering, calcium silicate w/cover 1" wall 1-1/4" diam	160.000	L.F.	446.40	771.20	1,217.60
Low water cut-off, quick hookup, in gage glass tappings	1.000	Ea.	166	28.50	194.50
TOTAL			5,927.40	8,348.70	14,276.10
COST PER S.F.			4.01	5.64	9.65

D3020 102	Small Heating Systems, Hydronic, Electric Boilers		COST PER S.F.		
			MAT.	INST.	TOTAL
1100	Small heating systems, hydronic, electric boilers				
1120	Steam, 1 floor, 1480 S.F., 61 M.B.H.		4	5.64	9.64
1160	3,000 S.F., 123 M.B.H.	R15500 -010	3	4.95	7.95
1200	5,000 S.F., 205 M.B.H.		2.57	4.56	7.13
1240	2 floors, 12,400 S.F., 512 M.B.H.	R15500 -020	2.04	4.53	6.57
1280	3 floors, 24,800 S.F., 1023 M.B.H.		2.16	4.46	6.62
1320	34,750 S.F., 1,433 M.B.H.	R15500 -030	1.95	4.34	6.29
1360	Hot water, 1 floor, 1,000 S.F., 41 M.B.H.		6.30	3.15	9.45
1400	2,500 S.F., 103 M.B.H.		4.53	5.65	10.18
1440	2 floors, 4,850 S.F., 205 M.B.H.		4.48	6.80	11.28
1480	3 floors, 9,700 S.F., 410 M.B.H.		4.24	7	11.24

Important: See the Reference Section for critical supporting data - Reference Numbers and City Cost Indexes

D3020 Heat Generating Systems

Boiler

Unit Heater

Large Electric Boiler System Considerations:

1. Terminal units are all unit heaters of the same size. Quantities are varied to accommodate total requirements.
2. All air is circulated through the heaters a minimum of three times per hour.
3. As the capacities are adequate for commercial use, floor levels are based on 10' ceiling heights.
4. All distribution piping is black steel pipe.

System Components	QUANTITY	UNIT	COST EACH		
			MAT.	INST.	TOTAL
SYSTEM D3020 104 1240					
LARGE HEATING SYSTEM, HYDRONIC, ELECTRIC BOILER					
9,280 S.F., 150 KW, 510 MBH, 1 FLOOR					
Boiler, electric hot water, std ctrls, trim, ftngs, valves, 150 KW, 510 MBH	1.000	Ea.	11,400	2,550	13,950
Expansion tank, painted steel, 60 Gal capacity ASME	1.000	Ea.	2,375	136	2,511
Circulating pump, CI, close cpld, 50 GPM, 2 HP, 2" pipe conn	1.000	Ea.	1,225	271	1,496
Unit heater, 1 speed propeller, horizontal, 200° EWT, 72.7 MBH	7.000	Ea.	4,410	1,043	5,453
Unit heater piping hookup with controls	7.000	Set	2,555	6,755	9,310
Pipe, steel, black, schedule 40, welded, 2-1/2" diam	380.000	L.F.	2,565	7,273.20	9,838.20
Pipe covering, calcium silicate w/cover, 1" wall, 2-1/2" diam	380.000	L.F.	1,368	1,888.60	3,256.60
TOTAL			25,898	19,916.80	45,814.80
COST PER S.F.			2.79	2.15	4.94

D3020 104	Large Heating Systems, Hydronic, Electric Boilers		COST PER S.F.		
			MAT.	INST.	TOTAL
1230	Large heating systems, hydronic, electric boilers				
1240	9,280 S.F., 150 K.W., 510 M.B.H., 1 floor		2.79	2.14	4.93
1280	14,900 S.F., 240 K.W., 820 M.B.H., 2 floors	R15500 -010	3.04	3.52	6.56
1320	18,600 S.F., 296 K.W., 1,010 M.B.H., 3 floors		3.01	3.82	6.83
1360	26,100 S.F., 420 K.W., 1,432 M.B.H., 4 floors	R15500 -020	2.97	3.75	6.72
1400	39,100 S.F., 666 K.W., 2,273 M.B.H., 4 floors		2.56	3.12	5.68
1440	57,700 S.F., 900 K.W., 3,071 M.B.H., 5 floors	R15500 -030	2.43	3.07	5.50
1480	111,700 S.F., 1,800 K.W., 6,148 M.B.H., 6 floors		2.21	2.64	4.85
1520	149,000 S.F., 2,400 K.W., 8,191 M.B.H., 8 floors		2.20	2.63	4.83
1560	223,300 S.F., 3,600 K.W., 12,283 M.B.H., 14 floors		2.33	3	5.33

SERVICES

D

Fossil Fuel Boiler System Considerations:

1. Terminal units are horizontal unit heaters. Quantities are varied to accommodate total heat loss per building.
2. Unit heater selection was determined by their capacity to circulate the building volume a minimum of three times per hour in addition to the BTU output.

3. Systems shown are forced hot water. Steam boilers cost slightly more than hot water boilers. However, this is compensated for by the smaller size or fewer terminal units required with steam.
4. Floor levels are based on 10′ story heights.
5. MBH requirements are gross boiler output.

Unit Heater

System Components	QUANTITY	UNIT	COST EACH MAT.	COST EACH INST.	COST EACH TOTAL
SYSTEM D3020 108 1280					
HEATING SYSTEM, HYDRONIC, FOSSIL FUEL, TERMINAL UNIT HEATERS					
CAST IRON BOILER, GAS, 80 MBH, 1,070 S.F. BUILDING					
Boiler, gas, hot water, CI, burner, controls, insulation, breeching, 80 MBH	1.000	Ea.	1,207.50	1,233.75	2,441.25
Pipe, steel, black, schedule 40, threaded, cplg & hngr 10′OC, 2″ diam	200.000	L.F.	700	2,540	3,240
Unit heater, 1 speed propeller, horizontal, 200° EWT, 72.7 MBH	2.000	Ea.	1,260	298	1,558
Unit heater piping hookup with controls	2.000	Set	730	1,930	2,660
Expansion tank, painted steel, ASME, 18 Gal capacity	1.000	Ea.	1,525	68	1,593
Circulating pump, CI, flange connection, 1/12 HP	1.000	Ea.	229	135	364
Pipe covering, calcium silicate w/cover, 1″ wall, 2″ diam	200.000	L.F.	674	994	1,668
TOTAL			6,325.50	7,198.75	13,524.25
COST PER S.F.			5.91	6.73	12.64

D3020 108	Heating Systems, Unit Heaters		COST PER S.F. MAT.	COST PER S.F. INST.	COST PER S.F. TOTAL
1260	Heating systems, hydronic, fossil fuel, terminal unit heaters,				
1280	Cast iron boiler, gas, 80 M.B.H., 1,070 S.F. bldg.		5.91	6.73	12.64
1320	163 M.B.H., 2,140 S.F. bldg.	R15500 -010	3.90	4.56	8.46
1360	544 M.B.H., 7,250 S.F. bldg.		2.81	3.28	6.09
1400	1,088 M.B.H., 14,500 S.F. bldg.	R15500 -020	2.47	3.12	5.59
1440	3,264 M.B.H., 43,500 S.F. bldg.		2.03	2.35	4.38
1480	5,032 M.B.H., 67,100 S.F. bldg.	R15500 -030	2.40	2.42	4.82
1520	Oil, 109 M.B.H., 1,420 S.F. bldg.		6	6	12
1560	235 M.B.H., 3,150 S.F. bldg.		3.94	4.35	8.29
1600	940 M.B.H., 12,500 S.F. bldg.		2.96	2.85	5.81
1640	1,600 M.B.H., 21,300 S.F. bldg.		2.86	2.74	5.60
1680	2,480 M.B.H., 33,100 S.F. bldg.		2.86	2.47	5.33
1720	3,350 M.B.H., 44,500 S.F. bldg.		2.50	2.52	5.02
1760	Coal, 148 M.B.H., 1,975 S.F. bldg.		5.05	3.92	8.97
1800	300 M.B.H., 4,000 S.F. bldg.		3.89	3.06	6.95
1840	2,360 M.B.H., 31,500 S.F. bldg.		2.80	2.50	5.30
1880	Steel boiler, gas, 72 M.B.H., 1,020 S.F. bldg.		5.90	4.98	10.88
1920	240 M.B.H., 3,200 S.F. bldg.		4.20	3.98	8.18
1960	480 M.B.H., 6,400 S.F. bldg.		3.39	2.96	6.35
2000	800 M.B.H., 10,700 S.F. bldg.		2.93	2.64	5.57

D3020 Heat Generating Systems

D3020 108	Heating Systems, Unit Heaters	COST PER S.F.		
		MAT.	INST.	TOTAL
2040	1,960 M.B.H., 26,100 S.F. bldg.	2.58	2.37	4.95
2080	3,000 M.B.H., 40,000 S.F. bldg.	2.56	2.42	4.98
2120	Oil, 97 M.B.H., 1,300 S.F. bldg.	5.55	5.65	11.20
2160	315 M.B.H., 4,550 S.F. bldg.	3.44	2.89	6.33
2200	525 M.B.H., 7,000 S.F. bldg.	3.71	2.96	6.67
2240	1,050 M.B.H., 14,000 S.F. bldg.	3.15	2.97	6.12
2280	2,310 M.B.H. 30,800 S.F. bldg.	2.77	2.56	5.33
2320	3,150 M.B.H., 42,000 S.F. bldg.	2.57	2.60	5.17

SERVICES

D

441

Fin Tube Radiator

Fossil Fuel Boiler System Considerations:

1. Terminal units are commercial steel fin tube radiation. Quantities are varied to accommodate total heat loss per building.
2. Systems shown are forced hot water. Steam boilers cost slightly more than hot water boilers. However, this is compensated for by the smaller size or fewer terminal units required with steam.
3. Floor levels are based on 10' story heights.
4. MBH requirements are gross boiler output.

System Components	QUANTITY	UNIT	COST EACH		
			MAT.	INST.	TOTAL
SYSTEM D3020 110 3240					
HEATING SYSTEM, HYDRONIC, FOSSIL FUEL, FIN TUBE RADIATION					
CAST IRON BOILER, GAS, 80 MBH, 1,070 S.F. BLDG.					
Boiler, gas, hot water, CI, burner, controls, insulation, breeching, 80 MBH	1.000	Ea.	1,207.50	1,233.75	2,441.25
Pipe, steel, black, schedule 40, threaded, 2" diam	200.000	L.F.	700	2,540	3,240
Radiation, steel 1-1/4" tube & 4-1/4" fin w/cover & damper, wall hung	80.000	L.F.	2,160	1,800	3,960
Rough-in, wall hung steel fin radiation with supply & balance valves	8.000	Set	1,288	4,000	5,288
Expansion tank, painted steel, ASME, 18 Gal capacity	1.000	Ea.	1,525	68	1,593
Circulating pump, CI flanged, 1/12 HP	1.000	Ea.	229	135	364
Pipe covering, calcium silicate w/cover, 1" wall, 2" diam	200.000	L.F.	674	994	1,668
TOTAL			7,783.50	10,770.75	18,554.25
COST PER S.F.			7.27	10.07	17.34

D3020 110	Heating System, Fin Tube Radiation		COST PER S.F.		
			MAT.	INST.	TOTAL
3230	Heating systems, hydronic, fossil fuel, fin tube radiation				
3240	Cast iron boiler, gas, 80 MBH, 1,070 S.F. bldg.		7.27	10.09	17.36
3280	169 M.B.H., 2,140 S.F. bldg.	R15500 -010	4.51	6.40	10.91
3320	544 M.B.H., 7,250 S.F. bldg.		3.81	5.55	9.36
3360	1,088 M.B.H., 14,500 S.F. bldg.	R15500 -020	3.54	5.45	8.99
3400	3,264 M.B.H., 43,500 S.F. bldg.		3.19	4.75	7.94
3440	5,032 M.B.H., 67,100 S.F. bldg.	R15500 -030	3.57	4.83	8.40
3480	Oil, 109 M.B.H., 1,420 S.F. bldg.		8.45	11	19.45
3520	235 M.B.H., 3,150 S.F. bldg.		4.94	6.55	11.49
3560	940 M.B.H., 12,500 S.F. bldg.		4.05	5.20	9.25
3600	1,600 M.B.H., 21,300 S.F. bldg.		4.03	5.15	9.18
3640	2,480 M.B.H., 33,100 S.F. bldg.		4.04	4.88	8.92
3680	3,350 M.B.H., 44,500 S.F. bldg.		3.65	4.93	8.58
3720	Coal, 148 M.B.H., 1,975 S.F. bldg.		6	6.10	12.10
3760	300 M.B.H., 4,000 S.F. bldg.		4.84	5.20	10.04
3800	2,360 M.B.H., 31,500 S.F. bldg.		3.94	4.87	8.81
3840	Steel boiler, gas, 72 M.B.H., 1,020 S.F. bldg.		8.30	9.55	17.85
3880	240 M.B.H., 3,200 S.F. bldg.		5.25	6.30	11.55
3920	480 M.B.H., 6,400 S.F. bldg.		4.42	5.20	9.62
3960	800 M.B.H., 10,700 S.F. bldg.		4.34	5.55	9.89
4000	1,960 M.B.H., 26,100 S.F. bldg.		3.70	4.74	8.44
4040	3,000 M.B.H., 40,000 S.F. bldg.		3.67	4.77	8.44

SERVICES D

D3020 Heat Generating Systems

D3020 110	Heating System, Fin Tube Radiation	COST PER S.F.		
		MAT.	INST.	TOTAL
4080	Oil, 97 M.B.H., 1,300 S.F. bldg.	6.25	9.05	15.30
4120	315 M.B.H., 4,550 S.F. bldg.	4.37	4.94	9.31
4160	525 M.B.H., 7,000 S.F. bldg.	4.83	5.30	10.13
4200	1,050 M.B.H., 14,000 S.F. bldg.	4.27	5.30	9.57
4240	2,310 M.B.H., 30,800 S.F. bldg.	3.89	4.91	8.80
4280	3,150 M.B.H., 42,000 S.F. bldg.	3.70	4.97	8.67

SERVICES

D

D3020 Heat Generating Systems

Pressure Relief Valve

Circulating Pump

Return Piping

Supply Piping

Expansion Tank

Electric Boiler, Hot Water

System Components	QUANTITY	UNIT	COST EACH		
			MAT.	INST.	TOTAL
SYSTEM D3020 126 1010					
BOILER, ELECTRIC, HOT WATER, 15 kW, 52 MBH					
Boilers, electric, ASME, hot water, 15 KW, 52 MBH	1.000	Ea.	2,650	975	3,625
Pipe, black steel, Sch 40, threaded, W/coupling & hangers, 10' OC, 3/4" dia	10.000	L.F.	18.63	85.10	103.73
Pipe, black steel, Sch 40, threaded, W/coupling & hangers, 10' OC, 1" dia	20.000	L.F.	44.51	182.75	227.26
Elbow, 90°, black, straight, 3/4" dia.	6.000	Ea.	9.12	192	201.12
Elbow, 90°, black, straight, 1" dia.	6.000	Ea.	16.20	207	223.20
Tee, black, straight, 3/4" dia.	1.000	Ea.	2.46	50	52.46
Tee, black, straight, 1" dia.	2.000	Ea.	8.40	113	121.40
Tee, black, reducing, 1" dia.	2.000	Ea.	9.60	113	122.60
Union, black with brass seat, 3/4" dia.	1.000	Ea.	6.90	34.50	41.40
Union, black with brass seat, 1" dia.	2.000	Ea.	19.80	75	94.80
Pipe nipples, 3/4" diam.	3.000	Ea.	2.43	11.10	13.53
Pipe nipples, 1" diam.	3.000	Ea.	3.11	12.75	15.86
Valves, bronze, gate, N.R.S., threaded, class 150, 3/4" size	2.000	Ea.	61	45	106
Valves, bronze, gate, N.R.S., threaded, class 150, 1" size	2.000	Ea.	79	48	127
Thermometer, stem type, 9" case, 8" stem, 3/4" NPT	1.000	Ea.	266	32.50	298.50
Tank, steel, liquid expansion, ASME, painted, 15 gallon capacity	1.000	Ea.	390	48	438
Pump, circulating, bronze, flange connection, 3/4" to 1-1/2" size, 1/8 HP	1.000	Ea.	605	135	740
Insulation, fiberglass pipe covering, 1" wall, 1" IPS	20.000	L.F.	33.60	72.20	105.80
TOTAL			4,225.76	2,431.90	6,657.66

D3020 126	Electric Boiler, Hot Water	COST EACH		
		MAT.	INST.	TOTAL
1010	Boiler, electric, hot water, 15 kW, 52 MBH	4,225	2,425	6,650
1020	30 kW, 103 MBH	5,050	2,550	7,600
1030	60 kW, 205 MBH	5,100	2,650	7,750
1040	120 kW, 410 MBH	8,550	3,175	11,725
1050	150 kW, 510 MBH	12,900	6,275	19,175
1060	210 kW, 716 MBH	15,800	6,850	22,650
1070	296 kW, 1010 MBH	16,700	7,050	23,750
1080	444 kW, 1515 MBH	20,300	9,525	29,825
1090	666 kW, 2273 MBH	24,300	10,300	34,600
1100	900 kW, 3071 MBH	29,700	11,900	41,600
1110	1320 kW, 4505 MBH	35,600	12,500	48,100
1120	2100 kW, 7167 MBH	54,500	15,100	69,600

Important: See the Reference Section for critical supporting data - Reference Numbers and City Cost Indexes

D30 HVAC

D3020 Heat Generating Systems

D3020 126	Electric Boiler, Hot Water	COST EACH		
		MAT.	INST.	TOTAL
1130	2610 kW, 8905 MBH	66,500	18,200	84,700
1140	3600 kW, 12,283 MBH	82,000	21,900	103,900

D3020 128	Electric Boiler, Steam	COST EACH		
		MAT.	INST.	TOTAL
1010	Boiler, electric, steam, 18 kW, 61.4 MBH	3,575	2,325	5,900
1020	36 kW, 123 MBH	4,475	2,475	6,950
1030	60 kW, 205 MBH	10,600	5,025	15,625
1040	120 kW, 409 MBH	12,200	5,550	17,750
1050	150 kW, 512 MBH	14,300	8,500	22,800
1060	210 kW, 716 MBH	18,900	9,075	27,975
1070	300 kW, 1023 MBH	22,100	9,950	32,050
1080	510 kW, 1740 MBH	27,300	13,000	40,300
1090	720 kW, 2456 MBH	30,800	13,600	44,400
1100	1080 kW, 3685 MBH	35,900	16,100	52,000
1110	1260 kW, 4300 MBH	37,900	17,100	55,000
1120	1620 kW, 5527 MBH	39,600	19,300	58,900
1130	2070 kW, 7063 MBH	53,500	20,300	73,800
1140	2340 kW, 7984 MBH	69,500	23,000	92,500

SERVICES

D

445

D3020 Heat Generating Systems

Cast Iron Boiler, Hot Water, Gas Fired

System Components	QUANTITY	UNIT	COST EACH		
			MAT.	INST.	TOTAL
SYSTEM D3020 130 1010					
BOILER, CAST IRON, GAS, HOT WATER, 100 MBH					
Boilers, gas fired, std controls, CI, insulated, HW, gross output 100 MBH	1.000	Ea.	1,325	1,275	2,600
Pipe, black steel, Sch 40, threaded, W/coupling & hangers, 10' OC, 3/4" dia	20.000	L.F.	36.45	166.50	202.95
Pipe, black steel, Sch 40, threaded, W/coupling & hangers, 10' OC, 1" dia.	20.000	L.F.	44.51	182.75	227.26
Elbow, 90°, black, straight, 3/4" dia.	9.000	Ea.	13.68	288	301.68
Elbow, 90°, black, straight, 1" dia.	6.000	Ea.	16.20	207	223.20
Tee, black, straight, 3/4" dia.	2.000	Ea.	4.92	100	104.92
Tee, black, straight, 1" dia.	2.000	Ea.	8.40	113	121.40
Tee, black, reducing, 1" dia.	2.000	Ea.	9.60	113	122.60
Pipe cap, black, 3/4" dia.	1.000	Ea.	1.75	14.10	15.85
Union, black with brass seat, 3/4" dia.	2.000	Ea.	13.80	69	82.80
Union, black with brass seat, 1" dia.	2.000	Ea.	19.80	75	94.80
Pipe nipples, black, 3/4" dia	5.000	Ea.	4.05	18.50	22.55
Pipe nipples, black, 1" dia	3.000	Ea.	3.11	12.75	15.86
Valves, bronze, gate, N.R.S., threaded, class 150, 3/4" size	2.000	Ea.	61	45	106
Valves, bronze, gate, N.R.S., threaded, class 150, 1" size	2.000	Ea.	79	48	127
Gas cock, brass, 3/4" size	1.000	Ea.	11.35	20.50	31.85
Thermometer, stem type, 9" case, 8" stem, 3/4" NPT	2.000	Ea.	266	32.50	298.50
Tank, steel, liquid expansion, ASME, painted, 15 gallon capacity	1.000	Ea.	390	48	438
Pump, circulating, bronze, flange connection, 3/4" to 1-1/2" size, 1/8 HP	1.000	Ea.	605	135	740
Vent chimney, all fuel, pressure tight, double wall, SS, 6" dia	20.000	L.F.	780	274	1,054
Vent chimney, elbow, 90° fixed, 6" dia	2.000	Ea.	440	55	495
Vent chimney, Tee, 6" dia	2.000	Ea.	290	69	359
Vent chimney, ventilated roof thimble, 6" dia	1.000	Ea.	210	31.50	241.50
Vent chimney, adjustable roof flashing, 6" dia	1.000	Ea.	51	27.50	78.50
Vent chimney, stack cap, 6" diameter	1.000	Ea.	154	17.90	171.90
Insulation, fiberglass pipe covering, 1" wall, 1" IPS	20.000	L.F.	33.60	72.20	105.80
TOTAL			4,872.22	3,510.70	8,382.92

D3020 130	Boiler, Cast Iron, Hot Water, Gas	COST EACH		
		MAT.	INST.	TOTAL
1010	Boiler, cast iron, gas, hot water, 100 MBH	4,875	3,500	8,375
1020	200 MBH	5,650	4,025	9,675
1030	320 MBH	6,025	4,400	10,425
1040	440 MBH	8,050	5,650	13,700

Important: See the Reference Section for critical supporting data - Reference Numbers and City Cost Indexes

D3020 Heat Generating Systems

D3020 130	Boiler, Cast Iron, Hot Water, Gas	COST EACH		
		MAT.	INST.	TOTAL
1050	544 MBH	12,100	8,975	21,075
1060	765 MBH	15,500	9,625	25,125
1070	1088 MBH	17,300	10,200	27,500
1080	1530 MBH	21,200	13,700	34,900
1090	2312 MBH	25,600	16,000	41,600
1100	2856 MBH	31,500	17,200	48,700
1110	3808 MBH	35,400	20,200	55,600
1120	4720 MBH	66,500	22,500	89,000
1130	6100 MBH	74,500	25,000	99,500
1140	6970 MBH	81,500	30,800	112,300

D3020 134	Boiler, Cast Iron, Steam, Gas	COST EACH		
		MAT.	INST.	TOTAL
1010	Boiler, cast iron, gas, steam, 100 MBH	4,050	3,400	7,450
1020	200 MBH	9,750	6,400	16,150
1030	320 MBH	11,000	7,200	18,200
1040	544 MBH	14,600	11,400	26,000
1050	765 MBH	16,600	11,900	28,500
1060	1275 MBH	19,600	13,000	32,600
1070	2675 MBH	28,300	19,000	47,300
1080	3570 MBH	33,500	21,300	54,800
1090	4720 MBH	37,600	22,800	60,400
1100	6100 MBH	72,500	26,300	98,800
1110	6970 MBH	79,500	30,700	110,200

D3020 136	Boiler, Cast Iron, Hot Water, Gas/Oil	COST EACH		
		MAT.	INST.	TOTAL
1010	Boiler, cast iron, gas & oil, hot water, 584 MBH	14,400	9,950	24,350
1020	876 MBH	19,600	10,700	30,300
1030	1168 MBH	21,000	12,000	33,000
1040	1460 MBH	23,800	14,600	38,400
1050	2044 MBH	26,600	15,100	41,700
1060	2628 MBH	34,800	18,200	53,000
1070	3210 MBH	39,400	19,500	58,900
1080	3796 MBH	42,300	20,100	62,400
1090	4672 MBH	48,800	21,000	69,800
1100	5256 MBH	60,500	25,400	85,900
1110	6000 MBH	70,500	25,800	96,300
1120	9800 MBH	88,000	44,000	132,000
1130	12,200 MBH	109,500	51,000	160,500
1140	13,500 MBH	118,500	54,500	173,000

D3020 138	Boiler, Cast Iron, Steam, Gas/Oil	COST EACH		
		MAT.	INST.	TOTAL
1010	Boiler, cast iron, gas & oil, steam, 810 MBH	18,500	12,900	31,400
1020	1360 MBH	21,300	13,700	35,000
1030	2040 MBH	27,000	17,700	44,700
1040	2700 MBH	29,300	19,600	48,900
1050	3270 MBH	33,900	21,900	55,800
1060	3770 MBH	59,000	22,400	81,400
1070	4650 MBH	66,000	25,300	91,300
1080	5520 MBH	76,000	26,600	102,600
1090	6100 MBH	77,000	28,000	105,000
1100	6970 MBH	84,000	34,000	118,000

SERVICES

D

D3020 Heat Generating Systems

Base Mounted End—Suction Pump

System Components	QUANTITY	UNIT	COST EACH		
			MAT.	INST.	TOTAL
SYSTEM D3020 330 1010					
PUMP, BASE MTD. WITH MOTOR, END-SUCTION, 2-1/2" SIZE, 3 HP, TO 150 GPM					
Pump, circulating, CI, base mounted, 2-1/2" size, 3 HP, to 150 GPM	1.000	Ea.	2,075	450	2,525
Pipe, black steel, Sch. 40, on yoke & roll hangers, 10' O.C., 2-1/2" dia	12.000	L.F.	81	229.68	310.68
Elbow, 90°, weld joint, steel, 2-1/2" pipe size	1.000	Ea.	10.20	112.80	123
Flange, weld neck, 150 LB, 2-1/2" pipe size	8.000	Ea.	176	451.20	627.20
Valve, iron body, gate, 125 lb., N.R.S., flanged, 2-1/2" size	1.000	Ea.	335	163	498
Strainer, Y type, iron body, flanged, 125 lb., 2-1/2" pipe size	1.000	Ea.	80	164	244
Multipurpose valve, CI body, 2" size	1.000	Ea.	385	57	442
Expansion joint, flanged spool, 6" F to F, 2-1/2" dia	2.000	Ea.	374	132	506
T-O-L, weld joint, socket, 1/4" pipe size, nozzle	2.000	Ea.	9.48	78.52	88
T-O-L, weld joint, socket, 1/2" pipe size, nozzle	2.000	Ea.	9.48	81.86	91.34
Control gauges, pressure or vacuum, 3-1/2" diameter dial	2.000	Ea.	46	28.40	74.40
Pressure/temperature relief plug, 316 SS, 3/4" OD, 7-1/2" insertion	2.000	Ea.	126	28.40	154.40
Insulation, fiberglass pipe covering, 1-1/2" wall, 2-1/2" IPS	12.000	L.F.	52.44	53.04	105.48
Pump control system	1.000	Ea.	945	420	1,365
Pump balancing	1.000	Ea.		200	200
TOTAL			4,704.60	2,649.90	7,354.50

D3020 330	Circulating Pump Systems, End Suction	COST EACH		
		MAT.	INST.	TOTAL
1010	Pump, base mtd with motor, end-suction, 2-1/2" size, 3 HP, to 150 GPM	4,700	2,650	7,350
1020	3" size, 5 HP, to 225 GPM	5,225	3,025	8,250
1030	4" size, 7-1/2 HP, to 350 GPM	6,400	3,650	10,050
1040	5" size, 15 HP, to 1000 GPM	8,900	5,525	14,425
1050	6" size, 25 HP, to 1550 GPM	11,900	6,700	18,600

D3020 340	Circulating Pump Systems, Double Suction	COST EACH		
		MAT.	INST.	TOTAL
1010	Pump, base mtd w/ motor, double suction, 6" size, 50 HP, to 1200 GPM	15,600	9,625	25,225
1020	8" size, 75 HP, to 2500 GPM	22,800	13,000	35,800
1030	8" size, 100 HP, to 3000 GPM	33,400	14,500	47,900
1040	10" size, 150 HP, to 4000 GPM	47,700	16,500	64,200

D3030 Cooling Generating Systems

Chilled Water Supply & Return Piping

Air Cooled Water Chiller Unit

Roof

Insulate
Return → Fan Coil Unit → Supply — Finish Ceiling

Design Assumptions: The chilled water, air cooled systems priced, utilize reciprocating hermetic compressors and propeller-type condenser fans. Piping with pumps and expansion tanks is included based on a two pipe system. No ducting is included and the fan-coil units are cooling only. Water treatment and balancing are not included. Chilled water piping is insulated. Area distribution is through the use of multiple fan coil units. Fewer but larger fan coil units with duct distribution would be approximately the same S.F. cost.

System Components	QUANTITY	UNIT	COST EACH MAT.	COST EACH INST.	COST EACH TOTAL
SYSTEM D3030 110 1200					
PACKAGED CHILLER, AIR COOLED, WITH FAN COIL UNIT					
APARTMENT CORRIDORS, 3,000 S.F., 5.50 TON					
Fan coil air conditioning unit, cabinet mounted & filters chilled water	1.000	Ea.	4,399.20	375.77	4,774.97
Water chiller,air conditioning unit, reciprocating, air cooled,	1.000	Ea.	8,250	1,416.25	9,666.25
Chilled water unit coil connections	1.000	Ea.	555	1,025	1,580
Chilled water distribution piping	440.000	L.F.	5,522	13,860	19,382
TOTAL			18,726.20	16,677.02	35,403.22
COST PER S.F.			6.24	5.56	11.80

*Cooling requirements would lead to choosing a water cooled unit

D3030 110	Chilled Water, Air Cooled Condenser Systems	COST PER S.F. MAT.	COST PER S.F. INST.	COST PER S.F. TOTAL
1180	Packaged chiller, air cooled, with fan coil unit			
1200	Apartment corridors, 3,000 S.F., 5.50 ton	6.23	5.57	11.80
1240	6,000 S.F., 11.00 ton	4.88	4.45	9.33
1280	10,000 S.F., 18.33 ton	4.41	3.38	7.79
1320	20,000 S.F., 36.66 ton	3.68	2.44	6.12
1360	40,000 S.F., 73.33 ton	3.86	2.46	6.32
1440	Banks and libraries, 3,000 S.F., 12.50 ton	9.65	6.55	16.20
1480	6,000 S.F., 25.00 ton	9.10	5.30	14.40
1520	10,000 S.F., 41.66 ton	7.65	3.72	11.37
1560	20,000 S.F., 83.33 ton	7.45	3.37	10.82
1600	40,000 S.F., 167 ton*			
1680	Bars and taverns, 3,000 S.F., 33.25 ton	19.40	7.75	27.15
1720	6,000 S.F., 66.50 ton	17.55	6	23.55
1760	10,000 S.F., 110.83 ton	15.45	1.92	17.37
1800	20,000 S.F., 220 ton*			
1840	40,000 S.F., 440 ton*			
1920	Bowling alleys, 3,000 S.F., 17.00 ton	12.45	7.35	19.80
1960	6,000 S.F., 34.00 ton	10.40	5.15	15.55
2000	10,000 S.F., 56.66 ton	9.20	3.75	12.95
2040	20,000 S.F., 113.33 ton	8.75	3.40	12.15
2080	40,000 S.F., 227 ton*			
2160	Department stores, 3,000 S.F., 8.75 ton	9.10	6.20	15.30
2200	6,000 S.F., 17.50 ton	6.75	4.80	11.55
2240	10,000 S.F., 29.17 ton	5.75	3.47	9.22

SERVICES

D

D3030 Cooling Generating Systems

D3030 110	Chilled Water, Air Cooled Condenser Systems	COST PER S.F.		
		MAT.	INST.	TOTAL
2280	20,000 S.F., 58.33 ton	5	2.56	7.56
2320	40,000 S.F., 116.66 ton	5.25	2.55	7.80
2400	Drug stores, 3,000 S.F., 20.00 ton	14.35	7.60	21.95
2440	6,000 S.F., 40.00 ton	12.25	5.65	17.90
2480	10,000 S.F., 66.66 ton	11.65	4.83	16.48
2520	20,000 S.F., 133.33 ton	10.80	3.59	14.39
2560	40,000 S.F., 267 ton*			
2640	Factories, 2,000 S.F., 10.00 ton	8.20	6.30	14.50
2680	6,000 S.F., 20.00 ton	7.65	5.05	12.70
2720	10,000 S.F., 33.33 ton	6.40	3.55	9.95
2760	20,000 S.F., 66.66 ton	6.25	3.19	9.44
2800	40,000 S.F., 133.33 ton	5.85	2.61	8.46
2880	Food supermarkets, 3,000 S.F., 8.50 ton	8.90	6.15	15.05
2920	6,000 S.F., 17.00 ton	6.60	4.77	11.37
2960	10,000 S.F., 28.33 ton	5.55	3.35	8.90
3000	20,000 S.F., 56.66 ton	4.85	2.50	7.35
3040	40,000 S.F., 113.33 ton	5.10	2.54	7.64
3120	Medical centers, 3,000 S.F., 7.00 ton	7.75	6	13.75
3160	6,000 S.F., 14.00 ton	5.75	4.61	10.36
3200	10,000 S.F., 23.33 ton	5.35	3.51	8.86
3240	20,000 S.F., 46.66 ton	4.50	2.57	7.07
3280	40,000 S.F., 93.33 ton	4.46	2.49	6.95
3360	Offices, 3,000 S.F., 9.50 ton	7.90	6.10	14
3400	6,000 S.F., 19.00 ton	7.40	5	12.40
3440	10,000 S.F., 31.66 ton	6.15	3.51	9.66
3480	20,000 S.F., 63.33 ton	6.15	3.25	9.40
3520	40,000 S.F., 126.66 ton	5.70	2.63	8.33
3600	Restaurants, 3,000 S.F., 15.00 ton	11.10	6.80	17.90
3640	6,000 S.F., 30.00 ton	9.55	5.15	14.70
3680	10,000 S.F., 50.00 ton	8.50	3.80	12.30
3720	20,000 S.F., 100.00 ton	8.65	3.55	12.20
3760	40,000 S.F., 200 ton*			
3840	Schools and colleges, 3,000 S.F., 11.50 ton	9.05	6.45	15.50
3880	6,000 S.F., 23.00 ton	8.55	5.20	13.75
3920	10,000 S.F., 38.33 ton	7.15	3.65	10.80
3960	20,000 S.F., 76.66 ton	7.10	3.40	10.50
4000	40,000 S.F., 153 ton*			

Important: See the Reference Section for critical supporting data - Reference Numbers and City Cost Indexes

D3030 Cooling Generating Systems

Reciprocating Package Chiller
Condenser Water
Cooling Tower
Chilled Water Supply & Return Piping
Cooling Tower Water Makeup
Roof Structure
Finish Ceiling
Insulate
Return
Supply
Fan Coil Unit

General: Water cooled chillers are available in the same sizes as air cooled units. They are also available in larger capacities.

Design Assumptions: The chilled water systems with water cooled condenser include reciprocating hermetic compressors, water cooling tower, pumps, piping and expansion tanks and are based on a two pipe system. Chilled water piping is insulated. No ducts are included and fan-coil units are cooling only. Area distribution is through use of multiple fan coil units. Fewer but larger fan coil units with duct distribution would be approximately the same S.F. cost. Water treatment and balancing are not included.

System Components	QUANTITY	UNIT	COST EACH		
			MAT.	INST.	TOTAL
SYSTEM D3030 115 1320					
PACKAGED CHILLER, WATER COOLED, WITH FAN COIL UNIT					
APARTMENT CORRIDORS, 4,000 S.F., 7.33 TON					
Fan coil air conditioner unit, cabinet mounted & filters, chilled water	2.000	Ea.	5,863.20	500.82	6,364.02
Water chiller, reciprocating, water cooled, 1 compressor semihermetic	1.000	Ea.	9,892.80	2,450.30	12,343.10
Cooling tower, draw thru single flow, belt drive	1.000	Ea.	634.05	103.72	737.77
Cooling tower pumps & piping	1.000	System	318.86	245.56	564.42
Chilled water unit coil connections	2.000	Ea.	1,110	2,050	3,160
Chilled water distribution piping	520.000	L.F.	6,526	16,380	22,906
TOTAL			24,344.91	21,730.40	46,075.31
COST PER S.F.			6.09	5.43	11.52
*Cooling requirements would lead to choosing a water cooled unit					

D3030 115	Chilled Water, Cooling Tower Systems	COST PER S.F.		
		MAT.	INST.	TOTAL
1300	Packaged chiller, water cooled, with fan coil unit			
1320	Apartment corridors, 4,000 S.F., 7.33 ton	6.08	5.43	11.51
1360	6,000 S.F., 11.00 ton	4.67	4.68	9.35
1400	10,000 S.F., 18.33 ton	4.35	3.47	7.82
1440	20,000 S.F., 26.66 ton	3.47	2.59	6.06
1480	40,000 S.F., 73.33 ton	3.18	2.65	5.83
1520	60,000 S.F., 110.00 ton	3.09	2.72	5.81
1600	Banks and libraries, 4,000 S.F., 16.66 ton	10.10	6	16.10
1640	6,000 S.F., 25.00 ton	8.20	5.25	13.45
1680	10,000 S.F., 41.66 ton	6.60	3.92	10.52
1720	20,000 S.F., 83.33 ton	5.85	3.74	9.59
1760	40,000 S.F., 166.66 ton	5.60	4.61	10.21
1800	60,000 S.F., 250.00 ton	5.50	4.88	10.38
1880	Bars and taverns, 4,000 S.F., 44.33 ton	16.20	7.55	23.75
1920	6,000 S.F., 66.50 ton	18.65	7.80	26.45
1960	10,000 S.F., 110.83 ton	12.80	6.15	18.95
2000	20,000 S.F., 221.66 ton	11.85	6.60	18.45
2040	40,000 S.F., 440 ton*			
2080	60,000 S.F., 660 ton*			
2160	Bowling alleys, 4,000 S.F., 22.66 ton	11.65	6.55	18.20
2200	6,000 S.F., 34.00 ton	9.85	5.80	15.65
2240	10,000 S.F., 56.66 ton	9.75	4.24	13.99

R15700 -020

SERVICES

D

D3030 Cooling Generating Systems

D3030 115	Chilled Water, Cooling Tower Systems	COST PER S.F.		
		MAT.	INST.	TOTAL
2280	20,000 S.F., 113.33 ton	6.95	3.94	10.89
2320	40,000 S.F., 226.66 ton	6.60	4.75	11.35
2360	60,000 S.F., 340 ton			
2440	Department stores, 4,000 S.F., 11.66 ton	6.65	5.90	12.55
2480	6,000 S.F., 17.50 ton	7.30	4.87	12.17
2520	10,000 S.F., 29.17 ton	5.30	3.63	8.93
2560	20,000 S.F., 58.33 ton	4.94	2.70	7.64
2600	40,000 S.F., 116.66 ton	4	2.80	6.80
2640	60,000 S.F., 175.00 ton	4.43	4.56	8.99
2720	Drug stores, 4,000 S.F., 26.66 ton	12.45	6.75	19.20
2760	6,000 S.F., 40.00 ton	10.45	5.80	16.25
2800	10,000 S.F., 66.66 ton	11.60	5.25	16.85
2840	20,000 S.F., 133.33 ton	8	4.12	12.12
2880	40,000 S.F., 266.67 ton	7.70	5.20	12.90
2920	60,000 S.F., 400 ton*			
3000	Factories, 4,000 S.F., 13.33 ton	8.45	5.70	14.15
3040	6,000 S.F., 20.00 ton	7.25	4.89	12.14
3080	10,000 S.F., 33.33 ton	5.90	3.74	9.64
3120	20,000 S.F., 66.66 ton	6.20	3.36	9.56
3160	40,000 S.F., 133.33 ton	4.46	2.88	7.34
3200	60,000 S.F., 200.00 ton	4.81	4.68	9.49
3280	Food supermarkets, 4,000 S.F., 11.33 ton	6.50	5.85	12.35
3320	6,000 S.F., 17.00 ton	6.35	4.77	11.12
3360	10,000 S.F., 28.33 ton	5.20	3.59	8.79
3400	20,000 S.F., 56.66 ton	5.10	2.70	7.80
3440	40,000 S.F., 113.33 ton	3.95	2.79	6.74
3480	60,000 S.F., 170.00 ton	4.37	4.53	8.90
3560	Medical centers, 4.000 S.F., 9.33 ton	5.60	5.40	11
3600	6,000 S.F., 14.00 ton	6.20	4.66	10.86
3640	10,000 S.F., 23.33 ton	4.83	3.47	8.30
3680	20,000 S.F., 46.66 ton	4.40	2.63	7.03
3720	40,000 S.F., 93.33 ton	3.60	2.73	6.33
3760	60,000 S.F., 140.00 ton	4	4.49	8.49
3840	Offices, 4,000 S.F., 12.66 ton	8.15	5.65	13.80
3880	6,000 S.F., 19.00 ton	7.10	5	12.10
3920	10,000 S.F., 31.66 ton	5.75	3.78	9.53
3960	20,000 S.F., 63.33 ton	6	3.36	9.36
4000	40,000 S.F., 126.66 ton	4.66	4.40	9.06
4040	60,000 S.F., 190.00 ton	4.63	4.61	9.24
4120	Restaurants, 4,000 S.F., 20.00 ton	10.40	6.10	16.50
4160	6,000 S.F., 30.00 ton	8.85	5.40	14.25
4200	10,000 S.F., 50.00 ton	8.85	4.07	12.92
4240	20,000 S.F., 100.00 ton	6.65	3.93	10.58
4280	40,000 S.F., 200.00 ton	5.85	4.46	10.31
4320	60,000 S.F., 300.00 ton	6.20	5.05	11.25
4400	Schools and colleges, 4,000 S.F., 15.33 ton	9.45	5.85	15.30
4440	6,000 S.F., 23.00 ton	7.70	5.15	12.85
4480	10,000 S.F., 38.33 ton	6.20	3.83	10.03
4520	20,000 S.F., 76.66 ton	5.60	3.69	9.29
4560	40,000 S.F., 153.33 ton	5.25	4.49	9.74
4600	60,000 S.F., 230.00 ton	5.05	4.69	9.74

Important: See the Reference Section for critical supporting data - Reference Numbers and City Cost Indexes

SERVICES D

D3030 Cooling Generating Systems

Chilled Water Piping   Reciprocating Chiller, Water Cooled   Condenser Water Piping

System Components	QUANTITY	UNIT	COST EACH		
			MAT.	INST.	TOTAL
SYSTEM D3030 130 1010					
CHILLER, RECIPROCATING, WATER COOLED, STD. CONTROLS, 60 TON					
Water chiller, reciprocating, water cooled, 60 ton cooling	1.000	Ea.	60,000	6,825	66,825
Pipe, black steel, welded, Sch. 40, on yoke & roll hgrs, 10' O.C., 4" dia	40.000	L.F.	396	973.60	1,369.60
Elbow, 90°, weld joint, butt, 4" pipe size	8.000	Ea.	164	1,442.40	1,606.40
Flange, welding neck, 150 lb., 4" pipe size	16.000	Ea.	456	1,442.40	1,898.40
Valves, iron body, butterfly, lug type, lever actuator, 4" size	4.000	Ea.	468	652	1,120
Expansion joints, bellows type, flange spool, 6" F to F, 4" dia	4.000	Ea.	908	390	1,298
T-O-L socket, weld joint, 1/4" pipe size, nozzle	4.000	Ea.	18.96	157.04	176
T-O-L socket, weld joint, 3/4" pipe size, nozzle	4.000	Ea.	22.20	170.44	192.64
Thermometers, stem type, 9" case, 8" stem, 3/4" NPT	4.000	Ea.	532	65	597
Gauges, pressure or vacuum, 3-1/2" diameter dial	4.000	Ea.	92	56.80	148.80
Insulation, fiberglass pipe covering, 1-1/2" wall, 4" IPS	20.000	L.F.	103	114	217
Chiller balancing	1.000	Ea.		385	385
TOTAL			63,160.16	12,673.68	75,833.84

D3030 130	Reciprocating Chiller, Water Cooled	COST EACH		
		MAT.	INST.	TOTAL
1010	Chiller, reciprocating, water cooled, std. controls, 60 Ton	63,000	12,700	75,700
1020	100 Ton	63,500	19,000	82,500
1030	150 Ton	84,000	20,800	104,800
1040	200 Ton	101,500	21,900	123,400

D3030 135	Reciprocating Chiller, Air Cooled	COST EACH		
		MAT.	INST.	TOTAL
1010	Chiller, reciprocating, air cooled, std. controls, 20 Ton	31,300	7,450	38,750
1020	30 Ton	38,000	7,825	45,825
1030	40 Ton	52,000	8,350	60,350
1040	60 Ton	75,500	9,400	84,900
1050	70 Ton	86,000	9,675	95,675
1060	80 Ton	92,000	9,850	101,850
1070	90 Ton	100,000	10,000	110,000
1080	100 Ton	111,000	10,200	121,200
1090	110 Ton	121,500	10,900	132,400
1100	130 Ton	134,500	11,100	145,600
1110	150 Ton	155,000	11,300	166,300
1120	175 Ton	165,000	11,400	176,400
1130	210 Ton	204,000	12,400	216,400

D SERVICES

D3030 Cooling Generating Systems

Chilled Water Piping

Centrifugal Chiller, Water Cooled, Hermetic

Condenser Water Piping

System Components

System Components	QUANTITY	UNIT	COST EACH MAT.	COST EACH INST.	COST EACH TOTAL
SYSTEM D3030 140 1010					
CHILLER, CENTRIF., WATER COOLED, PKGD. HERMETIC, STD. CONTROLS, 200 TON					
Water chiller, screw, water cooled, not incl tower, 200 ton	1.000	Ea.	70,500	13,600	84,100
Pipe, black steel, welded, Sch. 40, on yoke & roll hgrs, 10' o.c., 6" dia	40.000	L.F.	690	1,496	2,186
Elbow, 90°, weld joint, 6" pipe size	8.000	Ea.	384	2,162.40	2,546.40
Flange, welding neck, 150 lb., 6" pipe size	16.000	Ea.	704	2,154.40	2,858.40
Valves, iron body, butterfly, lug type, lever actuator, 6" size	4.000	Ea.	784	1,012	1,796
Expansion joints, bellows type, flange spool, 6" F to F, 6" dia	4.000	Ea.	1,148	480	1,628
T-O-L socket, weld jiont, 1/4" pipe size, nozzle	4.000	Ea.	18.96	157.04	176
T-O-L socket, weld joint, 3/4" pipe size, nozzle	4.000	Ea.	22.20	170.44	192.64
Thermometers, stem type, 9" case, 8" stem, 3/4" NPT	4.000	Ea.	532	65	597
Gauges, pressure or vacuum, 3-1/2" diameter dial	4.000	Ea.	92	56.80	148.80
Insulation, fiberglass pipe covering, 1-1/2" wall, 6" IPS	40.000	L.F.	236	290	526
Chiller balancing	1.000	Ea.		385	385
TOTAL			75,111.16	22,029.08	97,140.24

D3030 140	Centrifugal Chiller, Water Cooled	COST EACH MAT.	COST EACH INST.	COST EACH TOTAL
1010	Chiller, centrif., water cooled, pkgd. hermetic, std controls, 200 Ton	75,000	22,000	97,000
1020	400 Ton	139,000	30,300	169,300
1030	1,000 Ton	361,000	45,800	406,800
1040	1,500 Ton	531,500	56,000	587,500

D3030 Cooling Generating Systems

Chilled Water Piping
(Condenser Water Piping Similar)

Gas Absorption Chiller, Water Cooled

Gas Piping

System Components	QUANTITY	UNIT	COST EACH MAT.	COST EACH INST.	COST EACH TOTAL
SYSTEM D3030 145 1010					
CHILLER, GAS FIRED ABSORPTION, WATER COOLED, STD. CONTROLS, 800 TON					
Absorption water chiller, gas fired, water cooled, 800 ton	1.000	Ea.	555,000	20,600	575,600
Pipe, black steel, weld, Sch. 40, on yoke & roll hangers, 10' O.C., 10" dia	20.000	L.F.	950	1,122	2,072
Pipe, black steel, welded, Sch. 40, (no hgrs incl), 14" dia	44.000	L.F.	1,936	3,949	5,885
Pipe, blacksteel, welded, Sch. 40, on yoke & roll hangers, 10' O.C., 4" dia	20.000	L.F.	198	486.80	684.80
Elbow, 90°, weld joint, 10" pipe size	4.000	Ea.	664	1,796	2,460
Elbow, 90°, weld joint, 14" pipe size	6.000	Ea.	1,980	4,038	6,018
Elbow, 90°, weld joint, 4" pipe size	3.000	Ea.	61.50	540.90	602.40
Flange, welding neck, 150 lb., 10" pipe size	8.000	Ea.	992	1,803.20	2,795.20
Flange, welding neck, 150 lb., 14" pipe size	14.000	Ea.	3,640	4,202.80	7,842.80
Flange, welding neck, 150 lb., 4" pipe size	3.000	Ea.	85.50	270.45	355.95
Valves, iron body, butterfly, lug type, gear operated, 10" size	2.000	Ea.	1,120	630	1,750
Valves, iron body, butterfly, lug type, gear operated, 14" size	5.000	Ea.	4,475	2,750	7,225
Valves, semi-steel, lubricated plug valve, flanged, 4" pipe size	1.000	Ea.	365	271	636
Expansion joints, bellows type, flange spool, 6" F to F, 10" dia	2.000	Ea.	880	328	1,208
Expansion joints, bellows type, flange spool, 10" F to F, 14" dia	2.000	Ea.	1,430	430	1,860
T-O-L socket, weld joint, 1/4" pipe size, nozzle	4.000	Ea.	18.96	157.04	176
T-O-L socket, weld joint, 3/4" pipe size, nozzle	4.000	Ea.	22.20	170.44	192.64
Thermometers, stem type, 9" case, 8" stem, 3/4" NPT	4.000	Ea.	532	65	597
Gauges, pressure or vacuum, 3-1/2" diameter dial	4.000	Ea.	92	56.80	148.80
Insulation, fiberglass pipe covering, 2" wall, 10" IPS	20.000	L.F.	268	227	495
Vent chimney, all fuel, pressure tight, double wall, SS, 12" dia	20.000	L.F.	1,220	374	1,594
Chiller balancing	1.000	Ea.		385	385
TOTAL			575,930.16	44,653.43	620,583.59

D3030 145	Gas Absorption Chiller, Water Cooled	MAT.	INST.	TOTAL
1010	Chiller, gas fired absorption, water cooled, std. ctrls, 800 Ton	576,000	44,700	620,700
1020	1000 Ton	728,500	57,500	786,000

D3030 150	Steam Absorption Chiller, Water Cooled	COST EACH MAT.	COST EACH INST.	COST EACH TOTAL
1010	Chiller, steam absorption, water cooled, std. ctrls, 750 Ton	383,500	53,000	436,500
1020	955 Ton	445,500	63,000	508,500
1030	1465 Ton	643,000	73,000	716,000
1040	1660 Ton	769,000	82,500	851,500

SERVICES

D

D3030 Cooling Generating Systems

System Description: Rooftop single zone units are electric cooling and gas heat. Duct systems are low velocity, galvanized steel supply and return. Price variations between sizes are due to several factors. Jumps in the cost of the rooftop unit occur when the manufacturer shifts from the largest capacity unit on a small frame to the smallest capacity on the next larger frame, or changes from one compressor to two. As the unit capacity increases for larger areas the duct distribution grows in proportion. For most applications there is a tradeoff point where it is less expensive and more efficient to utilize smaller units with short simple distribution systems. Larger units also require larger initial supply and return ducts which can create a space problem. Supplemental heat may be desired in colder locations. The table below is based on one unit supplying the area listed. The 10,000 S.F. unit for bars and taverns is not listed because a nominal 110 ton unit would be required and this is above the normal single zone rooftop capacity.

System Components	QUANTITY	UNIT	COST EACH		
			MAT.	INST.	TOTAL
SYSTEM D3030 202 1280					
ROOFTOP, SINGLE ZONE, AIR CONDITIONER					
APARTMENT CORRIDORS, 500 S.F., .92 TON					
Rooftop air-conditioner, 1 zone, electric cool, standard controls, curb	1.000	Ea.	1,932	540.50	2,472.50
Ductwork package for rooftop single zone units	1.000	System	236.44	786.60	1,023.04
TOTAL			2,168.44	1,327.10	3,495.54
COST PER S.F.			4.34	2.65	6.99
*Size would suggest multiple units					

D3030 202	Rooftop Single Zone Unit Systems	COST PER S.F.		
		MAT.	INST.	TOTAL
1260	Rooftop, single zone, air conditioner			
1280	Apartment corridors, 500 S.F., .92 ton	4.35	2.65	7
1320	1,000 S.F., 1.83 ton	4.31	2.64	6.95
1360	1500 S.F., 2.75 ton	2.76	2.19	4.95
1400	3,000 S.F., 5.50 ton	2.55	2.10	4.65
1440	5,000 S.F., 9.17 ton	2.78	1.92	4.70
1480	10,000 S.F., 18.33 ton	3.33	1.81	5.14
1560	Banks or libraries, 500 S.F., 2.08 ton	9.80	6	15.80
1600	1,000 S.F., 4.17 ton	6.30	4.98	11.28
1640	1,500 S.F., 6.25 ton	5.80	4.77	10.57
1680	3,000 S.F., 12.50 ton	6.30	4.35	10.65
1720	5,000 S.F., 20.80 ton	7.55	4.10	11.65
1760	10,000 S.F., 41.67 ton	7.75	4.07	11.82
1840	Bars and taverns, 500 S.F. 5.54 ton	14	7.95	21.95
1880	1,000 S.F., 11.08 ton	15.40	6.85	22.25
1920	1,500 S.F., 16.62 ton	12.50	6.40	18.90
1960	3,000 S.F., 33.25 ton	19.65	6.10	25.75
2000	5,000 S.F., 55.42 ton	18.70	6.10	24.80
2040	10,000 S.F., 110.83 ton*			
2080	Bowling alleys, 500 S.F., 2.83 ton	8.55	6.75	15.30
2120	1,000 S.F., 5.67 ton	7.90	6.50	14.40
2160	1,500 S.F., 8.50 ton	8.60	5.95	14.55
2200	3,000 S.F., 17.00 ton	7.15	5.70	12.85
2240	5,000 S.F., 28.33 ton	10.80	5.55	16.35

Note: R15700 -020 (reference box appears near rows 1360-1400)

Important: See the Reference Section for critical supporting data - Reference Numbers and City Cost Indexes

D3030 Cooling Generating Systems

D3030 202	Rooftop Single Zone Unit Systems	COST PER S.F.		
		MAT.	INST.	TOTAL
2280	10,000 S.F., 56.67 ton	10.30	5.55	15.85
2360	Department stores, 500 S.F., 1.46 ton	6.85	4.21	11.06
2400	1,000 S.F., 2.92 ton	4.40	3.48	7.88
2440	1,500 S.F., 4.37 ton	4.06	3.34	7.40
2480	3,000 S.F., 8.75 ton	4.43	3.04	7.47
2520	5,000 S.F., 14.58 ton	3.67	2.92	6.59
2560	10,000 S.F., 29.17 ton	5.55	2.85	8.40
2640	Drug stores, 500 S.F., 3.33 ton	10.05	7.95	18
2680	1,000 S.F., 6.67 ton	9.25	7.65	16.90
2720	1,500 S.F., 10.00 ton	10.10	6.95	17.05
2760	3,000 S.F., 20.00 ton	12.10	6.55	18.65
2800	5,000 S.F., 33.33 ton	12.70	6.50	19.20
2840	10,000 S.F., 66.67 ton	12.10	6.50	18.60
2920	Factories, 500 S.F., 1.67 ton	7.85	4.81	12.66
2960	1,000 S.F., 3.33 ton	5.05	3.97	9.02
3000	1,500 S.F., 5.00 ton	4.64	3.82	8.46
3040	3,000 S.F., 10.00 ton	5.05	3.48	8.53
3080	5,000 S.F., 16.67 ton	4.19	3.34	7.53
3120	10,000 S.F., 33.33 ton	6.35	3.26	9.61
3200	Food supermarkets, 500 S.F., 1.42 ton	6.70	4.09	10.79
3240	1,000 S.F., 2.83 ton	4.23	3.37	7.60
3280	1,500 S.F., 4.25 ton	3.95	3.24	7.19
3320	3,000 S.F., 8.50 ton	4.30	2.96	7.26
3360	5,000 S.F., 14.17 ton	3.56	2.84	6.40
3400	10,000 S.F., 28.33 ton	5.40	2.77	8.17
3480	Medical centers, 500 S.F., 1.17 ton	5.50	3.37	8.87
3520	1,000 S.F., 2.33 ton	5.50	3.36	8.86
3560	1,500 S.F., 3.50 ton	3.52	2.79	6.31
3600	3,000 S.F., 7.00 ton	3.25	2.68	5.93
3640	5,000 S.F., 11.67 ton	3.54	2.44	5.98
3680	10,000 S.F., 23.33 ton	4.24	2.29	6.53
3760	Offices, 500 S.F., 1.58 ton	7.45	4.56	12.01
3800	1,000 S.F., 3.17 ton	4.77	3.78	8.55
3840	1,500 S.F., 4.75 ton	4.40	3.63	8.03
3880	3,000 S.F., 9.50 ton	4.80	3.31	8.11
3920	5,000 S.F., 15.83 ton	3.98	3.18	7.16
3960	10,000 S.F., 31.67 ton	6.05	3.10	9.15
4000	Restaurants, 500 S.F., 2.50 ton	11.80	7.20	19
4040	1,000 S.F., 5.00 ton	6.95	5.75	12.70
4080	1,500 S.F., 7.50 ton	7.60	5.25	12.85
4120	3,000 S.F., 15.00 ton	6.30	5	11.30
4160	5,000 S.F., 25.00 ton	9.10	4.93	14.03
4200	10,000 S.F., 50.00 ton	9.10	4.90	14
4240	Schools and colleges, 500 S.F., 1.92 ton	9.05	5.55	14.60
4280	1,000 S.F., 3.83 ton	5.75	4.56	10.31
4320	1,500 S.F., 5.75 ton	5.35	4.39	9.74
4360	3,000 S.F., 11.50 ton	5.80	4.01	9.81
4400	5,000 S.F., 19.17 ton	6.95	3.78	10.73
4440	10,000 S.F., 38.33 ton	7.10	3.75	10.85
5000				
5100				
5200				
9000	Components of ductwork packages for above systems, per ton of cooling:			
9010	Ductwork; galvanized steel, 120 pounds			
9020	Insulation; fiberglass 2″ thick, FRK faced, 52 SF			
9030	Diffusers; aluminum, 24″ x 12″, one			
9040	Registers; aluminum, return with dampers, one			

SERVICES

D

D3030 Cooling Generating Systems

System Description: Rooftop units are multizone with up to 12 zones, and include electric cooling, gas heat, thermostats, filters, supply and return fans complete. Duct systems are low velocity, galvanized steel supply and return with insulated supplies.

Multizone units cost more per ton of cooling than single zone. However, they offer flexibility where load conditions are varied due to heat generating areas or exposure to radiational heating. For example, perimeter offices on the "sunny side" may require cooling at the same

time "shady side" or central offices may require heating. It is possible to accomplish similar results using duct heaters in branches of the single zone unit. However, heater location could be a problem and total system operating energy efficiency could be lower.

System Components				COST EACH		
	QUANTITY	UNIT		MAT.	INST.	TOTAL
SYSTEM D3030 206 1280						
ROOFTOP, MULTIZONE, AIR CONDITIONER						
APARTMENT CORRIDORS, 3,000 S.F., 5.50 TON						
Rooftop multizone unit, standard controls, curb	1.000	Ea.		34,320	1,254	35,574
Ductwork package for rooftop multizone units	1.000	System		1,897.50	9,075	10,972.50
TOTAL				36,217.50	10,329	46,546.50
COST PER S.F.				12.07	3.44	15.51

Note A: Small single zone unit recommended
Note B: A combination of multizone units recommended

D3030 206	Rooftop Multizone Unit Systems		COST PER S.F.		
			MAT.	INST.	TOTAL
1240	Rooftop, multizone, air conditioner				
1260	Apartment corridors, 1,500 S.F., 2.75 ton. See Note A.				
1280	3,000 S.F., 5.50 ton	R15700 -020	12.07	3.43	15.50
1320	10,000 S.F., 18.30 ton		8.85	3.32	12.17
1360	15,000 S.F., 27.50 ton		7.55	3.31	10.86
1400	20,000 S.F., 36.70 ton		7.70	3.31	11.01
1440	25,000 S.F., 45.80 ton		6.60	3.33	9.93
1520	Banks or libraries, 1,500 S.F., 6.25 ton		27.50	7.85	35.35
1560	3,000 S.F., 12.50 ton		23	7.65	30.65
1600	10,000 S.F., 41.67 ton		15	7.60	22.60
1640	15,000 S.F., 62.50 ton		9.15	7.50	16.65
1680	20,000 S.F., 83.33 ton		8.55	7.50	16.05
1720	25,000 S.F., 104.00 ton		8.05	7.50	15.55
1800	Bars and taverns, 1,500 S.F., 16.62 ton		59.50	11.25	70.75
1840	3,000 S.F., 33.24 ton		44.50	10.85	55.35
1880	10,000 S.F., 110.83 ton		19.50	10.80	30.30
1920	15,000 S.F., 165 ton, See Note B				
1960	20,000 S.F., 220 ton, See Note B				
2000	25,000 S.F., 275 ton, See Note B				
2080	Bowling alleys, 1,500 S.F., 8.50 ton		37.50	10.65	48.15
2120	3,000 S.F., 17.00 ton		31.50	10.45	41.95
2160	10,000 S.F., 56.70 ton		20.50	10.30	30.80
2200	15,000 S.F., 85.00 ton		11.60	10.25	21.85
2240	20,000 S.F., 113.00 ton		10.90	10.15	21.05

D3030 Cooling Generating Systems

D3030 206	Rooftop Multizone Unit Systems	COST PER S.F.		
		MAT.	INST.	TOTAL
2280	25,000 S.F., 140.00 ton see Note B			
2360	Department stores, 1,500 S.F., 4.37 ton, See Note A.			
2400	3,000 S.F., 8.75 ton	19.20	5.50	24.70
2440	10,000 S.F., 29.17 ton	12.25	5.25	17.50
2480	15,000 S.F., 43.75 ton	10.50	5.30	15.80
2520	20,000 S.F., 58.33 ton	6.40	5.25	11.65
2560	25,000 S.F., 72.92 ton	5.95	5.25	11.20
2640	Drug stores, 1,500 S.F., 10.00 ton	44	12.50	56.50
2680	3,000 S.F., 20.00 ton	32	12.10	44.10
2720	10,000 S.F., 66.66 ton	14.65	12.05	26.70
2760	15,000 S.F., 100.00 ton	12.85	12	24.85
2800	20,000 S.F., 135 ton, See Note B			
2840	25,000 S.F., 165 ton, See Note B			
2920	Factories, 1,500 S.F., 5 ton, See Note A			
2960	3,000 S.F., 10.00 ton	22	6.25	28.25
3000	10,000 S.F., 33.33 ton	14	6	20
3040	15,000 S.F., 50.00 ton	12	6.05	18.05
3080	20,000 S.F., 66.66 ton	7.30	6	13.30
3120	25,000 S.F., 83.33 ton	6.80	6	12.80
3200	Food supermarkets, 1,500 S.F., 4.25 ton, See Note A			
3240	3,000 S.F., 8.50 ton	18.65	5.35	24
3280	10,000 S.F., 28.33 ton	11.70	5.10	16.80
3320	15,000 S.F., 42.50 ton	10.20	5.15	15.35
3360	20,000 S.F., 56.67 ton	6.20	5.10	11.30
3400	25,000 S.F., 70.83 ton	6.20	5.10	11.30
3480	Medical centers, 1,500 S.F., 3.5 ton, See Note A			
3520	3,000 S.F., 7.00 ton	15.35	4.38	19.73
3560	10,000 S.F., 23.33 ton	10.25	4.21	14.46
3600	15,000 S.F., 35.00 ton	9.80	4.21	14.01
3640	20,000 S.F., 46.66 ton	8.40	4.24	12.64
3680	25,000 S.F., 58.33 ton	5.10	4.21	9.31
3760	Offices, 1,500 S.F., 4.75 ton, See Note A			
3800	3,000 S.F., 9.50 ton	21	5.95	26.95
3840	10,000 S.F., 31.66 ton	13.30	5.70	19
3880	15,000 S.F., 47.50 ton	11.40	5.75	17.15
3920	20,000 S.F., 63.33 ton	6.95	5.70	12.65
3960	25,000 S.F., 79.16 ton	6.45	5.70	12.15
4000	Restaurants, 1,500 S.F., 7.50 ton	33	9.40	42.40
4040	3,000 S.F., 15.00 ton	27.50	9.20	36.70
4080	10,000 S.F., 50.00 ton	18	9.10	27.10
4120	15,000 S.F., 75.00 ton	11	9	20
4160	20,000 S.F., 100.00 ton	9.65	9	18.65
4200	25,000 S.F., 125 ton, See Note B			
4240	Schools and colleges, 1,500 S.F., 5.75 ton	25	7.20	32.20
4280	3,000 S.F., 11.50 ton	21	7.05	28.05
4320	10,000 S.F., 38.33 ton	13.80	6.95	20.75
4360	15,000 S.F., 57.50 ton	8.40	6.90	15.30
4400	20,000 S.F., 76.66 ton	7.85	6.90	14.75
4440	25,000 S.F., 95.83 ton	7.40	6.90	14.30
4450				
9000	Components of ductwork packages for above systems, per ton of cooling:			
9010	Ductwork; galvanized steel, 240 pounds			
9020	Insulation; fiberglass, 2" thick, FRK faced, 104 SF			
9030	Diffusers; aluminum, 24" x 12", two			
9040	Registers; aluminum, return with dampers, one			

SERVICES

D

System Description: Self-contained, single package water cooled units include cooling tower, pump, piping allowance. Systems for 1000 S.F. and up include duct and diffusers to provide for even distribution of air. Smaller units distribute

air through a supply air plenum, which is integral with the unit.

Returns are not ducted and supplies are not insulated.

Hot water or steam heating coils are included but piping to boiler and the boiler itself is not included.

Where local codes or conditions permit single pass cooling for the smaller units, deduct 10%.

System Components	QUANTITY	UNIT	COST EACH		
			MAT.	INST.	TOTAL
SYSTEM D3030 210 1300					
SELF-CONTAINED, WATER COOLED UNIT					
APARTMENT CORRIDORS, 500 S.F., .92 TON					
Self contained, water cooled, single package air conditioner unit	1.000	Ea.	1,730.40	428.40	2,158.80
Ductwork package for water or air cooled packaged units	1.000	System	55.66	602.60	658.26
Cooling tower, draw thru single flow, belt drive	1.000	Ea.	79.58	13.02	92.60
Cooling tower pumps & piping	1.000	System	40.02	30.82	70.84
TOTAL			1,905.66	1,074.84	2,980.50
COST PER S.F.			3.81	2.15	5.96

D3030 210	Self-contained, Water Cooled Unit Systems	COST PER S.F.		
		MAT.	INST.	TOTAL
1280	Self-contained, water cooled unit	3.81	2.16	5.97
1300	Apartment corridors, 500 S.F., .92 ton	3.80	2.15	5.95
1320	1,000 S.F., 1.83 ton	3.81	2.15	5.96
1360	3,000 S.F., 5.50 ton	3.02	1.80	4.82
1400	5,000 S.F., 9.17 ton	3.05	1.68	4.73
1440	10,000 S.F., 18.33 ton	2.60	1.51	4.11
1520	Banks or libraries, 500 S.F., 2.08 ton	8.40	2.14	10.54
1560	1,000 S.F., 4.17 ton	6.85	4.10	10.95
1600	3,000 S.F., 12.50 ton	6.95	3.81	10.76
1640	5,000 S.F., 20.80 ton	5.90	3.42	9.32
1680	10,000 S.F., 41.66 ton	5.10	3.42	8.52
1760	Bars and taverns, 500 S.F., 5.54 ton	17.45	3.62	21.07
1800	1,000 S.F., 11.08 ton	18	6.50	24.50
1840	3,000 S.F., 33.25 ton	13.70	5.15	18.85
1880	5,000 S.F., 55.42 ton	12.85	5.40	18.25
1920	10,000 S.F., 110.00 ton	12.55	5.30	17.85
2000	Bowling alleys, 500 S.F., 2.83 ton	11.45	2.92	14.37
2040	1,000 S.F., 5.66 ton	9.35	5.55	14.90
2080	3,000 S.F., 17.00 ton	8.05	4.66	12.71
2120	5,000 S.F., 28.33 ton	7.20	4.49	11.69
2160	10,000 S.F., 56.66 ton	6.75	4.61	11.36
2200	Department stores, 500 S.F., 1.46 ton	5.90	1.50	7.40
2240	1,000 S.F., 2.92 ton	4.81	2.87	7.68
2280	3,000 S.F., 8.75 ton	4.86	2.67	7.53

R15700 -020

Important: See the Reference Section for critical supporting data - Reference Numbers and City Cost Indexes

SERVICES D

D3030 Cooling Generating Systems

D3030 210	Self-contained, Water Cooled Unit Systems	COST PER S.F.		
		MAT.	INST.	TOTAL
2320	5,000 S.F., 14.58 ton	4.14	2.40	6.54
2360	10,000 S.F., 29.17 ton	3.72	2.31	6.03
2440	Drug stores, 500 S.F., 3.33 ton	13.45	3.43	16.88
2480	1,000 S.F., 6.66 ton	11	6.55	17.55
2520	3,000 S.F., 20.00 ton	9.45	5.50	14.95
2560	5,000 S.F., 33.33 ton	9.95	5.40	15.35
2600	10,000 S.F., 66.66 ton	7.95	5.45	13.40
2680	Factories, 500 S.F., 1.66 ton	6.70	1.72	8.42
2720	1,000 S.F. 3.37 ton	5.55	3.31	8.86
2760	3,000 S.F., 10.00 ton	5.55	3.05	8.60
2800	5,000 S.F., 16.66 ton	4.72	2.74	7.46
2840	10,000 S.F., 33.33 ton	4.25	2.64	6.89
2920	Food supermarkets, 500 S.F., 1.42 ton	5.70	1.45	7.15
2960	1,000 S.F., 2.83 ton	5.90	3.31	9.21
3000	3,000 S.F., 8.50 ton	4.67	2.78	7.45
3040	5,000 S.F., 14.17 ton	4.72	2.59	7.31
3080	10,000 S.F., 28.33 ton	3.62	2.25	5.87
3160	Medical centers, 500 S.F., 1.17 ton	4.70	1.20	5.90
3200	1,000 S.F., 2.33 ton	4.84	2.73	7.57
3240	3,000 S.F., 7.00 ton	3.84	2.29	6.13
3280	5,000 S.F., 11.66 ton	3.88	2.13	6.01
3320	10,000 S.F., 23.33 ton	3.30	1.92	5.22
3400	Offices, 500 S.F., 1.58 ton	6.40	1.63	8.03
3440	1,000 S.F., 3.17 ton	6.55	3.70	10.25
3480	3,000 S.F., 9.50 ton	5.25	2.89	8.14
3520	5,000 S.F., 15.83 ton	4.48	2.60	7.08
3560	10,000 S.F., 31.67 ton	4.03	2.51	6.54
3640	Restaurants, 500 S.F., 2.50 ton	10.10	2.58	12.68
3680	1,000 S.F., 5.00 ton	8.25	4.92	13.17
3720	3,000 S.F., 15.00 ton	8.30	4.58	12.88
3760	5,000 S.F., 25.00 ton	7.10	4.12	11.22
3800	10,000 S.F., 50.00 ton	4.21	3.79	8
3880	Schools and colleges, 500 S.F., 1.92 ton	7.75	1.97	9.72
3920	1,000 S.F., 3.83 ton	6.30	3.76	10.06
3960	3,000 S.F., 11.50 ton	6.40	3.50	9.90
4000	5,000 S.F., 19.17 ton	5.45	3.15	8.60
4040	10,000 S.F., 38.33 ton	4.68	3.15	7.83
4050				
9000	Components of ductwork packages for above systems, per ton of cooling:			
9010	Ductwork; galvanized steel, 108 pounds			
9020	Diffusers, aluminum, 24" x 12", two			

SERVICES

D

D3030 Cooling Generating Systems

System Description: Self-contained air cooled units with remote air cooled condenser and interconnecting tubing. Systems for 1000 S.F. and up include duct and diffusers. Smaller units distribute air directly.

Returns are not ducted and supplies are not insulated.

Potential savings may be realized by using a single zone rooftop system or through-the-wall unit, especially in the smaller capacities, if the application permits.

Hot water or steam heating coils are included but piping to boiler and the boiler itself is not included.

Condenserless models are available for 15% less where remote refrigerant source is available.

System Components	QUANTITY	UNIT	COST EACH		
			MAT.	INST.	TOTAL
SYSTEM D3030 214 1320					
SELF-CONTAINED, AIR COOLED UNIT					
APARTMENT CORRIDORS, 500 S.F., .92 TON					
Air cooled, package unit	1.000	Ea.	1,802	278.80	2,080.80
Ductwork package for water or air cooled packaged units	1.000	System	55.66	602.60	658.26
Refrigerant piping	1.000	System	110.40	377.20	487.60
Air cooled condenser, direct drive, propeller fan	1.000	Ea.	257.30	105.40	362.70
TOTAL			2,225.36	1,364	3,589.36
COST PER S.F.			4.45	2.73	7.18

D3030 214	Self-contained, Air Cooled Unit Systems		COST PER S.F.		
			MAT.	INST.	TOTAL
1300	Self-contained, air cooled unit				
1320	Apartment corridors, 500 S.F., .92 ton		4.45	2.75	7.20
1360	1,000 S.F., 1.83 ton	R15700	4.39	2.71	7.10
1400	3,000 S.F., 5.50 ton	-020	3.40	2.52	5.92
1440	5,000 S.F., 9.17 ton		3.02	2.41	5.43
1480	10,000 S.F., 18.33 ton		3.29	2.21	5.50
1560	Banks or libraries, 500 S.F., 2.08 ton		9.70	3.42	13.12
1600	1,000 S.F., 4.17 ton		7.70	5.75	13.45
1640	3,000 S.F., 12.50 ton		6.85	5.45	12.30
1680	5,000 S.F., 20.80 ton		7.45	5.05	12.50
1720	10,000 S.F., 41.66 ton		5.85	4.94	10.79
1800	Bars and taverns, 500 S.F., 5.54 ton		17.55	7.45	25
1840	1,000 S.F., 11.08 ton		17.90	10.90	28.80
1880	3,000 S.F., 33.25 ton		15.30	9.50	24.80
1920	5,000 S.F., 55.42 ton		14.80	9.50	24.30
1960	10,000 S.F., 110.00 ton		14.90	9.45	24.35
2040	Bowling alleys, 500 S.F., 2.83 ton		13.25	4.66	17.91
2080	1,000 S.F., 5.66 ton		10.50	7.80	18.30
2120	3,000 S.F., 17.00 ton		10.20	6.85	17.05
2160	5,000 S.F., 28.33 ton		8	6.70	14.70
2200	10,000 S.F., 56.66 ton		7.75	6.70	14.45
2240	Department stores, 500 S.F., 1.46 ton		6.80	2.41	9.21

D3030 Cooling Generating Systems

D3030 214	Self-contained, Air Cooled Unit Systems	COST PER S.F.		
		MAT.	INST.	TOTAL
2280	1,000 S.F., 2.92 ton	5.40	4.02	9.42
2320	3,000 S.F., 8.75 ton	4.80	3.83	8.63
2360	5,000 S.F., 14.58 ton	4.80	3.83	8.63
2400	10,000 S.F., 29.17 ton	4.12	3.45	7.57
2480	Drug stores, 500 S.F., 3.33 ton	15.55	5.50	21.05
2520	1,000 S.F., 6.66 ton	12.35	9.20	21.55
2560	3,000 S.F., 20.00 ton	12	8.10	20.10
2600	5,000 S.F., 33.33 ton	9.40	7.90	17.30
2640	10,000 S.F., 66.66 ton	9.15	7.95	17.10
2720	Factories, 500 S.F., 1.66 ton	7.90	2.77	10.67
2760	1,000 S.F., 3.33 ton	6.25	4.60	10.85
2800	3,000 S.F., 10.00 ton	5.50	4.38	9.88
2840	5,000 S.F., 16.66 ton	5.95	4.03	9.98
2880	10,000 S.F., 33.33 ton	4.70	3.95	8.65
2960	Food supermarkets, 500 S.F., 1.42 ton	6.65	2.33	8.98
3000	1,000 S.F., 2.83 ton	6.75	4.18	10.93
3040	3,000 S.F., 8.50 ton	5.25	3.90	9.15
3080	5,000 S.F., 14.17 ton	4.66	3.73	8.39
3120	10,000 S.F., 28.33 ton	3.99	3.36	7.35
3200	Medical centers, 500 S.F., 1.17 ton	5.50	1.94	7.44
3240	1,000 S.F., 2.33 ton	5.65	3.47	9.12
3280	3,000 S.F., 7.00 ton	4.33	3.22	7.55
3320	5,000 S.F., 16.66 ton	3.85	3.08	6.93
3360	10,000 S.F., 23.33 ton	4.19	2.83	7.02
3440	Offices, 500 S.F., 1.58 ton	7.40	2.61	10.01
3480	1,000 S.F., 3.16 ton	7.65	4.69	12.34
3520	3,000 S.F., 9.50 ton	5.25	4.16	9.41
3560	5,000 S.F., 15.83 ton	5.70	3.83	9.53
3600	10,000 S.F., 31.66 ton	4.47	3.74	8.21
3680	Restaurants, 500 S.F., 2.50 ton	11.65	4.10	15.75
3720	1,000 S.F., 5.00 ton	9.30	6.90	16.20
3760	3,000 S.F., 15.00 ton	8.20	6.55	14.75
3800	5,000 S.F., 25.00 ton	7	5.95	12.95
3840	10,000 S.F., 50.00 ton	7.05	5.90	12.95
3920	Schools and colleges, 500 S.F., 1.92 ton	8.95	3.16	12.11
3960	1,000 S.F., 3.83 ton	7.15	5.30	12.45
4000	3,000 S.F., 11.50 ton	6.30	5.05	11.35
4040	5,000 S.F., 19.17 ton	6.90	4.64	11.54
4080	10,000 S.F., 38.33 ton	5.40	4.54	9.94
4090				
9000	Components of ductwork packages for above systems, per ton of cooling:			
9010	Ductwork; galvanized steel, 102 pounds			
9020	Diffusers; aluminum, 24" x 12", two			

SERVICES

D

D3030 Cooling Generating Systems

Air Cooled Condensing Unit
Refrigerant Piping
Roof
Supply Duct
Fin. Ceiling
Return Grille
DX Air Handling Unit
Supply Diffuser

General: Split systems offer several important advantages which should be evaluated when a selection is to be made. They provide a greater degree of flexibility in component selection which permits an accurate match-up of the proper equipment size and type with the particular needs of the building. This allows for maximum use of modern energy saving concepts in heating and cooling. Outdoor installation of the air cooled condensing unit allows space savings in the building and also isolates the equipment operating sounds from building occupants.

Design Assumptions: The systems below are comprised of a direct expansion air handling unit and air cooled condensing unit with interconnecting copper tubing. Ducts and diffusers are also included for distribution of air. Systems are priced for cooling only. Heat can be added as desired either by putting hot water/steam coils into the air unit or into the duct supplying the particular area of need. Gas fired duct furnaces are also available. Refrigerant liquid line is insulated.

System Components	QUANTITY	UNIT	COST EACH		
			MAT.	INST.	TOTAL
SYSTEM D3030 218 1280					
SPLIT SYSTEM, AIR COOLED CONDENSING UNIT					
APARTMENT CORRIDORS, 1,000 S.F., 1.80 TON					
Fan coil AC unit, cabinet mntd & filters direct expansion air cool	1.000	Ea.	411.75	99.92	511.67
Ductwork package, for split system, remote condensing unit	1.000	System	52.16	571.88	624.04
Refrigeration piping	1.000	System	120.78	622.20	742.98
Condensing unit, air cooled, incls compressor & standard controls	1.000	Ea.	1,061.40	396.50	1,457.90
TOTAL			1,646.09	1,690.50	3,336.59
COST PER S.F.			1.65	1.69	3.34
*Cooling requirements would lead to choosing a water cooled unit					

D3030 218	Split Systems With Air Cooled Condensing Units	COST PER S.F.		
		MAT.	INST.	TOTAL
1260	Split system, air cooled condensing unit			
1280	Apartment corridors, 1,000 S.F., 1.83 ton	1.65	1.70	3.35
1320	2,000 S.F., 3.66 ton	1.22	1.71	2.93
1360	5,000 S.F., 9.17 ton	2.07	2.16	4.23
1400	10,000 S.F., 18.33 ton	2.18	2.36	4.54
1440	20,000 S.F., 36.66 ton	2.56	2.40	4.96
1520	Banks and libraries, 1,000 S.F., 4.17 ton	2.79	3.90	6.69
1560	2,000 S.F., 8.33 ton	4.72	4.90	9.62
1600	5,000 S.F., 20.80 ton	4.96	5.35	10.31
1640	10,000 S.F., 41.66 ton	5.85	5.45	11.30
1680	20,000 S.F., 83.32 ton	5.30	5.65	10.95
1760	Bars and taverns, 1,000 S.F., 11.08 ton	11.90	7.70	19.60
1800	2,000 S.F., 22.16 ton	15.30	9.20	24.50
1840	5,000 S.F., 55.42 ton	14.60	8.60	23.20
1880	10,000 S.F., 110.84 ton	12.80	8.95	21.75
1920	20,000 S.F., 220 ton*			
2000	Bowling alleys, 1,000 S.F., 5.66 ton	3.89	7.35	11.24
2040	2,000 S.F., 11.33 ton	6.40	6.65	13.05
2080	5,000 S.F., 28.33 ton	6.75	7.30	14.05
2120	10,000 S.F., 56.66 ton	7.95	7.40	15.35
2160	20,000 S.F., 113.32 ton	7.30	8	15.30
2320	Department stores, 1,000 S.F., 2.92 ton	1.97	2.67	4.64
2360	2,000 S.F., 5.83 ton	2.01	3.77	5.78
2400	5,000 S.F., 14.58 ton	3.31	3.42	6.73

Note: Row 1320 also shows reference box **R15700 -020**

D3030 Cooling Generating Systems

D3030 218	Split Systems With Air Cooled Condensing Units	COST PER S.F.		
		MAT.	INST.	TOTAL
2440	10,000 S.F., 29.17 ton	3.47	3.75	7.22
2480	20,000 S.F., 58.33 ton	4.09	3.80	7.89
2560	Drug stores, 1,000 S.F., 6.66 ton	4.59	8.60	13.19
2600	2,000 S.F., 13.32 ton	7.55	7.85	15.40
2640	5,000 S.F., 33.33 ton	9.35	8.70	18.05
2680	10,000 S.F., 66.66 ton	8.75	9.10	17.85
2720	20,000 S.F., 133.32 ton*			
2800	Factories, 1,000 S.F., 3.33 ton	2.25	3.05	5.30
2840	2,000 S.F., 6.66 ton	2.29	4.31	6.60
2880	5,000 S.F., 16.66 ton	3.97	4.28	8.25
2920	10,000 S.F., 33.33 ton	4.67	4.34	9.01
2960	20,000 S.F., 66.66 ton	4.37	4.54	8.91
3040	Food supermarkets, 1,000 S.F., 2.83 ton	1.92	2.59	4.51
3080	2,000 S.F., 5.66 ton	1.95	3.66	5.61
3120	5,000 S.F., 14.66 ton	3.22	3.33	6.55
3160	10,000 S.F., 28.33 ton	3.37	3.64	7.01
3200	20,000 S.F., 56.66 ton	3.97	3.70	7.67
3280	Medical centers, 1,000 S.F., 2.33 ton	1.61	2.10	3.71
3320	2,000 S.F., 4.66 ton	1.59	3.02	4.61
3360	5,000 S.F., 11.66 ton	2.64	2.74	5.38
3400	10,000 S.F., 23.33 ton	2.77	3	5.77
3440	20,000 S.F., 46.66 ton	3.27	3.04	6.31
3520	Offices, 1,000 S.F., 3.17 ton	2.14	2.90	5.04
3560	2,000 S.F., 6.33 ton	2.18	4.10	6.28
3600	5,000 S.F., 15.83 ton	3.58	3.73	7.31
3640	10,000 S.F., 31.66 ton	3.78	4.07	7.85
3680	20,000 S.F., 63.32 ton	4.15	4.32	8.47
3760	Restaurants, 1,000 S.F., 5.00 ton	3.45	6.50	9.95
3800	2,000 S.F., 10.00 ton	5.70	5.90	11.60
3840	5,000 S.F., 25.00 ton	5.95	6.45	12.40
3880	10,000 S.F., 50.00 ton	7	6.55	13.55
3920	20,000 S.F., 100.00 ton	6.45	7.10	13.55
4000	Schools and colleges, 1,000 S.F., 3.83 ton	2.56	3.58	6.14
4040	2,000 S.F., 7.66 ton	4.34	4.51	8.85
4080	5,000 S.F., 19.17 ton	4.56	4.93	9.49
4120	10,000 S.F., 38.33 ton	5.35	5	10.35
4160	20,000 S.F., 76.66 ton	5.05	5.20	10.25
5000				
5100				
5200				
9000	Components of ductwork packages for above systems, per ton of cooling:			
9010	Ductwork; galvanized steel, 102 pounds			
9020	Diffusers; aluminum, 24" x 12", two			

SERVICES

D

D30 HVAC

D3030 Cooling Generating Systems

Computer rooms impose special requirements on air conditioning systems. A prime requirement is reliability, due to the potential monetary loss that could be incurred by a system failure. A second basic requirement is the tolerance of control with which temperature and humidity are regulated, and dust eliminated. As the air conditioning system reliability is so vital, the additional cost of reserve capacity and redundant components is often justified.

System Descriptions: Computer areas may be environmentally controlled by one of three methods as follows:

1. Self-contained Units
These are units built to higher standards of performance and reliability.

They usually contain alarms and controls to indicate component operation failure, filter change, etc. It should be remembered that these units in the room will occupy space that is relatively expensive to build and that all alterations and service of the equipment will also have to be accomplished within the computer area.

2. Decentralized Air Handling Units In operation these are similar to the self-contained units except that their cooling capability comes from remotely located refrigeration equipment as refrigerant or chilled water. As no compressors or refrigerating equipment are required in the air units, they are smaller and require less service than

self-contained units. An added plus for this type of system occurs if some of the computer components themselves also require chilled water for cooling.

3. Central System Supply Cooling is obtained from a central source which, since it is not located within the computer room, may have excess capacity and permit greater flexibility without interfering with the computer components. System performance criteria must still be met.

Note: The costs shown below do not include an allowance for ductwork or piping.

D3030 222	Computer Room Cooling Units	COST EACH		
		MAT.	INST.	TOTAL
0560	Computer room unit, air cooled, includes remote condenser			
0580	3 ton	8,850	1,625	10,475
0600	5 ton	11,800	1,825	13,625
0620	8 ton	24,700	3,025	27,725
0640	10 ton	25,700	3,275	28,975
0660	15 ton	28,700	3,725	32,425
0680	20 ton	34,200	4,400	38,600
0700	23 ton	36,000	4,550	40,550
0800	Chilled water, for connection to existing chiller system			
0820	5 ton	9,975	1,100	11,075
0840	8 ton	14,300	1,625	15,925
0860	10 ton	15,800	1,675	17,475
0880	15 ton	17,000	1,700	18,700
0900	20 ton	17,400	1,775	19,175
0920	23 ton	18,300	1,950	20,250
1000	Glycol system, complete except for interconnecting tubing			
1020	3 ton	10,800	2,050	12,850
1040	5 ton	14,100	2,150	16,250
1060	8 ton	30,500	3,550	34,050
1080	10 ton	33,100	3,900	37,000
1100	15 ton	40,100	4,900	45,000
1120	20 ton	42,900	5,300	48,200
1140	23 ton	46,800	5,775	52,575
1240	Water cooled, not including condenser water supply or cooling tower			
1260	3 ton	8,000	1,325	9,325
1280	5 ton	11,300	1,525	12,825
1300	8 ton	23,200	2,475	25,675
1320	15 ton	28,400	3,025	31,425
1340	20 ton	33,500	3,350	36,850
1360	23 ton	35,500	3,750	39,250

SERVICES D

Important: See the Reference Section for critical supporting data - Reference Numbers and City Cost Indexes

D3030 Cooling Generating Systems

Gas Cock

Roof Curb

Gas Piping

Rooftop Air Conditioner Unit

System Components	QUANTITY	UNIT	COST EACH		
			MAT.	INST.	TOTAL
SYSTEM D3030 228 1010					
A/C, ROOFTOP, DX COOL, GAS HEAT, CURB, ECONOMIZER, FILTERS., 5 TON					
Roof top A/C, curb, economizer, sgl zone, elec cool, gas ht, 5 ton, 112 MBH	1.000	Ea.	5,675	1,450	7,125
Pipe, black steel, Sch 40, threaded, W/cplgs & hangers, 10' OC, 1" dia	20.000	L.F.	43.47	178.50	221.97
Elbow, 90°, black, straight, 3/4" dia.	3.000	Ea.	4.56	96	100.56
Elbow, 90°, black, straight, 1" dia.	3.000	Ea.	8.10	103.50	111.60
Tee, black, reducing, 1" dia.	1.000	Ea.	4.80	56.50	61.30
Union, black with brass seat, 1" dia.	1.000	Ea.	9.90	37.50	47.40
Pipe nipple, black, 3/4" dia	2.000	Ea.	3.24	14.80	18.04
Pipe nipple, black, 1" dia	2.000	Ea.	2.07	8.50	10.57
Cap, black, 1" dia.	1.000	Ea.	2.12	15.05	17.17
Gas cock, brass, 1" size	1.000	Ea.	13.85	24	37.85
Control system, pneumatic, Rooftop A/C unit	1.000	Ea.	2,050	1,600	3,650
Rooftop unit heat/cool balancing	1.000	Ea.		320	320
TOTAL			7,817.11	3,904.35	11,721.46

D3030 228	Rooftop Air Conditioner, Const. Volume	COST EACH		
		MAT.	INST.	TOTAL
1010	A/C, Rooftop, DX cool, gas heat, curb, economizer, fltrs, 5 Ton	7,825	3,900	11,725
1020	7-1/2 Ton	11,400	4,075	15,475
1030	12-1/2 Ton	16,900	4,475	21,375
1040	18 Ton	27,200	4,875	32,075
1050	25 Ton	40,200	5,575	45,775
1060	40 Ton	66,000	7,425	73,425

D3030 230	Rooftop Air Conditioner, Variable Air Volume	COST EACH		
		MAT.	INST.	TOTAL
1010	A/C, Rooftop, DX cool, gas heat, curb, ecmizr, fltrs, VAV, 12-1/2 Ton	19,000	4,775	23,775
1020	18 Ton	29,600	5,250	34,850
1030	25 Ton	43,800	6,025	49,825
1040	40 Ton	73,500	8,175	81,675
1050	60 Ton	102,500	11,100	113,600
1060	80 Ton	159,000	13,900	172,900

SERVICES

D

D3030 Cooling Generating Systems

D3030 232	Packaged A/C, Elec. Ht., Const. Volume	COST EACH		
		MAT.	INST.	TOTAL
1010	A/C Packaged, DX, air cooled, electric heat, constant vol., 5 Ton	9,100	2,850	11,950
1020	10 Ton	15,600	3,925	19,525
1030	20 Ton	32,200	5,175	37,375
1040	30 Ton	37,200	5,750	42,950
1050	40 Ton	48,500	6,600	55,100
1060	50 Ton	61,000	7,375	68,375

D3030 234	Packaged Air Conditioner, Elect. Heat, VAV	COST EACH		
		MAT.	INST.	TOTAL
1010	A/C Packaged, DX, air cooled, electric heat, VAV, 10 Ton	17,900	4,925	22,825
1020	20 Ton	36,200	6,525	42,725
1030	30 Ton	42,800	7,100	49,900
1040	40 Ton	56,500	8,125	64,625
1050	50 Ton	70,000	9,050	79,050
1060	60 Ton	80,000	10,600	90,600

D3030 236	Packaged A/C, Hot Wtr. Heat, Const. Volume	COST EACH		
		MAT.	INST.	TOTAL
1010	A/C Packaged, DX, air cooled, H/W heat, constant vol., 5 Ton	8,250	4,175	12,425
1020	10 Ton	13,600	5,500	19,100
1030	20 Ton	30,500	7,400	37,900
1040	30 Ton	32,600	9,300	41,900
1050	40 Ton	44,500	10,500	55,000
1060	50 Ton	54,500	11,300	65,800

D3030 238	Packaged A/C, Hot Wtr. Heat, VAV	COST EACH		
		MAT.	INST.	TOTAL
1010	A/C Packaged, DX, air cooled, H/W heat, VAV, 10 Ton	16,000	6,400	22,400
1020	20 Ton	34,500	8,425	42,925
1030	30 Ton	38,100	10,600	48,700
1040	40 Ton	52,500	11,400	63,900
1050	50 Ton	63,000	13,200	76,200
1060	60 Ton	76,000	14,800	90,800

Important: See the Reference Section for critical supporting data - Reference Numbers and City Cost Indexes

D3030 Cooling Generating Systems

Cold Water Fill

Drain To Waste Piping

Cooling Tower

Condenser Water Piping

SERVICES

D

System Components			COST EACH		
	QUANTITY	UNIT	MAT.	INST.	TOTAL
SYSTEM D3030 310 1010					
COOLING TOWER, GALVANIZED STEEL, PACKAGED UNIT, DRAW THRU, 60 TON					
Cooling towers, draw thru, single flow, belt drive, 60 tons	1.000	Ea.	5,190	849	6,039
Pipe, black steel, welded, Sch. 40, on yoke & roll hangers, 10' OC, 4" dia.	40.000	L.F.	396	973.60	1,369.60
Tubing, copper, Type L, couplings & hangers 10' OC, 1-1/2" dia.	20.000	L.F.	81	174	255
Elbow, 90°, copper, cu x cu, 1-1/2" dia.	3.000	Ea.	19.80	103.50	123.30
Elbow, 90°, weld joint, 4" pipe size	6.000	Ea.	123	1,081.80	1,204.80
Flange, welding neck, 150 lb., 4" pipe size	6.000	Ea.	171	540.90	711.90
Valves, bronze, gate, N.R.S., soldered, 125 psi, 1-1/2" size	1.000	Ea.	49.50	34.50	84
Valves, iron body, butterfly, lug type, lever actuator, 4" size	2.000	Ea.	234	326	560
Electric heat trace system, 400 degree, 115v, 5 watts per L.F.	160.000	L.F.	1,176	136	1,312
Insulation, fiberglass pipe covering, , 1-1/2" wall, 4" IPS	40.000	L.F.	206	228	434
Insulation, fiberglass pipe covering, 1" wall, 1-1/2" IPS	20.000	L.F.	41	75.80	116.80
Cooling tower condenser control system	1.000	Ea.	4,100	1,900	6,000
Cooling tower balancing	1.000	Ea.		295	295
TOTAL			11,787.30	6,718.10	18,505.40

D3030 310	Galvanized Draw Through Cooling Tower	COST EACH		
		MAT.	INST.	TOTAL
1010	Cooling tower, galvanized steel, packaged unit, draw thru, 60 Ton	11,800	6,725	18,525
1020	110 Ton	14,500	7,900	22,400
1030	300 Ton	28,300	11,800	40,100
1040	600 Ton	44,500	17,600	62,100
1050	1000 Ton	68,500	25,900	94,400

D3030 320	Fiberglass Draw Through Cooling Tower	COST EACH		
		MAT.	INST.	TOTAL
1010	Cooling tower, fiberglass, packaged unit, draw thru, 60 Ton	9,650	6,725	16,375
1020	125 Ton	14,000	7,900	21,900
1030	300 Ton	24,600	11,800	36,400
1040	600 Ton	41,600	17,600	59,200
1050	1000 Ton	67,000	25,900	92,900

D3030 330	Stainless Steel Draw Through Cooling Tower	COST EACH		
		MAT.	INST.	TOTAL
1010	Cooling tower, stainless steel, packaged unit, draw thru, 60 Ton	15,300	6,725	22,025
1020	110 Ton	19,700	7,900	27,600
1030	300 Ton	23,300	11,800	35,100
1040	600 Ton	28,500	17,600	46,100
1050	1000 Ton	108,500	17,400	125,900

D3040 Distribution Systems

Two-Way Valve Piping   Filters   Heating Coils   **Three-Way Valve Piping**

Service Lights (typ.)

Cooling Coil

18 Ga. Exterior with 7.5# per S.F. Insulation

Sound Absorptive 24 Ga. Perforated Inner Panels

Double Walled Access Doors

Built-Up Air Handling Unit

System Components	QUANTITY	UNIT	COST EACH		
			MAT.	INST.	TOTAL
SYSTEM D3040 106 1010					
AHU, FIELD FAB., BLT. UP, COOL/HEAT COILS, FLTRS., CONST. VOL., 40,000 CFM					
AHU, Built-up, cool/heat coils, filters, mix box, constant volume	1.000	Ea.	42,400	4,250	46,650
Pipe, black steel, welded, Sch. 40, on yoke & roll hgrs, 10' OC, 3" dia.	26.000	L.F.	211.90	543.66	755.56
Pipe, black steel, welded, Sch. 40, on yoke & roll hgrs, 10' OC, 4" dia	20.000	L.F.	198	486.80	684.80
Elbow, 90°, weld joint, 3" pipe size	5.000	Ea.	59.75	641.75	701.50
Elbow, 90°, weld joint, 4" pipe size	4.000	Ea.	82	721.20	803.20
Tee, welded, reducing on outlet, 3" by 2-1/2" pipe	3.000	Ea.	109.50	673.50	783
Tee, welded, reducing on outlet, 4" by 3" pipe	2.000	Ea.	93	600	693
Reducer, eccentric, weld joint, 3" pipe size	4.000	Ea.	84	451.20	535.20
Reducer, eccentric, weld joint, 4" pipe size	4.000	Ea.	110	597.60	707.60
Flange, weld neck, 150 LB, 2-1/2" pipe size	3.000	Ea.	66	169.20	235.20
Flange, weld neck, 150 lb., 3" pipe size	14.000	Ea.	336	898.10	1,234.10
Flange, weld neck, 150 lb., 4" pipe size	10.000	Ea.	285	901.50	1,186.50
Valves, iron body, butterfly, lug type, lever actuator, 3" size	2.000	Ea.	191	204	395
Valves, iron body, butterfly, lug type, lever actuator, 4" size	2.000	Ea.	234	326	560
Strainer, Y type, iron body, flanged, 125 lb., 3" pipe size	1.000	Ea.	93	182	275
Strainer, Y type, iron body, flanged, 125 lb., 4" pipe size	1.000	Ea.	170	273	443
Valve, electric motor actuated, iron body, 3 way, flanged, 2-1/2" pipe	1.000	Ea.	850	225	1,075
Valve, electric motor actuated, iron body, two way, flanged, 3" pipe size	1.000	Ea.	515	225	740
Expansion joints, bellows type, flange spool, 6" F to F, 3" dia	2.000	Ea.	424	144	568
Expansion joints, bellows type, flange spool, 6" F to F, 4" dia	2.000	Ea.	454	195	649
T-O-L socket, weld joint, 1/4" pipe size, nozzle	4.000	Ea.	18.96	157.04	176
T-O-L socket, weld joint, 1/2" pipe size, nozzle	12.000	Ea.	56.88	491.16	548.04
Thermometers, stem type, 9" case, 8" stem, 3/4" NPT	4.000	Ea.	532	65	597
Pressure/temperature relief plug, 316 SS, 3/4"OD, 7-1/2" insertion	4.000	Ea.	252	56.80	308.80
Insulation, fiberglass pipe covering, 1-1/2" wall, 3" IPS	26.000	L.F.	117	121.68	238.68
Insulation, fiberglass pipe covering, 1-1/2" wall, 4" IPS	20.000	L.F.	103	114	217
Fan coil control system	1.000	Ea.	925	420	1,345
Re-heat coil balancing	2.000	Ea.		168	168
Air conditioner balancing	1.000	Ea.		550	550
TOTAL			48,970.99	14,852.19	63,823.18

D3040 106	Field Fabricated Air Handling Unit	COST EACH		
		MAT.	INST.	TOTAL
1010	AHU, Field fab, blt up, cool/heat coils, fltrs, const vol, 40,000 CFM	49,000	14,800	63,800
1020	60,000 CFM	64,000	22,900	86,900
1030	75,000 CFM	78,000	24,100	102,100

Important: See the Reference Section for critical supporting data - Reference Numbers and City Cost Indexes

D3040 108	Field Fabricated VAV Air Handling Unit	COST EACH		
		MAT.	INST.	TOTAL
1010	AHU, Field fab, blt up, cool/heat coils, fltrs, VAV, 75,000 CFM	79,500	27,500	107,000
1020	100,000 CFM	125,000	36,500	161,500
1030	150,000 CFM	204,000	58,500	262,500

SERVICES

D

D3040 Distribution Systems

Chilled Water Piping **Central Station Air Handling Unit** **Hot Water Piping**

System Components	QUANTITY	UNIT	COST EACH		
			MAT.	INST.	TOTAL
SYSTEM D3040 110 1010					
AHU, CENTRAL STATION, COOL/HEAT COILS, CONST. VOL., FLTRS., 2,000 CFM					
Central station air handling unit, chilled water, 1900 CFM	1.000	Ea.	3,680	856.75	4,536.75
Pipe, black steel, Sch 40, threaded, W/cplgs & hangers, 10' OC, 1" dia.	26.000	L.F.	67.28	276.25	343.53
Pipe, black steel, Sch 40, threaded, W/cplg & hangers, 10' OC, 1-1/4" dia.	20.000	L.F.	63.96	237.90	301.86
Elbow, 90°, black, straight, 1" dia.	4.000	Ea.	10.80	138	148.80
Elbow, 90°, black, straight, 1-1/4" dia.	4.000	Ea.	17.68	148	165.68
Tee, black, reducing, 1" dia.	8.000	Ea.	38.40	452	490.40
Tee, black, reducing, 1-1/4" dia.	8.000	Ea.	66.40	464	530.40
Reducer, concentric, black, 1" dia.	4.000	Ea.	14.84	120	134.84
Reducer, concentric, black, 1-1/4" dia.	4.000	Ea.	21.80	126	147.80
Union, black with brass seat, 1" dia.	4.000	Ea.	39.60	150	189.60
Union, black with brass seat, 1-1/4" dia.	2.000	Ea.	26.40	77	103.40
Pipe nipple, black, 1" dia	13.000	Ea.	13.46	55.25	68.71
Pipe nipple, black, 1-1/4" dia	12.000	Ea.	14.76	54.90	69.66
Valves, bronze, gate, N.R.S., threaded, class 150, 1" size	2.000	Ea.	79	48	127
Valves, bronze, gate, N.R.S., threaded, class 150, 1-1/4" size	2.000	Ea.	106	60	166
Strainer, Y type, bronze body, screwed, 150 lb., 1" pipe size	1.000	Ea.	20.50	26.50	47
Strainer, Y type, bronze body, screwed, 150 lb., 1-1/4" pipe size	1.000	Ea.	46	30.50	76.50
Valve, electric motor actuated, brass, 3 way, screwed, 3/4" pipe	1.000	Ea.	261	25	286
Valve, electric motor actuated, brass, 2 way, screwed, 1" pipe size	1.000	Ea.	365	24	389
Thermometers, stem type, 9" case, 8" stem, 3/4" NPT	4.000	Ea.	532	65	597
Pressure/temperature relief plug, 316 SS, 3/4"OD, 7-1/2" insertion	4.000	Ea.	252	56.80	308.80
Insulation, fiberglass pipe covering, 1" wall, 1" IPS	26.000	L.F.	43.68	93.86	137.54
Insulation, fiberglass pipe covering, 1" wall, 1-1/4" IPS	20.000	L.F.	38	75.80	113.80
Fan coil control system	1.000	Ea.	925	420	1,345
Re-heat coil balancing	2.000	Ea.		168	168
Multizone A/C balancing	1.000	Ea.		274	274
TOTAL			6,743.56	4,523.51	11,267.07

D3040 110	Central Station Air Handling Unit	COST EACH		
		MAT.	INST.	TOTAL
1010	AHU, Central station, cool/heat coils, constant vol, fltrs, 2,000 CFM	6,750	4,525	11,275
1020	5,000 CFM	10,200	5,950	16,150
1030	10,000 CFM	13,600	8,475	22,075
1040	15,000 CFM	19,100	11,700	30,800
1050	20,000 CFM	22,800	13,700	36,500

Important: See the Reference Section for critical supporting data - Reference Numbers and City Cost Indexes

SERVICES D

D3040 112	Central Station Air Handling Unit, VAV	COST EACH		
		MAT.	INST.	TOTAL
1010	AHU, Central station, cool/heat coils, VAV, fltrs, 5,000 CFM	12,300	5,950	18,250
1020	10,000 CFM	16,100	8,325	24,425
1030	15,000 CFM	21,800	11,000	32,800
1040	20,000 CFM	27,000	13,300	40,300
1050	30,000 CFM	40,600	15,300	55,900

SERVICES

D

D3040 Distribution Systems

Cooling Coil

Two-Way Valve Piping

Heating Coil

Three-Way Valve Piping

Exhaust Section
Mixing Box
Filter Section
Coil Section

Return Air Fan Section

Roof Mounting Curb

Supply Air Fan Section

Rooftop Air Handling Unit

System Components	QUANTITY	UNIT	COST EACH		
			MAT.	INST.	TOTAL
SYSTEM D3040 114 1010					
AHU, ROOFTOP, COOL/HEAT COILS, FLTRS., CONST. VOL., 2,000 CFM					
AHU, Built-up, cool/heat coils, filter, mix box, rooftop, constant volume	1.000	Ea.	4,175	820	4,995
Pipe, black steel, Sch 40, threaded, W/cplgs & hangers, 10' OC, 1" dia.	26.000	L.F.	67.28	276.25	343.53
Pipe, black steel, Sch 40, threaded, W/cplg & hangers, 10' OC, 1-1/4" dia.	20.000	L.F.	63.96	237.90	301.86
Elbow, 90°, black, straight, 1" dia.	4.000	Ea.	10.80	138	148.80
Elbow, 90°, black, straight, 1-1/4" dia.	4.000	Ea.	17.68	148	165.68
Tee, black, reducing, 1" dia.	8.000	Ea.	38.40	452	490.40
Tee, black, reducing, 1-1/4" dia.	8.000	Ea.	66.40	464	530.40
Reducer, black, concentric, 1" dia.	4.000	Ea.	14.84	120	134.84
Reducer, black, concentric, 1-1/4" dia.	4.000	Ea.	21.80	126	147.80
Union, black with brass seat, 1" dia.	4.000	Ea.	39.60	150	189.60
Union, black with brass seat, 1-1/4" dia.	2.000	Ea.	26.40	77	103.40
Pipe nipples, 1" dia	13.000	Ea.	13.46	55.25	68.71
Pipe nipples, 1-1/4" dia	12.000	Ea.	14.76	54.90	69.66
Valves, bronze, gate, N.R.S., threaded, class 150, 1" size	2.000	Ea.	79	48	127
Valves, bronze, gate, N.R.S., threaded, class 150, 1-1/4" size	2.000	Ea.	106	60	166
Strainer, Y type, bronze body, screwed, 150 lb., 1" pipe size	1.000	Ea.	20.50	26.50	47
Strainer, Y type, bronze body, screwed, 150 lb., 1-1/4" pipe size	1.000	Ea.	46	30.50	76.50
Valve, electric motor actuated, brass, 3 way, screwed, 3/4" pipe	1.000	Ea.	261	25	286
Valve, electric motor actuated, brass, 2 way, screwed, 1" pipe size	1.000	Ea.	365	24	389
Thermometers, stem type, 9" case, 8" stem, 3/4" NPT	4.000	Ea.	532	65	597
Pressure/temperature relief plug, 316 SS, 3/4"OD, 7-1/2" insertion	4.000	Ea.	252	56.80	308.80
Insulation, fiberglass pipe covering, 1" wall, 1" IPS	26.000	L.F.	43.68	93.86	137.54
Insulation, fiberglass pipe covering, 1" wall, 1-1/4" IPS	20.000	L.F.	38	75.80	113.80
Insulation pipe covering .016" aluminum jacket finish, add	45.000	S.F.	20.70	178.65	199.35
Fan coil control system	1.000	Ea.	925	420	1,345
Re-heat coil balancing	2.000	Ea.		168	168
Rooftop unit heat/cool balancing	1.000	Ea.		320	320
TOTAL			7,259.26	4,711.41	11,970.67

D3040 114	Rooftop Air Handling Unit	COST EACH		
		MAT.	INST.	TOTAL
1010	AHU, Rooftop, cool/heat coils, constant vol., fltrs, 2,000 CFM	7,250	4,700	11,950
1020	5,000 CFM	11,100	5,975	17,075

D3040   Distribution Systems

D3040 114	Rooftop Air Handling Unit	COST EACH		
		MAT.	INST.	TOTAL
1030	10,000 CFM	14,800	8,450	23,250
1040	15,000 CFM	20,800	11,500	32,300
1050	20,000 CFM	24,800	13,600	38,400

D3040 116	Rooftop Air Handling Unit, VAV	COST EACH		
		MAT.	INST.	TOTAL
1010	AHU, Rooftop, cool/heat coils, VAV, fltrs, 5,000 CFM	12,900	6,200	19,100
1020	10,000 CFM	16,800	8,550	25,350
1030	15,000 CFM	22,900	11,400	34,300
1040	20,000 CFM	28,400	13,700	42,100
1050	30,000 CFM	43,000	15,700	58,700

SERVICES

D

Galvanized Steel Duct

Fiberglass Duct Insulation

Flexible Fiberglass Duct

Supply Piping

Horizontal Fan Air Conditioning System

Return Piping

System Components	QUANTITY	UNIT	COST EACH		
			MAT.	INST.	TOTAL
SYSTEM D3040 118 1010					
FAN COIL A/C SYSTEM, HORIZONTAL, W/HOUSING, CONTROLS, 2 PIPE, 1/2 TON					
Fan coil A/C, horizontal housing, filters, chilled water, 1/2 ton cooling	1.000	Ea.	945	102	1,047
Pipe, black steel, Sch 40, threaded, W/cplgs & hangers, 10' OC, 3/4" dia.	20.000	L.F.	34.83	159.10	193.93
Elbow, 90°, black, straight, 3/4" dia.	6.000	Ea.	9.12	192	201.12
Tee, black, straight, 3/4" dia.	2.000	Ea.	4.92	100	104.92
Union, black with brass seat, 3/4" dia.	2.000	Ea.	13.80	69	82.80
Pipe nipples, 3/4" diam	3.000	Ea.	2.43	11.10	13.53
Valves, bronze, gate, N.R.S., threaded, class 150, 3/4" size	2.000	Ea.	61	45	106
Circuit setter, balance valve, bronze body, threaded, 3/4" pipe size	1.000	Ea.	52	22.50	74.50
Insulation, fiberglass pipe covering, 1" wall, 3/4" IPS	20.000	L.F.	33	69.20	102.20
Ductwork, 12" x 8" fabricated, galvanized steel, 12 LF	55.000	Lb.	48.95	299.75	348.70
Insulation, ductwork, blanket type, fiberglass, 1" thk, 1-1/2 LB density	40.000	S.F.	20.40	90.80	111.20
Diffusers, aluminum, OB damper, ceiling, perf, 24"x24" panel size, 6"x6"	2.000	Ea.	150	57	207
Ductwork, flexible, fiberglass fabric, insulated, 1"thk, PE jacket, 6" dia	16.000	L.F.	25.92	50.72	76.64
Round volume control damper 6" dia.	2.000	Ea.	48	42	90
Fan coil unit balancing	1.000	Ea.		59.50	59.50
Re-heat coil balancing	1.000	Ea.		84	84
Diffuser/register, high, balancing	2.000	Ea.		165	165
TOTAL			1,449.37	1,618.67	3,068.04

D3040 118	Fan Coil A/C Unit, Two Pipe	COST EACH		
		MAT.	INST.	TOTAL
1010	Fan coil A/C system, cabinet mounted, controls, 2 pipe, 1/2 Ton	1,050	850	1,900
1020	1 Ton	1,275	955	2,230
1030	1-1/2 Ton	1,425	965	2,390
1040	2 Ton	2,125	1,175	3,300
1050	3 Ton	3,050	1,225	4,275

D3040 120	Fan Coil A/C Unit, Two Pipe, Electric Heat	COST EACH		
		MAT.	INST.	TOTAL
1010	Fan coil A/C system, cabinet mntd, elect. ht, controls, 2 pipe, 1/2 Ton	1,225	850	2,075
1020	1 Ton	1,475	955	2,430
1030	1-1/2 Ton	1,750	965	2,715
1040	2 Ton	2,600	1,200	3,800
1050	3 Ton	4,425	1,225	5,650

Important: See the Reference Section for critical supporting data - Reference Numbers and City Cost Indexes

D3040 Distribution Systems

D3040 122	Fan Coil A/C Unit, Four Pipe	COST EACH		
		MAT.	INST.	TOTAL
1010	Fan coil A/C system, cabinet mounted, controls, 4 pipe, 1/2 Ton	1,600	1,650	3,250
1020	1 Ton	1,875	1,750	3,625
1030	1-1/2 Ton	2,100	1,775	3,875
1040	2 Ton	2,650	1,850	4,500
1050	3 Ton	4,000	1,975	5,975

D3040 124	Fan Coil A/C, Horizontal, Duct Mount, 2 Pipe	COST EACH		
		MAT.	INST.	TOTAL
1010	Fan coil A/C system, horizontal w/housing, controls, 2 pipe, 1/2 Ton	1,450	1,625	3,075
1020	1 Ton	1,900	2,375	4,275
1030	1-1/2 Ton	2,375	3,075	5,450
1040	2 Ton	3,025	3,700	6,725
1050	3 Ton	3,650	4,900	8,550
1060	3-1/2 Ton	3,850	5,075	8,925
1070	4 Ton	3,950	5,775	9,725
1080	5 Ton	4,675	7,350	12,025
1090	6 Ton	4,750	7,875	12,625
1100	7 Ton	4,750	8,175	12,925
1110	8 Ton	4,750	8,350	13,100
1120	10 Ton	5,200	8,650	13,850

D3040 126	Fan Coil A/C, Horiz., Duct Mount, 2 Pipe, Elec. Ht.	COST EACH		
		MAT.	INST.	TOTAL
1010	Fan coil A/C system, horiz. hsng, elect. ht, ctrls, 2 pipe, 1/2 Ton	1,650	1,625	3,275
1020	1 Ton	2,150	2,375	4,525
1030	1-1/2 Ton	2,550	3,075	5,625
1040	2 Ton	3,325	3,700	7,025
1050	3 Ton	5,375	5,075	10,450
1060	3-1/2 Ton	6,075	5,075	11,150
1070	4 Ton	6,850	5,775	12,625
1080	5 Ton	7,850	7,350	15,200
1090	6 Ton	7,925	7,900	15,825
1100	7 Ton	8,000	7,850	15,850
1110	8 Ton	8,750	7,950	16,700
1120	10 Ton	9,325	8,175	17,500

D3040 128	Fan Coil A/C, Horizontal, Duct Mount, 4 Pipe	COST EACH		
		MAT.	INST.	TOTAL
1010	Fan coil A/C system, horiz. w/cabinet, controls, 4 pipe, 1/2 Ton	2,050	2,400	4,450
1020	1 Ton	2,575	3,250	5,825
1030	1-1/2 Ton	3,125	3,900	7,025
1040	2 Ton	3,600	4,375	7,975
1050	3 Ton	4,400	5,825	10,225
1060	3.5 Ton	5,050	7,150	12,200
1070	4 Ton	5,050	7,200	12,250
1080	5 Ton	6,075	9,925	16,000
1090	6 Ton	6,450	11,800	18,250
1100	7 Ton	7,300	13,300	20,600
1110	8 Ton	7,400	14,000	21,400
1120	10 Ton	8,900	17,200	26,100

SERVICES

D

D3040 Distribution Systems

Fiberglass Duct Insulation

Galvanized Steel Duct

Flexible Insulated Duct

VAV Terminal

Fan Powered VAV Terminal

System Components	QUANTITY	UNIT	COST EACH		
			MAT.	INST.	TOTAL
SYSTEM D3040 132 1010					
VAV TERMINAL, COOLING ONLY, WITH ACTUATOR / CONTROLS, 200 CFM					
Mixing box variable volume 300 to 600CFM, cool only	1.000	Ea.	335.50	82.50	418
Ductwork, 12" x 8" fabricated, galvanized steel, 12 LF	55.000	Lb.	48.95	299.75	348.70
Insulation, ductwork, blanket type, fiberglass, 1" thk, 1-1/2 LB density	40.000	S.F.	20.40	90.80	111.20
Diffusers, aluminum, OB damper, ceiling, perf, 24"x24" panel size, 6"x6"	2.000	Ea.	150	57	207
Ductwork, flexible, fiberglass fabric, insulated, 1"thk, PE jacket, 6" dia	16.000	L.F.	25.92	50.72	76.64
Round volume control damper 6" dia.	2.000	Ea.	48	42	90
VAV box balancing	1.000	Ea.		55	55
Diffuser/register, high, balancing	2.000	Ea.		165	165
TOTAL			628.77	842.77	1,471.54

D3040 132	VAV Terminal, Cooling Only	COST EACH		
		MAT.	INST.	TOTAL
1010	VAV Terminal, cooling only, with actuator / controls, 200 CFM	630	845	1,475
1020	400 CFM	810	1,500	2,310
1030	600 CFM	1,075	2,200	3,275
1040	800 CFM	1,150	2,675	3,825
1050	1000 CFM	1,350	3,100	4,450
1060	1250 CFM	1,475	3,950	5,425
1070	1500 CFM	1,575	4,650	6,225
1080	2000 CFM	2,075	6,600	8,675

D3040 134	VAV Terminal, Hot Water Reheat	COST EACH		
		MAT.	INST.	TOTAL
1010	VAV Terminal, cooling, HW reheat, with actuator / controls, 200 CFM	1,450	1,550	3,000
1020	400 CFM	1,650	2,125	3,775
1030	600 CFM	1,900	2,900	4,800
1040	800 CFM	2,000	3,400	5,400
1050	1000 CFM	2,150	3,825	5,975
1060	1250 CFM	2,350	4,700	7,050
1070	1500 CFM	2,575	5,550	8,125
1080	2000 CFM	3,125	7,550	10,675

Important: See the Reference Section for critical supporting data - Reference Numbers and City Cost Indexes

D3040 Distribution Systems

D3040 136	Fan Powered VAV Terminal, Cooling Only	COST EACH		
		MAT.	INST.	TOTAL
1010	VAV Terminal, cooling, fan powrd, with actuator / controls, 200 CFM	1,150	845	1,995
1020	400 CFM	1,325	1,375	2,700
1030	600 CFM	1,600	2,200	3,800
1040	800 CFM	1,675	2,625	4,300
1050	1000 CFM	1,825	3,050	4,875
1060	1250 CFM	2,125	4,000	6,125
1070	1500 CFM	2,225	4,675	6,900
1080	2000 CFM	2,775	6,800	9,575

D3040 138	Fan Powered VAV Terminal, Hot Water Reheat	COST EACH		
		MAT.	INST.	TOTAL
1010	VAV Terminal, cool, HW reht, fan powrd, with actuator/ctrls, 200 CFM	1,850	1,550	3,400
1020	400 CFM	2,050	2,125	4,175
1030	600 CFM	2,300	2,900	5,200
1040	800 CFM	2,375	3,350	5,725
1050	1000 CFM	2,550	3,775	6,325
1060	1250 CFM	2,925	4,725	7,650
1070	1500 CFM	3,150	5,600	8,750
1080	2000 CFM	3,725	7,750	11,475

SERVICES

D

Axial Flow Centrifugal Fan
Belt Drive

Belt Drive Utility Set

Centrifugal Roof Exhaust Fan
Direct Drive

System Components	QUANTITY	UNIT	COST EACH		
			MAT.	INST.	TOTAL
SYSTEM D3040 220 1010					
FAN SYSTEM, IN-LINE CENTRIFUGAL, 500 CFM					
Fans, in-line centrifugal, supply/exhaust, 500 CFM, 10" dia conn	1.000	Ea.	785	350	1,135
Louver, alum., W/screen, damper, mill fin, fxd blade, cont line, stormproof	4.050	S.F.	125.55	66.22	191.77
Motor for louver damper, electric or pneumatic	1.000	Ea.	365	32.50	397.50
Ductwork, 12" x 8" fabricated, galvanized steel, 60 LF	200.000	Lb.	145.60	1,352	1,497.60
Grilles, aluminum, air supply, single deflection, adjustable, 12" x 6"	1.000	Ea.	48.60	59.70	108.30
Fire damper, curtain type, vertical, 12" x 8"	1.000	Ea.	21.50	21	42.50
Duct access door, insulated, 10" x 10"	1.000	Ea.	17.90	41.50	59.40
Duct access door, insulated, 16" x 12"	1.000	Ea.	27	51	78
Pressure controller/switch, air flow controller	1.000	Ea.	121	38	159
Fan control switch and mounting base	1.000	Ea.	70.50	28.50	99
Fan balancing	1.000	Ea.		274	274
Diffuser/register balancing	3.000	Ea.		165	165
TOTAL			1,727.65	2,479.42	4,207.07

D3040 220	Centrifugal In Line Fan Systems	COST EACH		
		MAT.	INST.	TOTAL
1010	Fan system, in-line centrifugal, 500 CFM	1,725	2,475	4,200
1020	1300 CFM	2,150	4,300	6,450
1030	1500 CFM	2,475	4,975	7,450
1040	2500 CFM	3,725	15,200	18,925
1050	3500 CFM	4,675	20,500	25,175
1060	5000 CFM	5,950	30,400	36,350
1070	7500 CFM	7,150	31,100	38,250
1080	10,000 CFM	8,575	36,400	44,975

D3040 230	Utility Set Fan Systems	COST EACH		
		MAT.	INST.	TOTAL
1010	Utility fan set system, belt drive, 2000 CFM	1,600	8,050	9,650
1020	3500 CFM	2,700	14,200	16,900
1030	5000 CFM	3,400	19,700	23,100
1040	7500 CFM	4,175	22,900	27,075
1050	10,000 CFM	6,150	30,100	36,250
1060	15,000 CFM	6,725	30,700	37,425
1070	20,000 CFM	7,425	34,800	42,225

Important: See the Reference Section for critical supporting data - Reference Numbers and City Cost Indexes

D3040 Distribution Systems

D3040 240	Roof Exhaust Fan Systems	COST EACH		
		MAT.	INST.	TOTAL
1010	Roof vent. system, centrifugal, alum., galv curb, BDD, 500 CFM	695	1,625	2,320
1020	800 CFM	860	3,150	4,010
1030	1500 CFM	1,125	4,650	5,775
1040	2750 CFM	2,200	10,500	12,700
1050	3500 CFM	2,650	13,600	16,250
1060	5000 CFM	3,725	22,400	26,125
1070	8500 CFM	4,625	28,800	33,425
1080	13,800 CFM	6,525	41,800	48,325

D3040 Distribution Systems

Plate Heat Exchanger   Shell and Tube Heat Exchanger

System Components	QUANTITY	UNIT	COST EACH		
			MAT.	INST.	TOTAL
SYSTEM D3040 605 1010					
SHELL & TUBE HEAT EXCHANGER, 40 GPM					
Heat exchanger, 4 pass, 3/4" O.D. copper tubes, by steam at 10 psi, 40 GPM	1.000	Ea.	2,425	205	2,630
Pipe, black steel, Sch 40, threaded, W/couplings & hangers, 10' OC, 2" dia.	40.000	L.F.	176.75	641.35	818.10
Elbow, 90°, straight, 2" dia.	16.000	Ea.	160.80	720	880.80
Reducer, black steel, concentric, 2" dia.	2.000	Ea.	19.60	77	96.60
Valves, bronze, gate, rising stem, threaded, class 150, 2" size	6.000	Ea.	522	246	768
Valves, bronze, globe, class 150, rising stem, threaded, 2" size	2.000	Ea.	456	82	538
Strainers, Y type, bronze body, screwed, 150 lb., 2" pipe size	2.000	Ea.	118	70	188
Valve, electric motor actuated, brass, 2 way, screwed, 1-1/2" pipe size	1.000	Ea.	390	35.50	425.50
Union, black with brass seat, 2" dia.	8.000	Ea.	164	384	548
Tee, black, straight, 2" dia.	9.000	Ea.	112.95	666	778.95
Tee, black, reducing run and outlet, 2" dia.	4.000	Ea.	79.40	296	375.40
Pipe nipple, black, 2" dia.	21.000	Ea.	1.75	6.35	8.10
Thermometers, stem type, 9" case, 8" stem, 3/4" NPT	2.000	Ea.	266	32.50	298.50
Gauges, pressure or vacuum, 3-1/2" diameter dial	2.000	Ea.	46	28.40	74.40
Insulation, fiberglass pipe covering, 1" wall, 2" IPS	40.000	L.F.	89.20	158.80	248
Heat exchanger control system	1.000	Ea.	2,000	1,475	3,475
Coil balancing	1.000	Ea.		84	84
TOTAL			7,027.45	5,207.90	12,235.35

D3040 610	Heat Exchanger, Plate Type	COST EACH		
		MAT.	INST.	TOTAL
1010	Plate heat exchanger, 400 GPM	25,900	10,700	36,600
1020	800 GPM	40,600	13,600	54,200
1030	1200 GPM	53,000	18,300	71,300
1040	1800 GPM	77,500	22,800	100,300

D3040 620	Heat Exchanger, Shell & Tube	COST EACH		
		MAT.	INST.	TOTAL
1010	Shell & tube heat exchanger, 40 GPM	7,025	5,200	12,225
1020	96 GPM	12,500	9,475	21,975
1030	240 GPM	22,400	12,800	35,200
1040	600 GPM	42,000	17,800	59,800

D3050 Terminal & Package Units

Cabinet Unit Heater

Weather Cap Roof Vent →
Roof Flashing →
Vent Pipe ←
90° Elbow ←
Gas Cock →
Tee ←
Tee Cap ←

Gas Unit Heater

Hydronic Unit Heater

System Components	QUANTITY	UNIT	COST EACH		
			MAT.	INST.	TOTAL
SYSTEM D3050 120 1010					
SPACE HEATER, SUSPENDED, GAS FIRED, PROPELLER FAN, 20 MBH					
Space heater, propeller fan, 20 MBH output	1.000	Ea.	465	96	561
Pipe, black steel, Sch 40, threaded, W/coupling & hangers, 10' OC, 3/4" dia	20.000	L.F.	34.02	155.40	189.42
Elbow, 90°, black steel, straight, 1/2" dia.	3.000	Ea.	3.84	90	93.84
Elbow, 90°, black steel, straight, 3/4" dia.	3.000	Ea.	4.56	96	100.56
Tee, black steel, reducing, 3/4" dia.	1.000	Ea.	4.68	50	54.68
Union, black with brass seat, 3/4" dia.	1.000	Ea.	6.90	34.50	41.40
Pipe nipples, black, 1/2" dia	2.000	Ea.	2.90	14.30	17.20
Pipe nipples, black, 3/4" dia	2.000	Ea.	1.62	7.40	9.02
Pipe cap, black, 3/4" dia.	1.000	Ea.	1.75	14.10	15.85
Gas cock, brass, 3/4" size	1.000	Ea.	11.35	20.50	31.85
Thermostat, 1 set back, electric, timed	1.000	Ea.	89	57	146
Wiring, thermostat hook-up	1.000	Ea.	5.65	18.70	24.35
Vent chimney, prefab metal, U.L. listed, gas, double wall, galv. st, 4" dia	12.000	L.F.	57.48	145.20	202.68
Vent chimney, gas, double wall, galv. steel, elbow 90°, 4" dia.	2.000	Ea.	38.50	48	86.50
Vent chimney, gas, double wall, galv. steel, Tee, 4" dia.	1.000	Ea.	24.50	31.50	56
Vent chimney, gas, double wall, galv. steel, T cap, 4" dia.	1.000	Ea.	1.86	19.60	21.46
Vent chimney, gas, double wall, galv. steel, roof flashing, 4" dia.	1.000	Ea.	6.45	24	30.45
Vent chimney, gas, double wall, galv. steel, top, 4" dia.	1.000	Ea.	9.25	18.70	27.95
TOTAL			769.31	940.90	1,710.21

D3050 120	Unit Heaters, Gas	COST EACH		
		MAT.	INST.	TOTAL
1010	Space heater, suspended, gas fired, propeller fan, 20 MBH	770	940	1,710
1020	60 MBH	905	980	1,885
1030	100 MBH	1,125	1,025	2,150
1040	160 MBH	1,325	1,150	2,475
1050	200 MBH	1,600	1,250	2,850
1060	280 MBH	2,100	1,350	3,450
1070	320 MBH	2,800	1,525	4,325

D3050 130	Unit Heaters, Hydronic	COST EACH		
		MAT.	INST.	TOTAL
1010	Space heater, suspended, horiz. mount, HW, prop. fan, 20 MBH	1,325	1,000	2,325
1020	60 MBH	1,700	1,125	2,825

D3050 Terminal & Package Units

D3050 130	Unit Heaters, Hydronic	COST EACH		
		MAT.	INST.	TOTAL
1030	100 MBH	2,050	1,225	3,275
1040	150 MBH	2,575	1,350	3,925
1050	200 MBH	2,700	1,525	4,225
1060	300 MBH	2,950	1,800	4,750

D3050 140	Cabinet Unit Heaters, Hydronic	COST EACH		
		MAT.	INST.	TOTAL
1010	Unit heater, cabinet type, horizontal blower, hot water, 20 MBH	1,650	980	2,630
1020	60 MBH	2,350	1,075	3,425
1030	100 MBH	2,600	1,200	3,800
1040	120 MBH	2,675	1,225	3,900

Important: See the Reference Section for critical supporting data - Reference Numbers and City Cost Indexes

D3050 Terminal & Package Units

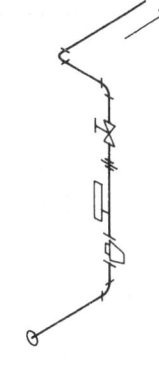

Condenser Supply Water Piping

Condenser Return Water Piping

Self—Contained Air Conditioner Unit, Water Cooled

System Components	QUANTITY	UNIT	COST EACH		
			MAT.	INST.	TOTAL
SYSTEM D3050 210 1010					
A/C, SELF CONTAINED, SINGLE PKG., WATER COOLED, ELECT. HEAT, 5 TON					
Self-contained, water cooled, elect. heat, not inc tower, 5 ton, const vol	1.000	Ea.	9,275	1,050	10,325
Pipe, black steel, Sch 40, threaded, W/cplg & hangers, 10' OC, 1-1/4" dia.	20.000	L.F.	52.89	196.73	249.62
Elbow, 90°, black, straight, 1-1/4" dia.	6.000	Ea.	26.52	222	248.52
Tee, black, straight, 1-1/4" dia.	2.000	Ea.	13.60	116	129.60
Tee, black, reducing, 1-1/4" dia.	2.000	Ea.	16.60	116	132.60
Thermometers, stem type, 9" case, 8" stem, 3/4" NPT	2.000	Ea.	266	32.50	298.50
Union, black with brass seat, 1-1/4" dia.	2.000	Ea.	26.40	77	103.40
Pipe nipple, black, 1-1/4" dia	3.000	Ea.	3.69	13.73	17.42
Valves, bronze, gate, N.R.S., threaded, class 150, 1-1/4" size	2.000	Ea.	106	60	166
Circuit setter, bal valve, bronze body, threaded, 1-1/4" pipe size	1.000	Ea.	99.50	30.50	130
Insulation, fiberglass pipe covering, 1" wall, 1-1/4" IPS	20.000	L.F.	38	75.80	113.80
Control system, pneumatic, A/C Unit with heat	1.000	Ea.	2,050	1,600	3,650
Re-heat coil balancing	1.000	Ea.		84	84
Rooftop unit heat/cool balancing	1.000	Ea.		320	320
TOTAL			11,974.20	3,994.26	15,968.46

D3050 210	AC Unit, Package, Elec. Ht., Water Cooled	COST EACH		
		MAT.	INST.	TOTAL
1010	A/C, Self contained, single pkg., water cooled, elect. heat, 5 Ton	12,000	4,000	16,000
1020	10 Ton	20,500	4,925	25,425
1030	20 Ton	27,700	6,350	34,050
1040	30 Ton	40,200	6,925	47,125
1050	40 Ton	51,500	7,650	59,150
1060	50 Ton	64,000	9,175	73,175

D3050 215	AC Unit, Package, Elec. Ht., Water Cooled, VAV	COST EACH		
		MAT.	INST.	TOTAL
1010	A/C, Self contained, single pkg., water cooled, elect. ht, VAV, 10 Ton	22,500	4,925	27,425
1020	20 Ton	31,300	6,350	37,650
1030	30 Ton	45,200	6,925	52,125
1040	40 Ton	59,000	7,650	66,650
1050	50 Ton	72,500	9,175	81,675
1060	60 Ton	83,500	11,300	94,800

SERVICES

D

D3050 Terminal & Package Units

D3050 220	AC Unit, Package, Hot Water Coil, Water Cooled	COST EACH		
		MAT.	INST.	TOTAL
1010	A/C, Self contn'd, single pkg., water cool, H/W ht, const. vol, 5 Ton	10,900	5,550	16,450
1020	10 Ton	18,400	6,625	25,025
1030	20 Ton	29,300	8,375	37,675
1040	30 Ton	38,400	10,800	49,200
1050	40 Ton	47,500	13,000	60,500
1060	50 Ton	58,500	16,000	74,500

D3050 225	AC Unit, Package, HW Coil, Water Cooled, VAV	COST EACH		
		MAT.	INST.	TOTAL
1010	A/C, Self contn'd, single pkg., water cool, H/W ht, VAV, 10 Ton	20,400	6,225	26,625
1020	20 Ton	28,900	8,150	37,050
1030	30 Ton	40,700	10,300	51,000
1040	40 Ton	55,000	11,200	66,200
1050	50 Ton	66,000	13,200	79,200
1060	60 Ton	78,500	14,400	92,900

Important: See the Reference Section for critical supporting data - Reference Numbers and City Cost Indexes

D3050 Terminal & Package Units

Heat Pump, Air/Air, Rooftop w/ Economizer

System Components	QUANTITY	UNIT	COST EACH		
			MAT.	INST.	TOTAL
SYSTEM D3050 245 1010					
HEAT PUMP, ROOF TOP, AIR/AIR, CURB., ECONOMIZER., SUP. ELECT. HEAT, 10 TON					
Heat pump, air to air, rooftop, curb, economizer, elect ht, 10 ton cooling	1.000	Ea.	39,800	3,175	42,975
Ductwork, 32" x 12" fabricated, galvanized steel, 16 LF	176.000	Lb.	156.64	959.20	1,115.84
Insulation, ductwork, blanket type, fiberglass, 1" thk, 1-1/2 LB density	117.000	S.F.	59.67	265.59	325.26
Rooftop unit heat/cool balancing	1.000	Ea.		320	320
TOTAL			40,016.31	4,719.79	44,736.10

D3050 230	Heat Pump, Water Source, Central Station	COST EACH		
		MAT.	INST.	TOTAL
1010	Heat pump, central station, water source, constant vol., 5 Ton	4,450	2,250	6,700
1020	10 Ton	14,400	3,150	17,550
1030	20 Ton	17,900	5,925	23,825
1040	30 Ton	31,100	8,575	39,675
1050	40 Ton	34,600	9,350	43,950
1060	50 Ton	47,900	12,600	60,500

D3050 235	Heat Pump, Water Source, Console	COST EACH		
		MAT.	INST.	TOTAL
1020	Heat pump, console, water source, 1 Ton	2,175	1,225	3,400
1030	1-1/2 Ton	2,525	1,275	3,800
1040	2 Ton	2,975	1,500	4,475
1050	3 Ton	3,250	1,600	4,850
1060	3-1/2 Ton	3,350	1,650	5,000

D3050 240	Heat Pump, Horizontal, Duct Mounted	COST EACH		
		MAT.	INST.	TOTAL
1020	Heat pump, horizontal, ducted, water source, 1 Ton	2,625	2,675	5,300
1030	1-1/2 Ton	3,200	3,375	6,575
1040	2 Ton	3,625	4,025	7,650
1050	3 Ton	4,300	5,300	9,600
1060	3-1/2 Ton	4,425	5,475	9,900

D **SERVICES**

D3050 Terminal & Package Units

D3050 245	Heat Pump, Roof Top, Air/Air	COST EACH		
		MAT.	INST.	TOTAL
1010	Heat pump, roof top, air/air, curb, economizer, sup. elect heat, 10 Ton	40,000	4,725	44,725
1020	20 Ton	57,000	8,000	65,000
1030	30 Ton	76,500	11,200	87,700
1040	40 Ton	85,500	12,900	98,400
1050	50 Ton	100,000	14,500	114,500
1060	60 Ton	105,500	16,400	121,900

Important: See the Reference Section for critical supporting data - Reference Numbers and City Cost Indexes

D3050 Terminal & Package Units

Thru-Wall Heat Pump   Thru-Wall Air Conditioner

System Components	QUANTITY	UNIT	COST EACH		
			MAT.	INST.	TOTAL
SYSTEM D3050 255 1010 **A/C UNIT, THRU-THE-WALL, SUP. ELECT. HEAT, CABINET, LOUVER, 1/2 TON** A/C unit, thru-wall, electric heat., cabinet, louver, 1/2 ton	1.000	Ea.	1,050	136	1,186
TOTAL			1,050	136	1,186

D3050 255	Thru-Wall A/C Unit	COST EACH		
		MAT.	INST.	TOTAL
1010	A/C Unit, thru-the-wall, sup. elect. heat, cabinet, louver, 1/2 Ton	1,050	136	1,186
1020	3/4 Ton	1,075	164	1,239
1030	1 Ton	1,125	205	1,330
1040	1-1/2 Ton	1,500	340	1,840
1050	2 Ton	1,500	430	1,930

D3050 260	Thru-Wall Heat Pump	COST EACH		
		MAT.	INST.	TOTAL
1010	Heat pump, thru-the-wall, cabinet, louver, 1/2 Ton	1,325	102	1,427
1020	3/4 Ton	1,425	136	1,561
1030	1 Ton	1,525	205	1,730
1040	Sup. elect. heat, 1-1/2 Ton	3,400	530	3,930
1050	2 Ton	3,475	545	4,020

D SERVICES

D3060 Controls & Instrumentation

Commercial/Industrial Dust Collection System

System Components	QUANTITY	UNIT	COST EACH		
			MAT.	INST.	TOTAL
SYSTEM D3060 410 1010					
COMMERCIAL/INDUSTRIAL VACUUM DUST COLLECTION SYSTEM, 500 CFM					
Dust collection central vac unit inc. stand, filter & shaker, 500 CFM 2 HP	1.000	Ea.	3,575	435	4,010
Galvanized tubing, 16 ga., 4″ OD	80.000	L.F.	331.20	183.20	514.40
Galvanized tubing, 90° ell, 4″ dia.	3.000	Ea.	70.50	49.35	119.85
Galvanized tubing, 45° ell, 4″ dia.	4.000	Ea.	82	65.80	147.80
Galvanized tubing, TY slip fit, 4″ dia.	4.000	Ea.	200	110	310
Flexible rubber hose, 4″ dia.	32.000	L.F.	264	73.28	337.28
Galvanized air gate valve, 4″ dia.	4.000	Ea.	600	144	744
Galvanized compression coupling, 4″ dia.	36.000	Ea.	707.40	900	1,607.40
Pipe hangers, steel, 4″ pipe size	16.000	Ea.	223.20	118.40	341.60
TOTAL			6,053.30	2,079.03	8,132.33

D3060 410	Commercial/Industrial Vacuum Dust Collection	COST EACH		
		MAT.	INST.	TOTAL
1010	Commercial / industrial, vacuum dust collection system, 500 CFM	6,050	2,075	8,125
1020	1000 CFM	7,800	2,350	10,150
1030	1500 CFM	9,475	2,625	12,100
1040	3000 CFM	19,000	3,675	22,675
1050	5000 CFM	28,400	5,500	33,900

Important: See the Reference Section for critical supporting data - Reference Numbers and City Cost Indexes

D3060 Controls & Instrumentation

Commercial Kitchen Exhaust/Make-up Air Rooftop System

Roof Fan →

Gas Fired Make-up Air Unit

Grease Filters

Make-up Air

Exhaust Air

Removable Grease Trough

System Components	QUANTITY	UNIT	COST EACH MAT.	COST EACH INST.	COST EACH TOTAL
SYSTEM D3060 420 1010					
COMMERCIAL KITCHEN EXHAUST/MAKE-UP AIR SYSTEM, ROOFTOP, GAS, 2000 CFM					
Make-up air rooftop unit, gas, 2090 CFM, 159 MBH	1.000	Ea.	5,130	459	5,589
Pipe, black steel, Sch 40, thrded, nipples, W/cplngs/hngrs, 10' OC, 1" dia.	20.000	L.F.	45.54	187	232.54
Elbow, 90°, black steel, straight, 3/4" dia.	2.000	Ea.	3.04	64	67.04
Elbow, 90°, black steel, straight, 1" dia.	3.000	Ea.	8.10	103.50	111.60
Tee, black steel, reducing, 1" dia.	1.000	Ea.	4.80	56.50	61.30
Union, black with brass seat, 1" dia.	1.000	Ea.	9.90	37.50	47.40
Pipe nipples, black, 3/4" dia	2.000	Ea.	3.24	14.80	18.04
Pipe cap, malleable iron, black, 1" dia.	1.000	Ea.	2.12	15.05	17.17
Gas cock, brass, 1" size	1.000	Ea.	13.85	24	37.85
Ductwork, 20" x 14" fabricated, galv. steel, 30 LF	220.000	Lb.	118.80	1,100	1,218.80
Ductwork, 18" x 10" fabricated, galv. steel, 12 LF	72.000	Lb.	38.88	360	398.88
Ductwork, 30" x 10" fabricated, galv. steel, 30 LF	260.000	Lb.	140.40	1,300	1,440.40
Ductwork, 16" x 10" fabricated, galv. steel, 12 LF	70.000	Lb.	37.80	350	387.80
Kitchen ventilation, island style, water wash, Stainless Steel	5.000	L.F.	8,500	900	9,400
Control system, Make-up Air Unit / Exhauster	1.000	Ea.	2,050	1,600	3,650
Rooftop unit heat/cool balancing	1.000	Ea.		320	320
Roof fan balancing	1.000	Ea.		183	183
TOTAL			16,106.47	7,074.35	23,180.82

D3060 420	Kitchen Exhaust/Make-Up Air	COST EACH MAT.	COST EACH INST.	COST EACH TOTAL
1010	Commercial kitchen exhaust/make-up air system, rooftop, gas, 2000 CFM	16,100	7,075	23,175
1020	3000 CFM	17,000	9,700	26,700
1030	5000 CFM	21,300	11,700	33,000
1040	8000 CFM	33,900	17,500	51,400
1050	12,000 CFM	49,900	21,200	71,100
1060	16,000 CFM	61,500	33,800	95,300

SERVICES

D3060   Controls & Instrumentation

System Components	QUANTITY	UNIT	COST EACH		
			MAT.	INST.	TOTAL
SYSTEM D3060 510 1010					
FUME HOOD EXHAUST SYSTEM, 3' LONG, 1000 CFM					
Fume hood	3.000	L.F.	4,050	990	5,040
Ductwork, plastic, FM approved for acid fume & smoke, straight, 10" dia	60.000	L.F.	1,350	582	1,932
Ductwork, plastic, FM approved for acid fume & smoke, elbow 90°, 10" dia	3.000	Ea.	348	318	666
Fan, corrosive fume resistant plastic, 1630 CFM, 1/2 HP	1.000	Ea.	3,400	262	3,662
Fan control switch & mounting base	1.000	Ea.	70.50	28.50	99
Roof fan balancing	1.000	Ea.		183	183
TOTAL			9,218.50	2,363.50	11,582

D3060 510	Fume Hood Exhaust Systems	COST EACH		
		MAT.	INST.	TOTAL
1010	Fume hood exhaust system, 3' long, 1000 CFM	9,225	2,375	11,600
1020	4' long, 2000 CFM	11,000	3,000	14,000
1030	6' long, 3500 CFM	15,400	4,450	19,850
1040	6' long, 5000 CFM	20,900	5,225	26,125
1050	10' long, 8000 CFM	28,200	7,175	35,375

Important: See the Reference Section for critical supporting data - Reference Numbers and City Cost Indexes

D3090-000 **Sample Worksheet**

Heating & Air Conditioning Estimate Worksheet

Equipment	Type	Table Number	Quantity	Unit Cost	Total Costs	Comments
Heat Source				$	$	
				$	$	
				$	$	
				$	$	
Pipe				$	$	
Duct				$	$	
Terminals				$	$	
Other				$	$	
				$	$	
				$	$	
				$	$	
Cold Source				$	$	
				$	$	
				$	$	
				$	$	
Pipe				$	$	
Duct				$	$	
Terminals				$	$	
Other				$	$	
SUBTOTAL				$	$	
Quality Complexity				%	$	
TOTAL					$	

D3090 Other HVAC Systems & Equipment

Vitrified Clay Garage Exhaust System

Dual Exhaust System

System Components	QUANTITY	UNIT	COST EACH		
			MAT.	INST.	TOTAL
SYSTEM D3090 320 1040					
GARAGE, EXHAUST, SINGLE OUTLET, 3" EXHAUST, CARS & LIGHT TRUCKS					
A Outlet top assy. for engine exhaust system with adapters and ftngs, 3" diam	1.000	Ea.	168	28.50	196.50
F Bullnose (guide) for engine exhaust system, 3" diam	1.000	Ea.	18.70		18.70
G Galvanized flexible tubing for engine exhaust system, 3" diam	8.000	L.F.	40		40
H Adapter for metal tubing end of engine exhaust system, 3" tail pipe	1.000	Ea.	37.50		37.50
J Pipe, sewer, vitrified clay, premium joint, 8" diam.	18.000	L.F.	78.48	128.70	207.18
Excavate utility trench w/chain trencher, 12 H.P., oper. walking	18.000	L.F.		23.94	23.94
Backfill utility trench by hand, incl. compaction, 8" wide 24" deep	18.000	L.F.		33.48	33.48
Stand for blower, concrete over polystyrene core, 6" high	1.000	Ea.	7.50	14.30	21.80
AC&V duct spiral reducer 10"x8"	1.000	Ea.	7.80	25.50	33.30
AC&V duct spiral reducer 12"x10"	1.000	Ea.	11.10	34.50	45.60
AC&V duct spiral preformed 45° elbow 8" diam	1.000	Ea.	3.64	29.50	33.14
AC&V utility fan, belt drive, 3 phase, 2000 CFM, 1 HP	2.000	Ea.	1,260	456	1,716
Safety switch, heavy duty fused, 240V, 3 pole, 30 amp	1.000	Ea.	126	140	266
Total			1,758.72	914.42	2,673.14

D3090 320	Garage Exhaust Systems	COST PER BAY		
		MAT.	INST.	TOTAL
1040	Garage, single exhaust, 3" outlet, cars & light trucks, one bay	1,750	915	2,665
1060	Additional bays up to seven bays	310	132	442
1500	4" outlet, trucks, one bay	1,775	915	2,690
1520	Additional bays up to six bays	315	132	447
1600	5" outlet, diesel trucks, one bay	2,000	915	2,915
1650	Additional single bays up to six	555	153	708
1700	Two adjoining bays	2,000	915	2,915
2000	Dual exhaust, 3" outlets, pair of adjoining bays	2,050	1,025	3,075
2100	Additional pairs of adjoining bays	575	153	728

For information about Means Estimating Seminars, see yellow pages 12 and 13 in back of book

SERVICES D

G BUILDING SITEWORK

G1030 Site Earthwork

Trenching Systems are shown on a cost per linear foot basis. The systems include: excavation; backfill and removal of spoil; and compaction for various depths and trench bottom widths. The backfill has been reduced to accommodate a pipe of suitable diameter and bedding.

The slope for trench sides varies from 0:1 to 2:1.

The Expanded System Listing shows Trenching Systems that range from 2' to 12' in width. Depths range from 2' to 25'.

System Components	QUANTITY	UNIT	COST PER L.F. EQUIP.	LABOR	TOTAL
SYSTEM G1030 805 1310					
TRENCHING, BACKHOE, 0 TO 1 SLOPE, 2' WIDE, 2' DP, 3/8 C.Y. BUCKET					
Excavation, trench, hyd. backhoe, track mtd., 3/8 C.Y. bucket	.174	C.Y.	.28	.82	1.10
Backfill and load spoil, from stockpile	.174	C.Y.	.10	.24	.34
Compaction by rammer tamper, 8" lifts, 4 passes	.014	C.Y.	.01	.03	.04
Remove excess spoil, 6 C.Y. dump truck, 2 mile roundtrip	.160	C.Y.	.54	.50	1.04
Total			.93	1.59	2.52

G1030 805	Trenching	COST PER L.F. EQUIP.	LABOR	TOTAL
1310	Trenching, backhoe, 0 to 1 slope, 2' wide, 2' deep, 3/8 C.Y. bucket	.93	1.59	2.52
1320	3' deep, 3/8 C.Y. bucket	1.16	2.32	3.48
1330	4' deep, 3/8 C.Y. bucket	1.38	3.06	4.44
1340	6' deep, 3/8 C.Y. bucket	1.71	3.90	5.61
1350	8' deep, 1/2 C.Y. bucket	2.02	4.96	6.98
1360	10' deep, 1 C.Y. bucket	3.24	6	9.24
1400	4' wide, 2' deep, 3/8 C.Y. bucket	1.92	3.20	5.12
1410	3' deep, 3/8 C.Y. bucket	2.73	4.76	7.49
1420	4' deep, 1/2 C.Y. bucket	2.72	4.77	7.49
1430	6' deep, 1/2 C.Y. bucket	3.55	7.25	10.80
1440	8' deep, 1/2 C.Y. bucket	5.80	9.55	15.35
1450	10' deep, 1 C.Y. bucket	6.70	11.85	18.55
1460	12' deep, 1 C.Y. bucket	8.55	15.20	23.75
1470	15' deep, 1-1/2 C.Y. bucket	7.40	12.95	20.35
1480	18' deep, 2-1/2 C.Y. bucket	9.60	18.55	28.15
1520	6' wide, 6' deep, 5/8 C.Y. bucket	7.20	10.25	17.45
1530	8' deep, 3/4 C.Y. bucket	8.90	12.40	21.30
1540	10' deep, 1 C.Y. bucket	9.95	14.95	24.90
1550	12' deep, 1-1/4 C.Y. bucket	8.80	11.70	20.50
1560	16' deep, 2 C.Y. bucket	12.75	13.25	26
1570	20' deep, 3-1/2 C.Y. bucket	14.80	24	38.80
1580	24' deep, 3-1/2 C.Y. bucket	24	40	64
1640	8' wide, 12' deep, 1-1/4 C.Y. bucket	13	16	29
1650	15' deep, 1-1/2 C.Y. bucket	15.30	19.70	35
1660	18' deep, 2-1/2 C.Y. bucket	19.65	19.80	39.45
1680	24' deep, 3-1/2 C.Y. bucket	33	53.50	86.50
1730	10' wide, 20' deep, 3-1/2 C.Y. bucket	27	40.50	67.50
1740	24' deep, 3-1/2 C.Y. bucket	41.50	67	108.50
1780	12' wide, 20' deep, 3-1/2 C.Y. bucket	32.50	48.50	81
1790	25' deep, bucket	55.50	90	145.50
1800	1/2 to 1 slope, 2' wide, 2' deep, 3/8 C.Y. bucket	1.23	2.03	3.26
1810	3' deep, 3/8 C.Y. bucket	1.73	3.35	5.08
1820	4' deep, 3/8 C.Y. bucket	2.23	4.91	7.14
1840	6' deep, 3/8 C.Y. bucket	3.24	7.55	10.79

Important: See the Reference Section for critical supporting data - Reference Numbers and City Cost Indexes

G1030 805	Trenching	COST PER L.F.		
		EQUIP.	LABOR	TOTAL
1860	8' deep, 1/2 C.Y. bucket	4.57	11.55	16.12
1880	10' deep, 1 C.Y. bucket	8.90	16.25	25.15
2300	4' wide, 2' deep, 3/8 C.Y. bucket	2.05	3.41	5.46
2310	3' deep, 3/8 C.Y. bucket	3.38	5.60	8.98
2320	4' deep, 1/2 C.Y. bucket	3.56	5.85	9.41
2340	6' deep, 1/2 C.Y. bucket	5.05	10.15	15.20
2360	8' deep, 1/2 C.Y. bucket	8.95	14.15	23.10
2380	10' deep, 1 C.Y. bucket	12.05	21	33.05
2400	12' deep, 1 C.Y. bucket	12.25	22.50	34.75
2430	15' deep, 1-1/2 C.Y. bucket	16.30	27.50	43.80
2460	18' deep, 2-1/2 C.Y. bucket	25	33	58
2840	6' wide, 6' deep, 5/8 C.Y. bucket	8.55	11.35	19.90
2860	8' deep, 3/4 C.Y. bucket	12.35	17.05	29.40
2880	10' deep, 1 C.Y. bucket	14.95	22.50	37.45
2900	12' deep, 1-1/4 C.Y. bucket	14.05	18.65	32.70
2940	16' deep, 2 C.Y. bucket	22	30	52
2980	20' deep, 3-1/2 C.Y. bucket	32	51.50	83.50
3020	24' deep, 3-1/2 C.Y. bucket	57.50	96	153.50
3100	8' wide, 12' deep, 1-1/4 C.Y. bucket	18.90	23	41.90
3120	15' deep, 1-1/2 C.Y. bucket	24	30.50	54.50
3140	18' deep, 2-1/2 C.Y. bucket	34	33.50	67.50
3180	24' deep, 3-1/2 C.Y. bucket	66.50	108	174.50
3270	10' wide, 20' deep, 3-1/2 C.Y. bucket	41.50	64	105.50
3280	24' deep, 3-1/2 C.Y. bucket	76	121	197
3370	12' wide, 20' deep, 3-1/2 C.Y. bucket	47	71.50	118.50
3380	25' deep, 3-1/2 C.Y. bucket	89	141	230
3500	1 to 1 slope, 2' wide, 2' deep, 3/8 C.Y. bucket	1.44	2.37	3.81
3520	3' deep, 3/8 C.Y. bucket	2.23	4.16	6.39
3540	4' deep, 3/8 C.Y. bucket	2.93	6.25	9.18
3560	6' deep, 3/8 C.Y. bucket	4.39	10	14.39
3580	8' deep, 1/2 C.Y. bucket	6.40	15.70	22.10
3600	10' deep, 1 C.Y. bucket	14.70	26	40.70
3800	4' wide, 2' deep, 3/8 C.Y. bucket	2.44	4.07	6.51
3820	3' deep, 3/8 C.Y. bucket	4.31	7.15	11.46
3840	4' deep, 1/2 C.Y. bucket	4.28	6.55	10.83
3860	6' deep, 1/2 C.Y. bucket	6.25	12.15	18.40
3880	8' deep, 1/2 C.Y. bucket	11.95	18.60	30.55
3900	10' deep, 1 C.Y. bucket	16	27	43
3920	12' deep, 1 C.Y. bucket	23	39	62
3940	15' deep, 1-1/2 C.Y. bucket	22.50	36	58.50
3960	18' deep, 2-1/2 C.Y. bucket	35.50	43	78.50
4030	6' wide, 6' deep, 5/8 C.Y. bucket	10.40	13.80	24.20
4040	8' deep, 3/4 C.Y. bucket	15.50	22	37.50
4050	10' deep, 1 C.Y. bucket	19.45	30.50	49.95
4060	12' deep, 1-1/4 C.Y. bucket	18.95	27.50	46.45
4070	16' deep, 2 C.Y. bucket	33.50	38.50	72
4080	20' deep, 3-1/2 C.Y. bucket	46.50	81.50	128
4090	24' deep, 3-1/2 C.Y. bucket	87.50	158	245.50
4500	8' wide, 12' deep, 1-1/4 C.Y. bucket	24.50	31	55.50
4550	15' deep, 1-1/2 C.Y. bucket	32	44	76
4600	18' deep, 2-1/2 C.Y. bucket	46.50	51	97.50
4650	24' deep, 3-1/2 C.Y. bucket	96	168	264
4800	10' wide, 20' deep, 3-1/2 C.Y. bucket	58.50	94	152.50
4850	24' deep, 3-1/2 C.Y. bucket	104	179	283
4950	12' wide, 20' deep, 3-1/2 C.Y. bucket	64	101	165
4980	25' deep, 3-1/2 C.Y. bucket	127	218	345
5000	1-1/2 to 1 slope, 2' wide, 2' deep, 3/8 C.Y. bucket	1.66	2.76	4.42
5020	3' deep, 3/8 C.Y. bucket	2.71	4.93	7.64

BUILDING SITEWORK

G

G1030 Site Earthwork

G1030 805	Trenching	COST PER L.F.		
		EQUIP.	LABOR	TOTAL
5040	4' deep, 3/8 C.Y. bucket	3.56	7.40	10.96
5060	6' deep, 3/8 C.Y. bucket	5.35	11.75	17.10
5080	8' deep, 1/2 C.Y. bucket	7.80	18.55	26.35
5100	10' deep, 1 C.Y. bucket	16.20	27	43.20
5300	4' wide, 2' deep, 3/8 C.Y. bucket	2.31	3.83	6.14
5320	3' deep, 3/8 C.Y. bucket	4.19	6.95	11.14
5340	4' deep, 1/2 C.Y. bucket	5.10	7.65	12.75
5360	6' deep, 1/2 C.Y. bucket	7.35	13.75	21.10
5380	8' deep, 1/2 C.Y. bucket	14.25	21	35.25
5400	10' deep, 1 C.Y. bucket	19.20	30.50	49.70
5420	12' deep, 1 C.Y. bucket	28	45	73
5450	15' deep, 1-1/2 C.Y. bucket	27.50	40	67.50
5480	18' deep, 2-1/2 C.Y. bucket	44	47	91
5660	6' wide, 6' deep, 5/8 C.Y. bucket	12	15.40	27.40
5680	8' deep, 3/4 C.Y. bucket	18.05	24.50	42.55
5700	10' deep, 1 C.Y. bucket	23	34.50	57.50
5720	12' deep, 1-1/4 C.Y. bucket	22	29	51
5760	16' deep, 2 C.Y. bucket	40	41.50	81.50
5800	20' deep, 3-1/2 C.Y. bucket	56	92.50	148.50
5840	24' deep, 3-1/2 C.Y. bucket	107	180	287
6020	8' wide, 12' deep, 1-1/4 C.Y. bucket	28.50	34	62.50
6050	15' deep, 1-1/2 C.Y. bucket	38	48.50	86.50
6080	18' deep, 2-1/2 C.Y. bucket	55.50	55	110.50
6140	24' deep, 3-1/2 C.Y. bucket	116	190	306
6300	10' wide, 20' deep, 3-1/2 C.Y. bucket	69	105	174
6350	24' deep, 3-1/2 C.Y. bucket	124	199	323
6450	12' wide, 20' deep, 3-1/2 C.Y. bucket	75.50	111	186.50
6480	25' deep, 3-1/2 C.Y. bucket	150	240	390
6600	2 to 1 slope, 2' wide, 2' deep, 3/8 C.Y. bucket	2.66	2.57	5.23
6620	3' deep, 3/8 C.Y. bucket	3.11	5.45	8.56
6640	4' deep, 3/8 C.Y. bucket	4.05	8.20	12.25
6660	6' deep, 3/8 C.Y. bucket	5.75	12.40	18.15
6680	8' deep, 1/2 C.Y. bucket	8.85	21	29.85
6700	10' deep, 1 C.Y. bucket	18.40	30.50	48.90
6900	4' wide, 2' deep, 3/8 C.Y. bucket	2.31	3.82	6.13
6920	3' deep, 3/8 C.Y. bucket	4.31	7.15	11.46
6940	4' deep, 1/2 C.Y. bucket	5.70	8.20	13.90
6960	6' deep, 1/2 C.Y. bucket	8.20	14.95	23.15
6980	8' deep, 1/2 C.Y. bucket	15.80	23.50	39.30
7000	10' deep, 1 C.Y. bucket	21.50	33.50	55
7020	12' deep, 1 C.Y. bucket	31.50	50.50	82
7050	15' deep, 1-1/2 C.Y. bucket	31	45	76
7080	18' deep, 2-1/2 C.Y. bucket	50	53	103
7260	6' wide, 6' deep, 5/8 C.Y. bucket	13.20	16.45	29.65
7280	8' deep, 3/4 C.Y. bucket	19.80	26.50	46.30
7300	10' deep, 3/4 C.Y. bucket	26.50	38	64.50
7320	12' deep, 1-1/4 C.Y. bucket	24.50	33	57.50
7360	16' deep, 2 C.Y. bucket	44.50	46	90.50
7400	20' deep, 3-1/2 C.Y. bucket	63	103	166
7440	24' deep, 3-1/2 C.Y. bucket	120	202	322
7620	8' wide, 12' deep, 1-1/4 C.Y. bucket	31.50	37	68.50
7650	15' deep, 1-1/2 C.Y. bucket	42	53	95
7680	18' deep, 2-1/2 C.Y. bucket	61.50	61	122.50
7740	24' deep, 3-1/2 C.Y. bucket	130	212	342
7920	10' wide, 20' deep, 3-1/2 C.Y. bucket	76.50	115	191.50
7940	24' deep, 3-1/2 C.Y. bucket	137	220	357
8060	12' wide, 20' deep, 3-1/2 C.Y. bucket	83.50	121	204.50
8080	25' deep, 3-1/2 C.Y. bucket	166	264	430

Important: See the Reference Section for critical supporting data - Reference Numbers and City Cost Indexes

G1030 Site Earthwork

The Pipe Bedding System is shown for various pipe diameters. Compacted bank sand is used for pipe bedding and to fill 12″ over the pipe. No backfill is included. Various side slopes are shown to accommodate different soil conditions. Pipe sizes vary from 6″ to 84″ diameter.

System Components	QUANTITY	UNIT	COST PER L.F.		
			MAT.	INST.	TOTAL
SYSTEM G1030 815 1440					
PIPE BEDDING, SIDE SLOPE 0 TO 1, 1′ WIDE, PIPE SIZE 6″ DIAMETER					
Borrow, bank sand, 2 mile haul, machine spread	.067	C.Y.	.29	.35	.64
Compaction, vibrating plate	.067	C.Y.		.12	.12
TOTAL			.29	.47	.76

G1030 815	Pipe Bedding	COST PER L.F.		
		MAT.	INST.	TOTAL
1440	Pipe bedding, side slope 0 to 1, 1′ wide, pipe size 6″ diameter	.29	.47	.76
1460	2′ wide, pipe size 8″ diameter	.64	1.05	1.69
1480	Pipe size 10″ diameter	.66	1.08	1.74
1500	Pipe size 12″ diameter	.67	1.11	1.78
1520	3′ wide, pipe size 14″ diameter	1.11	1.82	2.93
1540	Pipe size 15″ diameter	1.12	1.84	2.96
1560	Pipe size 16″ diameter	1.14	1.87	3.01
1580	Pipe size 18″ diameter	1.16	1.91	3.07
1600	4′ wide, pipe size 20″ diameter	1.68	2.74	4.42
1620	Pipe size 21″ diameter	1.69	2.78	4.47
1640	Pipe size 24″ diameter	1.74	2.85	4.59
1660	Pipe size 30″ diameter	1.78	2.91	4.69
1680	6′ wide, pipe size 32″ diameter	3.11	5.10	8.21
1700	Pipe size 36″ diameter	3.19	5.25	8.44
1720	7′ wide, pipe size 48″ diameter	4.15	6.80	10.95
1740	8′ wide, pipe size 60″ diameter	5.20	8.50	13.70
1760	10′ wide, pipe size 72″ diameter	7.50	12.25	19.75
1780	12′ wide, pipe size 84″ diameter	10.15	16.65	26.80
2140	Side slope 1/2 to 1, 1′ wide, pipe size 6″ diameter	.61	1.01	1.62
2160	2′ wide, pipe size 8″ diameter	1.01	1.65	2.66
2180	Pipe size 10″ diameter	1.10	1.80	2.90
2200	Pipe size 12″ diameter	1.17	1.92	3.09
2220	3′ wide, pipe size 14″ diameter	1.68	2.75	4.43
2240	Pipe size 15″ diameter	1.72	2.82	4.54
2260	Pipe size 16″ diameter	1.78	2.91	4.69
2280	Pipe size 18″ diameter	1.88	3.08	4.96
2300	4′ wide, pipe size 20″ diameter	2.48	4.06	6.54
2320	Pipe size 21″ diameter	2.54	4.17	6.71
2340	Pipe size 24″ diameter	2.72	4.45	7.17
2360	Pipe size 30″ diameter	3.05	5	8.05
2380	6′ wide, pipe size 32″ diameter	4.50	7.35	11.85
2400	Pipe size 36″ diameter	4.81	7.90	12.71
2420	7′ wide, pipe size 48″ diameter	6.60	10.80	17.40
2440	8′ wide, pipe size 60″ diameter	8.55	14.05	22.60
2460	10′ wide, pipe size 72″ diameter	12	19.65	31.65
2480	12′ wide, pipe size 84″ diameter	15.95	26	41.95
2620	Side slope 1 to 1, 1′ wide, pipe size 6″ diameter	.93	1.52	2.45
2640	2′ wide, pipe size 8″ diameter	1.40	2.28	3.68

BUILDING SITEWORK

G

G1030 Site Earthwork

G1030 815	Pipe Bedding	COST PER L.F.		
		MAT.	INST.	TOTAL
2660	Pipe size 10" diameter	1.53	2.50	4.03
2680	Pipe size 12" diameter	1.68	2.75	4.43
2700	3' wide, pipe size 14" diameter	2.25	3.69	5.94
2720	Pipe size 15" diameter	2.33	3.82	6.15
2740	Pipe size 16" diameter	2.41	3.96	6.37
2760	Pipe size 18" diameter	2.60	4.26	6.86
2780	4' wide, pipe size 20" diameter	3.28	5.40	8.68
2800	Pipe size 21" diameter	3.38	5.55	8.93
2820	Pipe size 24" diameter	3.70	6.05	9.75
2840	Pipe size 30" diameter	4.32	7.10	11.42
2860	6' wide, pipe size 32" diameter	5.90	9.65	15.55
2880	Pipe size 36" diameter	6.45	10.55	17
2900	7' wide, pipe size 48" diameter	9	14.75	23.75
2920	8' wide, pipe size 60" diameter	11.95	19.60	31.55
2940	10' wide, pipe size 72" diameter	16.50	27	43.50
2960	12' wide, pipe size 84" diameter	21.50	35.50	57
3000	Side slope 1-1/2 to 1, 1' wide, pipe size 6" diameter	1.25	2.05	3.30
3020	2' wide, pipe size 8" diameter	1.77	2.90	4.67
3040	Pipe size 10" diameter	1.96	3.22	5.18
3060	Pipe size 12" diameter	2.18	3.57	5.75
3080	3' wide, pipe size 14" diameter	2.82	4.62	7.44
3100	Pipe size 15" diameter	2.93	4.81	7.74
3120	Pipe size 16" diameter	3.05	5	8.05
3140	Pipe size 18" diameter	3.32	5.45	8.77
3160	4' wide, pipe size 20" diameter	4.10	6.70	10.80
3180	Pipe size 21" diameter	4.22	6.95	11.17
3200	Pipe size 24" diameter	4.68	7.65	12.33
3220	Pipe size 30" diameter	5.60	9.20	14.80
3240	6' wide, pipe size 32" diameter	7.30	11.95	19.25
3260	Pipe size 36" diameter	8.05	13.15	21.20
3280	7' wide, pipe size 48" diameter	11.40	18.70	30.10
3300	8' wide, pipe size 60" diameter	15.30	25	40.30
3320	10' wide, pipe size 72" diameter	21	34.50	55.50
3340	12' wide, pipe size 84" diameter	27.50	45.50	73
3400	Side slope 2 to 1, 1' wide, pipe size 6" diameter	1.60	2.63	4.23
3420	2' wide, pipe size 8" diameter	2.15	3.52	5.67
3440	Pipe size 10" diameter	2.40	3.93	6.33
3460	Pipe size 12" diameter	2.67	4.38	7.05
3480	3' wide, pipe size 14" diameter	3.38	5.55	8.93
3500	Pipe size 15" diameter	3.52	5.80	9.32
3520	Pipe size 16" diameter	3.69	6.05	9.74
3540	Pipe size 18" diameter	4.04	6.60	10.64
3560	4' wide, pipe size 20" diameter	4.89	8	12.89
3580	Pipe size 21" diameter	5.05	8.30	13.35
3600	Pipe size 24" diameter	5.65	9.30	14.95
3620	Pipe size 30" diameter	6.90	11.30	18.20
3640	6' wide, pipe size 32" diameter	8.65	14.20	22.85
3660	Pipe size 36" diameter	9.65	15.85	25.50
3680	7' wide, pipe size 48" diameter	13.85	22.50	36.35
3700	8' wide, pipe size 60" diameter	18.70	30.50	49.20
3720	10' wide, pipe size 72" diameter	25.50	41.50	67
3740	12' wide, pipe size 84" diameter	33.50	54.50	88

Important: See the Reference Section for critical supporting data - Reference Numbers and City Cost Indexes

G3010 Water Supply

G3010 110	Water Distribution Piping	COST PER L.F.		
		MAT.	INST.	TOTAL
2000	Piping, excav. & backfill excl., ductile iron class 250, mech. joint			
2130	4" diameter	10.65	7.25	17.90
2150	6" diameter	10.85	8.30	19.15
2160	8" diameter	14.95	9.70	24.65
2170	10" diameter	17.50	11.60	29.10
2180	12" diameter	22	19.85	41.85
2210	16" diameter	30	31	61
2220	18" diameter	39	34	73
3000	Tyton joint			
3130	4" diameter	7.65	6.75	14.40
3150	6" diameter	8.65	7.75	16.40
3160	8" diameter	12.35	9.10	21.45
3170	10" diameter	18.75	10.65	29.40
3180	12" diameter	19.75	13.40	33.15
3210	16" diameter	30	27.50	57.50
3220	18" diameter	33.50	33	66.50
3230	20" diameter	37	35	72
3250	24" diameter	48.50	35.50	84
4000	Copper tubing, type K			
4050	3/4" diameter	1.65	2.03	3.68
4060	1" diameter	2.19	2.54	4.73
4080	1-1/2" diameter	3.58	3.07	6.65
4090	2" diameter	5.55	3.53	9.08
4110	3" diameter	11.35	6.05	17.40
4130	4" diameter	18.80	8.55	27.35
4150	6" diameter	54	13.30	67.30
5000	Polyvinyl chloride class 160, S.D.R. 26			
5130	1-1/2" diameter	.65	3.48	4.13
5150	2" diameter	.98	4.18	5.16
5160	4" diameter	3.40	5.25	8.65
5170	6" diameter	7.35	5.80	13.15
5180	8" diameter	12.40	8.90	21.30
6000	Polyethylene 160 psi, S.D.R. 7			
6050	3/4" diameter	.28	1.99	2.27
6060	1" diameter	.35	2.15	2.50
6080	1-1/2" diameter	.64	2.32	2.96
6090	2" diameter	.98	2.86	3.84

BUILDING SITEWORK

G

G3020 Sanitary Sewer

G3020 110	Drainage & Sewage Piping	COST PER L.F.		
		MAT.	INST.	TOTAL
2000	Piping, excavation & backfill excluded, PVC, plain			
2130	4" diameter	1.71	2.79	4.50
2150	6" diameter	2.87	2.99	5.86
2160	8" diameter	4.84	3.12	7.96
2900	Box culvert, precast, 8' long			
3000	6' x 3'	169	23	192
3020	6' x 7'	256	25.50	281.50
3040	8' x 3'	233	24	257
3060	8' x 8'	315	32	347
3080	10' x 3'	345	29	374
3100	10' x 8'	395	40	435
3120	12' x 3'	335	32	367
3140	12' x 8'	580	48	628
4000	Concrete, nonreinforced			
4150	6" diameter	4.32	8.25	12.57
4160	8" diameter	4.76	9.70	14.46
4170	10" diameter	5.25	10.10	15.35
4180	12" diameter	6.50	10.90	17.40
4200	15" diameter	7.55	12.15	19.70
4220	18" diameter	9.30	15.15	24.45
4250	24" diameter	14.05	22	36.05
4400	Reinforced, no gasket			
4580	12" diameter	10.80	10.40	21.20
4600	15" diameter	13.80	12.45	26.25
4620	18" diameter	14.55	16.80	31.35
4650	24" diameter	21.50	22	43.50
4670	30" diameter	29.50	34.50	64
4680	36" diameter	42	42	84
4690	42" diameter	57	49.50	106.50
4700	48" diameter	71	56	127
4720	60" diameter	113	75	188
4730	72" diameter	160	90	250
4740	84" diameter	270	112	382
4800	With gasket			
4980	12" diameter	12.95	8.50	21.45
5000	15" diameter	15.55	8.90	24.45
5020	18" diameter	19.40	9.40	28.80
5050	24" diameter	29	10.50	39.50
5070	30" diameter	39	34.50	73.50
5080	36" diameter	58.50	42	100.50
5090	42" diameter	76	45	121
5100	48" diameter	95	56	151
5120	60" diameter	128	72	200
5130	72" diameter	246	90	336
5140	84" diameter	325	150	475
5700	Corrugated metal, alum. or galv. bit. coated			
5760	8" diameter	7.55	6.65	14.20
5770	10" diameter	9.10	8.35	17.45
5780	12" diameter	10.90	10.40	21.30
5800	15" diameter	13.25	10.90	24.15
5820	18" diameter	16.95	11.45	28.40
5850	24" diameter	20.50	13.65	34.15
5870	30" diameter	27.50	25.50	53
5880	36" diameter	40	25.50	65.50
5900	48" diameter	61	30	91
5920	60" diameter	79	48	127
5930	72" diameter	118	80	198
6000	Plain			

Important: See the Reference Section for critical supporting data - Reference Numbers and City Cost Indexes

G3020 Sanitary Sewer

G3020 110	Drainage & Sewage Piping	COST PER L.F.		
		MAT.	INST.	TOTAL
6060	8" diameter	5.85	6.15	12
6070	10" diameter	6.40	7.80	14.20
6080	12" diameter	7.35	9.95	17.30
6100	15" diameter	9.30	9.95	19.25
6120	18" diameter	12.35	10.60	22.95
6140	24" diameter	17.90	12.45	30.35
6170	30" diameter	23	23.50	46.50
6180	36" diameter	37.50	23.50	61
6200	48" diameter	50	27.50	77.50
6220	60" diameter	78.50	46	124.50
6230	72" diameter	116	102	218
6300	Steel or alum. oval arch, coated & paved invert			
6400	15" equivalent diameter	22.50	10.90	33.50
6420	18" equivalent diameter	29	14.55	43.50
6450	24" equivalent diameter	41.50	17.47	59
6470	30" equivalent diameter	50.50	21.80	72.50
6480	36" equivalent diameter	75	30.20	105
6490	42" equivalent diameter	90.50	33.50	124
6500	48" equivalent diameter	100	40.60	141
6600	Plain			
6700	15" equivalent diameter	12.10	9.72	22
6720	18" equivalent diameter	14.35	12.47	27
6750	24" equivalent diameter	23	14.55	37.50
6770	30" equivalent diameter	29	28.15	57
6780	36" equivalent diameter	47.50	28.15	75.50
6790	42" equivalent diameter	56.50	32.85	89.50
6800	48" equivalent diameter	49.50	40.60	90
8000	Polyvinyl chloride SDR 35			
8130	4" diameter	1.71	2.79	4.50
8150	6" diameter	2.87	2.99	5.86
8160	8" diameter	4.84	3.12	7.96
8170	10" diameter	7.30	4.33	11.63
8180	12" diameter	8.15	4.46	12.61
8200	15" diameter	12.30	7.50	19.80
9000	Vitrified clay, C-2000, premium joint			
9030	4" diameter	1.83	3.94	5.77
9050	6" diameter	3.06	5.25	8.31
9060	8" diameter	4.36	7.15	11.51
9070	10" diameter	7.30	7.50	14.80
9080	12" diameter	9.70	9.55	19.25
9100	15" diameter	17	13	30
9120	18" diameter	25.50	16.25	41.75
9150	24" diameter	46.50	32	78.50
9170	30" diameter	82	52	134
9180	36" diameter	135	81	216

BUILDING SITEWORK

G

G3030 Storm Sewer

Manhole   Catch Basin

The Manhole and Catch Basin System includes: excavation with a backhoe; a formed concrete footing; frame and cover; cast iron steps and compacted backfill.

The Expanded System Listing shows manholes that have a 4′, 5′ and 6′ inside diameter riser. Depths range from 4′ to 14′. Construction material shown is either concrete, concrete block, precast concrete, or brick.

BUILDING SITEWORK G

System Components	QUANTITY	UNIT	COST PER EACH		
			MAT.	INST.	TOTAL
SYSTEM G3030 210 1920					
MANHOLE/CATCH BASIN, BRICK, 4′ I.D. RISER, 4′ DEEP					
Excavation, hydraulic backhoe, 3/8 C.Y. bucket	14.815	C.Y.		84.59	84.59
Trim sides and bottom of excavation	64.000	S.F.		42.88	42.88
Forms in place, manhole base, 4 uses	20.000	SFCA	11.40	72	83.40
Reinforcing in place footings, #4 to #7	.019	Ton	11.12	17.29	28.41
Concrete, 3000 psi	.925	C.Y.	71.23		71.23
Place and vibrate concrete, footing, direct chute	.925	C.Y.		33.58	33.58
Catch basin or MH, brick, 4′ ID, 4′ deep	1.000	Ea.	310	700	1,010
Catch basin or MH steps; heavy galvanized cast iron	1.000	Ea.	10.50	9.95	20.45
Catch basin or MH frame and cover	1.000	Ea.	182	158	340
Fill, granular	12.954	C.Y.	112.70		112.70
Backfill, spread with wheeled front end loader	12.954	C.Y.		21.89	21.89
Backfill compaction, 12″ lifts, air tamp	12.954	C.Y.		79.67	79.67
TOTAL			708.95	1,219.85	1,928.80

G3030 210	Manholes & Catch Basins	COST PER EACH		
		MAT.	INST.	TOTAL
1920	Manhole/catch basin, brick, 4′ I.D. riser, 4′ deep	710	1,225	1,935
1940	6′ deep	950	1,700	2,650
1960	8′ deep	1,225	2,325	3,550
1980	10′ deep	1,575	2,875	4,450
3000	12′ deep	2,025	3,100	5,125
3020	14′ deep	2,525	4,375	6,900
3200	Block, 4′ I.D. riser, 4′ deep	660	990	1,650
3220	6′ deep	860	1,400	2,260
3240	8′ deep	1,100	1,925	3,025
3260	10′ deep	1,275	2,400	3,675
3280	12′ deep	1,575	3,025	4,600
3300	14′ deep	1,925	3,675	5,600
4620	Concrete, cast-in-place, 4′ I.D. riser, 4′ deep	790	1,700	2,490
4640	6′ deep	1,075	2,300	3,375
4660	8′ deep	1,475	3,325	4,800
4680	10′ deep	1,775	4,100	5,875
4700	12′ deep	2,150	5,100	7,250
4720	14′ deep	2,600	6,125	8,725
5820	Concrete, precast, 4′ I.D. riser, 4′ deep	875	920	1,795
5840	6′ deep	1,125	1,225	2,350

G3030 Storm Sewer

G3030 210	Manholes & Catch Basins	COST PER EACH		
		MAT.	INST.	TOTAL
5860	8' deep	1,375	1,750	3,125
5880	10' deep	1,675	2,125	3,800
5900	12' deep	2,075	2,600	4,675
5920	14' deep	2,525	3,350	5,875
6000	5' I.D. riser, 4' deep	940	995	1,935
6020	6' deep	1,250	1,400	2,650
6040	8' deep	1,575	1,825	3,400
6060	10' deep	2,000	2,375	4,375
6080	12' deep	2,450	3,025	5,475
6100	14' deep	2,950	3,675	6,625
6200	6' I.D. riser, 4' deep	1,325	1,325	2,650
6220	6' deep	1,725	1,750	3,475
6240	8' deep	2,175	2,425	4,600
6260	10' deep	2,725	3,075	5,800
6280	12' deep	3,350	3,875	7,225
6300	14' deep	4,000	4,725	8,725

G30 Site Mechanical Utilities

G3060 Fuel Distribution

G3060 110	Gas Service Piping	COST PER L.F.		
		MAT.	INST.	TOTAL
2000	Piping, excavation & backfill excluded, polyethylene			
2070	1-1/4" diam, SDR 10	.76	2.67	3.43
2090	2" diam, SDR 11	.94	2.98	3.92
2110	3" diam, SDR 11.5	1.98	3.56	5.54
2130	4" diam, SDR 11	4.55	6.85	11.40
2150	6" diam, SDR 21	14.15	7.35	21.50
2160	8" diam, SDR 21	19.30	8.85	28.15
3000	Steel, schedule 40, plain end, tarred & wrapped			
3060	1" diameter	2.33	6.05	8.38
3090	2" diameter	3.66	6.45	10.11
3110	3" diameter	6.05	6.95	13
3130	4" diameter	7.90	11.25	19.15
3140	5" diameter	11.45	13	24.45
3150	6" diameter	14	15.95	29.95
3160	8" diameter	22	20.50	42.50

Important: See the Reference Section for critical supporting data - Reference Numbers and City Cost Indexes

G3060 Fuel Distribution

Fiberglass Underground Fuel Storage Tank, Single Wall

System Components	QUANTITY	UNIT	COST EACH		
			MAT.	INST.	TOTAL
SYSTEM G3060 305 1010					
STORAGE TANK, FUEL, UNDERGROUND, SINGLE WALL FIBERGLASS, 550 GAL.					
Tank, fiberglass, underground, single wall, U.L. listed, 550 gal	1.000	Ea.	1,600	305	1,905
Tank, fiberglass, manways, add	1.000	Ea.	925		925
Tank, for hold-downs 500-4000 gal, add	1.000	Ea.	178	108	286
Foot valve, single poppet, 3/4" dia	1.000	Ea.	41.50	25	66.50
Tubing, copper, Type L, 3/4" dia	120.000	L.F.	234	714	948
Elbow 90°, copper, wrought, cu x cu, 3/4" dia.	8.000	Ea.	9.12	192	201.12
Fuel oil specialties, valve, ball chk, globe type, fusible, 3/4" dia	2.000	Ea.	78	45	123
Elbow 90°, black steel, straight, 2" dia.	3.000	Ea.	30.15	135	165.15
Union, black steel, with brass seat, 2" dia.	1.000	Ea.	20.50	48	68.50
Pipe and nipples, steel, Sch. 40, threaded, black, 2" dia	2.000	Ea.	133	482.60	615.60
Vent protector / breather, 2" dia.	1.000	Ea.	14.10	14.20	28.30
Pipe, black steel, welded, Sch. 40, 3" dia.	30.000	L.F.	244.50	627.30	871.80
Elbow 90°, steel, weld joint, butt, 3" dia.	3.000	Ea.	35.85	385.05	420.90
Flange, weld neck, 150 lb., 3" pipe size	1.000	Ea.	24	64.15	88.15
Fuel fill box, locking inner cover, 3" dia.	1.000	Ea.	112	91	203
Remote tank gauging system, 30', 5" pointer travel	1.000	Ea.	4,075	227	4,302
Tank leak detector system, 8 channel, external monitoring	1.000	Ea.	1,750		1,750
Tank leak detection system, probes, well monitoring, liquid phase detection	2.000	Ea.	1,570		1,570
Excavating, trench, NO sheeting or dewatering, 6'-10' D,3/4 CY hyd backhoe	18.000	C.Y.		102.78	102.78
Concrete hold down pad, 6' x 5' x 8" thick	30.000	C.Y.	61.20	31.50	92.70
Reinforcing in hold down pad, #3 to #7	120.000	Lb.	37.20	54	91.20
Stone back-fill, hand spread, pea gravel	9.000	C.Y.	301.50	306	607.50
Corrosion resistance, wrap & coat, small diam pipe, 1" diam, add	120.000	L.F.	139.20		139.20
Corrosion resistance, wrap & coat, small diameter pipe, 2" diam, add	36.000	L.F.	45.72		45.72
Corrosion resistance wrap & coat, 4" diam, add	30.000	L.F.	45.90		45.90
TOTAL			11,705.44	3,957.58	15,663.02

G3060 305	Fiberglass Fuel Tank, Single Wall	COST EACH		
		MAT.	INST.	TOTAL
1010	Storage tank, fuel, underground, single wall fiberglass, 550 Gal.	11,700	3,950	15,650
1020	2000 Gal.	13,900	5,400	19,300
1030	4000 Gal.	18,000	7,400	25,400
1040	6000 Gal.	19,100	8,250	27,350

G BUILDING SITEWORK

G3060 Fuel Distribution

G3060 305	Fiberglass Fuel Tank, Single Wall	COST EACH		
		MAT.	INST.	TOTAL
1050	8000 Gal.	23,300	10,100	33,400
1060	10,000 Gal.	25,400	11,000	36,400
1070	15,000 Gal.	29,100	13,600	42,700
1080	20,000 Gal.	36,600	16,600	53,200
1090	25,000 Gal.	46,400	19,300	65,700
1100	30,000 Gal.	53,500	20,900	74,400
1110	40,000 Gal.	67,000	26,300	93,300
1120	48,000 Gal.	93,000	30,400	123,400

G3060 310	Fiberglass Fuel Tank, Double Wall	COST EACH		
		MAT.	INST.	TOTAL
1010	Storage tank, fuel, underground, double wall fiberglass, 550 Gal.	13,900	4,525	18,425
1020	2500 Gal.	18,600	6,300	24,900
1030	4000 Gal.	22,700	7,850	30,550
1040	6000 Gal.	24,300	8,825	33,125
1050	8000 Gal.	30,100	10,700	40,800
1060	10,000 Gal.	32,300	12,300	44,600
1070	15,000 Gal.	42,100	14,200	56,300
1080	20,000 Gal.	50,500	17,300	67,800
1090	25,000 Gal.	60,500	20,000	80,500
1100	30,000 Gal.	69,000	21,300	90,300

Important: See the Reference Section for critical supporting data - Reference Numbers and City Cost Indexes

G30 Site Mechanical Utilities

G3060 Fuel Distribution

Steel Underground Fuel Storage Tank, Single Wall

System Components	QUANTITY	UNIT	COST EACH		
			MAT.	INST.	TOTAL
SYSTEM G3060 315 1010					
STORAGE TANK, FUEL, UNDERGROUND, SINGLE WALL STEEL, 550 GAL.					
Tank, steel, underground, sti-P3, set in place, 500 gal, 7 Ga. shell	1.000	Ea.	905	305	1,210
Tank, for manways, add	1.000	Ea.	650		650
Tanks, for hold-downs, 500-2000 gal, add	1.000	Ea.	310	108	418
Foot valve, single poppet, 3/4" dia	1.000	Ea.	41.50	25	66.50
Tubing, copper, Type L, 3/4" dia.	120.000	L.F.	234	714	948
Elbow 90°, copper, wrought, cu x cu, 3/4" dia.	8.000	Ea.	9.12	192	201.12
Fuel oil valve, ball chk, globe type, fusible, 3/4" dia.	2.000	Ea.	78	45	123
Elbow 90°, black steel, straight, 2" dia.	3.000	Ea.	30.15	135	165.15
Union, black steel, with brass seat, 3/4" dia.	1.000	Ea.	6.90	34.50	41.40
Pipe and nipples, steel, Sch. 40, threaded, black, 2" dia	2.000	Ea.	133	482.60	615.60
Vent protector / breather, 2" dia.	1.000	Ea.	14.10	14.20	28.30
Pipe, black steel, welded, Sch. 40, 3" dia.	30.000	L.F.	244.50	627.30	871.80
Elbow 90°, steel, weld joint, butt, 90° elbow, 3" dia.	3.000	Ea.	35.85	385.05	420.90
Flange, steel, weld neck, 150 lb., 3" pipe size	1.000	Ea.	24	64.15	88.15
Fuel fill box, locking inner cover, 3" dia.	1.000	Ea.	112	91	203
Remote tank gauging system, 30', 5" pointer travel	1.000	Ea.	4,075	227	4,302
Tank leak detector system, 8 channel, external monitoring	1.000	Ea.	1,750		1,750
Tank leak detection system, probes, well monitoring, liquid phase detection	2.000	Ea.	1,570		1,570
Excavation, NO sheeting or dewatering,6'-10' D,3/4 CY hyd backhoe	18.000	C.Y.		102.78	102.78
Concrete hold down pad, 6' x 5' x 8" thick	30.000	C.Y.	61.20	31.50	92.70
Reinforcing in hold down pad, #3 to #7	120.000	Lb.	37.20	54	91.20
Stone back-fill, hand spread, pea gravel	9.000	C.Y.	301.50	306	607.50
Corrosion resistance, wrap & coat, small diam pipe, 1" dia., add	120.000	L.F.	139.20		139.20
Corrosion resistance, wrap & coat, small diameter pipe, 2" dia., add	36.000	L.F.	45.72		45.72
Corrosion resistance wrap & coat, 4" dia., add	30.000	L.F.	45.90		45.90
TOTAL			10,853.84	3,944.08	14,797.92

G3060 315		Steel Fuel Tank, Single Wall	COST EACH		
			MAT.	INST.	TOTAL
1010	Storage tank, fuel, underground, single wall steel, 550 Gal.		10,900	3,950	14,850
1020	2000 Gal.		13,800	5,400	19,200
1030	5000 Gal.		19,500	7,675	27,175
1040	10,000 Gal.		28,400	11,700	40,1

G30 Site Mechanical Utilities

G3060 Fuel Distribution

G3060 315	Steel Fuel Tank, Single Wall	COST EACH		
		MAT.	INST.	TOTAL
1050	15,000 Gal.	33,400	13,600	47,000
1060	20,000 Gal.	40,600	16,400	57,000
1070	25,000 Gal.	48,700	19,200	67,900
1080	30,000 Gal.	56,000	20,600	76,600
1090	40,000 Gal.	69,000	26,000	95,000
1100	50,000 Gal.	83,000	31,500	114,500

G3060 320	Steel Fuel Tank, Double Wall	COST EACH		
		MAT.	INST.	TOTAL
1010	Storage tank, fuel, underground, double wall steel, 500 Gal.	13,600	4,625	18,225
1020	2000 Gal.	17,300	6,075	23,375
1030	4000 Gal.	22,800	7,950	30,750
1040	6000 Gal.	27,200	8,750	35,950
1050	8000 Gal.	31,800	10,800	42,600
1060	10,000 Gal.	35,000	12,500	47,500
1070	15,000 Gal.	41,200	14,500	55,700
1080	20,000 Gal.	49,100	17,500	66,600
1090	25,000 Gal.	66,000	20,000	86,000
1100	30,000 Gal.	75,500	21,400	96,900
1110	40,000 Gal.	92,000	26,900	118,900
1120	50,000 Gal.	109,000	32,400	141,400

G3060 325	Steel Tank, Above Ground, Single Wall	COST EACH		
		MAT.	INST.	TOTAL
1010	Storage tank, fuel, above ground, single wall steel, 550 Gal.	8,175	3,300	11,475
1020	2000 Gal.	10,300	3,350	13,650
1030	5000 Gal.	12,700	5,175	17,875
1040	10,000 Gal.	20,600	5,500	26,100
1050	15,000 Gal.	23,200	5,650	28,850
1060	20,000 Gal.	27,100	5,825	32,925
1070	25,000 Gal.	29,200	5,950	35,150
1080	30,000 Gal.	34,400	6,200	40,600

G3060 330	Steel Tank, Above Ground, Double Wall	COST EACH		
		MAT.	INST.	TOTAL
1010	Storage tank, fuel, above ground, double wall steel, 500 Gal.	12,000	3,325	15,325
1020	2000 Gal.	14,600	3,400	18,000
1030	4000 Gal.	19,300	3,475	22,775
1040	6000 Gal.	22,300	5,350	27,650
1050	8000 Gal.	24,400	5,500	29,900
1060	10,000 Gal.	25,700	5,600	31,300
1070	15,000 Gal.	33,000	5,775	38,775
1080	20,000 Gal.	36,100	5,950	42,050
1090	25,000 Gal.	41,600	6,125	47,725
1100	30,000 Gal.	44,500	6,350	50,850

For information about Means Estimating Seminars, see yellow pages 12 and 13 in back of book

BUILDING SITEWORK G

Reference Section

All the reference information is in one section, making it easy to find what you need to know ... and easy to use the book on a daily basis. This section is visually identified by a vertical gray bar on the edge of pages.

In the reference number information that follows, you'll see the background that relates to the "reference numbers" that appeared in the Unit Price Section. You'll find reference tables, explanations and estimating information that support how we arrived at the unit price data. Also included are alternate pricing methods, technical data and estimating procedures along with information on design and economy in construction.

Also in this Reference Section, we've included Change Orders, information on pricing changes to contract documents; Crew Listings, a full listing of all the crews, equipment and their costs; Historical Cost Indexes for cost comparisons over time; City Cost Indexes and Location Factors for adjusting costs to the region you are in, and an explanation of all abbreviations used in the book.

Table of Contents

R01100-005 Tips for Accurate Estimating

1. Use pre-printed or columnar forms for orderly sequence of dimensions and locations and for recording telephone quotations.

2. Use only the front side of each paper or form except for certain pre-printed summary forms.

3. Be consistent in listing dimensions: For example, length x width x height. This helps in rechecking to ensure that, the total length of partitions is appropriate for the building area.

4. Use printed (rather than measured) dimensions where given.

5. Add up multiple printed dimensions for a single entry where possible.

6. Measure all other dimensions carefully.

7. Use each set of dimensions to calculate multiple related quantities.

8. Convert foot and inch measurements to decimal feet when listing. Memorize decimal equivalents to .01 parts of a foot (1/8″ equals approximately .01′).

9. Do not "round off" quantities until the final summary.

10. Mark drawings with different colors as items are taken off.

11. Keep similar items together, different items separate.

12. Identify location and drawing numbers to aid in future checking for completeness.

13. Measure or list everything on the drawings or mentioned in the specifications.

14. It may be necessary to list items not called for to make the job complete.

15. Be alert for: Notes on plans such as N.T.S. (not to scale); changes in scale throughout the drawings; reduced size drawings; discrepancies between the specifications and the drawings.

16. Develop a consistent pattern of performing an estimate. For example:
    a. Start the quantity takeoff at the lower floor and move to the next higher floor.
    b. Proceed from the main section of the building to the wings.
    c. Proceed from south to north or vice versa, clockwise or counterclockwise.
    d. Take off floor plan quantities first, elevations next, then detail drawings.

17. List all gross dimensions that can be either used again for different quantities, or used as a rough check of other quantities for verification (exterior perimeter, gross floor area, individual floor areas, etc.).

18. Utilize design symmetry or repetition (repetitive floors, repetitive wings, symmetrical design around a center line, similar room layouts, etc.). Note: Extreme caution is needed here so as not to omit or duplicate an area.

19. Do not convert units until the final total is obtained. For instance, when estimating concrete work, keep all units to the nearest cubic foot, then summarize and convert to cubic yards.

20. When figuring alternatives, it is best to total all items involved in the basic system, then total all items involved in the alternates. Therefore you work with positive numbers in all cases. When adds and deducts are used, it is often confusing whether to add or subtract a portion of an item; especially on a complicated or involved alternate.

R01100-040 Builder's Risk Insurance

Builder's Risk Insurance is insurance on a building during construction. Premiums are paid by the owner or the contractor. Blasting, collapse and underground insurance would raise total insurance costs above those listed. Floater policy for materials delivered to the job runs $.75 to $1.25 per $100 value. Contractor equipment insurance runs $.50 to $1.50 per $100 value. Insurance for miscellaneous tools to $1,500 value runs from $3.00 to $7.50 per $100 value.

Tabulated below are New England Builder's Risk insurance rates in dollars per $100 value for $1,000 deductible. For $25,000 deductible, rates can be reduced 13% to 34%. On contracts over $1,000,000, rates may be lower than those tabulated. Policies are written annually for the total completed value in place. For "all risk" insurance (excluding flood, earthquake and certain other perils) add $.025 to total rates below.

Coverage	Frame Construction (Class 1) Range		Average	Brick Construction (Class 4) Range		Average	Fire Resistive (Class 6) Range		Average
Fire Insurance	$.350 to	$ .850	$.600	$.158 to	$.189	$.174	$.052 to	$.080	$.070
Extended Coverage	.115 to	.200	.158	.080 to	.105	.101	.081 to	.105	.100
Vandalism	.012 to	.016	.014	.008 to	.011	.011	.008 to	.011	.010
Total Annual Rate	$.477 to	$1.066	$.772	$.246 to	$.305	$.286	$.141 to	$.196	$.180

R01100-060 Workers' Compensation Insurance Rates by Trade

The table below tabulates the national averages for Workers' Compensation insurance rates by trade and type of building. The average "Insurance Rate" is multiplied by the "% of Building Cost" for each trade. This produces the "Workers' Compensation Cost" by % of total labor cost, to be added for each trade by building type to determine the weighted average Workers' Compensation rate for the building types analyzed.

Trade	Insurance Rate (% Labor Cost)			% of Building Cost			Workers' Compensation		
	Range		Average	Office Bldgs.	Schools & Apts.	Mfg.	Office Bldgs.	Schools & Apts.	Mfg.
Excavation, Grading, etc.	3.8 % to	24.7%	10.8%	4.8%	4.9%	4.5%	.52%	.53%	.49%
Piles & Foundations	6.8 to	62.6	25.2	7.1	5.2	8.7	1.79	1.31	2.19
Concrete	6.2 to	33.0	17.0	5.0	14.8	3.7	.85	2.52	.63
Masonry	5.3 to	34.8	16.1	6.9	7.5	1.9	1.11	1.21	.31
Structural Steel	6.8 to	112.0	41.0	10.7	3.9	17.6	4.39	1.60	7.22
Miscellaneous & Ornamental Metals	4.9 to	25.3	13.1	2.8	4.0	3.6	.37	.52	.47
Carpentry & Millwork	6.8 to	43.6	18.6	3.7	4.0	0.5	.69	.74	.09
Metal or Composition Siding	6.8 to	32.2	16.5	2.3	0.3	4.3	.38	.05	.71
Roofing	6.8 to	88.7	33.1	2.3	2.6	3.1	.76	.86	1.03
Doors & Hardware	4.1 to	25.5	11.4	0.9	1.4	0.4	.10	.16	.05
Sash & Glazing	5.7 to	44.9	14.2	3.5	4.0	1.0	.50	.57	.14
Lath & Plaster	5.2 to	61.9	15.9	3.3	6.9	0.8	.52	1.10	.13
Tile, Marble & Floors	3.7 to	23.3	9.9	2.6	3.0	0.5	.26	.30	.05
Acoustical Ceilings	3.5 to	26.7	11.5	2.4	0.2	0.3	.28	.02	.03
Painting	5.8 to	27.6	13.7	1.5	1.6	1.6	.21	.22	.22
Interior Partitions	6.8 to	43.6	18.6	3.9	4.3	4.4	.73	.80	.82
Miscellaneous Items	2.7 to	109.2	17.9	5.2	3.7	9.7	.93	.66	1.74
Elevators	1.8 to	19.6	7.7	2.1	1.1	2.2	.16	.08	.17
Sprinklers	2.7 to	21.2	9.1	0.5	—	2.0	.05	—	.18
Plumbing	3.0 to	16.9	8.3	4.9	7.2	5.2	.41	.60	.43
Heat., Vent., Air Conditioning	3.3 to	33.3	11.8	13.5	11.0	12.9	1.59	1.30	1.52
Electrical	2.3 to	10.5	6.6	10.1	8.4	11.1	.67	.55	.73
Total	2.3 % to	109.2%	—	100.0%	100.0%	100.0%	17.27%	15.70%	19.35%
				Overall Weighted Average		17.44%			

Workers' Compensation Insurance Rates by States

The table below lists the weighted average Workers' Compensation base rate for each state with a factor comparing this with the national average of 17.0%.

State	Weighted Average	Factor	State	Weighted Average	Factor	State	Weighted Average	Factor
Alabama	32.1%	189	Kentucky	19.9%	117	North Dakota	15.0%	88
Alaska	16.2	95	Louisiana	27.4	161	Ohio	19.3	114
Arizona	10.1	59	Maine	20.8	122	Oklahoma	22.1	130
Arkansas	15.5	91	Maryland	11.9	70	Oregon	13.9	82
California	18.7	110	Massachusetts	20.6	121	Pennsylvania	23.0	135
Colorado	26.6	156	Michigan	17.8	105	Rhode Island	21.1	124
Connecticut	25.3	149	Minnesota	27.9	164	South Carolina	13.9	82
Delaware	15.1	89	Mississippi	17.7	104	South Dakota	14.3	84
District of Columbia	21.9	129	Missouri	17.8	105	Tennessee	16.0	94
Florida	28.1	165	Montana	19.2	113	Texas	15.3	90
Georgia	25.0	147	Nebraska	15.8	93	Utah	13.0	76
Hawaii	16.8	99	Nevada	18.1	106	Vermont	18.0	106
Idaho	10.9	64	New Hampshire	22.7	134	Virginia	13.0	76
Illinois	22.3	131	New Jersey	9.6	56	Washington	8.9	52
Indiana	7.3	43	New Mexico	15.3	90	West Virginia	12.8	75
Iowa	14.2	84	New York	15.4	91	Wisconsin	14.7	86
Kansas	9.2	54	North Carolina	14.2	84	Wyoming	6.5	38
			Weighted Average for U.S. is	17.4% of payroll = 100%				

Rates in the following table are the base or manual costs per $100 of payroll for Workers' Compensation in each state. Rates are usually applied to straight time wages only and not to premium time wages and bonuses.

The weighted average skilled worker rate for 35 trades is 17.0%. For bidding purposes, apply the full value of Workers' Compensation directly to total labor costs, or if labor is 38%, materials 42% and overhead and profit 20% of total cost, carry 38/80 x 17.0% =8.1% of cost (before overhead and profit) into overhead. Rates vary not only from state to state but also with the experience rating of the contractor.

Rates are the most current available at the time of publication.

R01100-060 Workers' Compensation Insurance Rates by Trade and State (cont.)

State	Carpentry — 3 stories or less 5651	Carpentry — interior cab. work 5437	Carpentry — general 5403	Concrete Work — NOC 5213	Concrete Work — flat (flr., sdwk.) 5221	Electrical Wiring — inside 5190	Excavation — earth NOC 6217	Excavation — rock 6217	Glaziers 5462	Insulation Work 5479	Lathing 5443	Masonry 5022	Painting & Decorating 5474	Pile Driving 6003	Plastering 5480	Plumbing 5183	Roofing 5551	Sheet Metal Work (HVAC) 5538	Steel Erection — door & sash 5102	Steel Erection — inter., ornam. 5102	Steel Erection — structure 5040	Steel Erection — NOC 5057	Tile Work — (interior ceramic) 5348	Waterproofing 9014	Wrecking 5701
AL	21.76	13.96	32.26	17.91	14.04	9.73	18.64	18.64	44.93	20.78	15.43	34.76	27.25	36.98	61.88	12.17	88.73	33.28	24.00	24.00	55.03	52.56	15.68	6.26	55.03
AK	11.91	12.75	12.15	12.87	8.40	8.28	14.16	14.16	13.95	16.12	8.33	15.07	11.32	47.68	13.09	7.74	27.88	8.33	12.07	12.07	28.16	21.39	8.49	8.23	25.07
AZ	8.62	6.06	16.90	10.01	5.82	5.12	6.72	6.72	8.68	16.02	6.54	9.54	5.82	12.97	8.75	5.71	15.71	7.51	9.75	9.75	19.46	11.39	3.76	3.47	50.31
AR	15.74	10.25	15.82	14.78	7.19	7.57	10.06	10.06	14.17	22.24	14.44	10.14	11.36	16.81	13.94	6.90	29.53	12.32	8.87	8.87	33.72	32.80	7.13	4.61	33.72
CA	28.28	9.07	28.28	13.18	13.18	9.93	7.88	7.88	16.37	23.34	10.90	14.55	19.45	21.18	18.12	11.59	39.79	15.33	13.55	13.55	23.93	21.91	7.74	19.45	21.91
CO	32.22	13.54	20.95	20.96	15.13	9.20	15.99	15.99	15.98	21.31	14.68	31.06	21.85	41.53	37.40	14.50	59.10	12.54	11.83	11.83	73.46	43.49	15.04	12.26	73.46
CT	20.75	16.52	32.67	28.54	14.37	8.48	13.31	13.31	21.52	39.88	12.30	27.56	18.10	23.82	23.78	10.84	45.34	14.02	13.36	13.36	86.76	33.42	13.29	6.79	50.86
DE	14.72	14.72	12.47	12.67	9.21	6.23	10.00	10.00	12.72	12.47	13.30	12.17	15.53	21.07	13.30	7.85	28.41	11.01	13.46	13.46	29.88	13.46	10.48	12.17	29.88
DC	15.06	10.98	20.87	32.17	20.42	8.08	10.81	10.81	34.17	15.99	9.69	21.70	10.75	34.81	18.85	12.10	27.68	10.42	17.13	17.13	62.41	29.18	23.27	4.34	62.41
FL	30.50	25.48	28.80	33.02	15.37	10.50	14.27	14.27	20.62	27.82	16.71	26.13	23.62	62.53	36.48	10.75	53.17	16.89	15.61	15.61	51.50	42.80	12.40	9.83	51.50
GA	31.04	16.02	32.71	20.67	15.12	9.82	19.55	19.55	21.97	19.15	26.66	23.03	21.20	40.88	19.50	11.14	40.72	18.65	13.74	13.74	36.03	53.09	13.08	9.92	36.03
HI	17.38	11.62	28.20	13.74	12.27	7.10	7.73	7.73	20.55	21.19	10.94	17.87	11.33	20.15	15.61	6.06	33.24	8.02	11.11	11.11	31.71	21.70	9.78	11.54	31.71
ID	8.79	6.95	14.19	10.28	10.45	4.56	6.41	6.41	6.96	10.12	7.12	7.92	7.51	14.30	8.97	3.95	20.53	5.87	8.05	8.05	27.60	20.85	4.78	6.59	26.68
IL	20.54	14.78	27.06	32.94	13.02	9.50	10.70	10.70	19.97	20.23	13.08	20.05	11.79	30.85	16.31	11.62	35.07	16.56	19.55	19.55	56.07	36.68	15.66	5.40	56.07
IN	7.08	4.06	7.07	6.68	3.23	3.00	3.79	3.79	5.72	7.01	3.48	5.32	5.94	9.68	5.16	3.00	11.87	4.29	4.92	4.92	28.84	13.32	3.69	3.26	27.19
IA	11.69	5.98	13.84	14.77	7.26	4.38	6.66	6.66	13.73	11.80	6.77	9.91	9.31	14.56	9.53	5.56	20.95	7.27	13.52	13.52	60.41	31.50	6.20	4.98	60.41
KS	11.72	7.97	9.96	8.38	6.34	3.95	5.61	5.61	7.13	11.03	5.26	7.68	6.39	10.09	9.26	4.63	19.96	6.13	5.74	5.74	20.08	13.92	4.25	3.46	16.49
KY	16.47	15.26	24.34	22.12	9.53	8.95	13.67	13.67	15.96	14.68	10.62	22.00	19.83	30.44	13.94	10.38	32.83	15.92	13.21	13.21	45.21	33.73	10.60	8.85	45.21
LA	29.63	22.95	26.01	20.86	14.80	10.48	24.73	24.73	23.88	20.82	14.00	21.27	27.63	54.39	21.42	13.78	49.11	21.35	15.51	15.51	66.41	36.94	12.82	10.26	66.41
ME	13.43	10.42	43.59	24.29	11.12	5.87	11.62	11.62	13.35	15.39	14.03	17.26	16.07	31.44	17.95	7.52	32.07	8.82	15.08	15.08	37.84	61.81	11.66	6.73	37.84
MD	10.55	5.95	10.55	11.35	5.05	5.15	9.25	9.25	13.20	13.05	6.45	11.35	6.75	27.45	6.65	5.55	22.80	7.00	9.15	9.15	26.80	18.50	6.35	3.50	26.80
MA	12.21	9.11	18.28	31.54	13.45	4.92	8.59	8.59	13.44	22.23	11.23	23.25	11.97	23.70	11.02	6.80	54.04	10.57	12.87	12.87	64.68	46.56	13.18	5.30	50.36
MI	19.35	12.06	17.59	20.39	9.50	6.40	12.57	12.57	11.37	13.56	13.59	17.77	12.97	35.66	13.59	7.37	29.79	10.54	12.81	12.81	35.66	34.91	11.31	5.64	35.66
MN	18.94	20.38	41.30	20.33	14.86	6.79	15.72	15.72	18.94	22.75	23.83	25.13	18.89	29.83	23.83	10.56	65.54	10.92	16.66	16.66	112.03	36.34	17.83	6.43	132.92
MS	15.60	16.05	18.23	13.30	8.13	6.74	10.09	10.09	11.51	13.81	8.46	15.44	13.24	31.93	14.15	6.62	27.19	18.17	9.63	9.63	58.10	34.13	9.10	7.83	37.42
MO	20.97	8.42	13.26	14.87	11.94	6.48	12.38	12.38	10.05	22.39	10.83	15.20	13.18	18.91	15.93	7.95	28.83	10.93	15.48	15.48	53.74	35.80	6.53	5.96	53.74
MT	18.41	10.06	20.00	12.80	10.83	4.82	14.05	14.05	11.28	14.47	15.54	13.65	14.34	62.63	15.02	7.58	56.91	9.90	10.58	10.58	42.30	18.32	6.74	4.87	42.30
NE	16.19	9.23	15.94	17.31	12.73	5.43	10.19	10.19	12.36	16.79	9.34	18.07	11.58	22.01	14.62	9.06	33.91	13.86	10.75	10.75	25.03	26.96	7.88	5.47	25.03
NV	21.53	10.32	14.90	12.85	10.85	9.71	12.85	12.85	14.94	20.37	8.07	13.73	12.12	15.05	15.80	13.17	27.47	18.72	16.99	16.99	47.57	30.04	9.57	7.85	44.29
NH	13.36	13.42	17.35	28.68	10.00	6.89	15.81	15.81	13.97	38.85	9.95	23.46	17.06	21.14	17.68	11.41	52.99	12.14	16.19	16.19	56.75	44.76	14.21	7.29	56.75
NJ	10.22	7.20	10.20	9.37	6.76	3.52	5.65	5.65	7.04	9.49	8.45	11.22	7.73	11.79	8.45	5.00	26.87	5.93	8.77	8.77	14.23	8.74	4.61	4.24	25.28
NM	18.91	8.04	13.52	10.95	8.17	5.35	6.85	6.85	12.09	15.11	7.18	13.61	8.59	18.69	9.15	6.89	27.61	14.11	16.22	16.22	58.23	20.29	5.89	6.59	58.23
NY	13.69	6.52	14.65	19.32	12.87	6.06	8.60	8.60	11.62	10.81	20.44	20.50	12.89	18.88	10.81	8.40	33.69	14.96	11.47	11.47	18.74	24.53	9.71	5.99	29.99
NC	14.25	10.46	17.96	13.68	6.88	8.46	8.61	8.61	9.47	13.66	6.59	11.02	10.01	18.79	16.37	7.55	27.99	10.87	8.44	8.44	38.13	21.61	7.12	4.29	38.13
ND	11.64	11.64	11.64	7.53	7.53	4.71	7.02	7.02	6.69	11.64	11.18	7.53	7.47	25.26	11.18	6.68	27.33	6.68	25.26	25.26	25.26	25.26	8.56	27.33	17.90
OH	17.06	17.06	17.06	15.16	15.16	9.12	15.16	15.16	22.67	17.90	17.90	20.44	22.67	15.16	17.90	10.47	33.45	32.47	15.16	15.16	15.16	15.16	13.22	15.16	15.16
OK	26.16	11.79	19.33	15.51	11.50	7.41	18.45	18.45	13.55	18.00	12.05	16.80	17.10	36.26	18.29	8.72	42.26	12.45	11.17	11.17	69.45	46.76	11.42	8.20	69.45
OR	18.36	8.75	16.37	11.81	8.84	4.56	10.17	10.17	14.84	9.50	7.75	13.03	13.37	14.79	11.50	5.52	22.37	11.12	9.17	9.17	38.02	17.71	10.80	4.87	38.02
PA	15.80	15.80	19.85	28.18	11.84	7.80	12.74	12.74	15.69	19.85	18.94	18.96	22.05	32.12	18.94	11.57	43.30	12.58	24.71	24.71	59.05	24.71	11.83	18.96	81.34
RI	19.53	11.65	18.07	18.23	16.24	4.43	10.38	10.38	12.85	22.78	11.97	25.11	24.13	37.66	17.25	8.52	33.92	10.29	14.07	14.07	59.49	37.50	14.36	7.70	78.79
SC	19.10	13.03	19.24	12.88	6.19	7.02	8.32	8.32	12.22	10.69	7.27	9.55	11.19	16.64	18.89	6.88	28.98	11.42	9.49	9.49	21.80	23.66	6.23	4.65	21.80
SD	15.18	7.06	16.51	16.80	5.46	6.16	10.22	10.22	10.35	14.43	6.99	9.02	14.36	28.31	12.74	7.93	21.56	7.57	10.68	10.68	41.04	19.09	6.33	4.28	41.04
TN	15.52	11.50	14.25	14.34	8.85	5.88	10.07	10.07	9.46	17.08	9.49	13.44	12.46	16.80	14.75	16.86	32.77	10.46	9.28	9.28	53.04	19.67	7.22	6.14	53.04
TX	16.89	11.33	14.31	15.47	9.19	8.22	10.74	10.74	12.40	16.66	10.03	15.20	10.55	17.40	20.99	8.13	26.21	14.28	10.54	10.54	31.39	18.26	7.54	7.83	21.70
UT	11.17	11.17	11.17	17.68	7.99	7.05	6.38	6.38	9.55	10.95	12.32	13.75	15.29	17.24	10.60	6.41	26.27	6.30	8.99	8.99	23.51	23.51	6.06	5.83	26.53
VT	17.05	10.67	19.52	29.53	9.80	5.84	9.77	9.77	18.72	20.73	10.17	18.93	9.78	22.85	17.33	8.71	26.69	10.84	13.61	13.61	38.19	31.77	8.52	10.88	38.19
VA	10.65	8.00	11.99	12.17	6.45	4.59	8.52	8.52	8.12	7.68	11.40	9.17	10.02	19.91	8.03	6.35	22.51	9.22	16.59	16.59	30.83	24.72	9.28	3.66	30.83
WA	7.57	7.57	6.77	6.22	6.14	2.29	6.64	6.64	9.41	6.14	6.77	8.67	7.23	14.26	8.29	3.86	15.96	3.34	12.96	12.96	8.35	8.35	7.65	10.23	8.35
WV	12.00	12.00	12.00	23.80	23.80	5.13	7.74	7.74	6.08	6.08	16.88	11.57	12.20	12.89	12.20	4.77	13.44	6.08	14.98	14.98	9.84	13.15	16.02	3.88	13.27
WI	9.13	9.31	20.38	9.76	9.12	4.50	7.46	7.46	11.44	12.12	12.31	15.75	10.81	16.00	11.44	5.99	37.55	7.07	17.21	17.21	34.65	15.87	13.98	4.70	34.65
WY	6.76	6.76	6.76	6.76	6.76	6.76	6.76	6.76	6.76	6.76	6.76	6.76	6.76	6.76	6.76	6.76	6.76	6.76	6.76	6.76	6.76	6.76	6.76	6.76	6.76
AVG.	16.49	11.40	18.57	17.05	10.57	6.65	10.79	10.79	14.20	16.54	11.46	16.12	13.74	25.16	15.95	8.35	33.11	11.81	13.07	13.07	41.03	27.44	9.91	7.66	41.41

R01100-060 Workers' Compensation (cont.) (Canada in Canadian dollars)

Province		Alberta	British Columbia	Manitoba	Ontario	New Brunswick	Newfndld. & Labrador	Northwest Territories	Nova Scotia	Prince Edward Island	Quebec	Saskat-chewan	Yukon
Carpentry—3 stories or less	Rate	2.72	4.63	3.80	4.80	4.40	7.20	4.62	8.06	7.25	12.69	5.01	3.25
	Code	25401	721028	40102	723	422	403	4-41	4226	401	80110	B12-02	202
Carpentry—interior cab. work	Rate	1.17	4.85	3.80	4.80	3.48	7.20	4.62	5.69	3.49	12.69	3.25	3.25
	Code	42133	721021	40102	723	427	403	4-41	4274	402	80110	B11-27	202
CARPENTRY—general	Rate	2.72	4.63	3.80	4.80	4.40	7.20	4.62	8.06	7.25	12.69	5.01	3.25
	Code	25401	721028	40102	723	422	403	4-41	4226	401	80110	B12-02	202
CONCRETE WORK—NOC	Rate	4.69	6.48	5.76	18.00	4.40	7.20	4.62	4.97	7.25	13.89	6.50	3.25
	Code	42104	721010	40110	748	422	403	4-41	4224	401	80100	B13-14	203
CONCRETE WORK—flat (flr. sidewalk)	Rate	4.69	6.48	5.76	18.00	4.40	7.20	4.62	4.97	7.25	13.89	6.50	3.25
	Code	42104	721010	40110	748	422	403	4-41	4224	401	80100	B13-14	203
ELECTRICAL Wiring—inside	Rate	1.99	2.90	2.03	3.13	1.74	4.83	3.46	2.61	3.49	5.69	3.25	2.35
	Code	42124	721019	40203	704	426	400	4-46	4261	402	80170	B11-05	206
EXCAVATION—earth NOC	Rate	2.57	3.40	3.80	3.99	2.93	7.20	3.46	4.25	3.61	7.71	3.70	3.25
	Code	40604	721031	40706	711	421	403	4-43	4214	404	80030	R11-06	207
EXCAVATION—rock	Rate	2.57	3.40	3.80	3.99	2.93	7.20	3.46	4.25	3.61	7.71	3.70	3.25
	Code	40604	721031	40706	711	421	403	4-43	4214	404	80030	R11-06	207
GLAZIERS	Rate	2.79	3.21	3.77	8.22	6.01	5.12	4.62	8.06	3.49	14.40	6.50	2.35
	Code	42121	715020	40109	751	423	402	4-41	4233	402	80150	B13-04	212
INSULATION WORK	Rate	2.89	5.80	3.80	8.22	6.01	5.12	4.62	8.06	7.25	14.15	5.01	3.25
	Code	42184	721029	40102	751	423	402	4-41	4234	401	80120	B12-07	202
LATHING	Rate	5.76	8.29	3.80	4.80	3.48	5.12	4.62	5.69	3.49	14.15	6.50	3.25
	Code	42135	721033	40102	723	427	402	4-41	4271	402	80120	B13-16	202
MASONRY	Rate	4.69	8.29	3.80	11.66	6.01	7.20	4.62	8.06	7.25	13.89	6.50	3.25
	Code	42102	721037	40102	741	423	403	4-41	4231	401	80100	B13-18	202
PAINTING & DECORATING	Rate	4.07	5.80	4.20	7.60	3.28	5.12	4.62	5.69	3.49	12.69	5.01	3.25
	Code	42111	721041	40105	719	427	402	4-41	4275	402	80110	B12-01	202
PILE DRIVING	Rate	4.69	13.55	3.80	6.26	4.40	10.89	3.46	4.97	7.25	7.71	6.50	3.25
	Code	42159	722004	40706	732	422	404	4-43	4221	401	80030	B13-10	202
PLASTERING	Rate	5.76	8.29	4.46	7.60	3.28	5.12	4.62	5.69	3.49	12.69	5.01	3.25
	Code	42135	721042	40108	719	427	402	4-41	4271	402	80110	B12-21	202
PLUMBING	Rate	1.99	4.30	2.37	3.89	2.30	4.60	3.46	2.94	3.49	7.13	3.25	1.40
	Code	42122	721043	40204	707	424	401	4-46	4241	402	80160	B11-01	214
ROOFING	Rate	9.70	8.74	5.42	12.15	8.05	7.20	4.62	8.06	7.25	22.17	6.50	3.25
	Code	42118	721036	40403	728	430	403	4-41	4236	401	80130	B13-20	202
SHEET METAL WORK (HVAC)	Rate	1.99	4.30	5.14	3.89	2.30	4.60	3.46	2.94	3.49	7.13	3.25	2.35
	Code	42117	721043	40402	707	424	401	4-46	4244	402	80160	B11-07	208
STEEL ERECTION—door & sash	Rate	2.89	13.55	6.56	18.00	4.40	10.89	4.62	8.06	7.25	32.47	6.50	3.25
	Code	42106	722005	40502	748	422	404	4-41	4227	401	80080	B13-22	202
STEEL ERECTION—inter., ornam.	Rate	2.89	13.55	6.56	18.00	4.40	7.20	4.62	8.06	7.25	32.47	6.50	3.25
	Code	42106	722005	40502	748	422	403	4-41	4227	401	80080	B13-22	202
STEEL ERECTION—structure	Rate	2.89	13.55	6.56	18.00	4.40	10.89	4.62	8.06	7.25	32.47	6.50	3.25
	Code	42106	722005	40502	748	422	404	4-41	4227	401	80080	B13-22	202
STEEL ERECTION—NOC	Rate	2.89	13.55	6.56	18.00	4.40	10.89	4.62	8.06	7.25	32.47	6.50	3.25
	Code	42106	722005	40502	748	422	404	4-41	4227	401	80080	B13-22	202
TILE WORK—inter. (ceramic)	Rate	2.94	3.27	1.88	7.60	3.28	5.12	4.62	5.69	3.49	12.69	6.50	3.25
	Code	42113	721054	40103	719	427	402	4-41	4276	402	80110	B13-01	202
WATERPROOFING	Rate	4.07	5.80	3.80	4.80	6.01	7.20	4.62	8.06	3.49	22.17	5.01	3.25
	Code	42139	721016	40102	723	423	403	4-41	4239	402	80130	B12-17	202
WRECKING	Rate	2.57	6.08	5.15	18.00	2.93	7.20	3.46	4.25	7.25	12.69	6.50	3.25
	Code	40604	721005	40106	748	421	403	4-43	4211	401	80110	B13-09	202

R01100-070 Contractor's Overhead & Profit

Below are the **average** installing contractor's percentage mark-ups applied to base labor rates to arrive at typical billing rates.

Column A: Labor rates are based on union wages averaged for 30 major U.S. cities. Base rates including fringe benefits are listed hourly and daily. These figures are the sum of the wage rate and employer-paid fringe benefits such as vacation pay, employer-paid health and welfare costs, pension costs, plus appropriate training and industry advancement funds costs.

Column B: Workers' Compensation rates are the national average of state rates established for each trade.

Column C: Column C lists average fixed overhead figures for all trades. Included are Federal and State Unemployment costs set at 6.5%; Social Security Taxes (FICA) set at 7.65%; Builder's Risk Insurance costs set at 0.44%; and Public Liability costs set at 2.02%. All the percentages except those for Social Security Taxes vary from state to state as well as from company to company.

Columns D and E: Percentages in Columns D and E are based on the presumption that the installing contractor has annual billing of $4,000,000 and up. Overhead percentages may increase with smaller annual billing. The overhead percentages for any given contractor may vary greatly and depend on a number of factors, such as the contractor's annual volume, engineering and logistical support costs, and staff requirements. The figures for overhead and profit will also vary depending on the type of job, the job location, and the prevailing economic conditions. All factors should be examined very carefully for each job.

Column F: Column F lists the total of Columns B, C, D, and E.

Column G: Column G is Column A (hourly base labor rate) multiplied by the percentage in Column F (O&P percentage).

Column H: Column H is the total of Column A (hourly base labor rate) plus Column G (Total O&P).

Column I: Column I is Column H multiplied by eight hours.

Abbr.	Trade	A Base Rate Incl. Fringes Hourly	A Daily	B Workers' Comp. Ins.	C Average Fixed Overhead	D Over-head	E Profit	F Total Overhead & Profit %	G Amount	H Rate with O & P Hourly	I Daily
Skwk	Skilled Workers Average (35 trades)	$32.25	$258.00	17.0%	16.6%	13.0%	10.0%	56.6%	$18.25	$50.50	$404.00
	Helpers Average (5 trades)	23.70	189.60	18.7		11.0		56.3	$13.35	37.05	296.40
	Foreman Average, Inside ($.50 over trade)	32.75	262.00	17.0		13.0		56.6	18.55	51.30	410.40
	Foreman Average, Outside ($2.00 over trade)	34.25	274.00	17.0		13.0		56.6	19.40	53.65	429.20
Clab	Common Building Laborers	24.65	197.20	18.6		11.0		56.2	13.85	38.50	308.00
Asbe	Asbestos/Insulation Workers/Pipe Coverers	34.70	277.60	16.5		16.0		59.1	20.50	55.20	441.60
Boil	Boilermakers	38.70	309.60	14.5		16.0		57.1	22.10	60.80	486.40
Bric	Bricklayers	32.40	259.20	16.1		11.0		53.7	17.40	49.80	398.40
Brhe	Bricklayer Helpers	24.60	196.80	16.1		11.0		53.7	13.20	37.80	302.40
Carp	Carpenters	31.55	252.40	18.6		11.0		56.2	17.75	49.30	394.40
Cefi	Cement Finishers	30.20	241.60	10.6		11.0		48.2	14.55	44.75	358.00
Elec	Electricians	37.60	300.80	6.7		16.0		49.3	18.55	56.15	449.20
Elev	Elevator Constructors	38.75	310.00	7.7		16.0		50.3	19.50	58.25	466.00
Eqhv	Equipment Operators, Crane or Shovel	33.70	269.60	10.8		14.0		51.4	17.30	51.00	408.00
Eqmd	Equipment Operators, Medium Equipment	32.60	260.80	10.8		14.0		51.4	16.75	49.35	394.80
Eqlt	Equipment Operators, Light Equipment	31.10	248.80	10.8		14.0		51.4	16.00	47.10	376.80
Eqol	Equipment Operators, Oilers	28.30	226.40	10.8		14.0		51.4	14.55	42.85	342.80
Eqmm	Equipment Operators, Master Mechanics	34.20	273.60	10.8		14.0		51.4	17.60	51.80	414.40
Glaz	Glaziers	30.95	247.60	14.2		11.0		51.8	16.05	47.00	376.00
Lath	Lathers	29.65	237.20	11.5		11.0		49.1	14.55	44.20	353.60
Marb	Marble Setters	31.05	248.40	16.1		11.0		53.7	16.65	47.70	381.60
Mill	Millwrights	32.95	263.60	10.7		11.0		48.3	15.90	48.85	390.80
Mstz	Mosaic and Terrazzo Workers	30.40	243.20	9.9		11.0		47.5	14.45	44.85	358.80
Pord	Painters, Ordinary	28.05	224.40	13.7		11.0		51.3	14.40	42.45	339.60
Psst	Painters, Structural Steel	28.90	231.20	50.2		11.0		87.8	25.35	54.25	434.00
Pape	Paper Hangers	28.20	225.60	13.7		11.0		51.3	14.45	42.65	341.20
Pile	Pile Drivers	30.90	247.20	25.2		16.0		67.8	20.95	51.85	414.80
Plas	Plasterers	29.00	232.00	16.0		11.0		53.6	15.55	44.55	356.40
Plah	Plasterer Helpers	24.85	198.80	16.0		11.0		53.6	13.30	38.15	305.20
Plum	Plumbers	37.35	298.80	8.4		16.0		51.0	19.05	56.40	451.20
Rodm	Rodmen (Reinforcing)	35.55	284.40	27.4		14.0		68.0	24.15	59.70	477.60
Rofc	Roofers, Composition	27.65	221.20	33.1		11.0		70.7	19.55	47.20	377.60
Rots	Roofers, Tile and Slate	27.95	223.60	33.1		11.0		70.7	19.75	47.70	381.60
Rohe	Roofer Helpers (Composition)	20.55	164.40	33.1		11.0		70.7	14.55	35.10	280.80
Shee	Sheet Metal Workers	37.00	296.00	11.8		16.0		54.4	20.15	57.15	457.20
Spri	Sprinkler Installers	37.55	300.40	9.1		16.0		51.7	19.40	56.95	455.60
Stpi	Steamfitters or Pipefitters	37.60	300.80	8.4		16.0		51.0	19.20	56.80	454.40
Ston	Stone Masons	32.30	258.40	16.1		11.0		53.7	17.35	49.65	397.20
Sswk	Structural Steel Workers	35.65	285.20	41.0		14.0		81.6	29.10	64.75	518.00
Tilf	Tile Layers	30.40	243.20	9.9		11.0		47.5	14.45	44.85	358.80
Tilh	Tile Layer Helpers	23.90	191.20	9.9		11.0		47.5	11.35	35.25	282.00
Trlt	Truck Drivers, Light	24.90	199.20	15.1		11.0		52.7	13.10	38.00	304.00
Trhv	Truck Drivers, Heavy	25.65	205.20	15.1		11.0		52.7	13.50	39.15	313.20
Sswl	Welders, Structural Steel	35.65	285.20	41.0		14.0		81.6	29.10	64.75	518.00
Wrck	*Wrecking	24.65	197.20	41.4	▼	11.0	▼	79.0	19.45	44.10	352.80

*Not included in Averages.

R01100-080　Performance Bond

This table shows the cost of a Performance Bond for a construction job scheduled to be completed in 12 months. Add 1% of the premium cost per month for jobs requiring more than 12 months to complete. The rates are "standard" rates offered to contractors that the bonding company considers financially sound and capable of doing the work. Preferred rates are offered by some bonding companies based upon financial strength of the contractor. Actual rates vary from contractor to contractor and from bonding company to bonding company. Contractors should prequalify through a bonding agency before submitting a bid on a contract that requires a bond.

Contract Amount	Building Construction Class B Projects			Highways & Bridges					
				Class A New Construction			Class A-1 Highway Resurfacing		
First $ 100,000 bid	$25.00 per M			$15.00 per M			$9.40 per M		
Next 400,000 bid	$ 2,500	plus	$15.00 per M	$ 1,500	plus	$10.00 per M	$ 940	plus	$7.20 per M
Next 2,000,000 bid	8,500	plus	10.00 per M	5,500	plus	7.00 per M	3,820	plus	5.00 per M
Next 2,500,000 bid	28,500	plus	7.50 per M	19,500	plus	5.50 per M	15,820	plus	4.50 per M
Next 2,500,000 bid	47,250	plus	7.00 per M	33,250	plus	5.00 per M	28,320	plus	4.50 per M
Over 7,500,000 bid	64,750	plus	6.00 per M	45,750	plus	4.50 per M	39,570	plus	4.00 per M

R01100-090　Sales Tax by State

State sales tax on materials is tabulated below (5 states have no sales tax). Many states allow local jurisdictions, such as a county or city, to levy additional sales tax.

Some projects may be sales tax exempt, particularly those constructed with public funds.

State	Tax (%)	State	Tax (%)	State	Tax (%)	State	Tax (%)
Alabama	4	Illinois	6.25	Montana	0	Rhode Island	7
Alaska	0	Indiana	5	Nebraska	5	South Carolina	5
Arizona	5	Iowa	5	Nevada	6.5	South Dakota	4
Arkansas	4.625	Kansas	4.9	New Hampshire	0	Tennessee	6
California	6	Kentucky	6	New Jersey	6	Texas	6.25
Colorado	3	Louisiana	4	New Mexico	5	Utah	4.75
Connecticut	6	Maine	5	New York	4	Vermont	5
Delaware	0	Maryland	5	North Carolina	4	Virginia	3.5
District of Columbia	5.75	Massachusetts	5	North Dakota	5	Washington	6.5
Florida	6	Michigan	6	Ohio	5	West Virginia	6
Georgia	4	Minnesota	6.5	Oklahoma	4.5	Wisconsin	5
Hawaii	4	Mississippi	7	Oregon	0	Wyoming	4
Idaho	5	Missouri	4.225	Pennsylvania	6	Average	4.65 %

Sales Tax by Province (Canada)

GST - a value-added tax, which the government imposes on most goods and services provided in or imported into Canada.
PST - a retail sales tax, which five of the provinces impose on the price of most goods and some services.

QST - a value-added tax, similar to the federal GST, which Quebec imposes.
HST - Three provinces have combined their retail sales tax with the federal GST into one harmonized tax.

Province	PST (%)	QST (%)	GST(%)	HST(%)
Alberta	0	0	7	0
British Columbia	7.5	0	7	0
Manitoba	7	0	7	0
New Brunswick	0	0	0	15
Newfoundland	0	0	0	15
Northwest Territories	0	0	7	0
Nova Scotia	0	0	0	15
Ontario	8	0	7	0
Prince Edward Island	10	0	7	0
Quebec	0	7.5	7	0
Saskatchewan	6	0	7	0
Yukon	0	0	7	0

R01100-100　Unemployment Taxes and Social Security Taxes

Mass. State Unemployment tax ranges from 1.325% to 7.225% plus an experience rating assessment the following year, on the first $10,800 of wages. Federal Unemployment tax is 6.2% of the first $7,000 of wages. This is reduced by a credit for payment to the state. The minimum Federal Unemployment tax is 0.8% after all credits.

Combined rates in Mass. thus vary from 2.125% to 8.025% of the first $10,800 of wages. Combined average U.S. rate is about 6.5% of the first $7,000. Contractors with permanent workers will pay less since the average annual wages for skilled workers is $32.25 x 2,000 hours or about $64,500 per year. The average combined rate for U.S. would thus be 6.5% x $7,000 ÷ $64,500 = 0.7% of total wages for permanent employees.

Rates vary not only from state to state but also with the experience rating of the contractor.

Social Security (FICA) for 2003 is estimated at time of publication to be 7.65% of wages up to $84,900.

517

R01100-110 Overtime

One way to improve the completion date of a project or eliminate negative float from a schedule is to compress activity duration times. This can be achieved by increasing the crew size or working overtime with the proposed crew.

To determine the costs of working overtime to compress activity duration times, consider the following examples. Below is an overtime efficiency and

cost chart based on a five, six, or seven day week with an eight through twelve hour day. Payroll percentage increases for time and one half and double time are shown for the various working days.

Days per Week	Hours per Day	Production Efficiency					Payroll Cost Factors	
		1 Week	2 Weeks	3 Weeks	4 Weeks	Average 4 Weeks	@ 1-1/2 Times	@ 2 Times
5	8	100%	100%	100%	100%	100 %	100 %	100 %
	9	100	100	95	90	96.25	105.6	111.1
	10	100	95	90	85	91.25	110.0	120.0
	11	95	90	75	65	81.25	113.6	127.3
	12	90	85	70	60	76.25	116.7	133.3
6	8	100	100	95	90	96.25	108.3	116.7
	9	100	95	90	85	92.50	113.0	125.9
	10	95	90	85	80	87.50	116.7	133.3
	11	95	85	70	65	78.75	119.7	139.4
	12	90	80	65	60	73.75	122.2	144.4
7	8	100	95	85	75	88.75	114.3	128.6
	9	95	90	80	70	83.75	118.3	136.5
	10	90	85	75	65	78.75	121.4	142.9
	11	85	80	65	60	72.50	124.0	148.1
	12	85	75	60	55	68.75	126.2	152.4

R01100-710 Unit Gross Area Requirements

The figures in the table below indicate typical ranges in square feet as a function of the "occupant" unit. This table is best used in the preliminary design stages to help determine the probable size

requirement for the total project. See R17100-100 for the typical total size ranges for various types of buildings.

Building Type	Unit	Gross Area in S.F.		
		1/4	Median	3/4
Apartments	Unit	660	860	1,100
Auditorium & Play Theaters	Seat	18	25	38
Bowling Alleys	Lane		940	
Churches & Synagogues	Seat	20	28	39
Dormitories	Bed	200	230	275
Fraternity & Sorority Houses	Bed	220	315	370
Garages, Parking	Car	325	355	385
Hospitals	Bed	685	850	1,075
Hotels	Rental Unit	475	600	710
Housing for the elderly	Unit	515	635	755
Housing, Public	Unit	700	875	1,030
Ice Skating Rinks	Total	27,000	30,000	36,000
Motels	Rental Unit	360	465	620
Nursing Homes	Bed	290	350	450
Restaurants	Seat	23	29	39
Schools, Elementary	Pupil	65	77	90
Junior High & Middle		85	110	129
Senior High		102	130	145
Vocational		110	135	195
Shooting Ranges	Point		450	
Theaters & Movies	Seat		15	

R01100-720 Floor Area Ratios

Table below lists commonly used gross to net area and net to gross area ratios expressed in % for various building types.

Building Type	Gross to Net Ratio	Net to Gross Ratio	Building Type	Gross to Net Ratio	Net to Gross Ratio
Apartment	156	64	School Buildings (campus type)		
Bank	140	72	Administrative	150	67
Church	142	70	Auditorium	142	70
Courthouse	162	61	Biology	161	62
Department Store	123	81	Chemistry	170	59
Garage	118	85	Classroom	152	66
Hospital	183	55	Dining Hall	138	72
Hotel	158	63	Dormitory	154	65
Laboratory	171	58	Engineering	164	61
Library	132	76	Fraternity	160	63
Office	135	75	Gymnasium	142	70
Restaurant	141	70	Science	167	60
Warehouse	108	93	Service	120	83
			Student Union	172	59

The gross area of a building is the total floor area based on outside dimensions.

The net area of a building is the usable floor area for the function intended and excludes such items as stairways, corridors and mechanical rooms. In the case of a commercial building, it might be considered as the "leasable area."

R01100-730 Occupancy Determinations

Description		S.F. Required per Person		
		BOCA	SBC	UBC
Assembly Areas	Fixed Seats	**	6	7
	Movable Seats		15	15
	Concentrated	7		
	Unconcentrated	15		
	Standing Space	3		
Educational	Unclassified			
	Classrooms	20	40	20
	Shop Areas	50	100	50
Institutional	Unclassified		125	
	In-Patient Areas	240		
	Sleeping Areas	120		
Mercantile	Basement	30	30	20
	Ground Floor	30	30	30
	Upper Floors	60	60	50
Office		100	100	100

BOCA=Building Officials & Code Administrators
SBC=Southern Building Code
UBC=Uniform Building Code

** The occupancy load for assembly area with fixed seats shall be determined by the number of fixed seats installed.

R01107-030   Engineering Fees

Typical **Structural Engineering Fees** based on type of construction and total project size. These fees are included in Architectural Fees.

Type of Construction	Total Project Size (in thousands of dollars)			
	$500	$500-$1,000	$1,000-$5,000	Over $5000
Industrial buildings, factories & warehouses	Technical payroll times 2.0 to 2.5	1.60%	1.25%	1.00%
Hotels, apartments, offices, dormitories, hospitals, public buildings, food stores		2.00%	1.70%	1.20%
Museums, banks, churches and cathedrals		2.00%	1.75%	1.25%
Thin shells, prestressed concrete, earthquake resistive		2.00%	1.75%	1.50%
Parking ramps, auditoriums, stadiums, convention halls, hangars & boiler houses		2.50%	2.00%	1.75%
Special buildings, major alterations, underpinning & future expansion		Add to above 0.5%	Add to above 0.5%	Add to above 0.5%

For complex reinforced concrete or unusually complicated structures, add 20% to 50%.

Typical **Mechanical and Electrical Engineering Fees** are based on the size of the subcontract. The fee structure for Mechanical Engineering is shown below; Electrical Engineering fees range from 2 to 5%. These fees are included in Architectural Fees.

Type of Construction	Subcontract Size							
	$25,000	$50,000	$100,000	$225,000	$350,000	$500,000	$750,000	$1,000,000
Simple structures	6.4%	5.7%	4.8%	4.5%	4.4%	4.3%	4.2%	4.1%
Intermediate structures	8.0	7.3	6.5	5.6	5.1	5.0	4.9	4.8
Complex structures	12.0	9.0	9.0	8.0	7.5	7.5	7.0	7.0

For renovations, add 15% to 25% to applicable fee.

R01250-010 Repair and Remodeling

Cost figures are based on new construction utilizing the most cost-effective combination of labor, equipment and material with the work scheduled in proper sequence to allow the various trades to accomplish their work in an efficient manner.

The costs for repair and remodeling work must be modified due to the following factors that may be present in any given repair and remodeling project.

1. Equipment usage curtailment due to the physical limitations of the project, with only hand-operated equipment being used.

2. Increased requirement for shoring and bracing to hold up the building while structural changes are being made and to allow for temporary storage of construction materials on above-grade floors.

3. Material handling becomes more costly due to having to move within the confines of an enclosed building. For multi-story construction, low capacity elevators and stairwells may be the only access to the upper floors.

4. Large amount of cutting and patching and attempting to match the existing construction is required. It is often more economical to remove entire walls rather than create many new door and window openings. This sort of trade-off has to be carefully analyzed.

5. Cost of protection of completed work is increased since the usual sequence of construction usually cannot be accomplished.

6. Economies of scale usually associated with new construction may not be present. If small quantities of components must be custom fabricated due to job requirements, unit costs will naturally increase. Also, if only small work areas are available at a given time, job scheduling between trades becomes difficult and subcontractor quotations may reflect the excessive start-up and shut-down phases of the job.

7. Work may have to be done on other than normal shifts and may have to be done around an existing production facility which has to stay in production during the course of the repair and remodeling.

8. Dust and noise protection of adjoining non-construction areas can involve substantial special protection and alter usual construction methods.

9. Job may be delayed due to unexpected conditions discovered during demolition or removal. These delays ultimately increase construction costs.

10. Piping and ductwork runs may not be as simple as for new construction. Wiring may have to be snaked through walls and floors.

11. Matching "existing construction" may be impossible because materials may no longer be manufactured. Substitutions may be expensive.

12. Weather protection of existing structure requires additional temporary structures to protect building at openings.

13. On small projects, because of local conditions, it may be necessary to pay a tradesman for a minimum of four hours for a task that is completed in one hour.

All of the above areas can contribute to increased costs for a repair and remodeling project. Each of the above factors should be considered in the planning, bidding and construction stage in order to minimize the increased costs associated with repair and remodeling jobs.

R01540-100   Steel Tubular Scaffolding

On new construction, tubular scaffolding is efficient up to 60' high or five stories. Above this it is usually better to use a hung scaffolding if construction permits. Swing scaffolding operations may interfere with tenants. In this case, the tubular is more practical at all heights.

In repairing or cleaning the front of an existing building the cost of tubular scaffolding per S.F. of building front increases as the height increases above the first tier. The first tier cost is relatively high due to leveling and alignment.

The minimum efficient crew for erection is three workers. For heights over 50', a crew of four is more efficient. Use two or more on top and two at the bottom for handing up or hoisting. Four workers can erect and

dismantle about nine frames per hour up to five stories. From five to eight stories they will average six frames per hour. With 7' horizontal spacing this will run about 400 S.F. and 265 S.F. of wall surface, respectively. Time for placing planks must be added to the above. On heights above 50', five planks can be placed per labor-hour.

The table below shows the number of pieces required to erect tubular steel scaffolding for 1000 S.F. of building frontage. This area is made up of a scaffolding system that is 12 frames (11 bays) long by 2 frames high.

For jobs under twenty-five frames, add 50% to rental cost. Rental rates will be lower for jobs over three months duration. Large quantities for long periods can reduce rental rates by 20%.

Description of Component	CSI Line Item	Number of pieces for 1000 S.F. of Building Front	Unit
5' Wide Standard Frame, 6'-4" High	01540-750-2200	24	Ea.
Leveling Jack & Plate	01540-750-2650	24	
Cross Brace	01540-750-2500	44	
Side Arm Bracket, 21"	01540-750-2700	12	
Guardrail Post	01540-750-2550	12	
Guardrail, 7' section	01540-750-2600	22	
Stairway Section	01540-750-2900	2	
Stairway Starter Bar	01540-750-2910	1	
Stairway Inside Handrail	01540-750-2920	2	
Stairway Outside Handrail	01540-750-2930	2	
Walk-Thru Frame Guardrail	01540-750-2940	2	

Scaffolding is often used as falsework over 15' high during construction of cast-in-place concrete beams and slabs. Two foot wide scaffolding is generally used for heavy beam construction. The span between frames depends upon the load to be carried with a maximum span of 5'.

Heavy duty shoring frames with a capacity of 10,000#/leg can be spaced up to 10' O. C. depending upon form support design and loading.

Scaffolding used as horizontal shoring requires less than half the material required with conventional shoring.

On new construction, erection is done by carpenters.

Rolling towers supporting horizontal shores can reduce labor and speed the job. For maintenance work, catwalks with spans up to 70' can be supported by the rolling towers.

R01590-100   Contractor Equipment

Rental Rates shown in Division 01590 pertain to late model high quality machines in excellent working condition, rented from equipment dealers. Rental rates from contractors may be substantially lower than the rental rates from equipment dealers depending upon economic conditions. For older, less productive machines, reduce rates by a maximum of 15%. Any overtime must be added to the base rates. For shift work, rates are lower. Usual rule of thumb is 150% of one shift rate for two shifts; 200% for three shifts.

For periods of less than one week, operated equipment is usually more economical to rent than renting bare equipment and hiring an operator.

Costs to move equipment to a job site (mobilization) or from a job site (demobilization) are not included in rental rates in Division 01590, nor in any Equipment costs on any Unit Price line items or crew listings. These costs can be found in section 02305-250. If a piece of equipment is already at a job site, it is not appropriate to utilize mob/demob costs in an estimate again.

Rental rates vary throughout the country with larger cities generally having lower rates. Lease plans for new equipment are available for periods in excess of six months with a percentage of payments applying toward purchase.

Monthly rental rates vary from 2% to 5% of the cost of the equipment depending on the anticipated life of the equipment and its wearing parts. Weekly rates are about 1/3 the monthly rates and daily rental rates about 1/3 the weekly rate.

The hourly operating costs for each piece of equipment include costs to the user such as fuel, oil, lubrication, normal expendables for the equipment, and a percentage of mechanic's wages chargeable to maintenance. The hourly operating costs listed do not include the operator's wages.

The daily cost for equipment used in the standard crews is figured by dividing the weekly rate by five, then adding eight times the hourly operating cost to give the total daily equipment cost, not including the operator. This figure is in the right hand column of Division 01590 under Crew Equipment Cost/Day.

Pile Driving rates shown for pile hammer and extractor do not include leads, crane, boiler or compressor. Vibratory pile driving requires an added field specialist during set-up and pile driving operation for the electric model. The hydraulic model requires a field specialist for set-up only. Up to 125 reuses of sheet piling are possible using vibratory drivers. For normal conditions, crane capacity for hammer type and size are as follows.

Crane Capacity	Hammer Type and Size		
	Air or Steam	Diesel	Vibratory
25 ton	to 8,750 ft.-lb.		70 H.P.
40 ton	15,000 ft.-lb.	to 32,000 ft.-lb.	170 H.P.
60 ton	25,000 ft.-lb.		300 H.P.
100 ton		112,000 ft.-lb.	

Cranes should be specified for the job by size, building and site characteristics, availability, performance characteristics, and duration of time required.

Backhoes & Shovels rent for about the same as equivalent size cranes but maintenance and operating expense is higher. Crane operators rate must be adjusted for high boom heights. Average adjustments: for 150' boom add 2% per hour; over 185', add 4% per hour; over 210', add 6% per hour; over 250', add 8% per hour and over 295', add 12% per hour.

Tower Cranes of the climbing or static type have jibs from 50' to 200' and capacities at maximum reach range from 4,000 to 14,000 pounds. Lifting capacities increase up to maximum load as the hook radius decreases.

Typical rental rates, based on purchase price are about 2% to 3% per month.

Erection and dismantling runs between 500 and 2000 labor hours. Climbing operation takes 10 labor hours per 20' climb. Crane dead time is about 5 hours per 40' climb. If crane is bolted to side of the building add cost of ties and extra mast sections. Climbing cranes have from 80' to 180' of mast while static cranes have 80' to 800' of mast.

Truck Cranes can be converted to tower cranes by using tower attachments. Mast heights over 400' have been used. See Division 01590-600 for rental rates of high boom cranes.

A single 100' high material **Hoist and Tower** can be erected and dismantled in about 400 labor hours; a double 100' high hoist and tower in about 600 labor hours. Erection times for additional heights are 3 and 4 labor hours per vertical foot respectively up to 150', and 4 to 5 labor hours per vertical foot over 150' high. A 40' high portable Buck hoist takes about 160 labor hours to erect and dismantle. Additional heights take 2 labor hours per vertical foot to 80' and 3 labor hours per vertical foot for the next 100'. Most material hoists do not meet local code requirements for carrying personnel.

A 150' high **Personnel Hoist** requires about 500 to 800 labor hours to erect and dismantle. Budget erection time at 5 labor hours per vertical foot for all trades. Local code requirements or labor scarcity requiring overtime can add up to 50% to any of the above erection costs.

Earthmoving Equipment: The selection of earthmoving equipment depends upon the type and quantity of material, moisture content, haul distance, haul road, time available, and equipment available. Short haul cut and fill operations may require dozers only, while another operation may require excavators, a fleet of trucks, and spreading and compaction equipment. Stockpiled material and granular material are easily excavated with front end loaders. Scrapers are most economically used with hauls between 300' and 1-1/2 miles if adequate haul roads can be maintained. Shovels are often used for blasted rock and any material where a vertical face of 8' or more can be excavated. Special conditions may dictate the use of draglines, clamshells, or backhoes. Spreading and compaction equipment must be matched to the soil characteristics, the compaction required and the rate the fill is being supplied.

R02115-200 Underground Storage Tank Removal

Underground Storage Tank Removal can be divided into two categories: Non-Leaking and Leaking. Prior to removing an underground storage tank, tests should be made, with the proper authorities present, to determine whether a tank has been leaking or the surrounding soil has been contaminated.

To safely remove Liquid Underground Storage Tanks:
1. Excavate to the top of the tank.
2. Disconnect all piping.
3. Open all tank vents and access ports.
4. Remove all liquids and/or sludge.
5. Purge the tank with an inert gas.
6. Provide access to the inside of the tank and clean out the interior using proper personal protective equipment (PPE).
7. Excavate soil surrounding the tank using proper PPE for on-site personnel.
8. Pull and properly dispose of the tank.
9. Clean up the site of all contaminated material.
10. Install new tanks or close the excavation.

R02240-900 Wellpoints

A single stage wellpoint system is usually limited to dewatering an average 15' depth below normal ground water level. Multi-stage systems are employed for greater depth with the pumping equipment installed only at the lowest header level. Ejectors, with unlimited lift capacity, can be economical when two or more stages of wellpoints can be replaced or when horizontal clearance is restricted, such as in deep trenches or tunneling projects, and where low water flows are expected. Wellpoints are usually spaced on 2-1/2' to 10' centers along a header pipe. Wellpoint spacing, header size, and pump size are all determined by the expected flow, as dictated by soil conditions.

In almost all soils encountered in wellpoint dewatering, the wellpoints may be jetted into place. Cemented soils and stiff clays may require sand wicks about 12" in diameter around each wellpoint to increase efficiency and eliminate weeping into the excavation. These sand wicks require 1/2 to 3 C.Y. of washed filter sand and are installed by using a 12" diameter steel casing and hole puncher jetted into the ground 2' deeper than the wellpoint. Rock may require predrilled holes.

Labor required for the complete installation and removal of a single stage wellpoint system is in the range of 3/4 to 2 labor-hours per linear foot of header, depending upon jetting conditions, wellpoint spacing, etc.

Continuous pumping is necessary except in some free draining soil where temporary flooding is permissible (as in trenches which are backfilled after each day's work). Good practice requires provision of a stand-by pump during the continuous pumping operation.

Systems for continuous trenching below the water table should be installed three to four times the length of expected daily progress to insure uninterrupted digging, and header pipe size should not be changed during the job.

For pervious free draining soils, deep wells in place of wellpoints may be economical because of lower installation and maintenance costs. Daily production ranges between two to three wells per day, for 25' to 40' depths, to one well per day for depths over 50'.

Detailed analysis and estimating for any dewatering problem is available at no cost from wellpoint manufacturers. Major firms will quote "sufficient equipment" quotes or their affiliates offer lump sum proposals to cover complete dewatering responsibility.

	Description for 200' System with 8" Header	Quantities	1st Month Unit	1st Month Total	Thereafter Unit	Thereafter Total
Equipment & Material	Wellpoints 25' long, 2" diameter @ 5' O.C.	40 Each	$ 29.50	$ 1,180.00	$ 22.13	$ 885.20
	Header pipe, 8" diameter	200 L.F.	5.65	1,130.00	2.83	566.00
	Discharge pipe, 8" diameter	100 L.F.	3.78	378.00	2.46	246.00
	8" valves	3 Each	145.00	435.00	94.25	282.75
	Combination Jetting & Wellpoint pump (standby)	1 Each	2,350.00	2,350.00	1,645.00	1,645.00
	Wellpoint pump, 8" diameter	1 Each	2,300.00	2,300.00	1,610.00	1,610.00
	Transportation to and from site	1 Day	654.80	654.80	—	—
	Fuel 30 days x 60 gal./day	1800 Gallons	1.05	1,890.00	1.05	1,890.00
	Lubricants for 30 days x 16 lbs./day	480 Lbs.	1.00	480.00	1.00	480.00
	Sand for points	40 C.Y.	19.50	780.00	—	—
	Equipment & Materials Subtotal			$11,577.80		$ 7,604.95
Labor	Technician to supervise installation	1 Week	$ 34.20	$ 1,368.00	—	—
	Labor for installation and removal of system	300 Labor-hours	24.65	7,395.00	—	—
	4 Operators straight time 40 hrs./wk. for 4.33 wks.	693 Hrs.	31.10	21,552.30	$ 31.10	$21,552.30
	4 Operators overtime 2 hrs./wk. for 4.33 wks.	35 Hrs.	62.20	2,177.00	62.20	2,177.00
	Bare Labor Subtotal			$32,492.30		$23,729.30
	Total Bare Cost			$44,070.10		$31,334.25
	Monthly Bare Cost/L.F. Header			$ 220.35		$ 156.67

R02250-400 Wood Sheet Piling

Wood sheet piling may be used for depths to 20′ where there is no ground water. If moderate ground water is encountered Tongue & Groove sheeting will help to keep it out. When considerable ground water is present, steel sheeting must be used.

For estimating purposes on trench excavation, sizes are as follows:

Depth	Sheeting	Wales	Braces	B.F. per S.F.
To 8′	3 x 12's	6 x 8's, 2 line	6 x 8's, @ 10′	4.0 @ 8′
8′ x 12′	3 x 12's	10 x 10's, 2 line	10 x 10's, @ 9′	5.0 average
12′ to 20′	3 x 12's	12 x 12's, 3 line	12 x 12's, @ 8′	7.0 average

Sheeting to be toed in at least 2′ depending upon soil conditions. A five person crew with an air compressor and sheeting driver can drive and brace 440 SF/day at 8′ deep, 360 SF/day at 12′ deep, and 320 SF/day at 16′ deep. For normal soils, piling can be pulled in 1/3 the time to install. Pulling difficulty increases with the time in the ground. Production can be increased by high pressure jetting. Figures below assume 50% of lumber is salvaged and includes pulling costs. Some jurisdictions require an equipment operator in addition to Crew B-31.

Sheeting Pulled

Daily Cost Crew B-31	L.H./ Day	Hourly Cost	Daily Cost	8′ Depth, 440 S.F./Day		16′ Depth, 320 S.F./Day	
				To Drive (1 Day)	To Pull (1/3 Day)	To Drive (1 Day)	To Pull (1/3 Day)
1 Foreman	8	$26.65	$ 213.20	$ 213.20	$ 71.00	$ 213.20	$ 71.00
3 Laborers	24	24.65	591.60	$ 591.60	197.00	$ 591.60	197.00
1 Carpenter	8	31.35	250.80	$ 250.80	83.52	$ 250.80	83.52
1 Air Compressor			128.60	$ 128.60	42.82	$ 128.60	42.82
1 Sheeting Driver			8.80	$ 8.80	2.93	$ 8.80	2.93
2 -50 Ft. Air Hoses, 1-1/2″ Diam.			20.00	$ 20.00	6.66	$ 20.00	6.66
Lumber (50% salvage)				1.76 MBF 580.80		2.24 MBF 739.20	
Bare Total			$1,213.00	$1,793.80	$403.93	$1,952.20	$403.93
Bare Total/S.F.				$ 4.08	$ .92	$ 4.44	$ 1.26
Bare Total (Drive and Pull)/S.F.				$ 5.00		$ 5.70	

Sheeting Left in Place

Daily Cost	8′ Depth 440 S.F./Day		10′ Depth 400 S.F./Day		12′ Depth 360 S.F./Day		16′ Depth 320 S.F./Day		18′ Depth 305 S.F./Day		20′ Depth 280 S.F./Day	
Crew B-31		$1,204.60		$1,204.60		$1,204.60		$1,204.60		$1,204.60		$1,204.60
Lumber	1.76 MBF	1,161.60	1.8 MBF	1,188.00	1.8 MBF	1,188.00	2.24MBF	1,478.40	2.1 MBF	1,386.00	1.9 MBF	1,254.00
Bare Total in Place		$2,366.20		$2,392.60		$2,392.60		$2,683.00		$2,590.60		$2,458.60
Bare Total/S.F.		$ 5.38		$ 5.98		$ 6.65		$ 8.38		$ 8.49		$ 8.78
Bare Total/M.B.F.		$1,344.43		$1,329.22		$1,329.22		$1,197.77		$1,233.62		$1,294.00

R02250-450 Steel Sheet Piling

Limiting weights are 22 to 38#/S.F. of wall surface with 27#/S.F. average for usual types and sizes. (Weights of piles themselves are from 30.7#/L.F. to 57#/L.F. but they are 15″ to 21″ wide.) Lightweight sections 12″ to 28″ wide from 3 ga. to 12 ga. thick are also available for shallow excavations. Piles may be driven two at a time with an impact or vibratory hammer (use vibratory to pull) hung from a crane without leads. A reasonable estimate of the life of steel sheet piling is 10 uses with up to 125 uses possible if a vibratory hammer is used. Used piling costs from 50% to 80% of new piling depending on location and market conditions. Sheet piling and H piles

can be rented for about 30% of the delivered mill price for the first month and 5% per month thereafter. Allow 1 labor-hour per pile for cleaning and trimming after driving. These costs increase with depth and hydrostatic head. Vibratory drivers are faster in wet granular soils and are excellent for pile extraction. Pulling difficulty increases with the time in the ground and may cost more than driving. It is often economical to abandon the sheet piling, especially if it can be used as the outer wall form. Allow about 1/3 additional length or more, for toeing into ground. Add bracing, waler and strut costs. Waler costs can equal the cost per ton of sheeting.

Cost of Sheet Piling & Production Rate by Ton & S.F.						
Depth of Excavation	**15′ Depth**		**20′ Depth**		**25′ Depth**	
Description of Pile	**22 psf = 90.9 S.F./Ton**		**27 psf = 74 S.F./Ton**		**38 psf = 52.6 S.F./Ton**	
Type of operation:	Drive &	Drive &	Drive &	Drive &	Drive &	Drive &
Left in Place or Removed	Left	Extract*	Left	Extract*	Left	Extract*
Labor & Equip. to Drive 1 ton	$ 402.52	$725.20	$ 336.00	$664.31	$ 229.01	$414.40
Piling (75% Salvage)	800.00	200.00	800.00	200.00	800.00	200.00
Bare Cost/Ton in Place	$1,202.52	$925.20	$1,136.00	$864.31	$1,029.01	$614.40
Production Rate, Tons/Day	10.81	6.00	12.95	6.55	19.00	10.50
Bare Cost, S.F. in Place (incl. 33% toe in)	$ 13.23	$ 10.18	$ 15.35	$ 11.68	$ 19.56	$ 11.68
Production Rate, S.F./Day (incl. 33% toe in)	983	545	959	485	1000	553

	Bare Costs		Inc. Subs O&P		
Crew B-40	**Hr.**	**Daily**	**Hr.**	**Daily**	**Installation Cost per Ton**
1 Pile Driver Foreman (Out)	$32.90	$ 263.20	$55.20	$ 441.60	for 15′ Deep Excavation plus 33% toe-in
4 Pile Drivers	30.90	988.80	51.85	1,659.20	$4,351 per day/10.81 tons = $402.50 per ton
2 Equip. Oper. (crane)	33.70	539.20	51.00	816.00	**Installation Cost per S.F.**
1 Equip. Oper. Oiler	28.30	226.40	42.85	342.80	$\frac{2000\#}{22\#/S.F.}$ = 90.9 S.F. x 10.81 tons = 983 S.F. per day
1 Crane, 40 Ton		935.60		1,029.15	
Vibratory Hammer & Gen.		1,398.00		1,537.80	$4,351 / 983 S.F. per day = $4.43 per S.F.
64 L.H., Daily Totals		$4,351.20		$5,826.55	

*For driving & extracting, two mobilizations & demobilizations will be necessary

R13281-120   Asbestos Removal Process

Asbestos removal is accomplished by a specialty contractor who understands the federal and state regulations regarding the handling and disposal of the material. The process of asbestos removal is divided into many individual steps. An accurate estimate can be calculated only after all the steps have been priced.

The steps are generally as follows:

1. Obtain an asbestos abatement plan from an industrial hygienist.
2. Monitor the air quality in and around the removal area and along the path of travel between the removal area and transport area. This establishes the background contamination.
3. Construct a two part decontamination chamber at entrance to removal area.
4. Install a HEPA filter to create a negative pressure in the removal area.
5. Install wall, floor and ceiling protection as required by the plan, usually 2 layers of fireproof 6 mil polyethylene.

6. Industrial hygienist visually inspects work area to verify compliance with plan.
7. Provide temporary supports for conduit and piping affected by the removal process.
8. Proceed with asbestos removal and bagging process. Monitor air quality as described in Step #2. Discontinue operations when contaminate levels exceed applicable standards.
9. Document the legal disposal of materials in accordance with EPA standards.
10. Thoroughly clean removal area including all ledges, crevices and surfaces.
11. Post abatement inspection by industrial hygienist to verify plan compliance.
12. Provide a certificate from a licensed industrial hygienist attesting that contaminate levels are within acceptable standards before returning area to regular use.

R13600-610   Solar Heating (Space and Hot Water)

Collectors should face as close to due South as possible, however, variations of up to 20 degrees on either side of true South are acceptable. Local climate and collector type may influence the choice between east or west deviations. Obviously they should be located so they are not shaded from the sun's rays. Incline collectors at a slope of latitude minus 5 degrees for domestic hot water and latitude plus 15 degrees for space heating.

Flat plate collectors consist of a number of components as follows: Insulation to reduce heat loss through the bottom and sides of the collector. The enclosure which contains all the components in this assembly is usually weatherproof and prevents dust, wind and water from coming in contact with the absorber plate. The cover plate usually consists of one or more layers of a variety of glass or plastic and reduces the reradiation by creating an air space which traps the heat between the cover and the absorber plates.

The absorber plate must have a good thermal bond with the fluid passages. The absorber plate is usually metallic and treated with a surface coating which improves absorptivity. Black or dark paints or selective coatings are used for this purpose, and the design of this passage and plate combination helps determine a solar system's effectiveness.

Heat transfer fluid passage tubes are attached above and below or integral with an absorber plate for the purpose of transferring thermal energy from the absorber plate to a heat transfer medium. The heat exchanger is a device for transferring thermal energy from one fluid to another.

Piping and storage tanks should be well insulated to minimize heat losses.

Size domestic water heating storage tanks to hold 20 gallons of water per user, minimum, plus 10 gallons per dishwasher or washing machine. For domestic water heating an optimum collector size is approximately 3/4 square foot of area per gallon of water storage. For space heating of residences and small commercial applications the collector is commonly sized between 30% and 50% of the internal floor area. For space heating of large commercial applications, collector areas less than 30% of the internal floor area can still provide significant heat reductions.

A supplementary heat source is recommended for Northern states for December through February.

The solar energy transmission per square foot of collector surface varies greatly with the material used. Initial cost, heat transmittance and useful life are obviously interrelated.

SPECIAL CONSTRUCTION 13

REFERENCE NOS.

R15050-710 Subcontractors

On the unit cost pages of the R.S. Means Cost Data books, the last column is entitled "Total Incl. O&P". This is normally the cost of the installing contractor. In Division 15, this is the cost of the mechanical contractor. If the particular work being estimated is to be performed by a sub to the mechanical contractor, the mechanical's profit and handling charge (usually 10%) is added to the total of the last column.

R15050-720 Demolition (Selective vs. Removal for Replacement)

Demolition can be divided into two basic categories.

One type of demolition involves the removal of material with no concern for its replacement. The labor-hours to estimate this work are found in Div. 15055 under "Selective Demolition". It is selective in that individual items or all the material installed as a system or trade grouping such as plumbing or heating systems are removed. This may be accomplished by the easiest way possible, such as sawing, torch cutting, or sledge hammer as well as simple unbolting.

The second type of demolition is the removal of some item for repair or replacement. This removal may involve careful draining, opening of unions, disconnecting and tagging of electrical connections, capping of pipes/ducts to prevent entry of debris or leakage of the material contained as well as transport of the item away from its in-place location to a truck/dumpster. An approximation of the time required to accomplish this type of demolition is to use half of the time indicated as necessary to install a new unit. For example; installation of a new pump might be listed as requiring 6 labor-hours so if we had to estimate the removal of the old pump we would allow an additional 3 hours for a total of 9 hours. That is, the complete replacement of a defective pump with a new pump would be estimated to take 9 labor-hours.

R15100-050 Pipe Material Considerations

1. Malleable fittings should be used for gas service.
2. Malleable fittings are used where there are stresses/strains due to expansion and vibration.
3. Cast fittings may be broken as an aid to disassembling of heating lines frozen by long use, temperature and minerals.
4. Cast iron pipe is extensively used for underground and submerged service.
5. Type M (light wall) copper tubing is available in hard temper only and is used for nonpressure and less severe applications than K and L.
6. Type L (medium wall) copper tubing, available hard or soft for interior service.
7. Type K (heavy wall) copper tubing, available in hard or soft temper for use where conditions are severe. For underground and interior service.
8. Hard drawn tubing requires fewer hangers or supports but should not be bent. Silver brazed fittings are recommended, however soft solder is normally used.
9. Type DMV (very light wall) copper tubing designed for drainage, waste and vent plus other non-critical pressure services.

Domestic/Imported Pipe and Fittings Cost

The prices shown in this publication for steel/cast iron pipe and steel, cast iron, malleable iron fittings are based on domestic production sold at the normal trade discounts. The above listed items of foreign manufacture may be available at prices of 1/3 to 1/2 those shown. Some imported items after minor machining or finishing operations are being sold as domestic to further complicate the system.

Caution: Most pipe prices in this book also include a coupling and pipe hangers which for the larger sizes can add significantly to the per foot cost and should be taken into account when comparing "book cost" with quoted supplier's cost.

R15100-070 Piping to 10' High

When taking off pipe, it is important to identify the different material types and joining procedures, as well as distances between supports and components required for proper support.

During the takeoff, measure through all fittings. Do not subtract the lengths of the fittings, valves, or strainers, etc. This added length plus the final rounding of the totals will compensate for nipples and waste.

When rounding off totals always increase the actual amount to correspond with manufacturer's shipping lengths.

A. Both red brass and yellow brass pipe are normally furnished in 12' lengths, plain end. Division 15107-220 includes in the linear foot costs two field threads and one coupling per 10' length. A carbon steel clevis type hanger assembly every 10' is also prorated into the linear foot costs, including both material and labor.

B. Cast iron soil pipe is furnished in either 5' or 10' lengths. For pricing purposes, Division 15107-320 in *Means Plumbing Cost Data* features 10' lengths with a joint and a carbon steel clevis hanger assembly every 5' prorated into the per foot costs of both material and labor.

  Three methods of joining are considered in Division 15107-320, lead and oakum poured joints, or push-on gasket type joints for the bell and spigot pipe, and a joint clamp for the no-hub soil pipe. The labor and material costs for each of these individual joining procedures are also prorated into the linear costs per foot.

C. Copper tubing in Division 15107-420 covers types K, L, M, and DWV which are furnished in 20' lengths. Means pricing data is based on a tubing cut each length and a coupling and two soft soldered joints every 10'. A carbon steel, clevis type hanger assembly every 10' is also prorated into the per foot costs. The prices for refrigeration tubing are for materials only. Labor for full lengths may be based on the type L labor but short cut measures in tight areas can increase the installation labor-hours from 20 to 40%.

D. Corrosion-resistant piping does not lend itself to one particular standard of hanging or support assembly due to its diversity of application and placement. The several varieties of corrosion-resistant piping do not include any material or labor costs for hanger assemblies (See Division 15060-300 for appropriate selection). The estimator should refer to Division 15107-560 in *Means Plumbing Cost Data* for the compatible corrosion-resistant fittings and piping.

E. Glass pipe, as shown in *Means Plumbing Cost Data* Division 15106-120, is furnished in standard lengths either 5' or 10' long, beaded on one end. Special orders for diverse lengths beaded on both ends are also available. For pricing purposes, R.S. Means features 10' lengths with a coupling and a carbon steel band hanger assembly every 10' prorated into the per foot linear costs.

  Glass pipe is also available with conical ends and standard lengths ranging from 6" through 3' in 6" increments, then up to 10' in 12" increments. Special lengths can be customized for particular installation requirements.

  For pricing purposes, Means has based the labor and material pricing on 10' lengths. Included in these costs per linear foot are the prorated costs for a flanged assembly every 10' consisting of two flanges, a gasket, two insertable seals, and the required number of bolts and nuts. A carbon steel band hanger assembly based on 10' center lines has also been prorated into the costs per foot for labor and materials.

F. Plastic pipe of several compositions and joining methods are considered in Division 15108-520. Fiberglass reinforced pipe (FRP) is priced, based on 10' lengths (20' lengths are also available), with coupling and epoxy joints every 10'. FRP is furnished in both "General Service" and "High Strength." A carbon steel clevis hanger assembly, 3 for every 10', is built into the prorated labor and material costs on a per foot basis.

  The PVC and CPVC pipe schedules 40, 80 and 120, plus SDR ratings are all based on 20' lengths with a coupling installed every 10', as well as a carbon steel clevis hanger assembly every 3'. The PVC and ABS type DWV piping is based on 10' lengths with solvent weld couplings every 10', and with carbon steel clevis hanger assemblies, 3 for every 10'. The rest of the plastic piping in this section is based on flexible 100' coils and does not include any coupling or supports.

  This section ends with PVC drain and sewer piping based on 10' lengths with bell and spigot ends and 0-ring type, push-on joints. Material costs only are given, but the complete installation costs should be similar to those shown in line number 02620-270.

G. Stainless steel piping, Division 15107-900, includes both weld end and threaded piping, both in the type 304 and 316 specification and in the following schedules, 5, 10, 40, 80, and 160. Although this piping is usually furnished in 20' lengths, this cost grouping has a joint (either heli-arc butt-welded or threads and coupling) every 10'. A carbon steel clevis type hanger assembly is also included at 10' intervals and prorated into the linear foot costs.

H. Division 15107-620 covers carbon steel pipe, both black and galvanized. This section encompasses schedules 40 (standard) and 80 (extra heavy).

  Several common methods of joining steel pipe — such as thread and coupled, butt welded, and flanged (150 lb. weld neck flanges) are also included.

  For estimating purposes, it is assumed that the piping is purchased in 20' lengths and that a compatible joint is made up every 10'. These joints are prorated into the labor and material costs per linear foot. The following hanger and support assemblies every 10' are also included: carbon steel clevis for the T & C pipe, and single rod roll type for both the welded and flanged piping. All of these hangers are oversized to accommodate pipe insulation 3/4" thick through 5" pipe size and 1-1/2" thick from 6" through 12" pipe size.

I. Division 15107-690 prices grooved joint steel pipe, both black and galvanized, in schedules 10, 40, and 80, furnished in 20' lengths. This section describes two joining methods: cut groove and roll groove. The schedule 10 piping is roll-grooved, while the heavier schedules are cut-grooved. The labor and material costs are prorated into per linear foot prices, including a coupled joint every 10', as well as a carbon steel clevis hanger assembly.

Notes:

The pipe hanger assemblies mentioned in the preceding paragraphs include the described hanger; appropriately sized steel, box-type insert and nut; plus a threaded hanger rod. On average, the distance from the pipe center line to the insert face is 2'.

C clamps are used when the pipe is to be supported from steel shapes rather than anchored in the slab. C clamps are slightly less costly than inserts. However, to save time in estimating, it is advisable to use the given line number cost, rather than substituting a C clamp for the insert.

Add to piping labor for elevated installation:

10' to 15' high	10%	30' to 35' high	40%
15' to 20' high	20%	35' to 40' high	50%
20' to 25' high	25%	Over 40' high	55%
25' to 30' high	35%		

When using the percentage adds for elevated piping installations as shown above, bear in mind that the given heights are for the pipe supports, even though the insert, anchor, or clamp may be several feet higher than the pipe itself.

An allowance has been included in the piping installation time for testing and minor tightening of leaking joints, fittings, stuffing boxes, packing glands, etc. For extraordinary test requirements such as x-rays, prolonged pressure or demonstration tests, a percentage of the piping labor, based on the estimator's experience, must be added to the labor total. A testing service specializing in weld x-rays should be consulted for pricing if it is an estimate requirement. Equipment installation time includes start-up with associated adjustments.

529

R15100-080 Valve Materials

VALVE MATERIALS

Bronze:
Bronze is one of the oldest materials used to make valves. It is most commonly used in hot and cold water systems and other non-corrosive services. It is often used as a seating surface in larger iron body valves to ensure tight closure.

Carbon Steel:
Carbon steel is a high strength material. Therefore, valves made from this metal are used in higher pressure services, such as steam lines up to 600 psi at 850°F. Many steel valves are available with butt-weld ends for economy and are generally used in high pressure steam service as well as other higher pressure non-corrosive services.

Forged Steel:
Valves from tough carbon steel are used in service up to 2000 psi and temperatures up to 1000°F in Gate, Globe and Check valves.

Iron:
Valves are normally used in medium to large pipe lines to control non-corrosive fluid and gases, where pressures do not exceed 250 psi at 450° or 500 psi cold water, oil or gas.

Stainless Steel:
Developed steel alloys can be used in over 90% corrosive services.

Plastic PVC:
This is used in a great variety of valves generally in high corrosive service with lower temperatures and pressures.

VALVE SERVICE PRESSURES

Pressure ratings on valves provide an indication of the safe operating pressure for a valve at some elevated temperature. This temperature is dependent upon the materials used and the fabrication of the valve. When specific data is not available, a good "rule-of-thumb" to follow is the temperature of saturated steam on the primary rating indicated on the valve body. Example: The valve has the number 150S printed on the side indicating 150 psi and hence, a maximum operating temperature of 367°F (temperature of saturated steam and 150 psi).

DEFINITIONS

1. "WOG" – Water, oil, gas (cold working pressures).
2. "SWP" – Steam working pressure.
3. "100% area (full port) – means the area through the valve is equal to or greater than the area of standard pipe.
4. "Standard Opening" – means that the area through the valve is less than the area of standard pipe and therefore these valves should be used only where restriction of flow is unimportant.
5. "Round Port" – means the valve has a full round opening through the plug and body, of the same size and area as standard pipe.
6. "Rectangular Port" – valves have rectangular shaped ports through the plug body. The area of the port is either equal to 100% of the area of standard pipe, or restricted (standard opening). In either case it is clearly marked.
7. "ANSI" – American National Standards Institute.

R15100-090 Valve Selection Considerations

INTRODUCTION: In any piping application, valve performance is critical. Valves should be selected to give the best performance at the lowest cost.

The following is a list of performance characteristics generally expected of valves.
1. Stopping flow or starting it.
2. Throttling flow (Modulation).
3. Flow direction changing.
4. Checking backflow (Permitting flow in only one direction).
5. Relieving or regulating pressure.

In order to properly select the right valve, some facts must be determined.

A. What liquid or gas will flow through the valve?
B. Does the fluid contain suspended particles?
C. Does the fluid remain in liquid form at all times?
D. Which metals does fluid corrode?
E. What are the pressure and temperature limits? (As temperature and pressure rise, so will the price of the valve.)
F. Is there constant line pressure?
G. Is the valve merely an on-off valve?
H. Will checking of backflow be required?
I. Will the valve operate frequently or infrequently?

Valves are classified by design type into such classifications as Gate, Globe, Angle, Check, Ball, Butterfly and Plug. They are also classified by end connection, stem, pressure restrictions and material such as bronze, cast iron, etc. Each valve has a specific use. A quality valve used correctly will provide a lifetime of trouble- free service, but a high quality valve installed in the wrong service may require frequent attention.

STEM TYPES
(OS & Y)—Rising Stem-Outside Screw and Yoke

Offers a visual indication of whether the valve is open or closed. Recommended where high temperatures, corrosives, and solids in the line might cause damage to inside-valve stem threads. The stem threads are engaged by the yoke bushing so the stem rises through the hand wheel as it is turned.

(R.S.)—Rising Stem-Inside Screw

Adequate clearance for operation must be provided because both the hand wheel and the stem rise.
The valve wedge position is indicated by the position of the stem and hand wheel.

(N.R.S.)—Non-Rising Stem-Inside Screw

A minimum clearance is required for operating this type of valve. Excessive wear or damage to stem threads inside the valve may be caused by heat, corrosion, and solids. Because the hand wheel and stem do not rise, wedge position cannot be visually determined.

VALVE TYPES
Gate Valves

Provide full flow, minute pressure drop, minimum turbulence and minimum fluid trapped in the line.
They are normally used where operation is infrequent.

Globe Valves

Globe valves are designed for throttling and/or frequent operation with positive shut-off. Particular attention must be paid to the several types of seating materials available to avoid unnecessary wear. The seats must be compatible with the fluid in service and may be composition or metal. The configuration of the Globe valve opening causes turbulence which results in increased resistance. Most bronze Globe valves are rising stem-inside screw, but they are also available on O.S. & Y.

Angle Valves

The fundamental difference between the Angle valve and the Globe valve is the fluid flow through the Angle valve. It makes a 90° turn and offers less resistance to flow than the Globe valve while replacing an elbow. An Angle valve thus reduces the number of joints and installation time.

MECHANICAL 15

REFERENCE NOS.

Check Valves

Check valves are designed to prevent backflow by automatically seating when the direction of fluid is reversed.
Swing Check valves are generally installed with Gate-valves, as they provide comparable full flow. Usually recommended for lines where flow velocities are low and should not be used on lines with pulsating flow. Recommended for horizontal installation, or in vertical lines only where flow is upward.

Lift Check Valves

These are commonly used with Globe and Angle valves since they have similar diaphragm seating arrangements and are recommended for preventing backflow of steam, air, gas and water, and on vapor lines with high flow velocities. For horizontal lines, horizontal lift checks should be used and vertical lift checks for vertical lines.

Ball Valves

Ball valves are light and easily installed, yet because of modern elastomeric seats, provide tight closure. Flow is controlled by rotating up to 90° a drilled ball which fits tightly against resilient seals. This ball seats with flow in either direction, and valve handle indicates the degree of opening. Recommended for frequent operation readily adaptable to automation, ideal for installation where space is limited.

Butterfly Valves

Butterfly valves provide bubble-tight closure with excellent throttling characteristics. They can be used for full-open, closed and for throttling applications.

The Butterfly valve consists of a disc within the valve body which is controlled by a shaft. In its closed position, the valve disc seals against a resilient seat. The disc position throughout the full 90° rotation is visually indicated by the position of the operator.

A Butterfly valve is only a fraction of the weight of a Gate valve and requires no gaskets between flanges in most cases. Recommended for frequent operation and adaptable to automation where space is limited.

Wafer and Lug type bodies when installed between two pipe flanges, can be easily removed from the line. The pressure of the bolted flanges holds the valve in place.
Locating lugs makes installation easier.

Plug Valves

Lubricated plug valves, because of the wide range of service to which they are adapted, may be classified as all purpose valves. They can be safely used at all pressure and vacuums, and at all temperatures up to the limits of available lubricants. They are the most satisfactory valves for the handling of gritty suspensions and many other destructive, erosive, corrosive and chemical solutions.

15

MECHANICAL

REFERENCE NOS.

R15500-010 Heating Systems

Heating Systems

The basic function of a heating system is to bring an enclosed volume up to a desired temperature and then maintain that temperature within a reasonable range. To accomplish this, the selected system must have sufficient capacity to offset transmission losses resulting from the temperature difference on the interior and exterior of the enclosing walls in addition to losses due to cold air infiltration through cracks, crevices and around doors and windows. The amount of heat to be furnished is dependent upon the building size, construction, temperature difference, air leakage, use, shape, orientation and exposure. Air circulation is also an important consideration. Circulation will prevent stratification which could result in heat losses through uneven temperatures at various levels. For example, the most

efficient use of unit heaters can usually be achieved by circulating the space volume through the total number of units once every 20 minutes or 3 times an hour. This general rule must, of course, be adapted for special cases such as large buildings with low ratios of heat transmitting surface to cubical volume. The type of occupancy of a building will have considerable bearing on the number of heat transmitting units and the location selected. It is axiomatic, however, that the basis of any successful heating system is to provide the maximum amount of heat at the points of maximum heat loss such as exposed walls, windows, and doors. Large roof areas, wind direction, and wide doorways create problems of excessive heat loss and require special consideration and treatment.

Heat Transmission

Heat transfer is an important parameter to be considered during selection of the exterior wall style, material and window area. A high rate of transfer will permit greater heat loss during the wintertime with the resultant increase in heating energy costs and a greater rate of heat gain in the summer with proportionally greater cooling cost. Several terms are used to describe various aspects of heat transfer. However, for general estimating purposes this book lists U values for systems of construction materials. U is the "overall heat transfer coefficient." It is defined as the heat flow per hour through one square foot when the temperature difference in the air on either side of the structure wall, roof, ceiling or floor is one degree Fahrenheit. The structural segment may be a single homogeneous material or a composite.

Total heat transfer is found using the following equation:

$Q = AU(T_2 - T_1)$ where

Q = Heat flow, BTU per hour

A = Area, square feet

U = Overall heat transfer coefficient

$(T_2 - T_1)$ = Difference in temperature of air on each side of the construction component. (Also abbreviated TD)

Note that heat can flow through all surfaces of any building and this flow is in addition to heat gain or loss due to ventilation, infiltration and generation (appliances, machinery, people).

R15500-020 Heating Approximations for Quick Estimating

Oil Piping & Boiler Room Piping

Small System . . . . 20 to 30% of Boiler

Complex System

with Pumps, Headers, Etc. . . . . 80 to 110% of Boiler

Breeching With Insulation:

Small . . . . 10 to 15% of Boiler

Large . . . . 15 to 25% of Boiler

Coils: . . . . 15 to 30% of Containing Unit

Balancing (Independent) . . . . 1/2% of H.V.A.C. Estimate

Quality/Complexity Adjustment: For all heating installations add these adjustments to the estimate to more closely allow for the equipment and conditions of the particular job under consideration.

Economy installation, add . . . . 0 to 5% of System

Good quality, medium complexity, add . . . . 5 to 15% of System

Above average quality and complexity, add . . . . 15 to 25% of System

R15500-030 The Basics of a Heating System

The function of a heating system is to achieve and maintain a desired temperature in a room or building by replacing the amount of heat being dissipated. There are four kinds of heating systems: hot-water, steam, warm-air and electric resistance. Each has certain essential and similar elements with the exception of electric resistance heating which is covered in Section 16.

The basic elements of a heating system are:

A. A **combustion chamber** in which fuel is burned and heat transferred to a conveying medium.

B. The **"fluid"** used for conveying the heat (water, steam or air).

C. **Conductors** or pipes for transporting the fluid to specific desired locations.

D. A means of disseminating the heat, sometimes called **terminal units.**

A. The **combustion chamber** in a furnace heats air which is then distributed. This is called a warm-air system.

The combustion chamber in a boiler heats water which is either distributed as hot water or steam and this is termed a hydronic system.

The maximum allowable working pressures are limited by ASME "Code for Heating Boilers" to 15 PSI for steam and 160 PSI for hot water heating boilers, with a maximum temperature limitation of 250°F. Hot water boilers are generally rated for a working pressure of 30 PSI. High pressure boilers are governed by the ASME "Code for Power Boilers" which is used almost universally for boilers operating over 15 PSIG. High pressure boilers used for a combination of heating/process loads are usually designed for 150 PSIG.

Boiler ratings are usually indicated as either Gross or Net Output. The Gross Load is equal to the Net Load plus a piping and pickup allowance. When this allowance cannot be determined, divide the gross output rating by 1.25 for a value equal to or greater than the net heat loss requirement of the building.

B. Of the three **fluids** used, steam carries the greatest amount of heat per unit volume. This is due to the fact that it gives up its latent heat of vaporization at a temperature considerably above room temperature. Another advantage is that the pressure to produce a positive circulation is readily available. Piping conducts the steam to terminal units and returns condensate to the boiler.

The **steam system** is well adapted to large buildings because of its positive circulation, its comparatively economical installation and its ability to deliver large quantities of heat. Nearly all large office buildings, stores, hotels, and industrial buildings are so heated, in addition to many residences.

Hot water, when used as the heat carrying fluid, gives up a portion of its sensible heat and then returns to the boiler or heating apparatus for reheating. As the heat conveyed by each pound of water is about one-fiftieth of the heat conveyed by a pound of steam, it is necessary to circulate about fifty times as much water as steam by weight (although only one-thirtieth as much by volume). The hot water system is usually, although not necessarily, designed to operate at temperatures below that of the ordinary steam system and so the amount of heat transfer surface must be correspondingly greater. A temperature of 190°F to 200°F is normally the maximum. Circulation in small buildings may depend on the difference in density between hot water and the cool water returning to the boiler; circulating pumps are normally used to maintain a desired rate of flow. Pumps permit a greater degree of flexibility and better control.

In **warm-air** furnace systems, cool air is taken from one or more points in the building, passed over the combustion chamber and flue gas passages and then distributed through a duct system. A disadvantage of this system is that the ducts take up much more building volume than steam or hot water pipes. Advantages of this system are the relative ease with which humidification can be accomplished by the evaporation of water as the air circulates through the heater, and the lack of need for expensive disseminating units as the warm air simply becomes part of the interior atmosphere of the building.

C. Conductors (pipes and ducts) have been lightly treated in the discussion of conveying fluids. For more detailed information such as sizing and distribution methods, the reader is referred to technical publications such as the American Society of Heating, Refrigerating and Air-Conditioning Engineers "Handbook of Fundamentals."

D. Terminal units come in an almost infinite variety of sizes and styles, but the basic principles of operation are very limited. As previously mentioned, warm-air systems require only a simple register or diffuser to mix heated air with that present in the room. Special application items such as radiant coils and infrared heaters are available to meet particular conditions but are not usually considered for general heating needs. Most heating is accomplished by having air flow over coils or pipes containing the heat transporting medium (steam, hot-water, electricity). These units, while varied, may be separated into two general types, (1) radiator/convectors and (2) unit heaters.

Radiator/convectors may be cast, fin-tube or pipe assemblies. They may be direct, indirect, exposed, concealed or mounted within a cabinet enclosure, upright or baseboard style. These units are often collectively referred to as "radiators" or "radiation" although none gives off heat either entirely by radiation or by convection but rather a combination of both. The air flows over the units as a gravity "current." It is necessary to have one or more heat-emitting units in each room. The most efficient placement is low along an outside wall or under a window to counteract the cold coming into the room and achieve an even distribution.

In contrast to radiator/convectors which operate most effectively against the walls of smaller rooms, **unit heaters** utilize a fan to move air over heating coils and are very effective in locations of relatively large volume. Unit heaters, while usually suspended overhead, may be floor mounted. They also may take in fresh outside air for ventilation. The heat distributed by unit heaters may be from a remote source and conveyed by a fluid or it may be from the combustion of fuel in each individual heater. In the latter case the only piping required would be for fuel, however, a vent for the products of combustion would be necessary.

The following list gives may of the advantages of unit heaters for applications other than office or residential:

a. Large capacity so smaller number of units are required, **b.** Piping system simplified, **c.** Space saved where they are located overhead out of the way, **d.** Rapid heating directed where needed with effective wide distribution, **e.** Difference between floor and ceiling temperature reduced, **f.** Circulation of air obtained, and ventilation with introduction of fresh air possible, **g.** Heat output flexible and easily controlled.

R15500-050   Factor for Determining Heat Loss for Various Types of Buildings

General: While the most accurate estimates of heating requirements would naturally be based on detailed information about the building being considered, it is possible to arrive at a reasonable approximation using the following procedure:

1. Calculate the cubic volume of the room or building.
2. Select the appropriate factor from Table 1 below. Note that the factors apply only to inside temperatures listed in the first column and to 0°F outside temperature.

3. If the building has bad north and west exposures, multiply the heat loss factor by 1.1.
4. If the outside design temperature is other than 0°F, multiply the factor from Table 1 by the factor from Table 2.
5. Multiply the cubic volume by the factor selected from Table 1. This will give the estimated BTUH heat loss which must be made up to maintain inside temperature.

Table 1 — Building Type	Conditions	Qualifications	Loss Factor*
Factories & Industrial Plants General Office Areas at 70°F	One Story	Skylight in Roof	6.2
		No Skylight in Roof	5.7
	Multiple Story	Two Story	4.6
		Three Story	4.3
		Four Story	4.1
		Five Story	3.9
		Six Story	3.6
	All Walls Exposed	Flat Roof	6.9
		Heated Space Above	5.2
	One Long Warm Common Wall	Flat Roof	6.3
		Heated Space Above	4.7
	Warm Common Walls on Both Long Sides	Flat Roof	5.8
		Heated Space Above	4.1
Warehouses at 60°F	All Walls Exposed	Skylights in Roof	5.5
		No Skylight in Roof	5.1
		Heated Space Above	4.0
	One Long Warm Common Wall	Skylight in Roof	5.0
		No Skylight in Roof	4.9
		Heated Space Above	3.4
	Warm Common Walls on Both Long Sides	Skylight in Roof	4.7
		No Skylight in Roof	4.4
		Heated Space Above	3.0

*Note: This table tends to be conservative particularly for new buildings designed for minimum energy consumption.

Table 2 — Outside Design Temperature Correction Factor (for Degrees Fahrenheit)									
Outside Design Temperature	50	40	30	20	10	0	-10	-20	-30
Correction Factor	.29	.43	.57	.72	.86	1.00	1.14	1.28	1.43

MECHANICAL 15

REFERENCE NOS.

R15500-070 Transmission of Heat

R15500-070 and R15500-080 provide a way to calculate heat transmission of various construction materials from their U values and the TD (Temperature Difference).

1. From the Exterior Enclosure Division or elsewhere, determine U values for the construction desired.
2. Determine the coldest design temperature. The difference between this temperature and the desired interior temperature is the TD (temperature difference).

3. Enter R15500-070 or R15500-080 at correct U Value. Cross horizontally to the intersection with appropriate TD. Read transmission per square foot from bottom of figure.
4. Multiply this value of BTU per hour transmission per square foot of area by the total surface area of that type of construction.

R15500-080 Transmission of Heat (Low Rate)

R15700-010 Air Conditioning

General: The purpose of air conditioning is to control the environment of a space so that comfort is provided for the occupants and/or conditions are suitable for the processes or equipment contained therein. The several items which should be evaluated to define system objectives are:

Temperature Control
Humidity Control
Cleanliness
Odor, smoke and fumes
Ventilation

Efforts to control the above parameters must also include consideration of the degree or tolerance of variation, the noise level introduced, the velocity of air motion and the energy requirements to accomplish the desired results.

The variation in **temperature** and **humidity** is a function of the sensor and the controller. The controller reacts to a signal from the sensor and produces the appropriate suitable response in either the terminal unit, the conductor of the transporting medium (air, steam, chilled water, etc.), or the source (boiler, evaporating coils, etc.).

The **noise level** is a by-product of the energy supplied to moving components of the system. Those items which usually contribute the most noise are pumps, blowers, fans, compressors and diffusers. The level of noise can be partially controlled through use of vibration pads, isolators, proper sizing, shields, baffles and sound absorbing liners.

Some **air motion** is necessary to prevent stagnation and stratification. The maximum acceptable velocity varies with the degree of heating or cooling which is taking place. Most people feel air moving past them at velocities in excess of 25 FPM as an annoying draft. However, velocities up to 45 FPM may be acceptable in certain cases. Ventilation, expressed as air changes per hour and percentage of fresh air, is usually an item regulated by local codes.

Selection of the system to be used for a particular application is usually a trade-off. In some cases the building size, style, or room available for mechanical use limits the range of possibilities. Prime factors influencing the decision are first cost and total life (operating, maintenance and replacement costs). The accuracy with which each parameter is determined will be an important measure of the reliability of the decision and subsequent satisfactory operation of the installed system.

Heat delivery may be desired from an air conditioning system. Heating capability usually is added as follows: A gas fired burner or hot water/steam/electric coils may be added to the air handling unit directly and heat all air equally. For limited or localized heat requirements the water/steam/electric coils may be inserted into the duct branch supplying the cold areas. Gas fired duct furnaces are also available.

Note: When water or steam coils are used the cost of the piping and boiler must also be added. For a rough estimate use the cost per square foot of the appropriate sized hydronic system with unit heaters. This will provide a cost for the boiler and piping, and the unit heaters of the system would equate to the approximate cost of the heating coils.

R15700-020 Air Conditioning Requirements

BTU's per hour per S.F. of floor area and S.F. per ton of air conditioning.

Type of Building	BTU per S.F.	S.F. per Ton	Type of Building	BTU per S.F.	S.F. per Ton	Type of Building	BTU per S.F.	S.F. per Ton
Apartments, Individual	26	450	Dormitory, Rooms	40	300	Libraries	50	240
Corridors	22	550	Corridors	30	400	Low Rise Office, Exterior	38	320
Auditoriums & Theaters	40	300/18*	Dress Shops	43	280	Interior	33	360
Banks	50	240	Drug Stores	80	150	Medical Centers	28	425
Barber Shops	48	250	Factories	40	300	Motels	28	425
Bars & Taverns	133	90	High Rise Office—Ext. Rms.	46	263	Office (small suite)	43	280
Beauty Parlors	66	180	Interior Rooms	37	325	Post Office, Individual Office	42	285
Bowling Alleys	68	175	Hospitals, Core	43	280	Central Area	46	260
Churches	36	330/20*	Perimeter	46	260	Residences	20	600
Cocktail Lounges	68	175	Hotel, Guest Rooms	44	275	Restaurants	60	200
Computer Rooms	141	85	Corridors	30	400	Schools & Colleges	46	260
Dental Offices	52	230	Public Spaces	55	220	Shoe Stores	55	220
Dept. Stores, Basement	34	350	Industrial Plants, Offices	38	320	Shop'g. Ctrs., Supermarkets	34	350
Main Floor	40	300	General Offices	34	350	Retail Stores	48	250
Upper Floor	30	400	Plant Areas	40	300	Specialty	60	200

*Persons per ton
12,000 BTU = 1 ton of air conditioning

R15700-030 Psychrometric Table

Dewpoint or Saturation Temperature (F)

Relative humidity (%)															
100	32	35	40	45	50	55	60	65	70	75	80	85	90	95	100
90	30	33	37	42	47	52	57	62	67	72	77	82	87	92	97
80	27	30	34	39	44	49	54	58	64	68	73	78	83	88	93
70	24	27	31	36	40	45	50	55	60	64	69	74	79	84	88
60	20	24	28	32	36	41	46	51	55	60	65	69	74	79	83
50	16	20	24	28	33	36	41	46	50	55	60	64	69	73	78
40	12	15	18	23	27	31	35	40	45	49	53	58	62	67	71
30	8	10	14	18	21	25	29	33	37	42	46	50	54	59	62
20	6	7	8	9	13	16	20	24	28	31	35	40	43	48	52
10	4	4	5	5	6	8	9	10	13	17	20	24	27	30	34
	32	35	40	45	50	55	60	65	70	75	80	85	90	95	100

Dry bulb temperature (F)

This table shows the relationship between RELATIVE HUMIDITY, DRY BULB TEMPERATURE AND DEWPOINT.

As an example, assume that the thermometer in a room reads 75°F, and we know that the relative humidity is 50%. The chart shows the dewpoint temperature to be 55°. That is, any surface colder than 55°F will "sweat" or collect condensing moisture. This surface could be the outside of an uninsulated chilled water pipe in the summertime, or the inside surface of a wall or deck in the wintertime. After determining the extreme ambient parameters, the table at the left is useful in determining which surfaces need insulation or vapor barrier protection.

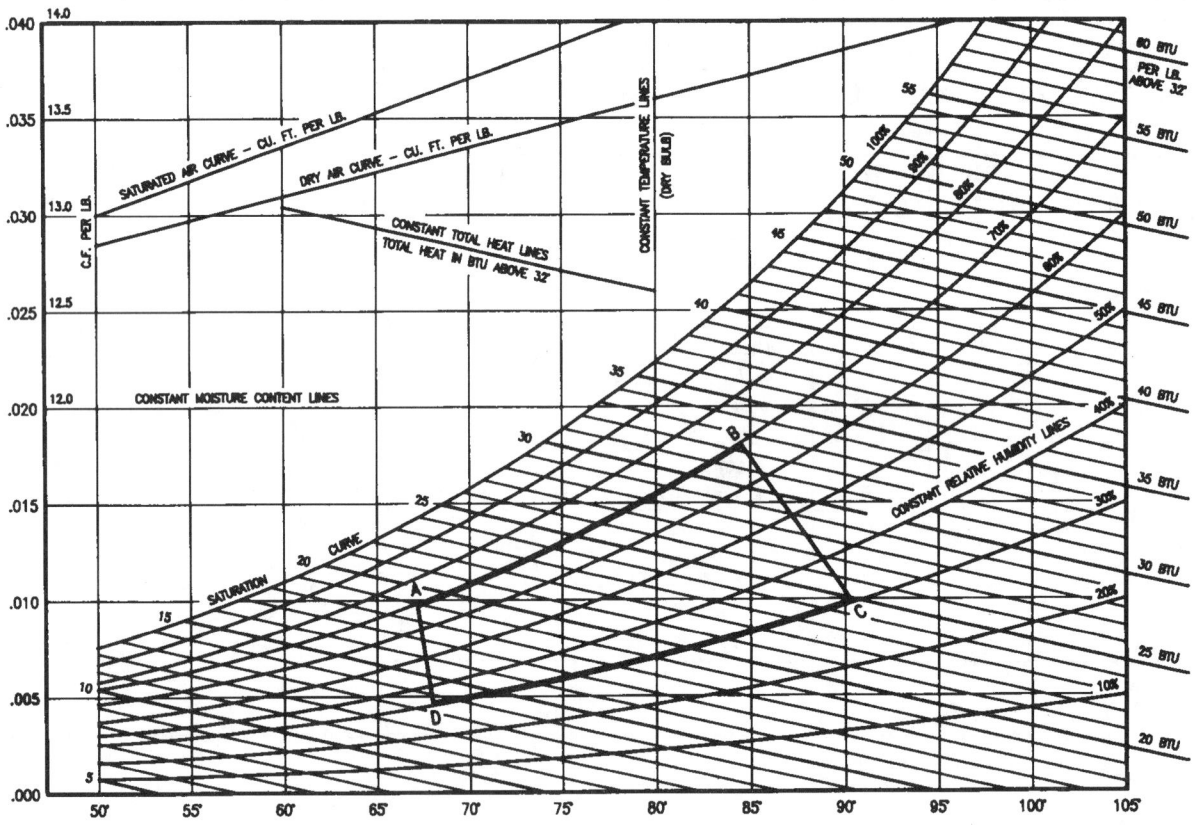

TEMPERATURE DEGREES FAHRENHEIT
TOTAL PRESSURE = 14.696 LB. PER SQ. IN. ABS.

Psychrometric chart showing different variables based on one pound of dry air. Space marked A B C D is temperature-humidity range which is most comfortable for majority of people.

R15700-040 Recommended Ventilation Air Changes

Table below lists range of time in minutes per change for various types of facilities.

Assembly Halls	2-10	Dance Halls	2-10	Laundries	1-3
Auditoriums	2-10	Dining Rooms	3-10	Markets	2-10
Bakeries	2-3	Dry Cleaners	1-5	Offices	2-10
Banks	3-10	Factories	2-5	Pool Rooms	2-5
Bars	2-5	Garages	2-10	Recreation Rooms	2-10
Beauty Parlors	2-5	Generator Rooms	2-5	Sales Rooms	2-10
Boiler Rooms	1-5	Gymnasiums	2-10	Theaters	2-8
Bowling Alleys	2-10	Kitchens-Hospitals	2-5	Toilets	2-5
Churches	5-10	Kitchens-Restaurant	1-3	Transformer Rooms	1-5

CFM air required for changes = Volume of room in cubic feet ÷ Minutes per change.

R15700-090 Quality/Complexity Adjustment for Air Conditioning Systems

Economy installation, add	0 to 5%
Good quality, medium complexity, add	5 to 15%
Above average quality and complexity, add	15 to 25%

Add the above adjustments to the estimate to more closely allow for the equipment and conditions of the particular job under consideration.

R15800-080 Loudness Levels for Moving Air thru Fans, Diffusers, Register, Etc. (Measured in Sones)

Recommended Loudness Levels (Sones)											
Area	Very Quiet	Quiet	Noisy	Area	Very Quiet	Quiet	Noisy	Area	Very Quiet	Quiet	Noisy
Auditoriums				**Hotels**				**Offices**			
Auditorium lobbies	3	4	6	Banquet Rooms	1.5	3	6	Conference rooms	1	1.7	3
Concert and opera halls	0.8	1	1.5	Individual rooms, suites	1	2	4	Drafting	2	4	8
Courtrooms	2	3	4	Kitchens and laundries	4	7	10	General open offices	2	4	8
Lecture halls	1.5	2	3	Lobbies	2	4	8	Halls and corridors	2.5	5	10
Movie theaters	1.5	2	3	**Indoor Sports**				Professional offices	1.5	3	6
Churches and Schools				Bowling alleys	3	4	6	Reception room	1.5	3	6
Kitchens	4	6	8	Gymnasiums	2	4	6	Tabulation & computation	3	6	12
Laboratories	2	4	6	Swimming pools	4	7	10	**Public Buildings**			
Libraries	1.5	2	3	**Manufacturing Areas**				Banks	2	4	6
Recreation halls	2	4	8	Assembly lines	5	12	30	Court houses	2.5	4	6
Sanctuaries	1	1.7	3	Foreman's office	3	5	8	Museums	2	3	4
Schools and classrooms	1.5	2.5	4	Foundries	10	20	40	Planetariums	1.5	2	3
Hospital and Clinics				General storage	5	10	20	Post offices	2.5	4	6
Laboratories	2	4	6	Heavy machinery	10	25	60	Public libraries	1.5	2	4
Lobbies, waiting rooms	2	4	6	Light machinery	5	12	30	Waiting rooms	3	5	8
Halls and corridors	2	4	6	Tool maintenance	4	7	10	**Restaurants**			
Operating rooms	1.5	2.5	4	**Stores**				Cafeterias	3	6	10
Private rooms	1	1.7	3	Department stores	3	6	8	Night clubs	2.5	4	6
Wards	1.5	2.5	4	Supermarkets	4	7	10	Restaurants	2	4	8

R15810-050 Ductwork

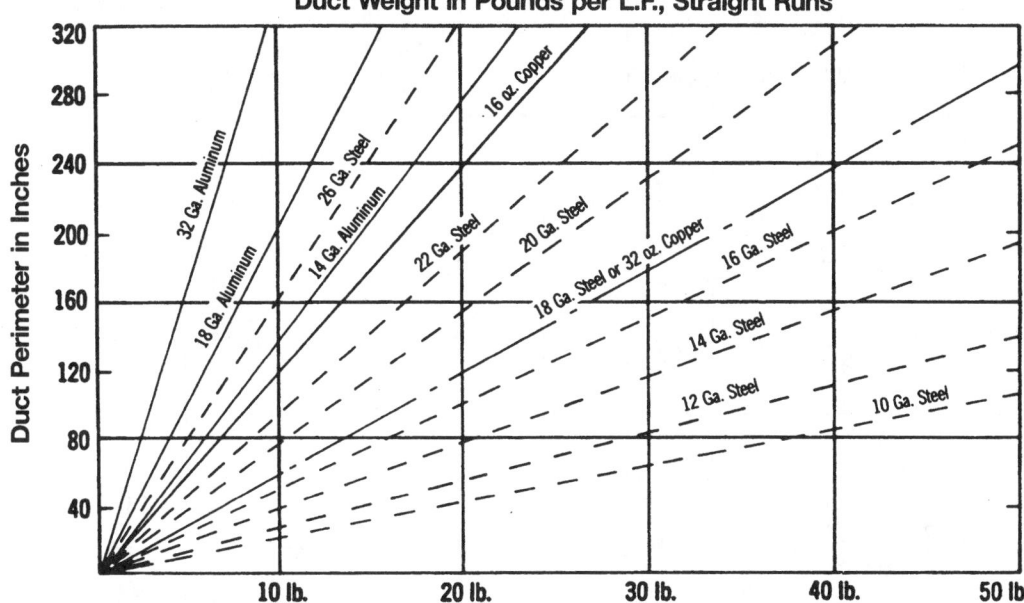

Duct Weight in Pounds per L.F., Straight Runs

Add to the above for fittings; 90° elbow is 3 L.F.; 45° elbow is 2.5 L.F.; offset is 4 L.F.; transition offset is 6 L.F.; square-to-round transition is 4 L.F.; 90° reducing elbow is 5 L.F. For bracing and waste, add 20% to aluminum and copper, 15% to steel.

R15810-070 Duct Fabrication/Installation

The labor cost for sheet metal duct using lines 15810-600-0070 thru 1060 includes both the cost of fabrication and installation of the duct. The split is approximately 40% for fabrication, 60% for installation. It is for this reason that the percentage add for elevated installation is less than the percentage add for prefabricated duct as found starting on line 15810-600-1283.

Example: assume a piece of duct cost $100 installed (labor only)

Sheet Metal Fabrication = 40% = $40

Installation = 60% = $60

The add for elevated installation is:

$$\frac{\text{Based on total labor}}{\text{(fabrication \& installation)}} = \$100 \times 6\% = \$6.00$$

$$\frac{\text{Based on installation cost only}}{\text{(Material purchased prefabricated)}} = \$60 \times 10\% = \$6.00$$

The $6.00 markup (10′ to 15′ high) is the same.

R15810-100 Sheet Metal Calculator (Weight in Lb./Ft. of Length)

Gauge	26	24	22	20	18	16	Gauge	26	24	22	20	18	16
Wt.-Lb./S.F.	.906	1.156	1.406	1.656	2.156	2.656	Wt.-Lb./S.F.	.906	1.156	1.406	1.656	2.156	2.656
SMACNA Max. Dimension – Long Side		30"	54"	84"	85" Up		SMACNA Max. Dimension – Long Side		30"	54"	84"	85" Up	
Sum-2 sides							Sum-2 Sides						
2	.3	.40	.50	.60	.80	.90	56	9.3	12.0	14.0	16.2	21.3	25.2
3	.5	.65	.80	.90	1.1	1.4	57	9.5	12.3	14.3	16.5	21.7	25.7
4	.7	.85	1.0	1.2	1.5	1.8	58	9.7	12.5	14.5	16.8	22.0	26.1
5	.8	1.1	1.3	1.5	1.9	2.3	59	9.8	12.7	14.8	17.1	22.4	26.6
6	1.0	1.3	1.5	1.7	2.3	2.7	60	10.0	12.9	15.0	17.4	22.8	27.0
7	1.2	1.5	1.8	2.0	2.7	3.2	61	10.2	13.1	15.3	17.7	23.2	27.5
8	1.3	1.7	2.0	2.3	3.0	3.6	62	10.3	13.3	15.5	18.0	23.6	27.9
9	1.5	1.9	2.3	2.6	3.4	4.1	63	10.5	13.5	15.8	18.3	24.0	28.4
10	1.7	2.2	2.5	2.9	3.8	4.5	64	10.7	13.7	16.0	18.6	24.3	28.8
11	1.8	2.4	2.8	3.2	4.2	5.0	65	10.8	13.9	16.3	18.9	24.7	29.3
12	2.0	2.6	3.0	3.5	4.6	5.4	66	11.0	14.1	16.5	19.1	25.1	29.7
13	2.2	2.8	3.3	3.8	4.9	5.9	67	11.2	14.3	16.8	19.4	25.5	30.2
14	2.3	3.0	3.5	4.1	5.3	6.3	68	11.3	14.6	17.0	19.7	25.8	30.6
15	2.5	3.2	3.8	4.4	5.7	6.8	69	11.5	14.8	17.3	20.0	26.2	31.1
16	2.7	3.4	4.0	4.6	6.1	7.2	70	11.7	15.0	17.5	20.3	26.6	31.5
17	2.8	3.7	4.3	4.9	6.5	7.7	71	11.8	15.2	17.8	20.6	27.0	32.0
18	3.0	3.9	4.5	5.2	6.8	8.1	72	12.0	15.4	18.0	20.9	27.4	32.4
19	3.2	4.1	4.8	5.5	7.2	8.6	73	12.2	15.6	18.3	21.2	27.7	32.9
20	3.3	4.3	5.0	5.8	7.6	9.0	74	12.3	15.8	18.5	21.5	28.1	33.3
21	3.5	4.5	5.3	6.1	8.0	9.5	75	12.5	16.1	18.8	21.8	28.5	33.8
22	3.7	4.7	5.5	6.4	8.4	9.9	76	12.7	16.3	19.0	22.0	28.9	34.2
23	3.8	5.0	5.8	6.7	8.7	10.4	77	12.8	16.5	19.3	22.3	29.3	34.7
24	4.0	5.2	6.0	7.0	9.1	10.8	78	13.0	16.7	19.5	22.6	29.6	35.1
25	4.2	5.4	6.3	7.3	9.5	11.3	79	13.2	16.9	19.8	22.9	30.0	35.6
26	4.3	5.6	6.5	7.5	9.9	11.7	80	13.3	17.1	20.0	23.2	30.4	36.0
27	4.5	5.8	6.8	7.8	10.3	12.2	81	13.5	17.3	20.3	23.5	30.8	36.5
28	4.7	6.0	7.0	8.1	10.6	12.6	82	13.7	17.5	20.5	23.8	31.2	36.9
29	4.8	6.2	7.3	8.4	11.0	13.1	83	13.8	17.8	20.8	24.1	31.5	37.4
30	5.0	6.5	7.5	8.7	11.4	13.5	84	14.0	18.0	21.0	24.4	31.9	37.8
31	5.2	6.7	7.8	9.0	11.8	14.0	85	14.2	18.2	21.3	24.7	32.3	38.3
32	5.3	6.9	8.0	9.3	12.2	14.4	86	14.3	18.4	21.5	24.9	32.7	38.7
33	5.5	7.1	8.3	9.6	12.5	14.9	87	14.5	18.6	21.8	25.2	33.1	39.2
34	5.7	7.3	8.5	9.9	12.9	15.3	88	14.7	18.8	22.0	25.5	33.4	39.6
35	5.8	7.5	8.8	10.2	13.3	15.8	89	14.8	19.0	22.3	25.8	33.8	40.1
36	6.0	7.8	9.0	10.4	13.7	16.2	90	15.0	19.3	22.5	26.1	34.2	40.5
37	6.2	8.0	9.3	10.7	14.1	16.7	91	15.2	19.5	22.8	26.4	34.6	41.0
38	6.3	8.2	9.5	11.0	14.4	17.1	92	15.3	19.7	23.0	26.7	35.0	41.4
39	6.5	8.4	9.8	11.3	14.8	17.6	93	15.5	19.9	23.3	27.0	35.3	41.9
40	6.7	8.6	10.0	11.6	15.2	18.0	94	15.7	20.1	23.5	27.3	35.7	42.3
41	6.8	8.8	10.3	11.9	15.6	18.5	95	15.8	20.3	23.8	27.6	36.1	42.8
42	7.0	9.0	10.5	12.2	16.0	18.9	96	16.0	20.5	24.0	27.8	36.5	43.2
43	7.2	9.2	10.8	12.5	16.3	19.4	97	16.2	20.8	24.3	28.1	36.9	43.7
44	7.3	9.5	11.0	12.8	16.7	19.8	98	16.3	21.0	24.5	28.4	37.2	44.1
45	7.5	9.7	11.3	13.1	17.1	20.3	99	16.5	21.2	24.8	28.7	37.6	44.6
46	7.7	9.9	11.5	13.3	17.5	20.7	100	16.7	21.4	25.0	29.0	38.0	45.0
47	7.8	10.1	11.8	13.6	17.9	21.2	101	16.8	21.6	25.3	29.3	38.4	45.5
48	8.0	10.3	12.0	13.9	18.2	21.6	102	17.0	21.8	25.5	29.6	38.8	45.9
49	8.2	10.5	12.3	14.2	18.6	22.1	103	17.2	22.0	25.8	29.9	39.1	46.4
50	8.3	10.7	12.5	14.5	19.0	22.5	104	17.3	22.3	26.0	30.2	39.5	46.8
51	8.5	11.0	12.8	14.8	19.4	23.0	105	17.5	22.5	26.3	30.5	39.9	47.3
52	8.7	11.2	13.0	15.1	19.8	23.4	106	17.7	22.7	26.5	30.7	40.3	47.7
53	8.8	11.4	13.3	15.4	20.1	23.9	107	17.8	22.9	26.8	31.0	40.7	48.2
54	9.0	11.6	13.5	15.7	20.5	24.3	108	18.0	23.1	27.0	31.3	41.0	48.6
55	9.2	11.8	13.8	16.0	20.9	24.8	109	18.2	23.3	27.3	31.6	41.4	49.1
							110	18.3	23.5	27.5	31.9	41.8	49.5

Example: If duct is 34" x 20" x 15' long, 34" is greater than 30" maximum, for 24 ga. so must be 22 ga. 34" + 20" = 54" going across from 54" find 13.5 lb. per foot. 13.5 x 15' = 202.5 lbs. For S.F. of surface area

202.5 ÷ 1.406 = 144 S.F.
Note: Figures include an allowance for scrap.

R15810-110   Ductwork Packages (per Ton of Cooling)

System	Sheet Metal	Insulation	Diffusers	Return Register
Roof Top Unit Single Zone	120 Lbs.	52 S.F.	1	1
Roof Top Unit Multizone	240 Lbs.	104 S.F.	2	1
Self-contained Air or Water Cooled	108 Lbs.	—	2	—
Split System Air Cooled	102 Lbs.	—	2	—

Systems reflect most common usage.
Refer to system graphics for duct layout.

R15850-060   Diffuser Evaluation

CFM = V × An × K where V = Outlet velocity in feet per minute. An = Neck area in square feet and K = Diffuser delivery factor. An undersized diffuser for a desired CFM will produce a high velocity and noise level. When air moves past people at a velocity in excess of 25 FPM, an annoying draft is felt. An oversized diffuser will result in low velocity with poor mixing. Consideration must be given to avoid vertical stratification or horizontal areas of stagnation.

R17100-100   Square Foot Project Size Modifier

One factor that affects the S.F. cost of a particular building is the size. In general, for buildings built to the same specifications in the same locality, the larger building will have the lower S.F. cost. This is due mainly to the decreasing contribution of the exterior walls plus the economy of scale usually achievable in larger buildings. The Area Conversion Scale shown below will give a factor to convert costs for the typical size building to an adjusted cost for the particular project.

The Square Foot Base Size lists the median costs, most typical project size in our accumulated data and the range in size of the projects.

The Size Factor for your project is determined by dividing your project area in S.F. by the typical project size for the particular Building Type. With this factor, enter the Area Conversion Scale at the appropriate Size Factor and determine the appropriate cost multiplier for your building size.

Example: Determine the cost per S.F. for a 100,000 S.F. Mid-rise apartment building.

$$\frac{\text{Proposed building area} = 100,000 \text{ S.F.}}{\text{Typical size form below} = 50,000 \text{ S.F.}} = 2.00$$

Enter Area Conversion scale at 2.0, intersect curve, read horizontally the appropriate cost multiplier of .94. Size adjusted cost becomes .94 x $74.30 = $69.85 based on national average costs.

Note: For Size Factors less than .50, the Cost Multiplier is 1.1
      For Size Factors greater than 3.5, the Cost Multiplier is .90

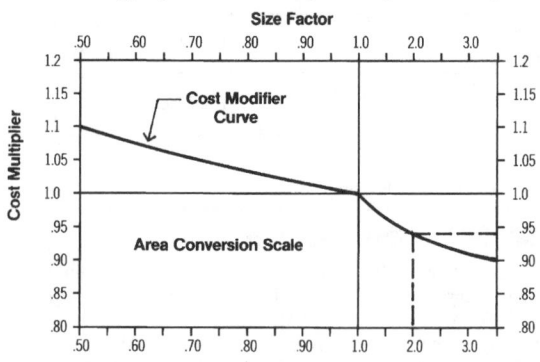

Building Type	Median Cost per S.F.	Typical Size Gross S.F.	Typical Range Gross S.F.	Building Type	Median Cost per S.F.	Typical Size Gross S.F.	Typical Range Gross S.F.
Apartments, Low Rise	$ 58.80	21,000	9,700 - 37,200	Jails	$180.00	40,000	5,500 - 145,000
Apartments, Mid Rise	74.30	50,000	32,000 - 100,000	Libraries	106.00	12,000	7,000 - 31,000
Apartments, High Rise	85.25	145,000	95,000 - 600,000	Medical Clinics	101.00	7,200	4,200 - 15,700
Auditoriums	98.40	25,000	7,600 - 39,000	Medical Offices	95.25	6,000	4,000 - 15,000
Auto Sales	60.85	20,000	10,800 - 28,600	Motels	72.95	40,000	15,800 - 120,000
Banks	132.00	4,200	2,500 - 7,500	Nursing Homes	95.05	23,000	15,000 - 37,000
Churches	88.85	17,000	2,000 - 42,000	Offices, Low Rise	79.45	20,000	5,000 - 80,000
Clubs, Country	88.60	6,500	4,500 - 15,000	Offices, Mid Rise	83.45	120,000	20,000 - 300,000
Clubs, Social	86.20	10,000	6,000 - 13,500	Offices, High Rise	107.00	260,000	120,000 - 800,000
Clubs, YMCA	86.40	28,300	12,800 - 39,400	Police Stations	133.00	10,500	4,000 - 19,000
Colleges (Class)	116.00	50,000	15,000 - 150,000	Post Offices	98.40	12,400	6,800 - 30,000
Colleges (Science Lab)	169.00	45,600	16,600 - 80,000	Power Plants	733.00	7,500	1,000 - 20,000
College (Student Union)	129.00	33,400	16,000 - 85,000	Religious Education	81.50	9,000	6,000 - 12,000
Community Center	92.60	9,400	5,300 - 16,700	Research	139.00	19,000	6,300 - 45,000
Court Houses	126.00	32,400	17,800 - 106,000	Restaurants	119.00	4,400	2,800 - 6,000
Dept. Stores	55.00	90,000	44,000 - 122,000	Retail Stores	58.45	7,200	4,000 - 17,600
Dormitories, Low Rise	94.90	25,000	10,000 - 95,000	Schools, Elementary	85.10	41,000	24,500 - 55,000
Dormitories, Mid Rise	124.00	85,000	20,000 - 200,000	Schools, Jr. High	86.70	92,000	52,000 - 119,000
Factories	53.30	26,400	12,900 - 50,000	Schools, Sr. High	86.70	101,000	50,500 - 175,000
Fire Stations	93.00	5,800	4,000 - 8,700	Schools, Vocational	86.35	37,000	20,500 - 82,000
Fraternity Houses	91.55	12,500	8,200 - 14,800	Sports Arenas	72.40	15,000	5,000 - 40,000
Funeral Homes	102.00	10,000	4,000 - 20,000	Supermarkets	58.65	44,000	12,000 - 60,000
Garages, Commercial	65.05	9,300	5,000 - 13,600	Swimming Pools	136.00	20,000	10,000 - 32,000
Garages, Municipal	83.15	8,300	4,500 - 12,600	Telephone Exchange	158.00	4,500	1,200 - 10,600
Garages, Parking	34.05	163,000	76,400 - 225,300	Theaters	86.80	10,500	8,800 - 17,500
Gymnasiums	86.00	19,200	11,600 - 41,000	Town Halls	95.45	10,800	4,800 - 23,400
Hospitals	159.00	55,000	27,200 - 125,000	Warehouses	39.35	25,000	8,000 - 72,000
House (Elderly)	80.50	37,000	21,000 - 66,000	Warehouse & Office	45.45	25,000	8,000 - 72,000
Housing (Public)	74.50	36,000	14,400 - 74,400				
Ice Rinks	82.85	29,000	27,200 - 33,600				

Square Foot Base Size (table title)

17 SQUARE FOOT REFERENCE NOS.

Change Orders

Change Order Considerations

A Change Order is a written document, usually prepared by the design professional, and signed by the owner, the architect/engineer and the contractor. A change order states the agreement of the parties to: an addition, deletion, or revision in the work; an adjustment in the contract sum, if any; or an adjustment in the contract time, if any. Change orders, or "extras" in the construction process occur after execution of the construction contract and impact architects/engineers, contractors and owners.

Change orders that are properly recognized and managed can ensure orderly, professional and profitable progress for all who are involved in the project. There are many causes for change orders and change order requests. In all cases, change orders or change order requests should be addressed promptly and in a precise and prescribed manner. The following paragraphs include information regarding change order pricing and procedures.

The Causes of Change Orders

Reasons for issuing change orders include:

- Unforeseen field conditions that require a change in the work
- Correction of design discrepancies, errors or omissions in the contract documents
- Owner-requested changes, either by design criteria, scope of work, or project objectives
- Completion date changes for reasons unrelated to the construction process
- Changes in building code interpretations, or other public authority requirements that require a change in the work
- Changes in availability of existing or new materials and products

Procedures

Properly written contract documents must include the correct change order procedures for all parties—owners, design professionals and contractors—to follow in order to avoid costly delays and litigation.

Being "in the right" is not always a sufficient or acceptable defense. The contract provisions requiring notification and documentation must be adhered to within a defined or reasonable time frame.

The appropriate method of handling change orders is by a written proposal and acceptance by all parties involved. Prior to starting work on a project, all parties should identify their authorized agents who may sign and accept change orders, as well as any limits placed on their authority.

Time may be a critical factor when the need for a change arises. For such cases, the contractor might be directed to proceed on a "time and materials" basis, rather than wait for all paperwork to be processed—a delay that could impede progress. In this situation, the contractor must still follow the prescribed change order procedures, including but not limited to, notification and documentation.

All forms used for change orders should be dated and signed by the proper authority. Lack of documentation can be very costly, especially if legal judgments are to be made and if certain field personnel are no longer available. For time and material change orders, the contractor should keep accurate daily records of all labor and material allocated to the change. Forms that can be used to document change order work are available in *Means Forms for Building Construction Professionals.*

Owners or awarding authorities who do considerable and continual building construction (such as the federal government) realize the inevitability of change orders for numerous reasons, both predictable and unpredictable. As a result, the federal government, the American Institute of Architects (AIA), the Engineers Joint Contract Documents Committee (EJCDC) and other contractor, legal and technical organizations have developed standards and procedures to be followed by all parties to achieve contract continuance and timely completion, while being financially fair to all concerned.

In addition to the change order standards put forth by industry associations, there are also many books available on the subject.

Pricing Change Orders

When pricing change orders, regardless of their cause, the most significant factor is *when* the change occurs. The need for a change may be perceived in the field or requested by the architect/engineer *before* any of the actual installation has begun, or may evolve or appear *during* construction when the item of work in question is partially installed. In the latter cases, the original sequence of construction is disrupted, along with all contiguous and supporting systems. Change orders cause the greatest impact when they occur *after* the installation has been completed and must be uncovered, or even replaced. Post-completion changes may be caused by necessary design changes, product failure, or changes in the owner's requirements that are not discovered until the building or the systems begin to function.

Specified procedures of notification and record keeping must be adhered to and enforced regardless of the stage of construction: *before, during,* or *after* installation. Some bidding documents anticipate change orders by requiring that unit prices including overhead and profit percentages—for additional as well as deductible changes—be listed. Generally these unit prices do not fully take into account the ripple effect, or impact on other trades, and should be used for general guidance only.

When pricing change orders, it is important to classify the time frame in which the change occurs. There are two basic time frames for change orders: *pre-installation change orders,* which occur before the start of construction, and *post-installation change orders,* which involve reworking after the original installation. Change orders that occur between these stages may be priced according to the extent of work completed using a combination of techniques developed for pricing *pre-* and *post-installation* changes.

The following factors are the basis for a check list to use when preparing a change order estimate.

Factors To Consider When Pricing Change Orders

As an estimator begins to prepare a change order, the following questions should be reviewed to determine their impact on the final price.

General

- Is the change order work *pre-installation* or *post-installation*?

Change order work costs vary according to how much of the installation has been completed. Once workers have the project scoped in their mind, even though they have not started, it can be difficult to refocus.

Consequently they may spend more than the normal amount of time understanding the change. Also, modifications to work in place such as trimming or refitting usually take more time than was initially estimated. The greater the amount of work in place, the more reluctant workers are to change it. Psychologically they may resent the change and as a result the rework takes longer than normal. Post-installation change order estimates must include demolition of existing work as required to accomplish the change. If the work is performed at a later time, additional obstacles such as building finishes may be present which must be

protected. Regardless of whether the change occurs pre-installation or post-installation, attempt to isolate the identifiable factors and price them separately. For example, add shipping costs that may be required pre-installation or any demolition required post-installation. Then analyze the potential impact on productivity of psychological and/or learning curve factors and adjust the output rates accordingly. One approach is to break down the typical workday into segments and quantify the impact on each segment. The following chart may be useful as a guide:

Task	Activities (Productivity) Expressed as Percentages of a Workday		
	Means Mechanical Cost Data (for New Construction)	Pre-Installation Change Orders	Post-Installation Change Orders
1. Study plans	3%	6%	6%
2. Material procurement	3%	3%	3%
3. Receiving and storing	3%	3%	3%
4. Mobilization	5%	5%	5%
5. Site movement	5%	5%	8%
6. Layout and marking	8%	10%	12%
7. Actual installation	64%	59%	54%
8. Clean-up	3%	3%	3%
9. Breaks—non-productive	6%	6%	6%
Total	100%	100%	100%

Change Order Installation Efficiency

The labor-hours expressed (for new construction) are based on average installation time, using an efficiency level of approximately 60-65%. For change order situations, adjustments to this efficiency level should reflect the daily labor-hour allocation for that particular occurrence.

If any of the specific percentages expressed in the above chart do not apply to a particular project situation, then those percentage points should be reallocated to the appropriate task(s). Example: Using data for new construction, assume there is no new material being utilized. The percentages for Tasks 2 and 3 would therefore be reallocated to other tasks. If the time required for Tasks 2 and 3 can now be applied to installation, we can add the time allocated for *Material Procurement* and *Receiving and Storing* to the *Actual Installation* time for new construction, thereby increasing the Actual Installation percentage.

This chart shows that, due to reduced productivity, labor costs will be higher than those for new construction by 5% to 15% for pre-installation change orders and by 15% to 25% for post-installation change orders. Each job and change order is unique and must be examined individually. Many factors, covered elsewhere in this section, can each have a significant impact on productivity and change order costs. All such factors should be considered in every case.

- Will the change substantially delay the original completion date?

  A significant change in the project may cause the original completion date to be extended. The extended schedule may subject the contractor to new wage rates dictated by relevant labor contracts. Project supervision and other project overhead must also be extended beyond the original completion date. The schedule extension may also put installation into a new weather season. For example, underground piping scheduled for October installation was delayed until January. As a result, frost penetrated the trench area, thereby changing the degree of difficulty of the task. Changes and delays may have a ripple effect throughout the project. This effect must be analyzed and negotiated with the owner.

- What is the net effect of a deduct change order?

  In most cases, change orders resulting in a deduction or credit reflect only bare costs. The contractor may retain the overhead and profit based on the original bid.

Materials

- Will you have to pay more or less for the new material, required by the change order, than you paid for the original purchase?

  The same material prices or discounts will usually apply to materials purchased for change orders as new construction. In some instances, however, the contractor may forfeit the advantages of competitive pricing for change orders. Consider the following example:

  A contractor purchased over $20,000 worth of fan coil units for an installation, and obtained the maximum discount. Some time later it was determined the project required an additional matching unit. The contractor has to purchase this unit from the original supplier to ensure a match. The supplier at this time may not discount the unit because of the small quantity, and the fact that he is no longer in a competitive situation. The impact of quantity on purchase can add between 0% and 25% to material prices and/or subcontractor quotes.

- If materials have been ordered or delivered to the job site, will they be subject to a cancellation charge or restocking fee?

  Check with the supplier to determine if ordered materials are subject to a cancellation charge. Delivered materials not used as result of a change order may be subject to a restocking fee if returned to the supplier. Common restocking charges run between 20% and 40%. Also, delivery charges to return the goods to the supplier must be added.

Labor

- How efficient is the existing crew at the actual installation?

  Is the same crew that performed the initial work going to do the change order? Possibly the change consists of the installation of a unit identical to one already installed; therefore the change should take less time. Be sure to consider this potential productivity increase and modify the productivity rates accordingly.

- If the crew size is increased, what impact will that have on supervision requirements?

  Under most bargaining agreements or management practices, there is a point at which a working foreman is replaced by a nonworking foreman. This replacement increases project overhead by adding a nonproductive worker. If additional workers are added to accelerate the project or to perform changes while maintaining the schedule, be sure to add additional supervision time if warranted. Calculate the hours involved and the additional cost directly if possible.

- What are the other impacts of increased crew size?

  The larger the crew, the greater the potential for productivity to decrease. Some of the factors that cause this productivity loss are: overcrowding (producing restrictive conditions in the working space), and possibly a shortage of any special tools and equipment required. Such factors affect not only the crew working on the elements directly involved in the change order, but other crews whose movement may also be hampered.

As the crew increases, check its basic composition for changes by the addition or deletion of apprentices or nonworking foreman and quantify the potential effects of equipment shortages or other logistical factors.

- As new crews, unfamiliar with the project, are brought onto the site, how long will it take them to become oriented to the project requirements?

  The orientation time for a new crew to become 100% effective varies with the site and type of project. Orientation is easiest at a new construction site, and most difficult at existing, very restrictive renovation sites. The type of work also affects orientation time. When all elements of the work are exposed, such as concrete or masonry work, orientation is decreased. When the work is concealed or less visible, such as existing electrical systems, orientation takes longer. Usually orientation can be accomplished in one day or less. Costs for added orientation should be itemized and added to the total estimated cost.

- How much actual production can be gained by working overtime?

  Short term overtime can be used effectively to accomplish more work in a day. However, as overtime is scheduled to run beyond several weeks, studies have shown marked decreases in output. The following chart shows the effect of long term overtime on worker efficiency. If the anticipated change requires extended overtime to keep the job on schedule, these factors can be used as a guide to predict the impact on time and cost. Add project overhead, particularly supervision, that may also be incurred.

Days per Week	Hours per Day	Production Efficiency					Payroll Cost Factors	
		1 Week	2 Weeks	3 Weeks	4 Weeks	Average 4 Weeks	@ 1-1/2 Times	@ 2 Times
5	8	100%	100%	100%	100%	100	100%	100%
	9	100	100	95	90	96.25	105.6	111.1
	10	100	95	90	85	91.25	110.0	120.0
	11	95	90	75	65	81.25	113.6	127.3
	12	90	85	70	60	76.25	116.7	133.3
6	8	100	100	95	90	96.25	108.3	116.7
	9	100	95	90	85	92.50	113.0	125.9
	10	95	90	85	80	87.50	116.7	133.3
	11	95	85	70	65	78.75	119.7	139.4
	12	90	80	65	60	73.75	122.2	144.4
7	8	100	95	85	75	88.75	114.3	128.6
	9	95	90	80	70	83.75	118.3	136.5
	10	90	85	75	65	78.75	121.4	142.9
	11	85	80	65	60	72.50	124.0	148.1
	12	85	75	60	55	68.75	126.2	152.4

Effects of Overtime

Caution: Under many labor agreements, Sundays and holidays are paid at a higher premium than the normal overtime rate.

The use of long-term overtime is counterproductive on almost any construction job; that is, the longer the period of overtime, the lower the actual production rate. Numerous studies have been conducted, and while they have resulted in slightly different numbers, all reach the same conclusion. The figure above tabulates the effects of overtime work on efficiency.

As illustrated, there can be a difference between the *actual* payroll cost per hour and the *effective* cost per hour for overtime work. This is due to the reduced production efficiency with the increase in weekly hours beyond 40. This difference between actual and effective cost results from overtime work over a prolonged period. Short-term overtime work does not result in as great a reduction in efficiency, and in such cases, effective cost may not vary significantly from the actual payroll cost. As the total hours per week are increased on a regular basis, more time is lost because of fatigue, lowered morale, and an increased accident rate.

As an example, assume a project where workers are working 6 days a week, 10 hours per day. From the figure above (based on productivity studies), the average effective productive hours over a four-week period are:

$$0.875 \times 60 = 52.5$$

Depending upon the locale and day of week, overtime hours may be paid at time and a half or double time. For time and a half, the overall (average) *actual* payroll cost (including regular and overtime hours) is determined as follows:

$$\frac{40 \text{ reg. hrs.} + (20 \text{ overtime hrs.} \times 1.5)}{60 \text{ hrs.}} = 1.167$$

Based on 60 hours, the payroll cost per hour will be 116.7% of the normal rate at 40 hours per week. However, because the effective production (efficiency) for 60 hours is reduced to the equivalent of 52.5 hours, the effective cost of overtime is calculated as follows:

For time and a half:

$$\frac{40 \text{ reg. hrs.} + (20 \text{ overtime hrs.} \times 1.5)}{52.5 \text{ hrs.}} = 1.33$$

Installed cost will be 133% of the normal rate (for labor).

Thus, when figuring overtime, the actual cost per unit of work will be higher than the apparent overtime payroll dollar increase, due to the reduced productivity of the longer workweek. These efficiency calculations are true only for those cost factors determined by hours worked. Costs that are applied weekly or monthly, such as equipment rentals, will not be similarly affected.

Equipment

• What equipment is required to complete the change order?

Change orders may require extending the rental period of equipment already on the job site, or the addition of special equipment brought in to accomplish the change work. In either case, the additional rental charges and operator labor charges must be added.

Summary

The preceding considerations and others you deem appropriate should be analyzed and applied to a change order estimate. The impact of each should be quantified and listed on the estimate to form an audit trail.

Change orders that are properly identified, documented, and managed help to ensure the orderly, professional and profitable progress of the work. They also minimize potential claims or disputes at the end of the project.

Crew No.	Bare Costs		Incl. Subs O & P		Cost Per Labor-Hour	
Crew A-1	Hr.	Daily	Hr.	Daily	Bare Costs	Incl. O&P
1 Building Laborer	$24.65	$197.20	$38.50	$308.00	$24.65	$38.50
1 Gas Eng. Power Tool		57.60		63.35	7.20	7.92
8 L.H., Daily Totals		$254.80		$371.35	$31.85	$46.42
Crew A-1A	Hr.	Daily	Hr.	Daily	Bare Costs	Incl. O&P
1 Skilled Worker	$32.25	$258.00	$50.50	$404.00	$32.25	$50.50
1 Shot Blaster, 20"		206.40		227.05	25.80	28.38
8 L.H., Daily Totals		$464.40		$631.05	$58.05	$78.88
Crew A-2	Hr.	Daily	Hr.	Daily	Bare Costs	Incl. O&P
2 Laborers	$24.65	$394.40	$38.50	$616.00	$24.73	$38.33
1 Truck Driver (light)	24.90	199.20	38.00	304.00		
1 Light Truck, 1.5 Ton		120.20		132.20	5.01	5.51
24 L.H., Daily Totals		$713.80		$1052.20	$29.74	$43.84
Crew A-2A	Hr.	Daily	Hr.	Daily	Bare Costs	Incl. O&P
2 Laborers	$24.65	$394.40	$38.50	$616.00	$24.73	$38.33
1 Truck Driver (light)	24.90	199.20	38.00	304.00		
1 Light Truck, 1.5 Ton		120.20		132.20		
1 Concrete Saw		108.20		119.00	9.52	10.47
24 L.H., Daily Totals		$822.00		$1171.20	$34.25	$48.80
Crew A-3	Hr.	Daily	Hr.	Daily	Bare Costs	Incl. O&P
1 Truck Driver (heavy)	$25.65	$205.20	$39.15	$313.20	$25.65	$39.15
1 Dump Truck, 12 Ton		306.00		336.60	38.25	42.08
8 L.H., Daily Totals		$511.20		$649.80	$63.90	$81.23
Crew A-3A	Hr.	Daily	Hr.	Daily	Bare Costs	Incl. O&P
1 Truck Driver (light)	$24.90	$199.20	$38.00	$304.00	$24.90	$38.00
1 Pickup Truck (4x4)		74.00		81.40	9.25	10.18
8 L.H., Daily Totals		$273.20		$385.40	$34.15	$48.18
Crew A-3B	Hr.	Daily	Hr.	Daily	Bare Costs	Incl. O&P
1 Equip. Oper. (medium)	$32.60	$260.80	$49.35	$394.80	$29.13	$44.25
1 Truck Driver (heavy)	25.65	205.20	39.15	313.20		
1 Dump Truck, 16 Ton		449.60		494.55		
1 F.E. Loader, 3 C.Y.		313.00		344.30	47.66	52.43
16 L.H., Daily Totals		$1228.60		$1546.85	$76.79	$96.68
Crew A-3C	Hr.	Daily	Hr.	Daily	Bare Costs	Incl. O&P
1 Equip. Oper. (light)	$31.10	$248.80	$47.10	$376.80	$31.10	$47.10
1 Wheeled Skid Steer Loader		194.80		214.30	24.35	26.79
8 L.H., Daily Totals		$443.60		$591.10	$55.45	$73.89
Crew A-3D	Hr.	Daily	Hr.	Daily	Bare Costs	Incl. O&P
1 Truck Driver, Light	$24.90	$199.20	$38.00	$304.00	$24.90	$38.00
1 Pickup Truck (4x4)		74.00		81.40		
1 Flatbed Trailer, 25 Ton		90.00		99.00	20.50	22.55
8 L.H., Daily Totals		$363.20		$484.40	$45.40	$60.55
Crew A-3E	Hr.	Daily	Hr.	Daily	Bare Costs	Incl. O&P
1 Equip. Oper. (crane)	$33.70	$269.60	$51.00	$408.00	$29.68	$45.08
1 Truck Driver (heavy)	25.65	205.20	39.15	313.20		
1 Pickup Truck (4x4)		74.00		81.40	4.63	5.09
16 L.H., Daily Totals		$548.80		$802.60	$34.31	$50.17

Crew No.	Bare Costs		Incl. Subs O & P		Cost Per Labor-Hour	
Crew A-3F	Hr.	Daily	Hr.	Daily	Bare Costs	Incl. O&P
1 Equip. Oper. (crane)	$33.70	$269.60	$51.00	$408.00	$29.68	$45.08
1 Truck Driver (heavy)	25.65	205.20	39.15	313.20		
1 Pickup Truck (4x4)		74.00		81.40		
1 Tractor, 6x2, 40 Ton Cap.		330.80		363.90		
1 Lowbed Trailer, 75 Ton		221.60		243.75	39.15	43.07
16 L.H., Daily Totals		$1101.20		$1410.25	$68.83	$88.15
Crew A-3G	Hr.	Daily	Hr.	Daily	Bare Costs	Incl. O&P
1 Equip. Oper. (crane)	$33.70	$269.60	$51.00	$408.00	$29.68	$45.08
1 Truck Driver (heavy)	25.65	205.20	39.15	313.20		
1 Pickup Truck (4x4)		74.00		81.40		
1 Tractor, 6x4, 45 Ton Cap.		376.00		413.60		
1 Lowbed Trailer, 75 Ton		221.60		243.75	41.98	46.17
16 L.H., Daily Totals		$1146.40		$1459.95	$71.66	$91.25
Crew A-4	Hr.	Daily	Hr.	Daily	Bare Costs	Incl. O&P
2 Carpenters	$31.55	$504.80	$49.30	$788.80	$30.38	$47.02
1 Painter, Ordinary	28.05	224.40	42.45	339.60		
24 L.H., Daily Totals		$729.20		$1128.40	$30.38	$47.02
Crew A-5	Hr.	Daily	Hr.	Daily	Bare Costs	Incl. O&P
2 Laborers	$24.65	$394.40	$38.50	$616.00	$24.68	$38.44
.25 Truck Driver (light)	24.90	49.80	38.00	76.00		
.25 Light Truck, 1.5 Ton		30.05		33.05	1.67	1.84
18 L.H., Daily Totals		$474.25		$725.05	$26.35	$40.28
Crew A-6	Hr.	Daily	Hr.	Daily	Bare Costs	Incl. O&P
1 Chief Of Party	$38.70	$309.60	$60.80	$486.40	$35.48	$55.65
1 Instrument Man	32.25	258.00	50.50	404.00		
16 L.H., Daily Totals		$567.60		$890.40	$35.48	$55.65
Crew A-7	Hr.	Daily	Hr.	Daily	Bare Costs	Incl. O&P
1 Chief Of Party	$38.70	$309.60	$60.80	$486.40	$33.97	$52.77
1 Instrument Man	32.25	258.00	50.50	404.00		
1 Rodman/Chainman	30.95	247.60	47.00	376.00		
24 L.H., Daily Totals		$815.20		$1266.40	$33.97	$52.77
Crew A-8	Hr.	Daily	Hr.	Daily	Bare Costs	Incl. O&P
1 Chief Of Party	$38.70	$309.60	$60.80	$486.40	$33.21	$51.33
1 Instrument Man	32.25	258.00	50.50	404.00		
2 Rodmen/Chainmen	30.95	495.20	47.00	752.00		
32 L.H., Daily Totals		$1062.80		$1642.40	$33.21	$51.33
Crew A-9	Hr.	Daily	Hr.	Daily	Bare Costs	Incl. O&P
1 Asbestos Foreman	$35.20	$281.60	$56.00	$448.00	$34.76	$55.30
7 Asbestos Workers	34.70	1943.20	55.20	3091.20		
64 L.H., Daily Totals		$2224.80		$3539.20	$34.76	$55.30
Crew A-10	Hr.	Daily	Hr.	Daily	Bare Costs	Incl. O&P
1 Asbestos Foreman	$35.20	$281.60	$56.00	$448.00	$34.76	$55.30
7 Asbestos Workers	34.70	1943.20	55.20	3091.20		
64 L.H., Daily Totals		$2224.80		$3539.20	$34.76	$55.30
Crew A-10A	Hr.	Daily	Hr.	Daily	Bare Costs	Incl. O&P
1 Asbestos Foreman	$35.20	$281.60	$56.00	$448.00	$34.87	$55.47
2 Asbestos Workers	34.70	555.20	55.20	883.20		
24 L.H., Daily Totals		$836.80		$1331.20	$34.87	$55.47

Crew No.	Bare Costs		Incl. Subs O & P		Cost Per Labor-Hour	
Crew A-10B	Hr.	Daily	Hr.	Daily	Bare Costs	Incl. O&P
1 Asbestos Foreman	$35.20	$281.60	$56.00	$448.00	$34.83	$55.40
3 Asbestos Workers	34.70	832.80	55.20	1324.80		
32 L.H., Daily Totals		$1114.40		$1772.80	$34.83	$55.40
Crew A-10C	Hr.	Daily	Hr.	Daily	Bare Costs	Incl. O&P
3 Asbestos Workers	$34.70	$832.80	$55.20	$1324.80	$34.70	$55.20
1 Flatbed Truck		120.20		132.20	5.01	5.51
24 L.H., Daily Totals		$953.00		$1457.00	$39.71	$60.71
Crew A-10D	Hr.	Daily	Hr.	Daily	Bare Costs	Incl. O&P
2 Asbestos Workers	$34.70	$555.20	$55.20	$883.20	$32.85	$51.06
1 Equip. Oper. (crane)	33.70	269.60	51.00	408.00		
1 Equip. Oper. Oiler	28.30	226.40	42.85	342.80		
1 Hydraulic Crane, 33 Ton		675.80		743.40	21.12	23.23
32 L.H., Daily Totals		$1727.00		$2377.40	$53.97	$74.29
Crew A-11	Hr.	Daily	Hr.	Daily	Bare Costs	Incl. O&P
1 Asbestos Foreman	$35.20	$281.60	$56.00	$448.00	$34.76	$55.30
7 Asbestos Workers	34.70	1943.20	55.20	3091.20		
2 Chipping Hammers		32.80		36.10	.51	.56
64 L.H., Daily Totals		$2257.60		$3575.30	$35.27	$55.86
Crew A-12	Hr.	Daily	Hr.	Daily	Bare Costs	Incl. O&P
1 Asbestos Foreman	$35.20	$281.60	$56.00	$448.00	$34.76	$55.30
7 Asbestos Workers	34.70	1943.20	55.20	3091.20		
1 Large Prod. Vac. Loader		490.25		539.30	7.66	8.43
64 L.H., Daily Totals		$2715.05		$4078.50	$42.42	$63.73
Crew A-13	Hr.	Daily	Hr.	Daily	Bare Costs	Incl. O&P
1 Equip. Oper. (light)	$31.10	$248.80	$47.10	$376.80	$31.10	$47.10
1 Large Prod. Vac. Loader		490.25		539.30	61.28	67.41
8 L.H., Daily Totals		$739.05		$916.10	$92.38	$114.51
Crew B-1	Hr.	Daily	Hr.	Daily	Bare Costs	Incl. O&P
1 Labor Foreman (outside)	$26.65	$213.20	$41.65	$333.20	$25.32	$39.55
2 Laborers	24.65	394.40	38.50	616.00		
24 L.H., Daily Totals		$607.60		$949.20	$25.32	$39.55
Crew B-2	Hr.	Daily	Hr.	Daily	Bare Costs	Incl. O&P
1 Labor Foreman (outside)	$26.65	$213.20	$41.65	$333.20	$25.05	$39.13
4 Laborers	24.65	788.80	38.50	1232.00		
40 L.H., Daily Totals		$1002.00		$1565.20	$25.05	$39.13
Crew B-3	Hr.	Daily	Hr.	Daily	Bare Costs	Incl. O&P
1 Labor Foreman (outside)	$26.65	$213.20	$41.65	$333.20	$26.64	$41.05
2 Laborers	24.65	394.40	38.50	616.00		
1 Equip. Oper. (med.)	32.60	260.80	49.35	394.80		
2 Truck Drivers (heavy)	25.65	410.40	39.15	626.40		
1 F.E. Loader, T.M., 2.5 C.Y.		778.40		856.25		
2 Dump Trucks, 16 Ton		899.20		989.10	34.95	38.45
48 L.H., Daily Totals		$2956.40		$3815.75	$61.59	$79.50
Crew B-3A	Hr.	Daily	Hr.	Daily	Bare Costs	Incl. O&P
4 Laborers	$24.65	$788.80	$38.50	$1232.00	$26.24	$40.67
1 Equip. Oper. (med.)	32.60	260.80	49.35	394.80		
1 Hyd. Excavator, 1.5 C.Y.		728.80		801.70	18.22	20.04
40 L.H., Daily Totals		$1778.40		$2428.50	$44.46	$60.71

Crew No.	Bare Costs		Incl. Subs O & P		Cost Per Labor-Hour	
Crew B-3B	Hr.	Daily	Hr.	Daily	Bare Costs	Incl. O&P
2 Laborers	$24.65	$394.40	$38.50	$616.00	$26.89	$41.38
1 Equip. Oper. (med.)	32.60	260.80	49.35	394.80		
1 Truck Driver (heavy)	25.65	205.20	39.15	313.20		
1 Backhoe Loader, 80 H.P.		244.20		268.60		
1 Dump Truck, 16 Ton		449.60		494.55	21.68	23.85
32 L.H., Daily Totals		$1554.20		$2087.15	$48.57	$65.23
Crew B-3C	Hr.	Daily	Hr.	Daily	Bare Costs	Incl. O&P
3 Laborers	$24.65	$591.60	$38.50	$924.00	$26.64	$41.21
1 Equip. Oper. (med.)	32.60	260.80	49.35	394.80		
1 F.E. Crawler Ldr, 4 C.Y.		1092.00		1201.20	34.13	37.54
32 L.H., Daily Totals		$1944.40		$2520.00	$60.77	$78.75
Crew B-4	Hr.	Daily	Hr.	Daily	Bare Costs	Incl. O&P
1 Labor Foreman (outside)	$26.65	$213.20	$41.65	$333.20	$25.15	$39.13
4 Laborers	24.65	788.80	38.50	1232.00		
1 Truck Driver (heavy)	25.65	205.20	39.15	313.20		
1 Tractor, 4 x 2, 195 H.P.		240.60		264.65		
1 Platform Trailer		143.00		157.30	7.99	8.79
48 L.H., Daily Totals		$1590.80		$2300.35	$33.14	$47.92
Crew B-5	Hr.	Daily	Hr.	Daily	Bare Costs	Incl. O&P
1 Labor Foreman (outside)	$26.65	$213.20	$41.65	$333.20	$27.21	$42.05
4 Laborers	24.65	788.80	38.50	1232.00		
2 Equip. Oper. (med.)	32.60	521.60	49.35	789.60		
1 Air Compr., 250 C.F.M.		128.60		141.45		
2 Air Tools & Accessories		29.20		32.10		
2-50 Ft. Air Hoses, 1.5" Dia.		10.00		11.00		
1 F.E. Loader, T.M., 2.5 C.Y.		778.40		856.25	16.90	18.59
56 L.H., Daily Totals		$2469.80		$3395.60	$44.11	$60.64
Crew B-5A	Hr.	Daily	Hr.	Daily	Bare Costs	Incl. O&P
1 Foreman	$26.65	$213.20	$41.65	$333.20	$26.85	$41.40
6 Laborers	24.65	1183.20	38.50	1848.00		
2 Equip. Oper. (med.)	32.60	521.60	49.35	789.60		
1 Equip. Oper. (light)	31.10	248.80	47.10	376.80		
2 Truck Drivers (heavy)	25.65	410.40	39.15	626.40		
1 Air Compr. 365 C.F.M.		147.60		162.35		
2 Pavement Breakers		29.20		32.10		
8 Air Hoses w/Coup.,1"		29.20		32.10		
2 Dump Trucks, 12 Ton		612.00		673.20	8.52	9.37
96 L.H., Daily Totals		$3395.20		$4873.75	$35.37	$50.77
Crew B-5B	Hr.	Daily	Hr.	Daily	Bare Costs	Incl. O&P
1 Powderman	$32.25	$258.00	$50.50	$404.00	$29.07	$44.44
2 Equip. Oper. (med.)	32.60	521.60	49.35	789.60		
3 Truck Drivers (heavy)	25.65	615.60	39.15	939.60		
1 F.E. Ldr. 2-1/2 CY		313.00		344.30		
3 Dump Trucks, 16 Ton		1348.80		1483.70		
1 Air Compr. 365 C.F.M.		147.60		162.35	37.70	41.47
48 L.H., Daily Totals		$3204.60		$4123.55	$66.77	$85.91

Crew B-5C

Crew No.	Bare Costs Hr.	Daily	Incl. Subs O&P Hr.	Daily	Cost Per Labor-Hour Bare Costs	Incl. O&P
3 Laborers	$24.65	$591.60	$38.50	$924.00	$27.48	$42.13
1 Equip. Oper. (medium)	32.60	260.80	49.35	394.80		
2 Truck Drivers (heavy)	25.65	410.40	39.15	626.40		
1 Equip. Oper. (crane)	33.70	269.60	51.00	408.00		
1 Equip. Oper. Oiler	28.30	226.40	42.85	342.80		
2 Dump Trucks, 16 Ton		899.20		989.10		
1 F.E. Crawler Ldr, 4 C.Y.		1092.00		1201.20		
1 Hyd. Crane, 25 Ton		657.80		723.60	41.39	45.53
64 L.H., Daily Totals		$4407.80		$5609.90	$68.87	$87.66

Crew B-6

Crew No.	Bare Costs Hr.	Daily	Incl. Subs O&P Hr.	Daily	Cost Per Labor-Hour Bare Costs	Incl. O&P
2 Laborers	$24.65	$394.40	$38.50	$616.00	$26.80	$41.37
1 Equip. Oper. (light)	31.10	248.80	47.10	376.80		
1 Backhoe Loader, 48 H.P.		218.80		240.70	9.12	10.03
24 L.H., Daily Totals		$862.00		$1233.50	$35.92	$51.40

Crew B-6A

Crew No.	Bare Costs Hr.	Daily	Incl. Subs O&P Hr.	Daily	Cost Per Labor-Hour Bare Costs	Incl. O&P
.5 Labor Foreman (outside)	$26.65	$106.60	$41.65	$166.60	$28.23	$43.47
1 Laborer	24.65	197.20	38.50	308.00		
1 Equip. Oper. (med.)	32.60	260.80	49.35	394.80		
1 Vacuum Trk., 5000 Gal.		301.30		331.45	15.07	16.57
20 L.H., Daily Totals		$865.90		$1200.85	$43.30	$60.04

Crew B-7

Crew No.	Bare Costs Hr.	Daily	Incl. Subs O&P Hr.	Daily	Cost Per Labor-Hour Bare Costs	Incl. O&P
1 Labor Foreman (outside)	$26.65	$213.20	$41.65	$333.20	$26.31	$40.83
4 Laborers	24.65	788.80	38.50	1232.00		
1 Equip. Oper. (med.)	32.60	260.80	49.35	394.80		
1 Chipping Machine		180.80		198.90		
1 F.E. Loader, T.M., 2.5 C.Y.		778.40		856.25		
2 Chain Saws, 36"		76.00		83.60	21.57	23.72
48 L.H., Daily Totals		$2298.00		$3098.75	$47.88	$64.55

Crew B-7A

Crew No.	Bare Costs Hr.	Daily	Incl. Subs O&P Hr.	Daily	Cost Per Labor-Hour Bare Costs	Incl. O&P
2 Laborers	$24.65	$394.40	$38.50	$616.00	$26.80	$41.37
1 Equip. Oper. (light)	31.10	248.80	47.10	376.80		
1 Rake w/Tractor		194.30		213.75		
2 Chain Saws, 18"		43.20		47.50	9.90	10.89
24 L.H., Daily Totals		$880.70		$1254.05	$36.70	$52.26

Crew B-8

Crew No.	Bare Costs Hr.	Daily	Incl. Subs O&P Hr.	Daily	Cost Per Labor-Hour Bare Costs	Incl. O&P
1 Labor Foreman (outside)	$26.65	$213.20	$41.65	$333.20	$27.59	$42.31
2 Laborers	24.65	394.40	38.50	616.00		
2 Equip. Oper. (med.)	32.60	521.60	49.35	789.60		
1 Equip. Oper. Oiler	28.30	226.40	42.85	342.80		
2 Truck Drivers (heavy)	25.65	410.40	39.15	626.40		
1 Hyd. Crane, 25 Ton		654.60		720.05		
1 F.E. Loader, T.M., 2.5 C.Y.		778.40		856.25		
2 Dump Trucks, 16 Ton		899.20		989.10	36.44	40.08
64 L.H., Daily Totals		$4098.20		$5273.40	$64.03	$82.39

Crew B-9

Crew No.	Bare Costs Hr.	Daily	Incl. Subs O&P Hr.	Daily	Cost Per Labor-Hour Bare Costs	Incl. O&P
1 Labor Foreman (outside)	$26.65	$213.20	$41.65	$333.20	$25.05	$39.13
4 Laborers	24.65	788.80	38.50	1232.00		
1 Air Compr., 250 C.F.M.		128.60		141.45		
2 Air Tools & Accessories		29.20		32.10		
2-50 Ft. Air Hoses, 1.5" Dia.		10.00		11.00	4.20	4.61
40 L.H., Daily Totals		$1169.80		$1749.75	$29.25	$43.74

Crew B-9A

Crew No.	Bare Costs Hr.	Daily	Incl. Subs O&P Hr.	Daily	Cost Per Labor-Hour Bare Costs	Incl. O&P
2 Laborers	$24.65	$394.40	$38.50	$616.00	$24.98	$38.72
1 Truck Driver (heavy)	25.65	205.20	39.15	313.20		
1 Water Tanker		144.80		159.30		
1 Tractor		240.60		264.65		
2-50 Ft. Disch. Hoses		9.10		10.00	16.44	18.08
24 L.H., Daily Totals		$994.10		$1363.15	$41.42	$56.80

Crew B-9B

Crew No.	Bare Costs Hr.	Daily	Incl. Subs O&P Hr.	Daily	Cost Per Labor-Hour Bare Costs	Incl. O&P
2 Laborers	$24.65	$394.40	$38.50	$616.00	$24.98	$38.72
1 Truck Driver (heavy)	25.65	205.20	39.15	313.20		
2-50 Ft. Disch. Hoses		9.10		10.00		
1 Water Tanker		144.80		159.30		
1 Tractor		240.60		264.65		
1 Pressure Washer		52.60		57.85	18.63	20.49
24 L.H., Daily Totals		$1046.70		$1421.00	$43.61	$59.21

Crew B-9C

Crew No.	Bare Costs Hr.	Daily	Incl. Subs O&P Hr.	Daily	Cost Per Labor-Hour Bare Costs	Incl. O&P
1 Labor Foreman (outside)	$26.65	$213.20	$41.65	$333.20	$25.05	$39.13
4 Laborers	24.65	788.80	38.50	1232.00		
1 Air Compr., 250 C.F.M.		128.60		141.45		
2-50 Ft. Air Hoses, 1.5" Dia.		10.00		11.00		
2 Breaker, Pavement, 60 lb.		29.20		32.10	4.20	4.61
40 L.H., Daily Totals		$1169.80		$1749.75	$29.25	$43.74

Crew B-10

Crew No.	Bare Costs Hr.	Daily	Incl. Subs O&P Hr.	Daily	Cost Per Labor-Hour Bare Costs	Incl. O&P
1 Equip. Oper. (med.)	$32.60	$260.80	$49.35	$394.80	$29.95	$45.73
.5 Laborer	24.65	98.60	38.50	154.00		
12 L.H., Daily Totals		$359.40		$548.80	$29.95	$45.73

Crew B-10A

Crew No.	Bare Costs Hr.	Daily	Incl. Subs O&P Hr.	Daily	Cost Per Labor-Hour Bare Costs	Incl. O&P
1 Equip. Oper. (med.)	$32.60	$260.80	$49.35	$394.80	$29.95	$45.73
.5 Laborer	24.65	98.60	38.50	154.00		
1 Roll. Compact., 2K Lbs.		77.80		85.60	6.48	7.13
12 L.H., Daily Totals		$437.20		$634.40	$36.43	$52.86

Crew B-10B

Crew No.	Bare Costs Hr.	Daily	Incl. Subs O&P Hr.	Daily	Cost Per Labor-Hour Bare Costs	Incl. O&P
1 Equip. Oper. (med.)	$32.60	$260.80	$49.35	$394.80	$29.95	$45.73
.5 Laborer	24.65	98.60	38.50	154.00		
1 Dozer, 200 H.P.		880.80		968.90	73.40	80.74
12 L.H., Daily Totals		$1240.20		$1517.70	$103.35	$126.47

Crew B-10C

Crew No.	Bare Costs Hr.	Daily	Incl. Subs O&P Hr.	Daily	Cost Per Labor-Hour Bare Costs	Incl. O&P
1 Equip. Oper. (med.)	$32.60	$260.80	$49.35	$394.80	$29.95	$45.73
.5 Laborer	24.65	98.60	38.50	154.00		
1 Dozer, 200 H.P.		880.80		968.90		
1 Vibratory Roller, Towed		203.40		223.75	90.35	99.39
12 L.H., Daily Totals		$1443.60		$1741.45	$120.30	$145.12

Crew B-10D

Crew No.	Bare Costs Hr.	Daily	Incl. Subs O&P Hr.	Daily	Cost Per Labor-Hour Bare Costs	Incl. O&P
1 Equip. Oper. (med.)	$32.60	$260.80	$49.35	$394.80	$29.95	$45.73
.5 Laborer	24.65	98.60	38.50	154.00		
1 Dozer, 200 H.P.		880.80		968.90		
1 Sheepsft. Roller, Towed		80.40		88.45	80.10	88.11
12 L.H., Daily Totals		$1320.60		$1606.15	$110.05	$133.84

Crew B-10E

Crew No.	Bare Costs Hr.	Daily	Incl. Subs O&P Hr.	Daily	Cost Per Labor-Hour Bare Costs	Incl. O&P
1 Equip. Oper. (med.)	$32.60	$260.80	$49.35	$394.80	$29.95	$45.73
.5 Laborer	24.65	98.60	38.50	154.00		
1 Tandem Roller, 5 Ton		107.00		117.70	8.92	9.81
12 L.H., Daily Totals		$466.40		$666.50	$38.87	$55.54

CREWS

Crew No.	Bare Costs		Incl. Subs O & P		Cost Per Labor-Hour	
Crew B-10F	Hr.	Daily	Hr.	Daily	Bare Costs	Incl. O&P
1 Equip. Oper. (med.)	$32.60	$260.80	$49.35	$394.80	$29.95	$45.73
.5 Laborer	24.65	98.60	38.50	154.00		
1 Tandem Roller, 10 Ton		183.40		201.75	15.28	16.81
12 L.H., Daily Totals		$542.80		$750.55	$45.23	$62.54
Crew B-10G	Hr.	Daily	Hr.	Daily	Bare Costs	Incl. O&P
1 Equip. Oper. (med.)	$32.60	$260.80	$49.35	$394.80	$29.95	$45.73
.5 Laborer	24.65	98.60	38.50	154.00		
1 Sheepsft. Roll., 130 H.P.		762.80		839.10	63.57	69.92
12 L.H., Daily Totals		$1122.20		$1387.90	$93.52	$115.65
Crew B-10H	Hr.	Daily	Hr.	Daily	Bare Costs	Incl. O&P
1 Equip. Oper. (med.)	$32.60	$260.80	$49.35	$394.80	$29.95	$45.73
.5 Laborer	24.65	98.60	38.50	154.00		
1 Diaphr. Water Pump, 2"		44.40		48.85		
1-20 Ft. Suction Hose, 2"		4.35		4.80		
2-50 Ft. Disch. Hoses, 2"		7.40		8.15	4.68	5.15
12 L.H., Daily Totals		$415.55		$610.60	$34.63	$50.88
Crew B-10I	Hr.	Daily	Hr.	Daily	Bare Costs	Incl. O&P
1 Equip. Oper. (med.)	$32.60	$260.80	$49.35	$394.80	$29.95	$45.73
.5 Laborer	24.65	98.60	38.50	154.00		
1 Diaphr. Water Pump, 4"		85.20		93.70		
1-20 Ft. Suction Hose, 4"		8.90		9.80		
2-50 Ft. Disch. Hoses, 4"		12.50		13.75	8.88	9.77
12 L.H., Daily Totals		$466.00		$666.05	$38.83	$55.50
Crew B-10J	Hr.	Daily	Hr.	Daily	Bare Costs	Incl. O&P
1 Equip. Oper. (med.)	$32.60	$260.80	$49.35	$394.80	$29.95	$45.73
.5 Laborer	24.65	98.60	38.50	154.00		
1 Centr. Water Pump, 3"		51.60		56.75		
1-20 Ft. Suction Hose, 3"		6.65		7.30		
2-50 Ft. Disch. Hoses, 3"		9.10		10.00	5.61	6.17
12 L.H., Daily Totals		$426.75		$622.85	$35.56	$51.90
Crew B-10K	Hr.	Daily	Hr.	Daily	Bare Costs	Incl. O&P
1 Equip. Oper. (med.)	$32.60	$260.80	$49.35	$394.80	$29.95	$45.73
.5 Laborer	24.65	98.60	38.50	154.00		
1 Centr. Water Pump, 6"		224.40		246.85		
1-20 Ft. Suction Hose, 6"		17.10		18.80		
2-50 Ft. Disch. Hoses, 6"		29.40		32.35	22.58	24.83
12 L.H., Daily Totals		$630.30		$846.80	$52.53	$70.56
Crew B-10L	Hr.	Daily	Hr.	Daily	Bare Costs	Incl. O&P
1 Equip. Oper. (med.)	$32.60	$260.80	$49.35	$394.80	$29.95	$45.73
.5 Laborer	24.65	98.60	38.50	154.00		
1 Dozer, 75 H.P.		336.40		370.05	28.03	30.84
12 L.H., Daily Totals		$695.80		$918.85	$57.98	$76.57
Crew B-10M	Hr.	Daily	Hr.	Daily	Bare Costs	Incl. O&P
1 Equip. Oper. (med.)	$32.60	$260.80	$49.35	$394.80	$29.95	$45.73
.5 Laborer	24.65	98.60	38.50	154.00		
1 Dozer, 300 H.P.		1130.00		1243.00	94.17	103.58
12 L.H., Daily Totals		$1489.40		$1791.80	$124.12	$149.31
Crew B-10N	Hr.	Daily	Hr.	Daily	Bare Costs	Incl. O&P
1 Equip. Oper. (med.)	$32.60	$260.80	$49.35	$394.80	$29.95	$45.73
.5 Laborer	24.65	98.60	38.50	154.00		
1 F.E. Loader, T.M., 1.5 C.Y		302.20		332.40	25.18	27.70
12 L.H., Daily Totals		$661.60		$881.20	$55.13	$73.43

Crew No.	Bare Costs		Incl. Subs O & P		Cost Per Labor-Hour	
Crew B-10O	Hr.	Daily	Hr.	Daily	Bare Costs	Incl. O&P
1 Equip. Oper. (med.)	$32.60	$260.80	$49.35	$394.80	$29.95	$45.73
.5 Laborer	24.65	98.60	38.50	154.00		
1 F.E. Loader, T.M., 2.25 C.Y.		554.00		609.40	46.17	50.78
12 L.H., Daily Totals		$913.40		$1158.20	$76.12	$96.51
Crew B-10P	Hr.	Daily	Hr.	Daily	Bare Costs	Incl. O&P
1 Equip. Oper. (med.)	$32.60	$260.80	$49.35	$394.80	$29.95	$45.73
.5 Laborer	24.65	98.60	38.50	154.00		
1 F.E. Loader, T.M., 2.5 C.Y.		778.40		856.25	64.87	71.35
12 L.H., Daily Totals		$1137.80		$1405.05	$94.82	$117.08
Crew B-10Q	Hr.	Daily	Hr.	Daily	Bare Costs	Incl. O&P
1 Equip. Oper. (med.)	$32.60	$260.80	$49.35	$394.80	$29.95	$45.73
.5 Laborer	24.65	98.60	38.50	154.00		
1 F.E. Loader, T.M., 5 C.Y.		1092.00		1201.20	91.00	100.10
12 L.H., Daily Totals		$1451.40		$1750.00	$120.95	$145.83
Crew B-10R	Hr.	Daily	Hr.	Daily	Bare Costs	Incl. O&P
1 Equip. Oper. (med.)	$32.60	$260.80	$49.35	$394.80	$29.95	$45.73
.5 Laborer	24.65	98.60	38.50	154.00		
1 F.E. Loader, W.M., 1 C.Y.		202.20		222.40	16.85	18.54
12 L.H., Daily Totals		$561.60		$771.20	$46.80	$64.27
Crew B-10S	Hr.	Daily	Hr.	Daily	Bare Costs	Incl. O&P
1 Equip. Oper. (med.)	$32.60	$260.80	$49.35	$394.80	$29.95	$45.73
.5 Laborer	24.65	98.60	38.50	154.00		
1 F.E. Loader, W.M., 1.5 C.Y.		252.60		277.85	21.05	23.16
12 L.H., Daily Totals		$612.00		$826.65	$51.00	$68.89
Crew B-10T	Hr.	Daily	Hr.	Daily	Bare Costs	Incl. O&P
1 Equip. Oper. (med.)	$32.60	$260.80	$49.35	$394.80	$29.95	$45.73
.5 Laborer	24.65	98.60	38.50	154.00		
1 F.E. Loader, W.M.,2.5 C.Y.		313.00		344.30	26.08	28.69
12 L.H., Daily Totals		$672.40		$893.10	$56.03	$74.42
Crew B-10U	Hr.	Daily	Hr.	Daily	Bare Costs	Incl. O&P
1 Equip. Oper. (med.)	$32.60	$260.80	$49.35	$394.80	$29.95	$45.73
.5 Laborer	24.65	98.60	38.50	154.00		
1 F.E. Loader, W.M., 5.5 C.Y.		717.60		789.35	59.80	65.78
12 L.H., Daily Totals		$1077.00		$1338.15	$89.75	$111.51
Crew B-10V	Hr.	Daily	Hr.	Daily	Bare Costs	Incl. O&P
1 Equip. Oper. (med.)	$32.60	$260.80	$49.35	$394.80	$29.95	$45.73
.5 Laborer	24.65	98.60	38.50	154.00		
1 Dozer, 700 H.P.		3190.00		3509.00	265.83	292.42
12 L.H., Daily Totals		$3549.40		$4057.80	$295.78	$338.15
Crew B-10W	Hr.	Daily	Hr.	Daily	Bare Costs	Incl. O&P
1 Equip. Oper. (med.)	$32.60	$260.80	$49.35	$394.80	$29.95	$45.73
.5 Laborer	24.65	98.60	38.50	154.00		
1 Dozer, 105 H.P.		438.40		482.25	36.53	40.19
12 L.H., Daily Totals		$797.80		$1031.05	$66.48	$85.92
Crew B-10X	Hr.	Daily	Hr.	Daily	Bare Costs	Incl. O&P
1 Equip. Oper. (med.)	$32.60	$260.80	$49.35	$394.80	$29.95	$45.73
.5 Laborer	24.65	98.60	38.50	154.00		
1 Dozer, 410 H.P.		1518.00		1669.80	126.50	139.15
12 L.H., Daily Totals		$1877.40		$2218.60	$156.45	$184.88

Crew B-10Y

Crew No.	Bare Costs Hr.	Daily	Incl. Subs O & P Hr.	Daily	Cost Per Labor-Hour Bare Costs	Incl. O&P
1 Equip. Oper. (med.)	$32.60	$260.80	$49.35	$394.80	$29.95	$45.73
.5 Laborer	24.65	98.60	38.50	154.00		
1 Vibratory Drum Roller		332.80		366.10	27.73	30.51
12 L.H., Daily Totals		$692.20		$914.90	$57.68	$76.24

Crew B-11A

Crew No.	Bare Costs Hr.	Daily	Incl. Subs O & P Hr.	Daily	Cost Per Labor-Hour Bare Costs	Incl. O&P
1 Equipment Oper. (med.)	$32.60	$260.80	$49.35	$394.80	$28.63	$43.92
1 Laborer	24.65	197.20	38.50	308.00		
1 Dozer, 200 H.P.		880.80		968.90	55.05	60.56
16 L.H., Daily Totals		$1338.80		$1671.70	$83.68	$104.48

Crew B-11B

Crew No.	Bare Costs Hr.	Daily	Incl. Subs O & P Hr.	Daily	Cost Per Labor-Hour Bare Costs	Incl. O&P
1 Equipment Oper. (med.)	$32.60	$260.80	$49.35	$394.80	$28.63	$43.92
1 Laborer	24.65	197.20	38.50	308.00		
1 Dozer, 200 H.P.		880.80		968.90		
1 Air Powered Tamper		21.10		23.20		
1 Air Compr. 365 C.F.M.		147.60		162.35		
2-50 Ft. Air Hoses, 1.5" Dia.		10.00		11.00	66.22	72.84
16 L.H., Daily Totals		$1517.50		$1868.25	$94.85	$116.76

Crew B-11C

Crew No.	Bare Costs Hr.	Daily	Incl. Subs O & P Hr.	Daily	Cost Per Labor-Hour Bare Costs	Incl. O&P
1 Equipment Oper. (med.)	$32.60	$260.80	$49.35	$394.80	$28.63	$43.92
1 Laborer	24.65	197.20	38.50	308.00		
1 Backhoe Loader, 48 H.P.		218.80		240.70	13.68	15.04
16 L.H., Daily Totals		$676.80		$943.50	$42.31	$58.96

Crew B-11K

Crew No.	Bare Costs Hr.	Daily	Incl. Subs O & P Hr.	Daily	Cost Per Labor-Hour Bare Costs	Incl. O&P
1 Equipment Oper. (med.)	$32.60	$260.80	$49.35	$394.80	$28.63	$43.92
1 Laborer	24.65	197.20	38.50	308.00		
1 Trencher, 8' D., 16" W.		1469.00		1615.90	91.81	100.99
16 L.H., Daily Totals		$1927.00		$2318.70	$120.44	$144.91

Crew B-11L

Crew No.	Bare Costs Hr.	Daily	Incl. Subs O & P Hr.	Daily	Cost Per Labor-Hour Bare Costs	Incl. O&P
1 Equipment Oper. (med.)	$32.60	$260.80	$49.35	$394.80	$28.63	$43.92
1 Laborer	24.65	197.20	38.50	308.00		
1 Grader, 30,000 Lbs.		457.20		502.90	28.58	31.43
16 L.H., Daily Totals		$915.20		$1205.70	$57.21	$75.35

Crew B-11M

Crew No.	Bare Costs Hr.	Daily	Incl. Subs O & P Hr.	Daily	Cost Per Labor-Hour Bare Costs	Incl. O&P
1 Equipment Oper. (med.)	$32.60	$260.80	$49.35	$394.80	$28.63	$43.92
1 Laborer	24.65	197.20	38.50	308.00		
1 Backhoe Loader, 80 H.P.		244.20		268.60	15.26	16.79
16 L.H., Daily Totals		$702.20		$971.40	$43.89	$60.71

Crew B-11N

Crew No.	Bare Costs Hr.	Daily	Incl. Subs O & P Hr.	Daily	Cost Per Labor-Hour Bare Costs	Incl. O&P
1 Labor Foreman	$26.65	$213.20	$41.65	$333.20	$27.31	$41.69
2 Equipment Operators (med.)	32.60	521.60	49.35	789.60		
6 Truck Drivers (hvy.)	25.65	1231.20	39.15	1879.20		
1 F.E. Loader, 5.5 C.Y.		717.60		789.35		
1 Dozer, 400 H.P.		1518.00		1669.80		
6 Off Hwy. Tks. 50 Ton		7038.00		7741.80	128.80	141.68
72 L.H., Daily Totals		$11239.60		$13202.95	$156.11	$183.37

Crew B-11Q

Crew No.	Bare Costs Hr.	Daily	Incl. Subs O & P Hr.	Daily	Cost Per Labor-Hour Bare Costs	Incl. O&P
1 Equipment Operator (med.)	$32.60	$260.80	$49.35	$394.80	$29.95	$45.73
.5 Laborer	24.65	98.60	38.50	154.00		
1 Dozer, 140 H.P.		547.20		601.90	45.60	50.16
12 L.H., Daily Totals		$906.60		$1150.70	$75.55	$95.89

Crew B-11R

Crew No.	Bare Costs Hr.	Daily	Incl. Subs O & P Hr.	Daily	Cost Per Labor-Hour Bare Costs	Incl. O&P
1 Equipment Operator (med.)	$32.60	$260.80	$49.35	$394.80	$29.95	$45.73
.5 Laborer	24.65	98.60	38.50	154.00		
1 Dozer, 215 H.P.		880.80		968.90	73.40	80.74
12 L.H., Daily Totals		$1240.20		$1517.70	$103.35	$126.47

Crew B-11S

Crew No.	Bare Costs Hr.	Daily	Incl. Subs O & P Hr.	Daily	Cost Per Labor-Hour Bare Costs	Incl. O&P
1 Equipment Operator	$32.60	$260.80	$49.35	$394.80	$29.95	$45.73
.5 Laborer	24.65	98.60	38.50	154.00		
1 Dozer, 285 H.P.		1130.00		1243.00	94.17	103.58
12 L.H., Daily Totals		$1489.40		$1791.80	$124.12	$149.31

Crew B-11T

Crew No.	Bare Costs Hr.	Daily	Incl. Subs O & P Hr.	Daily	Cost Per Labor-Hour Bare Costs	Incl. O&P
1 Equipment Operator (med.)	$32.60	$260.80	$49.35	$394.80	$29.95	$45.73
.5 Laborer	24.65	98.60	38.50	154.00		
1 Dozer, 370 H.P.		1518.00		1669.80	126.50	139.15
12 L.H., Daily Totals		$1877.40		$2218.60	$156.45	$184.88

Crew B-11U

Crew No.	Bare Costs Hr.	Daily	Incl. Subs O & P Hr.	Daily	Cost Per Labor-Hour Bare Costs	Incl. O&P
1 Equipment Operator (med.)	$32.60	$260.80	$49.35	$394.80	$29.95	$45.73
.5 Laborer	24.65	98.60	38.50	154.00		
1 Dozer, 520 H.P.		1984.00		2182.40	165.33	181.87
12 L.H., Daily Totals		$2343.40		$2731.20	$195.28	$227.60

Crew B-11V

Crew No.	Bare Costs Hr.	Daily	Incl. Subs O & P Hr.	Daily	Cost Per Labor-Hour Bare Costs	Incl. O&P
3 Laborer	$24.65	$591.60	$38.50	$924.00	$24.65	$38.50
1 Roll. Compact., 2K Lbs.		77.80		85.60	3.24	3.57
24 L.H., Daily Totals		$669.40		$1009.60	$27.89	$42.07

Crew B-12A

Crew No.	Bare Costs Hr.	Daily	Incl. Subs O & P Hr.	Daily	Cost Per Labor-Hour Bare Costs	Incl. O&P
1 Equip. Oper. (crane)	$33.70	$269.60	$51.00	$408.00	$31.00	$46.92
1 Equip. Oper. Oiler	28.30	226.40	42.85	342.80		
1 Hyd. Excavator, 1 C.Y.		523.60		575.95	32.73	36.00
16 L.H., Daily Totals		$1019.60		$1326.75	$63.73	$82.92

Crew B-12B

Crew No.	Bare Costs Hr.	Daily	Incl. Subs O & P Hr.	Daily	Cost Per Labor-Hour Bare Costs	Incl. O&P
1 Equip. Oper. (crane)	$33.70	$269.60	$51.00	$408.00	$31.00	$46.92
1 Equip. Oper. Oiler	28.30	226.40	42.85	342.80		
1 Hyd. Excavator, 1.5 C.Y.		728.80		801.70	45.55	50.11
16 L.H., Daily Totals		$1224.80		$1552.50	$76.55	$97.03

Crew B-12C

Crew No.	Bare Costs Hr.	Daily	Incl. Subs O & P Hr.	Daily	Cost Per Labor-Hour Bare Costs	Incl. O&P
1 Equip. Oper. (crane)	$33.70	$269.60	$51.00	$408.00	$31.00	$46.92
1 Equip. Oper. Oiler	28.30	226.40	42.85	342.80		
1 Hyd. Excavator, 2 C.Y.		951.20		1046.30	59.45	65.40
16 L.H., Daily Totals		$1447.20		$1797.10	$90.45	$112.32

Crew B-12D

Crew No.	Bare Costs Hr.	Daily	Incl. Subs O & P Hr.	Daily	Cost Per Labor-Hour Bare Costs	Incl. O&P
1 Equip. Oper. (crane)	$33.70	$269.60	$51.00	$408.00	$31.00	$46.92
1 Equip. Oper. Oiler	28.30	226.40	42.85	342.80		
1 Hyd. Excavator, 3.5 C.Y.		2018.00		2219.80	126.13	138.74
16 L.H., Daily Totals		$2514.00		$2970.60	$157.13	$185.66

Crew B-12E

Crew No.	Bare Costs Hr.	Daily	Incl. Subs O & P Hr.	Daily	Cost Per Labor-Hour Bare Costs	Incl. O&P
1 Equip. Oper. (crane)	$33.70	$269.60	$51.00	$408.00	$31.00	$46.92
1 Equip. Oper. Oiler	28.30	226.40	42.85	342.80		
1 Hyd. Excavator, .5 C.Y.		341.60		375.75	21.35	23.49
16 L.H., Daily Totals		$837.60		$1126.55	$52.35	$70.41

Crew No.	Bare Costs		Incl. Subs O & P		Cost Per Labor-Hour	
Crew B-12F	Hr.	Daily	Hr.	Daily	Bare Costs	Incl. O&P
1 Equip. Oper. (crane)	$33.70	$269.60	$51.00	$408.00	$31.00	$46.92
1 Equip. Oper. Oiler	28.30	226.40	42.85	342.80		
1 Hyd. Excavator, .75 C.Y.		484.00		532.40	30.25	33.28
16 L.H., Daily Totals		$980.00		$1283.20	$61.25	$80.20
Crew B-12G	Hr.	Daily	Hr.	Daily	Bare Costs	Incl. O&P
1 Equip. Oper. (crane)	$33.70	$269.60	$51.00	$408.00	$31.00	$46.92
1 Equip. Oper. Oiler	28.30	226.40	42.85	342.80		
1 Power Shovel, .5 C.Y.		266.95		293.65		
1 Clamshell Bucket, .5 C.Y.		39.00		42.90	19.12	21.03
16 L.H., Daily Totals		$801.95		$1087.35	$50.12	$67.95
Crew B-12H	Hr.	Daily	Hr.	Daily	Bare Costs	Incl. O&P
1 Equip. Oper. (crane)	$33.70	$269.60	$51.00	$408.00	$31.00	$46.92
1 Equip. Oper. Oiler	28.30	226.40	42.85	342.80		
1 Power Shovel, 1 C.Y.		744.00		818.40		
1 Clamshell Bucket, 1 C.Y.		54.00		59.40	49.88	54.86
16 L.H., Daily Totals		$1294.00		$1628.60	$80.88	$101.78
Crew B-12I	Hr.	Daily	Hr.	Daily	Bare Costs	Incl. O&P
1 Equip. Oper. (crane)	$33.70	$269.60	$51.00	$408.00	$31.00	$46.92
1 Equip. Oper. Oiler	28.30	226.40	42.85	342.80		
1 Power Shovel, .75 C.Y.		575.50		633.05		
1 Dragline Bucket, .75 C.Y.		19.00		20.90	37.16	40.87
16 L.H., Daily Totals		$1090.50		$1404.75	$68.16	$87.79
Crew B-12J	Hr.	Daily	Hr.	Daily	Bare Costs	Incl. O&P
1 Equip. Oper. (crane)	$33.70	$269.60	$51.00	$408.00	$31.00	$46.92
1 Equip. Oper. Oiler	28.30	226.40	42.85	342.80		
1 Gradall, 3 Ton, .5 C.Y.		846.40		931.05	52.90	58.19
16 L.H., Daily Totals		$1342.40		$1681.85	$83.90	$105.11
Crew B-12K	Hr.	Daily	Hr.	Daily	Bare Costs	Incl. O&P
1 Equip. Oper. (crane)	$33.70	$269.60	$51.00	$408.00	$31.00	$46.92
1 Equip. Oper. Oiler	28.30	226.40	42.85	342.80		
1 Gradall, 3 Ton, 1 C.Y.		988.80		1087.70	61.80	67.98
16 L.H., Daily Totals		$1484.80		$1838.50	$92.80	$114.90
Crew B-12L	Hr.	Daily	Hr.	Daily	Bare Costs	Incl. O&P
1 Equip. Oper. (crane)	$33.70	$269.60	$51.00	$408.00	$31.00	$46.92
1 Equip. Oper. Oiler	28.30	226.40	42.85	342.80		
1 Power Shovel, .5 C.Y.		266.95		293.65		
1 F.E. Attachment, .5 C.Y.		45.20		49.70	19.51	21.46
16 L.H., Daily Totals		$808.15		$1094.15	$50.51	$68.38
Crew B-12M	Hr.	Daily	Hr.	Daily	Bare Costs	Incl. O&P
1 Equip. Oper. (crane)	$33.70	$269.60	$51.00	$408.00	$31.00	$46.92
1 Equip. Oper. Oiler	28.30	226.40	42.85	342.80		
1 Power Shovel, .75 C.Y.		575.50		633.05		
1 F.E. Attachment, .75 C.Y.		51.00		56.10	39.16	43.07
16 L.H., Daily Totals		$1122.50		$1439.95	$70.16	$89.99
Crew B-12N	Hr.	Daily	Hr.	Daily	Bare Costs	Incl. O&P
1 Equip. Oper. (crane)	$33.70	$269.60	$51.00	$408.00	$31.00	$46.92
1 Equip. Oper. Oiler	28.30	226.40	42.85	342.80		
1 Power Shovel, 1 C.Y.		744.00		818.40		
1 F.E. Attachment, 1 C.Y.		57.80		63.60	50.11	55.12
16 L.H., Daily Totals		$1297.80		$1632.80	$81.11	$102.04
Crew B-12O	Hr.	Daily	Hr.	Daily	Bare Costs	Incl. O&P
1 Equip. Oper. (crane)	$33.70	$269.60	$51.00	$408.00	$31.00	$46.92
1 Equip. Oper. Oiler	28.30	226.40	42.85	342.80		
1 Power Shovel, 1.5 C.Y.		935.60		1029.15		
1 F.E. Attachment, 1.5 C.Y.		92.60		101.85	64.26	70.69
16 L.H., Daily Totals		$1524.20		$1881.80	$95.26	$117.61
Crew B-12P	Hr.	Daily	Hr.	Daily	Bare Costs	Incl. O&P
1 Equip. Oper. (crane)	$33.70	$269.60	$51.00	$408.00	$31.00	$46.92
1 Equip. Oper. Oiler	28.30	226.40	42.85	342.80		
1 Crawler Crane, 40 Ton		935.60		1029.15		
1 Dragline Bucket, 1.5 C.Y.		31.20		34.30	60.43	66.47
16 L.H., Daily Totals		$1462.80		$1814.25	$91.43	$113.39
Crew B-12Q	Hr.	Daily	Hr.	Daily	Bare Costs	Incl. O&P
1 Equip. Oper. (crane)	$33.70	$269.60	$51.00	$408.00	$31.00	$46.92
1 Equip. Oper. Oiler	28.30	226.40	42.85	342.80		
1 Hyd. Excavator, 5/8 C.Y.		440.00		484.00	27.50	30.25
16 L.H., Daily Totals		$936.00		$1234.80	$58.50	$77.17
Crew B-12R	Hr.	Daily	Hr.	Daily	Bare Costs	Incl. O&P
1 Equip. Oper. (crane)	$33.70	$269.60	$51.00	$408.00	$31.00	$46.92
1 Equip. Oper. Oiler	28.30	226.40	42.85	342.80		
1 Hyd. Excavator, 1.5 C.Y.		728.80		801.70	45.55	50.11
16 L.H., Daily Totals		$1224.80		$1552.50	$76.55	$97.03
Crew B-12S	Hr.	Daily	Hr.	Daily	Bare Costs	Incl. O&P
1 Equip. Oper. (crane)	$33.70	$269.60	$51.00	$408.00	$31.00	$46.92
1 Equip. Oper. Oiler	28.30	226.40	42.85	342.80		
1 Hyd. Excavator, 2.5 C.Y.		1450.00		1595.00	90.63	99.69
16 L.H., Daily Totals		$1946.00		$2345.80	$121.63	$146.61
Crew B-12T	Hr.	Daily	Hr.	Daily	Bare Costs	Incl. O&P
1 Equip. Oper. (crane)	$33.70	$269.60	$51.00	$408.00	$31.00	$46.92
1 Equip. Oper. Oiler	28.30	226.40	42.85	342.80		
1 Crawler Crane, 75 Ton		1191.00		1310.10		
1 F.E. Attachment, 3 C.Y.		147.40		162.15	83.65	92.02
16 L.H., Daily Totals		$1834.40		$2223.05	$114.65	$138.94
Crew B-12V	Hr.	Daily	Hr.	Daily	Bare Costs	Incl. O&P
1 Equip. Oper. (crane)	$33.70	$269.60	$51.00	$408.00	$31.00	$46.92
1 Equip. Oper. Oiler	28.30	226.40	42.85	342.80		
1 Crawler Crane, 75 Ton		1191.00		1310.10		
1 Dragline Bucket, 3 C.Y.		57.60		63.35	78.04	85.84
16 L.H., Daily Totals		$1744.60		$2124.25	$109.04	$132.76
Crew B-13	Hr.	Daily	Hr.	Daily	Bare Costs	Incl. O&P
1 Labor Foreman (outside)	$26.65	$213.20	$41.65	$333.20	$26.75	$41.36
4 Laborers	24.65	788.80	38.50	1232.00		
1 Equip. Oper. (crane)	33.70	269.60	51.00	408.00		
1 Equip. Oper. Oiler	28.30	226.40	42.85	342.80		
1 Hyd. Crane, 25 Ton		654.60		720.05	11.69	12.86
56 L.H., Daily Totals		$2152.60		$3036.05	$38.44	$54.22

Crews

Crew No.	Bare Costs		Incl. Subs O & P		Cost Per Labor-Hour	
Crew B-13A	Hr.	Daily	Hr.	Daily	Bare Costs	Incl. O&P
1 Foreman	$26.65	$213.20	$41.65	$333.20	$27.49	$42.24
2 Laborers	24.65	394.40	38.50	616.00		
2 Equipment Operators	32.60	521.60	49.35	789.60		
2 Truck Drivers (heavy)	25.65	410.40	39.15	626.40		
1 Crane, 75 Ton		1191.00		1310.10		
1 F.E. Lder, 3.75 C.Y.		1092.00		1201.20		
2 Dump Trucks, 12 Ton		612.00		673.20	51.70	56.87
56 L.H., Daily Totals		$4434.60		$5549.70	$79.19	$99.11

Crew No.	Bare Costs		Incl. Subs O & P		Cost Per Labor-Hour	
Crew B-13B	Hr.	Daily	Hr.	Daily	Bare Costs	Incl. O&P
1 Labor Foreman (outside)	$26.65	$213.20	$41.65	$333.20	$26.75	$41.36
4 Laborers	24.65	788.80	38.50	1232.00		
1 Equip. Oper. (crane)	33.70	269.60	51.00	408.00		
1 Equip. Oper. Oiler	28.30	226.40	42.85	342.80		
1 Hyd. Crane, 55 Ton		1156.00		1271.60	20.64	22.71
56 L.H., Daily Totals		$2654.00		$3587.60	$47.39	$64.07

Crew No.	Bare Costs		Incl. Subs O & P		Cost Per Labor-Hour	
Crew B-13C	Hr.	Daily	Hr.	Daily	Bare Costs	Incl. O&P
1 Labor Foreman (outside)	$26.65	$213.20	$41.65	$333.20	$26.75	$41.36
4 Laborers	24.65	788.80	38.50	1232.00		
1 Equip. Oper. (crane)	33.70	269.60	51.00	408.00		
1 Equip. Oper. Oiler	28.30	226.40	42.85	342.80		
1 Crawler Crane, 100 Ton		1546.00		1700.60	27.61	30.37
56 L.H., Daily Totals		$3044.00		$4016.60	$54.36	$71.73

Crew No.	Bare Costs		Incl. Subs O & P		Cost Per Labor-Hour	
Crew B-14	Hr.	Daily	Hr.	Daily	Bare Costs	Incl. O&P
1 Labor Foreman (outside)	$26.65	$213.20	$41.65	$333.20	$26.06	$40.46
4 Laborers	24.65	788.80	38.50	1232.00		
1 Equip. Oper. (light)	31.10	248.80	47.10	376.80		
1 Backhoe Loader, 48 H.P.		218.80		240.70	4.56	5.01
48 L.H., Daily Totals		$1469.60		$2182.70	$30.62	$45.47

Crew No.	Bare Costs		Incl. Subs O & P		Cost Per Labor-Hour	
Crew B-15	Hr.	Daily	Hr.	Daily	Bare Costs	Incl. O&P
1 Equipment Oper. (med)	$32.60	$260.80	$49.35	$394.80	$27.49	$41.97
.5 Laborer	24.65	98.60	38.50	154.00		
2 Truck Drivers (heavy)	25.65	410.40	39.15	626.40		
2 Dump Trucks, 16 Ton		899.20		989.10		
1 Dozer, 200 H.P.		880.80		968.90	63.57	69.93
28 L.H., Daily Totals		$2549.80		$3133.20	$91.06	$111.90

Crew No.	Bare Costs		Incl. Subs O & P		Cost Per Labor-Hour	
Crew B-16	Hr.	Daily	Hr.	Daily	Bare Costs	Incl. O&P
1 Labor Foreman (outside)	$26.65	$213.20	$41.65	$333.20	$25.40	$39.45
2 Laborers	24.65	394.40	38.50	616.00		
1 Truck Driver (heavy)	25.65	205.20	39.15	313.20		
1 Dump Truck, 16 Ton		449.60		494.55	14.05	15.46
32 L.H., Daily Totals		$1262.40		$1756.95	$39.45	$54.91

Crew No.	Bare Costs		Incl. Subs O & P		Cost Per Labor-Hour	
Crew B-17	Hr.	Daily	Hr.	Daily	Bare Costs	Incl. O&P
2 Laborers	$24.65	$394.40	$38.50	$616.00	$26.51	$40.81
1 Equip. Oper. (light)	31.10	248.80	47.10	376.80		
1 Truck Driver (heavy)	25.65	205.20	39.15	313.20		
1 Backhoe Loader, 48 H.P.		218.80		240.70		
1 Dump Truck, 12 Ton		306.00		336.60	16.40	18.04
32 L.H., Daily Totals		$1373.20		$1883.30	$42.91	$58.85

Crew No.	Bare Costs		Incl. Subs O & P		Cost Per Labor-Hour	
Crew B-18	Hr.	Daily	Hr.	Daily	Bare Costs	Incl. O&P
1 Labor Foreman (outside)	$26.65	$213.20	$41.65	$333.20	$25.32	$39.55
2 Laborers	24.65	394.40	38.50	616.00		
1 Vibrating Compactor		53.20		58.50	2.22	2.44
24 L.H., Daily Totals		$660.80		$1007.70	$27.54	$41.99

Crew No.	Bare Costs		Incl. Subs O & P		Cost Per Labor-Hour	
Crew B-19	Hr.	Daily	Hr.	Daily	Bare Costs	Incl. O&P
1 Pile Driver Foreman	$32.90	$263.20	$55.20	$441.60	$31.52	$50.93
4 Pile Drivers	30.90	988.80	51.85	1659.20		
2 Equip. Oper. (crane)	33.70	539.20	51.00	816.00		
1 Equip. Oper. Oiler	28.30	226.40	42.85	342.80		
1 Crane, 40 Ton & Access.		935.60		1029.15		
60 L.F. Leads, 15K Ft. Lbs.		105.00		115.50		
1 Hammer, 15K Ft. Lbs.		367.20		403.90		
1 Air Compr., 600 C.F.M.		249.20		274.10		
2-50 Ft. Air Hoses, 3" Dia.		35.40		38.95	26.44	29.09
64 L.H., Daily Totals		$3710.00		$5121.20	$57.96	$80.02

Crew No.	Bare Costs		Incl. Subs O & P		Cost Per Labor-Hour	
Crew B-19A	Hr.	Daily	Hr.	Daily	Bare Costs	Incl. O&P
1 Pile Driver Foreman	$32.90	$263.20	$55.20	$441.60	$31.52	$50.93
4 Pile Drivers	30.90	988.80	51.85	1659.20		
2 Equip. Oper. (crane)	33.70	539.20	51.00	816.00		
1 Equip. Oper. Oiler	28.30	226.40	42.85	342.80		
1 Crawler Crane, 75 Ton		1191.00		1310.10		
60 L.F. Leads, 25K Ft. Lbs.		144.00		158.40		
1 Air Compressor, 750 CFM		281.60		309.75		
4-50 Ft. Air Hoses, 3" Dia.		70.80		77.90	26.37	29.00
64 L.H., Daily Totals		$3705.00		$5115.75	$57.89	$79.93

Crew No.	Bare Costs		Incl. Subs O & P		Cost Per Labor-Hour	
Crew B-20	Hr.	Daily	Hr.	Daily	Bare Costs	Incl. O&P
1 Labor Foreman (out)	$26.65	$213.20	$41.65	$333.20	$27.85	$43.55
1 Skilled Worker	32.25	258.00	50.50	404.00		
1 Laborer	24.65	197.20	38.50	308.00		
24 L.H., Daily Totals		$668.40		$1045.20	$27.85	$43.55

Crew No.	Bare Costs		Incl. Subs O & P		Cost Per Labor-Hour	
Crew B-20A	Hr.	Daily	Hr.	Daily	Bare Costs	Incl. O&P
1 Labor Foreman	$26.65	$213.20	$41.65	$333.20	$29.64	$45.43
1 Laborer	24.65	197.20	38.50	308.00		
1 Plumber	37.35	298.80	56.40	451.20		
1 Plumber Apprentice	29.90	239.20	45.15	361.20		
32 L.H., Daily Totals		$948.40		$1453.60	$29.64	$45.43

Crew No.	Bare Costs		Incl. Subs O & P		Cost Per Labor-Hour	
Crew B-21	Hr.	Daily	Hr.	Daily	Bare Costs	Incl. O&P
1 Labor Foreman (out)	$26.65	$213.20	$41.65	$333.20	$28.69	$44.61
1 Skilled Worker	32.25	258.00	50.50	404.00		
1 Laborer	24.65	197.20	38.50	308.00		
.5 Equip. Oper. (crane)	33.70	134.80	51.00	204.00		
.5 S.P. Crane, 5 Ton		163.20		179.50	5.83	6.41
28 L.H., Daily Totals		$966.40		$1428.70	$34.52	$51.02

Crew No.	Bare Costs		Incl. Subs O & P		Cost Per Labor-Hour	
Crew B-21A	Hr.	Daily	Hr.	Daily	Bare Costs	Incl. O&P
1 Labor Foreman	$26.65	$213.20	$41.65	$333.20	$30.45	$46.54
1 Laborer	24.65	197.20	38.50	308.00		
1 Plumber	37.35	298.80	56.40	451.20		
1 Plumber Apprentice	29.90	239.20	45.15	361.20		
1 Equip. Oper. (crane)	33.70	269.60	51.00	408.00		
1 S.P. Crane, 12 Ton		493.40		542.75	12.34	13.57
40 L.H., Daily Totals		$1711.40		$2404.35	$42.79	$60.11

Crew No.	Bare Costs		Incl. Subs O & P		Cost Per Labor-Hour	
Crew B-22	Hr.	Daily	Hr.	Daily	Bare Costs	Incl. O&P
1 Labor Foreman (out)	$26.65	$213.20	$41.65	$333.20	$29.02	$45.04
1 Skilled Worker	32.25	258.00	50.50	404.00		
1 Laborer	24.65	197.20	38.50	308.00		
.75 Equip. Oper. (crane)	33.70	202.20	51.00	306.00		
.75 S.P. Crane, 5 Ton		244.80		269.30	8.16	8.98
30 L.H., Daily Totals		$1115.40		$1620.50	$37.18	$54.02

Crew No.	Bare Costs		Incl. Subs O & P		Cost Per Labor-Hour	

Crew B-22A

Crew B-22A	Hr.	Daily	Hr.	Daily	Bare Costs	Incl. O&P
1 Labor Foreman (out)	$26.65	$213.20	$41.65	$333.20	$28.10	$43.66
1 Skilled Worker	32.25	258.00	50.50	404.00		
2 Laborers	24.65	394.40	38.50	616.00		
.75 Equipment Oper. (crane)	33.70	202.20	51.00	306.00		
.75 Crane, 5 Ton		244.80		269.30		
1 Generator, 5 KW		39.40		43.35		
1 Butt Fusion Machine		426.40		469.05	18.70	20.57
38 L.H., Daily Totals		$1778.40		$2440.90	$46.80	$64.23

Crew B-22B

Crew B-22B	Hr.	Daily	Hr.	Daily	Bare Costs	Incl. O&P
1 Skilled Worker	$32.25	$258.00	$50.50	$404.00	$28.45	$44.50
1 Laborer	24.65	197.20	38.50	308.00		
1 Electro Fusion Machine		170.00		187.00	10.63	11.69
16 L.H., Daily Totals		$625.20		$899.00	$39.08	$56.19

Crew B-23

Crew B-23	Hr.	Daily	Hr.	Daily	Bare Costs	Incl. O&P
1 Labor Foreman (outside)	$26.65	$213.20	$41.65	$333.20	$25.05	$39.13
4 Laborers	24.65	788.80	38.50	1232.00		
1 Drill Rig, Wells		2295.00		2524.50		
1 Light Truck, 3 Ton		138.00		151.80	60.83	66.91
40 L.H., Daily Totals		$3435.00		$4241.50	$85.88	$106.04

Crew B-23A

Crew B-23A	Hr.	Daily	Hr.	Daily	Bare Costs	Incl. O&P
1 Labor Foreman (outside)	$26.65	$213.20	$41.65	$333.20	$27.97	$43.17
1 Laborer	24.65	197.20	38.50	308.00		
1 Equip. Operator (medium)	32.60	260.80	49.35	394.80		
1 Drill Rig, Wells		2295.00		2524.50		
1 Pickup Truck, 3/4 Ton		67.80		74.60	98.45	108.30
24 L.H., Daily Totals		$3034.00		$3635.10	$126.42	$151.47

Crew B-23B

Crew B-23B	Hr.	Daily	Hr.	Daily	Bare Costs	Incl. O&P
1 Labor Foreman (outside)	$26.65	$213.20	$41.65	$333.20	$27.97	$43.17
1 Laborer	24.65	197.20	38.50	308.00		
1 Equip. Operator (medium)	32.60	260.80	49.35	394.80		
1 Drill Rig, Wells		2295.00		2524.50		
1 Pickup Truck, 3/4 Ton		67.80		74.60		
1 Pump, Cntfgl, 6"		224.40		246.85	107.80	118.58
24 L.H., Daily Totals		$3258.40		$3881.95	$135.77	$161.75

Crew B-24

Crew B-24	Hr.	Daily	Hr.	Daily	Bare Costs	Incl. O&P
1 Cement Finisher	$30.20	$241.60	$44.75	$358.00	$28.80	$44.18
1 Laborer	24.65	197.20	38.50	308.00		
1 Carpenter	31.55	252.40	49.30	394.40		
24 L.H., Daily Totals		$691.20		$1060.40	$28.80	$44.18

Crew B-25

Crew B-25	Hr.	Daily	Hr.	Daily	Bare Costs	Incl. O&P
1 Labor Foreman	$26.65	$213.20	$41.65	$333.20	$27.00	$41.75
7 Laborers	24.65	1380.40	38.50	2156.00		
3 Equip. Oper. (med.)	32.60	782.40	49.35	1184.40		
1 Asphalt Paver, 130 H.P		1465.00		1611.50		
1 Tandem Roller, 10 Ton		183.40		201.75		
1 Roller, Pneumatic Wheel		247.00		271.70	21.54	23.69
88 L.H., Daily Totals		$4271.40		$5758.55	$48.54	$65.44

Crew B-25B

Crew B-25B	Hr.	Daily	Hr.	Daily	Bare Costs	Incl. O&P
1 Labor Foreman	$26.65	$213.20	$41.65	$333.20	$27.47	$42.38
7 Laborers	24.65	1380.40	38.50	2156.00		
4 Equip. Oper. (medium)	32.60	1043.20	49.35	1579.20		
1 Asphalt Paver, 130 H.P.		1465.00		1611.50		
2 Rollers, Steel Wheel		366.80		403.50		
1 Roller, Pneumatic Wheel		247.00		271.70	21.65	23.82
96 L.H., Daily Totals		$4715.60		$6355.10	$49.12	$66.20

Crew B-25C

Crew B-25C	Hr.	Daily	Hr.	Daily	Bare Costs	Incl. O&P
1 Labor Foreman	$26.65	$213.20	$41.65	$333.20	$27.63	$42.64
3 Laborers	24.65	591.60	38.50	924.00		
2 Equip. Oper. (medium)	32.60	521.60	49.35	789.60		
1 Asphalt Paver, 130 H.P.		1465.00		1611.50		
1 Rollers, Steel Wheel		183.40		201.75	34.34	37.78
48 L.H., Daily Totals		$2974.80		$3860.05	$61.97	$80.42

Crew B-26

Crew B-26	Hr.	Daily	Hr.	Daily	Bare Costs	Incl. O&P
1 Labor Foreman (outside)	$26.65	$213.20	$41.65	$333.20	$27.77	$43.25
6 Laborers	24.65	1183.20	38.50	1848.00		
2 Equip. Oper. (med.)	32.60	521.60	49.35	789.60		
1 Rodman (reinf.)	35.55	284.40	59.70	477.60		
1 Cement Finisher	30.20	241.60	44.75	358.00		
1 Grader, 30,000 Lbs.		457.20		502.90		
1 Paving Mach. & Equip.		1647.00		1811.70	23.91	26.30
88 L.H., Daily Totals		$4548.20		$6121.00	$51.68	$69.55

Crew B-27

Crew B-27	Hr.	Daily	Hr.	Daily	Bare Costs	Incl. O&P
1 Labor Foreman (outside)	$26.65	$213.20	$41.65	$333.20	$25.15	$39.29
3 Laborers	24.65	591.60	38.50	924.00		
1 Berm Machine		167.40		184.15	5.23	5.75
32 L.H., Daily Totals		$972.20		$1441.35	$30.38	$45.04

Crew B-28

Crew B-28	Hr.	Daily	Hr.	Daily	Bare Costs	Incl. O&P
2 Carpenters	$31.55	$504.80	$49.30	$788.80	$29.25	$45.70
1 Laborer	24.65	197.20	38.50	308.00		
24 L.H., Daily Totals		$702.00		$1096.80	$29.25	$45.70

Crew B-29

Crew B-29	Hr.	Daily	Hr.	Daily	Bare Costs	Incl. O&P
1 Labor Foreman (outside)	$26.65	$213.20	$41.65	$333.20	$26.75	$41.36
4 Laborers	24.65	788.80	38.50	1232.00		
1 Equip. Oper. (crane)	33.70	269.60	51.00	408.00		
1 Equip. Oper. Oiler	28.30	226.40	42.85	342.80		
1 Gradall, 3 Ton, 1/2 C.Y.		846.40		931.05	15.11	16.63
56 L.H., Daily Totals		$2344.40		$3247.05	$41.86	$57.99

Crew B-30

Crew B-30	Hr.	Daily	Hr.	Daily	Bare Costs	Incl. O&P
1 Equip. Oper. (med.)	$32.60	$260.80	$49.35	$394.80	$27.97	$42.55
2 Truck Drivers (heavy)	25.65	410.40	39.15	626.40		
1 Hyd. Excavator, 1.5 C.Y.		728.80		801.70		
2 Dump Trucks, 16 Ton		899.20		989.10	67.83	74.62
24 L.H., Daily Totals		$2299.20		$2812.00	$95.80	$117.17

Crew B-31

Crew B-31	Hr.	Daily	Hr.	Daily	Bare Costs	Incl. O&P
1 Labor Foreman (outside)	$26.65	$213.20	$41.65	$333.20	$26.43	$41.29
3 Laborers	24.65	591.60	38.50	924.00		
1 Carpenter	31.55	252.40	49.30	394.40		
1 Air Compr., 250 C.F.M.		128.60		141.45		
1 Sheeting Driver		8.80		9.70		
2-50 Ft. Air Hoses, 1.5" Dia.		10.00		11.00	3.69	4.05
40 L.H., Daily Totals		$1204.60		$1813.75	$30.12	$45.34

Crews

Crew B-32

Crew No.	Bare Costs Hr.	Daily	Incl. Subs O & P Hr.	Daily	Cost Per Labor-Hour Bare Costs	Incl. O&P
1 Laborer	$24.65	$197.20	$38.50	$308.00	$30.61	$46.64
3 Equip. Oper. (med.)	32.60	782.40	49.35	1184.40		
1 Grader, 30,000 Lbs.		457.20		502.90		
1 Tandem Roller, 10 Ton		183.40		201.75		
1 Dozer, 200 H.P.		880.80		968.90	47.54	52.30
32 L.H., Daily Totals		$2501.00		$3165.95	$78.15	$98.94

Crew B-32A

Crew No.	Bare Costs Hr.	Daily	Incl. Subs O & P Hr.	Daily	Cost Per Labor-Hour Bare Costs	Incl. O&P
1 Laborer	$24.65	$197.20	$38.50	$308.00	$29.95	$45.73
2 Equip. Oper. (medium)	32.60	521.60	49.35	789.60		
1 Grader, 30,000 Lbs.		457.20		502.90		
1 Roller, Vibratory, 29,000 Lbs.		447.60		492.35	37.70	41.47
24 L.H., Daily Totals		$1623.60		$2092.85	$67.65	$87.20

Crew B-32B

Crew No.	Bare Costs Hr.	Daily	Incl. Subs O & P Hr.	Daily	Cost Per Labor-Hour Bare Costs	Incl. O&P
1 Laborer	$24.65	$197.20	$38.50	$308.00	$29.95	$45.73
2 Equip. Oper. (medium)	32.60	521.60	49.35	789.60		
1 Dozer, 200 H.P.		880.80		968.90		
1 Roller, Vibratory, 29,000 Lbs.		447.60		492.35	55.35	60.89
24 L.H., Daily Totals		$2047.20		$2558.85	$85.30	$106.62

Crew B-32C

Crew No.	Bare Costs Hr.	Daily	Incl. Subs O & P Hr.	Daily	Cost Per Labor-Hour Bare Costs	Incl. O&P
1 Labor Foreman	$26.65	$213.20	$41.65	$333.20	$28.96	$44.45
2 Laborers	24.65	394.40	38.50	616.00		
3 Equip. Oper. (medium)	32.60	782.40	49.35	1184.40		
1 Grader, 30,000 Lbs.		457.20		502.90		
1 Roller, Steel Wheel		183.40		201.75		
1 Dozer, 200 H.P.		880.80		968.90	31.70	34.87
48 L.H., Daily Totals		$2911.40		$3807.15	$60.66	$79.32

Crew B-33A

Crew No.	Bare Costs Hr.	Daily	Incl. Subs O & P Hr.	Daily	Cost Per Labor-Hour Bare Costs	Incl. O&P
1 Equip. Oper. (med.)	$32.60	$260.80	$49.35	$394.80	$30.33	$46.25
.5 Laborer	24.65	98.60	38.50	154.00		
.25 Equip. Oper. (med.)	32.60	65.20	49.35	98.70		
1 Scraper, Towed, 7 C.Y.		139.55		153.50		
1.25 Dozer, 300 H.P.		1412.50		1553.75	110.86	121.95
14 L.H., Daily Totals		$1976.65		$2354.75	$141.19	$168.20

Crew B-33B

Crew No.	Bare Costs Hr.	Daily	Incl. Subs O & P Hr.	Daily	Cost Per Labor-Hour Bare Costs	Incl. O&P
1 Equip. Oper. (med.)	$32.60	$260.80	$49.35	$394.80	$30.33	$46.25
.5 Laborer	24.65	98.60	38.50	154.00		
.25 Equip. Oper. (med.)	32.60	65.20	49.35	98.70		
1 Scraper, Towed, 10 C.Y.		157.30		173.05		
1.25 Dozer, 300 H.P.		1412.50		1553.75	112.13	123.34
14 L.H., Daily Totals		$1994.40		$2374.30	$142.46	$169.59

Crew B-33C

Crew No.	Bare Costs Hr.	Daily	Incl. Subs O & P Hr.	Daily	Cost Per Labor-Hour Bare Costs	Incl. O&P
1 Equip. Oper. (med.)	$32.60	$260.80	$49.35	$394.80	$30.33	$46.25
.5 Laborer	24.65	98.60	38.50	154.00		
.25 Equip. Oper. (med.)	32.60	65.20	49.35	98.70		
1 Scraper, Towed, 12 C.Y.		157.30		173.05		
1.25 Dozer, 300 H.P.		1412.50		1553.75	112.13	123.34
14 L.H., Daily Totals		$1994.40		$2374.30	$142.46	$169.59

Crew B-33D

Crew No.	Bare Costs Hr.	Daily	Incl. Subs O & P Hr.	Daily	Cost Per Labor-Hour Bare Costs	Incl. O&P
1 Equip. Oper. (med.)	$32.60	$260.80	$49.35	$394.80	$30.33	$46.25
.5 Laborer	24.65	98.60	38.50	154.00		
.25 Equip. Oper. (med.)	32.60	65.20	49.35	98.70		
1 S.P. Scraper, 14 C.Y.		1537.00		1690.70		
.25 Dozer, 300 H.P.		282.50		310.75	129.96	142.96
14 L.H., Daily Totals		$2244.10		$2648.95	$160.29	$189.21

Crew B-33E

Crew No.	Bare Costs Hr.	Daily	Incl. Subs O & P Hr.	Daily	Cost Per Labor-Hour Bare Costs	Incl. O&P
1 Equip. Oper. (med.)	$32.60	$260.80	$49.35	$394.80	$30.33	$46.25
.5 Laborer	24.65	98.60	38.50	154.00		
.25 Equip. Oper. (med.)	32.60	65.20	49.35	98.70		
1 S.P. Scraper, 24 C.Y.		2378.00		2615.80		
.25 Dozer, 300 H.P.		282.50		310.75	190.04	209.04
14 L.H., Daily Totals		$3085.10		$3574.05	$220.37	$255.29

Crew B-33F

Crew No.	Bare Costs Hr.	Daily	Incl. Subs O & P Hr.	Daily	Cost Per Labor-Hour Bare Costs	Incl. O&P
1 Equip. Oper. (med.)	$32.60	$260.80	$49.35	$394.80	$30.33	$46.25
.5 Laborer	24.65	98.60	38.50	154.00		
.25 Equip. Oper. (med.)	32.60	65.20	49.35	98.70		
1 Elev. Scraper, 11 C.Y.		836.40		920.05		
.25 Dozer, 300 H.P.		282.50		310.75	79.92	87.91
14 L.H., Daily Totals		$1543.50		$1878.30	$110.25	$134.16

Crew B-33G

Crew No.	Bare Costs Hr.	Daily	Incl. Subs O & P Hr.	Daily	Cost Per Labor-Hour Bare Costs	Incl. O&P
1 Equip. Oper. (med.)	$32.60	$260.80	$49.35	$394.80	$30.33	$46.25
.5 Laborer	24.65	98.60	38.50	154.00		
.25 Equip. Oper. (med.)	32.60	65.20	49.35	98.70		
1 Elev. Scraper, 20 C.Y.		1658.00		1823.80		
.25 Dozer, 300 H.P.		282.50		310.75	138.61	152.47
14 L.H., Daily Totals		$2365.10		$2782.05	$168.94	$198.72

Crew B-33H

Crew No.	Bare Costs Hr.	Daily	Incl. Subs O & P Hr.	Daily	Cost Per Labor-Hour Bare Costs	Incl. O&P
.25 Laborer	$24.65	$49.30	$38.50	$77.00	$31.23	$47.48
1 Equipment Operator (med.)	32.60	260.80	49.35	394.80		
.2 Equipment Operator (med.)	32.60	52.16	49.35	78.96		
1 Scraper, 32-44 C.Y.		2896.00		3185.60		
.2 Dozer, 298 kw		303.60		333.95	275.83	303.41
11. L.H., Daily Totals		$3561.86		$4070.31	$307.06	$350.89

Crew B-33J

Crew No.	Bare Costs Hr.	Daily	Incl. Subs O & P Hr.	Daily	Cost Per Labor-Hour Bare Costs	Incl. O&P
1 Equipment Operator (med.)	$32.60	$260.80	$49.35	$394.80	$32.60	$49.35
1 Scraper 17 C.Y.		1537.00		1690.70	192.13	211.34
8 L.H., Daily Totals		$1797.80		$2085.50	$224.73	$260.69

Crew B-34A

Crew No.	Bare Costs Hr.	Daily	Incl. Subs O & P Hr.	Daily	Cost Per Labor-Hour Bare Costs	Incl. O&P
1 Truck Driver (heavy)	$25.65	$205.20	$39.15	$313.20	$25.65	$39.15
1 Dump Truck, 12 Ton		306.00		336.60	38.25	42.08
8 L.H., Daily Totals		$511.20		$649.80	$63.90	$81.23

Crew B-34B

Crew No.	Bare Costs Hr.	Daily	Incl. Subs O & P Hr.	Daily	Cost Per Labor-Hour Bare Costs	Incl. O&P
1 Truck Driver (heavy)	$25.65	$205.20	$39.15	$313.20	$25.65	$39.15
1 Dump Truck, 16 Ton		449.60		494.55	56.20	61.82
8 L.H., Daily Totals		$654.80		$807.75	$81.85	$100.97

Crew B-34C

Crew No.	Bare Costs Hr.	Daily	Incl. Subs O & P Hr.	Daily	Cost Per Labor-Hour Bare Costs	Incl. O&P
1 Truck Driver (heavy)	$25.65	$205.20	$39.15	$313.20	$25.65	$39.15
1 Truck Tractor, 40 Ton		330.80		363.90		
1 Dump Trailer, 16.5 C.Y.		106.60		117.25	54.68	60.14
8 L.H., Daily Totals		$642.60		$794.35	$80.33	$99.29

Crew B-34D

Crew No.	Bare Costs Hr.	Daily	Incl. Subs O & P Hr.	Daily	Cost Per Labor-Hour Bare Costs	Incl. O&P
1 Truck Driver (heavy)	$25.65	$205.20	$39.15	$313.20	$25.65	$39.15
1 Truck Tractor, 40 Ton		330.80		363.90		
1 Dump Trailer, 20 C.Y.		119.60		131.55	56.30	61.93
8 L.H., Daily Totals		$655.60		$808.65	$81.95	$101.08

557

Crew No.	Bare Costs		Incl. Subs O & P		Cost Per Labor-Hour	

Crew B-34E

	Hr.	Daily	Hr.	Daily	Bare Costs	Incl. O&P
1 Truck Driver (heavy)	$25.65	$205.20	$39.15	$313.20	$25.65	$39.15
1 Truck, Off Hwy., 25 Ton		914.00		1005.40	114.25	125.68
8 L.H., Daily Totals		$1119.20		$1318.60	$139.90	$164.83

Crew B-34F

	Hr.	Daily	Hr.	Daily	Bare Costs	Incl. O&P
1 Truck Driver (heavy)	$25.65	$205.20	$39.15	$313.20	$25.65	$39.15
1 Truck, Off Hwy., 22 C.Y.		940.40		1034.45	117.55	129.31
8 L.H., Daily Totals		$1145.60		$1347.65	$143.20	$168.46

Crew B-34G

	Hr.	Daily	Hr.	Daily	Bare Costs	Incl. O&P
1 Truck Driver (heavy)	$25.65	$205.20	$39.15	$313.20	$25.65	$39.15
1 Truck, Off Hwy., 34 C.Y.		1173.00		1290.30	146.63	161.29
8 L.H., Daily Totals		$1378.20		$1603.50	$172.28	$200.44

Crew B-34H

	Hr.	Daily	Hr.	Daily	Bare Costs	Incl. O&P
1 Truck Driver (heavy)	$25.65	$205.20	$39.15	$313.20	$25.65	$39.15
1 Truck, Off Hwy., 42 C.Y.		1257.00		1382.70	157.13	172.84
8 L.H., Daily Totals		$1462.20		$1695.90	$182.78	$211.99

Crew B-34J

	Hr.	Daily	Hr.	Daily	Bare Costs	Incl. O&P
1 Truck Driver (heavy)	$25.65	$205.20	$39.15	$313.20	$25.65	$39.15
1 Truck, Off Hwy., 60 C.Y.		1812.00		1993.20	226.50	249.15
8 L.H., Daily Totals		$2017.20		$2306.40	$252.15	$288.30

Crew B-34K

	Hr.	Daily	Hr.	Daily	Bare Costs	Incl. O&P
1 Truck Driver (heavy)	$25.65	$205.20	$39.15	$313.20	$25.65	$39.15
1 Truck Tractor, 240 H.P.		376.00		413.60		
1 Low Bed Trailer		221.60		243.75	74.70	82.17
8 L.H., Daily Totals		$802.80		$970.55	$100.35	$121.32

Crew B-34N

	Hr.	Daily	Hr.	Daily	Bare Costs	Incl. O&P
1 Truck Driver (heavy)	$25.65	$205.20	$39.15	$313.20	$25.65	$39.15
1 Dump Truck, 12 Ton		306.00		336.60		
1 Flatbed Trailer, 40 Ton		143.00		157.30	56.13	61.74
8 L.H., Daily Totals		$654.20		$807.10	$81.78	$100.89

Crew B-35

	Hr.	Daily	Hr.	Daily	Bare Costs	Incl. O&P
1 Laborer Foreman (out)	$26.65	$213.20	$41.65	$333.20	$30.48	$46.82
1 Skilled Worker	32.25	258.00	50.50	404.00		
1 Welder (plumber)	37.35	298.80	56.40	451.20		
1 Laborer	24.65	197.20	38.50	308.00		
1 Equip. Oper. (crane)	33.70	269.60	51.00	408.00		
1 Equip. Oper. Oiler	28.30	226.40	42.85	342.80		
1 Electric Welding Mach.		78.50		86.35		
1 Hyd. Excavator, .75 C.Y.		484.00		532.40	11.72	12.89
48 L.H., Daily Totals		$2025.70		$2865.95	$42.20	$59.71

Crew B-35A

	Hr.	Daily	Hr.	Daily	Bare Costs	Incl. O&P
1 Laborer Foreman (out)	$26.65	$213.20	$41.65	$333.20	$29.65	$45.63
2 Laborers	24.65	394.40	38.50	616.00		
1 Skilled Worker	32.25	258.00	50.50	404.00		
1 Welder (plumber)	37.35	298.80	56.40	451.20		
1 Equip. Oper. (crane)	33.70	269.60	51.00	408.00		
1 Equip. Oper. Oiler	28.30	226.40	42.85	342.80		
1 Welder, 300 amp		78.80		86.70		
1 Crane, 75 Ton		1191.00		1310.10	22.68	24.94
56 L.H., Daily Totals		$2930.20		$3952.00	$52.33	$70.57

Crew B-36

	Hr.	Daily	Hr.	Daily	Bare Costs	Incl. O&P
1 Labor Foreman (outside)	$26.65	$213.20	$41.65	$333.20	$28.23	$43.47
2 Laborers	24.65	394.40	38.50	616.00		
2 Equip. Oper. (med.)	32.60	521.60	49.35	789.60		
1 Dozer, 200 H.P.		880.80		968.90		
1 Aggregate Spreader		52.00		57.20		
1 Tandem Roller, 10 Ton		183.40		201.75	27.91	30.70
40 L.H., Daily Totals		$2245.40		$2966.65	$56.14	$74.17

Crew B-36A

	Hr.	Daily	Hr.	Daily	Bare Costs	Incl. O&P
1 Labor Foreman (outside)	$26.65	$213.20	$41.65	$333.20	$29.48	$45.15
2 Laborers	24.65	394.40	38.50	616.00		
4 Equip. Oper. (med.)	32.60	1043.20	49.35	1579.20		
1 Dozer, 200 H.P.		880.80		968.90		
1 Aggregate Spreader		52.00		57.20		
1 Roller, Steel Wheel		183.40		201.75		
1 Roller, Pneumatic Wheel		247.00		271.70	24.34	26.78
56 L.H., Daily Totals		$3014.00		$4027.95	$53.82	$71.93

Crew B-36B

	Hr.	Daily	Hr.	Daily	Bare Costs	Incl. O&P
1 Labor Foreman (outside)	$26.65	$213.20	$41.65	$333.20	$29.00	$44.40
2 Laborers	24.65	394.40	38.50	616.00		
4 Equip. Oper. (medium)	32.60	1043.20	49.35	1579.20		
1 Truck Driver, Heavy	25.65	205.20	39.15	313.20		
1 Grader, 30,000 Lbs.		457.20		502.90		
1 F.E. Loader, crl. 1.5 C.Y.		365.20		401.70		
1 Dozer, 300 H.P.		1130.00		1243.00		
1 Roller, Vibratory		447.60		492.35		
1 Truck, Tractor, 240 H.P.		376.00		413.60		
1 Water Tanker, 5000 Gal.		144.80		159.30	45.64	50.20
64 L.H., Daily Totals		$4776.80		$6054.45	$74.64	$94.60

Crew B-36C

	Hr.	Daily	Hr.	Daily	Bare Costs	Incl. O&P
1 Labor Foreman (outside)	$26.65	$213.20	$41.65	$333.20	$30.02	$45.77
3 Equip. Oper. (medium)	32.60	782.40	49.35	1184.40		
1 Truck Driver, Heavy	25.65	205.20	39.15	313.20		
1 Grader, 30,000 Lbs.		457.20		502.90		
1 Dozer, 300 H.P.		1130.00		1243.00		
1 Roller, Vibratory		447.60		492.35		
1 Truck, Tractor, 240 H.P.		376.00		413.60		
1 Water Tanker, 5000 Gal.		144.80		159.30	63.89	70.28
40 L.H., Daily Totals		$3756.40		$4641.95	$93.91	$116.05

Crew B-37

	Hr.	Daily	Hr.	Daily	Bare Costs	Incl. O&P
1 Labor Foreman (outside)	$26.65	$213.20	$41.65	$333.20	$26.06	$40.46
4 Laborers	24.65	788.80	38.50	1232.00		
1 Equip. Oper. (light)	31.10	248.80	47.10	376.80		
1 Tandem Roller, 5 Ton		107.00		117.70	2.23	2.45
48 L.H., Daily Totals		$1357.80		$2059.70	$28.29	$42.91

Crew B-38

	Hr.	Daily	Hr.	Daily	Bare Costs	Incl. O&P
1 Labor Foreman (outside)	$26.65	$213.20	$41.65	$333.20	$27.93	$43.02
2 Laborers	24.65	394.40	38.50	616.00		
1 Equip. Oper. (light)	31.10	248.80	47.10	376.80		
1 Equip. Oper. (medium)	32.60	260.80	49.35	394.80		
1 Backhoe Loader, 48 H.P.		218.80		240.70		
1 Hyd. Hammer, (1200 lb)		144.80		159.30		
1 F.E. Loader (170 H.P.)		464.40		510.85		
1 Pavt. Rem. Bucket		51.00		56.10	21.98	24.17
40 L.H., Daily Totals		$1996.20		$2687.75	$49.91	$67.19

Crew No.	Bare Costs		Incl. Subs O & P		Cost Per Labor-Hour	
Crew B-39	Hr.	Daily	Hr.	Daily	Bare Costs	Incl. O&P
1 Labor Foreman (outside)	$26.65	$213.20	$41.65	$333.20	$26.06	$40.46
4 Laborers	24.65	788.80	38.50	1232.00		
1 Equip. Oper. (light)	31.10	248.80	47.10	376.80		
1 Air Compr., 250 C.F.M.		128.60		141.45		
2 Air Tools & Accessories		29.20		32.10		
2-50 Ft. Air Hoses, 1.5" Dia.		10.00		11.00	3.50	3.85
48 L.H., Daily Totals		$1418.60		$2126.55	$29.56	$44.31

Crew No.	Bare Costs		Incl. Subs O & P		Cost Per Labor-Hour	
Crew B-40	Hr.	Daily	Hr.	Daily	Bare Costs	Incl. O&P
1 Pile Driver Foreman (out)	$32.90	$263.20	$55.20	$441.60	$31.52	$50.93
4 Pile Drivers	30.90	988.80	51.85	1659.20		
2 Equip. Oper. (crane)	33.70	539.20	51.00	816.00		
1 Equip. Oper. Oiler	28.30	226.40	42.85	342.80		
1 Crane, 40 Ton		935.60		1029.15		
1 Vibratory Hammer & Gen.		1398.00		1537.80	36.46	40.11
64 L.H., Daily Totals		$4351.20		$5826.55	$67.98	$91.04

Crew No.	Bare Costs		Incl. Subs O & P		Cost Per Labor-Hour	
Crew B-41	Hr.	Daily	Hr.	Daily	Bare Costs	Incl. O&P
1 Labor Foreman (outside)	$26.65	$213.20	$41.65	$333.20	$25.59	$39.84
4 Laborers	24.65	788.80	38.50	1232.00		
.25 Equip. Oper. (crane)	33.70	67.40	51.00	102.00		
.25 Equip. Oper. Oiler	28.30	56.60	42.85	85.70		
.25 Crawler Crane, 40 Ton		233.90		257.30	5.32	5.85
44 L.H., Daily Totals		$1359.90		$2010.20	$30.91	$45.69

Crew No.	Bare Costs		Incl. Subs O & P		Cost Per Labor-Hour	
Crew B-42	Hr.	Daily	Hr.	Daily	Bare Costs	Incl. O&P
1 Labor Foreman (outside)	$26.65	$213.20	$41.65	$333.20	$27.86	$44.28
4 Laborers	24.65	788.80	38.50	1232.00		
1 Equip. Oper. (crane)	33.70	269.60	51.00	408.00		
1 Equip. Oper. Oiler	28.30	226.40	42.85	342.80		
1 Welder	35.65	285.20	64.75	518.00		
1 Hyd. Crane, 25 Ton		654.60		720.05		
1 Gas Welding Machine		78.80		86.70		
1 Horz. Boring Csg. Mch.		461.60		507.75	18.67	20.54
64 L.H., Daily Totals		$2978.20		$4148.50	$46.53	$64.82

Crew No.	Bare Costs		Incl. Subs O & P		Cost Per Labor-Hour	
Crew B-43	Hr.	Daily	Hr.	Daily	Bare Costs	Incl. O&P
1 Labor Foreman (outside)	$26.65	$213.20	$41.65	$333.20	$27.10	$41.83
3 Laborers	24.65	591.60	38.50	924.00		
1 Equip. Oper. (crane)	33.70	269.60	51.00	408.00		
1 Equip. Oper. Oiler	28.30	226.40	42.85	342.80		
1 Drill Rig & Augers		2295.00		2524.50	47.81	52.59
48 L.H., Daily Totals		$3595.80		$4532.50	$74.91	$94.42

Crew No.	Bare Costs		Incl. Subs O & P		Cost Per Labor-Hour	
Crew B-44	Hr.	Daily	Hr.	Daily	Bare Costs	Incl. O&P
1 Pile Driver Foreman	$32.90	$263.20	$55.20	$441.60	$31.07	$50.39
4 Pile Drivers	30.90	988.80	51.85	1659.20		
2 Equip. Oper. (crane)	33.70	539.20	51.00	816.00		
1 Laborer	24.65	197.20	38.50	308.00		
1 Crane, 40 Ton, & Access.		935.60		1029.15		
45 L.F. Leads, 15K Ft. Lbs.		78.75		86.65	15.85	17.43
64 L.H., Daily Totals		$3002.75		$4340.60	$46.92	$67.82

Crew No.	Bare Costs		Incl. Subs O & P		Cost Per Labor-Hour	
Crew B-45	Hr.	Daily	Hr.	Daily	Bare Costs	Incl. O&P
1 Equip. Oper. (med.)	$32.60	$260.80	$49.35	$394.80	$29.13	$44.25
1 Truck Driver (heavy)	25.65	205.20	39.15	313.20		
1 Dist. Tank Truck, 3K Gal.		272.00		299.20		
1 Tractor, 4 x 2, 250 H.P.		331.60		364.75	37.73	41.50
16 L.H., Daily Totals		$1069.60		$1371.95	$66.86	$85.75

Crew No.	Bare Costs		Incl. Subs O & P		Cost Per Labor-Hour	
Crew B-46	Hr.	Daily	Hr.	Daily	Bare Costs	Incl. O&P
1 Pile Driver Foreman	$32.90	$263.20	$55.20	$441.60	$28.11	$45.73
2 Pile Drivers	30.90	494.40	51.85	829.60		
3 Laborers	24.65	591.60	38.50	924.00		
1 Chain Saw, 36" Long		38.00		41.80	.79	.87
48 L.H., Daily Totals		$1387.20		$2237.00	$28.90	$46.60

Crew No.	Bare Costs		Incl. Subs O & P		Cost Per Labor-Hour	
Crew B-47	Hr.	Daily	Hr.	Daily	Bare Costs	Incl. O&P
1 Blast Foreman	$26.65	$213.20	$41.65	$333.20	$27.47	$42.42
1 Driller	24.65	197.20	38.50	308.00		
1 Equip. Oper. (light)	31.10	248.80	47.10	376.80		
1 Crawler Type Drill, 4"		591.60		650.75		
1 Air Compr., 600 C.F.M.		249.20		274.10		
2-50 Ft. Air Hoses, 3" Dia.		35.40		38.95	36.51	40.16
24 L.H., Daily Totals		$1535.40		$1981.80	$63.98	$82.58

Crew No.	Bare Costs		Incl. Subs O & P		Cost Per Labor-Hour	
Crew B-47A	Hr.	Daily	Hr.	Daily	Bare Costs	Incl. O&P
1 Drilling Foreman	$26.65	$213.20	$41.65	$333.20	$29.55	$45.17
1 Equip. Oper. (heavy)	33.70	269.60	51.00	408.00		
1 Oiler	28.30	226.40	42.85	342.80		
1 Quarry Drill		818.40		900.25	34.10	37.51
24 L.H., Daily Totals		$1527.60		$1984.25	$63.65	$82.68

Crew No.	Bare Costs		Incl. Subs O & P		Cost Per Labor-Hour	
Crew B-47C	Hr.	Daily	Hr.	Daily	Bare Costs	Incl. O&P
1 Laborer	$24.65	$197.20	$38.50	$308.00	$27.88	$42.80
1 Equip. Oper. (light)	31.10	248.80	47.10	376.80		
1 Air Compressor, 750 CFM		281.60		309.75		
2-50' Air Hoses, 3"		35.40		38.95		
1 Air Track Drill, 4"		591.60		650.75	56.79	62.47
16 L.H., Daily Totals		$1354.60		$1684.25	$84.67	$105.27

Crew No.	Bare Costs		Incl. Subs O & P		Cost Per Labor-Hour	
Crew B-47E	Hr.	Daily	Hr.	Daily	Bare Costs	Incl. O&P
1 Laborer Foreman	$26.65	$213.20	$41.65	$333.20	$25.15	$39.29
3 Laborers	24.65	591.60	38.50	924.00		
1 Truck, Flatbed, 3 Ton		138.00		151.80	4.31	4.74
32 L.H., Daily Totals		$942.80		$1409.00	$29.46	$44.03

Crew No.	Bare Costs		Incl. Subs O & P		Cost Per Labor-Hour	
Crew B-48	Hr.	Daily	Hr.	Daily	Bare Costs	Incl. O&P
1 Labor Foreman (outside)	$26.65	$213.20	$41.65	$333.20	$27.67	$42.59
3 Laborers	24.65	591.60	38.50	924.00		
1 Equip. Oper. (crane)	33.70	269.60	51.00	408.00		
1 Equip. Oper. Oiler	28.30	226.40	42.85	342.80		
1 Equip. Oper. (light)	31.10	248.80	47.10	376.80		
1 Centr. Water Pump, 6"		224.40		246.85		
1-20 Ft. Suction Hose, 6"		17.10		18.80		
1-50 Ft. Disch. Hose, 6"		14.70		16.15		
1 Drill Rig & Augers		2295.00		2524.50	45.56	50.11
56 L.H., Daily Totals		$4100.80		$5191.10	$73.23	$92.70

Crew No.	Bare Costs		Incl. Subs O & P		Cost Per Labor-Hour	
Crew B-49	Hr.	Daily	Hr.	Daily	Bare Costs	Incl. O&P
1 Labor Foreman (outside)	$26.65	$213.20	$41.65	$333.20	$28.86	$45.06
3 Laborers	24.65	591.60	38.50	924.00		
2 Equip. Oper. (crane)	33.70	539.20	51.00	816.00		
2 Equip. Oper. Oilers	28.30	452.80	42.85	685.60		
1 Equip. Oper. (light)	31.10	248.80	47.10	376.80		
2 Pile Drivers	30.90	494.40	51.85	829.60		
1 Hyd. Crane, 25 Ton		654.60		720.05		
1 Centr. Water Pump, 6"		224.40		246.85		
1-20 Ft. Suction Hose, 6"		17.10		18.80		
1-50 Ft. Disch. Hose, 6"		14.70		16.15		
1 Drill Rig & Augers		2295.00		2524.50	36.43	40.07
88 L.H., Daily Totals		$5745.80		$7491.55	$65.29	$85.13

Crew No.	Bare Costs		Incl. Subs O & P		Cost Per Labor-Hour	
Crew B-50	Hr.	Daily	Hr.	Daily	Bare Costs	Incl. O&P
2 Pile Driver Foremen	$32.90	$526.40	$55.20	$883.20	$30.06	$48.70
6 Pile Drivers	30.90	1483.20	51.85	2488.80		
2 Equip. Oper. (crane)	33.70	539.20	51.00	816.00		
1 Equip. Oper. Oiler	28.30	226.40	42.85	342.80		
3 Laborers	24.65	591.60	38.50	924.00		
1 Crane, 40 Ton		935.60		1029.15		
60 L.F. Leads, 15K Ft. Lbs.		105.00		115.50		
1 Hammer, 15K Ft. Lbs.		367.20		403.90		
1 Air Compr., 600 C.F.M.		249.20		274.10		
2-50 Ft. Air Hoses, 3" Dia.		35.40		38.95		
1 Chain Saw, 36" Long		38.00		41.80	15.45	17.00
112 L.H., Daily Totals		$5097.20		$7358.20	$45.51	$65.70

Crew No.	Bare Costs		Incl. Subs O & P		Cost Per Labor-Hour	
Crew B-51	Hr.	Daily	Hr.	Daily	Bare Costs	Incl. O&P
1 Labor Foreman (outside)	$26.65	$213.20	$41.65	$333.20	$25.02	$38.94
4 Laborers	24.65	788.80	38.50	1232.00		
1 Truck Driver (light)	24.90	199.20	38.00	304.00		
1 Light Truck, 1.5 Ton		120.20		132.20	2.50	2.75
48 L.H., Daily Totals		$1321.40		$2001.40	$27.52	$41.69

Crew No.	Bare Costs		Incl. Subs O & P		Cost Per Labor-Hour	
Crew B-52	Hr.	Daily	Hr.	Daily	Bare Costs	Incl. O&P
1 Carpenter Foreman	$33.55	$268.40	$52.40	$419.20	$29.05	$45.21
1 Carpenter	31.55	252.40	49.30	394.40		
3 Laborers	24.65	591.60	38.50	924.00		
1 Cement Finisher	30.20	241.60	44.75	358.00		
.5 Rodman (reinf.)	35.55	142.20	59.70	238.80		
.5 Equip. Oper. (med.)	32.60	130.40	49.35	197.40		
.5 F.E. Ldr., T.M., 2.5 C.Y.		389.20		428.10	6.95	7.65
56 L.H., Daily Totals		$2015.80		$2959.90	$36.00	$52.86

Crew No.	Bare Costs		Incl. Subs O & P		Cost Per Labor-Hour	
Crew B-53	Hr.	Daily	Hr.	Daily	Bare Costs	Incl. O&P
1 Equip. Oper. (light)	$31.10	$248.80	$47.10	$376.80	$31.10	$47.10
1 Trencher, Chain, 12 H.P.		79.20		87.10	9.90	10.89
8 L.H., Daily Totals		$328.00		$463.90	$41.00	$57.99

Crew No.	Bare Costs		Incl. Subs O & P		Cost Per Labor-Hour	
Crew B-54	Hr.	Daily	Hr.	Daily	Bare Costs	Incl. O&P
1 Equip. Oper. (light)	$31.10	$248.80	$47.10	$376.80	$31.10	$47.10
1 Trencher, Chain, 40 H.P.		211.00		232.10	26.38	29.01
8 L.H., Daily Totals		$459.80		$608.90	$57.48	$76.11

Crew No.	Bare Costs		Incl. Subs O & P		Cost Per Labor-Hour	
Crew B-54A	Hr.	Daily	Hr.	Daily	Bare Costs	Incl. O&P
.17 Labor Foreman (outside)	$26.65	$36.24	$41.65	$56.64	$31.74	$48.23
1 Equipment Operator (med.)	32.60	260.80	49.35	394.80		
1 Wheel Trencher, 67 H.P.		685.60		754.15	73.25	80.57
9.36 L.H., Daily Totals		$982.64		$1205.59	$104.99	$128.80

Crew No.	Bare Costs		Incl. Subs O & P		Cost Per Labor-Hour	
Crew B-54B	Hr.	Daily	Hr.	Daily	Bare Costs	Incl. O&P
.25 Labor Foreman (outside)	$26.65	$53.30	$41.65	$83.30	$31.41	$47.81
1 Equipment Operator (med.)	32.60	260.80	49.35	394.80		
1 Wheel Trencher, 150 H.P.		1085.00		1193.50	108.50	119.35
10 L.H., Daily Totals		$1399.10		$1671.60	$139.91	$167.16

Crew No.	Bare Costs		Incl. Subs O & P		Cost Per Labor-Hour	
Crew B-55	Hr.	Daily	Hr.	Daily	Bare Costs	Incl. O&P
2 Laborers	$24.65	$394.40	$38.50	$616.00	$24.73	$38.33
1 Truck Driver (light)	24.90	199.20	38.00	304.00		
1 Auger, 4" to 36" Dia		616.40		678.05		
1 Flatbed 3 Ton Truck		138.00		151.80	31.43	34.58
24 L.H., Daily Totals		$1348.00		$1749.85	$56.16	$72.91

Crew No.	Bare Costs		Incl. Subs O & P		Cost Per Labor-Hour	
Crew B-56	Hr.	Daily	Hr.	Daily	Bare Costs	Incl. O&P
1 Laborer	$24.65	$197.20	$38.50	$308.00	$27.88	$42.80
1 Equip. Oper. (light)	31.10	248.80	47.10	376.80		
1 Crawler Type Drill, 4"		591.60		650.75		
1 Air Compr., 600 C.F.M.		249.20		274.10		
1-50 Ft. Air Hose, 3" Dia.		17.70		19.45	53.66	59.02
16 L.H., Daily Totals		$1304.50		$1629.10	$81.54	$101.82

Crew No.	Bare Costs		Incl. Subs O & P		Cost Per Labor-Hour	
Crew B-57	Hr.	Daily	Hr.	Daily	Bare Costs	Incl. O&P
1 Labor Foreman (outside)	$26.65	$213.20	$41.65	$333.20	$28.18	$43.27
2 Laborers	24.65	394.40	38.50	616.00		
1 Equip. Oper. (crane)	33.70	269.60	51.00	408.00		
1 Equip. Oper. (light)	31.10	248.80	47.10	376.80		
1 Equip. Oper. Oiler	28.30	226.40	42.85	342.80		
1 Power Shovel, 1 C.Y.		744.00		818.40		
1 Clamshell Bucket, 1 C.Y.		54.00		59.40		
1 Centr. Water Pump, 6"		224.40		246.85		
1-20 Ft. Suction Hose, 6"		17.10		18.80		
20-50 Ft. Disch. Hoses, 6"		294.00		323.40	27.78	30.56
48 L.H., Daily Totals		$2685.90		$3543.65	$55.96	$73.83

Crew No.	Bare Costs		Incl. Subs O & P		Cost Per Labor-Hour	
Crew B-58	Hr.	Daily	Hr.	Daily	Bare Costs	Incl. O&P
2 Laborers	$24.65	$394.40	$38.50	$616.00	$26.80	$41.37
1 Equip. Oper. (light)	31.10	248.80	47.10	376.80		
1 Backhoe Loader, 48 H.P.		218.80		240.70		
1 Small Helicopter, w/pilot		2069.00		2275.90	95.33	104.86
24 L.H., Daily Totals		$2931.00		$3509.40	$122.13	$146.23

Crew No.	Bare Costs		Incl. Subs O & P		Cost Per Labor-Hour	
Crew B-59	Hr.	Daily	Hr.	Daily	Bare Costs	Incl. O&P
1 Truck Driver (heavy)	$25.65	$205.20	$39.15	$313.20	$25.65	$39.15
1 Truck, 30 Ton		240.60		264.65		
1 Water tank, 5000 Gal.		144.80		159.30	48.18	52.99
8 L.H., Daily Totals		$590.60		$737.15	$73.83	$92.14

Crew No.	Bare Costs		Incl. Subs O & P		Cost Per Labor-Hour	
Crew B-59A	Hr.	Daily	Hr.	Daily	Bare Costs	Incl. O&P
2 Laborers	$24.65	$394.40	$38.50	$616.00	$24.98	$38.72
1 Truck Driver (heavy)	25.65	205.20	39.15	313.20		
1 Water tank, 5K w/pum		144.80		159.30		
1 Truck, 30 Ton		240.60		264.65	16.06	17.66
24 L.H., Daily Totals		$985.00		$1353.15	$41.04	$56.38

Crew No.	Bare Costs		Incl. Subs O & P		Cost Per Labor-Hour	
Crew B-60	Hr.	Daily	Hr.	Daily	Bare Costs	Incl. O&P
1 Labor Foreman (outside)	$26.65	$213.20	$41.65	$333.20	$28.59	$43.81
2 Laborers	24.65	394.40	38.50	616.00		
1 Equip. Oper. (crane)	33.70	269.60	51.00	408.00		
2 Equip. Oper. (light)	31.10	497.60	47.10	753.60		
1 Equip. Oper. Oiler	28.30	226.40	42.85	342.80		
1 Crawler Crane, 40 Ton		935.60		1029.15		
45 L.F. Leads, 15K Ft. Lbs.		78.75		86.65		
1 Backhoe Loader, 48 H.P.		218.80		240.70	22.02	24.22
56 L.H., Daily Totals		$2834.35		$3810.10	$50.61	$68.03

Crew No.	Bare Costs		Incl. Subs O & P		Cost Per Labor-Hour	
Crew B-61	Hr.	Daily	Hr.	Daily	Bare Costs	Incl. O&P
1 Labor Foreman (outside)	$26.65	$213.20	$41.65	$333.20	$26.34	$40.85
3 Laborers	24.65	591.60	38.50	924.00		
1 Equip. Oper. (light)	31.10	248.80	47.10	376.80		
1 Cement Mixer, 2 C.Y.		164.00		180.40		
1 Air Compr., 160 C.F.M.		86.40		95.05	6.26	6.89
40 L.H., Daily Totals		$1304.00		$1909.45	$32.60	$47.74

Crew B-62

Crew No.	Hr.	Daily	Hr.	Daily	Bare Costs	Incl. O&P
2 Laborers	$24.65	$394.40	$38.50	$616.00	$26.80	$41.37
1 Equip. Oper. (light)	31.10	248.80	47.10	376.80		
1 Loader, Skid Steer		153.00		168.30	6.38	7.01
24 L.H., Daily Totals		$796.20		$1161.10	$33.18	$48.38

Crew B-63

Crew No.	Hr.	Daily	Hr.	Daily	Bare Costs	Incl. O&P
4 Laborers	$24.65	$788.80	$38.50	$1232.00	$25.94	$40.22
1 Equip. Oper. (light)	31.10	248.80	47.10	376.80		
1 Loader, Skid Steer		153.00		168.30	3.83	4.21
40 L.H., Daily Totals		$1190.60		$1777.10	$29.77	$44.43

Crew B-64

Crew No.	Hr.	Daily	Hr.	Daily	Bare Costs	Incl. O&P
1 Laborer	$24.65	$197.20	$38.50	$308.00	$24.77	$38.25
1 Truck Driver (light)	24.90	199.20	38.00	304.00		
1 Power Mulcher (small)		117.00		128.70		
1 Light Truck, 1.5 Ton		120.20		132.20	14.83	16.31
16 L.H., Daily Totals		$633.60		$872.90	$39.60	$54.56

Crew B-65

Crew No.	Hr.	Daily	Hr.	Daily	Bare Costs	Incl. O&P
1 Laborer	$24.65	$197.20	$38.50	$308.00	$24.77	$38.25
1 Truck Driver (light)	24.90	199.20	38.00	304.00		
1 Power Mulcher (large)		220.40		242.45		
1 Light Truck, 1.5 Ton		120.20		132.20	21.29	23.42
16 L.H., Daily Totals		$737.00		$986.65	$46.06	$61.67

Crew B-66

Crew No.	Hr.	Daily	Hr.	Daily	Bare Costs	Incl. O&P
1 Equip. Oper. (light)	$31.10	$248.80	$47.10	$376.80	$31.10	$47.10
1 Backhoe Ldr. w/Attchmt.		178.60		196.45	22.33	24.56
8 L.H., Daily Totals		$427.40		$573.25	$53.43	$71.66

Crew B-67

Crew No.	Hr.	Daily	Hr.	Daily	Bare Costs	Incl. O&P
1 Millwright	$32.95	$263.60	$48.85	$390.80	$32.03	$47.97
1 Equip. Oper. (light)	31.10	248.80	47.10	376.80		
1 Forklift		241.40		265.55	15.09	16.60
16 L.H., Daily Totals		$753.80		$1033.15	$47.12	$64.57

Crew B-68

Crew No.	Hr.	Daily	Hr.	Daily	Bare Costs	Incl. O&P
2 Millwrights	$32.95	$527.20	$48.85	$781.60	$32.33	$48.27
1 Equip. Oper. (light)	31.10	248.80	47.10	376.80		
1 Forklift		241.40		265.55	10.06	11.06
24 L.H., Daily Totals		$1017.40		$1423.95	$42.39	$59.33

Crew B-69

Crew No.	Hr.	Daily	Hr.	Daily	Bare Costs	Incl. O&P
1 Labor Foreman (outside)	$26.65	$213.20	$41.65	$333.20	$27.10	$41.83
3 Laborers	24.65	591.60	38.50	924.00		
1 Equip Oper. (crane)	33.70	269.60	51.00	408.00		
1 Equip Oper. Oiler	28.30	226.40	42.85	342.80		
1 Truck Crane, 80 Ton		1093.00		1202.30	22.77	25.05
48 L.H., Daily Totals		$2393.80		$3210.30	$49.87	$66.88

Crew B-69A

Crew No.	Hr.	Daily	Hr.	Daily	Bare Costs	Incl. O&P
1 Labor Foreman	$26.65	$213.20	$41.65	$333.20	$27.23	$41.88
3 Laborers	24.65	591.60	38.50	924.00		
1 Equip. Oper. (medium)	32.60	260.80	49.35	394.80		
1 Concrete Finisher	30.20	241.60	44.75	358.00		
1 Curb Paver		567.60		624.35	11.83	13.01
48 L.H., Daily Totals		$1874.80		$2634.35	$39.06	$54.89

Crew B-69B

Crew No.	Hr.	Daily	Hr.	Daily	Bare Costs	Incl. O&P
1 Labor Foreman	$26.65	$213.20	$41.65	$333.20	$27.23	$41.88
3 Laborers	24.65	591.60	38.50	924.00		
1 Equip. Oper. (medium)	32.60	260.80	49.35	394.80		
1 Cement Finisher	30.20	241.60	44.75	358.00		
1 Curb/Gutter Paver		1015.00		1116.50	21.15	23.26
48 L.H., Daily Totals		$2322.20		$3126.50	$48.38	$65.14

Crew B-70

Crew No.	Hr.	Daily	Hr.	Daily	Bare Costs	Incl. O&P
1 Labor Foreman (outside)	$26.65	$213.20	$41.65	$333.20	$28.34	$43.60
3 Laborers	24.65	591.60	38.50	924.00		
3 Equip. Oper. (med.)	32.60	782.40	49.35	1184.40		
1 Motor Grader, 30,000 Lb.		457.20		502.90		
1 Grader Attach., Ripper		71.60		78.75		
1 Road Sweeper, S.P.		377.60		415.35		
1 F.E. Loader, 1-3/4 C.Y.		252.60		277.85	20.70	22.77
56 L.H., Daily Totals		$2746.20		$3716.45	$49.04	$66.37

Crew B-71

Crew No.	Hr.	Daily	Hr.	Daily	Bare Costs	Incl. O&P
1 Labor Foreman (outside)	$26.65	$213.20	$41.65	$333.20	$28.34	$43.60
3 Laborers	24.65	591.60	38.50	924.00		
3 Equip. Oper. (med.)	32.60	782.40	49.35	1184.40		
1 Pvmt. Profiler, 750 H.P.		4368.00		4804.80		
1 Road Sweeper, S.P.		377.60		415.35		
1 F.E. Loader, 1-3/4 C.Y.		252.60		277.85	89.25	98.18
56 L.H., Daily Totals		$6585.40		$7939.60	$117.59	$141.78

Crew B-72

Crew No.	Hr.	Daily	Hr.	Daily	Bare Costs	Incl. O&P
1 Labor Foreman (outside)	$26.65	$213.20	$41.65	$333.20	$28.87	$44.32
3 Laborers	24.65	591.60	38.50	924.00		
4 Equip. Oper. (med.)	32.60	1043.20	49.35	1579.20		
1 Pvmt. Profiler, 750 H.P.		4368.00		4804.80		
1 Hammermill, 250 H.P.		1291.00		1420.10		
1 Windrow Loader		799.40		879.35		
1 Mix Paver 165 H.P.		1670.00		1837.00		
1 Roller, Pneu. Tire, 12 T.		247.00		271.70	130.87	143.95
64 L.H., Daily Totals		$10223.40		$12049.35	$159.74	$188.27

Crew B-73

Crew No.	Hr.	Daily	Hr.	Daily	Bare Costs	Incl. O&P
1 Labor Foreman (outside)	$26.65	$213.20	$41.65	$333.20	$29.87	$45.67
2 Laborers	24.65	394.40	38.50	616.00		
5 Equip. Oper. (med.)	32.60	1304.00	49.35	1974.00		
1 Road Mixer, 310 H.P.		1625.00		1787.50		
1 Roller, Tandem, 12 Ton		183.40		201.75		
1 Hammermill, 250 H.P.		1291.00		1420.10		
1 Motor Grader, 30,000 Lb.		457.20		502.90		
.5 F.E. Loader, 1-3/4 C.Y.		126.30		138.95		
.5 Truck, 30 Ton		120.30		132.35		
.5 Water Tank 5000 Gal.		72.40		79.65	60.56	66.61
64 L.H., Daily Totals		$5787.20		$7186.40	$90.43	$112.28

Crew No.	Bare Costs		Incl. Subs O & P		Cost Per Labor-Hour	

Crew B-74	Hr.	Daily	Hr.	Daily	Bare Costs	Incl. O&P
1 Labor Foreman (outside)	$26.65	$213.20	$41.65	$333.20	$29.13	$44.48
1 Laborer	24.65	197.20	38.50	308.00		
4 Equip. Oper. (med.)	32.60	1043.20	49.35	1579.20		
2 Truck Drivers (heavy)	25.65	410.40	39.15	626.40		
1 Motor Grader, 30,000 Lb.		457.20		502.90		
1 Grader Attach., Ripper		71.60		78.75		
2 Stabilizers, 310 H.P.		2158.00		2373.80		
1 Flatbed Truck, 3 Ton		138.00		151.80		
1 Chem. Spreader, Towed		76.00		83.60		
1 Vibr. Roller, 29,000 Lb.		447.60		492.35		
1 Water Tank 5000 Gal.		144.80		159.30		
1 Truck, 30 Ton		240.60		264.65	58.34	64.17
64 L.H., Daily Totals		$5597.80		$6953.95	$87.47	$108.65

Crew B-75	Hr.	Daily	Hr.	Daily	Bare Costs	Incl. O&P
1 Labor Foreman (outside)	$26.65	$213.20	$41.65	$333.20	$29.62	$45.24
1 Laborer	24.65	197.20	38.50	308.00		
4 Equip. Oper. (med.)	32.60	1043.20	49.35	1579.20		
1 Truck Driver (heavy)	25.65	205.20	39.15	313.20		
1 Motor Grader, 30,000 Lb.		457.20		502.90		
1 Grader Attach., Ripper		71.60		78.75		
2 Stabilizers, 310 H.P.		2158.00		2373.80		
1 Dist. Truck, 3000 Gal.		272.00		299.20		
1 Vibr. Roller, 29,000 Lb.		447.60		492.35	60.83	66.91
56 L.H., Daily Totals		$5065.20		$6280.60	$90.45	$112.15

Crew B-76	Hr.	Daily	Hr.	Daily	Bare Costs	Incl. O&P
1 Dock Builder Foreman	$32.90	$263.20	$55.20	$441.60	$31.46	$51.03
5 Dock Builders	30.90	1236.00	51.85	2074.00		
2 Equip. Oper. (crane)	33.70	539.20	51.00	816.00		
1 Equip. Oper. Oiler	28.30	226.40	42.85	342.80		
1 Crawler Crane, 50 Ton		1222.00		1344.20		
1 Barge, 400 Ton		333.60		366.95		
1 Hammer, 15K Ft. Lbs.		367.20		403.90		
60 L.F. Leads, 15K Ft. Lbs.		105.00		115.50		
1 Air Compr., 600 C.F.M.		249.20		274.10		
2-50 Ft. Air Hoses, 3" Dia.		35.40		38.95	32.12	35.33
72 L.H., Daily Totals		$4577.20		$6218.00	$63.58	$86.36

Crew B-77	Hr.	Daily	Hr.	Daily	Bare Costs	Incl. O&P
1 Labor Foreman	$26.65	$213.20	$41.65	$333.20	$25.10	$39.03
3 Laborers	24.65	591.60	38.50	924.00		
1 Truck Driver (light)	24.90	199.20	38.00	304.00		
1 Crack Cleaner, 25 H.P.		53.60		58.95		
1 Crack Filler, Trailer Mtd.		144.80		159.30		
1 Flatbed Truck, 3 Ton		138.00		151.80	8.41	9.25
40 L.H., Daily Totals		$1340.40		$1931.25	$33.51	$48.28

Crew B-78	Hr.	Daily	Hr.	Daily	Bare Costs	Incl. O&P
1 Labor Foreman	$26.65	$213.20	$41.65	$333.20	$25.02	$38.94
4 Laborers	24.65	788.80	38.50	1232.00		
1 Truck Driver (light)	24.90	199.20	38.00	304.00		
1 Paint Striper, S.P.		167.80		184.60		
1 Flatbed Truck, 3 Ton		138.00		151.80		
1 Pickup Truck, 3/4 Ton		67.80		74.60	7.78	8.56
48 L.H., Daily Totals		$1574.80		$2280.20	$32.80	$47.50

Crew B-79	Hr.	Daily	Hr.	Daily	Bare Costs	Incl. O&P
1 Labor Foreman	$26.65	$213.20	$41.65	$333.20	$25.10	$39.03
3 Laborers	24.65	591.60	38.50	924.00		
1 Truck Driver (light)	24.90	199.20	38.00	304.00		
1 Thermo. Striper, T.M.		218.65		240.50		
1 Flatbed Truck, 3 Ton		138.00		151.80		
2 Pickup Trucks, 3/4 Ton		135.60		149.15	12.31	13.54
40 L.H., Daily Totals		$1496.25		$2102.65	$37.41	$52.57

Crew B-80	Hr.	Daily	Hr.	Daily	Bare Costs	Incl. O&P
1 Labor Foreman	$26.65	$213.20	$41.65	$333.20	$26.83	$41.31
1 Laborer	24.65	197.20	38.50	308.00		
1 Truck Driver (light)	24.90	199.20	38.00	304.00		
1 Equip. Oper. (light)	31.10	248.80	47.10	376.80		
1 Flatbed Truck, 3 Ton		138.00		151.80		
1 Fence Post Auger, T.M.		361.20		397.30	15.60	17.16
32 L.H., Daily Totals		$1357.60		$1871.10	$42.43	$58.47

Crew B-80A	Hr.	Daily	Hr.	Daily	Bare Costs	Incl. O&P
3 Laborers	$24.65	$591.60	$38.50	$924.00	$24.65	$38.50
1 Flatbed Truck, 3 Ton		138.00		151.80	5.75	6.33
24 L.H., Daily Totals		$729.60		$1075.80	$30.40	$44.83

Crew B-80B	Hr.	Daily	Hr.	Daily	Bare Costs	Incl. O&P
3 Laborers	$24.65	$591.60	$38.50	$924.00	$26.26	$40.65
1 Equip. Oper. (light)	31.10	248.80	47.10	376.80		
1 Crane, Flatbed Mnt.		203.80		224.20	6.37	7.01
32 L.H., Daily Totals		$1044.20		$1525.00	$32.63	$47.66

Crew B-81	Hr.	Daily	Hr.	Daily	Bare Costs	Incl. O&P
1 Laborer	$24.65	$197.20	$38.50	$308.00	$27.63	$42.33
1 Equip. Oper. (med.)	32.60	260.80	49.35	394.80		
1 Truck Driver (heavy)	25.65	205.20	39.15	313.20		
1 Hydromulcher, T.M.		218.20		240.00		
1 Tractor Truck, 4x2		240.60		264.65	19.12	21.03
24 L.H., Daily Totals		$1122.00		$1520.65	$46.75	$63.36

Crew B-82	Hr.	Daily	Hr.	Daily	Bare Costs	Incl. O&P
1 Laborer	$24.65	$197.20	$38.50	$308.00	$27.88	$42.80
1 Equip. Oper. (light)	31.10	248.80	47.10	376.80		
1 Horiz. Borer, 6 H.P.		62.60		68.85	3.91	4.30
16 L.H., Daily Totals		$508.60		$753.65	$31.79	$47.10

Crew B-83	Hr.	Daily	Hr.	Daily	Bare Costs	Incl. O&P
1 Tugboat Captain	$32.60	$260.80	$49.35	$394.80	$28.63	$43.92
1 Tugboat Hand	24.65	197.20	38.50	308.00		
1 Tugboat, 250 H.P.		446.00		490.60	27.88	30.66
16 L.H., Daily Totals		$904.00		$1193.40	$56.51	$74.58

Crew B-84	Hr.	Daily	Hr.	Daily	Bare Costs	Incl. O&P
1 Equip. Oper. (med.)	$32.60	$260.80	$49.35	$394.80	$32.60	$49.35
1 Rotary Mower/Tractor		228.60		251.45	28.58	31.43
8 L.H., Daily Totals		$489.40		$646.25	$61.18	$80.78

Crew No.	Bare Costs		Incl. Subs O & P		Cost Per Labor-Hour	
Crew B-85	Hr.	Daily	Hr.	Daily	Bare Costs	Incl. O&P
3 Laborers	$24.65	$591.60	$38.50	$924.00	$26.44	$40.80
1 Equip. Oper. (med.)	32.60	260.80	49.35	394.80		
1 Truck Driver (heavy)	25.65	205.20	39.15	313.20		
1 Aerial Lift Truck, 80'		492.80		542.10		
1 Brush Chipper, 130 H.P.		180.80		198.90		
1 Pruning Saw, Rotary		13.90		15.30	17.19	18.91
40 L.H., Daily Totals		$1745.10		$2388.30	$43.63	$59.71

Crew No.	Bare Costs		Incl. Subs O & P		Cost Per Labor-Hour	
Crew B-86	Hr.	Daily	Hr.	Daily	Bare Costs	Incl. O&P
1 Equip. Oper. (med.)	$32.60	$260.80	$49.35	$394.80	$32.60	$49.35
1 Stump Chipper, S.P.		104.50		114.95	13.06	14.37
8 L.H., Daily Totals		$365.30		$509.75	$45.66	$63.72

Crew B-86A	Hr.	Daily	Hr.	Daily	Bare Costs	Incl. O&P
1 Equip. Oper. (medium)	$32.60	$260.80	$49.35	$394.80	$32.60	$49.35
1 Grader, 30,000 Lbs.		457.20		502.90	57.15	62.87
8 L.H., Daily Totals		$718.00		$897.70	$89.75	$112.22

Crew B-86B	Hr.	Daily	Hr.	Daily	Bare Costs	Incl. O&P
1 Equip. Oper. (medium)	$32.60	$260.80	$49.35	$394.80	$32.60	$49.35
1 Dozer, 200 H.P.		880.80		968.90	110.10	121.11
8 L.H., Daily Totals		$1141.60		$1363.70	$142.70	$170.46

Crew B-87	Hr.	Daily	Hr.	Daily	Bare Costs	Incl. O&P
1 Laborer	$24.65	$197.20	$38.50	$308.00	$31.01	$47.18
4 Equip. Oper. (med.)	32.60	1043.20	49.35	1579.20		
2 Feller Bunchers, 50 H.P.		862.40		948.65		
1 Log Chipper, 22" Tree		1294.00		1423.40		
1 Dozer, 105 H.P.		336.40		370.05		
1 Chainsaw, Gas, 36" Long		38.00		41.80	63.27	69.60
40 L.H., Daily Totals		$3771.20		$4671.10	$94.28	$116.78

Crew B-88	Hr.	Daily	Hr.	Daily	Bare Costs	Incl. O&P
1 Laborer	$24.65	$197.20	$38.50	$308.00	$31.46	$47.80
6 Equip. Oper. (med.)	32.60	1564.80	49.35	2368.80		
2 Feller Bunchers, 50 H.P.		862.40		948.65		
1 Log Chipper, 22" Tree		1294.00		1423.40		
2 Log Skidders, 50 H.P.		1411.20		1552.30		
1 Dozer, 105 H.P.		336.40		370.05		
1 Chainsaw, Gas, 36" Long		38.00		41.80	70.39	77.43
56 L.H., Daily Totals		$5704.00		$7013.00	$101.85	$125.23

Crew B-89	Hr.	Daily	Hr.	Daily	Bare Costs	Incl. O&P
1 Equip. Oper. (light)	$31.10	$248.80	$47.10	$376.80	$28.00	$42.55
1 Truck Driver (light)	24.90	199.20	38.00	304.00		
1 Truck, Stake Body, 3 Ton		138.00		151.80		
1 Concrete Saw		108.20		119.00		
1 Water Tank, 65 Gal.		13.60		14.95	16.24	17.86
16 L.H., Daily Totals		$707.80		$966.55	$44.24	$60.41

Crew B-89A	Hr.	Daily	Hr.	Daily	Bare Costs	Incl. O&P
1 Skilled Worker	$32.25	$258.00	$50.50	$404.00	$28.45	$44.50
1 Laborer	24.65	197.20	38.50	308.00		
1 Core Drill (large)		93.05		102.35	5.82	6.40
16 L.H., Daily Totals		$548.25		$814.35	$34.27	$50.90

Crew No.	Bare Costs		Incl. Subs O & P		Cost Per Labor-Hour	
Crew B-89B	Hr.	Daily	Hr.	Daily	Bare Costs	Incl. O&P
1 Equip. Oper. (light)	$31.10	$248.80	$47.10	$376.80	$28.00	$42.55
1 Truck Driver, Light	24.90	199.20	38.00	304.00		
1 Wall Saw, Hydraulic, 10 H.P.		81.05		89.15		
1 Generator, Diesel, 100 KW		179.00		196.90		
1 Water Tank, 65 Gal.		13.60		14.95		
1 Flatbed Truck, 3 Ton		138.00		151.80	25.73	28.30
16 L.H., Daily Totals		$859.65		$1133.60	$53.73	$70.85

Crew B-90	Hr.	Daily	Hr.	Daily	Bare Costs	Incl. O&P
1 Labor Foreman (outside)	$26.65	$213.20	$41.65	$333.20	$26.76	$41.21
3 Laborers	24.65	591.60	38.50	924.00		
2 Equip. Oper. (light)	31.10	497.60	47.10	753.60		
2 Truck Drivers (heavy)	25.65	410.40	39.15	626.40		
1 Road Mixer, 310 H.P.		1625.00		1787.50		
1 Dist. Truck, 2000 Gal.		250.60		275.65	29.31	32.24
64 L.H., Daily Totals		$3588.40		$4700.35	$56.07	$73.45

Crew B-90A	Hr.	Daily	Hr.	Daily	Bare Costs	Incl. O&P
1 Labor Foreman	$26.65	$213.20	$41.65	$333.20	$29.48	$45.15
2 Laborers	24.65	394.40	38.50	616.00		
4 Equip. Oper. (medium)	32.60	1043.20	49.35	1579.20		
2 Graders, 30,000 Lbs.		914.40		1005.85		
1 Roller, Steel Wheel		183.40		201.75		
1 Roller, Pneumatic Wheel		247.00		271.70	24.01	26.42
56 L.H., Daily Totals		$2995.60		$4007.70	$53.49	$71.57

Crew B-90B	Hr.	Daily	Hr.	Daily	Bare Costs	Incl. O&P
1 Labor Foreman	$26.65	$213.20	$41.65	$333.20	$28.96	$44.45
2 Laborers	24.65	394.40	38.50	616.00		
3 Equip. Oper. (medium)	32.60	782.40	49.35	1184.40		
1 Roller, Steel Wheel		183.40		201.75		
1 Roller, Pneumatic Wheel		247.00		271.70		
1 Road Mixer, 310 H.P.		1625.00		1787.50	42.82	47.10
48 L.H., Daily Totals		$3445.40		$4394.55	$71.78	$91.55

Crew B-91	Hr.	Daily	Hr.	Daily	Bare Costs	Incl. O&P
1 Labor Foreman (outside)	$26.65	$213.20	$41.65	$333.20	$29.00	$44.40
2 Laborers	24.65	394.40	38.50	616.00		
4 Equip. Oper. (med.)	32.60	1043.20	49.35	1579.20		
1 Truck Driver (heavy)	25.65	205.20	39.15	313.20		
1 Dist. Truck, 3000 Gal.		272.00		299.20		
1 Aggreg. Spreader, S.P.		747.20		821.90		
1 Roller, Pneu. Tire, 12 Ton		247.00		271.70		
1 Roller, Steel, 10 Ton		183.40		201.75	22.65	24.92
64 L.H., Daily Totals		$3305.60		$4436.15	$51.65	$69.32

Crew B-92	Hr.	Daily	Hr.	Daily	Bare Costs	Incl. O&P
1 Labor Foreman (outside)	$26.65	$213.20	$41.65	$333.20	$25.15	$39.29
3 Laborers	24.65	591.60	38.50	924.00		
1 Crack Cleaner, 25 H.P.		53.60		58.95		
1 Air Compressor		56.60		62.25		
1 Tar Kettle, T.M.		42.30		46.55		
1 Flatbed Truck, 3 Ton		138.00		151.80	9.08	9.99
32 L.H., Daily Totals		$1095.30		$1576.75	$34.23	$49.28

Crew B-93	Hr.	Daily	Hr.	Daily	Bare Costs	Incl. O&P
1 Equip. Oper. (med.)	$32.60	$260.80	$49.35	$394.80	$32.60	$49.35
1 Feller Buncher, 50 H.P.		431.20		474.30	53.90	59.29
8 L.H., Daily Totals		$692.00		$869.10	$86.50	$108.64

Crew No.	Bare Costs		Incl. Subs O & P		Cost Per Labor-Hour	

Crew B-94A	Hr.	Daily	Hr.	Daily	Bare Costs	Incl. O&P
1 Laborer	$24.65	$197.20	$38.50	$308.00	$24.65	$38.50
1 Diaph. Water Pump, 2"		44.40		48.85		
1-20 Ft. Suction Hose, 2"		4.35		4.80		
2-50 Ft. Disch. Hoses, 2"		7.40		8.15	7.02	7.72
8 L.H., Daily Totals		$253.35		$369.80	$31.67	$46.22

Crew B-94B	Hr.	Daily	Hr.	Daily	Bare Costs	Incl. O&P
1 Laborer	$24.65	$197.20	$38.50	$308.00	$24.65	$38.50
1 Diaph. Water Pump, 4"		85.20		93.70		
1-20 Ft. Suction Hose, 4"		8.90		9.80		
2-50 Ft. Disch. Hoses, 4"		12.50		13.75	13.33	14.66
8 L.H., Daily Totals		$303.80		$425.25	$37.98	$53.16

Crew B-94C	Hr.	Daily	Hr.	Daily	Bare Costs	Incl. O&P
1 Laborer	$24.65	$197.20	$38.50	$308.00	$24.65	$38.50
1 Centr. Water Pump, 3"		51.60		56.75		
1-20 Ft. Suction Hose, 3"		6.65		7.30		
2-50 Ft. Disch. Hoses, 3"		9.10		10.00	8.42	9.26
8 L.H., Daily Totals		$264.55		$382.05	$33.07	$47.76

Crew B-94D	Hr.	Daily	Hr.	Daily	Bare Costs	Incl. O&P
1 Laborer	$24.65	$197.20	$38.50	$308.00	$24.65	$38.50
1 Centr. Water Pump, 6"		224.40		246.85		
1-20 Ft. Suction Hose, 6"		17.10		18.80		
2-50 Ft. Disch. Hoses, 6"		29.40		32.35	33.86	37.25
8 L.H., Daily Totals		$468.10		$606.00	$58.51	$75.75

Crew B-95A	Hr.	Daily	Hr.	Daily	Bare Costs	Incl. O&P
1 Equip. Oper. (crane)	$33.70	$269.60	$51.00	$408.00	$29.18	$44.75
1 Laborer	24.65	197.20	38.50	308.00		
1 Hyd. Excavator, 5/8 C.Y.		440.00		484.00	27.50	30.25
16 L.H., Daily Totals		$906.80		$1200.00	$56.68	$75.00

Crew B-95B	Hr.	Daily	Hr.	Daily	Bare Costs	Incl. O&P
1 Equip. Oper. (crane)	$33.70	$269.60	$51.00	$408.00	$29.18	$44.75
1 Laborer	24.65	197.20	38.50	308.00		
1 Hyd. Excavator, 1.5 C.Y.		728.80		801.70	45.55	50.11
16 L.H., Daily Totals		$1195.60		$1517.70	$74.73	$94.86

Crew B-95C	Hr.	Daily	Hr.	Daily	Bare Costs	Incl. O&P
1 Equip. Oper. (crane)	$33.70	$269.60	$51.00	$408.00	$29.18	$44.75
1 Laborer	24.65	197.20	38.50	308.00		
1 Hyd. Excavator, 2.5 C.Y.		1450.00		1595.00	90.63	99.69
16 L.H., Daily Totals		$1916.80		$2311.00	$119.81	$144.44

Crew C-1	Hr.	Daily	Hr.	Daily	Bare Costs	Incl. O&P
3 Carpenters	$31.55	$757.20	$49.30	$1183.20	$29.83	$46.60
1 Laborer	24.65	197.20	38.50	308.00		
32 L.H., Daily Totals		$954.40		$1491.20	$29.83	$46.60

Crew C-2	Hr.	Daily	Hr.	Daily	Bare Costs	Incl. O&P
1 Carpenter Foreman (out)	$33.55	$268.40	$52.40	$419.20	$30.73	$48.02
4 Carpenters	31.55	1009.60	49.30	1577.60		
1 Laborer	24.65	197.20	38.50	308.00		
48 L.H., Daily Totals		$1475.20		$2304.80	$30.73	$48.02

Crew C-2A	Hr.	Daily	Hr.	Daily	Bare Costs	Incl. O&P
1 Carpenter Foreman (out)	$33.55	$268.40	$52.40	$419.20	$30.51	$47.26
3 Carpenters	31.55	757.20	49.30	1183.20		
1 Cement Finisher	30.20	241.60	44.75	358.00		
1 Laborer	24.65	197.20	38.50	308.00		
48 L.H., Daily Totals		$1464.40		$2268.40	$30.51	$47.26

Crew C-3	Hr.	Daily	Hr.	Daily	Bare Costs	Incl. O&P
1 Rodman Foreman	$37.55	$300.40	$63.10	$504.80	$32.52	$53.25
4 Rodmen (reinf.)	35.55	1137.60	59.70	1910.40		
1 Equip. Oper. (light)	31.10	248.80	47.10	376.80		
2 Laborers	24.65	394.40	38.50	616.00		
3 Stressing Equipment		60.00		66.00		
.5 Grouting Equipment		88.97		97.85	2.33	2.56
64 L.H., Daily Totals		$2230.17		$3571.85	$34.85	$55.81

Crew C-4	Hr.	Daily	Hr.	Daily	Bare Costs	Incl. O&P
1 Rodman Foreman	$37.55	$300.40	$63.10	$504.80	$36.05	$60.55
3 Rodmen (reinf.)	35.55	853.20	59.70	1432.80		
3 Stressing Equipment		60.00		66.00	1.88	2.06
32 L.H., Daily Totals		$1213.60		$2003.60	$37.93	$62.61

Crew C-5	Hr.	Daily	Hr.	Daily	Bare Costs	Incl. O&P
1 Rodman Foreman	$37.55	$300.40	$63.10	$504.80	$34.54	$56.54
4 Rodmen (reinf.)	35.55	1137.60	59.70	1910.40		
1 Equip. Oper. (crane)	33.70	269.60	51.00	408.00		
1 Equip. Oper. Oiler	28.30	226.40	42.85	342.80		
1 Hyd. Crane, 25 Ton		654.60		720.05	11.69	12.86
56 L.H., Daily Totals		$2588.60		$3886.05	$46.23	$69.40

Crew C-6	Hr.	Daily	Hr.	Daily	Bare Costs	Incl. O&P
1 Labor Foreman (outside)	$26.65	$213.20	$41.65	$333.20	$25.91	$40.07
4 Laborers	24.65	788.80	38.50	1232.00		
1 Cement Finisher	30.20	241.60	44.75	358.00		
2 Gas Engine Vibrators		65.20		71.70	1.36	1.49
48 L.H., Daily Totals		$1308.80		$1994.90	$27.27	$41.56

Crew C-7	Hr.	Daily	Hr.	Daily	Bare Costs	Incl. O&P
1 Labor Foreman (outside)	$26.65	$213.20	$41.65	$333.20	$26.78	$41.23
5 Laborers	24.65	986.00	38.50	1540.00		
1 Cement Finisher	30.20	241.60	44.75	358.00		
1 Equip. Oper. (med.)	32.60	260.80	49.35	394.80		
1 Equip. Oper. (oiler)	28.30	226.40	42.85	342.80		
2 Gas Engine Vibrators		65.20		71.70		
1 Concrete Bucket, 1 C.Y.		20.40		22.45		
1 Hyd. Crane, 55 Ton		1156.00		1271.60	17.24	18.97
72 L.H., Daily Totals		$3169.60		$4334.55	$44.02	$60.20

Crew C-7A	Hr.	Daily	Hr.	Daily	Bare Costs	Incl. O&P
1 Labor Foreman (outside)	$26.65	$213.20	$41.65	$333.20	$25.15	$39.06
5 Laborers	24.65	986.00	38.50	1540.00		
2 Truck Drivers (Heavy)	25.65	410.40	39.15	626.40		
2 Conc. Transit Mixers		1513.60		1664.95	23.65	26.02
64 L.H., Daily Totals		$3123.20		$4164.55	$48.80	$65.08

Crew No.	Bare Costs		Incl. Subs O & P		Cost Per Labor-Hour	

Crew C-7B

	Hr.	Daily	Hr.	Daily	Bare Costs	Incl. O&P
1 Labor Foreman (outside)	$26.65	$213.20	$41.65	$333.20	$26.49	$41.00
5 Laborers	24.65	986.00	38.50	1540.00		
1 Equipment Operator (heavy)	33.70	269.60	51.00	408.00		
1 Equipment Oiler	28.30	226.40	42.85	342.80		
1 Conc. Bucket, 2 C.Y.		29.60		32.55		
1 Truck Crane, 165 Ton		1804.00		1984.40	28.65	31.52
64 L.H., Daily Totals		$3528.80		$4640.95	$55.14	$72.52

Crew C-7C

	Hr.	Daily	Hr.	Daily	Bare Costs	Incl. O&P
1 Labor Foreman (outside)	$26.65	$213.20	$41.65	$333.20	$26.89	$41.61
5 Laborers	24.65	986.00	38.50	1540.00		
2 Equipment Operator (medium)	32.60	521.60	49.35	789.60		
2 Wheel Loader, 4 C.Y.		928.80		1021.70	14.51	15.96
64 L.H., Daily Totals		$2649.60		$3684.50	$41.40	$57.57

Crew C-7D

	Hr.	Daily	Hr.	Daily	Bare Costs	Incl. O&P
1 Labor Foreman (outside)	$26.65	$213.20	$41.65	$333.20	$26.07	$40.50
5 Laborers	24.65	986.00	38.50	1540.00		
1 Equipment Operator (med.)	32.60	260.80	49.35	394.80		
1 Concrete Conveyer		154.80		170.30	2.76	3.04
56 L.H., Daily Totals		$1614.80		$2438.30	$28.83	$43.54

Crew C-8

	Hr.	Daily	Hr.	Daily	Bare Costs	Incl. O&P
1 Labor Foreman (outside)	$26.65	$213.20	$41.65	$333.20	$27.66	$42.29
3 Laborers	24.65	591.60	38.50	924.00		
2 Cement Finishers	30.20	483.20	44.75	716.00		
1 Equip. Oper. (med.)	32.60	260.80	49.35	394.80		
1 Concrete Pump (small)		737.20		810.90	13.16	14.48
56 L.H., Daily Totals		$2286.00		$3178.90	$40.82	$56.77

Crew C-8A

	Hr.	Daily	Hr.	Daily	Bare Costs	Incl. O&P
1 Labor Foreman (outside)	$26.65	$213.20	$41.65	$333.20	$26.83	$41.11
3 Laborers	24.65	591.60	38.50	924.00		
2 Cement Finishers	30.20	483.20	44.75	716.00		
48 L.H., Daily Totals		$1288.00		$1973.20	$26.83	$41.11

Crew C-8B

	Hr.	Daily	Hr.	Daily	Bare Costs	Incl. O&P
1 Labor Foreman (outside)	$26.65	$213.20	$41.65	$333.20	$26.64	$41.30
3 Laborers	24.65	591.60	38.50	924.00		
1 Equipment Operator	32.60	260.80	49.35	394.80		
1 Vibrating Screed		46.55		51.20		
1 Vibratory Roller		447.60		492.35		
1 Dozer, 200 H.P.		880.80		968.90	34.37	37.81
40 L.H., Daily Totals		$2440.55		$3164.45	$61.01	$79.11

Crew C-8C

	Hr.	Daily	Hr.	Daily	Bare Costs	Incl. O&P
1 Labor Foreman (outside)	$26.65	$213.20	$41.65	$333.20	$27.23	$41.88
3 Laborers	24.65	591.60	38.50	924.00		
1 Cement Finisher	30.20	241.60	44.75	358.00		
1 Equipment Operator (med.)	32.60	260.80	49.35	394.80		
1 Shotcrete Rig, 12 CY/hr		281.20		309.30	5.86	6.44
48 L.H., Daily Totals		$1588.40		$2319.30	$33.09	$48.32

Crew C-8D

	Hr.	Daily	Hr.	Daily	Bare Costs	Incl. O&P
1 Labor Foreman (outside)	$26.65	$213.20	$41.65	$333.20	$28.15	$43.00
1 Laborer	24.65	197.20	38.50	308.00		
1 Cement Finisher	30.20	241.60	44.75	358.00		
1 Equipment Operator (light)	31.10	248.80	47.10	376.80		
1 Compressor, 250 CFM		128.60		141.45		
2 Hoses, 1", 50'		7.30		8.05	4.25	4.67
32 L.H., Daily Totals		$1036.70		$1525.50	$32.40	$47.67

Crew C-8E

	Hr.	Daily	Hr.	Daily	Bare Costs	Incl. O&P
1 Labor Foreman (outside)	$26.65	$213.20	$41.65	$333.20	$28.15	$43.00
1 Laborer	24.65	197.20	38.50	308.00		
1 Cement Finisher	30.20	241.60	44.75	358.00		
1 Equipment Operator (light)	31.10	248.80	47.10	376.80		
1 Compressor, 250 CFM		128.60		141.45		
2 Hoses, 1", 50'		7.30		8.05		
1 Concrete Pump (small)		737.20		810.90	27.28	30.01
32 L.H., Daily Totals		$1773.90		$2336.40	$55.43	$73.01

Crew C-10

	Hr.	Daily	Hr.	Daily	Bare Costs	Incl. O&P
1 Laborer	$24.65	$197.20	$38.50	$308.00	$28.35	$42.67
2 Cement Finishers	30.20	483.20	44.75	716.00		
24 L.H., Daily Totals		$680.40		$1024.00	$28.35	$42.67

Crew C-11

	Hr.	Daily	Hr.	Daily	Bare Costs	Incl. O&P
1 Struc. Steel Foreman	$37.65	$301.20	$68.35	$546.80	$34.84	$61.19
6 Struc. Steel Workers	35.65	1711.20	64.75	3108.00		
1 Equip. Oper. (crane)	33.70	269.60	51.00	408.00		
1 Equip. Oper. Oiler	28.30	226.40	42.85	342.80		
1 Truck Crane, 150 Ton		1538.00		1691.80	21.36	23.50
72 L.H., Daily Totals		$4046.40		$6097.40	$56.20	$84.69

Crew C-12

	Hr.	Daily	Hr.	Daily	Bare Costs	Incl. O&P
1 Carpenter Foreman (out)	$33.55	$268.40	$52.40	$419.20	$31.09	$48.30
3 Carpenters	31.55	757.20	49.30	1183.20		
1 Laborer	24.65	197.20	38.50	308.00		
1 Equip. Oper. (crane)	33.70	269.60	51.00	408.00		
1 Hyd. Crane, 12 Ton		636.00		699.60	13.25	14.58
48 L.H., Daily Totals		$2128.40		$3018.00	$44.34	$62.88

Crew C-13

	Hr.	Daily	Hr.	Daily	Bare Costs	Incl. O&P
1 Struc. Steel Worker	$35.65	$285.20	$64.75	$518.00	$34.28	$59.60
1 Welder	35.65	285.20	64.75	518.00		
1 Carpenter	31.55	252.40	49.30	394.40		
1 Gas Welding Machine		78.80		86.70	3.28	3.61
24 L.H., Daily Totals		$901.60		$1517.10	$37.56	$63.21

Crew C-14

	Hr.	Daily	Hr.	Daily	Bare Costs	Incl. O&P
1 Carpenter Foreman (out)	$33.55	$268.40	$52.40	$419.20	$30.81	$48.61
5 Carpenters	31.55	1262.00	49.30	1972.00		
4 Laborers	24.65	788.80	38.50	1232.00		
4 Rodmen (reinf.)	35.55	1137.60	59.70	1910.40		
2 Cement Finishers	30.20	483.20	44.75	716.00		
1 Equip. Oper. (crane)	33.70	269.60	51.00	408.00		
1 Equip. Oper. Oiler	28.30	226.40	42.85	342.80		
1 Crane, 80 Ton, & Tools		1093.00		1202.30	7.59	8.35
144 L.H., Daily Totals		$5529.00		$8202.70	$38.40	$56.96

Crew C-14A

	Hr.	Daily	Hr.	Daily	Bare Costs	Incl. O&P
1 Carpenter Foreman (out)	$33.55	$268.40	$52.40	$419.20	$31.71	$50.04
16 Carpenters	31.55	4038.40	49.30	6310.40		
4 Rodmen (reinf.)	35.55	1137.60	59.70	1910.40		
2 Laborers	24.65	394.40	38.50	616.00		
1 Cement Finisher	30.20	241.60	44.75	358.00		
1 Equip. Oper. (med.)	32.60	260.80	49.35	394.80		
1 Gas Engine Vibrator		32.60		35.85		
1 Concrete Pump (small)		737.20		810.90	3.85	4.23
200 L.H., Daily Totals		$7111.00		$10855.55	$35.56	$54.27

Crews

Crew C-14B

Crew No.	Hr.	Daily	Hr.	Daily	Bare Costs	Incl. O&P
1 Carpenter Foreman (out)	$33.55	$268.40	$52.40	$419.20	$31.65	$49.84
16 Carpenters	31.55	4038.40	49.30	6310.40		
4 Rodmen (reinf.)	35.55	1137.60	59.70	1910.40		
2 Laborers	24.65	394.40	38.50	616.00		
2 Cement Finishers	30.20	483.20	44.75	716.00		
1 Equip. Oper. (med.)	32.60	260.80	49.35	394.80		
1 Gas Engine Vibrator		32.60		35.85		
1 Concrete Pump (small)		737.20		810.90	3.70	4.07
208 L.H., Daily Totals		$7352.60		$11213.55	$35.35	$53.91

Crew C-14C

Crew No.	Hr.	Daily	Hr.	Daily	Bare Costs	Incl. O&P
1 Carpenter Foreman (out)	$33.55	$268.40	$52.40	$419.20	$30.20	$47.60
6 Carpenters	31.55	1514.40	49.30	2366.40		
2 Rodmen (reinf.)	35.55	568.80	59.70	955.20		
4 Laborers	24.65	788.80	38.50	1232.00		
1 Cement Finisher	30.20	241.60	44.75	358.00		
1 Gas Engine Vibrator		32.60		35.85	.29	.32
112 L.H., Daily Totals		$3414.60		$5366.65	$30.49	$47.92

Crew C-14D

Crew No.	Hr.	Daily	Hr.	Daily	Bare Costs	Incl. O&P
1 Carpenter Foreman (out)	$33.55	$268.40	$52.40	$419.20	$31.39	$49.21
18 Carpenters	31.55	4543.20	49.30	7099.20		
2 Rodmen (reinf.)	35.55	568.80	59.70	955.20		
2 Laborers	24.65	394.40	38.50	616.00		
1 Cement Finisher	30.20	241.60	44.75	358.00		
1 Equip. Oper. (med.)	32.60	260.80	49.35	394.80		
1 Gas Engine Vibrator		32.60		35.85		
1 Concrete Pump (small)		737.20		810.90	3.85	4.23
200 L.H., Daily Totals		$7047.00		$10689.15	$35.24	$53.44

Crew C-14E

Crew No.	Hr.	Daily	Hr.	Daily	Bare Costs	Incl. O&P
1 Carpenter Foreman (out)	$33.55	$268.40	$52.40	$419.20	$31.18	$50.00
2 Carpenters	31.55	504.80	49.30	788.80		
4 Rodmen (reinf.)	35.55	1137.60	59.70	1910.40		
3 Laborers	24.65	591.60	38.50	924.00		
1 Cement Finisher	30.20	241.60	44.75	358.00		
1 Gas Engine Vibrator		32.60		35.85	.37	.41
88 L.H., Daily Totals		$2776.60		$4436.25	$31.55	$50.41

Crew C-14F

Crew No.	Hr.	Daily	Hr.	Daily	Bare Costs	Incl. O&P
1 Laborer Foreman (out)	$26.65	$213.20	$41.65	$333.20	$28.57	$43.02
2 Laborers	24.65	394.40	38.50	616.00		
6 Cement Finishers	30.20	1449.60	44.75	2148.00		
1 Gas Engine Vibrator		32.60		35.85	.45	.50
72 L.H., Daily Totals		$2089.80		$3133.05	$29.02	$43.52

Crew C-14G

Crew No.	Hr.	Daily	Hr.	Daily	Bare Costs	Incl. O&P
1 Laborer Foreman (out)	$26.65	$213.20	$41.65	$333.20	$28.11	$42.52
2 Laborers	24.65	394.40	38.50	616.00		
4 Cement Finishers	30.20	966.40	44.75	1432.00		
1 Gas Engine Vibrator		32.60		35.85	.58	.64
56 L.H., Daily Totals		$1606.60		$2417.05	$28.69	$43.16

Crew C-14H

Crew No.	Hr.	Daily	Hr.	Daily	Bare Costs	Incl. O&P
1 Carpenter Foreman (out)	$33.55	$268.40	$52.40	$419.20	$31.18	$48.99
2 Carpenters	31.55	504.80	49.30	788.80		
1 Rodman (reinf.)	35.55	284.40	59.70	477.60		
1 Laborer	24.65	197.20	38.50	308.00		
1 Cement Finisher	30.20	241.60	44.75	358.00		
1 Gas Engine Vibrator		32.60		35.85	.68	.75
48 L.H., Daily Totals		$1529.00		$2387.45	$31.86	$49.74

Crew C-15

Crew No.	Hr.	Daily	Hr.	Daily	Bare Costs	Incl. O&P
1 Carpenter Foreman (out)	$33.55	$268.40	$52.40	$419.20	$29.62	$46.19
2 Carpenters	31.55	504.80	49.30	788.80		
3 Laborers	24.65	591.60	38.50	924.00		
2 Cement Finishers	30.20	483.20	44.75	716.00		
1 Rodman (reinf.)	35.55	284.40	59.70	477.60		
72 L.H., Daily Totals		$2132.40		$3325.60	$29.62	$46.19

Crew C-16

Crew No.	Hr.	Daily	Hr.	Daily	Bare Costs	Incl. O&P
1 Labor Foreman (outside)	$26.65	$213.20	$41.65	$333.20	$29.41	$46.16
3 Laborers	24.65	591.60	38.50	924.00		
2 Cement Finishers	30.20	483.20	44.75	716.00		
1 Equip. Oper. (med.)	32.60	260.80	49.35	394.80		
2 Rodmen (reinf.)	35.55	568.80	59.70	955.20		
1 Concrete Pump (small)		737.20		810.90	10.24	11.26
72 L.H., Daily Totals		$2854.80		$4134.10	$39.65	$57.42

Crew C-17

Crew No.	Hr.	Daily	Hr.	Daily	Bare Costs	Incl. O&P
2 Skilled Worker Foremen	$34.25	$548.00	$53.65	$858.40	$32.65	$51.13
8 Skilled Workers	32.25	2064.00	50.50	3232.00		
80 L.H., Daily Totals		$2612.00		$4090.40	$32.65	$51.13

Crew C-17A

Crew No.	Hr.	Daily	Hr.	Daily	Bare Costs	Incl. O&P
2 Skilled Worker Foremen	$34.25	$548.00	$53.65	$858.40	$32.66	$51.13
8 Skilled Workers	32.25	2064.00	50.50	3232.00		
.125 Equip. Oper. (crane)	33.70	33.70	51.00	51.00		
.125 Crane, 80 Ton, & Tools		136.63		150.30	1.69	1.86
81 L.H., Daily Totals		$2782.33		$4291.70	$34.35	$52.99

Crew C-17B

Crew No.	Hr.	Daily	Hr.	Daily	Bare Costs	Incl. O&P
2 Skilled Worker Foremen	$34.25	$548.00	$53.65	$858.40	$32.68	$51.13
8 Skilled Workers	32.25	2064.00	50.50	3232.00		
.25 Equip. Oper. (crane)	33.70	67.40	51.00	102.00		
.25 Crane, 80 Ton, & Tools		273.25		300.60		
.25 Walk Behind Power Tools		8.85		9.75	3.44	3.78
82 L.H., Daily Totals		$2961.50		$4502.75	$36.12	$54.91

Crew C-17C

Crew No.	Hr.	Daily	Hr.	Daily	Bare Costs	Incl. O&P
2 Skilled Worker Foremen	$34.25	$548.00	$53.65	$858.40	$32.69	$51.13
8 Skilled Workers	32.25	2064.00	50.50	3232.00		
.375 Equip. Oper. (crane)	33.70	101.10	51.00	153.00		
.375 Crane, 80 Ton & Tools		409.88		450.85	4.94	5.43
83 L.H., Daily Totals		$3122.98		$4694.25	$37.63	$56.56

Crew C-17D

Crew No.	Hr.	Daily	Hr.	Daily	Bare Costs	Incl. O&P
2 Skilled Worker Foremen	$34.25	$548.00	$53.65	$858.40	$32.70	$51.12
8 Skilled Workers	32.25	2064.00	50.50	3232.00		
.5 Equip. Oper. (crane)	33.70	134.80	51.00	204.00		
.5 Crane, 80 Ton & Tools		546.50		601.15	6.51	7.16
84 L.H., Daily Totals		$3293.30		$4895.55	$39.21	$58.28

Crew C-17E

Crew No.	Hr.	Daily	Hr.	Daily	Bare Costs	Incl. O&P
2 Skilled Worker Foremen	$34.25	$548.00	$53.65	$858.40	$32.65	$51.13
8 Skilled Workers	32.25	2064.00	50.50	3232.00		
1 Hyd. Jack with Rods		76.10		83.70	.95	1.05
80 L.H., Daily Totals		$2688.10		$4174.10	$33.60	$52.18

CREWS

Crew No.	Bare Costs Hr.	Daily	Incl. Subs O & P Hr.	Daily	Bare Costs	Incl. O&P
Crew C-18	**Hr.**	**Daily**	**Hr.**	**Daily**	**Bare Costs**	**Incl. O&P**
.125 Labor Foreman (out)	$26.65	$26.65	$41.65	$41.65	$24.87	$38.85
1 Laborer	24.65	197.20	38.50	308.00		
1 Concrete Cart, 10 C.F.		51.40		56.55	5.71	6.28
9 L.H., Daily Totals		$275.25		$406.20	$30.58	$45.13

Crew No.	Bare Costs Hr.	Daily	Incl. Subs O & P Hr.	Daily	Bare Costs	Incl. O&P
Crew C-19	**Hr.**	**Daily**	**Hr.**	**Daily**	**Bare Costs**	**Incl. O&P**
.125 Labor Foreman (out)	$26.65	$26.65	$41.65	$41.65	$24.87	$38.85
1 Laborer	24.65	197.20	38.50	308.00		
1 Concrete Cart, 18 C.F.		77.80		85.60	8.64	9.51
9 L.H., Daily Totals		$301.65		$435.25	$33.51	$48.36

Crew No.	Bare Costs Hr.	Daily	Incl. Subs O & P Hr.	Daily	Bare Costs	Incl. O&P
Crew C-20	**Hr.**	**Daily**	**Hr.**	**Daily**	**Bare Costs**	**Incl. O&P**
1 Labor Foreman (outside)	$26.65	$213.20	$41.65	$333.20	$26.59	$41.03
5 Laborers	24.65	986.00	38.50	1540.00		
1 Cement Finisher	30.20	241.60	44.75	358.00		
1 Equip. Oper. (med.)	32.60	260.80	49.35	394.80		
2 Gas Engine Vibrators		65.20		71.70		
1 Concrete Pump (small)		737.20		810.90	12.54	13.79
64 L.H., Daily Totals		$2504.00		$3508.60	$39.13	$54.82

Crew No.	Bare Costs Hr.	Daily	Incl. Subs O & P Hr.	Daily	Bare Costs	Incl. O&P
Crew C-21	**Hr.**	**Daily**	**Hr.**	**Daily**	**Bare Costs**	**Incl. O&P**
1 Labor Foreman (outside)	$26.65	$213.20	$41.65	$333.20	$26.59	$41.03
5 Laborers	24.65	986.00	38.50	1540.00		
1 Cement Finisher	30.20	241.60	44.75	358.00		
1 Equip. Oper. (med.)	32.60	260.80	49.35	394.80		
2 Gas Engine Vibrators		65.20		71.70		
1 Concrete Conveyer		154.80		170.30	3.44	3.78
64 L.H., Daily Totals		$1921.60		$2868.00	$30.03	$44.81

Crew No.	Bare Costs Hr.	Daily	Incl. Subs O & P Hr.	Daily	Bare Costs	Incl. O&P
Crew C-22	**Hr.**	**Daily**	**Hr.**	**Daily**	**Bare Costs**	**Incl. O&P**
1 Rodman Foreman	$37.55	$300.40	$63.10	$504.80	$35.71	$59.74
4 Rodmen (reinf.)	35.55	1137.60	59.70	1910.40		
.125 Equip. Oper. (crane)	33.70	33.70	51.00	51.00		
.125 Equip. Oper. Oiler	28.30	28.30	42.85	42.85		
.125 Hyd. Crane, 25 Ton		81.83		90.00	1.95	2.14
42 L.H., Daily Totals		$1581.83		$2599.05	$37.66	$61.88

Crew No.	Bare Costs Hr.	Daily	Incl. Subs O & P Hr.	Daily	Bare Costs	Incl. O&P
Crew C-23	**Hr.**	**Daily**	**Hr.**	**Daily**	**Bare Costs**	**Incl. O&P**
2 Skilled Worker Foremen	$34.25	$548.00	$53.65	$858.40	$32.40	$50.42
6 Skilled Workers	32.25	1548.00	50.50	2424.00		
1 Equip. Oper. (crane)	33.70	269.60	51.00	408.00		
1 Equip. Oper. Oiler	28.30	226.40	42.85	342.80		
1 Crane, 90 Ton		1203.00		1323.30	15.04	16.54
80 L.H., Daily Totals		$3795.00		$5356.50	$47.44	$66.96

Crew No.	Bare Costs Hr.	Daily	Incl. Subs O & P Hr.	Daily	Bare Costs	Incl. O&P
Crew C-24	**Hr.**	**Daily**	**Hr.**	**Daily**	**Bare Costs**	**Incl. O&P**
2 Skilled Worker Foremen	$34.25	$548.00	$53.65	$858.40	$32.40	$50.42
6 Skilled Workers	32.25	1548.00	50.50	2424.00		
1 Equip. Oper. (crane)	33.70	269.60	51.00	408.00		
1 Equip. Oper. Oiler	28.30	226.40	42.85	342.80		
1 Truck Crane, 150 Ton		1538.00		1691.80	19.23	21.15
80 L.H., Daily Totals		$4130.00		$5725.00	$51.63	$71.57

Crew No.	Bare Costs Hr.	Daily	Incl. Subs O & P Hr.	Daily	Bare Costs	Incl. O&P
Crew C-25	**Hr.**	**Daily**	**Hr.**	**Daily**	**Bare Costs**	**Incl. O&P**
2 Rodmen (reinf.)	$35.55	$568.80	$59.70	$955.20	$28.05	$47.40
2 Rodmen Helpers	20.55	328.80	35.10	561.60		
32 L.H., Daily Totals		$897.60		$1516.80	$28.05	$47.40

Crew No.	Bare Costs Hr.	Daily	Incl. Subs O & P Hr.	Daily	Bare Costs	Incl. O&P
Crew D-1	**Hr.**	**Daily**	**Hr.**	**Daily**	**Bare Costs**	**Incl. O&P**
1 Bricklayer	$32.40	$259.20	$49.80	$398.40	$28.50	$43.80
1 Bricklayer Helper	24.60	196.80	37.80	302.40		
16 L.H., Daily Totals		$456.00		$700.80	$28.50	$43.80

Crew No.	Bare Costs Hr.	Daily	Incl. Subs O & P Hr.	Daily	Bare Costs	Incl. O&P
Crew D-2	**Hr.**	**Daily**	**Hr.**	**Daily**	**Bare Costs**	**Incl. O&P**
3 Bricklayers	$32.40	$777.60	$49.80	$1195.20	$29.49	$45.39
2 Bricklayer Helpers	24.60	393.60	37.80	604.80		
.5 Carpenter	31.55	126.20	49.30	197.20		
44 L.H., Daily Totals		$1297.40		$1997.20	$29.49	$45.39

Crew No.	Bare Costs Hr.	Daily	Incl. Subs O & P Hr.	Daily	Bare Costs	Incl. O&P
Crew D-3	**Hr.**	**Daily**	**Hr.**	**Daily**	**Bare Costs**	**Incl. O&P**
3 Bricklayers	$32.40	$777.60	$49.80	$1195.20	$29.39	$45.20
2 Bricklayer Helpers	24.60	393.60	37.80	604.80		
.25 Carpenter	31.55	63.10	49.30	98.60		
42 L.H., Daily Totals		$1234.30		$1898.60	$29.39	$45.20

Crew No.	Bare Costs Hr.	Daily	Incl. Subs O & P Hr.	Daily	Bare Costs	Incl. O&P
Crew D-4	**Hr.**	**Daily**	**Hr.**	**Daily**	**Bare Costs**	**Incl. O&P**
1 Bricklayer	$32.40	$259.20	$49.80	$398.40	$28.18	$43.13
2 Bricklayer Helpers	24.60	393.60	37.80	604.80		
1 Equip. Oper. (light)	31.10	248.80	47.10	376.80		
1 Grout Pump, 50 C.F./hr		120.15		132.15		
1 Hoses & Hopper		22.40		24.65		
1 Accessories		11.50		12.65	4.81	5.30
32 L.H., Daily Totals		$1055.65		$1549.45	$32.99	$48.43

Crew No.	Bare Costs Hr.	Daily	Incl. Subs O & P Hr.	Daily	Bare Costs	Incl. O&P
Crew D-5	**Hr.**	**Daily**	**Hr.**	**Daily**	**Bare Costs**	**Incl. O&P**
1 Bricklayer	$32.40	$259.20	$49.80	$398.40	$32.40	$49.80
8 L.H., Daily Totals		$259.20		$398.40	$32.40	$49.80

Crew No.	Bare Costs Hr.	Daily	Incl. Subs O & P Hr.	Daily	Bare Costs	Incl. O&P
Crew D-6	**Hr.**	**Daily**	**Hr.**	**Daily**	**Bare Costs**	**Incl. O&P**
3 Bricklayers	$32.40	$777.60	$49.80	$1195.20	$28.62	$44.02
3 Bricklayer Helpers	24.60	590.40	37.80	907.20		
.25 Carpenter	31.55	63.10	49.30	98.60		
50 L.H., Daily Totals		$1431.10		$2201.00	$28.62	$44.02

Crew No.	Bare Costs Hr.	Daily	Incl. Subs O & P Hr.	Daily	Bare Costs	Incl. O&P
Crew D-7	**Hr.**	**Daily**	**Hr.**	**Daily**	**Bare Costs**	**Incl. O&P**
1 Tile Layer	$30.40	$243.20	$44.85	$358.80	$27.15	$40.05
1 Tile Layer Helper	23.90	191.20	35.25	282.00		
16 L.H., Daily Totals		$434.40		$640.80	$27.15	$40.05

Crew No.	Bare Costs Hr.	Daily	Incl. Subs O & P Hr.	Daily	Bare Costs	Incl. O&P
Crew D-8	**Hr.**	**Daily**	**Hr.**	**Daily**	**Bare Costs**	**Incl. O&P**
3 Bricklayers	$32.40	$777.60	$49.80	$1195.20	$29.28	$45.00
2 Bricklayer Helpers	24.60	393.60	37.80	604.80		
40 L.H., Daily Totals		$1171.20		$1800.00	$29.28	$45.00

Crew No.	Bare Costs Hr.	Daily	Incl. Subs O & P Hr.	Daily	Bare Costs	Incl. O&P
Crew D-9	**Hr.**	**Daily**	**Hr.**	**Daily**	**Bare Costs**	**Incl. O&P**
3 Bricklayers	$32.40	$777.60	$49.80	$1195.20	$28.50	$43.80
3 Bricklayer Helpers	24.60	590.40	37.80	907.20		
48 L.H., Daily Totals		$1368.00		$2102.40	$28.50	$43.80

Crew No.	Bare Costs Hr.	Daily	Incl. Subs O & P Hr.	Daily	Bare Costs	Incl. O&P
Crew D-10	**Hr.**	**Daily**	**Hr.**	**Daily**	**Bare Costs**	**Incl. O&P**
1 Bricklayer Foreman	$34.40	$275.20	$52.85	$422.80	$29.94	$45.85
1 Bricklayer	32.40	259.20	49.80	398.40		
2 Bricklayer Helpers	24.60	393.60	37.80	604.80		
1 Equip. Oper. (crane)	33.70	269.60	51.00	408.00		
1 Truck Crane, 12.5 Ton		493.40		542.75	12.34	13.57
40 L.H., Daily Totals		$1691.00		$2376.75	$42.28	$59.42

Crews

Crew No.	Bare Costs Hr.	Daily	Incl. Subs O & P Hr.	Daily	Cost Per Labor-Hour Bare Costs	Incl. O&P
Crew D-11	**Hr.**	**Daily**	**Hr.**	**Daily**	**Bare Costs**	**Incl. O&P**
1 Bricklayer Foreman	$34.40	$275.20	$52.85	$422.80	$30.47	$46.82
1 Bricklayer	32.40	259.20	49.80	398.40		
1 Bricklayer Helper	24.60	196.80	37.80	302.40		
24 L.H., Daily Totals		$731.20		$1123.60	$30.47	$46.82
Crew D-12	**Hr.**	**Daily**	**Hr.**	**Daily**	**Bare Costs**	**Incl. O&P**
1 Bricklayer Foreman	$34.40	$275.20	$52.85	$422.80	$29.00	$44.56
1 Bricklayer	32.40	259.20	49.80	398.40		
2 Bricklayer Helpers	24.60	393.60	37.80	604.80		
32 L.H., Daily Totals		$928.00		$1426.00	$29.00	$44.56
Crew D-13	**Hr.**	**Daily**	**Hr.**	**Daily**	**Bare Costs**	**Incl. O&P**
1 Bricklayer Foreman	$34.40	$275.20	$52.85	$422.80	$30.21	$46.42
1 Bricklayer	32.40	259.20	49.80	398.40		
2 Bricklayer Helpers	24.60	393.60	37.80	604.80		
1 Carpenter	31.55	252.40	49.30	394.40		
1 Equip. Oper. (crane)	33.70	269.60	51.00	408.00		
1 Truck Crane, 12.5 Ton		493.40		542.75	10.28	11.31
48 L.H., Daily Totals		$1943.40		$2771.15	$40.49	$57.73
Crew E-1	**Hr.**	**Daily**	**Hr.**	**Daily**	**Bare Costs**	**Incl. O&P**
1 Welder Foreman	$37.65	$301.20	$68.35	$546.80	$34.80	$60.07
1 Welder	35.65	285.20	64.75	518.00		
1 Equip. Oper. (light)	31.10	248.80	47.10	376.80		
1 Gas Welding Machine		78.80		86.70	3.28	3.61
24 L.H., Daily Totals		$914.00		$1528.30	$38.08	$63.68
Crew E-2	**Hr.**	**Daily**	**Hr.**	**Daily**	**Bare Costs**	**Incl. O&P**
1 Struc. Steel Foreman	$37.65	$301.20	$68.35	$546.80	$34.61	$60.17
4 Struc. Steel Workers	35.65	1140.80	64.75	2072.00		
1 Equip. Oper. (crane)	33.70	269.60	51.00	408.00		
1 Equip. Oper. Oiler	28.30	226.40	42.85	342.80		
1 Crane, 90 Ton		1203.00		1323.30	21.48	23.63
56 L.H., Daily Totals		$3141.00		$4692.90	$56.09	$83.80
Crew E-3	**Hr.**	**Daily**	**Hr.**	**Daily**	**Bare Costs**	**Incl. O&P**
1 Struc. Steel Foreman	$37.65	$301.20	$68.35	$546.80	$36.32	$65.95
1 Struc. Steel Worker	35.65	285.20	64.75	518.00		
1 Welder	35.65	285.20	64.75	518.00		
1 Gas Welding Machine		78.80		86.70	3.28	3.61
24 L.H., Daily Totals		$950.40		$1669.50	$39.60	$69.56
Crew E-4	**Hr.**	**Daily**	**Hr.**	**Daily**	**Bare Costs**	**Incl. O&P**
1 Struc. Steel Foreman	$37.65	$301.20	$68.35	$546.80	$36.15	$65.65
3 Struc. Steel Workers	35.65	855.60	64.75	1554.00		
1 Gas Welding Machine		78.80		86.70	2.46	2.71
32 L.H., Daily Totals		$1235.60		$2187.50	$38.61	$68.36
Crew E-5	**Hr.**	**Daily**	**Hr.**	**Daily**	**Bare Costs**	**Incl. O&P**
2 Struc. Steel Foremen	$37.65	$602.40	$68.35	$1093.60	$35.12	$61.90
5 Struc. Steel Workers	35.65	1426.00	64.75	2590.00		
1 Equip. Oper. (crane)	33.70	269.60	51.00	408.00		
1 Welder	35.65	285.20	64.75	518.00		
1 Equip. Oper. Oiler	28.30	226.40	42.85	342.80		
1 Crane, 90 Ton		1203.00		1323.30		
1 Gas Welding Machine		78.80		86.70	16.02	17.62
80 L.H., Daily Totals		$4091.40		$6362.40	$51.14	$79.52

Crew No.	Bare Costs Hr.	Daily	Incl. Subs O & P Hr.	Daily	Cost Per Labor-Hour Bare Costs	Incl. O&P
Crew E-6	**Hr.**	**Daily**	**Hr.**	**Daily**	**Bare Costs**	**Incl. O&P**
3 Struc. Steel Foremen	$37.65	$903.60	$68.35	$1640.40	$35.16	$62.09
9 Struc. Steel Workers	35.65	2566.80	64.75	4662.00		
1 Equip. Oper. (crane)	33.70	269.60	51.00	408.00		
1 Welder	35.65	285.20	64.75	518.00		
1 Equip. Oper. Oiler	28.30	226.40	42.85	342.80		
1 Equip. Oper. (light)	31.10	248.80	47.10	376.80		
1 Crane, 90 Ton		1203.00		1323.30		
1 Gas Welding Machine		78.80		86.70		
1 Air Compr., 160 C.F.M.		86.40		95.05		
2 Impact Wrenches		36.80		40.50	10.98	12.07
128 L.H., Daily Totals		$5905.40		$9493.55	$46.14	$74.16
Crew E-7	**Hr.**	**Daily**	**Hr.**	**Daily**	**Bare Costs**	**Incl. O&P**
1 Struc. Steel Foreman	$37.65	$301.20	$68.35	$546.80	$35.12	$61.90
4 Struc. Steel Workers	35.65	1140.80	64.75	2072.00		
1 Equip. Oper. (crane)	33.70	269.60	51.00	408.00		
1 Equip. Oper. Oiler	28.30	226.40	42.85	342.80		
1 Welder Foreman	37.65	301.20	68.35	546.80		
2 Welders	35.65	570.40	64.75	1036.00		
1 Crane, 90 Ton		1203.00		1323.30		
2 Gas Welding Machines		157.60		173.35	17.01	18.71
80 L.H., Daily Totals		$4170.20		$6449.05	$52.13	$80.61
Crew E-8	**Hr.**	**Daily**	**Hr.**	**Daily**	**Bare Costs**	**Incl. O&P**
1 Struc. Steel Foreman	$37.65	$301.20	$68.35	$546.80	$34.89	$61.20
4 Struc. Steel Workers	35.65	1140.80	64.75	2072.00		
1 Welder Foreman	37.65	301.20	68.35	546.80		
4 Welders	35.65	1140.80	64.75	2072.00		
1 Equip. Oper. (crane)	33.70	269.60	51.00	408.00		
1 Equip. Oper. Oiler	28.30	226.40	42.85	342.80		
1 Equip. Oper. (light)	31.10	248.80	47.10	376.80		
1 Crane, 90 Ton		1203.00		1323.30		
4 Gas Welding Machines		315.20		346.70	14.60	16.06
104 L.H., Daily Totals		$5147.00		$8035.20	$49.49	$77.26
Crew E-9	**Hr.**	**Daily**	**Hr.**	**Daily**	**Bare Costs**	**Incl. O&P**
2 Struc. Steel Foremen	$37.65	$602.40	$68.35	$1093.60	$35.16	$62.09
5 Struc. Steel Workers	35.65	1426.00	64.75	2590.00		
1 Welder Foreman	37.65	301.20	68.35	546.80		
5 Welders	35.65	1426.00	64.75	2590.00		
1 Equip. Oper. (crane)	33.70	269.60	51.00	408.00		
1 Equip. Oper. Oiler	28.30	226.40	42.85	342.80		
1 Equip. Oper. (light)	31.10	248.80	47.10	376.80		
1 Crane, 90 Ton		1203.00		1323.30		
5 Gas Welding Machines		394.00		433.40	12.48	13.72
128 L.H., Daily Totals		$6097.40		$9704.70	$47.64	$75.81
Crew E-10	**Hr.**	**Daily**	**Hr.**	**Daily**	**Bare Costs**	**Incl. O&P**
1 Welder Foreman	$37.65	$301.20	$68.35	$546.80	$36.65	$66.55
1 Welder	35.65	285.20	64.75	518.00		
1 Gas Welding Machine		78.80		86.70		
1 Truck, 3 Ton		138.00		151.80	13.55	14.91
16 L.H., Daily Totals		$803.20		$1303.30	$50.20	$81.46
Crew E-11	**Hr.**	**Daily**	**Hr.**	**Daily**	**Bare Costs**	**Incl. O&P**
2 Painters, Struc. Steel	$28.90	$462.40	$54.25	$868.00	$28.39	$48.53
1 Building Laborer	24.65	197.20	38.50	308.00		
1 Equip. Oper. (light)	31.10	248.80	47.10	376.80		
1 Air Compressor 250 C.F.M.		128.60		141.45		
1 Sand Blaster		22.40		24.65		
1 Sand Blasting Accessories		11.50		12.65	5.08	5.59
32 L.H., Daily Totals		$1070.90		$1731.55	$33.47	$54.12

CREWS

Crew No.	Bare Costs		Incl. Subs O & P		Cost Per Labor-Hour	
Crew E-12	Hr.	Daily	Hr.	Daily	Bare Costs	Incl. O&P
1 Welder Foreman	$37.65	$301.20	$68.35	$546.80	$34.38	$57.72
1 Equip. Oper. (light)	31.10	248.80	47.10	376.80		
1 Gas Welding Machine		78.80		86.70	4.93	5.42
16 L.H., Daily Totals		$628.80		$1010.30	$39.31	$63.14

Crew No.	Bare Costs		Incl. Subs O & P		Cost Per Labor-Hour	
Crew E-13	Hr.	Daily	Hr.	Daily	Bare Costs	Incl. O&P
1 Welder Foreman	$37.65	$301.20	$68.35	$546.80	$35.47	$61.27
.5 Equip. Oper. (light)	31.10	124.40	47.10	188.40		
1 Gas Welding Machine		78.80		86.70	6.57	7.22
12 L.H., Daily Totals		$504.40		$821.90	$42.04	$68.49

Crew No.	Bare Costs		Incl. Subs O & P		Cost Per Labor-Hour	
Crew E-14	Hr.	Daily	Hr.	Daily	Bare Costs	Incl. O&P
1 Welder Foreman	$37.65	$301.20	$68.35	$546.80	$37.65	$68.35
1 Gas Welding Machine		78.80		86.70	9.85	10.84
8 L.H., Daily Totals		$380.00		$633.50	$47.50	$79.19

Crew No.	Bare Costs		Incl. Subs O & P		Cost Per Labor-Hour	
Crew E-16	Hr.	Daily	Hr.	Daily	Bare Costs	Incl. O&P
1 Welder Foreman	$37.65	$301.20	$68.35	$546.80	$36.65	$66.55
1 Welder	35.65	285.20	64.75	518.00		
1 Gas Welding Machine		78.80		86.70	4.93	5.42
16 L.H., Daily Totals		$665.20		$1151.50	$41.58	$71.97

Crew No.	Bare Costs		Incl. Subs O & P		Cost Per Labor-Hour	
Crew E-17	Hr.	Daily	Hr.	Daily	Bare Costs	Incl. O&P
1 Structural Steel Foreman	$37.65	$301.20	$68.35	$546.80	$36.65	$66.55
1 Structural Steel Worker	35.65	285.20	64.75	518.00		
1 Power Tool		4.00		4.40	.25	.28
16 L.H., Daily Totals		$590.40		$1069.20	$36.90	$66.83

Crew No.	Bare Costs		Incl. Subs O & P		Cost Per Labor-Hour	
Crew E-18	Hr.	Daily	Hr.	Daily	Bare Costs	Incl. O&P
1 Structural Steel Foreman	$37.65	$301.20	$68.35	$546.80	$35.44	$62.39
3 Structural Steel Workers	35.65	855.60	64.75	1554.00		
1 Equipment Operator (med.)	32.60	260.80	49.35	394.80		
1 Crane, 20 Ton		681.30		749.45	17.03	18.74
40 L.H., Daily Totals		$2098.90		$3245.05	$52.47	$81.13

Crew No.	Bare Costs		Incl. Subs O & P		Cost Per Labor-Hour	
Crew E-19	Hr.	Daily	Hr.	Daily	Bare Costs	Incl. O&P
1 Structural Steel Worker	$35.65	$285.20	$64.75	$518.00	$34.80	$60.07
1 Structural Steel Foreman	37.65	301.20	68.35	546.80		
1 Equip. Oper. (light)	31.10	248.80	47.10	376.80		
1 Power Tool		4.00		4.40		
1 Crane, 20 Ton		681.30		749.45	28.55	31.41
24 L.H., Daily Totals		$1520.50		$2195.45	$63.35	$91.48

Crew No.	Bare Costs		Incl. Subs O & P		Cost Per Labor-Hour	
Crew E-20	Hr.	Daily	Hr.	Daily	Bare Costs	Incl. O&P
1 Structural Steel Foreman	$37.65	$301.20	$68.35	$546.80	$34.74	$60.74
5 Structural Steel Workers	35.65	1426.00	64.75	2590.00		
1 Equip. Oper. (crane)	33.70	269.60	51.00	408.00		
1 Oiler	28.30	226.40	42.85	342.80		
1 Power Tool		4.00		4.40		
1 Crane, 40 Ton		817.60		899.35	12.84	14.12
64 L.H., Daily Totals		$3044.80		$4791.35	$47.58	$74.86

Crew No.	Bare Costs		Incl. Subs O & P		Cost Per Labor-Hour	
Crew E-22	Hr.	Daily	Hr.	Daily	Bare Costs	Incl. O&P
1 Skilled Worker Foreman	$34.25	$274.00	$53.65	$429.20	$32.92	$51.55
2 Skilled Workers	32.25	516.00	50.50	808.00		
24 L.H., Daily Totals		$790.00		$1237.20	$32.92	$51.55

Crew No.	Bare Costs		Incl. Subs O & P		Cost Per Labor-Hour	
Crew E-24	Hr.	Daily	Hr.	Daily	Bare Costs	Incl. O&P
3 Structural Steel Workers	$35.65	$855.60	$64.75	$1554.00	$34.89	$60.90
1 Equipment Operator (medium)	32.60	260.80	49.35	394.80		
1-25 Ton Crane		654.60		720.05	20.46	22.50
32 L.H., Daily Totals		$1771.00		$2668.85	$55.35	$83.40

Crew No.	Bare Costs		Incl. Subs O & P		Cost Per Labor-Hour	
Crew E-25	Hr.	Daily	Hr.	Daily	Bare Costs	Incl. O&P
1 Welder Foreman	$37.65	$301.20	$68.35	$546.80	$37.65	$68.35
1 Cutting Torch		24.00		26.40		
1 Gases		36.00		39.60	7.50	8.25
8 L.H., Daily Totals		$361.20		$612.80	$45.15	$76.60

Crew No.	Bare Costs		Incl. Subs O & P		Cost Per Labor-Hour	
Crew F-3	Hr.	Daily	Hr.	Daily	Bare Costs	Incl. O&P
4 Carpenters	$31.55	$1009.60	$49.30	$1577.60	$31.98	$49.64
1 Equip. Oper. (crane)	33.70	269.60	51.00	408.00		
1 Hyd. Crane, 12 Ton		636.00		699.60	15.90	17.49
40 L.H., Daily Totals		$1915.20		$2685.20	$47.88	$67.13

Crew No.	Bare Costs		Incl. Subs O & P		Cost Per Labor-Hour	
Crew F-4	Hr.	Daily	Hr.	Daily	Bare Costs	Incl. O&P
4 Carpenters	$31.55	$1009.60	$49.30	$1577.60	$31.37	$48.51
1 Equip. Oper. (crane)	33.70	269.60	51.00	408.00		
1 Equip. Oper. Oiler	28.30	226.40	42.85	342.80		
1 Hyd. Crane, 55 Ton		1156.00		1271.60	24.08	26.49
48 L.H., Daily Totals		$2661.60		$3600.00	$55.45	$75.00

Crew No.	Bare Costs		Incl. Subs O & P		Cost Per Labor-Hour	
Crew F-5	Hr.	Daily	Hr.	Daily	Bare Costs	Incl. O&P
1 Carpenter Foreman	$33.55	$268.40	$52.40	$419.20	$32.05	$50.08
3 Carpenters	31.55	757.20	49.30	1183.20		
32 L.H., Daily Totals		$1025.60		$1602.40	$32.05	$50.08

Crew No.	Bare Costs		Incl. Subs O & P		Cost Per Labor-Hour	
Crew F-6	Hr.	Daily	Hr.	Daily	Bare Costs	Incl. O&P
2 Carpenters	$31.55	$504.80	$49.30	$788.80	$29.22	$45.32
2 Building Laborers	24.65	394.40	38.50	616.00		
1 Equip. Oper. (crane)	33.70	269.60	51.00	408.00		
1 Hyd. Crane, 12 Ton		636.00		699.60	15.90	17.49
40 L.H., Daily Totals		$1804.80		$2512.40	$45.12	$62.81

Crew No.	Bare Costs		Incl. Subs O & P		Cost Per Labor-Hour	
Crew F-7	Hr.	Daily	Hr.	Daily	Bare Costs	Incl. O&P
2 Carpenters	$31.55	$504.80	$49.30	$788.80	$28.10	$43.90
2 Building Laborers	24.65	394.40	38.50	616.00		
32 L.H., Daily Totals		$899.20		$1404.80	$28.10	$43.90

Crew No.	Bare Costs		Incl. Subs O & P		Cost Per Labor-Hour	
Crew G-1	Hr.	Daily	Hr.	Daily	Bare Costs	Incl. O&P
1 Roofer Foreman	$29.65	$237.20	$50.60	$404.80	$25.91	$44.23
4 Roofers, Composition	27.65	884.80	47.20	1510.40		
2 Roofer Helpers	20.55	328.80	35.10	561.60		
1 Application Equipment		148.00		162.80		
1 Tar Kettle/Pot		52.50		57.75		
1 Crew Truck		81.50		89.65	5.04	5.54
56 L.H., Daily Totals		$1732.80		$2787.00	$30.95	$49.77

Crew No.	Bare Costs		Incl. Subs O & P		Cost Per Labor-Hour	
Crew G-2	Hr.	Daily	Hr.	Daily	Bare Costs	Incl. O&P
1 Plasterer	$29.00	$232.00	$44.55	$356.40	$26.17	$40.40
1 Plasterer Helper	24.85	198.80	38.15	305.20		
1 Building Laborer	24.65	197.20	38.50	308.00		
1 Grouting Equipment		120.15		132.15	5.01	5.51
24 L.H., Daily Totals		$748.15		$1101.75	$31.18	$45.91

CREWS

Crew G-3

Crew No.	Bare Costs Hr.	Bare Costs Daily	Incl. Subs O & P Hr.	Incl. Subs O & P Daily	Cost Per Labor-Hour Bare Costs	Cost Per Labor-Hour Incl. O&P
2 Sheet Metal Workers	$37.00	$592.00	$57.15	$914.40	$30.83	$47.83
2 Building Laborers	24.65	394.40	38.50	616.00		
32 L.H., Daily Totals		$986.40		$1530.40	$30.83	$47.83

Crew G-4

Crew No.	Bare Costs Hr.	Bare Costs Daily	Incl. Subs O & P Hr.	Incl. Subs O & P Daily	Bare Costs	Incl. O&P
1 Labor Foreman (outside)	$26.65	$213.20	$41.65	$333.20	$25.32	$39.55
2 Building Laborers	24.65	394.40	38.50	616.00		
1 Light Truck, 1.5 Ton		120.20		132.20		
1 Air Compr., 160 C.F.M.		86.40		95.05	8.61	9.47
24 L.H., Daily Totals		$814.20		$1176.45	$33.93	$49.02

Crew G-5

Crew No.	Bare Costs Hr.	Bare Costs Daily	Incl. Subs O & P Hr.	Incl. Subs O & P Daily	Bare Costs	Incl. O&P
1 Roofer Foreman	$29.65	$237.20	$50.60	$404.80	$25.21	$43.04
2 Roofers, Composition	27.65	442.40	47.20	755.20		
2 Roofer Helpers	20.55	328.80	35.10	561.60		
1 Application Equipment		148.00		162.80	3.70	4.07
40 L.H., Daily Totals		$1156.40		$1884.40	$28.91	$47.11

Crew G-6A

Crew No.	Bare Costs Hr.	Bare Costs Daily	Incl. Subs O & P Hr.	Incl. Subs O & P Daily	Bare Costs	Incl. O&P
2 Roofers Composition	$27.65	$442.40	$47.20	$755.20	$27.65	$47.20
1 Small Compressor		17.85		19.65		
2 Pneumatic Nailers		38.40		42.25	3.52	3.87
16 L.H., Daily Totals		$498.65		$817.10	$31.17	$51.07

Crew G-7

Crew No.	Bare Costs Hr.	Bare Costs Daily	Incl. Subs O & P Hr.	Incl. Subs O & P Daily	Bare Costs	Incl. O&P
1 Carpenter	$31.55	$252.40	$49.30	$394.40	$31.55	$49.30
1 Small Compressor		17.85		19.65		
1 Pneumatic Nailer		19.20		21.10	4.63	5.09
8 L.H., Daily Totals		$289.45		$435.15	$36.18	$54.39

Crew H-1

Crew No.	Bare Costs Hr.	Bare Costs Daily	Incl. Subs O & P Hr.	Incl. Subs O & P Daily	Bare Costs	Incl. O&P
2 Glaziers	$30.95	$495.20	$47.00	$752.00	$33.30	$55.88
2 Struc. Steel Workers	35.65	570.40	64.75	1036.00		
32 L.H., Daily Totals		$1065.60		$1788.00	$33.30	$55.88

Crew H-2

Crew No.	Bare Costs Hr.	Bare Costs Daily	Incl. Subs O & P Hr.	Incl. Subs O & P Daily	Bare Costs	Incl. O&P
2 Glaziers	$30.95	$495.20	$47.00	$752.00	$28.85	$44.17
1 Building Laborer	24.65	197.20	38.50	308.00		
24 L.H., Daily Totals		$692.40		$1060.00	$28.85	$44.17

Crew H-3

Crew No.	Bare Costs Hr.	Bare Costs Daily	Incl. Subs O & P Hr.	Incl. Subs O & P Daily	Bare Costs	Incl. O&P
1 Glazier	$30.95	$247.60	$47.00	$376.00	$27.33	$42.03
1 Helper	23.70	189.60	37.05	296.40		
16 L.H., Daily Totals		$437.20		$672.40	$27.33	$42.03

Crew J-1

Crew No.	Bare Costs Hr.	Bare Costs Daily	Incl. Subs O & P Hr.	Incl. Subs O & P Daily	Bare Costs	Incl. O&P
3 Plasterers	$29.00	$696.00	$44.55	$1069.20	$27.34	$41.99
2 Plasterer Helpers	24.85	397.60	38.15	610.40		
1 Mixing Machine, 6 C.F.		88.00		96.80	2.20	2.42
40 L.H., Daily Totals		$1181.60		$1776.40	$29.54	$44.41

Crew J-2

Crew No.	Bare Costs Hr.	Bare Costs Daily	Incl. Subs O & P Hr.	Incl. Subs O & P Daily	Bare Costs	Incl. O&P
3 Plasterers	$29.00	$696.00	$44.55	$1069.20	$27.73	$42.36
2 Plasterer Helpers	24.85	397.60	38.15	610.40		
1 Lather	29.65	237.20	44.20	353.60		
1 Mixing Machine, 6 C.F.		88.00		96.80	1.83	2.02
48 L.H., Daily Totals		$1418.80		$2130.00	$29.56	$44.38

Crew J-3

Crew No.	Bare Costs Hr.	Bare Costs Daily	Incl. Subs O & P Hr.	Incl. Subs O & P Daily	Bare Costs	Incl. O&P
1 Terrazzo Worker	$30.40	$243.20	$44.85	$358.80	$27.55	$40.65
1 Terrazzo Helper	24.70	197.60	36.45	291.60		
1 Terrazzo Grinder, Electric		61.85		68.05		
1 Terrazzo Mixer		118.40		130.25	11.27	12.39
16 L.H., Daily Totals		$621.05		$848.70	$38.82	$53.04

Crew J-4

Crew No.	Bare Costs Hr.	Bare Costs Daily	Incl. Subs O & P Hr.	Incl. Subs O & P Daily	Bare Costs	Incl. O&P
1 Tile Layer	$30.40	$243.20	$44.85	$358.80	$27.15	$40.05
1 Tile Layer Helper	23.90	191.20	35.25	282.00		
16 L.H., Daily Totals		$434.40		$640.80	$27.15	$40.05

Crew K-1

Crew No.	Bare Costs Hr.	Bare Costs Daily	Incl. Subs O & P Hr.	Incl. Subs O & P Daily	Bare Costs	Incl. O&P
1 Carpenter	$31.55	$252.40	$49.30	$394.40	$28.23	$43.65
1 Truck Driver (light)	24.90	199.20	38.00	304.00		
1 Truck w/Power Equip.		138.00		151.80	8.63	9.49
16 L.H., Daily Totals		$589.60		$850.20	$36.86	$53.14

Crew K-2

Crew No.	Bare Costs Hr.	Bare Costs Daily	Incl. Subs O & P Hr.	Incl. Subs O & P Daily	Bare Costs	Incl. O&P
1 Struc. Steel Foreman	$37.65	$301.20	$68.35	$546.80	$32.73	$57.03
1 Struc. Steel Worker	35.65	285.20	64.75	518.00		
1 Truck Driver (light)	24.90	199.20	38.00	304.00		
1 Truck w/Power Equip.		138.00		151.80	5.75	6.33
24 L.H., Daily Totals		$923.60		$1520.60	$38.48	$63.36

Crew L-1

Crew No.	Bare Costs Hr.	Bare Costs Daily	Incl. Subs O & P Hr.	Incl. Subs O & P Daily	Bare Costs	Incl. O&P
1 Electrician	$37.60	$300.80	$56.15	$449.20	$37.47	$56.28
1 Plumber	37.35	298.80	56.40	451.20		
16 L.H., Daily Totals		$599.60		$900.40	$37.47	$56.28

Crew L-2

Crew No.	Bare Costs Hr.	Bare Costs Daily	Incl. Subs O & P Hr.	Incl. Subs O & P Daily	Bare Costs	Incl. O&P
1 Carpenter	$31.55	$252.40	$49.30	$394.40	$27.63	$43.17
1 Carpenter Helper	23.70	189.60	37.05	296.40		
16 L.H., Daily Totals		$442.00		$690.80	$27.63	$43.17

Crew L-3

Crew No.	Bare Costs Hr.	Bare Costs Daily	Incl. Subs O & P Hr.	Incl. Subs O & P Daily	Bare Costs	Incl. O&P
1 Carpenter	$31.55	$252.40	$49.30	$394.40	$34.42	$52.98
.5 Electrician	37.60	150.40	56.15	224.60		
.5 Sheet Metal Worker	37.00	148.00	57.15	228.60		
16 L.H., Daily Totals		$550.80		$847.60	$34.42	$52.98

Crew L-3A

Crew No.	Bare Costs Hr.	Bare Costs Daily	Incl. Subs O & P Hr.	Incl. Subs O & P Daily	Bare Costs	Incl. O&P
1 Carpenter Foreman (outside)	$33.55	$268.40	$52.40	$419.20	$34.70	$53.98
.5 Sheet Metal Worker	37.00	148.00	57.15	228.60		
12 L.H., Daily Totals		$416.40		$647.80	$34.70	$53.98

Crew L-4

Crew No.	Bare Costs Hr.	Bare Costs Daily	Incl. Subs O & P Hr.	Incl. Subs O & P Daily	Bare Costs	Incl. O&P
2 Skilled Workers	$32.25	$516.00	$50.50	$808.00	$29.40	$46.02
1 Helper	23.70	189.60	37.05	296.40		
24 L.H., Daily Totals		$705.60		$1104.40	$29.40	$46.02

Crew L-5

Crew No.	Bare Costs Hr.	Bare Costs Daily	Incl. Subs O & P Hr.	Incl. Subs O & P Daily	Bare Costs	Incl. O&P
1 Struc. Steel Foreman	$37.65	$301.20	$68.35	$546.80	$35.66	$63.30
5 Struc. Steel Workers	35.65	1426.00	64.75	2590.00		
1 Equip. Oper. (crane)	33.70	269.60	51.00	408.00		
1 Hyd. Crane, 25 Ton		654.60		720.05	11.69	12.86
56 L.H., Daily Totals		$2651.40		$4264.85	$47.35	$76.16

Crew L-5A

Crew No.	Bare Costs Hr.	Daily	Incl. Subs O & P Hr.	Daily	Cost Per Labor-Hour Bare Costs	Incl. O&P
1 Structural Steel Foreman	$37.65	$301.20	$68.35	$546.80	$35.66	$62.21
2 Structural Steel Workers	35.65	570.40	64.75	1036.00		
1 Equip. Oper. (crane)	33.70	269.60	51.00	408.00		
1 Crane, SP, 25 Ton		657.80		723.60	20.56	22.61
32 L.H., Daily Totals		$1799.00		$2714.40	$56.22	$84.82

Crew L-6

Crew No.	Bare Costs Hr.	Daily	Incl. Subs O & P Hr.	Daily	Cost Per Labor-Hour Bare Costs	Incl. O&P
1 Plumber	$37.35	$298.80	$56.40	$451.20	$37.43	$56.32
.5 Electrician	37.60	150.40	56.15	224.60		
12 L.H., Daily Totals		$449.20		$675.80	$37.43	$56.32

Crew L-7

Crew No.	Bare Costs Hr.	Daily	Incl. Subs O & P Hr.	Daily	Cost Per Labor-Hour Bare Costs	Incl. O&P
2 Carpenters	$31.55	$504.80	$49.30	$788.80	$30.44	$47.19
1 Building Laborer	24.65	197.20	38.50	308.00		
.5 Electrician	37.60	150.40	56.15	224.60		
28 L.H., Daily Totals		$852.40		$1321.40	$30.44	$47.19

Crew L-8

Crew No.	Bare Costs Hr.	Daily	Incl. Subs O & P Hr.	Daily	Cost Per Labor-Hour Bare Costs	Incl. O&P
2 Carpenters	$31.55	$504.80	$49.30	$788.80	$32.71	$50.72
.5 Plumber	37.35	149.40	56.40	225.60		
20 L.H., Daily Totals		$654.20		$1014.40	$32.71	$50.72

Crew L-9

Crew No.	Bare Costs Hr.	Daily	Incl. Subs O & P Hr.	Daily	Cost Per Labor-Hour Bare Costs	Incl. O&P
1 Labor Foreman (inside)	$25.15	$201.20	$39.30	$314.40	$28.64	$46.47
2 Building Laborers	24.65	394.40	38.50	616.00		
1 Struc. Steel Worker	35.65	285.20	64.75	518.00		
.5 Electrician	37.60	150.40	56.15	224.60		
36 L.H., Daily Totals		$1031.20		$1673.00	$28.64	$46.47

Crew L-10

Crew No.	Bare Costs Hr.	Daily	Incl. Subs O & P Hr.	Daily	Cost Per Labor-Hour Bare Costs	Incl. O&P
1 Structural Steel Foreman	$37.65	$301.20	$68.35	$546.80	$35.67	$61.37
1 Structural Steel Worker	35.65	285.20	64.75	518.00		
1 Equip. Oper. (crane)	33.70	269.60	51.00	408.00		
1 Hyd. Crane, 12 Ton		636.00		699.60	26.50	29.15
24 L.H., Daily Totals		$1492.00		$2172.40	$62.17	$90.52

Crew M-1

Crew No.	Bare Costs Hr.	Daily	Incl. Subs O & P Hr.	Daily	Cost Per Labor-Hour Bare Costs	Incl. O&P
3 Elevator Constructors	$38.75	$930.00	$58.25	$1398.00	$36.81	$55.34
1 Elevator Apprentice	31.00	248.00	46.60	372.80		
5 Hand Tools		83.50		91.85	2.61	2.87
32 L.H., Daily Totals		$1261.50		$1862.65	$39.42	$58.21

Crew M-3

Crew No.	Bare Costs Hr.	Daily	Incl. Subs O & P Hr.	Daily	Cost Per Labor-Hour Bare Costs	Incl. O&P
1 Electrician Foreman (out)	$39.60	$316.80	$59.10	$472.80	$33.45	$50.54
1 Common Laborer	24.65	197.20	38.50	308.00		
.25 Equipment Operator, Medium	32.60	65.20	49.35	98.70		
1 Elevator Constructor	38.75	310.00	58.25	466.00		
1 Elevator Apprentice	31.00	248.00	46.60	372.80		
.25 Crane, SP, 4 x 4, 20 ton		146.85		161.55	4.32	4.75
34 L.H., Daily Totals		$1284.05		$1879.85	$37.77	$55.29

Crew M-4

Crew No.	Bare Costs Hr.	Daily	Incl. Subs O & P Hr.	Daily	Cost Per Labor-Hour Bare Costs	Incl. O&P
1 Electrician Foreman (out)	$39.60	$316.80	$59.10	$472.80	$33.22	$50.20
1 Common Laborer	24.65	197.20	38.50	308.00		
.25 Equipment Operator, Crane	33.70	67.40	51.00	102.00		
.25 Equipment Operator, Oiler	28.30	56.60	42.85	85.70		
1 Elevator Constructor	38.75	310.00	58.25	466.00		
1 Elevator Apprentice	31.00	248.00	46.60	372.80		
.25 Crane, Hyd, SP, 4WD, 40 Ton		272.50		299.75	7.57	8.33
36 L.H., Daily Totals		$1468.50		$2107.05	$40.79	$58.53

Crew Q-1

Crew No.	Bare Costs Hr.	Daily	Incl. Subs O & P Hr.	Daily	Cost Per Labor-Hour Bare Costs	Incl. O&P
1 Plumber	$37.35	$298.80	$56.40	$451.20	$33.63	$50.78
1 Plumber Apprentice	29.90	239.20	45.15	361.20		
16 L.H., Daily Totals		$538.00		$812.40	$33.63	$50.78

Crew Q-1C

Crew No.	Bare Costs Hr.	Daily	Incl. Subs O & P Hr.	Daily	Cost Per Labor-Hour Bare Costs	Incl. O&P
1 Plumber	$37.35	$298.80	$56.40	$451.20	$33.28	$50.30
1 Plumber Apprentice	29.90	239.20	45.15	361.20		
1 Equip. Oper. (medium)	32.60	260.80	49.35	394.80		
1 Trencher, Chain		1469.00		1615.90	61.21	67.33
24 L.H., Daily Totals		$2267.80		$2823.10	$94.49	$117.63

Crew Q-2

Crew No.	Bare Costs Hr.	Daily	Incl. Subs O & P Hr.	Daily	Cost Per Labor-Hour Bare Costs	Incl. O&P
2 Plumbers	$37.35	$597.60	$56.40	$902.40	$34.87	$52.65
1 Plumber Apprentice	29.90	239.20	45.15	361.20		
24 L.H., Daily Totals		$836.80		$1263.60	$34.87	$52.65

Crew Q-3

Crew No.	Bare Costs Hr.	Daily	Incl. Subs O & P Hr.	Daily	Cost Per Labor-Hour Bare Costs	Incl. O&P
1 Plumber Foreman (inside)	$37.85	$302.80	$57.15	$457.20	$35.61	$53.78
2 Plumbers	37.35	597.60	56.40	902.40		
1 Plumber Apprentice	29.90	239.20	45.15	361.20		
32 L.H., Daily Totals		$1139.60		$1720.80	$35.61	$53.78

Crew Q-4

Crew No.	Bare Costs Hr.	Daily	Incl. Subs O & P Hr.	Daily	Cost Per Labor-Hour Bare Costs	Incl. O&P
1 Plumber Foreman (inside)	$37.85	$302.80	$57.15	$457.20	$35.61	$53.78
1 Plumber	37.35	298.80	56.40	451.20		
1 Welder (plumber)	37.35	298.80	56.40	451.20		
1 Plumber Apprentice	29.90	239.20	45.15	361.20		
1 Electric Welding Mach.		78.50		86.35	2.45	2.70
32 L.H., Daily Totals		$1218.10		$1807.15	$38.06	$56.48

Crew Q-5

Crew No.	Bare Costs Hr.	Daily	Incl. Subs O & P Hr.	Daily	Cost Per Labor-Hour Bare Costs	Incl. O&P
1 Steamfitter	$37.60	$300.80	$56.80	$454.40	$33.85	$51.13
1 Steamfitter Apprentice	30.10	240.80	45.45	363.60		
16 L.H., Daily Totals		$541.60		$818.00	$33.85	$51.13

Crew Q-6

Crew No.	Bare Costs Hr.	Daily	Incl. Subs O & P Hr.	Daily	Cost Per Labor-Hour Bare Costs	Incl. O&P
2 Steamfitters	$37.60	$601.60	$56.80	$908.80	$35.10	$53.02
1 Steamfitter Apprentice	30.10	240.80	45.45	363.60		
24 L.H., Daily Totals		$842.40		$1272.40	$35.10	$53.02

Crew Q-7

Crew No.	Bare Costs Hr.	Daily	Incl. Subs O & P Hr.	Daily	Cost Per Labor-Hour Bare Costs	Incl. O&P
1 Steamfitter Foreman (inside)	$38.10	$304.80	$57.55	$460.40	$35.85	$54.15
2 Steamfitters	37.60	601.60	56.80	908.80		
1 Steamfitter Apprentice	30.10	240.80	45.45	363.60		
32 L.H., Daily Totals		$1147.20		$1732.80	$35.85	$54.15

Crew Q-8

Crew No.	Bare Costs Hr.	Daily	Incl. Subs O & P Hr.	Daily	Cost Per Labor-Hour Bare Costs	Incl. O&P
1 Steamfitter Foreman (inside)	$38.10	$304.80	$57.55	$460.40	$35.85	$54.15
1 Steamfitter	37.60	300.80	56.80	454.40		
1 Welder (steamfitter)	37.60	300.80	56.80	454.40		
1 Steamfitter Apprentice	30.10	240.80	45.45	363.60		
1 Electric Welding Mach.		78.50		86.35	2.45	2.70
32 L.H., Daily Totals		$1225.70		$1819.15	$38.30	$56.85

Crew Q-9

Crew No.	Bare Costs Hr.	Daily	Incl. Subs O & P Hr.	Daily	Cost Per Labor-Hour Bare Costs	Incl. O&P
1 Sheet Metal Worker	$37.00	$296.00	$57.15	$457.20	$33.30	$51.43
1 Sheet Metal Apprentice	29.60	236.80	45.70	365.60		
16 L.H., Daily Totals		$532.80		$822.80	$33.30	$51.43

CREWS

Crew No.	Bare Costs		Incl. Subs O & P		Cost Per Labor-Hour	

Crew Q-10

Crew Q-10	Hr.	Daily	Hr.	Daily	Bare Costs	Incl. O&P
2 Sheet Metal Workers	$37.00	$592.00	$57.15	$914.40	$34.53	$53.33
1 Sheet Metal Apprentice	29.60	236.80	45.70	365.60		
24 L.H., Daily Totals		$828.80		$1280.00	$34.53	$53.33

Crew Q-11

Crew Q-11	Hr.	Daily	Hr.	Daily	Bare Costs	Incl. O&P
1 Sheet Metal Foreman (inside)	$37.50	$300.00	$57.90	$463.20	$35.28	$54.48
2 Sheet Metal Workers	37.00	592.00	57.15	914.40		
1 Sheet Metal Apprentice	29.60	236.80	45.70	365.60		
32 L.H., Daily Totals		$1128.80		$1743.20	$35.28	$54.48

Crew Q-12

Crew Q-12	Hr.	Daily	Hr.	Daily	Bare Costs	Incl. O&P
1 Sprinkler Installer	$37.55	$300.40	$56.95	$455.60	$33.80	$51.28
1 Sprinkler Apprentice	30.05	240.40	45.60	364.80		
16 L.H., Daily Totals		$540.80		$820.40	$33.80	$51.28

Crew Q-13

Crew Q-13	Hr.	Daily	Hr.	Daily	Bare Costs	Incl. O&P
1 Sprinkler Foreman (inside)	$38.05	$304.40	$57.70	$461.60	$35.80	$54.30
2 Sprinkler Installers	37.55	600.80	56.95	911.20		
1 Sprinkler Apprentice	30.05	240.40	45.60	364.80		
32 L.H., Daily Totals		$1145.60		$1737.60	$35.80	$54.30

Crew Q-14

Crew Q-14	Hr.	Daily	Hr.	Daily	Bare Costs	Incl. O&P
1 Asbestos Worker	$34.70	$277.60	$55.20	$441.60	$31.23	$49.68
1 Asbestos Apprentice	27.75	222.00	44.15	353.20		
16 L.H., Daily Totals		$499.60		$794.80	$31.23	$49.68

Crew Q-15

Crew Q-15	Hr.	Daily	Hr.	Daily	Bare Costs	Incl. O&P
1 Plumber	$37.35	$298.80	$56.40	$451.20	$33.63	$50.78
1 Plumber Apprentice	29.90	239.20	45.15	361.20		
1 Electric Welding Mach.		78.50		86.35	4.91	5.40
16 L.H., Daily Totals		$616.50		$898.75	$38.54	$56.18

Crew Q-16

Crew Q-16	Hr.	Daily	Hr.	Daily	Bare Costs	Incl. O&P
2 Plumbers	$37.35	$597.60	$56.40	$902.40	$34.87	$52.65
1 Plumber Apprentice	29.90	239.20	45.15	361.20		
1 Electric Welding Mach.		78.50		86.35	3.27	3.60
24 L.H., Daily Totals		$915.30		$1349.95	$38.14	$56.25

Crew Q-17

Crew Q-17	Hr.	Daily	Hr.	Daily	Bare Costs	Incl. O&P
1 Steamfitter	$37.60	$300.80	$56.80	$454.40	$33.85	$51.13
1 Steamfitter Apprentice	30.10	240.80	45.45	363.60		
1 Electric Welding Mach.		78.50		86.35	4.91	5.40
16 L.H., Daily Totals		$620.10		$904.35	$38.76	$56.53

Crew Q-17A

Crew Q-17A	Hr.	Daily	Hr.	Daily	Bare Costs	Incl. O&P
1 Steamfitter	$37.60	$300.80	$56.80	$454.40	$33.80	$51.08
1 Steamfitter Apprentice	30.10	240.80	45.45	363.60		
1 Equip. Oper. (crane)	33.70	269.60	51.00	408.00		
1 Truck Crane, 12 Ton		636.00		699.60		
1 Electric Welding Mach.		78.50		86.35	29.77	32.75
24 L.H., Daily Totals		$1525.70		$2011.95	$63.57	$83.83

Crew Q-18

Crew Q-18	Hr.	Daily	Hr.	Daily	Bare Costs	Incl. O&P
2 Steamfitters	$37.60	$601.60	$56.80	$908.80	$35.10	$53.02
1 Steamfitter Apprentice	30.10	240.80	45.45	363.60		
1 Electric Welding Mach.		78.50		86.35	3.27	3.60
24 L.H., Daily Totals		$920.90		$1358.75	$38.37	$56.62

Crew Q-19

Crew Q-19	Hr.	Daily	Hr.	Daily	Bare Costs	Incl. O&P
1 Steamfitter	$37.60	$300.80	$56.80	$454.40	$35.10	$52.80
1 Steamfitter Apprentice	30.10	240.80	45.45	363.60		
1 Electrician	37.60	300.80	56.15	449.20		
24 L.H., Daily Totals		$842.40		$1267.20	$35.10	$52.80

Crew Q-20

Crew Q-20	Hr.	Daily	Hr.	Daily	Bare Costs	Incl. O&P
1 Sheet Metal Worker	$37.00	$296.00	$57.15	$457.20	$34.16	$52.37
1 Sheet Metal Apprentice	29.60	236.80	45.70	365.60		
.5 Electrician	37.60	150.40	56.15	224.60		
20 L.H., Daily Totals		$683.20		$1047.40	$34.16	$52.37

Crew Q-21

Crew Q-21	Hr.	Daily	Hr.	Daily	Bare Costs	Incl. O&P
2 Steamfitters	$37.60	$601.60	$56.80	$908.80	$35.72	$53.80
1 Steamfitter Apprentice	30.10	240.80	45.45	363.60		
1 Electrician	37.60	300.80	56.15	449.20		
32 L.H., Daily Totals		$1143.20		$1721.60	$35.72	$53.80

Crew Q-22

Crew Q-22	Hr.	Daily	Hr.	Daily	Bare Costs	Incl. O&P
1 Plumber	$37.35	$298.80	$56.40	$451.20	$33.63	$50.78
1 Plumber Apprentice	29.90	239.20	45.15	361.20		
1 Truck Crane, 12 Ton		636.00		699.60	39.75	43.73
16 L.H., Daily Totals		$1174.00		$1512.00	$73.38	$94.51

Crew Q-22A

Crew Q-22A	Hr.	Daily	Hr.	Daily	Bare Costs	Incl. O&P
1 Plumber	$37.35	$298.80	$56.40	$451.20	$31.40	$47.76
1 Plumber Apprentice	29.90	239.20	45.15	361.20		
1 Laborer	24.65	197.20	38.50	308.00		
1 Equip. Oper. (crane)	33.70	269.60	51.00	408.00		
1 Truck Crane, 12 Ton		636.00		699.60	19.88	21.86
32 L.H., Daily Totals		$1640.80		$2228.00	$51.28	$69.62

Crew Q-23

Crew Q-23	Hr.	Daily	Hr.	Daily	Bare Costs	Incl. O&P
1 Plumber Foreman	$39.35	$314.80	$59.40	$475.20	$36.43	$55.05
1 Plumber	37.35	298.80	56.40	451.20		
1 Equip. Oper. (medium)	32.60	260.80	49.35	394.80		
1 Power Tools		4.00		4.40		
1 Crane, 20 Ton		681.30		749.45	28.55	31.41
24 L.H., Daily Totals		$1559.70		$2075.05	$64.98	$86.46

Crew R-1

Crew R-1	Hr.	Daily	Hr.	Daily	Bare Costs	Incl. O&P
1 Electrician Foreman	$38.10	$304.80	$56.90	$455.20	$33.05	$49.91
3 Electricians	37.60	902.40	56.15	1347.60		
2 Helpers	23.70	379.20	37.05	592.80		
48 L.H., Daily Totals		$1586.40		$2395.60	$33.05	$49.91

Crew R-1A

Crew R-1A	Hr.	Daily	Hr.	Daily	Bare Costs	Incl. O&P
1 Electrician	$37.60	$300.80	$56.15	$449.20	$30.65	$46.60
1 Helper	23.70	189.60	37.05	296.40		
16 L.H., Daily Totals		$490.40		$745.60	$30.65	$46.60

Crew R-2

Crew R-2	Hr.	Daily	Hr.	Daily	Bare Costs	Incl. O&P
1 Electrician Foreman	$38.10	$304.80	$56.90	$455.20	$33.14	$50.06
3 Electricians	37.60	902.40	56.15	1347.60		
2 Helpers	23.70	379.20	37.05	592.80		
1 Equip. Oper. (crane)	33.70	269.60	51.00	408.00		
1 S.P. Crane, 5 Ton		326.40		359.05	5.83	6.41
56 L.H., Daily Totals		$2182.40		$3162.65	$38.97	$56.47

Crew No.	Bare Costs		Incl. Subs O & P		Cost Per Labor-Hour	

Crew R-3	Hr.	Daily	Hr.	Daily	Bare Costs	Incl. O&P
1 Electrician Foreman	$38.10	$304.80	$56.90	$455.20	$37.02	$55.42
1 Electrician	37.60	300.80	56.15	449.20		
.5 Equip. Oper. (crane)	33.70	134.80	51.00	204.00		
.5 S.P. Crane, 5 Ton		163.20		179.50	8.16	8.98
20 L.H., Daily Totals		$903.60		$1287.90	$45.18	$64.40

Crew R-4	Hr.	Daily	Hr.	Daily	Bare Costs	Incl. O&P
1 Struc. Steel Foreman	$37.65	$301.20	$68.35	$546.80	$36.44	$63.75
3 Struc. Steel Workers	35.65	855.60	64.75	1554.00		
1 Electrician	37.60	300.80	56.15	449.20		
1 Gas Welding Machine		78.80		86.70	1.97	2.17
40 L.H., Daily Totals		$1536.40		$2636.70	$38.41	$65.92

Crew R-5	Hr.	Daily	Hr.	Daily	Bare Costs	Incl. O&P
1 Electrician Foreman	$38.10	$304.80	$56.90	$455.20	$32.59	$49.27
4 Electrician Linemen	37.60	1203.20	56.15	1796.80		
2 Electrician Operators	37.60	601.60	56.15	898.40		
4 Electrician Groundmen	23.70	758.40	37.05	1185.60		
1 Crew Truck		81.50		89.65		
1 Tool Van		113.70		125.05		
1 Pickup Truck, 3/4 Ton		67.80		74.60		
.2 Crane, 55 Ton		231.20		254.30		
.2 Crane, 12 Ton		127.20		139.90		
.2 Auger, Truck Mtd.		459.00		504.90		
1 Tractor w/Winch		243.20		267.50	15.04	16.55
88 L.H., Daily Totals		$4191.60		$5791.90	$47.63	$65.82

Crew R-6	Hr.	Daily	Hr.	Daily	Bare Costs	Incl. O&P
1 Electrician Foreman	$38.10	$304.80	$56.90	$455.20	$32.59	$49.27
4 Electrician Linemen	37.60	1203.20	56.15	1796.80		
2 Electrician Operators	37.60	601.60	56.15	898.40		
4 Electrician Groundmen	23.70	758.40	37.05	1185.60		
1 Crew Truck		81.50		89.65		
1 Tool Van		113.70		125.05		
1 Pickup Truck, 3/4 Ton		67.80		74.60		
.2 Crane, 55 Ton		231.20		254.30		
.2 Crane, 12 Ton		127.20		139.90		
.2 Auger, Truck Mtd.		459.00		504.90		
1 Tractor w/Winch		243.20		267.50		
3 Cable Trailers		471.60		518.75		
.5 Tensioning Rig		154.88		170.35		
.5 Cable Pulling Rig		898.00		987.80	32.36	35.60
88 L.H., Daily Totals		$5716.08		$7468.80	$64.95	$84.87

Crew R-7	Hr.	Daily	Hr.	Daily	Bare Costs	Incl. O&P
1 Electrician Foreman	$38.10	$304.80	$56.90	$455.20	$26.10	$40.36
5 Electrician Groundmen	23.70	948.00	37.05	1482.00		
1 Crew Truck		81.50		89.65	1.70	1.87
48 L.H., Daily Totals		$1334.30		$2026.85	$27.80	$42.23

Crew R-8	Hr.	Daily	Hr.	Daily	Bare Costs	Incl. O&P
1 Electrician Foreman	$38.10	$304.80	$56.90	$455.20	$33.05	$49.91
3 Electrician Linemen	37.60	902.40	56.15	1347.60		
2 Electrician Groundmen	23.70	379.20	37.05	592.80		
1 Pickup Truck, 3/4 Ton		67.80		74.60		
1 Crew Truck		81.50		89.65	3.11	3.42
48 L.H., Daily Totals		$1735.70		$2559.85	$36.16	$53.33

Crew R-9	Hr.	Daily	Hr.	Daily	Bare Costs	Incl. O&P
1 Electrician Foreman	$38.10	$304.80	$56.90	$455.20	$30.71	$46.69
1 Electrician Lineman	37.60	300.80	56.15	449.20		
2 Electrician Operators	37.60	601.60	56.15	898.40		
4 Electrician Groundmen	23.70	758.40	37.05	1185.60		
1 Pickup Truck, 3/4 Ton		67.80		74.60		
1 Crew Truck		81.50		89.65	2.33	2.57
64 L.H., Daily Totals		$2114.90		$3152.65	$33.04	$49.26

Crew R-10	Hr.	Daily	Hr.	Daily	Bare Costs	Incl. O&P
1 Electrician Foreman	$38.10	$304.80	$56.90	$455.20	$35.37	$53.09
4 Electrician Linemen	37.60	1203.20	56.15	1796.80		
1 Electrician Groundman	23.70	189.60	37.05	296.40		
1 Crew Truck		81.50		89.65		
3 Tram Cars		334.95		368.45	8.68	9.54
48 L.H., Daily Totals		$2114.05		$3006.50	$44.05	$62.63

Crew R-11	Hr.	Daily	Hr.	Daily	Bare Costs	Incl. O&P
1 Electrician Foreman	$38.10	$304.80	$56.90	$455.20	$33.84	$51.01
4 Electricians	37.60	1203.20	56.15	1796.80		
1 Helper	23.70	189.60	37.05	296.40		
1 Common Laborer	24.65	197.20	38.50	308.00		
1 Crew Truck		81.50		89.65		
1 Crane, 12 Ton		636.00		699.60	12.81	14.09
56 L.H., Daily Totals		$2612.30		$3645.65	$46.65	$65.10

Crew R-12	Hr.	Daily	Hr.	Daily	Bare Costs	Incl. O&P
1 Carpenter Foreman	$32.05	$256.40	$50.05	$400.40	$29.55	$46.85
4 Carpenters	31.55	1009.60	49.30	1577.60		
4 Common Laborers	24.65	788.80	38.50	1232.00		
1 Equip. Oper. (med.)	32.60	260.80	49.35	394.80		
1 Steel Worker	35.65	285.20	64.75	518.00		
1 Dozer, 200 H.P.		880.80		968.90		
1 Pickup Truck, 3/4 Ton		67.80		74.60	10.78	11.86
88 L.H., Daily Totals		$3549.40		$5166.30	$40.33	$58.71

Crew R-13	Hr.	Daily	Hr.	Daily	Bare Costs	Incl. O&P
1 Electrician Foreman	$38.10	$304.80	$56.90	$455.20	$35.74	$53.51
3 Electricians	37.60	902.40	56.15	1347.60		
.25 Equip. Oper. (crane)	33.70	67.40	51.00	102.00		
1 Equipment Oiler	28.30	226.40	42.85	342.80		
.25-1 Hyd. Crane, 33 Ton		168.95		185.85	4.02	4.42
42 L.H., Daily Totals		$1669.95		$2433.45	$39.76	$57.93

Crew R-15	Hr.	Daily	Hr.	Daily	Bare Costs	Incl. O&P
1 Electrician Foreman	$38.10	$304.80	$56.90	$455.20	$36.60	$54.77
4 Electricians	37.60	1203.20	56.15	1796.80		
1 Equipment Operator	31.10	248.80	47.10	376.80		
1 Aerial Lift Truck		239.60		263.55	4.99	5.49
48 L.H., Daily Totals		$1996.40		$2892.35	$41.59	$60.26

Crew R-18	Hr.	Daily	Hr.	Daily	Bare Costs	Incl. O&P
.25 Electrician Foreman	$38.10	$76.20	$56.90	$113.80	$29.08	$44.45
1 Electrician	37.60	300.80	56.15	449.20		
2 Helpers	23.70	379.20	37.05	592.80		
26 L.H., Daily Totals		$756.20		$1155.80	$29.08	$44.45

Crew R-19	Hr.	Daily	Hr.	Daily	Bare Costs	Incl. O&P
.5 Electrician Foreman	$38.10	$152.40	$56.90	$227.60	$37.70	$56.30
2 Electricians	37.60	601.60	56.15	898.40		
20 L.H., Daily Totals		$754.00		$1126.00	$37.70	$56.30

Crew No.	Bare Costs		Incl. Sub O & P		Cost Per Labor-Hour	
Crew R-21	Hr.	Daily	Hr.	Daily	Bare Costs	Incl. O&P
1 Electrician Foreman	$38.10	$304.80	$56.90	$455.20	$37.60	$56.17
3 Electricians	37.60	902.40	56.15	1347.60		
.1 Equip. Oper. (med.)	32.60	26.08	49.35	39.48		
.1 Hyd. Crane 25 Ton		65.78		72.35	2.01	2.21
32. L.H., Daily Totals		$1299.06		$1914.63	$39.61	$58.38
Crew R-22	Hr.	Daily	Hr.	Daily	Bare Costs	Incl. O&P
.66 Electrician Foreman	$38.10	$201.17	$56.90	$300.43	$31.71	$48.06
2 Helpers	23.70	379.20	37.05	592.80		
2 Electricians	37.60	601.60	56.15	898.40		
37.28 L.H., Daily Totals		$1181.97		$1791.63	$31.71	$48.06
Crew R-30	Hr.	Daily	Hr.	Daily	Bare Costs	Incl. O&P
.25 Electrician	$39.60	$79.20	$59.10	$118.20	$29.78	$45.52
1 Electrician	37.60	300.80	56.15	449.20		
2 Laborers, (Semi-Skilled)	24.65	394.40	38.50	616.00		
26 L.H., Daily Totals		$774.40		$1183.40	$29.78	$45.52
Crew W-41E	Hr.	Daily	Hr.	Daily	Bare Costs	Incl. O&P
1 Laborer, (Semi-Skilled)	$24.65	$197.20	$38.50	$308.00	$32.67	$49.84
1 Plumber	37.35	298.80	56.40	451.20		
.5 Plumber	39.35	157.40	59.40	237.60		
20 L.H., Daily Totals		$653.40		$996.80	$32.67	$49.84

Historical Cost Indexes

The table below lists both the Means Historical Cost Index based on Jan. 1, 1993 = 100 as well as the computed value of an index based on Jan. 1, 2003 costs. Since the Jan. 1, 2003 figure is estimated, space is left to write in the actual index figures as they become available through either the quarterly "Means Construction Cost Indexes" or as printed in the "Engineering News-Record." To compute the actual index based on Jan. 1, 2003 = 100, divide the Historical Cost Index for a particular year by the actual Jan. 1, 2003 Construction Cost Index. Space has been left to advance the index figures as the year progresses.

Year	Historical Cost Index Jan. 1, 1993 = 100		Current Index Based on Jan. 1, 2003 = 100		Year	Historical Cost Index Jan. 1, 1993 = 100	Current Index Based on Jan. 1, 2003 = 100			Year	Historical Cost Index Jan. 1, 1993 = 100	Current Index Based on Jan. 1, 2003 = 100	
	Est.	Actual	Est.	Actual		Actual	Est.	Actual			Actual	Est.	Actual
Oct 2003					July 1988	89.9	69.0		July 1970	28.7	22.0		
July 2003					1987	87.7	67.3		1969	26.9	20.7		
April 2003					1986	84.2	64.7		1968	24.9	19.1		
Jan 2003	130.2		100.0	100.0	1985	82.6	63.5		1967	23.5	18.0		
July 2002		128.7	98.8		1984	82.0	62.9		1966	22.7	17.4		
2001		125.1	96.1		1983	80.2	61.6		1965	21.7	16.7		
2000		120.9	92.9		1982	76.1	58.5		1964	21.2	16.3		
1999		117.6	90.3		1981	70.0	53.8		1963	20.7	15.9		
1998		115.1	88.4		1980	62.9	48.3		1962	20.2	15.5		
1997		112.8	86.6		1979	57.8	44.4		1961	19.8	15.2		
1996		110.2	84.6		1978	53.5	41.1		1960	19.7	15.1		
1995		107.6	82.6		1977	49.5	38.0		1959	19.3	14.8		
1994		104.4	80.2		1976	46.9	36.0		1958	18.8	14.4		
1993		101.7	78.1		1975	44.8	34.4		1957	18.4	14.1		
1992		99.4	76.4		1974	41.4	31.8		1956	17.6	13.5		
1991		96.8	74.4		1973	37.7	29.0		1955	16.6	12.7		
1990		94.3	72.4		1972	34.8	26.7		1954	16.0	12.3		
1989		92.1	70.8		1971	32.1	24.7		1953	15.8	12.1		

Adjustments to Costs

The Historical Cost Index can be used to convert National Average building costs at a particular time to the approximate building costs for some other time.

Example:

Estimate and compare construction costs for different years in the same city.

To estimate the National Average construction cost of a building in 1970, knowing that it cost $900,000 in 2003:

INDEX in 1970 = 28.7

INDEX in 2003 = 130.2

Note: The City Cost Indexes for Canada can be used to convert U.S. National averages to local costs in Canadian dollars.

Time Adjustment using the Historical Cost Indexes:

$$\frac{\text{Index for Year A}}{\text{Index for Year B}} \times \text{Cost in Year B} = \text{Cost in Year A}$$

$$\frac{\text{INDEX 1970}}{\text{INDEX 2003}} \times \text{Cost 2003} = \text{Cost 1970}$$

$$\frac{28.7}{130.2} \times \$900,000 = .220 \times \$900,000 = \$198,000$$

The construction cost of the building in 1970 is $198,000.

How to Use the City Cost Indexes

What you should know before you begin

Means City Cost Indexes (CCI) are an extremely useful tool to use when you want to compare costs from city to city and region to region.

This publication contains average construction cost indexes for 316 major U.S. and Canadian cities and Location Factors covering over 930 three-digit zip code locations.

Keep in mind that a City Cost Index number is a *percentage ratio* of a specific city's cost to the national average cost of the same item at a stated time period.

In other words, these index figures represent relative construction *factors* (or, if you prefer, multipliers) for Material and Installation costs, as well as the weighted average for Total In Place costs for each CSI MasterFormat division. Installation costs include both labor and equipment rental costs.

The 30 City Average Index is the average of 30 major U.S. cities and serves as a National Average.

Index figures for both material and installation are based on the 30 major city average of 100 and represent the cost relationship as of July 1, 2002. The index for each division is computed from representative material and labor quantities for that division. The weighted average for each city is a weighted total of the components listed above it, but does not include relative productivity between trades or cities.

As changes occur in local material prices, labor rates and equipment rental rates, the impact of these changes should be accurately measured by the change in the City Cost Index for each particular city (as compared to the 30 City Average).

> *Therefore, if you know (or have estimated) building costs in one city today, you can easily convert those costs to expected building costs in another city.*

> *In addition, by using the Historical Cost Index, you can easily convert National Average building costs at a particular time to the approximate building costs for some other time. The City Cost Indexes can then be applied to calculate the costs for a particular city.*

Quick Calculations

Location Adjustment Using the City Cost Indexes:

$$\frac{\text{Index for City A}}{\text{Index for City B}} \times \text{Cost in City B} = \text{Cost in City A}$$

Time Adjustment for the National Average Using the Historical Cost Index:

$$\frac{\text{Index for Year A}}{\text{Index for Year B}} \times \text{Cost in Year B} = \text{Cost in Year A}$$

Adjustment from the National Average:

$$\frac{\text{Index for City A}}{100} \times \text{National Average Cost} = \text{Cost in City A}$$

Since each of the other R.S. Means publications contains many different items, any *one* item multiplied by the particular city index may give incorrect results. However, the larger the number of items compiled, the closer the results should be to actual costs for that particular city.

The City Cost Indexes for Canadian cities are calculated using Canadian material and equipment prices and labor rates, in Canadian dollars. Therefore, indexes for Canadian cities can be used to convert U.S. National Average prices to local costs in Canadian dollars.

How to use this section

1. Compare costs from city to city.

In using the Means Indexes, remember that an index number is not a fixed number but a *ratio:* It's a percentage ratio of a building component's cost at any stated time to the National Average cost of that same component at the same time period. Put in the form of an equation:

$$\frac{\text{Specific City Cost}}{\text{National Average Cost}} \times 100 = \text{City Index Number}$$

Therefore, when making cost comparisons between cities, do not subtract one city's index number from the index number of another city and read the result as a percentage difference. Instead, divide one city's index number by that of the other city. The resulting number may then be used as a multiplier to calculate cost differences from city to city.

The formula used to find cost differences between cities for the purpose of comparison is as follows:

$$\frac{\text{City A Index}}{\text{City B Index}} \times \text{City B Cost (Known)} = \text{City A Cost (Unknown)}$$

In addition, you can use *Means CCI* to calculate and compare costs division by division between cities using the same basic formula. (Just be sure that you're comparing similar divisions.)

2. Compare a specific city's construction costs with the National Average.

When you're studying construction location feasibility, it's advisable to compare a prospective project's cost index with an index of the National Average cost.

For example, divide the weighted average index of construction costs of a specific city by that of the 30 City Average, which = 100.

$$\frac{\text{City Index}}{100} = \% \text{ of National Average}$$

As a result, you get a ratio that indicates the relative cost of construction in that city in comparison with the National Average.

3. Convert U.S. National Average to actual costs in Canadian City

$$\frac{\text{Index for Canadian City}}{100} \times \text{National Average Cost} = \text{Cost in Canadian City in \$ CAN}$$

4. Adjust construction cost data based on a National Average.

When you use a source of construction cost data which is based on a National Average (such as *Means cost data publications*), it is necessary to adjust those costs to a specific location.

$$\frac{\text{City Index}}{100} \quad x \quad \begin{array}{c}\text{"Book" Cost Based on}\\\text{National Average Costs}\end{array} = \begin{array}{c}\text{City Cost}\\\text{(Unknown)}\end{array}$$

5. When applying the City Cost Indexes to demolition projects, use the appropriate division index. For example, for removal of existing doors and windows, use the Division 8 index.

What you might like to know about how we developed the Indexes

To create a reliable index, R.S. Means researched the building type most often constructed in the United States and Canada. Because it was concluded that no one type of building completely represented the building construction industry, nine different types of buildings were combined to create a composite model.

The exact material, labor and equipment quantities are based on detailed analysis of these nine building types, then each quantity is weighted in proportion to expected usage. These various material items, labor hours, and equipment rental rates are thus combined to form a composite building representing as closely as possible the actual usage of materials, labor and equipment used in the North American Building Construction Industry.

The following structures were chosen to make up that composite model:

1. Factory, 1 story
2. Office, 2–4 story
3. Store, Retail
4. Town Hall, 2–3 story
5. High School, 2–3 story
6. Hospital, 4–8 story
7. Garage, Parking
8. Apartment, 1–3 story
9. Hotel/Motel, 2–3 story

For the purposes of ensuring the timeliness of the data, the components of the index for the composite model have been streamlined. They currently consist of:

- specific quantities of 66 commonly used construction materials;
- specific labor-hours for 21 building construction trades; and
- specific days of equipment rental for 6 types of construction equipment (normally used to install the 66 material items by the 21 trades.)

A sophisticated computer program handles the updating of all costs for each city on a quarterly basis. Material and equipment price quotations are gathered quarterly from over 316 cities in the United States and Canada. These prices and the latest negotiated labor wage rates for 21 different building trades are used to compile the quarterly update of the City Cost Index.

The 30 major U.S. cities used to calculate the National Average are:

Atlanta, GA	Memphis, TN
Baltimore, MD	Milwaukee, WI
Boston, MA	Minneapolis, MN
Buffalo, NY	Nashville, TN
Chicago, IL	New Orleans, LA
Cincinnati, OH	New York, NY
Cleveland, OH	Philadelphia, PA
Columbus, OH	Phoenix, AZ
Dallas, TX	Pittsburgh, PA
Denver, CO	St. Louis, MO
Detroit, MI	San Antonio, TX
Houston, TX	San Diego, CA
Indianapolis, IN	San Francisco, CA
Kansas City, MO	Seattle, WA
Los Angeles, CA	Washington, DC

F.Y.I.: The CSI MasterFormat Divisions

1. General Requirements
2. Site Work
3. Concrete
4. Masonry
5. Metals
6. Wood & Plastics
7. Thermal & Moisture Protection
8. Doors & Windows
9. Finishes
10. Specialties
11. Equipment
12. Furnishings
13. Special Construction
14. Conveying Systems
15. Mechanical
16. Electrical

The information presented in the CCI is organized according to the Construction Specifications Institute (CSI) MasterFormat.

What the CCI does not indicate

The weighted average for each city is a total of the components listed above weighted to reflect typical usage, but it does *not* include the productivity variations between trades or cities.

In addition, the CCI does not take into consideration factors such as the following:

- managerial efficiency
- competitive conditions
- automation
- restrictive union practices
- Unique local requirements
- regional variations due to specific building codes

DIVISION		UNITED STATES 30 CITY AVERAGE			BIRMINGHAM			HUNTSVILLE			MOBILE			MONTGOMERY			TUSCALOOSA		
		MAT.	INST.	TOTAL	MAT.	INST.	TOTAL	MAT.	INST.	TOTAL	MAT.	INST.	TOTAL	MAT.	INST.	TOTAL	MAT.	INST.	TOTAL
01590	EQUIPMENT RENTAL	.0	100.0	100.0	.0	101.4	101.4	.0	101.3	101.3	.0	97.7	97.7	.0	97.7	97.7	.0	101.3	101.3
02	SITE CONSTRUCTION	100.0	100.0	100.0	85.9	93.6	91.7	84.0	93.1	90.8	95.2	86.5	88.6	95.6	87.0	89.1	84.5	92.4	90.4
03100	CONCRETE FORMS & ACCESSORIES	100.0	100.0	100.0	91.9	77.8	79.6	93.7	73.8	76.3	93.7	57.4	62.0	92.2	51.9	57.0	93.6	42.8	49.2
03200	CONCRETE REINFORCEMENT	100.0	100.0	100.0	91.2	86.7	88.6	91.2	82.6	86.1	94.1	57.5	72.5	94.1	85.2	88.8	91.2	85.5	87.8
03300	CAST-IN-PLACE CONCRETE	100.0	100.0	100.0	93.1	72.7	84.7	88.1	70.2	80.7	93.0	58.6	78.9	94.7	52.8	77.5	91.7	50.4	74.8
03	CONCRETE	100.0	100.0	100.0	89.6	78.8	84.2	87.2	75.3	81.2	90.0	59.4	74.6	90.7	60.3	75.4	89.0	55.5	72.1
04	MASONRY	100.0	100.0	100.0	85.5	80.5	82.4	85.4	70.0	75.9	86.0	56.8	67.9	86.6	39.4	57.4	85.7	41.9	58.6
05	METALS	100.0	100.0	100.0	98.4	96.4	97.7	98.1	93.5	96.4	96.7	82.8	91.5	96.7	93.1	95.3	97.1	93.7	95.8
06	WOOD & PLASTICS	100.0	100.0	100.0	93.0	77.6	84.9	92.2	74.6	83.0	92.2	57.5	74.0	90.5	52.7	70.6	92.2	41.8	65.7
07	THERMAL & MOISTURE PROTECTION	100.0	100.0	100.0	96.3	84.0	90.3	96.0	78.6	87.6	96.0	72.9	84.8	95.7	65.9	81.3	95.9	64.6	80.8
08	DOORS & WINDOWS	100.0	100.0	100.0	98.4	80.2	94.0	98.4	70.1	91.6	98.4	57.6	88.4	98.4	61.1	89.2	98.4	60.9	89.2
09200	PLASTER & GYPSUM BOARD	100.0	100.0	100.0	105.5	77.4	86.7	102.1	74.4	83.5	102.1	56.8	71.7	102.1	51.8	68.4	102.1	40.6	60.9
095,098	CEILINGS & ACOUSTICAL TREATMENT	100.0	100.0	100.0	96.3	77.4	83.5	96.3	74.4	81.4	96.3	56.8	69.4	96.3	51.8	66.1	96.3	40.6	58.5
09600	FLOORING	100.0	100.0	100.0	103.1	54.5	90.9	103.1	47.1	89.1	111.7	59.0	98.5	111.7	30.5	91.3	103.1	44.8	88.5
097,099	WALL FINISHES, PAINTS & COATINGS	100.0	100.0	100.0	94.2	92.9	93.4	94.2	69.9	79.8	94.2	61.9	75.2	94.2	60.9	74.6	94.2	52.4	69.6
09	FINISHES	100.0	100.0	100.0	98.7	74.8	86.0	98.1	68.1	82.2	102.1	57.7	78.5	102.2	47.7	73.3	98.0	42.6	68.6
10-14	TOTAL DIV. 10000 - 14000	100.0	100.0	100.0	100.0	84.2	96.6	100.0	82.6	96.2	100.0	76.7	95.0	100.0	74.5	94.5	100.0	71.7	93.9
15	MECHANICAL	100.0	100.0	100.0	99.9	71.2	87.0	99.9	68.3	85.6	99.9	66.5	84.8	99.9	45.2	75.3	99.9	39.8	72.8
16	ELECTRICAL	100.0	100.0	100.0	96.8	68.7	77.0	97.2	72.5	79.8	97.2	55.8	68.0	96.8	69.9	77.8	97.2	68.7	77.1
01-16	WEIGHTED AVERAGE	100.0	100.0	100.0	96.5	78.6	87.7	96.0	75.5	85.9	96.9	65.0	81.3	97.0	61.4	79.5	96.1	59.4	78.1

ALABAMA (BIRMINGHAM, HUNTSVILLE, MOBILE, MONTGOMERY, TUSCALOOSA)

DIVISION		ANCHORAGE			FAIRBANKS			JUNEAU			FLAGSTAFF			MESA/TEMPE			PHOENIX		
		MAT.	INST.	TOTAL	MAT.	INST.	TOTAL	MAT.	INST.	TOTAL	MAT.	INST.	TOTAL	MAT.	INST.	TOTAL	MAT.	INST.	TOTAL
01590	EQUIPMENT RENTAL	.0	118.6	118.6	.0	118.6	118.6	.0	118.6	118.6	.0	94.8	94.8	.0	97.5	97.5	.0	98.1	98.1
02	SITE CONSTRUCTION	141.0	134.7	136.3	124.3	134.8	132.2	136.6	134.7	135.2	85.6	101.5	97.6	84.8	104.1	99.3	85.2	104.9	100.1
03100	CONCRETE FORMS & ACCESSORIES	133.3	120.7	122.3	135.1	125.4	126.6	134.8	120.7	122.5	104.0	66.1	71.3	101.5	66.9	71.3	102.8	72.3	76.1
03200	CONCRETE REINFORCEMENT	140.5	110.2	122.6	118.1	110.2	113.5	104.5	110.2	107.9	104.3	76.6	87.9	102.6	74.2	85.8	100.7	77.3	86.9
03300	CAST-IN-PLACE CONCRETE	192.5	118.5	162.2	160.5	117.2	142.7	193.4	118.5	162.6	97.4	80.5	90.5	103.6	73.2	91.1	103.7	80.7	94.3
03	CONCRETE	152.8	117.2	134.9	129.2	118.9	124.0	149.1	117.2	133.1	121.1	72.9	96.8	100.5	70.3	85.3	100.2	75.9	88.0
04	MASONRY	210.8	124.3	157.3	211.6	124.3	157.6	213.5	124.3	158.3	103.1	58.8	75.7	108.8	55.0	75.5	96.5	70.1	80.2
05	METALS	130.0	103.8	120.3	130.1	104.1	120.5	130.3	103.8	120.5	98.0	69.9	87.6	98.4	70.0	87.9	99.7	72.7	89.7
06	WOOD & PLASTICS	115.1	119.9	117.6	115.4	126.0	120.9	115.1	119.9	117.6	108.4	65.4	85.8	103.0	73.0	87.2	104.1	73.1	87.8
07	THERMAL & MOISTURE PROTECTION	199.6	119.6	161.0	195.5	121.9	160.0	196.2	119.6	159.3	109.6	70.0	90.5	107.5	67.2	88.1	107.4	73.0	90.8
08	DOORS & WINDOWS	127.0	113.6	123.7	124.1	116.9	122.3	124.1	113.6	121.5	102.2	68.8	93.9	99.0	68.5	91.5	100.2	73.0	93.5
09200	PLASTER & GYPSUM BOARD	135.3	120.4	125.3	135.3	126.7	129.5	135.3	120.4	125.3	97.5	64.4	75.3	100.9	72.1	81.6	101.5	72.2	81.9
095,098	CEILINGS & ACOUSTICAL TREATMENT	123.4	120.4	121.4	123.4	126.7	125.6	123.4	120.4	121.4	103.3	64.4	76.9	101.6	72.1	81.6	101.6	72.2	81.6
09600	FLOORING	165.4	129.9	156.5	165.4	129.9	156.5	165.4	129.9	156.5	97.1	55.9	86.8	100.7	69.7	92.9	101.0	72.8	94.0
097,099	WALL FINISHES, PAINTS & COATINGS	170.6	109.1	134.4	170.6	125.1	143.7	170.6	109.1	134.4	96.5	53.1	70.9	108.3	57.8	78.5	108.3	65.7	83.2
09	FINISHES	156.4	121.5	137.9	154.0	126.9	139.6	154.9	121.5	137.2	97.8	62.2	78.9	100.1	66.2	82.1	100.3	71.3	84.9
10-14	TOTAL DIV. 10000 - 14000	100.0	116.3	103.5	100.0	117.1	103.7	100.0	116.3	103.5	100.0	79.8	95.7	100.0	76.0	94.8	100.0	81.2	96.0
15	MECHANICAL	100.5	109.6	104.6	100.5	115.9	107.5	100.5	109.6	104.6	100.2	79.7	90.9	100.2	72.5	87.7	100.2	79.7	91.0
16	ELECTRICAL	161.8	112.6	127.1	164.7	112.6	127.9	164.7	112.6	127.9	100.7	52.3	66.5	96.6	65.6	74.7	105.3	65.3	77.1
01-16	WEIGHTED AVERAGE	134.9	116.5	125.9	131.0	118.9	125.1	134.1	116.5	125.5	102.6	70.3	86.7	99.8	71.1	85.7	100.1	76.0	88.3

ALASKA (ANCHORAGE, FAIRBANKS, JUNEAU); ARIZONA (FLAGSTAFF, MESA/TEMPE, PHOENIX)

DIVISION		PRESCOTT			TUCSON			FORT SMITH			JONESBORO			LITTLE ROCK			PINE BLUFF		
		MAT.	INST.	TOTAL	MAT.	INST.	TOTAL	MAT.	INST.	TOTAL	MAT.	INST.	TOTAL	MAT.	INST.	TOTAL	MAT.	INST.	TOTAL
01590	EQUIPMENT RENTAL	.0	94.8	94.8	.0	97.5	97.5	.0	85.6	85.6	.0	107.1	107.1	.0	85.6	85.6	.0	85.6	85.6
02	SITE CONSTRUCTION	72.6	101.0	94.1	81.9	104.6	99.0	76.5	83.7	81.9	99.9	98.9	99.1	76.3	83.7	81.9	78.5	83.7	82.4
03100	CONCRETE FORMS & ACCESSORIES	99.2	60.7	65.5	102.1	71.8	75.6	99.2	45.4	52.2	84.5	51.9	56.0	93.2	63.3	67.1	76.5	63.2	64.8
03200	CONCRETE REINFORCEMENT	104.3	73.9	86.3	99.6	76.6	86.0	97.4	77.0	85.4	93.2	53.0	69.4	97.6	76.1	84.9	97.6	76.1	84.9
03300	CAST-IN-PLACE CONCRETE	97.3	68.3	85.4	103.7	80.5	94.2	90.5	66.5	80.6	86.2	60.7	75.8	90.5	66.6	80.7	83.0	66.6	76.3
03	CONCRETE	105.8	65.9	85.7	100.0	75.5	87.6	88.1	59.5	73.6	84.7	56.9	70.7	87.7	67.2	77.4	85.6	67.1	76.3
04	MASONRY	103.5	63.8	78.9	96.9	58.7	73.3	96.2	57.8	72.5	90.9	50.8	66.1	94.5	57.8	71.8	115.3	57.8	79.8
05	METALS	98.0	69.7	87.5	99.0	70.7	88.5	96.7	73.6	88.1	90.9	79.5	86.7	96.3	73.6	87.8	95.4	73.4	87.2
06	WOOD & PLASTICS	103.9	59.4	80.5	103.3	73.1	87.4	103.9	44.1	72.5	89.4	53.1	70.3	100.8	67.9	83.5	82.2	67.9	74.7
07	THERMAL & MOISTURE PROTECTION	107.9	66.1	87.7	107.9	66.4	87.9	98.7	52.0	76.2	108.6	55.5	83.0	97.5	54.4	76.7	97.3	54.4	76.6
08	DOORS & WINDOWS	102.2	62.9	92.5	96.0	73.0	90.4	97.1	48.8	85.2	98.7	51.7	87.1	97.1	62.3	88.5	92.4	62.3	85.0
09200	PLASTER & GYPSUM BOARD	95.4	58.2	70.5	101.8	72.2	82.0	90.3	43.3	58.8	96.7	52.1	66.8	90.3	67.8	75.2	83.2	67.8	72.9
095,098	CEILINGS & ACOUSTICAL TREATMENT	103.3	58.2	72.7	103.0	72.2	82.1	99.3	43.3	61.2	100.0	52.1	67.4	99.3	67.8	77.9	95.2	67.8	76.6
09600	FLOORING	95.4	55.7	85.4	100.2	55.9	89.0	118.1	75.6	107.4	78.9	49.2	71.4	119.4	75.6	108.4	106.8	75.6	98.9
097,099	WALL FINISHES, PAINTS & COATINGS	96.5	53.1	70.9	105.8	53.1	74.8	96.8	63.4	77.1	85.1	57.0	68.6	96.8	65.1	78.1	96.8	65.1	78.1
09	FINISHES	95.7	58.4	75.9	100.0	66.7	82.3	100.6	52.5	75.0	90.9	52.0	70.2	101.0	66.6	82.7	95.3	66.6	80.1
10-14	TOTAL DIV. 10000 - 14000	100.0	78.7	95.4	100.0	81.2	96.0	100.0	70.1	93.6	100.0	60.6	91.5	100.0	73.2	94.2	100.0	73.2	94.2
15	MECHANICAL	100.2	78.8	90.6	100.1	73.2	88.0	100.0	46.9	76.1	100.2	46.9	76.2	100.0	58.1	81.1	100.0	49.6	77.3
16	ELECTRICAL	100.3	50.3	65.0	100.0	62.4	73.5	94.8	65.7	74.3	99.7	52.0	66.0	95.8	68.1	76.2	93.7	68.1	75.6
01-16	WEIGHTED AVERAGE	100.0	68.4	84.5	99.0	72.2	85.8	96.5	60.1	78.6	95.3	58.7	77.3	96.3	66.3	81.6	95.7	64.6	80.5

ARIZONA (PRESCOTT, TUCSON); ARKANSAS (FORT SMITH, JONESBORO, LITTLE ROCK, PINE BLUFF)

City Cost Indexes

DIVISION		ARKANSAS TEXARKANA MAT.	INST.	TOTAL	CALIFORNIA ANAHEIM MAT.	INST.	TOTAL	BAKERSFIELD MAT.	INST.	TOTAL	FRESNO MAT.	INST.	TOTAL	LOS ANGELES MAT.	INST.	TOTAL	OAKLAND MAT.	INST.	TOTAL
01590	EQUIPMENT RENTAL	.0	86.3	86.3	.0	102.2	102.2	.0	99.5	99.5	.0	99.5	99.5	.0	97.9	97.9	.0	102.3	102.3
02	SITE CONSTRUCTION	95.2	84.5	87.1	99.4	109.6	107.1	104.2	106.6	106.0	105.7	106.6	106.4	95.0	108.1	104.9	141.2	104.4	113.4
03100	CONCRETE FORMS & ACCESSORIES	85.1	43.0	48.3	104.3	120.8	118.7	97.4	120.5	117.6	100.9	118.7	116.4	106.8	120.5	118.8	108.0	135.5	132.1
03200	CONCRETE REINFORCEMENT	97.1	51.4	70.1	106.2	110.9	109.0	105.7	110.9	108.8	106.3	111.1	109.1	108.7	111.2	110.2	98.9	111.8	106.5
03300	CAST-IN-PLACE CONCRETE	90.5	47.4	72.8	105.5	119.2	111.1	100.7	118.1	107.8	110.3	116.2	112.7	86.5	116.6	98.9	129.4	120.3	125.7
03	CONCRETE	84.5	47.1	65.7	110.6	117.4	114.0	107.7	116.9	112.3	112.8	115.5	114.1	103.7	116.4	110.1	124.1	124.7	124.4
04	MASONRY	96.2	35.6	58.7	88.2	114.0	104.1	108.2	114.1	111.9	110.7	113.8	112.6	98.3	118.6	110.8	148.5	125.1	134.0
05	METALS	87.7	62.2	78.3	110.5	99.1	106.3	105.0	98.5	102.6	106.9	99.1	104.0	111.3	97.6	106.2	101.7	104.8	102.9
06	WOOD & PLASTICS	91.2	46.0	67.5	91.9	120.3	106.8	84.0	120.4	103.2	97.4	117.3	107.9	90.2	119.9	105.8	100.0	137.3	119.6
07	THERMAL & MOISTURE PROTECTION	98.2	45.4	72.7	120.6	115.4	118.1	105.6	110.1	107.8	101.6	110.8	106.0	111.7	116.8	114.2	112.2	129.0	120.3
08	DOORS & WINDOWS	97.5	44.3	84.4	102.8	115.0	105.8	101.6	112.4	104.2	101.9	111.2	104.2	96.4	114.8	100.9	106.3	126.4	111.3
09200	PLASTER & GYPSUM BOARD	87.0	45.2	59.0	97.6	121.0	113.3	97.6	121.0	113.3	97.1	117.8	111.0	82.7	121.0	108.3	101.4	138.1	126.0
095,098	CEILINGS & ACOUSTICAL TREATMENT	100.7	45.2	63.0	113.8	121.0	118.6	113.8	121.0	118.6	113.8	117.8	116.5	109.3	121.0	117.2	112.9	138.1	130.0
09600	FLOORING	109.6	46.9	93.8	121.6	105.3	117.5	117.0	68.2	104.8	133.4	141.2	135.4	115.8	105.3	113.2	111.1	126.0	114.9
097,099	WALL FINISHES, PAINTS & COATINGS	96.8	40.0	63.3	108.5	111.5	110.3	111.6	100.6	105.1	135.6	98.2	113.5	103.3	115.8	110.7	116.6	139.1	129.9
09	FINISHES	98.2	43.5	69.1	112.0	117.0	114.7	113.0	109.5	111.2	120.4	121.0	120.7	106.1	117.2	112.0	114.8	135.0	125.5
10-14	TOTAL DIV. 10000 - 14000	100.0	43.3	87.8	100.0	114.8	103.2	100.0	128.8	106.2	100.0	130.0	106.5	100.0	113.6	102.9	100.0	134.4	107.4
15	MECHANICAL	100.0	41.4	73.6	100.2	110.3	104.8	100.2	105.8	102.7	100.2	111.2	105.2	100.0	110.2	104.6	100.2	132.7	114.9
16	ELECTRICAL	96.8	43.3	59.0	95.2	107.4	103.8	95.6	99.0	98.0	94.4	99.6	98.1	107.5	112.8	111.2	113.2	133.4	127.4
01-16	WEIGHTED AVERAGE	95.1	48.4	72.2	103.9	111.4	107.6	103.5	107.9	105.7	105.3	110.2	107.7	102.8	112.4	107.5	110.8	125.9	118.2

DIVISION		CALIFORNIA OXNARD MAT.	INST.	TOTAL	REDDING MAT.	INST.	TOTAL	RIVERSIDE MAT.	INST.	TOTAL	SACRAMENTO MAT.	INST.	TOTAL	SAN DIEGO MAT.	INST.	TOTAL	SAN FRANCISCO MAT.	INST.	TOTAL
01590	EQUIPMENT RENTAL	.0	98.1	98.1	.0	99.2	99.2	.0	100.7	100.7	.0	101.9	101.9	.0	97.7	97.7	.0	107.6	107.6
02	SITE CONSTRUCTION	105.7	104.3	104.6	110.7	106.0	107.1	97.2	107.2	104.8	107.0	110.7	109.8	101.9	101.6	101.6	143.3	110.8	118.8
03100	CONCRETE FORMS & ACCESSORIES	103.3	120.8	118.6	102.7	131.8	128.1	105.4	120.7	118.8	106.4	131.5	128.3	107.0	110.3	109.9	108.4	136.5	132.9
03200	CONCRETE REINFORCEMENT	105.7	110.8	108.7	102.5	111.0	107.6	104.8	110.9	108.4	100.0	111.1	106.5	108.7	110.6	109.8	112.6	112.2	112.4
03300	CAST-IN-PLACE CONCRETE	106.9	118.3	111.6	121.1	117.2	119.5	104.5	119.2	110.5	110.8	116.6	113.2	112.2	103.2	108.5	129.3	121.8	126.2
03	CONCRETE	111.2	117.1	114.2	121.7	121.6	121.6	110.0	117.4	113.7	114.8	121.2	118.0	116.6	107.2	111.9	125.7	125.6	125.7
04	MASONRY	113.2	110.5	111.5	114.6	114.5	114.5	85.9	113.5	103.0	119.2	112.8	115.3	102.7	108.9	106.5	148.7	129.1	136.6
05	METALS	104.4	98.5	102.2	108.9	99.4	105.4	110.5	99.0	106.2	97.8	98.6	98.1	110.7	98.0	106.0	107.1	106.0	106.7
06	WOOD & PLASTICS	91.6	120.4	106.7	92.5	134.0	114.3	91.9	120.3	106.8	94.2	134.2	115.2	98.6	106.7	102.9	100.0	137.6	119.7
07	THERMAL & MOISTURE PROTECTION	110.8	112.2	111.5	111.5	114.4	112.9	119.7	113.2	116.6	122.8	114.5	118.8	114.1	104.8	109.6	112.2	131.9	121.7
08	DOORS & WINDOWS	100.4	115.1	104.0	103.2	122.6	108.0	102.8	115.0	105.8	117.8	122.7	119.0	104.5	106.8	105.1	110.8	127.3	114.9
09200	PLASTER & GYPSUM BOARD	97.6	121.0	113.3	97.3	134.9	122.5	97.3	121.0	113.2	99.5	134.9	123.3	110.7	106.8	108.0	103.2	138.1	126.6
095,098	CEILINGS & ACOUSTICAL TREATMENT	113.8	121.0	118.6	119.3	134.9	129.9	112.4	121.0	118.2	119.8	134.9	130.1	110.7	106.8	108.0	118.0	138.1	131.6
09600	FLOORING	117.0	105.3	114.1	116.3	108.3	114.9	121.5	105.3	117.4	113.9	115.2	114.3	113.8	111.6	113.2	111.1	126.0	114.9
097,099	WALL FINISHES, PAINTS & COATINGS	111.0	104.3	107.1	111.0	122.1	117.6	108.5	111.5	110.3	115.1	122.1	119.3	110.8	111.5	111.2	116.6	147.3	134.7
09	FINISHES	112.8	116.3	114.7	114.0	129.2	122.1	111.4	117.0	114.4	114.9	129.1	122.4	110.6	110.5	110.5	116.1	136.3	126.8
10-14	TOTAL DIV. 10000 - 14000	100.0	115.1	103.2	100.0	132.9	107.1	100.0	114.7	103.2	100.0	132.7	107.0	100.0	113.5	102.9	100.0	135.1	107.6
15	MECHANICAL	100.2	110.3	104.8	100.2	114.6	106.7	100.1	110.3	104.7	100.1	115.4	107.0	100.1	110.3	104.7	100.2	157.0	125.8
16	ELECTRICAL	99.4	104.5	103.0	104.7	95.3	98.1	95.4	101.9	100.0	103.6	105.5	104.9	96.3	97.1	96.9	108.9	152.8	139.9
01-16	WEIGHTED AVERAGE	104.5	109.8	107.1	107.6	112.8	110.1	103.6	110.1	106.8	107.3	114.8	111.0	105.6	105.3	105.5	112.2	135.5	123.6

DIVISION		CALIFORNIA SAN JOSE MAT.	INST.	TOTAL	SANTA BARBARA MAT.	INST.	TOTAL	STOCKTON MAT.	INST.	TOTAL	VALLEJO MAT.	INST.	TOTAL	COLORADO COLORADO SPRINGS MAT.	INST.	TOTAL	DENVER MAT.	INST.	TOTAL
01590	EQUIPMENT RENTAL	.0	99.9	99.9	.0	99.5	99.5	.0	99.2	99.2	.0	102.6	102.6	.0	95.8	95.8	.0	101.0	101.0
02	SITE CONSTRUCTION	143.5	100.0	110.7	105.6	106.6	106.4	103.6	105.9	105.3	103.9	110.8	109.2	101.4	97.7	98.6	99.9	107.1	105.4
03100	CONCRETE FORMS & ACCESSORIES	105.5	135.7	131.9	104.0	120.8	118.7	98.1	118.6	116.0	107.6	134.2	130.9	89.9	80.5	81.7	99.3	83.1	85.1
03200	CONCRETE REINFORCEMENT	102.8	111.9	108.2	105.7	110.8	108.7	106.3	111.0	109.1	98.6	111.9	106.6	101.3	84.2	91.2	100.7	85.1	91.5
03300	CAST-IN-PLACE CONCRETE	125.3	120.1	123.2	106.5	118.2	111.3	106.4	116.1	110.4	116.1	117.9	116.8	104.8	88.0	97.9	98.0	88.7	94.2
03	CONCRETE	120.2	124.7	122.5	111.0	117.0	114.1	110.7	115.4	113.1	117.3	123.0	120.2	112.4	84.0	98.1	105.5	85.5	95.5
04	MASONRY	148.7	125.2	134.2	108.7	111.6	110.4	112.8	113.6	113.3	84.8	127.4	111.2	108.7	66.4	82.5	108.3	85.0	93.9
05	METALS	111.1	106.0	109.2	104.2	98.6	102.1	106.5	99.1	103.7	101.1	100.4	100.8	101.3	86.9	96.0	104.9	88.4	98.8
06	WOOD & PLASTICS	99.7	137.2	119.4	91.6	120.4	106.7	87.7	117.3	103.3	92.2	137.0	115.7	93.0	81.9	87.2	100.8	83.8	91.9
07	THERMAL & MOISTURE PROTECTION	107.3	130.9	118.7	106.8	110.6	108.7	110.9	110.1	110.5	125.6	127.0	126.3	105.1	79.6	92.8	104.4	84.7	94.9
08	DOORS & WINDOWS	92.7	127.1	101.2	101.6	115.1	104.9	100.8	113.6	104.0	119.2	127.0	121.2	98.3	87.4	95.6	99.7	88.4	96.9
09200	PLASTER & GYPSUM BOARD	98.7	138.1	125.1	97.6	121.0	113.3	98.7	117.8	111.5	100.0	138.1	125.5	95.0	81.4	85.9	104.8	83.6	90.6
095,098	CEILINGS & ACOUSTICAL TREATMENT	105.5	138.1	127.6	113.8	121.0	118.6	113.8	117.8	116.5	119.9	138.1	132.2	101.6	81.4	87.9	100.2	83.6	88.9
09600	FLOORING	114.1	126.0	117.1	117.0	108.3	114.9	117.0	114.8	116.5	118.1	126.0	120.1	108.3	96.1	105.2	109.2	96.1	105.9
097,099	WALL FINISHES, PAINTS & COATINGS	112.7	137.4	127.3	111.0	104.3	107.1	111.0	97.5	103.1	113.1	127.3	121.5	107.4	58.3	78.5	107.4	77.0	89.5
09	FINISHES	112.3	134.8	124.3	113.0	116.8	115.1	113.0	116.2	114.7	113.5	133.3	124.0	99.8	80.0	89.3	101.0	84.6	92.3
10-14	TOTAL DIV. 10000 - 14000	100.0	134.0	107.3	100.0	115.1	103.2	100.0	130.1	106.5	100.0	132.8	107.1	100.0	87.6	97.3	100.0	88.0	97.4
15	MECHANICAL	100.2	145.2	120.5	100.2	110.3	104.8	100.2	109.1	104.2	100.1	124.0	110.9	100.1	82.2	92.0	100.0	88.3	94.7
16	ELECTRICAL	110.0	146.0	135.4	92.2	108.1	103.4	103.6	110.3	108.3	99.5	107.6	105.2	97.7	87.2	90.3	99.9	92.1	94.4
01-16	WEIGHTED AVERAGE	109.5	130.3	119.7	103.8	110.8	107.2	105.0	111.0	108.0	105.8	119.7	112.7	102.0	83.6	93.0	102.1	89.5	95.9

DIVISION		COLORADO												CONNECTICUT					
		FORT COLLINS			GRAND JUNCTION			GREELEY			PUEBLO			BRIDGEPORT			BRISTOL		
		MAT.	INST.	TOTAL	MAT.	INST.	TOTAL	MAT.	INST.	TOTAL	MAT.	INST.	TOTAL	MAT.	INST.	TOTAL	MAT.	INST.	TOTAL
01590	EQUIPMENT RENTAL	.0	97.3	97.3	.0	99.9	99.9	.0	97.3	97.3	.0	96.8	96.8	.0	102.6	102.6	.0	102.6	102.6
02	SITE CONSTRUCTION	112.3	100.6	103.5	127.1	101.9	108.1	97.9	99.3	98.9	118.4	96.8	102.1	103.7	105.1	104.7	102.8	105.0	104.5
03100	CONCRETE FORMS & ACCESSORIES	98.7	79.4	81.8	109.6	48.3	56.1	96.0	52.8	58.2	104.3	80.8	83.7	103.4	112.8	111.6	103.4	112.5	111.3
03200	CONCRETE REINFORCEMENT	105.5	84.0	92.8	109.5	83.7	94.3	105.1	83.3	92.2	102.1	84.1	91.5	110.6	118.4	115.2	110.6	118.4	115.2
03300	CAST-IN-PLACE CONCRETE	113.9	80.4	100.1	120.2	48.3	90.7	95.1	64.7	82.7	107.7	88.9	100.0	106.5	115.3	110.1	99.8	115.2	106.1
03	CONCRETE	120.0	80.8	100.2	119.1	56.0	87.3	103.0	63.6	83.1	107.7	84.5	96.0	108.3	114.6	111.5	104.9	114.4	109.7
04	MASONRY	122.5	66.1	87.6	148.5	51.9	88.8	116.1	43.8	71.4	107.3	64.4	80.8	107.6	118.3	114.2	97.5	118.3	110.3
05	METALS	100.7	84.4	94.6	101.3	82.4	94.3	100.7	82.6	94.0	102.9	87.9	97.3	99.1	117.7	106.0	99.1	117.4	105.9
06	WOOD & PLASTICS	100.6	81.7	90.6	108.5	49.1	77.3	98.1	52.6	74.2	104.1	82.2	92.6	100.8	110.8	106.0	100.8	110.8	106.0
07	THERMAL & MOISTURE PROTECTION	105.1	74.9	90.5	105.9	59.9	83.7	104.2	63.3	84.5	105.3	78.4	92.3	102.2	117.3	109.5	102.4	114.3	108.1
08	DOORS & WINDOWS	95.7	87.3	93.6	100.7	69.7	93.1	95.7	71.5	89.7	95.2	87.6	93.3	107.1	121.2	110.6	107.1	114.3	108.9
09200	PLASTER & GYPSUM BOARD	103.6	81.4	88.7	115.1	47.7	69.9	102.6	51.4	68.3	91.4	81.4	84.7	104.3	110.1	108.2	104.3	110.1	108.2
095,098	CEILINGS & ACOUSTICAL TREATMENT	94.7	81.4	85.7	97.8	47.7	63.8	94.7	51.4	65.3	106.1	81.4	89.3	107.9	110.1	109.4	107.9	110.1	109.4
09600	FLOORING	109.1	77.2	101.1	118.3	77.2	107.9	107.7	77.2	100.0	114.5	96.1	109.9	101.6	116.6	105.4	101.6	116.6	105.4
097,099	WALL FINISHES, PAINTS & COATINGS	107.4	57.8	78.2	119.2	33.9	68.9	107.4	36.0	65.3	119.2	54.0	80.8	94.4	109.2	103.1	94.4	109.7	103.4
09	FINISHES	100.4	76.9	87.9	108.7	51.7	78.4	99.1	54.5	75.4	105.5	80.8	92.4	103.1	112.8	108.2	103.1	112.8	108.3
10 - 14	TOTAL DIV. 10000 - 14000	100.0	68.9	93.3	100.0	59.4	91.3	100.0	60.4	91.5	100.0	88.5	97.5	100.0	114.8	103.2	100.0	114.8	103.2
15	MECHANICAL	100.0	84.9	93.2	100.0	37.1	71.7	100.0	78.3	90.2	100.0	74.4	88.5	100.1	103.2	101.5	100.1	103.2	101.5
16	ELECTRICAL	92.3	64.3	72.5	94.6	71.2	78.1	92.3	64.3	72.5	96.0	78.4	83.6	104.4	101.4	102.3	104.4	102.2	102.9
01 - 16	WEIGHTED AVERAGE	103.4	78.9	91.3	107.0	61.4	84.6	100.3	68.7	84.8	102.2	80.6	91.6	103.0	109.9	106.4	102.0	109.7	105.8

DIVISION		CONNECTICUT																	
		HARTFORD			NEW BRITAIN			NEW HAVEN			NORWALK			STAMFORD			WATERBURY		
		MAT.	INST.	TOTAL	MAT.	INST.	TOTAL	MAT.	INST.	TOTAL	MAT.	INST.	TOTAL	MAT.	INST.	TOTAL	MAT.	INST.	TOTAL
01590	EQUIPMENT RENTAL	.0	102.6	102.6	.0	102.6	102.6	.0	103.1	103.1	.0	102.6	102.6	.0	102.6	102.6	.0	102.6	102.6
02	SITE CONSTRUCTION	103.1	105.0	104.6	103.0	105.0	104.5	102.8	105.9	105.2	103.5	105.1	104.7	104.1	105.1	104.8	103.2	105.1	104.6
03100	CONCRETE FORMS & ACCESSORIES	102.2	112.5	111.2	103.7	112.5	111.4	103.2	112.7	111.5	103.4	113.0	111.8	103.4	113.2	112.0	103.4	112.7	111.5
03200	CONCRETE REINFORCEMENT	110.6	118.4	115.2	110.6	118.4	115.2	110.6	118.4	115.2	110.6	118.7	115.4	110.6	118.8	115.4	110.6	118.4	115.2
03300	CAST-IN-PLACE CONCRETE	99.5	115.2	105.9	101.4	115.2	107.1	103.1	115.3	108.1	104.8	123.3	112.4	106.5	123.4	113.4	106.5	115.3	110.1
03	CONCRETE	104.7	114.4	109.6	105.7	114.4	110.1	121.4	114.5	117.9	107.4	117.5	112.5	108.3	117.6	113.0	108.3	114.5	111.4
04	MASONRY	97.4	118.3	110.3	97.6	118.3	110.4	97.7	118.3	110.4	97.8	123.3	113.6	98.0	123.3	113.6	98.0	118.3	110.5
05	METALS	99.7	117.4	106.3	95.8	117.4	103.8	96.0	117.6	104.0	99.1	118.3	106.2	99.1	118.5	106.3	99.1	117.6	106.0
06	WOOD & PLASTICS	100.8	110.8	106.0	100.8	110.8	106.0	100.8	110.8	106.0	100.8	110.8	106.0	100.8	110.8	106.0	100.8	110.8	106.0
07	THERMAL & MOISTURE PROTECTION	101.0	114.3	107.4	102.4	114.3	108.1	102.5	114.3	108.2	102.4	119.8	110.8	102.4	119.8	110.8	102.4	114.3	108.1
08	DOORS & WINDOWS	107.1	114.3	108.9	107.1	114.3	108.9	107.1	121.2	110.6	107.1	121.2	110.6	107.1	121.2	110.6	107.1	121.2	110.6
09200	PLASTER & GYPSUM BOARD	104.3	110.1	108.2	104.3	110.1	108.2	104.3	110.1	108.2	104.3	110.1	108.2	104.3	110.1	108.2	104.3	110.1	108.2
095,098	CEILINGS & ACOUSTICAL TREATMENT	107.9	110.1	109.4	107.9	110.1	109.4	107.9	110.1	109.4	107.9	110.1	109.4	107.9	110.1	109.4	107.9	110.1	109.4
09600	FLOORING	101.6	116.6	105.4	101.6	116.6	105.4	101.6	116.6	105.4	101.6	116.6	105.4	101.6	116.6	105.4	101.6	116.6	105.4
097,099	WALL FINISHES, PAINTS & COATINGS	94.4	109.7	103.4	94.4	109.7	103.4	94.4	107.1	101.9	94.4	109.2	103.1	94.4	109.2	103.1	94.4	109.7	103.4
09	FINISHES	103.1	112.8	108.2	103.1	112.8	108.3	103.2	112.5	108.1	103.1	112.8	108.2	103.2	112.8	108.3	103.0	112.8	108.2
10 - 14	TOTAL DIV. 10000 - 14000	100.0	114.8	103.2	100.0	114.8	103.2	100.0	114.8	103.2	100.0	114.8	103.2	100.0	114.9	103.2	100.0	114.8	103.2
15	MECHANICAL	100.1	103.2	101.5	100.1	103.2	101.5	100.1	103.2	101.5	100.1	103.2	101.5	100.1	103.3	101.5	100.1	103.2	101.5
16	ELECTRICAL	103.6	98.5	100.0	104.4	102.2	102.9	104.4	102.2	102.9	104.4	101.4	102.3	104.4	125.1	119.0	103.6	101.4	102.0
01 - 16	WEIGHTED AVERAGE	101.9	109.0	105.4	101.6	109.7	105.6	103.6	110.0	106.7	102.3	110.9	106.6	102.5	115.1	108.6	102.4	109.8	106.0

DIVISION		D.C.			DELAWARE			FLORIDA											
		WASHINGTON			WILMINGTON			DAYTONA BEACH			FORT LAUDERDALE			JACKSONVILLE			MELBOURNE		
		MAT.	INST.	TOTAL	MAT.	INST.	TOTAL	MAT.	INST.	TOTAL	MAT.	INST.	TOTAL	MAT.	INST.	TOTAL	MAT.	INST.	TOTAL
01590	EQUIPMENT RENTAL	.0	103.0	103.0	.0	118.8	118.8	.0	97.7	97.7	.0	89.3	89.3	.0	97.7	97.7	.0	97.7	97.7
02	SITE CONSTRUCTION	102.7	89.7	92.9	85.9	112.5	106.0	117.6	87.0	94.5	103.0	73.4	80.7	117.7	87.6	95.0	125.6	87.4	96.7
03100	CONCRETE FORMS & ACCESSORIES	100.0	82.4	84.6	96.6	103.3	102.4	94.3	67.6	70.9	92.2	66.6	69.8	93.9	57.8	62.3	88.6	70.3	72.6
03200	CONCRETE REINFORCEMENT	99.6	90.4	94.2	97.2	100.0	98.9	94.1	85.0	89.2	94.1	70.5	80.8	95.2	85.1	89.2	95.2	85.1	89.2
03300	CAST-IN-PLACE CONCRETE	112.0	86.3	101.5	77.2	97.6	85.6	90.2	71.9	82.7	94.7	65.4	82.7	91.1	62.9	79.5	104.8	75.5	92.8
03	CONCRETE	105.6	86.7	96.1	98.3	101.7	100.0	88.7	73.7	81.2	90.8	68.4	79.5	89.1	60.7	74.8	98.3	76.2	87.2
04	MASONRY	91.6	82.6	86.0	107.8	89.9	96.7	86.3	66.3	73.9	86.6	57.5	68.6	85.9	55.8	67.2	84.0	72.5	76.9
05	METALS	97.1	110.5	102.0	97.9	117.0	105.0	99.0	96.5	98.0	98.3	90.5	95.4	98.5	83.2	92.8	106.9	96.9	103.2
06	WOOD & PLASTICS	97.6	82.1	89.4	95.8	104.8	100.5	92.8	68.9	80.2	88.8	70.1	79.0	92.8	56.8	73.9	87.5	68.9	77.7
07	THERMAL & MOISTURE PROTECTION	95.0	83.9	89.7	100.0	107.8	103.7	96.0	71.0	83.9	96.0	65.6	81.3	96.3	61.6	79.6	96.2	75.4	86.2
08	DOORS & WINDOWS	102.0	91.5	99.4	94.4	106.8	97.5	100.8	68.5	92.9	98.4	65.7	90.4	100.8	53.7	89.2	100.0	72.5	93.2
09200	PLASTER & GYPSUM BOARD	112.1	81.5	91.6	103.9	104.8	104.5	102.1	68.5	79.6	101.6	69.8	80.2	102.1	56.1	71.2	98.6	68.5	78.4
095,098	CEILINGS & ACOUSTICAL TREATMENT	102.1	81.5	88.1	102.3	104.8	104.0	96.3	68.5	77.4	96.3	69.8	78.3	96.3	56.1	69.0	92.1	68.5	76.1
09600	FLOORING	102.9	96.0	101.1	80.2	96.5	84.3	117.3	72.9	106.2	117.3	65.8	104.4	117.3	54.0	101.4	114.0	72.9	103.7
097,099	WALL FINISHES, PAINTS & COATINGS	109.6	92.8	99.7	84.9	101.9	94.9	110.1	78.4	91.4	106.5	54.3	75.7	110.1	52.3	76.1	110.1	94.3	100.8
09	FINISHES	99.8	85.8	92.4	99.6	101.7	100.7	107.1	69.7	87.2	105.0	65.1	83.8	107.1	56.0	79.9	105.3	73.0	88.1
10 - 14	TOTAL DIV. 10000 - 14000	100.0	96.2	99.2	100.0	109.4	102.0	100.0	77.5	95.2	100.0	80.8	95.9	100.0	73.9	94.4	100.0	80.1	95.7
15	MECHANICAL	100.0	90.8	95.9	100.3	108.9	104.2	99.9	66.0	84.6	99.9	65.1	84.2	99.9	55.9	80.0	99.9	69.6	86.3
16	ELECTRICAL	97.1	97.6	97.5	97.6	104.4	102.4	97.2	60.7	71.4	97.2	76.3	82.4	96.7	65.9	75.0	96.9	69.2	77.4
01 - 16	WEIGHTED AVERAGE	99.7	91.6	95.7	98.7	105.4	102.0	98.6	71.8	85.4	97.8	70.2	84.3	98.5	64.2	81.7	100.7	75.7	88.4

COST INDEXES

FLORIDA

DIVISION		MIAMI MAT.	INST.	TOTAL	ORLANDO MAT.	INST.	TOTAL	PANAMA CITY MAT.	INST.	TOTAL	PENSACOLA MAT.	INST.	TOTAL	ST. PETERSBURG MAT.	INST.	TOTAL	TALLAHASSEE MAT.	INST.	TOTAL
01590	EQUIPMENT RENTAL	.0	89.3	89.3	.0	97.7	97.7	.0	97.7	97.7	.0	97.7	97.7	.0	97.7	97.7	.0	97.7	97.7
02	SITE CONSTRUCTION	102.3	73.2	80.3	118.2	86.7	94.4	132.7	84.3	96.1	130.2	86.7	97.4	118.5	86.2	94.1	119.0	86.2	94.2
03100	CONCRETE FORMS & ACCESSORIES	91.9	66.6	69.8	94.0	57.7	62.2	93.0	30.1	38.1	83.6	55.4	59.0	91.6	51.4	56.4	93.9	42.4	48.9
03200	CONCRETE REINFORCEMENT	94.1	70.6	80.2	94.1	81.1	86.4	98.3	53.5	71.8	100.7	53.9	73.0	97.5	61.6	76.3	94.1	54.1	70.5
03300	CAST-IN-PLACE CONCRETE	92.2	65.6	81.3	98.1	71.7	87.3	95.8	37.1	71.7	95.7	58.8	80.6	102.1	59.3	84.5	94.4	51.9	77.0
03	CONCRETE	89.5	68.4	78.9	90.4	68.6	79.4	97.5	38.8	67.9	96.4	58.1	77.1	94.9	57.8	76.2	90.8	50.0	70.2
04	MASONRY	84.8	57.9	68.2	87.5	66.3	74.4	90.7	29.9	53.1	87.8	54.1	66.9	134.3	53.4	84.3	87.4	42.1	59.4
05	METALS	98.8	89.9	95.5	108.2	94.7	103.1	97.2	68.3	86.5	97.1	82.7	91.8	101.0	84.0	94.7	98.9	81.6	92.5
06	WOOD & PLASTICS	88.8	70.1	79.0	92.8	56.0	73.5	91.7	30.3	59.5	82.9	56.0	68.8	90.4	51.4	69.9	91.0	40.7	64.6
07	THERMAL & MOISTURE PROTECTION	99.0	67.0	83.5	96.3	69.7	83.5	96.6	33.6	66.2	96.3	56.6	77.1	95.9	53.1	75.3	96.3	49.8	73.9
08	DOORS & WINDOWS	98.4	66.3	90.5	100.8	59.8	90.7	98.4	28.8	81.3	98.4	54.2	87.5	99.4	50.2	87.3	99.4	44.3	85.9
09200	PLASTER & GYPSUM BOARD	101.6	69.8	80.2	105.5	55.2	71.8	100.3	28.8	52.4	96.2	55.3	68.8	99.8	50.5	66.8	102.1	39.5	60.1
095,098	CEILINGS & ACOUSTICAL TREATMENT	96.3	69.8	78.3	96.3	55.2	68.4	90.7	28.8	48.7	90.7	55.3	66.7	90.7	50.5	63.4	96.3	39.5	57.7
09600	FLOORING	124.8	66.5	110.1	117.3	72.9	106.2	116.7	20.7	92.6	110.9	57.4	97.5	115.7	57.2	101.0	117.3	42.0	98.4
097,099	WALL FINISHES, PAINTS & COATINGS	106.5	54.3	75.7	110.1	60.1	80.6	110.1	26.7	61.0	110.1	60.8	81.1	110.1	50.6	75.1	110.1	43.1	70.6
09	FINISHES	107.4	65.2	85.0	107.5	60.0	82.3	106.8	27.5	64.7	104.2	56.3	78.7	105.2	52.2	77.0	107.2	41.5	72.3
10-14	TOTAL DIV. 10000 - 14000	100.0	80.8	95.9	100.0	75.7	94.8	100.0	51.2	89.5	100.0	57.3	90.8	100.0	59.9	91.4	100.0	66.9	92.9
15	MECHANICAL	99.9	66.1	84.7	99.9	59.2	81.5	99.9	27.4	67.2	99.9	54.5	79.4	99.9	54.3	79.3	99.9	43.4	74.4
16	ELECTRICAL	97.7	76.8	83.0	97.8	48.5	63.0	95.5	36.3	53.7	100.9	58.7	71.1	97.6	52.4	65.7	97.8	44.9	60.5
01-16	WEIGHTED AVERAGE	98.0	70.6	84.5	100.3	65.8	83.4	99.7	40.5	70.6	99.3	61.5	80.7	102.1	59.6	81.2	98.8	52.3	76.0

FLORIDA / GEORGIA

DIVISION		TAMPA (FL) MAT.	INST.	TOTAL	ALBANY MAT.	INST.	TOTAL	ATLANTA MAT.	INST.	TOTAL	AUGUSTA MAT.	INST.	TOTAL	COLUMBUS MAT.	INST.	TOTAL	MACON MAT.	INST.	TOTAL
01590	EQUIPMENT RENTAL	.0	97.7	97.7	.0	90.1	90.1	.0	93.0	93.0	.0	92.3	92.3	.0	90.1	90.1	.0	102.6	102.6
02	SITE CONSTRUCTION	118.8	86.2	94.2	103.2	76.1	82.7	105.0	95.2	97.6	101.3	92.5	94.6	103.2	76.3	82.9	103.9	94.1	96.5
03100	CONCRETE FORMS & ACCESSORIES	95.3	51.7	57.2	93.6	46.4	52.4	92.4	81.9	83.2	89.6	55.9	60.1	94.9	59.4	63.8	92.5	60.0	64.1
03200	CONCRETE REINFORCEMENT	94.1	61.6	74.9	94.1	91.5	92.6	100.3	93.3	96.1	106.0	78.4	89.7	94.1	91.6	92.7	96.5	91.7	93.7
03300	CAST-IN-PLACE CONCRETE	99.8	59.5	83.3	96.4	44.6	75.1	99.3	77.2	90.2	93.8	52.4	76.8	96.0	45.3	75.2	94.7	48.8	75.9
03	CONCRETE	93.6	58.0	75.7	91.7	56.0	73.7	94.7	82.0	88.4	90.9	59.4	75.0	91.6	62.0	76.7	91.1	63.5	77.2
04	MASONRY	87.4	53.4	66.4	88.5	34.6	55.1	88.3	73.6	79.2	88.5	43.6	60.7	88.6	38.1	57.4	102.6	41.6	64.9
05	METALS	101.8	84.4	95.4	96.2	90.3	94.0	92.2	81.0	88.1	90.9	71.3	83.6	96.6	91.5	94.7	91.5	91.9	91.6
06	WOOD & PLASTICS	94.0	51.4	71.6	92.2	46.4	68.1	90.6	84.6	87.5	88.1	58.1	72.4	93.7	63.9	78.0	100.8	62.7	80.8
07	THERMAL & MOISTURE PROTECTION	96.2	54.1	75.9	96.1	54.2	75.9	92.8	78.5	85.9	92.2	53.4	73.5	95.7	56.9	77.0	94.5	60.9	78.3
08	DOORS & WINDOWS	100.8	50.5	88.4	98.4	52.8	87.2	100.2	80.5	95.3	94.3	56.6	85.0	98.4	62.5	89.6	96.9	62.7	88.5
09200	PLASTER & GYPSUM BOARD	102.1	50.5	67.5	102.1	45.3	64.0	116.8	84.6	95.2	115.8	57.3	76.6	102.1	63.4	76.1	111.0	62.1	78.2
095,098	CEILINGS & ACOUSTICAL TREATMENT	96.3	50.5	65.2	96.3	45.3	61.7	95.9	84.6	88.2	97.5	57.3	70.2	96.3	63.4	73.9	88.6	62.1	70.6
09600	FLOORING	117.3	57.2	102.2	117.3	35.1	96.7	85.6	78.9	83.9	84.4	44.6	74.4	117.3	40.9	98.1	91.5	41.2	78.9
097,099	WALL FINISHES, PAINTS & COATINGS	110.1	50.6	75.1	106.5	45.0	70.2	90.1	81.2	84.8	90.1	42.8	62.2	106.5	43.1	69.1	108.2	52.7	75.5
09	FINISHES	107.1	52.2	77.9	105.0	42.9	72.0	92.8	81.5	86.8	92.4	52.4	71.1	104.9	54.2	78.0	91.2	55.5	72.2
10-14	TOTAL DIV. 10000 - 14000	100.0	75.7	94.8	100.0	73.1	94.2	100.0	83.5	96.5	100.0	75.2	94.7	100.0	75.2	94.7	100.0	76.6	95.0
15	MECHANICAL	99.9	54.6	79.5	99.9	49.2	77.1	100.0	80.6	91.3	100.0	46.6	75.9	99.9	40.3	73.0	99.9	45.4	75.3
16	ELECTRICAL	96.7	52.4	65.4	91.3	65.8	73.3	95.2	85.8	88.5	97.2	49.4	63.5	92.6	42.6	57.3	90.6	64.9	72.5
01-16	WEIGHTED AVERAGE	99.7	60.2	80.3	97.4	57.8	77.9	96.3	82.6	89.6	94.9	57.1	76.4	97.5	55.4	76.8	95.8	62.8	79.6

GEORGIA / HAWAII / IDAHO

DIVISION		SAVANNAH MAT.	INST.	TOTAL	VALDOSTA MAT.	INST.	TOTAL	HONOLULU MAT.	INST.	TOTAL	BOISE MAT.	INST.	TOTAL	LEWISTON MAT.	INST.	TOTAL	POCATELLO MAT.	INST.	TOTAL
01590	EQUIPMENT RENTAL	.0	91.2	91.2	.0	90.1	90.1	.0	99.3	99.3	.0	101.6	101.6	.0	94.4	94.4	.0	101.6	101.6
02	SITE CONSTRUCTION	103.8	77.4	83.9	113.9	76.7	85.8	135.6	107.2	114.2	81.2	103.4	97.9	87.7	97.2	94.9	82.3	103.3	98.2
03100	CONCRETE FORMS & ACCESSORIES	93.4	56.8	61.4	80.6	49.2	53.1	106.7	151.2	145.6	98.5	85.3	87.0	112.4	71.8	76.9	98.5	84.9	86.6
03200	CONCRETE REINFORCEMENT	99.7	78.7	87.3	99.9	47.0	68.6	105.7	125.7	117.5	100.8	79.1	88.0	114.2	96.5	103.7	101.1	78.6	87.8
03300	CAST-IN-PLACE CONCRETE	93.0	54.6	77.2	94.3	55.2	78.3	196.5	130.9	169.5	101.8	93.4	98.4	110.4	88.2	101.3	101.0	93.2	97.8
03	CONCRETE	90.6	61.6	76.0	96.1	52.6	74.2	156.3	137.7	146.9	106.8	86.8	96.7	115.3	82.4	98.7	103.0	86.4	94.6
04	MASONRY	91.7	57.1	70.3	94.8	51.8	68.2	133.4	132.9	133.1	137.8	78.6	101.2	138.4	85.0	105.4	133.0	71.1	94.7
05	METALS	96.9	86.5	93.1	96.4	75.6	88.7	116.3	110.6	114.2	113.0	79.2	100.5	96.6	88.3	93.5	113.3	78.2	100.3
06	WOOD & PLASTICS	106.4	54.7	79.3	80.0	45.2	61.7	95.1	156.5	127.3	96.5	84.5	90.2	106.1	65.8	84.9	96.5	84.5	90.2
07	THERMAL & MOISTURE PROTECTION	96.1	56.3	76.9	95.8	61.0	79.0	113.3	133.9	123.2	97.9	83.2	90.8	169.5	81.5	127.1	98.2	74.2	86.6
08	DOORS & WINDOWS	99.6	54.1	88.4	93.8	41.8	81.0	105.9	143.6	115.2	94.7	79.7	91.0	114.8	71.0	104.0	94.7	73.2	89.4
09200	PLASTER & GYPSUM BOARD	102.1	54.0	69.8	95.2	44.1	61.0	118.0	158.1	144.9	89.8	83.9	85.9	160.0	64.8	96.1	89.8	83.9	85.9
095,098	CEILINGS & ACOUSTICAL TREATMENT	96.3	54.0	67.5	90.7	44.1	59.1	113.8	158.1	143.9	106.1	83.9	91.0	153.0	64.8	93.1	106.1	83.9	91.0
09600	FLOORING	117.3	52.6	101.1	109.1	42.0	92.3	168.0	135.4	159.8	101.7	60.7	91.4	140.1	89.2	127.3	102.0	60.7	91.6
097,099	WALL FINISHES, PAINTS & COATINGS	106.5	53.4	75.2	106.5	39.0	66.7	105.5	154.7	134.5	105.7	56.4	76.6	131.9	77.9	100.1	105.7	59.1	78.2
09	FINISHES	105.2	55.4	78.8	101.0	46.1	71.8	134.4	150.3	142.9	98.8	77.8	87.7	168.4	74.5	118.5	100.0	77.9	87.9
10-14	TOTAL DIV. 10000 - 14000	100.0	78.0	95.3	100.0	76.6	95.0	100.0	127.3	105.9	100.0	87.9	97.4	100.0	96.9	99.3	100.0	87.9	97.4
15	MECHANICAL	99.9	51.3	78.0	99.9	45.2	75.2	100.2	123.7	110.8	100.0	82.5	92.2	101.3	89.5	95.9	100.0	82.5	92.1
16	ELECTRICAL	93.9	57.2	68.0	89.5	32.8	49.5	114.4	125.1	122.0	81.9	79.4	80.1	84.2	86.9	86.1	84.9	77.6	79.7
01-16	WEIGHTED AVERAGE	98.0	61.3	80.0	97.4	51.5	74.9	117.9	128.7	123.2	102.3	83.3	93.0	113.2	85.4	99.6	101.8	81.7	91.9

COST INDEXES

DIVISION		IDAHO TWIN FALLS			ILLINOIS CHICAGO			DECATUR			EAST ST. LOUIS			JOLIET			PEORIA		
		MAT.	INST.	TOTAL	MAT.	INST.	TOTAL	MAT.	INST.	TOTAL	MAT.	INST.	TOTAL	MAT.	INST.	TOTAL	MAT.	INST.	TOTAL
01590	EQUIPMENT RENTAL	.0	101.6	101.6	.0	93.0	93.0	.0	102.3	102.3	.0	108.9	108.9	.0	90.8	90.8	.0	101.5	101.5
02	SITE CONSTRUCTION	89.4	101.3	98.4	92.5	91.2	91.5	86.1	95.1	92.9	102.8	95.6	97.4	92.5	90.1	90.7	95.0	94.2	94.4
03100	CONCRETE FORMS & ACCESSORIES	101.5	38.2	46.2	104.0	138.2	133.9	98.3	111.6	109.9	91.6	111.1	108.6	104.9	135.1	131.3	96.3	115.0	112.7
03200	CONCRETE REINFORCEMENT	109.9	52.5	75.9	97.0	148.5	127.5	98.6	98.8	98.7	104.3	102.3	103.1	97.0	130.5	116.8	98.6	105.4	102.6
03300	CAST-IN-PLACE CONCRETE	103.6	47.9	80.8	111.8	134.3	121.1	99.3	100.3	99.7	93.0	114.0	101.6	111.8	125.5	117.4	97.1	111.1	102.8
03	CONCRETE	112.1	45.3	78.4	108.6	137.9	123.3	96.2	105.5	100.9	87.2	111.4	99.4	108.6	129.9	119.3	94.9	112.0	103.5
04	MASONRY	136.4	42.3	78.2	94.7	136.3	120.5	68.1	99.5	87.5	73.5	117.0	100.0	96.2	124.4	113.6	112.6	111.7	112.0
05	METALS	113.2	65.1	95.4	96.1	126.5	107.4	95.4	107.3	99.8	92.9	121.2	103.4	94.2	116.1	102.3	95.4	112.3	101.7
06	WOOD & PLASTICS	99.3	38.2	67.2	104.9	137.0	121.8	100.5	111.3	106.2	97.1	108.9	103.3	106.9	137.6	123.0	100.6	113.7	107.5
07	THERMAL & MOISTURE PROTECTION	99.3	50.0	75.5	100.4	131.8	115.5	97.7	99.9	98.8	92.6	107.9	100.0	100.2	125.9	112.6	97.6	108.9	103.1
08	DOORS & WINDOWS	98.0	38.9	83.4	103.7	140.9	112.8	96.3	107.5	99.0	86.0	114.5	93.0	101.6	136.4	110.2	96.3	110.6	99.8
09200	PLASTER & GYPSUM BOARD	89.9	36.2	53.9	97.7	137.8	124.6	101.4	111.4	108.1	98.2	109.0	105.4	95.1	138.6	124.2	101.4	113.9	109.8
095,098	CEILINGS & ACOUSTICAL TREATMENT	101.9	36.2	57.3	100.7	137.8	125.9	98.8	111.4	107.4	93.3	109.0	104.0	100.7	138.6	126.4	98.8	113.9	109.0
09600	FLOORING	103.4	60.7	92.6	84.0	132.2	96.1	99.1	91.3	97.2	108.4	112.7	109.5	83.6	125.4	94.1	99.1	107.6	101.3
097,099	WALL FINISHES, PAINTS & COATINGS	105.7	34.8	63.9	78.3	134.9	111.6	90.8	101.6	97.2	98.8	98.3	98.5	76.1	125.7	105.3	90.8	104.8	99.1
09	FINISHES	99.2	41.7	68.6	90.5	137.0	115.2	96.9	107.4	102.5	96.7	108.8	103.1	89.9	133.1	112.8	97.0	112.9	105.4
10-14	TOTAL DIV. 10000-14000	100.0	53.6	90.0	100.0	123.8	105.1	100.0	103.9	100.8	100.0	105.4	101.2	100.0	120.3	104.4	100.0	103.4	100.7
15	MECHANICAL	100.0	39.5	72.8	100.1	125.5	111.6	100.1	99.1	99.7	100.0	99.6	99.8	100.3	121.0	109.6	100.1	106.3	102.9
16	ELECTRICAL	87.7	34.2	50.0	100.3	130.6	121.6	102.9	88.7	92.9	99.1	99.8	99.6	99.9	115.8	111.1	101.8	96.6	98.1
01-16	WEIGHTED AVERAGE	104.0	48.6	76.8	99.5	128.2	113.6	96.0	100.2	98.1	93.6	106.6	100.0	99.1	120.6	109.7	98.6	106.3	102.4

DIVISION		ILLINOIS ROCKFORD			SPRINGFIELD			INDIANA ANDERSON			BLOOMINGTON			EVANSVILLE			FORT WAYNE		
		MAT.	INST.	TOTAL	MAT.	INST.	TOTAL	MAT.	INST.	TOTAL	MAT.	INST.	TOTAL	MAT.	INST.	TOTAL	MAT.	INST.	TOTAL
01590	EQUIPMENT RENTAL	.0	101.5	101.5	.0	102.3	102.3	.0	97.7	97.7	.0	86.4	86.4	.0	122.0	122.0	.0	97.7	97.7
02	SITE CONSTRUCTION	94.1	94.2	94.2	92.4	95.2	94.5	89.2	97.6	95.5	78.5	96.2	91.9	83.6	130.9	119.3	89.9	97.8	95.8
03100	CONCRETE FORMS & ACCESSORIES	101.3	118.4	116.3	100.6	112.4	110.9	92.2	81.2	82.6	103.7	82.1	84.8	92.3	81.6	82.9	91.7	81.3	82.6
03200	CONCRETE REINFORCEMENT	98.6	139.0	122.5	98.6	98.9	98.8	95.4	86.7	90.2	88.2	86.5	87.2	96.6	83.9	89.1	95.4	81.2	87.0
03300	CAST-IN-PLACE CONCRETE	99.3	104.2	101.3	92.3	107.0	98.3	102.1	84.1	94.7	101.5	80.5	92.9	96.9	94.2	95.8	108.4	81.7	97.5
03	CONCRETE	96.4	117.5	107.0	92.8	108.2	100.6	97.6	83.8	90.6	105.9	82.0	93.9	105.7	86.6	96.1	100.7	82.0	91.3
04	MASONRY	83.9	118.0	105.0	67.5	100.6	87.9	93.8	82.4	86.8	91.7	80.3	84.7	87.8	85.4	86.3	91.1	84.3	86.9
05	METALS	95.4	127.6	107.4	95.4	107.7	100.0	102.0	92.1	98.3	97.3	78.1	90.2	90.7	88.6	89.9	102.0	90.2	97.6
06	WOOD & PLASTICS	100.5	116.8	109.1	103.5	111.3	107.6	111.1	80.7	95.2	122.9	81.4	101.1	97.5	78.7	87.6	110.6	80.2	94.6
07	THERMAL & MOISTURE PROTECTION	97.6	115.4	106.2	97.2	105.5	101.2	100.7	80.7	91.0	92.1	85.5	88.9	95.4	88.5	92.1	100.4	86.4	93.6
08	DOORS & WINDOWS	96.3	121.3	102.4	96.2	107.5	99.0	99.3	83.8	95.5	103.1	84.1	98.5	95.6	81.0	92.0	99.3	79.5	94.4
09200	PLASTER & GYPSUM BOARD	101.4	117.0	111.9	101.4	111.4	108.1	104.1	80.8	88.5	98.6	81.6	87.2	94.5	77.1	82.8	99.0	80.2	86.4
095,098	CEILINGS & ACOUSTICAL TREATMENT	98.8	117.0	111.2	98.8	111.4	107.4	98.0	80.8	86.3	84.9	81.6	82.7	96.4	77.1	83.3	98.0	80.2	85.9
09600	FLOORING	99.1	100.7	99.5	99.3	91.4	97.4	86.4	95.6	88.7	102.9	95.6	101.1	96.3	87.3	94.0	86.4	86.1	86.3
097,099	WALL FINISHES, PAINTS & COATINGS	90.8	118.3	107.0	90.8	98.1	95.1	90.2	73.6	80.4	90.0	93.8	92.3	96.8	91.9	93.9	90.2	81.3	84.9
09	FINISHES	97.0	115.1	106.6	97.0	107.3	102.5	91.3	83.0	86.9	96.0	85.8	90.5	95.2	83.3	88.9	90.5	82.1	86.0
10-14	TOTAL DIV. 10000-14000	100.0	111.9	102.6	100.0	104.4	100.9	100.0	88.1	97.4	100.0	85.9	97.0	100.0	91.3	98.1	100.0	92.9	98.5
15	MECHANICAL	100.1	107.2	103.3	100.1	103.7	101.8	99.9	83.0	92.3	99.7	87.3	94.1	100.0	89.4	95.2	99.9	85.5	93.4
16	ELECTRICAL	101.8	109.1	106.9	102.9	96.2	98.2	81.6	94.4	90.6	101.1	87.7	91.6	95.8	81.8	85.9	82.1	84.1	83.5
01-16	WEIGHTED AVERAGE	97.1	112.5	104.7	95.8	103.0	99.3	97.2	87.2	92.3	99.2	85.5	92.5	96.7	90.0	93.4	97.4	85.7	91.7

DIVISION		INDIANA GARY			INDIANAPOLIS			MUNCIE			SOUTH BEND			TERRE HAUTE			IOWA CEDAR RAPIDS		
		MAT.	INST.	TOTAL	MAT.	INST.	TOTAL	MAT.	INST.	TOTAL	MAT.	INST.	TOTAL	MAT.	INST.	TOTAL	MAT.	INST.	TOTAL
01590	EQUIPMENT RENTAL	.0	97.7	97.7	.0	92.5	92.5	.0	96.6	96.6	.0	107.8	107.8	.0	122.0	122.0	.0	93.0	93.0
02	SITE CONSTRUCTION	89.8	99.6	96.2	89.4	101.2	98.3	78.4	96.3	91.9	89.3	97.5	95.5	85.1	130.9	119.7	87.3	95.1	93.2
03100	CONCRETE FORMS & ACCESSORIES	92.7	108.5	106.5	92.8	86.1	86.9	89.9	81.0	82.1	97.6	84.2	85.0	93.7	83.8	85.0	104.4	80.0	83.1
03200	CONCRETE REINFORCEMENT	95.4	106.4	101.9	94.8	84.3	88.6	97.6	86.6	91.1	95.4	81.6	87.3	96.6	86.2	90.5	98.2	85.2	90.5
03300	CAST-IN-PLACE CONCRETE	106.6	107.8	107.1	98.2	86.5	93.4	106.6	83.6	97.2	99.3	88.5	94.8	93.8	96.3	94.8	108.4	84.8	98.7
03	CONCRETE	99.9	107.9	103.9	100.1	85.5	92.7	103.6	83.5	93.5	91.8	86.7	89.2	108.9	88.7	98.7	99.2	83.1	91.1
04	MASONRY	95.3	99.0	97.6	101.1	84.7	90.9	93.9	82.4	86.8	88.3	84.0	85.6	94.8	83.7	87.9	106.7	80.1	90.3
05	METALS	102.0	106.0	103.5	104.1	79.2	94.8	98.9	92.1	96.4	102.0	105.8	103.4	91.5	90.0	90.9	89.4	90.2	89.7
06	WOOD & PLASTICS	110.1	109.9	110.0	107.6	85.9	96.2	112.3	80.6	95.7	111.3	83.9	96.9	99.7	83.0	90.9	111.1	78.2	93.8
07	THERMAL & MOISTURE PROTECTION	99.7	103.0	101.3	99.5	87.5	93.7	93.7	80.9	87.6	99.5	88.5	94.2	95.5	85.2	90.5	99.5	81.2	90.6
08	DOORS & WINDOWS	99.3	110.8	102.1	107.0	85.9	101.8	98.1	83.7	94.5	92.6	82.8	90.2	96.1	84.8	93.3	99.4	82.6	95.3
09200	PLASTER & GYPSUM BOARD	100.0	110.9	107.3	99.9	85.9	90.5	93.2	80.8	84.9	103.6	84.1	90.5	94.5	81.6	85.8	103.0	78.0	86.2
095,098	CEILINGS & ACOUSTICAL TREATMENT	98.0	110.9	106.7	96.6	85.9	89.3	86.3	80.8	82.6	98.0	84.1	88.5	96.4	81.6	86.4	111.0	78.0	88.6
09600	FLOORING	86.4	107.2	91.6	86.1	95.6	88.5	95.5	95.6	95.5	86.4	83.5	85.6	96.3	94.2	95.8	126.1	62.7	110.2
097,099	WALL FINISHES, PAINTS & COATINGS	90.2	110.7	102.3	90.2	93.9	92.4	90.0	73.6	80.3	90.2	85.5	87.4	96.8	90.0	92.8	110.2	82.9	94.1
09	FINISHES	90.7	108.7	100.3	90.8	88.8	89.8	92.7	82.9	87.5	91.2	84.2	87.5	95.2	86.2	90.4	113.1	76.1	93.4
10-14	TOTAL DIV. 10000-14000	100.0	104.7	101.0	100.0	90.1	97.9	100.0	87.7	97.3	100.0	86.3	97.0	100.0	92.5	98.4	100.0	87.5	97.3
15	MECHANICAL	99.9	100.4	100.1	99.9	87.4	94.3	99.7	82.9	92.1	99.9	81.4	91.6	100.0	84.7	93.1	100.4	81.6	91.9
16	ELECTRICAL	88.7	111.7	105.0	99.5	94.3	95.8	89.5	83.2	85.1	94.6	90.8	91.9	93.0	87.7	89.3	94.9	87.0	89.3
01-16	WEIGHTED AVERAGE	98.0	105.3	101.6	100.3	88.8	94.6	97.5	85.1	91.4	96.2	88.4	92.4	97.5	90.8	94.2	99.5	84.1	91.9

COST INDEXES

City Cost Indexes

IOWA

DIVISION		COUNCIL BLUFFS MAT.	INST.	TOTAL	DAVENPORT MAT.	INST.	TOTAL	DES MOINES MAT.	INST.	TOTAL	DUBUQUE MAT.	INST.	TOTAL	SIOUX CITY MAT.	INST.	TOTAL	WATERLOO MAT.	INST.	TOTAL
01590	EQUIPMENT RENTAL	.0	92.9	92.9	.0	97.1	97.1	.0	99.0	99.0	.0	91.6	91.6	.0	97.1	97.1	.0	97.1	97.1
02	SITE CONSTRUCTION	91.9	91.6	91.7	85.9	99.2	95.9	78.8	100.7	95.3	85.4	92.1	90.5	95.6	96.2	96.0	86.5	95.3	93.1
03100	CONCRETE FORMS & ACCESSORIES	80.4	63.1	65.2	104.0	93.7	95.0	106.2	78.6	82.1	82.3	70.8	72.2	104.4	67.2	71.9	105.0	53.3	59.8
03200	CONCRETE REINFORCEMENT	100.3	76.3	86.1	98.2	98.7	98.5	98.2	81.2	88.2	96.9	84.9	89.8	98.2	69.7	81.3	98.2	84.6	90.2
03300	CAST-IN-PLACE CONCRETE	112.8	79.3	99.0	104.4	107.7	105.8	108.7	82.7	98.0	106.2	93.3	100.9	107.5	61.0	88.4	108.4	51.7	85.1
03	CONCRETE	101.1	72.1	86.5	97.2	99.6	98.4	98.3	81.1	89.6	95.6	81.9	88.7	98.7	66.3	82.4	99.2	59.8	79.4
04	MASONRY	106.4	72.9	85.7	104.2	95.0	98.5	100.3	84.4	90.5	107.8	73.4	86.5	99.4	64.6	77.8	101.0	57.5	74.1
05	METALS	94.7	85.3	91.2	89.4	99.9	93.3	89.3	90.4	89.7	87.9	89.6	88.5	89.4	82.0	86.6	89.4	88.1	88.9
06	WOOD & PLASTICS	87.3	59.3	72.6	111.1	90.9	100.5	112.5	77.0	93.9	89.4	68.1	78.2	111.1	66.0	87.4	111.7	53.7	81.2
07	THERMAL & MOISTURE PROTECTION	98.6	69.9	84.8	98.8	91.5	95.3	99.7	81.1	90.7	99.0	72.8	86.3	98.8	61.8	81.0	98.5	55.4	77.7
08	DOORS & WINDOWS	98.4	64.5	90.1	99.4	93.5	97.9	99.4	82.2	95.2	98.4	78.6	93.6	99.4	66.6	91.3	95.0	65.7	87.7
09200	PLASTER & GYPSUM BOARD	93.3	58.5	70.0	103.0	90.7	94.8	99.3	76.4	84.0	93.8	67.6	76.2	103.0	65.1	77.6	103.0	52.4	69.1
095,098	CEILINGS & ACOUSTICAL TREATMENT	106.9	58.5	74.0	111.0	90.7	97.2	109.6	76.4	87.1	106.9	67.6	80.2	111.0	65.1	79.8	111.0	52.4	71.2
09600	FLOORING	100.8	51.2	88.3	111.9	85.9	105.4	111.7	46.8	95.4	115.2	43.7	97.2	112.5	66.1	100.8	113.7	66.4	101.8
097,099	WALL FINISHES, PAINTS & COATINGS	101.6	76.0	86.6	105.9	100.0	102.5	105.9	79.5	90.4	109.0	81.7	92.9	107.2	67.5	83.8	107.2	40.1	67.6
09	FINISHES	101.9	60.9	80.1	108.0	92.4	99.7	106.3	71.8	88.0	107.3	65.6	85.1	109.4	66.8	86.7	108.7	53.8	79.5
10-14	TOTAL DIV. 10000-14000	100.0	80.3	95.7	100.0	92.9	98.5	100.0	87.6	97.3	100.0	84.6	96.7	100.0	84.1	96.6	100.0	77.9	95.2
15	MECHANICAL	100.4	80.9	91.6	100.4	99.9	100.2	100.4	86.3	94.0	100.4	85.2	93.5	100.4	80.5	91.4	100.4	50.9	78.1
16	ELECTRICAL	103.2	84.8	90.2	92.4	86.4	88.1	95.7	87.1	89.7	100.9	76.5	83.7	94.9	84.5	87.6	94.9	66.2	74.7
01-16	WEIGHTED AVERAGE	99.6	77.4	88.7	98.3	95.3	96.9	98.1	85.2	91.8	98.3	80.0	89.3	98.8	76.6	87.9	98.1	64.6	81.7

KANSAS / KENTUCKY

DIVISION		DODGE CITY MAT.	INST.	TOTAL	KANSAS CITY MAT.	INST.	TOTAL	SALINA MAT.	INST.	TOTAL	TOPEKA MAT.	INST.	TOTAL	WICHITA MAT.	INST.	TOTAL	BOWLING GREEN MAT.	INST.	TOTAL
01590	EQUIPMENT RENTAL	.0	103.4	103.4	.0	100.0	100.0	.0	103.4	103.4	.0	101.5	101.5	.0	103.4	103.4	.0	96.6	96.6
02	SITE CONSTRUCTION	109.7	92.1	96.4	89.7	90.5	90.3	99.2	92.4	94.1	91.8	89.8	90.3	93.1	92.8	92.9	70.6	100.9	93.5
03100	CONCRETE FORMS & ACCESSORIES	94.6	47.1	53.1	100.4	99.5	99.6	90.1	49.8	54.9	99.9	48.7	55.1	96.7	62.9	67.2	83.0	84.0	83.9
03200	CONCRETE REINFORCEMENT	106.7	62.5	80.6	101.4	87.7	93.3	106.1	82.7	92.3	98.6	89.8	93.4	98.6	83.8	89.8	87.5	73.2	79.1
03300	CAST-IN-PLACE CONCRETE	115.3	61.6	93.3	90.4	98.1	93.6	100.0	52.5	80.5	91.4	57.2	77.4	87.2	61.7	76.7	88.5	84.3	86.8
03	CONCRETE	113.4	56.7	84.8	94.3	97.2	95.7	100.2	58.6	79.2	92.3	61.0	76.6	90.0	67.7	78.8	97.1	82.2	89.6
04	MASONRY	104.1	45.8	68.0	103.5	88.9	94.5	118.9	44.0	72.6	97.1	55.1	71.1	91.6	70.0	78.2	95.3	79.3	85.4
05	METALS	92.7	82.2	88.8	97.6	98.6	98.0	92.5	91.0	91.9	95.5	95.9	95.6	95.5	93.6	94.8	94.7	79.8	89.2
06	WOOD & PLASTICS	93.6	47.3	69.3	100.1	100.6	100.4	89.8	50.9	69.4	97.4	45.0	69.9	95.4	61.9	77.8	93.8	82.9	88.1
07	THERMAL & MOISTURE PROTECTION	98.7	52.8	76.5	96.6	95.1	95.8	98.0	58.4	78.9	97.9	72.8	85.8	97.6	70.0	84.3	85.6	79.8	82.8
08	DOORS & WINDOWS	96.1	50.8	85.0	95.0	95.5	95.1	96.1	56.0	86.2	96.3	62.4	87.9	96.3	66.5	88.9	95.3	76.9	90.7
09200	PLASTER & GYPSUM BOARD	98.9	45.5	63.1	94.6	100.4	98.5	97.8	49.2	65.2	101.4	43.2	62.4	101.4	60.6	74.0	89.9	82.7	85.1
095,098	CEILINGS & ACOUSTICAL TREATMENT	92.0	45.5	60.4	98.8	100.4	99.9	92.0	49.2	62.9	98.8	43.2	61.0	98.8	60.6	72.8	92.3	82.7	85.8
09600	FLOORING	98.5	55.2	87.6	88.9	58.9	81.4	96.5	40.3	82.4	100.0	51.3	87.7	99.1	83.3	95.1	92.3	68.0	86.2
097,099	WALL FINISHES, PAINTS & COATINGS	90.8	54.3	69.3	98.8	97.4	98.0	90.8	42.8	62.5	90.8	71.4	79.4	90.8	67.0	76.8	96.8	75.9	84.5
09	FINISHES	96.2	49.0	71.1	95.4	91.9	93.5	94.8	46.5	69.1	97.2	49.5	71.8	97.0	66.9	81.0	92.6	80.2	86.0
10-14	TOTAL DIV. 10000-14000	100.0	55.9	90.5	100.0	93.7	98.7	100.0	71.6	93.9	100.0	77.8	95.2	100.0	76.6	95.0	100.0	69.9	93.5
15	MECHANICAL	100.1	52.1	78.5	100.0	95.9	98.1	100.1	39.4	72.8	100.1	73.6	88.2	100.1	71.5	87.2	100.0	83.7	92.6
16	ELECTRICAL	97.4	54.8	67.3	106.9	104.9	105.5	96.9	73.1	80.1	103.8	79.0	86.3	101.6	73.1	81.5	93.3	84.4	87.0
01-16	WEIGHTED AVERAGE	100.0	58.6	79.7	98.1	96.1	97.2	98.6	60.7	80.0	97.4	71.0	84.5	96.7	74.3	85.7	95.5	83.3	89.5

KENTUCKY / LOUISIANA

DIVISION		LEXINGTON MAT.	INST.	TOTAL	LOUISVILLE MAT.	INST.	TOTAL	OWENSBORO MAT.	INST.	TOTAL	ALEXANDRIA MAT.	INST.	TOTAL	BATON ROUGE MAT.	INST.	TOTAL	LAKE CHARLES MAT.	INST.	TOTAL
01590	EQUIPMENT RENTAL	.0	104.1	104.1	.0	96.6	96.6	.0	122.0	122.0	.0	86.3	86.3	.0	87.4	87.4	.0	87.4	87.4
02	SITE CONSTRUCTION	78.3	103.0	96.9	67.8	100.7	92.6	83.1	130.6	119.0	102.4	85.2	89.4	116.9	86.2	93.7	118.6	85.9	93.9
03100	CONCRETE FORMS & ACCESSORIES	94.1	72.4	75.2	92.0	82.9	84.0	88.1	70.2	72.5	79.6	47.5	51.5	101.0	52.3	58.4	101.2	53.2	59.2
03200	CONCRETE REINFORCEMENT	96.6	96.8	96.7	96.6	97.8	97.3	87.6	97.0	93.2	98.9	66.1	79.5	100.9	54.5	73.5	100.9	54.4	73.4
03300	CAST-IN-PLACE CONCRETE	95.6	84.4	91.0	92.5	81.9	88.1	91.3	82.4	87.6	94.6	46.4	74.8	88.0	56.0	74.9	93.2	57.6	78.6
03	CONCRETE	98.7	81.5	90.1	97.0	85.4	91.2	107.3	79.9	93.5	89.7	51.7	70.6	96.9	54.9	75.7	99.5	55.8	77.5
04	MASONRY	91.8	52.1	67.3	92.2	81.9	85.8	92.3	56.4	70.1	113.8	46.7	72.3	99.3	55.8	72.4	98.2	56.1	72.2
05	METALS	96.2	90.1	94.0	96.2	90.4	94.1	86.2	90.0	87.6	86.6	74.6	82.2	101.1	69.0	89.2	100.6	69.1	88.9
06	WOOD & PLASTICS	99.7	74.9	86.7	102.4	82.9	92.2	93.7	69.4	80.9	85.9	47.7	65.8	110.3	52.6	80.0	108.3	53.4	79.5
07	THERMAL & MOISTURE PROTECTION	95.8	80.8	88.6	85.8	81.3	83.6	95.4	72.2	84.2	98.8	54.9	77.6	97.4	58.2	78.5	100.3	58.1	80.0
08	DOORS & WINDOWS	96.1	82.2	92.7	96.1	85.8	93.6	93.5	80.4	90.3	99.2	53.6	88.0	104.9	52.6	92.0	104.9	54.7	92.5
09200	PLASTER & GYPSUM BOARD	94.5	73.3	80.3	97.9	82.7	87.7	89.3	67.5	74.7	84.3	46.9	59.3	105.6	51.8	69.5	105.6	52.6	70.1
095,098	CEILINGS & ACOUSTICAL TREATMENT	96.4	73.3	80.7	96.4	82.7	87.1	82.7	67.5	72.4	98.0	46.9	63.3	100.5	51.8	67.5	100.5	52.6	68.0
09600	FLOORING	97.2	47.2	84.6	95.7	72.4	89.8	94.6	68.0	87.9	107.0	54.2	93.7	114.2	65.2	101.9	114.5	57.0	100.0
097,099	WALL FINISHES, PAINTS & COATINGS	96.8	65.0	78.0	96.8	78.3	85.9	96.8	103.9	101.0	96.8	45.8	66.7	101.1	52.5	72.5	101.1	48.4	70.0
09	FINISHES	95.6	66.3	80.0	95.2	80.2	87.2	91.5	72.8	81.6	97.0	48.0	71.0	105.8	54.6	78.6	105.8	53.2	77.9
10-14	TOTAL DIV. 10000-14000	100.0	90.4	97.9	100.0	92.8	98.5	100.0	91.6	98.2	100.0	79.2	95.5	100.0	72.9	94.2	100.0	73.2	94.2
15	MECHANICAL	100.0	52.9	78.7	100.0	81.4	91.6	100.0	58.1	81.1	100.0	41.7	73.8	100.0	47.6	76.4	100.0	58.5	81.3
16	ELECTRICAL	94.0	55.2	66.6	94.0	84.4	87.3	92.4	81.2	84.5	94.7	58.0	68.8	98.6	54.8	67.7	98.6	62.8	73.3
01-16	WEIGHTED AVERAGE	96.7	69.7	83.5	95.9	85.4	90.8	95.6	78.4	87.1	96.7	55.9	76.7	101.3	58.1	80.1	101.6	61.7	82.0

COST INDEXES

DIVISION		LOUISIANA								MAINE								
		MONROE			NEW ORLEANS			SHREVEPORT			AUGUSTA			BANGOR			LEWISTON		
		MAT.	INST.	TOTAL	MAT.	INST.	TOTAL	MAT.	INST.	TOTAL	MAT.	INST.	TOTAL	MAT.	INST.	TOTAL	MAT.	INST.	TOTAL
01590	EQUIPMENT RENTAL	.0	86.3	86.3	.0	88.6	88.6	.0	86.3	86.3	.0	102.6	102.6	.0	102.6	102.6	.0	102.6	102.6
02	SITE CONSTRUCTION	102.4	85.0	89.3	121.3	88.4	96.5	101.0	84.9	88.8	84.5	102.4	98.0	84.4	102.1	97.8	82.6	102.1	97.3
03100	CONCRETE FORMS & ACCESSORIES	78.9	48.3	52.2	100.3	69.7	73.5	100.3	51.2	57.4	101.9	70.0	74.0	95.6	87.1	88.1	102.0	87.1	89.0
03200	CONCRETE REINFORCEMENT	97.9	66.1	79.1	100.9	67.6	81.2	97.4	65.9	78.8	90.0	112.4	103.2	90.0	112.8	103.5	110.6	112.8	111.9
03300	CAST-IN-PLACE CONCRETE	94.6	51.0	76.7	92.2	62.6	80.1	93.3	55.0	77.5	82.6	67.6	76.4	82.5	69.9	77.3	92.0	69.9	82.9
03	CONCRETE	89.5	53.7	71.5	99.0	67.3	83.0	89.5	56.3	72.8	98.4	77.1	87.7	97.5	85.3	91.4	100.9	85.4	93.1
04	MASONRY	108.5	48.4	71.3	100.1	64.2	77.9	101.4	43.6	65.7	94.2	56.8	71.1	111.8	65.2	83.0	95.2	65.2	76.7
05	METALS	86.6	74.3	82.0	103.5	76.8	93.6	87.4	73.8	82.4	95.1	85.9	91.7	94.9	84.5	91.1	98.2	84.5	93.1
06	WOOD & PLASTICS	85.1	48.2	65.7	104.6	72.2	87.6	105.8	52.4	77.8	99.7	69.1	83.6	93.8	90.0	91.8	99.7	90.0	94.6
07	THERMAL & MOISTURE PROTECTION	98.8	57.2	78.7	101.1	66.3	84.3	97.7	56.4	77.8	102.7	56.5	80.4	102.6	64.0	84.0	102.4	64.0	83.9
08	DOORS & WINDOWS	99.2	58.3	89.1	105.5	70.8	97.0	97.1	56.2	87.0	103.9	65.5	94.5	103.8	78.7	97.7	107.1	78.7	100.1
09200	PLASTER & GYPSUM BOARD	83.8	47.4	59.4	104.5	71.9	82.7	90.3	51.8	64.5	103.0	67.1	78.9	100.1	88.6	92.4	104.8	88.6	94.0
095,098	CEILINGS & ACOUSTICAL TREATMENT	98.0	47.4	63.6	99.0	71.9	80.6	99.3	51.8	67.0	99.7	67.1	77.5	98.3	88.6	91.7	107.9	88.6	94.8
09600	FLOORING	106.6	47.1	91.6	114.9	55.9	100.1	117.9	49.4	100.7	101.6	47.3	87.9	99.1	59.1	89.0	101.6	59.1	90.9
097,099	WALL FINISHES, PAINTS & COATINGS	96.8	51.0	69.8	102.9	67.5	82.0	96.8	45.8	66.7	94.4	39.6	62.1	94.4	36.5	60.3	94.4	36.5	60.3
09	FINISHES	96.8	47.6	70.6	106.0	67.1	85.4	101.7	49.7	74.1	100.2	62.0	79.9	98.8	77.0	87.2	101.6	77.0	88.5
10-14	TOTAL DIV. 10000 - 14000	100.0	70.3	93.6	100.0	77.6	95.2	100.0	72.6	94.1	100.0	71.0	93.7	100.0	92.2	98.3	100.0	92.2	98.3
15	MECHANICAL	100.0	54.6	79.5	100.0	68.0	85.6	100.0	59.4	81.7	100.1	58.6	81.4	100.1	77.5	89.9	100.1	77.5	89.9
16	ELECTRICAL	97.7	58.0	69.7	99.0	68.4	77.4	95.7	72.4	79.3	104.4	82.4	88.9	101.1	82.4	87.9	104.4	84.8	90.6
01-16	WEIGHTED AVERAGE	96.5	58.8	78.0	102.3	70.5	86.7	96.5	62.2	79.7	99.2	72.5	86.1	99.6	81.2	90.6	100.5	81.7	91.2

DIVISION		MAINE			MARYLAND						MASSACHUSETTS								
		PORTLAND			BALTIMORE			HAGERSTOWN			BOSTON			BROCKTON			FALL RIVER		
		MAT.	INST.	TOTAL	MAT.	INST.	TOTAL	MAT.	INST.	TOTAL	MAT.	INST.	TOTAL	MAT.	INST.	TOTAL	MAT.	INST.	TOTAL
01590	EQUIPMENT RENTAL	.0	102.6	102.6	.0	102.4	102.4	.0	98.3	98.3	.0	109.5	109.5	.0	104.7	104.7	.0	105.9	105.9
02	SITE CONSTRUCTION	81.6	102.1	97.1	96.8	91.4	92.8	86.4	87.2	87.0	89.7	109.6	104.7	87.2	105.5	101.0	86.2	105.6	100.9
03100	CONCRETE FORMS & ACCESSORIES	101.1	87.1	88.8	104.4	75.5	79.1	91.8	76.3	78.3	107.9	131.5	128.5	107.6	114.8	113.9	107.6	108.5	108.4
03200	CONCRETE REINFORCEMENT	110.6	112.8	111.9	100.0	88.4	93.1	88.4	75.1	80.5	109.4	140.5	127.8	110.6	139.9	127.9	110.6	124.0	118.6
03300	CAST-IN-PLACE CONCRETE	85.5	69.9	79.1	95.1	80.1	88.9	80.9	63.5	73.7	106.5	141.8	121.0	101.4	129.3	112.9	98.1	138.2	114.5
03	CONCRETE	97.6	85.4	91.4	99.0	80.7	89.8	84.6	72.8	78.7	111.4	135.9	123.7	108.4	124.0	116.3	106.7	121.3	114.1
04	MASONRY	93.2	65.2	75.9	91.4	74.2	80.7	95.7	74.6	82.7	114.7	141.3	131.1	110.5	127.6	121.0	110.2	136.6	126.5
05	METALS	98.3	84.5	93.2	96.0	99.9	97.4	95.0	92.4	94.1	101.1	124.0	109.6	98.1	119.8	106.2	98.1	114.3	104.1
06	WOOD & PLASTICS	99.7	90.0	94.6	102.5	76.7	88.9	90.4	76.5	83.1	103.6	131.0	118.0	102.5	112.1	107.5	102.5	103.8	103.2
07	THERMAL & MOISTURE PROTECTION	102.2	64.0	83.8	95.0	79.7	87.6	94.2	73.2	84.1	102.9	137.7	119.7	102.7	126.2	114.0	102.7	121.9	111.9
08	DOORS & WINDOWS	107.1	78.7	100.1	93.8	84.9	91.6	91.1	77.2	87.7	103.0	133.6	110.6	101.0	119.6	105.6	101.0	110.7	103.4
09200	PLASTER & GYPSUM BOARD	104.8	88.6	94.0	106.0	76.4	86.1	100.9	76.4	84.4	108.2	130.9	123.4	101.4	111.4	108.1	101.4	102.5	102.1
095,098	CEILINGS & ACOUSTICAL TREATMENT	107.9	88.6	94.8	90.7	76.4	81.0	92.3	76.4	81.5	106.4	130.9	123.0	109.5	111.4	110.8	109.5	102.5	104.7
09600	FLOORING	101.6	59.1	90.9	94.8	79.8	91.0	89.6	78.2	86.8	102.2	150.6	114.4	102.7	150.6	114.7	102.5	150.6	114.6
097,099	WALL FINISHES, PAINTS & COATINGS	94.4	36.5	60.3	98.1	77.3	85.9	98.1	44.7	66.7	97.5	142.3	123.9	96.9	120.1	110.6	96.9	120.1	110.6
09	FINISHES	101.7	77.0	88.6	94.7	76.0	84.8	92.0	73.4	82.1	102.0	136.4	120.3	101.8	121.8	112.5	101.8	117.0	109.9
10-14	TOTAL DIV. 10000 - 14000	100.0	92.2	98.3	100.0	85.0	96.8	100.0	86.9	97.2	100.0	122.1	104.8	100.0	118.0	103.9	100.0	117.9	103.9
15	MECHANICAL	100.1	77.5	89.9	100.0	85.5	93.4	100.0	85.4	93.4	100.1	123.8	110.8	100.1	99.4	99.8	100.1	99.4	99.8
16	ELECTRICAL	104.6	84.8	90.7	101.7	89.0	92.7	97.8	80.4	85.5	99.4	121.3	114.8	99.1	93.2	94.9	98.5	93.2	94.7
01-16	WEIGHTED AVERAGE	99.9	81.7	91.0	97.4	84.8	91.2	94.4	80.5	87.6	102.8	127.7	115.0	101.4	111.5	106.3	101.1	110.4	105.6

DIVISION		MASSACHUSETTS																	
		HYANNIS			LAWRENCE			LOWELL			NEW BEDFORD			PITTSFIELD			SPRINGFIELD		
		MAT.	INST.	TOTAL	MAT.	INST.	TOTAL	MAT.	INST.	TOTAL	MAT.	INST.	TOTAL	MAT.	INST.	TOTAL	MAT.	INST.	TOTAL
01590	EQUIPMENT RENTAL	.0	104.7	104.7	.0	104.7	104.7	.0	102.6	102.6	.0	105.9	105.9	.0	102.6	102.6	.0	102.6	102.6
02	SITE CONSTRUCTION	83.9	105.5	100.2	87.8	105.5	101.2	86.9	105.4	100.9	85.8	105.6	100.8	88.0	104.2	100.3	87.1	104.3	100.1
03100	CONCRETE FORMS & ACCESSORIES	97.4	108.0	106.7	107.4	112.1	111.5	103.7	112.4	111.3	107.6	108.5	108.4	103.7	97.9	98.7	103.9	103.1	103.2
03200	CONCRETE REINFORCEMENT	88.7	118.2	106.1	109.7	126.3	119.5	110.6	126.6	120.1	110.6	124.0	118.5	91.9	107.3	101.0	110.6	114.3	112.8
03300	CAST-IN-PLACE CONCRETE	92.5	137.4	110.9	102.2	137.8	116.8	92.9	137.9	111.4	95.2	138.2	112.8	101.4	108.0	104.1	95.2	109.7	101.1
03	CONCRETE	98.6	119.7	109.3	108.7	123.2	116.0	99.1	123.2	111.2	105.3	121.3	113.4	101.1	102.8	101.9	100.2	107.0	103.6
04	MASONRY	109.5	136.7	126.3	109.6	130.2	122.3	95.3	135.1	119.9	110.0	136.6	126.5	95.9	109.8	104.5	95.5	109.8	104.3
05	METALS	94.7	111.3	100.8	95.7	114.7	102.7	95.6	112.1	101.7	98.1	114.3	104.1	95.5	99.0	96.8	98.0	102.2	99.6
06	WOOD & PLASTICS	93.2	103.4	98.6	102.5	108.3	105.5	101.7	108.3	105.2	102.5	103.8	103.2	101.7	96.5	99.0	101.7	102.9	102.3
07	THERMAL & MOISTURE PROTECTION	102.1	122.9	112.2	102.7	127.6	114.7	102.4	128.9	115.2	102.7	121.9	111.9	102.5	106.8	104.6	102.4	107.7	104.9
08	DOORS & WINDOWS	97.1	107.2	99.6	101.0	113.8	104.1	107.1	113.8	108.7	101.0	110.7	103.4	107.1	101.6	105.7	107.1	106.9	107.0
09200	PLASTER & GYPSUM BOARD	95.7	102.5	100.2	104.8	107.5	106.6	104.8	107.5	106.6	101.4	102.5	102.1	104.8	95.3	98.5	104.8	101.9	102.9
095,098	CEILINGS & ACOUSTICAL TREATMENT	99.8	102.5	101.6	107.9	107.5	107.6	107.9	107.5	107.6	109.5	102.5	104.7	107.9	95.3	99.4	107.9	101.9	103.8
09600	FLOORING	98.7	150.6	111.8	101.6	150.6	113.9	101.6	150.6	113.9	102.5	150.6	114.6	101.8	120.7	106.5	101.5	120.7	106.3
097,099	WALL FINISHES, PAINTS & COATINGS	96.9	119.0	109.9	94.5	120.1	109.6	94.4	119.0	108.9	96.9	119.0	109.9	94.4	93.2	93.7	96.2	93.2	94.5
09	FINISHES	97.8	116.6	107.8	101.4	119.6	111.1	101.4	119.5	111.0	101.7	116.9	109.8	101.5	101.6	101.5	101.5	105.4	103.6
10-14	TOTAL DIV. 10000 - 14000	100.0	116.9	103.6	100.0	117.6	103.8	100.0	117.6	103.8	100.0	117.9	103.9	100.0	106.4	101.4	100.0	107.2	101.6
15	MECHANICAL	100.1	99.2	99.7	100.1	104.4	102.1	100.1	114.0	106.4	100.1	99.3	99.8	100.1	83.9	92.8	100.1	94.9	97.8
16	ELECTRICAL	94.3	93.2	93.5	103.6	102.1	102.5	104.4	102.1	102.8	99.4	93.2	95.0	104.4	80.7	87.6	104.4	86.9	92.1
01-16	WEIGHTED AVERAGE	98.2	109.7	103.8	101.3	113.2	107.1	100.0	115.3	107.5	100.9	110.4	105.5	100.3	95.7	98.1	100.5	100.6	100.5

COST INDEXES

City Cost Indexes

MASSACHUSETTS / MICHIGAN

DIVISION		WORCESTER MAT.	WORCESTER INST.	WORCESTER TOTAL	ANN ARBOR MAT.	ANN ARBOR INST.	ANN ARBOR TOTAL	DEARBORN MAT.	DEARBORN INST.	DEARBORN TOTAL	DETROIT MAT.	DETROIT INST.	DETROIT TOTAL	FLINT MAT.	FLINT INST.	FLINT TOTAL	GRAND RAPIDS MAT.	GRAND RAPIDS INST.	GRAND RAPIDS TOTAL
01590	EQUIPMENT RENTAL	.0	102.6	102.6	.0	111.1	111.1	.0	111.1	111.1	.0	96.7	96.7	.0	111.1	111.1	.0	105.9	105.9
02	SITE CONSTRUCTION	86.9	105.4	100.9	81.9	93.1	90.4	81.7	93.3	90.4	97.4	95.0	95.6	71.4	92.4	87.2	83.7	88.6	87.4
03100	CONCRETE FORMS & ACCESSORIES	104.3	111.5	110.6	97.6	118.7	116.0	97.4	124.8	121.4	98.9	125.0	121.7	98.8	98.3	98.4	97.2	85.9	87.3
03200	CONCRETE REINFORCEMENT	110.6	139.4	127.6	94.2	120.9	110.0	94.2	121.5	110.3	93.6	121.5	110.1	94.2	120.7	109.9	96.6	92.6	94.2
03300	CAST-IN-PLACE CONCRETE	92.9	126.5	106.7	94.0	118.2	103.9	91.9	122.8	104.6	100.9	122.8	109.9	94.6	100.6	97.1	95.8	107.4	100.5
03	CONCRETE	99.1	121.3	110.3	91.1	119.4	105.3	90.0	123.8	107.0	94.5	122.3	108.5	91.5	104.3	97.9	99.0	93.9	96.4
04	MASONRY	95.3	135.1	119.9	99.4	110.6	106.3	99.2	123.1	114.0	98.2	123.1	113.6	99.5	100.4	100.0	96.4	60.8	74.4
05	METALS	98.1	116.2	104.8	100.1	126.7	109.9	100.1	128.1	110.5	100.6	105.3	102.4	100.1	124.6	109.2	97.3	84.2	92.4
06	WOOD & PLASTICS	102.1	108.3	105.4	101.5	119.6	111.0	101.5	125.3	114.0	102.2	125.3	114.3	101.5	98.5	100.0	98.0	86.9	92.2
07	THERMAL & MOISTURE PROTECTION	102.4	121.0	111.4	96.6	113.7	104.9	95.5	125.1	109.8	94.0	125.1	109.0	94.8	98.9	96.8	92.1	67.1	80.0
08	DOORS & WINDOWS	107.1	121.0	110.5	96.4	114.8	100.9	96.4	117.9	101.7	97.9	121.1	103.6	96.4	102.7	97.9	94.2	76.8	89.9
09200	PLASTER & GYPSUM BOARD	104.8	107.5	106.6	103.1	118.6	113.5	103.1	124.5	117.4	103.1	124.5	117.4	103.1	96.9	99.0	94.5	81.0	85.5
095,098	CEILINGS & ACOUSTICAL TREATMENT	107.9	107.5	107.6	100.9	118.6	112.9	100.9	124.5	116.9	102.4	124.5	117.4	100.9	96.9	98.2	96.4	81.0	86.0
09600	FLOORING	101.6	133.8	109.7	89.1	113.6	95.2	88.6	125.3	97.9	88.7	125.3	97.9	88.9	80.8	86.8	95.9	47.7	83.8
097,099	WALL FINISHES, PAINTS & COATINGS	94.4	113.1	105.4	88.5	112.4	102.6	88.5	115.9	104.7	90.3	115.9	105.4	88.5	87.9	88.1	96.8	46.4	67.1
09	FINISHES	101.4	115.3	108.8	94.8	117.5	106.8	94.6	124.5	110.5	95.8	124.5	111.0	94.1	93.4	93.7	95.1	74.5	84.1
10-14	TOTAL DIV. 10000-14000	100.0	111.1	102.4	100.0	113.6	102.9	100.0	115.7	103.4	100.0	115.7	103.4	100.0	102.4	100.5	100.0	108.8	101.9
15	MECHANICAL	100.1	106.9	103.2	100.0	107.4	103.4	100.0	115.6	107.1	100.0	120.2	109.1	100.0	101.6	100.7	100.0	61.9	82.8
16	ELECTRICAL	104.4	92.5	96.0	94.8	95.6	95.4	94.8	118.1	111.2	96.2	118.0	111.6	94.8	96.8	96.2	93.5	66.7	74.6
01-16	WEIGHTED AVERAGE	100.4	111.8	106.0	96.9	109.6	103.2	96.8	118.4	107.4	98.2	117.3	107.6	96.6	101.2	98.8	96.9	75.0	86.1

MICHIGAN / MINNESOTA

DIVISION		KALAMAZOO MAT.	KALAMAZOO INST.	KALAMAZOO TOTAL	LANSING MAT.	LANSING INST.	LANSING TOTAL	MUSKEGON MAT.	MUSKEGON INST.	MUSKEGON TOTAL	SAGINAW MAT.	SAGINAW INST.	SAGINAW TOTAL	DULUTH MAT.	DULUTH INST.	DULUTH TOTAL	MINNEAPOLIS MAT.	MINNEAPOLIS INST.	MINNEAPOLIS TOTAL
01590	EQUIPMENT RENTAL	.0	105.9	105.9	.0	111.1	111.1	.0	105.9	105.9	.0	111.1	111.1	.0	97.6	97.6	.0	102.0	102.0
02	SITE CONSTRUCTION	84.3	88.8	87.7	88.7	92.3	91.4	82.3	88.7	87.2	73.6	92.1	87.6	84.4	102.9	98.3	84.1	109.3	103.1
03100	CONCRETE FORMS & ACCESSORIES	96.3	92.5	93.0	101.1	98.6	98.9	97.2	89.8	90.7	97.6	96.3	96.5	100.5	113.1	111.5	101.2	142.3	137.1
03200	CONCRETE REINFORCEMENT	96.6	94.3	95.3	94.2	120.5	109.8	97.2	94.1	95.4	94.2	120.1	109.5	95.9	106.9	102.4	96.1	132.4	117.6
03300	CAST-IN-PLACE CONCRETE	97.4	107.1	101.4	94.0	100.9	96.8	95.2	104.8	99.1	92.9	99.6	95.6	109.5	104.1	107.3	107.9	130.2	117.0
03	CONCRETE	102.4	97.0	99.7	91.3	104.5	97.9	96.8	95.0	95.9	90.5	103.0	96.8	97.7	109.2	103.5	98.5	136.0	117.4
04	MASONRY	99.1	90.0	93.5	93.3	100.9	98.0	96.2	85.5	89.6	100.9	91.7	95.2	107.4	112.0	110.3	107.4	141.1	128.3
05	METALS	97.5	88.2	94.0	99.3	124.2	108.6	95.5	87.8	92.6	100.1	123.3	108.7	96.4	118.1	104.5	98.4	134.5	111.8
06	WOOD & PLASTICS	99.7	92.1	95.7	104.3	97.7	100.8	96.4	89.1	92.5	97.3	96.7	97.0	112.0	112.7	112.4	112.4	142.4	128.2
07	THERMAL & MOISTURE PROTECTION	91.6	89.7	90.7	96.3	98.6	97.4	90.8	83.6	87.3	95.4	94.4	94.9	100.7	114.3	107.3	100.6	135.1	117.2
08	DOORS & WINDOWS	90.9	86.8	89.9	96.4	102.2	97.8	90.1	86.6	89.2	95.9	101.7	97.3	97.2	116.8	102.0	100.3	146.0	111.5
09200	PLASTER & GYPSUM BOARD	94.5	86.4	89.1	106.5	96.0	99.5	87.3	83.3	84.6	103.1	95.0	97.7	99.5	113.7	109.0	99.5	144.1	129.4
095,098	CEILINGS & ACOUSTICAL TREATMENT	96.4	86.4	89.6	100.9	96.0	97.6	101.9	83.3	89.3	100.9	95.0	96.9	86.1	113.7	104.9	86.1	144.1	125.5
09600	FLOORING	95.9	75.7	90.8	96.6	89.5	94.8	95.1	87.9	93.2	89.1	64.2	82.8	109.8	109.3	109.7	107.1	122.5	110.9
097,099	WALL FINISHES, PAINTS & COATINGS	96.8	84.3	89.4	100.8	102.7	101.9	96.0	67.3	79.1	88.5	88.2	88.3	96.9	112.1	105.8	103.3	117.8	111.9
09	FINISHES	95.1	88.2	91.4	99.2	97.0	98.0	94.4	86.7	90.3	94.5	88.9	91.6	100.9	112.7	107.2	100.4	136.7	119.7
10-14	TOTAL DIV. 10000-14000	100.0	110.3	102.2	100.0	103.5	100.8	100.0	109.5	102.1	100.0	102.2	100.5	100.0	101.0	100.2	100.0	113.5	102.9
15	MECHANICAL	100.0	93.1	96.9	100.0	95.8	98.1	99.8	94.2	97.3	100.0	95.1	97.8	100.0	100.9	100.4	100.0	116.8	107.6
16	ELECTRICAL	93.2	78.1	82.6	93.6	96.0	95.3	93.4	73.3	79.2	97.3	90.6	92.5	100.7	102.5	102.0	101.9	117.2	112.7
01-16	WEIGHTED AVERAGE	97.1	89.4	93.3	97.1	100.3	98.7	95.7	87.6	91.7	96.7	96.9	96.8	99.1	107.6	103.3	99.9	126.9	113.2

MINNESOTA / MISSISSIPPI

DIVISION		ROCHESTER MAT.	ROCHESTER INST.	ROCHESTER TOTAL	SAINT PAUL MAT.	SAINT PAUL INST.	SAINT PAUL TOTAL	ST. CLOUD MAT.	ST. CLOUD INST.	ST. CLOUD TOTAL	BILOXI MAT.	BILOXI INST.	BILOXI TOTAL	GREENVILLE MAT.	GREENVILLE INST.	GREENVILLE TOTAL	JACKSON MAT.	JACKSON INST.	JACKSON TOTAL
01590	EQUIPMENT RENTAL	.0	97.6	97.6	.0	97.6	97.6	.0	99.2	99.2	.0	98.3	98.3	.0	98.3	98.3	.0	98.3	98.3
02	SITE CONSTRUCTION	83.4	102.5	97.8	86.5	103.7	99.5	78.7	105.6	99.0	105.2	86.6	91.1	109.4	86.3	92.0	101.5	86.3	90.1
03100	CONCRETE FORMS & ACCESSORIES	101.0	110.0	108.9	91.5	140.0	133.9	83.3	107.4	104.3	96.1	46.1	52.4	76.8	38.1	43.0	91.3	43.5	49.6
03200	CONCRETE REINFORCEMENT	95.9	131.7	117.1	92.6	132.4	116.1	101.5	131.6	119.3	94.1	65.2	77.1	102.0	48.0	70.1	94.1	52.0	69.3
03300	CAST-IN-PLACE CONCRETE	105.8	102.0	104.3	108.9	129.3	117.3	98.4	125.7	109.6	104.9	49.6	82.2	105.3	43.5	79.9	102.9	46.2	79.6
03	CONCRETE	95.9	112.0	104.0	100.1	134.7	117.5	89.4	119.0	104.3	96.1	52.9	74.3	98.8	44.0	71.2	94.8	48.1	71.3
04	MASONRY	106.7	115.4	112.1	119.6	141.1	132.9	109.9	122.0	117.4	89.1	41.3	59.5	132.2	42.8	76.9	91.9	42.8	61.5
05	METALS	96.3	132.0	109.6	95.1	134.2	109.6	96.3	133.1	109.9	97.9	86.0	93.5	96.3	78.1	89.5	97.9	80.0	91.3
06	WOOD & PLASTICS	112.4	109.6	111.0	103.1	139.5	122.2	94.3	99.1	96.8	96.3	46.2	70.0	77.2	37.5	56.3	91.4	44.5	66.8
07	THERMAL & MOISTURE PROTECTION	100.5	108.4	104.3	100.5	134.6	116.9	98.1	117.9	107.7	95.9	50.1	73.8	95.9	45.4	71.6	95.6	46.8	72.0
08	DOORS & WINDOWS	97.2	128.3	104.8	94.4	144.4	106.7	91.6	122.6	99.3	98.4	52.0	87.0	97.8	42.4	84.2	98.9	47.3	86.2
09200	PLASTER & GYPSUM BOARD	99.5	110.6	106.9	94.8	141.3	126.0	87.7	99.8	95.8	104.1	45.1	64.6	95.2	36.2	55.6	104.1	43.4	63.4
095,098	CEILINGS & ACOUSTICAL TREATMENT	86.1	110.6	102.7	83.4	141.3	122.7	69.8	99.8	90.2	96.3	45.1	61.5	90.7	36.2	53.7	96.3	43.4	60.4
09600	FLOORING	109.6	88.5	104.3	102.5	122.5	107.5	100.1	122.5	105.7	117.3	44.6	99.1	107.8	38.1	90.3	117.3	44.1	98.9
097,099	WALL FINISHES, PAINTS & COATINGS	99.3	103.8	101.9	103.3	116.9	111.3	105.8	117.8	112.9	106.5	39.7	67.1	106.5	39.0	66.7	106.5	39.0	66.7
09	FINISHES	101.0	105.8	103.5	98.0	134.9	117.6	94.0	109.9	102.4	105.2	44.7	73.1	100.2	37.9	67.1	105.2	43.3	72.3
10-14	TOTAL DIV. 10000-14000	100.0	101.7	100.4	100.0	112.7	102.7	100.0	100.7	100.1	100.0	61.3	91.7	100.0	59.4	91.3	100.0	60.3	91.5
15	MECHANICAL	100.0	92.3	96.5	100.0	116.4	107.4	99.7	113.5	105.9	99.9	60.8	82.3	99.9	38.7	72.3	99.9	38.4	72.2
16	ELECTRICAL	100.7	91.8	94.4	99.8	109.9	106.9	97.7	109.9	106.3	97.2	49.7	63.7	96.1	41.4	57.5	97.8	41.4	58.0
01-16	WEIGHTED AVERAGE	98.8	105.5	102.1	99.2	124.5	111.6	96.1	115.0	105.4	98.7	57.9	78.7	100.5	49.0	75.2	98.6	50.6	75.0

City Cost Indexes

		MISSISSIPPI			MISSOURI																
	DIVISION	MERIDIAN			CAPE GIRARDEAU			COLUMBIA			JOPLIN			KANSAS CITY			SPRINGFIELD				
		MAT.	INST.	TOTAL	MAT.	INST.	TOTAL	MAT.	INST.	TOTAL	MAT.	INST.	TOTAL	MAT.	INST.	TOTAL	MAT.	INST.	TOTAL		
01590	EQUIPMENT RENTAL	.0	98.3	98.3	.0	108.5	108.5	.0	108.9	108.9	.0	104.8	104.8	.0	102.8	102.8	.0	102.6	102.6		
02	SITE CONSTRUCTION	100.4	86.3	89.8	94.5	91.8	92.5	100.8	93.7	95.4	102.0	95.7	97.3	95.1	94.8	94.9	94.2	92.7	93.1		
03100	CONCRETE FORMS & ACCESSORIES	75.4	37.0	41.9	83.8	79.6	80.2	85.1	77.4	78.4	103.4	68.0	72.5	102.7	103.8	103.7	101.7	71.1	74.9		
03200	CONCRETE REINFORCEMENT	100.7	51.4	71.6	103.4	89.9	95.4	102.6	106.8	105.1	111.0	78.6	91.9	105.5	108.2	107.1	98.6	106.9	103.5		
03300	CAST-IN-PLACE CONCRETE	99.3	46.2	77.5	86.5	88.3	87.2	88.9	72.6	82.2	102.0	73.4	90.3	94.8	104.6	98.9	100.5	68.1	87.2		
03	CONCRETE	91.3	45.1	68.1	85.9	86.3	86.1	83.8	83.2	83.5	97.1	73.2	85.1	93.2	105.4	99.3	97.0	78.0	87.4		
04	MASONRY	88.7	31.9	53.6	113.7	78.0	91.6	131.8	72.9	95.4	97.0	57.6	72.6	101.8	103.8	103.0	87.2	70.9	77.1		
05	METALS	96.2	79.0	89.8	93.8	112.6	100.8	92.9	119.7	102.8	95.8	94.1	95.2	99.3	112.8	104.3	96.2	105.7	99.8		
06	WOOD & PLASTICS	75.9	36.0	54.9	80.7	78.9	79.8	91.4	74.5	82.5	103.0	69.2	85.3	103.0	103.5	103.2	100.6	72.6	85.9		
07	THERMAL & MOISTURE PROTECTION	95.4	43.1	70.1	96.8	82.3	89.8	92.8	81.4	87.3	98.4	64.7	82.2	97.6	105.0	101.2	96.9	73.5	85.6		
08	DOORS & WINDOWS	97.8	41.0	83.8	93.8	80.3	90.5	90.8	94.9	91.8	91.6	76.7	87.9	97.7	104.5	99.4	96.3	80.7	92.4		
09200	PLASTER & GYPSUM BOARD	94.7	34.6	54.4	91.9	78.0	82.5	96.6	73.5	81.1	108.8	68.0	81.4	103.2	103.4	103.3	101.4	71.5	81.4		
095,098	CEILINGS & ACOUSTICAL TREATMENT	90.7	34.6	52.6	92.0	78.0	82.4	93.3	73.5	79.9	96.6	68.0	77.2	102.1	103.4	102.9	98.8	71.5	80.3		
09600	FLOORING	106.7	31.0	87.6	95.6	84.3	92.8	105.3	70.2	96.5	116.1	56.7	101.2	94.9	91.8	94.1	108.0	56.7	95.1		
097,099	WALL FINISHES, PAINTS & COATINGS	106.5	38.5	66.4	100.6	85.4	91.6	98.8	81.4	88.6	92.9	50.4	67.9	97.7	110.3	105.1	93.2	72.2	80.8		
09	FINISHES	99.2	35.4	65.3	92.4	80.3	86.0	95.3	74.5	84.3	104.3	63.7	82.8	99.2	101.8	100.6	100.2	68.0	83.1		
10 - 14	TOTAL DIV. 10000 - 14000	100.0	59.4	91.2	100.0	76.5	94.9	100.0	95.8	99.1	100.0	85.2	96.8	100.0	98.9	99.8	100.0	87.5	97.3		
15	MECHANICAL	99.9	38.5	72.2	100.2	97.3	98.9	100.0	99.3	99.7	100.1	58.8	81.5	100.0	104.6	102.0	100.1	73.5	88.1		
16	ELECTRICAL	95.6	61.4	71.5	99.3	108.0	105.4	96.0	78.9	83.9	92.3	69.3	76.0	106.9	107.4	107.3	102.6	71.5	80.6		
01 - 16	WEIGHTED AVERAGE	96.7	51.1	74.3	96.2	92.8	94.5	96.7	88.4	92.6	97.8	71.2	84.8	99.1	104.4	101.7	97.9	78.1	88.2		

		MISSOURI						MONTANA													
	DIVISION	ST. JOSEPH			ST. LOUIS			BILLINGS			BUTTE			GREAT FALLS			HELENA				
		MAT.	INST.	TOTAL	MAT.	INST.	TOTAL	MAT.	INST.	TOTAL	MAT.	INST.	TOTAL	MAT.	INST.	TOTAL	MAT.	INST.	TOTAL		
01590	EQUIPMENT RENTAL	.0	101.4	101.4	.0	109.7	109.7	.0	96.0	96.0	.0	95.7	95.7	.0	95.7	95.7	.0	95.7	95.7		
02	SITE CONSTRUCTION	96.6	91.3	92.6	94.2	95.6	95.3	84.9	96.9	94.0	91.6	95.6	94.6	95.2	96.2	95.9	97.1	96.2	96.4		
03100	CONCRETE FORMS & ACCESSORIES	102.6	89.8	91.4	99.6	108.5	107.4	98.1	70.7	74.2	84.2	68.3	70.3	104.3	69.1	73.6	104.4	69.9	74.2		
03200	CONCRETE REINFORCEMENT	104.2	103.9	104.0	94.6	111.8	104.8	98.2	77.9	86.2	106.6	78.0	89.7	98.2	77.9	86.2	101.7	70.3	83.2		
03300	CAST-IN-PLACE CONCRETE	94.8	100.3	97.0	86.5	112.0	97.0	119.7	73.4	100.7	123.2	73.5	102.8	130.2	60.7	101.7	132.7	71.8	107.7		
03	CONCRETE	93.0	96.8	94.9	85.9	111.4	98.7	104.5	73.6	88.9	104.2	72.6	88.3	110.1	68.6	89.2	111.8	71.3	91.4		
04	MASONRY	101.2	87.8	92.9	95.3	110.7	104.8	120.3	72.0	90.4	120.3	73.0	91.1	124.0	74.0	93.1	121.2	71.2	90.3		
05	METALS	97.9	108.6	101.9	96.3	125.5	107.1	98.1	85.2	93.3	97.5	85.1	92.9	98.5	85.0	93.5	97.7	80.7	91.4		
06	WOOD & PLASTICS	103.7	89.1	96.0	95.3	107.0	101.4	103.2	70.9	86.2	91.0	69.3	79.6	110.9	69.0	88.9	111.1	69.0	89.0		
07	THERMAL & MOISTURE PROTECTION	98.2	94.2	96.2	96.8	106.4	101.4	99.4	72.3	86.3	98.6	72.5	86.0	99.6	70.6	85.6	99.6	71.8	86.2		
08	DOORS & WINDOWS	96.5	95.7	96.3	92.8	113.4	97.8	99.1	68.4	91.6	95.9	67.4	88.9	99.4	67.4	91.5	98.9	65.5	90.6		
09200	PLASTER & GYPSUM BOARD	109.7	88.5	95.5	99.0	106.9	104.3	103.0	70.4	81.1	94.1	68.7	77.1	103.0	68.5	79.9	102.4	68.5	79.7		
095,098	CEILINGS & ACOUSTICAL TREATMENT	100.7	88.5	92.4	96.1	106.9	103.4	111.0	70.4	83.4	108.2	68.7	81.4	111.0	68.5	82.1	108.2	68.5	81.2		
09600	FLOORING	97.4	90.9	95.7	103.1	102.7	103.0	110.4	55.2	96.5	101.8	44.0	87.3	110.4	59.7	97.7	110.4	57.8	97.2		
097,099	WALL FINISHES, PAINTS & COATINGS	93.5	87.2	89.8	100.6	110.2	106.2	101.6	67.3	81.4	101.6	45.8	68.7	101.6	51.6	72.1	101.6	57.3	75.5		
09	FINISHES	100.3	89.2	94.4	96.7	107.3	102.4	107.1	67.2	85.9	102.4	60.8	80.3	107.1	65.1	84.8	106.7	66.0	85.1		
10 - 14	TOTAL DIV. 10000 - 14000	100.0	94.2	98.8	100.0	105.7	101.2	100.0	79.8	95.6	100.0	78.5	95.4	100.0	79.6	95.6	100.0	65.3	92.5		
15	MECHANICAL	100.1	93.2	97.0	100.2	105.6	102.6	100.4	80.2	91.3	100.4	76.7	89.7	100.4	80.3	91.4	100.4	79.5	91.0		
16	ELECTRICAL	106.3	90.8	95.3	103.4	108.0	106.6	93.8	78.4	83.0	105.1	75.8	84.4	93.7	76.0	81.2	93.7	76.0	81.2		
01 - 16	WEIGHTED AVERAGE	98.9	93.5	96.3	96.2	108.6	102.3	101.3	77.9	89.8	101.1	75.7	88.7	102.7	76.6	89.9	102.6	75.8	89.4		

		MONTANA			NEBRASKA												NEVADA				
	DIVISION	MISSOULA			GRAND ISLAND			LINCOLN			NORTH PLATTE			OMAHA			CARSON CITY				
		MAT.	INST.	TOTAL	MAT.	INST.	TOTAL	MAT.	INST.	TOTAL	MAT.	INST.	TOTAL	MAT.	INST.	TOTAL	MAT.	INST.	TOTAL		
01590	EQUIPMENT RENTAL	.0	95.7	95.7	.0	101.5	101.5	.0	101.5	101.5	.0	101.5	101.5	.0	90.6	90.6	.0	101.6	101.6		
02	SITE CONSTRUCTION	74.8	95.4	90.4	97.3	89.9	91.7	88.4	89.9	89.5	98.2	89.3	91.5	78.6	88.6	86.1	63.5	104.2	94.2		
03100	CONCRETE FORMS & ACCESSORIES	88.5	64.0	67.1	97.9	54.9	60.3	103.0	50.5	57.1	97.8	51.3	57.2	97.3	76.1	78.7	98.3	102.3	101.8		
03200	CONCRETE REINFORCEMENT	108.4	82.0	92.8	107.3	72.7	86.9	98.6	73.7	83.9	108.8	72.5	87.3	103.2	74.2	86.1	109.0	116.6	113.5		
03300	CAST-IN-PLACE CONCRETE	90.7	69.7	82.1	116.1	62.6	94.1	104.8	65.3	88.6	116.1	56.3	91.5	108.8	81.0	97.4	118.5	89.8	106.7		
03	CONCRETE	82.1	70.1	76.1	107.1	62.3	84.5	99.2	61.4	80.2	107.2	58.7	82.8	100.5	77.5	88.9	112.6	100.4	106.5		
04	MASONRY	146.1	65.6	96.3	105.6	54.1	73.7	94.3	72.2	80.7	90.9	43.9	61.8	99.9	81.7	88.7	136.4	79.8	101.4		
05	METALS	97.6	85.7	93.1	92.8	84.1	89.6	95.8	85.6	92.0	93.0	87.4	90.9	98.4	76.4	90.2	103.8	102.1	103.2		
06	WOOD & PLASTICS	95.5	65.1	79.5	97.3	49.6	72.2	101.8	43.3	71.1	97.2	49.6	72.2	95.3	76.9	85.6	95.8	104.6	100.4		
07	THERMAL & MOISTURE PROTECTION	97.8	74.9	86.7	97.4	61.7	80.2	97.8	63.5	81.3	97.3	50.9	74.9	92.9	75.2	84.4	104.2	90.0	97.4		
08	DOORS & WINDOWS	95.9	67.1	88.8	90.0	52.5	80.7	95.4	52.1	84.8	89.3	57.5	81.4	98.8	69.2	91.5	94.3	110.2	98.2		
09200	PLASTER & GYPSUM BOARD	96.2	64.5	74.9	97.8	47.8	64.3	101.4	41.4	61.2	98.2	47.8	64.4	108.2	76.7	87.1	88.9	104.6	99.4		
095,098	CEILINGS & ACOUSTICAL TREATMENT	108.2	64.5	78.5	92.0	47.8	62.0	98.8	41.4	59.8	93.3	47.8	62.4	136.1	76.7	95.7	101.9	104.6	103.7		
09600	FLOORING	103.9	68.7	95.1	97.3	39.2	82.7	99.1	47.2	86.1	97.3	40.8	83.1	126.8	49.0	107.3	103.1	72.0	95.3		
097,099	WALL FINISHES, PAINTS & COATINGS	101.6	51.6	72.1	90.8	46.6	64.7	90.8	47.0	65.0	90.8	56.4	70.5	146.3	77.6	105.8	105.7	88.8	95.7		
09	FINISHES	102.5	63.1	81.6	95.2	49.7	71.0	97.4	47.7	71.0	95.5	49.2	70.9	122.9	71.1	95.4	97.7	95.2	96.4		
10 - 14	TOTAL DIV. 10000 - 14000	100.0	61.2	91.6	100.0	80.1	95.7	100.0	79.3	95.5	100.0	61.3	91.7	100.0	81.9	96.1	100.0	115.2	103.3		
15	MECHANICAL	100.4	72.9	88.0	100.1	84.3	93.0	100.1	84.3	93.0	100.1	79.9	91.0	99.8	83.5	92.5	100.0	94.2	97.4		
16	ELECTRICAL	101.8	78.7	85.5	93.1	69.4	76.4	102.9	69.4	79.2	96.4	47.9	62.2	90.4	85.6	87.0	93.2	94.2	93.9		
01 - 16	WEIGHTED AVERAGE	99.2	74.3	87.0	97.8	69.9	84.1	98.0	71.4	84.9	97.2	63.5	80.7	100.4	80.3	90.6	101.8	96.5	99.2		

NEVADA / NEW HAMPSHIRE / NEW JERSEY

DIVISION	LAS VEGAS MAT.	INST.	TOTAL	RENO MAT.	INST.	TOTAL	MANCHESTER MAT.	INST.	TOTAL	NASHUA MAT.	INST.	TOTAL	PORTSMOUTH MAT.	INST.	TOTAL	CAMDEN MAT.	INST.	TOTAL
01590 EQUIPMENT RENTAL	.0	101.6	101.6	.0	101.6	101.6	.0	102.6	102.6	.0	102.6	102.6	.0	102.6	102.6	.0	100.4	100.4
02 SITE CONSTRUCTION	63.4	106.0	95.5	63.8	104.2	94.3	87.6	101.5	98.1	89.1	101.5	98.4	83.3	100.6	96.3	92.4	105.2	102.0
03100 CONCRETE FORMS & ACCESSORIES	95.0	107.9	106.3	98.7	102.3	101.8	103.3	68.0	72.4	103.9	68.0	72.5	90.1	63.7	67.0	103.8	115.4	114.0
03200 CONCRETE REINFORCEMENT	100.6	117.0	110.3	100.6	116.6	110.1	110.6	88.8	97.8	110.6	88.8	97.8	88.4	88.7	88.6	110.6	104.6	107.1
03300 CAST-IN-PLACE CONCRETE	111.0	110.6	110.8	120.7	89.8	108.0	100.8	86.4	94.8	93.9	86.4	90.8	89.1	80.4	85.5	81.7	124.3	99.2
03 CONCRETE	107.7	110.2	108.9	112.8	100.4	106.6	105.4	78.5	91.9	102.0	78.5	90.2	93.6	74.5	84.0	95.9	115.4	105.7
04 MASONRY	128.2	93.3	106.6	134.3	79.8	100.5	97.2	94.7	95.7	95.7	94.7	95.1	91.2	85.1	87.4	89.3	118.9	107.6
05 METALS	104.5	105.2	104.7	104.3	102.2	103.5	98.0	86.7	93.8	98.0	86.7	93.8	94.6	84.4	90.9	97.9	96.8	97.5
06 WOOD & PLASTICS	91.8	106.4	99.5	96.2	104.6	100.6	101.7	61.0	80.3	101.7	61.0	80.3	88.3	61.0	74.0	101.7	114.1	108.2
07 THERMAL & MOISTURE PROTECTION	103.7	99.2	101.5	104.2	90.0	97.4	102.2	91.3	96.9	102.6	91.3	97.1	102.1	90.0	96.3	102.0	118.4	109.9
08 DOORS & WINDOWS	94.7	111.2	98.7	94.7	111.3	98.8	107.1	68.3	97.5	107.1	68.3	97.5	108.0	61.9	96.7	107.1	109.3	107.6
09200 PLASTER & GYPSUM BOARD	88.8	106.5	100.6	89.8	104.6	99.7	104.8	58.7	73.9	104.8	58.7	73.9	96.5	58.7	71.2	104.8	113.4	110.6
095,098 CEILINGS & ACOUSTICAL TREATMENT	106.1	106.5	106.3	106.1	104.6	105.1	107.9	58.7	74.5	107.9	58.7	74.5	98.3	58.7	71.4	107.9	113.4	111.7
09600 FLOORING	103.1	85.7	98.7	103.1	72.0	95.3	101.8	105.1	102.6	101.6	105.1	102.5	96.2	105.1	98.5	101.6	126.2	107.8
097,099 WALL FINISHES, PAINTS & COATINGS	105.7	114.4	110.8	105.7	88.8	95.7	94.4	95.6	95.1	94.4	95.6	95.1	94.4	43.9	64.6	94.4	130.2	115.5
09 FINISHES	98.3	104.0	101.4	98.5	95.2	96.8	101.6	76.6	88.3	101.8	76.6	88.4	96.8	68.3	81.7	102.1	119.0	111.1
10-14 TOTAL DIV. 10000 - 14000	100.0	113.0	102.8	100.0	115.2	103.3	100.0	88.1	97.4	100.0	88.1	97.4	100.0	84.2	96.6	100.0	113.6	102.9
15 MECHANICAL	100.0	112.4	105.6	100.0	94.2	97.4	100.1	87.2	94.3	100.1	87.2	94.3	100.1	82.1	92.0	100.1	111.6	105.3
16 ELECTRICAL	95.2	107.8	104.1	93.2	94.2	93.9	104.6	76.4	84.7	104.4	76.4	84.6	101.1	76.3	83.6	104.4	116.7	113.1
01-16 WEIGHTED AVERAGE	101.0	106.8	103.8	102.0	96.6	99.3	101.3	84.0	92.8	100.8	84.0	92.6	98.1	79.9	89.2	99.8	112.8	106.2

NEW JERSEY / NEW MEXICO

DIVISION	ELIZABETH MAT.	INST.	TOTAL	JERSEY CITY MAT.	INST.	TOTAL	NEWARK MAT.	INST.	TOTAL	PATERSON MAT.	INST.	TOTAL	TRENTON MAT.	INST.	TOTAL	ALBUQUERQUE MAT.	INST.	TOTAL
01590 EQUIPMENT RENTAL	.0	102.6	102.6	.0	100.4	100.4	.0	102.6	102.6	.0	102.6	102.6	.0	99.9	99.9	.0	116.2	116.2
02 SITE CONSTRUCTION	107.1	106.2	106.4	92.4	105.7	102.4	111.1	106.2	107.4	105.3	105.6	105.5	93.6	106.0	103.0	82.0	112.0	104.6
03100 CONCRETE FORMS & ACCESSORIES	114.8	117.5	117.1	103.9	118.7	116.8	101.9	117.8	115.8	102.5	117.8	115.8	102.7	115.3	113.8	96.6	70.7	74.0
03200 CONCRETE REINFORCEMENT	85.2	120.5	106.0	110.6	120.5	116.5	110.6	120.5	116.5	110.6	120.5	116.5	110.6	114.8	113.1	101.4	66.9	81.0
03300 CAST-IN-PLACE CONCRETE	90.0	124.3	104.1	81.7	126.1	99.9	92.0	124.4	105.3	104.1	124.4	112.4	91.9	117.0	102.2	110.9	76.6	96.8
03 CONCRETE	99.5	119.5	109.6	95.9	120.5	108.3	100.9	119.7	110.4	107.0	119.7	113.4	100.9	114.8	107.9	107.8	73.1	90.3
04 MASONRY	110.6	119.0	115.8	89.3	121.9	109.5	98.2	119.0	111.1	93.4	119.0	109.2	90.2	114.7	105.4	122.3	69.1	89.4
05 METALS	94.4	105.8	98.6	97.9	103.0	99.8	97.9	104.8	100.5	97.9	106.1	101.0	98.7	100.4	99.4	105.5	88.7	99.3
06 WOOD & PLASTICS	116.0	116.9	116.5	101.7	116.7	109.5	103.5	116.9	110.6	103.5	116.9	110.6	101.7	113.8	108.0	96.5	72.6	84.0
07 THERMAL & MOISTURE PROTECTION	103.0	116.0	109.3	102.0	117.3	109.4	102.3	116.1	108.9	102.8	116.7	109.5	100.7	115.5	107.8	103.0	74.5	89.3
08 DOORS & WINDOWS	108.1	116.2	110.1	107.1	116.0	109.3	113.2	115.2	113.7	113.2	116.2	113.9	107.1	113.1	108.6	94.6	72.9	89.3
09200 PLASTER & GYPSUM BOARD	107.9	116.4	113.6	104.8	116.1	112.4	104.8	116.4	112.6	104.8	116.4	112.6	104.8	113.2	110.5	93.2	71.3	78.5
095,098 CEILINGS & ACOUSTICAL TREATMENT	98.3	116.4	110.6	107.9	116.1	113.5	107.9	116.4	113.7	107.9	116.4	113.7	107.9	113.2	111.5	106.1	71.3	82.5
09600 FLOORING	106.2	126.2	111.3	101.6	126.2	107.8	101.8	126.2	107.9	101.6	126.2	107.8	101.8	113.4	104.7	103.1	68.9	94.5
097,099 WALL FINISHES, PAINTS & COATINGS	94.3	130.2	115.5	94.4	130.2	115.5	94.3	130.2	115.5	94.3	130.2	115.5	94.4	130.2	115.5	105.7	59.9	78.7
09 FINISHES	103.1	120.7	112.4	102.1	121.3	112.3	102.7	120.7	112.3	102.3	120.7	112.1	102.2	116.2	109.7	99.4	69.3	83.4
10-14 TOTAL DIV. 10000 - 14000	100.0	92.5	98.4	100.0	93.7	98.6	100.0	92.6	98.4	100.0	92.5	98.4	100.0	113.4	102.9	100.0	79.1	95.5
15 MECHANICAL	100.1	107.5	103.4	100.1	121.3	109.7	100.1	119.0	108.6	100.1	119.7	108.9	100.1	118.3	108.3	100.0	72.0	87.4
16 ELECTRICAL	99.2	104.9	103.2	106.1	125.0	119.4	105.8	125.0	119.4	106.1	120.8	116.5	104.6	122.0	116.9	87.9	76.8	80.0
01-16 WEIGHTED AVERAGE	101.4	111.4	106.3	99.9	117.7	108.7	102.4	116.9	109.5	102.7	116.5	109.5	100.6	114.7	107.5	101.0	77.9	89.7

NEW MEXICO / NEW YORK

DIVISION	FARMINGTON MAT.	INST.	TOTAL	LAS CRUCES MAT.	INST.	TOTAL	ROSWELL MAT.	INST.	TOTAL	SANTA FE MAT.	INST.	TOTAL	ALBANY MAT.	INST.	TOTAL	BINGHAMTON MAT.	INST.	TOTAL
01590 EQUIPMENT RENTAL	.0	116.2	116.2	.0	86.8	86.8	.0	116.2	116.2	.0	116.2	116.2	.0	116.4	116.4	.0	116.0	116.0
02 SITE CONSTRUCTION	87.7	112.0	106.0	93.7	87.1	88.7	92.5	112.0	107.2	81.2	112.0	104.5	71.3	107.4	98.5	93.3	90.2	90.9
03100 CONCRETE FORMS & ACCESSORIES	96.5	70.7	74.0	93.8	68.9	72.1	96.5	70.6	73.9	96.5	70.7	74.0	97.5	87.4	88.7	102.7	75.5	78.9
03200 CONCRETE REINFORCEMENT	110.7	66.9	84.8	104.3	59.9	78.0	109.9	60.4	80.6	108.7	66.9	84.0	100.6	91.1	95.0	99.6	91.6	94.9
03300 CAST-IN-PLACE CONCRETE	111.5	76.6	97.2	97.0	66.5	84.5	102.7	76.5	92.0	104.8	76.6	93.2	87.6	98.9	92.2	104.6	92.9	99.8
03 CONCRETE	111.3	73.1	92.0	87.7	67.1	77.3	111.6	71.8	91.5	105.6	73.1	89.2	101.6	93.2	97.4	100.7	86.8	93.7
04 MASONRY	127.4	69.1	91.3	111.2	63.7	81.8	127.5	69.1	91.4	115.7	69.1	86.9	90.1	94.3	92.7	106.1	85.6	93.5
05 METALS	104.7	88.7	98.8	98.8	76.4	90.5	104.6	85.6	97.6	104.6	88.7	98.7	100.2	106.7	102.6	95.3	118.3	103.8
06 WOOD & PLASTICS	96.5	72.6	84.0	87.5	71.1	78.9	96.5	72.6	84.0	96.5	72.6	84.0	97.6	84.7	90.8	106.5	71.6	88.2
07 THERMAL & MOISTURE PROTECTION	104.0	74.5	89.7	89.5	67.1	78.7	104.3	74.5	89.9	103.3	74.5	89.4	89.6	92.7	91.1	100.1	85.2	92.9
08 DOORS & WINDOWS	97.5	72.9	91.5	87.0	70.0	82.8	93.5	70.9	87.9	93.7	72.9	88.6	95.8	80.1	91.9	90.9	73.4	86.6
09200 PLASTER & GYPSUM BOARD	88.3	71.3	76.9	90.1	71.3	77.5	88.3	71.3	76.9	88.3	71.3	76.9	96.6	83.9	88.1	103.4	70.2	81.1
095,098 CEILINGS & ACOUSTICAL TREATMENT	99.2	71.3	80.2	98.2	71.3	79.9	99.2	71.3	80.2	99.2	71.3	80.2	97.6	83.9	88.3	97.6	70.2	79.0
09600 FLOORING	103.1	68.9	94.5	135.9	70.4	119.4	103.1	68.9	94.5	103.1	68.9	94.5	84.7	96.5	87.6	95.1	86.5	93.0
097,099 WALL FINISHES, PAINTS & COATINGS	105.7	59.9	78.7	99.1	59.9	76.0	105.7	59.9	78.7	105.7	59.9	78.7	83.2	82.1	82.5	87.8	79.4	82.9
09 FINISHES	97.8	69.3	82.7	114.9	68.5	90.2	98.3	69.3	82.9	97.6	69.3	82.6	94.6	88.1	91.1	95.5	77.0	85.7
10-14 TOTAL DIV. 10000 - 14000	100.0	79.1	95.5	100.0	74.8	94.6	100.0	79.1	95.5	100.0	79.1	95.5	100.0	93.9	98.7	100.0	92.7	98.4
15 MECHANICAL	100.0	72.0	87.4	100.3	71.5	87.3	100.0	71.9	87.4	100.0	72.0	87.4	100.3	92.3	96.7	100.5	87.3	94.5
16 ELECTRICAL	85.1	76.8	79.2	86.2	62.3	69.3	86.3	76.8	79.6	87.9	76.8	80.0	100.8	89.7	93.0	101.8	81.9	87.7
01-16 WEIGHTED AVERAGE	101.8	77.9	90.1	97.4	70.0	84.0	101.7	77.3	89.7	100.0	77.9	89.1	97.6	93.8	95.7	98.3	87.3	92.9

		NEW YORK																
	DIVISION	BUFFALO			HICKSVILLE			NEW YORK			RIVERHEAD			ROCHESTER			SCHENECTADY		
		MAT.	INST.	TOTAL	MAT.	INST.	TOTAL	MAT.	INST.	TOTAL	MAT.	INST.	TOTAL	MAT.	INST.	TOTAL	MAT.	INST.	TOTAL
01590	EQUIPMENT RENTAL	.0	92.7	92.7	.0	118.3	118.3	.0	118.7	118.7	.0	118.3	118.3	.0	115.5	115.5	.0	116.4	116.4
02	SITE CONSTRUCTION	94.7	93.0	93.4	112.3	131.2	126.6	135.6	129.2	130.8	112.5	131.5	126.8	73.0	107.2	98.8	70.8	107.4	98.4
03100	CONCRETE FORMS & ACCESSORIES	103.8	113.1	111.9	88.8	156.5	148.0	111.0	171.9	164.3	93.1	158.7	150.4	103.7	98.1	98.8	102.3	87.4	89.3
03200	CONCRETE REINFORCEMENT	101.1	102.5	101.9	101.0	184.1	150.1	105.5	184.3	152.1	102.9	184.1	150.9	102.0	90.3	95.1	99.3	91.1	94.4
03300	CAST-IN-PLACE CONCRETE	116.2	118.1	117.0	97.8	150.8	119.5	125.0	162.6	140.4	96.0	153.9	119.8	110.6	104.0	107.9	96.0	98.8	97.2
03	CONCRETE	107.5	111.9	109.7	104.9	158.1	131.7	122.8	169.0	146.1	103.8	160.2	132.2	113.6	99.6	106.5	106.0	93.2	99.5
04	MASONRY	102.4	116.1	110.9	108.1	155.6	137.5	105.6	165.0	142.3	114.8	161.3	143.6	98.5	100.5	99.7	90.8	94.3	92.9
05	METALS	100.3	93.9	97.9	109.8	138.2	120.3	109.8	139.3	120.7	109.8	138.2	120.4	103.6	108.8	105.5	100.2	106.7	102.6
06	WOOD & PLASTICS	109.7	112.7	111.3	88.8	160.4	126.4	108.3	174.7	143.2	92.6	160.4	128.2	106.0	98.7	102.2	102.9	84.7	93.4
07	THERMAL & MOISTURE PROTECTION	99.3	106.0	102.6	103.4	149.7	125.8	104.7	158.4	130.6	103.5	152.2	127.0	95.9	96.8	96.3	89.7	92.7	91.1
08	DOORS & WINDOWS	93.3	101.8	95.4	88.5	159.2	105.9	97.3	166.9	114.4	88.5	159.2	105.9	96.6	90.9	95.2	95.8	80.1	91.9
09200	PLASTER & GYPSUM BOARD	96.8	112.7	107.5	92.0	162.1	139.0	106.8	176.5	153.5	93.0	162.1	139.4	98.5	98.6	98.5	96.6	83.9	88.1
095,098	CEILINGS & ACOUSTICAL TREATMENT	96.3	112.7	107.4	82.5	162.1	136.6	108.9	176.5	155.1	82.5	162.1	136.6	100.5	98.6	99.2	97.6	83.9	88.3
09600	FLOORING	93.4	116.3	99.2	94.8	97.2	95.4	96.9	160.0	112.8	96.2	97.2	96.4	83.2	92.6	85.5	84.7	96.5	87.6
097,099	WALL FINISHES, PAINTS & COATINGS	94.5	117.7	108.2	110.2	152.7	135.2	98.0	152.7	130.2	110.2	152.7	135.2	92.1	97.9	95.5	83.2	82.1	82.5
09	FINISHES	94.4	114.8	105.3	105.8	144.5	126.4	108.0	168.5	140.1	106.3	145.9	127.4	95.6	97.1	96.4	94.3	88.1	91.0
10 - 14	TOTAL DIV. 10000 - 14000	100.0	107.3	101.6	100.0	130.8	106.6	100.0	146.0	109.9	100.0	133.2	107.1	100.0	97.9	99.5	100.0	93.8	98.7
15	MECHANICAL	99.8	97.7	98.8	99.8	144.7	120.1	100.3	158.7	126.6	99.8	147.8	121.4	99.8	92.1	96.3	100.3	92.3	96.7
16	ELECTRICAL	98.7	102.0	101.0	101.6	153.1	137.9	109.7	167.7	150.6	102.7	154.4	139.1	103.6	93.7	96.6	101.2	89.7	93.1
01 - 16	WEIGHTED AVERAGE	99.5	104.1	101.8	102.1	147.7	124.5	107.1	159.1	132.6	102.6	149.6	125.7	100.6	98.0	99.3	98.2	93.8	96.0

		NEW YORK															NORTH CAROLINA		
	DIVISION	SYRACUSE			UTICA			WATERTOWN			WHITE PLAINS			YONKERS			ASHEVILLE		
		MAT.	INST.	TOTAL	MAT.	INST.	TOTAL	MAT.	INST.	TOTAL	MAT.	INST.	TOTAL	MAT.	INST.	TOTAL	MAT.	INST.	TOTAL
01590	EQUIPMENT RENTAL	.0	116.4	116.4	.0	116.4	116.4	.0	116.4	116.4	.0	118.5	118.5	.0	118.5	118.5	.0	93.3	93.3
02	SITE CONSTRUCTION	92.1	107.5	103.7	68.8	106.2	97.0	77.1	107.9	100.4	125.2	126.5	126.2	135.0	126.3	128.4	102.6	72.1	79.6
03100	CONCRETE FORMS & ACCESSORIES	101.6	84.8	86.9	102.9	77.5	80.7	84.0	78.2	78.9	111.7	141.7	137.9	110.8	141.6	137.7	96.1	46.7	52.9
03200	CONCRETE REINFORCEMENT	100.6	93.9	96.6	100.6	85.6	91.8	101.2	63.9	79.1	98.8	182.5	148.3	102.7	182.5	149.9	94.2	42.2	63.4
03300	CAST-IN-PLACE CONCRETE	97.1	95.2	96.3	88.8	90.1	89.3	103.4	90.7	98.1	110.1	127.5	117.2	123.2	127.4	125.0	93.7	50.6	76.0
03	CONCRETE	104.5	91.3	97.8	102.5	84.8	93.6	116.4	81.5	98.8	110.2	143.2	126.8	121.8	143.1	132.5	96.6	49.2	72.7
04	MASONRY	95.6	91.3	92.9	89.4	84.0	86.0	90.5	92.8	91.9	100.1	127.3	116.9	105.2	127.3	118.9	78.7	40.2	54.9
05	METALS	100.0	106.7	102.5	98.3	103.3	100.1	98.3	101.0	99.3	99.6	133.5	112.2	106.6	133.4	116.6	94.0	78.2	88.1
06	WOOD & PLASTICS	102.9	81.8	91.8	102.9	75.4	88.5	85.0	74.1	79.3	110.4	142.3	127.1	109.2	142.3	126.6	97.0	48.5	71.5
07	THERMAL & MOISTURE PROTECTION	97.9	93.0	95.5	89.7	89.2	89.4	90.0	90.1	90.0	105.6	139.1	121.7	106.0	137.3	121.1	95.5	45.4	71.3
08	DOORS & WINDOWS	93.8	80.3	90.4	95.8	74.6	90.6	95.8	72.3	90.0	91.8	149.4	106.0	95.4	149.4	108.7	91.9	44.7	80.3
09200	PLASTER & GYPSUM BOARD	96.6	80.9	86.1	96.6	74.4	81.7	89.9	73.0	78.6	103.1	143.0	129.9	106.5	143.0	131.0	104.3	46.8	65.7
095,098	CEILINGS & ACOUSTICAL TREATMENT	97.6	80.9	86.3	97.6	74.4	81.8	97.6	73.0	80.9	90.5	143.0	126.2	108.4	143.0	131.9	87.7	46.8	59.9
09600	FLOORING	86.2	81.6	85.0	84.7	81.6	83.9	77.2	79.8	77.9	93.2	160.0	110.0	92.8	160.0	109.7	97.1	44.6	83.9
097,099	WALL FINISHES, PAINTS & COATINGS	88.4	88.6	88.5	83.2	86.8	85.3	83.2	79.9	81.2	95.6	152.7	129.2	95.6	152.7	129.2	108.1	42.6	69.5
09	FINISHES	95.7	83.9	89.5	94.3	78.6	85.9	91.5	77.8	84.2	101.9	146.8	125.7	106.1	146.8	127.7	95.3	45.9	69.1
10 - 14	TOTAL DIV. 10000 - 14000	100.0	94.8	98.9	100.0	90.0	97.8	100.0	89.6	97.8	100.0	128.9	106.2	100.0	137.2	108.0	100.0	69.3	93.4
15	MECHANICAL	100.3	91.6	96.4	100.3	84.3	93.1	100.3	79.2	90.8	100.6	129.2	113.5	100.6	125.7	111.9	100.0	46.9	76.1
16	ELECTRICAL	101.2	84.9	89.7	101.2	84.9	89.7	101.2	85.9	90.4	94.4	142.3	128.2	107.2	142.3	132.0	99.3	43.9	60.2
01 - 16	WEIGHTED AVERAGE	99.0	91.8	95.5	97.3	87.3	92.4	98.9	86.6	92.9	101.2	136.6	118.6	106.0	136.0	120.7	95.9	51.5	74.1

		NORTH CAROLINA																	
	DIVISION	CHARLOTTE			DURHAM			FAYETTEVILLE			GREENSBORO			RALEIGH			WILMINGTON		
		MAT.	INST.	TOTAL	MAT.	INST.	TOTAL	MAT.	INST.	TOTAL	MAT.	INST.	TOTAL	MAT.	INST.	TOTAL	MAT.	INST.	TOTAL
01590	EQUIPMENT RENTAL	.0	93.3	93.3	.0	99.6	99.6	.0	99.6	99.6	.0	99.6	99.6	.0	99.6	99.6	.0	93.3	93.3
02	SITE CONSTRUCTION	102.8	72.1	79.7	102.5	82.0	87.0	100.5	82.0	86.6	102.4	82.0	87.0	103.6	82.0	87.3	103.6	72.2	79.9
03100	CONCRETE FORMS & ACCESSORIES	104.4	46.7	54.0	98.4	47.0	53.5	93.5	47.0	52.9	98.4	47.0	53.5	101.3	47.0	53.9	97.8	47.0	53.4
03200	CONCRETE REINFORCEMENT	94.6	42.2	63.6	94.6	58.5	73.2	93.6	58.5	72.8	94.6	58.5	73.2	94.6	58.5	73.2	94.9	58.5	73.4
03300	CAST-IN-PLACE CONCRETE	96.0	50.7	77.4	93.9	50.8	76.2	90.6	50.8	74.3	93.1	50.8	75.7	99.4	50.8	79.4	93.2	50.8	75.8
03	CONCRETE	97.6	49.3	73.3	96.2	52.5	74.2	93.8	52.5	73.0	95.8	52.5	74.0	99.1	52.5	75.6	96.4	52.5	74.3
04	MASONRY	80.8	40.2	55.7	79.2	40.2	55.1	83.0	40.2	56.5	79.0	40.2	55.0	83.7	40.2	56.8	69.3	40.2	51.3
05	METALS	95.3	78.2	88.9	95.3	85.1	91.5	94.3	85.1	90.9	96.1	85.1	92.0	95.4	85.1	91.5	94.3	85.1	90.9
06	WOOD & PLASTICS	105.9	48.5	75.8	99.1	48.5	72.6	94.1	48.5	70.2	99.1	48.5	72.6	102.4	48.5	74.1	98.6	48.5	72.3
07	THERMAL & MOISTURE PROTECTION	95.5	46.4	71.8	96.1	46.0	71.9	95.8	46.0	71.7	96.1	46.0	71.9	95.9	46.0	71.9	95.5	46.0	71.6
08	DOORS & WINDOWS	96.0	44.7	83.4	96.0	50.9	84.9	92.0	50.9	81.9	96.0	50.9	84.9	92.8	50.9	82.5	92.1	50.9	81.9
09200	PLASTER & GYPSUM BOARD	110.3	46.8	67.7	110.3	46.8	67.7	104.6	46.8	65.8	110.3	46.8	67.7	113.7	46.8	68.9	105.1	46.8	66.0
095,098	CEILINGS & ACOUSTICAL TREATMENT	91.8	46.8	61.2	91.8	46.8	61.2	89.0	46.8	60.3	91.8	46.8	61.2	91.8	46.8	61.2	89.0	46.8	60.3
09600	FLOORING	100.7	44.6	86.6	100.9	44.6	86.7	97.2	44.6	84.0	100.9	44.6	86.7	100.9	44.6	86.7	98.1	44.6	84.6
097,099	WALL FINISHES, PAINTS & COATINGS	108.1	42.6	69.5	108.1	42.6	69.5	108.1	42.6	69.5	108.1	42.6	69.5	108.1	42.6	69.5	108.1	42.6	69.5
09	FINISHES	98.0	45.9	70.3	98.2	45.9	70.4	95.6	45.9	69.2	98.2	45.9	70.4	98.7	45.9	70.6	96.0	45.9	69.4
10 - 14	TOTAL DIV. 10000 - 14000	100.0	69.3	93.4	100.0	72.3	94.0	100.0	72.3	94.0	100.0	69.3	93.4	100.0	72.3	94.0	100.0	72.3	94.0
15	MECHANICAL	100.0	46.9	76.1	100.0	47.1	76.2	100.0	47.1	76.2	100.0	47.1	76.2	100.0	47.1	76.2	100.0	47.1	76.2
16	ELECTRICAL	99.3	39.5	57.1	97.2	42.0	58.3	92.5	42.0	56.9	98.3	42.0	58.6	97.9	42.0	58.5	99.1	42.0	58.8
01 - 16	WEIGHTED AVERAGE	97.2	50.8	74.4	96.8	53.6	75.5	95.4	53.6	74.8	96.9	53.5	75.6	97.2	53.6	75.7	95.5	52.6	74.5

City Cost Indexes

		NORTH CAROLINA			NORTH DAKOTA												OHIO		
	DIVISION	WINSTON-SALEM			BISMARCK			FARGO			GRAND FORKS			MINOT			AKRON		
		MAT.	INST.	TOTAL	MAT.	INST.	TOTAL	MAT.	INST.	TOTAL	MAT.	INST.	TOTAL	MAT.	INST.	TOTAL	MAT.	INST.	TOTAL
01590	EQUIPMENT RENTAL	.0	99.6	99.6	.0	95.7	95.7	.0	95.7	95.7	.0	95.7	95.7	.0	95.7	95.7	.0	98.6	98.6
02	SITE CONSTRUCTION	102.8	82.0	87.1	88.2	96.5	94.4	87.7	96.5	94.3	96.1	95.0	95.2	93.7	96.5	95.8	102.2	107.9	106.5
03100	CONCRETE FORMS & ACCESSORIES	99.9	46.8	53.5	96.8	54.9	60.1	97.5	55.3	60.6	93.2	47.9	53.6	87.8	62.4	65.6	100.5	99.0	99.2
03200	CONCRETE REINFORCEMENT	94.6	42.2	63.6	106.6	77.3	89.3	98.2	76.3	85.3	105.0	56.4	76.3	108.6	77.3	90.1	96.1	97.8	97.1
03300	CAST-IN-PLACE CONCRETE	96.1	50.7	77.4	105.5	59.9	86.8	110.6	62.3	90.7	108.0	57.9	87.4	108.0	58.4	87.6	98.6	108.7	102.8
03	CONCRETE	97.4	49.3	73.1	98.2	61.6	79.8	109.2	62.5	85.7	104.4	53.9	79.0	102.8	64.4	83.5	101.3	101.1	101.2
04	MASONRY	79.3	40.2	55.1	105.9	72.4	85.2	106.2	54.1	74.0	106.1	55.7	74.9	107.1	72.5	85.7	89.8	103.2	98.1
05	METALS	95.3	78.3	89.0	95.3	78.4	89.0	95.2	76.9	88.4	95.3	68.2	85.2	95.6	78.4	89.2	91.7	83.2	88.5
06	WOOD & PLASTICS	99.1	48.5	72.6	89.2	50.2	68.7	89.2	51.0	69.1	85.7	46.2	65.0	80.8	60.4	70.1	94.5	97.2	95.9
07	THERMAL & MOISTURE PROTECTION	96.1	45.0	71.9	100.0	59.4	80.4	100.5	55.3	78.9				100.5	60.3	81.1	105.4	101.8	103.7
08	DOORS & WINDOWS	96.0	46.5	83.8	99.6	53.8	88.3	99.5	54.2	88.4	99.6	43.7	85.8	99.7	59.3	89.8	104.2	97.8	102.6
09200	PLASTER & GYPSUM BOARD	110.3	46.8	67.7	120.0	49.1	72.5	120.0	49.9	73.0	117.9	45.0	69.0	116.4	59.6	78.3	99.7	96.5	97.6
095,098	CEILINGS & ACOUSTICAL TREATMENT	91.8	46.8	61.2	143.4	49.1	79.4	143.4	49.9	79.9	143.4	45.0	76.5	143.4	59.6	86.5	100.9	96.5	97.9
09600	FLOORING	100.9	44.6	86.7	110.9	84.6	104.3	110.7	50.5	95.6	109.2	50.5	94.5	106.5	84.8	101.1	105.1	94.9	102.5
097,099	WALL FINISHES, PAINTS & COATINGS	108.1	42.6	69.5	101.6	42.3	66.7	101.6	80.1	88.9	101.6	32.3	60.8	101.6	35.4	62.6	105.3	121.0	114.5
09	FINISHES	98.2	45.9	70.4	115.5	58.3	85.1	115.3	56.0	83.8	115.2	46.2	78.6	113.9	63.5	87.2	102.6	100.3	101.4
10-14	TOTAL DIV. 10000 - 14000	100.0	69.3	93.4	100.0	82.2	96.2	100.0	82.3	96.2	100.0	53.4	90.0	100.0	83.5	96.5	100.0	100.7	100.2
15	MECHANICAL	100.0	46.9	76.1	100.6	63.8	84.0	100.6	69.8	86.7	100.6	50.7	78.2	100.6	63.8	84.1	99.9	104.0	101.7
16	ELECTRICAL	98.3	42.0	58.6	95.7	72.6	79.4	95.6	66.0	74.7	100.4	62.9	73.9	105.4	74.0	83.2	97.1	96.1	96.4
01-16	WEIGHTED AVERAGE	97.0	52.2	75.0	100.4	69.3	85.1	101.7	67.3	84.8	101.6	58.7	80.6	101.7	71.0	86.6	99.1	99.8	99.4

		OHIO																
	DIVISION	CANTON			CINCINNATI			CLEVELAND			COLUMBUS			DAYTON			LORAIN		
		MAT.	INST.	TOTAL	MAT.	INST.	TOTAL	MAT.	INST.	TOTAL	MAT.	INST.	TOTAL	MAT.	INST.	TOTAL	MAT.	INST.	TOTAL
01590	EQUIPMENT RENTAL	.0	98.6	98.6	.0	103.1	103.1	.0	99.0	99.0	.0	97.3	97.3	.0	97.3	97.3	.0	98.6	98.6
02	SITE CONSTRUCTION	102.3	108.0	106.6	74.6	105.9	100.9	102.2	107.7	106.3	87.6	103.3	99.5	73.3	109.3	100.4	101.6	107.1	105.7
03100	CONCRETE FORMS & ACCESSORIES	100.5	87.6	89.2	99.4	89.4	90.7	100.6	108.0	107.1	95.3	89.6	90.3	99.3	86.2	87.9	100.5	88.4	89.9
03200	CONCRETE REINFORCEMENT	96.1	82.5	88.0	94.6	89.4	91.5	96.5	98.2	97.5	102.7	89.2	94.7	94.6	84.9	88.8	96.1	97.9	97.2
03300	CAST-IN-PLACE CONCRETE	99.6	105.2	101.9	86.8	94.9	90.1	96.7	116.5	104.9	88.1	97.5	92.0	79.1	89.7	83.4	93.9	109.2	100.2
03	CONCRETE	101.8	92.0	96.8	94.8	91.3	93.0	100.4	107.9	104.2	96.5	91.8	94.1	90.9	86.7	88.8	98.9	96.6	97.8
04	MASONRY	90.6	92.3	91.7	86.4	97.5	93.3	94.1	112.5	105.5	96.4	97.6	97.1	85.8	91.0	89.0	86.8	102.2	96.3
05	METALS	91.7	76.4	86.0	94.8	89.2	92.8	93.1	86.1	90.5	96.2	83.4	91.5	94.2	79.3	88.7	92.3	83.6	89.0
06	WOOD & PLASTICS	94.5	86.1	90.1	97.9	86.8	92.1	93.6	105.2	99.7	100.4	87.9	93.8	99.0	85.2	91.8	94.5	83.9	88.9
07	THERMAL & MOISTURE PROTECTION	106.0	97.3	101.8	90.3	98.0	94.0	104.2	115.5	109.6	99.9	99.8	99.8	95.8	92.2	94.1	105.9	106.1	106.0
08	DOORS & WINDOWS	98.6	79.5	93.9	95.8	84.9	93.1	95.0	103.0	96.9	99.6	86.1	96.3	96.0	83.0	92.8	98.6	91.5	96.8
09200	PLASTER & GYPSUM BOARD	99.7	85.1	89.9	97.6	86.8	90.3	98.9	104.7	102.8	100.0	87.3	91.5	97.6	85.1	89.2	99.7	82.9	88.4
095,098	CEILINGS & ACOUSTICAL TREATMENT	100.9	85.1	90.2	95.0	86.8	89.4	99.5	104.7	103.1	96.3	87.3	90.2	96.6	85.1	88.8	100.9	82.9	88.6
09600	FLOORING	105.3	87.7	100.9	98.1	104.8	99.8	104.9	112.5	106.8	99.7	95.1	98.5	100.7	90.6	98.2	105.3	110.8	106.7
097,099	WALL FINISHES, PAINTS & COATINGS	105.3	93.4	98.3	104.1	97.8	100.4	105.3	124.0	116.3	99.9	103.6	102.1	104.1	94.6	98.5	105.3	124.0	116.3
09	FINISHES	102.7	87.2	94.5	95.2	92.8	93.9	102.3	110.4	106.6	98.4	91.7	94.8	96.3	87.6	91.7	102.7	95.5	98.9
10-14	TOTAL DIV. 10000 - 14000	100.0	86.8	97.2	100.0	92.3	98.3	100.0	107.7	101.7	100.0	93.1	98.5	100.0	90.5	98.0	100.0	101.5	100.3
15	MECHANICAL	99.9	88.2	94.6	99.8	96.5	98.3	99.9	110.9	104.9	100.0	94.6	97.5	100.8	88.2	95.1	99.9	93.7	97.1
16	ELECTRICAL	95.6	98.9	98.0	97.2	83.6	87.6	96.2	109.2	105.4	101.9	91.4	94.5	95.8	87.9	90.2	95.6	94.4	94.8
01-16	WEIGHTED AVERAGE	98.4	91.5	95.0	95.5	93.2	94.4	98.2	107.5	102.7	98.4	93.1	95.8	95.3	89.2	92.3	97.9	95.9	96.9

		OHIO									OKLAHOMA								
	DIVISION	SPRINGFIELD			TOLEDO			YOUNGSTOWN			ENID			LAWTON			MUSKOGEE		
		MAT.	INST.	TOTAL	MAT.	INST.	TOTAL	MAT.	INST.	TOTAL	MAT.	INST.	TOTAL	MAT.	INST.	TOTAL	MAT.	INST.	TOTAL
01590	EQUIPMENT RENTAL	.0	97.3	97.3	.0	100.4	100.4	.0	98.6	98.6	.0	76.6	76.6	.0	77.6	77.6	.0	86.3	86.3
02	SITE CONSTRUCTION	73.8	108.4	99.9	86.9	104.5	100.1	102.1	108.1	106.6	108.7	89.0	93.8	103.8	90.6	93.8	94.1	85.0	87.2
03100	CONCRETE FORMS & ACCESSORIES	99.3	88.0	89.4	95.3	105.6	104.3	100.5	92.0	93.0	94.8	43.6	50.0	99.1	59.9	64.8	100.7	39.6	47.3
03200	CONCRETE REINFORCEMENT	94.6	84.9	88.8	102.7	92.5	96.6	96.1	91.8	93.5	97.2	85.2	90.1	97.4	85.2	90.2	97.2	42.0	64.5
03300	CAST-IN-PLACE CONCRETE	82.9	90.1	85.8	88.1	109.8	97.0	97.7	107.3	101.6	93.3	54.9	77.5	90.2	54.9	75.7	83.9	42.2	66.8
03	CONCRETE	92.8	87.6	90.2	96.5	103.9	100.2	100.8	96.4	98.6	91.5	55.5	73.4	87.9	62.7	75.2	83.3	42.1	62.6
04	MASONRY	86.1	91.7	89.5	106.6	105.2	105.8	89.9	99.9	96.1	101.4	62.4	77.3	96.2	62.4	75.3	110.7	46.9	71.2
05	METALS	94.2	79.2	88.6	96.1	88.9	93.4	91.8	80.9	87.7	93.4	68.8	84.2	95.4	68.9	85.6	93.3	61.8	81.6
06	WOOD & PLASTICS	100.3	87.4	93.5	100.4	105.9	103.3	94.5	89.3	91.8	100.4	40.8	69.1	103.9	62.7	82.3	105.6	40.3	71.3
07	THERMAL & MOISTURE PROTECTION	95.7	92.8	94.3	101.7	110.3	105.8	106.2	101.3	103.8	99.1	64.0	82.2	98.8	66.2	83.1	98.6	49.0	74.7
08	DOORS & WINDOWS	94.0	84.8	91.7	97.6	98.9	97.9	98.6	95.2	97.7	95.5	54.8	85.5	97.1	66.6	89.6	95.5	39.5	81.7
09200	PLASTER & GYPSUM BOARD	97.6	87.3	90.7	100.0	105.9	104.0	99.7	88.4	92.1	87.5	40.0	55.6	90.3	62.5	71.7	89.0	39.4	55.7
095,098	CEILINGS & ACOUSTICAL TREATMENT	96.6	87.3	90.3	96.3	105.9	102.8	100.9	88.4	92.4	91.1	40.0	56.4	99.3	62.5	74.3	91.1	39.4	56.0
09600	FLOORING	100.7	90.6	98.2	98.8	102.7	99.8	105.3	102.8	104.6	115.1	57.9	100.7	118.1	57.9	103.0	119.4	46.7	101.1
097,099	WALL FINISHES, PAINTS & COATINGS	104.1	94.6	98.5	100.0	115.9	109.4	105.3	106.3	105.9	96.8	62.2	76.4	96.8	62.2	76.4	96.8	39.5	63.0
09	FINISHES	96.3	89.0	92.4	98.1	106.4	102.5	102.8	94.9	98.6	99.7	46.4	71.3	102.0	59.2	79.3	100.1	40.9	68.7
10-14	TOTAL DIV. 10000 - 14000	100.0	91.1	98.1	100.0	102.8	100.6	100.0	100.0	100.0	100.0	70.7	93.7	100.0	73.6	94.3	100.0	69.8	93.5
15	MECHANICAL	100.8	88.5	95.3	100.0	104.1	101.8	99.9	92.4	96.5	100.0	67.0	85.1	100.0	67.0	85.2	100.0	32.8	69.7
16	ELECTRICAL	95.8	88.0	90.3	102.6	103.6	103.3	95.6	92.5	93.4	94.1	70.6	77.5	96.8	69.7	77.7	93.6	38.2	54.5
01-16	WEIGHTED AVERAGE	95.4	89.7	92.6	98.7	103.0	100.8	98.3	94.9	96.6	97.3	64.6	81.3	97.4	68.0	83.0	96.5	46.5	71.9

COST INDEXES

589

Oklahoma / Oregon

	DIVISION	OKLAHOMA CITY MAT.	INST.	TOTAL	TULSA MAT.	INST.	TOTAL	EUGENE MAT.	INST.	TOTAL	MEDFORD MAT.	INST.	TOTAL	PORTLAND MAT.	INST.	TOTAL	SALEM MAT.	INST.	TOTAL
01590	EQUIPMENT RENTAL	.0	77.9	77.9	.0	86.3	86.3	.0	99.5	99.5	.0	99.5	99.5	.0	99.5	99.5	.0	99.5	99.5
02	SITE CONSTRUCTION	105.2	91.2	94.6	100.7	86.0	89.6	104.9	105.5	105.3	114.5	105.4	107.7	106.0	105.5	105.6	105.0	105.5	105.3
03100	CONCRETE FORMS & ACCESSORIES	100.3	51.5	57.6	100.3	52.2	58.3	103.4	108.5	107.8	98.1	108.3	107.0	104.8	108.8	108.3	104.7	108.7	108.2
03200	CONCRETE REINFORCEMENT	97.4	85.2	90.2	97.4	85.2	90.2	108.6	98.1	102.4	100.3	98.0	99.0	103.8	98.5	100.7	109.8	98.5	103.1
03300	CAST-IN-PLACE CONCRETE	97.6	58.5	81.6	91.4	53.7	75.9	104.3	106.9	105.4	107.7	106.8	107.4	112.0	107.0	110.0	108.2	107.0	107.7
03	CONCRETE	91.7	60.3	75.9	88.6	59.9	74.2	110.3	105.5	107.9	117.4	105.4	111.3	113.6	105.8	109.7	112.4	105.7	109.0
04	MASONRY	98.6	64.7	77.6	95.4	64.8	76.5	113.0	101.3	105.7	109.7	101.3	104.5	114.2	108.6	110.7	115.0	108.6	111.0
05	METALS	95.6	68.9	85.7	96.0	84.3	91.6	95.2	96.2	95.6	94.7	96.0	95.2	93.9	97.1	95.1	95.8	97.0	96.2
06	WOOD & PLASTICS	105.8	50.2	76.6	104.6	51.7	76.8	91.7	108.6	100.5	86.5	108.6	98.1	93.4	108.6	101.3	93.4	108.6	101.3
07	THERMAL & MOISTURE PROTECTION	97.9	66.0	82.5	98.6	63.6	81.7	110.5	92.9	102.0	111.4	90.6	101.4	110.3	102.0	106.3	110.3	99.6	105.2
08	DOORS & WINDOWS	97.1	59.9	87.9	97.1	60.2	88.0	98.8	107.6	100.9	101.5	107.6	103.0	96.3	107.6	99.0	98.1	107.6	100.4
09200	PLASTER & GYPSUM BOARD	90.3	49.7	63.1	90.3	51.1	64.0	97.6	108.7	105.1	94.7	108.7	104.1	97.1	108.7	104.9	98.6	108.7	105.4
095,098	CEILINGS & ACOUSTICAL TREATMENT	99.3	49.7	65.6	99.3	51.1	66.6	113.8	108.7	110.3	112.4	108.7	109.9	113.8	108.7	110.3	120.6	108.7	112.6
09600	FLOORING	118.1	57.9	103.0	117.9	60.3	103.4	115.6	100.5	111.8	113.1	100.5	110.0	115.6	100.5	111.8	115.6	100.5	111.8
097,099	WALL FINISHES, PAINTS & COATINGS	96.8	62.2	76.4	96.8	58.3	74.1	116.6	73.7	91.3	116.6	68.3	88.1	116.6	73.7	91.3	116.6	73.7	91.3
09	FINISHES	102.1	52.5	75.7	101.5	53.4	76.0	112.8	103.5	107.8	112.0	102.9	107.1	112.7	103.5	107.8	114.1	103.5	108.4
10-14	TOTAL DIV. 10000 - 14000	100.0	72.9	94.2	100.0	73.6	94.3	100.0	108.6	101.8	100.0	108.5	101.8	100.0	108.6	101.9	100.0	108.6	101.8
15	MECHANICAL	100.0	68.3	85.7	100.0	68.0	85.6	100.2	111.8	105.4	100.2	111.7	105.4	100.2	111.8	105.4	100.2	111.8	105.4
16	ELECTRICAL	95.7	70.6	78.0	96.8	49.8	63.7	104.4	98.7	100.4	110.1	90.4	96.2	104.2	107.3	106.4	104.0	98.7	100.3
01-16	WEIGHTED AVERAGE	98.0	67.1	82.8	97.4	64.4	81.2	103.2	103.9	103.5	104.7	102.3	103.5	103.2	106.5	104.8	103.7	104.9	104.3

Pennsylvania

	DIVISION	ALLENTOWN MAT.	INST.	TOTAL	ALTOONA MAT.	INST.	TOTAL	ERIE MAT.	INST.	TOTAL	HARRISBURG MAT.	INST.	TOTAL	PHILADELPHIA MAT.	INST.	TOTAL	PITTSBURGH MAT.	INST.	TOTAL
01590	EQUIPMENT RENTAL	.0	116.4	116.4	.0	116.4	116.4	.0	116.4	116.4	.0	115.6	115.6	.0	94.8	94.8	.0	117.3	117.3
02	SITE CONSTRUCTION	90.9	107.3	103.3	95.2	107.4	104.4	91.8	108.0	104.0	81.3	105.8	99.8	99.9	94.6	95.9	98.5	110.4	107.5
03100	CONCRETE FORMS & ACCESSORIES	101.4	109.5	108.5	82.0	93.0	91.6	100.0	100.1	100.1	93.3	85.6	86.5	99.8	127.3	123.8	96.3	101.3	100.6
03200	CONCRETE REINFORCEMENT	100.6	105.3	103.4	97.6	97.7	97.7	99.6	98.1	98.7	100.6	97.4	98.7	98.8	131.7	118.3	103.3	111.1	107.9
03300	CAST-IN-PLACE CONCRETE	87.9	98.0	92.1	98.0	92.8	95.8	96.4	90.7	94.0	97.0	89.0	93.7	95.2	121.5	106.0	95.1	98.5	96.5
03	CONCRETE	98.8	105.8	102.3	93.3	95.2	94.3	92.2	97.8	95.0	100.8	90.6	95.7	107.7	125.7	116.8	93.6	103.4	98.5
04	MASONRY	94.2	94.3	94.3	96.6	86.4	90.3	86.8	91.5	89.7	92.5	83.3	86.8	96.5	126.1	114.8	89.6	102.3	97.5
05	METALS	98.2	122.7	107.3	91.8	118.5	101.7	92.0	119.6	102.2	99.9	118.4	106.8	100.2	123.0	108.7	98.3	127.0	109.0
06	WOOD & PLASTICS	102.9	112.5	107.9	80.9	96.1	88.9	98.3	101.0	99.7	96.2	85.0	90.3	98.5	127.4	113.7	93.8	100.9	97.6
07	THERMAL & MOISTURE PROTECTION	97.9	111.4	104.4	96.9	93.7	95.4	97.0	98.8	97.9	101.6	103.0	102.3	100.5	127.6	113.6	94.5	102.6	98.4
08	DOORS & WINDOWS	93.8	109.2	97.6	88.2	101.1	91.4	88.4	98.0	90.8	93.8	95.1	94.1	96.8	131.8	105.4	94.1	111.5	98.4
09200	PLASTER & GYPSUM BOARD	96.6	112.5	107.3	88.5	95.6	93.3	96.6	100.6	99.3	96.6	84.2	88.3	95.5	128.1	117.3	86.1	100.6	95.8
095,098	CEILINGS & ACOUSTICAL TREATMENT	97.6	112.5	107.7	93.5	95.6	94.9	97.6	100.6	99.7	97.6	84.2	88.5	101.0	128.1	119.4	93.3	100.6	98.2
09600	FLOORING	86.2	89.9	87.1	80.5	64.2	76.4	88.1	84.1	87.1	86.4	84.6	85.9	82.3	127.0	93.5	95.7	104.8	98.0
097,099	WALL FINISHES, PAINTS & COATINGS	88.4	101.5	96.1	84.1	104.2	96.0	94.6	87.3	90.3	88.4	86.8	87.4	86.7	129.7	112.1	93.7	108.4	102.4
09	FINISHES	95.7	105.3	100.8	92.3	88.6	90.3	97.3	95.7	96.5	94.6	84.9	89.5	98.9	127.4	114.0	96.0	102.4	99.4
10-14	TOTAL DIV. 10000 - 14000	100.0	103.9	100.8	100.0	99.3	99.8	100.0	103.4	100.7	100.0	96.0	99.1	100.0	121.8	104.7	100.0	103.5	100.7
15	MECHANICAL	100.3	103.7	101.8	99.8	83.0	92.2	99.8	93.4	96.9	100.3	88.2	94.8	100.1	122.6	110.3	99.9	97.1	98.6
16	ELECTRICAL	101.2	90.0	93.3	86.9	97.2	94.1	89.2	84.1	85.6	100.2	78.0	84.5	95.3	130.7	120.3	95.5	97.2	96.7
01-16	WEIGHTED AVERAGE	97.9	103.5	100.7	94.2	95.1	94.6	94.3	96.8	95.6	97.9	91.1	94.6	100.0	123.3	111.4	96.7	103.9	100.2

Pennsylvania / Puerto Rico / Rhode Island / South Carolina

	DIVISION	READING MAT.	INST.	TOTAL	SCRANTON MAT.	INST.	TOTAL	YORK MAT.	INST.	TOTAL	SAN JUAN MAT.	INST.	TOTAL	PROVIDENCE MAT.	INST.	TOTAL	CHARLESTON MAT.	INST.	TOTAL
01590	EQUIPMENT RENTAL	.0	118.7	118.7	.0	116.4	116.4	.0	115.6	115.6	.0	90.1	90.1	.0	104.4	104.4	.0	99.2	99.2
02	SITE CONSTRUCTION	97.4	111.8	108.3	91.4	107.3	103.4	80.6	105.9	99.7	120.1	93.5	100.0	81.7	104.6	99.0	95.7	81.1	84.7
03100	CONCRETE FORMS & ACCESSORIES	99.3	87.4	88.9	101.5	89.6	91.1	80.9	86.0	85.3	101.0	21.9	31.9	106.3	108.5	108.2	98.4	41.7	48.8
03200	CONCRETE REINFORCEMENT	97.2	100.0	98.8	100.6	103.3	102.2	99.6	97.3	98.3	198.8	13.4	89.2	110.6	115.3	113.4	94.6	67.1	78.4
03300	CAST-IN-PLACE CONCRETE	69.6	93.0	79.2	91.8	92.1	91.9	86.5	89.7	87.8	106.5	33.8	76.7	87.2	114.4	98.5	79.8	52.8	68.7
03	CONCRETE	94.6	93.3	94.0	100.7	94.6	97.6	99.2	91.0	95.1	109.9	25.1	67.2	101.2	111.6	106.5	89.1	52.2	70.5
04	MASONRY	98.0	89.0	92.4	94.5	93.3	93.7	93.8	83.0	87.1	210.1	18.8	91.8	105.4	112.1	109.5	86.9	39.2	57.4
05	METALS	97.2	120.0	105.7	100.0	121.5	108.0	97.3	118.3	105.1	129.5	32.8	93.6	98.0	107.8	101.7	95.3	84.2	91.2
06	WOOD & PLASTICS	98.6	85.1	91.5	102.9	87.3	94.7	87.8	85.0	86.3	92.4	22.2	55.5	102.4	106.8	104.7	99.1	41.0	68.6
07	THERMAL & MOISTURE PROTECTION	100.5	105.5	102.9	97.7	102.5	100.0	96.2	102.7	99.4	170.6	25.9	100.8	101.2	107.3	104.1	95.7	47.1	72.2
08	DOORS & WINDOWS	94.4	92.8	94.1	93.8	98.1	94.8	90.6	95.1	91.7	154.4	18.5	120.9	101.0	110.2	103.2	96.0	45.3	83.5
09200	PLASTER & GYPSUM BOARD	103.9	84.5	90.9	96.6	86.5	89.9	92.0	84.2	86.8	265.1	19.6	100.5	101.4	106.0	104.5	110.3	39.0	62.5
095,098	CEILINGS & ACOUSTICAL TREATMENT	102.3	84.5	90.2	97.6	86.5	90.1	90.7	84.2	86.3	389.0	19.6	138.1	109.5	106.0	107.1	91.8	39.0	55.9
09600	FLOORING	80.4	91.2	83.1	86.2	88.6	86.8	82.2	84.6	82.8	203.9	18.5	157.3	102.5	120.6	107.0	100.9	44.2	86.6
097,099	WALL FINISHES, PAINTS & COATINGS	84.9	96.7	91.8	88.4	96.7	93.3	88.4	83.1	85.3	205.2	17.2	94.3	96.9	113.3	106.6	108.1	43.0	69.7
09	FINISHES	99.5	88.3	93.5	95.7	89.5	92.4	91.5	84.8	88.0	250.4	21.0	128.5	101.7	111.4	106.9	98.4	41.7	68.3
10-14	TOTAL DIV. 10000 - 14000	100.0	98.2	99.6	100.0	99.8	100.0	100.0	96.5	99.2	100.0	23.4	83.5	100.0	109.2	102.0	100.0	70.8	93.7
15	MECHANICAL	100.3	99.8	100.1	100.3	90.2	95.7	100.3	88.8	95.1	105.2	17.3	65.5	100.1	99.1	99.6	100.0	48.5	76.8
16	ELECTRICAL	97.9	75.2	81.9	101.2	84.7	89.5	91.1	78.0	81.8	144.9	17.1	54.7	98.9	93.0	94.7	98.2	36.1	54.4
01-16	WEIGHTED AVERAGE	98.0	94.9	96.4	98.5	95.3	96.9	95.8	91.2	93.6	140.5	27.9	85.2	99.9	104.7	102.3	96.2	51.7	74.3

SOUTH CAROLINA / SOUTH DAKOTA

DIVISION		COLUMBIA			FLORENCE			GREENVILLE			SPARTANBURG			ABERDEEN			PIERRE		
		MAT.	INST.	TOTAL	MAT.	INST.	TOTAL	MAT.	INST.	TOTAL	MAT.	INST.	TOTAL	MAT.	INST.	TOTAL	MAT.	INST.	TOTAL
01590	EQUIPMENT RENTAL	.0	99.2	99.2	.0	99.2	99.2	.0	99.2	99.2	.0	99.2	99.2	.0	95.7	95.7	.0	95.7	95.7
02	SITE CONSTRUCTION	95.4	81.1	84.6	107.2	81.0	87.5	102.0	80.7	85.9	101.8	80.7	85.8	83.9	94.5	91.9	82.1	94.5	91.4
03100	CONCRETE FORMS & ACCESSORIES	103.4	44.1	51.6	82.5	44.1	48.9	98.1	43.8	50.7	102.2	43.8	51.2	97.2	43.5	50.3	95.4	45.1	51.5
03200	CONCRETE REINFORCEMENT	94.6	67.1	78.4	94.3	67.1	78.2	94.2	42.7	63.8	94.2	42.7	63.8	105.2	52.5	74.1	104.7	63.4	80.3
03300	CAST-IN-PLACE CONCRETE	78.1	52.4	67.6	70.6	52.7	63.2	70.6	52.6	63.2	70.6	52.6	63.2	102.9	54.8	83.2	100.0	50.2	79.5
03	CONCRETE	88.6	53.2	70.8	91.2	53.3	72.1	89.8	48.6	69.0	90.1	48.6	69.2	97.5	50.3	73.7	95.2	51.5	73.2
04	MASONRY	85.8	40.2	57.6	72.7	39.2	52.0	70.6	39.2	51.2	72.7	39.2	52.0	108.2	61.2	79.1	104.6	60.9	77.6
05	METALS	95.3	84.1	91.1	94.2	83.9	90.4	94.1	75.1	87.0	94.1	75.1	87.1	107.1	69.9	93.3	107.0	74.3	94.9
06	WOOD & PLASTICS	104.9	44.6	73.2	83.2	44.6	62.9	98.8	44.6	70.3	103.1	44.6	72.4	103.1	41.5	70.8	101.3	43.9	71.1
07	THERMAL & MOISTURE PROTECTION	95.6	46.9	72.1	96.0	47.5	72.6	96.0	47.5	72.6	96.0	47.5	72.6	98.4	55.7	77.8	98.6	53.8	77.0
08	DOORS & WINDOWS	96.0	47.2	84.0	92.0	47.2	81.0	92.0	41.5	79.5	92.0	41.5	79.5	95.1	44.4	82.6	98.4	48.8	86.2
09200	PLASTER & GYPSUM BOARD	110.3	42.7	65.0	99.4	42.7	61.4	105.3	42.7	63.3	107.4	42.7	64.0	104.3	40.2	61.3	103.0	42.6	62.5
095,098	CEILINGS & ACOUSTICAL TREATMENT	91.8	42.7	58.4	89.0	42.7	57.6	87.7	42.7	57.1	87.7	42.7	57.1	107.8	40.2	61.9	106.4	42.6	63.1
09600	FLOORING	100.7	44.2	86.5	91.5	44.2	79.6	98.4	45.1	85.0	100.1	45.1	86.3	111.3	63.3	99.2	110.4	45.1	94.0
097,099	WALL FINISHES, PAINTS & COATINGS	108.1	43.0	69.7	108.1	43.0	69.7	108.1	43.0	69.7	108.1	43.0	69.7	101.6	38.8	64.6	101.6	44.2	67.8
09	FINISHES	98.4	43.8	69.4	94.4	43.8	67.5	96.9	43.9	68.8	97.8	43.9	69.2	106.7	46.2	74.5	105.9	44.5	73.2
10-14	TOTAL DIV. 10000 - 14000	100.0	71.3	93.8	100.0	71.2	93.8	100.0	71.3	93.8	100.0	71.3	93.8	100.0	63.7	92.2	100.0	73.9	94.4
15	MECHANICAL	100.0	40.9	73.4	100.0	40.9	73.4	100.0	40.8	73.3	100.0	40.8	73.3	100.3	43.6	74.7	100.3	41.7	73.9
16	ELECTRICAL	99.3	38.1	56.1	95.2	21.7	43.4	98.6	34.6	53.5	98.6	34.6	53.5	101.2	66.1	76.5	94.9	55.3	67.0
01-16	WEIGHTED AVERAGE	96.2	51.1	74.1	94.6	48.2	71.8	94.8	48.8	72.2	95.0	48.8	72.3	101.0	58.3	80.0	100.3	56.8	78.9

SOUTH DAKOTA / TENNESSEE

DIVISION		RAPID CITY			SIOUX FALLS			CHATTANOOGA			JACKSON			JOHNSON CITY			KNOXVILLE		
		MAT.	INST.	TOTAL	MAT.	INST.	TOTAL	MAT.	INST.	TOTAL	MAT.	INST.	TOTAL	MAT.	INST.	TOTAL	MAT.	INST.	TOTAL
01590	EQUIPMENT RENTAL	.0	95.7	95.7	.0	97.0	97.0	.0	104.4	104.4	.0	104.8	104.8	.0	97.4	97.4	.0	97.4	97.4
02	SITE CONSTRUCTION	82.3	94.2	91.3	83.4	96.4	93.2	103.9	97.1	98.8	102.4	96.1	97.6	113.8	85.9	92.7	90.5	85.9	87.1
03100	CONCRETE FORMS & ACCESSORIES	108.5	39.9	48.6	95.5	46.1	52.3	94.6	51.2	56.6	85.0	41.7	47.2	79.5	52.0	55.5	93.4	52.1	57.3
03200	CONCRETE REINFORCEMENT	98.2	63.2	77.6	98.0	47.7	66.1	92.7	47.7	66.1	98.0	47.4	68.1	96.7	47.9	67.8	92.7	47.9	66.2
03300	CAST-IN-PLACE CONCRETE	99.2	47.3	77.9	103.4	53.3	82.8	101.5	52.9	81.6	100.9	45.8	78.3	81.6	57.9	71.9	95.3	53.1	77.0
03	CONCRETE	94.9	48.2	71.4	94.9	53.0	73.8	93.5	53.0	73.1	93.9	46.4	70.0	99.6	55.1	77.2	90.6	53.5	71.9
04	MASONRY	104.6	52.7	72.5	101.8	60.9	76.5	98.3	48.2	67.3	111.5	37.4	65.7	110.6	47.9	71.8	76.0	47.9	58.6
05	METALS	109.1	74.3	96.2	109.6	74.8	96.7	96.9	80.8	90.9	95.0	79.1	89.1	94.5	80.3	89.3	97.1	80.7	91.0
06	WOOD & PLASTICS	110.9	39.2	73.3	101.3	44.6	71.5	98.0	52.4	74.1	83.6	41.7	61.6	73.5	53.6	63.0	86.4	53.6	69.2
07	THERMAL & MOISTURE PROTECTION	99.1	48.6	74.7	98.5	53.4	76.7	101.2	55.4	79.1	100.1	43.6	72.8	96.4	52.8	75.4	94.0	53.1	74.2
08	DOORS & WINDOWS	99.4	46.3	86.3	99.4	49.1	87.0	101.3	52.1	89.2	102.1	43.9	87.8	97.1	53.3	86.3	93.6	53.3	83.7
09200	PLASTER & GYPSUM BOARD	104.2	37.8	59.7	104.2	43.3	63.4	88.0	51.6	63.6	91.5	40.5	57.3	96.0	52.8	67.1	102.9	52.8	69.3
095,098	CEILINGS & ACOUSTICAL TREATMENT	111.9	37.8	61.6	111.9	43.3	65.3	102.4	51.6	67.9	96.9	40.5	58.6	93.8	52.8	66.0	99.3	52.8	67.7
09600	FLOORING	110.4	72.3	100.8	110.4	77.8	102.2	102.2	55.7	90.5	91.9	27.0	75.5	95.4	57.2	85.8	101.1	57.2	90.1
097,099	WALL FINISHES, PAINTS & COATINGS	101.6	44.2	67.8	101.6	44.2	67.8	108.4	51.6	74.9	95.5	35.0	59.8	106.3	60.6	79.3	106.3	60.6	79.3
09	FINISHES	107.1	46.0	74.6	107.1	51.5	77.6	99.0	52.1	74.1	95.5	38.3	65.1	99.1	53.8	75.1	93.9	53.8	72.6
10-14	TOTAL DIV. 10000 - 14000	100.0	71.2	93.8	100.0	74.1	94.4	100.0	61.6	91.7	100.0	58.1	91.0	100.0	70.7	93.7	100.0	70.8	93.7
15	MECHANICAL	100.3	38.9	72.6	100.3	40.6	73.4	100.0	50.6	77.7	100.0	62.7	83.2	99.8	63.3	83.3	99.8	65.0	84.1
16	ELECTRICAL	95.6	55.3	67.2	94.6	66.1	74.5	103.7	58.8	72.0	99.9	57.0	69.6	91.1	57.0	67.0	100.5	64.3	74.9
01-16	WEIGHTED AVERAGE	100.9	54.8	78.3	100.7	59.7	80.6	99.1	59.8	79.8	98.9	57.2	78.4	98.7	61.5	80.4	95.1	63.0	79.3

TENNESSEE / TEXAS

DIVISION		MEMPHIS			NASHVILLE			ABILENE			AMARILLO			AUSTIN			BEAUMONT		
		MAT.	INST.	TOTAL	MAT.	INST.	TOTAL	MAT.	INST.	TOTAL	MAT.	INST.	TOTAL	MAT.	INST.	TOTAL	MAT.	INST.	TOTAL
01590	EQUIPMENT RENTAL	.0	102.2	102.2	.0	105.5	105.5	.0	86.3	86.3	.0	86.3	86.3	.0	86.8	86.8	.0	87.7	87.7
02	SITE CONSTRUCTION	95.4	91.3	92.3	96.5	100.3	99.4	101.1	85.2	89.1	101.5	86.1	89.9	86.3	85.9	86.0	90.8	86.5	87.6
03100	CONCRETE FORMS & ACCESSORIES	94.7	65.1	68.8	93.9	65.1	68.7	97.2	47.5	53.7	100.8	57.2	62.7	96.9	61.1	65.6	103.5	58.8	64.4
03200	CONCRETE REINFORCEMENT	89.5	69.7	77.8	95.4	68.5	79.5	97.4	57.1	73.6	97.4	56.4	73.2	95.5	57.0	72.7	93.5	48.7	67.1
03300	CAST-IN-PLACE CONCRETE	90.8	66.8	81.0	93.8	62.4	80.9	94.8	46.3	74.9	99.7	52.8	80.5	90.4	54.4	75.6	86.7	59.9	75.7
03	CONCRETE	87.8	68.3	78.0	90.9	66.4	78.6	90.1	49.9	69.9	92.8	56.4	74.4	81.1	58.6	69.8	86.2	58.1	72.0
04	MASONRY	85.5	64.0	72.2	86.8	62.9	72.0	98.9	57.3	73.2	102.7	43.4	66.0	93.4	48.9	65.9	102.6	62.4	77.7
05	METALS	96.5	95.4	96.1	99.4	93.1	97.1	96.7	70.6	87.0	96.7	69.5	86.6	95.9	68.4	85.7	95.8	68.3	85.6
06	WOOD & PLASTICS	93.2	67.9	79.9	93.2	66.6	79.2	101.0	47.1	72.7	103.9	60.5	81.1	97.7	65.1	80.6	113.1	59.9	85.2
07	THERMAL & MOISTURE PROTECTION	97.6	69.8	84.2	97.2	66.3	82.3	98.7	54.4	77.4	101.3	49.9	76.5	96.1	55.6	76.6	101.6	63.1	83.0
08	DOORS & WINDOWS	100.7	70.8	93.3	98.6	69.1	91.3	92.8	50.1	82.3	92.8	54.8	83.4	93.8	64.0	86.5	97.3	55.1	86.9
09200	PLASTER & GYPSUM BOARD	97.3	67.3	77.2	93.2	66.1	75.1	90.3	46.4	60.9	90.3	60.2	70.1	92.9	64.8	74.1	91.9	59.6	70.2
095,098	CEILINGS & ACOUSTICAL TREATMENT	94.9	67.3	76.2	94.7	66.1	75.3	99.3	46.4	63.4	99.3	60.2	72.7	90.1	64.8	72.9	104.0	59.6	73.8
09600	FLOORING	94.0	50.8	83.2	106.7	73.5	98.4	118.1	71.7	106.4	117.9	48.0	100.3	98.7	55.2	87.8	114.7	75.3	104.8
097,099	WALL FINISHES, PAINTS & COATINGS	96.9	63.5	77.2	112.9	57.0	80.0	95.6	61.6	75.5	95.6	42.7	64.4	93.7	47.9	66.7	91.7	57.5	71.6
09	FINISHES	93.3	61.9	76.6	103.2	65.7	83.3	101.5	53.1	75.8	101.6	54.0	76.3	92.9	58.8	74.8	95.6	61.7	77.6
10-14	TOTAL DIV. 10000 - 14000	100.0	76.2	94.9	100.0	74.8	94.6	100.0	71.2	93.8	100.0	70.9	93.7	100.0	71.7	93.9	100.0	76.0	94.8
15	MECHANICAL	100.0	71.8	87.3	99.8	74.9	88.6	100.0	50.9	77.9	100.0	57.5	80.8	99.8	61.9	82.7	100.0	65.0	84.2
16	ELECTRICAL	98.8	80.2	85.6	102.3	68.6	78.5	96.8	54.4	66.9	97.8	64.2	74.1	94.8	72.3	78.9	94.4	71.1	78.0
01-16	WEIGHTED AVERAGE	96.2	74.7	85.7	98.0	73.9	86.2	97.3	57.8	77.9	98.1	60.5	79.6	94.3	64.6	79.7	96.6	66.6	81.9

COST INDEXES

City Cost Indexes

TEXAS

DIVISION		CORPUS CHRISTI MAT.	INST.	TOTAL	DALLAS MAT.	INST.	TOTAL	EL PASO MAT.	INST.	TOTAL	FORT WORTH MAT.	INST.	TOTAL	HOUSTON MAT.	INST.	TOTAL	LAREDO MAT.	INST.	TOTAL
01590	EQUIPMENT RENTAL	.0	95.4	95.4	.0	94.8	94.8	.0	86.3	86.3	.0	86.3	86.3	.0	98.0	98.0	.0	86.8	86.8
02	SITE CONSTRUCTION	119.1	81.1	90.4	130.0	83.8	95.1	101.6	84.9	89.0	101.0	85.5	89.3	117.5	84.7	92.7	86.4	85.6	85.8
03100	CONCRETE FORMS & ACCESSORIES	100.6	43.0	50.2	97.8	60.2	65.0	97.7	52.6	58.3	99.6	59.8	64.8	91.2	71.4	73.9	89.3	43.1	48.9
03200	CONCRETE REINFORCEMENT	93.6	51.2	68.5	99.1	60.2	76.1	97.4	54.1	71.8	97.4	60.1	75.3	96.2	65.9	78.3	95.5	51.5	69.5
03300	CAST-IN-PLACE CONCRETE	107.1	50.4	83.8	96.6	59.4	81.3	97.2	40.6	73.9	93.3	55.0	77.6	89.4	72.5	82.5	82.5	57.9	72.4
03	CONCRETE	91.5	49.1	70.2	88.6	61.6	75.0	91.3	49.5	70.3	89.5	59.0	74.1	85.5	72.4	78.9	80.9	50.7	65.7
04	MASONRY	84.5	47.9	61.9	99.3	63.0	76.8	97.7	53.6	70.4	93.6	62.9	74.6	97.8	69.2	80.1	93.0	53.9	68.8
05	METALS	95.4	80.8	90.0	96.3	86.3	92.6	96.5	66.6	85.4	96.4	72.0	87.4	100.2	93.6	97.7	96.8	63.1	84.3
06	WOOD & PLASTICS	111.7	42.8	75.5	103.0	61.1	81.0	102.2	56.7	78.3	109.8	61.1	84.2	97.8	72.2	84.4	88.5	42.7	64.5
07	THERMAL & MOISTURE PROTECTION	99.0	50.0	75.3	95.0	62.7	79.4	98.3	56.0	77.9	99.2	58.1	79.4	93.2	71.3	82.7	95.0	54.3	75.4
08	DOORS & WINDOWS	100.9	43.7	86.8	105.6	58.3	94.0	92.8	51.3	82.6	87.2	58.4	80.1	107.0	70.2	97.9	94.5	42.5	81.7
09200	PLASTER & GYPSUM BOARD	94.1	41.7	59.0	93.5	60.8	71.6	90.3	56.2	67.4	90.3	60.8	70.5	89.9	72.0	77.9	90.3	41.7	57.7
095,098	CEILINGS & ACOUSTICAL TREATMENT	95.6	41.7	59.0	99.2	60.8	73.1	99.3	56.2	70.0	99.3	60.8	73.1	103.6	72.0	82.2	90.1	41.7	57.2
09600	FLOORING	111.1	54.5	96.9	106.8	66.9	96.7	118.1	71.7	106.4	153.1	53.5	128.1	102.3	73.6	95.1	95.5	54.5	85.2
097,099	WALL FINISHES, PAINTS & COATINGS	107.3	42.8	69.3	102.5	61.0	78.0	95.6	41.0	63.4	96.8	60.8	75.6	100.1	67.4	80.8	93.7	52.1	69.2
09	FINISHES	99.6	44.5	70.4	100.3	61.4	80.9	101.5	55.5	77.1	113.6	58.7	84.4	100.8	71.4	85.2	91.6	45.5	67.1
10-14	TOTAL DIV. 10000 - 14000	100.0	72.4	94.1	100.0	75.0	94.6	100.0	69.0	93.3	100.0	75.1	94.6	100.0	80.9	95.9	100.0	69.4	93.4
15	MECHANICAL	99.8	52.0	78.2	99.8	68.9	85.9	100.0	38.8	72.4	100.0	61.6	82.7	100.0	75.0	88.7	99.7	46.7	75.8
16	ELECTRICAL	91.7	53.5	64.7	97.7	69.7	77.9	95.7	59.6	70.2	95.6	67.9	76.0	96.2	72.0	79.1	94.8	65.3	74.0
01-16	WEIGHTED AVERAGE	97.5	55.9	77.1	99.6	69.0	84.6	97.3	55.9	77.0	97.5	65.4	81.7	99.0	75.5	87.5	94.2	56.6	75.7

TEXAS / UTAH

DIVISION		LUBBOCK MAT.	INST.	TOTAL	ODESSA MAT.	INST.	TOTAL	SAN ANTONIO MAT.	INST.	TOTAL	WACO MAT.	INST.	TOTAL	WICHITA FALLS MAT.	INST.	TOTAL	LOGAN MAT.	INST.	TOTAL
01590	EQUIPMENT RENTAL	.0	96.4	96.4	.0	86.3	86.3	.0	89.8	89.8	.0	86.3	86.3	.0	86.3	86.3	.0	100.6	100.6
02	SITE CONSTRUCTION	131.8	83.3	95.2	101.1	85.6	89.4	86.1	90.2	89.2	99.9	85.5	89.0	100.7	85.2	89.0	93.7	101.5	99.6
03100	CONCRETE FORMS & ACCESSORIES	98.5	46.5	53.0	97.2	44.4	51.0	89.4	59.2	63.0	99.2	44.8	51.6	99.2	47.6	54.1	101.8	66.4	70.9
03200	CONCRETE REINFORCEMENT	96.8	56.3	72.9	99.6	56.2	73.9	95.5	54.5	71.2	97.4	56.8	73.4	97.4	55.5	72.6	101.1	78.9	88.0
03300	CAST-IN-PLACE CONCRETE	95.0	53.0	77.8	94.8	48.6	75.9	80.9	63.5	73.8	85.6	56.8	73.8	91.5	52.5	75.5	93.7	77.3	87.0
03	CONCRETE	88.8	52.7	70.6	90.4	49.2	69.6	80.1	60.5	70.2	85.6	52.3	68.8	88.5	51.9	70.1	111.1	72.9	91.8
04	MASONRY	98.3	40.1	62.3	98.9	40.1	62.5	92.9	61.0	73.2	95.9	52.4	69.0	96.3	58.5	72.9	121.6	67.5	88.1
05	METALS	100.7	85.0	94.9	96.8	68.7	86.4	97.4	69.8	87.1	96.6	69.6	86.6	96.5	70.9	87.0	105.2	78.2	95.2
06	WOOD & PLASTICS	102.7	47.2	73.6	101.0	44.8	71.5	88.5	59.3	73.2	108.7	41.6	73.5	108.7	47.1	76.3	88.2	64.9	76.0
07	THERMAL & MOISTURE PROTECTION	90.7	52.0	72.0	98.7	47.1	73.8	95.0	66.8	81.4	99.5	51.8	76.5	99.5	56.0	78.5	103.6	73.4	89.0
08	DOORS & WINDOWS	104.2	48.1	90.4	92.8	46.3	81.4	96.4	58.6	87.1	87.2	43.3	76.4	87.2	50.0	78.0	89.2	65.0	83.3
09200	PLASTER & GYPSUM BOARD	90.6	46.3	60.9	90.3	43.9	59.2	90.3	58.7	69.1	90.3	40.7	57.1	90.3	46.3	60.8	89.8	63.7	72.3
095,098	CEILINGS & ACOUSTICAL TREATMENT	100.7	46.3	63.8	99.3	43.9	61.7	90.1	58.7	68.8	99.3	40.7	59.5	99.3	46.3	63.3	106.1	63.7	77.3
09600	FLOORING	109.2	45.4	93.2	118.1	44.8	99.7	95.5	58.7	86.3	153.3	41.6	125.2	153.8	73.7	133.7	103.1	67.2	94.1
097,099	WALL FINISHES, PAINTS & COATINGS	107.3	38.3	66.6	95.6	38.3	61.8	93.7	52.1	69.2	96.8	39.5	63.0	98.6	62.6	77.4	105.7	54.0	75.2
09	FINISHES	103.0	44.7	72.0	101.5	43.2	70.6	91.6	57.9	73.7	113.6	42.5	75.8	113.9	53.7	81.9	100.2	64.6	81.3
10-14	TOTAL DIV. 10000 - 14000	100.0	70.9	93.7	100.0	68.2	93.1	100.0	74.6	94.5	100.0	72.5	94.1	100.0	69.2	93.4	100.0	83.8	96.5
15	MECHANICAL	99.7	53.0	78.7	100.0	42.5	74.1	99.7	71.5	87.0	100.0	60.6	82.3	100.0	50.0	77.5	100.0	76.7	89.5
16	ELECTRICAL	95.3	51.2	64.2	96.8	47.6	62.1	94.8	65.4	74.0	96.8	65.0	74.4	101.6	62.5	74.0	93.2	75.0	80.4
01-16	WEIGHTED AVERAGE	99.7	56.3	78.4	97.4	51.4	74.8	94.4	67.2	81.1	97.2	59.6	78.7	98.0	59.4	79.0	101.4	75.5	88.7

UTAH / VERMONT / VIRGINIA

DIVISION		OGDEN MAT.	INST.	TOTAL	PROVO MAT.	INST.	TOTAL	SALT LAKE CITY MAT.	INST.	TOTAL	BURLINGTON MAT.	INST.	TOTAL	RUTLAND MAT.	INST.	TOTAL	ALEXANDRIA MAT.	INST.	TOTAL
01590	EQUIPMENT RENTAL	.0	100.6	100.6	.0	99.3	99.3	.0	100.6	100.6	.0	102.6	102.6	.0	102.6	102.6	.0	101.1	101.1
02	SITE CONSTRUCTION	81.3	101.5	96.6	90.1	99.5	97.2	80.9	101.5	96.4	80.5	100.5	95.6	80.5	100.5	95.6	113.4	85.6	92.4
03100	CONCRETE FORMS & ACCESSORIES	101.8	66.4	70.9	103.4	66.5	71.2	100.0	66.6	70.8	92.9	56.2	60.9	104.3	56.3	62.4	94.7	76.6	78.9
03200	CONCRETE REINFORCEMENT	101.1	78.9	88.0	109.4	78.9	91.3	99.0	78.9	87.1	110.6	59.8	80.6	110.6	59.8	80.6	84.0	84.0	84.0
03300	CAST-IN-PLACE CONCRETE	95.1	77.3	87.8	93.8	77.4	87.0	102.8	77.4	92.4	98.4	69.7	86.6	93.8	69.7	83.9	96.9	81.3	90.5
03	CONCRETE	100.2	72.9	86.4	110.4	72.9	91.5	119.6	72.9	96.1	103.5	62.1	82.7	102.0	62.1	81.9	99.6	80.8	90.1
04	MASONRY	114.7	67.5	85.5	127.4	67.5	90.4	133.3	67.5	92.6	96.4	67.3	78.4	84.2	67.3	73.7	85.4	75.2	79.1
05	METALS	105.6	78.2	95.4	102.6	78.3	93.5	104.4	78.3	94.7	98.7	70.7	88.3	98.0	70.8	87.9	95.3	98.3	96.4
06	WOOD & PLASTICS	88.2	64.9	76.0	89.8	64.9	76.7	88.4	64.9	76.1	89.6	54.6	71.2	102.1	54.6	77.2	98.1	76.4	86.7
07	THERMAL & MOISTURE PROTECTION	102.0	73.4	88.2	106.1	73.4	90.3	105.3	73.4	89.9	102.0	61.3	82.4	101.7	62.5	82.8	95.9	81.4	88.9
08	DOORS & WINDOWS	89.2	65.0	83.3	93.6	65.0	86.5	89.2	65.0	83.3	107.1	50.7	93.2	107.1	50.7	93.2	96.0	78.1	91.6
09200	PLASTER & GYPSUM BOARD	89.8	63.7	72.3	89.1	63.7	72.1	89.8	63.7	72.3	104.8	52.2	69.5	104.8	52.2	69.5	110.3	75.5	87.0
095,098	CEILINGS & ACOUSTICAL TREATMENT	106.1	63.7	77.3	106.1	63.7	77.3	106.1	63.7	77.3	107.9	52.2	71.3	107.9	52.2	71.3	91.8	75.5	80.7
09600	FLOORING	103.1	67.2	94.1	103.9	67.2	94.7	102.7	67.2	93.8	101.6	77.6	95.6	101.6	77.6	95.6	100.9	85.1	96.9
097,099	WALL FINISHES, PAINTS & COATINGS	105.7	54.0	75.2	105.7	67.1	82.9	105.7	67.1	82.9	94.4	44.7	65.1	94.4	44.7	65.1	119.1	87.5	100.5
09	FINISHES	99.1	64.6	80.7	99.7	66.1	81.9	99.4	66.1	81.7	100.8	58.5	78.3	100.7	58.5	78.3	98.9	79.2	88.4
10-14	TOTAL DIV. 10000 - 14000	100.0	83.8	96.5	100.0	83.8	96.5	100.0	83.9	96.5	100.0	92.2	98.3	100.0	92.2	98.3	100.0	86.8	97.2
15	MECHANICAL	100.0	76.7	89.5	100.0	76.8	89.5	100.2	76.8	89.6	100.1	70.1	86.6	100.1	70.1	86.6	100.0	85.6	93.5
16	ELECTRICAL	93.2	75.0	80.4	93.7	75.0	80.5	96.5	81.6	86.0	105.6	70.7	81.0	104.4	70.7	80.6	98.4	94.5	95.6
01-16	WEIGHTED AVERAGE	99.2	75.5	87.6	101.8	75.5	88.9	102.9	76.8	90.1	100.8	69.7	85.5	99.8	69.8	85.1	98.0	85.4	91.8

VIRGINIA

DIVISION		ARLINGTON			NEWPORT NEWS			NORFOLK			PORTSMOUTH			RICHMOND			ROANOKE		
		MAT.	INST.	TOTAL	MAT.	INST.	TOTAL	MAT.	INST.	TOTAL	MAT.	INST.	TOTAL	MAT.	INST.	TOTAL	MAT.	INST.	TOTAL
01590	EQUIPMENT RENTAL	.0	99.6	99.6	.0	104.7	104.7	.0	105.4	105.4	.0	104.6	104.6	.0	104.7	104.7	.0	99.6	99.6
02	SITE CONSTRUCTION	124.4	82.6	92.8	106.5	84.0	89.6	105.8	85.2	90.2	105.0	83.7	88.9	107.1	84.8	90.2	104.2	80.8	86.5
03100	CONCRETE FORMS & ACCESSORIES	91.7	73.5	75.8	98.4	51.4	57.3	103.1	51.5	58.0	85.1	51.4	55.6	99.2	62.0	66.7	98.1	39.9	47.3
03200	CONCRETE REINFORCEMENT	94.8	69.2	79.7	94.6	71.9	81.2	94.6	71.9	81.2	94.3	71.9	81.0	94.6	72.2	81.4	94.6	68.7	79.3
03300	CAST-IN-PLACE CONCRETE	94.3	77.1	87.2	94.9	57.4	79.5	97.7	57.5	81.2	94.0	59.3	79.7	100.9	59.8	84.1	106.6	51.0	83.8
03	CONCRETE	103.9	75.1	89.4	96.7	59.1	77.8	98.4	59.2	78.7	95.3	59.7	77.4	99.8	64.7	82.1	102.6	51.2	76.7
04	MASONRY	98.1	68.2	79.6	88.1	55.6	68.0	96.2	55.6	71.1	93.2	55.5	69.9	84.3	55.1	66.3	88.5	37.2	56.8
05	METALS	94.2	91.1	93.1	95.3	91.5	93.1	94.6	91.7	93.5	94.5	90.8	93.2	95.3	92.3	94.2	95.2	85.4	91.5
06	WOOD & PLASTICS	94.5	76.4	85.0	99.1	51.7	74.2	104.6	51.7	76.8	85.9	51.7	67.9	100.3	64.8	81.7	99.1	39.7	67.9
07	THERMAL & MOISTURE PROTECTION	96.7	75.2	86.3	95.7	51.3	74.3	95.5	51.3	74.2	95.6	51.4	74.3	95.2	56.1	76.3	95.6	45.1	71.3
08	DOORS & WINDOWS	94.0	73.9	89.0	96.0	54.1	85.7	96.0	56.9	86.4	96.1	56.9	86.4	96.0	60.2	87.2	96.0	45.9	83.7
09200	PLASTER & GYPSUM BOARD	106.6	75.5	85.8	110.3	49.1	69.3	110.3	49.1	69.3	101.6	49.1	66.4	110.3	62.7	78.4	110.3	37.6	61.6
095,098	CEILINGS & ACOUSTICAL TREATMENT	89.0	75.5	79.8	91.8	49.1	62.8	91.8	49.1	62.8	91.8	49.1	62.8	91.8	62.7	72.0	91.8	37.6	55.0
09600	FLOORING	99.3	76.1	93.5	100.9	44.4	86.7	100.7	44.4	86.5	92.5	44.4	80.4	100.7	70.5	93.1	100.9	39.8	85.5
097,099	WALL FINISHES, PAINTS & COATINGS	119.1	87.5	100.5	108.1	47.3	72.3	108.1	61.2	80.4	108.1	41.5	68.8	108.1	61.9	80.9	108.1	36.9	66.2
09	FINISHES	98.4	75.7	86.3	98.2	49.1	72.1	98.2	50.6	72.9	94.3	48.4	69.9	98.1	63.9	79.9	98.1	39.1	66.7
10 - 14	TOTAL DIV. 10000 - 14000	100.0	80.0	95.7	100.0	75.6	94.7	100.0	75.6	94.7	100.0	73.6	94.3	100.0	76.9	95.0	100.0	69.8	93.5
15	MECHANICAL	100.0	81.5	91.7	100.0	63.4	83.5	100.0	65.7	84.5	100.0	65.6	84.5	100.0	67.7	85.5	100.0	44.3	74.9
16	ELECTRICAL	94.9	89.7	91.3	98.3	59.6	71.0	98.3	61.7	72.5	95.5	61.7	71.6	99.3	69.3	78.1	98.3	38.1	55.8
01 - 16	WEIGHTED AVERAGE	98.9	80.5	89.8	97.5	63.6	80.9	98.1	64.9	81.8	96.8	64.4	80.9	97.8	69.3	83.8	98.1	50.6	74.8

WASHINGTON

DIVISION		EVERETT			RICHLAND			SEATTLE			SPOKANE			TACOMA			VANCOUVER		
		MAT.	INST.	TOTAL	MAT.	INST.	TOTAL	MAT.	INST.	TOTAL	MAT.	INST.	TOTAL	MAT.	INST.	TOTAL	MAT.	INST.	TOTAL
01590	EQUIPMENT RENTAL	.0	105.2	105.2	.0	89.8	89.8	.0	104.9	104.9	.0	89.8	89.8	.0	105.2	105.2	.0	98.1	98.1
02	SITE CONSTRUCTION	96.1	116.5	111.5	102.0	90.0	92.9	100.5	114.3	110.9	102.8	90.0	93.1	99.1	116.6	112.3	113.0	102.0	104.7
03100	CONCRETE FORMS & ACCESSORIES	109.9	101.0	102.1	116.1	83.8	87.9	93.8	101.4	100.4	123.5	83.9	88.9	93.7	101.4	100.4	98.2	99.8	99.6
03200	CONCRETE REINFORCEMENT	105.8	91.0	97.0	100.5	87.1	92.6	99.8	91.0	94.6	101.1	87.0	92.8	99.8	91.0	94.6	102.7	90.8	95.7
03300	CAST-IN-PLACE CONCRETE	100.2	108.3	103.5	115.6	85.2	103.1	105.2	108.6	106.6	119.7	85.2	105.6	103.1	108.6	105.3	115.1	102.5	109.9
03	CONCRETE	100.4	101.0	100.7	108.6	84.8	96.6	102.7	101.4	102.1	111.3	84.8	97.9	101.7	101.3	101.5	113.3	98.7	105.9
04	MASONRY	129.1	104.1	113.6	114.3	86.4	97.0	124.7	104.1	112.0	116.0	83.2	95.7	124.5	104.1	111.9	125.4	102.6	111.3
05	METALS	99.2	86.8	94.6	93.2	82.9	89.4	100.7	89.3	96.5	93.0	82.7	89.2	100.2	87.5	95.5	98.7	89.1	95.2
06	WOOD & PLASTICS	100.2	100.0	100.1	97.7	83.5	90.2	85.1	100.0	92.9	106.3	83.5	94.3	83.9	100.0	92.4	81.7	99.5	91.1
07	THERMAL & MOISTURE PROTECTION	108.0	99.0	103.7	173.1	82.1	129.2	107.8	100.2	104.2	170.4	81.8	127.7	107.6	100.6	104.2	110.7	93.7	102.5
08	DOORS & WINDOWS	98.3	96.9	98.0	112.1	78.1	103.7	100.5	96.9	99.6	111.8	77.9	103.5	98.9	96.9	98.4	95.8	96.0	95.9
09200	PLASTER & GYPSUM BOARD	103.6	99.8	101.1	128.8	82.9	98.0	97.8	99.8	99.2	131.9	82.9	99.0	100.9	99.8	100.2	102.5	99.8	100.7
095,098	CEILINGS & ACOUSTICAL TREATMENT	109.8	99.8	103.0	115.1	82.9	93.2	110.8	99.8	103.4	115.1	82.9	93.2	112.5	99.8	103.9	121.0	99.8	106.6
09600	FLOORING	119.9	99.2	114.7	110.7	48.7	95.1	111.6	99.2	108.5	114.3	77.9	105.1	112.3	105.3	110.5	118.0	89.7	110.9
097,099	WALL FINISHES, PAINTS & COATINGS	110.6	88.5	97.6	114.1	77.7	92.6	110.6	96.3	102.2	114.1	77.7	92.6	110.6	96.3	102.2	118.3	69.7	89.7
09	FINISHES	114.9	99.3	106.6	126.9	76.0	99.8	111.8	100.1	105.6	128.5	82.0	103.8	112.6	101.4	106.6	113.4	94.6	103.4
10 - 14	TOTAL DIV. 10000 - 14000	100.0	102.5	100.5	100.0	96.9	99.3	100.0	102.6	100.6	100.0	96.8	99.3	100.0	102.6	100.6	100.0	93.7	98.6
15	MECHANICAL	100.0	99.3	99.7	100.6	97.7	99.3	100.0	107.0	103.2	100.6	88.1	94.9	100.1	101.0	100.5	100.3	106.2	102.9
16	ELECTRICAL	106.0	92.1	96.2	104.6	90.1	94.4	105.6	104.9	105.1	94.9	85.0	88.0	105.6	100.9	102.3	118.3	104.9	108.9
01 - 16	WEIGHTED AVERAGE	103.5	99.2	101.4	107.7	87.6	97.8	103.6	103.1	103.4	107.6	85.2	96.6	103.2	101.5	102.4	105.6	100.2	102.9

WASHINGTON / WEST VIRGINIA / WISCONSIN

DIVISION		WASHINGTON YAKIMA			WEST VIRGINIA CHARLESTON			HUNTINGTON			PARKERSBURG			WHEELING			WISCONSIN EAU CLAIRE		
		MAT.	INST.	TOTAL	MAT.	INST.	TOTAL	MAT.	INST.	TOTAL	MAT.	INST.	TOTAL	MAT.	INST.	TOTAL	MAT.	INST.	TOTAL
01590	EQUIPMENT RENTAL	.0	105.2	105.2	.0	99.6	99.6	.0	99.6	99.6	.0	99.6	99.6	.0	99.6	99.6	.0	98.0	98.0
02	SITE CONSTRUCTION	102.7	115.4	112.3	101.9	85.3	89.3	103.3	85.8	90.1	108.2	85.2	90.9	108.8	85.1	90.9	84.4	102.6	98.2
03100	CONCRETE FORMS & ACCESSORIES	96.1	96.1	96.1	106.5	92.4	94.1	99.3	93.2	94.0	87.9	90.4	90.1	90.0	89.3	89.4	99.8	97.7	98.0
03200	CONCRETE REINFORCEMENT	102.2	89.9	94.9	94.6	83.1	87.8	94.6	92.5	93.4	93.2	91.3	92.1	92.6	93.7	93.2	101.8	101.5	101.6
03300	CAST-IN-PLACE CONCRETE	110.1	85.0	99.8	92.8	106.7	98.5	100.2	107.5	103.2	94.6	92.2	93.0	93.6	100.1	96.3	100.7	97.5	99.4
03	CONCRETE	107.3	90.5	98.8	96.2	96.3	96.2	99.4	98.7	99.1	99.5	92.0	95.7	99.6	94.7	97.1	96.4	98.4	97.4
04	MASONRY	116.3	79.4	93.5	86.5	90.4	88.9	88.0	94.1	91.8	76.2	85.9	82.2	99.3	90.4	93.8	92.3	97.1	95.3
05	METALS	99.1	82.5	92.9	95.3	99.6	96.9	95.4	103.3	98.3	94.2	102.0	97.1	94.3	103.7	97.8	93.8	99.2	95.8
06	WOOD & PLASTICS	86.3	100.0	93.5	107.6	91.5	99.1	99.1	91.6	95.2	88.3	89.1	88.7	90.2	88.4	89.2	115.1	97.5	105.9
07	THERMAL & MOISTURE PROTECTION	107.9	80.7	94.8	95.6	90.5	93.1	95.9	92.6	94.3	96.1	86.5	91.4	96.2	87.9	92.2	98.0	90.5	94.4
08	DOORS & WINDOWS	98.5	84.6	95.1	97.2	83.5	93.8	96.0	85.7	93.5	96.6	82.4	93.1	97.5	88.1	95.2	100.9	97.0	99.9
09200	PLASTER & GYPSUM BOARD	101.1	99.8	100.2	110.3	91.0	97.4	110.3	91.2	97.5	103.5	88.6	93.5	104.0	87.8	93.1	113.6	97.8	103.0
095,098	CEILINGS & ACOUSTICAL TREATMENT	108.4	99.8	102.6	91.8	91.0	91.3	91.8	91.2	91.4	89.0	88.6	88.7	89.0	87.8	88.2	106.9	97.8	100.7
09600	FLOORING	113.6	73.3	103.5	100.7	104.9	101.7	100.7	106.2	102.1	95.7	91.1	94.5	96.7	106.6	99.2	92.8	69.4	86.9
097,099	WALL FINISHES, PAINTS & COATINGS	110.6	77.7	91.2	108.1	92.9	99.1	108.1	94.8	100.3	108.1	100.7	103.7	108.1	89.4	97.1	91.2	76.0	82.2
09	FINISHES	112.5	90.4	100.8	98.2	95.0	96.5	98.0	95.9	96.9	95.5	91.7	93.4	95.9	92.4	94.0	100.8	89.9	95.0
10 - 14	TOTAL DIV. 10000 - 14000	100.0	98.9	99.8	100.0	100.4	100.1	100.0	100.9	100.2	100.0	100.1	100.0	100.0	99.2	99.8	100.0	93.9	98.7
15	MECHANICAL	100.1	97.6	99.0	100.0	87.4	94.3	100.0	87.9	94.6	100.0	91.7	96.3	100.0	89.1	95.1	100.3	90.7	96.0
16	ELECTRICAL	110.5	90.1	96.1	98.3	90.4	92.7	98.3	83.0	87.5	99.1	92.8	94.7	94.4	92.2	92.9	103.6	87.9	92.5
01 - 16	WEIGHTED AVERAGE	103.7	92.2	98.0	97.5	91.4	94.5	97.7	91.6	94.7	96.8	91.3	94.1	98.0	92.0	95.1	98.3	94.0	96.2

WISCONSIN

DIVISION		GREEN BAY MAT.	INST.	TOTAL	KENOSHA MAT.	INST.	TOTAL	LA CROSSE MAT.	INST.	TOTAL	MADISON MAT.	INST.	TOTAL	MILWAUKEE MAT.	INST.	TOTAL	RACINE MAT.	INST.	TOTAL
01590	EQUIPMENT RENTAL	.0	95.7	95.7	.0	97.9	97.9	.0	98.0	98.0	.0	100.1	100.1	.0	86.4	86.4	.0	100.1	100.1
02	SITE CONSTRUCTION	86.9	98.7	95.8	91.6	102.0	99.4	78.5	102.6	96.7	86.2	105.3	100.6	87.4	93.6	92.1	86.5	105.8	101.1
03100	CONCRETE FORMS & ACCESSORIES	118.4	97.4	100.1	113.8	102.2	103.7	82.7	97.7	95.8	102.0	95.6	96.4	104.8	109.5	108.9	101.6	102.2	102.1
03200	CONCRETE REINFORCEMENT	98.2	91.6	94.3	107.2	100.5	103.3	100.3	90.4	94.4	100.4	90.7	94.7	98.7	100.6	99.8	100.4	100.5	100.5
03300	CAST-IN-PLACE CONCRETE	104.1	99.8	102.4	116.5	101.9	110.5	90.5	97.1	93.2	105.0	103.1	104.3	105.0	107.7	106.1	105.0	101.6	103.6
03	CONCRETE	98.0	97.2	97.6	106.3	101.8	104.0	87.3	96.1	91.8	98.9	97.4	98.2	99.2	106.3	102.8	98.9	101.7	100.3
04	MASONRY	123.6	98.7	108.2	98.9	110.0	105.8	91.5	99.8	96.7	102.7	102.9	102.9	102.7	111.7	108.3	102.7	110.0	107.2
05	METALS	96.3	95.3	95.9	96.5	101.4	98.3	93.7	94.7	94.1	96.8	96.5	96.7	98.4	91.9	96.0	96.8	101.4	98.5
06	WOOD & PLASTICS	126.2	97.5	111.1	121.6	99.9	110.2	98.1	97.5	97.8	116.2	93.4	104.2	119.0	108.7	113.6	116.4	99.9	107.7
07	THERMAL & MOISTURE PROTECTION	100.5	89.5	95.4	96.7	104.0	100.2	97.3	90.9	94.2	96.0	99.3	97.6	95.3	106.7	100.7	96.7	103.5	100.0
08	DOORS & WINDOWS	99.2	90.4	97.0	96.7	102.1	98.0	100.8	86.8	97.4	101.6	95.5	100.1	103.9	106.9	104.6	101.6	102.1	101.7
09200	PLASTER & GYPSUM BOARD	98.3	97.8	98.0	99.2	100.3	99.9	106.2	97.8	100.6	111.0	93.7	99.4	111.0	109.2	109.8	111.0	100.3	103.8
095,098	CEILINGS & ACOUSTICAL TREATMENT	101.8	97.8	99.1	88.4	100.3	96.5	101.4	97.8	99.0	97.5	93.7	94.9	97.5	109.2	105.4	97.5	100.3	99.4
09600	FLOORING	111.4	108.6	110.7	110.9	112.8	111.4	85.5	110.1	91.7	94.1	114.4	99.2	96.9	113.6	101.1	94.1	112.8	98.8
097,099	WALL FINISHES, PAINTS & COATINGS	101.6	82.5	90.3	104.2	100.0	101.7	91.2	76.0	82.2	93.8	95.5	94.8	96.2	107.2	102.7	93.8	95.7	94.9
09	FINISHES	105.4	98.4	101.6	102.6	104.0	103.4	96.0	98.2	97.2	99.3	98.9	99.1	100.5	110.5	105.8	99.3	103.5	101.6
10 - 14	TOTAL DIV. 10000 - 14000	100.0	95.8	99.1	100.0	98.2	99.6	100.0	93.9	98.7	100.0	93.3	98.5	100.0	100.5	100.1	100.0	98.2	99.6
15	MECHANICAL	100.6	89.4	95.6	100.3	98.5	99.5	100.3	90.4	95.8	100.0	94.5	97.5	100.0	98.7	99.4	100.0	98.5	99.4
16	ELECTRICAL	94.9	82.9	86.4	92.3	93.5	93.2	104.1	83.2	89.4	93.2	92.8	92.9	92.0	99.7	97.4	90.1	97.7	95.5
01 - 16	WEIGHTED AVERAGE	100.7	93.0	96.9	99.5	100.7	100.1	96.3	93.3	94.9	98.9	97.2	98.1	99.5	102.2	100.8	98.7	101.7	100.2

WYOMING / CANADA

DIVISION		CASPER MAT.	INST.	TOTAL	CHEYENNE MAT.	INST.	TOTAL	ROCK SPRINGS MAT.	INST.	TOTAL	CALGARY, ALBERTA MAT.	INST.	TOTAL	EDMONTON, ALBERTA MAT.	INST.	TOTAL	HALIFAX, NOVA SCOTIA MAT.	INST.	TOTAL
01590	EQUIPMENT RENTAL	.0	101.6	101.6	.0	101.6	101.6	.0	101.6	101.6	.0	100.3	100.3	.0	100.3	100.3	.0	98.6	98.6
02	SITE CONSTRUCTION	82.2	101.0	96.4	82.0	101.0	96.3	85.7	100.5	96.9	108.7	97.2	100.0	110.1	97.2	100.4	97.5	96.3	96.6
03100	CONCRETE FORMS & ACCESSORIES	99.1	46.9	53.5	99.2	46.9	53.5	104.4	33.1	42.1	123.2	75.2	81.2	120.7	75.2	80.9	90.1	72.0	74.3
03200	CONCRETE REINFORCEMENT	107.8	52.1	74.9	101.1	52.3	72.2	109.9	51.9	75.6	161.4	62.9	103.1	161.4	62.9	103.1	148.2	53.1	91.9
03300	CAST-IN-PLACE CONCRETE	105.5	61.5	87.4	105.5	61.5	87.4	106.0	46.2	81.4	162.8	85.1	130.9	163.6	85.1	131.4	178.4	73.1	135.2
03	CONCRETE	106.0	53.7	79.7	105.2	53.8	79.3	106.9	42.4	74.4	146.4	77.0	111.4	146.6	77.0	111.5	152.1	69.5	110.5
04	MASONRY	108.4	42.6	67.7	107.4	44.9	68.8	175.4	34.8	88.4	177.9	75.6	114.6	177.8	75.6	114.6	159.9	75.5	107.7
05	METALS	104.9	64.4	89.9	106.2	64.6	90.8	105.0	63.1	89.4	100.9	82.9	94.2	100.9	82.9	94.2	100.8	75.7	91.5
06	WOOD & PLASTICS	97.1	45.4	70.0	96.4	45.4	69.6	103.2	32.2	65.9	114.0	74.2	93.1	110.4	74.2	91.4	84.8	71.4	77.8
07	THERMAL & MOISTURE PROTECTION	103.9	58.1	81.8	103.9	58.7	82.1	104.6	44.2	75.5	120.5	77.2	99.6	120.5	77.2	99.6	103.4	72.7	88.6
08	DOORS & WINDOWS	93.5	47.1	82.0	94.2	47.1	82.6	100.6	39.9	85.6	91.7	72.3	87.0	91.7	72.3	87.0	82.4	66.0	78.4
09200	PLASTER & GYPSUM BOARD	88.6	43.7	58.5	89.8	43.7	58.9	93.7	30.0	51.0	153.3	73.2	99.6	145.5	73.2	97.1	187.7	70.6	109.2
095,098	CEILINGS & ACOUSTICAL TREATMENT	100.5	43.7	61.9	106.1	43.7	63.7	100.5	30.0	52.6	122.7	73.2	89.1	99.3	73.2	81.6	99.3	70.6	79.8
09600	FLOORING	103.1	46.5	88.9	103.1	74.6	95.9	105.4	45.6	90.4	132.7	77.6	118.9	132.7	77.6	118.9	110.1	67.5	99.4
097,099	WALL FINISHES, PAINTS & COATINGS	105.7	61.9	79.9	105.6	61.9	79.8	105.7	37.6	65.6	109.3	74.0	88.5	109.4	74.0	88.6	109.4	65.7	83.6
09	FINISHES	98.1	47.6	71.3	99.2	53.3	74.8	99.5	35.0	65.3	124.4	75.7	98.5	119.3	75.7	96.1	115.9	70.6	91.9
10 - 14	TOTAL DIV. 10000 - 14000	100.0	76.8	95.0	100.0	76.8	95.0	100.0	55.9	90.5	140.0	92.8	129.8	140.0	92.8	129.8	140.0	69.1	124.7
15	MECHANICAL	100.0	64.2	83.9	100.0	48.7	76.9	100.0	34.4	70.4	100.1	67.8	85.5	100.1	67.8	85.5	99.6	75.0	88.5
16	ELECTRICAL	93.1	65.3	73.5	93.1	57.2	67.8	92.1	49.3	61.9	113.9	73.6	85.5	113.8	73.6	85.4	123.6	76.2	90.2
01 - 16	WEIGHTED AVERAGE	100.1	61.7	81.2	100.3	58.2	79.6	105.1	47.9	77.0	117.3	76.9	97.5	116.8	76.9	97.2	114.4	75.5	95.3

CANADA

DIVISION		HAMILTON, ONTARIO MAT.	INST.	TOTAL	KITCHENER, ONTARIO MAT.	INST.	TOTAL	LAVAL, QUEBEC MAT.	INST.	TOTAL	LONDON, ONTARIO MAT.	INST.	TOTAL	MONTREAL, QUEBEC MAT.	INST.	TOTAL	OSHAWA, ONTARIO MAT.	INST.	TOTAL
01590	EQUIPMENT RENTAL	.0	103.1	103.1	.0	103.0	103.0	.0	98.6	98.6	.0	103.1	103.1	.0	98.6	98.6	.0	103.0	103.0
02	SITE CONSTRUCTION	112.0	104.1	106.0	99.3	103.5	102.5	92.4	97.3	96.1	112.0	103.8	105.8	92.1	97.3	96.0	113.6	103.7	106.1
03100	CONCRETE FORMS & ACCESSORIES	122.6	94.3	97.9	116.6	86.4	90.2	131.9	88.2	93.7	128.3	89.5	94.4	131.9	88.2	93.7	124.2	90.4	94.7
03200	CONCRETE REINFORCEMENT	174.6	95.2	127.7	110.3	95.1	101.3	156.9	89.5	117.0	123.4	93.5	105.7	159.5	89.5	118.1	174.6	95.8	128.0
03300	CAST-IN-PLACE CONCRETE	157.2	101.6	134.4	146.7	81.8	120.1	140.1	95.3	121.7	157.2	99.0	133.3	137.5	95.3	120.2	169.7	87.6	136.0
03	CONCRETE	145.1	96.9	120.8	125.2	86.6	105.7	135.1	91.0	112.8	139.4	93.6	116.3	134.1	91.0	112.3	151.5	90.5	120.8
04	MASONRY	161.5	98.9	122.7	159.6	94.5	119.3	158.6	88.3	115.1	161.5	95.2	120.9	158.5	88.3	115.1	162.3	98.1	122.6
05	METALS	101.0	93.2	98.1	100.8	92.7	97.8	101.1	89.0	96.6	99.3	92.8	96.9	101.1	89.0	96.6	100.8	93.0	97.9
06	WOOD & PLASTICS	114.8	93.2	103.5	108.8	84.8	96.2	126.9	88.4	106.7	114.8	87.8	100.6	126.9	88.4	106.7	115.7	88.7	101.5
07	THERMAL & MOISTURE PROTECTION	106.8	94.7	101.0	105.3	90.0	97.9	104.8	91.8	98.5	115.2	92.7	104.4	104.8	91.8	98.6	106.8	90.6	99.0
08	DOORS & WINDOWS	91.7	92.6	92.0	83.2	86.3	83.9	91.7	78.0	88.4	92.6	88.6	91.6	91.7	78.0	88.4	90.8	89.8	90.6
09200	PLASTER & GYPSUM BOARD	208.9	93.1	131.3	163.0	84.4	110.3	157.9	88.1	111.1	208.9	87.5	127.5	157.9	88.1	111.1	170.5	88.4	115.5
095,098	CEILINGS & ACOUSTICAL TREATMENT	99.3	93.1	95.1	99.3	84.4	89.2	99.3	88.1	91.7	99.3	87.5	91.3	99.3	88.1	91.7	99.3	88.4	91.9
09600	FLOORING	132.7	99.2	124.3	128.3	99.2	121.0	132.7	99.9	124.5	134.8	99.2	125.9	132.7	99.9	124.5	132.7	101.7	124.9
097,099	WALL FINISHES, PAINTS & COATINGS	109.4	108.0	108.6	109.4	98.9	103.2	109.4	93.9	100.3	109.4	108.0	108.6	109.4	93.9	100.3	109.4	113.5	111.8
09	FINISHES	126.9	96.8	110.9	118.9	90.1	103.5	119.7	91.4	104.7	127.6	93.2	109.3	119.7	91.4	104.7	121.9	95.1	107.7
10 - 14	TOTAL DIV. 10000 - 14000	140.0	101.5	131.7	140.0	99.1	131.2	140.0	87.6	128.7	140.0	100.1	131.4	140.0	87.6	128.7	140.0	100.6	131.5
15	MECHANICAL	100.1	93.1	96.9	99.6	89.1	94.9	99.6	81.2	91.3	100.1	89.3	95.2	100.1	81.2	91.5	99.6	90.9	95.7
16	ELECTRICAL	127.1	102.3	109.6	127.0	98.7	107.1	123.6	79.8	92.7	127.1	98.9	107.2	123.6	79.8	92.7	127.1	99.5	107.6
01 - 16	WEIGHTED AVERAGE	117.1	97.5	107.4	112.0	92.9	102.6	114.1	86.8	100.7	116.5	94.6	105.8	114.1	86.8	100.7	117.2	95.1	106.3

City Cost Indexes

		CANADA																	
	DIVISION	OTTAWA, ONTARIO			QUEBEC, QUEBEC			REGINA, SASKATCHEWAN			SASKATOON, SASKATCHEWAN			ST CATHARINES, ONTARIO			ST JOHNS, NEWFOUNDLAND		
		MAT.	INST.	TOTAL	MAT.	INST.	TOTAL	MAT.	INST.	TOTAL	MAT.	INST.	TOTAL	MAT.	INST.	TOTAL	MAT.	INST.	TOTAL
01590	EQUIPMENT RENTAL	.0	103.0	103.0	.0	98.6	98.6	.0	98.6	98.6	.0	98.6	98.6	.0	100.4	100.4	.0	98.6	98.6
02	SITE CONSTRUCTION	112.3	103.7	105.8	93.9	97.3	96.4	110.6	95.3	99.1	105.5	95.4	97.9	99.8	99.4	99.5	113.5	95.0	99.6
03100	CONCRETE FORMS & ACCESSORIES	120.9	90.4	94.3	131.9	88.3	93.8	102.4	60.8	66.1	102.4	60.7	66.0	114.0	88.3	91.5	100.5	61.8	66.7
03200	CONCRETE REINFORCEMENT	174.6	93.5	126.7	148.2	89.5	113.5	113.9	70.3	88.1	119.0	70.3	90.2	111.3	95.2	101.7	148.2	55.0	93.1
03300	CAST-IN-PLACE CONCRETE	159.6	98.9	134.7	151.5	95.4	128.5	162.8	70.3	124.8	148.0	70.2	116.1	140.1	86.6	118.1	179.6	69.2	134.3
03	CONCRETE	146.2	94.0	119.9	139.7	91.0	115.2	132.6	66.5	99.3	125.7	66.5	95.9	121.8	89.1	105.3	166.0	64.0	114.6
04	MASONRY	161.7	95.4	120.7	159.5	88.3	115.5	166.6	65.8	104.2	166.8	65.8	104.3	159.1	91.7	117.4	160.0	65.1	101.3
05	METALS	101.0	92.7	98.0	101.1	89.2	96.7	98.2	76.0	90.0	98.5	75.9	90.1	98.0	92.8	96.0	102.2	73.3	91.5
06	WOOD & PLASTICS	114.0	89.6	101.1	126.9	88.4	106.7	94.5	59.4	76.1	92.9	59.4	75.3	106.3	89.4	97.4	93.9	60.9	76.6
07	THERMAL & MOISTURE PROTECTION	106.8	91.3	99.3	104.8	91.8	98.6	104.8	66.3	86.3	103.8	65.3	85.2	105.3	91.3	98.6	106.3	63.8	85.8
08	DOORS & WINDOWS	91.7	89.5	91.2	91.7	82.8	89.5	86.7	58.1	79.6	85.6	58.1	78.9	82.7	89.2	84.3	98.0	58.1	88.2
09200	PLASTER & GYPSUM BOARD	254.6	89.3	143.8	215.6	88.1	130.1	145.3	58.3	86.9	145.3	58.3	86.9	147.2	89.1	108.3	194.0	59.8	104.1
095,098	CEILINGS & ACOUSTICAL TREATMENT	99.3	89.3	92.5	99.3	88.1	91.7	96.6	58.3	70.5	96.6	58.3	70.5	96.6	89.1	91.5	102.1	59.8	73.4
09600	FLOORING	132.7	97.7	123.9	132.7	99.9	124.5	120.2	63.5	105.9	120.2	63.5	105.9	126.7	99.2	119.8	115.0	57.4	100.5
097,099	WALL FINISHES, PAINTS & COATINGS	109.4	101.0	104.4	109.4	93.9	100.3	109.4	68.4	85.2	109.4	58.4	79.3	109.4	101.1	104.5	109.4	63.3	82.2
09	FINISHES	132.7	93.0	111.6	127.1	91.4	108.1	114.6	61.8	86.6	114.3	60.7	85.8	115.7	92.0	103.1	120.0	61.1	88.7
10 - 14	TOTAL DIV. 10000 - 14000	140.0	97.9	130.9	140.0	87.6	128.7	140.0	66.6	124.2	140.0	66.6	124.2	140.0	75.4	126.1	140.0	67.4	124.4
15	MECHANICAL	100.1	89.6	95.4	100.1	81.2	91.6	99.6	67.5	85.1	99.6	67.5	85.1	99.6	87.3	94.1	99.6	63.9	83.5
16	ELECTRICAL	123.6	100.0	107.0	123.6	79.8	92.7	128.8	69.0	86.6	128.9	69.0	86.6	127.0	100.1	108.0	123.7	64.1	81.6
01 - 16	WEIGHTED AVERAGE	117.6	94.8	106.4	115.6	87.0	101.6	113.3	69.6	91.8	112.1	69.5	91.2	110.8	92.3	101.7	119.2	67.3	93.7

		CANADA																	
	DIVISION	THUNDER BAY, ONTARIO			TORONTO, ONTARIO			VANCOUVER, B C			WINDSOR, ONTARIO			WINNIPEG, MANITOBA					
		MAT.	INST.	TOTAL	MAT.	INST.	TOTAL	MAT.	INST.	TOTAL	MAT.	INST.	TOTAL	MAT.	INST.	TOTAL	MAT.	INST.	TOTAL
01590	EQUIPMENT RENTAL	.0	100.4	100.4	.0	103.1	103.1	.0	107.5	107.5	.0	100.4	100.4	.0	101.5	101.5	.0	.0	.0
02	SITE CONSTRUCTION	105.7	99.5	101.0	113.2	104.5	106.6	110.9	107.6	108.4	95.5	99.7	98.7	110.4	100.3	102.8	.0	.0	.0
03100	CONCRETE FORMS & ACCESSORIES	124.0	89.4	93.8	124.2	101.4	104.2	115.7	91.6	94.6	124.0	90.1	94.3	123.3	72.5	78.9	.0	.0	.0
03200	CONCRETE REINFORCEMENT	99.4	94.4	96.5	170.9	96.2	126.8	161.4	87.2	117.5	109.0	93.6	99.9	161.4	60.3	101.6	.0	.0	.0
03300	CAST-IN-PLACE CONCRETE	154.2	99.1	131.6	163.2	109.2	141.1	158.4	97.6	133.4	143.5	100.3	125.8	151.2	74.0	119.5	.0	.0	.0
03	CONCRETE	131.6	93.7	112.5	147.8	102.9	125.2	148.0	93.2	120.4	123.9	94.3	109.0	140.3	71.4	105.6	.0	.0	.0
04	MASONRY	160.0	92.6	118.3	185.4	107.5	137.2	175.2	96.1	126.3	159.4	99.5	122.3	165.5	68.6	105.6	.0	.0	.0
05	METALS	97.4	91.8	95.3	101.0	95.1	98.9	101.2	94.0	98.5	97.5	93.1	95.9	100.9	78.2	92.5	.0	.0	.0
06	WOOD & PLASTICS	115.7	89.8	102.1	115.7	99.9	107.4	113.5	89.5	100.9	115.7	88.2	101.3	113.8	73.5	92.6	.0	.0	.0
07	THERMAL & MOISTURE PROTECTION	105.8	90.5	98.4	107.1	101.9	104.6	114.0	94.3	104.5	105.4	94.5	100.1	104.6	73.6	89.7	.0	.0	.0
08	DOORS & WINDOWS	81.7	88.7	83.4	90.8	98.2	92.6	94.3	87.4	92.6	81.4	88.4	83.1	91.7	65.9	85.4	.0	.0	.0
09200	PLASTER & GYPSUM BOARD	179.8	89.5	119.3	170.5	100.0	123.2	152.1	89.2	109.9	173.0	87.9	116.0	157.7	72.8	100.8	.0	.0	.0
095,098	CEILINGS & ACOUSTICAL TREATMENT	95.2	89.5	91.4	99.3	100.0	99.8	99.3	89.2	92.5	95.2	87.9	90.3	99.3	72.8	81.3	.0	.0	.0
09600	FLOORING	132.7	57.1	113.7	132.7	105.1	125.8	132.7	96.2	123.5	132.7	99.9	124.5	132.7	70.5	117.1	.0	.0	.0
097,099	WALL FINISHES, PAINTS & COATINGS	109.4	103.3	105.8	112.3	113.5	113.0	109.3	107.4	108.2	109.4	102.8	105.5	109.4	60.1	80.4	.0	.0	.0
09	FINISHES	122.1	85.8	102.8	122.3	103.6	112.3	120.6	93.6	106.3	120.8	93.3	106.2	120.7	71.1	94.3	.0	.0	.0
10 - 14	TOTAL DIV. 10000 - 14000	140.0	76.6	126.3	140.0	104.0	132.3	140.0	97.3	130.8	140.0	77.9	126.6	140.0	69.2	124.7	.0	.0	.0
15	MECHANICAL	99.6	88.1	94.4	100.1	99.5	99.8	100.1	91.9	96.4	99.6	91.0	95.7	100.1	74.4	88.5	.0	.0	.0
16	ELECTRICAL	127.0	98.1	106.6	127.4	103.0	110.2	131.1	90.1	102.2	127.1	100.0	108.0	128.8	76.6	92.0	.0	.0	.0
01 - 16	WEIGHTED AVERAGE	112.8	91.9	102.6	118.3	102.0	110.3	118.4	94.0	106.4	111.3	94.7	103.2	116.1	75.7	96.2	.0	.0	.0

Location Factors

Costs shown in *Means cost data publications* are based on National Averages for materials and installation. To adjust these costs to a specific location, simply multiply the base cost by the factor and divide by 100 for that city. The data is arranged alphabetically by state and postal zip code numbers. For a city not listed, use the factor for a nearby city with similar economic characteristics.

STATE/ZIP	CITY	MAT.	INST.	TOTAL
ALABAMA				
350-352	Birmingham	96.5	78.6	87.7
354	Tuscaloosa	96.1	59.4	78.1
355	Jasper	97.3	55.0	76.5
356	Decatur	96.1	61.5	79.1
357-358	Huntsville	96.0	75.5	85.9
359	Gadsden	96.9	63.0	80.2
360-361	Montgomery	97.0	61.4	79.5
362	Anniston	95.9	49.8	73.3
363	Dothan	96.8	53.7	75.6
364	Evergreen	96.1	55.0	75.9
365-366	Mobile	96.9	65.0	81.3
367	Selma	96.4	55.6	76.4
368	Phenix City	97.2	60.5	79.2
369	Butler	96.5	53.7	75.5
ALASKA				
995-996	Anchorage	134.9	116.5	125.9
997	Fairbanks	131.0	118.9	125.1
998	Juneau	134.1	116.5	125.5
999	Ketchikan	146.1	116.5	131.6
ARIZONA				
850,853	Phoenix	100.1	76.0	88.3
852	Mesa/Tempe	99.8	71.1	85.7
855	Globe	100.7	66.3	83.8
856-857	Tucson	99.0	72.2	85.8
859	Show Low	100.8	66.8	84.1
860	Flagstaff	102.6	70.3	86.7
863	Prescott	100.0	68.4	84.5
864	Kingman	98.5	68.9	84.0
865	Chambers	98.6	66.1	82.7
ARKANSAS				
716	Pine Bluff	95.7	64.6	80.5
717	Camden	94.0	43.0	69.0
718	Texarkana	95.1	48.4	72.2
719	Hot Springs	93.1	42.4	68.2
720-722	Little Rock	96.3	66.3	81.6
723	West Memphis	95.4	58.7	77.4
724	Jonesboro	95.3	58.7	77.3
725	Batesville	94.1	53.4	74.1
726	Harrison	95.5	53.4	74.8
727	Fayetteville	92.6	37.0	65.3
728	Russellville	94.0	51.5	73.2
729	Fort Smith	96.5	60.1	78.6
CALIFORNIA				
900-902	Los Angeles	102.8	112.4	107.5
903-905	Inglewood	99.0	109.2	104.0
906-908	Long Beach	100.7	109.2	104.9
910-912	Pasadena	101.1	109.5	105.2
913-916	Van Nuys	105.1	109.0	107.0
917-918	Alhambra	103.7	109.5	106.6
919-921	San Diego	105.6	105.3	105.5
922	Palm Springs	101.5	106.2	103.8
923-924	San Bernardino	99.0	107.4	103.1
925	Riverside	103.6	110.1	106.8
926-927	Santa Ana	101.3	108.9	105.0
928	Anaheim	103.9	111.4	107.6
930	Oxnard	104.5	109.8	107.1
931	Santa Barbara	103.8	110.8	107.2
932-933	Bakersfield	103.5	107.9	105.7
934	San Luis Obispo	105.4	108.6	107.0
935	Mojave	102.0	105.6	103.7
936-938	Fresno	105.3	110.2	107.7
939	Salinas	106.3	116.0	111.1
940-941	San Francisco	112.2	135.5	123.6
942,956-958	Sacramento	107.3	114.8	111.0
943	Palo Alto	105.6	128.7	116.9
944	San Mateo	108.8	127.4	117.9
945	Vallejo	105.8	119.7	112.7
946	Oakland	110.8	125.9	118.2
947	Berkeley	110.3	125.2	117.6
948	Richmond	110.0	123.6	116.7
949	San Rafael	111.7	125.6	118.5
950	Santa Cruz	110.4	116.1	113.2

STATE/ZIP	CITY	MAT.	INST.	TOTAL
CALIFORNIA (CONT'D)				
951	San Jose	109.5	130.3	119.7
952	Stockton	105.0	111.0	108.0
953	Modesto	105.1	110.8	107.9
954	Santa Rosa	106.4	125.1	115.6
955	Eureka	107.6	105.9	106.7
959	Marysville	106.6	110.3	108.4
960	Redding	107.6	112.8	110.1
961	Susanville	107.6	112.8	110.2
COLORADO				
800-802	Denver	102.1	89.5	95.9
803	Boulder	100.0	68.7	84.6
804	Golden	102.6	86.2	94.6
805	Fort Collins	103.4	78.9	91.3
806	Greeley	100.3	68.7	84.8
807	Fort Morgan	100.9	81.5	91.4
808-809	Colorado Springs	102.0	83.6	93.0
810	Pueblo	102.2	80.6	91.6
811	Alamosa	104.9	68.6	87.1
812	Salida	104.9	68.6	87.1
813	Durango	105.5	67.0	86.6
814	Montrose	103.8	64.6	84.6
815	Grand Junction	107.0	61.4	84.6
816	Glenwood Springs	104.8	81.0	93.1
CONNECTICUT				
060	New Britain	101.6	109.7	105.6
061	Hartford	101.9	109.0	105.4
062	Willimantic	102.3	107.3	104.8
063	New London	99.0	109.7	104.3
064	Meriden	101.6	110.1	105.7
065	New Haven	103.6	110.0	106.7
066	Bridgeport	103.0	109.9	106.4
067	Waterbury	102.4	109.8	106.0
068	Norwalk	102.3	110.9	106.6
069	Stamford	102.5	115.1	108.6
D.C.				
200-205	Washington	99.7	91.6	95.7
DELAWARE				
197	Newark	99.3	105.4	102.3
198	Wilmington	98.7	105.4	102.0
199	Dover	99.3	105.4	102.3
FLORIDA				
320,322	Jacksonville	98.5	64.2	81.7
321	Daytona Beach	98.6	71.8	85.4
323	Tallahassee	98.8	52.3	76.0
324	Panama City	99.7	40.5	70.6
325	Pensacola	99.3	61.5	80.7
326,344	Gainesville	100.1	60.1	80.5
327-328,347	Orlando	100.3	65.8	83.4
329	Melbourne	100.7	75.7	88.4
330-332,340	Miami	98.0	70.6	84.5
333	Fort Lauderdale	97.8	70.2	84.3
334,349	West Palm Beach	96.7	66.9	82.1
335-336,346	Tampa	99.7	60.2	80.3
337	St. Petersburg	102.1	59.6	81.2
338	Lakeland	98.8	59.7	79.6
339,341	Fort Myers	98.1	57.1	78.0
342	Sarasota	99.8	57.4	79.0
GEORGIA				
300-303,399	Atlanta	96.3	82.6	89.6
304	Statesboro	96.5	47.8	72.6
305	Gainesville	95.2	61.7	78.7
306	Athens	94.4	66.5	80.7
307	Dalton	96.4	34.3	65.9
308-309	Augusta	94.9	57.1	76.4
310-312	Macon	95.8	62.8	79.6
313-314	Savannah	98.0	61.3	80.0
315	Waycross	97.8	46.3	72.5
316	Valdosta	97.4	51.5	74.9
317	Albany	97.4	57.8	77.9
318-319	Columbus	97.5	55.4	76.8

Location Factors

STATE/ZIP	CITY	MAT.	INST.	TOTAL
HAWAII				
967	Hilo	116.3	128.7	122.4
968	Honolulu	117.9	128.7	123.2
STATES & POSS.				
969	Guam	200.3	56.9	129.9
IDAHO				
832	Pocatello	101.8	81.7	91.9
833	Twin Falls	104.0	48.6	76.8
834	Idaho Falls	101.1	59.6	80.8
835	Lewiston	113.2	85.4	99.6
836-837	Boise	102.3	83.3	93.0
838	Coeur d'Alene	112.9	59.7	86.8
ILLINOIS				
600-603	North Suburban	99.1	121.5	110.1
604	Joliet	99.1	120.6	109.7
605	South Suburban	99.1	121.4	110.0
606	Chicago	99.5	128.2	113.6
609	Kankakee	95.5	110.5	102.9
610-611	Rockford	97.1	112.5	104.7
612	Rock Island	95.2	101.6	98.3
613	La Salle	96.6	105.1	100.8
614	Galesburg	96.3	105.3	100.7
615-616	Peoria	98.6	106.3	102.4
617	Bloomington	95.6	106.2	100.8
618-619	Champaign	99.5	104.6	102.0
620-622	East St. Louis	93.6	106.6	100.0
623	Quincy	95.6	99.9	97.7
624	Effingham	94.9	103.2	98.9
625	Decatur	96.0	100.2	98.1
626-627	Springfield	95.8	103.0	99.3
628	Centralia	92.9	104.0	98.4
629	Carbondale	92.6	102.0	97.2
INDIANA				
460	Anderson	97.2	87.2	92.3
461-462	Indianapolis	100.3	88.8	94.6
463-464	Gary	98.0	105.3	101.6
465-466	South Bend	96.2	88.4	92.4
467-468	Fort Wayne	97.4	85.7	91.7
469	Kokomo	95.6	86.9	91.3
470	Lawrenceburg	93.4	84.4	89.0
471	New Albany	95.1	79.9	87.6
472	Columbus	97.4	85.3	91.4
473	Muncie	97.5	85.1	91.4
474	Bloomington	99.2	85.5	92.5
475	Washington	95.7	88.9	92.4
476-477	Evansville	96.7	90.0	93.4
478	Terre Haute	97.5	90.8	94.2
479	Lafayette	97.1	84.0	90.7
IOWA				
500-503,509	Des Moines	98.1	85.2	91.8
504	Mason City	96.8	66.7	82.0
505	Fort Dodge	97.0	62.5	80.1
506-507	Waterloo	98.1	64.6	81.7
508	Creston	97.5	68.2	83.1
510-511	Sioux City	98.8	76.6	87.9
512	Sibley	97.9	55.9	77.3
513	Spencer	100.1	53.9	77.4
514	Carroll	96.5	59.4	78.3
515	Council Bluffs	99.6	77.4	88.7
516	Shenandoah	96.7	54.3	75.8
520	Dubuque	98.3	80.0	89.3
521	Decorah	97.8	57.6	78.0
522-524	Cedar Rapids	99.5	84.1	91.9
525	Ottumwa	97.8	73.0	85.6
526	Burlington	96.7	75.1	86.1
527-528	Davenport	98.3	95.3	96.9
KANSAS				
660-662	Kansas City	98.1	96.1	97.2
664-666	Topeka	97.4	71.0	84.5
667	Fort Scott	97.4	70.2	84.0
668	Emporia	97.2	63.4	80.6
669	Belleville	99.2	58.6	79.3
670-672	Wichita	96.7	74.3	85.7
673	Independence	98.9	56.2	77.9
674	Salina	98.6	60.7	80.0
675	Hutchinson	93.9	54.7	74.7
676	Hays	98.4	58.6	78.9
677	Colby	99.2	58.6	79.3

STATE/ZIP	CITY	MAT.	INST.	TOTAL
KANSAS (CONT'D)				
678	Dodge City	100.0	58.6	79.7
679	Liberal	98.3	48.7	73.9
KENTUCKY				
400-402	Louisville	95.9	85.4	90.8
403-405	Lexington	96.7	69.7	83.5
406	Frankfort	96.0	71.3	83.9
407-409	Corbin	94.7	47.3	71.4
410	Covington	95.6	95.4	95.5
411-412	Ashland	93.9	99.6	96.7
413-414	Campton	95.6	47.7	72.1
415-416	Pikeville	96.8	66.2	81.8
417-418	Hazard	94.9	47.4	71.6
420	Paducah	93.7	89.3	91.5
421-422	Bowling Green	95.5	83.3	89.5
423	Owensboro	95.6	78.4	87.1
424	Henderson	93.2	90.6	91.9
425-426	Somerset	92.5	49.1	71.2
427	Elizabethtown	92.1	85.4	88.8
LOUISIANA				
700-701	New Orleans	102.3	70.5	86.7
703	Thibodaux	102.6	66.8	85.1
704	Hammond	99.3	65.0	82.5
705	Lafayette	101.5	57.6	79.9
706	Lake Charles	101.6	61.7	82.0
707-708	Baton Rouge	101.3	58.1	80.1
710-711	Shreveport	96.5	62.2	79.7
712	Monroe	96.5	58.8	78.0
713-714	Alexandria	96.7	55.9	76.7
MAINE				
039	Kittery	97.3	73.6	85.7
040-041	Portland	99.9	81.7	91.0
042	Lewiston	100.5	81.7	91.2
043	Augusta	99.2	72.5	86.1
044	Bangor	99.6	81.2	90.6
045	Bath	98.7	75.2	87.2
046	Machias	98.0	72.2	85.4
047	Houlton	98.2	76.3	87.5
048	Rockland	97.1	70.9	84.2
049	Waterville	98.6	67.5	83.3
MARYLAND				
206	Waldorf	97.8	75.3	86.7
207-208	College Park	97.7	81.8	89.9
209	Silver Spring	97.0	79.9	88.6
210-212	Baltimore	97.4	84.8	91.2
214	Annapolis	96.9	81.1	89.2
215	Cumberland	94.2	79.4	87.0
216	Easton	95.9	48.0	72.4
217	Hagerstown	94.4	80.5	87.6
218	Salisbury	96.4	55.2	76.2
219	Elkton	93.2	70.9	82.2
MASSACHUSETTS				
010-011	Springfield	100.5	100.6	100.5
012	Pittsfield	100.3	95.7	98.1
013	Greenfield	98.5	98.2	98.4
014	Fitchburg	96.9	107.7	102.2
015-016	Worcester	100.4	111.8	106.0
017	Framingham	96.5	113.3	104.8
018	Lowell	100.0	115.3	107.5
019	Lawrence	101.3	113.2	107.1
020-022, 024	Boston	102.8	127.7	115.0
023	Brockton	101.4	111.5	106.3
025	Buzzards Bay	95.8	109.7	102.6
026	Hyannis	98.2	109.7	103.8
027	New Bedford	100.9	110.4	105.5
MICHIGAN				
480,483	Royal Oak	94.5	109.9	102.1
481	Ann Arbor	96.9	109.6	103.2
482	Detroit	98.2	117.3	107.6
484-485	Flint	96.6	101.2	98.8
486	Saginaw	96.7	96.9	96.8
487	Bay City	96.4	97.0	96.7
488-489	Lansing	97.1	100.3	98.7
490	Battle Creek	96.7	92.3	94.5
491	Kalamazoo	97.1	89.4	93.3
492	Jackson	95.3	96.7	96.0
493,495	Grand Rapids	96.9	75.0	86.1
494	Muskegon	95.7	87.6	91.7

STATE/ZIP	CITY	MAT.	INST.	TOTAL
MICHIGAN (CONT'D)				
496	Traverse City	94.7	78.0	86.5
497	Gaylord	96.2	82.2	89.3
498-499	Iron Mountain	98.2	91.4	94.9
MINNESOTA				
550-551	Saint Paul	99.2	124.5	111.6
553-555	Minneapolis	99.9	126.9	113.2
556-558	Duluth	99.1	107.6	103.3
559	Rochester	98.8	105.5	102.1
560	Mankato	97.3	101.4	99.3
561	Windom	96.0	83.6	89.9
562	Willmar	95.3	90.6	93.0
563	St. Cloud	96.1	115.0	105.4
564	Brainerd	97.2	101.3	99.2
565	Detroit Lakes	99.4	97.2	98.3
566	Bemidji	98.6	98.6	98.6
567	Thief River Falls	97.6	90.2	94.0
MISSISSIPPI				
386	Clarksdale	96.8	32.0	65.0
387	Greenville	100.5	49.0	75.2
388	Tupelo	98.5	40.7	70.1
389	Greenwood	98.3	34.4	66.9
390-392	Jackson	98.6	50.6	75.0
393	Meridian	96.7	51.1	74.3
394	Laurel	98.4	35.5	67.5
395	Biloxi	98.7	57.9	78.7
396	McComb	96.7	55.8	76.7
397	Columbus	98.3	40.1	69.8
MISSOURI				
630-631	St. Louis	96.2	108.6	102.3
633	Bowling Green	96.0	92.5	94.3
634	Hannibal	94.8	89.1	92.0
635	Kirksville	97.4	80.9	89.3
636	Flat River	97.1	95.8	96.5
637	Cape Girardeau	96.2	92.8	94.5
638	Sikeston	94.8	89.8	92.3
639	Poplar Bluff	94.3	89.2	91.8
640-641	Kansas City	99.1	104.4	101.7
644-645	St. Joseph	98.9	93.5	96.3
646	Chillicothe	96.3	73.9	85.3
647	Harrisonville	95.8	101.5	98.6
648	Joplin	97.8	71.2	84.8
650-651	Jefferson City	95.1	86.9	91.1
652	Columbia	96.7	88.4	92.6
653	Sedalia	96.2	87.4	91.9
654-655	Rolla	94.9	80.3	87.7
656-658	Springfield	97.9	78.1	88.2
MONTANA				
590-591	Billings	101.3	77.9	89.8
592	Wolf Point	102.1	74.2	88.4
593	Miles City	99.9	74.8	87.6
594	Great Falls	102.7	76.6	89.9
595	Havre	100.8	74.4	87.8
596	Helena	102.6	75.8	89.4
597	Butte	101.1	75.7	88.7
598	Missoula	99.2	74.3	87.0
599	Kalispell	98.6	73.6	86.3
NEBRASKA				
680-681	Omaha	100.4	80.3	90.6
683-685	Lincoln	98.0	71.4	84.9
686	Columbus	96.9	54.6	76.1
687	Norfolk	99.3	66.8	83.3
688	Grand Island	97.8	69.9	84.1
689	Hastings	97.6	62.1	80.2
690	Mccook	97.3	50.7	74.4
691	North Platte	97.2	63.5	80.7
692	Valentine	100.9	42.8	72.4
693	Alliance	100.3	38.9	70.2
NEVADA				
889-891	Las Vegas	101.0	106.8	103.8
893	Ely	102.2	86.0	94.3
894-895	Reno	102.0	96.6	99.3
897	Carson City	101.8	96.5	99.2
898	Elko	100.8	85.6	93.3
NEW HAMPSHIRE				
030	Nashua	100.8	84.0	92.6
031	Manchester	101.3	84.0	92.8

STATE/ZIP	CITY	MAT.	INST.	TOTAL
NEW HAMPSHIRE (CONT'D)				
032-033	Concord	98.8	84.0	91.6
034	Keene	98.0	55.8	77.3
035	Littleton	98.1	62.5	80.6
036	Charleston	97.5	52.6	75.4
037	Claremont	96.6	52.6	75.0
038	Portsmouth	98.1	79.9	89.2
NEW JERSEY				
070-071	Newark	102.4	116.9	109.5
072	Elizabeth	101.4	111.4	106.3
073	Jersey City	99.9	117.7	108.7
074-075	Paterson	102.7	116.5	109.5
076	Hackensack	99.9	117.2	108.4
077	Long Branch	99.5	113.0	106.1
078	Dover	100.2	117.0	108.4
079	Summit	100.3	110.2	105.2
080,083	Vineland	97.7	113.7	105.6
081	Camden	99.8	112.8	106.2
082,084	Atlantic City	98.7	114.1	106.3
085-086	Trenton	100.6	114.7	107.5
087	Point Pleasant	100.1	112.6	106.3
088-089	New Brunswick	100.5	116.1	108.2
NEW MEXICO				
870-872	Albuquerque	101.0	77.9	89.7
873	Gallup	101.8	77.9	90.1
874	Farmington	101.8	77.9	90.1
875	Santa Fe	100.0	77.9	89.1
877	Las Vegas	100.0	77.9	89.1
878	Socorro	99.5	77.9	88.9
879	Truth/Consequences	99.3	72.5	86.2
880	Las Cruces	97.4	70.0	84.0
881	Clovis	100.2	77.3	88.9
882	Roswell	101.7	77.3	89.7
883	Carrizozo	102.5	77.9	90.4
884	Tucumcari	101.1	77.3	89.4
NEW YORK				
100-102	New York	107.1	159.1	132.6
103	Staten Island	103.3	156.8	129.6
104	Bronx	100.6	156.9	128.3
105	Mount Vernon	101.5	136.9	118.9
106	White Plains	101.2	136.6	118.6
107	Yonkers	106.0	136.0	120.7
108	New Rochelle	102.2	136.9	119.2
109	Suffern	102.0	126.0	113.8
110	Queens	102.2	157.0	129.1
111	Long Island City	104.1	157.0	130.1
112	Brooklyn	104.3	156.5	129.9
113	Flushing	105.0	157.0	130.5
114	Jamaica	103.0	157.0	129.5
115,117,118	Hicksville	102.1	147.7	124.5
116	Far Rockaway	105.1	157.0	130.6
119	Riverhead	102.6	149.6	125.7
120-122	Albany	97.6	93.8	95.7
123	Schenectady	98.2	93.8	96.0
124	Kingston	102.7	116.4	109.4
125-126	Poughkeepsie	101.8	120.7	111.1
127	Monticello	100.9	114.9	107.8
128	Glens Falls	92.8	89.5	91.2
129	Plattsburgh	98.4	85.4	92.0
130-132	Syracuse	99.0	91.8	95.5
133-135	Utica	97.3	87.3	92.4
136	Watertown	98.9	86.6	92.9
137-139	Binghamton	98.3	87.3	92.9
140-142	Buffalo	99.5	104.1	101.8
143	Niagara Falls	97.8	106.4	102.0
144-146	Rochester	100.6	98.0	99.3
147	Jamestown	97.0	89.7	93.5
148-149	Elmira	96.3	87.1	91.8
NORTH CAROLINA				
270,272-274	Greensboro	96.9	53.5	75.6
271	Winston-Salem	97.0	52.2	75.0
275-276	Raleigh	97.2	53.6	75.7
277	Durham	96.8	53.6	75.5
278	Rocky Mount	95.5	38.6	67.5
279	Elizabeth City	96.2	41.0	69.1
280	Gastonia	97.0	50.5	74.1
281-282	Charlotte	97.2	50.8	74.4
283	Fayetteville	95.4	53.6	74.8
284	Wilmington	95.5	52.6	74.5
285	Kinston	94.0	38.1	66.6

STATE/ZIP	CITY	MAT.	INST.	TOTAL
NORTH CAROLINA (CONT'D)				
286	Hickory	94.2	37.3	66.3
287-288	Asheville	95.9	51.5	74.1
289	Murphy	95.6	36.3	66.5
NORTH DAKOTA				
580-581	Fargo	101.7	67.3	84.8
582	Grand Forks	101.6	58.7	80.6
583	Devils Lake	101.8	62.9	82.7
584	Jamestown	101.7	57.5	80.0
585	Bismarck	100.4	69.3	85.1
586	Dickinson	102.4	66.0	84.5
587	Minot	101.7	71.0	86.6
588	Williston	100.8	65.1	83.3
OHIO				
430-432	Columbus	98.4	93.1	95.8
433	Marion	95.5	92.5	94.0
434-436	Toledo	98.7	103.0	100.8
437-438	Zanesville	95.7	87.8	91.8
439	Steubenville	97.6	96.8	97.2
440	Lorain	97.9	95.9	96.9
441	Cleveland	98.2	107.5	102.7
442-443	Akron	99.1	99.8	99.4
444-445	Youngstown	98.3	94.9	96.6
446-447	Canton	98.4	91.5	95.0
448-449	Mansfield	96.0	91.2	93.6
450	Hamilton	95.3	91.9	93.6
451-452	Cincinnati	95.5	93.2	94.4
453-454	Dayton	95.3	89.2	92.3
455	Springfield	95.4	89.7	92.6
456	Chillicothe	94.9	96.6	95.7
457	Athens	98.3	80.5	89.6
458	Lima	98.9	88.9	94.0
OKLAHOMA				
730-731	Oklahoma City	98.0	67.1	82.8
734	Ardmore	95.7	66.4	81.3
735	Lawton	97.4	68.0	83.0
736	Clinton	97.3	64.7	81.3
737	Enid	97.3	64.6	81.3
738	Woodward	95.8	64.7	80.6
739	Guymon	97.1	35.8	67.0
740-741	Tulsa	97.4	64.4	81.2
743	Miami	94.5	69.2	82.1
744	Muskogee	96.5	46.5	71.9
745	Mcalester	94.0	58.1	76.4
746	Ponca City	94.7	65.1	80.1
747	Durant	94.7	65.3	80.2
748	Shawnee	96.4	63.3	80.2
749	Poteau	93.7	66.3	80.2
OREGON				
970-972	Portland	103.2	106.5	104.8
973	Salem	103.7	104.9	104.3
974	Eugene	103.2	103.9	103.5
975	Medford	104.7	102.3	103.5
976	Klamath Falls	105.7	102.3	104.0
977	Bend	104.5	104.0	104.3
978	Pendleton	98.4	101.8	100.1
979	Vale	95.8	95.6	95.7
PENNSYLVANIA				
150-152	Pittsburgh	96.7	103.9	100.2
153	Washington	94.1	103.1	98.5
154	Uniontown	94.3	100.7	97.4
155	Bedford	95.4	93.3	94.4
156	Greensburg	95.5	100.9	98.1
157	Indiana	94.2	98.8	96.4
158	Dubois	95.7	95.2	95.5
159	Johnstown	95.3	97.7	96.5
160	Butler	92.4	101.2	96.7
161	New Castle	92.4	101.5	96.8
162	Kittanning	93.0	102.8	97.8
163	Oil City	92.3	95.8	94.0
164-165	Erie	94.3	96.8	95.6
166	Altoona	94.2	95.1	94.6
167	Bradford	95.9	95.2	95.5
168	State College	95.6	95.3	95.4
169	Wellsboro	96.6	88.6	92.7
170-171	Harrisburg	97.9	91.1	94.6
172	Chambersburg	96.2	89.4	92.9
173-174	York	95.8	91.2	93.6
175-176	Lancaster	94.8	88.4	91.6

STATE/ZIP	CITY	MAT.	INST.	TOTAL
PENNSYLVANIA (CONT'D)				
177	Williamsport	93.0	83.7	88.5
178	Sunbury	95.5	89.9	92.8
179	Pottsville	94.5	90.0	92.3
180	Lehigh Valley	96.2	105.6	100.8
181	Allentown	97.9	103.5	100.7
182	Hazleton	95.5	95.6	95.5
183	Stroudsburg	95.4	100.7	98.0
184-185	Scranton	98.5	95.3	96.9
186-187	Wilkes-Barre	95.1	94.0	94.6
188	Montrose	94.7	94.7	94.7
189	Doylestown	95.0	115.3	105.0
190-191	Philadelphia	100.0	123.3	111.4
193	Westchester	97.1	114.4	105.6
194	Norristown	95.8	118.3	106.9
195-196	Reading	98.0	94.9	96.4
PUERTO RICO				
009	San Juan	140.5	27.9	85.2
RHODE ISLAND				
028	Newport	99.6	104.7	102.1
029	Providence	99.9	104.7	102.3
SOUTH CAROLINA				
290-292	Columbia	96.2	51.1	74.1
293	Spartanburg	95.0	48.8	72.3
294	Charleston	96.2	51.7	74.3
295	Florence	94.6	48.2	71.8
296	Greenville	94.8	48.8	72.2
297	Rock Hill	94.9	35.7	65.9
298	Aiken	96.0	72.8	84.6
299	Beaufort	96.6	39.6	68.7
SOUTH DAKOTA				
570-571	Sioux Falls	100.7	59.7	80.6
572	Watertown	99.8	54.6	77.6
573	Mitchell	98.7	54.8	77.1
574	Aberdeen	101.0	58.3	80.0
575	Pierre	100.3	56.8	78.9
576	Mobridge	99.3	54.7	77.4
577	Rapid City	100.9	54.8	78.3
TENNESSEE				
370-372	Nashville	98.0	73.9	86.2
373-374	Chattanooga	99.1	59.8	79.8
375,380-381	Memphis	96.2	74.7	85.7
376	Johnson City	98.7	61.5	80.4
377-379	Knoxville	95.1	63.0	79.3
382	Mckenzie	97.3	53.3	75.7
383	Jackson	98.9	57.2	78.4
384	Columbia	95.7	60.6	78.5
385	Cookeville	97.1	55.0	76.4
TEXAS				
750	Mckinney	99.4	60.6	80.4
751	Waxahackie	99.4	61.6	80.8
752-753	Dallas	99.6	69.0	84.6
754	Greenville	99.6	43.7	72.1
755	Texarkana	98.8	51.5	75.6
756	Longview	99.2	45.1	72.7
757	Tyler	99.9	57.9	79.3
758	Palestine	95.6	45.4	70.9
759	Lufkin	96.5	51.2	74.3
760-761	Fort Worth	97.5	65.4	81.7
762	Denton	97.7	56.5	77.5
763	Wichita Falls	98.0	59.4	79.0
764	Eastland	97.1	44.3	71.1
765	Temple	95.6	53.1	74.7
766-767	Waco	97.2	59.6	78.7
768	Brownwood	98.0	41.1	70.1
769	San Angelo	97.6	49.8	74.1
770-772	Houston	99.0	75.5	87.5
773	Huntsville	97.8	44.3	71.5
774	Wharton	99.3	49.2	74.7
775	Galveston	97.1	73.7	85.6
776-777	Beaumont	96.6	66.6	81.9
778	Bryan	94.3	68.1	81.4
779	Victoria	99.5	52.6	76.5
780	Laredo	94.2	56.6	75.7
781-782	San Antonio	94.4	67.2	81.1
783-784	Corpus Christi	97.5	55.9	77.1
785	Mc Allen	98.0	49.6	74.3
786-787	Austin	94.3	64.6	79.7

STATE/ZIP	CITY	MAT.	INST.	TOTAL
TEXAS (CONT'D)				
788	Del Rio	97.4	34.4	66.4
789	Giddings	94.4	44.3	69.8
790-791	Amarillo	98.1	60.5	79.6
792	Childress	97.6	55.3	76.8
793-794	Lubbock	99.7	56.3	78.4
795-796	Abilene	97.3	57.8	77.9
797	Midland	100.6	52.9	77.2
798-799,885	El Paso	97.3	55.9	77.0
UTAH				
840-841	Salt Lake City	102.9	76.8	90.1
842,844	Ogden	99.2	75.5	87.6
843	Logan	101.4	75.5	88.7
845	Price	102.2	54.6	78.8
846-847	Provo	101.8	75.5	88.9
VERMONT				
050	White River Jct.	99.7	50.7	75.6
051	Bellows Falls	97.9	52.9	75.8
052	Bennington	98.3	53.3	76.2
053	Brattleboro	98.8	53.5	76.5
054	Burlington	100.8	69.7	85.5
056	Montpelier	97.9	69.8	84.1
057	Rutland	99.8	69.8	85.1
058	St. Johnsbury	99.9	49.8	75.3
059	Guildhall	98.2	49.3	74.2
VIRGINIA				
220-221	Fairfax	98.1	81.4	89.9
222	Arlington	98.9	80.5	89.8
223	Alexandria	98.0	85.4	91.8
224-225	Fredericksburg	96.6	69.9	83.4
226	Winchester	97.4	58.1	78.1
227	Culpeper	97.1	59.5	78.7
228	Harrisonburg	97.5	51.2	74.8
229	Charlottesville	97.8	63.7	81.1
230-232	Richmond	97.8	69.3	83.8
233-235	Norfolk	98.1	64.9	81.8
236	Newport News	97.5	63.6	80.9
237	Portsmouth	96.8	64.4	80.9
238	Petersburg	97.5	69.3	83.7
239	Farmville	96.9	45.5	71.7
240-241	Roanoke	98.1	50.6	74.8
242	Bristol	96.6	50.7	74.1
243	Pulaski	96.3	44.9	71.1
244	Staunton	97.1	49.6	73.8
245	Lynchburg	97.2	53.7	75.8
246	Grundy	96.5	44.9	71.2
WASHINGTON				
980-981,987	Seattle	103.6	103.1	103.4
982	Everett	103.5	99.2	101.4
983-984	Tacoma	103.2	101.5	102.4
985	Olympia	103.1	101.3	102.2
986	Vancouver	105.6	100.2	102.9
988	Wenatchee	104.4	87.7	96.2
989	Yakima	103.7	92.2	98.0
990-992	Spokane	107.6	85.2	96.6
993	Richland	107.7	87.6	97.8
994	Clarkston	106.4	86.1	96.4
WEST VIRGINIA				
247-248	Bluefield	95.4	81.4	88.5
249	Lewisburg	97.0	84.8	91.0
250-253	Charleston	97.5	91.4	94.5
254	Martinsburg	96.9	69.5	83.4
255-257	Huntington	97.7	91.6	94.7
258-259	Beckley	95.2	90.6	92.9
260	Wheeling	98.0	92.0	95.0
261	Parkersburg	96.8	91.3	94.1
262	Buckhannon	96.8	90.8	93.9
263-264	Clarksburg	97.3	91.2	94.3
265	Morgantown	97.4	91.1	94.3
266	Gassaway	96.6	92.1	94.4
267	Romney	96.5	84.8	90.8
268	Petersburg	96.5	86.6	91.6
WISCONSIN				
530,532	Milwaukee	99.5	102.2	100.8
531	Kenosha	99.5	100.7	100.1
534	Racine	98.7	101.7	100.2
535	Beloit	98.6	98.3	98.5
537	Madison	98.9	97.2	98.1

STATE/ZIP	CITY	MAT.	INST.	TOTAL
WISCONSIN (CONT'D)				
538	Lancaster	97.0	90.7	93.9
539	Portage	95.3	97.2	96.2
540	New Richmond	96.8	95.9	96.4
541-543	Green Bay	100.7	93.0	96.9
544	Wausau	95.8	93.2	94.5
545	Rhinelander	99.4	92.9	96.2
546	La Crosse	96.3	93.3	94.9
547	Eau Claire	98.3	94.0	96.2
548	Superior	96.6	99.0	97.8
549	Oshkosh	96.4	93.6	95.0
WYOMING				
820	Cheyenne	100.3	58.2	79.6
821	Yellowstone Nat'l Park	99.8	51.9	76.3
822	Wheatland	101.5	51.2	76.8
823	Rawlins	103.0	47.4	75.7
824	Worland	100.6	47.8	74.7
825	Riverton	101.8	50.7	76.7
826	Casper	100.1	61.7	81.2
827	Newcastle	100.4	47.4	74.4
828	Sheridan	101.5	55.9	79.1
829-831	Rock Springs	105.1	47.9	77.0
CANADIAN FACTORS (reflect Canadian currency)				
ALBERTA				
	Calgary	117.3	76.9	97.5
	Edmonton	116.8	76.9	97.2
	Fort McMurray	114.7	76.9	96.2
	Lethbridge	114.8	76.9	96.2
	Lloydminster	114.7	76.9	96.2
	Medicine Hat	114.8	76.9	96.2
	Red Deer	115.0	76.9	96.3
BRITISH COLUMBIA				
	Kamloops	115.1	90.0	102.8
	Prince George	118.3	90.0	104.4
	Vancouver	118.4	94.0	106.4
	Victoria	118.7	90.6	104.9
MANITOBA				
	Brandon	115.5	75.5	95.9
	Portage la Prairie	114.8	75.5	95.5
	Winnipeg	116.1	75.7	96.2
NEW BRUNSWICK				
	Bathurst	113.6	67.4	90.9
	Dalhousie	113.6	67.4	90.9
	Fredericton	114.1	72.3	93.6
	Moncton	113.2	67.4	90.7
	Newcastle	113.6	67.4	90.9
	Saint John	115.5	72.3	94.3
NEWFOUNDLAND				
	Corner Brook	119.6	67.3	93.9
	St. John's	119.2	67.3	93.7
NORTHWEST TERRITORIES				
	Yellowknife	112.0	69.2	91.0
NOVA SCOTIA				
	Dartmouth	114.9	74.1	94.9
	Halifax	114.4	75.5	95.3
	New Glasgow	114.7	74.1	94.8
	Sydney	111.9	74.1	93.3
	Yarmouth	114.6	74.1	94.7
ONTARIO				
	Barrie	119.3	91.6	105.7
	Brantford	117.6	96.1	107.1
	Cornwall	117.4	91.9	104.9
	Hamilton	117.1	97.5	107.4
	Kingston	117.9	93.9	106.1
	Kitchener	112.0	92.9	102.6
	London	116.5	94.6	105.8
	North Bay	117.6	91.6	104.8
	Oshawa	117.2	95.1	106.3
	Ottawa	117.6	94.8	106.4
	Owen Sound	119.4	91.5	105.7
	Peterborough	117.7	91.7	104.9
	Sarnia	117.7	97.5	107.8
	St. Catharines	110.8	92.3	101.7
	Sudbury	111.2	91.8	101.7

Location Factors

STATE/ZIP	CITY	MAT.	INST.	TOTAL
ONTARIO (CONT'D)				
	Thunder Bay	112.8	91.9	102.6
	Toronto	118.3	102.0	110.3
	Windsor	111.3	94.7	103.2
PRINCE EDWARD ISLAND				
	Charlottetown	116.9	62.8	90.3
	Summerside	117.0	62.8	90.4
QUEBEC				
	Cap-de-la-Madeleine	114.9	87.0	101.2
	Charlesbourg	114.9	87.0	101.2
	Chicoutimi	113.6	86.7	100.4
	Gatineau	112.2	86.8	99.7
	Laval	114.1	86.8	100.7
	Montreal	114.1	86.8	100.7
	Quebec	115.6	87.0	101.6
	Sherbrooke	114.4	86.8	100.8
	Trois Rivieres	114.9	87.0	101.2
SASKATCHEWAN				
	Moose Jaw	112.4	69.6	91.4
	Prince Albert	111.7	69.4	90.9
	Regina	113.3	69.6	91.8
	Saskatoon	112.1	69.5	91.2
YUKON				
	Whitehorse	112.0	68.4	90.6

A	Area Square Feet; Ampere	Cab.	Cabinet	d.f.u.	Drainage Fixture Units
ABS	Acrylonitrile Butadiene Stryrene; Asbestos Bonded Steel	Cair.	Air Tool Laborer	D.H.	Double Hung
		Calc	Calculated	DHW	Domestic Hot Water
A.C.	Alternating Current; Air-Conditioning; Asbestos Cement; Plywood Grade A & C	Cap.	Capacity	Diag.	Diagonal
		Carp.	Carpenter	Diam.	Diameter
		C.B.	Circuit Breaker	Distrib.	Distribution
		C.C.A.	Chromate Copper Arsenate	Dk.	Deck
A.C.I.	American Concrete Institute	C.C.F.	Hundred Cubic Feet	D.L.	Dead Load; Diesel
AD	Plywood, Grade A & D	cd	Candela	DLH	Deep Long Span Bar Joist
Addit.	Additional	cd/sf	Candela per Square Foot	Do.	Ditto
Adj.	Adjustable	CD	Grade of Plywood Face & Back	Dp.	Depth
af	Audio-frequency	CDX	Plywood, Grade C & D, exterior glue	D.P.S.T.	Double Pole, Single Throw
A.G.A.	American Gas Association			Dr.	Driver
Agg.	Aggregate	Cefi.	Cement Finisher	Drink.	Drinking
A.H.	Ampere Hours	Cem.	Cement	D.S.	Double Strength
A hr.	Ampere-hour	CF	Hundred Feet	D.S.A.	Double Strength A Grade
A.H.U.	Air Handling Unit	C.F.	Cubic Feet	D.S.B.	Double Strength B Grade
A.I.A.	American Institute of Architects	CFM	Cubic Feet per Minute	Dty.	Duty
AIC	Ampere Interrupting Capacity	c.g.	Center of Gravity	DWV	Drain Waste Vent
Allow.	Allowance	CHW	Chilled Water; Commercial Hot Water	DX	Deluxe White, Direct Expansion
alt.	Altitude			dyn	Dyne
Alum.	Aluminum	C.I.	Cast Iron	e	Eccentricity
a.m.	Ante Meridiem	C.I.P.	Cast in Place	E	Equipment Only; East
Amp.	Ampere	Circ.	Circuit	Ea.	Each
Anod.	Anodized	C.L.	Carload Lot	E.B.	Encased Burial
Approx.	Approximate	Clab.	Common Laborer	Econ.	Economy
Apt.	Apartment	C.L.F.	Hundred Linear Feet	EDP	Electronic Data Processing
Asb.	Asbestos	CLF	Current Limiting Fuse	EIFS	Exterior Insulation Finish System
A.S.B.C.	American Standard Building Code	CLP	Cross Linked Polyethylene	E.D.R.	Equiv. Direct Radiation
Asbe.	Asbestos Worker	cm	Centimeter	Eq.	Equation
A.S.H.R.A.E.	American Society of Heating, Refrig. & AC Engineers	CMP	Corr. Metal Pipe	Elec.	Electrician; Electrical
		C.M.U.	Concrete Masonry Unit	Elev.	Elevator; Elevating
A.S.M.E.	American Society of Mechanical Engineers	CN	Change Notice	EMT	Electrical Metallic Conduit; Thin Wall Conduit
		Col.	Column		
A.S.T.M.	American Society for Testing and Materials	CO_2	Carbon Dioxide	Eng.	Engine, Engineered
		Comb.	Combination	EPDM	Ethylene Propylene Diene Monomer
Attchmt.	Attachment	Compr.	Compressor		
Avg.	Average	Conc.	Concrete	EPS	Expanded Polystyrene
A.W.G.	American Wire Gauge	Cont.	Continuous; Continued	Eqhv.	Equip. Oper., Heavy
AWWA	American Water Works Assoc.	Corr.	Corrugated	Eqlt.	Equip. Oper., Light
Bbl.	Barrel	Cos	Cosine	Eqmd.	Equip. Oper., Medium
B. & B.	Grade B and Better; Balled & Burlapped	Cot	Cotangent	Eqmm.	Equip. Oper., Master Mechanic
		Cov.	Cover	Eqol.	Equip. Oper., Oilers
B. & S.	Bell and Spigot	C/P	Cedar on Paneling	Equip.	Equipment
B. & W.	Black and White	CPA	Control Point Adjustment	ERW	Electric Resistance Welded
b.c.c.	Body-centered Cubic	Cplg.	Coupling	E.S.	Energy Saver
B.C.Y.	Bank Cubic Yards	C.P.M.	Critical Path Method	Est.	Estimated
BE	Bevel End	CPVC	Chlorinated Polyvinyl Chloride	esu	Electrostatic Units
B.F.	Board Feet	C.Pr.	Hundred Pair	E.W.	Each Way
Bg. cem.	Bag of Cement	CRC	Cold Rolled Channel	EWT	Entering Water Temperature
BHP	Boiler Horsepower; Brake Horsepower	Creos.	Creosote	Excav.	Excavation
		Crpt.	Carpet & Linoleum Layer	Exp.	Expansion, Exposure
B.I.	Black Iron	CRT	Cathode-ray Tube	Ext.	Exterior
Bit.; Bitum.	Bituminous	CS	Carbon Steel, Constant Shear Bar Joist	Extru.	Extrusion
Bk.	Backed			f.	Fiber stress
Bkrs.	Breakers	Csc	Cosecant	F	Fahrenheit; Female; Fill
Bldg.	Building	C.S.F.	Hundred Square Feet	Fab.	Fabricated
Blk.	Block	CSI	Construction Specifications Institute	FBGS	Fiberglass
Bm.	Beam			F.C.	Footcandles
Boil.	Boilermaker	C.T.	Current Transformer	f.c.c.	Face-centered Cubic
B.P.M.	Blows per Minute	CTS	Copper Tube Size	f'c.	Compressive Stress in Concrete; Extreme Compressive Stress
BR	Bedroom	Cu	Copper, Cubic		
Brg.	Bearing	Cu. Ft.	Cubic Foot	F.E.	Front End
Brhe.	Bricklayer Helper	cw	Continuous Wave	FEP	Fluorinated Ethylene Propylene (Teflon)
Bric.	Bricklayer	C.W.	Cool White; Cold Water		
Brk.	Brick	Cwt.	100 Pounds	F.G.	Flat Grain
Brng.	Bearing	C.W.X.	Cool White Deluxe	F.H.A.	Federal Housing Administration
Brs.	Brass	C.Y.	Cubic Yard (27 cubic feet)	Fig.	Figure
Brz.	Bronze	C.Y./Hr.	Cubic Yard per Hour	Fin.	Finished
Bsn.	Basin	Cyl.	Cylinder	Fixt.	Fixture
Btr.	Better	d	Penny (nail size)	Fl. Oz.	Fluid Ounces
BTU	British Thermal Unit	D	Deep; Depth; Discharge	Flr.	Floor
BTUH	BTU per Hour	Dis.;Disch.	Discharge	F.M.	Frequency Modulation; Factory Mutual
B.U.R.	Built-up Roofing	Db.	Decibel		
BX	Interlocked Armored Cable	Dbl.	Double	Fmg.	Framing
c	Conductivity, Copper Sweat	DC	Direct Current	Fndtn.	Foundation
C	Hundred; Centigrade	DDC	Direct Digital Control	Fori.	Foreman, Inside
C/C	Center to Center, Cedar on Cedar	Demob.	Demobilization	Foro.	Foreman, Outside

Abbreviations

Fount.	Fountain	J.I.C.	Joint Industrial Council	M.C.P.	Motor Circuit Protector
FPM	Feet per Minute	K	Thousand; Thousand Pounds;	MD	Medium Duty
FPT	Female Pipe Thread		Heavy Wall Copper Tubing, Kelvin	M.D.O.	Medium Density Overlaid
Fr.	Frame	K.A.H.	Thousand Amp. Hours	Med.	Medium
F.R.	Fire Rating	KCMIL	Thousand Circular Mils	MF	Thousand Feet
FRK	Foil Reinforced Kraft	KD	Knock Down	M.F.B.M.	Thousand Feet Board Measure
FRP	Fiberglass Reinforced Plastic	K.D.A.T.	Kiln Dried After Treatment	Mfg.	Manufacturing
FS	Forged Steel	kg	Kilogram	Mfrs.	Manufacturers
FSC	Cast Body; Cast Switch Box	kG	Kilogauss	mg	Milligram
Ft.	Foot; Feet	kgf	Kilogram Force	MGD	Million Gallons per Day
Ftng.	Fitting	kHz	Kilohertz	MGPH	Thousand Gallons per Hour
Ftg.	Footing	Kip.	1000 Pounds	MH, M.H.	Manhole; Metal Halide; Man-Hour
Ft. Lb.	Foot Pound	KJ	Kiljoule	MHz	Megahertz
Furn.	Furniture	K.L.	Effective Length Factor	Mi.	Mile
FVNR	Full Voltage Non-Reversing	K.L.F.	Kips per Linear Foot	MI	Malleable Iron; Mineral Insulated
FXM	Female by Male	Km	Kilometer	mm	Millimeter
Fy.	Minimum Yield Stress of Steel	K.S.F.	Kips per Square Foot	Mill.	Millwright
g	Gram	K.S.I.	Kips per Square Inch	Min., min.	Minimum, minute
G	Gauss	kV	Kilovolt	Misc.	Miscellaneous
Ga.	Gauge	kVA	Kilovolt Ampere	ml	Milliliter, Mainline
Gal.	Gallon	K.V.A.R.	Kilovar (Reactance)	M.L.F.	Thousand Linear Feet
Gal./Min.	Gallon per Minute	KW	Kilowatt	Mo.	Month
Galv.	Galvanized	KWh	Kilowatt-hour	Mobil.	Mobilization
Gen.	General	L	Labor Only; Length; Long;	Mog.	Mogul Base
G.F.I.	Ground Fault Interrupter		Medium Wall Copper Tubing	MPH	Miles per Hour
Glaz.	Glazier	Lab.	Labor	MPT	Male Pipe Thread
GPD	Gallons per Day	lat	Latitude	MRT	Mile Round Trip
GPH	Gallons per Hour	Lath.	Lather	ms	Millisecond
GPM	Gallons per Minute	Lav.	Lavatory	M.S.F.	Thousand Square Feet
GR	Grade	lb.; #	Pound	Mstz.	Mosaic & Terrazzo Worker
Gran.	Granular	L.B.	Load Bearing; L Conduit Body	M.S.Y.	Thousand Square Yards
Grnd.	Ground	L. & E.	Labor & Equipment	Mtd.	Mounted
H	High; High Strength Bar Joist;	lb./hr.	Pounds per Hour	Mthe.	Mosaic & Terrazzo Helper
	Henry	lb./L.F.	Pounds per Linear Foot	Mtng.	Mounting
H.C.	High Capacity	lbf/sq.in.	Pound-force per Square Inch	Mult.	Multi; Multiply
H.D.	Heavy Duty; High Density	L.C.L.	Less than Carload Lot	M.V.A.	Million Volt Amperes
H.D.O.	High Density Overlaid	Ld.	Load	M.V.A.R.	Million Volt Amperes Reactance
Hdr.	Header	LE	Lead Equivalent	MV	Megavolt
Hdwe.	Hardware	LED	Light Emitting Diode	MW	Megawatt
Help.	Helpers Average	L.F.	Linear Foot	MXM	Male by Male
HEPA	High Efficiency Particulate Air	Lg.	Long; Length; Large	MYD	Thousand Yards
	Filter	L & H	Light and Heat	N	Natural; North
Hg	Mercury	LH	Long Span Bar Joist	nA	Nanoampere
HIC	High Interrupting Capacity	L.H.	Labor Hours	NA	Not Available; Not Applicable
HM	Hollow Metal	L.L.	Live Load	N.B.C.	National Building Code
H.O.	High Output	L.L.D.	Lamp Lumen Depreciation	NC	Normally Closed
Horiz.	Horizontal	L-O-L	Lateralolet	N.E.M.A.	National Electrical Manufacturers
H.P.	Horsepower; High Pressure	lm	Lumen		Assoc.
H.P.F.	High Power Factor	lm/sf	Lumen per Square Foot	NEHB	Bolted Circuit Breaker to 600V.
Hr.	Hour	lm/W	Lumen per Watt	N.L.B.	Non-Load-Bearing
Hrs./Day	Hours per Day	L.O.A.	Length Over All	NM	Non-Metallic Cable
HSC	High Short Circuit	log	Logarithm	nm	Nanometer
Ht.	Height	L.P.	Liquefied Petroleum; Low Pressure	No.	Number
Htg.	Heating	L.P.F.	Low Power Factor	NO	Normally Open
Htrs.	Heaters	LR	Long Radius	N.O.C.	Not Otherwise Classified
HVAC	Heating, Ventilation & Air-	L.S.	Lump Sum	Nose.	Nosing
	Conditioning	Lt.	Light	N.P.T.	National Pipe Thread
Hvy.	Heavy	Lt. Ga.	Light Gauge	NQOD	Combination Plug-on/Bolt on
HW	Hot Water	L.T.L.	Less than Truckload Lot		Circuit Breaker to 240V.
Hyd.;Hydr.	Hydraulic	Lt. Wt.	Lightweight	N.R.C.	Noise Reduction Coefficient
Hz.	Hertz (cycles)	L.V.	Low Voltage	N.R.S.	Non Rising Stem
I.	Moment of Inertia	M	Thousand; Material; Male;	ns	Nanosecond
I.C.	Interrupting Capacity		Light Wall Copper Tubing	nW	Nanowatt
ID	Inside Diameter	M²CA	Meters Squared Contact Area	OB	Opposing Blade
I.D.	Inside Dimension; Identification	m/hr; M.H.	Man-hour	OC	On Center
I.F.	Inside Frosted	mA	Milliampere	OD	Outside Diameter
I.M.C.	Intermediate Metal Conduit	Mach.	Machine	O.D.	Outside Dimension
In.	Inch	Mag. Str.	Magnetic Starter	ODS	Overhead Distribution System
Incan.	Incandescent	Maint.	Maintenance	O.G.	Ogee
Incl.	Included; Including	Marb.	Marble Setter	O.H.	Overhead
Int.	Interior	Mat; Mat'l.	Material	O & P	Overhead and Profit
Inst.	Installation	Max.	Maximum	Oper.	Operator
Insul.	Insulation/Insulated	MBF	Thousand Board Feet	Opng.	Opening
I.P.	Iron Pipe	MBH	Thousand BTU's per hr.	Orna.	Ornamental
I.P.S.	Iron Pipe Size	MC	Metal Clad Cable	OSB	Oriented Strand Board
I.P.T.	Iron Pipe Threaded	M.C.F.	Thousand Cubic Feet	O. S. & Y.	Outside Screw and Yoke
I.W.	Indirect Waste	M.C.F.M.	Thousand Cubic Feet per Minute	Ovhd.	Overhead
J	Joule	M.C.M.	Thousand Circular Mils	OWG	Oil, Water or Gas

603

Oz.	Ounce	SCFM	Standard Cubic Feet per Minute	THW.	Insulated Strand Wire
P.	Pole; Applied Load; Projection	Scaf.	Scaffold	THWN;	Nylon Jacketed Wire
p.	Page	Sch.; Sched.	Schedule	T.L.	Truckload
Pape.	Paperhanger	S.C.R.	Modular Brick	T.M.	Track Mounted
P.A.P.R.	Powered Air Purifying Respirator	S.D.	Sound Deadening	Tot.	Total
PAR	Parabolic Reflector	S.D.R.	Standard Dimension Ratio	T-O-L	Threadolet
Pc., Pcs.	Piece, Pieces	S.E.	Surfaced Edge	T.S.	Trigger Start
P.C.	Portland Cement; Power Connector	Sel.	Select	Tr.	Trade
P.C.F.	Pounds per Cubic Foot	S.E.R.;	Service Entrance Cable	Transf.	Transformer
P.C.M.	Phase Contract Microscopy	S.E.U.	Service Entrance Cable	Trhv.	Truck Driver, Heavy
P.E.	Professional Engineer;	S.F.	Square Foot	Trlr	Trailer
	Porcelain Enamel;	S.F.C.A.	Square Foot Contact Area	Trlt.	Truck Driver, Light
	Polyethylene; Plain End	S.F.G.	Square Foot of Ground	TV	Television
Perf.	Perforated	S.F. Hor.	Square Foot Horizontal	T.W.	Thermoplastic Water Resistant
Ph.	Phase	S.F.R.	Square Feet of Radiation		Wire
P.I.	Pressure Injected	S.F. Shlf.	Square Foot of Shelf	UCI	Uniform Construction Index
Pile.	Pile Driver	S4S	Surface 4 Sides	UF	Underground Feeder
Pkg.	Package	Shee.	Sheet Metal Worker	UGND	Underground Feeder
Pl.	Plate	Sin.	Sine	U.H.F.	Ultra High Frequency
Plah.	Plasterer Helper	Skwk.	Skilled Worker	U.L.	Underwriters Laboratory
Plas.	Plasterer	SL	Saran Lined	Unfin.	Unfinished
Pluh.	Plumbers Helper	S.L.	Slimline	URD	Underground Residential
Plum.	Plumber	Sldr.	Solder		Distribution
Ply.	Plywood	SLH	Super Long Span Bar Joist	US	United States
p.m.	Post Meridiem	S.N.	Solid Neutral	USP	United States Primed
Pntd.	Painted	S-O-L	Socketolet	UTP	Unshielded Twisted Pair
Pord.	Painter, Ordinary	sp	Standpipe	V	Volt
pp	Pages	S.P.	Static Pressure; Single Pole; Self-	V.A.	Volt Amperes
PP; PPL	Polypropylene		Propelled	V.C.T.	Vinyl Composition Tile
P.P.M.	Parts per Million	Spri.	Sprinkler Installer	VAV	Variable Air Volume
Pr.	Pair	spwg	Static Pressure Water Gauge	VC	Veneer Core
P.E.S.B.	Pre-engineered Steel Building	Sq.	Square; 100 Square Feet	Vent.	Ventilation
Prefab.	Prefabricated	S.P.D.T.	Single Pole, Double Throw	Vert.	Vertical
Prefin.	Prefinished	SPF	Spruce Pine Fir	V.F.	Vinyl Faced
Prop.	Propelled	S.P.S.T.	Single Pole, Single Throw	V.G.	Vertical Grain
PSF; psf	Pounds per Square Foot	SPT	Standard Pipe Thread	V.H.F.	Very High Frequency
PSI; psi	Pounds per Square Inch	Sq. Hd.	Square Head	VHO	Very High Output
PSIG	Pounds per Square Inch Gauge	Sq. In.	Square Inch	Vib.	Vibrating
PSP	Plastic Sewer Pipe	S.S.	Single Strength; Stainless Steel	V.L.F.	Vertical Linear Foot
Pspr.	Painter, Spray	S.S.B.	Single Strength B Grade	Vol.	Volume
Psst.	Painter, Structural Steel	sst	Stainless Steel	VRP	Vinyl Reinforced Polyester
P.T.	Potential Transformer	Sswk.	Structural Steel Worker	W	Wire; Watt; Wide; West
P. & T.	Pressure & Temperature	Sswl.	Structural Steel Welder	w/	With
Ptd.	Painted	St.; Stl.	Steel	W.C.	Water Column; Water Closet
Ptns.	Partitions	S.T.C.	Sound Transmission Coefficient	W.F.	Wide Flange
Pu	Ultimate Load	Std.	Standard	W.G.	Water Gauge
PVC	Polyvinyl Chloride	STK	Select Tight Knot	Wldg.	Welding
Pvmt.	Pavement	STP	Standard Temperature & Pressure	W. Mile	Wire Mile
Pwr.	Power	Stpi.	Steamfitter, Pipefitter	W-O-L	Weldolet
Q	Quantity Heat Flow	Str.	Strength; Starter; Straight	W.R.	Water Resistant
Quan.; Qty.	Quantity	Strd.	Stranded	Wrck.	Wrecker
Q.C.	Quick Coupling	Struct.	Structural	W.S.P.	Water, Steam, Petroleum
r	Radius of Gyration	Sty.	Story	WT., Wt.	Weight
R	Resistance	Subj.	Subject	WWF	Welded Wire Fabric
R.C.P.	Reinforced Concrete Pipe	Subs.	Subcontractors	XFER	Transfer
Rect.	Rectangle	Surf.	Surface	XFMR	Transformer
Reg.	Regular	Sw.	Switch	XHD	Extra Heavy Duty
Reinf.	Reinforced	Swbd.	Switchboard	XHHW; XLPE	Cross-Linked Polyethylene Wire
Req'd.	Required	S.Y.	Square Yard		Insulation
Res.	Resistant	Syn.	Synthetic	XLP	Cross-linked Polyethylene
Resi.	Residential	S.Y.P.	Southern Yellow Pine	Y	Wye
Rgh.	Rough	Sys.	System	yd	Yard
RGS	Rigid Galvanized Steel	t.	Thickness	yr	Year
R.H.W.	Rubber, Heat & Water Resistant;	T	Temperature; Ton	Δ	Delta
	Residential Hot Water	Tan	Tangent	%	Percent
rms	Root Mean Square	T.C.	Terra Cotta	~	Approximately
Rnd.	Round	T & C	Threaded and Coupled	Ø	Phase
Rodm.	Rodman	T.D.	Temperature Difference	@	At
Rofc.	Roofer, Composition	T.E.M.	Transmission Electron Microscopy	#	Pound; Number
Rofp.	Roofer, Precast	TFE	Tetrafluoroethylene (Teflon)	<	Less Than
Rohe.	Roofer Helpers (Composition)	T. & G.	Tongue & Groove;	>	Greater Than
Rots.	Roofer, Tile & Slate		Tar & Gravel		
R.O.W.	Right of Way	Th.; Thk.	Thick		
RPM	Revolutions per Minute	Thn.	Thin		
R.S.	Rapid Start	Thrded	Threaded		
Rsr	Riser	Tilf.	Tile Layer, Floor		
RT	Round Trip	Tilh.	Tile Layer, Helper		
S.	Suction; Single Entrance; South	THHN	Nylon Jacketed Wire		

Index

Notes

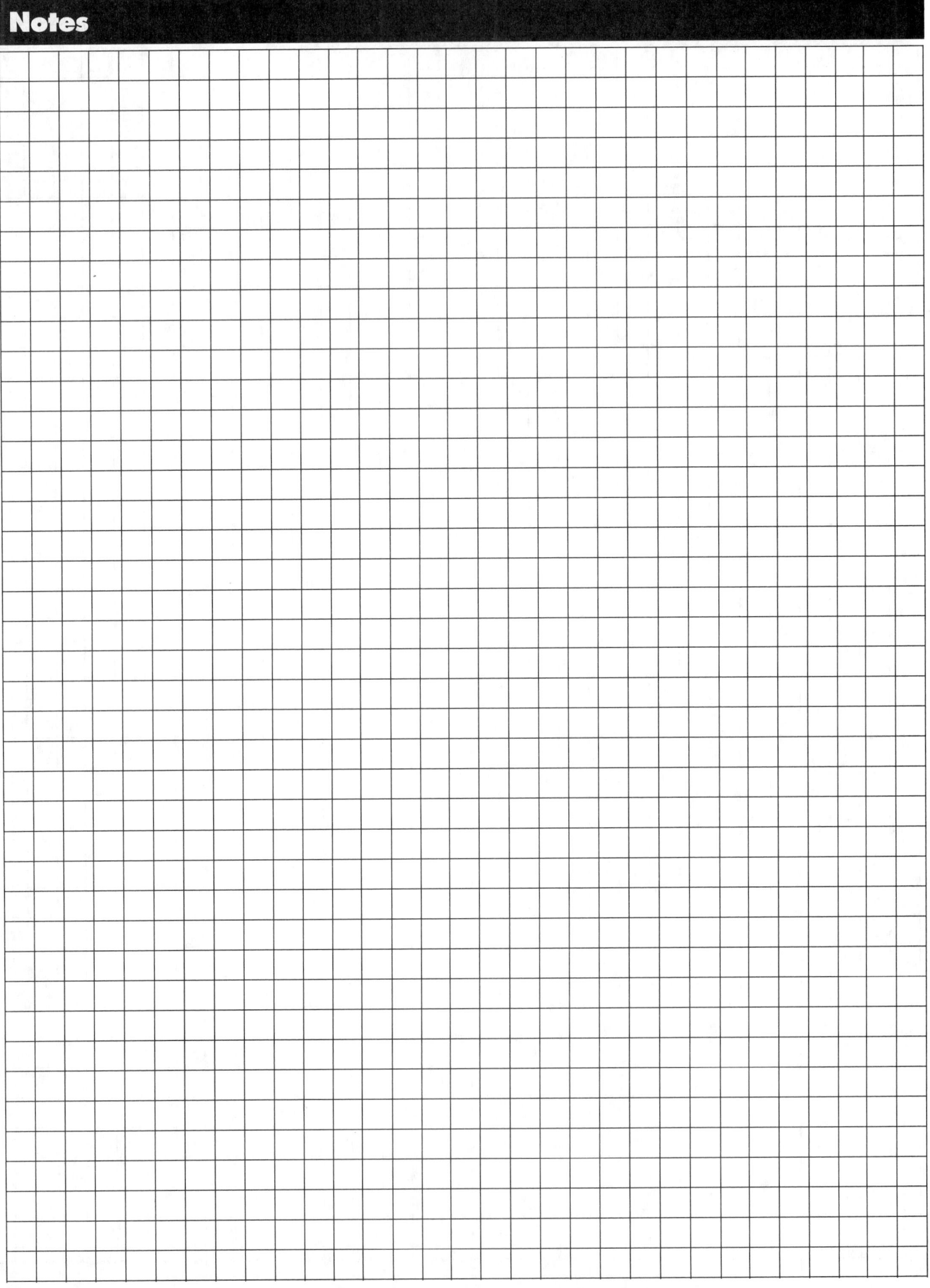

Division Notes

	CREW	DAILY OUTPUT	LABOR-HOURS	UNIT	2003 BARE COSTS				TOTAL INCL O&P
					MAT.	LABOR	EQUIP.	TOTAL	

	CREW	DAILY OUTPUT	LABOR-HOURS	UNIT	2003 BARE COSTS				TOTAL INCL O&P
					MAT.	LABOR	EQUIP.	TOTAL	
	CREW	DAILY OUTPUT	LABOR-HOURS	UNIT	MAT.	LABOR	EQUIP.	TOTAL	TOTAL INCL O&P

Division Notes

	CREW	DAILY OUTPUT	LABOR-HOURS	UNIT	2003 BARE COSTS				TOTAL INCL O&P
					MAT.	LABOR	EQUIP.	TOTAL	

	CREW	DAILY OUTPUT	LABOR-HOURS	UNIT	2003 BARE COSTS				TOTAL INCL O&P
					MAT.	LABOR	EQUIP.	TOTAL	

	CREW	DAILY OUTPUT	LABOR-HOURS	UNIT	2003 BARE COSTS				TOTAL INCL O&P
					MAT.	LABOR	EQUIP.	TOTAL	

Division Notes

		CREW	DAILY OUTPUT	LABOR-HOURS	UNIT	2003 BARE COSTS				TOTAL INCL O&P
						MAT.	LABOR	EQUIP.	TOTAL	

Division Notes

	CREW	DAILY OUTPUT	LABOR-HOURS	UNIT	2003 BARE COSTS				TOTAL INCL O&P
					MAT.	LABOR	EQUIP.	TOTAL	

Division Notes

	CREW	DAILY OUTPUT	LABOR-HOURS	UNIT	2003 BARE COSTS				TOTAL INCL O&P
					MAT.	LABOR	EQUIP.	TOTAL	

Reed Contruction Data

Reed Construction Data, a leading worldwide provider of total construction information solutions, is comprised of three main product groups designed specifically to help construction professionals advance their businesses with timely, accurate and actionable project, product, and cost data. Reed Construction Data is a division of Reed Business Information, a member of the Reed Elsevier plc group of companies.

The *Project, Product, and Cost & Estimating* divisions offer a variety of innovative products and services designed for the full spectrum of design, construction, and manufacturing professionals. Through it's *International* companies, Reed Construction Data's reputation for quality construction market data is growing worldwide.

Cost Information
R.S. Means, the undisputed market leader and authority on construction costs, publishes current cost and estimating information in annual cost books and on the CostWorks CD-ROM. R.S. Means furnishes the construction industry with a rich library of complementary reference books and a series of professional seminars that are designed to sharpen professional skills and maximize the effective use of cost estimating and management tools. R.S. Means also provides construction cost consulting for Owners, Manufacturers, Designers, and Contractors.

Project Data
Reed Construction Data provides complete, accurate and relevant project information through all stages of construction. Customers are supplied industry data through leads, project reports, contact lists, plans and specifications surveys, market penetration analyses and sales evaluation reports. Any of these products can pinpoint a county, look at a state, or cover the country. Data is delivered via paper, e-mail, CD-ROM or the Internet.

Building Product Information
The First Source suite of products is the only integrated building product information system offered to the commercial construction industry for comparing and specifying building products. These print and online resources include *CMD First Source,* CSI's SPEC-DATA™, CSI's MANU-SPEC™, CADBlocks®, and Manufacturer Catalogs. Written by industry professionals and organized using CSI's MasterFormat™, construction professionals use this information to make better design decisions.

CMD FirstSource.com combines Reed Construction Data's project, product and cost data with news and information from Reed Business Information's *Building Design & Construction* and *Consulting-Specifying Engineer,* this industry-focused site offers easy and unlimited access to vital information for all construction professionals.

International
BIMSA/Mexico provides construction project news, product information, cost-data, seminars and consulting services to construction professionals in Mexico. Its subsidiary, PRISMA, provides job costing software.

Byggfakta Scandinavia AB, founded in 1936, is the parent company for the leaders of customized construction market data for Denmark, Estonia, Finland, Norway and Sweden. Each company fully covers the local construction market and provides information across several platforms including subscription, ad-hoc basis, electronically and on paper.

Reed Construction Data Canada serves the Canadian construction market with reliable and comprehensive project and product information services that cover all facets of construction. Core services include: *BuildSource, BuildSpec, BuildSelect,* product selection and specification tools available in print and on the Internet; Building Reports, a national construction project lead service; CanaData, statistical and forecasting information; *Daily Commercial News,* a construction newspaper reporting on news and projects in Ontario; and *Journal of Commerce,* reporting news in British Columbia and Alberta.

Cordell Building Information Services, with its complete range of project and cost and estimating services, is Australia's specialist in the construction information industry. Cordell provides in-depth and historical information on all aspects of construction projects and estimation, including several customized reports, construction and sales leads, and detailed cost information among others.

For more information, please visit our Web site at www.cmdg.com.

Reed Construction Data Corporate Office
30 Technology Parkway South
Norcross, GA 30092-2912
(800) 322-6996
(800) 895-8661 (fax)
info@cmdg.com
www.cmdg.com

 Reed Construction Data

Means Project Cost Report

By filling out and returning the Project Description, you can receive a discount of $20.00 off any one of the Means products advertised in the following pages. The cost information required includes all items marked (✔) except those where no costs occurred. The sum of all major items should equal the Total Project Cost.

$20.00 Discount per product for each report you submit.

DISCOUNT PRODUCTS AVAILABLE—FOR U.S. CUSTOMERS ONLY—STRICTLY CONFIDENTIAL

Project Description (No remodeling projects, please.)

✔ Type Building _____

✔ Location _____

  Capacity _____

✔ Frame _____

✔ Exterior _____

✔ Basement: full ☐ partial ☐ none ☐ crawl ☐

✔ Height in Stories _____

✔ Total Floor Area _____

  Ground Floor Area _____

✔ Volume in C.F. _____

  % Air Conditioned _____ Tons _____

  Comments _____

  Owner _____

  Architect _____

  General Contractor _____

✔ Bid Date _____

  Typical Bay Size _____

✔ Labor Force: _____ % Union _____ % Non-Union

✔ Project Description (Circle one number in each line)
- 1. Economy 2. Average 3. Custom 4. Luxury
- 1. Square 2. Rectangular 3. Irregular 4. Very Irregular

					$
	✔ **Total Project Cost**				$
A	✔ **General Conditions**				$
B	✔ **Site Work**				$
BS	Site Clearing & Improvement				
BE	Excavation	(	C.Y.)		
BF	Caissons & Piling	(	L.F.)		
BU	Site Utilities				
BP	Roads & Walks Exterior Paving	(	S.Y.)		
C	✔ **Concrete**				$
C	Cast in Place	(	C.Y.)		
CP	Precast	(	S.F.)		
D	✔ **Masonry**				$
DB	Brick	(	M)		
DC	Block	(	M)		
DT	Tile	(	S.F.)		
DS	Stone	(	S.F.)		
E	✔ **Metals**				$
ES	Structural Steel	(	Tons)		
EM	Misc. & Ornamental Metals				
F	✔ **Wood & Plastics**				$
FR	Rough Carpentry	(	MBF)		
FF	Finish Carpentry				
FM	Architectural Millwork				
G	✔ **Thermal & Moisture Protection**				$
GW	Waterproofing-Dampproofing	(	S.F.)		
GN	Insulation	(	S.F.)		
GR	Roofing & Flashing	(	S.F.)		
GM	Metal Siding/Curtain Wall	(	S.F.)		
H	✔ **Doors and Windows**				$
HD	Doors	(	Ea.)		
HW	Windows	(	S.F.)		
HH	Finish Hardware				
HG	Glass & Glazing	(	S.F.)		
HS	Storefronts	(	S.F.)		

					$
J	✔ **Finishes**				$
JL	Lath & Plaster	(	S.Y.)		
JD	Drywall	(	S.F.)		
JM	Tile & Marble	(	S.F.)		
JT	Terrazzo	(	S.F.)		
JA	Acoustical Treatment	(	S.F.)		
JC	Carpet	(	S.Y.)		
JF	Hard Surface Flooring	(	S.F.)		
JP	Painting & Wall Covering	(	S.F.)		
K	✔ **Specialties**				$
KB	Bathroom Partitions & Access.	(	S.F.)		
KF	Other Partitions	(	S.F.)		
KL	Lockers	(	Ea.)		
L	✔ **Equipment**				$
LK	Kitchen				
LS	School				
LO	Other				
M	✔ **Furnishings**				$
MW	Window Treatment				
MS	Seating	(	Ea.)		
N	✔ **Special Construction**				$
NA	Acoustical	(	S.F.)		
NB	Prefab. Bldgs.	(	S.F.)		
NO	Other				
P	✔ **Conveying Systems**				$
PE	Elevators	(	Ea.)		
PS	Escalators	(	Ea.)		
PM	Material Handling				
Q	✔ **Mechanical**				$
QP	Plumbing	(No. of fixtures	)		
QS	Fire Protection (Sprinklers)				
QF	Fire Protection (Hose Standpipes)				
QB	Heating, Ventilating & A.C.				
QH	Heating & Ventilating	(BTU Output	)		
QA	Air Conditioning	(	Tons)		
R	✔ **Electrical**				$
RL	Lighting	(	S.F.)		
RP	Power Service				
RD	Power Distribution				
RA	Alarms				
RG	Special Systems				
S	✔ **Mech./Elec. Combined**				$

Product Name _____

Product Number _____

Your Name _____

Title _____

Company _____

☐ Company

☐ Home Street Address _____

City, State, Zip _____

☐ Please send _____ forms.

Please specify the Means product you wish to receive. Complete the address information as requested and return this form with your check (product cost less $20.00) to address below.

R.S. Means Company, Inc.,
Square Foot Costs Department
P.O. Box 800
Kingston, MA 02364-9988

R.S. Means Company, Inc. . . . a tradition of excellence in Construction Cost Information and Services since 1942.

For more information visit Means Web Site at www.rsmeans.com

Table of Contents

Book Selection Guide

The following table provides definitive information on the content of each cost data publication. The number of lines of data provided in each unit price or assemblies division, as well as the number of reference tables and crews is listed for each book. The presence of other elements such as an historical cost index, city cost indexes, square foot models or cross-referenced index is also indicated. You can use the table to help select the Means' book that has the quantity and type of information you most need in your work.

Unit Cost Divisions	Building Construction Costs	Mechanical	Electrical	Repair & Remodel.	Square Foot	Site Work Landsc.	Assemblies	Interior	Concrete Masonry	Open Shop	Heavy Construc.	Light Commercial	Facil. Construc.	Plumbing	Western Construction Costs	Residential
1	971	785	870	976		1008		810	977	1066	1014	716	1491	837	1065	673
2	2745	1270	459	1633		7834		777	1455	2700	5011	840	4555	1585	2719	921
3	1374	75	56	691		1210		174	1750	1369	1370	189	1276	34	1372	220
4	782	22	0	587		640		535	1014	756	561	345	1002	0	766	274
5	1764	213	153	888		699		893	629	1732	967	736	1745	285	1746	714
6	1194	82	78	1142		422		1186	311	1185	551	1331	1269	47	1533	1398
7	1230	155	71	1191		435		472	405	1229	317	921	1282	165	1229	717
8	1762	47	0	1783		258		1620	636	1744	7	1123	1953	0	1763	1092
9	1470	47	0	1312		142		1568	290	1422	97	1247	1686	47	1466	1135
10	847	47	25	468		194		686	169	849	0	380	867	221	847	215
11	1010	347	174	528		121		805	28	920	80	217	1069	298	917	103
12	291	0	0	37		210		1404	27	282	0	62	1404	0	282	63
13	1083	950	359	445		367		845	75	1067	258	445	1696	889	1049	193
14	323	36	0	236		22		291	0	323	30	12	322	14	321	6
15	1929	12662	616	1758		1544		1130	59	1940	1746	1191	10584	9248	1960	826
16	1282	470	9606	1003		744		1108	55	1300	765	1081	9373	415	1231	552
17	428	354	428	0		0		0	0	428	0	0	428	356	428	0
Totals	20465	17562	12895	14678		15850		14304	7880	20312	12774	10836	42002	14441	20694	9102

Assembly Divisions	Building Construction Costs	Mechanical	Electrical	Repair & Remodel.	Square Foot	Site Work Landsc.	Assemblies	Interior	Concrete Masonry	Open Shop	Heavy Construc.	Light Commercial	Facil. Construc.	Plumbing	Western Construction Costs	Asm Div	Residential
A		19	0	192	150	540	612	0	550		542	149	24	0		1	374
B		0	0	820	2480	0	5591	333	1915		0	2030	145	0		2	217
C		0	0	700	869	0	1292	1573	146		0	769	266	0		3	588
D		1034	810	729	1853	0	2471	748	0		0	1358	1031	895		4	871
E		0	0	85	255	0	292	5	0		0	255	5	0		5	393
F		0	0	0	123	0	126	0	0		0	123	3	0		6	363
G		475	148	370	87	1858	576	0	492		432	86	87	540		7	299
																8	760
																9	80
																10	0
																11	0
																12	0
Totals		1528	958	2896	5817	2398	10959	2659	3103		974	4770	1561	1435			3945

Reference Section	Building Construction Costs	Mechanical	Electrical	Repair & Remodel.	Square Foot	Site Work Landsc.	Assemblies	Interior	Concrete Masonry	Open Shop	Heavy Construc.	Light Commercial	Facil. Construc.	Plumbing	Western Construction Costs	Residential
Tables	129	45	85	69	4	80	219	46	71	130	61	57	104	51	131	42
Models					102							43				32
Crews	410	410	410	391		410		410	410	393	410	393	391	410	410	393
City Cost Indexes	yes	yes	yes	yes	yes	yes	yes	yes	yes	yes	yes	yes	yes	yes	yes	yes
Historical Cost Indexes	yes	yes	yes	yes	yes	yes	yes	yes	yes	yes	yes	yes	yes	yes	yes	no
Index	yes	yes	yes	yes	no	yes	yes	yes	yes	yes	yes	yes	yes	yes	yes	yes

1

Annual Cost Guides

For more information
visit Means Web Site
at www.rsmeans.com

Means
Building Construction
Cost Data 2003

Available in Both Softbound and Looseleaf Editions

The "Bible" of the industry comes in the standard softcover edition or the looseleaf edition.

Many customers enjoy the convenience and flexibility of the looseleaf binder, which increases the usefulness of *Means Building Construction Cost Data 2003* by making it easy to add and remove pages. You can insert your own cost information pages, so everything is in one place. Copying pages for faxing is easier also. Whichever edition you prefer, softbound or the convenient looseleaf edition, you'll be eligible to receive *The Change Notice* FREE.

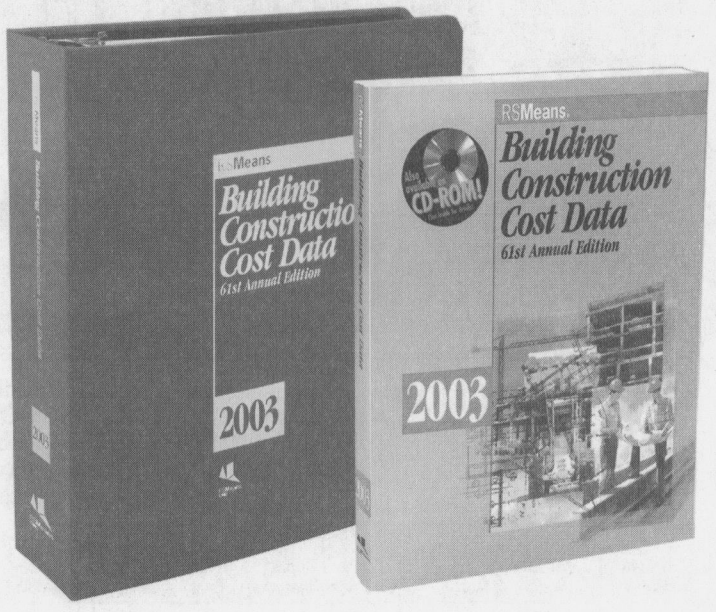

$104.95 per copy, Softbound
Catalog No. 60013

$131.95 per copy, Looseleaf
Catalog No. 61013

Means Building Construction Cost Data 2003

Offers you unchallenged unit price reliability in an easy-to-use arrangement. Whether used for complete, finished estimates or for periodic checks, it supplies more cost facts better and faster than any comparable source. Over 23,000 unit prices for 2003. The City Cost Indexes cover over 930 areas, for indexing to any project location in North America. Order and get *The Change Notice* FREE. You'll have year-long access to the Means Estimating **HOTLINE** FREE with your subscription. Expert assistance when using Means data is just a phone call away.

$104.95 per copy
Over 650 pages, illustrated, available Oct. 2002
Catalog No. 60013

Means Building Construction Cost Data 2003

Metric Version

The Federal Government has stated that all federal construction projects must now use metric documentation. The *Metric Version* of *Means Building Construction Cost Data 2003* is presented in metric measurements covering all construction areas. Don't miss out on these billion dollar opportunities. Make the switch to metric today.

$104.95 per copy
Over 650 pages, illus., available Nov. 2002
Catalog No. 63013

For more information
visit Means Web Site
at www.rsmeans.com

Annual Cost Guides

Means Mechanical Cost Data 2003

• HVAC • Controls

Total unit and systems price guidance for mechanical construction...materials, parts, fittings, and complete labor cost information. Includes prices for piping, heating, air conditioning, ventilation, and all related construction.

Plus new 2003 unit costs for:
• Over 2500 installed HVAC/controls assemblies
• "On Site" Location Factors for over 930 cities and towns in the U.S. and Canada
• Crews, labor and equipment

$104.95 per copy
Over 600 pages, illustrated, available Oct. 2002
Catalog No. 60023

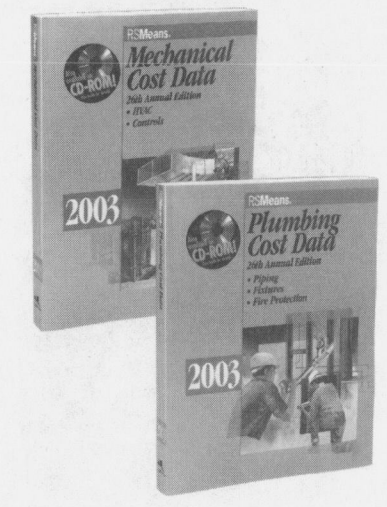

Means Plumbing Cost Data 2003

Comprehensive unit prices and assemblies for plumbing, irrigation systems, commercial and residential fire protection, point-of-use water heaters, and the latest approved materials. This publication and its companion, *Means Mechanical Cost Data*, provide full-range cost estimating coverage for all the mechanical trades.

$104.95 per copy
Over 500 pages, illustrated, available Nov. 2002
Catalog No. 60213

Means Electrical Cost Data 2003

Pricing information for every part of electrical cost planning: More than 15,000 unit and systems costs with design tables; clear specifications and drawings; engineering guides and illustrated estimating procedures; complete labor-hour and materials costs for better scheduling and procurement; the latest electrical products and construction methods.
• A Variety of Special Electrical Systems including Cathodic Protection
• Costs for maintenance, demolition, HVAC/mechanical, specialties, equipment, and more

$104.95 per copy
Over 450 pages, illustrated, available Oct. 2002
Catalog No. 60033

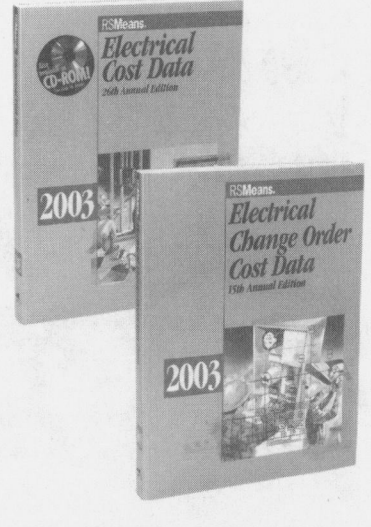

Means Electrical Change Order Cost Data 2003

You are provided with electrical unit prices exclusively for pricing change orders based on the recent, direct experience of contractors and suppliers. Analyze and check your own change order estimates against the experience others have had doing the same work. It also covers productivity analysis and change order cost justifications. With useful information for calculating the effects of change orders and dealing with their administration.

$104.95 per copy
Over 450 pages, available Oct. 2002
Catalog No. 60233

Means Facilities Maintenance & Repair Cost Data 2003

Published in a looseleaf format, *Means Facilities Maintenance & Repair Cost Data* gives you a complete system to manage and plan your facility repair and maintenance costs and budget efficiently. Guidelines for auditing a facility and developing an annual maintenance plan. Budgeting is included, along with reference tables on cost and management and information on frequency and productivity of maintenance operations.

The only nationally recognized source of maintenance and repair costs. Developed in cooperation with the Army Corps of Engineers.

$230.95 per copy
Over 600 pages, illustrated, available Dec. 2002
Catalog No. 60303

Means Square Foot Costs 2003

It's Accurate and Easy To Use!

• **Updated 2003 price information,** based on nationwide figures from suppliers, estimators, labor experts and contractors.
• **"How-to-Use"** Sections, with **clear examples** of commercial, residential, industrial, and institutional structures.
• Realistic graphics, offering true-to-life illustrations of building projects.
• Extensive information on using square foot cost data, including **sample estimates** and **alternate pricing methods.**

$115.95 per copy
Over 450 pages, illustrated, available Nov. 2002
Catalog No. 60053

Means Repair & Remodeling Cost Data 2003

Commercial/Residential

You can use this valuable tool to estimate commercial and residential renovation and remodeling.

Includes: New costs for hundreds of unique methods, materials and conditions that only come up in repair and remodeling. PLUS:

- Unit costs for over 16,000 construction components
- Installed costs for over 90 assemblies
- Costs for 300+ construction crews
- Over 930 "On Site" localization factors for the U.S. and Canada.

$92.95 per copy
Over 600 pages, illustrated, available Nov. 2002
Catalog No. 60043

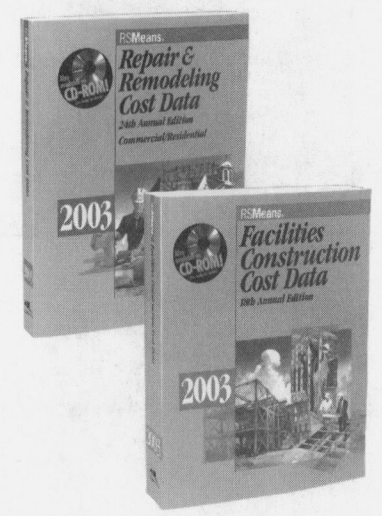

Means Facilities Construction Cost Data 2003

For the maintenance and construction of commercial, industrial, municipal, and institutional properties. Costs are shown for new and remodeling construction and are broken down into materials, labor, equipment, overhead, and profit. Special emphasis is given to sections on mechanical, electrical, furnishings, site work, building maintenance, finish work, and demolition. More than 45,000 unit costs, plus assemblies and reference sections are included.

$253.95 per copy
Over 1150 pages, illustrated, available Nov. 2002
Catalog No. 60203

Means Residential Cost Data 2003

Contains square foot costs for 30 basic home models with the look of today, plus hundreds of custom additions and modifications you can quote right off the page. With costs for the 100 residential systems you're most likely to use in the year ahead. Complete with blank estimating forms, sample estimates and step-by-step instructions.

$92.95 per copy
Over 550 pages, illustrated, available Nov. 2002
Catalog No. 60173

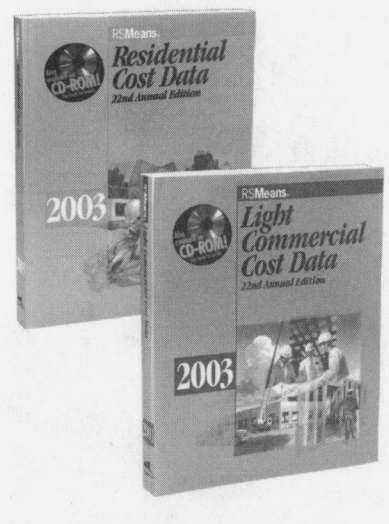

Means Light Commercial Cost Data 2003

Specifically addresses the light commercial market, which is an increasingly specialized niche in the industry. Aids you, the owner/ designer/contractor, in preparing all types of estimates, from budgets to detailed bids. Includes new advances in methods and materials. Assemblies section allows you to evaluate alternatives in the early stages of design/planning.

Over 13,000 unit costs for 2003 ensure you have the prices you need...when you need them.

$92.95 per copy
Over 600 pages, illustrated, available Dec. 2002
Catalog No. 60183

Means Assemblies Cost Data 2003

Means Assemblies Cost Data 2003 takes the guesswork out of preliminary or conceptual estimates. Now you don't have to try to calculate the assembled cost by working up individual components costs. We've done all the work for you.

Presents detailed illustrations, descriptions, specifications and costs for every conceivable building assembly—240 types in all—arranged in the easy-to-use UNIFORMAT II system. Each illustrated "assembled" cost includes a complete grouping of materials and associated installation costs including the installing contractor's overhead and profit.

$173.95 per copy
Over 550 pages, illustrated, available Oct. 2002
Catalog No. 60063

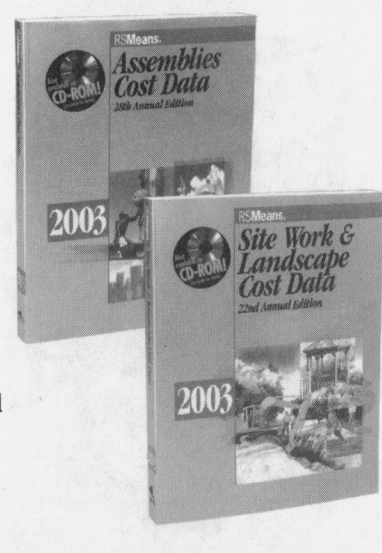

Means Site Work & Landscape Cost Data 2003

Means Site Work & Landscape Cost Data 2003 is organized to assist you in all your estimating needs. Hundreds of fact-filled pages help you make accurate cost estimates efficiently.

Updated for 2003!

- Demolition features—including ceilings, doors, electrical, flooring, HVAC, millwork, plumbing, roofing, walls and windows
- State-of-the-art segmental retaining walls
- Flywheel trenching costs and details
- Updated Wells section
- Landscape materials, flowers, shrubs and trees

$104.95 per copy
Over 600 pages, illustrated, available Oct. 2002
Catalog No. 60283

For more information
visit Means Web Site
at www.rsmeans.com

Annual Cost Guides

Means Open Shop Building Construction Cost Data 2003

The latest costs for accurate budgeting and estimating of new commercial and residential construction... renovation work... change orders... cost engineering. *Means Open Shop BCCD* will assist you to...
• Develop benchmark prices for change orders
• Plug gaps in preliminary estimates, budgets
• Estimate complex projects
• Substantiate invoices on contracts
• Price ADA-related renovations

$104.95 per copy
Over 650 pages, illustrated, available Nov. 2002
Catalog No. 60153

Means Building Construction Cost Data 2003
Western Edition

This regional edition provides more precise cost information for western North America. Labor rates are based on union rates from 13 western states and western Canada. Included are western practices and materials not found in our national edition: tilt-up concrete walls, glu-lam structural systems, specialized timber construction, seismic restraints, landscape and irrigation systems.

$104.95 per copy
Over 600 pages, illustrated, available Dec. 2002
Catalog No. 60223

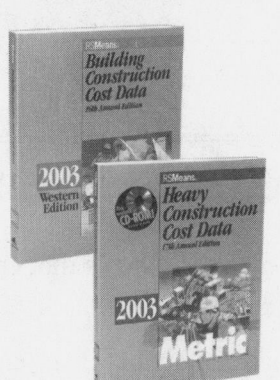

Means Construction Cost Indexes 2003

Who knows what 2003 holds? What materials and labor costs will change unexpectedly? By how much?
• Breakdowns for 305 major cities.
• National averages for 30 key cities.
• Expanded five major city indexes.
• Historical construction cost indexes.

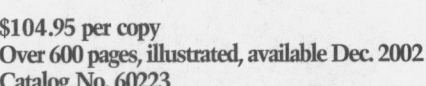

$228.95 per year
$57.25 individual quarters
Catalog No. 60143 A,B,C,D

Means Concrete & Masonry Cost Data 2003

Provides you with cost facts for virtually all concrete/masonry estimating needs, from complicated formwork to various sizes and face finishes of brick and block, all in great detail. The comprehensive unit cost section contains more than 8,500 selected entries. Also contains an assemblies cost section, and a detailed reference section which supplements the cost data.

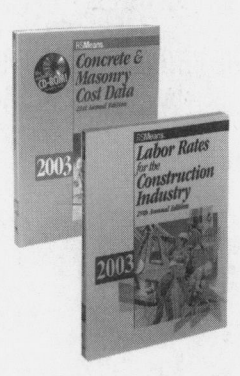

$97.95 per copy
Over 450 pages, illustrated, available Nov. 2002
Catalog No. 60113

Means Heavy Construction Cost Data 2003

A comprehensive guide to heavy construction costs. Includes costs for highly specialized projects such as tunnels, dams, highways, airports, and waterways. Information on different labor rates, equipment, and material costs is included. Has unit price costs, systems costs, and numerous reference tables for costs and design. Valuable not only to contractors and civil engineers, but also to government agencies and city/town engineers.

$104.95 per copy
Over 450 pages, illustrated, available Nov. 2002
Catalog No. 60163

Means Heavy Construction Cost Data 2003
Metric Version

Make sure you have the Means industry standard metric costs for the federal, state, municipal and private marketplace. With thousands of up-to-date metric unit prices in tables by CSI standard divisions. Supplies you with assemblies costs using the metric standard for reliable cost projections in the design stage of your project. Helps you determine sizes, material amounts, and has tips for handling metric estimates.

$104.95 per copy
Over 450 pages, illustrated, available Dec. 2002
Catalog No. 63163

Means Interior Cost Data 2003

Provides you with prices and guidance needed to make accurate interior work estimates. Contains costs on materials, equipment, hardware, custom installations, furnishings, labor costs ... every cost factor for new and remodel commercial and industrial interior construction, including updated information on office furnishings, plus more than 50 reference tables. For contractors, facility managers, owners.

$104.95 per copy
Over 550 pages, illustrated, available Oct. 2002
Catalog No. 60093

Means Labor Rates for the Construction Industry 2003

Complete information for estimating labor costs, making comparisons and negotiating wage rates by trade for over 300 cities (United States and Canada). With 46 construction trades listed by local union number in each city, and historical wage rates included for comparison. No similar book is available through the trade.

Each city chart lists the county and is alphabetically arranged with handy visual flip tabs for quick reference.

$230.95 per copy
Over 300 pages, available Dec. 2002
Catalog No. 60123

Reference Books

For more information
visit Means Web Site
at www.rsmeans.com

Preventive Maintenance for Higher Education Facilities

By Applied Management Engineering, Inc.

An easy-to-use system to help facilities professionals establish the value of PM, and to develop and budget for an appropriate PM program for their college or university. Features interactive campus building models typical of those found in different-sized higher education facilities, such as dormitories, classroom buildings, laboratories, athletic facilities, and more. Includes PM checklists linked to each piece of equipment or system in hard copy and electronic format, along with required labor hours to complete the PM tasks. Helps users select and develop the best possible PM plan within budget.

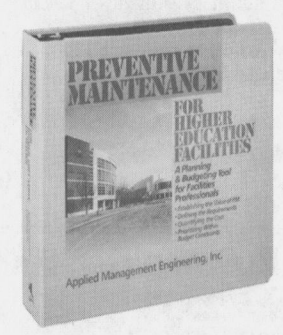

$149.95 per copy
150 pages, Hardcover
Catalog No. 67337

Preventive Maintenance Guidelines for School Facilities

By John C. Maciha

This new publication is a complete PM program for K-12 schools that ensures sustained security, safety, property integrity, user satisfaction, and reasonable ongoing expenditures.

Includes schedules for weekly, monthly, semiannual, and annual maintenance. Comes as a 3-ring binder with hard copy and electronic forms available at Means website, and a laminated wall chart.

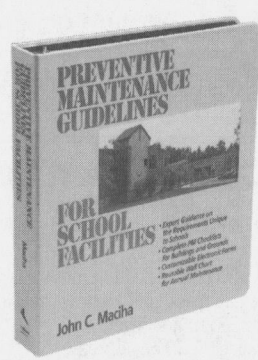

$149.95 per copy
Over 225 pages, Hardcover
Catalog No. 67326

Historic Preservation: Project Planning & Estimating

By Swanke Hayden Connell Architects

- *Managing Historic Restoration, Rehabilitation, and Preservation Building Projects and*
- *Determining and Controlling Their Costs*

The authors explain:
- How to determine whether a structure qualifies as historic
- Where to obtain funding and other assistance
- How to evaluate and repair more than 75 historic building materials
- How to properly research, document, and manage the project to meet code, agency, and other special requirements
- How to approach the upgrade of major building systems

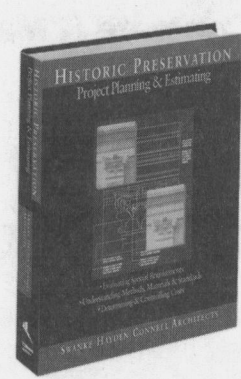

$99.95 per copy
Over 675 pages, Hardcover
Catalog No. 67323

6

For more information
visit Means Web Site
at www.rsmeans.com

Reference Books

Value Engineering: Practical Applications

. . . For Design, Construction, Maintenance & Operations

By Alphonse Dell'Isola, PE

A tool for immediate application—for engineers, architects, facility managers, owners, and contractors. Includes: Making the Case for VE—The Management Briefing, Integrating VE into Planning and Budgeting, Conducting Life Cycle Costing, Integrating VE into the Design Process, Using VE Methodology in Design Review and Consultant Selection, Case Studies, VE Workbook, and a Life Cycle Costing program on disk.

$79.95 per copy
Over 450 pages, illustrated, Hardcover
Catalog No. 67319

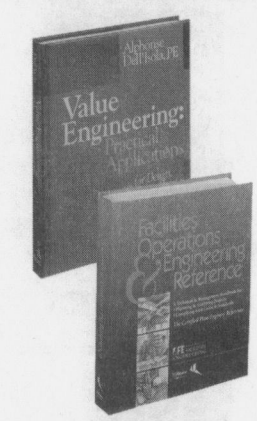

Facilities Operations & Engineering Reference

By the Association for Facilities Engineering and R.S. Means

An all-in-one technical referance for planning and managing facility projects and solving day-to-day operations problems. Selected as the official Certified Plant Engineer reference, this handbook covers financial analysis, maintenance, HVAC and energy efficiency, and more.

$109.95 per copy
Over 700 pages, illustrated, Hardcover
Catalog No. 67318

Builder's Essentials: Advanced Framing Methods

By Scot Simpson

A highly illustrated, "framer-friendly" approach to advanced framing elements. Provides expert, but easy-to-interpret, instruction for laying out and framing complex walls, roofs, and stairs, and special requirements for earthquake and hurricane protection. Also helps bring framers up to date on the latest building code changes, and provides tips on the lead framer's role and responsibilities, how to prepare for a job, and how to get the crew started.

$24.95 per copy
250 pages, illustrated, Softcover
Catalog No. 67330

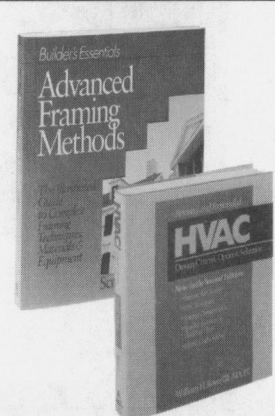

HVAC: Design Criteria, Options, Selection

Expanded 2nd Edition

By William H. Rowe III, AIA, PE

Includes Indoor Air Quality, CFC Removal, Energy Efficient Systems, and Special Systems by Building Type. Helps you solve a wide range of HVAC system design and selection problems effectively and economically. Gives you clear explanations of the latest ASHRAE standards.

$84.95 per copy
Over 600 pages, illustrated, Hardcover
Catalog No. 67306

Total Productive Facilities Management

By Richard W. Sievert, Jr.

This book provides a comprehensive program to:
• Achieve business goals by optimizing facility resources
• Implement best practices through benchmarking, evaluation & project management
• Increase your value to the organization

$39.98 per copy
Over 270 pages, illustrated, Hardcover
Catalog No. 67321

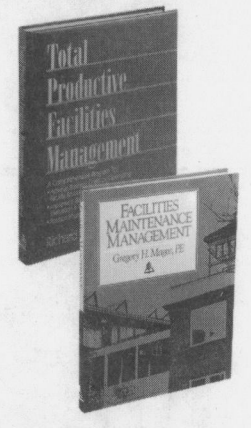

Facilities Maintenance Management

By Gregory H. Magee, PE

Now you can get successful management methods and techniques for all aspects of facilities maintenance. This comprehensive reference explains and demonstrates successful management techniques for all aspects of maintenance, repair, and improvements for buildings, machinery, equipment, and grounds. Plus, guidance for outsourcing and managing internal staffs.

$86.95 per copy
Over 280 pages with illustrations, Hardcover
Catalog No. 67249

Cost Planning & Estimating for Facilities Maintenance

In this unique book, a team of facilities management authorities shares their expertise at:
• Evaluating and budgeting maintenance operations
• Maintaining & repairing key building components
• Applying *Means Facilities Maintenance & Repair Cost Data* to your estimating

Covers special maintenance requirements of the 10 major building types.

$89.95 per copy
Over 475 pages, Hardcover
Catalog No. 67314

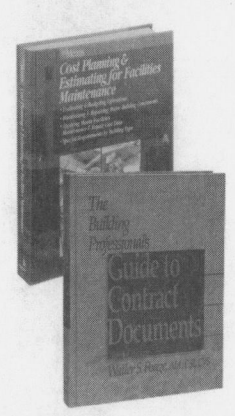

The Building Professional's Guide to Contract Documents

New 3rd Edition

By Waller S. Poage, AIA, CSI, CVS

This comprehensive treatment of Contract Documents is an important reference for owners, design professionals, contractors, and students.

• Structure your Documents for Maximum Efficiency
• Effectively communicate construction requirements to all concerned
• Understand the Roles and Responsibilities of Construction Professionals
• Improve Methods of Project Delivery

$64.95 per copy, 400 pages
Diagrams and construction forms, Hardcover
Catalog No. 67261A

Reference Books

For more information
visit Means Web Site
at www.rsmeans.com

Builder's Essentials: Plan Reading & Material Takeoff

By Wayne J. DelPico

For Residential and Light Commercial Construction

A valuable tool for understanding plans and specs, and accurately calculating material quantities.
Step-by-step instructions and takeoff procedures based on a full set of working drawings.

$35.95 per copy
Over 420 pages, Softcover
Catalog No. 67307

Planning & Managing Interior Projects

2nd Edition
By Carol E. Farren, CFM

Addresses changes in technology and business, guiding you through commercial design and construction from initial client meetings to post-project administration. Includes: evaluating space requirements, alternative work models, telecommunications and data management, and environmental issues.

$69.95 per copy
Over 400 pages, illustrated, Hardcover
Catalog No. 67245A

Builder's Essentials: Best Business Practices for Builders & Remodelers:

An Easy-to-Use Checklist System
By Thomas N. Frisby

A comprehensive guide covering all aspects of running a construction business, with more than 40 user–friendly checklists. This book provides expert guidance on: increasing your revenue and keeping more of your profit; planning for long-term growth; keeping good employees and managing subcontractors.

$29.95 per copy
Over 220 pages, Softcover
Catalog No. 67329

Builder's Essentials: Framing & Rough Carpentry 2nd Edition

By Scot Simpson

A complete training manual for apprentice and experienced carpenters. Develop and improve your skills with "framer-friendly," easy-to-follow instructions, and step-by-step illustrations. Learn proven techniques for framing walls, floors, roofs, stairs, doors, and windows. Updated guidance on standards, building codes, safety requirements, and more. Also available in Spanish!

$24.95 per copy
Over 150 pages, Softcover
Catalog No. 67298A   Spanish Catalog No. 67298AS

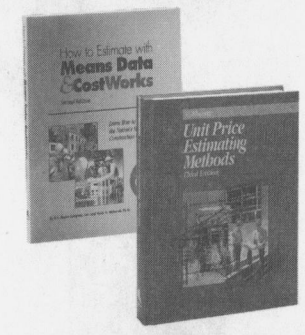

How to Estimate with Means Data & CostWorks

By R.S. Means and Saleh Mubarak
New 2nd Edition!

Learn estimating techniques using Means cost data. Includes an instructional version of Means CostWorks CD–ROM with Sample Building Plans.
The step-by-step guide takes you through all the major construction items. Over 300 sample estimating problems are included.

$59.95 per copy
Over 190 pages, Softcover
Catalog No. 67324A

Unit Price Estimating Methods

New 3rd Edition

This new edition includes up-to-date cost data and estimating examples, updated to reflect changes to the CSI numbering system and new features of Means cost data. It describes the most productive, universally accepted ways to estimate, and uses checklists and forms to illustrate shortcuts and timesavers. A model estimate demonstrates procedures. A new chapter explores computer estimating alternatives.

$59.95 per copy
Over 350 pages, illustrated, Hardcover
Catalog No. 67303A

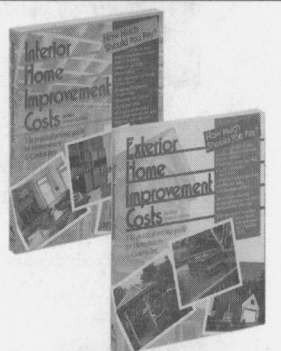

Interior Home Improvement Costs, 8th Edition

Provides 66 updated estimates for the most popular projects including estimates for home offices, in-law apartments, and remodeling for disabled residents. Includes guidance for remodeling older homes as well as attic and basement conversions, kitchen and bath remodeling, lighting and security; fireplaces; storage; stairs, new floors, and walls and ceilings.

$19.95 per copy
250 pages, illustrated, Softcover
Catalog No. 67308D

Exterior Home Improvement Costs, 8th Edition

Provides quick estimates for 64 projects including room additions and garages; windows and doors; dormers, roofs, and skylights; decks and pergolas; paving and patios; painting and siding; walls, fences, and porches; landscaping and driveways. It also contains price comparison sheets for material and equipment.

$19.95 per copy
270 pages, illustrated, Softcover
Catalog No. 67309D

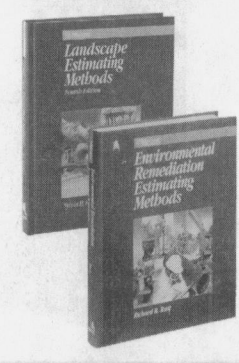

Means Landscape Estimating Methods

4th Edition
By Sylvia H. Fee

This revised edition offers expert guidance for preparing accurate estimates for new landscape construction and grounds maintenance. Includes a complete project estimate featuring the latest equipment and methods, and **two chapters on Life Cycle Costing, and Landscape Maintenance Estimating.**

$62.95 per copy
Over 300 pages, illustrated, Hardcover
Catalog No. 67295B

Means Environmental Remediation Estimating Methods

By Richard R. Rast

Guidelines for estimating 50 standard remediation technologies. Use it to prepare preliminary budgets, develop estimates, compare costs and solutions, estimate liability, review quotes, negotiate settlements.

A valuable support tool for *Means Environmental Remediation Unit Price* and *Assemblies* books.

$99.95 per copy
Over 600 pages, illustrated, Hardcover
Catalog No. 64777

For more information
visit Means Web Site
at www.rsmeans.com

Reference Books

Means Estimating Handbook

This comprehensive reference covers a full spectrum of technical data for estimating, with information on sizing, productivity, equipment requirements, codes, design standards, and engineering factors.
Means Estimating Handbook will help you: evaluate architectural plans and specifications, prepare accurate quantity takeoffs, prepare estimates from conceptual to detailed, and evaluate change orders.

$99.95 per copy
Over 900 pages, Hardcover
Catalog No. 67276

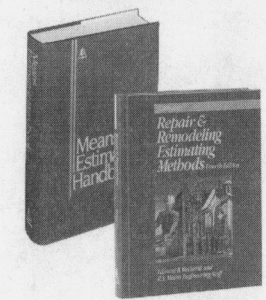

Means Repair and Remodeling Estimating
New 4th Edition
By Edward B. Wetherill & R.S. Means

Focuses on the unique problems of estimating renovations of existing structures. It helps you determine the true costs of remodeling through careful evaluation of architectural details and a site visit.
New section on disaster restoration costs.

$69.95 per copy
Over 450 pages, illustrated, Hardcover
Catalog No. 67265B

Facilities Planning & Relocation
New, lower price and user-friendly format.
By David D. Owen

A complete system for planning space needs and managing relocations. Includes step-by-step manual, over 50 forms, and extensive reference section on materials and furnishings.

$89.95 per copy
Over 450 pages, Softcover
Catalog No. 67301

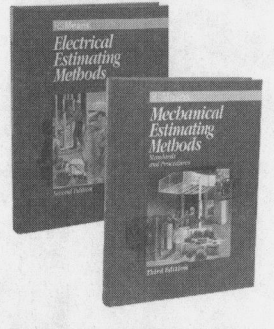

Means Square Foot & Assemblies Estimating Methods
New 3rd Edition!

Develop realistic Square Foot and Assemblies Costs for budgeting and construction funding. The new edition features updated guidance on square foot and assemblies estimating using UNIFORMAT II. An essential reference for anyone who performs conceptual estimates.

$69.95 per copy
Over 300 pages, illustrated, Hardcover
Catalog No. 67145B

Means Electrical Estimating Methods
2nd Edition

Expanded version includes sample estimates and cost information in keeping with the latest version of the CSI MasterFormat. Contains new coverage of Fiber Optic and Uninterruptible Power Supply electrical systems, broken down by components and explained in detail. A practical companion to *Means Electrical Cost Data*.

$64.95 per copy
Over 325 pages, Hardcover
Catalog No. 67230A

Means Mechanical Estimating Methods
New 3rd Edition

This guide assists you in making a review of plans, specs, and bid packages with suggestions for takeoff procedures, listings, substitutions and pre-bid scheduling. Includes suggestions for budgeting labor and equipment usage. Compares materials and construction methods to allow you to select the best option.

$64.95 per copy
Over 350 pages, illustrated, Hardcover
Catalog No. 67294A

Successful Estimating Methods:
From Concept to Bid
By John D. Bledsoe, PhD, PE

A highly practical, all-in-one guide to the tips and practices of today's successful estimator. Presents techniques for all types of estimates, and advanced topics such as life cycle cost analysis, value engineering, and automated estimating. Estimate spreadsheets available at Means Web site.

$32.48 per copy
Over 300 pages, illustrated, Hardcover
Catalog No. 67287

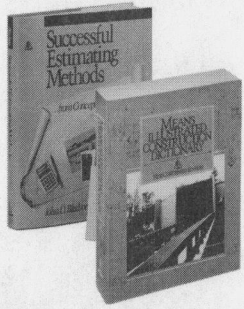

Illustrated Construction Dictionary, Condensed, 2nd Edition
New for 2003!

Updated, expanded edition coming in 2003! Based on Means *Illustrated Construction Dictionary*, this condensed version features 9,000 construction terms—precise definitions for every area of construction, from civil engineering to home improvement. The new edition features hundreds of new terms and illustrations.

$59.95 per copy
Over 500 pages, illustrated, Softcover
Catalog No. 67282A

Means Productivity Standards for Construction
Expanded Edition (*Formerly Man-Hour Standards*)

Here is the working encyclopedia of labor productivity information for construction professionals, with labor requirements for thousands of construction functions in CSI MasterFormat.
Completely updated, with over 3,000 new work items.

$69.98 per copy
Over 800 pages, Hardcover
Catalog No. 67236A

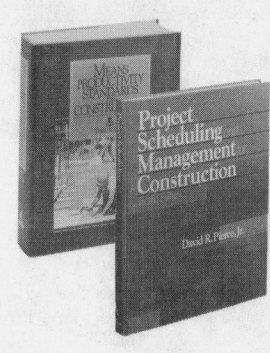

Project Scheduling & Management for Construction
New 2nd Edition
By David R. Pierce, Jr.

A comprehensive yet easy-to-follow guide to construction project scheduling and control—from vital project management principles through the latest scheduling, tracking, and controlling techniques. The author is a leading authority on scheduling with years of field and teaching experience at leading academic institutions. Spend a few hours with this book and come away with a solid understanding of this essential management topic.

$64.95 per copy
Over 250 pages, illustrated, Hardcover
Catalog No. 67247A

Reference Books

For more information
visit Means Web Site
at www.rsmeans.com

Cyberplaces: The Internet Guide for AECs & Facility Managers
By Paul Doherty, AIA, CSI
2nd Edition
$59.95 per copy
Catalog No. 67317A

Concrete Repair and Maintenance Illustrated
By Peter H. Emmons
$69.95 per copy
Catalog No. 67146

Maintenance Management Audit
Now $32.48 per copy, limited quantity
Catalog No. 67299

Superintending for Contractors:
How to Bring Jobs in On-Time, On-Budget
By Paul J. Cook
$35.95 per copy
Catalog No. 67233

HVAC Systems Evaluation
By Harold R. Colen, PE
$84.95 per copy
Catalog No. 67281

Basics for Builders: How to Survive and Prosper in Construction
By Thomas N. Frisby
$34.95 per copy
Catalog No. 67273

Successful Interior Projects Through Effective Contract Documents
By Joel Downey & Patricia K. Gilbert
Now $24.98 per copy, limited quantity
Catalog No. 67313

Building Spec Homes Profitably
By Kenneth V. Johnson
$29.95 per copy
Catalog No. 67312

Estimating for Contractors
How to Make Estimates that Win Jobs
By Paul J. Cook
$35.95 per copy
Catalog No. 67160

Means Forms for Building Construction Professionals
$47.48 per copy
Catalog No. 67231

Understanding Building Automation Systems
By Reinhold A. Carlson, PE & Robert Di Giandomenico
$29.98 per copy, limited quantity
Catalog No. 67284

Managing Construction Purchasing
By John G. McConville, CCC, CPE
Now $19.98 per copy, limited quantity
Catalog No. 67302

Fundamentals of the Construction Process
By Kweku K. Bentil, AIC
Now $34.98 per copy, limited quantity
Catalog No. 67260

Means ADA Compliance Pricing Guide
$59.98 per copy
Catalog No. 67310

Means Facilities Maintenance Standards
By Roger W. Liska, PE, AIC
Now $79.95 per copy
Catalog No. 67246

How to Estimate with Metric Units
Now $9.98 per copy, limited quantity
Catalog No. 67304

For more information
visit Means Web Site
at www.rsmeans.com

Reference Books

From Model Codes to the IBC: A Transitional Guide
By Rolf Jensen & Associates, Inc.

A time–saving resource for Architects, Engineers, Building Officials and Authorities Having Jurisdiction (AHJs), Contractors, Manufacturers, Building Owners, and Facility Managers.

Provides comprehensive, user-friendly guidance on making the transition to the International Building Code® from the model codes you're familiar with. Includes side-by-side code comparison of the IBC to the UBC, NBC, SBC, and NFPA 101®. Also features professional code commentary, quick-find indexes, and a Web site with regular code updates.

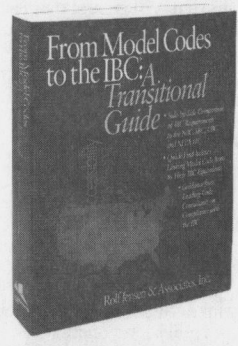

Also contains illustrations, abbreviations key, and an extensive resource section.

$114.95 per copy
880 pages, Softcover
Catalog No. 67328

Means Illustrated Construction Dictionary, 3rd Edition

Long regarded as the Industry's finest, the Means *Illustrated Construction Dictionary* is now even better. With the addition of over 1,000 new terms and hundreds of new illustrations, it is the clear choice for the most comprehensive and current information.

The companion CD-ROM that comes with this new edition adds many extra features: larger graphics, expanded definitions, and links to both CSI MasterFormat numbers and product information.

• 19,000 construction words, terms, phrases, symbols, weights, measures, and equivalents
• 1,000 new entries
• 1,200 helpful illustrations
• Easy-to-use format, with thumbtabs

$99.95 per copy
Over 790 pages, illustrated Hardcover
Catalog No. 67292A

Means Spanish/English Construction Dictionary
By R.S. Means, The International Conference of Building Officials (ICBO), and Rolf Jensen & Associates (RJA)

Designed to facilitate communication among Spanish- and English-speaking construction personnel—improving performance and job-site safety. Features the most common words and phrases used in the construction industry, with easy-to-follow pronunciations. Includes extensive building systems and tools illustrations.

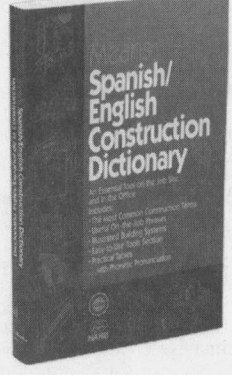

$22.95 per copy
250 pages, illustrated, Softcover
Catalog No. 67327

For more information
visit Means Web Site
at www.rsmeans.com

Seminars

Means CostWorks Training

This one-day seminar course has been designed with the intention of assisting both new and existing users to become more familiar with CostWorks program. The class is broken into two unique sections: (1) A one-half day presentation on the function of each icon; and each student will be shown how to use the software to develop a cost estimate. (2) Hands-on estimating exercises that will ensure that each student thoroughly understands how to use CostWorks.

CostWorks Benefits/Features:
- Estimate in your own spreadsheet format
- Power of Means National Database
- Database automatically regionalized
- Save time with keyword searches
- Save time by establishing common estimate items in "Bookmark" files
- Customize your spreadsheet template
- Hot key to Product Manufacturers' listings and specs
- Merge capability for networking environments
- View crews and assembly components
- AutoSave capability
- Enhanced sorting capability

Advanced Project Management

This two-day seminar will teach you how to effectively manage and control the entire design-build process and allow you to take home tangible skills that will be immediately applicable on existing projects.

Some Of What You'll Learn:
- Value engineering, bonding, fast-tracking and bid package creation
- How estimates and schedules can be integrated to provide advanced project management tools
- Cost engineering, quality control, productivity measurement and improvement
- Front loading a project and predicting its cash flow

Who Should Attend: Owners, project managers, architectural and engineering managers, construction managers, contractors...and anyone else who is responsible for the timely design and completion of construction projects.

Unit Price Estimating

This interactive two-day seminar teaches attendees how to interpret project information and process it into final, detailed estimates with the greatest accuracy level.

The single most important credential an estimator can take to the job is the ability to visualize construction in the mind's eye, and thereby estimate accurately.

Some Of What You'll Learn:
- Interpreting the design in terms of cost
- The most detailed, time tested methodology for accurate "pricing"
- Key cost drivers—material, labor, equipment, staging and subcontracts
- Understanding direct and indirect costs for accurate job cost accounting and change order management

Who Should Attend: Corporate and government estimators and purchasers, architects, engineers...and others needing to produce accurate project estimates.

Mechanical and Electrical Estimating

This two-day course teaches attendees how to prepare more accurate and complete mechanical/electrical estimates, avoiding the pitfalls of omission and double-counting, while understanding the composition and rationale within the Means Mechanical/Electrical database.

Some Of What You'll Learn:
- The unique way mechanical and electrical systems are interrelated
- M&E estimates, conceptual, planning, budgeting and bidding stages
- Order of magnitude, square foot, assemblies and unit price estimating
- Comparative cost analysis of equipment and design alternatives

Who Should Attend: Architects, engineers, facilities managers, mechanical and electrical contractors...and others needing a highly reliable method for developing, understanding and evaluating mechanical and electrical contracts.

Square Foot and Assemblies Cost Estimating

This two-day course teaches attendees how to quickly deliver accurate square foot estimates using limited budget and design information.

Some Of What You'll Learn:
- How square foot costing gets the estimate done faster
- Taking advantage of a "systems" or "assemblies" format
- The Means "building assemblies/square foot cost approach"
- How to create a very reliable preliminary and systems estimate using bare-bones design information

Who Should Attend: Facilities managers, facilities engineers, estimators, planners, developers, construction finance professionals...and others needing to make quick, accurate construction cost estimates at commercial, government, educational and medical facilities.

Facilities Maintenance and Repair Estimating

This two-day course teaches attendees how to plan, budget, and estimate the cost of ongoing and preventive maintenance and repair for existing buildings and grounds.

Some Of What You'll Learn:
- The most financially favorable maintenance, repair and replacement scheduling and estimating
- Auditing and value engineering facilities
- Preventive planning and facilities upgrading
- Determining both in-house and contract-out service costs; annual, asset-protecting M&R plan

Who Should Attend: Facility managers, maintenance supervisors, buildings and grounds superintendents, plant managers, planners, estimators...and others involved in facilities planning and budgeting.

Repair and Remodeling Estimating

This two-day seminar emphasizes all the underlying considerations unique to repair/remodeling estimating and presents the correct methods for generating accurate, reliable R&R project costs using the unit price and assemblies methods.

Some Of What You'll Learn:
- Estimating considerations—like labor-hours, building code compliance, working within existing structures, purchasing materials in smaller quantities, unforeseen deficiencies
- Identifies problems and provides solutions to estimating building alterations
- Rules for factoring in minimum labor costs, accurate productivity estimates and allowances for project contingencies
- R&R estimating examples are calculated using unite prices and assemblies data

Who Should Attend: Facilities managers, plant engineers, architects, contractors, estimators, builders...and others who are concerned with the proper preparation and/or evaluation of repair and remodeling estimates.

Plan Reading and Material Takeoff

This two-day program teaches attendees to read and understand construction documents and to use them in the preparation of material takeoffs.

Some of What You'll Learn:
- Skills necessary to read and understand typical contract documents—blueprints and specifications
- Details and symbols used by architects and engineers
- Construction specifications' importance in conjunction with blueprints
- Accurate takeoff of construction materials and industry-accepted takeoff methods

Who Should Attend: Facilities managers, construction supervisors, office managers...and other responsible for the execution and administration of a construction project including government, medical, commercial, educational or retail facilities.

Scheduling and Project Management

This two-day course teaches attendees the most current and proven scheduling and management techniques needed to bring projects in on time and on budget.

Some Of What You'll Learn:
- Crucial phases of planning and scheduling
- How to establish project priorities, develop realistic schedules and management techniques
- Critical Path and Precedence Methods
- Special emphasis on cost control

Who Should Attend: Construction project managers, supervisors, engineers, estimators, contractors...and others who want to improve their project planning, scheduling and management skills.

Seminars

Seminars

2003 Means Seminar Schedule

For more information visit Means Web Site at www.rsmeans.com

Location	Dates
Las Vegas, NV	March TBD
Washington, DC	April 7-10
Denver, CO	May 19-22
San Francisco, CA	June 9-12
Cape Cod, MA	September 8-11
Washington, DC	September 22-25
Las Vegas, NV	October TBD
Orlando, FL	November TBD
Atlantic City, NJ	November TBD
San Diego, CA	December TBD

Note: Call for exact dates and details.

Registration Information

Register Early... Save up to $125! Register 30 days before the start date of a seminar and save $125 off your total fee. *Note: This discount can be applied only once per order. It cannot be applied to team discount registrations or any other special offer.*

How to Register Register by phone today! Means toll-free number for making reservations is: **1-800-334-3509, ext. 5115.**

Individual Seminar Registration Fee $895. Individual CostWorks Training Registration Fee $349. To register by mail, complete the registration form and return with your full fee to: Seminar Division, R.S. Means Company, Inc., 63 Smiths Lane, Kingston, MA 02364.

Federal Government Pricing All Federal Government employees save 25% off regular seminar price. Other promotional discounts cannot be combined with Federal Government discount.

Team Discount Program Two to four seminar registrations: $760 per person– Five or more seminar registrations: $710 per person–Ten or more seminar registrations: Call for pricing.

Consecutive Seminar Offer One individual signing up for two separate courses at the same location during the designated time period pays only $1,400. You get the second course for only $525 **(a 40% discount).** Payment must be received at least ten days prior to seminar dates to confirm attendance.

Refund Policy Cancellations will be accepted up to ten days prior to the seminar start. There are no refunds for cancellations received later than ten working days prior to the first day of the seminar. A $150 processing fee will be applied for all cancellations. Written notice of cancellation is required . Substitutions can be made at anytime before the session starts. **No-shows are subject to the full seminar fee.**

AACE Approved Courses The R.S. Means Construction Estimating and Management Seminars described and offered to you here have each been approved for 14 hours (1.4 recertification credits) of credit by the AACE International Certification Board toward meeting the continuing education requirements for re-certification as a Certified Cost Engineer/Certified Cost Consultant.

AIA Continuing Education R.S. Means is registered with the AIA Continuing Education System (AIA/CES) and is committed to developing quality learning activities in accordance with the CES criteria. R.S. Means seminars meet the AIA/CES criteria for Quality Level 2. AIA members will receive (28) learning units (LUs) for each two-day R.S. Means Course.

Daily Course Schedule The first day of each seminar session begins at 8:30 A.M. and ends at 4:30 P.M. The second day is 8:00 A.M.–4:00 P.M. Participants are urged to bring a hand-held calculator since many actual problems will be worked out in each session.

Continental Breakfast Your registration includes the cost of a continental breakfast, a morning coffee break, and an afternoon break. These informal segments will allow you to discuss topics of mutual interest with other members of the seminar. (You are free to make your own lunch and dinner arrangements.)

Hotel/Transportation Arrangements R.S. Means has arranged to hold a block of rooms at each hotel hosting a seminar. To take advantage of special group rates when making your reservation, be sure to mention that you are attending the Means Seminar. You are, of course, free to stay at the lodging place of your choice. (**Hotel reservations and transportation arrangements should be made directly by seminar attendees.**)

Important Class sizes are limited, so please register as soon as possible.

Note: Pricing subject to change.

Registration Form Call 1-800-334-3509 x5115 to register or FAX 1-800-632-6732. Visit our Web site www.rsmeans.com

Please register the following people for the Means Construction Seminars as shown here. Full payment or deposit is enclosed, and we understand that we must make our own hotel reservations if overnight stays are necessary.

☐ Full payment of $ _____ enclosed.

☐ Bill me

Name of Registrant(s)
(To appear on certificate of completion)

_____

_____

_____

P.O. #: _____

GOVERNMENT AGENCIES MUST SUPPLY PURCHASE ORDER NUMBER

Firm Name _____

Address _____

City/State/Zip _____

Telephone No. fax No. _____

E-mail Address _____

Charge our registration(s) to: ☐ MasterCard ☐ VISA ☐ American Express ☐ Discover

Account No. _____ Exp. Date _____

Cardholder's Signature _____

Seminar Name	City	Dates

Please mail check to: Seminar Division, R.S. Means Company, Inc., 63 Smiths Lane, P.O.Box 800, Kingston, MA 02364 USA

13

MeansData™

CONSTRUCTION COSTS FOR SOFTWARE APPLICATIONS

Your construction estimating software is only as good as your cost data.

A proven construction cost database is a mandatory part of any estimating package. The following list of software providers can offer you MeansData™ as an added feature for their estimating systems. See the table below for what types of products and services they offer (match their numbers). Visit them online at **www.rsmeans.com/demo/** for more information and free demos. Or call their numbers listed below.

1. **3D International**
713-871-7000
venegas@3di.com

2. **4Clicks-Solutions, LLC**
719-574-7721
mbrown@4clicks-solutions.com

3. **Aepco, Inc.**
301-670-4642
blueworks@aepco.com

4. **Applied Flow Technology**
800-589-4943
info@aft.com

5. **ArenaSoft Estimating**
888-370-8806
info@arenasoft.com

6. **ARES Corporation**
925-299-6700
sales@arescorporation.com

7. **Benchmark Inc.**
319-393-9100
sales@benchmark-inc.com

8. **Beneco Enterprises, Inc.**
877-719-4748
inquire@beneco.com

9. **BSD - Building Systems Design, Inc.**
888-273-7638
bsd@bsdsoftlink.com

10. **Cadafis, Inc.**
614-444-9851
cadafis@rrohio.com

11. **Caelix, Inc.**
602-787-8600
kwebster@caelix.com

12. **CMS - Computerized Micro Solutions**
800-255-7407
cms@proest.com

13. **Conac Software**
800-663-2338
sales@conac.com

14. **Estimating Systems, Inc.**
800-967-8572
esipulsar@adelphia.net

15. **eCaliper Estimating Software, A Marathon Systems Service Business**
877-469-3700
sales@ecaliper.com

16. **G2 Estimator A Divi. of Valli Info. Systems, Inc.**
800-627-3283
info@g2estimator.com

17. **G/C Emuni**
514-633-8339
rpa@gcei.com

18. **Geac Computer Corp., LTD.**
800-554-9865
info@geac.com

19. **HardDollar**
800-637-7496
sales@harddollar.com

20. **LUQS International**
888-682-5573
info@luqs.com

21. **MAESTRO Estimator Schwaab Technology Solutions, Inc.**
281-578-3039
Stefan@schwaabtech.com

22. **Magellan K-12**
936-447-1744
sam.wilson@magellan-K12.com

23. **MC2 - Management Computer Controls**
800-225-5622
vkeys@mc2-ice.com

24. **Maximus Asset Solutions**
800-659-9001
assetsolutions@maximus.com

25. **Prism Computer Corp.**
800-774-7622
famis@prismcc.com

26. **Quest Solutions, Inc.**
800-452-2342
info@questsolutions.com

27. **Sanders Software, Inc.**
800-280-9760
hsander@vallnet.com

28. **Timberline Software Corp.** 800-628-6583
pruduct.info@timberline.com

29. **TMA SYSTEMS, Inc.**
800-862-1130
sales@tmasys.com

30. **US Cost, Inc.**
800-372-4003
sales@uscost.com

31. **Vanderweil Facility Advisors**
617-451-5100
info@VFA.com

32. **Vertigraph, Inc.**
800-989-4243
info-request@vertigraph.com

33. **Wendleware**
714-895-7222
sales@corecon.com

34. **WinEstimator, Inc.**
800-950-2374
sales@winest.com

TYPE:	1	2	3	4	5	6	7	8	9	10	11	12	13	14	15	16	17	18	19	20	21	22	23	24	25	26	27	28	29	30	31	32	33	34
BID									•				•				•		•				•			•		•				•		•
Estimating		•			•	•		•	•	•	•		•	•	•		•	•	•	•			•			•		•		•	•			•
DOC/JOC/SABER		•						•						•										•										•
IDIQ		•						•																										•
Asset Mgmt										•								•				•		•	•									
Facility Mgmt	•		•							•											•	•							•		•			
Project Mgmt	•					•	•				•		•		•				•		•							•				•	•	
TAKE-OFF					•						•					•	•		•	•			•			•		•						
EARTHWK																			•	•			•			•								
Pipe flow				•																														
HVAC/Plumbing																				•							•							
Roofing							•			•																								
Design	•							•																				•						•
Other Offers/Links:																																		
Accounting/HR								•					•						•									•						
Scheduling										•	•								•		•							•	•	•				•
CAD										•	•					•												•						•
PDA									•																			•		•				
Lt Versions																												•						
Consulting		•	•						•					•		•								•				•						•
Training		•						•						•		•								•				•						•

Reseller applications now being accepted. Call Carol Polio Ext. 5107.

FOR MORE INFORMATION
CALL 1-800-448-8182, EXT. 5107 OR FAX 1-800-632-6732

14

For more information
visit Means Web Site
at www.rsmeans.com

New Titles

Residential & Light Commercial Construction Standards, 2nd Edition
By R.S. Means and Contributing Authors

New, updated second edition of this unique collection of industry standards that define quality construction. For contractors, subcontractors, owners, developers, architects, engineers, attorneys, and insurance personnel, this book provides authoritative requirements and recommendations compiled from the nation's leading professional associations, industry publications, and building code organizations. This one-stop reference is enhanced by helpful commentary from respected practitioners, including identification of items most frequently targeted for construction defect claims. The new second edition provides the latest building code requirements.

$59.95 per copy
600 pages, illustrated, Softcover
Catalog No. 67322A

Green Building: Project Planning & Cost Estimating
By R.S. Means and Contributing Authors

Written by a team of leading experts in sustainable design, this new book is a complete guide to planning and estimating green building projects, a growing trend in building design and construction – commercial, industrial, institutional and residential. It explains:

- All the different criteria for "green-ness"
- What criteria your building needs to meet to get a LEED, Energy Star, or other recognized rating for green buildings
- How the project team works differently on a green versus a traditional building project
- How to select and specify green products
- How to evaluate the cost and value of green products versus conventional ones — not only for their first (installation) cost, but their cost over time (in maintenance and operation).

Features an extensive Green Building Cost Data section, which details the available products, how they are specified, and how much they cost.

$89.95 per copy
350 pages, illustrated, Hardcover
Catalog No. 67338

Coming Soon!
Building Security: Strategies & Costs
By R.S. Means and David Owen

- Assessing Risks Unique to Your Facility
- Developing a Rational Response Plan
- Incorporating Protection into Design & Space Planning
- Cost Data for Security Systems & Building Materials

For commercial, industrial, and institutional buildings, this new book provides guidance to owners, facility managers, architects, and interior designers on how to identify security risks—both the likelihood of an event (including terrorist, criminal, or natural disaster), and the monetary risk (the price of a potential loss). Provides guidance on:

- How to develop a plan to minimize risk, and create a plan of action to be used in case of an event.
- How to design buildings and plan interior space in a security-conscious way to minimize risk.
- What building systems, materials, and equipment are available to promote security.

Features a cost data section for building materials and systems, as well as security devices and equipment, including line items with labor, equipment, and materials costs.

$89.95 per copy
Over 250 pages, illustrated, Hardcover
Catalog No. 67339

2003 Order Form

ORDER TOLL FREE 1-800-334-3509
OR FAX 1-800-632-6732.

Qty.	Book No.	COST ESTIMATING BOOKS	Unit Price	Total
	60063	Assemblies Cost Data 2003	$173.95	
	60013	Building Construction Cost Data 2003	104.95	
	61013	Building Const. Cost Data–Looseleaf Ed. 2003	124.95	
	63013	Building Const. Cost Data–Metric Version 2003	104.95	
	60223	Building Const. Cost Data–Western Ed. 2003	104.95	
	60113	Concrete & Masonry Cost Data 2003	97.95	
	60143	Construction Cost Indexes 2003	228.95	
	60143A	Construction Cost Index–January 2003	57.25	
	60143B	Construction Cost Index–April 2003	57.25	
	60143C	Construction Cost Index–July 2003	57.25	
	60143D	Construction Cost Index–October 2003	57.25	
	60343	Contr. Pricing Guide: Resid. R & R Costs 2003	38.95	
	60333	Contr. Pricing Guide: Resid. Detailed 2003	38.95	
	60323	Contr. Pricing Guide: Resid. Sq. Ft. 2003	39.95	
	64023	ECHOS Assemblies Cost Book 2003	172.95	
	64013	ECHOS Unit Cost Book 2003	115.95	
	54003	ECHOS (Combo set of both books)	241.95	
	60233	Electrical Change Order Cost Data 2003	104.95	
	60033	Electrical Cost Data 2003	104.95	
	60203	Facilities Construction Cost Data 2003	253.95	
	60303	Facilities Maintenance & Repair Cost Data 2003	230.95	
	60163	Heavy Construction Cost Data 2003	104.95	
	63163	Heavy Const. Cost Data–Metric Version 2003	104.95	
	60093	Interior Cost Data 2003	104.95	
	60123	Labor Rates for the Const. Industry 2003	230.95	
	60183	Light Commercial Cost Data 2003	92.95	
	60023	Mechanical Cost Data 2003	104.95	
	60153	Open Shop Building Const. Cost Data 2003	104.95	
	60213	Plumbing Cost Data 2003	104.95	
	60043	Repair and Remodeling Cost Data 2003	92.95	
	60173	Residential Cost Data 2003	92.95	
	60283	Site Work & Landscape Cost Data 2003	104.95	
	60053	Square Foot Costs 2003	115.95	
		REFERENCE BOOKS		
	67147A	ADA in Practice	59.98	
	67310	ADA Pricing Guide	59.98	
	67273	Basics for Builders: How to Survive and Prosper	34.95	
	67330	Bldrs Essentials: Adv. Framing Methods	24.95	
	67329	Bldrs Essentials: Best Bus. Practices for Bldrs	29.95	
	67298A	Bldrs Essentials: Framing/Carpentry 2nd Ed.	24.95	
	67298AS	Bldrs Essentials: Framing/Carpentry Spanish	24.95	
	67307	Bldrs Essentials: Plan Reading & Takeoff	35.95	
	67261A	Bldg. Prof. Guide to Contract Documents 3rd Ed.	64.95	
	67312	Building Spec Homes Profitably	29.95	
	67146	Concrete Repair & Maintenance Illustrated	69.95	
	67278	Construction Delays	29.48	
	67314	Cost Planning & Est. for Facil. Maint.	89.95	
	67317A	Cyberplaces: The Internet Guide 2nd Ed.	59.95	
	67230A	Electrical Estimating Methods 2nd Ed.	64.95	

Qty.	Book No.	REFERENCE BOOKS (Cont.)	Unit Price	Total
	64777	Environmental Remediation Est. Methods	$ 99.95	
	67160	Estimating for Contractors	35.95	
	67276	Estimating Handbook	99.95	
	67249	Facilities Maintenance Management	86.95	
	67246	Facilities Maintenance Standards	79.95	
	67318	Facilities Operations & Engineering Reference	109.95	
	67301	Facilities Planning & Relocation	89.95	
	67231	Forms for Building Const. Professional	47.48	
	67328	From Model Codes to IBC: Transitional Guide	114.95	
	67260	Fundamentals of the Construction Process	34.98	
	67323	Historic Preservation: Proj. Planning & Est.	99.95	
	67308D	Home Improvement Costs–Int. Projects 8th Ed.	19.95	
	67309D	Home Improvement Costs–Ext. Projects 8th Ed.	19.95	
	67324A	How to Est. w/Means Data & CostWorks, 2nd Ed.	59.95	
	67304	How to Estimate with Metric Units	9.98	
	67306	HVAC: Design Criteria, Options, Select. 2nd Ed.	84.95	
	67281	HVAC Systems Evaluation	84.95	
	67282	Illustrated Construction Dictionary, Condensed	59.95	
	67292A	Illustrated Construction Dictionary, w/CD-ROM	99.95	
	67295B	Landscape Estimating 4th Ed.	62.95	
	67299	Maintenance Management Audit	32.48	
	67302	Managing Construction Purchasing	19.98	
	67294A	Mechanical Estimating 3rd Ed.	64.95	
	67245A	Planning and Managing Interior Projects 2nd Ed.	69.95	
	67283A	Plumbing Estimating Methods 2nd Ed.	59.95	
	67326	Preventive Maint. Guidelines for School Facil.	149.95	
	67337	Preventive Maint. for Higher Education Facilities	149.95	
	67236A	Productivity Standards for Constr.–3rd Ed.	69.98	
	67247A	Project Scheduling & Management for Constr.	64.95	
	67265B	Repair & Remodeling Estimating 4th Ed.	69.95	
	67322A	Resi. & Light Commercial Const. Stds.2nd Ed.	59.95	
	67254	Risk Management for Building Professionals	15.98	
	67327	Spanish/English Construction Dictionary	22.95	
	67145B	Sq. Ft. & Assem. Estimating Methods 3rd Ed.	69.95	
	67287	Successful Estimating Methods	32.48	
	67313	Successful Interior Projects	24.98	
	67233	Superintending for Contractors	35.95	
	67321	Total Productive Facilities Management	39.98	
	67284	Understanding Building Automation Systems	29.98	
	67303A	Unit Price Estimating Methods 3rd Ed.	59.95	
	67319	Value Engineering: Practical Applications	79.95	

MA residents add 5% state sales tax	
Shipping & Handling**	
Total (U.S. Funds)*	

Prices are subject to change and are for U.S. delivery only. *Canadian customers may call for current prices. **Shipping & handling charges: Add 7% of total order for check and credit card payments. Add 9% of total order for invoiced orders.

Send Order To: ADDV-1001

Name (Please Print) _____

Company _____

☐ **Company**
☐ **Home**   Address _____

City/State/Zip _____

Phone # _____ P.O. # _____

Mail To: **R.S. Means Company, Inc.,** P.O. Box 800, Kingston, MA 02364-0800   (Must accompany all orders being billed)